Common Graphs

Linear Function
$y = mx + b$

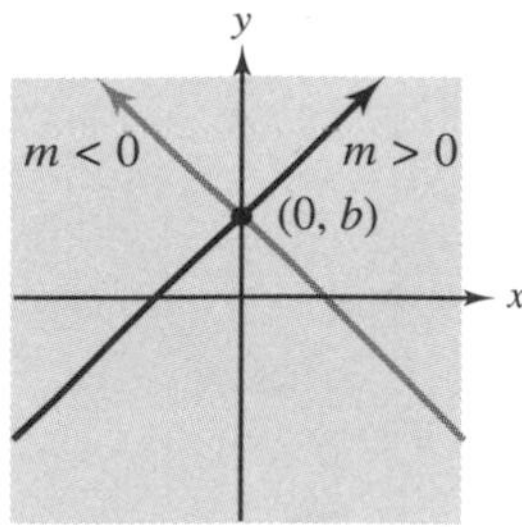

Quadratic Function
$y = ax^2$

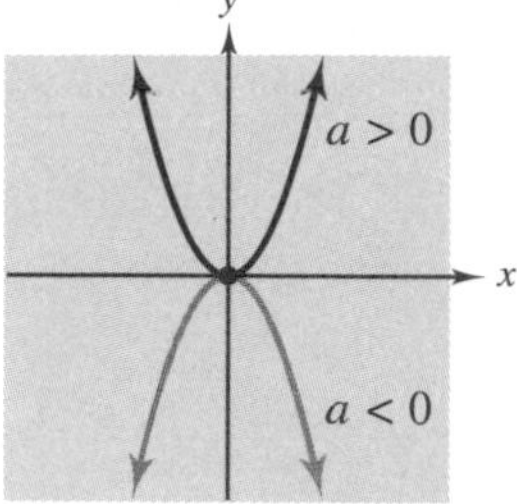

Absolute Value Function
$y = |x|$

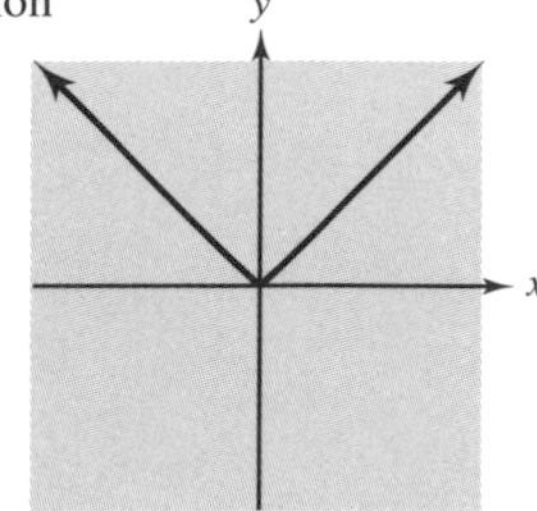

Square Root Function
$y = \sqrt{x}$

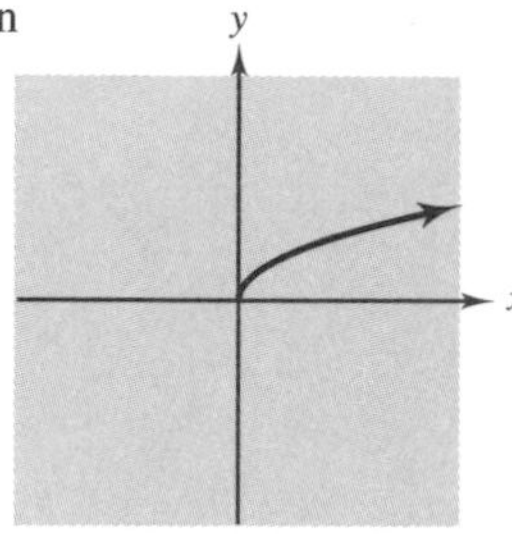

Exponential Function
$y = b^x, b > 1$

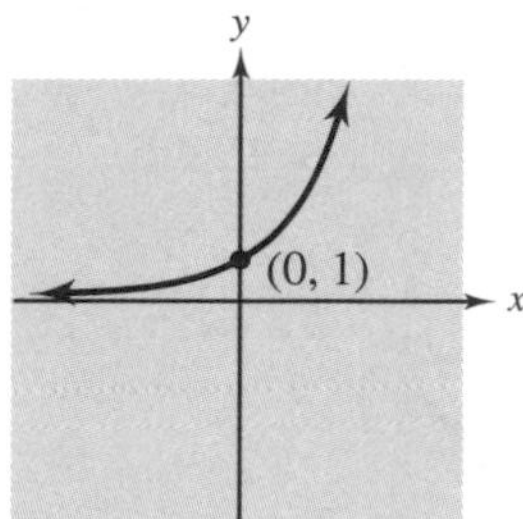

Logarithmic Function
$y = \log_b x, b > 1$

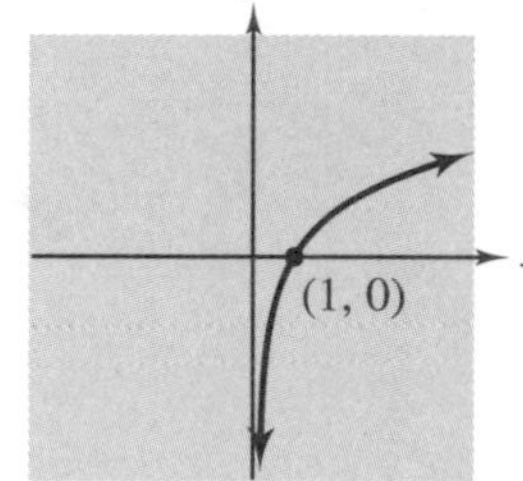

Cubic Function
$y = x^3$

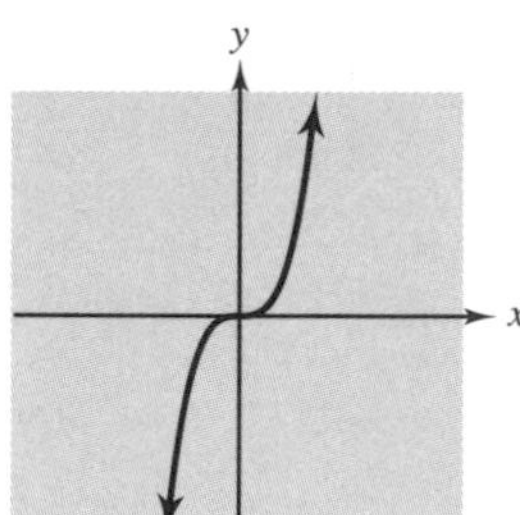

Circle
$x^2 + y^2 = r^2$

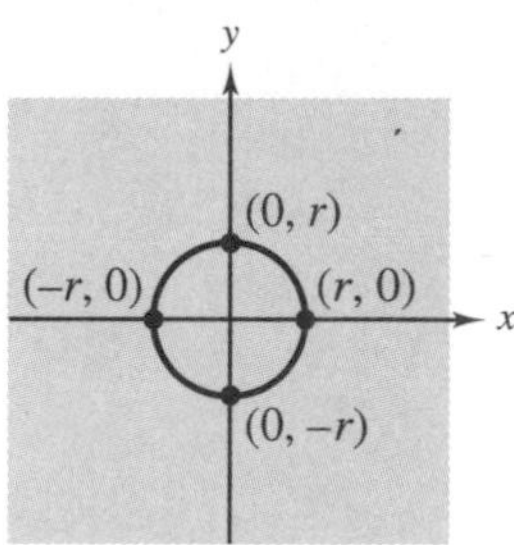

Ellipse
$\frac{x^2}{a^2} + \frac{y^2}{b^2} = 1$

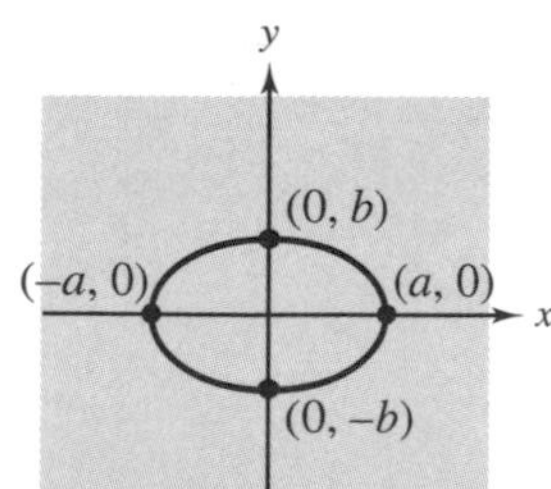

Hyperbola
$\frac{x^2}{a^2} - \frac{y^2}{b^2} = 1$

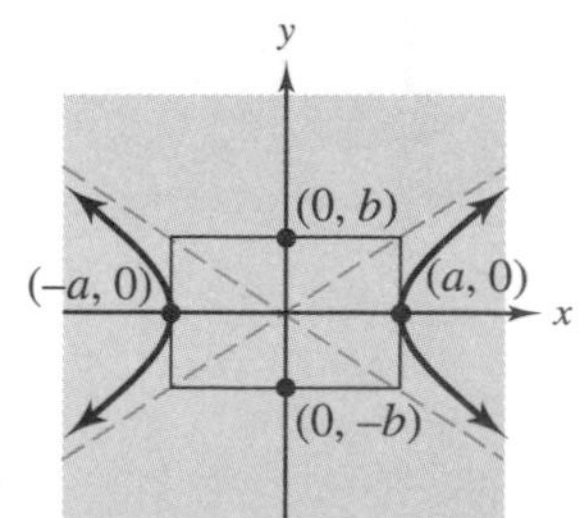

INTERMEDIATE ALGEBRA

Second Edition

Elaine Hubbard

KENNESAW STATE UNIVERSITY

Ronald D. Robinson

HOUGHTON MIFFLIN COMPANY BOSTON NEW YORK

Editor-in-Chief: Charles Hartford
Associate Editor: Mary Beckwith
Senior Project Editor: Maria Morelli
Editorial Assistant: Lauren M. Gagliardi
Senior Production/Design Coordinator: Jennifer Waddell
Senior Manufacturing Coordinator: Sally Culler
Marketing Manager: Roslyn Kane

Cover design: Stoltze Design, Wing Ngan

Cover image: SuperStock, Inc.

Photo credits: Chapter 1: Galen Rowell/Corbis; AP Photo/Amy Sancetta; Chapter 2: Official U.S. Navy Photo by PH2 Dante DeAngelis; © Dick Hemingway; Chapter 3: Raymond Gehman/Corbis; Gary Carter/Corbis; Chapter 4: AP Photo/David Longstreath; Joseph Sohm/ChromoSohm Inc./Corbis; Chapter 5: Ed Young/Corbis; Raymond Gehman/Corbis; Chapter 6: Philip Gould/Corbis; Phil Schermeister/© Corbis; Chapter 7: Jennie Woodcock, Reflections Photolibrary/Corbis; Jim Sugar Photography/Corbis; Chapter 8: Judy Griesedieck/© Corbis; The Purcell Team/Corbis; Chapter 9: AP Photo/Gazette Telegraph, Mark Reis; AP Photo/Enric F. Marti; Chapter 10: AP Photo/Natasha Lane, File; AP Photo/Salt Lake Tribune, Al Hartmann.

Printed in the U.S.A.

Library of Congress Catalog Card Number: 98-72043

ISBN
Student text: 0-395-90113-8
Instructor's Annotated Edition: 0-395-93332-3

23456789-DW-01 00 99

Contents

Chapter 12 SEQUENCES AND SERIES 741

Preface

The Approach

Every learning theory emphasizes the value of widening the sensory spectrum. To accomplish this in teaching mathematics, we believe that one must give students a visual connection that allows students to "see" mathematics in context.

In this second edition, we continue to use our extensive teaching experience to create a balance between traditional approaches and the use of a graphing calculator. We have developed an approach to teaching with a graphing calculator that works successfully for us, for our colleagues, and for students.

The Exploration/Discovery format of the first edition has been revised to "Exploring the Concept." In this new feature, we begin at a visual, concrete level at which concepts and relationships are illustrated and outcomes are suggested. By experimenting and asking "What if?" questions, the instructor can help students to become active participants in the learning process.

As in the previous edition, we strive to maintain a proper perspective. The focus is on mathematics, with the graphing calculator serving as a tool for better understanding.

Modeling and Real-Life Applications

Each chapter begins with a presentation of real data and background information on a topic of interest. This topic is pursued further in a Group Project within the chapter and in a Chapter Project at the end of the chapter. In addition to this thread, a large number of modeling problems have been added to this edition. These exercises contain actual data (with sources), tables, graphs, and guidance in modeling, analyzing, and interpreting the data.

Most exercise sets contain real-life applications that we have written in the student's experiential context. Some sections are devoted exclusively to such applications.

Other New Features

As was true for our first edition, we have written this book with careful attention given to AMATYC's *Crossroads in Mathematics* standards. Two major elements of this document are the use of a graphing calculator and the inclusion of real-data examples and exercises. We have expanded and improved in both of these areas.

In addition, we have added the critical-thinking feature, "Think About It," to each section. In the area of written and verbal communication skills, we have continued to include a generous number of writing exercises.

In the following pages, the special features of the text are highlighted and discussed in detail.

The goals of this book are to provide an effective instructional framework for the classroom and to engage students in a better understanding of the nature of mathematics in their personal and career lives. We earnestly hope that our work will promote achievement and success for all.

Supplements

Instructor's Annotated Edition This version of the text includes answers to the odd- and even-numbered exercises listed consecutively, as well as the answers to Think About It and Looking Ahead. Instructor annotations are highlighted in red. To help instructors better plan students' homework assignments in accordance with their teaching styles and strategies, graphing calculator icons are now used in the Instructor's Annotated Edition to identify problems best completed using a graphing calculator. Similarly, Concept Extension exercises are identified with the notation CE. The graphing calculator icons and CE notation do *not* appear in the student text.

Student Solutions Manual All solutions to the odd-numbered exercises are worked out in this manual.

Instructor's Resource Manual All solutions to the even-numbered exercises and three forms of sample tests per chapter are included.

Test Item File This printed manual contains over 2500 multiple-choice and free-response test questions.

Computerized Testing A computerized test bank of multiple-choice and free-response test questions for Windows or Macintosh is available.

Tutorial Software This tutorial software is organized according to the text and provides students with the opportunity to solve problems in the areas in which they need additional practice. Versions for Macintosh and IBM-Windows are available.

Graphing Calculator Keystroke Guide This guide offers valuable support and keystroke instruction for several calculator models, including the TI-83. A Key Word icon in the text alerts students to specific keystroke information in this supplement.

Videotapes A series of videotapes for instructors provides a thorough review of concepts and worked-out examples to reinforce lessons within the text.

Acknowledgments

We would like to thank the following colleagues who reviewed the second edition manuscript and made many helpful suggestions:

Randall Allbritton, *Daytona Beach Community College*
Linda A. Britton, *Kellogg Community College*
Mitzi Chaffer, *Central Michigan University*
Peggy Clifton, *Redlands Community College*
Karen Driskell, *Calhoun State Community College*
Wallace Etterbeek, *California State University—Sacramento*
Gene Garza, *University of Montevallo*
Virginia Hamilton, *Shawnee State University*
Nancy Krzesicki, *St. Petersburg Junior College*
Martha Ann Larkin, *Southern Utah University*
Lou Ann Mahaney, *Tarrant County Junior College—Northeast*
Gael Mericle, *Mankato State University*
Claudinna Rowley, *Johnson County Community College*
Barbara D. Sehr, *Indiana University—Kokomo*
Fred Schineller, *Arizona State University*
Patricia P. Taylor, *Thomas Nelson Community College*

Elaine Hubbard
Ronald D. Robinson

Opening Features

Chapter Opener

Each chapter begins with a short introduction to a real-data application. This opening topic is continued as a Group Project later in the chapter and as a Chapter Project at the end of the chapter. The chapter opener also includes a helpful overview of the topics that will be covered in the chapter and a list of the section titles.

Chapter 2

The Coordinate Plane and Functions

2.1 The Coordinate Plane
2.2 The Graph of an Expression
2.3 Relations and Functions
2.4 Functions: Notation and Evaluation
2.5 Analysis of Functions

As more and more states enacted helmet laws for bicycle riders, the sales of helmets increased dramatically. The bar graph shows the sales (in millions of dollars) of bicycle helmets for selected years in the period 1990–1995. (Source: U.S. Bicycle Federation.)

We can write a **function** to model the trend in helmet sales during this time, and we can graph the function in a **coordinate system.** Both the function and the graph can be used, for example, to estimate the helmet sales for 1991 or to predict sales after 1995. (For more on this topic, see Exercises 79–82 at the end of Section 2.4, and see the Chapter Project.)

In this chapter we introduce the coordinate plane for associating ordered pairs of real numbers with points. We learn how to graph an expression in the coordinate plane, and we introduce the basic features of the graphing calculator. The remainder of the chapter is devoted to the important topic of functions, including graphs, notation, evaluation, and analysis.

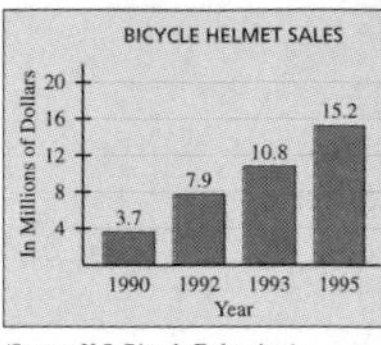

(Source: U.S. Bicycle Federation.)

65

Section Opener

Each section begins with a list of subsection titles that provides a brief outline of the material that follows.

96 CHAPTER 2 The Coordinate Plane and Functions

2.4 FUNCTIONS: NOTATION AND EVALUATION

Function Notation • Evaluating Functions

Function Notation

Consider the function defined by the equation $y = 2x + 1$. For convenience, we often use **function notation** to name the function and to indicate the value of the function.

$$f(x) = 2x + 1$$

The notation $f(x)$ is read "f of x." In this notation, the name of the function is f, the variable (domain element) is x, and the value of the function (range element) corresponding to x is $f(x)$.

Note: The parentheses in the notation $f(x)$ does *not* indicate multiplication. We read $f(x)$ as "f of x," *not* "f times x."

Just as we can use symbols other than x as a variable, we can use letters other than f to name a function. Each of the following defines the same function.

$$f(x) = x^2 - 3x + 4 \qquad g(t) = t^2 - 3t + 4 \qquad g(w) = w^2 - 3w + 4$$

Evaluating Functions

Determining the value of $f(x)$ for a specific value of x is called **evaluating the function.**

EXPLORING THE CONCEPT

Using function notation for evaluating functions usually helps to sell students on the value of the notation.

Point out that the x-coordinate is the value of the variable and the y-coordinate is the value of the function.

LEARNING TIP

Keep relating an expression and its value to points of a graph. If $f(2) = 6$, then (2, 6) is a point of the graph of f. As you study mathematics, go slowly and take the time to visualize.

Evaluating Functions

To indicate the value of the function $f(x) = 3x^2 - 5x + 4$ when x has a value of 2, we write $f(2)$. To evaluate $f(2)$, replace each occurrence of the variable with 2.

$$f(x) = 3x^2 - 5x + 4$$
$$f(2) = 3(2)^2 - 5(2) + 4 = 12 - 10 + 4 = 6$$

We also can use a graph to estimate the value of a function. Figure 2.34 shows the graph of $f(x) = 3x^2 - 5x + 4$ with the tracing cursor on (2, 6). Note that the first coordinate is the value of x, and the second coordinate is $f(2) = 6$.

Figure 2.34

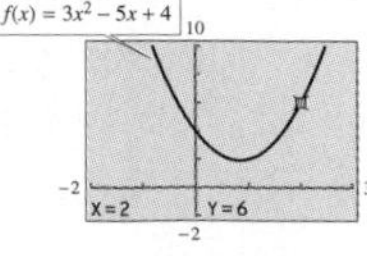

Modeling with Real Data

In accordance with the current standards of professional organizations such as AMATYC and NCTM, we have added an abundance of exercises in the general category of Modeling with Real Data. Students are taught how to organize and interpret data, how to model it with mathematical functions, and how to use such models to extrapolate and predict. Sourced data from a variety of subjects are used in dedicated blocks of modeling exercises, Group Projects, and Chapter Projects.

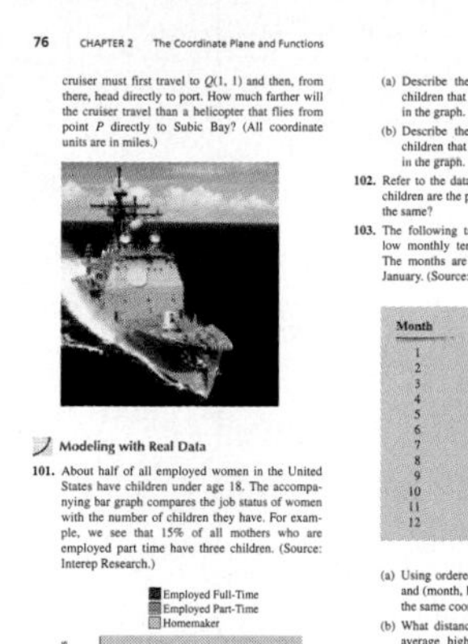

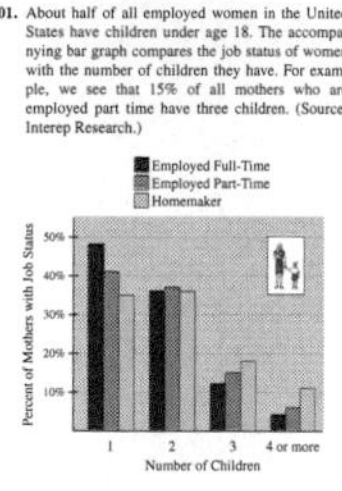

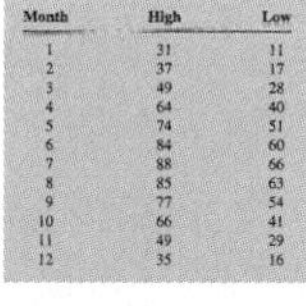

Month	High	Low
1	31	11
2	37	17
3	49	28
4	64	40
5	74	51
6	84	60
7	88	66
8	85	63
9	77	54
10	66	41
11	49	29
12	35	16

Real Data and Real Life

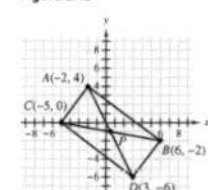

Real-Life Applications

The majority of application problems are written in the context of real-life knowledge and experience. Some sections are devoted exclusively to examples and exercises involving real-life applications, and most other sections contain dedicated blocks of such problems. Connecting mathematics with students' view of the world is in accordance with national standards and leads students to a better understanding of the practical nature of the discipline.

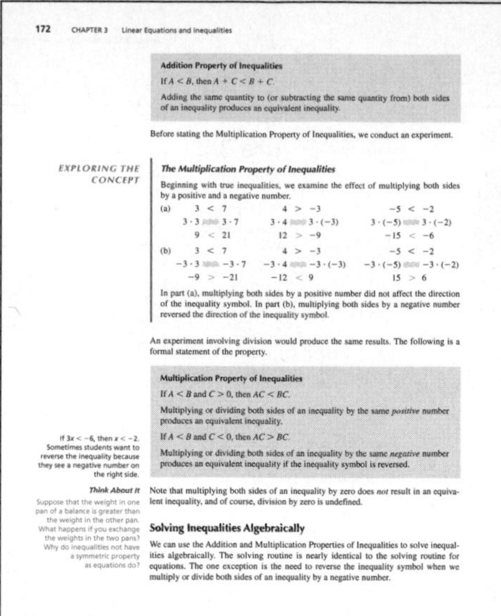

172 CHAPTER 3 Linear Equations and Inequalities

Addition Property of Inequalities

If $A < B$, then $A + C < B + C$.

Adding the same quantity to (or subtracting the same quantity from) both sides of an inequality produces an equivalent inequality.

Before stating the Multiplication Property of Inequalities, we conduct an experiment.

EXPLORING THE CONCEPT

The Multiplication Property of Inequalities

Beginning with true inequalities, we examine the effect of multiplying both sides by a positive and a negative number.

(a)	$3 < 7$	$4 > -3$	$-5 < -2$
	$3 \cdot 3 \;\square\; 3 \cdot 7$	$3 \cdot 4 \;\square\; 3 \cdot (-3)$	$3 \cdot (-5) \;\square\; 3 \cdot (-2)$
	$9 < 21$	$12 > -9$	$-15 < -6$
(b)	$3 < 7$	$4 > -3$	$-5 < -2$
	$-3 \cdot 3 \;\square\; -3 \cdot 7$	$-3 \cdot 4 \;\square\; -3 \cdot (-3)$	$-3 \cdot (-5) \;\square\; -3 \cdot (-2)$
	$-9 > -21$	$-12 < 9$	$15 > 6$

In part (a), multiplying both sides by a positive number did not affect the direction of the inequality symbol. In part (b), multiplying both sides by a negative number reversed the direction of the inequality symbol.

An experiment involving division would produce the same results. The following is a formal statement of the property.

Multiplication Property of Inequalities

If $A < B$ and $C > 0$, then $AC < BC$.

Multiplying or dividing both sides of an inequality by the same *positive* number produces an equivalent inequality.

If $A < B$ and $C < 0$, then $AC > BC$.

Multiplying or dividing both sides of an inequality by the same *negative* number produces an equivalent inequality if the inequality symbol is reversed.

If $3x < -6$, then $x < -2$. Sometimes students want to reverse the inequality because they see a negative number on the right side.

Think About It

Suppose that the weight in one pan of a balance is greater than the weight in the other pan. What happens if you exchange the weights in the two pans? Why do inequalities not have a symmetric property as equations do?

Note that multiplying both sides of an inequality by zero does *not* result in an equivalent inequality, and of course, division by zero is undefined.

Solving Inequalities Algebraically

We can use the Addition and Multiplication Properties of Inequalities to solve inequalities algebraically. The solving routine is nearly identical to the solving routine for equations. The one exception is the need to reverse the inequality symbol when we multiply or divide both sides of an inequality by a negative number.

Exploring the Concept

This new pedagogical device guides students from concrete experiences to generalizations and formal rules. Students become active participants in the learning process by experimenting and asking "what if" questions.

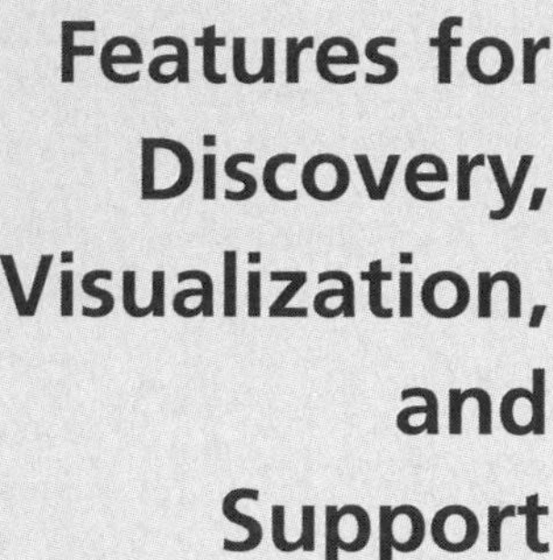

Features for Discovery, Visualization, and Support

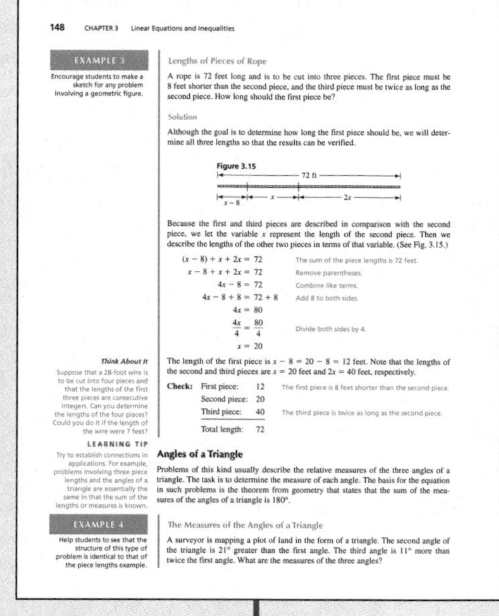

148 CHAPTER 3 Linear Equations and Inequalities

EXAMPLE 3

Encourage students to make a sketch for any problem involving a geometric figure.

Lengths of Pieces of Rope

A rope is 72 feet long and is to be cut into three pieces. The first piece must be 8 feet shorter than the second piece, and the third piece must be twice as long as the second piece. How long should the first piece be?

Solution

Although the goal is to determine how long the first piece should be, we will determine all three lengths so that the results can be verified.

Figure 3.15

Because the first and third pieces are described in comparison with the second piece, we let the variable x represent the length of the second piece. Then we describe the lengths of the other two pieces in terms of that variable. (See Fig. 3.15.)

$$(x - 8) + x + 2x = 72 \quad \text{The sum of the piece lengths is 72 feet.}$$
$$x - 8 + x + 2x = 72 \quad \text{Remove parentheses.}$$
$$4x - 8 = 72 \quad \text{Combine like terms.}$$
$$4x - 8 + 8 = 72 + 8 \quad \text{Add 8 to both sides.}$$
$$4x = 80$$
$$\frac{4x}{4} = \frac{80}{4} \quad \text{Divide both sides by 4.}$$
$$x = 20$$

Think About It

Suppose that a 28-foot wire is to be cut into four pieces and that the lengths of the first three pieces are consecutive integers. Can you determine the lengths of the four pieces? Could you do it if the length of the wire were 7 feet?

The length of the first piece is $x - 8 = 20 - 8 = 12$ feet. Note that the lengths of the second and third pieces are $x = 20$ feet and $2x = 40$ feet, respectively.

Check:

	First piece:	12	The first piece is 8 feet shorter than the second piece.
	Second piece:	20	
	Third piece:	40	The third piece is twice as long as the second piece.
	Total length:	72	

LEARNING TIP

Try to establish connections in applications. For example, problems involving three piece lengths and the angles of a triangle are essentially the same in that the sum of the lengths or measures is known.

Angles of a Triangle

Problems of this kind usually describe the relative measures of the three angles of a triangle. The task is to determine the measure of each angle. The basis for the equation in such problems is the theorem from geometry that states that the sum of the measures of the angles of a triangle is 180°.

EXAMPLE 4

Help students to see that the structure of this type of problem is identical to that of the piece lengths example.

The Measures of the Angles of a Triangle

A surveyor is mapping a plot of land in the form of a triangle. The second angle of the triangle is 21° greater than the first angle. The third angle is 11° more than twice the first angle. What are the measures of the three angles?

Examples

All sections contain numerous, titled Examples, many with multiple parts graded by difficulty. These Examples illustrate concepts, procedures, and techniques, and they reinforce the reasoning and critical thinking needed for problem solving. Detailed solutions include helpful comments that justify the steps taken and explain their purpose.

Features for Discovery, Visualization, and Support

Figure 2.19

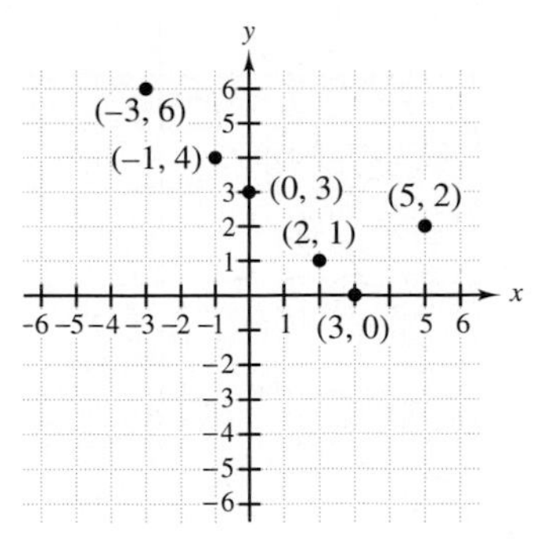

Figure 3.1

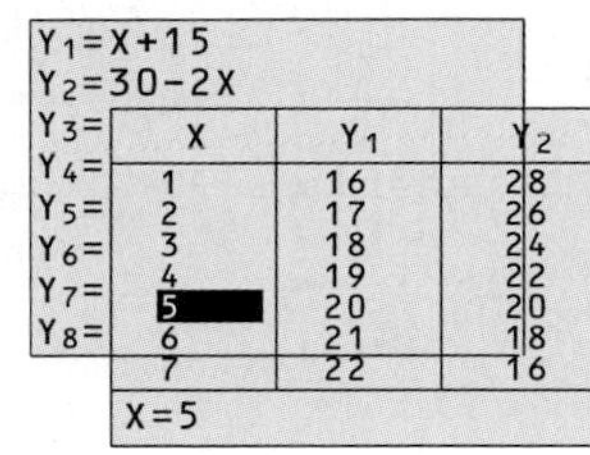

Figure 3.2

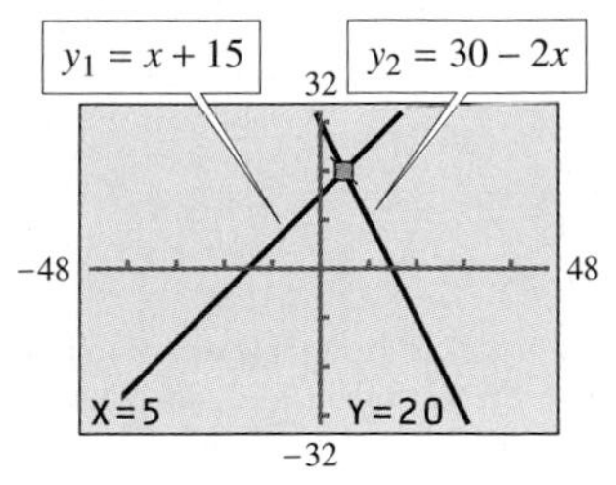

Graphs

Both traditional and calculator graphs are used throughout the exposition and exercises to assist students in visualizing concepts. Calculator displays are representative and are intended to resemble what students typically obtain on their own calculators.

Decimal Usually you can set the number of decimal places that you want your calculator to display.

The set I of decimal numbers that neither terminate nor repeat is called the set of **irrational numbers.** Examples of irrational numbers are π and most square roots, such as $\sqrt{6}$, $\sqrt{15}$, and $-\sqrt{7}$. Note that $\sqrt{25} = 5$, and so $\sqrt{25}$ is a rational number.

Square Root Most calculators have a key for calculating square roots.

The rational numbers and the irrational numbers are two distinct sets of numbers. Taken together, the two sets form the set **R** of **real numbers.** We can think of the real numbers as all numbers with decimal representations. Figure 1.1 shows the relationships of the various sets of numbers that we have discussed.

Key Words

We indicate the appropriate use of a calculator with a Key Word and a short description of the pertinent calculator function. These Key Words appear at the initial point of use. Each Key Word references the accompanying *Graphing Calculator Keystroke Guide,* where specific keystroke information for several popular calculator models can be found, including the TI-83. Selected keys from typical graphing calculators are inside the back cover.

Function Notation

Consider the function defined by the equation $y = 2x + 1$. For convenience, we often use **function notation** to name the function and to indicate the value of the function.

$$f(x) = 2x + 1$$

The notation $f(x)$ is read "f of x." In this notation, the name of the function is f, the variable (domain element) is x, and the value of the function (range element) corresponding to x is $f(x)$.

Note: The parentheses in the notation $f(x)$ does *not* indicate multiplication. We read $f(x)$ as "f of x," *not* "f times x."

Just as we can use symbols other than x as a variable, we can use letters other than f to name a function. Each of the following defines the same function.

$$f(x) = x^2 - 3x + 4 \qquad g(t) = t^2 - 3t + 4 \qquad g(w) = w^2 - 3w + 4$$

Notes

Special remarks and cautionary notes that offer additional insight appear throughout the text.

In Chapter 1 we briefly considered equations and methods for testing whether numbers were solutions of equations. Now we present a more formal treatment of these topics along with some related vocabulary.

Definition of Equation

An **equation** is a statement that two algebraic expressions have the same value.

The expressions to the left and right of the equality symbol are called the *left side* and

15 ← right side

variable, we try to find replacements for that variable that ... call such replacements **solutions** of the equation.

Note: Because a table or a graph may or may not reveal an exact solution, we generally refer to solutions obtained in these ways as *estimates*. All solutions should be verified by substitution.

Here is a summary of the graphing approach to solving an equation.

Equation Solving: The Graphing Method

1. Graph y_1 = left side of the original equation.
2. Graph y_2 = right side of the original equation.
3. Trace to the point of intersection.
4. The x-coordinate of that point is the solution of the equation.
5. The y-coordinate is the value of both the left side and the right side of the original equation when x is replaced with the solution.

Definitions, Properties, and Procedures

Important definitions, properties, and procedures are shaded and titled for easy reference.

LEARNING TIP

Consider writing a word equation before you attempt to write an algebraic equation. For example, if you know or can represent the three quantities, then writing "original price − discount = sale price" gives you a guide for writing the actual equation.

When an item is placed on sale, the amount by which the price is reduced is called the **discount,** which is expressed as a percentage of the **original price.**

Retail store owners buy their goods at **wholesale prices.** Then, to cover overhead and the profit they want to make, they add an amount called **markup,** which is expressed as a percentage of the wholesale price. The result is the **retail price.**

EXAMPLE 2

Wholesale Price Plus Markup Equals Retail Price

All the goods at a hobby store are marked up 40%. If the retail price of one item is \$32.13, what was the wholesale cost to the store owner?

Help the students translate this information into equations.

Original price − discount = sale price

Wholesale price + markup = retail price

Solution

Let w = the wholesale cost.

$\text{Markup} = 0.40w$ — Markup = 40% of wholesale cost

$w + 0.40w = 32.13$ — Wholesale cost + markup = retail price

$1.00w + 0.40w = 32.13$

$1.40w = 32.13$

$\frac{1.40w}{1.40} = \frac{32.13}{1.40}$

$w = 22.95$

The wholesale cost was \$22.95.

Think About It

Suppose that the wholesale price of an item is w and that the retail price reflects a 30% markup. If the item is later sold at a 20% discount, will the seller make money or lose money?

Fixed and Variable Costs

The cost of doing business can be separated into two components. One is the **fixed cost,** which is a cost that is incurred even if no products are sold or no service is rendered. The other component is the **variable cost,** which is a cost that depends on the number of products produced or sold or on the amount of service rendered.

Learning Tip

Every section now has at least one Learning Tip that offers students helpful strategies and alternative ways of thinking about concepts.

Think About It

New to the Second Edition, Think About It is a question or series of questions that requires critical thinking and reasoning in order to broaden and extend concepts. These questions are designed to spark students' imagination and interest and appear in the margin of each section. Answers appear in the Instructor's Annotated Edition.

End-of-Section Features

A typical section ends with Quick Reference and Section Exercises.

3.2 QUICK REFERENCE

Properties of Equations

- **Equivalent equations** are equations that have exactly the same solution sets.
- An equivalent equation results when
  1. the same quantity is added to or subtracted from both sides of an equation (Addition Property of Equations).
  2. both sides of an equation are multiplied or divided by the same nonzero quantity (Multiplication Property of Equations).
  3. the two sides of an equation are swapped (Symmetric Property of Equations).
  4. an expression in an equation is replaced with an equivalent expression (Substitution Property of Equations).

Applying the Properties

- The Addition Property of Equations can be used to eliminate a variable term from one side of an equation and to isolate a variable term.
- The Multiplication Property of Equations can be used to isolate a variable by eliminating the coefficient of the isolated variable term.

Quick Reference

Quick Reference appears at the end of all sections except those dealing exclusively with applications. These detailed summaries of the important rules, properties, and procedures are grouped by subsection for a handy reference and review tool.

Exercises

A typical exercise set in each section includes exercises from each of the following groups: Concepts and Skills (including Writing and Concept Extension), Real-Life Applications and Modeling with Real Data, Group Project, and Challenge.

Concepts and Skills

Most exercise sets begin with the basic skills and concepts discussed within the text. These include Writing Exercises, designed to help students gain confidence in their ability to communicate, and Concept Extension exercises, which go slightly beyond the text examples. The Concept Extension exercises are identified in the Instructor's Annotated as CE.

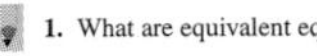

3.2 EXERCISES

Concepts and Skills

1. What are equivalent equations?
2. What is an important advantage of the algebraic method over the graphing method for solving equations?
3. Consider the following two equations.
   (i) $x + 2 = 3$ (ii) $-2x = 6$
   When we solve these equations, explain why we *add* -2 to both sides of the first equation and *divide* both sides by -2 in the second equation.
4. The following are possible first steps in solving the equation $\frac{2}{3}x = 10$. Explain why each approach is correct.
   (i) Multiply both sides by $\frac{3}{2}$.
   (ii) Divide both sides by $\frac{2}{3}$.
   (iii) Multiply both sides by 3 and then divide both sides by 2.

In Exercises 5–20, solve the equation algebraically and verify the solution. For special cases, write the solution set.

5. $5x + 8 = 23$
6. $4n - 7 = -35$
7. $7x - 5 = 8x + 7$
8. $7 + 8x = 4x - 13$
9. $3(3 - 2x) = 33 - 2x$
10. $2(3y + 8) = -2(2 - y)$
11. $7w - 5 = 11w - 5 - 4w$
12. $9w + 7 = 16w + 7(1 - w)$
13. $9x + 5 = 5x + 3(x - 1)$
14. $11x + 2(4 - 3x) = 3(3 - x) + 15$
15. $16 + 7(6 - x) = 15 - 4(x + 2)$
16. $7 - 4(t - 3) = 6t - 3(t + 3)$
17. $2(3a + 2) = 2(a + 1) + 4a$
18. $8t - 7 - 3t = 5t + 2$
19. $-2(x + 5) = 5(1 - x) + 3(7 - x)$
20. $2y - 3(2y - 3) = 2y - 5(3y - 4)$

In Exercises 5–20, encourage the use of a calculator for verifying solutions.

21. Consider solving the equation $\frac{1}{3} - 2x = \frac{2}{5}$. To clear fractions, we multiply both sides by 15. Why is it necessary to multiply $2x$ by 15?
22. What is the solution set of $3x = 2x$? Now divide both sides of $3x = 2x$ by x. What is the solution set of the resulting equation? Why are the equations not equivalent?

In Exercises 23–32, solve the given equation and use your calculator to verify the solution.

23. $2x - \frac{3}{4} = -\frac{5}{6}$
24. $\frac{2}{15}y + \frac{3}{5} = 2 - \frac{2}{3}y$
25. $\frac{y}{2} + \frac{y}{4} = \frac{7}{8} - \frac{y}{8}$
26. $\frac{5 + 3x}{2} + 7 = 2x$
27. $\frac{5}{4}(x + 2) = \frac{x}{2}$
28. $\frac{1}{4}(x - 4) = \frac{1}{3}(x + 6)$
29. $x + \frac{3x - 1}{9} = 4 + \frac{3x + 1}{3}$
30. $\frac{2x}{5} - \frac{2x - 1}{2} = \frac{x}{5}$
31. $\frac{1}{4} + \frac{1}{6}(4a + 5) = 2(a - 3) + \frac{5}{12}$
32. $\frac{5}{6}(1 - t) - \frac{t}{2} = -\frac{1}{3}(1 - 3t)$

Using a calculator to verify solutions is particularly helpful for more complicated equations.

In Exercises 33–36, solve the given equation and use your calculator to verify the solution.

33. $2.6 = 0.4z + 1$
34. $1 - 0.3c = -0.5$
35. $0.08x + 0.15(x + 200) = 7$
36. $0.6(2x + 1) - 0.4(x - 2) = 1$

CE In Exercises 37–42, determine whether the given equations are equivalent equations.

37. $x + 4 = 7$ and $3x - 5 = 4$
38. $\frac{1}{3}x = 2$ and $x + 3 = 9$
39. $x + 3 = 4 + x$ and $2x - 3 = 4 + 2x$
40. $x + 3 = 3 + x$ and $2x - 5 = -(5 - 2x)$

End-of-Section Features

Group Project

Group Projects now appear in the exercises for many sections. These series of exercises focus on real data and allow students to work together to solve problems. An index of Group Projects is inside the back cover.

Group Project: ATM Cards

Research indicates that about 60% of American adults have automatic teller machine (ATM) cards. The accompanying pie chart shows the percentages of those cardholders who use their cards a certain number of times per month. (Source: Research Partnership.)

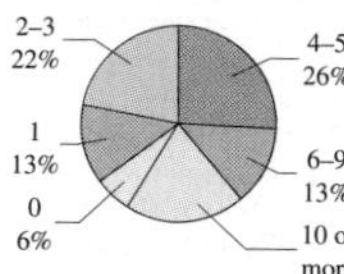

91. What percent of the cardholders use their cards fewer than four times per month?

92. What percent of the cardholders use their cards at least ten times per month?

93. Why does the data not allow you to determine the actual number of cardholders in any of the categories?

94. Bank fees for ATM use have increased dramatically. Do you think that higher fees might affect the percentages shown in the pie chart? If so, in what way?

Graphing Calculator Icon

Exercises best completed with a graphing calculator are now identified by icon in the Instructor's Annotated Edition. The icons do not appear in the student text to allow instructors flexibility in planning assignments and to help students determine appropriate use.

69. If the population continues to grow according to the function P, what is the projected population in the year 2000?

70. Does the shape of the graph suggest that the gray wolf population will increase indefinitely? What natural factors might change the shape of the graph in the future?

Challenge

In Exercises 71–76, use the integer setting to graph each function. Then write the set of x-values for which

(a) $y \le 0$. (b) $y \ge 0$.

71. $y = x - 3$ **72.** $y = x + 4$

73. $y = x^2 - 9$ **74.** $y = 9 - x^2$

75. $y = \sqrt{x}$ **76.** $y = |x| + 1$

77. (a) Draw a graph of the following piece-wise function.

$$f(x) = \begin{cases} x + 3, & x < 0 \\ 5, & x = 0 \\ -x, & x > 0 \end{cases}$$

(b) What is the absolute maximum of the function?

(c) In words, how would you describe your estimate of the range of the function?

78. Consider the **constant function** $g(x) = 5$.

(a) What are the domain and range of g?

(b) We say that a function f is an **increasing function** if $f(x_2) \ge f(x_1)$ when $x_2 > x_1$ for all x in the domain of f. Explain why function g is an increasing function.

Challenge

These problems appear at the end of most exercise sets and offer more challenging work than the standard and Concept Extension problems.

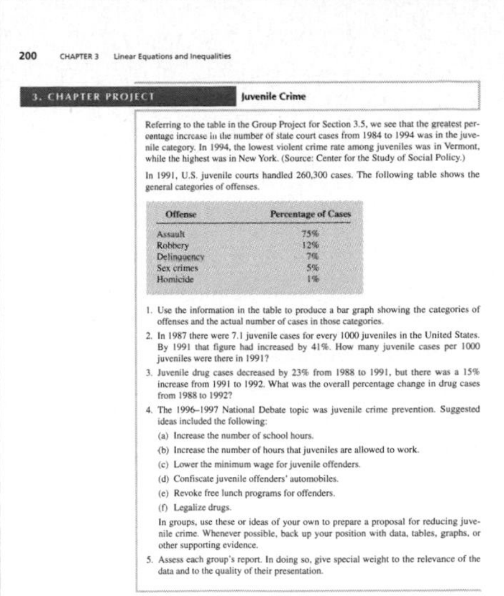
200 CHAPTER 3 Linear Equations and Inequalities

3. CHAPTER PROJECT **Juvenile Crime**

Referring to the table in the Group Project for Section 3.5, we see that the greatest percentage increase in the number of state court cases from 1984 to 1994 was in the juvenile category. In 1994, the lowest violent crime rate among juveniles was in Vermont, while the highest was in New York. (Source: Center for the Study of Social Policy.)

In 1991, U.S. juvenile courts handled 260,300 cases. The following table shows the general categories of offenses.

Offense	Percentage of Cases
Assault	75%
Robbery	12%
Delinquency	7%
Sex crimes	5%
Homicide	1%

1. Use the information in the table to produce a bar graph showing the categories of offenses and the actual number of cases in those categories.
2. In 1987 there were 7.1 juvenile cases for every 1000 juveniles in the United States. By 1991 that figure had increased by 41%. How many juvenile cases per 1000 juveniles were there in 1991?
3. Juvenile drug cases decreased by 23% from 1988 to 1991, but there was a 15% increase from 1991 to 1992. What was the overall percentage change in drug cases from 1988 to 1992?
4. The 1996–1997 National Debate topic was juvenile crime prevention. Suggested ideas included the following:
   (a) Increase the number of school hours.
   (b) Increase the number of hours that juveniles are allowed to work.
   (c) Lower the minimum wage for juvenile offenders.
   (d) Confiscate juvenile offenders' automobiles.
   (e) Revoke free lunch programs for offenders.
   (f) Legalize drugs.
   In groups, use these or ideas of your own to prepare a proposal for reducing juvenile crime. Whenever possible, back up your position with data, tables, graphs, or other supporting evidence.
5. Assess each group's report. In doing so, give special weight to the relevance of the data and to the quality of their presentation.

Chapter Project

Each Chapter Project is the culmination of the real-data application introduced in the chapter opener and continued in a Group Project. These more in-depth projects emphasize the organization, modeling, and interpretation of data, and they provide opportunities for students to report their findings. Topics are listed in the Contents. Answers appear in the Instructor's Annotated Edition.

Chapter Review Exercises

Each chapter ends with a set of review exercises. These exercises include helpful section references that direct students to the appropriate sections for review. The answers to the odd-numbered review exercises are included at the back of the text.

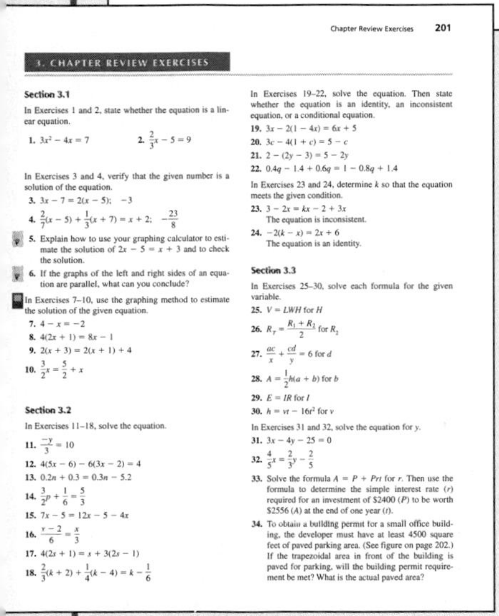

End-of-Chapter Features

At the end of each chapter, these features appear in the following order: Chapter Project, Chapter Review Exercises, Looking Ahead (except Chapter 12), Chapter Test, Cumulative Test (at the end of selected chapters).

Looking Ahead

New to the second edition, this short list of review exercises focuses on previously discussed skills and concepts that will be needed in the upcoming chapter. Answers are included in the Instructor's Annotated Edition.

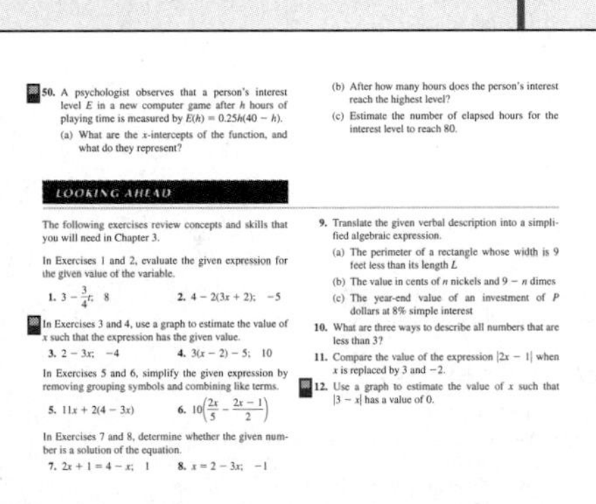
50. A psychologist observes that a person's interest level E in a new computer game after h hours of playing time is measured by $E(h) = 0.25h(40 - h)$.
   (a) What are the x-intercepts of the function, and what do they represent?
   (b) After how many hours does the person's interest reach the highest level?
   (c) Estimate the number of elapsed hours for the interest level to reach 80.

LOOKING AHEAD

The following exercises review concepts and skills that you will need in Chapter 3.

In Exercises 1 and 2, evaluate the given expression for the given value of the variable.

1. $3 - \frac{3}{4}t$; 8
2. $4 - 2(3x + 2)$; -5

In Exercises 3 and 4, use a graph to estimate the value of x such that the expression has the given value.

3. $2 - 3x$; -4
4. $3(x - 2) - 5$; 10

In Exercises 5 and 6, simplify the given expression by removing grouping symbols and combining like terms.

5. $11x + 2(4 - 3x)$
6. $10\left(\frac{2x}{5} - \frac{2x - 1}{2}\right)$

In Exercises 7 and 8, determine whether the given number is a solution of the equation.

7. $2x + 1 = 4 - x$; 1
8. $x = 2 - 3x$; -1

9. Translate the given verbal description into a simplified algebraic expression.
   (a) The perimeter of a rectangle whose width is 9 feet less than its length L
   (b) The value in cents of n nickels and $9 - n$ dimes
   (c) The year-end value of an investment of P dollars at 8% simple interest
10. What are three ways to describe all numbers that are less than 3?
11. Compare the value of the expression $|2x - 1|$ when x is replaced by 3 and -2.
12. Use a graph to estimate the value of x such that $|3 - x|$ has a value of 0.

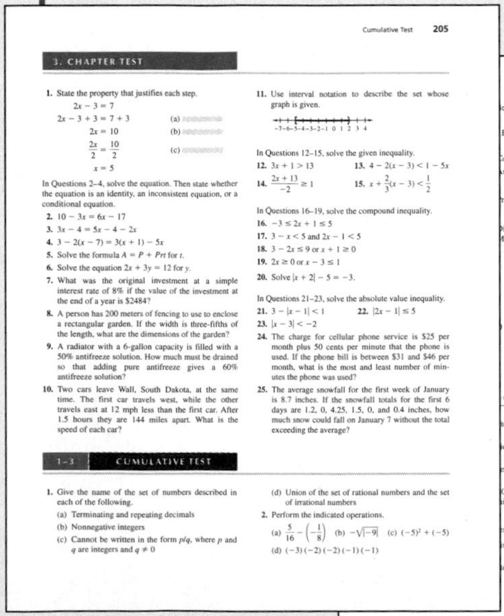

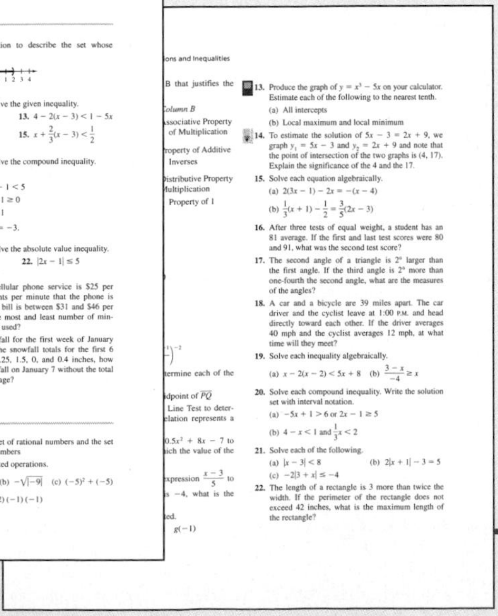

Chapter Test

A Chapter Test follows each chapter review. The answers to all the test questions, with the appropriate section references, are included at the back of the text.

Cumulative Test

A Cumulative Test appears at the end of Chapters 3, 5, 7, 10 and 12. The answers to all the test questions, with section references, are included at the back of the text.

Chapter 1

The Real Number System

In 1996 the American space shuttle docked with the Russian space station Mir. Shannon Lucid, an American astronaut, stayed on Mir, where she set an endurance record in space.

The mathematics of space exploration typically involves very large numbers, which can be represented with **exponents** and **scientific notation**. Equally large are the costs of NASA programs. The accompanying bar graph shows the proposed budget cuts for 1996–2000. (For more on this topic, see Exercises 95–98 at the end of Section 1.7, and see the Chapter Project.)

Chapter 1 begins with a review of the real numbers, the operations that we perform with them, and the properties of the real numbers. Next, we learn how to evaluate and simplify algebraic expressions. Finally, we introduce integer exponents and the basic rules for working with exponential expressions.

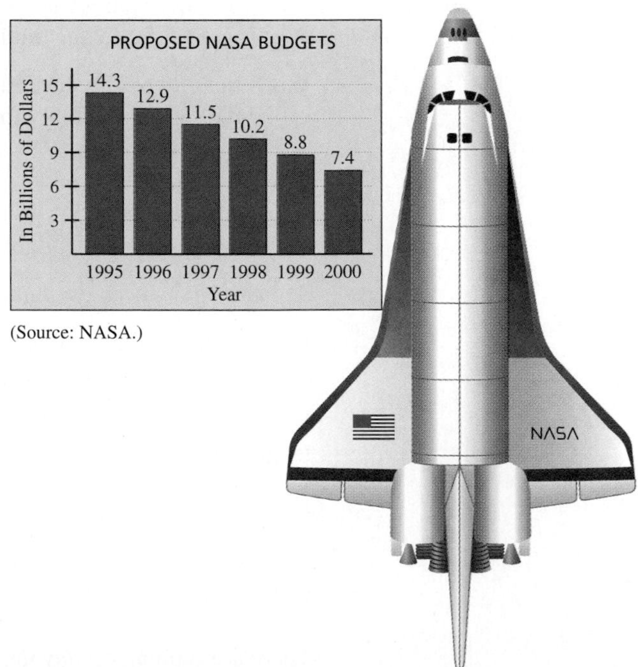

(Source: NASA.)

1.1 THE REAL NUMBERS

Structure of the Real Numbers • The Number Line • Order of the Real Numbers • Opposites and Absolute Value

Structure of the Real Numbers

It is convenient to organize the numbers of mathematics into sets. A **set** is *any* collection of objects, but our focus will be on sets of numbers.

We use braces to enclose the **members** or **elements** of a set, and we usually assign a letter name to the set. For example, the set N of **natural,** or **counting, numbers,** is written as

$$N = \{1, 2, 3, 4, \ldots\}$$

The dots . . . mean that the list continues without end. Because N has infinitely many elements, it is an example of an **infinite set.**

A set with no elements is called the **empty set** and is named Ø.

Note: The set $\{0\}$, which contains *one element*, 0, is not the same as the empty set, which contains *no elements*.

The number of elements in $A = \{2, 4, 6, 8\}$ is the counting number 4. If the number of elements of a set is 0 (empty set) or a counting number, the set is a **finite set.** To indicate that 8 is an element of set A, we write $8 \in A$.

We wrote sets N and A by listing all the elements of the set. This method of describing a set is called the **roster method.**

When the number 0 is included with the natural numbers, the collection of numbers is called the set W of **whole numbers.**

$$W = \{0, 1, 2, 3, \ldots\}$$

A **variable** is a symbol, usually a letter, that represents an unknown number. In the following, we use the variable n to describe set A in **set-builder notation.**

$$A = \{n \mid n \text{ is an even number between 2 and 8, inclusive}\}$$

In this context, the symbol | means "such that." Thus we say "the set of all numbers n such that n is an even number between 2 and 8, inclusive."

When the temperature falls to 7° below zero, we use a negative number to describe the temperature: $-7°$. The whole numbers and their negatives form the set J of **integers.**

$$J = \{\ldots, -3, -2, -1, 0, 1, 2, 3, \ldots\}$$

Negative

On some calculators you must use a negative key (rather than the minus key) for entering negative numbers.

There are infinitely many numbers that are not integers. For example, fractions such as

$\frac{1}{2}$, $\frac{2}{3}$, and $-\frac{3}{8}$ are called **rational numbers.** We write the set Q of rational numbers with set-builder notation.

$$Q = \left\{ \frac{p}{q} \mid p \text{ and } q \text{ are integers, } q \neq 0 \right\}$$

Note that the integers are also rational numbers because we can write any integer n as $\frac{n}{1}$.

It can be shown that the decimal representation of any rational number is either a terminating or a repeating decimal. For example, $\frac{3}{4} = 0.75$, which is a terminating decimal, and $\frac{3}{11} = 0.272727\ldots$, which is a repeating decimal, usually written $0.\overline{27}$.

Decimal

Usually you can set the number of decimal places that you want your calculator to display.

The set I of decimal numbers that neither terminate nor repeat is called the set of **irrational numbers.** Examples of irrational numbers are π and most square roots, such as $\sqrt{6}$, $\sqrt{15}$, and $-\sqrt{7}$. Note that $\sqrt{25} = 5$, and so $\sqrt{25}$ is a rational number.

Square Root

Most calculators have a key for calculating square roots.

The rational numbers and the irrational numbers are two distinct sets of numbers. Taken together, the two sets form the set **R** of **real numbers.** We can think of the real numbers as all numbers with decimal representations. Figure 1.1 shows the relationships of the various sets of numbers that we have discussed.

Figure 1.1

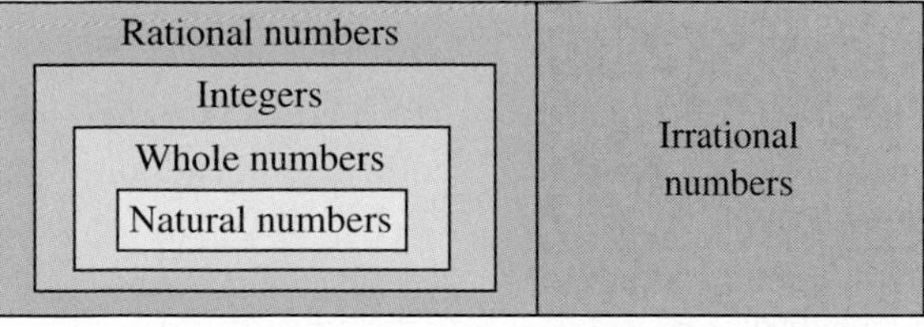

EXAMPLE 1

Classifying the Real Numbers

Indicate every set to which the following real numbers belong.

	3.2	0	−4	$3\frac{5}{6}$	$-\frac{3}{4}$	$\sqrt{9}$	$\sqrt{6}$
Irrational							✓
Rational	✓	✓	✓	✓	✓	✓	
Integer		✓	✓			✓	
Whole		✓				✓	
Natural						✓	

The Number Line

We use a **number line** to visualize real numbers and their relation to each other.

To construct a number line, we choose a point corresponding to the number 0. Points at equally spaced intervals are then associated with the integers. The positive integers are to the right of 0, and the negative integers are to the left of 0. All other real numbers are associated with intervening points.

The number associated with a point is called the **coordinate** of that point. The point associated with 0 is called the **origin.**

Using a number line to highlight points corresponding to the numbers in a set is called **plotting** the points. The resulting set of plotted points is called the **graph** of the set of numbers.

EXAMPLE 2

Graphing a Set of Real Numbers

Graph the set $\{-4, -1.5, 2, \pi, 4\frac{1}{2}, \sqrt{36}\}$.

Solution

Figure 1.2

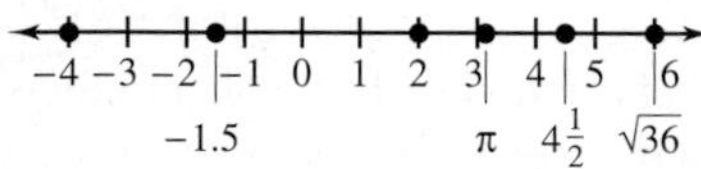

Order of the Real Numbers

We can use a number line to represent the **order** of the real numbers.

LEARNING TIP

Visualizing or drawing a number line is especially helpful when you work with inequalities. For example, by observing that −4.5 is to the left of −4 on the number line, you can easily conclude that $-4.5 < -4$.

If two numbers a and b are associated with the same point, we say that the numbers are **equal,** and we write $a = b$.

The **inequality** symbols $\neq$, $<$, and $>$ indicate that two numbers are not equal. If a number a is to the left of another number b on the number line, we say that a is **less than** b, and we write $a < b$. Alternatively, we say that b is **greater than** a, and we write $b > a$. Writing $a \neq b$, which is read "a is not equal to b," indicates that either $a > b$ or $a < b$.

Test

If you enter an equation on your home screen, the reported value is 1 if the equation is true and 0 if the equation is false. Inequalities can be tested in the same way.

EXAMPLE 3

Writing Inequalities

Insert < or > to make the statement true.

(a) $0 \quad \square \quad -5$ (b) $\frac{1}{3} \quad \square \quad \frac{1}{4}$ (c) $-7 \quad \square \quad -6$

Solution

(a) $0 > -5$ (b) $\frac{1}{3} > \frac{1}{4}$ (c) $-7 < -6$

Inequality symbols can be combined with the equality symbol. The symbol $\leq$ means "less than or equal to," and the symbol $\geq$ means "greater than or equal to."

To describe all the numbers in the set that are less than or equal to 2, we write $\{x \mid x \leq 2\}$. Figure 1.3 shows the graph of this set.

Figure 1.3

$x \leq 2$

−4 −3 −2 −1 0 1 2 3 4

We use a right bracket to indicate that 2 is included in the set. The same notation is used to write the set in **interval notation:** $(-\infty, 2]$. The symbol $-\infty$ indicates that the interval extends to the left without end.

Note: The infinity symbols ∞ and $-\infty$ do not represent numbers. They are used with parentheses in interval notation to indicate that the interval extends without end.

Figure 1.4

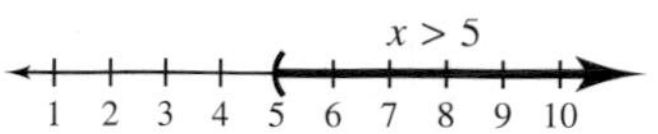

Note the difference between the graph in Fig. 1.3 and the graph of $\{x \mid x > 5\}$ in Fig. 1.4. We use a left parenthesis to indicate that 5 is not included in the set. The corresponding interval notation is $(5, \infty)$. Here, the infinity symbol ∞ indicates that the interval extends to the right without end.

Think About It

What do you think $(-\infty, \infty)$ represents?

EXAMPLE 4

Graphs and Interval Notation

Graph the given set and write the set in interval notation.

(a) $\{x \mid 0 < x < 7\}$ (b) $\{x \mid -3 < x \leq 5\}$

Solution

(a) We use parentheses to indicate that neither 0 nor 7 is included in the set. In interval notation, the set is $(0, 7)$. (See Fig. 1.5.)

(b) We use a left parenthesis at -3 and a right bracket at 5. The interval notation is $(-3, 5]$. (See Fig. 1.6.)

Figure 1.5

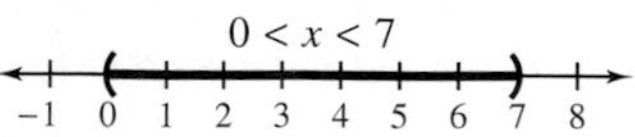

Figure 1.6

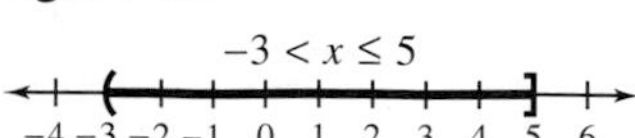

Opposites and Absolute Value

On the number line, two different numbers that are on opposite sides of the origin and that are the same distance from the origin are called **opposites** (or **additive inverses**). The numbers 4 and -4 are examples of opposites. (See Fig. 1.7.)

Figure 1.7

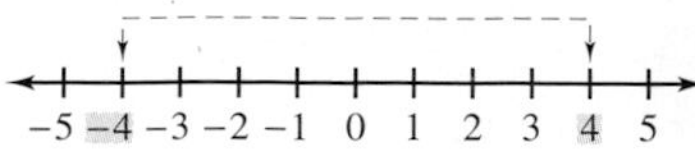

So that every real number has an opposite, we shall agree that the opposite of 0 is 0. For any real number x, we use the symbol $-x$ to represent "the opposite of x."

Note: It is incorrect to read $-x$ as "minus x" because "minus" refers to subtraction. It is also incorrect to read $-x$ as "negative x" because $-x$ does not necessarily represent a negative number. The symbol $-x$ refers to the *opposite* of the number represented by x.

Figure 1.7 also shows that the opposite of -4 is 4. We write this fact as $-(-4) = 4$, and we read the statement, "The opposite of negative 4 is 4."

The Opposite of the Opposite of a Number

For any real number a, $-(-a) = a$.

On the number line, the distance between any number and 0 is called the **absolute value** of the number. For a real number a, we denote its absolute value by $|a|$.

 Absolute Value

The typical key for calculating absolute value is ABS.

We also can state an algebraic definition of absolute value.

Algebraic Definition of Absolute Value

For any real number a, $|a| = \begin{cases} a & \text{if } a \geq 0 \\ -a & \text{if } a < 0 \end{cases}$

Note: In the definition of absolute value, if $a < 0$, then $-a$ represents a *positive* number. The absolute value of a number is never negative.

A **numerical expression** is any combination of numbers and arithmetic operations. When we **evaluate** a numerical expression, we determine the value of the expression by performing the indicated operations.

EXAMPLE 5

Evaluating Absolute Value Expressions

(a) $|10| = 10$

(b) $|-5| = 5$

(c) $-|8| = -8$

(d) $-|-7| = -7$

(e) $|8| + |-7| = 8 + 7 = 15$

(f) $|-12| - |-9| = 12 - 9 = 3$

1.1 QUICK REFERENCE

Structure of the Real Numbers

- A **set** is any collection of objects, such as numbers. The objects are called **members** or **elements** of the set.
- The set $N = \{1, 2, 3, \ldots\}$ is called the set of **natural,** or **counting, numbers.**

- A set with no elements is called the **empty set** $\emptyset$. If the number of elements in a set is 0 or a counting number, the set is a **finite set.** If a set has infinitely many elements, it is an **infinite set.**
- The set W of **whole numbers** is $W = \{0, 1, 2, 3, \dots\}$.
- Listing the elements of a set is called the **roster method.**
- A **variable** is a symbol that represents an unknown number. A variable is used to write a set in **set-builder notation.**
- The set J of **integers** is $J = \{\dots, -3, -2, -1, 0, 1, 2, 3, \dots\}$.
- The set Q of **rational numbers** is $Q = \left\{\frac{p}{q} \mid p \text{ and } q \text{ are integers, } q \neq 0\right\}$. The decimal representation of a rational number is either a terminating or repeating decimal.
- The set I of decimal numbers that neither terminate nor repeat is called the set of **irrational numbers.**
- The set **R** of **real numbers** is the set of rational numbers together with the set of irrational numbers.

The Number Line

- Each point of a **number line** is associated with a real number, called the **coordinate** of the point. The point whose coordinate is 0 is the **origin.**
- Using a number line to highlight points corresponding to the numbers in a set is called **plotting** the points. The resulting set of plotted points is called the **graph** of the set of numbers.

Order of the Real Numbers

- Two numbers a and b that are associated with the same point of a number line are **equal,** and we write $a = b$. Writing $a \neq b$ indicates that a and b are not equal.
- **Inequality** symbols $<$ (*less than*) and $>$ (*greater than*) indicate that two numbers are not equal. We also use the combined symbols $\leq$ (*less than or equal to*) and $\geq$ (*greater than or equal to*).
- The real numbers that satisfy a given inequality can be described with set-builder notation, with **interval notation,** or with a number line graph.

Opposite and Absolute Value

- On a number line, two different numbers that are on opposite sides of 0 and the same distance from 0 are called **opposites** (or **additive inverses**).
- The opposite of any real number x is denoted by $-x$.
- For any real number a, $-(-a) = a$.
- On the number line, the distance between any number and 0 is called the **absolute value** of the number. For a real number a, we denote its absolute value by $|a|$.
- For any real number a, $|a| = \begin{cases} a & \text{if } a \geq 0 \\ -a & \text{if } a < 0 \end{cases}$
- A **numerical expression** is any combination of numbers and arithmetic operations. When we **evaluate** a numerical expression, we determine the value of the expression by performing the indicated operations.

1.1 EXERCISES

Concepts and Skills

Use the following for all the exercises in this section.

$N = \{\text{natural numbers}\}$

$W = \{\text{whole numbers}\}$

$J = \{\text{integers}\}$

$Q = \{\text{rational numbers}\}$

$I = \{\text{irrational numbers}\}$

$\mathbf{R} = \{\text{real numbers}\}$

1. Write a sentence to describe the decimal name of a rational number.

2. Write a sentence to describe the decimal name of an irrational number.

In Exercises 3–6, state whether the given number has a terminating or repeating decimal name.

3. $\frac{3}{8}$ **4.** $-2\frac{23}{100}$ **5.** $\frac{35}{99}$ **6.** $\frac{5}{11}$

In Exercises 7–14, determine whether the statement is true or false.

7. $-7 \in N$ **8.** $-12 \in J$

9. $-\frac{7}{4} \in Q$ **10.** $\sqrt{5} \in Q$

11. $\sqrt{7} \in I$ **12.** $0 \in W$

13. $0 \in I$ **14.** $\frac{\pi}{2} \in I$

In Exercises 15–18, use the roster method to write the given set.

15. $\{w \mid w \in W \text{ and } w \text{ is less than } 7\}$

16. $\{i \mid i \in J \text{ and } i \text{ is between } -3 \text{ and } 3\text{, inclusive}\}$

17. $\{x \mid x \in W \text{ and } 0 < x < 1\}$

18. $\{n \mid n \in N \text{ and } n \leq 1\}$

In Exercises 19–22, describe, in words, numbers that satisfy the given conditions.

19. Integers that are not whole numbers

20. Rational numbers that are not integers

21. Rational numbers that are integers

22. Integers that are neither negative nor natural numbers

In Exercises 23–28, determine whether the statement is true or false.

23. Every integer is also a whole number.

24. Every integer is a rational number.

25. Every whole number is also an integer.

26. Every natural number is also a whole number.

27. Every nonterminating decimal repeats.

28. The number 0 is a whole number but not an integer.

29. Explain the difference between 1.75 and $1.\overline{75}$.

30. Explain the difference between the graphs of $x > 2$ and $x \geq 2$.

In Exercises 31–38, write the given number in decimal form. What does the decimal name suggest as to whether the number is rational or irrational?

31. $\frac{35}{37}$ **32.** $\frac{8}{11}$ **33.** $-\frac{\pi}{3}$ **34.** $\frac{\pi}{4}$

35. $\frac{\sqrt{2}}{\sqrt{8}}$ **36.** $-\frac{\sqrt{3}}{\sqrt{27}}$ **37.** $-\frac{\sqrt{5}}{2}$ **38.** $\frac{\sqrt{7}}{5}$

In Exercises 39 and 40, list all the numbers in the given set that are members of the following sets.

(a) N (b) W (c) J (d) Q (e) I

39. $\left\{-4, 0, \frac{3}{5}, \sqrt{7}, 0.25, -17, 0.\overline{63}, \pi, \sqrt{16}\right\}$

40. $\left\{-24, -\sqrt{5}, -4\frac{7}{8}, 0.\overline{36}, \frac{\sqrt{3}}{2}, \frac{\pi}{7}, 0.2863\right\}$

In Exercises 41–46, graph the given set.

41. $\{1, 2, 3, -2, -4\}$ **42.** $\{-1, -3, 5, 3\}$

43. $\left\{\frac{1}{2}, 0.75, -\frac{4}{3}, 0, -1.2\right\}$ **44.** $\left\{-2\frac{2}{3}, -\frac{3}{2}, -1, 3\frac{4}{5}\right\}$

45. $\left\{-3, -\frac{1}{2}, 3, -\pi, 4.3, \sqrt{4}, -\sqrt{4}, -2\frac{1}{3}\right\}$

46. $\left\{5, -4, 2.5, \frac{\pi}{2}, -\sqrt{9}, -\sqrt{6}, 3\frac{2}{3}, \frac{7}{2}\right\}$

In Exercises 47–56, insert $<$ or $>$ to make the statement true.

47. -12 ▭ -5 **48.** 11 ▭ 5

49. $-\frac{15}{7}$ ▭ $-\frac{7}{15}$ **50.** $\frac{22}{23}$ ▭ $\frac{23}{24}$

51. $\frac{22}{7}$ ▢ π

52. 3.14 ▢ π

53. $\sqrt{7}$ ▢ $2.\overline{645}$

54. $-\sqrt{30}$ ▢ $-5.4\overline{7}$

55. $|-4|$ ▢ -4

56. $-|-3|$ ▢ $-(-3)$

In Exercises 57–60, write an inequality with the same meaning but with the inequality symbol reversed.

57. $x < 10$

58. $5 \geq -1$

59. $-14 > -14.2$

60. $y \leq 0$

In Exercises 61–68, write the given information as an inequality.

61. 3 is less than y.

62. a is greater than -3.

63. x is between -2 and 3, inclusive.

64. n is between -4 and 3, including 3 but not -4.

65. a is negative.

66. b is nonnegative.

67. x is between -2 and 2.

68. y is greater than or equal to 0.

In Exercises 69–78, graph the given set on a number line. Then describe the set with interval notation.

69. $\{x \mid x < -4\}$

70. $\{x \mid x \leq 6\}$

71. $\{x \mid x \geq -5\}$

72. $\{x \mid x > 4\}$

73. $\{x \mid -3 < x < -1\}$

74. $\{x \mid -2 < x < 8\}$

75. $\{x \mid 2 \leq x \leq 5\}$

76. $\{x \mid -2 \leq x \leq 6\}$

77. $\{x \mid -3 < x \leq 7\}$

78. $\{x \mid 3 \leq x < 8\}$

In Exercises 79–82, a set of numbers is described with interval notation. Write the set in set-builder notation.

79. $[-3, 7)$

80. $[0, 5]$

81. $(-\infty, 0]$

82. $(-1, \infty)$

In Exercises 83–86, write the opposite of each number.

83. -6

84. 13

85. 0

86. $|-4|$

87. Using the word *distance*, explain why $|7| = |-7|$.

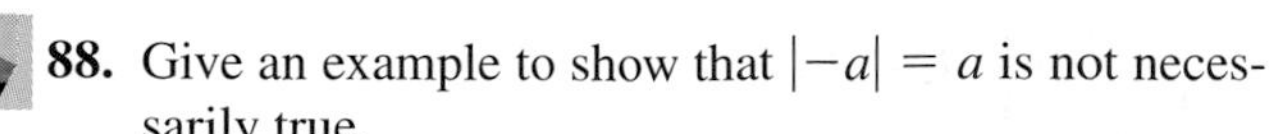

88. Give an example to show that $|-a| = a$ is not necessarily true.

In Exercises 89–92, a number n is described in reference to a number line. State all possible values of n.

89. The number n is 5 units from 0.

90. The number n is 3 units from -2.

91. When increased by 2, n is 4 units from 0.

92. The number n has an absolute value of 8.

In Exercises 93–102, evaluate the given expression.

93. $|-6|$

94. $|-4|$

95. $-|7|$

96. $-|-10|$

97. $|0|$

98. $-|12|$

99. $|-2|-|2|$

100. $|-4| + |5|$

101. $|-6| - |-5|$

102. $|-12| - |-10|$

Real-Life Applications

103. Before your trip to England, you converted \$1440 to pounds at an exchange rate of 1 pound per \$1.60. While in England, you spent 600 pounds. When you returned and converted pounds to dollars, the exchange rate had fallen to \$1.50 per pound. How many dollars did you receive?

104. A total of \$9500 is to be divided among four people. Two people are each to receive 30% of the total, and one person is to receive \$2000. How much will the fourth person receive?

105. As a produce manager, you have purchased 50 pounds of grapes at a wholesale price of \$75. If you want to make a profit of 40 cents per pound, at what retail price should you sell the grapes?

106. As a store owner, you must collect a 5% state sales tax on all goods sold. If your total receipts (including tax) for the month were \$21,000, how much tax must you remit to your state?

Group Project: Frozen Desserts

In recent years, Americans have consumed an average of about 5 gallons of frozen desserts per person per year. The accompanying figure indicates the popularity of the four basic categories. (Source: U.S. Department of Agriculture.)

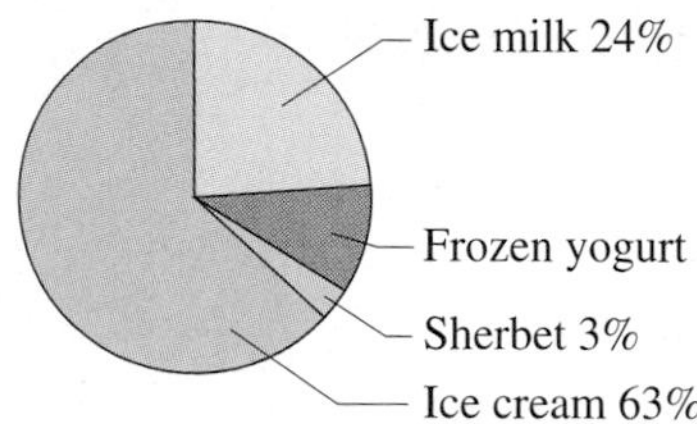

107. On average, how many gallons of ice cream does each American consume annually?

108. What percent of the total consumption is frozen yogurt?

109. On average, how many ounces of sherbet does each American consume annually? (There are 128 ounces in a gallon.)

110. The cost of compiling statistics such as these is borne by the taxpayer. How, if at all, do businesses and consumers benefit?

Challenge

In Exercises 111–114, fill in the blanks with $<$, $>$, or $=$.

111. If $a < 0$, then $-a$ ____ 0.

112. If $a > 0$, then $|a|$ ____ $-a$.

113. If $a < 0$, then $|a|$ ____ a.

114. If $a < 0$, then $-(-a)$ ____ a.

In Exercises 115–118, write the given number as the simplified quotient of two integers.

115. 2.7 **116.** 0.352 **117.** $0.\overline{3}$ **118.** $1.\overline{6}$

1.2 OPERATIONS WITH REAL NUMBERS

Addition and Subtraction • Multiplication and Division • Exponents • Square Roots

Addition and Subtraction

In addition, the numbers being added are called **addends.** The indicated addition and the result are both called a **sum.**

The following is a summary of the rules for adding two nonzero real numbers.

Rules for Adding Two Nonzero Real Numbers

1. If the addends have *like signs*, add the absolute values of the addends. The sign of the sum is the same as the sign of the addends.
2. If the addends have *unlike signs*, take the absolute value of each addend and subtract the smaller absolute value from the larger absolute value. The sign of the sum is the same as the sign of the addend with the larger absolute value.

Add

Use the + key for addition.

EXAMPLE 1

Adding Real Numbers

(a) $-12 + 7 = -5$

(b) $-7 + (-15) = -22$

(c) $20 + (-8) = 12$

(d) $\frac{9}{10} + \left(-\frac{3}{4}\right) = \frac{18}{20} + \left(-\frac{15}{20}\right) = \frac{3}{20}$

(e) $-362 + 287 = -75$

(f) $-3 + \frac{3}{5} = -\frac{15}{5} + \frac{3}{5} = -\frac{12}{5}$

(g) $-12.48 + (-23.4) = -35.88$

Fraction

On some calculators, you can request results in simplified fraction form.

LEARNING TIP

When a sum consists of more than two addends, try adding all the positive numbers first and then all the negative numbers. Then add the two results.

When addition involves more than two addends, grouping symbols, such as parentheses, brackets, or braces, indicate which addends are added first. When grouping symbols are not present, addition is performed from left to right.

EXAMPLE 2

Adding Three or More Addends

Determine the sum.

(a) $-4 + (-2) + 6 + (-7)$ (b) $-|-8| + (-6 + 2)$

Solution

(a) $-4 + (-2) + 6 + (-7) = -6 + 6 + (-7) = 0 + (-7) = -7$

(b) $-|-8| + (-6 + 2) = -8 + (-4) = -12$

In the indicated subtraction $8 - 5$, the number 8 is called the **minuend,** and the number 5 is called the **subtrahend.** Both the indicated subtraction and the result of performing the subtraction are called the **difference.**

Subtraction is defined in terms of addition.

Definition of Subtraction

For any real numbers a and b, $a - b = a + (-b)$.

Subtract

If your calculator has both a minus key and a negative key, be sure to use the minus key for subtraction.

EXAMPLE 3

Subtracting Real Numbers

(a) $9 - 17 = 9 + (-17) = -8$

(b) $-5 - 14 = -5 + (-14) = -19$

(c) $8 - (-12) = 8 + 12 = 20$

(d) $-10 - (-5) = -10 + 5 = -5$

(e) $\frac{1}{2} - \frac{7}{5} = \frac{5}{10} - \frac{14}{10} = \frac{5 + (-14)}{10} = -\frac{9}{10}$

(f) $-7.6 - (-9.37) = -7.6 + 9.37 = 1.77$

(g) $-735 - 579 = -735 + (-579) = -1314$

LEARNING TIP

Try not to rush subtraction. You will make fewer sign errors if you actually write the conversion step: $-2 - 9 = -2 + (-9)$.

We use subtraction to determine the distance between two points of a number line.

Distance Between Two Points of a Number Line

On a number line, if the coordinates of points A and B are a and b, respectively, then the distance d between the points is

$$d = |a - b| = |b - a|$$

EXAMPLE 4

Finding the Distance Between Points of a Number Line

The coordinates of points P and Q are given. Determine the distance d between the points.

(a) -16 and -8 (b) -6 and 9

Solution

(a) $d = |-16 - (-8)| = |-8| = 8$

(b) $d = |-6 - 9| = |-15| = 15$

Think About It

Suppose that you use the number line as a ruler to measure the distance between two points. Does the positioning of the ruler affect the result? Why?

Multiplication and Division

In multiplication, the numbers that are multiplied are called **factors.** Both the indicated multiplication and the result are called the **product.**

We know that the product of 0 and any number is 0.

Multiplying Two Nonzero Real Numbers

For any nonzero real numbers a and b,

1. if a and b have like signs, the product is positive; if a and b have unlike signs, the product is negative.
2. the absolute value of the product ab is the product of the absolute values of a and b.

Multiply

Although the x key may be used for multiplication, the symbol * may be displayed on the screen.

EXAMPLE 5

Multiplying Real Numbers

(a) $(-6)(-7) = 42$ (b) $9(-5) = -45$

(c) $-9 \cdot 8 = -72$ (d) $\frac{3}{5} \cdot \left(-\frac{4}{9}\right) = -\frac{4}{15}$

(e) $10 \cdot \left(-\frac{2}{5}\right) = -4$ (f) $(-36)7 = -252$

LEARNING TIP

The sign rules for multiplication are quite different from the sign rules for addition. As you do the multiplication exercises, go back and rework some addition exercises to make sure that the different rules are clear in your mind.

When a product consists of more than two nonzero factors,

1. if there is an even number of negative factors, then the product is positive.
2. if there is an odd number of negative factors, then the product is negative.

EXAMPLE 6

Multiplying Three or More Factors

(a) $(-1)(-2)(3)(-4) = -24$ There is an odd number of negative factors.

(b) $2 \cdot (-3) \cdot (-5) = 30$ There is an even number of negative factors.

(c) $(-0.16)(74)(0)(-3.67) = 0$ One factor is 0.

In the indicated division $20 \div 4$ or $\frac{20}{4}$, the number 20 is called the **dividend,** and the number 4 is called the **divisor.** Both the indicated division and the result are called the **quotient.**

Because division is defined in terms of multiplication, the sign rules for division are the same as the sign rules for multiplication.

Definition of Division

For real numbers a, b, and c, with $b \neq 0$,

$$a \div b = c \quad \text{if} \quad c \cdot b = a.$$

The definition of division requires that the divisor be nonzero. If we attempt to evaluate $7 \div 0$, we must find a number c such that $c \cdot 0 = 7$, which is impossible. To evaluate $0 \div 0$, we must find a number c such that $c \cdot 0 = 0$. Because this is true for any number c, we say that $0 \div 0$ is *indeterminate*. In short, division by 0 is undefined.

To perform a division that involves fractions, multiply the dividend by the reciprocal of the divisor.

Divide

Although the $\div$ key may be used for division, the symbol / may be displayed on the screen.

EXAMPLE 7

Dividing Real Numbers

(a) $\dfrac{24}{-6} = -4$

(b) $-14 \div (-4) = 3.5$

(c) $\dfrac{0}{-7} = 0$

(d) $5 \div 0$ is undefined

(e) $\left(-\dfrac{15}{2}\right) \div (-12) = \left(-\dfrac{15}{2}\right) \cdot \left(-\dfrac{1}{12}\right) = \dfrac{5}{8}$

(f) $\dfrac{\frac{3}{5}}{-\frac{15}{4}} = \dfrac{3}{5} \div \left(-\dfrac{15}{4}\right) = \dfrac{3}{5} \cdot \left(-\dfrac{4}{15}\right) = -\dfrac{4}{25}$

The following summarizes the equivalence of quotients with opposite signs.

Opposite Signs in Quotients

For real numbers a and b ($b \neq 0$),

$$-\frac{a}{b} = \frac{-a}{b} = \frac{a}{-b} \quad \text{and} \quad \frac{-a}{-b} = \frac{a}{b}$$

LEARNING TIP

Knowing that $-(-a) = a$ and $\frac{-a}{-b} = \frac{a}{b}$ allows you to remove pairs of negative (or opposite) signs. Eliminating signs, when possible, simplifies the appearance of the expression and reduces sign errors.

As always, grouping symbols indicate the order in which operations are to be performed. For example,

$$-30 \div [(-8) \div 4] = -30 \div (-2) = 15$$

Exponents

We use an exponent to indicate repeated multiplication.

$$3^4 = 3 \cdot 3 \cdot 3 \cdot 3 = 81$$

The number 3 is called the **base,** and the number 4 is called the **exponent.** The expression is read as "three to the fourth power." For exponents of 2 and 3, the words *squared* and *cubed* are sometimes used rather than *to the second power* and *to the third power.*

The expression 3^4 is in **exponential form,** and the expression $3 \cdot 3 \cdot 3 \cdot 3$ is in **product** or **factored form.** When the exponent is a natural number, the base can be any real number.

Exponent

Your calculator may have special keys for squaring and cubing, but typically, the general exponent key is ^.

EXAMPLE 8

Evaluating Exponential Expressions

(a) $5^3 = 5 \cdot 5 \cdot 5 = 125$

(b) $(-3)^4 = (-3) \cdot (-3) \cdot (-3) \cdot (-3) = 81$

(c) $-3^4 = -(3 \cdot 3 \cdot 3 \cdot 3) = -81$

(d) $\left(-\frac{3}{2}\right)^3 = \left(-\frac{3}{2}\right) \cdot \left(-\frac{3}{2}\right) \cdot \left(-\frac{3}{2}\right) = \frac{-27}{8}$

(e) $(0.3)^2 = (0.3)(0.3) = 0.09$

Square Roots

Because 16 can be written as 4^2, 4 is called a *square root* of 16. Similarly, 16 can be written as $(-4)^2$, so -4 is also a square root of 16.

Definition of Square Root

For real numbers a and b, if $b^2 = a$, then b is a **square root** of a.

In the system of real numbers, b^2 cannot be negative. Therefore, a negative number cannot have a square root. Also, because 0 is the only number that can be squared to obtain 0, 0 has only one square root. All positive numbers have two square roots, one positive and one negative.

The positive square root (or **principal** square root) of a number is denoted by a **radical** symbol $\sqrt{\ }$. If $b^2 = a$, where $b \geq 0$, then $\sqrt{a} = b$. The number or expression under the radical symbol is called the **radicand.**

EXAMPLE 9

Evaluating Square Roots

(a) $\sqrt{25} = 5$ — $5^2 = 25$

(b) $\sqrt{\frac{4}{9}} = \frac{2}{3}$ — $\left(\frac{2}{3}\right)^2 = \frac{4}{9}$

(c) $\sqrt{0} = 0$ — $0^2 = 0$

(d) $-\sqrt{121} = -11$ — Note that the radical symbol is also a grouping symbol.

(e) $\sqrt{56} \approx 7.48$ — The $\approx$ symbol means "approximately equal to."

(f) $\sqrt{-9}$ is not a real number.

Because $2^3 = 8$, 2 is called a **cube root** of 8, and we write $\sqrt[3]{8} = 2$. Because $3^4 = 81$, 3 is called a **fourth root** of 81, and we write $\sqrt[4]{81} = 3$. The concept can be extended to higher and higher roots.

1.2 QUICK REFERENCE

Addition and Subtraction

- A **sum** is an indicated addition. The numbers being added are called **addends.**
- Rules for adding two nonzero real numbers:
  1. If the addends have *like signs*, add the absolute values of the addends. The sign of the sum is the same as the sign of the addends.
  2. If the addends have *unlike signs*, take the absolute value of each addend and subtract the smaller absolute value from the larger absolute value. The sign of the sum is the same as the sign of the addend with the larger absolute value.
- A **difference** is an indicated subtraction. For $a - b$, a is called the **minuend,** and b is called the **subtrahend.**
- For any real numbers a and b, $a - b = a + (-b)$.
- On a number line, if the coordinates of points A and B are a and b, respectively, then the distance d between the points is $d = |a - b| = |b - a|$.

Multiplication and Division

- A **product** is an indicated multiplication. The numbers being multiplied are called **factors.**
- Rules for multiplying two real numbers:
  1. If at least one factor is 0, then the product is 0.
  2. If neither factor is 0, then

(a) if the factors have like signs, the product is positive; if the factors have unlike signs, the product is negative.

(b) the absolute value of the product is the product of the absolute values of the factors.

- A **quotient** is an indicated division. For $a \div b$, a is called the **dividend,** and b is called the **divisor.**
- If $a \neq 0$, $\frac{0}{a} = 0$. However, division by 0 is undefined.
- For real numbers a and b $(b \neq 0)$,

$$-\frac{a}{b} = \frac{-a}{b} = \frac{a}{-b} \quad \text{and} \quad \frac{-a}{-b} = \frac{a}{b}$$

Exponents

- An exponent is used to indicate repeated multiplication. In the **exponential form** b^n, b is called the **base,** and n is called the **exponent.**

Square Roots

- For real numbers a and b, if $b^2 = a$, then b is a **square root** of a. The **principal** square root of a is written $\sqrt{a}$, where a is the **radicand.**

1.2 EXERCISES

Concepts and Skills

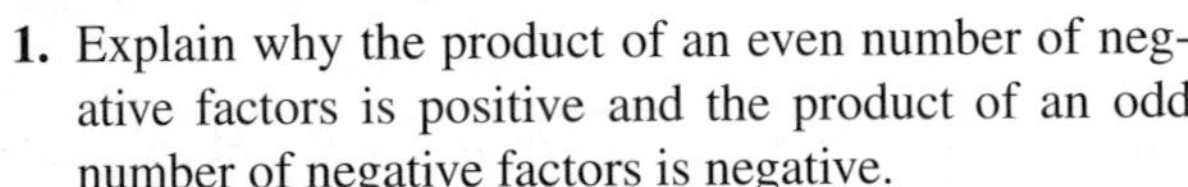

1. Explain why the product of an even number of negative factors is positive and the product of an odd number of negative factors is negative.

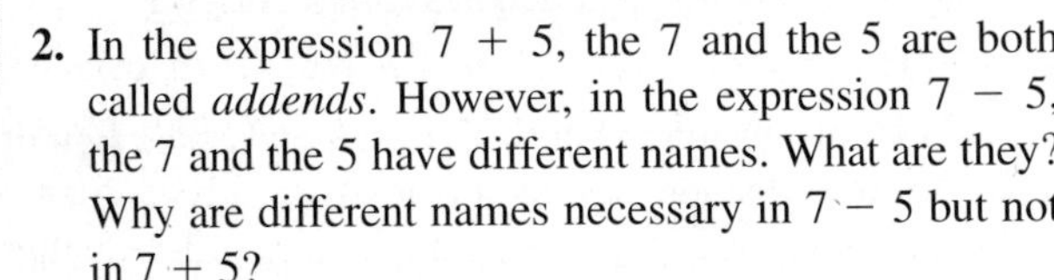

2. In the expression $7 + 5$, the 7 and the 5 are both called *addends*. However, in the expression $7 - 5$, the 7 and the 5 have different names. What are they? Why are different names necessary in $7 - 5$ but not in $7 + 5$?

In Exercises 3–18, perform the indicated operations.

3. $-6 + (-2)$ **4.** $-7 + 5$ **5.** $12 + (-4)$

6. $8 + (-14)$ **7.** $-2 - (-8)$ **8.** $-12 - 9$

9. $4 - 15$ **10.** $13 - (-5)$ **11.** $-\frac{2}{3} + \frac{3}{5}$

12. $-\frac{3}{4} - \left(-\frac{2}{3}\right)$ **13.** $-3 + [6 + (-9)]$

14. $(-2) + [(-5) + (-3)] + 9$

15. $12.3 + (-23.7)$ **16.** $-9.7 - 19.5$

17. $-|3| + |-10|$ **18.** $|5| - |-11|$

In Exercises 19–22, the coordinates of points A and B are given. Determine the distance from A to B.

19. $-7, -1$ **20.** $-4, 6$ **21.** $\frac{7}{2}, \frac{3}{4}$ **22.** $4, -\frac{1}{5}$

In Exercises 23–34, perform the indicated operations.

23. $-6 \cdot (-10)$ **24.** $11(-4)$

25. $(-1)(-1)(2)(-3)(-5)$

26. $(-1)(-1)(2)(0)(-5)(-3)$

27. $-\frac{2}{5} \cdot \left(-\frac{5}{8}\right)$ **28.** $-\frac{7}{8} \cdot 16$

29. $\frac{-24}{6}$ **30.** $-40 \div (-5)$

31. $\frac{-2}{0}$ **32.** $0 \div (-35)$

33. $\left(-\frac{3}{4}\right) \div 8$ **34.** $-\frac{2}{3} \div \frac{8}{27}$

In Exercises 35–42, evaluate the exponential expression.

35. 2^4 **36.** 5^3 **37.** $(-3)^3$ **38.** -4^2

39. -7^2 **40.** $-(-4)^2$ **41.** $\left(-\frac{3}{4}\right)^3$ **42.** $-\left(-\frac{2}{3}\right)^4$

43. For each of the following, propose a rule that states the condition under which the statement is true when $b < 0$.

(a) b^n is positive. (b) b^n is negative.

44. Explain how to verify that $\sqrt{2809} = 53$.

In Exercises 45–52, evaluate the square root, if possible.

45. $\sqrt{4}$ 46. $\sqrt{16}$ 47. $-\sqrt{36}$ 48. $-\sqrt{100}$

49. $\sqrt{-64}$ 50. $\sqrt{-1}$ 51. $\sqrt{\frac{4}{9}}$ 52. $\sqrt{0.81}$

In Exercises 53–66, perform the indicated operations.

53. $15 + (-1) + 2 + (-6)$
54. $15 - 20 - 5$
55. $10 - (6 - 9)$
56. $(-8 + 10) - (-7 + 2)$
57. $\sqrt{16} + |-3|$
58. $|4| - |-12|$
59. $-5(7)(-2)$
60. $-2 \cdot (\sqrt{25})$
61. $-9 \cdot (-6)$
62. $\frac{|-42|}{|7|}$
63. -6^2
64. $\left(-\frac{2}{3}\right)^2$
65. $(-1)^{10}$
66. $4(2^3)$

In Exercises 67–72, write a numerical expression that models the given information and then evaluate the expression.

67. The difference of 6 and -8
68. 12 increased by -4
69. Two-thirds of -21
70. The quotient of $-\frac{8}{5}$ and 12
71. The opposite of the square of -6
72. The principal square root of one-hundredth

In Exercises 73–78, determine the number.

73. What number subtracted from -2 results in 5?
74. If the sum of three numbers is 0, and two of the numbers are opposites, what is the third number?
75. The quotient of what number and 2 is $-\frac{1}{4}$?
76. If the product of three factors is 1, and two of the factors are 1 and $-\frac{3}{2}$, what is the third factor?
77. What number raised to the third power is 64?
78. The principal square root of what number is 6?

Real-Life Applications

79. After writing an electric bill check for \$76.27, a person notices that his new checkbook balance is $-\$26.84$. He decides not to mail the electric bill until he makes a deposit of \$35.25. What is the checkbook balance after the deposit and the payment of the electric bill?

80. The **average** or **mean** of n numbers is found by adding the n numbers and dividing the sum by n. During a particularly active week, the stock prices for a utility company changed by the following amounts: $-\frac{3}{8}$, $+\frac{1}{2}$, $+\frac{3}{4}$, $-\frac{5}{16}$, and $+\frac{1}{8}$. What was the average change in price for the week?

81. In the state of California, the highest elevation is Mount Whitney with an altitude of 14,495 feet. The lowest altitude is Death Valley at 280 feet below sea level. What is the difference in the two altitudes?

82. With a balance of \$564.28 in her checkbook, a person wrote a check for \$834.75 to purchase a new stereo system. What is the new checkbook balance? Will a quick deposit of \$300 rescue her credit rating?

83. A car buyer has obtained a loan for \$1075 from a credit union. He has an automatic payroll deduction of \$225.39 each month for the next 5 months to pay off his loan. At the end of the loan period, how much interest will he have paid?

84. A dog owner buys a 50-pound bag of dog food for her golden retriever. If the dog eats 12 ounces of food each day, how many pounds of dog food does the owner have left after 3 weeks? (There are 16 ounces in 1 pound.)

85. To discourage guessing, a test-grading scheme is as follows.

Correct answer	$+1$
Answer left blank	0
Incorrect answer	-1

On a 100-question test, a student left 13 answers blank and answered 50 questions correctly. What was his test score?

86. The weather service reported that during a certain 1-week period, the daily high temperature differed from the average high by 12°, −6°, −2°, 5°, −10°, −9°, and 8°. What was the average deviation from the normal high temperature?

Modeling with Real Data

87. The accompanying table shows the amount (in dollars) by which the average hourly pay for production workers in selected countries compared with the U.S. average of $16.73 in 1996. Complete the table.

Country	Average Hourly Pay (in dollars)	Difference from United States
Germany	25.71	(a)
Hong Kong	(b)	−12.44
Japan	(c)	2.88
Mexico	(d)	−14.14
Spain	11.50	(e)

(Source: Bureau of Labor Statistics.)

88. The accompanying graph shows the worldwide distribution of Coca-Cola products in 1995. (Source: Coca-Cola Company.)

North America	Latin America	Greater Europe	Middle and Far East	Africa
32%	24%	21%	18%	5%

(a) Of the 12.7 billion cases sold in 1995, how many were sold in the Western Hemisphere?

(b) Group the five regions into two categories, each accounting for 50% of the sales.

89. The accompanying table shows the number of teachers in certain city school systems in the 1995–1996 school year. The table also shows the projected need for additional teachers in the 1996–1997 school year.

City	Number of Teachers	Projected need for 1996–1997
Los Angeles	28,000	1600
Chicago	28,000	1000
Houston	12,000	1000
Memphis	6,000	500

(Source: U.S. Department of Education.)

(a) Which two school systems had the same percent increase projected for 1996–1997?

(b) In Chicago, 73% of the teachers were over 40 years old. Assuming retirement at age 60, how many teachers would need to be replaced over the next 20 years?

90. From 1995 to 1996, the average cost of owning and operating an automobile rose to 42.6 cents per mile per year. The costliest car to own was the Chevy Blazer, at 49.9 cents per mile. (Source: Runzheimer.) Based on driving 12,000 miles per year, how much above average was the annual cost of driving a Blazer?

Group Project: ATM Cards

Research indicates that about 60% of American adults have automatic teller machine (ATM) cards. The accompanying pie chart shows the percentages of those cardholders who use their cards a certain number of times per month. (Source: Research Partnership.)

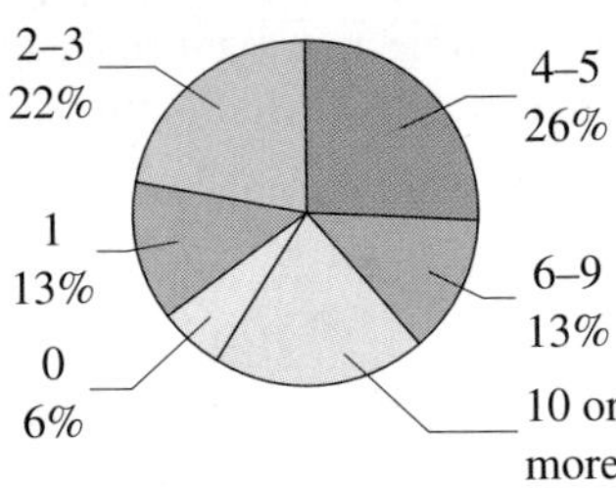

91. What percent of the cardholders use their cards fewer than four times per month?

92. What percent of the cardholders use their cards at least ten times per month?

93. Why does the data not allow you to determine the actual number of cardholders in any of the categories?

94. Bank fees for ATM use have increased dramatically. Do you think that higher fees might affect the percentages shown in the pie chart? If so, in what way?

Challenge

In Exercises 95–98, let a represent a positive number. Determine whether the given expression is always positive, always negative, always 0, or none of these. If none of these, give values of a to support your answer.

95. $a^2 - a$ **96.** $|a| - a$ **97.** $a - \sqrt{a}$ **98.** $a - \frac{1}{a}$

99. The value of the following expression is 5.

$$1 - 2 + 3 - 4 + 5 - 6 + 7 - 8 + 9$$

Insert parentheses (as many as needed) so that the value of the expression is 7.

100. The value of the following expression is 45.

$$1 + 2 + 3 + 4 + 5 + 6 + 7 + 8 + 9$$

Is it possible to replace some of the addition symbols with subtraction symbols so that the value of the expression is 0? Why?

In Exercises 101 and 102, for what value(s) of x is the given expression (a) positive? (b) 0? (c) negative?

101. $-x^2$ **102.** $(-x)^2$

1.3 PROPERTIES OF THE REAL NUMBERS

Commutative Properties • Associative Properties • Identity Properties • Inverse Properties • Special Multiplication Properties • Distributive Property

Commutative Properties

In preceding sections we informally used certain rules to perform operations with real numbers. In this section we summarize and formally state these rules and other important properties of real numbers.

In addition, the order of the addends does not affect the sum. In multiplication, the order of the factors does not affect the product.

$$5 + (-8) = -8 + 5 \qquad -7 \cdot 4 = 4(-7)$$

Commutative Properties of Addition and Multiplication

For all real numbers a and b,

$a + b = b + a$	Commutative Property of Addition
$ab = ba$	Commutative Property of Multiplication

Note that subtraction and division are *not* commutative operations.

$$5 - 3 \neq 3 - 5$$

$$12 \div 6 \neq 6 \div 12$$

Associative Properties

In addition, the grouping of two or more addends does not affect the sum. In multiplication, the grouping of two or more factors does not affect the product.

$$-6 + [2 + (-5)] = (-6 + 2) + (-5)$$
$$2 \cdot [(-3)(-7)] = [2(-3)] \cdot (-7)$$

Associative Properties of Addition and Multiplication

For all real numbers a, b, and c,

$(a + b) + c = a + (b + c)$ Associative Property of Addition

$(ab)c = a(bc)$ Associative Property of Multiplication

LEARNING TIP

Your instructor may or may not ask you to know the names of these properties. Try to use everyday words to assist you. For example, *commute* might suggest travel back and forth, as in $a + b = b + a$, and *associate* might refer to joining or grouping, as in $a(bc) = (ab)c$.

Subtraction and division are *not* associative operations.

$(12 - 8) - 2 = 2,$ but $12 - (8 - 2) = 6$

$(24 \div 4) \div 2 = 3,$ but $24 \div (4 \div 2) = 12$

The associative properties are used to *regroup* addends and factors, and the commutative properties are used to *reorder* addends and factors.

Although subtraction is neither commutative nor associative, we frequently wish to reorder and regroup an expression involving subtraction.

EXAMPLE 1

Reordering and Regrouping a Numerical Expression

Simplify $5 - 3 + (-7) - (-8) + 2$.

Solution

By first converting subtractions into additions, we can take advantage of the commutative and associative properties to group the positive numbers and the negative numbers.

$5 - 3 + (-7) - (-8) + 2$

$= 5 + (-3) + (-7) + 8 + 2$ Definition of Subtraction

$= 5 + 8 + 2 + (-3) + (-7)$ Commutative Property of Addition

$= (5 + 8 + 2) + [(-3) + (-7)]$ Associative Property of Addition

$= 15 + (-10)$

$= 5$

Identity Properties

The sum of a number and 0 is the original number, and the product of a number and 1 is the original number. We call 0 the **additive identity,** and we call 1 the **multiplicative identity.**

Identity Properties

For any real number a,

$a + 0 = 0 + a = a$ Additive Identity Property

$a \cdot 1 = 1 \cdot a = a$ Multiplicative Identity Property

Inverse Properties

The sum of a number and its opposite is 0, which is the additive identity. The formal statement of this property usually uses *additive inverse* rather than *opposite*.

Property of Additive Inverses

For every real number a, there is a unique number $-a$ such that

$$a + (-a) = -a + a = 0$$

A corresponding rule exists for multiplication. Each *nonzero* real number a has a **multiplicative inverse** or **reciprocal,** represented by the symbol $\frac{1}{a}$. The product of a nonzero number and its reciprocal is 1, which is the multiplicative identity.

Property of Multiplicative Inverses

For every nonzero real number a, there is a unique number $\frac{1}{a}$ such that

$$a \cdot \frac{1}{a} = \frac{1}{a} \cdot a = 1$$

 Reciprocal

Usually the reciprocal key is $1/x$ or x^{-1}.

From the sign rules for multiplication, we can see that a number and its reciprocal always have the same sign.

For a nonzero rational number $\frac{p}{q}$, the reciprocal is $\frac{q}{p}$. For example, because $\frac{3}{4} \cdot \frac{4}{3} = 1$, the reciprocal of $\frac{3}{4}$ is $\frac{4}{3}$.

EXAMPLE 2

Inverses of Real Numbers

Write the additive inverse (opposite) and the multiplicative inverse (reciprocal) of each number.

(a) -12 (b) $\frac{9}{5}$

Solution

	Additive Inverse	*Multiplicative Inverse*	
(a) -12	12	$-\frac{1}{12}$	$\frac{1}{-12} = -\frac{1}{12}$
(b) $\frac{9}{5}$	$-\frac{9}{5}$	$\frac{5}{9}$	

We defined the quotient $a \div b$ $(b \neq 0)$ as the number c such that $c \cdot b = a$. Division is sometimes defined as follows.

Alternative Definition of Division

For real numbers a and b, $b \neq 0$,

$$a \div b = a \cdot \frac{1}{b}$$

In words, to divide a by b, multiply a by the reciprocal of b.

Special Multiplication Properties

There are two useful multiplication properties that involve 0 and –1.

Multiplication Properties of 0 and –1

For any real number a,

$a \cdot 0 = 0 \cdot a = 0$ Multiplication Property of 0

$a \cdot (-1) = -1 \cdot a = -a$ Multiplication Property of -1

Distributive Property

All the properties summarized so far have involved either addition or multiplication. The Distributive Property involves both operations.

Distributive Property

For all real numbers a, b, and c,

$$a(b + c) = ab + ac$$

More precisely, this property is called the Distributive Property of Multiplication Over Addition. There is also a Distributive Property of Multiplication Over Subtraction.

$$a(b - c) = ab - ac$$

We will refer to both versions simply as the Distributive Property. The Distributive Property can be read in either direction.

1. $a(b + c) = ab + ac$
2. $ab + ac = a(b + c)$

In (1), we change an indicated product into a sum. In this case we are using the Distributive Property to *multiply*. In (2), we change an indicated sum into a product. In this case we are using the Distributive Property to *factor*.

EXAMPLE 3

Using the Distributive Property to Multiply and Factor

Use the Distributive Property to rewrite each indicated product as a sum and each indicated sum as a product.

(a) $-7(x - 3)$ (b) $3x + 3y$

(c) $a(y - z)$ (d) $-4a - 4b$

Solution

(a) $-7(x - 3) = -7 \cdot x - (-7) \cdot 3 = -7x + 21$ Multiply.

(b) $3x + 3y = 3 \cdot x + 3 \cdot y = 3(x + y)$ Factor.

(c) $a(y - z) = ay - az$

(d) $-4a - 4b = -4a + (-4)b = -4(a + b)$

Think About It

Consider the process of evaluating 3(7 + 2) and simplifying $3(x + 2)$. Although the Distributive Property can be applied in both expressions, why is it essential in one case but not in the other?

Some properties of real numbers can be deduced from previously stated properties. The following is an example.

$$\begin{aligned}
-(a - b) &= (-1)(a - b) && \text{Multiplication Property of } -1\\
&= (-1)a - (-1)b && \text{Distributive Property}\\
&= -a - (-b) && \text{Multiplication Property of } -1\\
&= -a + b && \text{Definition of Subtraction}\\
&= b + (-a) && \text{Commutative Property of Addition}\\
&= b - a && \text{Definition of Subtraction}
\end{aligned}$$

We have proven the following important property.

Property of the Opposite of a Difference

For any real numbers a and b,

$$-(a - b) = b - a$$

EXAMPLE 4

Identifying the Properties of Real Numbers

Identify the property that justifies the given statement.

(a) $x + (2 + x) = x + (x + 2)$ Commutative Property of Addition

(b) $1 \cdot (y - 2) = y - 2$ Multiplicative Identity Property

(c) $(x + 5) - 1 = x + (5 - 1)$ Associative Property of Addition

(d) $-3(2 - y) = -6 + 3y$ Distributive Property

(e) $\frac{1}{5} \cdot 5 = 1$ Property of Multiplicative Inverses

(f) $0 \cdot (5x) = 0$ Multiplication Property of 0

(g) $-(3b) + (3b) = 0$ Property of Additive Inverses

(h) $-(3 - w) = w - 3$ Property of the Opposite of a Difference

EXAMPLE 5

Using the Properties to Rewrite Expressions

Use the given property to rewrite the expression.

(a) Commutative Property of Multiplication: $x \cdot 3 =$ ______

(b) Distributive Property: $2a + 3a =$ ______

(c) Associative Property of Multiplication: $\frac{1}{2}(6x) =$ ______

(d) Additive Identity Property: $0 + y =$ ______

Solution

(a) $x \cdot 3 = 3x$

(b) $2a + 3a = (2 + 3)a$

(c) $\frac{1}{2}(6x) = \left(\frac{1}{2} \cdot 6\right) \cdot x$

(d) $0 + y = y$

1.3 QUICK REFERENCE

The following is a summary of the properties of real numbers, where a, b, and c represent any real number.

Commutative Properties

- $a + b = b + a$ Commutative Property of Addition
- $ab = ba$ Commutative Property of Multiplication
- Subtraction and division are not commutative.
- The commutative properties are used to reorder addends and factors.

Associative Properties

- $(a + b) + c = a + (b + c)$ Associative Property of Addition
- $(ab)c = a(bc)$ Associative Property of Multiplication
- Subtraction and division are not associative.
- The associative properties are used to regroup addends and factors.

Identity Properties	• $a + 0 = 0 + a = a$	Additive Identity Property
	• $a \cdot 1 = 1 \cdot a = a$	Multiplicative Identity Property
Inverse Properties	• $a + (-a) = -a + a = 0$	Property of Additive Inverses
	• $a \cdot \frac{1}{a} = \frac{1}{a} \cdot a = 1 \quad (a \neq 0)$	Property of Multiplicative Inverses
	• For any real number a, $-a$ is unique; for any nonzero real number a, $\frac{1}{a}$ is unique.	
	• An alternative definition of division is $a \div b = a \cdot \frac{1}{b}$, where $b \neq 0$.	
Special Multiplication Properties	• $a \cdot 0 = 0 \cdot a = 0$	Multiplication Property of 0
	• $a \cdot (-1) = -1 \cdot a = -a$	Multiplication Property of -1
Distributive Property	• $a(b + c) = ab + ac$ and $a(b - c) = ab - ac$	
	• We use $a(b + c) = ab + ac$ to *multiply*; we use $ab + ac = a(b + c)$ to *factor*.	
	• The Distributive Property can be used to prove the Property of the Opposite of a Difference: $-(a - b) = b - a$.	

1.3 EXERCISES

Concepts and Skills

1. Although subtraction is not commutative, writing $-9 - 4x + x^2$ as $x^2 - 4x - 9$ is correct. Why?

2. (a) Name a number that is its own opposite. Show how your answer satisfies the Property of Additive Inverses.

   (b) Name two numbers that are their own reciprocals. Show how your answers satisfy the Property of Multiplicative Inverses.

In Exercises 3–16, name the property that is illustrated.

3. $3 + (2 + b) = (3 + 2) + b$
4. $-x + 2 = 2 - x$
5. $4x + 0 = 4x$
6. $-6x + 6x = 0$
7. $4(x - 5) = 4x - 20$
8. $(x^2 + 1) \cdot \frac{1}{x^2 + 1} = 1$
9. $\frac{2}{3}x - \frac{2}{3} = \frac{2}{3}(x - 1)$
10. $10 \cdot \left(\frac{1}{2}x\right) = \left(10 \cdot \frac{1}{2}\right)x$
11. $5 \cdot 1 = 5$
12. $(2x - 7) \cdot 0 = 0$
13. $-1 \cdot (2y) = -2y$
14. $-3 - x = -1 \cdot (3 + x)$
15. $5x - 3x = (5 - 3)x$
16. $(y + 5)(y - 3) = (y - 3)(y + 5)$

In Exercises 17–24, match the expression in column A with the corresponding expression in column B.

Column A	*Column B*
17. $x \cdot (-7)$	(a) $y - x$
18. $-x + y$	(b) $(y + 2) + 9$
19. $y + (2 + 9)$	(c) $2x - y$
20. $3(x + 7)$	(d) 1
21. $1(2x - y)$	(e) 0
22. $-(x - 2)$	(f) $(2 \cdot 3)x$
23. $2(3x)$	(g) $3x + 21$
24. $2x \cdot \frac{1}{2x}\ (x \neq 0)$	(h) $-1(x - 2)$
	(i) $-7x$

In Exercises 25–34, use the given property to write the expression in an alternative form.

25. Associative Property of Addition:

$(x + 4) + 3 =$ ______

26. Commutative Property of Multiplication:

$(x + 6) \cdot 3 =$ ______

27. Associative Property of Multiplication:

$6(2x) =$ ______

28. Commutative Property of Addition:

$8 - x =$ ______

29. Additive Identity Property:

$(x + 0) + 6 =$ ______

30. Property of Additive Inverses:

$-(y + 3) + (y + 3) =$ ______

31. Property of Multiplicative Inverses:

$\left(3 \cdot \frac{1}{3}\right)x =$ ______

32. Multiplicative Identity Property:

$1(n - 4) =$ ______

33. Distributive Property:

$12z + 15 =$ ______

34. Distributive Property:

$-3[a + (-6)] =$ ______

35. Explain why 0 does not have a reciprocal.

36. Use the definition of reciprocal to show why the reciprocal of $\frac{p}{q}$ is $\frac{q}{p}$, where $p \neq 0$ and $q \neq 0$.

In Exercises 37–42, fill in the blank and state the property that justifies your answer.

37. $5 \cdot$ ______ $= 1$

38. $-7 +$ ______ $= 0$

39. $-\frac{2}{3} \cdot$ ______ $= -\frac{2}{3}$

40. $8 + (-x + x) = 8 +$ ______

41. ______ $(2x - 5) = -2x + 5$

42. $-x + 5 = -1 \cdot$ ______

In Exercises 43–46, name the property or definition that justifies each step.

43. $(y + 0) + (-y)$

$= (0 + y) + (-y)$ (a) ______

$= 0 + [y + (-y)]$ (b) ______

$= 0 + 0$ (c) ______

$= 0$ (d) ______

44. $-(2x - 3)$

$= -1(2x - 3)$ (a) ______

$= -2x + 3$ (b) ______

$= 3 - 2x$ (c) ______

45. $(a - b) + b$

$= [a + (-b)] + b$ (a) ______

$= a + [(-b) + b]$ (b) ______

$= a + 0$ (c) ______

$= a$ (d) ______

46. $5 \cdot \left(\frac{1}{5}x\right)$

$= \left(5 \cdot \frac{1}{5}\right)x$ (a) ______

$= 1x$ (b) ______

$= x$ (c) ______

In Exercises 47–54, use the associative properties to rewrite the given expression in simpler form or in a form with an exponent.

47. $(x + 3) + 2$

48. $-4 + (8 + y)$

49. $(2z + 7) - 2$

50. $(3x - 5) + 7$

51. $-5(3x)$

52. $(2y)y$

53. $-\frac{5}{6} \cdot \left(-\frac{9}{10}x\right)$

54. $\frac{3}{4}(2xy)$

In Exercises 55–70, use the Distributive Property to rewrite each expression.

55. $5(4 + 3b)$

56. $-2(3a - 7)$

57. $-\frac{3}{5}(5x + 20)$

58. $\frac{1}{3}(12x - 3)$

59. $-3(5 - 2x)$

60. $-4(-3x + 5)$

61. $x(y + z)$

62. $s(t + 3)$

63. $5x + 5y$

64. $11a - 11b$

65. $3x + 12$

66. $7y - 14$

67. $5y + 5$

68. $2 + 2b$

69. $6 - 3x$

70. $2x - 2$

In Exercises 71–74, write the given expression without grouping symbols.

71. $-(x - 4)$ **72.** $-(3 + 5x)$

73. $-(2x + 5y - 3)$ **74.** $-(-4x - y + 7)$

In Exercises 75–82, use the commutative and associative properties to rewrite the given expression in simpler form or in a form with an exponent.

75. $(3x) \cdot 4$ **76.** $y(3y)$

77. $(-6x)(2x)$ **78.** $3x \cdot 4y$

79. $(5 + x) + 3$ **80.** $x + [7 + (-x)]$

81. $-5x + (3 + 5x)$ **82.** $-7 + (2x - 3)$

In Exercises 83–86, use the commutative and associative properties to write the given expression so that the indicated operations can be performed easily. Then evaluate the expression.

83. $\frac{2}{9} + \frac{3}{7} + \frac{5}{9} + \frac{4}{7} + \frac{2}{9}$

84. $77 + 40 + 23 + 30 + 70 + 60$

85. $2 \cdot (-87) \cdot 5 \cdot (-1)$ **86.** $5 \cdot \frac{1}{3} \cdot \frac{1}{5} \cdot 3$

The Distributive Property can be useful in performing multiplication mentally. The following is an example.

$$\begin{aligned} 25 \cdot 103 &= 25(100 + 3) \\ &= 25 \cdot 100 + 25 \cdot 3 \\ &= 2500 + 75 \\ &= 2575 \end{aligned}$$

In Exercises 87–90, show how you would use the Distributive Property to determine the given product mentally.

87. $7 \cdot 108$ **88.** $30 \cdot 109$

89. $15 \cdot 98$ **90.** $4 \cdot 49$

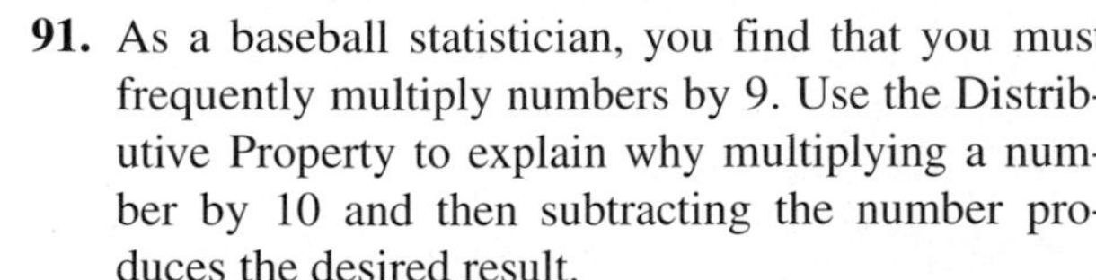

91. As a baseball statistician, you find that you must frequently multiply numbers by 9. Use the Distributive Property to explain why multiplying a number by 10 and then subtracting the number produces the desired result.

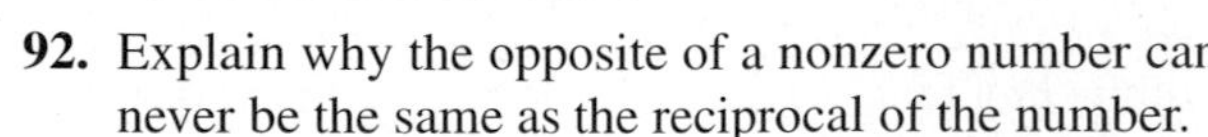

92. Explain why the opposite of a nonzero number can never be the same as the reciprocal of the number.

In Exercises 93–100, give the additive inverse (opposite) and the multiplicative inverse (reciprocal), if any, of the given number or expression.

93. $\frac{5}{7}$ **94.** -8 **95.** $\frac{2}{3}x$ **96.** $\frac{2}{y}$

97. $x + 3$ **98.** $y - 5$ **99.** $4 - 3x$ **100.** $2y$

Real-Life Applications

101. Suppose that during the month of June you wrote checks totaling the exact amount of your monthly paycheck, which you deposited on June 1. Which property of real numbers guarantees that the sum of your monthly income and expenses is 0?

102. For a theater-in-the-round production, the stage designer decides to use a triangular stage. To purchase construction material, the designer must calculate the stage area, which is found by multiplying 0.5 times the base of the triangle times the height of the triangle. Which property of real numbers guarantees that the calculated area will be the same no matter which two of the three numbers are multiplied first?

103. A tax auditor adds a column of numbers from top to bottom. She then checks her work by adding the column from bottom to top. Which property of real numbers guarantees that the sum is the same either way?

104. At the Two-Dollar Shop, all merchandise is sold for \$2. During one month, 20 key chains and 35 birthday balloons were sold. Write two numerical expressions, one a product and the other a sum, that represent the total income from these items. Then identify the property that guarantees that the expressions have the same value.

Group Project: Living Expenses

The majority of a person's time on the job is spent earning the money to pay taxes and living expenses. The accompanying table shows the number of minutes in an 8-hour workday that are used by the average American to pay for selected categories of expenses.

Expenses	Minutes per Day
Federal tax	110
Housing	82
Health care	58
State/local taxes	57
Food	50
Transportation	35
Clothing	20

(Source: Tax Foundation.)

105. (a) If a person goes to work at 8:00 A.M. and has no break until all taxes are paid, at what time is the first break?

(b) How many hours per week must a person work to pay for housing and food?

106. Create a similar table to show the days per week and weeks per year for each category. (Assume a 40-hour, 5-day work week and 50 work weeks per year.)

107. (a) Assuming that the expenses shown in the table are a person's only expenses, what percent of the workday produces disposable income?

(b) According to these figures, if your annual salary were $30,000, how much per month would you have after expenses?

108. How is it possible that an economy can grow vigorously at the same time that individuals struggle financially?

Challenge

In Exercises 109–114, for nonzero numbers a and b, indicate whether the given statement is always true, always false, or sometimes true. If sometimes, give examples of a and b for which the statement is true and for which the statement is false.

109. $a - b = b - a$

110. $\frac{a}{b} = \frac{b}{a}$

111. $|a + b| = |a| + |b|$

112. $|a - b| = |a| - |b|$

113. $(a + b)^2 = a^2 + b^2$

114. $\sqrt{a + b} = \sqrt{a} + \sqrt{b}$

115. Show how the Distributive Property can be used to prove that

$$\frac{a + b}{3} = \frac{a}{3} + \frac{b}{3}$$

1.4 ALGEBRAIC EXPRESSIONS AND FORMULAS

Order of Operations • Evaluating Algebraic Expressions • Equations and Formulas

Order of Operations

When a numerical expression contains more than one operation, the order in which those operations are performed can affect the result. We perform operations in the following order.

The Order of Operations

1. Perform all operations within grouping symbols according to the order in steps 2–4. For grouping symbols within grouping symbols, start with the innermost group and work outward.
2. Evaluate exponents and roots.
3. Perform multiplication and division from left to right.
4. Perform addition and subtraction from left to right.

Common grouping symbols are parentheses, brackets, and braces. However, fraction bars, absolute value symbols, and radical symbols also serve as grouping symbols.

EXAMPLE 1

Using the Order of Operations to Evaluate a Numerical Expression

Evaluate the given numerical expression.

(a) $8 - 5 \cdot 4$

(b) $5 + 2 \cdot 3^2 - (3 - 7)$

(c) $\dfrac{8 - 3 \cdot 2^2}{5 - (6 - 9)}$

(d) $7 - 3\left[(5 - 3) \div \dfrac{1}{3}\right]$

(e) $-32 \div 4 \cdot 3 - 2^3$

(f) $-(-5) - \sqrt{6^2 - 4 \cdot 5}$

LEARNING TIP

For each problem, work through the Order of Operations in a disciplined, step-by-step manner. Trying to do too much in each step is a major contributor to errors.

Solution

(a) $8 - 5 \cdot 4 = 8 - 20 = -12$ Multiply, then subtract.

(b) $5 + 2 \cdot 3^2 - (3 - 7)$

$= 5 + 2 \cdot 3^2 - (-4)$ Grouping symbols first

$= 5 + 2 \cdot 9 - (-4)$ Then exponents

$= 5 + 18 - (-4)$ Then multiply.

$= 27$ Add and subtract last.

(c) $\dfrac{8 - 3 \cdot 2^2}{5 - (6 - 9)}$ The fraction bar is a grouping symbol.

$= \dfrac{8 - 3 \cdot 4}{5 - (-3)}$ Grouping symbols, then exponent

$= \dfrac{8 - 12}{5 - (-3)}$ Then multiply.

$= \dfrac{-4}{8}$ Then subtract.

$= -\dfrac{1}{2}$ Simplify the fraction.

(d) $7 - 3\left[(5 - 3) \div \dfrac{1}{3}\right]$ Start with the innermost grouping symbols.

$= 7 - 3\left[2 \div \dfrac{1}{3}\right]$ Now work inside the brackets.

$= 7 - 3 \cdot 6$ Then multiply.

$= 7 - 18$ Then subtract.

$= -11$

(e) $-32 \div 4 \cdot 3 - 2^3$

$= -32 \div 4 \cdot 3 - 8$ Exponent first

$= -8 \cdot 3 - 8$ Then divide.

$= -24 - 8$ Then multiply.

$= -32$ Subtract last.

(f) $-(-5) - \sqrt{6^2 - 4 \cdot 5}$	Radical symbol is a grouping symbol.
$= -(-5) - \sqrt{36 - 4 \cdot 5}$	Exponent first
$= -(-5) - \sqrt{36 - 20}$	Then multiply.
$= -(-5) - \sqrt{16}$	Then subtract.
$= -(-5) - 4$	Now take the square root.
$= 5 - 4$	$-(-x) = x$
$= 1$	

Evaluating Algebraic Expressions

We can use variables to represent numbers in expressions. An **algebraic expression** is any combination of numbers, variables, grouping symbols, and operation symbols. The following are examples of algebraic expressions.

$$2t^2 - 5t + 1 \qquad \frac{a^2 + 4}{6} \qquad 5(x + 6y) - 4 \qquad r - \sqrt{2t}$$

When we replace the variables in an algebraic expression with specific values, the expression becomes a numerical expression. Performing all indicated operations to determine the value of the expression is called **evaluating** the expression.

In Example 2(b), we include two methods for evaluating an expression with a calculator.

Store

We assign a value to a variable by *storing* the value in the variable.

Evaluate

When we enter an algebraic expression on the home screen, the calculator will return the value of the expression for the currently stored value(s) of the variable(s).

Alpha

Special keystrokes are usually needed to enter variables other than x.

EXAMPLE 2

Evaluating Algebraic Expressions

Evaluate each expression for the given value(s) of the variables(s).

(a) $x^2 + 3x - 4, \quad x = -2$

(b) $5x + 2(1 - x), \quad x = -3$

(c) $3x - |y - 2x|, \quad x = 3, y = 1$

(d) $\dfrac{a + 2b}{2a^2b}, \quad a = -1, b = 3$

Solution

(a) Replace each occurrence of x with -2. Then evaluate the resulting numerical expression.

$$x^2 + 3x - 4 = (-2)^2 + 3(-2) - 4 = 4 - 6 - 4 = -6$$

(b) Replace each occurrence of x with -3.

$$5x + 2(1 - x) = 5(-3) + 2[1 - (-3)]$$

You can then enter the resulting *numerical* expression in your calculator to evaluate the expression. Figure 1.8 shows that the value of the expression is -7.

Figure 1.8

```
5(-3)+2(1-(-3))
                -7
```

Figure 1.9

```
-3→X
            -3
5X+2(1-X)
            -7
```

An alternative method begins with storing -3 in the variable x. Then we enter $5x + 2(1 - x)$ to evaluate the *algebraic* expression. (See Fig. 1.9.)

(c) Replace x with 3 and y with 1.

$$3x - |y - 2x| = 3(3) - |1 - 2(3)| = 9 - |-5| = 4$$

If you use your calculator to evaluate this expression, you will need to store values for both x and y.

(d) $\dfrac{a + 2b}{2a^2b} = \dfrac{-1 + 2 \cdot 3}{2(-1)^2 \cdot 3} = \dfrac{-1 + 6}{2 \cdot 1 \cdot 3} = \dfrac{5}{6}$

Note: The two calculator methods illustrated in Example 2 work well if an expression is to be evaluated just once or twice. For repeated evaluation of an expression, we are better served by other techniques, which we will introduce later.

Equations and Formulas

An **equation** is a statement that two expressions have the same value. If both expressions are numerical, then the truth of the equation can be determined simply by evaluating the expressions.

If one or both sides of an equation are algebraic expressions, then the equation may be true or false, depending on the replacement for the variable. A value of the variable that makes the equation *true* is called a **solution** of the equation.

EXAMPLE 3

Testing Solutions of an Equation

For each equation, determine whether the given number is a solution.

(a) $5 - x^2 = 3x - (2x + 1)$; 2 (b) $\sqrt{4 - x} - \sqrt{x + 6} = 2$; 3

Solution

(a) $5 - x^2 = 3x - (2x + 1)$

$5 - 2^2 = 3 \cdot 2 - (2 \cdot 2 + 1)$ Replace x with 2.

$1 = 1$ True

The last equation is true, so 2 is a solution.

(b) $\sqrt{4 - x} - \sqrt{x + 6} = 2$

$\sqrt{4 - 3} - \sqrt{3 + 6} = 2$ Replace x with 3.

$1 - 3 = 2$

$-2 = 2$ False

The last equation is false, so 3 is not a solution.

A **formula** is an equation that uses an expression to represent some specific quantity. For example, if the base of a triangle is b and the height is h, then the formula $A = \frac{1}{2}bh$ tells us how to determine the area A of the triangle.

An important formula is contained in the **Pythagorean Theorem.**

The Pythagorean Theorem

A triangle is a right triangle if and only if the sum of the squares of the **legs** is equal to the square of the **hypotenuse.** Symbolically, $a^2 + b^2 = c^2$.

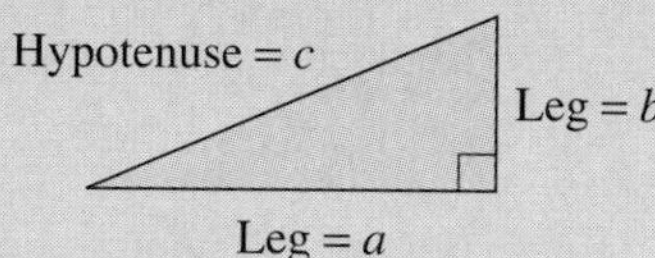

Note: The legs of a right triangle form the right angle; the side opposite the right angle is the hypotenuse.

We can use the formula in the Pythagorean Theorem to determine whether a triangle is a right triangle.

EXAMPLE 4

LEARNING TIP

When you work with the Pythagorean Theorem, it is helpful to begin by identifying the hypotenuse. It is the longest side, and it is located opposite the right angle.

Determining Whether a Triangle Is a Right Triangle

The three numbers given in each part are the lengths of the sides of a triangle. Determine whether the triangle is a right triangle.

(a) 2, 4, 5 (b) 5, 13, 12

Solution

(a) We must determine whether the Pythagorean Theorem formula is true for the given lengths. The longest side is 5, so $c = 5$. It does not matter how we assign the values for a and b.

$$a^2 + b^2 = c^2$$

$$2^2 + 4^2 = 5^2 \qquad \text{Let } a = 2,\ b = 4, \text{ and } c = 5.$$

$$4 + 16 = 25$$

$$20 = 25 \qquad \text{False}$$

The triangle is not a right triangle.

(b) The longest side is 13, so $c = 13$.

$$a^2 + b^2 = c^2$$

$$5^2 + 12^2 = 13^2 \qquad \text{Let } a = 5,\ b = 12, \text{ and } c = 13.$$

$$25 + 144 = 169$$

$$169 = 169 \qquad \text{True}$$

The triangle is a right triangle.

Think About It

Is it necessary to use the Pythagorean Theorem to test 2, 3, and 7 to determine whether these numbers could be the lengths of the sides of a right triangle? Why?

1.4 QUICK REFERENCE

Order of Operations

- To evaluate a numerical expression, we perform the indicated operations in the following order.
  1. If grouping symbols are present, perform all operations inside of them according to the order shown in steps 2–4. For grouping symbols within grouping symbols, start with the innermost group and work outward.
  2. Evaluate exponents and roots.
  3. Perform multiplication and division from left to right.
  4. Perform addition and subtraction from left to right.
- Grouping symbols can be parentheses, brackets, braces, radical symbols, and fraction bars.

Evaluating Algebraic Expressions

- An **algebraic expression** is any combination of numbers, variables, grouping symbols, and operation symbols.
- To **evaluate** an algebraic expression, we replace each variable with a specific value and then perform all indicated operations.

Equations and Formulas

- An **equation** is a statement that two expressions have the same value.
- If an equation contains a variable, a value of the variable that makes the equation true is called a **solution.**
- A **formula** is an equation that uses an algebraic expression to represent some specific quantity.
- In a right triangle, the sides forming the right angle are called **legs;** the side opposite the right angle is called the **hypotenuse.**
- For a right triangle with legs a and b and hypotenuse c, the **Pythagorean Theorem** states that $a^2 + b^2 = c^2$.
- If the lengths of the three sides of a triangle satisfy the formula $a^2 + b^2 = c^2$, where c is the length of the longest side, then the triangle is a right triangle.

1.4 EXERCISES

Concepts and Skills

1. Explain the difference between a numerical expression and an algebraic expression.

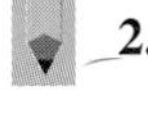
2. For the following two expressions, describe the difference in the order in which the operations are to be performed.

(a) $5 - 3^2$ (b) $(5 - 3)^2$

In Exercises 3–14, perform the indicated operations. Use your calculator to verify results.

3. $-3 - 2^3$

4. $30 \div 5 \cdot 6$

5. $-2|-4| + |5|$

6. $-2\sqrt{16} - \sqrt{6 + 3}$

7. $6 - 4 \cdot 2 + 20 \div 5$

8. $6 - (5 - 3)^2$

9. $\dfrac{5^2 + 7}{11 - 3^2}$

10. $\dfrac{5 \cdot 6 - 2(-3)}{6 - (-6)}$

11. $12 - (5 + 3) \div 2 + 6$

12. $1 - (2 + 7) \div (5 + 4)$

13. $5 - [3(2 - 1) - 4(3 - 5)]$

14. $6 - 4[6 - 4(6 - 4)]$

In Exercises 15–18, write the verbal description as a numerical expression. Then evaluate the expression.

15. Four less than the product of -2 and 7

16. Six more than the square of -8, with the result divided by 10

17. Add -3 to 7 and divide the sum by 2. Then subtract -4 from the quotient.

18. The square of the difference of 3 and -2

In Exercises 19 and 20, insert grouping symbols so that the numerical expression has the values given in parts (a), (b), and (c).

19. $-3 \cdot 4 - 5$

(a) 3

(b) -17

(c) -7

20. $7 - 5^2 + 1$

(a) 5

(b) -19

(c) -17

In Exercises 21–28, use your calculator to perform the operations. Round decimal answers to the nearest hundredth.

21. $-2^4 + 5^1 \cdot 2^3$

22. $2 \cdot 3^2 - 4 \cdot 3 + 1$

23. $(5 - 2)^2 + (-3 - 1)^2$

24. $\sqrt{(7 - 1)^2 + (2 - 5)^2}$

25. $(-3)^2 - 4(2)(-1)$

26. $\sqrt{6^2 - 4(3)(1)}$

27. $\dfrac{5 - \sqrt{5^2 - 4(2)(1)}}{2(2)}$

28. $\dfrac{-3 + \sqrt{(-3)^2 - 4(1)(2)}}{2(1)}$

29. Explain why the expression $\dfrac{x + 1}{x - 2}$ cannot be evaluated for $x = 2$.

30. The expression $\dfrac{x - 7}{7 - x}$ has a value of -1 for any replacement of x except 7. Why?

In Exercises 31–38, evaluate the expression for the given value of the variable. Use your calculator only to verify results.

31. $3x - 2$ for $x = 4$

32. $x - 2(x - 1)$ for $x = -3$

33. $2x^2 - 4x - 9$ for $x = 3$

34. $x^2 - 5x$ for $x = -2$

35. $5 - |t + 3|$ for $t = -5$

36. $\sqrt{7x + 4}$ for $x = 3$

37. $\dfrac{2x - 3}{3 - 2x}$ for $x = 1$

38. $\dfrac{x^2 - 3}{x - 4}$ for $x = -1$

In Exercises 39–46, evaluate the expression for the given values of the variables. Use your calculator only to verify results.

39. $x - y$ for $x = 3$ and $y = -2$

40. $2x - (x - y)$ for $x = 0$ and $y = -5$

41. $|2s - t|$ for $s = -4$ and $t = 2$

42. $5t - 2|t - s|$ for $s = 5$ and $t = 2$

43. $\dfrac{2x + y}{2x - y}$ for $x = 3$ and $y = 6$

44. $\dfrac{xy}{xy - 12}$ for $x = 4$ and $y = 3$

45. $\dfrac{2x + 3y}{x^2 + y^2}$ for $x = -6$ and $y = 4$

46. $\dfrac{4y^2}{9} + \dfrac{x^2}{4}$ for $x = -2$ and $y = 6$

In Exercises 47–50, evaluate $\dfrac{-b + \sqrt{b^2 - 4ac}}{2a}$ for the given values of the variables.

47. $a = 2, b = -3, c = -2$

48. $a = 3, b = -5, c = 2$

49. $a = 9, b = 12, c = 4$

50. $a = 2, b = 3, c = -14$

In Exercises 51–54, evaluate $\dfrac{y_2 - y_1}{x_2 - x_1}$ for the given values of the variables.

51. $x_1 = -2, x_2 = -4, y_1 = 3, y_2 = -1$

52. $x_1 = -12, x_2 = 4, y_1 = 5, y_2 = 5$

53. $x_1 = 0, x_2 = 5, y_1 = 2, y_2 = 0$

54. $x_1 = -3, x_2 = -3, y_1 = 4, y_2 = -2$

In Exercises 55–60, use your calculator to evaluate the expression for the given value(s) of the variable(s). Round decimal answers to the nearest hundredth.

55. $3x - 2$ for $x = -2.47$

56. $t^2 - 3t$ for $t = -4.56$

57. $-y + 3z$ for $y = 2.1$ and $z = -3.7$

58. $m^2 - 2n^2$ for $m = -2.14$ and $n = 7.12$

59. $\sqrt{5a} - |-25b|$ for $a = 2$ and $b = 10$

60. $\sqrt{x^2 - 3x + 2}$ for $x = 2$

61. Suppose that you store 5 in t and then you evaluate $\sqrt{4 - t}$. Explain the result.

62. Suppose that someone turns on your calculator and presses the X key and the ENTER key. How does the calculator respond? Why?

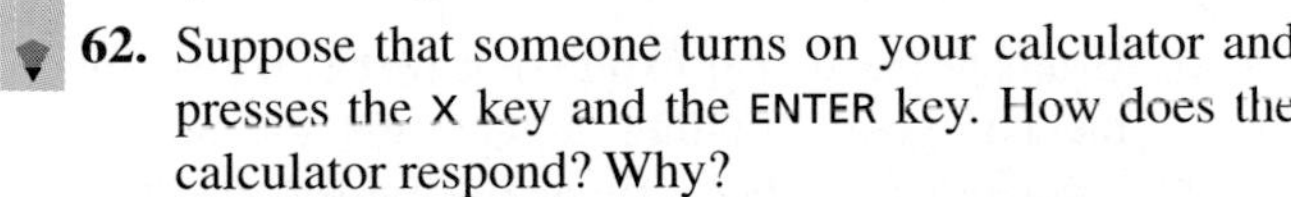

In Exercises 63–68, suppose that you store 5 for x. What value would you need to store for y in order for the following expressions to have a value of 9? Use your calculator to verify your answers.

63. $x + y$ **64.** $x - y$ **65.** xy

66. $\dfrac{y}{x}$ **67.** $|y - x|$ **68.** y

69. Explain the difference between evaluating an expression and verifying that a number is a solution of an equation.

70. Explain how to determine which member of the set $A = \{-8, -1, 0, 2.3, 8, 37.92\}$ is a solution of the equation $3 \cdot (2x) = (3 \cdot 2)x$ without substituting any of the numbers from set A.

In Exercises 71–78, determine whether the given number is a solution of the equation.

71. $3x - 12 = -3(4 - x);\quad 0$

72. $x - 7 = 7 - x;\quad 2$

73. $x(x - 3) = 2;\quad 2$

74. $(x - 5)(x + 2) = 0;\quad -2$

75. $x^2 - x - 6 = 0;\quad -2$

76. $2x^2 - 7x + 3 = 0;\quad 1$

77. $\dfrac{x}{2} - 3 = 5;\quad 4$ **78.** $\dfrac{x}{3} + \dfrac{x}{6} = \dfrac{3}{2};\quad 3$

Geometric Models

In Exercises 79–86, the three given numbers are the lengths of the sides of a triangle. Determine whether the triangle is a right triangle.

79. 5, 7, 4

80. 12, 13, 17

81. 17, 8, 15 **82.** 24, 25, 7

83. 9, 12, 15 **84.** 25, 24, 10

85. $1, 1, \sqrt{2}$ **86.** $1, 2, \sqrt{3}$

87. Some community volunteers have laid out a softball field at the local park. The distance between the bases is 60 feet. The distance from home plate to second base is 90 feet. (See figure.) Is the infield perfectly square? If not, what should be the distance from home plate to second base (to the nearest tenth)?

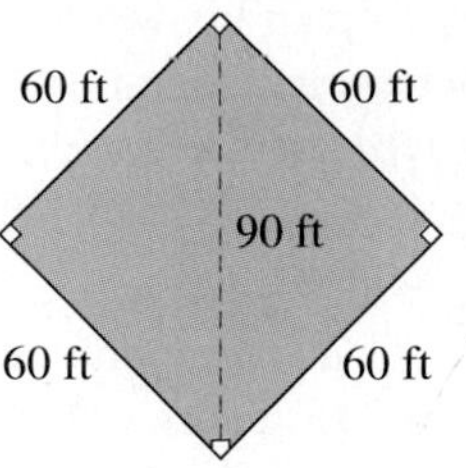

88. Standard letter paper is 8.5 inches wide and 11 inches long. To the nearest tenth, what is the diagonal distance d across the paper? (See figure.)

Figure for 88

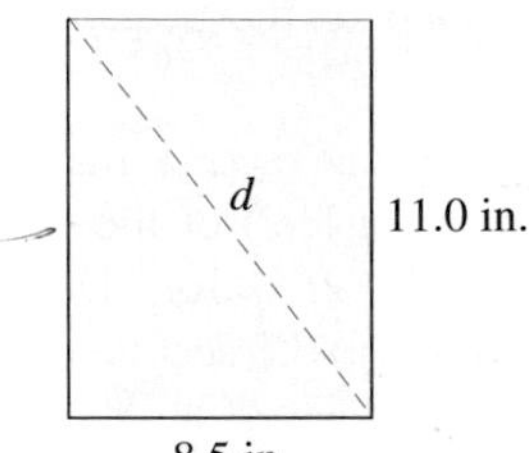

Figure for 89

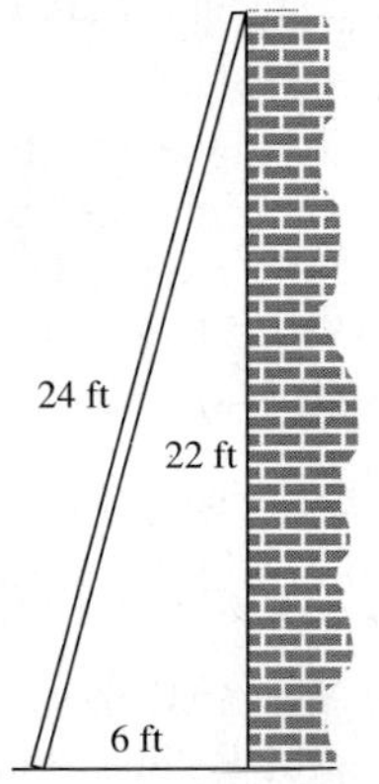

89. A building inspector is checking on an apartment building that is under construction. She finds that a window ledge, exactly 22 feet from the level ground, can be reached with a ladder that is 24 feet long. (See figure.) The bottom of the ladder is 6 feet away from the base of the wall, and the top of the ladder just reaches the window ledge. Is the wall perfectly vertical? Explain.

90. **Perpendicular lines** are lines that intersect to form right angles. A surveyor places pins at three corners of a property. (See figure.) The distance along one side of the property is 80 feet, and the distance along the other side is 150 feet. Explain what measurement the surveyor can use to determine if the two sides of the property are perpendicular. What should the measurement be?

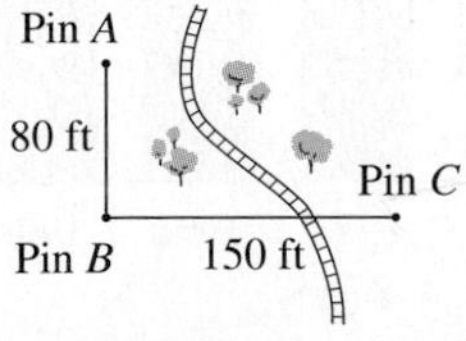

Real-Life Applications

91. A tourist is planning a January trip from his home in London to San Francisco. To plan his wardrobe, he learns that the average January temperature is 50°F. To convert to Celsius, the temperature scale in England, he uses the formula $C = \frac{5}{9}(F - 32)$. What is the corresponding Celsius temperature?

92. A radio station in Toronto, Canada, reports that the current temperature is $-10°$. Listeners who live in Buffalo, New York, know that this is a Celsius temperature. To convert to Fahrenheit, they use the formula $F = \frac{9}{5}C + 32$. What is the Fahrenheit temperature?

93. If a college fund of $1000 is invested at 6% simple interest, what will be the value of the investment after 5 years?

94. To signal the beginning of a yacht race, a flare is shot into the air. The height h (in feet) of the flare after t seconds is given by $h = 112t - 16t^2$. Determine the height of the flare after (a) 3.5 seconds and (b) 6 seconds.

Modeling with Real Data

95. We frequently model data with an expression in which the variable represents a year. For convenience, we usually select a base year and let the variable represent the number of years after the base year.

The accompanying bar graph shows the number of countries in which McDonald's restaurants were established during the period 1980–1996. (Source: McDonald's Corporation.)

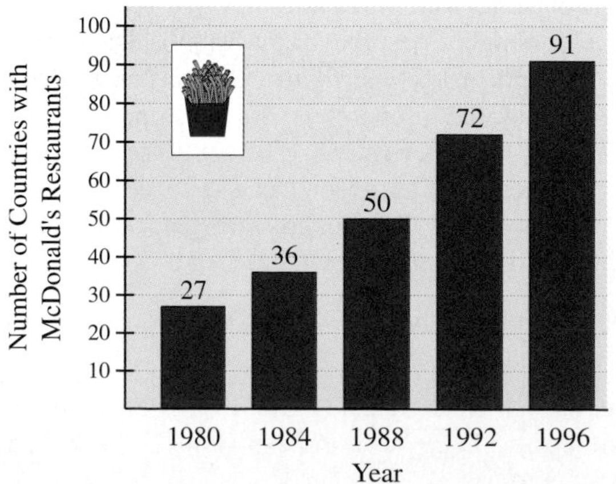

We can model the number y of countries with the expression $4.1x + 22.4$, where x is the number of years since 1980.

(a) Evaluate the model expression for each year in the bar graph. Letting A represent the actual number of countries, make a table with the headings *Year*, x, A, y, and $A - y$.

(b) In what year is the model expression most accurate? least accurate?

(c) Determine the average of the entries in the column $A - y$. What does the result indicate, if anything, about the accuracy of the model expression?

96. The number of letter carriers who were bitten by dogs decreased from 7000 in 1983 to 2787 in 1994. (Source: U.S. Postal Service.) With Y representing the number of years since 1980, the expression $-0.38Y + 8.15$ is a model for the number of thousands of letter carriers who were bitten in year Y during this time interval. Use the model to estimate the number of letter carriers who were bitten in 1990.

Group Project: Single-Parent Families

During the past 10 years, the number of families with women as the sole parent has increased. The accompanying graph shows the number of such families (in millions) for selected years. (Source: U.S. Department of Commerce.)

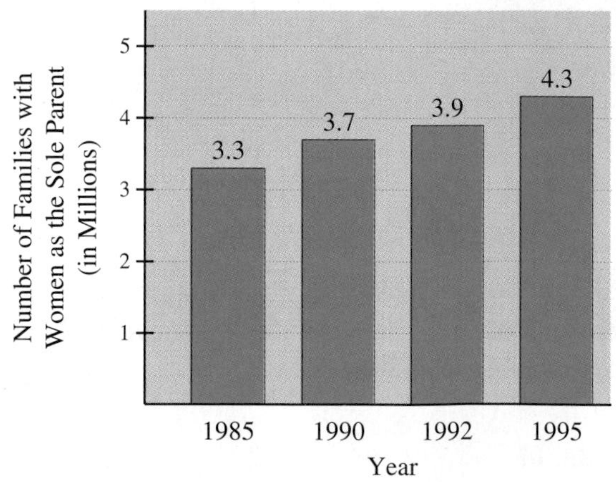

The data can be modeled by either of the following expressions.

Expression 1 (E_1): $0.003t^2 + 0.07t + 3.3$

Expression 2 (E_2): $0.08t + 3.3$

In both models, t is the number of years since 1985.

97. To the nearest tenth, evaluate each expression for each of the given 4 years in the graph. How accurate are the models for these years?

98. Evaluate each expression to estimate the number of such families in 2005. What do the results suggest about the validity of both models beyond 1995?

99. How can the two models appear to be so different but both model the data? (Hint: Consider the value of the term $0.003t^2$ for small values of t.)

100. What is the apparent trend in the number of families with women as the sole parent? What do you think are the social and economic implications of this trend on society in general?

Challenge

101. For the following expressions, $a > 0$, $b > 0$, and $c < 0$. Determine whether the given expression represents a positive number, a negative number, or either.

(a) $a - c$ (b) $a + c$ (c) ab

(d) ac (e) $bc - a$ (f) a^2c

(g) ac^2 (h) $\dfrac{a+b}{c}$

102. The value of the following expression is 362,880.

$$1 \cdot 2 \cdot 3 \cdot 4 \cdot 5 \cdot 6 \cdot 7 \cdot 8 \cdot 9$$

Replace some of the multiplication symbols with $+$, $-$, and/or $\div$ so that the value of the resulting expression is 100.

103. One way to write 3 with three 2's is $2 + 2 \div 2$. Write 30 with three 3's.

104. Using four 4's, we can write 1 as $\frac{44}{44}$ and 2 as $\frac{4}{4} + \frac{4}{4}$. Write the numbers 3 through 10 with four 4's.

105. The following equations illustrate sets of three integers that satisfy the Pythagorean Theorem formula $a^2 + b^2 = c^2$.

$$3^2 + 4^2 = 5^2$$
$$5^2 + 12^2 = 13^2$$
$$7^2 + 24^2 = 25^2$$
$$9^2 + 40^2 = ?$$

What pattern is revealed by these equations? Write the next two equations in the list and use your calculator to verify that they satisfy the Pythagorean Theorem formula.

106. In the accompanying figure, ΔABC is a right triangle. Squares I, II, and III are drawn so that one side of each square is also a side of the triangle. Explain how the area of Square I is related to the areas of the other two squares. (Hint: Consider the Pythagorean Theorem.)

Figure for 106

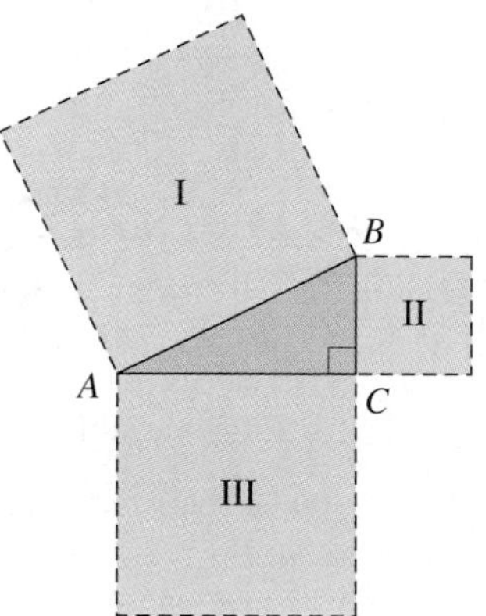

1.5 SIMPLIFYING EXPRESSIONS

Terms and Coefficients • Simplifying Algebraic Expressions • Translations

Terms and Coefficients

We refer to the addends of an algebraic expression as **terms.** A term with no variable is called a **constant term.**

The expression $3y^2 - 5y + 1$ can be written as $3y^2 + (-5y) + 1$. Thus the terms are $3y^2$, $-5y$, and 1, and the constant term is 1. Because the expression $x^2 + 7y^2$ can be written as $x^2 + 7y^2 + 0$, the constant term is 0.

Grouping symbols can affect the number of terms of an expression.

Expression	*Terms*	
$x^2 + 5x + 3$	$x^2, 5x, 3$	Three terms
$x^2 + 5(x + 3)$	$x^2, 5(x + 3)$	Two terms

Think About It

In the expression $5[(x + y + 2) - 7]$, how many terms are inside the parentheses? inside the brackets? in the entire expression?

Note: An easy way to identify terms of an expression is to determine the parts of the expression that are separated by plus or minus symbols that are not inside grouping symbols.

The numerical factor in a term is called the **numerical coefficient,** or simply the **coefficient.** For instance, the coefficient of $3a$ is 3, the coefficient of $-xy$ is -1, and the coefficient of y is 1.

EXAMPLE 1

Identifying Terms and Coefficients

Identify the terms and coefficients in each expression.

Expression	*Terms*	*Coefficients*
(a) $4a^2 - 3a + 5$	$4a^2, -3a, 5$	$4, -3, 5$
(b) $y - (x - 3) - 2$	$y, -(x - 3), -2$	$1, -1, -2$
(c) $3y - x + \dfrac{y - 5}{2}$	$3y, -x, \dfrac{y - 5}{2}$	$3, -1, \dfrac{1}{2}$

Note that $\dfrac{y - 5}{2} = \dfrac{1}{2}(y - 5)$. Therefore, the coefficient is $\dfrac{1}{2}$.

Two terms are called **like terms** if they are both constant terms or if they have the same variable factors with the same exponents.

Like Terms	*Unlike Terms*
$2x, 7x$	$2x, 7$
$5y, 3y$	$5y, 3y^2$
$3a^2b, 6a^2b$	$3a^2b, 6ab^2$

Simplifying Algebraic Expressions

Equivalent expressions are expressions that have the same value for all replacements for the variables for which both expressions are defined. For instance, because $\dfrac{x}{x}$ and 1 have the same value for all values of x except 0, they are equivalent expressions.

To **simplify** an algebraic expression, we remove grouping symbols and combine like terms. The result is an expression that is equivalent to the original expression.

Combining like terms is accomplished by applying the Distributive Property.

$$3a + 5a = a(3 + 5) = a \cdot 8 = 8a$$

Because the Commutative Property of Multiplication allows us to write factors in any order, we can save a step when we combine like terms.

$$6n^2 - 7n^2 = (6 - 7)n^2 = -1n^2 = -n^2$$

EXAMPLE 2

Combining Like Terms

(a) $2y - 3 - 5y + 1 = (2 - 5)y + (-3 + 1) = -3y - 2$

(b) $2a^2 + b - 2b - a^2 = (2 - 1)a^2 + (1 - 2)b = a^2 - b$

(c) $a^2b - ab^2 + 3a^2b - 5ab^2 = 4a^2b - 6ab^2$

We also can use the Distributive Property to remove certain grouping symbols. An expression of the form $a(b + c)$ is equivalent to the expression $ab + ac$.

An expression of the form $-(b + c)$ is equivalent to the expression $-1(b + c)$, which is equivalent to $-b - c$. In effect, the opposite sign in front of the group has the effect of changing the sign of each term inside the group.

EXAMPLE 3

Simplifying Algebraic Expressions

(a) $(3 + 4x) - (5x - 1) = 3 + 4x - 5x + 1$ Remove parentheses.
$= -x + 4$ Combine like terms.

(b) $2(3x^2 - 2) - 3(x + 5) = 6x^2 - 4 - 3x - 15$ Distributive Property
$= 6x^2 - 3x - 19$ Combine like terms.

(c) $3x + x(5 - x) - 7x = 3x + 5x - x^2 - 7x$ Distributive Property
$= -x^2 + x$ Combine like terms.

(d) $7 - 3[4 - (x - 2)] = 7 - 3[4 - x + 2]$ Remove parentheses.
$= 7 - 3[6 - x]$ Combine like terms.
$= 7 - 18 + 3x$ Remove brackets.
$= -11 + 3x$ Combine like terms.

Translations

We can think of a **model** as a representation of an object, an idea, or information. When we use algebra to solve problems, it is often necessary to write the given information as an algebraic expression or equation that symbolically models the conditions of the problem and the operations to be performed.

The following are some typical translations involving the four basic operations.

Addition

Phrase	*Expression*
5 added to a number	$5 + x$
a number increased by 8	$x + 8$
7 more than a number	$x + 7$
the sum of two numbers	$x + y$

Subtraction

Phrase	*Expression*
-3 subtracted from a number	$x - (-3)$
a number decreased by 7	$x - 7$
6 less than a number	$x - 6$
the difference between 8 and a number	$8 - x$

Multiplication

Phrase	*Expression*
a number multiplied by -4	$-4x$
two-thirds of a number	$\frac{2}{3}x$
the product of two numbers	xy
twice a number	$2x$
6% of a number	$0.06x$

Division

Phrase	*Expression*
the quotient of -3 and a number	$\frac{-3}{x}$
the ratio of two numbers	$\frac{x}{y}$
a number divided by 5	$\frac{x}{5}$

When we translate information into an equation, we watch for words such as *is*, *equals*, *obtain*, and *results in*, which correspond to the equality symbol.

EXAMPLE 4

Translating into Equations

Translate the given information into an equation.

(a) The ratio of a number and 5 is 3.

(b) If a number is subtracted from -7, the result is 9.

(c) If you increase the product of two numbers by 1, you obtain 11.

LEARNING TIP

Avoid trying to translate all the given information at once. Break the information down into words and phrases, and build your expression or equation a little bit at a time.

Solution

(a) Letting x represent the number, $\frac{x}{5} = 3$.

(b) Letting n represent the number, $-7 - n = 9$.

(c) Letting a and b represent the numbers, $ab + 1 = 11$.

In addition to knowing how to translate such typical phrases, some common knowledge is sometimes needed to write expressions.

EXAMPLE 5

Translating Word Phrases into Algebraic Expressions

Translate the given verbal expressions into algebraic expressions.

(a) The value in cents of n nickels and d dimes.

(b) The difference between two numbers divided by the sum of their squares.

(c) The reciprocal of a number x increased by two-thirds of the number.

(d) The year-end value of y dollars invested at a 7% simple interest rate.

Solution

(a) $5n + 10d$

(b) $\dfrac{x - y}{x^2 + y^2}$

(c) $\dfrac{1}{x} + \dfrac{2}{3}x$

(d) $y + 0.07y = 1y + 0.07y = 1.07y$

1.5 QUICK REFERENCE

Terms and Coefficients

- The addends of an algebraic expression are called **terms.** A term with no variable is a **constant term.**
- The numerical factor in a term is called the **coefficient.**
- Two terms are called **like terms** if they are both constant terms or if they have the same variable factors with the same exponents.

Simplifying Algebraic Expressions

- **Equivalent expressions** are expressions that have the same defined value regardless of the replacements for the variables.
- To **simplify** an algebraic expression, remove grouping symbols and combine like terms.
- We use the Distributive Property to combine like terms. The procedure is to combine the coefficients of the terms and retain the common variable factors.
- We also use the Distributive Property to remove grouping symbols.
  1. $a(b + c) = ab + ac$
  2. $-(b + c) = -b - c$

1.5 EXERCISES

Concepts and Skills

1. Use the Additive Identity Property to explain why every algebraic expression has a constant term.

2. Use the Multiplicative Identity Property to explain why every term has a numerical coefficient.

In Exercises 3–10, identify the terms and coefficients in each expression.

3. $2x - y + 5$

4. $6a - 5b + c - 2d$

5. $3a^3 - 4b^2 + 5c - 6d$

6. $3x^2 + 5x - 6$

7. $7y^2 - 2(3x - 4) + \frac{2}{3}$ **8.** $\frac{5x}{3} - 2(y - 3)$

9. $\frac{x + 4}{5} - 5x + 2y^3$

10. $\frac{2x^2}{5} - y^2 - 8x + 2$

11. Explain why $2x + 3$ has two terms but $2(x + 3)$ is one term.

12. Explain why $5x^3y^2$ and $5x^2y^3$ are not like terms.

In Exercises 13–28, simplify the expressions by combining like terms.

13. $2x + 1 - 3x + 2$ **14.** $3 - x + 3x - 5$

15. $2x - 3y + x - 5$ **16.** $x - y - 3y + 4$

17. $2x + 4x - 7y$ **18.** $5y - 8y + 3x$

19. $-x + 3 - 4x + 1 + 5x$

20. $4 + 2x + x - 3 - x - 7$

21. $\frac{3}{2}x + \frac{2}{3}y - \left(-\frac{1}{3}\right)y - \frac{5}{2}x$

22. $\frac{3}{7}m - \frac{3}{8}n + \frac{4}{7}m + \frac{5}{8}n$

23. $8x^2 + 9x^2 - 7x + 10x$

24. $17x^3 - 4x^2 + 5x^3 - 7x^2$

25. $3ac^2 - 7ac + 7ac^2 - 6ac$

26. $10ax^2 + 10a^2x - 8ax^2 - 2a^2x$

27. $5.23x^2 - 5.23x^3 + 7.98x^3 - 9.63x^2$

28. $7.36a^2 - 8.97b^2 + 4.86a^2$

29. Explain why we can remove the parentheses in $-(2x - 5)$ and in $3 - (4a + 1)$ by changing the sign of each term inside the grouping symbols.

30. Do the expressions $2x^2z$ and $2xyz$ have the same number of factors? Why?

In Exercises 31–38, simplify each expression by removing the grouping symbols.

31. $-(a - b)$ **32.** $-(3 + n)$

33. $5(2x - 3)$ **34.** $-3(4 - x)$

35. $-(2a - 3b + 7d)$ **36.** $-(3h - 7k - 8g)$

37. $3(x - 2y) - (z - 2)$

38. $-3(2x + 4y) - 7(z - 2)$

In Exercises 39–44, match each expression in column A with its equivalent simplified expression in column B.

Column A	*Column B*
39. $2(x - 3) - 5x$	(a) $-(x - 1)$
40. $-(x - y) + x$	(b) $x^2y + xy^2$
41. $-x^2 + (-x)^2$	(c) $4x + 3$
42. $\frac{1}{2}(8x + 6)$	(d) $-3x - 6$
43. $-[-(1 - x)]$	(e) y
44. $xy^2 + x^2y$	(f) 0

In Exercises 45–68, simplify the expressions by removing grouping symbols and combining like terms.

45. $2x - (3x + 1)$ **46.** $5 + (3x - 4)$

47. $3(x - 4) - 2x$ **48.** $-2(3 - x) - 4$

49. $-(2a + 4b) - (5a - 2b)$

50. $-(3a - 7b) - (8b - a)$

51. $3x + 2(x - 4) - 5x$

52. $4 - 2(2x - 5) - x + 3$

53. $3(x + 5) + 2(1 - 3x)$

54. $5 - (x + 3) - 4$

55. $-(x - 1) + 2x + 1$

56. $5x - 1 + (x - 5)$

57. $(x + 1) + (2x - 1) + 5$

58. $-x - y - (x + y) + x$

59. $5(2t + 1) - 2(t - 4)$

60. $-(t + 1) - (2t - 1)$

61. $3x - 5y + 2x - (3x - 5y + 4)$

62. $7y - 6x + 9y - (7y - 3x - 8y)$

63. $6c - 5[6c - 5(c - 5)]$

64. $4a - 3[4a - 3(2a - 7)]$

65. $2(3x - 5) - 3(4x - 7) - 8(x + 6)$

66. $3(4x - 7) - 10(2x - 3) - 9(4 - x)$

67. $7 + 3[-(x - 2) - (3 - x)]$

68. $6 + 4[-2(5 - 3x) - (5x - 3)]$

In Exercises 69–72, simplify the expressions by removing grouping symbols and combining like terms.

69. $2x^2y^3 - 5x^3y^2 + 8x^3y^2 - x^2y^3$

70. $3x^4y - (x^2y^2 + x^4y)$

71. $-(x^5y^2 - x^2y^5) - (x^2y^5 - x^5y^2)$

72. $xyz + xy^2 - y(xz - xy)$

Modeling Conditions

In Exercises 73–76, translate the given verbal expressions into algebraic expressions. Let n represent the number. Simplify the expression, if possible.

73. The quotient of twice the number and 12

74. The product of the number and the number increased by 4

75. Three times the number plus the number itself

76. Eight more than the number less twice the number

In Exercises 77–80, column A contains a word phrase describing a number n. Match the word phrase with the algebraic expression in column B.

Column A	*Column B*
77. Five less than a number	(a) $5 - n$
78. The difference between 5 and a number	(b) $n - 5$
79. The difference between a number and 5	
80. Five more than the opposite of a number	

In Exercises 81–90, translate the given information into an equation. Let x represent the unknown number and let y represent any second unknown number.

81. Five less than a number is 3.

82. One more than twice a number is 7.

83. The product of 8 and a number is 11.

84. The quotient of a number and 4 is 3 times the number.

85. Twice a number is 3 less than the number.

86. If 25 is added to the square of a number, the result is 10 times the number.

87. Decreasing a number by 40% of the number results in 50.

88. If one number is increased by half of another number, the result is 16.

89. If you take the difference between a number and twice another number, you obtain -5.

90. If the numerator of the ratio $\frac{x}{3}$ is increased by 2, the value of the ratio becomes 7.

In Exercises 91–104, translate the verbal description into an algebraic expression.

91. The perimeter of a rectangle whose width is half its length L

92. The area of a triangle whose base is twice its height h

93. The value (in cents) of q quarters and d dimes

94. The distance (in miles) traveled in h hours at an average speed of 55 mph

95. The cost of online service for x hours if the monthly charge is \$15 plus \$2.50 per hour

96. The portion of a job done in 1 hour if it takes h hours to do the entire job

97. The sum of three consecutive integers, the smallest being n

98. The quotient of two consecutive odd integers, the smallest being n

99. The annual interest earned on d dollars if the simple interest rate is 8%

100. The amount needed to pay off a 1-year loan of d dollars if the simple interest rate is 9%

101. The hypotenuse of a right triangle if one leg is 2 units less than the other leg, whose length is a

102. The selling price of a television set purchased at a 35% discount off the original price c

103. A person's salary after a 5% raise if the old salary was s

104. The amount of acid in L liters of a 20% solution

105. Which expression is the correct translation of the phrase "the product of 3 and a number increased by 7": $3x + 7$ or $3(x + 7)$? Explain.

106. Evaluate the expression $\sqrt{a + b}$ and the expression $\sqrt{a} + \sqrt{b}$ for $a = 1$ and $b = 0$. Are these equivalent expressions? Use the definition of equivalent expressions to explain your answer.

Modeling with Real Data

107. The 8–17 age group averages about 12 shopping trips per month, and the average amount spent is about \$25 per trip. The 8–12 age group averages \$18.50 per trip, while the 13–17 age group averages \$31.20 per trip. (Source: International Mass Retail Association.)

(a) The expression $2.54A - 6.9$ can be used as a model to estimate the amount spent per trip by a person A years old. Create an appropriate table and estimate the amount spent per trip by persons aged 10, 12, 14, and 16.

(b) Use your table to estimate the age at which spending exceeds the average for the entire age group.

(c) Using the average number of shopping trips per month, estimate the average amount spent per month for ages 10, 12, 14, and 16.

108. In 1992, the number of motor vehicle thefts per 100,000 people was 632. (Source: Federal Bureau of Investigation.) Write an algebraic expression that models the number of motor vehicle thefts in a city with population P.

Group Project: Altitude and Temperature

Temperature decreases with altitude. Using a reference temperature of 60°F at sea level, the temperature is approximately −67°F at an altitude of 36,000 feet (Source: American Meteorological Society.)

109. Using the given reference point, what is the total temperature change from sea level to an altitude of 36,000 feet?

110. Assuming a constant decrease in temperature, what negative number represents the temperature decrease for each altitude increase of 1000 feet? (Round your answer to the nearest tenth of a degree.)

111. Using the assumption and the result in Exercise 110, what temperature would you predict at an altitude of 22,000 feet?

112. Using the assumption and the result in Exercise 110, at what altitude would you be flying if the outside temperature is at the freezing point?

Challenge

In Exercises 113–118, determine whether the given expression is equivalent to (a) x^2 or (b) $-x^2$.

113. $(-x)^2$ **114.** $-(-x)^2$ **115.** $|x^2|$

116. $|-x^2|$ **117.** $-|x^2|$ **118.** $|x|^2$

In Exercises 119 and 120, simplify the expression.

119. $x \cdot |x| + x^2$, where $x > 0$

120. $x \cdot |x| + x^2$, where $x < 0$

1.6 INTEGER EXPONENTS

Bases • Product and Quotient Rules for Exponents • Zero and Negative Exponents

Bases

In Section 1.2 we used a positive integer *exponent* to indicate repeated multiplication. The exponent indicates the number of times the *base* is used as a factor.

$$6 \cdot 6 \cdot 6 \cdot 6 = 6^4 \qquad t^5 = t \cdot t \cdot t \cdot t \cdot t$$

In the expression 6^4, called an **exponential expression,** the exponent is 4 and the base is 6. When no exponent is indicated, we assume it to be 1: $b = b^1$.

LEARNING TIP

Writing the understood exponent 1 helps as you simplify an expression in which no exponent is indicated.

Think About It

What repeated exponential operation would result in exponents in the following sequence?

1, 2, 4, 8, 16, 32, 64, . . .

When we work with exponential expressions, identifying the base is important.

$(-5)^2 = (-5)(-5) = 25$ The base is −5.

$-5^2 = -1 \cdot 5^2 = -1 \cdot 5 \cdot 5 = -25$ The base is 5.

$3a^4 = 3 \cdot a \cdot a \cdot a \cdot a$ The base for the exponent 4 is a.

$(3a)^4 = (3a)(3a)(3a)(3a)$ The base for the exponent 4 is $3a$.

Product and Quotient Rules for Exponents

Consider the product $b^3 \cdot b^5$. Using the fact that a positive integer exponent indicates repeated multiplication, we can rewrite the product.

$$b^3 \cdot b^5 = \underbrace{b \cdot b \cdot b}_{b^3} \cdot \underbrace{b \cdot b \cdot b \cdot b \cdot b}_{b^5} = b^8$$

This result suggests that we can simplify a product of exponential expressions with like bases by adding the exponents and retaining the base.

Product Rule for Exponents

For any real number b, and for any positive integers m and n,

$$b^m \cdot b^n = b^{m+n}$$

Note: In the Product Rule for Exponents, the bases can be numbers, variables, or expressions, but they must be the same.

EXAMPLE 1

Using the Product Rule for Exponents

(a) $3^7 \cdot 3^9 = 3^{7+9} = 3^{16}$

(b) $x^{18} \cdot x^{13} = x^{18+13} = x^{31}$

(c) $(2x^2y)(5x^5y^4) = (2 \cdot 5)(x^2 \cdot x^5)(y^1 \cdot y^4) = 10x^{2+5}y^{1+4} = 10x^7y^5$

(d) $(x + 3)^2(x + 3)^5 = (x + 3)^{2+5} = (x + 3)^7$

We can use the Product Rule for Exponents to simplify a quotient such as $\frac{b^5}{b^2}$.

$$\frac{b^5}{b^2} = \frac{b^{2+3}}{b^2} = \frac{b^2 \cdot b^3}{b^2} = \frac{b^2}{b^2} \cdot \frac{b^3}{1} = 1 \cdot b^3 = b^3$$

This result suggests that we can simplify a quotient of exponential expressions with like bases by subtracting the exponents and retaining the base.

Quotient Rule for Exponents

For any real number b, and for any positive integers m and n,

$$\frac{b^m}{b^n} = b^{m-n}$$

EXAMPLE 2

Using the Quotient Rule for Exponents

Use the Quotient Rule for Exponents to perform the indicated operations. Assume that no divisor has a value of 0.

(a) $\frac{20^{12}}{20^8} = 20^{12-8} = 20^4 = 160{,}000$

(b) $\frac{6x^9}{9x^6} = \frac{3}{3} \cdot \frac{2}{3} \cdot \frac{x^9}{x^6} = 1 \cdot \frac{2}{3} \cdot x^{9-6} = \frac{2}{3}x^3$

(c) $\frac{(x-2)^9}{(x-2)^5} = (x-2)^{9-5} = (x-2)^4$

(d) $\dfrac{5a^7b^6}{10a^6b^2} = \dfrac{5}{5} \cdot \dfrac{1}{2} \cdot a^{7-6}b^{6-2} = \dfrac{1}{2}ab^4$

Zero and Negative Exponents

Applying the Quotient Rule for Exponents may lead to exponents other than natural numbers. For instance,

$$\frac{b^3}{b^3} = b^{3-3} = b^0$$

Because any nonzero number divided by itself is 1, we also know that $\dfrac{b^3}{b^3} = 1$. This observation is the basis for defining a zero exponent.

Definition of the Zero Exponent

For any nonzero real number b, $b^0 = 1$.

Note: The base can be any quantity except 0; that is, 0^0 is undefined. In all expressions involving a variable base and zero exponent, we will assume that the base is not zero.

EXAMPLE 3

Exponential Expressions with Zero Exponents

(a) $5^0 = 1$ — The base is 5.

(b) $(-5)^0 = 1$ — The base is -5.

(c) $-5^0 = -1 \cdot 5^0 = -1 \cdot 1 = -1$ — The base is 5.

(d) $(4y)^0 = 1$ — The base is $4y$.

(e) $4y^0 = 4 \cdot y^0 = 4 \cdot 1 = 4$ — The base of the exponent 0 is y.

We can use the Quotient Rule for Exponents to write $\dfrac{b^4}{b^7} = b^{4-7} = b^{-3}$.

Another approach to simplifying the expression uses the Product Rule for Exponents.

$$\frac{b^4}{b^7} = \frac{1 \cdot b^4}{b^3 \cdot b^4} = \frac{1}{b^3} \cdot \frac{b^4}{b^4} = \frac{1}{b^3} \cdot 1 = \frac{1}{b^3}$$

This result suggests the following definition.

Definition of an Integer Exponent

For any nonzero real number b and any integer n,

$$b^{-n} = \frac{1}{b^n}$$

Note: A base with a negative exponent cannot be 0; that is, 0^n is undefined if $n < 0$. In all expressions involving a variable base and a negative exponent, we will assume that the base is not zero.

Because n can be any integer, the definition of a negative exponent may be expressed in either of the following ways.

$$b^{-n} = \frac{1}{b^n} \quad \text{or} \quad b^n = \frac{1}{b^{-n}}$$

According to the definition of an integer exponent, each of the following is true.

$$7^{-4} = \frac{1}{7^4} \qquad y^{-1} = \frac{1}{y} \qquad (-5)^{-3} = \frac{1}{(-5)^3} \qquad \left(\frac{5}{8}\right)^{-2} = \left(\frac{8}{5}\right)^2 \qquad \frac{1}{c^{-7}} = c^7$$

 Exponent

You can use your calculator to evaluate an exponential expression with a negative exponent.

EXAMPLE 4

Exponential Expressions with Negative Exponents

(a) $(-4)^{-2} = \dfrac{1}{(-4)^2} = \dfrac{1}{16} = 0.0625$ — The sign of the exponent changes, but the sign of the base does not change.

(b) $-4^{-2} = -1 \cdot 4^{-2} = -1 \cdot \dfrac{1}{4^2} = -\dfrac{1}{16} = -0.0625$

(c) $-(-5)^{-1} = -1 \cdot (-5)^{-1} = \dfrac{-1}{(-5)^1} = \dfrac{1}{5}$

(d) $3^{-1} + 2^{-1} = \dfrac{1}{3} + \dfrac{1}{2} = \dfrac{2}{6} + \dfrac{3}{6} = \dfrac{5}{6}$

(e) $(3 + 2)^{-1} = 5^{-1} = \dfrac{1}{5}$

(f) $\left(\dfrac{2}{5}\right)^{-2} = \left(\dfrac{5}{2}\right)^2 = \dfrac{25}{4}$ — $\left(\dfrac{a}{b}\right)^{-n} = \left(\dfrac{b}{a}\right)^n$

EXAMPLE 5

Writing Exponential Expressions with Positive Exponents

(a) $x^{-5} = \dfrac{1}{x^5}$

(b) $(3k)^{-5} = \dfrac{1}{(3k)^5}$ — The base is $3k$.

(c) $3k^{-5} = 3 \cdot \dfrac{1}{k^5} = \dfrac{3}{k^5}$ — The base is k.

(d) $\dfrac{1}{b^{-3}} = b^3$

(e) $\dfrac{3}{p^{-8}} = 3 \cdot \dfrac{1}{p^{-8}} = 3p^8$

(f) $\dfrac{7^{-2}}{2^{-5}} = \dfrac{7^{-2}}{1} \cdot \dfrac{1}{2^{-5}} = \dfrac{1}{7^2} \cdot \dfrac{2^5}{1} = \dfrac{2^5}{7^2}$

Example 5 suggests that we can move exponential factors in a fraction from the numerator to the denominator or from the denominator to the numerator if we change the sign of the exponent. This method applies only to exponential *factors*, not exponential terms.

EXAMPLE 6

Writing Exponential Expressions with Positive Exponents

(a) $\frac{s^{-6}}{t^{-9}} = \frac{t^9}{s^6}$

(b) $\frac{c^2d^{-3}}{5k^{-2}} = \frac{c^2k^2}{5d^3}$ — The factors c^2 and 5 have positive exponents and are not moved.

(c) $6x^{-1}y = \frac{6x^{-1}y}{1} = \frac{6y}{x}$ — The factors 6 and y have positive exponents and are not moved.

(d) $x^{-2}y + xy^{-2} = \frac{y}{x^2} + \frac{x}{y^2}$ — Apply the definition of negative exponent to each term.

The product and quotient rules for exponents apply to zero and negative exponents as well as positive exponents.

EXAMPLE 7

Applying the Product and Quotient Rules to Exponential Expressions with Negative Exponents

(a) $(3x^{-3}y)(2x^5y^{-4}) = (3 \cdot 2)(x^{-3} \cdot x^5)(y^1 \cdot y^{-4}) = 6x^2y^{-3} = \frac{6x^2}{y^3}$

(b) $\frac{a^{-3}b^7}{a^2b^{-4}} = a^{-3-2}b^{7-(-4)} = a^{-5}b^{11} = \frac{b^{11}}{a^5}$

1.6 QUICK REFERENCE

Bases

- An **exponential expression** is an expression that involves exponents.
- For natural number exponents, the **base** is the number being multiplied, and the **exponent** indicates the number of factors of the base.

Product Rule for Exponents

- For any real number b and for any positive integers m and n, $b^m \cdot b^n = b^{m+n}$.
- In words, to multiply exponential expressions with like bases, add the exponents.

Quotient Rule for Exponents

- For any nonzero real number b and natural numbers m and n, $\frac{b^m}{b^n} = b^{m-n}$.
- In words, to divide exponential expressions with like bases, subtract the exponents.

Zero Exponent

- For any nonzero real number b, $b^0 = 1$.

Negative Exponents

- For any nonzero real number b and any integer n, $b^{-n} = \frac{1}{b^n}$ and $b^n = \frac{1}{b^{-n}}$.
- The product and quotient rules for exponents hold for any integer exponents.

1.6 EXERCISES

Concepts and Skills

1. Explain why $2x^0$ and $(2x)^0$ have different values.

2. In the definition of b^{-n}, why must b be a nonzero number?

In Exercises 3–24, evaluate the expression.

3. 4^0
4. $\frac{1}{9^0}$
5. -6^0
6. $(7y)^0$
7. $3t^0$
8. $-7x^0$
9. $4x^0 - y^0$
10. $-s^0 \cdot 2t^0$
11. -2^{-3}
12. 5^{-2}
13. $\frac{1}{15^{-1}}$
14. $\frac{1}{-3^{-2}}$
15. $\left(\frac{1}{4}\right)^{-1}$
16. $\left(-\frac{2}{3}\right)^{-3}$
17. $5^0 \cdot 3^{-2} + 3 \cdot 2^{-1}$
18. $2^{-1} - 5^{-1}$
19. $\frac{2^{-2}}{2}$
20. $\frac{3^0}{4^{-2}}$
21. $\frac{2^{-2} + 3^{-1}}{2^{-2} - 3^{-1}}$
22. $\frac{8 \cdot 4^{-2}}{3^{-3}}$
23. $(4^{-1} + 2^{-2})^{-2}$
24. $\frac{3^{-1}}{4} + \frac{2^{-1}}{6}$

25. Which is the correct way to write $3x^{-5}$ with a positive exponent: $\frac{1}{3x^5}$ or $\frac{3}{x^5}$? Why?

26. Explain why we can multiply $2x^2$ by $3x$ but we cannot combine the terms in the sum $2x^2 + 3x$.

In Exercises 27–42, write the expression with positive exponents.

27. x^{-7}
28. $\frac{1}{y^{-3}}$
29. $4x^{-4}$
30. $-5y^{-1}$
31. $2^{-5}x^3$
32. $4u^{-3}v^2$
33. $-x^{-3}y^{-4}$
34. yz^{-3}
35. x^0y^{-2}
36. $-10x^0y^{-2}$
37. $\frac{1}{-5k^{-6}}$
38. $\frac{1}{3t^{-5}}$
39. $(2 + y)^{-3}$
40. $\frac{3}{(x+1)^{-5}}$
41. $\frac{-3a^{-1}}{2b^{-1}}$
42. $\frac{-2a^{-4}}{b^{-3}}$

In Exercises 43–52, use the Product Rule for Exponents and express each result in exponential form with positive exponents.

43. $2^4 \cdot 2^7$
44. $t \cdot t^4$
45. $(6 + t)^5(6 + t)^4$
46. $(3x^3y^2)(-5x^6)$
47. $x^5 \cdot x^{-5}$
48. $(-4x^{-4})(3x)$
49. $(x^{-4})(-3x^{-6})(-x^7)$
50. $(x^{15}y^{-12})(x^{-1}y^5)$
51. $7x^{-4}y^5(2x^{-3}y^{-4})$
52. $(x^{-4}y^7)(x^8y^{-10})$

53. Explain why both of the following methods are correct.

$$\frac{x^2}{x^{-3}} = x^2 \cdot x^3 = x^5 \qquad \frac{x^2}{x^{-3}} = x^{2-(-3)} = x^5$$

54. If $b > 0$ and $b \neq 1$, is it possible for b^2 to be less than b? Illustrate with an example.

In Exercises 55–66, use the Quotient Rule for Exponents and express each result in exponential form with positive exponents.

55. $\frac{2^7}{2^4}$
56. $\frac{5^3}{5^6}$
57. $\frac{x^{12}}{x}$
58. $\frac{z^{-2}}{z^5}$
59. $\frac{y^{-8}}{y^{-3}}$
60. $\frac{x^6}{x^{-7}}$
61. $\frac{(x+1)^5}{(x+1)^3}$
62. $\frac{4x^8}{2x^5}$
63. $\frac{-5x^4}{10x^3}$
64. $\frac{-3z^5}{-4z^2}$
65. $\frac{-2t^5}{t^{-7}}$
66. $\frac{21t^4}{7t^{10}}$

In Exercises 67–70, determine the unknown expression.

67. The quotient of the expression and x^{-3} is x^{-5}.
68. The product of the expression and x^5 is x.
69. The product of the expression and x^{-4} is 1.
70. The quotient of the expression and x^{-2} is 3.

In Exercises 71 and 72, list the given expressions in order from least to greatest.

71. $(-10)^0$, 0, $-|8|^0$, $-1 - 4^0$, $4^0 + x^0$
72. $-x^0 + 1$, $(-x)^0 + 1$, $|-6|^0$, $(-2)(-x)^0$

In Exercises 73–80, insert grouping symbols so that the expression has the given value.

73. $4 - 3^{-9}$; 1 **74.** $7 - 2^2$; 25

75. $2^5 + 1^0$; 1 **76.** $5 + 3^2 \div 4$; 16

77. $8 \cdot 5^2$; 1600 **78.** $5 \cdot 2^{-1}$; 0.1

79. $5 \div 3^{-1}$; $\frac{3}{5}$ **80.** $9 \div 3^2$; 9

In Exercises 81–94, use the Quotient Rule for Exponents and express each result in exponential form with positive exponents.

81. $\frac{3z^{-8}}{2z^{-5}}$ **82.** $\frac{-(3u)^0}{v^{-5}}$

83. $\frac{c^{-12}d^{15}}{c^{-6}d^5}$ **84.** $\frac{a^4b^5}{b^2}$

85. $\frac{x^{-2}y^{-3}}{x^{-3}y^{-1}}$ **86.** $\frac{x^{21}y^{-22}}{x^{-18}y^{26}}$

87. $\frac{3^{-2}y^{-5}}{3^{-1}y^{-7}}$ **88.** $\frac{4^{-1}x^{-3}}{-4^2x^{-2}}$

89. $\frac{10x^{-5}y^3}{5x^{-7}y^7}$ **90.** $\frac{3^{-2}x^{-5}y^{-7}}{4x^5y^{-9}}$

91. $\frac{x^{-3}y^8}{x^2y^{-2}}$ **92.** $\frac{a^4b^5}{a^2b^4}$

93. $\frac{8x^{-3}y^0}{24x^2y^{-5}}$ **94.** $\frac{6x^{12}y^8}{12x^6y^4}$

Real-Life Applications

95. The planner for a new municipal transportation system estimated that t years after establishment of the system, the number of buses in the fleet would be $4t$ and the average number of passengers transported per day would be $160t^2$. Write a simplified expression for the average number of passengers per day that a bus would transport.

96. The distribution manager for a morning newspaper estimates that the number C of daily subscriber complaints from any newspaper route depends on the number of months m that the carrier has been on that route: $C = 18m^{-2}$. How many daily complaints does the manager anticipate receiving at the end of a carrier's first 3 months on a route?

97. The number of new cases of a certain strain of flu reported by residents of a certain town d days after the first reported case is modeled by $8500d^{-2}$. How many new cases are estimated 10 days after the first report?

98. A sociologist uses an index I to measure the frustration level of a certain group of people in a restaurant. The formula for the index is $I = W \cdot E$, where W is the tension level caused by having to wait for a table, and E is the moderating influence of the availability of entertainment while waiting. With x representing the number of 5-minute intervals, $W = 6x^3$ and $E = 2x^{-1}$. Write a simplified model expression for I.

Modeling with Real Data

99. The accompanying bar graph shows the sales of lottery tickets for selected years in the period 1980–1994. From the graph we see that lottery ticket sales rose from 2.4 billion tickets in 1980 to 28.5 billion tickets in 1994. The total sales include "lotto" ticket sales, which rose from 52 million in 1980 to 10 billion in 1994. Proceeds from all ticket sales rose from \$1 billion to \$10 billion over the same time period. (Source: *1995 World Lottery Almanac*.)

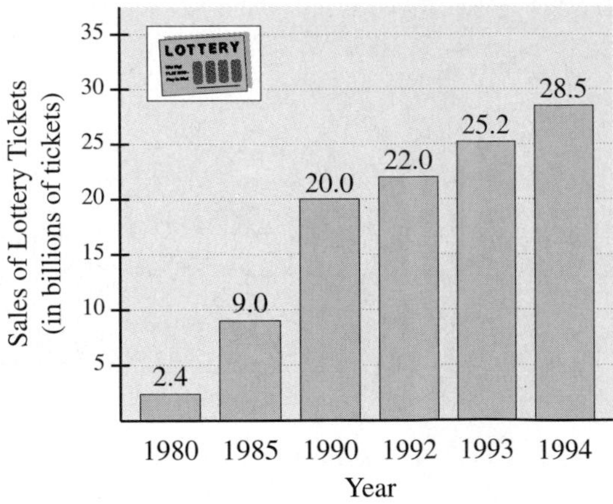

Exponential expressions model the total sales (in billions of tickets), lotto sales (in billions of tickets), and total proceeds (in billions of dollars), where t is the number of years since 1975.

Total sales: $0.08t^2$

Lotto sales: $0.0001t^4$

Total proceeds: $0.03t^2$

(a) Estimate the total ticket sales for 2002.

(b) Write a simplified exponential expression to model the portion of total ticket sales that were lotto sales.

100. Use the information in Exercise 99 to write a simplified exponential expression for the average proceeds per ticket.

101. The percentage of people who participate in sports decreases from a high of 59% in the 18–24-year-old age group to 18% in the 65–74-year-old group. (Source: U.S. National Endowment for the Arts.) The percentage of people of age a who participate in sports as a leisure activity can be modeled by the exponential expression $1553a^{-1}$.

(a) Write the expression with a positive exponent.

(b) Estimate the percentage of people in your age group who participate in sports.

102. The use of telegrams has declined dramatically, and consequently, the operating revenue for telegraph companies declined from $1.4 billion in 1985 to $367 million in 1993. (Source: Federal Communications Commission.) The operating revenue (in billions of dollars) can be modeled by $45y^{-2}$, where y is the number of years since 1980.

(a) Write the expression with a positive exponent.

(b) In 1987 there were approximately 25 million messages delivered. If those messages had been the only source of revenue, estimate the cost per message for that year.

Challenge

In Exercises 103–104, show that the statement is true.

103. $(-x)^2 + (-x^2) - x^2 + (x^1)^2 = 0$

104. $\dfrac{x^{-4}}{x^{-5}} - \dfrac{x^{-3}}{x^{-4}} + \dfrac{x^{-2}}{x^{-3}} - \dfrac{x^{-1}}{x^{-2}} + \dfrac{x^0}{x^{-1}} - x = 0$

In Exercises 105–106, determine the values of m and n that make the statement true.

105. $\dfrac{a^3b^n}{a^5b^2} \cdot \dfrac{a^4b^3}{a^mb^{-5}} = 1$

106. $\dfrac{c^{-2}d^{-3}}{c^{-9}d^n} \cdot \dfrac{c^md^5}{c^4d^{-1}} = 1$

In Exercises 107–108, show that the statement is true.

107. $\dfrac{b^{m+n}}{b^{m-n}} = b^{2n}$

108. $\dfrac{x^{-m}}{y^{-n}} - \dfrac{y^n}{x^m} = 0$

1.7 FURTHER PROPERTIES OF EXPONENTS

Power to a Power Rule • Products and Quotients Raised to a Power • Scientific Notation • Real-Life Applications

Power to a Power Rule

To simplify a power to a power, such as $(x^5)^3$, we can apply the Product Rule for Exponents.

$$(x^5)^3 = x^5 \cdot x^5 \cdot x^5 = x^{5+5+5} = x^{15}$$

Note that the exponent of the result is the product of the exponents in the original expression.

Power to a Power Rule

For any real number b and any integers m and n for which the expressions are defined,

$$(b^m)^n = b^{mn}$$

EXAMPLE 1

Using the Power to a Power Rule

Use the Power to a Power Rule to simplify the given expression. Leave the result in exponential form.

(a) $(6^3)^7 = 6^{3 \cdot 7} = 6^{21}$

(b) $(y^5)^{-4} = y^{5 \cdot (-4)} = y^{-20} = \dfrac{1}{y^{20}}$

(c) $(a^{-5})^3 \cdot (a^{-4})^{-2} = a^{-15} \cdot a^8 = a^{-7} = \dfrac{1}{a^7}$

Products and Quotients Raised to a Power

To raise a product or quotient to a power, we can apply the definition of exponent.

Product to a Power

$$\begin{aligned}(3x)^4 &= (3x)(3x)(3x)(3x)\\ &= 3 \cdot 3 \cdot 3 \cdot 3 \cdot x \cdot x \cdot x \cdot x\\ &= 3^4x^4\end{aligned}$$

Quotient to a Power

$$\begin{aligned}\left(\frac{x}{5}\right)^3 &= \frac{x}{5} \cdot \frac{x}{5} \cdot \frac{x}{5}\\ &= \frac{x \cdot x \cdot x}{5 \cdot 5 \cdot 5}\\ &= \frac{x^3}{5^3}\end{aligned}$$

Think About It

Show why $\left(\frac{a}{b}\right)^{-2} = \left(\frac{b}{a}\right)^2$.

Note that each factor of the product $3x$ is raised to the power 4 and that both the numerator and the denominator of the quotient $\frac{x}{5}$ is raised to the power 3.

Products and Quotients Raised to a Power

For any real numbers a and b and any integer n for which the expressions are defined,

$(ab)^n = a^nb^n$ Product to a Power Rule

and

$\left(\dfrac{a}{b}\right)^n = \dfrac{a^n}{b^n}$ Quotient to a Power Rule

EXAMPLE 2

Raising Products and Quotients to a Power

Simplify the given expression.

(a) $(2r^{-3}t^2)^{-3}$ (b) $\left(\dfrac{2m^2}{n^{-3}}\right)^{-4}$ (c) $\dfrac{2cd^{-1}}{(2cd)^{-1}}$

LEARNING TIP

When more than one exponent rule is needed, there is often more than one way to simplify an expression. To become aware of the choices that you have, practice varying your approaches, and confirm that the results are the same.

Solution

(a) $(2r^{-3}t^2)^{-3} = 2^{-3}(r^{-3})^{-3}(t^2)^{-3} = 2^{-3}r^9t^{-6} = \dfrac{r^9}{2^3t^6} = \dfrac{r^9}{8t^6}$

(b) $\left(\dfrac{2m^2}{n^{-3}}\right)^{-4} = \dfrac{2^{-4}(m^2)^{-4}}{(n^{-3})^{-4}} = \dfrac{2^{-4}m^{-8}}{n^{12}} = \dfrac{1}{2^4m^8n^{12}} = \dfrac{1}{16m^8n^{12}}$

(c) $\dfrac{2cd^{-1}}{(2cd)^{-1}} = \dfrac{2c(2cd)}{d} = \dfrac{4c^2d}{d} = 4c^2 \cdot \dfrac{d}{d} = 4c^2 \cdot 1 = 4c^2$

Scientific Notation

The value of 50^9 is 1,953,125,000,000,000, and the value of 5^{-8} is 0.00000256. Very large and very small numbers are awkward to write and cannot be displayed by your calculator.

A convenient way to write such numbers is based on multiplication of a number by integer powers of 10. If we multiply a number n by 10^p, the decimal point moves p places to the right if p is positive and $|p|$ places to the left if p is negative. By simply counting decimal places, you can easily verify the following.

$1.3 \cdot 10^6 = 1{,}300{,}000$ Move the decimal point 6 places to the right.

$7.92 \cdot 10^{-5} = 0.0000792$ Move the decimal point 5 places to the left.

The numbers $1.3 \cdot 10^6$ and $7.92 \cdot 10^{-5}$ are said to be in **scientific notation.**

LEARNING TIP

When numbers are written in the form $n \cdot 10^p$, at least be aware that the number is large when p is positive and small when p is negative, Also try to develop a number sense by being familiar with certain powers of 10: 10^3 means 1000, 10^6 means 1 million, and so on.

General Scientific Notation

The **scientific notation** for a number is $n \cdot 10^p$, where $1 \le n < 10$ and p is an integer.

To convert a number that is written in scientific notation to decimal form, we simply shift the decimal point right or left, as illustrated previously. To convert a number that is written in decimal form to scientific notation $n \cdot 10^p$, we must determine n and p.

Writing a Number in Scientific Notation

1. To obtain n, place the decimal point after the first nonzero digit.
2. The number of places from the decimal point to the original position is $|p|$.
3. If the original position is to the right, p is positive; if the original position is to the left, p is negative.

EXAMPLE 3

Writing Numbers in Scientific Notation

Write each number in scientific notation.

(a) 47,900,000 (b) 0.00089

Solution

(a) $n = 4.79$ Place the decimal point after the first nonzero digit, 4.

$p = 7$ 47,900,000 The original location of the decimal point is 7 places to the right.

$47{,}900{,}000 = 4.79 \cdot 10^7$

(b) $n = 8.9$ Place the decimal point after the first nonzero digit, 8.

$p = -4$ 0.00089 The original location of the decimal point is 4 places to the left.

$0.00089 = 8.9 \cdot 10^{-4}$

Scientific

Your calculator will display results in scientific notation when they are very large or very small. Moreover, you can set your calculator to display *all* results in scientific notation or to enter numbers in scientific notation.

EXAMPLE 4

Performing Operations with Large and Small Numbers

Use a calculator to perform the following operations. Write the result in scientific notation $n \cdot 10^p$ with n rounded to the nearest tenth.

(a) 500^{20} (b) $0.00005 \div 600{,}000$ (c) $857{,}000 \cdot 26{,}984$

Figure 1.10

```
500^20
                9.5E53
.00005/600000
                8.3E-11
857000*26984
                2.3E10
```

Solution

Figure 1.10 is a typical display of the results. The letter E means the exponent on 10.

(a) $500^{20} \approx 9.5 \cdot 10^{53}$

(b) $0.00005 \div 600{,}000 \approx 8.3 \cdot 10^{-11}$

(c) $857{,}000 \cdot 26{,}984 \approx 2.3 \cdot 10^{10}$

EXAMPLE 5

Calculations with Numbers in Scientific Notation

Set your calculator to enter numbers and to report results in scientific notation. Round results to the nearest hundredth.

(a) $(4.2 \cdot 10^5)(7.23 \cdot 10^{-9})$

(b) $\dfrac{4.59 \cdot 10^{-5}}{1.3 \cdot 10^4}$

(c) $(8.45 \cdot 10^3)^6$

Figure 1.11

```
(4.2E5)(7.23E-9)
                3.04E-3
(4.59E-5)/(1.3E4)
                3.53E-9
(8.45E3)^6
                3.64E23
```

Solution

Figure 1.11 shows a typical screen display of the results.

Real-Life Applications

Scientific notation is useful when only approximations are needed in calculations involving large or small numbers.

EXAMPLE 6

Water Use in the United States

In the United States, an estimated 1620 gallons of water per day per person are withdrawn from ground and surface sources.

(a) Based on a population of 260 million people, how many gallons of water per day are withdrawn?

(b) If the volume of fresh water in the world is approximately $9.9 \cdot 10^{18}$ gallons, what percentage of the world's freshwater supply does the United States withdraw each year?

Solution

(a) Each day, $(1.62 \cdot 10^3)(2.60 \cdot 10^8) \approx 4.21 \cdot 10^{11}$ gallons are withdrawn in the United States.

(b) Each year, $365 \cdot (4.21 \cdot 10^{11}) \approx 1.54 \cdot 10^{14}$ gallons are withdrawn in the United States.

$$\frac{1.54 \cdot 10^{14}}{9.9 \cdot 10^{18}} \approx 1.56 \cdot 10^{-5} = 0.0000156 \approx 0.00002$$

The United States uses approximately 0.002% of the world's freshwater supply each year.

EXAMPLE 7

The Doomsday Gap

If the sun were suddenly to burn out, how long would it take for earthlings to find out about it? (Use 93,000,000 miles as the distance from Earth to the sun and 186,000 miles per second as the speed of light.)

Solution

$$\begin{aligned}\text{Time} &= \frac{\text{distance from sun to Earth}}{\text{speed of light}}\\ &= \frac{9.3 \cdot 10^7 \text{ miles}}{1.86 \cdot 10^5 \dfrac{\text{miles}}{\text{second}}}\\ &= \frac{9.3}{1.86} \cdot 10^{7-5} \text{ seconds}\\ &= 5 \cdot 10^2 \text{ seconds}\\ &= 500 \text{ seconds}\end{aligned}$$

It would take 500 seconds, or approximately 8.3 minutes, before disaster would strike the Earth.

1.7 QUICK REFERENCE

Power to a Power Rule

- For any real number b and any integers m and n for which the expressions are defined, $(b^m)^n = b^{mn}$.
- To raise a power to a power, multiply the exponents.

Products and Quotients Raised to a Power

- For any real numbers a and b and any integer n for which the expressions are defined,

1. $(ab)^n = a^n b^n$ Product to a Power Rule
2. $\left(\frac{a}{b}\right)^n = \frac{a^n}{b^n}$ Quotient to a Power Rule

- To raise a product to a power, raise each factor of the product to the power.
- To raise a quotient to a power, raise both the numerator and the denominator to the power.

Scientific Notation

- The **scientific notation** for a number is $n \cdot 10^p$, where $1 \leq n < 10$ and p is an integer.
- To convert a number from scientific notation $n \cdot 10^p$ to decimal form, shift the decimal point p places to the right if p is positive and $|p|$ places to the left if p is negative.
- To write a number in scientific notation,

1. place the decimal point after the first nonzero digit to obtain n.
2. The number of places from the decimal point to the original position is $|p|$.
3. If the original position is to the right, p is positive; if the original position is to the left, p is negative.

1.7 EXERCISES

Concepts and Skills

1. When performing operations with exponential expressions, sometimes we multiply exponents, sometimes we add exponents, and sometimes we do nothing with the exponents. Use the following expressions to explain the differences.

$x^3 \cdot x^3$, $x^3 + x^3$, and $(x^3)^3$

2. From the following list, identify the two expressions that are equivalent to $(2x^3)^5$ and explain why the third expression is not equivalent to $(2x^3)^5$.

(i) $2^5(x^3)^5$ (ii) $2x^{15}$ (iii) $32x^{15}$

In Exercises 3–24, simplify and express each result with positive exponents.

3. $(x^3)^4$

4. $(x^{-6})^5$

5. $(2a^{-5})^{-4}$

6. $(4t)^{-3}$

7. $[(x-3)^{-3}]^{-4}$

8. $[(2-5x)^6]^2$

9. $(a^5b^9)^6$

10. $(-2x^{-2}y^4)^3$

11. $(a^{-4}b^5)^{-6}$

12. $(2x^6y^{-8})^{-2}$

13. $(2^{-2}x^3y^{-4})^{-2}$

14. $(-3a^{-3}b^5c^{-2})^3$

15. $[x(x+2)^3]^4$

16. $[y^{-2}(4-y)^{-1}]^{-1}$

17. $\left(\frac{7x^3}{5y^5}\right)^2$

18. $\left(\frac{3x^2}{2y^4}\right)^3$

19. $\left(-\frac{2}{k}\right)^{-5}$

20. $\left(\frac{5a^5}{-2b^6}\right)^4$

21. $\left(\frac{4^{-1}x^6}{y^5}\right)^2$

22. $\left(\frac{x^3}{y^{-4}}\right)^{-3}$

23. $\left(\frac{5x^{-3}}{3y^2}\right)^{-2}$

24. $\left(\frac{-4x^{-3}}{5y^4}\right)^3$

25. We know that $(cd)^2 = c^2d^2$ because of the Product to a Power Rule. However, there is no corresponding exponent rule for a *sum* raised to a power. For example, how can you show that $(c + d)^2$ is *not* the same as $c^2 + d^2$?

26. Explain how you can determine the signs of the results in advance when you simplify $(-x^4y^6)^7$ and $(-x^3y^5)^6$.

In Exercises 27–42, perform the indicated operations and express each result with positive exponents.

27. $(-5x^{-2}y)^{-1}$

28. $(-2ab^7c^4)^5$

29. $\dfrac{-4x^2y^{-2}}{8y^5}$

30. $\dfrac{2^4y^8z^4}{2y^4z}$

31. $3x^5(2x^3)^3$

32. $(2x^{-2}y^3)(-3x^4y^{-7})$

33. $\dfrac{(x^{-2})^0\,x^{-3}}{x^5}$

34. $\dfrac{(t^{-6})^2}{t^4(t^3)^{-4}}$

35. $(-2x^3)^4(3x^5)$

36. $(3x^{-2})^0(2x^{-3})^{-4}$

37. $(2y^3)^4(-4y^2)^{-2}$

38. $(a^4b^{-3})^{-2}(a^{-3}b^{-2})^4$

39. $\dfrac{(2x^3)^2}{10x^{10}}$

40. $\left(\dfrac{-2x^{-2}}{y^5}\right)^{-3}$

41. $\dfrac{(a^2b^{-3})^{-4}}{(a^{-3}b^5)^2}$

42. $\dfrac{8x^4y^{-2}}{(2x^2y^{-3})^{-3}}$

In Exercises 43–46, determine the number or expression that satisfies the given conditions.

43. The square of the number is 5^{-4}.

44. The number raised to the -2 power is $\dfrac{9}{16}$.

45. The expression raised to the -2 power is $\dfrac{x^6}{9}$.

46. When the expression is squared and then multiplied by $3x^{-2}$, the result is $3x^6$.

In Exercises 47–50, insert grouping symbols in the expression on the left side of the equation to make the statement true.

47. $2xy^{-1} = \dfrac{2}{xy}$

48. $2xy^{-2} = \dfrac{2}{x^2y^2}$

49. $x^{-1} + y^{-1} = \dfrac{1}{\dfrac{1}{x} + y}$

50. $x^5y^{-3}z^0 = x^5$

51. A number is in scientific notation when it is written in the form $n \cdot 10^p$. What must be true about the numbers n and p?

52. From the following list, identify the one number that is in scientific notation. Explain why the other two numbers are not in scientific notation.

(i) $35 \cdot 10^{-4}$ (ii) $1 \cdot 10^{-9}$ (iii) $10 \cdot 10^3$

In Exercises 53–60, write the given number in scientific notation.

53. 125,000

54. 2570.4

55. $-537{,}600{,}000$

56. 456,700,000,000

57. 0.025

58. 0.0000003749

59. 0.00000645

60. -0.00000001

In Exercises 61–68, write the given number in decimal form.

61. $1.34 \cdot 10^6$

62. $7.358 \cdot 10^{-5}$

63. $-4.214 \cdot 10^{-7}$

64. $6.87 \cdot 10^4$

65. $-5.72 \cdot 10^8$

66. $6.3 \cdot 10^{-9}$

67. $8.9 \cdot 10^{-4}$

68. $-4.83 \cdot 10^7$

In Exercises 69–72, use a calculator to perform the operations. Write the result in scientific notation $n \cdot 10^p$ with n rounded to the nearest tenth.

69. 428^{20}

70. -376^{-9}

71. $0.0000006 \div 50{,}000$

72. $8{,}765{,}432 \cdot 654{,}321$

In Exercises 73–78, estimate the value of the expression. Then verify your estimate by performing the operations on your calculator.

73. $(2 \cdot 10^2)(3 \cdot 10^1)$

74. $(2.5 \cdot 10^{-5})(1.2 \cdot 10^3)$

75. $\dfrac{4.2 \cdot 10^8}{2.1 \cdot 10^6}$

76. $\dfrac{1}{10^{-3}}$

77. $(2 \cdot 10^{-1})^{-1}$

78. $(4 \cdot 10^{-2})^{-1}$

In Exercises 79–82, perform the indicated operations. Express each result in scientific notation.

79. $(3.72 \cdot 10^7)(9.8 \cdot 10^3)$

80. $\dfrac{4.23 \cdot 10^8}{5.2 \cdot 10^{-5}}$

81. $(5.7 \cdot 10^3)^4$

82. $\dfrac{(4.9 \cdot 10^5)(3 \cdot 10^{-7})}{3.79 \cdot 10^8}$

Real-Life Applications

83. The distance between an electron and a proton in a hydrogen atom is $5.3 \cdot 10^{-11}$ meters. Write this distance in decimal form.

84. The radius of the nucleus of a heavy atom is 0.000000000007 millimeters. Write the radius in scientific notation.

85. A computer can do one addition problem in exactly $6 \cdot 10^{-6}$ seconds. How many such problems can the computer do in $\frac{1}{2}$ minute?

86. The mass of one hydrogen atom, in grams, is approximately $1.67339 \cdot 10^{-24}$. If this number were written in decimal notation, how many 0's would there be between the decimal point and the first nonzero digit?

87. Property tax rates are expressed in mils, where one mil is one dollar for each thousand dollars of assessed property value. If the millage rate for a county is 43 mils and the total property value in the county is $5.3 \cdot 10^7$ dollars, what will the annual tax revenue be for the year? Express your answer both in decimal notation and in scientific notation.

88. One way to convert feet into miles is to multiply the number of feet by $1.894 \cdot 10^{-4}$. To the nearest thousandth of a mile, how long is a 100-yard football field?

 Modeling with Real Data

In Exercises 89–94, write your answers in scientific notation.

89. In 1996, AIDS and measles each caused the deaths of 1 million people worldwide. There were 110% more deaths from malaria than from AIDS, and 210% more people died from tuberculosis than from measles. (Source: *World Health Report.*) What was the total number of deaths due to these four diseases?

90. Americans make 30 billion visits to fast-food restaurants annually. (Source: *USA Today.*) If the average sale per visit is $5, what are the total annual sales?

91. About 60 million people in the 26–44 age group commute to work and average 22 minutes of travel. (Source: Bureau of Census.) What is the number of person-hours per week spent commuting?

92. The typical American family watches television for an average of 7 hours and 42 minutes each day. (Source: Nielsen Media Research.) What is the number of person-hours per year that the 97 million households in America spend watching television?

93. The hummingbird produces the smallest bird egg. A total of 78 eggs would weigh only 1 ounce. (Source: *A Guide to Birds.*) What is the weight in pounds of a single hummingbird egg?

94. The most accurate clock is the Hewlett-Packard atomic clock, which is accurate to within 1 second in 1.6 million years. (Source: *Guinness Book of Records.*) How much time each year does the clock lose?

Group Project: NASA Budget

In 1995, the total budget of the National Aeronautics and Space Administration (NASA) was $14.3 billion. The budget for the space shuttle program was $3,324,000,000. (Source: NASA.)

95. Write the total budget and the shuttle budget figures in scientific notation.

96. Using the numbers in scientific notation, write the ratio of the space shuttle budget to the total NASA budget. Then simplify the expression and determine the percentage of the total budget that is represented by the shuttle budget.

97. If the long-range federal budget included a 7% increase in the NASA budget over each of the 5 years from 1995 to 2000, what was the projected NASA budget in 1998? Perform your calculations with scientific notation, but express your result in decimal form.

98. The total NASA budget is close to the sum of the welfare budgets of all 50 states. To reduce taxes, which program would you target for reduction? Why?

Challenge

In Exercises 99 and 100, show that the statement is true.

99. $(x^3)^4 + x^7 \cdot x^5 + \dfrac{x^{20}}{x^8} = 3x^{12}$

100. $\left(\frac{a}{b}\right)^{-n}\left(\frac{b}{a}\right)^{n} = \left(\frac{b}{a}\right)^{2n}$

In Exercises 101 and 102, determine the values of m and n that make the statement true.

101. $(xy^{-3})^2(x^ny^m) = \frac{x}{y}$

102. $\left(\frac{x^{-2}y}{x^2y^{-4}}\right)(x^my^n) = x^{12}$

In Exercises 103 and 104, determine the value of n that makes the statement true.

103. $\frac{(3 \cdot 10^4)(8 \cdot 10^n)}{12 \cdot 10^{-2}} = 2$

104. $\frac{(18 \cdot 10^5)(5 \cdot 10^{-3})}{(8 \cdot 10^n)(15 \cdot 10^7)} = \frac{3}{4}$

1. CHAPTER PROJECT

NASA Facts

The National Aeronautics and Space Administration (NASA) came into existence in 1958. Among its personnel are mathematicians, computer scientists, physicists, and engineers who have developed the mathematics needed for space flight and exploration.

1. To remain in a near-earth orbit, an object must travel at a velocity of about 4.9 miles per second. Convert this velocity to miles per hour, and write the result in both scientific and decimal notation.
2. The *period* of an orbit is the time required to travel one orbit. Use the formula $d = rt$ and the result in Exercise 1 to calculate the period (in minutes) of a near-earth orbit of 24,900 miles.
3. What would be the result if an object in orbit reduced its velocity to less than 4.9 miles per second? What would happen if the velocity were increased to more than 4.9 miles per second?
4. In 1996 the American space shuttle joined with the Russian Mir space station. Shannon Lucid, an American astronaut, stayed with Mir and set an American endurance record in space. Using both scientific notation and decimal notation, write the total number of miles that she traveled during one 30-day period.
5. In Section 1.7 we assumed increases in the NASA budget. However, attempts to balance the budget have resulted in budget cuts for NASA. In 1995 the president's budget plan called for a 35% reduction in the NASA budget. Then Congress proposed an additional 20% reduction. (Source: The Planetary Society.) If the 1995 NASA budget was \$14.3 billion, what would have been the funding level for 1996 if both proposals had been adopted?
6. Mission to Planet Earth is a NASA program designed to provide data in the areas of climate forecasting, global rainfall, ice sheets, air pollution, and ozone layers. NASA's Space Science programs are responsible for planetary exploration and the Hubble Space Telescope. If scarce budgetary resources required the elimination of one of these two programs, which one would you choose? Why?

1. CHAPTER REVIEW EXERCISES

Section 1.1

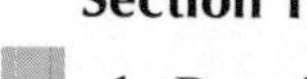

1. Describe the decimal names of the rational numbers and of the irrational numbers.

In Exercises 2–5, indicate whether the statement is true or false.

2. All square roots are irrational numbers.
3. The numbers 0, -13, $\frac{2}{5}$, and $2.\overline{41}$ are all rational numbers.
4. If $b > a$, then a is to the left of b on the number line.
5. The set-builder notation $\{x \mid 2 < x \leq 5\}$ and the interval notation $[2, 5)$ both represent the same set of numbers.

In Exercises 6 and 7, graph the given inequality.

6. $x \geq -2$
7. $0 \leq x < 3$

8. Referring to a number line, explain why $|a| = |-a|$.
9. Evaluate $-|-(-3)|$.
10. Write an inequality to model the given information.
    (a) a is positive.
    (b) x is between -1 and 5.

Section 1.2

In Exercises 11–16, perform the indicated operations.

11. $\frac{3}{4} - \left(-\frac{5}{8}\right)$
12. $-\sqrt{\frac{9}{4}}$
13. $(-1)(-2)(-3)(-4)(-5)$
14. $-|-8| + (-8)$
15. -3^3
16. $0 \div (-6)$
17. If the coordinates of points A and B of a number line are -4 and 7, find the distance AB.
18. Translate the following information into a numerical expression and evaluate it: "The dividend is -7 and the divisor is -28."
19. Your March 1 checkbook balance was \$852.39. During the month, you wrote checks totaling \$485.11, and you deposited \$500.00. On March 31 your account statement shows a service fee of \$3.00 and interest to you in the amount of \$2.00. What is your March 31 closing balance?
20. From the following list, identify the two numerical expressions that are undefined and explain why.
    (i) $6 \div 0$ (ii) $\sqrt{-9}$ (iii) $\sqrt{7}$

Section 1.3

21. What property allows us to change the grouping of the addends of a sum?
22. What property allows us to change the order of the factors of a product?

In Exercises 23–26, name the property that justifies the statement.

23. The sum of x and $-x$ is the additive identity.
24. $3x + 9x = 12x$
25. $2a(b + c) = 2ab + 2ac$
26. $-x = -1x$

In Exercises 27 and 28, write the expression without grouping symbols.

27. $-(7 - 8x - x^2)$
28. $\frac{a}{b}\left[\frac{b}{a}(a + b)\right]$
29. If b is the reciprocal of a, what is the value of ab?
30. If $ab = a$, where $a \neq 0$, what do we call b?

Section 1.4

In Exercises 31–35, evaluate the expression.

31. $18 - 5 \cdot 2 + 8^2 \div 4$
32. $-5^2 + (-5)^2$
33. Six less than the quotient of 20 and 4
34. $\sqrt{3 - 2x}$ when $x = -11$
35. $b^2 - 4ac$ when $a = 2$, $b = -1$, and $c = -3$
36. Translate the following into an equation: "The sum of a number and its square is 4 more than twice the number."
37. Which of the following is a solution of the equation $\frac{x}{2} + 3 = 3x - 17$?
    (a) 8 (b) 0 (c) -3
38. If the lengths of the sides of a triangle are 3, 5, and $\sqrt{34}$, show whether the triangle is a right triangle.

39. Which of the following are grouping symbols?

(a) parentheses

(b) radical signs

(c) fraction bars

(d) absolute value symbols

40. Evaluate $\sqrt{y - x} + x^2$ for $x = 1.8$ and $y = 2.6$. Round your result to the nearest hundredth.

Section 1.5

41. For the expression $3x^2 - 5(x + 1)$, state how many terms there are and list the coefficients.

42. Why do we say that $5x + 3x$ and $8x$ are equivalent expressions?

In Exercises 43–47, simplify.

43. $2x - 3y - y + 3x^2$

44. $-(x + 2a) + 2(x - 3a)$

45. $3 + 2[7x - 4(x - 1)]$

46. $ab - a(a + b) + a^2 + a(b + 1)$

47. $-xy^2z + xyz^2 + x^2yz + 2xy^2z$

In Exercises 48 and 49, translate the verbal description into an algebraic expression.

48. Three less than the product of a number n and 2

49. The retail price of an item if the wholesale price w is marked up by 20%

50. Name two properties needed to simplify the expression $-(m + n)$.

Section 1.6

51. How is b restricted in b^0 and in b^n when $n < 0$?

52. List the following numbers in order from least to greatest.

$$(-4)^0, \quad -4^0, \quad 4^{-1}, \quad (-4)^{-1}$$

In Exercises 53–58, simplify and express the results with positive exponents. Assume all expressions are defined.

53. $(x^{-8}y^3)(2x^4y)$

54. $\dfrac{21x^{-5}}{28x}$

55. $(x + 3)^3(x + 3)^{-3}$

56. $(-2x^5y^{-4})(x^{-2}y^{-1})$

57. $(xy^{-1}z)(x^{-1}yz)(xyz^{-1})$

58. $\dfrac{-5^{-1}x^{-2}y^{-3}}{(-5)^{-1}x^0y^4}$

59. Explain why $(-2)^{-1} = -\dfrac{1}{2}$, not $\dfrac{1}{2}$.

60. Explain why $-3x^{-1} = -\dfrac{3}{x}$, not $-\dfrac{1}{3x}$.

Section 1.7

61. How can we simplify $(x^5)^2$ with

(a) the Product Rule for Exponents?

(b) the Power to a Power Rule?

62. Insert grouping symbols on the left side so that $2x^{-1}y^2 = \dfrac{y^2}{2x}$.

In Exercises 63–66, simplify and express the results with positive exponents. Assume all expressions are defined.

63. $(a^{-3}b^2)^{-2}(ab^{-3})^2$

64. $(-2a^{-1}b^2)^2(-3ab)^{-2}$

65. $\dfrac{(-3x^{-2})^3}{6x}$

66. $\left(\dfrac{2x^{-3}}{x^2}\right)^{-4}$

67. Write 0.000029 in scientific notation.

68. Write $6 \cdot 10^{-7}$ in decimal form.

69. Calculate 23^{64} and express the result in scientific notation.

70. NASA's *Mars Observer* made a 450-million-mile journey at a cost of about \$1 billion.

(a) Write the distance and cost in scientific notation.

(b) What was the cost per mile?

LOOKING AHEAD

1. What is the coordinate of the point of the number line that is
   (a) 4 units to the right of -3?
   (b) 7 units to the left of 2?
2. Determine the distance between the numbers -7 and 1 on a number line.
3. Explain why $(3 - 7)^2$ and $|3 - 7|^2$ have the same value.

In Exercises 4 and 5, evaluate the expression for the given value of the variable.

4. $6 - x$; -9
5. $x^2 - 2x - 7$; -3

In Exercises 6 and 7, evaluate the expression for the given values of the variables.

6. $\dfrac{a+b}{2}$; $a = 4, b = -8$
7. $\sqrt{(a-b)^2 + (c-d)^2}$; $a = -3, b = 5, c = -2, d = -8$
8. Suppose that $ab > 0$. What can you conclude about the signs of a and b?
9. What is the perimeter of a triangle whose sides have lengths (in inches) 4, 5, and 2?
10. Determine the area of a rectangular lot that is 70 feet long and 40 feet wide.

In Exercises 11 and 12, simplify the given expression.

11. $3(5 + h) - 4$
12. $(-3t)^2 + 4(-3t) - 8$

1. CHAPTER TEST

1. Which elements of
   $\{-3.7, -2, -0.\overline{14}, 0, \frac{2}{3}, \pi, \sqrt{7}, 6, \sqrt{9}\}$
   also belong to the following sets?
   (a) integers (b) irrational numbers
   (c) rational numbers
2. Use the roster method to write the set
   $\{n \mid n \text{ is an integer and } -3 < n \leq 5\}$.
3. Graph the set $\{x \mid -2 < x \leq 3\}$ on a number line.
4. Evaluate each of the following.
   (a) $-7 - (-3)$ (b) $\left(-\dfrac{3}{10}\right)\left(\dfrac{2}{9}\right)$
   (c) $\dfrac{2 - (-3)}{3(7) - 2(-8)}$ (d) $3 - 2[4 - (2 - 3)]$
   (e) $\dfrac{-3 + \sqrt{3^2 - 4(2)}}{2}$ (f) $|6 - 2(5)|$
   (g) $-\dfrac{3}{4} + \left(-\dfrac{1}{2}\right)$ (h) $-2.5 - 6.7$
   (i) $(-2)(5)(-4)(0)(6)$ (j) $-32 \div (-4)$
5. Use the indicated property to complete the following statements.
   (a) Distributive Property: $3(x - 4) =$ _____
   (b) Commutative Property of Addition: $4 - x =$ _____
   (c) Associative Property of Addition: $(3 + t) - 3r =$ _____
6. The sum of the numbers 4 and -9 is multiplied by 2. The result is squared and then divided by 3 less than -17. Write the numerical expression and evaluate it.
7. A certain pharmaceutical is manufactured under very controlled temperature conditions. If the temperature in the lab is $-2.45°$C and is decreased by $1.04°$, what is the new temperature?
8. Identify the elements of $\{-2, 0, 2, 3\}$ that are solutions of the equation $3x - 4 = -10$.
9. Identify the terms and coefficients of the expression $5x^2 - 4x + 2$.
10. Explain the difference in the order in which the operations are performed for the expressions -2^2 and $(-2)^2$. Show that the expressions are not equal.
11. Simplify the following expressions and express your answers with positive exponents. Assume all expressions are defined.
    (a) $(x^3y^2)(xy^4)$ (b) $t^3 \cdot t^{-2} \cdot t$
    (c) $(2a^{-5}b^4)^2$ (d) $\left(\dfrac{a^2}{b^{-3}}\right)^{-2}$
    (e) $\dfrac{c^{-3}d^4}{c^{-5}d^{-2}}$

12. Evaluate each of the following expressions.

(a) $\dfrac{(-2)^{-3}}{(-2)^2}$ (b) $3(-x)^0$ (c) $(-2)^{-3}$

13. Write the following in decimal notation.

(a) $2.7 \cdot 10^7$ (b) $4.56 \cdot 10^{-5}$

14. The wavelength of sunlight is approximately 0.00000057 meters. Write this number in scientific notation.

15. Evaluate each of the following expressions.

(a) $3 - 5x$ for $x = 1.4$

(b) $-x^2 + 1$ for $x = 2$

(c) $\frac{1}{2}bh$ for $b = 5.32$ and $h = 7.6$

16. Simplify each of the following.

(a) $(3x - 2y) + 7y$

(b) $(2s - 5t) + 2(s + 2t)$

(c) $4 - 2[3x - (x - 1)]$

17. Determine whether the numbers 5, 12, and 13 are the lengths of the sides of a right triangle.

18. Write an algebraic expression for the width of a rectangle if the width is 3 less than half the length. (Let L represent the length.)

19. Translate the following information into an algebraic expression: the sum of a number, 3 times the number, and twice 1 less than the number. (Let n represent the number.)

20. It takes approximately 5.53 hours for light to travel from the sun to Pluto. How far is Pluto from the sun? Express the answer in scientific notation. The speed of light is $1.86 \cdot 10^5$ miles per second. (Hint: Use distance = rate · time and convert hours to seconds.)

Chapter 2

The Coordinate Plane and Functions

As more and more states enacted helmet laws for bicycle riders, the sales of helmets increased dramatically. The bar graph shows the sales (in millions of dollars) of bicycle helmets for selected years in the period 1990–1995. (Source: U.S. Bicycle Federation.)

We can write a **function** to model the trend in helmet sales during this time, and we can graph the function in a **coordinate system.** Both the function and the graph can be used, for example, to estimate the helmet sales for 1991 or to predict sales after 1995. (For more on this topic, see Exercises 79–82 at the end of Section 2.4, and see the Chapter Project.)

In this chapter we introduce the coordinate plane for associating ordered pairs of real numbers with points. We learn how to graph an expression in the coordinate plane, and we introduce the basic features of the graphing calculator. The remainder of the chapter is devoted to the important topic of functions, including graphs, notation, evaluation, and analysis.

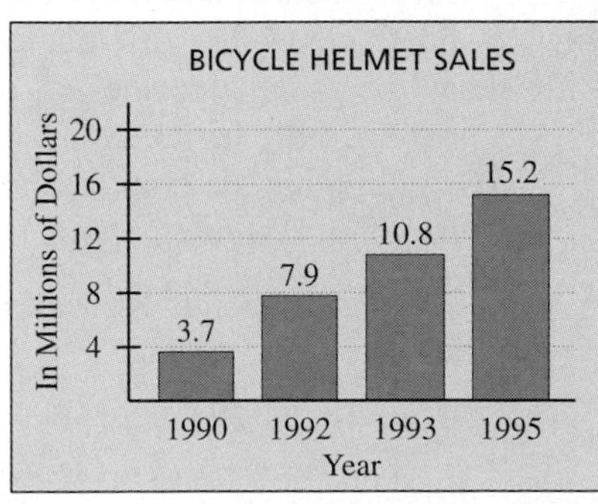

(Source: U.S. Bicycle Federation.)

2.1 THE COORDINATE PLANE

Rectangular Coordinate System • The Graphing Calculator • Distance Between Points • Midpoints of Line Segments • Real-Life Applications

Rectangular Coordinate System

Just as we use a number line to visualize real numbers, we use a system of two perpendicular number lines to visualize **ordered pairs** (x, y) of real numbers. The horizontal number line is called the ***x*-axis,** and the vertical number line is called the ***y*-axis.** The plane containing the axes is called the **coordinate plane.** The axes intersect at a point called the **origin,** and they divide the coordinate plane into four regions called **quadrants.** Figure 2.1 shows the arrangement of the **rectangular** (or Cartesian) **coordinate system.**

Figure 2.1

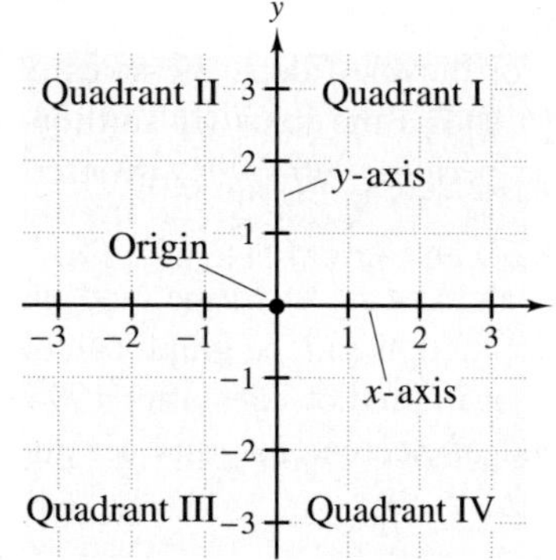

Figure 2.2

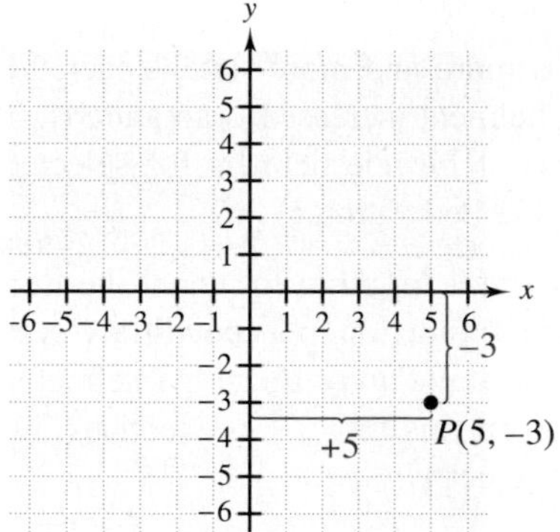

LEARNING TIP

Think of the coordinate plane as a city laid out in blocks. The directions from the center of the city (origin) are given as a number of blocks east/west followed by a number of blocks north/south.

Each ordered pair (x, y) of real numbers is associated with a unique point in the coordinate plane. The first number x, called the ***x*-coordinate** of the point, indicates how far the point is to the left or right of the y-axis. The second number y, called the ***y*-coordinate** of the point, indicates how far the point is above or below the x-axis. Highlighting the point associated with an ordered pair is called **plotting** or **graphing** the point. Figure 2.2 shows the point $P(5, -3)$ plotted in a coordinate plane.

Plotting the points associated with all the ordered pairs of a set is called **graphing the set.**

EXAMPLE 1

Graphing Sets of Ordered Pairs

Graph the following set:

$$\{(2, 5), (-3, -4), (4, -3), (-5, 4), (5, 0), (0, 4)\}$$

Solution

Figure 2.3

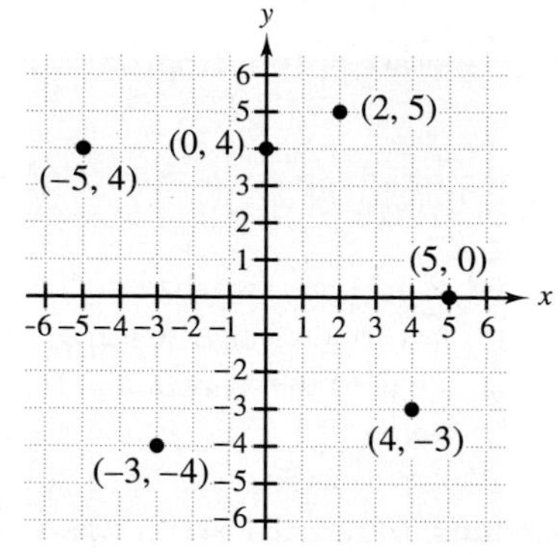

The results in Example 1 suggest the following sign patterns for points in the four quadrants:

Quadrant I	*Quadrant II*	*Quadrant III*	*Quadrant IV*
$(+, +)$	$(-, +)$	$(-, -)$	$(+, -)$

Example 1 also suggests the following generalizations about points of the axes:

1. Points of the x-axis have a y-coordinate of 0.
2. Points of the y-axis have an x-coordinate of 0.

EXAMPLE 2

Determining the Coordinates of a Point

Determine the coordinates of each point in Fig. 2.4.

Figure 2.4

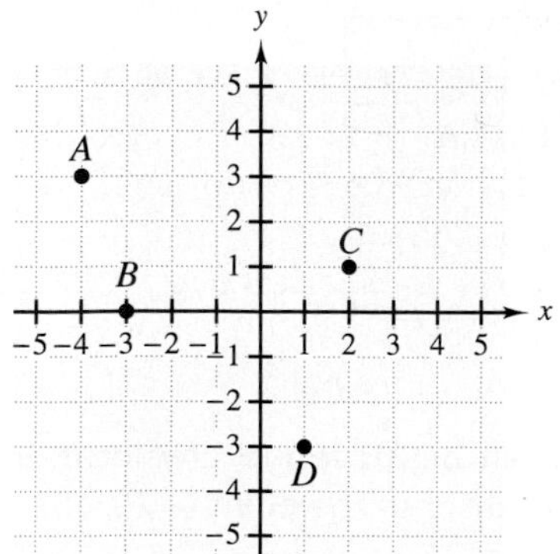

Think About It

Suppose that you wanted to represent ordered triples rather than ordered pairs. What kind of coordinate system could you use?

Solution

(a) $A(-4, 3)$ (b) $B(-3, 0)$

(c) $C(2, 1)$ (d) $D(1, -3)$

The Graphing Calculator

A graphing calculator can be used to produce any portion of a coordinate system. You should become familiar with the following basic features.

Clear The clear key is used to erase entries from various calculator screens.

Y Screen For purposes to be described later, we enter algebraic expressions on the Y screen. (See Fig. 2.5.)

Figure 2.5

```
Y1=
Y2=
Y3=
Y4=
Y5=
Y6=
Y7=
Y8=
```

Figure 2.6

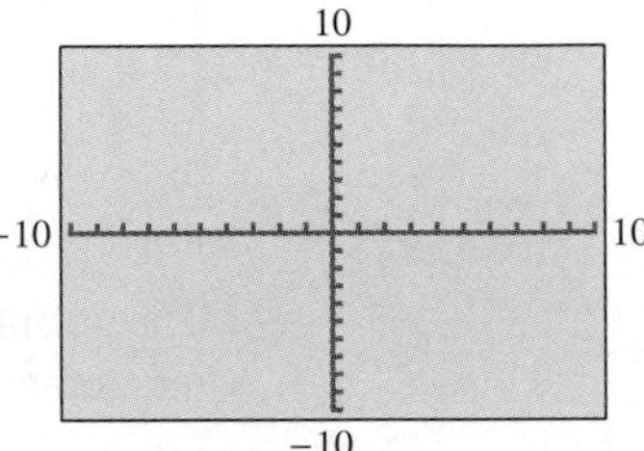

Graph Coordinate systems are displayed in the graph screen.

Default For most calculators, the *default* or *standard* setting places the origin at the center of the graph screen. (See Fig. 2.6.)

Integer The integer setting differs from the default setting in that at least one of the displayed coordinates is an integer. (See Fig. 2.7.)

Figure 2.7

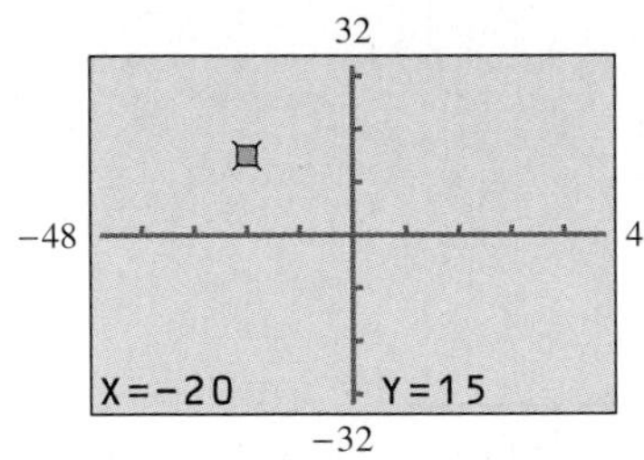

Cursor A *general cursor* can be moved to any location on the graph screen. Figure 2.7 shows the general cursor at the point (−20, 15).

Window The window screen is used to display particular portions of the coordinate plane. You can choose the minimum and maximum values to be displayed on each axis as well as the spacing between tick marks.

In Fig. 2.8(a), the x-axis is set from -8 to 12, with a tick-mark spacing of 4. The y-axis is set from -6 to 18, with a tick-mark spacing of 3. Figure 2.8(b) shows the display of the resulting coordinate system.

Figure 2.8

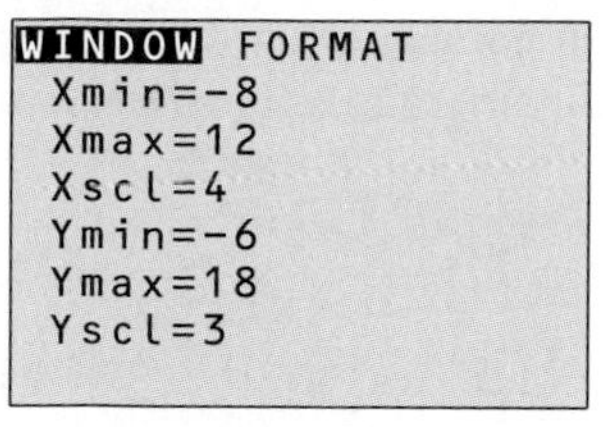

(a)

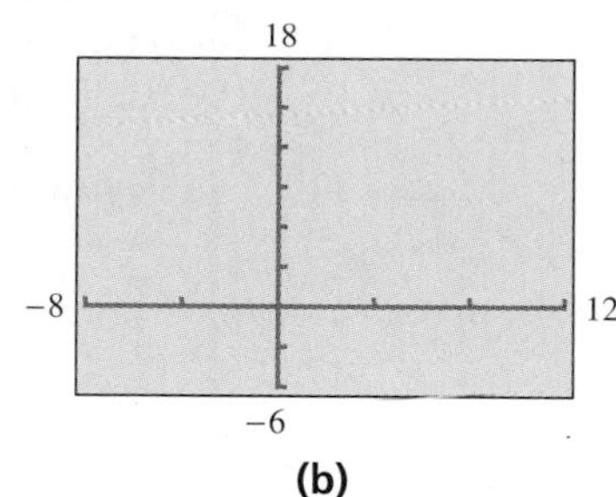

(b)

Distance Between Points

We have seen that if the coordinates of two points of a number line are a and b, then the distance between the points can be found by calculating $|a - b|$ or $|b - a|$.

To find the distance between two points in a rectangular coordinate system, we use the same principle along with the Pythagorean Theorem.

EXPLORING THE CONCEPT

The Distance Formula

Figure 2.9 shows two points $A(-3, 2)$ and $B(5, 8)$. Suppose we need to know the distance AB between the points.

Figure 2.9

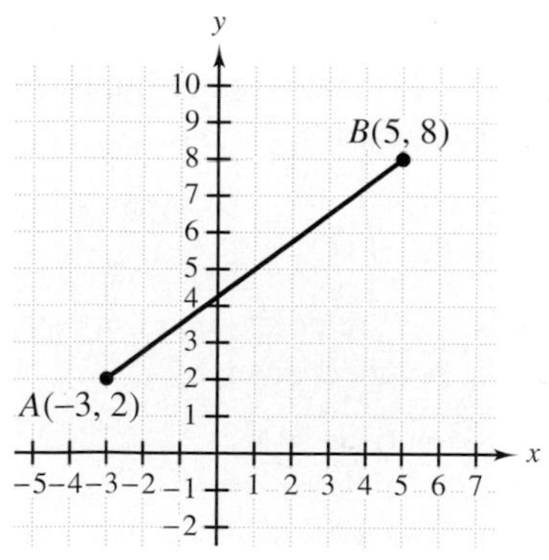

Figure 2.10

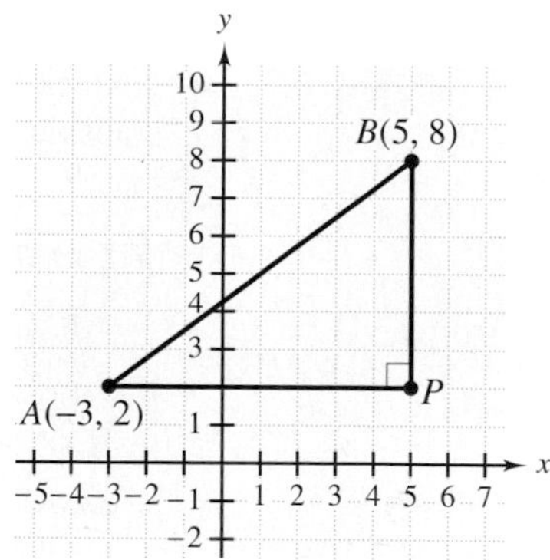

We begin by adding a third point P, which is on a horizontal line containing point A and on a vertical line containing point B. (See Fig. 2.10.) Because point P has the same first coordinate as point B and the same second coordinate as point A, the coordinates of point P are $(5, 2)$.

Because $\angle P$ is a right angle, ΔAPB is a right triangle, and we can use the Pythagorean Theorem to calculate the length AB of the hypotenuse.

$$AP = |-3 - 5| = |-8| = 8 \quad \text{Length of one leg}$$

$$BP = |8 - 2| = |6| = 6 \quad \text{Length of the other leg}$$

$$(AB)^2 = (AP)^2 + (BP)^2 \quad \text{Pythagorean Theorem}$$

$$= 8^2 + 6^2 = 64 + 36 = 100$$

Therefore, $AB = 10$.

We can use the preceding technique to obtain a general method for finding the distance d between any two points $A(x_1, y_1)$ and $B(x_2, y_2)$. Figure 2.11 shows points A and B along with a third point $P(x_2, y_1)$ located so that ΔAPB is a right triangle.

Figure 2.11

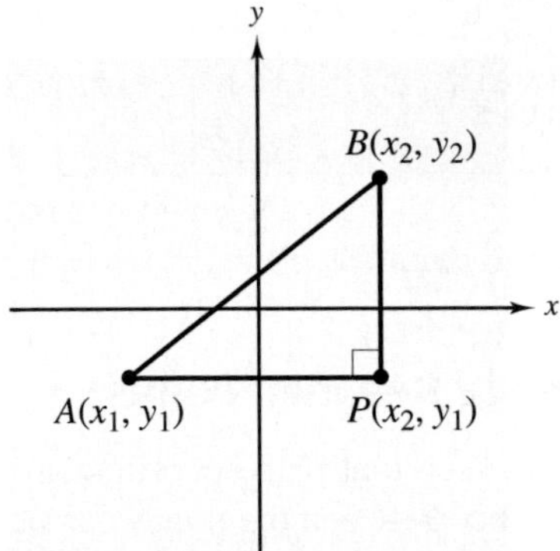

The horizontal distance between A and B is $|x_2 - x_1|$, and the vertical distance is $|y_2 - y_1|$. According to the Pythagorean Theorem,

$$d^2 = |x_2 - x_1|^2 + |y_2 - y_1|^2$$

Therefore,

$$d = \sqrt{|x_2 - x_1|^2 + |y_2 - y_1|^2}$$

This formula is called the **distance formula** for the distance between two points in the Cartesian coordinate system. Because the differences are being squared, the order in which we subtract the coordinates is not important, and it is not necessary to take the absolute value of each difference.

The Distance Formula

If $A(x_1, y_1)$ and $B(x_2, y_2)$ are points in a Cartesian coordinate system, then the distance d between the points is given by

$$d = \sqrt{(x_2 - x_1)^2 + (y_2 - y_1)^2}$$

EXAMPLE 3

Using the Distance Formula

Find the distance between the points $E(-2, 1)$ and $F(3, 13)$.

Solution

$$d = \sqrt{(x_2 - x_1)^2 + (y_2 - y_1)^2}$$
$$d = \sqrt{(-2 - 3)^2 + (1 - 13)^2}$$
$$= \sqrt{(-5)^2 + (-12)^2}$$
$$= \sqrt{25 + 144}$$
$$= \sqrt{169}$$
$$= 13$$

Note: We can use a calculator to determine d in one keying sequence. Remember, though, that the differences are quantities and must be enclosed in parentheses. Also, the square root symbol serves as a grouping symbol, so the entire quantity under the square root symbol also must be enclosed in parentheses.

Midpoints of a Line Segments

The **midpoint** of a line segment is the point of the segment that is equidistant (the same distance) from the endpoints of the segment. It can be shown that the coordinates of the midpoint are the averages of the endpoint coordinates.

The Midpoint of a Line Segment

For the line segment with endpoints $A(x_1, y_1)$ and $B(x_2, y_2)$, the midpoint M has coordinates

$$x_m = \frac{x_1 + x_2}{2} \quad \text{and} \quad y_m = \frac{y_1 + y_2}{2}$$

EXAMPLE 4

Finding the Midpoint of a Line Segment

(a) Determine the midpoint M of the line segment whose endpoints are $P(4, 5)$ and $Q(-8, 9)$.

(b) Use the distance formula to verify that $MP = MQ$.

Solution

(a) $x_m = \dfrac{4 + (-8)}{2} = -2 \qquad y_m = \dfrac{5 + 9}{2} = 7$

The coordinates of midpoint M are $(-2, 7)$. (See Fig. 2.12.)

Figure 2.12

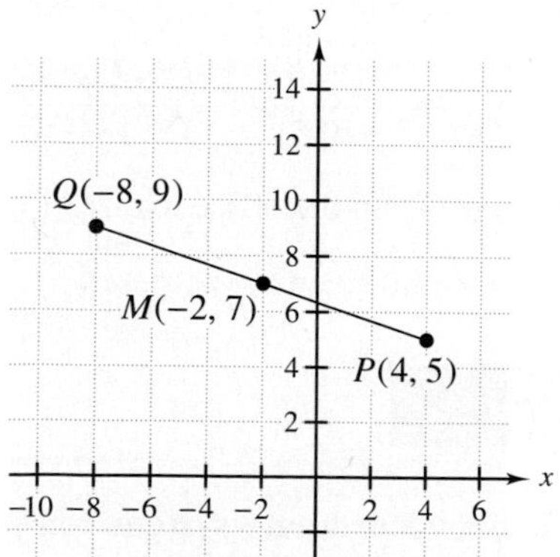

(b) $MP = \sqrt{(-2 - 4)^2 + (7 - 5)^2} = \sqrt{36 + 4} = \sqrt{40}$

$MQ = \sqrt{[-2 - (-8)]^2 + (9 - 7)^2} = \sqrt{36 + 4} = \sqrt{40}$

Therefore, $MP = MQ$.

Real-Life Applications

We can apply the distance and midpoint formulas to problems involving geometric figures drawn in a coordinate plane.

In Example 5 we use the following definition: If two line segments have the same midpoint, then the line segments **bisect** each other.

EXAMPLE 5

Figure 2.13

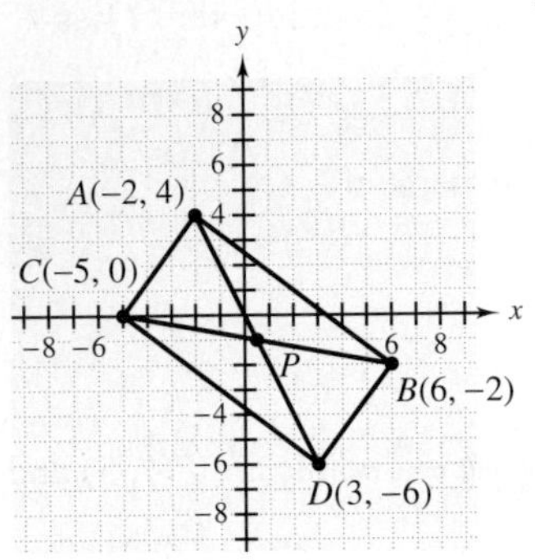

The Diagonals of a Rectangle Bisect Each Other

Figure 2.13 shows a grid map of forest ranger facilities. Observation towers are located at points A, B, C, and D, which are the vertices of a rectangle. The main office is located at point P, where the diagonals of the rectangle intersect. Each tick mark represents 1 mile.

(a) Show that the main office is equidistant from the four observation towers by proving that the diagonals of the rectangle bisect each other.

(b) Find the distance from the main office to any one of the four towers.

Solution

(a) For the diagonal with endpoints A and D, the coordinates of the midpoint are as follows:

$$x_m = \frac{-2+3}{2} = \frac{1}{2} \qquad y_m = \frac{4+(-6)}{2} = -1$$

For the diagonal with endpoints B and C, the coordinates of the midpoint are as follows:

$$x_m = \frac{-5+6}{2} = \frac{1}{2} \qquad y_m = \frac{0+(-2)}{2} = -1$$

Because $\left(\frac{1}{2}, -1\right)$ is the midpoint of both diagonals, the diagonals bisect each other. Therefore, the main office is equidistant from all four towers.

(b) The distance from the main office at $P\left(\frac{1}{2}, -1\right)$ to the tower at $A(-2, 4)$ is calculated with the distance formula.

$$d = \sqrt{(-2-0.5)^2 + [4-(-1)]^2} \approx 5.6 \text{ miles}$$

The distance from the office to any of the other towers is the same.

2.1 QUICK REFERENCE

Rectangular Coordinate System

- The **rectangular coordinate system** consists of two perpendicular number lines called the **x-axis** and the **y-axis,** which intersect at the **origin.** The axes divide the **coordinate plane** into four regions called **quadrants.**
- Each ordered pair of real numbers (x, y) is associated with a point of the coordinate plane. We call the numbers x and y the **x-coordinate** and the **y-coordinate** of the point, respectively.

- Highlighting the point associated with an ordered pair is called **plotting** or **graphing** the point. Plotting the points associated with all the ordered pairs of a set is called **graphing the set.**
- The following are the sign patterns for the coordinates in the four quadrants.

  Quadrant I: (+, +) Quadrant II: (−, +)

  Quadrant III: (−, −) Quadrant IV: (+, −)
- Every point of the x-axis has a y-coordinate of 0. Every point of the y-axis has an x-coordinate of 0.

The Graphing Calculator

- We can produce a coordinate system on a graphing calculator with viewing windows automatically set or customized.
- A *general cursor* is available for highlighting points in the graph screen and displaying their coordinates.

Distance Between Points

- If $A(x_1, y_1)$ and $B(x_2, y_2)$ are points in a coordinate system, then the distance d between them is given by the **distance formula.**

$$d = \sqrt{(x_2 - x_1)^2 + (y_2 - y_1)^2}$$

Midpoints of Line Segments

- For the line segment with endpoints $A(x_1, y_1)$ and $B(x_2, y_2)$, the midpoint $M(x_m, y_m)$ has coordinates that are the averages of the endpoint coordinates.

$$x_m = \frac{x_1 + x_2}{2} \qquad y_m = \frac{y_1 + y_2}{2}$$

2.1 EXERCISES

Concepts and Skills

1. Explain how to plot the point $A(-2, 4)$.

2. Give an example of a point that is not in any quadrant, and describe its location.

In Exercises 3–6, plot the given points.

3. $A(4, -6)$, $B(-6, 4)$, $C(3, 5)$
4. $A(-2, -3)$, $B(-3, -2)$, $C(-5, 6)$
5. $A(0, -3)$, $B(-5, 0)$, $C(-2, -1)$
6. $A(3, 1)$, $B(3, 7)$, $C(4, -5)$

In Exercises 7–12, plot the ordered pairs in the given set. From the resulting pattern, predict the next ordered pair.

7. $\{(-2, -2), (0, -1), (2, 0), (4, 1), (6, 2), \ldots\}$
8. $\{(-5, 6), (-2, 4), (1, 2), (4, 0), (7, -2), \ldots\}$
9. $\{(-3, 5), (-2, 5), (-1, 5), (0, 5), (1, 5), \ldots\}$
10. $\{(4, 2), (4, 1), (4, 0), (4, -1), (4, -2), \ldots\}$
11. $\{(-4, 0), (-2, 3), (0, 0), (2, 3), (4, 0), \ldots\}$
12. $\{(-1, 1), (0, 0), (1, -1), (2, 0), (3, 1), \ldots\}$

In Exercises 13–18, without plotting the given point, name the quadrant in which the point lies.

13. $P\left(-2, -\frac{4}{5}\right)$
14. $B\left(\frac{3}{2}, -9\right)$
15. $P(2.4, 6.5)$
16. $B(-\sqrt{2}, -\sqrt{3})$
17. $A(\sqrt{5}, -7)$
18. $P(-\sqrt{6}, 10)$

In Exercises 19–24, determine the coordinates of the points in the figure on the next page.

19. A
20. B
21. C
22. D
23. E
24. F

Figure for 19 – 24

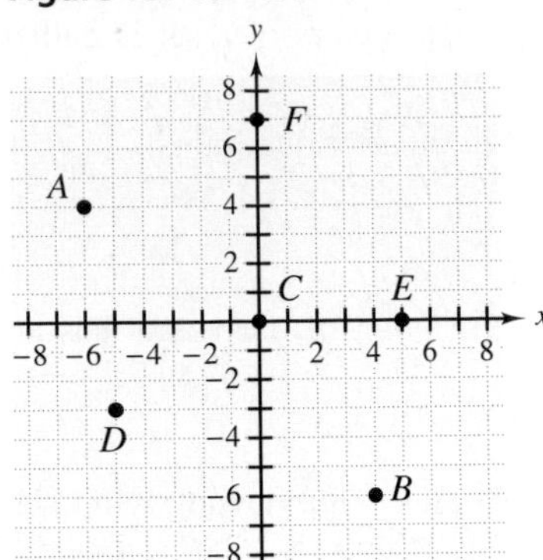

25. Describe the set of points whose x-coordinates are 0.

26. Describe the set of points whose y-coordinates are 0.

In Exercises 27–32, graph the given set. Assume that x and y represent integers.

27. $\{(x, y) \mid -3 \le x \le 3, y = 2\}$

28. $\{(x, y) \mid -2 \le y \le 5, x = -2\}$

29. $\{(x, y) \mid -4 \le y \le 2, x = -y\}$

30. $\{(x, y) \mid -3 \le x \le 4, y = -x + 1\}$

31. $\{(x, y) \mid -2 \le x \le 1, 1 \le y \le 3\}$

32. $\{(x, y) \mid 1 \le x \le 4, -3 \le y \le 2\}$

In Exercises 33–42, in which *possible* quadrants or on which axes could the point lie?

33. $(x, y), \quad x > 0$

34. $(x, y), \quad x < 0$

35. $(x, y), \quad y < 0$

36. $(x, y), \quad y > 0$

37. $(x, 0)$

38. $(0, y)$

39. $(-3, y)$

40. $(x, 5)$

41. $(x, y), \quad xy < 0$

42. $(x, y), \quad xy > 0$

43. Suppose that $P(x, y)$ is a point of the x-axis to the left of the origin. Describe the location of the given points.
(a) $Q(y, x)$ (b) $R(-x, y)$ (c) $S(y, -x)$

44. If $P(x, y)$ is in Quadrant II, describe the location of the given points.
(a) $Q(-x, y)$ (b) $R(-x, -y)$ (c) $S(x, -y)$

A graph is *reflected across the x-axis* if each ordered pair (x, y) is replaced with $(x, -y)$. A graph is *reflected across the y-axis* if each ordered pair (x, y) is replaced with $(-x, y)$.

In Exercises 45–48, sketch the graph of the geometric figure with the given vertices. Then sketch the graph of the reflection of that figure across (a) the x-axis and (b) the y-axis. In each case, give the vertices of the reflected figure.

45. $(2, 1), (3, 7), (5, 4)$

46. $(-10, 3), (-5, 9), (-2, 6)$

47. $(1, -5), (5, -1), (1, -11), (11, -1)$

48. $(-6, -2), (-8, -5), (-1, -5), (-3, -2)$

In Exercises 49–52, sketch the graph of the geometric figure with the given vertices. Then interchange the coordinates of those vertices and sketch the graph of the new figure.

49. $(2, 6), (5, 1), (7, 3)$

50. $(-3, 2), (-1, 4), (-2, 1)$

51. $(-1, -2), (2, -3), (4, -1), (5, 4)$

52. $(-6, 2), (-4, 4), (2, 6), (2, -2)$

In Exercises 53–56, sketch the geometric figure with the given vertices. Then shift the vertices of that figure as indicated and sketch the resulting figure. Give the vertices of the resulting figure.

53. $(-4, 6), (1, 3), (-6, 0)$; down 6, right 4

54. $(0, 0), (1, 7), (-3, 1), (5, 4)$; up 3, left 1

55. $(-2, 5), (4, 2), (3, -4), (-4, -1)$; up 1, right 2

56. $(0, -2), (8, -1), (5, 3)$; down 2, left 3

In Exercises 57–60, sketch the geometric figure with the given vertices. Then shift the figure horizontally and vertically so that vertex A is located at the origin and sketch the resulting figure. Give the vertices of the resulting figure.

57. $A(-3, -4), (-2, -1), (1, -1), (2, -4)$

58. $A(5, 6), (3, -2), (8, 4)$

59. $A(-8, 3), (1, 5), (1, -2)$

60. $A(3, -2), (5, 3), (5, -4), (8, -2)$

61. In your own words, explain how to determine the distance between two points in a coordinate plane.

62. In your own words, explain how to determine the coordinates of the midpoint of a line segment.

In Exercises 63–72, determine the distance (to the nearest hundredth) between the given points P and Q.

63. $P(5, 2), Q(-12, 2)$

64. $P(7, -4), Q(-17, -4)$

65. $P(3, 8), Q(3, -8)$

66. $P(-5, -14), Q(-5, 13)$

67. $P(-3, 2), Q(5, 8)$

68. $P(-5, -6), Q(4, 3)$

69. $P(-7, -2), Q(-3, -5)$

70. $P(-8, 5), Q(7, -9)$

71. $P(-4, 3), Q(-6, -7)$

72. $P(-5, -7), Q(5, -4)$

In Exercises 73–76, use the distance formula to determine whether the three given points are collinear. (Points are **collinear** if they are points of a straight line.)

73. $A(2, 4), B(0, -2), C(-1, -5)$

74. $A(-4, 3), B(2, 0), C(4, -1)$

75. $A(-4, 1), B(-1, 2), C(4, -1)$

76. $A(5, -4), B(3, -2), C(-1, 2)$

In Exercises 77–82, determine the midpoint of the line segment whose endpoints are given.

77. $P(4, -3), Q(4, -1)$

78. $P(-5, 6), Q(5, -6)$

79. $P(3, 7), Q(-1, 3)$

80. $P(3, -2), Q(-5, 5)$

81. $P\left(2, \frac{1}{2}\right), Q(-2, 3)$

82. $P\left(\frac{1}{3}, -\frac{11}{3}\right), Q\left(\frac{5}{3}, \frac{2}{3}\right)$

Geometric Models

In Exercises 83–86, determine (a) the area and (b) the perimeter of the geometric figure. Round results to the nearest hundredth.

83. Square; the endpoints of one side are $A(-2, -5)$ and $B(-4, 1)$.

84. Rectangle; the vertices are $A(-7, -2)$, $B(1, 6)$, $C(5, 2)$, and $D(-3, -6)$.

85. Trapezoid; the vertices are $A(-4, 2)$, $B(0, 4)$, $C(4, 1)$, and $D(-2, -2)$. (*Hint*: $\overline{AD}$ is the altitude of the trapezoid.)

86. Right triangle; the endpoints of the hypotenuse are $A(2, 5)$ and $C(6, -1)$; the legs are parallel to the coordinate axes.

In Exercises 87 and 88, the given points are the vertices of a rectangle. Show that the diagonals bisect each other.

87. $A(2, 5), B(8, 1), C(2, -8), D(-4, -4)$

88. $A(2, 4), B(6, -4), C(0, -7), D(-4, 1)$

In Exercises 89 and 90, show that the given points are vertices of a parallelogram; that is, show that both pairs of opposite sides are equal in length.

89. $A(-2, 5), B(4, 3), C(-2, -3), D(-8, -1)$

90. $A(0, 7), B(8, 3), C(4, -1), D(-4, 3)$

In Exercises 91–94, determine whether the three given points are vertices of a right triangle.

91. $A(-2, 4), B(0, 9), C(3, 2)$

92. $A(-2, -1), B(4, -9), C(2, 2)$

93. $A(2, 4), B(-2, 1), C(3, -5)$

94. $A(-5, 0), B(2, 6), C(1, -7)$

In Exercises 95 and 96, the vertices of an isosceles triangle are given. If M is the midpoint of base $\overline{BC}$, then $\overline{AM}$ is the altitude of the triangle. Determine (a) the area and (b) the perimeter of the triangle. Round results to the nearest hundredth.

95. $A(-4, -6), B(1, 3), C(5, -1)$

96. $A(-4, 13), B(4, -1), C(-6, -3)$

Real-Life Applications

97. A train carrying hazardous waste derails and leaks poisonous gas. Authorities use a grid map to help them develop evacuation plans. They issue evacuation orders for all people living within a radius of 6 miles of the derailment. On the grid map, the scene of the accident is located at $(3, -2)$, and your house is located at $(-1, 4)$. Do you need to evacuate? (All coordinate units are in miles.)

98. The weather service is tracking Hurricane Jeffrey on a grid map with coordinate units representing 10 miles. At 10:00 A.M., the center of the hurricane is located at $(3, -4)$. Hurricane-force winds extend 70 miles out from the center. It is predicted that the eye of Jeffrey will be located at $(-1, -2)$ in 2 hours. Your cabin by Lake Helen is located at $(-7, -5)$ on the grid map. Will you feel the force of hurricane winds at noon? What is the predicted rate (in miles per hour) of the movement of the eye of Jeffrey?

99. When a rectangular coordinate system is superimposed on an aerial photograph, a church is at the origin and a school is at $(3, 2)$. Coordinate units are in miles. A cartographer uses the grid photograph to produce a map whose scale is 6 inches = 1 mile. To the nearest tenth of an inch, how far apart are the church and the school on the map?

100. On a grid map, a Navy cruiser is at $P(-20, 5)$ and is steaming toward Subic Bay located at $S(30, -15)$. To remain in the required sea lane, the

cruiser must first travel to $Q(1, 1)$ and then, from there, head directly to port. How much farther will the cruiser travel than a helicopter that flies from point P directly to Subic Bay? (All coordinate units are in miles.)

Modeling with Real Data

101. About half of all employed women in the United States have children under age 18. The accompanying bar graph compares the job status of women with the number of children they have. For example, we see that 15% of all mothers who are employed part time have three children. (Source: Interep Research.)

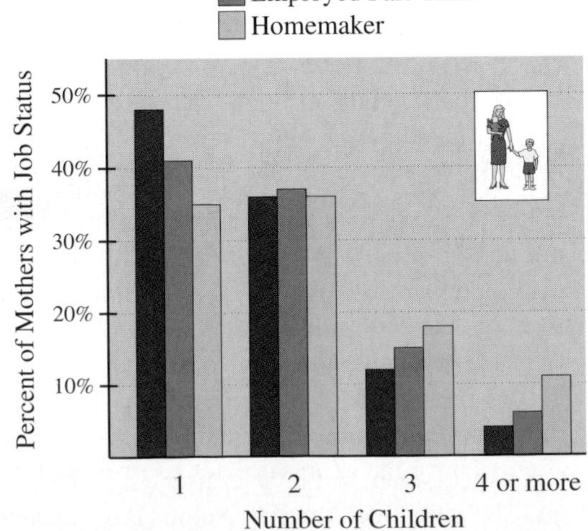

(a) Describe the job status and the number of children that are represented by the highest bar in the graph.

(b) Describe the job status and the number of children that are represented by the lowest bar in the graph.

102. Refer to the data in Exercise 101. For how many children are the percentages in each category about the same?

103. The following table gives the average high and low monthly temperatures in Omaha, Nebraska. The months are numbered beginning with 1 for January. (Source: National Climatic Data Center.)

Month	High	Low
1	31	11
2	37	17
3	49	28
4	64	40
5	74	51
6	84	60
7	88	66
8	85	63
9	77	54
10	66	41
11	49	29
12	35	16

(a) Using ordered pairs of the form (month, high) and (month, low), plot these two sets of data in the same coordinate plane.

(b) What distance indicates the difference in the average high and low temperatures for any month?

104. The graphs in Exercise 103 indicate some gradual and some abrupt average temperature changes. During what 2-month period does the greatest change in the average high or low temperature occur?

Group Project: Population Trends

The figure on the next page shows a bar graph of population trends in the Midwest and South. (Source: U.S. Bureau of the Census.)

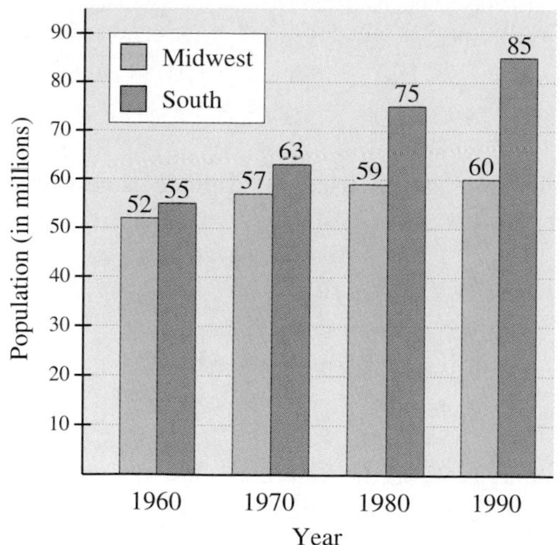

105. Write the four ordered pairs represented in the graph for the South.

106. Determine the coordinates of the midpoint of the line segment whose endpoints are (1960, 55) and (1990, 85).

107. What is a reasonable interpretation of your result in Exercise 106?

108. The land that is now Ohio was once part of the Western Reserve. In fact, Ohio is still considered a Midwestern state. Why is this geographically odd but historically logical?

Challenge

With most calculators, you can plot individual points. The steps for doing this are as follows:

 Erase Data

(a) Erase unwanted graphs and set an appropriate window.

(b) Erase any previously stored ordered pairs.

 Enter Data

(c) Enter your data in the form of ordered pairs.

 Plot Data

(d) Plot the points.

109. Use your calculator to plot the points associated with the following ordered pairs.

$A(-4, -3)$ $B(-3, 4)$ $C(2, 5)$
$D(3, 0)$ $E(5, -2)$

 **110.** In Exercise 109, why is point D not visible?

 Line Graph

You also may be able to connect plotted points to create a **line graph.**

111. Use your calculator to produce a line graph of the data in Exercise 109.

 Histogram

A **histogram** is a special kind of bar graph. You may be able to use your calculator to produce a histogram.

112. Select a suitable window and use your calculator to draw a histogram of the following data.

x	y
1	25
2	112
3	258
4	173
5	81

2.2 THE GRAPH OF AN EXPRESSION

Repeated Evaluation • Graphing Expressions

Repeated Evaluation

Consider the expression $|3 - x|$, where $x = -3, -1, 0, 2, 3$, and 5. We can use a **table of values** to summarize the results of evaluating the expression for each of the replacements for x. (See Fig. 2.14.)

Figure 2.14

x	-3	-1	0	2	3	5
$\lvert 3 - x \rvert$	6	4	3	1	0	2

In Chapter 1 we saw how a calculator can be used to evaluate an expression on the home screen. When an expression must be evaluated repeatedly, more efficient methods are available.

Evaluate Y

We can evaluate an expression by entering it on the Y screen, storing the given values of x on the home screen, and determining the corresponding values of the expression.

Figure 2.15 shows the entry of $|3 - x|$ as Y_1 on the Y screen. On the home screen, we stored -3 in X and then retrieved the value of Y_1, which is the value of $|3 - x|$ when $x = -3$.

Figure 2.15

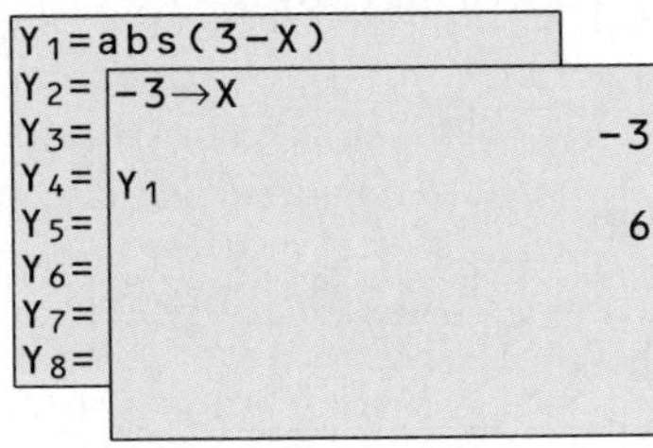

The advantage of this method is that with calculator assistance we can rapidly complete the second row of the table in Fig. 2.14. An even more efficient method is to have your calculator create the entire table for you.

Table

A calculator can create a table of values for an expression that has been entered on the Y screen.

Figure 2.16

X	Y_1	
-3	6	
-2	5	
-1	4	
0	3	
1	2	
2	1	
3	0	
Y1=ABS(3-X)		

Figure 2.16 is a typical display of a table of values for the expression $|3 - x|$, which was entered previously as Y_1. The x-values are in increments of 1, but you can set the increments as you wish.

By scrolling through the table, you will observe that all the entries of the table in Fig. 2.14 are shown along with many more.

EXAMPLE 1

Repeated Evaluation of an Expression

Evaluate the expression $3x - x^2$ for $x = -4, -2, 0, 8$, and 15 and create a table of values.

Solution

The calculator table in Fig. 2.17 shows the values of $3x - x^2$ for $x = -4, -2$, and 0. By scrolling, we also can find the values of the expression for $x = 8$ and 15.

Figure 2.17

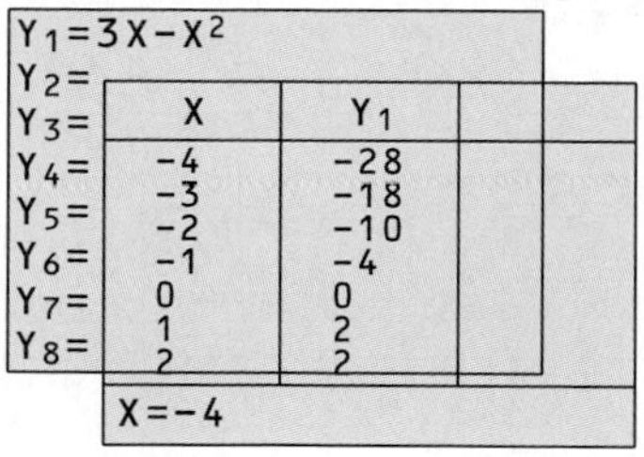

```
Y1=3X-X²
Y2=
Y3=
Y4=
Y5=
Y6=
Y7=
Y8=
```

X	Y_1
-4	-28
-3	-18
-2	-10
-1	-4
0	0
1	2
2	2

X=-4

By scrolling the calculator table, we can easily create the table of values.

x	-4	-2	0	8	15
$3x - x^2$	-28	-10	0	-40	-180

Graphing Expressions

Referring again to the table in Fig. 2.14, we can add a third row by pairing the entries in the first two rows. In general, the pairs have the form (value of the variable, value of the expression). The table in Fig. 2.18 shows the particular case of ordered pairs of the form $(x, |3 - x|)$.

Figure 2.18

x	-3	-1	0	2	3	5
$\|3 - x\|$	6	4	3	1	0	2
$(x, \|3 - x\|)$	$(-3, 6)$	$(-1, 4)$	$(0, 3)$	$(2, 1)$	$(3, 0)$	$(5, 2)$

In Fig. 2.19 we have plotted the points corresponding to these ordered pairs.

Figure 2.19

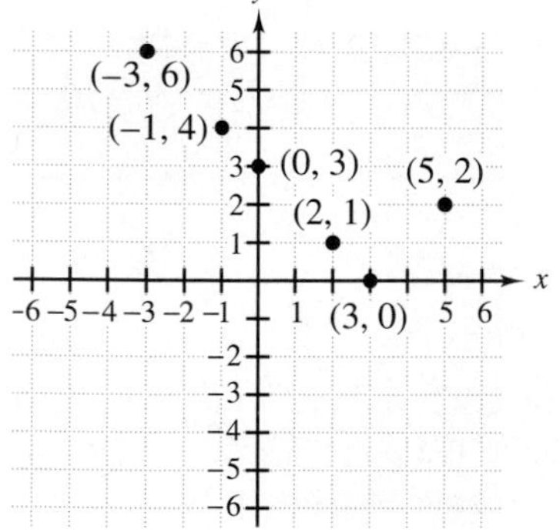

Because x can be replaced with infinitely many values, there are infinitely many pairs of the form $(x, |3 - x|)$, and so there are infinitely many points that we could add to the graph in Fig. 2.19. We call the graph of the set of all such pairs the **graph of the expression.**

A graphing calculator can do the work of plotting many points in a very short time. In Fig. 2.20 we entered the expression $|3 - x|$ on the Y screen and used the integer setting to produce the graph of the expression.

Figure 2.20

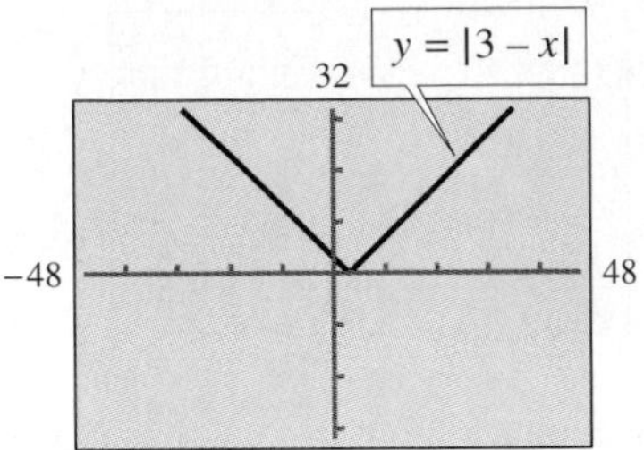

Figure 2.21

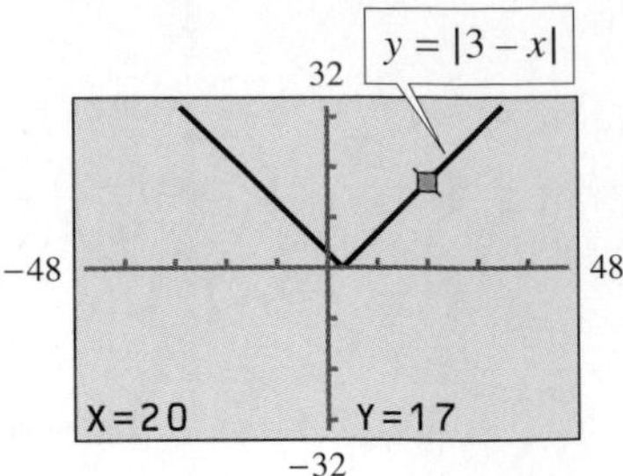

Trace

The *tracing cursor* is another kind of cursor that can be moved along a graph on the graph screen.

Figure 2.21 shows the tracing cursor on the point (20, 17). The coordinates of each point of the graph have the form (value of the variable, value of the expression). Thus, when the value of x is 20, the value of the expression $|3 - x|$ is 17.

Associated with each point of the graph of an expression is an ordered pair whose first coordinate is the value of the variable and whose second coordinate is the corresponding value of the expression. Using a graph or a table of values, we can determine either of the following:

1. Given the value of the variable, we can determine the value of the expression.
2. Given the value of the expression, we can determine the value of the variable.

EXAMPLE 2

Using the Graph of an Expression

Use the graph of $2x - 5$ to complete the following table.

x	-3				0
$2x - 5$			7		
$(x, 2x - 5)$		(2,)		(, −3)	

Solution

Figure 2.22

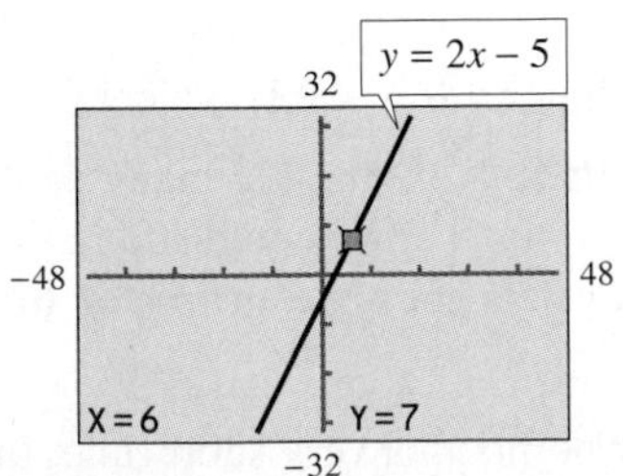

In Fig. 2.22 we have used a calculator to produce the graph of $2x - 5$, and we have traced to the point whose second coordinate (value of the expression) is 7. We see that the ordered pair $(x, 2x - 5)$ is (6, 7) and that the first coordinate (value of the variable) is 6.

By continuing to trace the graph, we can complete the table.

x	-3	2	6	1	0
$2x - 5$	-11	-1	7	-3	-5
$(x, 2x - 5)$	$(-3, -11)$	$(2, -1)$	$(6, 7)$	$(1, -3)$	$(0, -5)$

EXAMPLE 3

Using the Graph of an Expression

For the expression $x^2 + 2x - 5$, use a graph to

(a) evaluate the expression for $x = -5, 2$, and 6.

(b) determine the value(s) of x for which the expression has a value of -2.

Solution

Use a calculator to produce the graph of the expression, and trace the graph to determine the values in parts (a) and (b).

(a) Figure 2.23 shows the graph of $y = x^2 + 2x - 5$ with the tracing cursor on $(-5, 10)$. Because $(-5, 10)$ belongs to the graph, the expression $x^2 + 2x - 5$ has a value of 10 when x is -5. Similarly, for $x = 2$, $x^2 + 2x - 5 = 3$ and for $x = 6$, $x^2 + 2x - 5 = 43$.

Figure 2.23

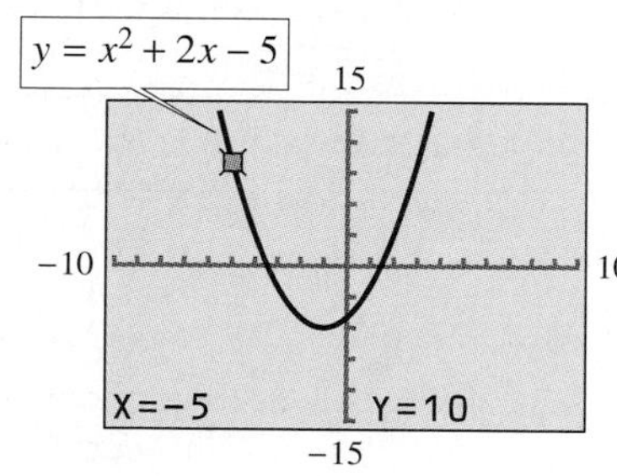

Figure 2.24

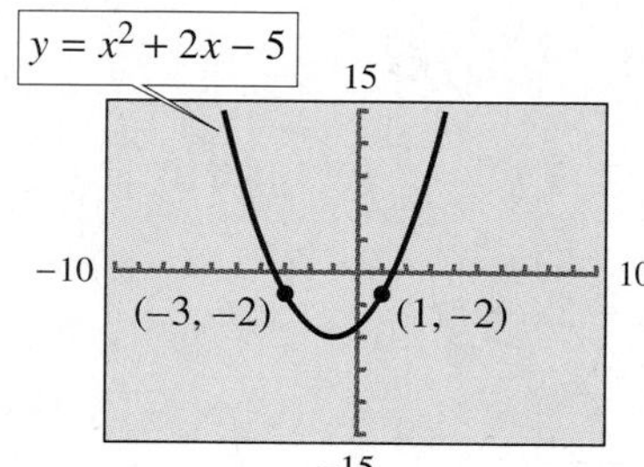

(b) Trace to a point whose y-coordinate is -2. The two such points of the graph are $(-3, -2)$ and $(1, -2)$. (See Fig. 2.24.) Replacing x with either -3 or 1 results in a value of -2 for the expression.

LEARNING TIP

You can think of the tracing cursor as an electronic finger that glides along a graph, just as you might move your own finger along the graph. As you trace, keep your eye on the known coordinate, and stop when you reach it. Then you can read the other coordinate.

Think About It

Suppose that you have graphed an expression and moved the tracing cursor to a point where the value of the expression is 0. Where is your cursor?

2.2 QUICK REFERENCE

Repeated Evaluation

- A **table of values** can be used to summarize the results of repeatedly evaluating an expression.
- A calculator's Y screen and table feature are efficient methods for evaluating an expression repeatedly.

Graphing Expressions

- From a table of values, we can form ordered pairs (value of the variable, value of the expression) and plot the associated points. The graph of all such possible pairs is called the **graph of the expression.**
- A calculator can be used to graph an expression and to trace the graph to display the coordinates of selected points.
- Using a graph or a table of values, we can determine either of the following:
  1. Given the value of the variable, we can determine the value of the expression.
  2. Given the value of the expression, we can determine the value of the variable.

2.2 EXERCISES

Concepts and Skills

1. To use the Y screen to evaluate an expression for $x = 2$, we begin by entering the expression on the Y screen. What are the other two steps that are needed?

2. The headings of the two columns in a calculator table are X and Y_1. What are the meanings of the entries in these two columns?

In Exercises 3–10, use the Y screen to determine the value of the expression for each of the given values of the variable.

3. $4x - 7;\quad \frac{3}{4}, -1.5, 3, -1$

4. $\frac{1}{2} + 2x;\quad -\frac{3}{4}, 0.25, -4, 0$

5. $5 + 3x - 2x^2;\quad -\frac{1}{2}, 0, -3, 1$

6. $x^2 - x - 1;\quad -2, 1, -5, 6$

7. $\sqrt{x - 5};\quad 6, 9, 5, 10$

8. $\sqrt{1 - 2x};\quad -4, 0, -12, -8$

9. $\frac{3 - x}{2x + 1};\quad -1, 2, -3, 3$

10. $\frac{4}{x^2 + 2};\quad 2, -1, 0, 4$

In Exercises 11–18, use a calculator table to determine the value of the expression for each of the given values of the variable.

11. $\frac{3}{4}x + 5;\quad -4, 2, 5$

12. $7 - 0.2x;\quad -5, -1, 9$

13. $x^2 - 5x + 15;\quad -1, 2, 3$

14. $-x^2 - x + 1;\quad -5, 0, 4$

15. $3 + \sqrt{x};\quad 1, 4, 100$

16. $\sqrt{x + 1} + x + 1;\quad -1, 3, 15$

17. $\frac{x}{1 - x^2};\quad 0, 1, -3$

18. $x^3 + 2x^2 - x + 1;\quad -8, 2, 5$

19. Suppose that you trace to a point of the graph of $2x + 1$ with a first coordinate of 3. What is the second coordinate of the point? Why?

20. Suppose that you trace the graph of an expression to a point with a y-coordinate of -4. What is the significance of the first coordinate?

In Exercises 21–24, use a graph to complete the table of values.

21.

x	-2			0
$3x$		3	-12	
$(x, 3x)$				

22.

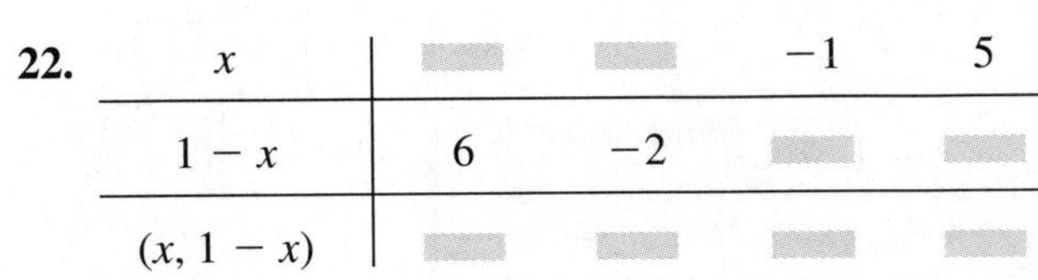

x			-1	5
$1 - x$	6	-2		
$(x, 1 - x)$				

23.

x		-2	2	-4
$2x^2$	0			
$(x, 2x^2)$				

24.

x	-3	0	3	
$\lvert x\rvert + 1$				1
$(x, \lvert x\rvert + 1)$				

In Exercises 25 and 26, the graph of an expression is given. Which expression does it represent?

25. (i) $3x$ (ii) $x + 3$ (iii) $3 - x$

Figure for 25

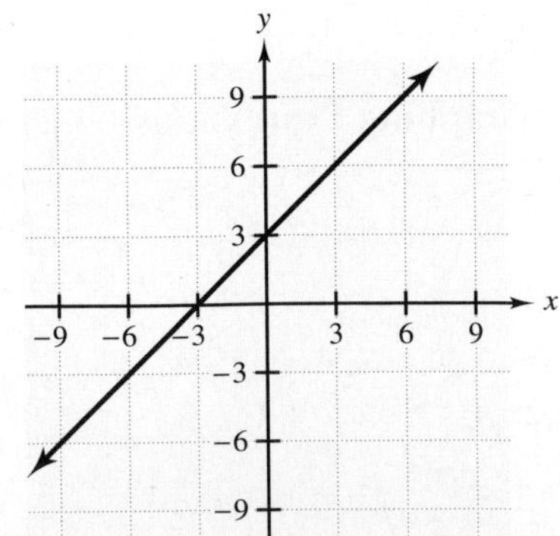

26. (i) $x^2 - 1$ (ii) $1 - x^2$ (iii) x^2

Figure for 26

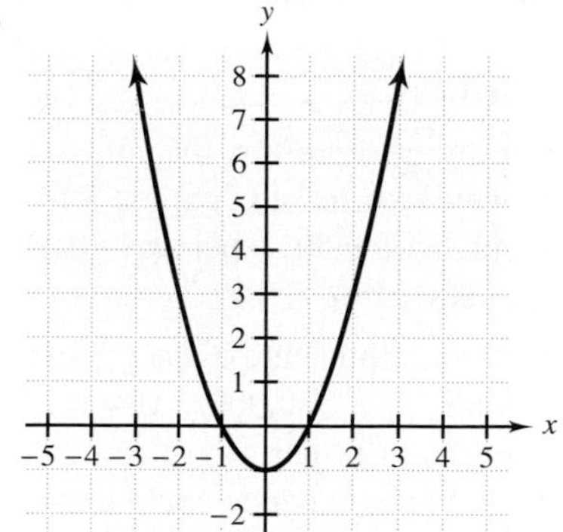

In Exercises 27–34, use a graph to estimate the value of the expression for the given values of the variable.

27. $\frac{2}{3}x + 1$; $1, 9, -10, -6$

28. $-2(1 - x)$; $-5, 1, 0, 4$

29. $1 - 2x - x^2$; $-4, 0, -2, 2$

30. $2x(x - 3)$; $3, 5, -1, -5$

31. $\frac{1}{x + 2}$; $-7, 1, 2, -4$

32. $\frac{x}{1 + x^2}$; $3, 0, -2, -7$

33. $|x| + |x - 3|$; $-4, 1, 4, 7$

34. $x + 2|x|$; -5; $3, -10, 4$

In Exercises 35–40, use a graph to estimate the value of x so that the expression has the given value.

35. $7 + \frac{3}{2}x$; $22, -11, -3.5, 4$

36. $-\frac{1}{4}x - 8$; $0, -1.25, -10, 1.5$

37. $x|x|$; $1, 16, -9, -25$

38. $x^3 + 2x + 1$; $34, -11, 1, 73$

39. $\sqrt{x + 9}$; $3, 1, 6, -1$

40. $3 - \sqrt{x}$; $-3, 2, 3, -1$

In Exercises 41–44, use a graph to estimate all values of x so that the expression has the given value.

41. $x(6 + x)$; $0, -9, -15, 16$

42. $8x - x^2$; $25, 16, -20, 0$

43. $|2x + 1|$; $15, -4, 1, 0$

44. $12 - \frac{|x|}{2}$; $0, 15, -4.5, 12$

45. Suppose that you use the graph of an expression to estimate the value of the expression when the variable has a value of 0. Would you trace to a point of the x-axis or the y-axis. Why?

46. Suppose that the graph of an expression intersects the x-axis. Explain how you know the value of the expression that is represented by the point of intersection?

In Exercises 47–54, use a graph to estimate (a) the value of the expression when x has a value of 0 and (b) the value of x when the expression has a value of 0.

47. $12 - 3x$ **48.** $2x + 10$

49. $-x^2 + 9$ **50.** $1 + x^2$

51. $|2 + x|$ **52.** $|x - 5|$

53. $x^4 - 16$ **54.** $x^3 + 8$

In Exercises 55–60, use a table or graph to estimate the value of x so that the expressions have the same value.

55. $x - 15, 9 - 2x$ **56.** $\frac{1}{2}x + 12, 22 - \frac{1}{3}x$

57. $x^2 + 3x + 9, 9 - 2x$ **58.** $(x - 4)^2, 2x - 5$

59. $-\frac{1}{2}x + 3, |x + 3|$ **60.** $12 - |x|, -\frac{1}{3}x + 4$

In Exercises 61–64, how many times do the graphs of the given expressions intersect?

61. $3x - 7, 5 - x$

62. $|x|, 5 - x^2$

63. $x^3 + 2x^2 - 7x - 5, x^2 + x + 5$

64. $\sqrt{x + 5}, 0.5x + 1$

In Exercises 65–68, which ordered pairs belong to the graph of the given expression?

65. $\frac{-2x}{x^2 + 1}$

(i) $\left(-3, \frac{3}{5}\right)$ (ii) $(1, -1)$ (iii) $\left(2, \frac{4}{5}\right)$

66. $\frac{2x + 1}{3}$

(i) $\left(-1, \frac{1}{3}\right)$ (ii) $(4, 3)$ (iii) $(1, 1)$

67. $\sqrt{1 - 2x}$

(i) $(-4, 3)$ (ii) $(0, -1)$ (iii) $(-12, 5)$

68. $|x| - |x + 3|$
(i) $(2, -3)$ (ii) $(6, 3)$ (iii) $(1, -3)$

In Exercises 69–72, use the integer setting to produce the graph of the given expression. Then move the tracing cursor along the graph to determine the amount by which the y-coordinate changes for each unit of change in the x-coordinate.

69. $2x - 11$

70. $-\frac{1}{2}x + 15$

71. $8 + 0.6x$

72. $-x + 20$

In Exercises 73–76, use the integer setting to produce the graph of the given expression. Starting at the given point A, move your *general* cursor as indicated to the destination point B. Verify that the y-coordinate of point B is the value of the expression when the value of x is the x-coordinate of point B.

73. $\frac{2}{3}x + 7$; $A(9, 13)$; up 4, right 6

74. $2x - 11$; $A(-4, -19)$; up 14, right 7

75. $12 - x$; $A(8, 4)$; down 10, right 10

76. $-\frac{1}{2}x - 5$; $A(-18, 4)$; up 3, left 6

Real-Life Applications

77. Suppose that you plan to expand your construction business by adding additional workers. You estimate that for each additional worker, your annual revenue will increase by \$40,000 from the current million dollar annual revenue.

(a) Write an expression for the annual revenue.

(b) Use a graph to estimate the number of workers to be added for revenue to grow by 16%.

78. In addition to a monthly salary of \$1500, an automobile sales agent receives a bonus of \$500 for each car he sells during the month.

(a) If the expression $1500 + 500c$ models the agent's monthly income, what does c represent?

(b) Suppose that the agent sells three cars in one month. Use a graph to estimate the agent's monthly income.

(c) How many cars must the agent sell per month for an annual income of \$60,000?

79. A law enforcement agency expects that the number of callers per day who offer information on a crime will be 12 plus 4 additional callers for each \$5000 in reward money.

(a) Letting x represent the amount of the reward money, write an expression to model the number of callers per day that the agency can expect.

(b) Suppose that the agency decides to offer a reward of \$20,000. Use a graph of the expression in part (a) to estimate the number of calls that the agency could expect.

(c) The agency would like to receive 60 calls per day. Use a graph to estimate the reward that would be required.

80. RenTool rents a trencher for \$60 per day plus \$6.50 per hour.

(a) Write an expression to model the cost of a rental.

(b) RenTool charged a customer \$99. Use a graph to estimate the number of hours for which the customer was charged.

(c) Use a table or graph to estimate the charge for
(i) 3 hours. (ii) 7 hours. (iii) 12 hours.

Modeling with Real Data

81. From 1985 to 1995 there was a decrease in the percentage of people who believed that driving anything but a new car would hurt their image among friends and co-workers. The accompanying table shows the percentages of image-conscious drivers for three selected years in the period. (Source: CNW Marketing/Research.)

Year	Percentage
1985	31%
1990	22%
1995	15%

The expression $-1.6x + 31$ models the percentage of image-conscious drivers, where x is the number of years since 1985; that is, $x = 0$ for 1985, $x = 5$ for 1990, and $x = 10$ for 1995. Use the integer setting to produce the graph of the model expression. Then trace the graph to the points that represent the data in the table. For which years does the model appear to be completely accurate?

82. Trace the graph in Exercise 81 to estimate the year when the model projects that only 7% of the drivers will be image conscious.

83. The accompanying table shows the change in consumers' perceptions of celebrity endorsements during the 1990s. The first column shows the percentage of consumers who are favorable toward celebrity endorsements and the second column shows the percentage who say that such endorsements lack credibility. (Source: Video Storyboard Tests.)

Year	Favorable	Lack Credibility
1992	27%	35%
1993	28%	35%
1994	24%	45%
1995	21%	52%
1996	17%	53%

Expressions that model the percentages t years after 1990 are as follows.

Favorable: $-0.64t^2 + 2.44t + 25.2$

Lack credibility: $22.25\sqrt{t}$

For what years do the two expressions most accurately model the actual data?

84. Refer to the model expressions in Exercise 83.

(a) Use a graph to estimate the percentage of consumers who had a favorable view of celebrity endorsements in 1997.

(b) According to the model, in what year would the unfavorable reaction reach 70%?

Group Project: Christmas Trees

The accompanying table shows the numbers of both real and artificial Christmas trees sold during the years 1992–1995. (Source: National Christmas Tree Growers Association.)

Year	Real Trees (in millions)	Artificial Trees (in millions)
1992	34.4	38
1993	35.2	41.6
1994	33	41
1995	37.1	39.3

The following expressions model the number (in millions) of trees sold x years after 1990.

Real: $0.9x + 32$

Artificial: $-1.15x + 45$

85. Produce the graph of each expression. Compare the trend in sales.

86. Use a graph to estimate the year in which the model predicts that 40 million real trees will be sold.

87. Use a table to predict the first year after 1990 in which the sales of real trees surpassed the sales of artificial trees.

88. What are some reasons that consumers prefer each type of tree?

Challenge

In Exercises 89–92, use a graph or table to estimate the value(s) of the variable so that the expression has the given value.

89. $|x| + |x - 4|$; 8, 4, 1

90. $|x + 2| - |x - 2|$; $-4, 0, 4$

91. $\frac{2x + 1}{x - 3}$; 1, 2, 9

92. $\sqrt{x^2 - 6x + 8}$; 0

In Exercises 93 and 94, use the integer setting and your tracing cursor.

93. If the value of $|x - 12|$ is 5, what are the possible values of $x^2 - 3$?

94. If the value of $\frac{70}{x}$ is 14, what is the value of $\frac{70}{x - 5}$?

2.3 RELATIONS AND FUNCTIONS

Relations • Functions • Graphs of Functions

Relations

In everyday life we encounter situations in which one item is associated or paired with another. For instance, a student's tuition cost may be associated with the number of credit hours, or one person may be paired with another. In mathematics, we call sets of such pairs a **relation.**

Definition of a Relation

A **relation** is a set of ordered pairs. The set of all first coordinates is the **domain** of the relation, and the set of all second coordinates is the **range** of the relation.

EXAMPLE 1

Determining the Domain and Range of a Relation

State the domain and range of each relation.

(a) $\{(3, -2), (7, 4), (7, 1), (6, 3), (9, 3), (9, 2)\}$

(b) $\{(1, 3), (2, 4), (3, 5), (4, 6), \ldots\}$

(c) $\{(x, y) \mid y = |x|\}$

Solution

(a) Domain: $\{3, 7, 6, 9\}$ Range: $\{-2, 4, 1, 3, 2\}$

(b) Domain: $\{1, 2, 3, 4, \ldots\}$ Range: $\{3, 4, 5, 6, \ldots\}$

(c) Because the first coordinate x can be any real number, the domain is **R.** The second coordinate y is the value of the algebraic expression $|x|$, which is never negative. Thus the range is $\{y \mid y \geq 0\}$.

As illustrated in Example 1, a relation can be a finite set or an infinite set. A relation can be described by listing the ordered pairs, as in parts (a) and (b), or with an algebraic expression, as in part (c).

Another method for describing a relation is a **mapping diagram,** which illustrates the correspondence between the elements of the domain and the elements of the range. For the relation in Example 1(a), the mapping diagram in Fig 2.25 shows the domain elements in one row and the range elements in another row, with arrows indicating the correspondence.

Figure 2.25

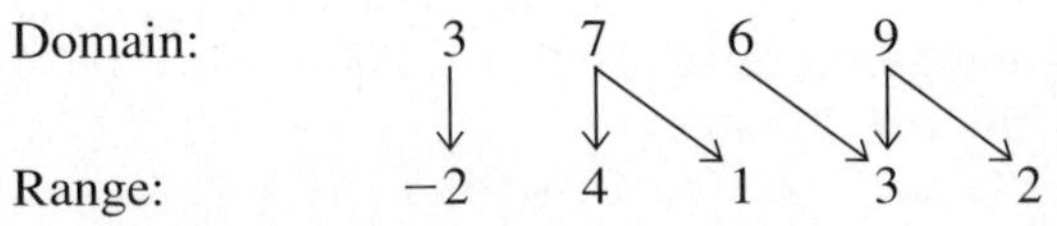

Domain: 3 7 6 9

Range: −2 4 1 3 2

The result of plotting the pairs of a relation is called the **graph of the relation.** When a relation is defined by an algebraic expression, the graph of the relation is the same as the graph of the expression.

EXAMPLE 2

Graphing a Relation

Graph the given relation.

(a) $\{(1, -1), (2, -2), (3, -3, (4, -4), \ldots\}$
(b) $\{(x, y) \mid y = x^2 + 2\}$

Solution

(a) In Fig. 2.26 the arrow indicates that the pattern of the graph extends forever.

Figure 2.26

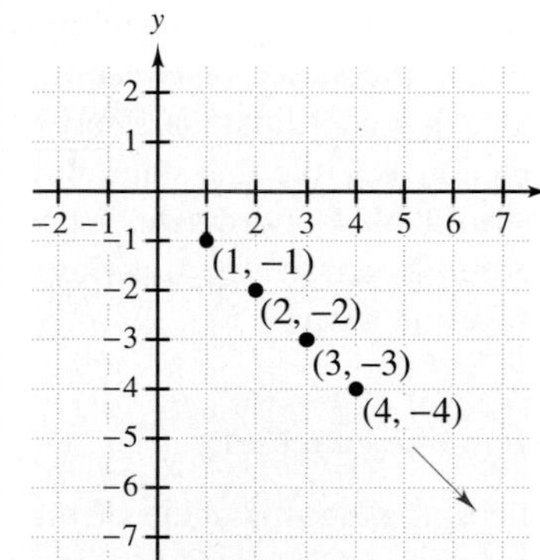

Figure 2.27

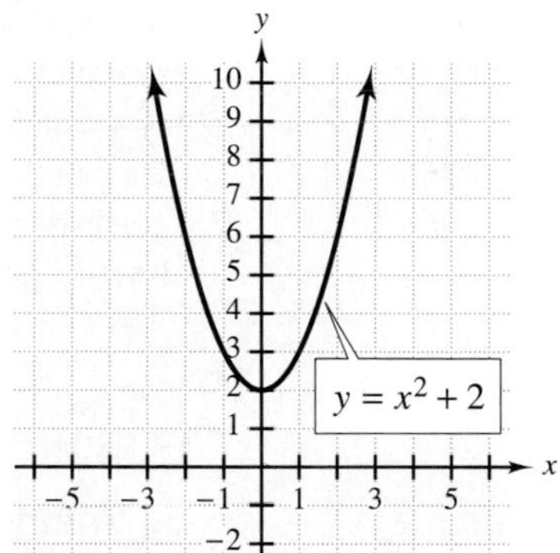

(b) The graph of the relation $\{(x, y) \mid y = x^2 + 2\}$ is the same as the graph of the expression $x^2 + 2$. (See Fig. 2.27.)

Note that a graph of a relation provides a way to visually estimate the domain and range of the relation. For instance, the graph of $y = x^2 + 2$ in Example 2(b) appears to extend forever to the left and right. Thus the domain of the relation is **R.** Because the point with the smallest y-coordinate is (0, 2) and from that point the graph rises forever, the range is $\{y \mid y \geq 2\}$.

Functions

A **function** is a special kind of relation.

Definition of a Function

A **function** is a relation in which no two ordered pairs have the same first coordinate.

EXAMPLE 3

Recognizing Functions

Which of the following relations is a function?

(a) $A = \{(-2, 3), (1, 4), (4, 3), (5, -2)\}$

(b) $B = \{(1, 3), (2, 5), (-2, 5), (2, -1)\}$

(c) $C = \{(-1, 1), (-2, 2), (-3, 3), \dots\}$

Solution

Think About It

Consider a relation whose ordered pairs are of the form (person, name of person). Is this relation a function? If the pairs were reversed, would the relation be a function?

(a) Relation A is a function because no two ordered pairs have the same first coordinate. There are two pairs, $(-2, 3)$ and $(4, 3)$, with the same second coordinate, but that does not violate the definition of a function.

(b) Relation B is not a function because two pairs, $(2, 5)$ and $(2, -1)$, have the same first coordinate.

(c) Assuming that the indicated pattern continues, relation C is a function.

In addition to listing the ordered pairs, we can describe a function with mapping diagrams, graphs, algebraic expressions, or equations. For example, $\{(x, y) \mid y = x + 5\}$ is a function in which each y-coordinate is 5 more than the x-coordinate. In this case we can think of the function as a rule that states how to determine the y-coordinate that corresponds to each permissible x-coordinate. When a rule such as $y = x + 5$ indicates how to generate the ordered pairs of the function, we simply say that the function is $y = x + 5$ rather than $\{(x, y) \mid y = x + 5\}$.

Alternative Definition of a Function

A **function** is a rule that assigns a value of the variable x to a unique value y. The **domain** of the function is the set of all permissible values of x. The **range** of the function is the set of all possible values of y.

In the alternative definition, the word *unique* means that each permissible replacement for x must result in *exactly one* value of y. Informally, we can say that each *input* value of x results in exactly one *output* value of y.

EXAMPLE 4

Determining Whether a Relation Represents a Function

Determine whether the given relation represents a function.

(a) $y = |x|$ (b) $x = |y|$

Solution

LEARNING TIP

To judge whether an equation represents a function, ask yourself, "Can I choose any x-value that will result in more than one y-value?" If the answer is yes, then the equation does not represent a function.

(a) For each real number x, the expression $|x|$ has a unique (exactly one) value. Thus $y = |x|$ represents a function.

(b) For $x = 5$, for example, $5 = |y|$ is true for two values of y: 5 and -5. That is, replacing x with 5 results in two different values of y. Therefore, $x = |y|$ does not represent a function.

Graphs of Functions

Because a function is a relation, we graph a function by plotting the ordered pairs in a coordinate system. For a function that is described by a rule, such as $y = 2x - 1$, we can use a calculator to produce the graph of the function by entering $2x - 1$ on the Y screen. (See Fig. 2.28.)

Figure 2.28

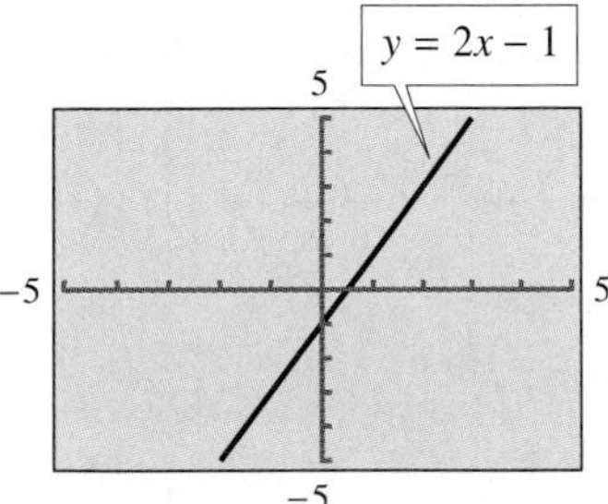

An easy way to determine whether a relation is a function is to examine its graph. Figure 2.29 shows the graph of a relation. A vertical line intersects two points of the graph, and these points have the same first coordinate. Thus the relation is not a function because no two ordered pairs of a function can have the same first coordinate. Figure 2.30 shows the graph of a relation for which a vertical line intersects the graph at most once. Thus this graph represents a function.

Figure 2.29

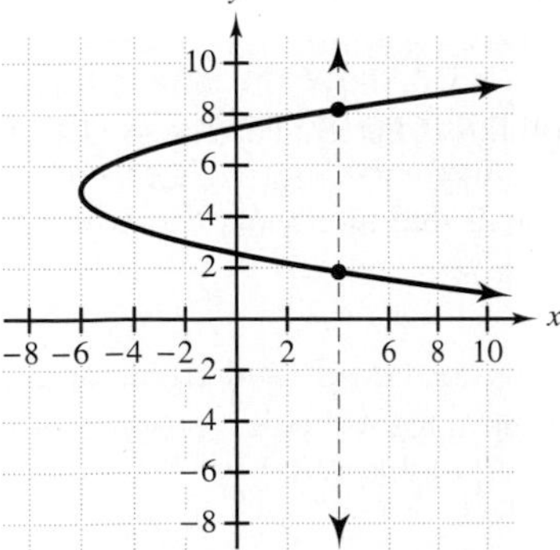

Figure 2.30

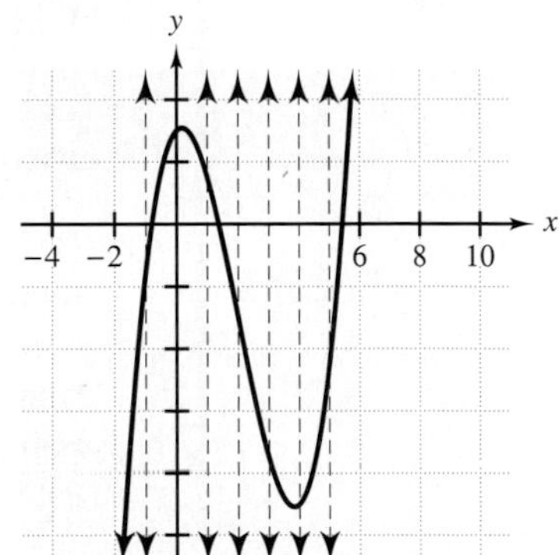

> **The Vertical Line Test**
>
> A graph of a relation represents a function if and only if no vertical line intersects the graph at more than one point.

EXAMPLE 5

Using the Vertical Line Test

Determine whether the relation whose graph is given is a function. Then estimate the domain and range.

(a) **Figure 2.31** (b) **Figure 2.32** (c) **Figure 2.33**

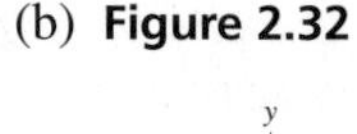

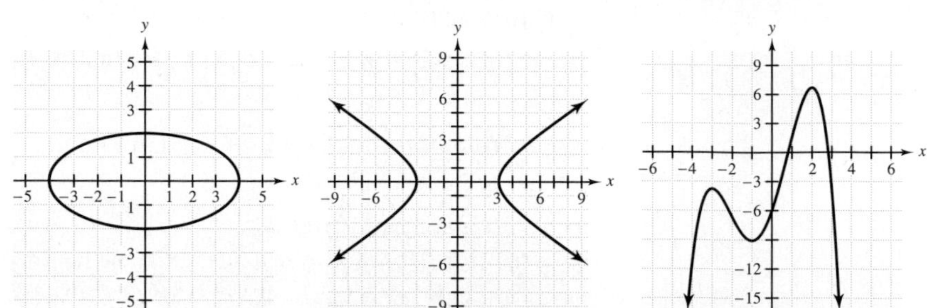

Solution

(a) In Fig. 2.31 we can draw a vertical line that would intersect the graph at more than one point. Thus the graph does not represent a function.

The first coordinates of the points of the graph extend from -4 to 4. The second coordinates extend from -2 to 2. Thus the domain is $\{x \mid -4 \leq x \leq 4\}$ and the range is $\{y \mid -2 \leq y \leq 2\}$.

(b) Applying the Vertical Line Test reveals that the graph does not represent a function.

From the point $(-3, 0)$, the graph extends forever to the left, and from the point $(3, 0)$, the graph extends forever to the right. The domain is $\{x \mid x \leq -3 \text{ or } x \geq 3\}$. Above the x-axis, the graph rises forever, and below the x-axis, the graph falls forever. The range is **R.**

(c) Because no vertical line intersects the graph at more than one point, the graph represents a function.

The graph widens to the left and right, and so the domain is **R.** The highest point of the graph appears to have a y-coordinate of approximately 6.6. The estimated range is $\{y \mid y \leq 6.6\}$.

Note: In Example 5(c), an exact description of the range, $\left\{y \mid y \leq \frac{20}{3}\right\}$, involves the use of more advanced methods. Remember that a graph is only suggestive and that conclusions based on the graph are estimates.

2.3 QUICK REFERENCE

Relations

- A **relation** is a set of ordered pairs.
- The set of all first coordinates is the **domain** of the relation, and the set of all second coordinates is the **range** of the relation.
- A **mapping diagram** depicts the correspondence between the elements of the domain of a relation and the elements of the range.
- The result of plotting the pairs of a relation is called the **graph of the relation.**

Functions

- A **function** is defined as a relation in which no two ordered pairs have the same first coordinate.
- Alternatively, a **function** is a rule that assigns a value of the variable x to a unique value y. The **domain** of the function is the set of all permissible values of x. The **range** of the function is the set of all possible values of y.

Graphs of Functions

- The Vertical Line Test: A graph of a relation represents a function if and only if no vertical line intersects the graph at more than one point.
- The graph of a function can be used to make a visual estimate of the domain and range of the function.

2.3 EXERCISES

Concepts and Skills

1. In your own words, explain the difference between a relation and a function.

2. Explain the meaning of *domain* and *range*.

In Exercises 3–8, use the given description to write a set of ordered pairs. Then determine the domain and range of the relation.

3. The first coordinate is an integer between -3 and 3, inclusive; the second coordinate is the square of the first coordinate.

4. The first coordinate is a positive integer less than 8; the second coordinate is the sum of the first coordinate and -4.

5. The first coordinate is the length of the side of a square, where the length is a positive integer not exceeding 5; the second coordinate is the perimeter of the square.

6. The first coordinate is the radius of a circle, where the radius is a positive integer not exceeding 4; the second coordinate is the area of the circle.

7. The first coordinate is the number of hours worked (10, 20, 30, 40); the second coordinate is the salary at $7 per hour.

8. The first coordinate is the number of toppings on a pizza (up to 5); the second coordinate is the price of a pizza, which is $7 plus $1 per topping.

In Exercises 9–18, sketch the graph of each relation and give the domain and range. Identify any relations that are functions.

9. $\{(1, 2), (3, 4), (5, 6), (7, 8), (9, 10)\}$

10. $\{(-1, 2), (-3, 4), (-5, 6), (-7, 8), (-9, 10)\}$

11. $\{(3, 1), (4, 1), (5, 1)\}$

12. $\{(2, 3), (5, 3), (6, 3)\}$

13. $\{(-2, 4), (-2, 5), (-2, 7)\}$

14. $\{(-1, -3), (-1, -5), (-1, -7)\}$

15. $\{(-2, 2), (-2, -3), (3, 4), (2, -3)\}$

16. $\{(4, -4), (-2, 3), (4, 3), (5, 0)\}$

17. $\{(-3, -4), (-4, -5), (-5, -6), \ldots\}$

18. $\{(-3, 4), (-4, 5), (-5, 6), \ldots\}$

19. Describe how to use the graph of a relation to estimate the domain.

20. Describe how to use the graph of a relation to estimate the range.

21. In your own words, state the Vertical Line Test and explain the reasoning behind it.

22. Are all functions relations? Are all relations functions? Explain.

In Exercises 23–26, a relation is described by a table of values. Give the domain and range. Identify any relation that is a function.

23.

Input: Cost of Mailing (in dollars)	Output: Weight of Package (in pounds)
3.00	2.7
4.50	3.8
3.00	1.2
6.00	5.3
4.50	4.1

24.

Input: Super Bowl Winner	Output: Year
Washington	1992
Dallas	1993
Dallas	1994
San Francisco	1995
Dallas	1996

25.

Input: State	Output: Capital
Maine	Augusta
Michigan	Lansing
Montana	Helena
Texas	Austin
Oregon	Salem

26.

Input: Census Year	Output: U.S. Population (in millions)
1950	151
1960	179
1970	203
1980	227
1990	249

In Exercises 27–38, determine whether the relation whose graph is given is a function. Estimate the domain and range.

27. 28.

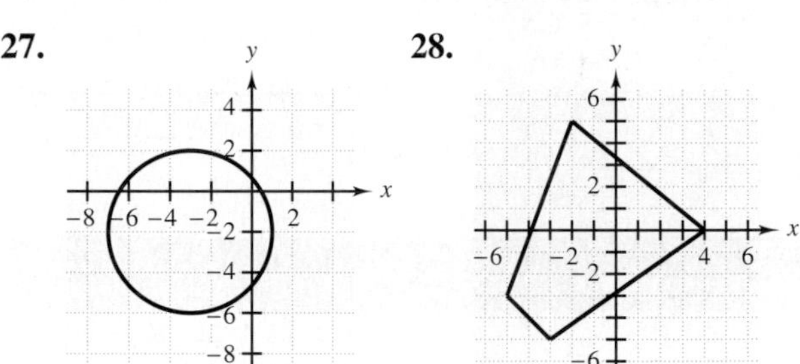

29. 30.

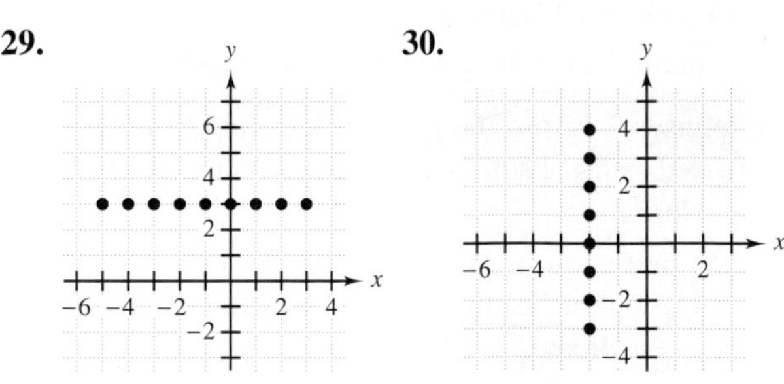

31. 32.

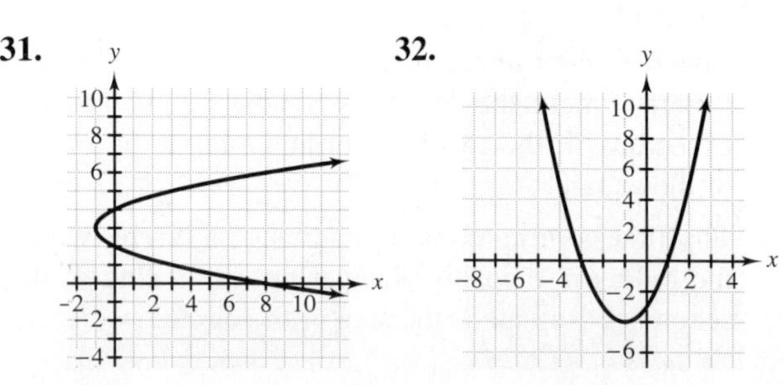

33.

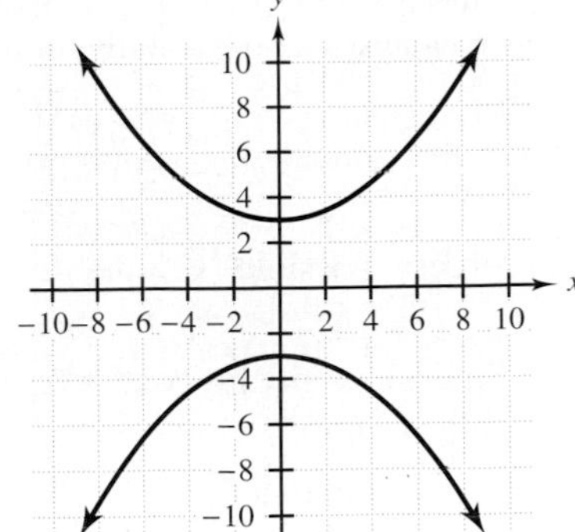

34.

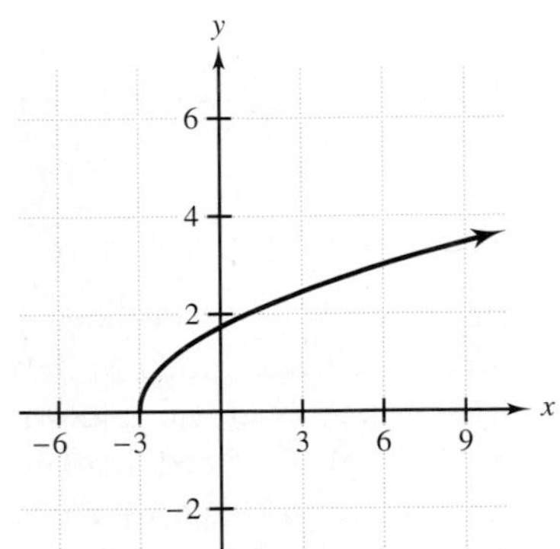

35.

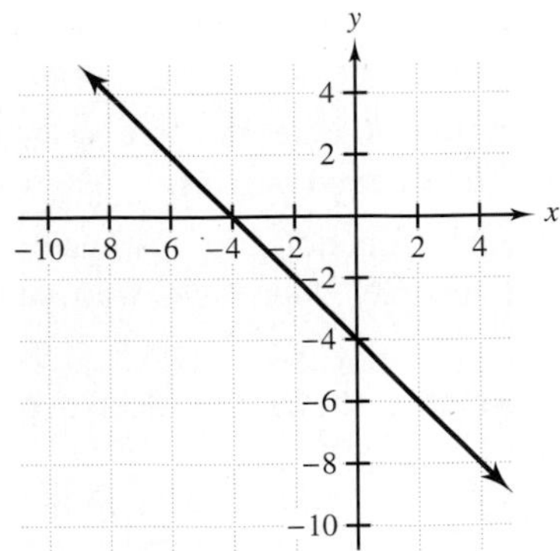

36.

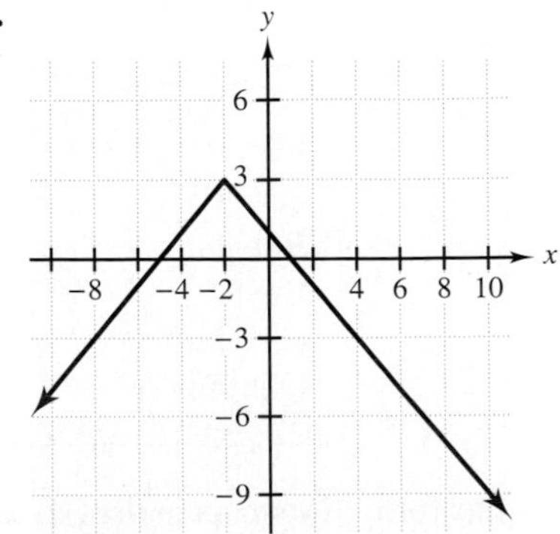

37.

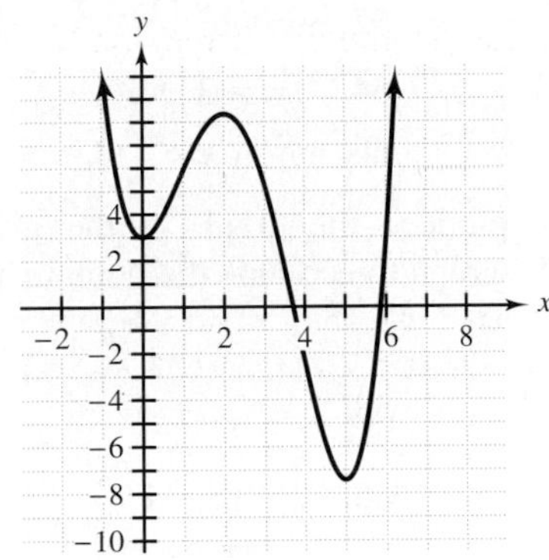

38.

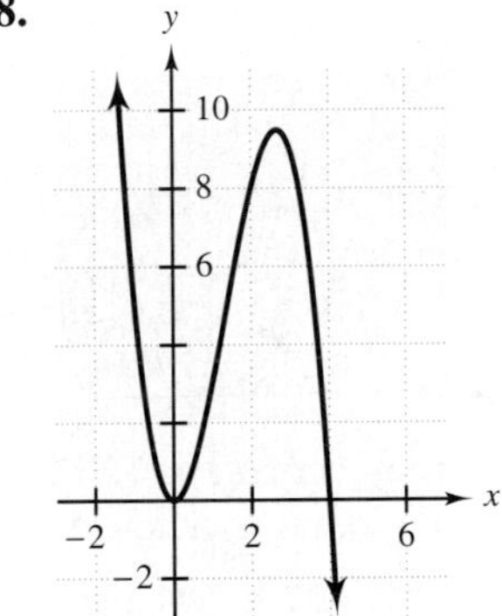

In Exercises 39–56, determine whether the given rule defines a function.

39. $x - 3 = y$

40. $y = x - 12$

41. $x = y^2$

42. $y = x^2$

43. $y = x^2 - 2$

44. $x + 1 = y^2$

45. $y^2 = x + 9$

46. $y = (x - 4)^2$

47. $x = y^3$

48. $y = x^3$

49. $y^2 = 9 - x^2$

50. $y^2 = x^2 - 4$

51. $|x| = |y|$ **52.** $y^2 = x^2$

53. $y = \sqrt{x+5}$ **54.** $\sqrt{y} = 2 - x$

55. $xy = 1$ **56.** $y = \sqrt[3]{x}$

In Exercises 57–68, produce the graph of the given function. Then use the graph to estimate the domain and range.

57. $y = 1 - 2x$ **58.** $y = \frac{1}{2}x + 3$

59. $y = x^2 + 2x - 3$ **60.** $y = 4x - x^2$

61. $y = 4 - |x + 1|$ **62.** $y = |3 - x|$

63. $y = \sqrt{x+9} - 5$ **64.** $y = 1 - \sqrt{4 - x}$

65. $y = |x| + x$ **66.** $y = \frac{|x|}{x}$

67. $y = \sqrt{x^2}$ **68.** $y = \frac{1}{4}x^4 - \frac{1}{3}x^3 - 6x^2$

Real-Life Applications

69. A carpet installer agrees to lay carpet for \$50 plus \$140 per room for up to seven rooms in a house.

(a) Write a relation that models this cost plan for installing carpet.

(b) What are the domain and range of the relation?

(c) Is the relation a function?

70. A group of construction volunteers plans to travel to a certain location to repair storm-damaged homes. The weekly cost to rent a van is \$480, and the volunteers have agreed to share the cost equally.

(a) Write a relation that models the per-person cost of renting the van for up to six people.

(b) What are the domain and range of the relation?

(c) Is the relation a function?

71. A florist charges $12 - x$ dollars per arrangement to deliver x floral arrangements.

(a) Write a function that models the florist's income for delivering x floral arrangements.

(b) What is the greatest income from delivery charges that the florist can expect?

(c) How many floral arrangements result in the greatest income?

(d) What is the domain of the function?

(e) What is the range of the function?

(f) How many floral arrangements are delivered if the income is \$32?

72. The height of a produce shipping container with a square base is $6 - x$, where x is the length (in feet) of a side of the base.

(a) Write a function that models the volume of the container.

(b) What is the largest possible volume of the container?

(c) What are the dimensions of the container with the largest possible volume?

(d) What is the domain of the function?

(e) What is the range of the function?

(f) Is it possible for the container to have a volume of 35 cubic feet?

Modeling with Real Data

73. In 1995 the number of movie-goers declined from a high of 82% in the 18–24 age group to 58% in the 45–54 age group. (Source: National Endowment for the Arts.) The percentage P of people of age a who attended movies can be modeled by the function $P = -1.06a + 106$.

(a) Produce the graph of the function and use it to describe the trend in movie-going with respect to age.

(b) Use the graph to estimate the age at which approximately 70% attend movies.

(c) Use the model function to estimate the percentage of people your age who attend movies.

74. The accompanying table shows the elected presidents in the period 1968–1996.

Year	President
1968	Nixon
1972	Nixon
1976	Carter
1980	Reagan
1984	Reagan
1988	Bush
1992	Clinton
1996	Clinton

Let A be a set of the form {(year, president)}, and let B be a set of the form {(president, year)}.

(a) Which of the two sets, if either, is a relation? Why?

(b) Which of the two sets, if either, is a function? Why?

75. The accompanying bar graph shows that the number of political action committees (PACs) increased dramatically from 1980 to 1990. (Source: U.S. Federal Election Commission.)

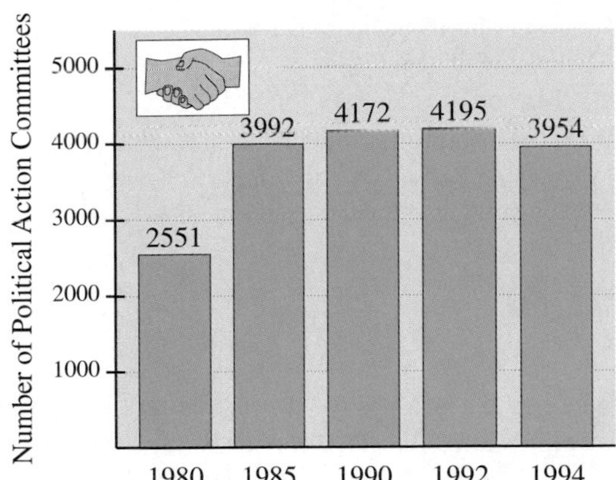

The number P of political action committees y years after 1975 can be modeled by $P = 1620\sqrt[3]{y}$.

(a) Write the ordered pairs associated with the data in the bar graph, and write the corresponding pairs that result from evaluating the model function. Then use these two relations to determine the year in which the model function is most accurate.

(b) In which year is the model function least accurate?

76. Six popular trees in the Southeast are red oak, silver maple, pink dogwood, silver oak, red maple, and white dogwood. (Source: *Landscape Plants of the Southeast*.)

(a) Construct a mapping diagram of the relation A whose form is {(color, tree)} and use it to determine whether set A is a function.

(b) If the coordinates of set A were reversed, would the new relation be a function? Why?

Group Project: Day-Care Costs

Based on a monthly fee for a 3-year-old in a for-profit center 5 days per week, 8 hours per day, the median cost of day care nationwide is $388. The accompanying bar graph shows the metropolitan areas that had the cheapest day care in 1996. (Source: Runzheimer International.)

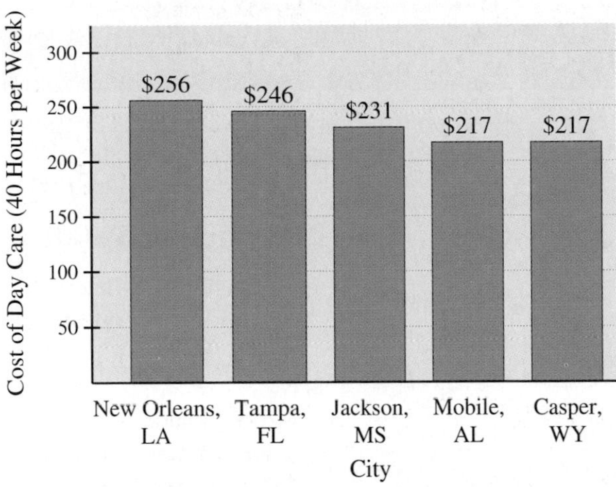

77. Write the information shown in the bar graph as a set A of ordered pairs of the form (city, cost). Is set A a function?

78. Consider the set of ordered pairs (city, cost), where the city is any city in the United States, and the cost is the average cost of day care for that city. Is this set a function?

79. Suppose that you selected a city and made a table of the costs at all day-care centers in that city. Would the relation {(city, cost)} be a function?

80. In 1996 the minimum wage was $4.25 per hour. Given the national median cost of day care, why might a person with children have been disinclined to take a full-time job that paid minimum wage?

Challenge

In Exercises 81–86, the variables x and y represent human beings. Determine whether the given relation is a function.

81. $A = \{(x, y) \mid y \text{ is an ancestor of } x\}$

82. $B = \{(x, y) \mid x \text{ is the mother of } y\}$

83. $C = \{(x, y) \mid y \text{ is the mother of } x\}$

84. $D = \{(x, y) \mid x \text{ is a brother of } y\}$

85. $E = \{(x, y) \mid y \text{ is an aunt of } x\}$

86. $F = \{(x, y) \mid x \text{ is a grandfather of } y\}$

In Exercises 87–90, use a graph to estimate the domain and range of the function.

87. $y = \dfrac{x^2 - 9}{x + 3}$

88. $y = \dfrac{x + 3}{x - 2}$

89. $y = \dfrac{1}{1 + x^2}$

90. $y = \dfrac{1}{x^2 - 1}$

2.4 FUNCTIONS: NOTATION AND EVALUATION

Function Notation • Evaluating Functions

Function Notation

Consider the function defined by the equation $y = 2x + 1$. For convenience, we often use **function notation** to name the function and to indicate the value of the function.

$$f(x) = 2x + 1$$

The notation $f(x)$ is read "f of x." In this notation, the name of the function is f, the variable (domain element) is x, and the value of the function (range element) corresponding to x is $f(x)$.

Note: The parentheses in the notation $f(x)$ does *not* indicate multiplication. We read $f(x)$ as "f of x," *not* "f times x."

Just as we can use symbols other than x as a variable, we can use letters other than f to name a function. Each of the following defines the same function.

$$f(x) = x^2 - 3x + 4 \qquad g(t) = t^2 - 3t + 4 \qquad g(w) = w^2 - 3w + 4$$

Evaluating Functions

Determining the value of $f(x)$ for a specific value of x is called **evaluating the function.**

EXPLORING THE CONCEPT

Evaluating Functions

To indicate the value of the function $f(x) = 3x^2 - 5x + 4$ when x has a value of 2, we write $f(2)$. To evaluate $f(2)$, replace each occurrence of the variable with 2.

$$f(x) = 3x^2 - 5x + 4$$
$$f(2) = 3(2)^2 - 5(2) + 4 = 12 - 10 + 4 = 6$$

We also can use a graph to estimate the value of a function. Figure 2.34 shows the graph of $f(x) = 3x^2 - 5x + 4$ with the tracing cursor on (2, 6). Note that the first coordinate is the value of x, and the second coordinate is $f(2) = 6$.

LEARNING TIP

Keep relating an expression and its value to points of a graph. If $f(2) = 6$, then (2, 6) is a point of the graph of f. As you study mathematics, go slowly and take the time to visualize.

Figure 2.34

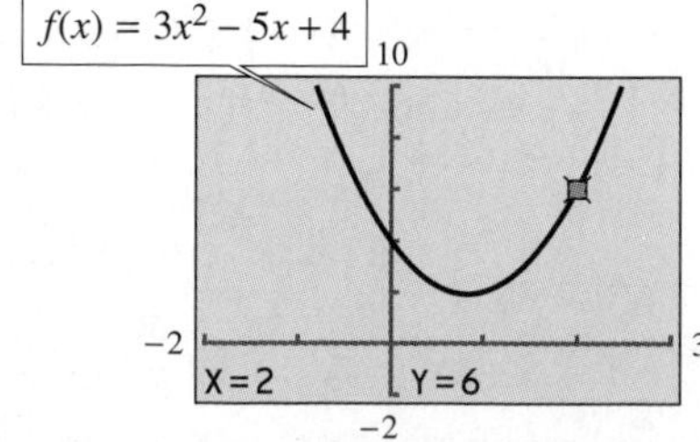

EXAMPLE 1

Evaluating a Function

Let $g(t) = -t^2 + 2t + 6$. Evaluate $g(-4)$.

Solution

$$g(t) = -t^2 + 2t + 6$$
$$g(-4) = -(-4)^2 + 2(-4) + 6 \quad \text{Replace } t \text{ with } -4.$$
$$= -16 - 8 + 6 \quad -(-4)^2 = -1 \cdot (-4)^2 = -1 \cdot 16 = -16$$
$$= -18$$

EXAMPLE 2

Evaluating Functions

Let $h(x) = |3 - 2x|$ and $r(x) = \dfrac{12}{x}$. Evaluate each of the following.

(a) $h(5)$ (b) $r(4)$ (c) $h(0) - r(-4)$

Solution

(a) $h(x) = |3 - 2x|$

$$h(5) = |3 - 2 \cdot 5| \quad \text{Replace } x \text{ with } 5.$$
$$= |3 - 10|$$
$$= |-7|$$
$$= 7$$

(b) $r(x) = \dfrac{12}{x}$

$$r(4) = \frac{12}{4} \quad \text{Replace } x \text{ with } 4.$$
$$= 3$$

(c) $h(0) = |3 - 2 \cdot 0| = |3 - 0| = |3| = 3$

$$r(-4) = \frac{12}{-4} = -3$$

Therefore, $h(0) - r(-4) = 3 - (-3) = 6$.

Recall that we can evaluate an expression with a calculator by storing the value of the variable and then entering the expression on the home screen or on the Y screen.

Evaluate Y

Some calculator models allow us to evaluate a function with function notation. The advantage of this feature is that we do not explicitly have to store the value of the variable.

EXAMPLE 3

Using a Calculator to Evaluate a Function

Let $f(x) = \sqrt{x - 5}$ and $g(x) = 4x^2 - x$.

(a) Enter $f(x)$ on the Y screen, and evaluate $f(30)$ by storing 30 in X.

(b) Use the calculator's function notation feature to determine $g(-1)$ and $g(3)$.

Solution

(a) We begin by entering $\sqrt{x-5}$ on the Y screen. (See Fig. 2.35.)

Figure 2.35

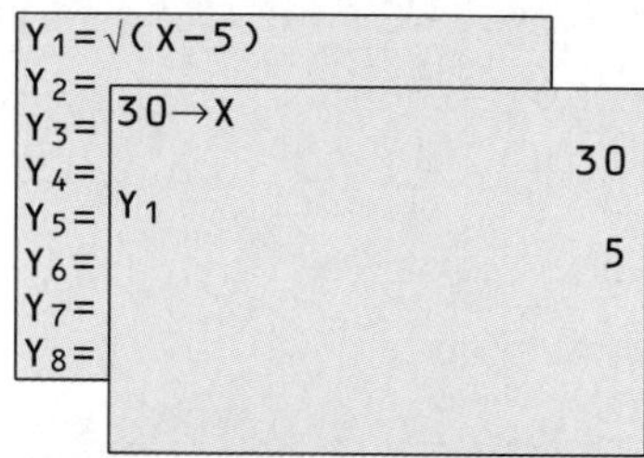

Then we store 30 in X and retrieve Y_1. Figure 2.35 shows that $f(30) = 5$.

(b) Enter $4x^2 - x$ as Y_1 on the Y screen. Then enter $Y_1(-1)$ and $Y_1(3)$ on the home screen. (See Fig. 2.36.)

Figure 2.36

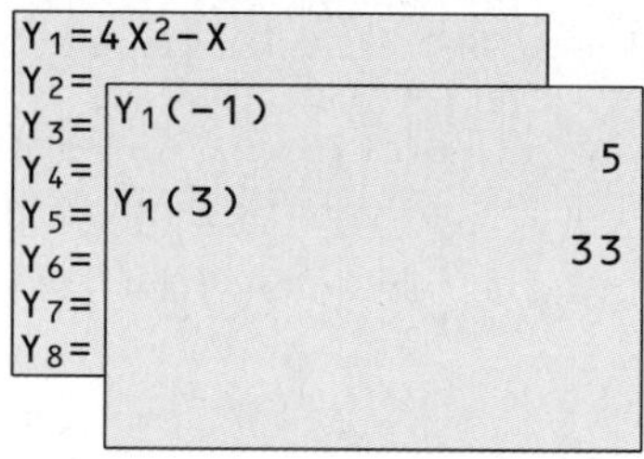

The values are $g(-1) = 5$ and $g(3) = 33$.

As the next two examples illustrate, a function can be evaluated for a variable expression instead of a specific value.

EXAMPLE 4

Evaluating a Function for a Variable Expression

Let $g(x) = \sqrt{x^2 - 4}$. Find $g(t)$ and $g(3t)$.

Think About It

If $f(x) = x^2 - 5$, you can use your calculator to evaluate $f(-5)$. Can you also use your calculator to evaluate $f(4t)$?

Solution

$g(t) = \sqrt{(t)^2 - 4} = \sqrt{t^2 - 4}$ Replace x with t.

$g(3t) = \sqrt{(3t)^2 - 4} = \sqrt{9t^2 - 4}$ Replace x with $3t$.

EXAMPLE 5

Operations with Functions

Let $f(x) = 3x - 4$. Determine each of the following.

(a) $f(5)$ (b) $f(5 + h)$ (c) $f(5 + h) - f(5)$ (d) $\dfrac{f(5 + h) - f(5)}{h}$

Solution

(a) $f(5) = 3(5) - 4 = 15 - 4 = 11$

(b) $f(5 + h) = 3(5 + h) - 4$ Substitute $5 + h$ for x.

$= 15 + 3h - 4$

$= 11 + 3h$

(c) $f(5 + h) - f(5) = (11 + 3h) - 11$ Use the results of parts (a) and (b).

$= 3h$

(d) $\dfrac{f(5 + h) - f(5)}{h} = \dfrac{3h}{h}$ Use the result of part (c).

$= 3$

2.4 QUICK REFERENCE

Function Notation

- A special notation called **function notation** can be used to represent a function and to represent its value for a given value of the variable.
- The notation $f(x)$ is read "f of x" and indicates that the name of the function is f and that the variable in the function is x.

Evaluating Functions

- Function notation is especially convenient when we want to indicate the value of a function for a particular value of the variable.
- If $f(x)$ represents a certain function and $f(3) = 5$, then (3, 5) is a point of the graph of the function.
- Several methods are available for using a calculator to evaluate functions.

2.4 EXERCISES

Concepts and Skills

1. If $f(x) = 2x + 1$, explain how to evaluate $f(3)$.

2. If $g(x) = 1 - 3x$, explain how to evaluate $g(t + 4)$.

In Exercises 3 and 4, fill in the blank and simplify.

3. $f(x) = -x^2 + 3x$
   (a) $f(-2) = -(\quad)^2 + 3(\quad)$
   (b) $f(2t) = -(\quad)^2 + 3(\quad)$
   (c) $f(a + 1) = -(\quad)^2 + 3(\quad)$

4. $g(x) = |2 - 5x|$
   (a) $g(3) = |2 - 5(\quad)|$
   (b) $g(-b) = |2 - 5(\quad)|$
   (c) $g(3 - y) = |2 - 5(\quad)|$

In Exercises 5–12, evaluate each function as indicated.

5. $f(x) = 4 - 9x$
   $f(0), f(2)$

6. $g(x) = 3x + 1$
   $g(-1), g(3)$

7. $h(t) = 3t - t^2$
   $h(-2), h(4)$

8. $f(y) = y^3 + 1$
   $f(-1), f(0)$

9. $g(t) = t - |2 - t|$
   $g(-3), g(3)$

10. $h(x) = -3\sqrt{x + 2}$
    $h(7), h(-1)$

11. $f(x) = \dfrac{x + 2}{x}$
    $f(-1), f(4)$

12. $g(x) = \dfrac{x^2}{4 - x}$
    $g(3), g(0)$

Piece-Wise

A **piece-wise function** is a function that is defined by two or more rules, each for a specified domain.

In Exercises 13–16, evaluate each piece-wise function as indicated.

13. $f(x) = \begin{cases} 2x + 1, & x < 0 \\ -2x + 1, & x \geq 0 \end{cases}$

(a) $f(-2)$ (b) $f(5)$

14. $g(t) = \begin{cases} 4t, & t > -3 \\ 2 - t, & t \leq -3 \end{cases}$

(a) $g(2)$ (b) $g(-3)$

15. $h(t) = \begin{cases} |t|, & t \geq 4 \\ -t, & t < 4 \end{cases}$

(a) $h(4)$ (b) $h(-3)$

16. $f(x) = \begin{cases} x^2, & x \leq 1 \\ x - x^2, & x > 1 \end{cases}$

(a) $f(0)$ (b) $f(2)$

17. Suppose that a certain expression is represented by $f(x)$ and that $f(4) = -7$. What is one point of the graph of f? Explain how you know.

18. Suppose that the graph of a function contains the point $(-3, 5)$. If the function is represented by $g(x)$, what is $g(-3)$? Explain how you know.

In Exercises 19–24, evaluate the given functions as indicated.

19. $f(x) = 7x - 2$ and $g(x) = 2x - 5$

(a) $g(-4)$ (b) $f(-3)$

(c) $g(3) - f(2)$ (d) $f(0) + g(1)$

20. $P(x) = x^2 - 3x + 5$ and $Q(x) = x^2 + 4x - 8$

(a) $P(-1)$ (b) $Q(1)$

(c) $Q(2) + P(0)$ (d) $P(2) - Q(0)$

21. $g(x) = -x^2 - 3x$ and $h(x) = x^4 - x$

(a) $g(-1)$ (b) $h(-1)$

(c) $h(-2)$ (d) $g(-3)$

22. $s(x) = -x^3 - 5x + 3$ and $T(x) = x^3 + 2x^2 - 5$

(a) $s(1)$ (b) $T(1)$

(c) $T(-2)$ (d) $s(-2)$

23. $h(x) = \sqrt{3x + 4}$ and $g(x) = |3 - 2x|$

(a) $h(-1)$ (b) $h(4)$

(c) $g(-2)$ (d) $g(2)$

24. $f(x) = |2x - 1|$ and $g(x) = \sqrt{5 - 4x}$

(a) $f(0)$ (b) $f(-1)$

(c) $g(1)$ (d) $g(-5)$

25. Explain how to use your calculator to evaluate a function with the Y screen.

26. Explain how to use your calculator to evaluate a function with only the home screen.

In Exercises 27–30, evaluate the given functions as indicated.

27. $g(x) = 3x - 5$ and $f(x) = 8x - 3$

(a) $g(-9.3)$ (b) $g\left(\frac{3}{4}\right)$

(c) $f\left(\frac{5}{8}\right)$ (d) $f(50)$

28. $f(x) = 7 - 4x$ and $g(x) = \frac{3}{4}x + 1$

(a) $f(-2.7)$ (b) $f\left(\frac{2}{3}\right)$

(c) $g(25)$ (d) $g(0.8)$

29. $p(x) = -6x^2 - 8x + 3$ and $h(x) = -4x^2 + 3x - 5$

(a) $h\left(\frac{2}{5}\right)$ (b) $h(-11.4)$

(c) $p(25.7)$ (d) $p\left(-\frac{1}{7}\right)$

30. $f(x) = 2x^3 - 4x - 15$ and $g(x) = 4x^3 + 3x - 1$

(a) $f(-2.3)$ (b) $f(-1.9)$

(c) $g\left(\frac{3}{7}\right)$ (d) $g\left(-\frac{1}{7}\right)$

31. Suppose that $h(x) = \sqrt{3x - 6} \div 1 - x$ and that $k(x) = \sqrt{3x - 6} \div (1 - x)$.

(a) Evaluate $h(5)$ and $k(5)$.

(b) Explain the role of the parentheses.

32. Suppose that $B(x) = 4x + 8 \div 2 - x$ and that $C(x) = (4x + 8) \div (2 - x)$.

(a) Evaluate $B(1)$ and $C(1)$.

(b) Explain the role of the parentheses.

In Exercises 33–40, evaluate the given function as indicated. Verify the result with your calculator.

33. $f(x) = x^2 - (4 - x);\quad f(-2)$

34. $g(z) = -z^2 - (3 - z);\quad g(-2)$

35. $h(c) = 5 - 5(5 - c);\quad h(5)$

36. $f(t) = 3 - 3(8 - t);\ \ f(1)$

37. $k(m) = 7(3m - 1) - (8 - m);\ \ k(-2)$

38. $f(p) = 6(2p - 3) - (1 - 3p);\ \ f(2)$

39. $m(s) = [4 - (s - 1)] + 8s;\ \ m(3)$

40. $n(t) = -[4 - (t - 2)] + 8t;\ \ n(3)$

41. Suppose $G(x) = 3\{x - [x - (x - 1)]\} - 3x$.

(a) Evaluate $G(1)$, $G(-4)$, and $G(2.3)$.

(b) What is your conjecture about $G(x)$ for any value of x?

(c) Show that your conjecture is correct by simplifying the expression.

42. Suppose $F(x) = 2[(x - 1) - 2] - 2(x - 2)$.

(a) Evaluate $F(3)$, $F(1)$, and $F(-3.7)$.

(b) What is your conjecture about $F(x)$ for any value of x?

(c) Show that your conjecture is correct by simplifying the expression.

43. Suppose $f(x) = 2x - 3(x - \sqrt{x - 1})$.

(a) Determine $f(1)$ without your calculator.

(b) Insert any parentheses necessary to evaluate f with your calculator.

2X − 3X − √X − 1

(c) Verify the result in part (a) by using your calculator.

44. Consider the following function.

$$g(x) = (5x + 2)^2 - \sqrt{12 - x} - (x - 1)$$

(a) Determine $g(3)$ without your calculator.

(b) Insert any parentheses necessary to evaluate g with your calculator.

5X + 2 ^ 2 − √ 12 − X − X − 1

(c) Verify the result in part (a) by using your calculator.

In Exercises 45 and 46, insert any parentheses necessary to evaluate the function with your calculator. Then use your calculator to evaluate the function as indicated.

45. $K(x) = 3 - 5 \cdot |x - 3|$

(a) 3 − 5 * ABS X − 3

(b) $K(9)$

46. $n(x) = |x + 2| - |x - 2|$

(a) ABS X + 2 − ABS X − 2

(b) $n(5)$

In Exercises 47–50, evaluate each function as indicated. Round results to the nearest hundredth.

47. $h(x) = \sqrt{3x + 4}$

(a) $h(20.8)$ (b) $h(0.9)$

48. $g(x) = \sqrt{5 - 4x}$

(a) $g(-2.4)$ (b) $g(-4.9)$

49. $k(x) = \sqrt{x^2 - 2x - 15}$

(a) $k(11.3)$ (b) $k(-4.5)$

50. $A(x) = \sqrt{2x^2 - x - 1}$

(a) $A(3.4)$ (b) $A(-5.7)$

In Exercises 51–54, evaluate each function as indicated.

51. $c(x) = |x^2 + 2x - 7|$

(a) $c(106)$ (b) $c(-89)$

52. $d(x) = |2 - x - x^2|$

(a) $d(57)$ (b) $d(-104)$

53. $f(x) = \dfrac{x + 1}{2x - x^2}$

(a) $f(-1)$ (b) $f(0.7)$

54. $g(x) = \dfrac{x^2 + x}{3x + 2}$

(a) $g(1)$ (b) $g(-0.3)$

55. Given $f(x) = -x^2$, $g(x) = x^2$, and $h(x) = (-x)^2$, determine the values in parts (a) and (b). Then answer the questions in parts (c)–(e).

(a) $f(7)$, $g(7)$, and $h(7)$

(b) $f(-7)$, $g(-7)$, and $h(-7)$

(c) Let A represent the step "multiply by -1"; let B represent the step "square." Using A and B, write the sequence to be followed in evaluating each of the three functions.

(d) Will any of the three functions have the same graph? Explain.

(e) Verify your prediction by graphing the functions.

56. Given $f(x) = -x^3$, $g(x) = x^3$, and $h(x) = (-x)^3$, determine the values in parts (a) and (b). Then answer the questions in parts (c) and (d).
(a) $f(2)$, $g(2)$, and $h(2)$
(b) $f(-2)$, $g(-2)$, and $h(-2)$
(c) Which of the following is/are suggested by the results? Explain your answer.
(i) For all x, $f(x) = g(x)$.
(ii) For all x, $g(x) = h(x)$.
(iii) For all x, $f(x) = h(x)$.
(d) Verify your prediction by graphing the functions.

57. Suppose $B(x) = \sqrt{x^2 + 25}$ and $C(x) = x + 5$.
(a) Evaluate $B(0)$ and $C(0)$.
(b) Evaluate $B(3)$ and $C(3)$.
(c) Evaluate $B(-6)$ and $C(-6)$.
(d) Would you expect the two functions to have the same graph? Explain.

58. Suppose $f(x) = \sqrt{x^2}$, $g(x) = x$, and $h(x) = |x|$.
(a) Evaluate $f(3)$, $g(3)$, and $h(3)$.
(b) Evaluate $f(-5)$, $g(-5)$, and $h(-5)$.
(c) Which two of these three functions do you think have the same graph? Why?

59. Suppose $f(x) = |x + 7|$ and $g(x) = |x| + 7$.
(a) If we replace x with only positive values, would the graphs of the two functions be the same?
(b) If we replace x with any real number, would the graphs of the two functions be the same?
(c) Test your conclusions in parts (a) and (b) by producing the graphs of f and g on your calculator.

60. Suppose $A(x) = (\sqrt{x - 5})^2$ and $B(x) = |x - 5|$.
(a) Evaluate $A(14)$ and $B(14)$.
(b) Evaluate $A(1)$ and $B(1)$.
(c) Would you expect the two functions to have the same graph?
(d) Test your conclusion in part (c) by producing the graphs of A and B.

In Exercises 61–68, evaluate the given function as indicated.

61. $f(x) = 5x - 7$
(a) $f(2a)$ (b) $f(a + 2)$

62. $g(x) = 6x + 2$
(a) $g(3 + z)$ (b) $g(3z)$

63. $h(x) = x^2 - 5x + 6$
(a) $h(t)$ (b) $h(2t)$

64. $g(x) = x^2 + 4x - 8$
(a) $g(-3t)$ (b) $g(t)$

65. $f(x) = \sqrt{x^2 + 5}$
(a) $f(a)$ (b) $f(-2a)$

66. $g(x) = \sqrt{6 - x^2}$
(a) $g(-3b)$ (b) $g(m)$

67. $f(x) = |x^3 - 2|$
(a) $f(-s)$ (b) $f(3s)$

68. $h(x) = |x^3 + 4|$
(a) $h(2r)$ (b) $h(-r)$

69. Let $f(x) = 2x - 5$. Evaluate f as indicated.
(a) $f(3)$ (b) $f(3 + h)$
(c) $f(3 + h) - f(3)$
(d) $\dfrac{f(3 + h) - f(3)}{h}$

70. Let $f(x) = 5 - 3x$. Evaluate f as indicated.
(a) $f(-2)$ (b) $f(-2 + h)$
(c) $f(-2 + h) - f(-2)$
(d) $\dfrac{f(-2 + h) - f(-2)}{h}$

71. Let $g(x) = 7 + 4x$. Evaluate g as indicated.
(a) $g(-8)$ (b) $g(-8 + h)$
(c) $g(-8 + h) - g(-8)$
(d) $\dfrac{g(-8 + h) - g(-8)}{h}$

72. Let $g(x) = 5x + 7$. Evaluate g as indicated.
(a) $g(4)$ (b) $g(4 + h)$
(c) $g(4 + h) - g(4)$
(d) $\dfrac{g(4 + h) - g(4)}{h}$

Real-Life Applications

73. A person walking a treadmill finds that her number of paces p per minute can be modeled by the function $p(m) = 60 - \sqrt{2m}$, where m is the number of minutes. To the nearest whole number, what is her pace rate halfway through a 30-minute workout?

74. A "square" of shingles covers 100 square feet. For a roof of x square feet, the function $N(x) = \frac{x}{100}$ models the number N of squares that will be needed. Evaluate N to determine the number of squares required for a roof of 5100 square feet.

75. The mosquito population M (in thousands) in a certain area d days after spraying the area with insecticide is given by $M(d) = 2d^2 - 20d + 125$. Use a graph of M to estimate the answers to the following questions.

(a) In how many days after spraying will the mosquito population be at its lowest? What is the population of mosquitos at that time?

(b) In how many days after spraying will the mosquito population be the same as it was when spraying began?

76. A certain insurance company found that the probability that a driver would be involved in an automobile accident could be predicted by the function $P(x) = \frac{2}{x + 6}$, where x is the difference between the driver's age and 16. What is the probability that a driver will be involved in an accident at

(a) age 20? (b) age 34?

Modeling with Real Data

77. From 1990 to 1995, the number of semiprivate golf courses (clubs that also open to the public for a daily fee) rose, while the number of private clubs declined. (Source: National Golf Foundation.)

Year	Semiprivate	Private
1990	6500	5500
1995	8142	4759

With x representing the number of years since 1990, the following functions model the data, where S is the number of semiprivate clubs and P is the number of private clubs.

$$S(x) = 328.4x + 6500 \quad P(x) = -148.2x + 5500$$

Produce graphs of S and P and trace them to complete the following table.

Year		1995		2001
x				
Semiprivate			9127	
Private	5204			

78. Veterinarians use the drug lufenuron for control of fleas in cats. The dosage depends on the weight of the cat. For example, a 15-pound cat receives 270 milligrams of lufenuron. (Source: Ciba Animal Health.) You might use $D(w) = 13.5w + 67.5$ to estimate the dosage D (in milligrams) for a cat that weighs w pounds.

(a) If you owned a 12-pound cat, what dosage of lufenuron would you administer?

(b) Produce and trace the graph of function D to estimate the weight (to the nearest pound) of a cat that receives 202 milligrams of lufenuron.

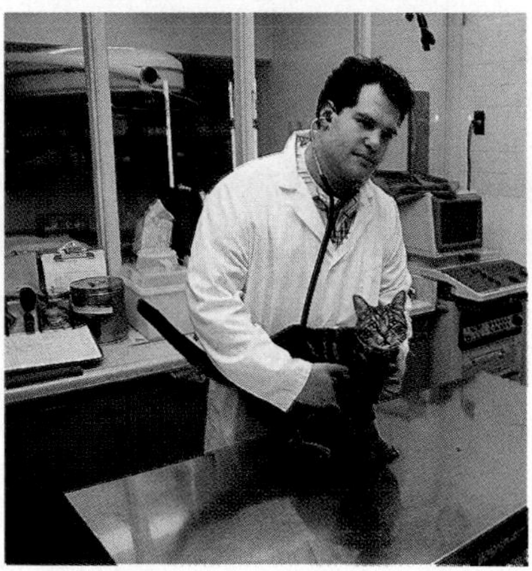

Group Project: Bicycle Helmet Sales

As more and more states enacted helmet laws for bicycle riders, the sales of helmets increased dramatically. The bar graph on the next page shows sales (in millions of dollars) of bicycle helmets for selected years in the period 1990–1995. (Source: U.S. Bicycle Federation.)

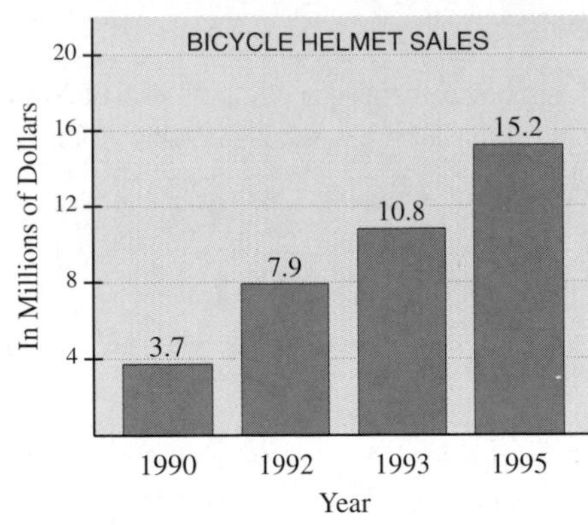

If we let t represent the number of years since 1990, then the sales S (in millions of dollars) can be modeled by the following function.

$$S(t) = 2.33t + 3.59$$

79. What values should be substituted for t in order to obtain approximations of the sales in the bar graph?

80. If you evaluate the model expression for $t = 10$, what is your interpretation of the result?

81. What value of t refers to 1988? Use this value of t to estimate sales in 1988. What does the result suggest about the validity of the model prior to 1990?

82. How might legislators have profited from enacting helmet laws if hearings on the matter had been closed to the public?

Challenge

83. Suppose $f(x) = \frac{3}{x}$.

(a) Evaluate each of the following.

(i) $f(1)$

(ii) $f(0.1)$

(iii) $f(0.01)$

(iv) $f(0.001)$

(v) $f(0.0001)$

(b) As the x-values become closer to zero, what is happening to the values of $f(x)$?

(c) Use a calculator to evaluate $f(0)$. Explain the result.

84. Suppose $g(x) = x^2$.

(a) Evaluate each of the following.

(i) $g(11)$

(ii) $g(111)$

(iii) $g(1111)$

(iv) $g(11{,}111)$

(b) Based on the pattern in part (a), predict the value of $g(11{,}111{,}111)$.

(c) Can you confirm your prediction by evaluating $g(11{,}111{,}111)$ with your calculator? Why?

2.5 ANALYSIS OF FUNCTIONS

Goals of Analysis • Maximums, Minimums, and Intercepts • Estimating Domain and Range • Points of Intersection

Goals of Analysis

Roughly speaking, **analyzing a function** refers to learning how a function behaves. To analyze a function, we ask questions such as the following.

1. What are the highest and lowest points of the graph?
2. What are the domain and range of the function?
3. Where does the graph intersect the x- and y-axes, if at all?

Analyzing a function often involves methods that are beyond the scope of this book. However, we can estimate the answers to such questions by exploring graphs and visually observing the behavior of the function.

Maximums, Minimums, and Intercepts

In the following discussion we consider the flight of a thrown ball. It would be very difficult to make physical measurements of times and distances while the ball is in the air. Instead, we use a function to model the flight of the ball.

EXPLORING THE CONCEPT

Maximums, Minimums, and Intercepts

Suppose that a person standing at the top of a 20-story building throws a ball upward and outward. (See Fig. 2.37.)

Figure 2.37

The flight of the ball is described by the function $h(x) = -16x^2 + 100x + 200$, where h is the height of the ball (in feet) from the ground after x seconds. Because both h and x are nonnegative numbers, we restrict the graph of h to the first quadrant and to the axes. (See Fig. 2.38.)

Figure 2.38

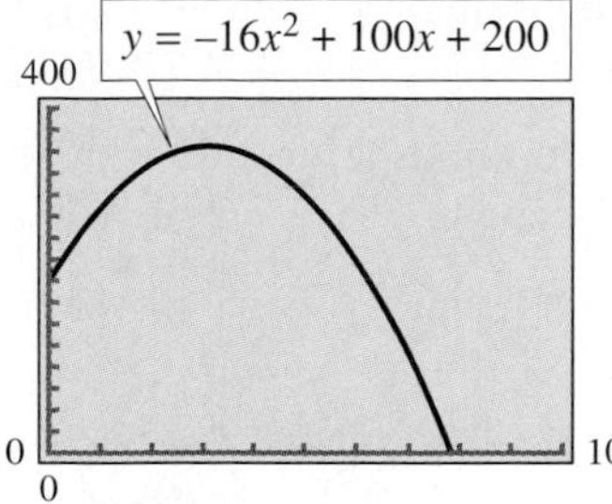

LEARNING TIP

As you produce graphs of various functions, you may need to adjust the window. Be willing to experiment. There is usually no one correct setting.

When the person releases the ball, the time x is 0. By tracing the graph to the point whose x-coordinate is 0, we see that $h(0) = 200$. (See Fig. 2.39.) Thus the ball was released from a height of 200 feet. A point whose x-coordinate is 0 is called a **y-intercept.**

Figure 2.39

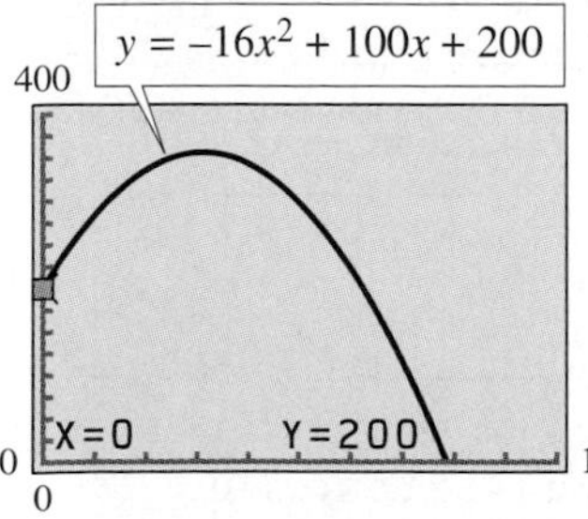

Figure 2.40

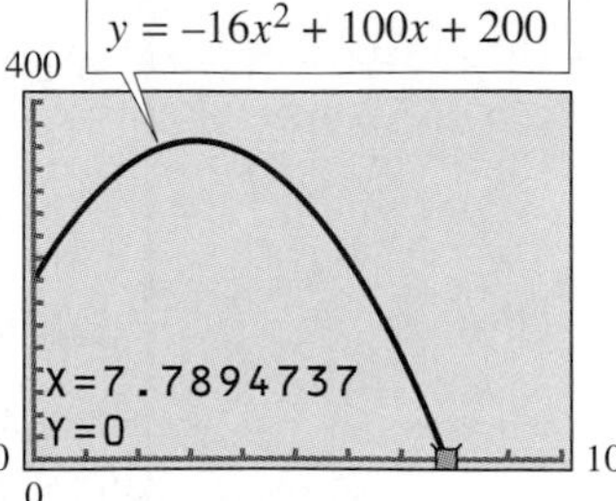

When the ball reaches the ground, the height y is 0. Figure 2.40 shows that the y-coordinate is 0 when $x \approx 7.8$. Thus the ball reaches the ground in about 7.8 seconds. A point whose y-coordinate is 0 is called an **x-intercept.**

The maximum height of the ball corresponds to the y-coordinate of the highest point of the graph. By tracing to the top of the curve, we estimate that the ball's maximum height is approximately 356 feet. (See Fig. 2.41.)

Figure 2.41

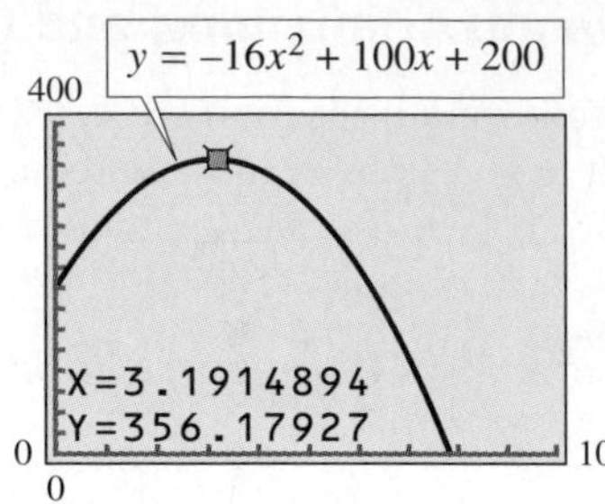

The y-coordinate of the highest point of a graph is called the **absolute maximum** of the function. Similarly, the y-coordinate of the lowest point of a graph is called the **absolute minimum** of the function.

Summary of Definitions

The **absolute maximum** of a function is the y-coordinate of the highest point of the graph of the function.

The **absolute minimum** of a function is the y-coordinate of the lowest point of the graph of the function.

An ***x*-intercept** of a graph is a point at which the graph intersects the x-axis. At this point, $y = 0$.

A ***y*-intercept** of a graph is a point at which the graph intersects the y-axis. At this point, $x = 0$.

Sometimes the graph of a function has peaks and dips that are of interest as we analyze the function.

EXPLORING THE CONCEPT

Local Maximums and Minimums

Figure 2.42 shows a graph with two peaks and one dip.

Figure 2.42

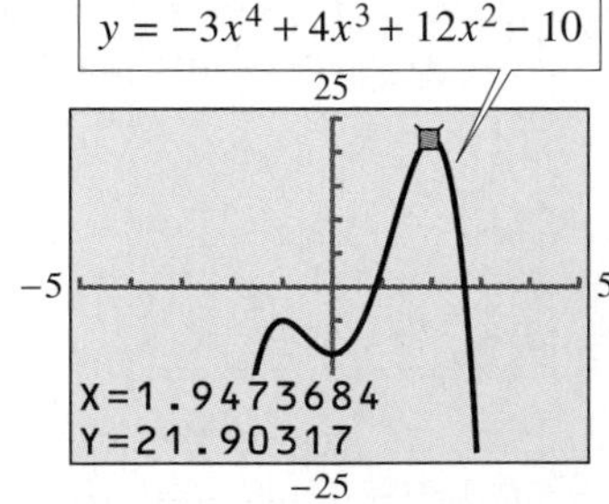

Note that the tracing cursor is on the highest point of the graph. The point is approximately (2, 22), so we estimate that the absolute maximum of the function is about 22. The function has no absolute minimum.

In Fig. 2.43, observe that the tracing cursor is on the relatively high point $(-1, -5)$. Informally, we say that this point is the highest point in its *neighborhood*, and we call the y-coordinate of such a point a **local maximum.**

Figure 2.43

Figure 2.44

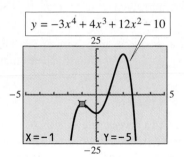

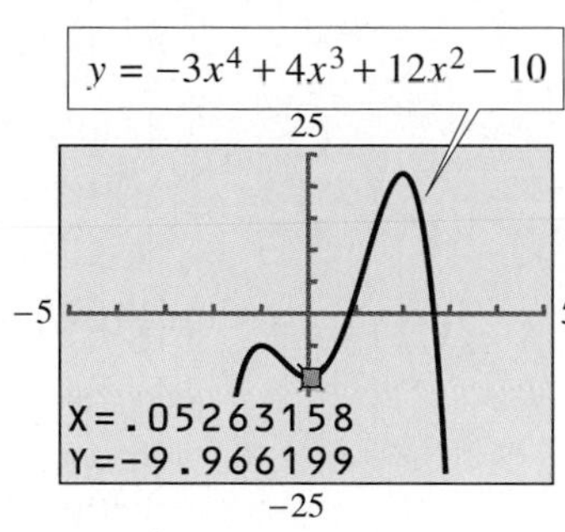

Similarly, the relatively low point in Fig. 2.44 is approximately (0, −10). The y-coordinate of such a point is called a **local minimum.**

Think About It
Using a window with a large maximum x-value, produce the graph of $f(x) = \frac{x}{x+1}$. Might you conclude from the graph that the absolute maximum of the function is 1? How can you show that this conclusion is false?

Note: When speaking of maximums and minimums, we often refer to the point itself as a maximum or minimum rather than the y-coordinate.

An absolute maximum (minimum) is also a local maximum (minimum) because it is the highest (lowest) point in its neighborhood as well as the highest (lowest) point of the graph.

Estimating Domain and Range

We can use estimates of absolute maximums and minimums to approximate the range of a function. As we saw in Section 2.3, we also can use a graph to approximate the domain of a function.

EXAMPLE 1

Estimating the Domain and Range

(a) Use a suitable window setting to display the graph of $f(x) = \sqrt{x+3}$.

(b) Estimate the absolute minimum and the absolute maximum. Use these estimates to approximate the range of the function.

(c) Estimate the least and greatest x-coordinates. Use these estimates to approximate the domain of the function.

Solution

LEARNING TIP
Remember that x-values correspond to the left and right, so domains involve x-values. Also, y-values correspond to up and down, so ranges, maximums, and minimums involve y-values.

(a) **Figure 2.45**

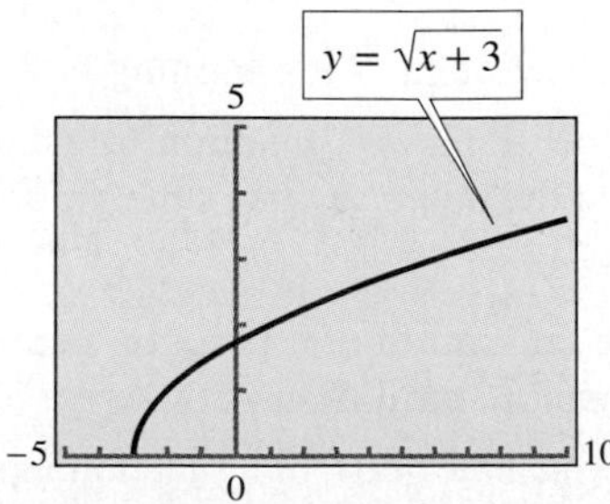

(b) The lowest point of the graph shown in Fig. 2.45 is the x-intercept, where the y-coordinate (absolute minimum) is 0. There is no absolute maximum. The range appears to be $\{y \mid y \geq 0\}$.

(c) The smallest x-coordinate occurs at the x-intercept and is approximately -3. There is no greatest x-coordinate. The domain appears to be $\{x \mid x \geq -3\}$.

Points of Intersection

It is frequently useful to display the graphs of two functions on the same set of axes. In most calculators we can store two functions and produce the graph of either one or both of them.

Activate

Deactivate

A function whose graph is displayed is called an *active function*. A function that has been entered but whose graph is not displayed is called an *inactive function*.

EXAMPLE 2

Two Graphs on One Set of Axes

Let $f(x) = x^3$ and $g(x) = -x + 10$. Produce the graphs of both functions on the same coordinate system and determine the x-coordinate of the point at which $f(x) = g(x)$.

Solution

Figure 2.46

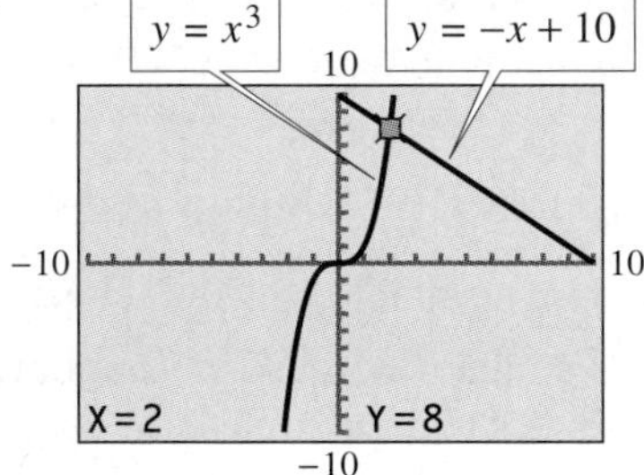

The **point of intersection** of the two graphs is the point at which $f(x) = g(x)$. The estimated coordinates are (2, 8). (See Fig. 2.46.)

2.5 QUICK REFERENCE

Goals of Analysis

- Analyzing a function involves learning how a function behaves.
- We can use the graph of a function to estimate its high and low points, its domain and range, and the points at which the graph intersects the axes.

Maximums, Minimums, and Intercepts

- The **absolute maximum** of a function is the y-coordinate of the highest point of the graph; the **absolute minimum** is the y-coordinate of the lowest point of the graph.
- The point at which the graph of a function intersects the x-axis is called the ***x*-intercept;** the point at which the graph intersects the y-axis is called the ***y*-intercept.**

- Informally, we call the y-coordinate of the point that is highest in the neighborhood of a peak of a graph a **local maximum.** The y-coordinate of the point that is lowest in the neighborhood of a dip is called a **local minimum.**

Estimating Domain and Range

- We can use estimates of absolute maximums and minimums to approximate the range of a function.
- We can use the graph of a function to estimate the domain by determining the smallest and greatest x-coordinates.

Points of Intersection

- We can store two or more functions in a calculator and display their graphs simultaneously.
- Having stored two functions, we can *activate* one so that its graph will be displayed, and we can *deactivate* the other so that its graph will not be displayed.
- By displaying the graphs of two different functions, we can estimate the coordinates of their **point(s) of intersection.**

2.5 EXERCISES

Concepts and Skills

Many of the exercises in this section require estimates. Round all decimal answers to the nearest hundredth.

1. Explain how you can estimate or determine the y-intercept of a graph (a) with your calculator and (b) algebraically.

2. Explain how to estimate the x-intercept of a graph with your calculator.

3. Explain why a function has at most one y-intercept.

4. In parts (a)–(d), sketch an example of a graph having the given number of x-intercepts. Then answer the question in part (e).

(a) 0 (b) 1

(c) 2 (d) 3

(e) Use the definition of function to explain why the graph of a function can have more than one x-intercept.

In Exercises 5–12, use the integer setting to graph the given function. Then use the graph to estimate (a) the x-value for the given value of $f(x)$ and (b) the x- and y-intercepts.

5. $f(x) = x - 4;\ \ f(x) = 6$

6. $f(x) = 4 - x;\ \ f(x) = -3$

7. $f(x) = \frac{1}{3}x - 8;\ \ f(x) = -5$

8. $f(x) = -\frac{1}{3}x - 7;\ \ f(x) = -8$

9. $f(x) = x^2 + 2x - 3;\ \ f(x) = 2$ and $x < 0$

10. $f(x) = x^2 - 2x - 8;\ \ f(x) = -3$ and $x > 0$

11. $f(x) = 8 - 2x - x^2;\ \ f(x) = 3$ and $x > 0$

12. $f(x) = 3 + 2x - x^2;\ \ f(x) = -4$ and $x < 0$

In Exercises 13–20, use the integer setting to graph the given function. Then use your graph to estimate the values in parts (a)–(c).

13. $h(x) = \frac{2}{5}x - 9$

(a) $h(-15)$ (b) x, if $h(x) = -5$

(c) x, if $h(x) = 11$

14. $c(x) = -\frac{1}{3}x + 5$

(a) $c(-9)$ (b) x, if $c(x) = -5$

(c) x, if $c(x) = 16$

15. $g(x) = x^2 - 2x - 8$

(a) $g(7)$ (b) x, if $g(x) = 7$

(c) x, if $g(x) = -10$

16. $k(x) = -x^2 - x + 56$

(a) $k(-2)$ (b) x, if $k(x) = 14$

(c) x, if $k(x) = 50$

17. $f(x) = 10 - \sqrt{8x}$

(a) $f(8)$ (b) x, if $f(x) = -2$

(c) x, if $f(x) = 10$

18. $h(x) = \sqrt{12x + 3}$

(a) $h(12)$ (b) x, if $h(x) = 0$

(c) x, if $h(x) = 3$

19. $f(x) = |x| + 5$

(a) $f(6)$ (b) x, if $f(x) = 19$

(c) x, if $f(x) = 4$

20. $p(x) = 15 - |2x|$

(a) $p(17)$ (b) x, if $p(x) = 7$

(c) x, if $p(x) = 16$

In Exercises 21–26, use the integer setting to graph the given function on your calculator. Trace the graph to estimate the function's absolute maximum, if any, and the function's absolute minimum, if any.

21. $f(x) = x^2 + 10x + 25$

22. $g(x) = -x^2 + 8x - 20$

23. $h(x) = 9 - |x|$ **24.** $m(x) = |x| - 9$

25. $p(x) = 6 + \sqrt{10x}$ **26.** $r(x) = 6 - \sqrt{10x}$

In Exercises 27–30, use a graph to estimate (a) the coordinates of the local minimum in Quadrant IV and (b) the coordinates of the local maximum in Quadrant II.

27. $f(x) = x^3 - 2x$

28. $h(x) = x^3 - 7x$

29. $g(x) = x^3 - 3x^2 - 5x + 6$

30. $h(x) = x^3 + 2x^2 - 5x - 6$

In Exercises 31 and 32, estimate the coordinates of the point representing the local minimum or local maximum.

31. $f(x) = x^3 - x^2 - 9x + 9$

32. $f(x) = x^3 - x^2 - 10x - 8$

33. Explain how you can determine the number of x-intercepts of $f(x) = -(2 + |x|)$. How many are there?

34. Suppose the range of function g is $\{y \mid y \geq 3\}$. What can you conclude about the absolute minimum and absolute maximum, if any, of function g?

In Exercises 35–38, sketch an example of a graph that has the given characteristics.

35. The graph has an absolute maximum and an absolute minimum.

36. The graph has an absolute maximum but no absolute minimum.

37. The graph has an absolute minimum but no absolute maximum.

38. The graph has no absolute maximum and no absolute minimum.

In Exercises 39 and 40, sketch an example of a graph that meets the given conditions.

39. The graph has three x-intercepts: $(-3, 0)$, $(1, 0)$, and $(5, 0)$.

The graph has local maximums at $(-1, 4)$ and at $(3, 7)$.

The graph has a local minimum at $(1, 0)$.

40. The graph has an absolute minimum at $(-3, -5)$.

The graph has an x-intercept at the origin.

The graph has a local maximum at $(2, 4)$.

The graph has a local minimum at $(5, 3)$.

In Exercises 41 and 42, use a graph to determine (a) the number of x- and y-intercepts, (b) the number of local minimum and local maximum points, (c) the approximate domain and range, and (d) the approximate coordinates of the x-intercept that is the farthest to the right.

41. $f(x) = x^4 + 3x^3 - 2x^2 - 5x + 3$

42. $g(x) = 4x^2 - 2x^3 - x^4 + 3x - 5$

43. Explain why -5 is not in the domain of the function $g(x) = \sqrt{2x + 3}$.

44. Explain why -2 is not in the range of the function $h(x) = |x + 2|$.

In Exercises 45–48, use the integer setting to graph the given function on your calculator. Then answer the following questions.

(a) What are the x- and y-intercepts?

(b) What is the absolute minimum and the absolute maximum?

(c) What are your estimates of the domain and range?

45. $f(x) = |x + 2| + 3$ **46.** $g(x) = 5 - |x - 2|$

47. $h(x) = \sqrt{x - 1}$ **48.** $h(x) = 5 - \sqrt{x + 4}$

In Exercises 49–52, use the integer setting to graph the two given functions on the same coordinate axes. Then name the graphs to which the given points belong.

49. $f(x) = x + 1$; $g(x) = 2x - 3$

(a) $(3, 4)$ (b) $(0, -3)$ (c) $(4, 5)$

50. $f(x) = \frac{1}{2}x - 2$; $g(x) = -x + 1$

(a) $(-5, 6)$ (b) $(6, 1)$ (c) $(2, -1)$

51. $f(x) = \frac{1}{2}x$; $g(x) = |x|$

(a) $(10, 5)$ (b) $(10, 10)$ (c) $(0, 0)$

52. $f(x) = \frac{1}{10}x^2$; $g(x) = 2x - 10$

(a) $(8, 6)$ (b) $(8, 6.4)$ (c) $(10, 10)$

Dot

A function can be entered in the calculator and graphed for a specified domain.

In Exercises 53–56, use the integer setting and dot mode to graph the function $y = x + 17$ for the specified domain.

53. $x \geq -25$ **54.** $x \leq 6$

55. $-23 \leq x \leq -8$ **56.** $x \leq -15$ or $x \geq -5$

Piece-Wise

A **piece-wise function** is described by two or more rules, each for a specified domain.

In Exercises 57 and 58, graph the given piece-wise function.

57. $y = \begin{cases} x + 17, & x \leq -15 \\ -x + 8, & x > -15 \end{cases}$

58. $y = \begin{cases} x + 17, & x \leq -15 \\ -x + 8, & -15 < x \leq 20 \end{cases}$

Real-Life Applications

59. Visual Graphics, Inc., is designing an advertisement that will be placed on a rectangular board. The customer has specified that the length L (in feet) must be twice the width W, and the area and the perimeter of the board must be the same. The designer makes the following calculations.

$$\text{Area} = LW = (2W) \cdot W = 2W^2$$

$$\text{Perimeter} = 2L + 2W = 2(2W) + 2W = 4W + 2W = 6W$$

Enter these two functions in your calculator and use the default setting to graph the functions on the same axes.

(a) To which graph does (2, 8) belong? Interpret the meaning of that ordered pair.

(b) To which graph does (2, 12) belong? Interpret the meaning of that ordered pair.

(c) What is the point of intersection?

(d) How wide and how long should the board be?

(e) What is the area of the board and what is the perimeter?

60. At the local hardware store, the cash register is programmed to add a 5% sales tax to the price x of all goods purchased. The total cost C is then computed as follows.

$$C(x) = x + 0.05x = 1.05x$$

Use the integer setting to graph function C on your calculator. Then answer the following questions.

(a) Describe the apparent shape of the graph.

(b) What is $C(26)$?

(c) What is $C(11)$?

(d) What do your results in parts (b) and (c) represent?

(e) Calculate $\dfrac{C(26) - C(11)}{26 - 11}$.

(f) Compare your result in part (e) to $C(x) = 1.05x$. What do you observe?

61. Use the window settings indicated in Fig. 2.38 to produce the graph of the function that models the flight path of the thrown ball. Then trace the graph to estimate the time(s) when the ball is 300 feet from the ground.

62. Referring to Figs. 2.39, 2.40, and 2.41, estimate the domain and range of the function that models the flight path of the thrown ball.

Modeling with Real Data

63. The accompanying bar graph shows the annual funding for the Department of Housing and Urban Development (HUD) in recent years. (Source: U.S. Department of Housing and Urban Development.)

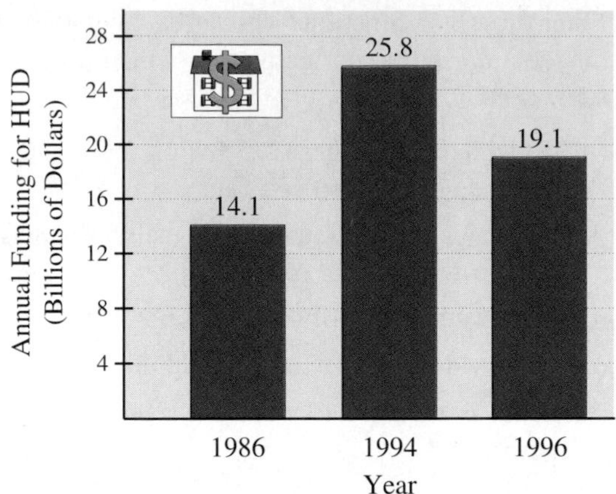

The function $F(x) = -0.48x^2 + 11x - 35$ can be used to model HUD funding F (in billions of dollars), where x is the number of years since 1980.

(a) Use a graph of F to estimate the year during this period when HUD received its largest funding.

(b) Estimate the funding for that year.

64. Observe that the graph of function F in Exercise 63 has two x-intercepts.

(a) Estimate and interpret these intercepts.

(b) What do your interpretations suggest about the validity of the model function before 1986 and after 1996?

65. The category C of a hurricane can be modeled with a piece-wise function $C(w)$, where w is the maximum wind speed in miles per hour. (Source: National Hurricane Center.)

$$C(w) = \begin{cases} 1, & 74 \leq W \leq 95 \\ 2, & 95 < W \leq 110 \\ 3, & 110 < W \leq 130 \\ 4, & 130 < W \leq 155 \\ 5, & 155 < W \end{cases}$$

Evaluate $C(w)$ to complete the "Category" column in the following table.

Hurricane	Maximum Wind Speed (mph)	Category
Opal	150	
Erin	90	
Roxanne	115	
Camille	172	
David	95	
Humberto	104	
Carla	145	

66. For the function $C(w)$ in Exercise 65, determine (a) the domain and (b) the range.

Group Project: Gray Wolf Population

Since the Endangered Species Act of 1973, the gray wolf population in the continental United States has increased. The figure shows the estimated populations of gray wolves between 1960 and 1990. (Source: U.S. Fish and Wildlife Service.) If we let x represent the number of years since 1960, a function that approximates the population P of gray wolves during this 30-year period is $P(x) = 0.74x^2 + 7.75x + 800$.

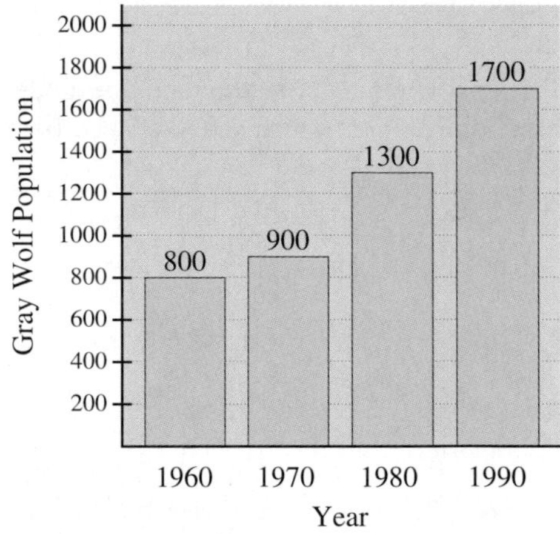

67. Select a suitable window and graph function P on your calculator. During the period 1960–1990, when was the gray wolf population at its lowest (the absolute minimum of the function)? Estimate the population that year.

68. During this period, when was the gray wolf population at its highest (the absolute maximum of the function)? Estimate the population that year.

69. If the population continues to grow according to the function P, what is the projected population in the year 2000?

70. Does the shape of the graph suggest that the gray wolf population will increase indefinitely? What natural factors might change the shape of the graph in the future?

Challenge

In Exercises 71–76, use the integer setting to graph each function. Then write the set of x-values for which

(a) $y \leq 0$. (b) $y \geq 0$.

71. $y = x - 3$ **72.** $y = x + 4$

73. $y = x^2 - 9$ **74.** $y = 9 - x^2$

75. $y = \sqrt{x}$ **76.** $y = |x| + 1$

77. (a) Draw a graph of the following piece-wise function.

$$f(x) = \begin{cases} x + 3, & x < 0 \\ 5, & x = 0 \\ -x, & x > 0 \end{cases}$$

(b) What is the absolute maximum of the function?

(c) In words, how would you describe your estimate of the range of the function?

78. Consider the **constant function** $g(x) = 5$.

(a) What are the domain and range of g?

(b) We say that a function f is an **increasing function** if $f(x_2) \geq f(x_1)$ when $x_2 > x_1$ for all x in the domain of f. Explain why function g is an increasing function.

2. CHAPTER PROJECT Bicycle Helmet Sales

In Section 2.4, Exercises 79–82, we modeled the sales of bicycle helmets with the function $S = 2.33t + 3.59$, where S is the sales (in millions of dollars) and t is the number of years since 1990.

1. Complete the following table.

t	Actual Sales A	Modeled Sales S	$S - A$	$(S - A)^2$
0	3.7	3.59	−0.11	0.0121
2	7.9	8.25	+0.35	0.1225
3	10.8			
5	15.2			

2. The columns headed "$S - A$" and "$(S - A)^2$" give some indication of how accurate the model is for a given value of t. Add the entries in the "$(S - A)^2$" column and take the square root of the result. Call the resulting number D.
3. The following are four other models that could be used.

   (a) $S = 2.30t + 3.70$ (b) $S = 2.90t + 2.10$

   (c) $S = 2.37t + 3.70$ (d) $S = 2.43t + 3.04$

   Form four groups, with each group selecting one of these alternative models.
4. For each model, create a table similar to that shown in part (1). Then calculate the value of D as described in part (2).
5. Along with the other groups, report your values of D. Assess whether you think (a) your model is better or worse than the original model and (b) your model is better or worse than the alternative models. Explain your reasoning.

2. CHAPTER REVIEW EXERCISES

Section 2.1

1. Determine the coordinates of the points in the accompanying figure.

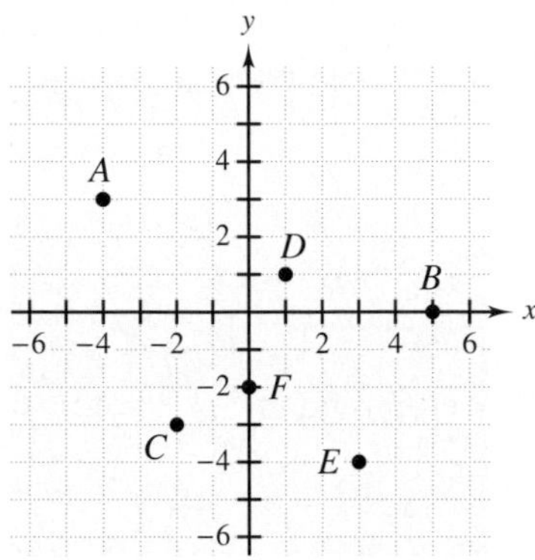

(a) A (b) B (c) C
(d) D (e) E (f) F

In Exercises 2–4, state the quadrant in which the given point lies.

2. $P(-\pi, \pi)$
3. $Q(5, -6)$
4. $R(a, b)$ if $a < 0$ and $b > 0$
5. On which axis is $P(c, d)$ if $c = 0$?
6. In which quadrant is $P(a, b)$ if $Q(b, a)$ is in Quadrant IV? Why?
7. Graph the following set.

$$\{(-3, 5), (0, 6), (2, -4), (-5, -3), (7, 0)\}$$

8. If point P is directly to the right of $(-3, 4)$ and directly above $(2, -7)$, what are the coordinates of P?

In Exercises 9 and 10, determine each of the following.

(a) The distance (to the nearest hundredth) between the points A and B

(b) The midpoint of the line segment whose endpoints are A and B

9. $A(-5, 4)$ and $B(-8, -5)$
10. $A(7, -1)$ and $B(13, -9)$
11. Find the perimeter of the parallelogram $ABCD$ if the vertices are $A(-6, 1)$, $B(-3, 5)$, $C(9, 0)$, and $D(6, -4)$.
12. Determine whether points $A(-3, -2)$, $B(-1, 8)$, and $C(3, 4)$ are vertices of a right triangle.

Section 2.2

13. Evaluate $3 + 2x - x^2$ for $x = -5, -1, 0$, and 6.
14. Use a graph to estimate the value of $5\sqrt{8 - x}$ for $x = -1, -8, 8$, and 10.
15. Use a graph to estimate the value of x for which the expression $12 - \dfrac{x}{2}$ has a value of 14.
16. Use a graph to estimate the values of x for which $8 + 2x - x^2$ has each of the following values.

(a) -7 (b) 9 (c) 12

In Exercises 17 and 18, use a table or graph to estimate the value(s) of x for which the expressions have the same value.

17. $x - 5, \quad 3x + 8$
18. $|x|, \quad \dfrac{1}{2}x + 9$
19. Suppose that you know that the value of an expression is 0 and you trace to a point of the graph of the expression to estimate the value of the variable. Explain how you know which axis contains the point.
20. For the expression $|2x - 9| - 15$, estimate (a) the value of the expression when x is 0 and (b) the value of the variable if the expression has a value of 0.
21. Which of the given ordered pairs belong to the graph of the expression $\dfrac{8x + 3}{x - 5}$?

(i) $(-5, 3.7)$ (ii) $(0, -0.6)$
(iii) $(6, 51)$ (iv) $(5, 0)$

22. A person pays an annual distributor fee of \$50 for the right to sell plastic kitchenware and receives \$1.75 commission for each item sold. Write an expression to model the possible annual profit, and use a graph of the expression to estimate each of the following.

(a) The profit if 20 items are sold
(b) The profit if 56 items are sold
(c) The number of items that must be sold to earn a profit of \$251

Section 2.3

In Exercises 23 and 24, state the domain and range of the given relation. Identify any relation that is a function.

23. $\{(0, 1), (3, 5), (-4, 7), (5, -6)\}$

24.

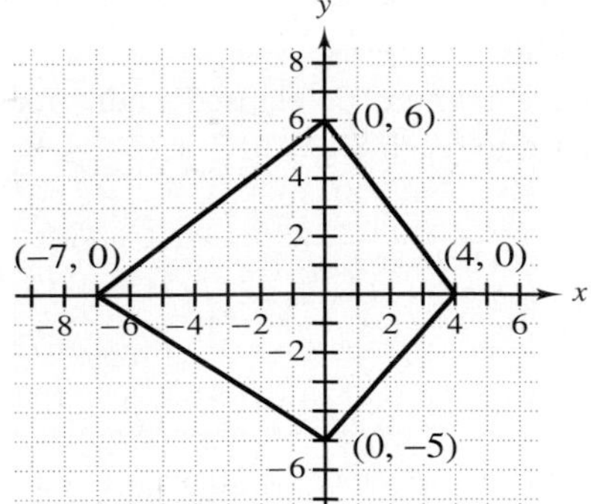

In Exercises 25–27, determine whether the given relation defines a function.

25. $y = |x + 2|$ **26.** $x = \sqrt{y + 1}$ **27.** $x = y^2 - 4$

28. Which of the following statements is true? Why?

(a) Every relation is a function.

(b) Every function is a relation.

In Exercises 29 and 30, use a graph to estimate the domain and range of the given function.

29. $y = \sqrt{9 - x}$ **30.** $y = x^2 + 4x - 5$

31. Which of the following graphs, if either, represents a function?

(i) A half-circle centered at the origin with endpoints on the y-axis

(ii) A half-circle centered at the origin with endpoints on the x-axis

32. The sum of the lengths of the base and height of a triangular poster is 14 inches.

(a) Write a function for the area of the poster.

(b) What is the greatest possible area of the poster?

(c) What is the domain of the function?

(d) What is the range of the function?

Section 2.4

In Exercises 33–40, evaluate the given function as indicated.

33. $f(x) = -x^2 - 5x + 2;\quad f(-2)$

34. $g(a) = a^3 - 6a;\quad g(-1)$

35. $h(c) = \sqrt{c^2 + 4c};\quad h(3)$

36. $A(x) = |2x - 3|;\quad A(-5)$

37. $B(r) = \dfrac{r^{-2} + r^{-3}}{5r - r^{-1}};\quad B(2)$

38. $Q(a) = 5a - 3(5a - 3) - a;\quad Q(-1)$

39. $P(c) = c^3 - 3 \cdot |2c^{-1} - 5|;\quad P(2)$

40. $R(d) = \sqrt{5d + 4} - d^2;\quad R(-0.23)$

41. Let $f(x) = |2x - 5|$ and $g(x) = \dfrac{3x}{x^2 + 1}$. Evaluate the given expression.

(a) $f(1) + g(0)$ (b) $g(1) - f(-2)$

42. Determine $f(2 + h)$ if $f(x) = 7 - 3x$.

43. Suppose that function f is defined as follows.

$$f(x) = \frac{\sqrt{x^2 - 3} - 4}{|x^{-1} - 2|}$$

Show how you would insert parentheses to enter f in your calculator.

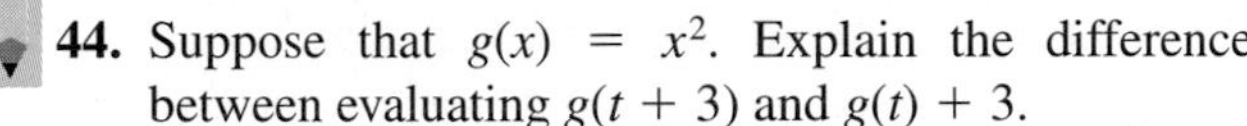

44. Suppose that $g(x) = x^2$. Explain the difference between evaluating $g(t + 3)$ and $g(t) + 3$.

Section 2.5

In Exercises 45–48, use the integer setting to produce the graph of the given function. Use the graph to answer the questions in each part.

45. $f(x) = 20 - \sqrt{2x + 4}$

(a) Estimate the domain and range.

(b) What is $f(16)$?

(c) For what value of x is $f(x) = 16$?

46. $g(x) = -0.2x^2 + 3.6x + 8$

(a) What are the x- and y-intercepts?

(b) What is $g(3)$?

(c) For what value of x is $g(x) = 25$?

47. $h(x) = x + 3$

(a) What are the x- and y-intercepts?

(b) What is $h(2)$?

(c) For what value of x is $h(x) = 7$?

48. $k(x) = 0.1x^3 - 7x$

(a) Estimate the local maximum and the local minimum.

(b) Estimate the domain and range.

(c) What is the y-intercept?

49. For what points of the graph of a function f is $f(x) = 0$?

50. A psychologist observes that a person's interest level E in a new computer game after h hours of playing time is measured by $E(h) = 0.25h(40 - h)$.

(a) What are the x-intercepts of the function, and what do they represent?

(b) After how many hours does the person's interest reach the highest level?

(c) Estimate the number of elapsed hours for the interest level to reach 80.

LOOKING AHEAD

The following exercises review concepts and skills that you will need in Chapter 3.

In Exercises 1 and 2, evaluate the given expression for the given value of the variable.

1. $3 - \frac{3}{4}t$; 8

2. $4 - 2(3x + 2)$; -5

In Exercises 3 and 4, use a graph to estimate the value of x such that the expression has the given value.

3. $2 - 3x$; -4

4. $3(x - 2) - 5$; 10

In Exercises 5 and 6, simplify the given expression by removing grouping symbols and combining like terms.

5. $11x + 2(4 - 3x)$

6. $10\left(\frac{2x}{5} - \frac{2x - 1}{2}\right)$

In Exercises 7 and 8, determine whether the given number is a solution of the equation.

7. $2x + 1 = 4 - x$; 1

8. $x = 2 - 3x$; -1

9. Translate the given verbal description into a simplified algebraic expression.

(a) The perimeter of a rectangle whose width is 9 feet less than its length L

(b) The value in cents of n nickels and $9 - n$ dimes

(c) The year-end value of an investment of P dollars at 8% simple interest

10. What are three ways to describe all numbers that are less than 3?

11. Compare the value of the expression $|2x - 1|$ when x is replaced by 3 and -2.

12. Use a graph to estimate the value of x such that $|3 - x|$ has a value of 0.

2. CHAPTER TEST

1. Determine the coordinates of the points in the accompanying figure.

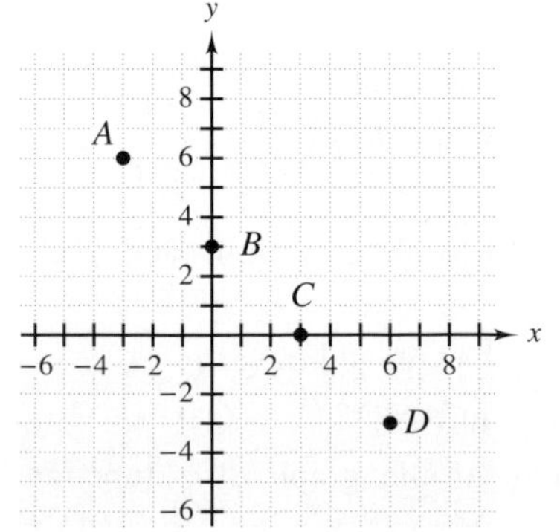

2. Use the given conditions to determine in which quadrant(s) or on which axis the point $P(a, b)$ could be located.

(a) $a > 0$

(b) $Q(b, a)$ is in Quadrant IV

3. Consider the points $A(-2, 1)$, $B(3, -2)$, and $C(2, 4)$.

(a) Determine the midpoint of the line segment $\overline{BC}$.

(b) If $\overline{AB}$ is a side of a triangle, how long is that side?

4. Suppose that a weather radar system can detect a tornado within 75 miles of the radar location. The radar is located 30 miles east and 20 miles north of your home, and a tornado is located 40 miles west and 10 miles south of your home.

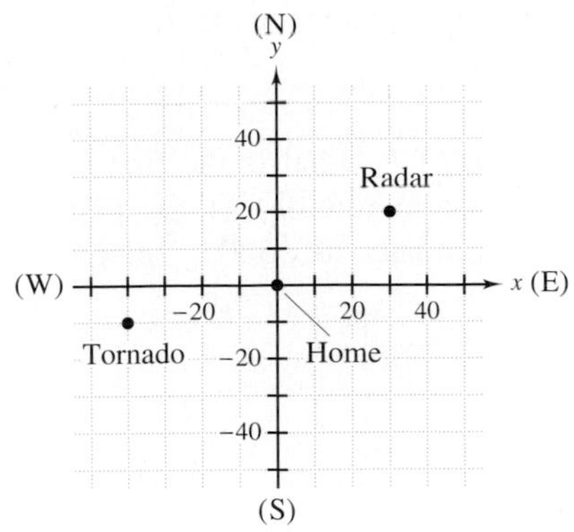

(a) How far is the tornado from the radar site? Is it within range of the radar?

(b) Show whether your home is on the line connecting the tornado and the radar location.

5. Suppose that you trace the graph of an expression to a point of the y-axis. What does the second coordinate of that point represent?

6. Use a graph or a table to evaluate the expression $\frac{5x}{x+1}$ for $x = -3.5, 0, 3,$ and 7.

7. Use a graph of the expression $5 - |x|$ to estimate each of the following.

(a) The value of the expression for $x = -3$.

(b) The value of x if $5 - |x|$ has a value of -3.

8. Use a graph to estimate the value(s) of x for which the expressions $x^2 + x$ and $10 - 2x$ have the same value.

In Questions 9–11, evaluate the function as indicated.

9. $g(x) = \frac{-x^{-2} + x^{-1}}{|x - 3|}$; $g(2)$

10. $f(x) = \sqrt{2x - 1}$; $f(5)$

11. $h(x) = 3x + 2$; $h(1 + t) - h(1)$

12. Determine the domain and range for the given relation.

(a)

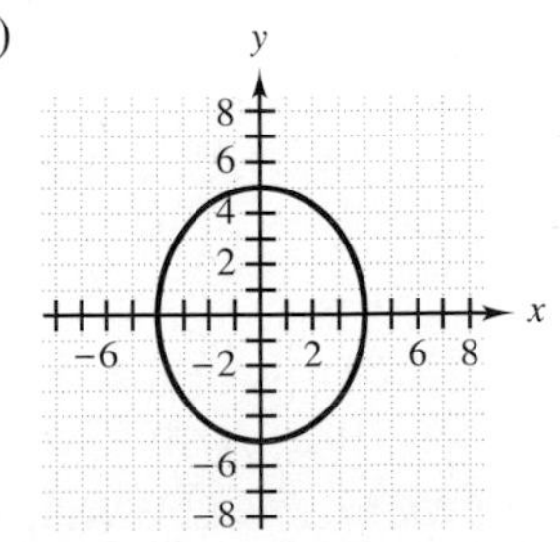

(b) $\{(1, 3), (2, 7), (3, -1), (5, 3)\}$

13. Determine whether the given graph represents a function and explain your answer.

(a)

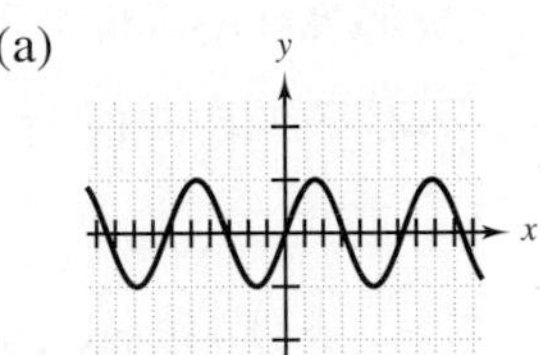

(b)

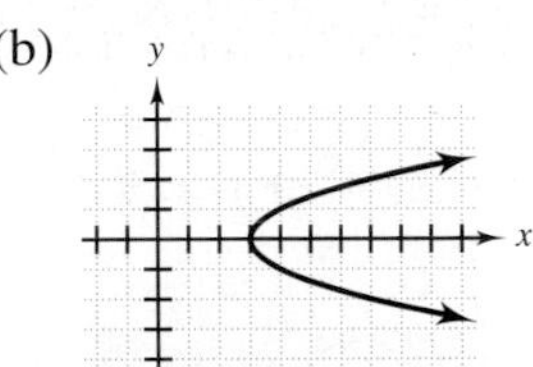

14. Use the integer setting to produce the graph of the function $f(x) = x^2 - 9x - 10$. Then use the graph to answer each part.

(a) Estimate the range of f.

(b) Determine the x-intercept(s).

15. Use the integer setting to graph $g(x) = \sqrt{4 - x}$. Then use the graph to answer each part.

(a) For what value of x is $g(x) = 3$?

(b) Estimate the domain of g.

16. Use the integer setting to graph $f(x) = |x - 2|$ and $g(x) = -\frac{1}{2}x + 4$ on the same coordinate axes. Then use the graphs to answer each part.

(a) Determine the point(s) of intersection of the graphs.

(b) What is the y-intercept of f?

17. In the accompanying figure, what name is given to the following points?

(a) A

(b) B

(c) C

(d) D

(e) E

(f) F

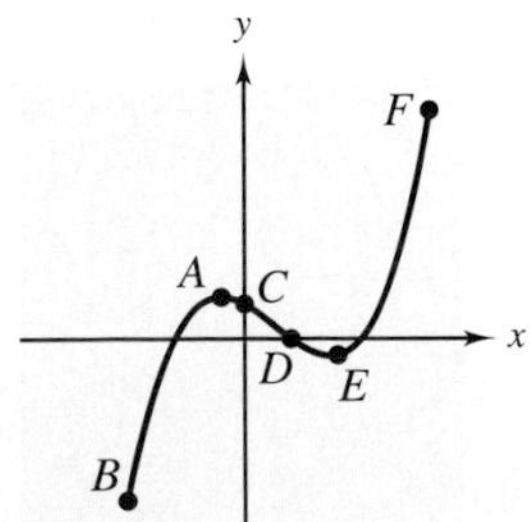

18. The daily profit P at a theater is modeled by the function $P(a) = 500a - 50a^2 - 450$, where a is the price of admission and P and a are in dollars. Use a graph of function P to estimate the answers to each part.

(a) To the nearest whole number, what is the absolute maximum of function P? Interpret your answer.

(b) On the graph of P, what is the first coordinate (to the nearest hundredth) of the point whose second coordinate is 500? Interpret your answer.

Chapter 3

Linear Equations and Inequalities

In 1994, state courts handled 98% of the volume of all court cases. The accompanying bar graph shows the percentage changes in various categories of court cases from 1984 to 1994. (Source: *USA Today*.)

Equations and **formulas** involving percentages are used commonly in application problems. For example, if we know that there were 14.3 million civil cases in 1994, we can write an equation to determine the number of civil cases (in millions) in 1984. (For more on this topic, see Exercises 67–70 at the end of Section 3.5, and see the Chapter Project.)

In this chapter we introduce linear equations and formulas as models, and we learn methods for solving and using them in various applications. We then consider solution techniques and applications for linear and compound inequalities, as well as absolute value equations and inequalities.

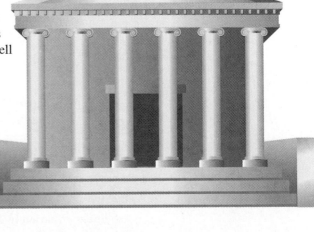

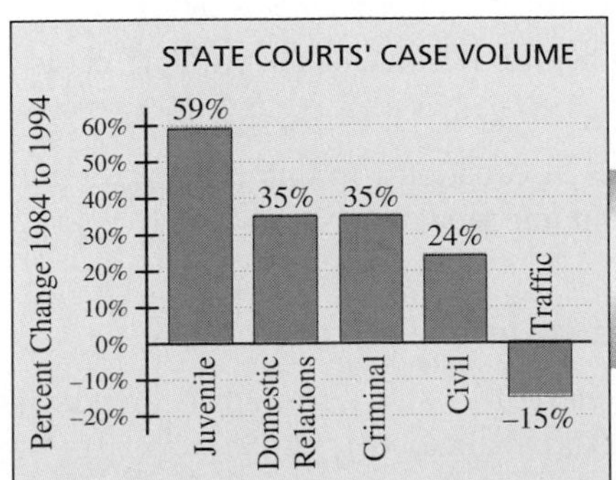

(Source: *USA Today*.)

3.1 INTRODUCTION TO LINEAR EQUATIONS

Equations • The Graphing Method

Equations

In Chapter 1 we briefly considered equations and methods for testing whether numbers were solutions of equations. Now we present a more formal treatment of these topics along with some related vocabulary.

Definition of Equation

An **equation** is a statement that two algebraic expressions have the same value.

The expressions to the left and right of the equality symbol are called the *left side* and *right side* of the equation.

left side → $2x - 5 = 15$ ← right side

If an equation contains a variable, we try to find replacements for that variable that make the equation true. We call such replacements **solutions** of the equation.

Definitions of Solution and Solution Set

A **solution** of an equation is a replacement for the variable that makes the equation true. A **solution set** is the set of all solutions of an equation.

To **solve** an equation means to determine each value of the variable that is a solution.

Some equations, called **identities,** are true for all permissible replacements of the variable. Other equations, called **inconsistent equations** (or **contradictions**), are false for all replacements of the variable.

Identities	*Contradictions*
$2(3x + 4) = 6x + 8$	$0 \cdot (x + 1) = 2$
$x + 7 = 7 + x$	$x + 7 = x + 1$

An equation that is true for at least one value of the variable but is not an identity is a **conditional equation.**

There are many different types of equations. In this chapter we will focus on just one type called a **linear equation in one variable.**

Definition of a Linear Equation in One Variable

A **linear equation in one variable** is an equation that can be written in the form $Ax + B = 0$, where A and B are real numbers and $A \neq 0$.

Similarly, a **linear function** is a function of the form $f(x) = ax + b$, where $a \neq 0$.

The following are examples of linear equations.

$$3x + 5 = 0 \qquad 5(x + 3) = 4 - x$$

The first equation is in the form $Ax + B = 0$. It can be shown that the other equation also can be written in that form.

The following are examples of equations that are *not* linear equations.

$\frac{3}{x} = 9 + x$ The variable cannot appear in a denominator.

$x + 2 = \sqrt{2x} - 1$ The variable cannot appear under a radical symbol.

$x^2 - 25 = 0$ The variable cannot have an exponent other than 1.

The Graphing Method

Verifying a solution involves testing whether a replacement for the variable makes the equation true. *Determining* a solution requires other methods. One approach is to estimate the solution of an equation by using a table or a graph.

EXPLORING THE CONCEPT

Estimating Solutions of Linear Equations

Consider the equation $x + 15 = 30 - 2x$. In Fig. 3.1, the left side is entered as Y_1 and the right side is entered as Y_2. We have then produced a table of values.

Figure 3.1

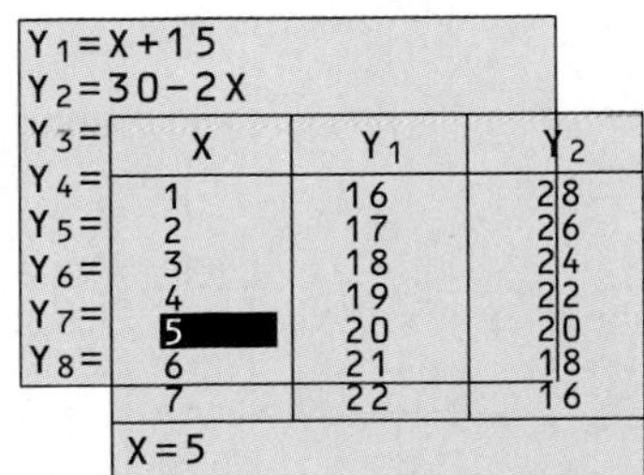

X	Y1	Y2
1	16	28
2	17	26
3	18	24
4	19	22
5	20	20
6	21	18
7	22	16

X=5

The table reveals that Y_1 and Y_2 both have a value of 20 when $x = 5$. This means that $x + 15$ and $30 - 2x$ are equal when x is 5, so 5 is a solution of the equation.

Figure 3.2 shows the graphs of the left and right sides of the same equation. Observe that the tracing cursor is on the point of intersection (5, 20).

Figure 3.2

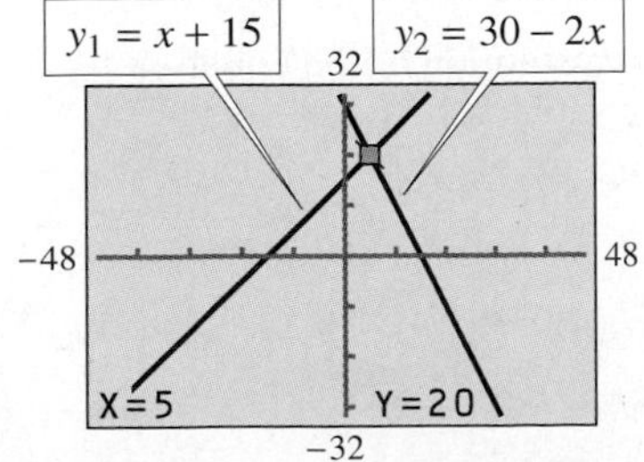

From the graph, we see that the expressions $x + 15$ and $30 - 2x$ both have a value of 20 when $x = 5$. Thus $x + 15 = 30 - 2x$ is true when $x = 5$, so 5 is a solution of the equation.

Note: Because a table or a graph may or may not reveal an exact solution, we generally refer to solutions obtained in these ways as *estimates*. All solutions should be verified by substitution.

Here is a summary of the graphing approach to solving an equation.

Equation Solving: The Graphing Method

1. Graph y_1 = left side of the original equation.
2. Graph y_2 = right side of the original equation.
3. Trace to the point of intersection.
4. The x-coordinate of that point is the solution of the equation.
5. The y-coordinate is the value of both the left side and the right side of the original equation when x is replaced with the solution.

EXAMPLE 1

Estimating Solutions with the Graphing Method

Use the graphing method to estimate the solution of $\frac{1}{2}x - 19 = -2x + 6$.

Solution

We use the integer setting to produce the graphs of $y_1 = \frac{1}{2}x - 19$ and $y_2 = -2x + 6$. (See Fig. 3.3.)

Figure 3.3

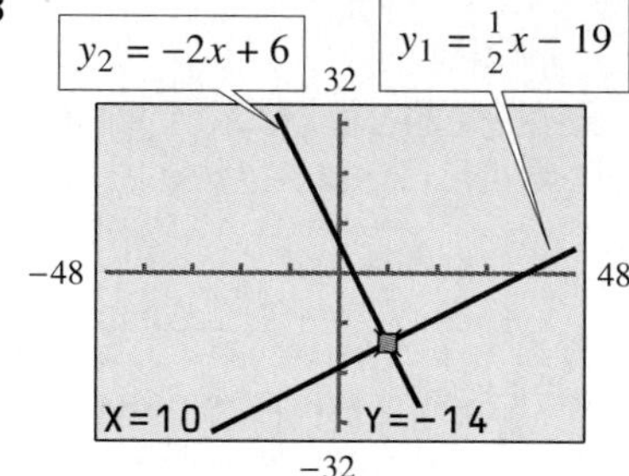

Tracing to the point of intersection, we find that the x-coordinate is 10, which is the estimated solution of the equation.

Verify that 10 is a solution of the equation.

$$\frac{1}{2}x - 19 = -2x + 6$$

$$\frac{1}{2}(10) - 19 = -2(10) + 6 \qquad \text{Replace } x \text{ with 10.}$$

$$5 - 19 = -20 + 6 \qquad \text{Simplify.}$$

$$-14 = -14 \qquad \text{True}$$

Figure 3.4

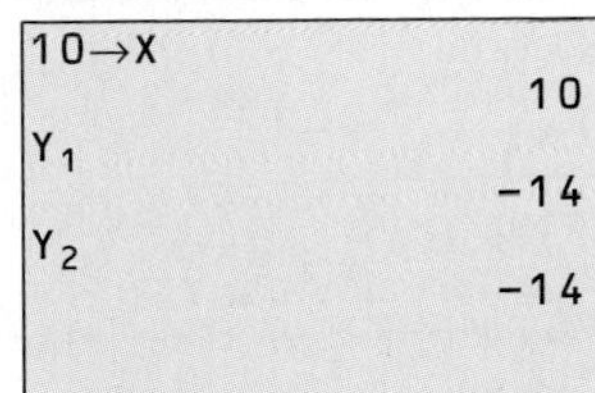

This verifies that 10 is a solution.

Having entered the two sides of the equation as y_1 and y_2 in order to produce their graphs, it is easy to verify the solution by storing 10 in X and evaluating y_1 and y_2. (See Fig. 3.4.)

The equation in Example 1 is a conditional equation with one solution. In Example 2 we learn how inconsistent equations and identities are revealed when we use the graphing method.

EXAMPLE 2

Inconsistent Equations and Identities

Use the graphing method to estimate the solution(s) of each of the following equations.

(a) $x + 3 = 1 + x$ (b) $x + 3 = 3 + x$

Solution

Figures 3.5 and 3.6 show the graphs of the left and right sides of the equations.

Figure 3.5

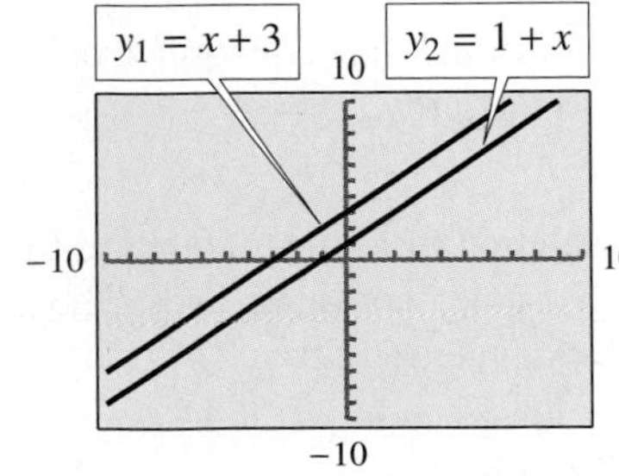

Figure 3.6

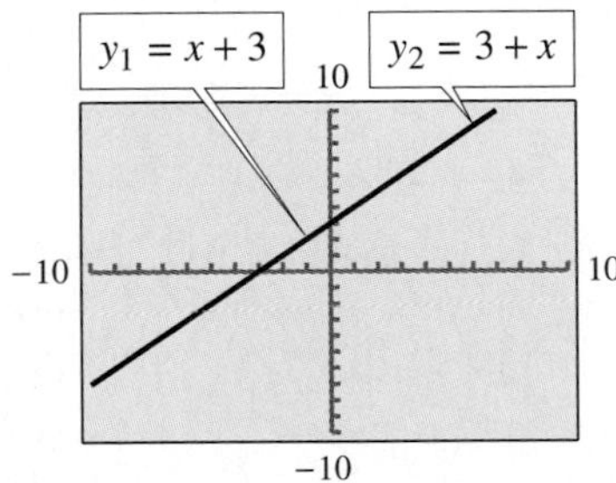

(a) In Fig. 3.5 the graphs appear to be parallel lines. Because there is no point of intersection corresponding to a solution, the equation has no solution and is an inconsistent equation. The solution set is Ø.

(b) In Fig. 3.6 the graphs appear to coincide. Because every point of the coinciding lines is a point of intersection corresponding to a solution, every real number is a solution, and the equation is an identity. The solution set is **R.**

Think About It

Suppose that you use the graphing method to estimate the solution of $0.51x - 3 = 0.5x + 3$. What is a reasonable conclusion? Now evaluate each side for $x = 600$. Do the results cause you to revise your conclusion?

We can now summarize what we have learned about the three types of equations discussed in this section.

1. An equation is called an **identity** if every permissible replacement for the variable is a solution. The graphs of the left and right sides coincide. The solution set is **R.**
2. An equation with no solution is called an **inconsistent equation.** The graphs of the left and right sides never intersect, and the solution set is Ø.
3. A **conditional equation** is an equation that has at least one solution but is not an identity. The point of intersection of the graphs of the left and right sides corresponds to a solution. A linear equation in one variable is an example of a conditional equation.

LEARNING TIP

Try to keep visualizing lines as you think about the three types of equations: conditional equations (intersecting lines), inconsistent equations (parallel lines), and identities (coinciding lines). Mental pictures promote understanding.

Although the graphing method is visually helpful in understanding the concept of solving an equation, algebraic techniques are often more useful when an exact solution is required. We will learn about such techniques in the next section.

3.1 QUICK REFERENCE

Equations

- An **equation** is a statement that two algebraic expressions have the same value.
- A **solution** of an equation is a replacement for the variable that makes the equation true.
- The **solution set** of an equation is the set of all solutions of the equation.
- We verify a solution by replacing every occurrence of the variable by the proposed solution. If the replacement makes the equation true, it is a solution.
- A **linear equation in one variable** is an equation that can be written in the form $Ax + B = 0$, where A and B are real numbers and $A \neq 0$.
- A **linear function** is a function of the form $f(x) = ax + b$, where $a \neq 0$.
- An **identity** is an equation that is true for every permissible replacement for the variable.
- An **inconsistent equation** (or **contradiction**) is false for all replacements of the variable.
- A **conditional equation** is an equation that is true for at least one value of the variable but is not an identity.

The Graphing Method

- We estimate the solution of a linear equation by graphing the left and right sides of the equation on the same coordinate axes. The x-coordinate of the point of intersection, if any, is the estimated solution.
- We verify the estimated solution by testing it in the equation to see if the replacement makes the equation true.
- The graphs of the left and right sides of an inconsistent equation never intersect, and the solution set is Ø.
- The graphs of the left and right sides of an identity coincide, and the solution set is **R**.

3.1 EXERCISES

Concepts and Skills

1. State the definition of an equation.

2. State the definition of a linear equation in one variable. Give two examples of equations that are not linear equations in one variable.

In Exercises 3–12, determine whether each of the following is an equation.

3. $x - 3 + 2x = 5$

4. $10 = x$

5. $4(x - 5) - 7(2x + 3)$

6. $\frac{9}{5} \cdot 5 - 6 + 7x$

7. $0 = -7$

8. $3 + 4 - 5 = 6$

9. $3x - 5 + 7x + 12$

10. $5t - 6 = 7$

11. $x^2 = 5x + 6$

12. $t^2 - 5t + 6$

In Exercises 13–20, state whether each equation is a linear equation in one variable.

13. $\frac{4}{7}x + 3 = 0$ **14.** $3x - 2 = 7$

15. $x^3 - 2x = 4$ **16.** $2x^2 + 5x = 6$

17. $2\pi^2 + 5x = 6$ **18.** $3\sqrt{2} - 7x = 0$

19. $\sqrt{x} = 5$ **20.** $6x = \sqrt{5}$

21. Although the equation $|x| = x$ has infinitely many solutions, it is a conditional equation. Why?

22. The solution of $2x - 3 = x + 4$ is 7. The graphs of the left and right sides intersect at a point. How can you know the coordinates of that point without actually producing the graphs?

In Exercises 23–32, verify that the given number is a solution of the given equation.

23. $3x - 5 = 7x + 7;\quad -3$

24. $15 - 2x = 3x;\quad 3$

25. $4(x - 2) + 5 = 5(3 - x) + 3(x + 4);\quad 5$

26. $29 + 4(3x - 5) = 2(6x + 7) - 5;\quad \frac{2}{3}$

27. $-2[2x - 3(1 - x)] = -5(x + 3) + 11;\quad 2$

28. $\frac{2}{3}(x - 4) + \frac{3}{4}(x + 5) = 2(x - 1);\quad \frac{37}{7}$

29. $\frac{3}{5}y = -2;\quad -\frac{10}{3}$ **30.** $\frac{x}{2} - \frac{2x}{5} + \frac{1}{2} = \frac{7}{10}x - \frac{2}{5};\quad \frac{3}{2}$

31. $6x - (5x - 8) = 1;\quad -7$

32. $-0.2t = 5.4;\quad -27$

In Exercises 33–40, determine which of the given numbers is a solution of the given equation.

33. $3x - 4 = 11;\quad 3, 5, 7$

34. $4x + 7 = 17 - x;\quad -2, 1, 2$

35. $\frac{x}{2} + \frac{1}{3} = \frac{x}{3} - \frac{1}{2};\quad -1, -4, -5$

36. $\frac{t}{4} - 2 = 3 - \frac{3}{4}t;\quad 2, 5, 8$

37. $3(x - 2) + 4 = 3x - 2;\quad -3, 0, 2$

38. $4 - 2(3x + 2) = -4x - 2x;\quad -5, -1, 4$

39. $5(t + 2) = 3(t - 4);\quad -11, -1, 0$

40. $2y - 5 = 5(y - 1) - y;\quad -2, 0, 2$

41. Describe the graph and solution set of an inconsistent equation.

42. Describe the graph and solution set of an identity.

In Exercises 43–46, the figure shows the graphs of the left and right sides of an equation. From the figure estimate the solution of the equation.

43.

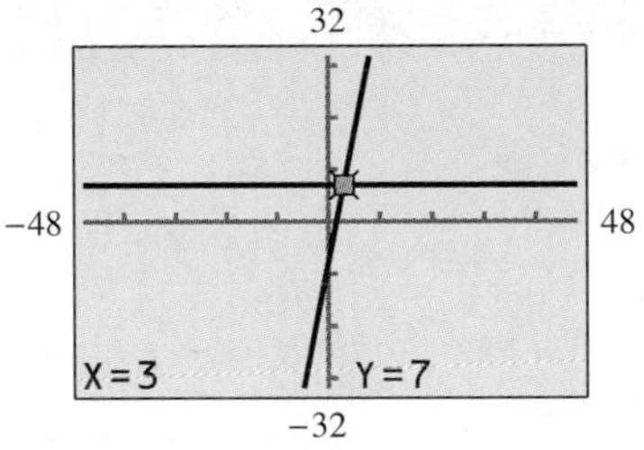

44.

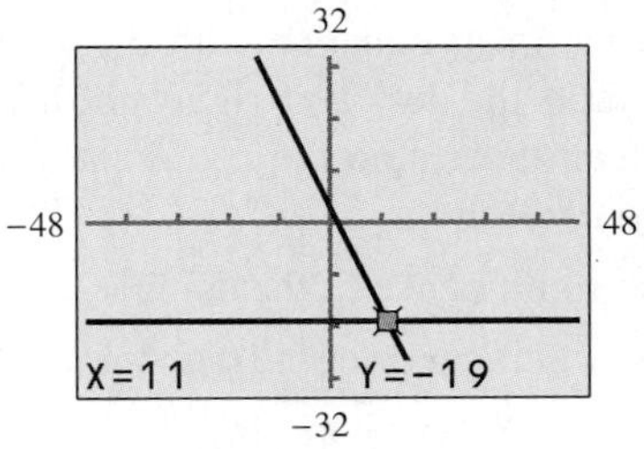

45.

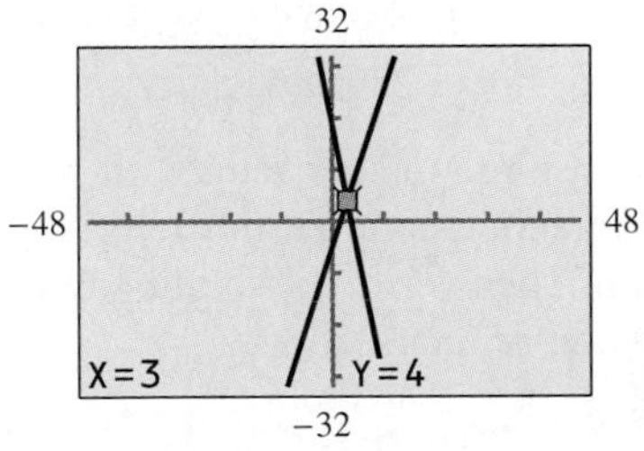

46.

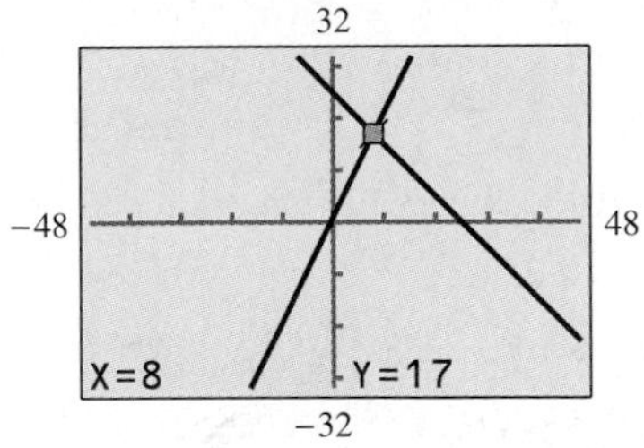

In Exercises 47–66, use the graphing method to estimate the solution(s) of the given equation. Then verify the solution. For identities and inconsistent equations, state the solution set.

47. $4x - 3 = 5$ **48.** $4x + 13 = -3$

49. $2 - 3x = -4$ **50.** $5x = 3x - 4$

51. $2x + 3 = 3 + 2x$ **52.** $x - 2 = -(2 - x)$

53. $4 - x = 3x + 12$ **54.** $2x - 3 = 7 - 3x$

55. $2(x + 3) = 6 + 5x$ **56.** $2(x - 5) = 8$

57. $2 - 3x = -3(x + 4)$ **58.** $2x + 5 - x = 7 + x$

59. $58 - (4 - 3x) = 0$ **60.** $3(x - 2) - 5 = 10$

61. $1 - 2(x - 1) = -(2x - 3)$

62. $x - (2x - 5 - x) = 5$

63. $\frac{1}{2}x + 2 = -\frac{1}{6}x + 6$

64. $1 - \frac{t}{3} = 5$

65. $\frac{t - 3}{3} = \frac{t}{6}$

66. $x - 5 = \frac{9 - x}{3}$

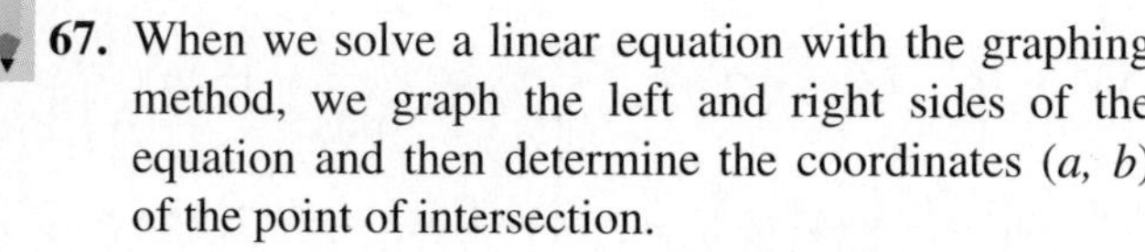

67. When we solve a linear equation with the graphing method, we graph the left and right sides of the equation and then determine the coordinates (a, b) of the point of intersection.

(a) What is the significance of the number a?

(b) What is the significance of the number b?

68. Use the Distributive Property to verify that

(a) $-(x + 1) = -x + 1$ is an inconsistent equation.

(b) $-(3 - x) + 5 = x + 2$ is an identity.

In Exercises 69–76, use the integer setting on your calculator and the graphing method to estimate the solution of the given equation. Then verify the solution. Write the coordinates of the point of intersection.

69. $2x + 3 = 3x - 8$

70. $-2x - 2 = 12 - x$

71. $\frac{3}{2}(x + 7) = 3$

72. $-(x - 3) = 2(10 - x)$

73. $3.75 - 0.5x = -0.25x$

74. $4x - 7 = 3x$

75. $\frac{5}{6}x + 3 = -7$

76. $\frac{1}{3}x + \frac{5}{3} = \frac{x}{3} - 1$

Real-Life Applications

77. An Internet provider charges a flat fee of \$20 per month plus \$2 per hour of on-line time. An alternative plan has a \$10 flat fee per month plus \$2.25 per hour on-line.

(a) Letting h represent the number of on-line hours, write expressions for the total monthly cost of each plan.

(b) Equate your expressions in part (a) and use the graphing method to estimate the solution of the equation.

(c) What does the estimated solution represent?

(d) What does the second coordinate of the solution point represent?

78. A *price-sensitive* item is one for which a small change in price results in a large change in demand. Suppose that the expression $-5x + 500$ represents the number of calculators that can be sold per month when the price is x dollars.

(a) Write an equation indicating monthly sales of 100 calculators. Use the graphing method to estimate the solution of your equation.

(b) Write an equation indicating monthly sales of 50 calculators. Use the graphing method to estimate the solution of your equation.

(c) What price change caused the sales to be cut in half?

(d) According to this model, at what price would there be no calculators sold at all?

Modeling with Real Data

79. Passports are valid for 10 years. The line graphs in the accompanying figure compare new applications with renewals for 1994 and 1996. (Source: U.S. State Department.)

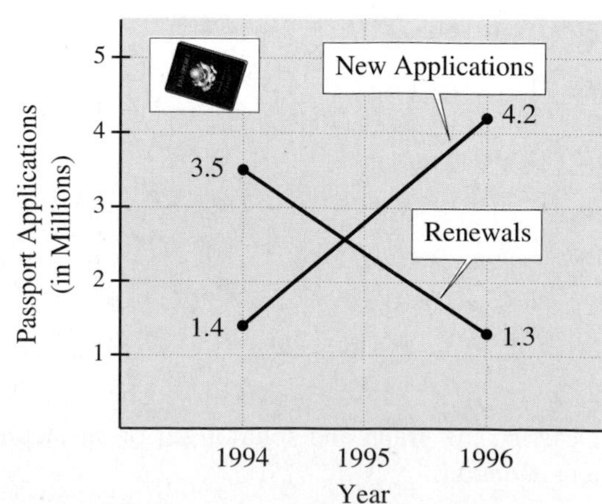

With t representing the number of years since 1990, the following expressions model the number (in millions) of passports issued.

New applications:	$1.4t - 4.2$
Renewals:	$-1.1t + 7.9$

(a) Write an equation indicating that the number of new applications and the number of renewals is the same. Use the graphing method to estimate the solution of your equation.

(b) Interpret both coordinates of the point that represents the estimated solution.

(c) What visual evidence suggests that the model for renewals will not be valid indefinitely?

80. Referring to the data in Exercise 79, write an equation to predict the year when the total number of new applications and renewals would be 7 million.

81. In 1996, the percentage of people who paid off all credit card bills every month increased with the age of the card holder. For instance, 49% of those in the 30–45 age group paid their bills monthly, whereas 87% of those age 60 and over paid off monthly. (Source: *USA Today*.)

The function $P(a) = 1.16a + 9.27$ models the percentage P of people of age a who paid off their bills monthly. Write an equation to estimate the age at which 40% of card holders paid their bills monthly. Estimate the solution of your equation.

82. At what age does the model in Exercise 81 predict that all people of that age paid off their credit card bills monthly?

Challenge

In Exercises 83–90, use the graphing method to estimate the solution of the given equation. When necessary, round your estimates to the nearest tenth.

83. $3x + 40 = 3.5x - 20$

84. $2(x + 15) = x - 40$

85. $\sqrt{3}x - 12 = \pi x$

86. $\dfrac{x}{\sqrt{2}} = 8 - x$

87. $|x + 2| = \dfrac{1}{2}x + 7$

88. $x - 3 = |x - 3|$

89. $\dfrac{x}{x^2 + 1} = 2$

90. $\dfrac{6 - x}{x} = x$

91. If B and C are constants, is $0 \cdot x + B = C$ a linear equation in one variable? Why? Under what condition would the equation be an identity?

92. Show that a linear equation in one variable cannot be an inconsistent equation.

3.2 SOLVING LINEAR EQUATIONS

Properties of Equations • Applying the Properties • Simplifying Both Sides • Clearing Fractions • Special Cases • The Algebraic Solving Routine

Properties of Equations

To obtain exact solutions of equations, we use algebraic methods that are based on the basic properties of equations.

Definition of Equivalent Equations

Equivalent equations are equations that have exactly the same solution sets.

The following equations are examples of equivalent equations because the only solution of each equation is -4.

$$5 - 3x = 17 \qquad -3x = 12 \qquad x = -4$$

The following properties of equations may be applied to produce equivalent equations.

Properties of Equations

Suppose A, B, and C represent algebraic expressions.

1. If $A = B$, then $A + C = B + C$. Addition Property of Equations

   Adding the same quantity to (or subtracting the same quantity from) both sides of an equation produces an equivalent equation.

2. If $A = B$ and $C \neq 0$, then $AC = BC$. Multiplication Property of Equations

   Multiplying (or dividing) both sides of an equation by the same nonzero quantity produces an equivalent equation.

3. If $A = B$, then $B = A$. Symmetric Property of Equations

   The two sides of an equation can be exchanged to produce an equivalent equation.

4. If $A = B$, then A may be replaced with B in any equation. Substitution Property of Equations

   Any expression in an equation may be replaced with an equivalent expression.

LEARNING TIP

You can think of the properties of equations in terms of a pan balance. For example, you can add the same weight to each pan, or you can swap the contents of the two pans, or you can replace any weight with an equivalent weight—all these actions maintain the balance.

Applying the Properties

To solve an equation algebraically, we apply the properties of equations to produce simpler equivalent equations. Usually the goal is to find an equivalent equation in which the variable stands alone (is isolated) on one side of the equation so that the solution is obvious. To accomplish this, the Addition Property of Equations is used to isolate the variable term, and the Multiplication Property of Equations is used to isolate the variable itself.

Our first example requires only these two properties. However, a full summary of the algebraic method can be found at the end of this section.

EXAMPLE 1

Applying the Addition and Multiplication Properties

Solve the following equations and verify the solutions by substitution.

(a) $2x + 3 = 11$ (b) $7x + 8 = -22 - 3x$

Solution

(a)

$$2x + 3 = 11$$

$$2x + 3 - 3 = 11 - 3 \quad \text{Addition Property of Equations}$$

$$2x = 8$$

$$\frac{2x}{2} = \frac{8}{2} \quad \text{Multiplication Property of Equations}$$

$$x = 4$$

We verify that 4 is the solution.

$$2x + 3 = 11$$

$$2(4) + 3 = 11 \quad \text{Replace } x \text{ with 4.}$$

$$11 = 11 \quad \text{True}$$

(b)

$$
\begin{aligned}
7x + 8 &= -22 - 3x \\
7x + 8 + 3x &= -22 - 3x + 3x && \text{Add } 3x \text{ to both sides of the equation.} \\
10x + 8 &= -22 && \text{Combine like terms.} \\
10x + 8 - 8 &= -22 - 8 && \text{Subtract 8 from both sides of the equation.} \\
10x &= -30 \\
\frac{10x}{10} &= \frac{-30}{10} && \text{Divide both sides of the equation by 10.} \\
x &= -3
\end{aligned}
$$

To verify by substitution, replace x with -3 in the original equation.

$$
\begin{aligned}
7x + 8 &= -22 - 3x \\
7(-3) + 8 &= -22 - 3(-3) && \text{Replace } x \text{ with } -3. \\
-21 + 8 &= -22 + 9 \\
-13 &= -13 && \text{True}
\end{aligned}
$$

Simplifying Both Sides

The next example illustrates how the expressions on the two sides of an equation may need to be simplified. Recall that the Distributive Property allows us to remove parentheses and combine like terms.

EXAMPLE 2

Simplifying Both Sides of an Equation

Solve the following equations.

(a) $4 - 2(x - 5) = 3x - 4(x + 2)$

(b) $1.09 - 0.3t = 0.4t - 1.7 + 0.2t$

Solution

(a)

$$
\begin{aligned}
4 - 2(x - 5) &= 3x - 4(x + 2) \\
4 - 2x + 10 &= 3x - 4x - 8 && \text{Remove parentheses.} \\
14 - 2x &= -x - 8 && \text{Combine like terms.} \\
14 - 2x + x &= -x - 8 + x && \text{Add } x \text{ to both sides of the equation.} \\
14 - x &= -8 && \text{Combine like terms.} \\
14 - x - 14 &= -8 - 14 && \text{Subtract 14 from both sides of the equation.} \\
-1x &= -22 \\
\frac{-1x}{-1} &= \frac{-22}{-1} && \text{Divide both sides of the equation by } -1. \\
x &= 22
\end{aligned}
$$

(b)

$$
\begin{aligned}
1.09 - 0.3t &= 0.4t - 1.7 + 0.2t \\
1.09 - 0.3t &= 0.6t - 1.7 && \text{Combine like terms.} \\
1.09 - 0.3t - 0.6t &= 0.6t - 0.6t - 1.7 && \text{Subtract } 0.6t \text{ from both sides.}
\end{aligned}
$$

$$1.09 - 0.9t = -1.7$$ Combine like terms.

$$1.09 - 1.09 - 0.9t = -1.7 - 1.09$$ Subtract 1.09 from both sides.

$$-0.9t = -2.79$$

$$\frac{-0.9t}{-0.9} = \frac{-2.79}{-0.9}$$ Divide both sides by -0.9.

$$t = 3.1$$

Think About It

By what number would you multiply both sides of each of the following equations in order to clear the decimals?

(a) $2 - 0.1x = 3$
(b) $4x + 0.237 = 1.58$

Write a general rule for determining the multiplier to clear decimals from an equation.

In part (b) of Example 2 we could have eliminated (or *cleared*) the decimals from the original equation by multiplying both sides by 100.

$$1.09 - 0.3t = 0.4t - 1.7 + 0.2t$$
$$100(1.09 - 0.3t) = 100(0.4t - 1.7 + 0.2t)$$
$$109 - 30t = 40t - 170 + 20t$$

This makes the resulting equivalent equation somewhat easier to manage, although the use of a calculator to do the routine arithmetic makes this step unnecessary.

Clearing Fractions

When equations involve fractions, it is often helpful to clear the fractions by using the Distributive Property to multiply both sides of the equation by the least common denominator (LCD).

Note: Remember to multiply each term by the LCD, even terms that are not fractions.

EXAMPLE 3

Clearing Fractions from an Equation

Solve the following equations.

(a) $\frac{3}{5}x - \frac{2}{3} = \frac{9}{10} + \frac{1}{15}x$ (b) $\frac{1}{2} - \frac{1}{3}(x - 2) = 2(x - 1) + \frac{2}{3}$

Solution

(a) $\frac{3}{5}x - \frac{2}{3} = \frac{9}{10} + \frac{1}{15}x$

Multiply each term of both sides by the LCD, 30.

$$30 \cdot \frac{3}{5}x - 30 \cdot \frac{2}{3} = 30 \cdot \frac{9}{10} + 30 \cdot \frac{1}{15}x$$

$$\left(30 \cdot \frac{3}{5}\right)x - 30 \cdot \frac{2}{3} = 30 \cdot \frac{9}{10} + \left(30 \cdot \frac{1}{15}\right)x$$ Associative Property of Multiplication

$$18x - 20 = 27 + 2x$$

$$18x - 20 - 2x = 27 + 2x - 2x$$ Subtract $2x$ from both sides.

$$16x - 20 = 27 \quad \text{Combine like terms.}$$
$$16x - 20 + 20 = 27 + 20 \quad \text{Add 20 to both sides.}$$
$$16x = 47$$
$$\frac{16x}{16} = \frac{47}{16} \quad \text{Divide both sides by 16.}$$
$$x = \frac{47}{16}$$

The solution can be verified with a calculator. (See Fig. 3.7.) Note the essential use of the parentheses.

Figure 3.7

```
47/16→X
              2.9375
(3/5)X-2/3
         1.095833333
9/10+(1/15)X
         1.095833333
```

Replacing x with $\frac{47}{16}$ (or 2.9375) makes both the left and right sides of the equation equal to 1.095833333. The solution is verified.

(b)
$$\frac{1}{2} - \frac{1}{3}(x - 2) = 2(x - 1) + \frac{2}{3} \quad \text{The LCD is 6.}$$
$$6 \cdot \frac{1}{2} - 6 \cdot \frac{1}{3}(x - 2) = 6 \cdot 2(x - 1) + 6 \cdot \frac{2}{3} \quad \text{Multiply both sides by 6.}$$
$$6 \cdot \frac{1}{2} - \left(6 \cdot \frac{1}{3}\right)(x - 2) = (6 \cdot 2)(x - 1) + 6 \cdot \frac{2}{3} \quad \text{Associative Property of Multiplication}$$
$$3 - 2(x - 2) = 12(x - 1) + 4$$
$$3 - 2x + 4 = 12x - 12 + 4 \quad \text{Distributive Property}$$
$$7 - 2x = 12x - 8 \quad \text{Combine like terms.}$$
$$7 - 2x - 12x = 12x - 8 - 12x \quad \text{Subtract } 12x \text{ from both sides.}$$
$$7 - 14x = -8$$
$$7 - 14x - 7 = -8 - 7 \quad \text{Subtract 7 from both sides.}$$
$$-14x = -15$$
$$\frac{-14x}{-14} = \frac{-15}{-14} \quad \text{Divide both sides by } -14.$$
$$x = \frac{15}{14}$$

We verify this solution with a calculator in the usual way. (See Fig. 3.8.)

Figure 3.8

```
15/14→X
         1.071428571
1/2-(1/3)(X-2)
         .8095238095
2(X-1)+2/3
         .8095238095
```

Special Cases

In Section 3.1 we observed that the equation $x + 3 = 1 + x$ appeared to have no solution, whereas every real number appeared to be a solution of $x + 3 = 3 + x$. These observations can be confirmed algebraically.

$x + 3 = 1 + x$	$x + 3 = 3 + x$
$x + 3 - x = 1 + x - x$	$x + 3 - x = 3 + x - x$
$3 = 1$ False	$3 = 3$ True
Solution set: Ø	Solution set: **R**

EXAMPLE 4

Special Cases

Solve each equation.

(a) $5 - 4x = -2(3 + 2x)$ (b) $3(x - 2) = -1 - (5 - 3x)$

Solution

(a)

$$5 - 4x = -2(3 + 2x)$$

$$5 - 4x = -6 - 4x \quad \text{Distributive Property}$$

$$5 - 4x + 4x = -6 - 4x + 4x \quad \text{Eliminate the variable term from one side.}$$

$$5 = -6 \quad \text{False}$$

The resulting equation is never true. Thus the original equation is a contradiction, and the solution set is Ø.

(b)

$$3(x - 2) = -1 - (5 - 3x)$$

$$3x - 6 = -1 - 5 + 3x \quad \text{Distributive Property}$$

$$3x - 6 = -6 + 3x \quad \text{Combine like terms.}$$

$$3x - 6 - 3x = -6 + 3x - 3x \quad \text{Eliminate the variable term from one side.}$$

$$-6 = -6 \quad \text{True}$$

The resulting equation is always true. Thus the original equation is an identity, and the solution set is **R**.

As we saw in Example 4, applying the properties of equations may lead to an equivalent equation with no variable.

1. If the resulting equivalent equation is always false, then the original equation is a contradiction, and the solution set is Ø.
2. If the resulting equivalent equation is always true, then the original equation is an identity, and the solution set is **R**.

The Algebraic Solving Routine

The following is a general summary of the steps for solving a linear equation in one variable algebraically.

Algebraic Routine for Solving a Linear Equation in One Variable

1. If necessary, clear fractions by multiplying every term of both sides of the equation by the LCD of all the fractions.
2. Simplify both sides of the equation by
   (a) removing grouping symbols and/or
   (b) combining like terms.
3. If the variable is on both sides, use the Addition Property of Equations to eliminate the variable term from one side or the other.
4. Use the Addition Property of Equations to isolate the variable term.
5. Use the Multiplication Property of Equations to isolate the variable itself.
6. Verify the solution.

The algebraic method of solving an equation has the advantage of producing an exact solution.

Nevertheless, graphic interpretations are extremely valuable in the development of a conceptual knowledge of equations and their solutions. Before computers and graphing calculators, the task of producing graphs was time-consuming and laborious. Now the graphing calculator can be an excellent tool for promoting a visual understanding of the relationship between equations and graphs.

3.2 QUICK REFERENCE

Properties of Equations

- **Equivalent equations** are equations that have exactly the same solution sets.
- An equivalent equation results when
  1. the same quantity is added to or subtracted from both sides of an equation (Addition Property of Equations).
  2. both sides of an equation are multiplied or divided by the same nonzero quantity (Multiplication Property of Equations).
  3. the two sides of an equation are swapped (Symmetric Property of Equations).
  4. an expression in an equation is replaced with an equivalent expression (Substitution Property of Equations).

Applying the Properties

- The Addition Property of Equations can be used to eliminate a variable term from one side of an equation and to isolate a variable term.
- The Multiplication Property of Equations can be used to isolate a variable by eliminating the coefficient of the isolated variable term.

Simplifying Both Sides

- Before applying the properties of equations, it is usually best to simplify both sides of the equation by removing parentheses and combining like terms.

Clearing Fractions

- Remove (clear) fractions from an equation by multiplying every term of both sides of the equation by the LCD.

Special Cases

- The algebraic routine for solving an equation will reveal special cases.
  1. If a resulting equivalent equation is a false statement, the original equation is inconsistent, and the solution set is Ø.
  2. If a resulting equivalent equation is a true statement, the original equation is an identity, and the solution set is the **R.**

Algebraic Solving Routine

- The summary at the end of this section lists the suggested steps for solving a linear equation in one variable algebraically.

3.2 EXERCISES

Concepts and Skills

1. What are equivalent equations?

2. What is an important advantage of the algebraic method over the graphing method for solving equations?

3. Consider the following two equations.

(i) $x + 2 = 3$ (ii) $-2x = 6$

When we solve these equations, explain why we *add* -2 to both sides of the first equation and *divide* both sides by -2 in the second equation.

4. The following are possible first steps in solving the equation $\frac{2}{3}x = 10$. Explain why each approach is correct.

(i) Multiply both sides by $\frac{3}{2}$.
(ii) Divide both sides by $\frac{2}{3}$.
(iii) Multiply both sides by 3 and then divide both sides by 2.

In Exercises 5–20, solve the equation algebraically and verify the solution. For special cases, write the solution set.

5. $5x + 8 = 23$
6. $4n - 7 = -35$
7. $7x - 5 = 8x + 7$
8. $7 + 8x = 4x - 13$
9. $3(3 - 2x) = 33 - 2x$
10. $2(3y + 8) = -2(2 - y)$
11. $7w - 5 = 11w - 5 - 4w$
12. $9w + 7 = 16w + 7(1 - w)$
13. $9x + 5 = 5x + 3(x - 1)$
14. $11x + 2(4 - 3x) = 3(3 - x) + 15$
15. $16 + 7(6 - x) = 15 - 4(x + 2)$
16. $7 - 4(t - 3) = 6t - 3(t + 3)$
17. $2(3a + 2) = 2(a + 1) + 4a$
18. $8t - 7 - 3t = 5t + 2$
19. $-2(x + 5) = 5(1 - x) + 3(7 - x)$
20. $2y - 3(2y - 3) = 2y - 5(3y - 4)$

21. Consider solving the equation $\frac{1}{3} - 2x = \frac{2}{5}$. To clear fractions, we multiply both sides by 15. Why is it necessary to multiply $2x$ by 15?

22. What is the solution set of $3x = 2x$? Now divide both sides of $3x = 2x$ by x. What is the solution set of the resulting equation? Why are the equations not equivalent?

In Exercises 23–32, solve the given equation and use your calculator to verify the solution.

23. $2x - \frac{3}{4} = -\frac{5}{6}$
24. $\frac{2}{15}y + \frac{3}{5} = 2 - \frac{2}{3}y$
25. $\frac{y}{2} + \frac{y}{4} = \frac{7}{8} - \frac{y}{8}$
26. $\frac{5 + 3x}{2} + 7 = 2x$
27. $\frac{5}{4}(x + 2) = \frac{x}{2}$
28. $\frac{1}{4}(x - 4) = \frac{1}{3}(x + 6)$
29. $x + \frac{3x - 1}{9} = 4 + \frac{3x + 1}{3}$
30. $\frac{2x}{5} - \frac{2x - 1}{2} = \frac{x}{5}$
31. $\frac{1}{4} + \frac{1}{6}(4a + 5) = 2(a - 3) + \frac{5}{12}$
32. $\frac{5}{6}(1 - t) - \frac{t}{2} = -\frac{1}{3}(1 - 3t)$

In Exercises 33–36, solve the given equation and use your calculator to verify the solution.

33. $2.6 = 0.4z + 1$
34. $1 - 0.3c = -0.5$
35. $0.08x + 0.15(x + 200) = 7$
36. $0.6(2x + 1) - 0.4(x - 2) = 1$

In Exercises 37–42, determine whether the given equations are equivalent equations.

37. $x + 4 = 7$ and $3x - 5 = 4$
38. $\frac{1}{3}x = 2$ and $x + 3 = 9$
39. $x + 3 = 4 + x$ and $2x - 3 = 4 + 2x$
40. $x + 3 = 3 + x$ and $2x - 5 = -(5 - 2x)$

41. $\frac{1}{4}x = -2$ and $8 - x = 0$

42. $3x = 0$ and $x - 3 = 0$

In Exercises 43–50, solve the equation. State whether the equation is an identity, an inconsistent equation, or a conditional equation.

43. $x - 6 = 6 - x$

44. $2x - 7 = x + 4$

45. $x + 6 = 6 + x$

46. $x + 3 = 5 + x$

47. $2x - 5 = 7 + 3x - (x - 3)$

48. $x + 3 = 5x - [2x - (3 - 2x)]$

49. $\frac{1}{5}x - 2 = \frac{1}{3}x - 4$

50. $-\frac{1}{3}x + 8 = 0$

In Exercises 51–54, determine a value of c so that the equations are equivalent.

51. $5 - 3x = c + 2x$ and $2x - 3 = 5$

52. $7 - 2x = 4x - c$ and $3x - 8 = 7x$

53. $2x + c = 3x + 2$ and $4x + 7 = 6x - 5$

54. $3x - c = 5x - 3$ and $6 - 5x = 3x + 14$

In Exercises 55–58, determine a value of k so that the equation is inconsistent.

55. $3x - 2 = 4 + kx$

56. $5 - kx = 7x + 2$

57. $2x - 4 = x - 1 + kx$

58. $kx - 3x = 2x + 1$

In Exercises 59–62, determine a value of c so that the equation is an identity.

59. $2x + 3 = c + 2x$

60. $x - 4 = -(c - x)$

61. $cx + 2 = 2 + 5x$

62. $4 - 2x = 3x - (cx - 4)$

In Exercises 63–66, determine a value of m so that the solution of the equation is 7.

63. $3x - 5 = m$

64. $8 - 2x = m + 4$

65. $3x + m - 2 = 4 - x$

66. $mx - 7 = 13$

In Exercises 67–84, solve the equations.

67. $2x - 5 = 5 - 2x$

68. $19 - 5z = 6z + 41$

69. $17 - 4z = 3z + 52$

70. $3y + 2 = 4 + 3y$

71. $3(x + 1) - 2 = 2(x + 1) + x - 1$

72. $x + 3(2 - x) = 6 - 2x$

73. $5 + 3[t - (1 - 4t)] = 7$

74. $13 - 3(x + 2) = 5x - 9 + 2(x - 7)$

75. $3(2x + 1) = 8x - 2(x - 2)$

76. $3 + 2[3 - (7t - 4)] = 2t - [3 - (3t + 5)]$

77. $\frac{4}{9}t - \frac{1}{6} = \frac{1}{3}t + 1$

78. $\frac{t}{5} + \frac{7}{15} = 2t + \frac{2}{3}$

79. $\frac{3}{4}(x - 3) - \frac{1}{2}(3x - 5) = 2(3 - x)$

80. $\frac{3t - 8}{2} = -8 - \frac{t}{3}$

81. $0.12 - 0.05y = 0.04y - 0.15$

82. $1 - 0.3x = 0.6x - 1.7$

83. $x(x + 6) = 5(x + 1) + x^2$

84. $x^2 - 4 = x(x + 1)$

Real-Life Applications

85. The volume (in decibels) of a digital alarm clock depends on the loudness setting, where a setting of 1 produces 30 decibels and each increment from 1 to 9 produces an increase of 8 decibels.
   (a) Write an expression that models the volume of the alarm for any setting s.
   (b) Write an equation that indicates a volume of 62 decibels, and solve the equation to determine the setting.

86. At a medical facility the parking lot is designed so that one-third of the spaces are for small cars, one-half of the spaces are for standard-sized cars, and the remaining 35 spaces are reserved for professional staff.
   (a) Letting x represent the total number of parking spaces, write an expression that models the sum of the three categories.
   (b) Write an equation showing that the sum of the three categories is equal to the total number of spaces.
   (c) Solve the equation in part (b), and determine the number of parking spaces for small cars.

Modeling with Real Data

87. The accompanying bar graph shows the number of registered shareholders (in thousands) of Exxon stock at the end of each year between 1986 and 1990. (Source: Exxon Corporation.)

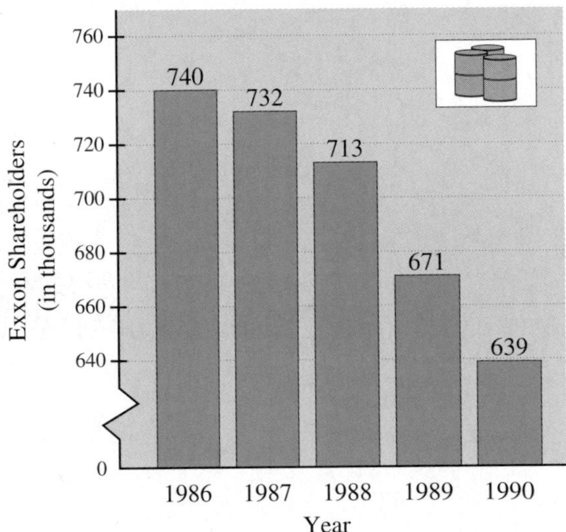

The number of shareholders s (in thousands) can be modeled by the function $s(x) = 778 - 26x$, where x is the number of years since 1985.

(a) For which year is the model most accurate?

(b) If the model were to remain valid indefinitely, in what year would there be no shareholders?

88. In 1996, 1.2 million students received a bachelor's degree. (Source: National Center for Educational Statistics.) With t representing the number of years since 1990, the function $D(t) = 0.02t + 1.09$ models the number of degrees D (in millions).

(a) What is the predicted number of graduates in 2001?

(b) Write and solve an equation that describes the year in which 1.25 million degrees will be granted.

89. The Social Security Act was passed in 1935. The accompanying table shows that the combined tax rate for employers and employees for Social Security and Medicare has risen steadily. (Source: Department of Health and Human Services.)

Year	Combined tax rate
1945	2.00%
1955	4.00%
1965	7.25%
1975	11.70%
1985	14.10%
1995	15.30%

The function $T(x) = 0.28x - 1.07$ models the tax rate T, where x is the number of years since 1935.

(a) Write an equation for estimating the year in which the tax rate will reach 10 times the rate in 1945.

(b) Determine and interpret the solution of the equation.

90. Referring to the data in Exercise 89, during the first 60 years of its existence, the Social Security tax rate climbed to 15.3%. What tax rate does the model function project after another 60 years?

Group Project: Production Jobs

The number of production-line jobs in the manufacturing sector of the United States has decreased significantly in recent years. In 1979 there were 21 million people employed as production workers. By 1995 that number had decreased to 18 million. (Source: AFL-CIO.) If we let x represent the number of years since 1979, the number (in millions) of production workers can be modeled by the expression $21 - 0.1875x$.

91. Use the integer setting to produce the graph of the model expression. Then trace the graph to the points whose x-coordinates are 0 and 16. What do these points represent? Do they accurately reflect the given data?

92. Use the graphing method to estimate the solution of the equation $21 - 0.1875x = 18.75$. Then solve the equation algebraically.

93. What is your interpretation of the coordinates of the point corresponding to the solution of the equation in Exercise 92?

94. What are several factors that you think have contributed to the decline in production workers over this period?

Challenge

In Exercises 95–98, solve for x.

95. $b - ax = c$

96. $a(x - b) = c$

97. $ax + b = cx + d$

98. $\dfrac{x + a}{3} = b$

In Exercises 99 and 100, solve the equation.

99. $\dfrac{x}{2} + \dfrac{x+1}{3} + \dfrac{x+2}{4} + \dfrac{x+3}{5} = 1$

100. $x - 2\{(x - 1) - 3[(x - 2) - 4(x - 3)]\} = 19$

In Exercises 101 and 102, determine whether the given equations are equivalent.

101. $h^2 = 25$ and $h = 5$

102. $\dfrac{3}{x} = 0$ and $x = 0$

3.3 FORMULAS

Solving Formulas • Real-Life Applications

Solving Formulas

In Section 1.4 we described a **formula** as an equation that uses an expression to represent a specific quantity.

The formula for the perimeter P of a rectangle is $P = 2L + 2W$, where L represents the length of the rectangle and W represents the width. For given values of L and W, it is easy to calculate P. On the other hand, if the values of P and W are given, the task of calculating L involves solving an equation. If we must perform this task repeatedly, it is easier if we have a formula for L in terms of P and W.

The process of isolating a variable in a formula is called *solving the formula* for the variable. Because a formula is simply an equation, the procedure for solving a formula is exactly the same as the procedure for solving equations.

EXAMPLE 1

Solving the Perimeter Formula

(a) Solve the formula $P = 2W + 2L$ for L.

(b) Use a calculator and the formula from part (a) to find the length of a rectangle whose perimeter is 110.4 inches and whose width is 20.7 inches.

Solution

(a) Because the task is to solve for L, we treat L as the variable and W and P as constants. (For convenience, we swap the sides of the formula so that L appears on the left side.)

LEARNING TIP

When you solve a formula for a particular variable, consider writing the variable in color or circling it in order to distinguish it from the other letters.

$$2W + 2L = P$$

$$2W - 2W + 2L = P - 2W \quad \text{Subtract } 2W \text{ from both sides.}$$

$$2L = P - 2W$$

$$\frac{2L}{2} = \frac{P - 2W}{2} \quad \text{Divide both sides by 2.}$$

$$L = \frac{P - 2W}{2}$$

In the last step we used the Multiplication Property of Equations to divide both *sides* by 2. We might choose to divide *each term* by 2. Therefore, an alternative way to write the formula is

$$L = \frac{P}{2} - \frac{2W}{2} = \frac{P}{2} - W$$

(b) We store the given values for P and W and use the derived formula to evaluate L. (See Fig. 3.9.)

In Fig. 3.10 we verify that the alternate formula gives the same value for L.

Figure 3.9

```
110.4→P
                110.4
20.7→W
                 20.7
(P-2W)/2
                 34.5
```

Figure 3.10

```
110.4→P
                110.4
20.7→W
                 20.7
P/2-W
                 34.5
```

In Example 1, solving the formula for L, in general, involves exactly the same steps as solving the equation for L using specific values of P and W. By solving the formula for L, we have to go through the steps only once, and the new formula provides an arithmetic means of calculating L.

A **trapezoid** is a four-sided figure with one pair of sides parallel. (See Fig. 3.11.) The formula for calculating the area of a trapezoid is

$$A = \frac{1}{2}(b_1 + b_2)h$$

Figure 3.11

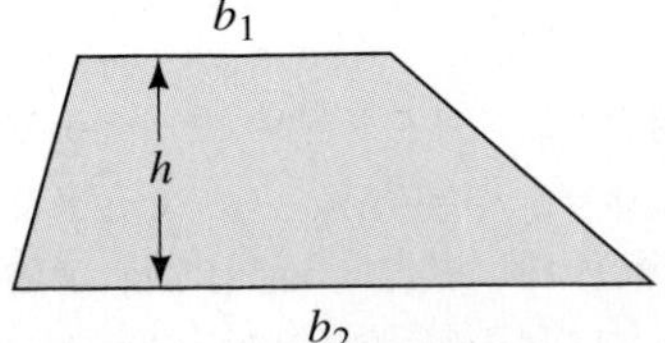

EXAMPLE 2

A Formula for the Height of a Trapezoid

Solve the formula for the area of a trapezoid for h.

Solution

In the formula $\frac{1}{2}(b_1 + b_2)h = A$, treat h as the variable and regard all other variables as constants.

$$2 \cdot \frac{1}{2}(b_1 + b_2)h = 2 \cdot A$$ Multiply both sides by 2 to clear the fraction.

$$(b_1 + b_2)h = 2A$$ The coefficient of h is $(b_1 + b_2)$.

$$\frac{(b_1 + b_2)h}{(b_1 + b_2)} = \frac{2A}{(b_1 + b_2)}$$ Divide both sides by $(b_1 + b_2)$.

$$h = \frac{2A}{b_1 + b_2}$$

Think About It

Suppose the circumference of any circle is increased by 1 foot. By how much is the radius increased?

The formula for converting Fahrenheit temperatures F to Celsius temperatures C is $C = \frac{5}{9}(F - 32)$.

EXAMPLE 3

Converting Celsius Temperatures to Fahrenheit Temperatures

(a) Solve the formula $C = \frac{5}{9}(F - 32)$ for F.

(b) Use a calculator and the formula for F to convert 0°C and 100°C to Fahrenheit temperatures.

Solution

(a)

$$\frac{5}{9}(F - 32) = C$$ Swap sides to place F on the left.

$$\frac{9}{5} \cdot \frac{5}{9}(F - 32) = \frac{9}{5} \cdot C$$ Multiply both sides by $\frac{9}{5}$ to clear the fraction.

$$F - 32 = \frac{9}{5}C$$

$$F = \frac{9}{5}C + 32$$ Add 32 to both sides.

Figure 3.12

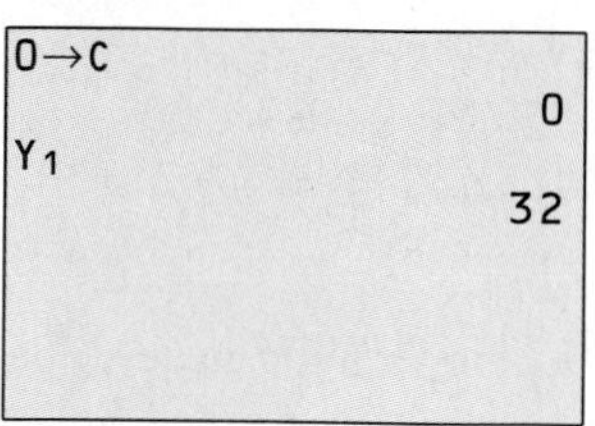

Figure 3.13

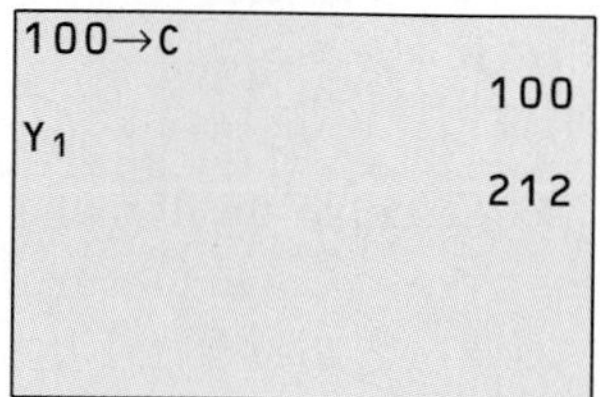

(b) To avoid having to enter the expression for F each time, we enter it as Y_1. Then we can evaluate the expression for both of the given values of C. (See Figs. 3.12 and 3.13.)

Note that 32°F, the freezing temperature of water, corresponds to 0°C and that 212°F, the boiling temperature of water, corresponds to 100°C.

When we use a calculator to graph a function, the function must be in the form $y =$ expression. In other words, the equation must be solved for y.

EXAMPLE 4

Solving an Equation for y

Solve the equation $3x - y = 4$ for y.

Solution

We treat x as if it were a constant and isolate the variable y.

$$3x - y = 4$$

$$3x - y - 3x = -3x + 4 \qquad \text{Subtract } 3x \text{ from both sides.}$$

$$-1y = -3x + 4$$

$$\frac{-1y}{-1} = \frac{-3x}{-1} + \frac{4}{-1} \qquad \text{Divide both sides by } -1.$$

$$y = 3x - 4$$

Real-Life Applications

Example 5 involves the volume V of a rectangular solid. The formula is $V = LWH$, where L, W, and H are the length, width, and height, respectively.

EXAMPLE 5

A Concrete Slab in the Form of a Rectangular Solid

In order to install an underground electric cable underneath an existing parking lot, a strip of concrete 200 feet long and 4 feet wide had to be removed. (See Fig. 3.14.) After the installation, the contractor ordered 8 cubic yards of concrete to repair the parking lot.

Figure 3.14

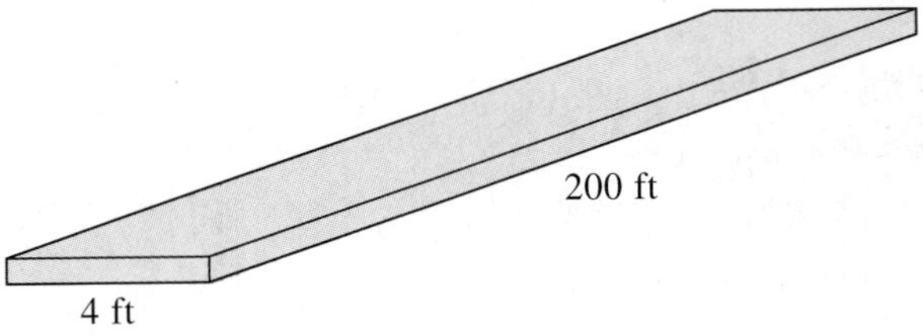

(a) Given the formula $V = LWH$, for which variable does the formula need to be solved in order to determine the thickness of the new slab of concrete? Solve the formula for that variable.

(b) Use the new formula to determine the thickness of the slab of concrete.

Solution

(a) The thickness of the concrete is represented by H, so we solve the formula $V = LWH$ for H.

$$LWH = V \qquad \text{The coefficient of } H \text{ is } LW.$$

$$\frac{LWH}{LW} = \frac{V}{LW} \qquad \text{Divide both sides by } LW.$$

$$H = \frac{V}{LW}$$

(b) Because the length and width are expressed in feet, the volume of 8 cubic yards of concrete ordered must be expressed in cubic feet.

$$8 \text{ cubic yards} \cdot 27 \frac{\text{cubic feet}}{\text{cubic yard}} = 216 \text{ cubic feet}$$

$$H = \frac{216}{(200)(4)} = 0.27 \text{ feet} \qquad V = 216, L = 200, W = 4$$

The concrete slab will be 0.27 feet, or about 3 inches, thick.

It is impossible to be a good equation solver and a poor formula solver. The procedures are identical in both cases.

No matter what your current or future line of work may be, you may very well find that being able to solve a formula is a valuable skill to have.

3.3 QUICK REFERENCE

Solving Formulas

- We previously described a **formula** as an equation that uses an expression to represent a specific quantity.
- Formulas are written in the form in which they are most commonly used. However, we sometimes need to know the value of another variable in the formula. If we need to calculate that variable repeatedly, it is usually best to **solve the formula** for that variable.
- When solving a formula for a particular variable, treat all other variables as constants. The methods for solving a formula are identical to the algebraic methods for solving any other equation.

3.3 EXERCISES

Concepts and Skills

1. In your own words, explain what is meant by solving a formula for a given variable.

2. If we solve the formula $I = Prt$ for r, which letter do we regard as the variable? How do we treat the other letters?

In Exercises 3–32, solve the formula for the given variable.

3. $F = ma$ for m

4. $C = \pi d$ for d

5. $I = Prt$ for r

6. $V = LWH$ for W

7. $v = \frac{s}{t}$ for t

8. $r = \frac{d}{t}$ for t

9. $A = \frac{1}{2}bh$ for b

10. $V = \frac{1}{3}Bh$ for h

11. $A = \frac{1}{2}h(a + b)$ for a

12. $A = \frac{1}{2}h(a + b)$ for h

13. $A = \pi r^2$ for π

14. $x^2 = 4py$ for y

15. $E = mc^2$ for m

16. $V = \frac{4}{3}\pi r^3$ for π

17. $a = \frac{v - w}{t}$ for v

18. $z = \frac{x - u}{s}$ for u

19. $A = P + Prt$ for r

20. $IR + Ir = E$ for r

21. $P = \frac{nRT}{V}$ for R

22. $P = \frac{nRT}{V}$ for V

23. $R_T = \frac{R_1 + R_2}{2}$ for R_1

24. $A = \frac{r^2\theta}{2}$ for θ

25. $ax + by + c = 0$ for y

26. $y = mx + b$ for m

27. $s = \frac{1}{2}gt^2 + vt$ for v

28. $s = \frac{1}{2}gt^2 + vt$ for g

29. $A = a + (n - 1)d$ for n

30. $A = a + (n - 1)d$ for d

31. $S = \frac{n}{2}[2a + (n - 1)d]$ for a

32. $S = \frac{n}{2}[2a + (n - 1)d]$ for d

33. The following shows two different ways of solving the equation $2x + 3y = 6$ for y. Explain why both methods are correct.

$$\begin{aligned} 2x + 3y &= 6 \\ 3y &= -2x + 6 \\ y &= -\frac{2}{3}x + 2 \end{aligned} \qquad \begin{aligned} 2x + 3y &= 6 \\ 3y &= -2x + 6 \\ y &= \frac{-2x + 6}{3} \end{aligned}$$

34. In terms of graphing, what might be our motivation for solving the equation $2x + 3y = 6$ for y?

In Exercises 35–52, solve the given equation for y.

35. $2x + 5y = 10$ **36.** $3x - y = 9$

37. $x - y + 5 = 0$ **38.** $x - 5y + 10 = 0$

39. $2x + 3y - 9 = 0$ **40.** $5x - 6y + 12 = 0$

41. $3x = 4y + 11$ **42.** $5x = 7y - 15$

43. $3x - y = 5x + 3y$

44. $3(2x - y) = 4x - y - 2$

45. $0.4x - 0.3y + 12 = 0$

46. $10 = 0.4y - 0.5x$

47. $y - 4 = \frac{1}{2}(x + 6)$

48. $y + 7 = -\frac{3}{4}(x - 12)$

49. $\frac{4}{5}x - \frac{2}{3}y + 8 = 0$ **50.** $\frac{y}{8} - \frac{x}{4} = 0$

51. $\frac{y - 2}{3} = \frac{x + 3}{4}$ **52.** $\frac{3 - x}{2} + 5 = \frac{y + 1}{3}$

In Exercises 53–54, round all decimal results to the nearest hundredth.

53. The surface area A of a right circular cylinder is given by the formula $A = 2\pi r^2 + 2\pi rh$, where r is the radius of the circular base and h is the height of the cylinder.

(a) What is the surface area of a right circular cylinder if the radius is 2.7 inches and the height is 5.83 inches?

(b) Solve the formula for h.

(c) What is the height of a right circular cylinder if its surface area is 2858.85 square centimeters and its radius is 13 centimeters?

54. The volume V of a right circular cone is given by the formula $V = \frac{1}{3}\pi r^2 h$, where r is the radius of the circular base and h is the height of the cone.

(a) What is the volume of a right circular cone if the radius of the base is 2.7 meters and the height is 11.3 meters?

(b) Solve the formula for h.

(c) What is the height of a right circular cone if the volume is 1256.64 cubic feet and the radius is 10 feet?

55. Which of the following formulas are equivalent to the formula $A = \frac{1}{2}bh$?

(i) $b = \frac{2A}{h}$ (ii) $b = 2A \cdot \frac{1}{h}$

(iii) $b = \frac{2}{h} \cdot A$ (iv) $b = \frac{A}{\frac{1}{2}h}$

(v) $b = \frac{A}{2h}$

56. A student drove 20 miles from home to college to take her math test. Her average speed was 40 mph. When she arrived at the college, she realized she had forgotten her calculator. Traveling at an average speed of 50 mph, she returned home to get her calculator. Finally, driving at an average speed of 60 mph, she drove back to the college to take the test. Because her speeds were 40, 50, and 60 mph, was her average speed 50 mph? Explain.

Geometric Models

57. A runner jogged 4 miles in ten laps around a circular track. What is the greatest distance straight across the track?

58. A person needs to buy a tablecloth for a circular table that has a diameter of 5 feet. The tablecloth is to hang over the edge of the table by 6 inches. A store advertises circular tablecloths with an area of 28.3 square feet. Is the tablecloth the proper size?

59. The perimeter of a rectangular swimming pool is 180 feet. If the length is twice the width, how far is it across the diagonal of the pool? (Hint: Use the Pythagorean Theorem.)

60. Suppose you want to put fringe around a circular tablecloth whose diameter is 6 feet. If you already have 6 yards of fringe, do you need to buy more? Explain.

61. Suppose you have two fish tanks. One tank is 5 feet by 2 feet, and the depth of the water is 18 inches. Will the water be deeper if you pour it into the other tank, which is 6 feet by 1.5 feet? Explain.

62. A regulation baseball has a circumference between 9 and $9\frac{1}{4}$ inches. Between what two values is the radius of the baseball?

63. The front of a concrete dam is in the shape of a trapezoid and has dimensions as shown in the accompanying figure. If the surface area of the front of the dam is 482,850 square feet, what is the height of the dam?

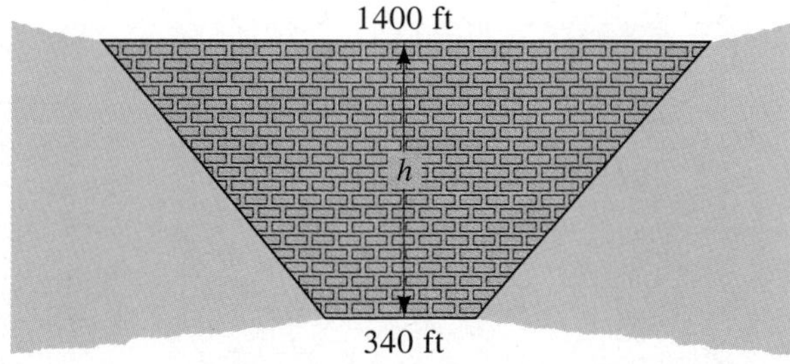

64. A garden in the shape of a trapezoid with the dimensions shown in the accompanying figure has an area of 1386 square feet. What is the length of the side perpendicular to the bases?

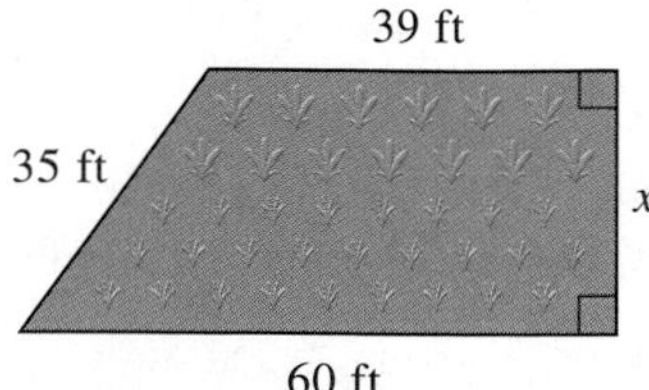

65. Determine a formula for the surface area A of the rectangular box shown in the accompanying figure. (The surface area is the sum of areas of the six sides.)

Figure for 65

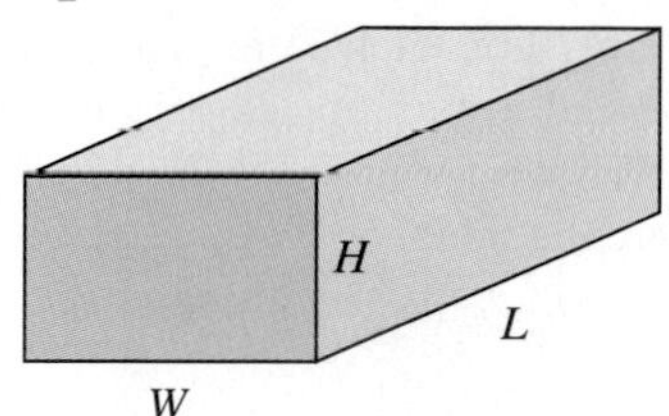

66. If a couple decides to paddle their canoe directly across (through the center of) a circular pond whose circumference is 200 meters, how far will they travel?

67. A city's building code requires a certain building to have 36,000 square feet of parking. The developer plans a trapezoidal parking lot. Complete the accompanying table to show some possible dimensions of the lot. (Dimensions in the table are in feet.)

b_1	b_2	h
330	120	(a)
(b)	170	240
(c)	210	180
300	(d)	150

68. The organizers of a weekend festival need a cylindrical tank that holds 300 cubic feet of water. To the nearest hundredth of a foot, what must be the height for the given radius of the base? (The volume of a right circular cylinder is $V = \pi r^2 h$.)

(a) 4 feet (b) 30 inches (c) 1 yard

Real-Life Applications

69. A consumer affairs investigator is checking advertised prices and discounts in area stores. If the discount rate is r, write a formula for the discount price D for an item whose regular price was R. Then use the formula to complete the table.

Discount Price	Regular Price	Discount Rate
(a)	$150.25	32%
$ 42.94	$ 56.50	(b)
$ 29.47	(c)	65%
$187.17	(d)	15%
$257.04	$459.00	(e)

70. A stock manager for a department store is responsible for pricing items. For each item, she uses the wholesale price W and a markup rate r to establish the retail price R. Write a formula for R, and use the formula to complete the table.

Retail Price	Wholesale Price	Markup Rate
(a)	$123.50	18%
$ 81.13	(b)	22%
$442.54	$406.00	(c)
$319.29	$220.20	(d)
$ 20.77	(e)	34%

In Exercises 71 and 72, solve the formula for the appropriate variable. Use your calculator for the repeated evaluations.

71. A satellite travels in a circular orbit of the earth. The circumference of the earth is approximately 25,000 miles. (The formula for the circumference C of a circle with radius r is $C = 2\pi r$.) How high above the earth is the satellite if the length of one orbit is

(a) 25,800 miles?

(b) 26,000 miles?

(c) 26,500 miles?

72. Simple interest I on an investment P is calculated by the formula $I = Prt$, where r is the interest rate and t is the time in years. As a financial advisor, you have gathered the information shown in the following table to assist a client in deciding on an investment. Which investment has the best interest rate?

Amount Invested	Time	Interest
(i) $10,000	6 months	$375.00
(ii) $ 5,000	5 months	$170.83
(iii) $ 5,000	3 months	$100.00
(iv) $ 7,500	8 months	$350.00
(v) $ 8,000	1 year	$640.00

Modeling with Real Data

73. The liquid hydrogen tank of the space shuttle has a volume of approximately 1450 cubic meters and a diameter of approximately 8.4 meters. (Source: NASA.)

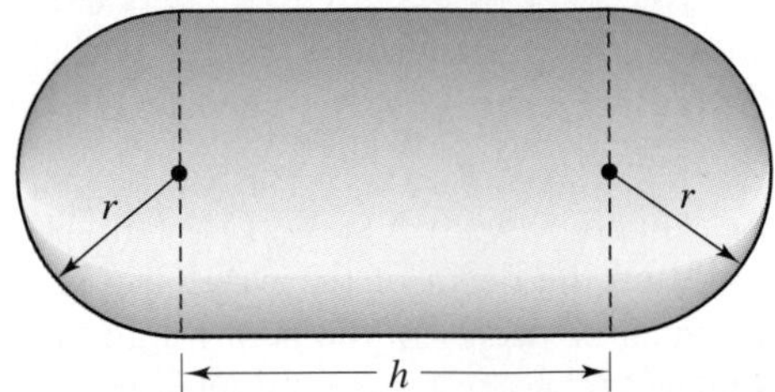

Use the formula $V = \frac{4}{3}\pi r^3 + \pi r^2 h$ to estimate the height h of the cylindrical part of the tank.

74. The wind speed of a storm is often given in *nautical* miles per hour. The accompanying table shows some notable storms with wind speeds given in either nautical or statute miles per hour. (Source: National Oceanic and Atmospheric Administration.)

Storm	Nautical mph	Statute mph
Beulah		109
Carla	126	
Donna		130
Juan	55	

The formula $N \approx 0.869M$ relates statute miles M to nautical miles N. Use this formula to complete the table.

Group Project: Wal-Mart Stores

From a modest beginning, Wal-Mart Stores has risen to become one of the largest retailing chains in America. The bar graph on the next page shows the number of stores at the end of selected years in the period 1985–1995. (Source: Wal-Mart Stores, Inc.) The solid line in the figure shows the graph of a model function $y = 127x + 750$, where y represents the number of stores at the end of year x, and x is the number of years since 1985.

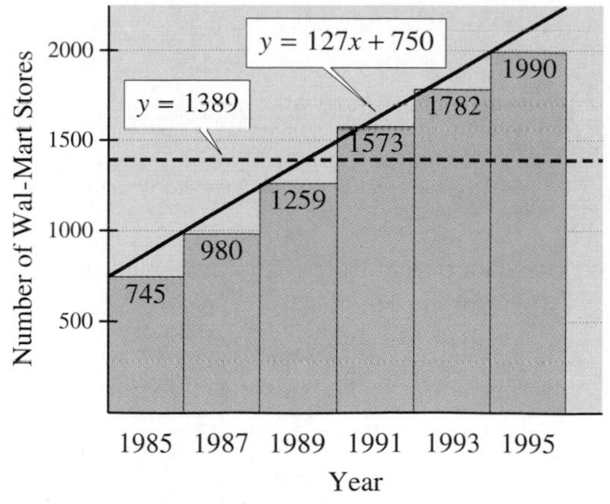

$y = 127x + 750$
$y = 1389$
745 980 1259 1573 1782 1990
Number of Wal-Mart Stores: 500 1000 1500 2000
1985 1987 1989 1991 1993 1995
Year

75. Based on just a visual inspection of the figure, how accurate do you judge the model function to be?

76. For each of the years shown in the figure, calculate the quantity (actual data − model data). Then calculate the average of these six differences. Does the small result indicate that the model is accurate?

77. Suppose that we model the data with the function $y = 1389$. The graph of this function is shown as a dashed line in the figure. Using this new model, repeat the calculations in Exercise 76. Is the average of the differences small? Would you conclude that $y = 1389$ is a good model of the data?

78. Consider the impact chain stores such as Wal-Mart have had on large urban department stores. How would the trend in the number of such urban stores compare with the trend indicated in the figure?

Challenge

In Exercises 79 and 80, solve the given equation for x.

79. $a(x + c) = b(x - c)$

80. $\dfrac{2x - 1}{x + 3} = y$

In Exercises 81–86, solve the given formula for the indicated variable.

81. $S = \dfrac{a}{1 - r}$ for r

82. $F = \dfrac{mv^2}{r}$ for r

83. $\dfrac{P_1V_1}{T_1} = \dfrac{P_2V_2}{T_2}$ for V_2

84. $E = IR + Ir$ for I

85. $A = 2LW + 2LH + 2WH$ for L

86. $A = P + Prt$ for P

3.4 MODELING AND PROBLEM SOLVING

A General Approach to Problem Solving • Numbers • Piece Lengths • Angles of a Triangle • Rectangles • Things of Value

A General Approach to Problem Solving

From a practical standpoint, it makes little sense to learn mathematics and not apply it to anything. This section and the next provide the opportunity to bring together several skills and to use them in solving various kinds of application problems.

Many of the problems involve common, everyday situations. Others, while not as realistic, are important because they help you obtain a conceptual understanding of how mathematics can be used to model a problem and lead to the development of problem-solving strategies.

Although we will illustrate various types of application problems, real-life problems rarely fall into a specific category. It is extremely important to concentrate on the general approach and to develop skills and strategies that can be applied to any problem.

The following is a summary of a general approach to problem solving. We will elaborate on these strategies as we discuss examples.

A General Approach to Problem Solving

1. On your first reading of the problem, make sure you understand what the problem is about and what question(s) you need to answer.
2. If appropriate, draw a figure or diagram or make a chart or table to organize information.
3. Assign a variable to the unknown quantity. Be specific about what the variable represents. If there are other unknown quantities, represent them in terms of the same variable.
4. Write an equation that describes, in symbols, the information given in the problem.
5. Solve the equation.
6. Use the solution to answer the question(s) asked in the problem.
7. Check the answer(s) by confirming that the conditions stated in the problem are met.

The following examples illustrate how this general approach can be used in a variety of applications. Naturally, there is no way to include every possible type of problem that could ever be encountered. However, these examples will provide a good start toward developing methods that generally will work well for any problem.

Numbers

An important step in writing an equation to model information is translating relationships into mathematical terms. Number problems provide an introduction to these basic relationships.

EXAMPLE 1

Numbers of Friends and Relatives

Of the 20 friends and relatives who attended a holiday party, half the friends and one-quarter of the relatives remained to help clean up. If 7 people helped, how many relatives left without helping?

Solution

Let $x =$ the number of relatives who attended the party. Then $20 - x =$ the number of friends who attended the party.

$\frac{1}{4}x + \frac{1}{2}(20 - x) = 7$	Half the friends and a fourth of the relatives remained.
$x + 2(20 - x) = 28$	Multiply by 4 to clear fractions.
$x + 40 - 2x = 28$	Distributive Property
$-x + 40 = 28$	Combine like terms.
$-x = -12$	
$x = 12$	

Of the 12 relatives who attended, only one-quarter helped. Thus 9 relatives left without helping.

Two or more integers that follow immediately after one another on the number line are called **consecutive integers.** If we let n represent the first integer, the next integers can be described in terms of n.

	Example	*Representation*
Consecutive integers:	4, 5, 6	$n, n + 1, n + 2, \dots$
Consecutive even integers:	$-6, -4, -2$	$n, n + 2, n + 4, \dots$
Consecutive odd integers:	7, 9, 11	$n, n + 2, n + 4, \dots$

Note that consecutive even integers and consecutive odd integers are represented the same way. However, in one case, n represents an even integer, and in the other, n represents an odd integer.

EXAMPLE 2

Consecutively Numbered Lockers

The lockers on the right side of the hall are numbered consecutively with even integers; the lockers on the left side of the hall are numbered consecutively with odd integers. Two people have adjoining lockers on the left side of the hall. They note that their locker numbers have a sum of 60. What are their locker numbers?

Solution

The goal is to determine the two locker numbers. Because the lockers are on the left side of the hall, the numbers are consecutive odd integers.

LEARNING TIP

When you assign a variable, always be precise. If the unknown quantity is the price of a rose, writing x = rose is not meaningful and can be misleading. Remember that a variable represents a number, not a thing.

Let x = the first odd integer and $x + 2$ = the second odd integer.

$$x + (x + 2) = 60 \quad \text{The sum of the locker numbers is 60.}$$
$$x + x + 2 = 60 \quad \text{Remove parentheses.}$$
$$2x + 2 = 60 \quad \text{Combine like terms.}$$
$$2x + 2 - 2 = 60 - 2 \quad \text{Subtract 2 from both sides.}$$
$$2x = 58$$
$$\frac{2x}{2} = \frac{58}{2} \quad \text{Divide both sides by 2.}$$
$$x = 29$$

Because x (the first locker number) is 29, $x + 2$ (the second locker number) is 31. The locker numbers are 29 and 31.

Check: The locker numbers are consecutive odd integers, and their sum is $29 + 31 = 60$.

Piece Lengths

In this type of problem, material of some known length is to be cut into two or more pieces. We are given the relative sizes of the pieces, and the task is to determine the actual length of each piece.

EXAMPLE 3

Lengths of Pieces of Rope

A rope is 72 feet long and is to be cut into three pieces. The first piece must be 8 feet shorter than the second piece, and the third piece must be twice as long as the second piece. How long should the first piece be?

Solution

Although the goal is to determine how long the first piece should be, we will determine all three lengths so that the results can be verified.

Figure 3.15

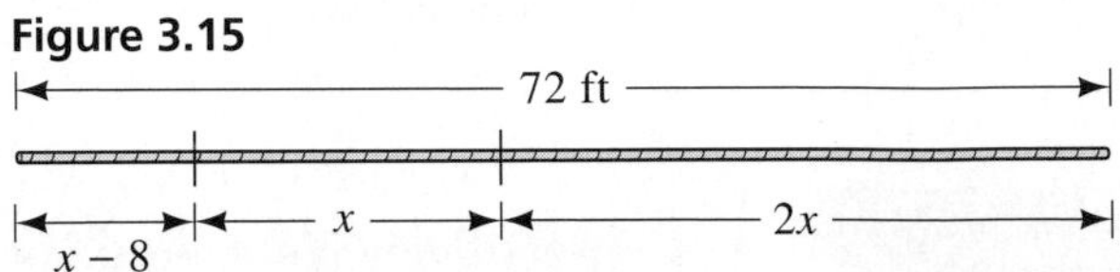

Because the first and third pieces are described in comparison with the second piece, we let the variable x represent the length of the second piece. Then we describe the lengths of the other two pieces in terms of that variable. (See Fig. 3.15.)

$(x - 8) + x + 2x = 72$	The sum of the piece lengths is 72 feet.
$x - 8 + x + 2x = 72$	Remove parentheses.
$4x - 8 = 72$	Combine like terms.
$4x - 8 + 8 = 72 + 8$	Add 8 to both sides.
$4x = 80$	
$\frac{4x}{4} = \frac{80}{4}$	Divide both sides by 4.
$x = 20$	

The length of the first piece is $x - 8 = 20 - 8 = 12$ feet. Note that the lengths of the second and third pieces are $x = 20$ feet and $2x = 40$ feet, respectively.

Check:		
First piece:	12	The first piece is 8 feet shorter than the second piece.
Second piece:	20	
Third piece:	40	The third piece is twice as long as the second piece.
Total length:	72	

Think About It

Suppose that a 28-foot wire is to be cut into four pieces and that the lengths of the first three pieces are consecutive integers. Can you determine the lengths of the four pieces? Could you do it if the length of the wire were 7 feet?

LEARNING TIP

Try to establish connections in applications. For example, problems involving three piece lengths and the angles of a triangle are essentially the same in that the sum of the lengths or measures is known.

Angles of a Triangle

Problems of this kind usually describe the relative measures of the three angles of a triangle. The task is to determine the measure of each angle. The basis for the equation in such problems is the theorem from geometry that states that the sum of the measures of the angles of a triangle is 180°.

EXAMPLE 4

The Measures of the Angles of a Triangle

A surveyor is mapping a plot of land in the form of a triangle. The second angle of the triangle is 21° greater than the first angle. The third angle is 11° more than twice the first angle. What are the measures of the three angles?

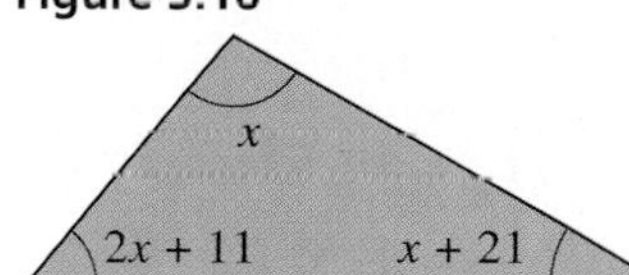

Figure 3.16

Solution

The goal is to determine the measures of all three angles of the triangle.

Because the measures of the second and third angles are described in comparison with the measure of the first angle, we assign the variable x to the measure of the first angle. Figure 3.16 shows how the measures of the other two angles are represented.

$$x + (x + 21) + (2x + 11) = 180$$ The sum of the measures of the angles of a triangle is 180°.

$$x + x + 21 + 2x + 11 = 180$$ Remove parentheses.

$$4x + 32 = 180$$ Combine like terms.

$$4x + 32 - 32 = 180 - 32$$ Subtract 32 from both sides.

$$4x = 148$$

$$\frac{4x}{4} = \frac{148}{4}$$ Divide both sides by 4.

$$x = 37$$

The measure of the first angle is 37°. The second angle is $x + 21 = 37 + 21$, or 58°. The third angle is $2x + 11 = 2 \cdot 37 + 11$, or 85°.

Check:			
	First angle:	37°	
	Second angle:	58°	The second angle is 21° greater than the first angle.
	Third angle:	85°	The third angle is 11° more than twice the first angle.
	Total:	180°	

Rectangles

The basic quantities involving a rectangle are the length, width, perimeter, and area. Rectangle problems typically describe the relationship between the length and width and provide either the perimeter or the area.

EXAMPLE 5

Carpet for a Rectangular Room

A carpet layer has been to a home to provide an estimate of the cost of carpeting a rectangular dining room. When he returns to his office, he finds that he has misplaced the dimensions. He remembers that the perimeter is 52 feet and that the width is 4 feet less than the length. If the carpet costs $12.95 per square yard, how can he still provide the estimate, and what is it?

Figure 3.17

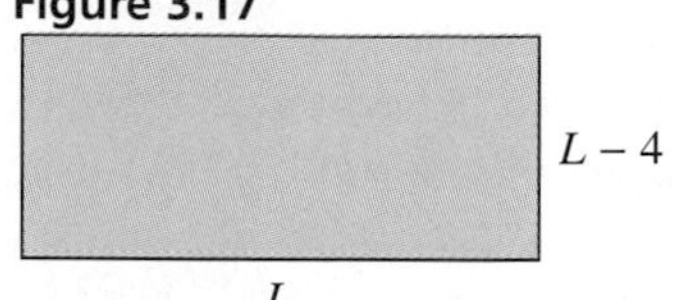

Solution

To provide a cost estimate, the carpet layer must determine the dimensions of the room.

As shown in Fig. 3.17, we let L represent the length and $L - 4$ the width. Note that we could have let W represent the width and $W + 4$ the length.

$$2L + 2W = P$$ Formula for the perimeter of a rectangle

$$2L + 2(L - 4) = 52$$ Replace W with $L - 4$. The perimeter P is 52 feet.

$$2L + 2L - 8 = 52$$ Distributive Property

$$4L - 8 = 52 \quad \text{Combine like terms.}$$
$$4L - 8 + 8 = 52 + 8 \quad \text{Add 8 to both sides.}$$
$$4L = 60$$
$$\frac{4L}{4} = \frac{60}{4} \quad \text{Divide both sides by 4.}$$
$$L = 15$$

The length of the dining room is 15 feet, and the width is $15 - 4$, or 11, feet. The area A is $A = LW = 15(11) = 165$ square feet, or about 18.3 square yards. At \$12.95 per square yard, the cost estimate is 12.95(18.3) or about \$237.

Check: Length: 15

Width: 11 The width is 4 feet less than the length.

Perimeter: $2(15) + 2(11) = 52$

Things of Value

There are many applications involving things of value. The objects may be coins, tickets to a play, or pounds of bananas.

The problems usually involve more than one of an item. Typically, the per-unit value is given, as is the total value of all the items. The total value for a collection of items is determined by multiplying the number of items times their per-unit value. To organize the information in the problem, a table is particularly useful.

EXAMPLE 6

Two Kinds of Tickets with Different Values

A teacher organized a field trip to a zoo for students and their parents. The admission was \$2.00 for students and \$4.75 for adults. If there were six fewer adults than children on the trip and the total cost was \$93.00, how many adults went on the trip?

Solution

The question asks for the number of adults who went on the trip, but to check our results, we also will determine the number of children who went.

	Number of Tickets	Per-Ticket Cost	Cost of Tickets
Children	x	2.00	$2.00x$
Adults	$x - 6$	4.75	$4.75(x - 6)$
Total			93.00

The table shows how we have assigned the variable and how we have represented other unknown quantities.

$2.00x + 4.75(x - 6) = 93.00$	The total cost of the tickets is \$93.00.
$2.00x + 4.75x - 28.50 = 93.00$	Distributive Property
$6.75x - 28.50 = 93.00$	Combine like terms.
$6.75x - 28.50 + 28.50 = 93.00 + 28.50$	Add 28.50 to both sides.
$6.75 = 121.50$	
$\dfrac{6.75x}{6.75} = \dfrac{121.50}{6.75}$	Divide both sides by 6.75.
$x = 18$	

The number of children who went to the zoo is 18. The number of adults is $18 - 6 = 12$.

Check: There were six fewer adults (12) than children (18).

Cost of children's tickets:	18(2.00) =	\$36.00
Cost of adults' tickets:	12(4.75) =	\$57.00
Total cost:		\$93.00

Note that the monetary units used in Example 6 were dollars. The units could have been cents to avoid the decimals, but the arithmetic involved in solving the equation is virtually the same either way. When solving applications involving monetary units, just be sure your units are the same in all expressions.

An application problem can have explicit or implied conditions that cannot be met. The following are some examples of applications with solutions that must be disqualified.

Application Problem	*Disqualified Solution*
Consecutive integers	Not an integer
Consecutive even integers	Odd integer
Consecutive odd integers	Even integer
Dimensions of a figure	Not a positive number
Number of things of value	Not a whole number
Time, rate, or distance	Not a positive number

Remember that a solution of an equation is not necessarily the answer to the question. Make sure your answer meets all the conditions of the problem.

3.4 EXERCISES

Real-Life Applications

Numbers

1. Suppose that the total of two numbers is T. If one of the numbers is n, how can you represent the other number in terms of n? Show that the sum of these representations is T.

2. Explain why we should check the answer to an application problem by verifying that the answer meets the stated conditions of the problem rather than by substituting into the original equation.

3. Increasing a number by 5 and multiplying the result by 7 is the same as multiplying the number by -5 and subtracting 1 from the result. What is the number?

4. The difference of three times a number and seven less than the number is one less than the number. What is the number?

5. A bakery offers 32 types of cakes and cookies. If there are 6 more types of cookies than cakes, how many types of cakes are available?

6. On a placement test, a student correctly answered the same number of one-point questions as two-point questions. If the total score was 57 points, how many two-point questions did the student answer correctly?

7. A history class has 12 more men than women. The same number of men and women made an A on the first exam. If half the women and one-third of the men made an A on the first exam, how many men were in the class?

8. A car dealer began the month with the same number of used and new cars. After buying three more used cars and selling two of the new cars, the number of new cars was $\frac{4}{9}$ the number of used cars. How many cars did the dealer have at the beginning of the month?

9. Explain why x and $x + 2$ can be used to represent either two consecutive *even* integers or two consecutive *odd* integers.

10. For three consecutive integers, the sum of the smallest and largest integers equals twice the middle integer. What are the integers? Explain the result.

11. A student's scores on two 50-point quizzes were consecutive integers, and the total number of points was 95. What was the score on the second quiz?

12. For a four-number lottery, a person's first three picks were consecutive odd integers, and the fourth was the sum of the first three numbers. If the fourth number was 45, what was the third number?

13. The number of red, white, and yellow flowers available for floral arrangements are consecutive even integers. All the red flowers and half the yellow flowers are in one container, and all the white flowers are in a second container. The number of flowers in the first container exceeds the number in the second container by 11. How many yellow flowers are in the first container?

14. The number of the grades D, C, B, and A in a class were consecutive integers. The number of D's was the same as the sum of one-third the number of A's and half the number of C's. How many students received a B?

Piece Lengths

15. A 45-foot rope is to be cut into three pieces. The second piece must be twice as long as the first piece, and the third piece must be 9 feet longer than three times the length of the second piece. How long should each of the three pieces be?

16. A 54-foot log is to be cut into three smaller logs whose lengths are consecutive even integers. How long should each piece be?

17. A 105-yard bolt of cloth is to be cut into three lengths. The second piece is to be 5 yards shorter than the first piece, and the third piece is to be twice the length of the second piece. What are the lengths of the three pieces?

18. A 32-foot pipe is cut into three pieces. The first two lengths are consecutive integers, and the third piece is 6 feet longer than the second. How long are the three pieces?

Angles of a Triangle

19. One angle of a triangular banner is 20° greater than the first angle. The third angle is twice as large as the first angle. What are the measures of the three angles?

20. The second angle of a triangular tarpaulin is 7° more than three times the first angle. The third angle is 12° greater than the second angle. What are the measures of the three angles?

21. Three walkways intersect to enclose a triangular grassy area. If the second angle of the triangular area is twice the first angle, and the third angle is one and one-half times the second angle, determine whether the triangle is a right triangle.

22. An **isosceles triangle** is a triangle with at least two angles having the same measure. (See figure.) If the first angle of a triangular highway warning sign is 30° smaller than the third angle, and the third angle is 20° less than twice the second angle, determine whether the triangle is an isosceles triangle.

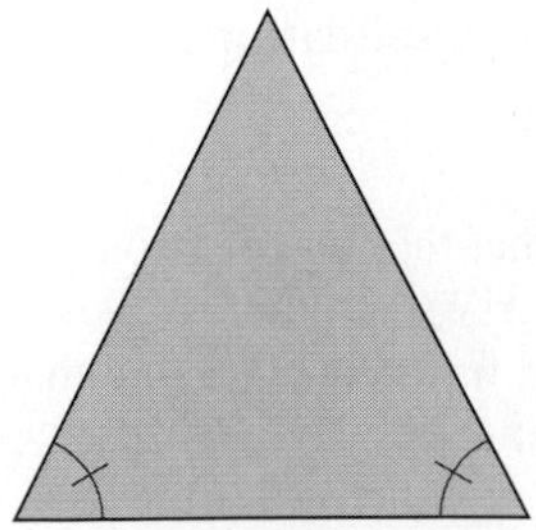

Rectangles

23. The width of a rectangular garage is 8 feet less than the length. If the perimeter of the garage is 64 feet, what are the dimensions?

24. The width of a rectangular real estate sign is 16 inches less than the length. If the perimeter of the sign is 88 inches, what are the dimensions?

25. The length of a rectangular piece of carpet is 4 yards greater than the width. If 60 yards of fringe border is required for the edge of the carpet, what are the dimensions of the carpet?

26. Suppose that 134 meters of fencing are required to enclose a rectangular animal feed lot. If the length of the lot is 7 meters greater than the width, what are the dimensions of the lot?

Things of Value

27. A Salvation Army collection kettle contains \$2.25 in nickels and dimes. If the kettle contains nine more nickels than dimes, how many nickels are in the kettle?

28. The coin box of a gum vending machine, which accepts only dimes and quarters, contains \$5.20. If the number of quarters is four more than the number of dimes, how many quarters are in the box?

29. If a change dispenser contains 32 coins consisting of only dimes and quarters worth \$5.15, how many dimes are in the dispenser?

30. If a Lions Club mint box contains 36 coins consisting of only nickels and dimes worth \$2.50, how many nickels are in the box?

Geometric Models

31. A supporting guy wire is attached to the top of a utility pole and is fastened at the ground 10 feet from the base of the pole. (See figure.) If the angle between the guy wire and the ground is eight times the angle between the guy wire and the pole, what is the measure of the angle between the guy wire and the ground?

Figure for 31

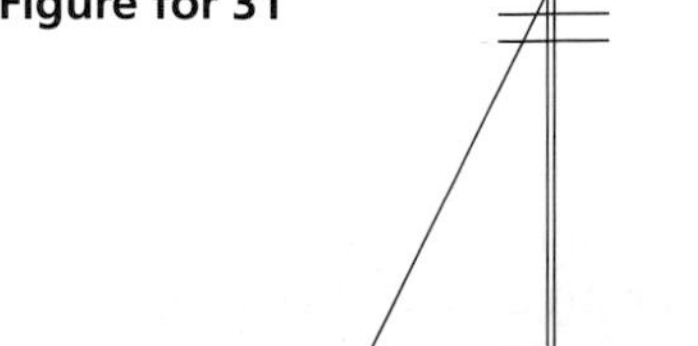

32. An Atlanta Braves pennant is a triangle with two of the angles having the same measure. The third angle is one-fifth the sum of the two equal angles. What is the measure of the smallest angle?

33. The height of a book is 4 inches greater than its width. If the perimeter of the book is 48 inches, can the book be placed upright on a bookcase shelf 12 inches high?

34. The width of one poster is 7 inches more than half the length. The width of a second poster is 9 inches less than the length of the first poster, and the length of the second poster is 10 inches greater than its width. If the two posters have the same perimeter, which poster is wider?

35. An art student drew a preliminary sketch for a painting on a small rectangular pad that was 5 inches longer than it was wide. To improve the scale of her drawing, she switched to another sketch pad that was twice as long as the smaller pad and 8 inches wider. If the perimeter of the larger pad was 60 inches, how long was the smaller pad?

36. The width of a sheet of paper is 2.5 inches less than its length. A photocopy is made that is 90% of the original dimensions. The reduced copy has a perimeter of 35.1 inches. How wide was the original paper?

37. Curtains 40 inches long are to be hung on a rectangular window. The height of the window is one and a half times its width, and the perimeter of the window is 150 inches. Will the curtains cover the full height of the window?

38. A computer is 2 inches taller than it is wide. The perimeter of the front of the computer is 60 inches. Will the computer fit under a shelf that is 18 inches above the top of the desk?

39. An **obtuse triangle** is a triangle with one angle having a measure that is greater than 90°. (See figure.) If one angle of a triangle is four times the second angle, and the third angle is 2° less than twice the second angle, determine whether the triangle is an obtuse triangle.

Figure for 39

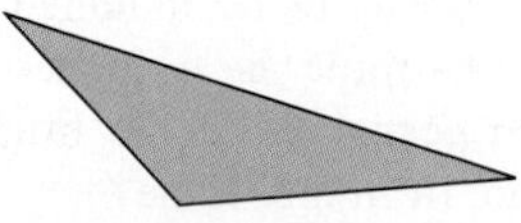

40. An **acute triangle** is a triangle with all three angles having measures that are less than 90°. (See figure.) If the second angle of a triangle is six-sevenths as large as the third angle, and the first angle is five-sixths as large as the second angle, determine whether the triangle is an acute triangle.

Figure for 40

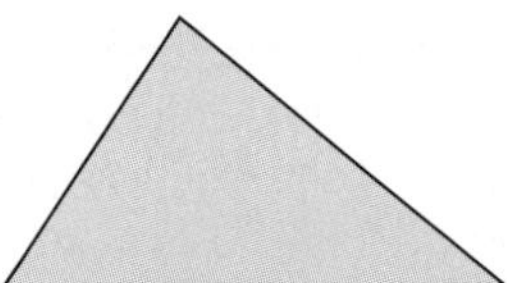

41. Two angles are **supplementary angles** if the sum of their measures is 180°. What is the measure of the smaller of two supplementary angles if the measure of the larger is 10.8° more than twice the measure of the smaller?

42. Two angles are **complementary angles** if the sum of their measures is 90°. What is the measure of the larger of two complementary angles if the measure of the smaller angle is 15° less than half the measure of the larger angle?

43. A ladder rests against a wall. (See figure.) The angle made by the ladder and the ground is 6° less than seven times the angle made by the ladder and the wall. What is the measure of the angle made by the ladder and the wall?

Figure for 43

44. A 9-inch by 12-inch picture is to be mounted in a frame. The width of the frame is 5 inches less than the length, and the perimeter of the frame is 42 inches. Will the picture fit in the frame?

Miscellaneous Applications

45. For any three consecutive integers, show that the average of the smallest and largest integers is the middle integer.

46. Show that the absolute value of the difference between any two consecutive integers is always 1.

47. The sum of three consecutive even integers is 19. What are the integers? Could you have anticipated your answer without solving an equation?

48. The sum of three consecutive odd integers is 19. What are the integers? Can the conditions of the problem be met? Explain.

49. A couple decided to keep a record of the number of miles they walked during the year. By July 1, the woman had walked four times as many miles as the man. If each walks an additional 80 miles by the end of the year, the woman will have walked only twice as many miles as the man. How many miles had the woman walked by July 1?

50. A small library had 5 times as many nonfiction books as fiction books. After adding 140 volumes of each type book, the library now has three times as many nonfiction books as fiction books. What is the total number of volumes in the expanded collection?

51. Two couples took a business associate to dinner. The first couple agreed to pay $\frac{3}{5}$ of the bill if the second couple paid $\frac{2}{5}$ of the bill and the tip of 15%. If the second couple paid $74.58, how much did the first couple pay?

52. A child saved $3.55 in nickels and dimes. If the child has ten fewer nickels than dimes, how many nickels does the child have?

53. A utility pole rises 10 feet above the trees. Half the entire length of the pole is hidden by the trees, and one-sixth of the entire length of the pole is in the ground. How long is the pole?

54. The first angle of a triangular stained-glass window ornament is 10° smaller than the second angle. The third angle is 10° less than twice the second angle. Show that the shape of the ornament is a right triangle.

55. A 21-inch strip of metal is bent in the shape of an isosceles triangle to make a frame for a stained-glass window decoration. If the length of the base is 3 inches less than twice the length of one of the other two sides, what are the dimensions of the frame?

56. The police use 88 feet of crime tape to enclose a rectangular crime scene. If the width of the location is 2 feet more than half the length, what is the width of the crime scene?

57. On a certain day, the volume of first-class mail at a small post office was seven times the volume of all other mail. The next day, when the amount of first-class mail decreased by 80 and the volume of other mail increased by 40, the volume of first-class mail was only four times that of all other mail. How many pieces of mail did the post office handle the first day?

58. One weekend, the Saturday attendance at an outdoor craft show was three times the Sunday attendance. On the next weekend, rain caused the Saturday attendance to drop by 1300 people, but the Sunday attendance rose by 400. Nevertheless, the Saturday attendance was twice the Sunday attendance. What was the total weekend attendance on the second weekend?

59. There are 39 students in a political science class. The instructor notices that there are half as many men as women. How many women are in the class?

60. An uncle wills $3000 to his two nieces and one nephew. The older niece is to receive twice what the younger niece receives, and the nephew is to receive an amount equal to the sum of the nieces' amounts. How much will each person receive?

61. In a poker game two people began with equal sums of money. When the loser lost $50 more than a quarter of his money, the winner had twice as much money as the loser. How much money did each person have at the end of the game?

62. An appliance company had four more washers than dryers in its inventory. On Friday half the dryers and one-third of the washers were shipped out. If a total of eight appliances were shipped, how many were dryers?

63. A retail store ordered some sofas and chairs from a manufacturer. The number of chairs ordered was four more than the number of sofas. The store received a partial shipment consisting of half the sofas and one-third of the chairs that were ordered. If the total number of pieces was the same as the number of sofas originally ordered, how many sofas did the store receive?

64. Some guests had arrived at a party by 8:00. At 8:30, two more arrived. By 9:00, the 21 guests in attendance were three times the number who were there at 8:30. How many had arrived by 8:00?

65. By 10:00 A.M., the Dow-Jones average had dropped from the opening average. By noon, the Dow-Jones average slipped another 8 points. At the close of trading, the 50-point drop was five times the amount by which the average had dropped at noon. By how much had the average dropped at 10:00 A.M.?

66. The government's farm price support program pays a farmer not to plant any crops this year, so he must decide how not to plant his 100 acres. He decides not to plant corn on three times as many acres as he will not plant strawberries. The number of acres on which he will not harvest hay is equal to the total number of acres on which he will plant neither corn nor strawberries. On how many acres will the farmer not plant corn?

67. At the sports club, two people have adjacent lockers, and the sum of their locker numbers is 434. Show how you know that the lockers are not numbered consecutively?

68. A precinct worker issues voting passes to three voters consecutively. The serial numbers on the passes total 237. What was the serial number of the second voter's pass?

69. After three college basketball games, a forward, center, and guard have a total of 143 points. The forward has scored the most points, and the guard has scored the fewest. The forward's and center's points are consecutive integers. The guard's and center's points are consecutive even integers. How many points has the forward scored?

70. A bookmark is placed in a book between two pages whose sum is 193. On what page did the reader stop reading? (There are two possible answers.)

71. A scout leader is on a camping trip with his troop. He wants to cut a 40-meter rope into three lengths. The second length must be 2 meters longer than the first length. The third length must be 6 meters longer than twice the first length. What are the three lengths of rope?

72. A fabric store worker has a 106-yard bolt of cloth that is to be cut into three lengths. The second piece is to be 4 yards longer than the first. The third piece is to be three times as long as the second piece. What is the length of each piece?

73. The sponsors of a Boys and Girls Club took some members to a play and bought seven more tickets for members than for adults. The members' tickets cost \$1.25, and the adults' tickets cost \$2.75. If a total of \$28.75 was paid for all the tickets, how many members went to the play?

74. The total receipts for a college basketball game were \$674, with a student ticket selling for \$1.25 and a nonstudent ticket selling for \$2.75. If there were 136 more students than nonstudents in attendance, how many people were at the game?

75. Hamburgers sell for \$2 each and hot dogs sell for \$1.25 each. On a certain day, the number of hamburgers sold was twice the number of hot dogs sold. Determine the number of hot dogs sold if the total income was \$304.50.

76. Mama's Pizzeria sells individual pizzas for \$8 each and an order of bread sticks for \$1 each. During a 2-week period, the number of pizzas sold was four times the number of bread stick orders sold. How many pizzas were sold if the total revenue was \$2970?

77. A person invested one-fourth of an inheritance in bank stock, one-fifth in bonds, and one-half in a mutual fund. If the sum of her investments was \$38,000, how much did she inherit?

78. A commercial developer reads the zoning laws on how he must allot his 50 acres of land. The number of acres of lawn and landscaping must equal the number of acres used for buildings. The number of acres used for parking must be two-fifths the number of acres used for buildings. The number of acres for roads and utilities must be one-tenth the number of acres for lawn and landscaping. How many acres of parking lot must the developer provide?

79. There are 266 men, women, and children at the company picnic. There are four times as many men as children and twice as many women as children. How many of each came to the picnic?

80. Three adjacent offices have office numbers that are consecutive odd integers. The manager is in a fourth office with a number that is 13 more than the number of the middle office. If the sum of the four office numbers is 97, what is the manager's office number?

81. A member of a three-person bowling team brags to his wife that he bowled over 200 last night, but he will not give the actual score. His wife asks another member of the team about it, but she learns only that the team's three scores were consecutive odd integers. His wife then checks with the third member of the team, who says that the team score was 597. Could her husband have bowled over 200? Can she know for sure? Explain.

82. Soft drinks sell for 75 cents each, and milk shakes sell for \$1.25 each. On a certain day, the number of soft drinks sold was three times the number of milk shakes sold. How many soft drinks were sold if the total income was \$280?

Modeling with Real Data

83. The accompanying figure shows that the average debt for law school graduates rose to \$37,700 in 1995. (Source: National Law Journal.)

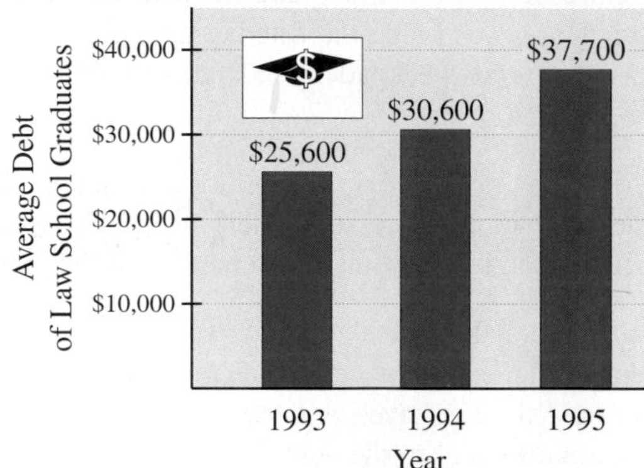

(a) Assuming that the average debt increases by \$6000 for all future years, write an expression to describe the average debt y years after 1995.

(b) Write and solve an equation that can be used to estimate the year in which the average debt will reach \$61,700.

84. By the year 2020, the number of people over age 65 who will be living in New York City is projected to be 3,028,000. This represents about a 25% increase from 1993. (Source: U.S. Census Bureau.) Letting p represent the population of New Yorkers who were over age 65 in 1993, write and solve an equation that models the given information.

Group Project: Carbon Monoxide Emissions

Carbon monoxide emissions rose from 80.3 metric tons in 1940 to a 1970 high of 123.6 metric tons. By 1980 the emissions level had declined to 100 metric tons and by

1990 to 67.7 metric tons. (Source: U.S. Environmental Protection Agency.) The number C of metric tons of emissions can be modeled by $C(x) = -3x + 127$, where x is the number of years since 1970.

85. Write an equation that you can use to predict the year in which emissions are expected to be half the 1980 level. Then estimate the solution by graphing.

86. Solve the equation in Exercise 85 algebraically.

87. Use the graph to estimate the average decrease in carbon monoxide emissions per year.

88. Why do you think carbon monoxide emissions rose from 1940 to 1970 but then decreased significantly from 1970 to 1990?

Challenge

89. George P. Burdell enjoys entertaining friends with number games. In one game he asks a friend to choose any number. Next, he tells his friend to subtract one from the number, multiply the result by three, and then subtract twice the number increased by one. Finally, he tells his friend to add five to the result. George asks what the resulting number is and tells his friend that it is the chosen number. Show that George's trick works for any number.

90. George P. Burdell enjoys entertaining friends with number games. In one game he asks a friend to choose any number and add eight to it; multiply the result by three and then subtract three from the result; multiply by two and then divide by six; and finally, subtract the original number. George knows that the result is 7. Show that George's trick works for any number.

3.5 APPLICATIONS

Percentages • Fixed and Variable Costs • Simple Interest • Distance, Rate, and Time • Mixtures • Liquid Solutions

For most of the application problems in the preceding section, equations were based on information provided by the problem itself.

This section deals with problems for which certain common knowledge is needed. In a sense, the use of this common knowledge makes the applications in this section more practical.

Although checking answers continues to be an important step in the general approach to problem solving, we will leave this step to you in the examples that follow.

Percentages

Ad valorem taxes are taxes based on the value of property. The valuation of the property, for tax purposes, is usually some percentage of the market or list value of the property. The ad valorem tax is then some percentage of the valuation.

EXAMPLE 1 Ad Valorem Tax on an Automobile

Suppose the valuation of an automobile for tax purposes is 40% of the list value of the car, and the ad valorem tax rate is 2%. If the tax due on an automobile is \$27, what is the list value of the car?

Solution

Let x = the list value of the car.

$$\text{Valuation} = 0.40x \qquad \text{Valuation} = 40\% \text{ of list value}$$

$$\text{Tax due} = (0.02)(0.40x) = 0.008x \qquad \text{Tax due} = 2\% \text{ of the valuation}$$

$$0.008x = 27.00 \qquad \text{The tax due is \$27.}$$

$$\frac{0.008x}{0.008} = \frac{27.00}{0.008}$$

$$x = 3375$$

The list value of the car is $3375.00.

LEARNING TIP

Consider writing a word equation before you attempt to write an algebraic equation. For example, if you know or can represent the three quantities, then writing "original price − discount = sale price" gives you a guide for writing the actual equation.

When an item is placed on sale, the amount by which the price is reduced is called the **discount,** which is expressed as a percentage of the **original price.**

Retail store owners buy their goods at **wholesale prices.** Then, to cover overhead and the profit they want to make, they add an amount called **markup,** which is expressed as a percentage of the wholesale price. The result is the **retail price.**

EXAMPLE 2

Wholesale Price Plus Markup Equals Retail Price

All the goods at a hobby store are marked up 40%. If the retail price of one item is $32.13, what was the wholesale cost to the store owner?

Solution

Let w = the wholesale cost.

$$\text{Markup} = 0.40w \qquad \text{Markup} = 40\% \text{ of wholesale cost}$$

$$w + 0.40w = 32.13 \qquad \text{Wholesale cost + markup = retail price}$$

$$1.00w + 0.40w = 32.13$$

$$1.40w = 32.13$$

$$\frac{1.40w}{1.40} = \frac{32.13}{1.40}$$

$$w = 22.95$$

The wholesale cost was $22.95.

Think About It

Suppose that the wholesale price of an item is w and that the retail price reflects a 30% markup. If the item is later sold at a 20% discount, will the seller make money or lose money?

Fixed and Variable Costs

The cost of doing business can be separated into two components. One is the **fixed cost,** which is a cost that is incurred even if no products are sold or no service is rendered. The other component is the **variable cost,** which is a cost that depends on the number of products produced or sold or on the amount of service rendered.

EXAMPLE 3

The Cost of an Auto Rental

An employee submits an expense account for a 1-day round trip to a town 40 miles away. The expense account includes a car rental for \$82.00. The fixed daily rate is \$28.00 and the mileage rate is \$0.20 per mile. Is there reason to question the expense of the car rental?

Solution

Let M = the number of miles driven.

$\text{Fixed cost} = 28.00$ — Fixed daily rate is \$28.00.

$\text{Variable cost} = 0.20M$ — Mileage rate is \$0.20 per mile.

$$28.00 + 0.20M = 82.00 \quad \text{Fixed cost + variable cost = total cost}$$

$$28.00 + 0.20M - 28.00 = 82.00 - 28.00$$

$$0.20M = 54.00$$

$$\frac{0.20M}{0.20} = \frac{54.00}{0.20}$$

$$M = 270$$

The employee has been charged for 270 miles when the trip should have been only 80 miles. Either the employee or the car rental agency has made a mistake.

Simple Interest

When money is deposited in a bank account, the bank pays for the use of the money. The amount deposited is called the **principal** (P). The amount paid by the bank is called the **interest** (I). Interest is expressed as a percentage (r) of the principal.

Simple interest is interest paid just once at the end of a given time period. To calculate the interest that accumulates over a period of t years, we use the formula $I = Prt$.

EXAMPLE 4

Loan Plus Interest Equals Loan Payoff

A used-car buyer borrows money for 1 year at a simple interest rate of 12%. If the payoff of the loan is \$5040, what was the amount of the original loan?

Solution

Let P = the amount of the loan.

$\text{Interest} = 0.12P$ — Interest is 12% of the loan amount.

$$P + 0.12P = 5040 \quad \text{Loan amount + interest = loan payoff}$$

$$1.00P + 0.12P = 5040$$

$$1.12P = 5040$$

$$\frac{1.12P}{1.12} = \frac{5040}{1.12}$$

$$P = 4500$$

The original loan was \$4500.

In a *dual-investment* problem, a given amount of money is divided into two different investments, each yielding a given interest percentage. With the total amount of interest earned given, the problem is to determine the individual investment amounts.

A table can be very useful in arranging the information for such problems.

EXAMPLE 5

Investments in Two Bond Funds

A total of \$10,000 was invested in two bond mutual funds, a junk bond fund and a government bond fund. The junk bond fund is risky and yields 11% interest. The safer government bond fund yields only 5%. The year's total income from the two investments was \$740. How much was invested in each fund?

Solution

Let x = the amount of money invested in the junk bond fund. The remaining money, $10{,}000 - x$, is the amount invested in the government bond fund.

	Amount Invested	Interest Rate	Interest Earned
Junk bonds	x	0.11	$0.11x$
Government bonds	$10{,}000 - x$	0.05	$0.05(10{,}000 - x)$
Totals	10,000		740

$$
\begin{aligned}
0.11x + 0.05(10{,}000 - x) &= 740 \\
0.11x + 500 - 0.05x &= 740 \\
0.06x + 500 &= 740 \\
0.06x + 500 - 500 &= 740 - 500 \\
0.06x &= 240 \\
\frac{0.06x}{0.06} &= \frac{240}{0.06} \\
x &= 4000
\end{aligned}
$$

The total interest earned is the sum of the interest earned from each fund.

The amount invested in the junk bond fund was \$4000, and $10{,}000 - 4000$, or \$6000, was invested in the government bond fund.

Distance, Rate, and Time

Distance d is related to the speed or rate r and time t by the formula $d = rt$. Again, a table is often useful for organizing the information.

EXAMPLE 6

Distance, Rate, and Time

Two friends drove from Deadwood, South Dakota, to Cody, Wyoming, a distance of 370 miles. The average driving speed of one driver was 16 mph slower than that of the other driver. The faster driver drove for 2.5 hours and the other driver half an hour longer. What was the average speed of each driver?

Solution

Let $r =$ the average speed for the faster driver.

	Rate	Time	Distance
Fast driver	r	2.5	2.5r
Slow driver	$r - 16$	3	$3(r - 16)$
Total			370

$$2.5r + 3(r - 16) = 370 \quad \text{The sum of the distances is 370 miles.}$$
$$2.5r + 3r - 48 = 370$$
$$5.5r - 48 = 370$$
$$5.5r = 418$$
$$\frac{5.5r}{5.5} = \frac{418}{5.5}$$
$$r = 76$$

The faster driver averaged 76 mph, and the slower driver averaged 60 mph.

Mixtures

Mixture problems usually involve the mixing or blending of two items, each with its own unit price or cost. The resulting mixture has a unit value that is *between* the unit prices of the individual items.

Note: The unit price of a mixture is *not* the sum of the unit prices of the items that are mixed.

EXAMPLE 7

A Mixture of Nursery Stock

The owner of a small nursery specializes in iris and day lily starter plants. He has been selling irises at a price of $3.00 per dozen and day lilies at a price of $8.00 per dozen.

A landscaper wants to buy 15 dozen of a mixture of irises and day lilies, and she is willing to pay $5.00 per dozen for the mixture. How many dozen of each type should the nursery owner include in the mixture?

Solution

Let $x =$ the number of dozen irises to include. Then $15 - x =$ the number of dozen day lilies.

The table summarizes the information in the problem.

	Number of Dozen	Price per Dozen	Total Cost
Iris	x	3.00	$3.00x$
Day lily	$15 - x$	8.00	$8.00(15 - x)$
Totals	15	5.00	75.00

$3x + 8(15 - x) = 75$ The total cost of the mixture is $75.

This time we use the graphing method to estimate the solution of the equation. We use the integer setting to produce the graph of $y_1 = 3x + 8(15 - x)$. Then we trace to the point whose y-coordinate is 75. (See Fig. 3.18.)

Figure 3.18

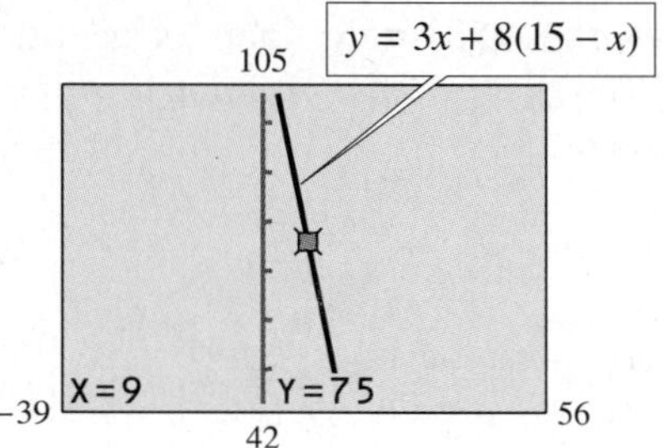

Because $x = 9$ when $y = 75$, we conclude that the mixture should consist of 9 dozen irises and $15 - 9 = 6$ dozen day lilies. This result can be confirmed by solving the equation $3x + 8(15 - x) = 75$.

By tracing to other points along the graph, we can estimate the cost of *any* iris and day lily combination.

Liquid Solutions

When two liquids are mixed together, the result is a **liquid solution.**

Figure 3.19

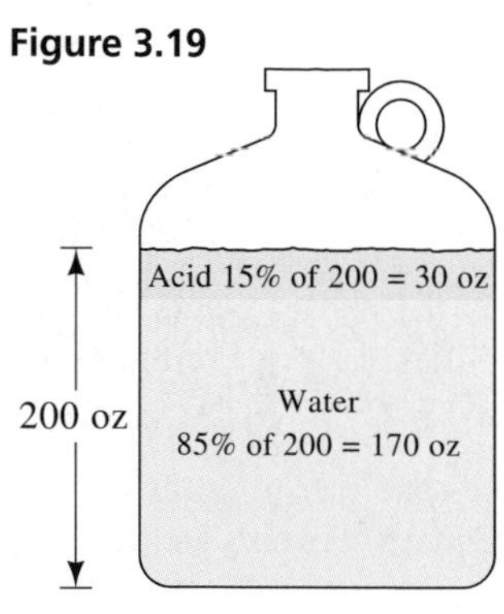

Figure 3.19 shows a bottle containing 200 ounces of a 15% acid solution, meaning that 15% of the volume is pure acid. We say that the **concentration** of acid in the solution is 15%.

Because 15% of the total volume is acid, the number of ounces of acid in the bottle is $(0.15)(200) = 30$ ounces. The remaining 85% of the solution is water, which means that $(0.85)(200) = 170$ ounces of water are in the bottle. Of course, the total is $30 + 170 = 200$ ounces.

Liquid solution problems often involve the mixing of two solutions of different concentrations. The result is a solution whose concentration is between the concentrations of the original two solutions.

Note: When two solutions are mixed, the concentration of the result is *not* the sum of the concentrations of the two solutions.

In Fig. 3.20, 50 liters of a 30% acid solution are mixed with 100 liters of a 75% solution. The result is a 150-liter solution.

Figure 3.20

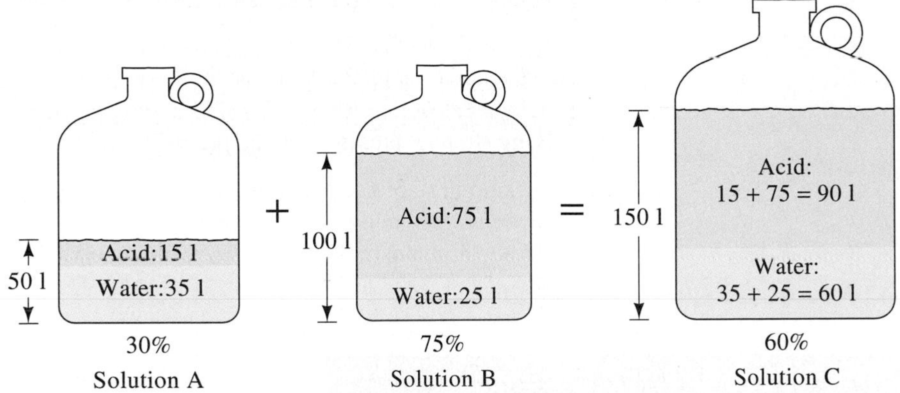

Solution A has (0.30)(50) = 15 liters of acid and solution B has (0.75)(100) = 75 liters of acid. Thus, the resulting solution C will have 15 + 75 = 90 liters of acid. The resulting concentration is 90 ÷ 150 = 0.60 or 60%.

EXAMPLE 8

Mixing Two Acid Solutions

An embossing shop etches metal with acid solutions. The owner is discussing a certain job with the supervisor, and they decide to use 30 ounces of a 40% acid solution.

Upon checking the acid solution supplies, the supervisor has found some 25% acid solution and some 50% acid solution, but there is no 40% acid solution. The owner explains to the supervisor how the two existing solutions can be mixed to obtain the 40% solution. How much of each solution is needed?

Solution

Let x = the number of ounces of the 25% solution. The remaining $30 - x$ ounces will be 50% solution.

The table summarizes the information in the problem.

	Ounces of Solution	Concentration of Acid	Ounces of Acid
25% solution	x	0.25	$0.25x$
50% solution	$30 - x$	0.50	$0.50(30 - x)$
Mixed solution	30	0.40	12

$$
\begin{aligned}
0.25x + 0.5(30 - x) &= 12 \\
0.25x + 15 - 0.5x &= 12 \\
-0.25x + 15 &= 12 \\
-0.25x + 15 - 15 &= 12 - 15 \\
-0.25x &= -3 \\
\frac{-0.25x}{-0.25} &= \frac{-3}{-0.25} \\
x &= 12
\end{aligned}
$$

The total amount of acid in the mixture is the sum of the amounts of acid in each individual solution.

By mixing 12 ounces of the 25% solution and 30 − 12 or 18 ounces of the 50% solution, the supervisor obtains the required 30 ounces of 40% solution.

Although the sample problems in these sections have been organized into typical problem categories, they are offered only as illustrations of the general problem-solving strategy presented in the preceding section.

The examples are not meant to suggest that every problem can be wedged into some known category and solved according to a fixed routine. It is more important to *think* about the problem rather than to seek out quick, ready-made prescriptions.

3.5 EXERCISES

Real-Life Applications

Percentages

1. How do we translate the word *of* in mathematics? Translate the phrase *15% of 30* into a mathematical expression.

2. Suppose there is a 5% sales tax in your area. In which equation below does x represent the cost of an item *before* the sales tax? In which equation does x represent the cost *including* the sales tax?

   (i) $(1.05)(32.00) = x$ (ii) $1.05x = 32.00$

3. A boy delivers newspapers for The Daily News, Inc., and he keeps 12% of all the money he collects. One week he kept $28. How much money did he turn in to The Daily News, Inc.?

4. Washington County collects a 6% sales tax. If the tax on a new car is $929.88, what is the total amount paid for the car? (Ignore all other fees associated with the purchase.)

5. On Monday an investor bought some stock. By Wednesday the stock had increased in value by 4%. However, on Friday the stock dropped 4% from Wednesday's value. The investor immediately sold the stock for $349.44. What did the investor pay for the stock? Did she make money, lose money, or break even?

6. A student correctly answered 15 test questions of equal value. If his percentage grade was 83% (rounded off), how many questions were on the test?

7. Each month a worker's salary is reduced by 7% for Social Security, 20% for federal taxes, 6% for her pension, and 7% for state taxes. If the worker's monthly take-home pay is $1200, what is her monthly salary?

8. A person's weekly wages are reduced by 36% for taxes and other deductions. He decides to spend his entire week's take-home pay on some stereo equipment costing $220 plus a 4% sales tax. What is this person's weekly pay?

9. A $60 blouse is on sale for $48. What is the discount percentage?

10. A salesperson receives a base salary of $360 a week and a commission of 4% on all sales. What was the amount of his sales during the first week of December if his weekly check was for $500?

11. A teapot is on sale for $17.50. If the teapot is marked down 30%, what was the original price?

12. A coffee maker is on sale for $30.60. If the price of the coffee maker is marked down 15%, what was the original price?

13. A sign in a store window reads, "Everything marked down 20%." A customer found a suit on sale for $239.20. What was the price of the suit before the markdown?

14. Cardigans Galore and The Sweater Shop both claim that their goods are priced exactly the same. When The Sweater Shop held a "20% off" sale, a shopper found a sweater on sale for $68.00. Cardigans Galore sells the same sweater for $80.00. Was the original price at The Sweater Shop the same as the current price at Cardigans Galore?

Fixed and Variable Costs

15. Classify each of the following costs of running a business as a fixed cost or a variable cost.
    (a) Expenditures for materials
    (b) Annual insurance
    (c) Electric bill
    (d) Wages of hourly employees
    (e) Annual salary of the manager
16. If the cost C of producing x items is given by the function $C(x) = ax + b$, which term represents the fixed cost? Which term represents the variable cost?
17. Ace Construction Company rents a bulldozer for \$100 per day plus \$50 for each hour that the bulldozer is in use. If the bulldozer rental bill on April 18 was \$480, how many hours was the bulldozer used?

18. Sport Car Rentals rents a car for \$35 per day and 12 cents per mile. After 2 days, a bill (before tax) was \$127.84. How many miles were driven?
19. EMH Corporation has leased a photocopier. Because the company must pay \$300 per month plus 3 cents per copy made, it makes a policy that no more than 3600 copies can be made per month. At the end of the first month, the accounting department receives a photocopying bill for \$414. By how many copies has the company exceeded its limit?
20. Last year, Evans Company produced furnace vents with a variable cost of \$5.00. This year, the fixed cost has risen by 11%. If the cost of producing 100 furnace vents is now \$508.88, what was the fixed cost last year?

Simple Interest

21. If P represents the amount of an investment at 7% simple interest, which of the following represents the investment value after 1 year?
    (i) $P + 0.07$ (ii) $0.07P$
    (iii) $P + 0.07P$ (iv) $0.07 + 0.07P$
22. A big spender has two credit cards, one that carries an 18% finance charge and another that carries a 20% finance charge. This person uses only one of the cards to avoid paying 38% in finance charges. What is your opinion of this reasoning?
23. Suppose that you plan to save for 1 year to buy a sofa that costs \$600. What amount must you invest at 5% simple interest in order to have enough to buy the sofa?
24. Your father-in-law will sell his rider mower to you, but you do not have to pay him now. Instead, he will charge 5.5% simple interest, and you will owe him \$1000 at the end of 1 year. For how much is he selling the mower?
25. One year ago a college sophomore received a grant that she did not need until this year. She invested the money at 7% simple interest, and she now has \$1177. How much was the grant?
26. Last year a friend borrowed money from you at 9% simple interest, and he promised to repay the loan plus interest a year later. Now he has paid you \$1000, but he could not pay the remaining \$362.50 that he still owes. How much did he borrow from you?
27. The owner of a sporting goods store had \$15,000 cash that he could apply toward the cost of expanding his store. He needed to borrow the rest of the cost at 8% simple interest for 1 year. If his total cash outlay at the end of the year was \$36,600, what was the amount of his loan?
28. A homeowner took out an 8% simple interest equity loan on his home. At the end of the first year of the loan, he paid the interest due plus \$3800 toward the principal. At the end of the second year, before making any payment, he still owed \$33,000. What was the amount of the equity loan?
29. In 1996 you invested money in low-income mortgage loans paying 9% simple interest, while your more conservative sister placed her money in a safe 6% simple interest investment. If the total of your

investments was \$6500, and the total of your interest incomes 1 year later was \$465, how much did each of you invest?

30. Two sons borrowed a total of \$26,000 from their father for 1 year. One son was charged 6% simple interest, but the less reliable son was charged 10% simple interest. If both paid off their loans on time and the father received a total of \$2000 in interest, how much did each son borrow?

31. To purchase a car costing \$10,000, the buyer borrowed part of the money from the bank at 9% simple interest and the rest from her mother-in-law at 12% simple interest. If her total interest for the year was \$1080, how much did she borrow from the bank?

32. The owner of a small business, hoping to maintain good relations with two banks, invested a total of \$5000. Part of the money went to UltraBank, which paid 8% simple interest. The remainder went to Capital Savings, which paid 8% compounded continuously, yielding the equivalent of 8.3% simple interest. At the end of 1 year, the owner received a combined income of \$408.10. How much was invested at each bank?

Distance, Rate, and Time

33. A boy left home at 8:30 A.M. to return his friend's bike. He rode at 15 mph and arrived at his friend's house at 9:10 A.M. After chatting awhile, he began his walk home at 9:30 A.M. How fast did he walk if he arrived home at noon?

34. Two travelers left a restaurant in Salina, Kansas, and traveled in opposite directions on Interstate 70. If one driver averaged 65 mph and the other averaged 60 mph, how long was it before they were 400 miles apart?

35. A student left her college campus by bus to travel home, a distance of 472 miles. At the same time, her father left home in his car to meet the bus. Her father met the bus in 4 hours and his average speed was 6 mph faster than that of the bus. Determine the speeds of the bus and the car.

36. Two cars left the Ranch Motel at the same time and traveled in opposite directions around a circular scenic loop. The first car averaged 40 mph and the other car averaged 35 mph. Determine the distance around the loop if the first car returned to the Ranch Motel 30 minutes ahead of the second car.

37. One car left Valdosta averaging 55 mph and another car left Atlanta averaging 70 mph. They traveled toward each other on Interstate 75. Atlanta is north of Valdosta by 300 miles. When did they meet if they both left at noon? How far south of Atlanta did they meet?

38. A mother jogs at a rate of 4.5 mph, and her daughter jogs at a rate of 5 mph. The mother begins her jog, and 15 minutes later her daughter begins her jog following the same route. How long will it take the daughter to overtake her mother, and how far will they have jogged?

Mixtures

39. The owner of The Nuttery wishes to mix 20 pounds of pecans selling for \$1.80 per pound with some cashews selling for \$2.40 per pound to make a mixture that will sell for \$2 per pound. How many pounds of cashews should be used?

40. A seed company has fescue grass seed worth 55 cents per pound and bluegrass seed worth 95 cents per pound. How many pounds of each should be mixed to produce 100 pounds of seed mix that will sell for 70 cents per pound?

41. The Nature Center sells bird seed for 75 cents per pound and sunflower seed for \$1.10 per pound. To the nearest pound, how many pounds of sunflower seed should be used to create a 50-pound mixture that sells for \$0.90 per pound?

42. The Indian Tea Company makes a tea blend of cinnamon tea worth \$2.50 per kilogram and black tea worth \$2.00 per kilogram. How many kilograms of each should be used to produce a 20-kilogram blend whose total worth is \$46.00?

43. A special grade of lime worth 90 cents per pound is to be mixed with 5 pounds of standard lime worth 50 cents per pound. If the mixture is to be worth 65 cents per pound, how many pounds of the special grade of lime should be used?

44. Two parts of pea gravel worth $10.00 per ton are mixed with one part of sand worth $8.00 per ton to form a base for road paving. If the bill for a project is $728.00, what is the approximate price per ton for the mixture?

Liquid Solutions

45. Cagle's Dairy mixed two grades of milk containing 3% and 4.5% butterfat, respectively, to obtain 150 gallons of milk that contained 4% butterfat. How many gallons of each were used in the mixture?

46. A precious metals firm received an order for 160 pounds of an alloy that is 35% silver. The firm melted together two blocks of alloy, one that was 50% silver and one that was 25% silver. What was the weight of each original block?

47. A pharmacist has two decongestants, one containing 0.8% pseudoephedrine HCl, the other containing 1.4% of the same substance. A prescription calls for a decongestant containing 1.1% pseudoephedrine HCl. How much of each should be used to produce 4 ounces of a 1.1% solution?

48. A biologist wishes to have 15 gallons of an 80% formaldehyde solution. In her inventory, she has pure formaldehyde and some 50% formaldehyde solution. How many gallons of each should she mix to obtain the desired solution?

49. A worker has 64 ounces of a trisodium phosphate (TSP) solution to clean a wooden deck. However, the 20% concentration is too strong, so he must add water to dilute the solution to a 10% concentration. How many ounces of water should he add?

50. A leather finisher determines that his 8-ounce solution of 12% boric acid is too weak. How many ounces of pure boric acid must be added to increase the concentration to 20%?

Miscellaneous Applications

51. A person wanting to buy a computer decided to save for 1 year by investing money at 6% simple interest. However, at the end of the year, the price of the computer had risen to $1800, which was $236.50 more than the value of the investment. What was the original amount of the person's investment?

52. A furniture store advertises that no payment is due on any purchase for 1 year. However, customers must pay 11% simple interest on the amount of their purchases. If a certain customer buys a bedroom suite and pays $1470.75 one year later, what was the cost of the furniture?

53. A college student bought his textbooks at the bookstore for $119. He spent $4 more on his biology book than on his English book. His math book cost $16 more than his English book. How much did he spend on each book?

54. The cash register of Soft Hardware contains $6.65 in just nickels and dimes. If there is a total of 85 coins, how many nickels are in the drawer?

55. Party Ideas has two grades of confetti. How many pounds of plain confetti worth $4.50 per pound should be mixed with 6 pounds of sparkling confetti worth $6 per pound in order to obtain a mixture worth $5.40 per pound?

56. Candy Corner has two kinds of candy creams and wishes to make a mixture of 100 pounds to sell at $3 per pound. If the two types are priced at $2.50 per pound and $3.75 per pound, how many pounds of each must be mixed to produce the desired amount?

57. In the accompanying figure, Elm and Maple are parallel streets that are intersected by 3rd Avenue. By a theorem in geometry, $\angle A$ and $\angle B$ are supplementary. The measure of the smaller angle is 20° less than one-fourth the measure of the larger angle. What is the measure of the angle between Elm Street and 3rd Avenue?

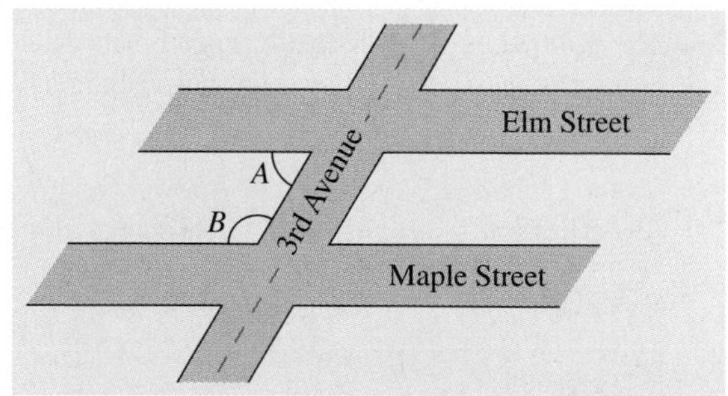

58. The fire-control computer on an M1-A1 tank plots a solution based on a right triangle in which the acute angles are complementary. The measure of one angle is 30° less than twice the measure of the other angle. What is the measure of each of the acute angles?

59. A chef has a 12% vinegar solution and a 25% vinegar solution. How much of each should he use to make 2 cups of a 15% vinegar solution?

60. A chemist added a solution that was 90% acid to an 18-gallon solution that was 60% acid. She obtained a solution that was 66% acid. How many gallons of 90% acid did she add?

61. A college student leaves her dorm at 10:00 A.M. and walks to the campus bookstore 2 miles away. She arrives at 10:30 A.M., buys her books, and leaves at 11:15 A.M. She then returns to the dorm, but she passes it and keeps walking an additional mile to the sorority house. What time does she arrive?

62. The small digital computers used on road bikes measure time and distance, but the clock is active only when the bike is actually moving. The computer also calculates and displays the average speed for the entire trip. A cyclist rides 15 miles from 2:00 to 3:00 and stops for a sundae. At 3:30, she leaves for home and arrives at 4:20. To the nearest tenth, what average speed will the computer display?

63. To obtain an $80,000 mortgage on their home, a couple had to pay 2.5 discount points. A **discount point** is 1% of the mortgage amount. If the points were added to the mortgage amount, how much was the total loan?

64. To be able to buy a home, the purchaser had to borrow 90% of the purchase price. If her mortgage loan was $108,000, what was the purchase price of the home?

Modeling with Real Data

65. The following table lists the speeds (in miles per hour) that various animals can run. (Source: World Almanac.)

Animal	Speed (mph)
Rabbit	35
Cat	30
Elephant	25
Coyote	43
Tortoise	0.17
Snail	0.03

Use the formula $d = rt$ to determine the time for each animal to complete a 100-yard dash.

66. In 1995, the average starting salary for graduates with a bachelor's degree in nursing was $33,500, which was a 1.7% increase from 1994. (Source: Employment Research Institute.) What was the dollar amount of increase in average salaries from 1994 to 1995?

Group Project: State Court Cases

State courts handle 98% of all court cases and reported 86.5 million new cases filed in 1994. The accompanying table shows the cases by type along with the percent change since 1984. (Source: *USA Today*.)

Type	1994 Cases (millions)	Percent Change (1984–1994)
Traffic	52.1	−15%
Civil	14.3	+24%
Criminal	13.5	+35%
Domestic relations	4.7	+35%
Juvenile	1.9	+59%

67. Use the information in the table to determine the number of cases of each type for 1984.

68. In 1994, how many cases were handled by courts other than state courts?

69. What is the percent change in the total number of state court cases from 1984 to 1994?

70. Tort reform proposals are designed to reduce the volume of cases in the courts and to discourage "frivolous" law suits. Which category in the table would be most affected by such reforms, and what is the most common argument against the proposals?

Challenge

71. You have decided to sell your business. Your buyer has agreed to purchase your inventory at what it cost you to buy (wholesale price). The buyer asks, "What was your markup on all of your goods?" You reply that you marked up everything 25%. The buyer says, "Just give me a 25% discount on everything, and we'll be even." Will you agree to this plan? Why or why not?

72. Big Ed's Car Rental rents cars for $25 a day plus $0.40 per mile driven. Pop's Car Rental rents cars for $33 a day plus $0.30 per mile driven.

(a) Which agency provides the lowest cost for a 100-mile trip?

(b) Let f represent the total cost of renting a car from Big Ed's, and let g represent the total cost of renting a car from Pop's. Let m represent the number of miles driven. Write f and g as functions of m.

(c) Use your calculator to display both graphs. What is the significance of the point of intersection?

(d) Set the two functions equal to each other, and solve the resulting equation for m.

(e) What are the values of the left and right sides of the equation in part (d) for the solution? What do these values represent?

(f) From parts (d) and (e), what are the coordinates of the point of intersection of the graphs in part (c)?

(g) How would you use the information obtained in parts (d) and (e) to make decisions about which car rental agency to use for *any* trip?

3.6 LINEAR INEQUALITIES

Linear Inequalities in One Variable

When we write an equation, we use an equality symbol to indicate that two algebraic expressions have the same value. In this section we use an inequality symbol to indicate that two algebraic expressions do *not* have the same value, and we call the relation an **inequality.**

As is true for equations, an inequality may be true or false.

$-2 > -5$ True

$x < 9$ False if, for example, $x = 15$

$z \geq 0$ True if, for example $z = 0$

The last inequality is true if z is replaced with 0 or any number greater than 0. We say that these replacements **satisfy** the inequality.

There are many different types of inequalities. In this section we discuss one type called a **linear inequality in one variable.**

Definition of Linear Inequality in One Variable

A **linear inequality in one variable** is an inequality that can be written in the form $Ax + B < 0$, where A and B are real numbers and $A \neq 0$.

This definition, and all others in this section, is also valid for the symbols $>$, $\leq$, and $\geq$.

The following are examples of linear inequalities.

$3x + 6 \geq 0$ $\qquad t > -3$ $\qquad 3(x - 5) \leq 6 - 3x$

The first example is written in the form $Ax + B \geq 0$. It can be shown that the other two also can be written in the defined form. An inequality written with the symbol $<$ or $>$ is called a **strict inequality.**

The following are examples of inequalities that are *not* linear inequalities.

$x^2 - 5 < 0$ No exponent can exceed 1.

$\frac{3}{x} + 2x > 5$ No variable can be in a denominator.

$\sqrt{x} - 9 > 0$ There can be no variable in a radicand.

The definitions of **solution** and **solution set** for inequalities are similar to those for equations.

Definition of Solution and Solution Set of an Inequality

A **solution** of an inequality is any replacement for the variable that makes the inequality true. The **solution set** of an inequality is the set of all solutions of the inequality.

A solution of an inequality can be verified by substitution or with the same calculator techniques used for verifying solutions of equations.

The solutions of the inequality $x \geq 3$ are 3 and all numbers greater than 3. We can use a number line graph to help us visualize these infinitely many solutions. (See Fig. 3.21.)

LEARNING TIP

Be aware of the visual advantages of representing sets of numbers with a number-line graph. Even when you are not asked to do so, sketch these number line graphs so that you can see what the numbers are.

Figure 3.21

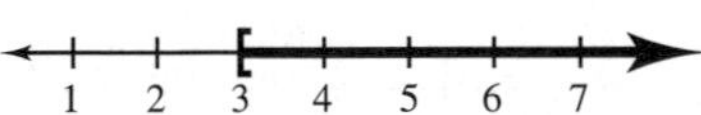

We also can describe the solution set with the set-builder notation $\{x \mid x \geq 3\}$ or with the interval notation $[3, \infty)$. Recall that the bracket indicates that 3 is included in the set, and the ∞ symbol indicates that the interval extends without end. For $x > 3$, we write $(3, \infty)$, where the left parenthesis indicates that 3 is not included in the solution set.

Solving Inequalities by Graphing

We can use a calculator to estimate solutions of inequalities. The techniques are very similar to those used in estimating solutions of equations.

EXPLORING THE CONCEPT

Estimating Solutions of Linear Inequalities

Consider the inequality $15 - 2x < -11$. Letting $y_1 = 15 - 2x$ and $y_2 = -11$, we can write the inequality in a new way: $y_1 < y_2$. Figure 3.22 shows the graphs of y_1 and y_2 with the tracing cursor on the point of intersection $(13, -11)$.

Figure 3.22

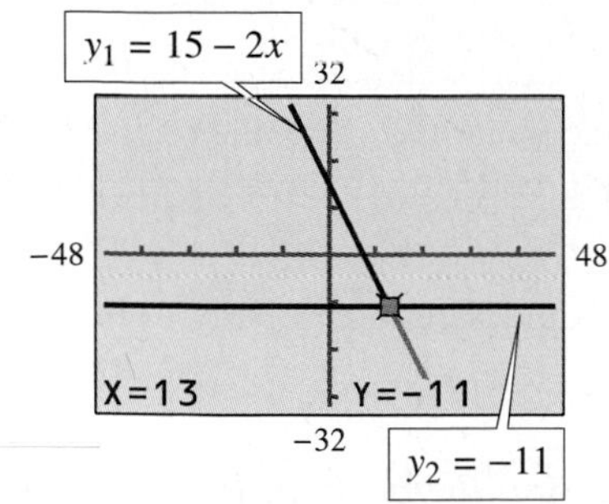

The inequality $y_1 < y_2$ implies that the graph of y_1 must be *below* the graph of y_2. We can see from the graph that this requirement is met by all points of y_1 that are to the right of the point of intersection. Because the x-coordinates of those points are solutions, the solution set of the inequality is $(13, \infty)$.

Note: The point of intersection is the point for which $y_1 = y_2$, and so 13 is not a solution of this inequality. If the inequality symbol had been $\leq$, then 13 would have been included in the solution set.

The following summary describes the graphing method for estimating solutions of an inequality.

Estimating Solutions of Inequalities Graphically

To use a graph to estimate the solution set of an inequality, perform the following steps:

1. In the same coordinate system, produce the graphs of the left-side expression y_1 and the right-side expression y_2.
2. By tracing to the point of intersection, estimate the x-coordinate of that point. For inequalities with $\leq$ or $\geq$, this x-value is a solution.
3. Trace the graph of y_1 to find the following:
   (a) For an inequality in the form $y_1 < y_2$ or $y_1 \leq y_2$, determine the x-values for which the graph of y_1 is below the graph of y_2.
   (b) For an inequality in the form $y_1 > y_2$ or $y_1 \geq y_2$, determine the x-values for which the graph of y_1 is above the graph of y_2.

   These x-values are all solutions of the inequality.

As always, graphing methods provide only estimates of solution sets. For accuracy, we must use algebraic methods. Before we discuss such methods, we consider some properties of inequalities.

Properties of Inequalities

As was true for equations, inequalities are **equivalent** if they have the same solution set. The Addition and Multiplication Properties of Inequalities can be applied to produce equivalent inequalities.

Addition Property of Inequalities

If $A < B$, then $A + C < B + C$.

Adding the same quantity to (or subtracting the same quantity from) both sides of an inequality produces an equivalent inequality.

Before stating the Multiplication Property of Inequalities, we conduct an experiment.

EXPLORING THE CONCEPT

The Multiplication Property of Inequalities

Beginning with true inequalities, we examine the effect of multiplying both sides by a positive and a negative number.

(a)	$3 < 7$	$4 > -3$	$-5 < -2$
	$3 \cdot 3 \quad 3 \cdot 7$	$3 \cdot 4 \quad 3 \cdot (-3)$	$3 \cdot (-5) \quad 3 \cdot (-2)$
	$9 < 21$	$12 > -9$	$-15 < -6$
(b)	$3 < 7$	$4 > -3$	$-5 < -2$
	$-3 \cdot 3 \quad -3 \cdot 7$	$-3 \cdot 4 \quad -3 \cdot (-3)$	$-3 \cdot (-5) \quad -3 \cdot (-2)$
	$-9 > -21$	$-12 < 9$	$15 > 6$

In part (a), multiplying both sides by a positive number did not affect the direction of the inequality symbol. In part (b), multiplying both sides by a negative number reversed the direction of the inequality symbol.

An experiment involving division would produce the same results. The following is a formal statement of the property.

Multiplication Property of Inequalities

If $A < B$ and $C > 0$, then $AC < BC$.

Multiplying or dividing both sides of an inequality by the same *positive* number produces an equivalent inequality.

If $A < B$ and $C < 0$, then $AC > BC$.

Multiplying or dividing both sides of an inequality by the same *negative* number produces an equivalent inequality if the inequality symbol is reversed.

Think About It

Suppose that the weight in one pan of a balance is greater than the weight in the other pan. What happens if you exchange the weights in the two pans? Why do inequalities not have a symmetric property as equations do?

Note that multiplying both sides of an inequality by zero does *not* result in an equivalent inequality, and of course, division by zero is undefined.

Solving Inequalities Algebraically

We can use the Addition and Multiplication Properties of Inequalities to solve inequalities algebraically. The solving routine is nearly identical to the solving routine for equations. The one exception is the need to reverse the inequality symbol when we multiply or divide both sides of an inequality by a negative number.

EXAMPLE 1

Solving an Inequality Algebraically

Use algebraic methods to solve the inequality $15 - 2x < -11$.

Solution

$$15 - 2x < -11$$

$$15 - 2x - 15 < -11 - 15 \quad \text{Addition Property of Inequalities}$$

$$-2x < -26$$

$$\frac{-2x}{-2} > \frac{-26}{-2} \quad \text{Multiplication Property of Inequalities—Dividing both sides by } -2 \text{ reverses the inequality symbol.}$$

$$x > 13$$

The solution set is $(13, \infty)$, which confirms the result obtained earlier. (See Fig. 3.23.)

Figure 3.23

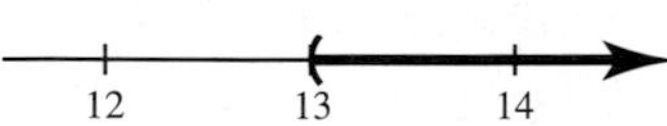

As with equations, it may be necessary to simplify both sides of an inequality by removing grouping symbols and combining like terms.

EXAMPLE 2

Simplifying Both Sides of an Inequality

Estimate the solution of $2 + 3(x - 4) > x - 10$ graphically. Then solve the inequality.

Solution

Figure 3.24

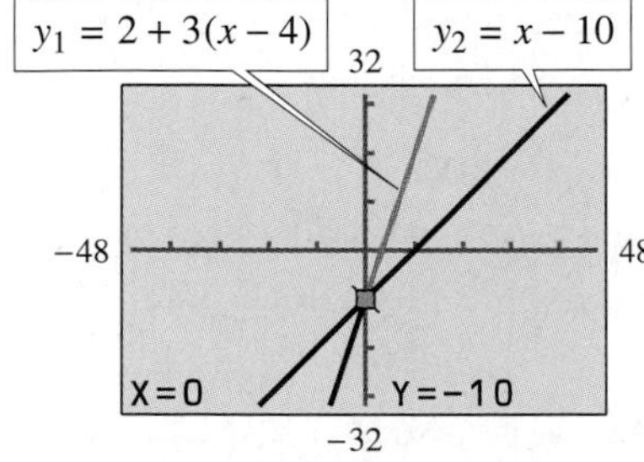

From Fig. 3.24, we see that the graph of y_1 is above the graph of y_2 for all x-values greater than 0. Thus our estimate of the solution set is $(0, \infty)$.

To solve algebraically, begin by simplifying the left side.

$$2 + 3(x - 4) > x - 10$$

$$2 + 3x - 12 > x - 10 \quad \text{Distributive Property}$$

$$3x - 10 > x - 10$$

$$3x - 10 - x > x - 10 - x \quad \text{Addition Property of Inequalities}$$
$$2x - 10 > -10 \quad \text{Combine like terms.}$$
$$2x - 10 + 10 > -10 + 10 \quad \text{Addition Property of Inequalities}$$
$$2x > 0$$
$$\frac{2x}{2} > \frac{0}{2} \quad \text{Multiplication Property of Inequalities}$$
$$x > 0$$

Figure 3.25

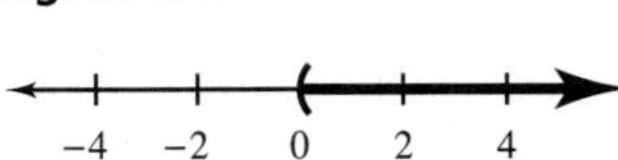

Note that dividing both sides by 2 did not affect the inequality symbol. The solution set is $(0, \infty)$. (See Fig. 3.25.)

When solving equations, it is often useful to clear fractions. The same is true when solving inequalities.

EXAMPLE 3

Clearing Fractions from an Inequality

Solve $\frac{1}{2}x + \frac{2}{3} < \frac{5}{6}x + 1$.

Solution

The LCD is 6. To clear the fractions, multiply every term of both sides by 6.

$$\frac{6}{1} \cdot \frac{1}{2}x + \frac{6}{1} \cdot \frac{2}{3} < \frac{6}{1} \cdot \frac{5}{6}x + 6 \cdot 1 \quad \text{Multiplication Property of Inequalities}$$
$$3x + 4 < 5x + 6$$
$$3x + 4 - 5x < 5x + 6 - 5x \quad \text{Addition Property of Inequalities}$$
$$-2x + 4 < 6 \quad \text{Combine like terms.}$$
$$-2x + 4 - 4 < 6 - 4 \quad \text{Addition Property of Inequalities}$$
$$-2x < 2$$
$$\frac{-2x}{-2} > \frac{2}{-2} \quad \text{Multiplication Property of Inequalities—Dividing by } -2 \text{ reverses the inequality symbol.}$$
$$x > -1$$

Figure 3.26

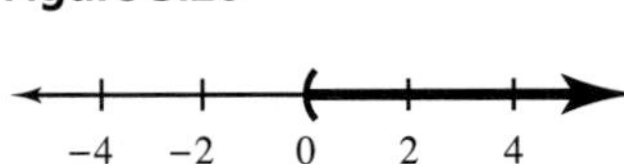

The solution set is $(-1, \infty)$. (See Fig. 3.26.)

Special Cases

When the graphs of the left and right sides of an equation do not intersect, the equation has no solution. A similar special case can arise with inequalities.

EXPLORING THE CONCEPT

Special Cases

Figure 3.27 shows the parallel graphs of $y_1 = x + 12$ and $y_2 = x - 9$.

Figure 3.27

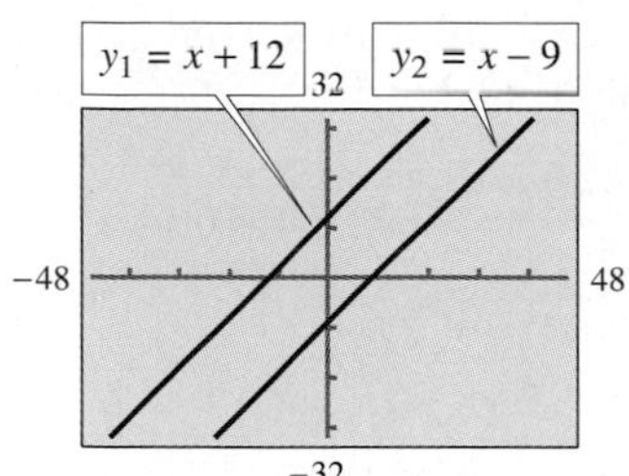

Because the graph of y_1 is *never* below the graph of y_2, there are no x-values for which $y_1 < y_2$. Thus the solution set of the inequality $x + 12 < x - 9$ is $\emptyset$.

Similarly, the graph of y_1 is *always* above the graph of y_2, so every x-value satisfies the inequality $y_1 > y_2$. Thus the solution set of the inequality $x + 12 > x - 9$ is **R.**

These same results also can be obtained with algebraic methods.

EXAMPLE 4

Detecting Special Cases Algebraically

Solve the following inequalities.

(a) $x + 12 < x - 9$ (b) $x + 12 > x - 9$

Solution

(a) $x + 12 < x - 9$

$x - x + 12 < x - x - 9$ Addition Property of Inequalities

$12 < -9$ False inequality

The resulting false inequality is equivalent to the original inequality, so we conclude that there is no replacement that makes the original inequality true. The solution set is $\emptyset$.

(b) $x + 12 > x - 9$

$x - x + 12 > x - x - 9$ Addition Property of Inequalities

$12 > -9$ True inequality

The resulting true inequality is equivalent to the original inequality, so we conclude that every replacement makes the original inequality true. The solution set is **R.**

Real-Life Applications

Example 5 illustrates an application that can be solved with a linear inequality.

EXAMPLE 5

Minimum Bowling Average

To qualify for the semifinals in a bowling tournament, a bowler must have at least a 160 average for five games. Suppose the scores for your first four games were 172, 150, 148, and 162. What is the lowest score you can bowl in your fifth game and qualify for the semifinals?

Solution

Let $x =$ the fifth score. Your final average will be determined by adding your five scores and dividing by 5.

$$\frac{172 + 150 + 148 + 162 + x}{5} \geq 160 \qquad \text{Your average must be at least 160.}$$

$$\frac{632 + x}{5} \geq 160$$

$$\frac{5}{1} \cdot \frac{632 + x}{5} \geq 5 \cdot 160 \qquad \text{Multiply both sides by 5.}$$

$$632 + x \geq 800$$

$$632 + x - 632 \geq 800 - 632 \qquad \text{Subtract 632 from both sides.}$$

$$x \geq 168$$

You must bowl *at least* 168 for your fifth game in order to qualify.

3.6 QUICK REFERENCE

Linear Inequalities in One Variable

- A **linear inequality in one variable** is an inequality that can be written in the form $Ax + B < 0$, where A and B are real numbers and $A \neq 0$. The definitions involving $>$, $\geq$, and $\leq$ are similar.
- A **solution** of an inequality is any replacement for the variable that makes the inequality true. The **solution set** of an inequality is the set of all solutions.
- Set-builder notation, number-line graphs, and interval notation can all be used to represent numbers that satisfy an inequality.

Solving Inequalities by Graphing

- To use a graph to estimate the solution set of an inequality, perform the following steps:
  1. Produce the graphs of the left-side expression y_1 and the right-side expression y_2.
  2. By tracing to the point of intersection, estimate the x-coordinate of that point. For inequalities with $\leq$ or $\geq$, this x-value is a solution.
  3. Trace the graph of y_1 to find the following:
     (a) For an inequality in the form $y_1 < y_2$ or $y_1 \leq y_2$, determine the x-values for which the graph of y_1 is below the graph of y_2.
     (b) For an inequality in the form $y_1 > y_2$ or $y_1 \geq y_2$, determine the x-values for which the graph of y_1 is above the graph of y_2.

  These x-values are all solutions of the inequality.

Properties of Inequalities

- Inequalities are **equivalent** if they have the same solution set.
- Addition Property of Inequalities: If $A < B$, then $A + C < B + C$. Adding the same quantity to (or subtracting the same quantity from) both sides of an inequality produces an equivalent inequality.

- Multiplication Property of Inequalities:
  1. If $A < B$ and $C > 0$, then $AC < BC$. Multiplying or dividing both sides of an inequality by the same *positive* number produces an equivalent inequality.
  2. If $A < B$ and $C < 0$, then $AC > BC$. Multiplying or dividing both sides of an inequality by the same *negative* number produces an equivalent inequality if the inequality symbol is reversed.

Solving Inequalities Algebraically

- The routine for solving inequalities is nearly identical to that for solving equations. The exception is the need to reverse the inequality symbol when we multiply or divide both sides by a negative number.
- As with equations, simplifying both sides and clearing fractions are steps in the solving process.

Special Cases

- If the graphs of the left and right sides of an inequality do not intersect, the solution set is either Ø or **R.**
- These special cases can be detected algebraically when we obtain an equivalent inequality that is false (in which case the solution set is Ø) or true (in which case the solution set is **R**).

3.6 EXERCISES

Concepts and Skills

1. Explain the difference between $[a, b]$ and $(a, b]$.

2. For $\{x \mid x \geq 2\}$, is it correct to write $[2, \infty]$ or $[2, \infty)$?

In Exercises 3–8, represent the solution set with interval notation, set notation, and a number-line graph.

3. $x < 2$
4. $x > -5$
5. $x \geq 3$
6. $x \leq -4$
7. $-3 < x \leq 2$
8. $-2 \leq x < 4$

9. Compare the graphing method for solving equations with the graphing method for solving inequalities.

10. When you use the graphing method to solve an inequality, under what condition does the point of intersection represent a solution?

In Exercises 11 and 12, the given calculator display shows the graphs of the left and right sides of the given inequality. Write the solution set of the inequality with interval notation.

11. (a) $y_1 \leq y_2$
(b) $y_1 > y_2$

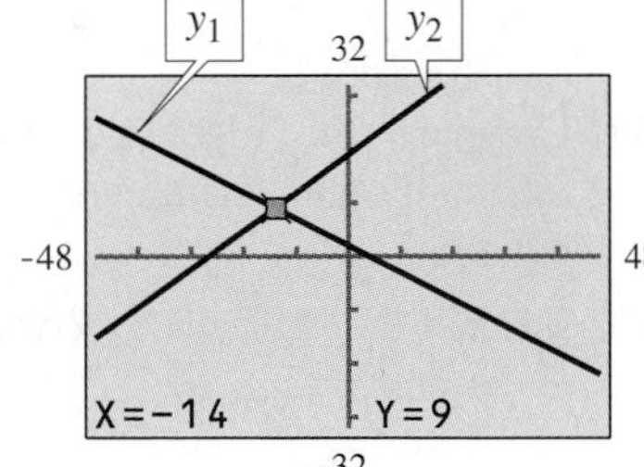

12. (a) $y_1 < y_2$
(b) $y_1 \geq y_2$

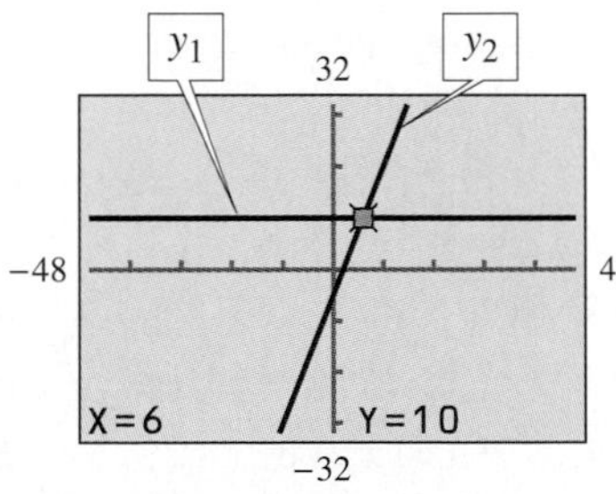

In Exercises 13–22, use the graphing method to estimate the solution set of the given inequality.

13. $3x - 16 < 8$
14. $10 - 2x < 5x - 11$
15. $3x + 9 \geq -2x - 16$
16. $-x \geq 2x - 12$
17. $3x - 8 \leq 3(2 + x)$
18. $6(1 + x) > 2(3x - 5)$

19. $20 - 3x \geq 2x$

20. $2x - 7 \geq 11 - x$

21. $2(x - 5) > 2x + 1$

22. $2 - 3x \geq 15 - 3x$

23. Describe the graphs of the left and right sides of an inequality whose solution set is

(a) the empty set.

(b) the set of all real numbers.

24. If $-x < -z$, what is true about x? Explain.

In Exercises 25–32, assume that $a < b$ and insert the correct inequality symbol.

25. $2a$ ▒ $2b$

26. $a - 6$ ▒ $b - 6$

27. b ▒ a

28. $-\frac{1}{2}a$ ▒ $-\frac{1}{2}b$

29. $-a$ ▒ $-b$

30. $\frac{a}{4}$ ▒ $\frac{b}{4}$

31. $-3 + a$ ▒ $-3 + b$

32. $a \div (-13)$ ▒ $b \div (-13)$

In Exercises 33–38, assume that $a < b$ and determine the values of c for which the inequality is true.

33. $ac > bc$

34. $ac < bc$

35. $a + c < b + c$

36. $a - c < b - c$

37. $\frac{a}{c} < \frac{b}{c}$

38. $\frac{a}{c} > \frac{b}{c}$

39. Compare the algebraic method for solving an inequality with the algebraic method for solving an equation.

40. Explain the difference between the solution sets of $x < a$ and $x \leq a$.

In Exercises 41–74, solve the inequality algebraically. Except for special cases, write the solution set with interval notation.

41. $3x - 4 < 14$

42. $7x + 1 \geq 15$

43. $7 - 3t \leq -8$

44. $5 - y > 0$

45. $5x + 3 < 7 + 5x$

46. $3 - 2x \geq -2x + 1$

47. $4x - 4 \leq 7x - 13$

48. $2x + 7 < 7 + 5x$

49. $8 - x > x - 8$

50. $5 - t < t + 5$

51. $3 - x \geq 4x + 1$

52. $6 + 4x < 11 - 2x$

53. $3x - x < 2x$

54. $1 + 2x \leq 2x + 1$

55. $-(x - 2) \leq 2(x + 1)$

56. $3(x + 4) > 2x - 3$

57. $5 - 4x \geq 4(2 - x)$

58. $2 - 5x < -5(1 + x)$

59. $x - 4(x - 4) - 4x > x - 4$

60. $3(x - 1) - 4(2x + 3) > 0$

61. $5 + 4(x - 3) \geq 2x + 3(x - 1) + 4$

62. $5 - 5(5 - x) \leq 5(x - 1) + 2x$

63. $\frac{2}{3}x - 5 < \frac{5}{6}$

64. $\frac{5}{4}x - \frac{1}{2} \geq x - \frac{5}{4}$

65. $\frac{2x + 1}{-3} < 6$

66. $\frac{5x - 1}{-5} > \frac{3}{10}$

67. $x - \frac{1}{3} + \frac{5}{6}x \leq \frac{1}{2} - 3x$

68. $\frac{5}{12}x + \frac{1}{3} - x \leq \frac{7}{12} - \frac{x}{2}$

69. $\frac{3}{4}(x + 3) + 2x < 1$

70. $\frac{1}{4}(x - 2) < \frac{1}{2}x - \frac{2}{3}$

71. $0.2x + 1.7 \geq 2.94 - 5.3x$

72. $0.75x + 0.25 \leq x - 1.5$

73. $2.8(1.3x - 0.9) \leq 1.2 - 9.97x$

74. $0.5(0.6x + 3.1) \geq x - 1.73$

Real-Life Applications

75. The lengths of the sides of a triangle are represented by a, b, and c. What must be true about any two sides with respect to the third side? Write three true inequalities involving a, b, and c.

76. The length of a rectangle is 6 inches more than the width. The perimeter of the rectangle can be no more than 48 inches. What is the maximum width?

77. A pipe is at least 21 feet long, and you want to cut it into three pieces. The second piece is to be twice as long as the first piece, and the third piece is to be 1 foot longer than the second piece. What is the minimum length of the first piece?

78. You know that a rope is no more than 100 feet long. You need to cut the rope into three pieces. The second piece is to be three times as long as the first piece, and the third piece must be 18 feet long. What is the maximum length of the second piece?

79. Cross-country runners are awarded these quality points according to how the runners finish in a race.

Finish Position	Points
1	10
2	7
3	5
4	3
5	1

To win a trophy at the end of ten races, a runner must have an average of at least 7 points.

Here are your finish positions for your first nine races.

1 4 1 2 2 1 2 2 3

What is the minimum position in which you can finish the tenth race in order to win a trophy?

80. Your grades on tests 1, 2, and 4 are 82, 76, and 90. Unfortunately, you cut the third test and received a 0. If you have one test left to take, and if the passing grade for the course is 70, can you still pass the course?

81. A car rental agency rents cars for \$26.20 per day plus \$0.22 per mile driven. If your travel budget is \$200, what is the maximum number of miles you can drive during a 1-day rental?

82. Suppose that you are running a concession stand when a person gives you \$18 and asks for six soft drinks and as many hot dogs as the remaining money will buy. If soft drinks are \$1.00 and hot dogs are \$1.75, what is the maximum number of hot dogs the person can buy?

83. A rich, eccentric aunt leaves some money in her will for Calvin, John, and Carrie. Calvin is to receive the least; Carrie is to receive the most. The aunt requires the three relatives to solve a puzzle before they can receive their inheritance. The will reads as follows:

> Calvin and John are given consecutive even integers. Carrie's number is 5 more than Calvin's number. If one-seventh of Calvin's number, one-third of John's number, and one-sixth of Carrie's number are added together, the sum exceeds 5. Determine the smallest numbers that satisfy these conditions and multiply them by \$1000. The result is your inheritance.

How much will Calvin inherit?

84. Suppose you have a gift certificate worth \$20 for one long-distance phone call. If the charge is \$1.10 for the first minute and \$0.42 for each additional minute, what is the longest that you can talk?

Modeling with Real Data

85. From 1986 to 1996, the number of nonincumbent candidates for the House of Representatives more than doubled. The accompanying bar graph shows the trend. (Source: Federal Elections Commission.)

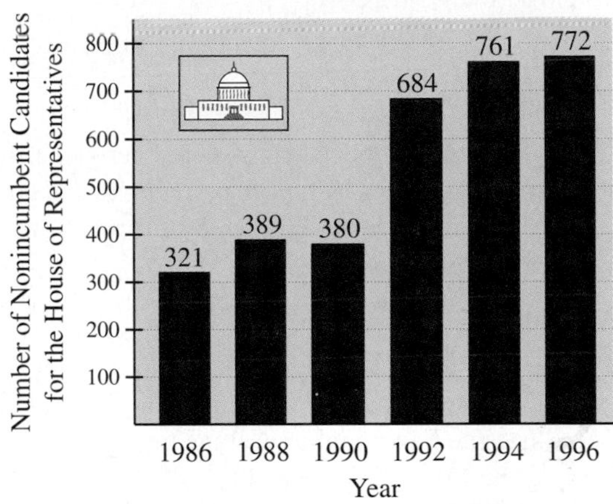

The number C of nonincumbent candidates can be modeled by the function $C(t) = \frac{105}{2}t - \frac{79}{3}$, where t is the number of years since 1980.

(a) Assuming that the trend continues, write an inequality indicating an average of at least two nonincumbent candidates for each seat in the House of Representatives. (There are 435 members in the House.)

(b) Determine and interpret the solution set of your inequality in part (a).

86. The average daily cost of a car vacation for a family of four on a 300-mile trip approximately doubled from 1985 to 1996. This average cost includes such expenses as meals, lodging, and gasoline. (Source: American Automobile Association.) If x is the number of years since 1985, then the daily average cost C can be modeled by the function $C(x) = \frac{69}{11}x + 164$.

(a) Write an inequality to describe the years in which the average daily cost does not exceed \$300.

(b) Determine and interpret the solution set of your inequality in part (a).

87. The following table shows the average number of runs scored per game in the National and American

Leagues during the period 1992–1996. (Source: Elias Sports Bureau.)

Year	National League	American League
1992	7.76	8.64
1993	8.98	9.41
1994	9.24	10.45
1995	9.26	10.12
1996	9.66	11.03

The average number of runs scored per game for the National League (N) and the American League (A) can be modeled by the following functions, where x is the number of years since 1990.

$N(x) = 0.408x + 7.348$

$A(x) = 0.549x + 7.734$

(a) For the period shown in the table, the American League had a higher average number of runs per game. Write an inequality to determine the year when the average run production in the National League will exceed that of the American League.

(b) Determine and interpret the solution set of your inequality in part (a).

88. Produce the graphs of the functions in Exercise 87 to explain your interpretation in part (b).

Challenge

In Exercises 89–92, solve the inequality for x.

89. $5x + a \le b$

90. $cx < 10$, where $c < 0$

91. $a^2x \ge 8, a \ne 0$

92. $b - x > a$

93. Your instructor offers a one-million-dollar prize to anyone who can solve this puzzle. There are three consecutive even integers, and at least two of them are positive. The sum of the first and third integers is no more than the second integer. Show why your instructor will never have to pay.

94. Suppose you graph the left and right sides of a linear inequality and you find that the lines coincide. Explain your answers to the following.

(a) If you correctly conclude that the solution set is empty, what is the inequality symbol? (There are two possibilities.)

(b) If you correctly conclude that the solution set is the set of real numbers, what is the inequality symbol? (There are two possibilities.)

3.7 COMPOUND INEQUALITIES

Double Inequalities • Conjunctions • Disjunctions • Real-Life Applications

Double Inequalities

A **compound inequality** consists of two inequalities connected with the words *and* or *or*.

(i) $3 - x < 4$ *and* $2x - 1 \le 9$ (ii) $x + 2 \ge 6$ *or* $x - 2 \le 0$

In (i), the word *and* means that both inequalities must be true, and in (ii), the word *or* means that at least one inequality must be true.

A **double inequality** is a special kind of compound inequality in which the value of an expression lies between two values. For example, $-26 \le 3x - 8 \le 16$ means that the value of $3x - 8$ is between -26 and 16, inclusive. The implied connective in a double inequality is *and*: $-26 \le 3x - 8$ *and* $3x - 8 \le 16$.

EXPLORING THE CONCEPT

Double Inequalities

If we let $y_1 = -26$, $y_2 = 3x - 8$, and $y_3 = 16$, we can write $-26 \le 3x - 8 \le 16$ as $y_1 \le y_2 \le y_3$, and we can estimate the solution set graphically. Figure 3.28 shows that the point of intersection of the graphs of y_1 and y_2 is $(-6, -26)$. Similarly, the point of intersection of the graphs of y_2 and y_3 is $(8, 16)$. (See Fig. 3.29.)

Figure 3.28

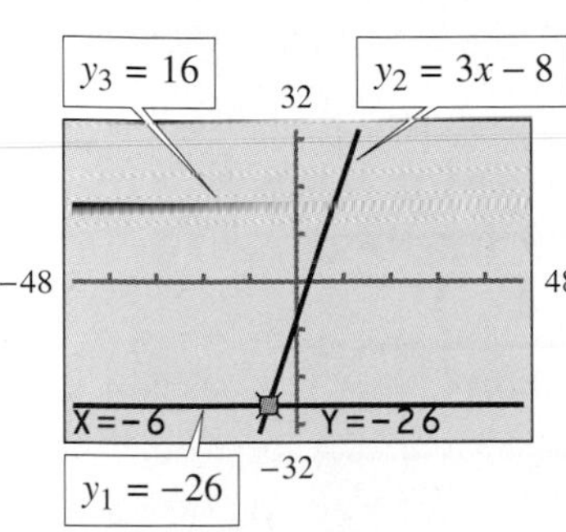

Figure 3.29

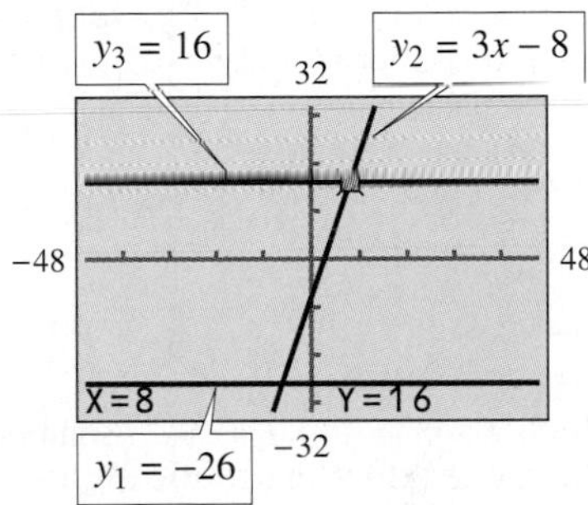

The solutions of the double inequality are represented by the points of the graph of y_2 that lie between the graphs of y_1 and y_3. These are the points whose x-coordinates are between -6 and 8, inclusive. Thus the solution set of $-26 \le 3x - 8 \le 16$ is $[-6, 8]$.

In Example 1 we solve the same double inequality algebraically. When we use algebraic methods, our goal is to isolate the variable in the middle.

EXAMPLE 1

Solving a Double Inequality Algebraically

Solve $-26 \le 3x - 8 \le 16$ algebraically.

Solution

$$-26 \le 3x - 8 \le 16$$

$$-26 + 8 \le 3x - 8 + 8 \le 16 + 8 \qquad \text{Addition Property of Inequalities}$$

$$-18 \le 3x \le 24$$

$$\frac{-18}{3} \le \frac{3x}{3} \le \frac{24}{3} \qquad \text{Multiplication Property of Inequalities}$$

$$-6 \le x \le 8$$

As before, the solution set is all numbers between -6 and 8, inclusive. The solution set is $[-6, 8]$. (See Fig. 3.30.)

Figure 3.30

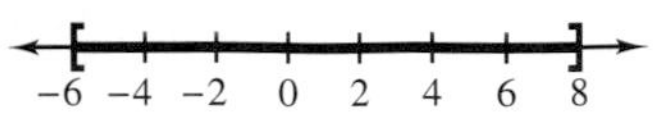

−6 −4 −2 0 2 4 6 8

Conjunctions

A compound inequality whose connective is *and* is called a **conjunction.**

Suppose that A and B are the solution sets for the two inequalities of a conjunction. Then the solution set for the conjunction is the set S of numbers that belong to both A and B. We call the set S the **intersection** of sets A and B. Symbolically, we write $A \cap B$.

EXAMPLE 2

LEARNING TIP

When you use number-line graphs to assist you in forming unions and intersections, align the tick marks. This allows you to run your pencil vertically through the graphs to see where numbers are included or overlap.

Intersection of Solution Sets

Describe the solution set of the compound inequality

$$x < 4 \quad \textit{and} \quad x > -3$$

Solution

The solution set is the intersection of the individual sets. The graph of the solution set is as shown in Fig. 3.31. In interval notation the solution set is $(-3, 4)$.

Figure 3.31

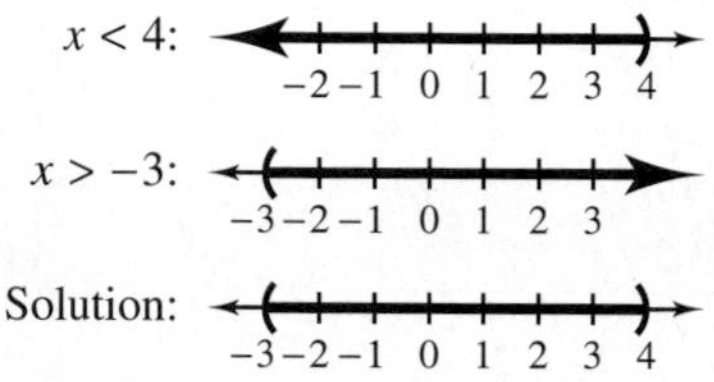

$x < 4$: −2 −1 0 1 2 3 4

$x > -3$: −3 −2 −1 0 1 2 3

Solution: −3 −2 −1 0 1 2 3 4

To solve a conjunction, we solve each individual inequality and then form the intersection of the individual solution sets.

EXAMPLE 3

Solving a Conjunction

Solve the conjunction $3 - x < 5$ and $2x - 3 \le 7$.

Solution

Solve each inequality individually.

$$\begin{aligned} 3 - x &< 5 && \text{and} & 2x - 3 &\le 7 \\ 3 - x - 3 &< 5 - 3 && \text{and} & 2x - 3 + 3 &\le 7 + 3 \\ -x &< 2 && \text{and} & 2x &\le 10 \\ x &> -2 && \text{and} & x &\le 5 \end{aligned}$$

The connective is *and*, so the solution set for the compound inequality is the intersection of the two sets. The solution set is $(-2, 5]$. (See Fig. 3.32.)

Figure 3.32

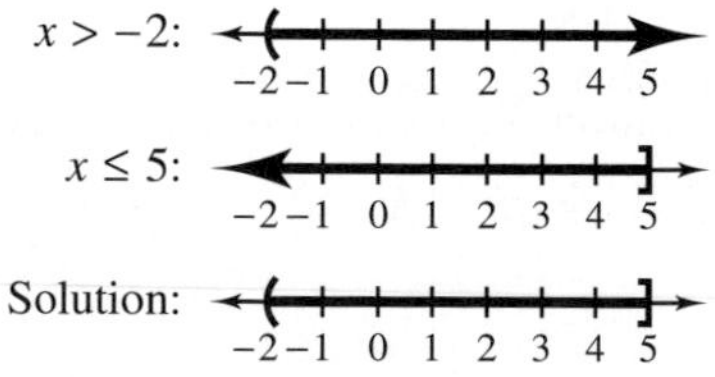

$x > -2$: −2 −1 0 1 2 3 4 5

$x \le 5$: −2 −1 0 1 2 3 4 5

Solution: −2 −1 0 1 2 3 4 5

EXAMPLE 4

Solving a Conjunction

Solve the conjunction $x - 4 \ge -3$ and $2x - 6 > 0$.

Solution

Solve each inequality individually.

$$
\begin{array}{rcr}
x - 4 \geq -3 & \text{and} & 2x - 6 > 0 \\
x - 4 + 4 \geq -3 + 4 & \text{and} & 2x - 6 + 6 > 0 + 6 \\
x \geq 1 & \text{and} & 2x > 6 \\
x \geq 1 & \text{and} & x > 3
\end{array}
$$

From the graphs we can visually determine the intersection by noting the points common to both graphs. The solution set is $(3, \infty)$. (See Fig. 3.33.)

Figure 3.33

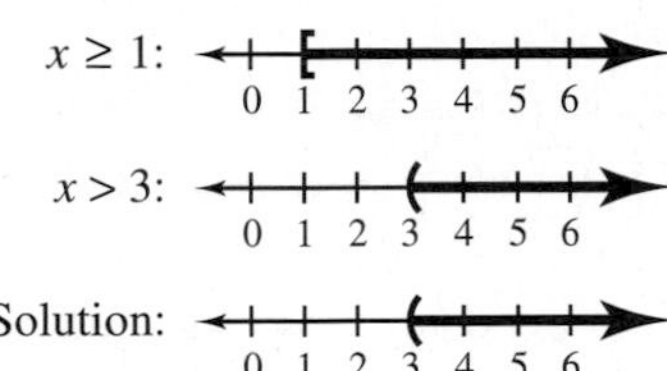

EXAMPLE 5

Solving a Conjunction

Solve the conjunction $x + 2 \leq 0$ and $5 - x \leq 3$.

Solution

Solve each inequality individually.

$$
\begin{array}{rcr}
x + 2 \leq 0 & \text{and} & 5 - x \leq 3 \\
x + 2 - 2 \leq 0 - 2 & \text{and} & 5 - x - 5 \leq 3 - 5 \\
x \leq -2 & \text{and} & -x \leq -2 \\
x \leq -2 & \text{and} & x \geq 2
\end{array}
$$

Because there are no points common to both graphs in Fig. 3.34, there are no solutions of the compound inequality. The solution set is Ø.

Figure 3.34

$x \leq -2$:
−2 −1 0 1 2 3 4

$x \geq 2$:
−3 −2 −1 0 1 2 3

Disjunctions

A compound inequality whose connective is *or* is called a **disjunction.**

Suppose that A and B are the solution sets for the two inequalities of a disjunction. Then the solution set for the disjunction is the set S of numbers that belong to A or B (or to both). We call the set S the **union** of sets A and B. Symbolically, we write $A \cup B$.

EXAMPLE 6

Union of Solution Sets

Describe the solution set of the compound inequality

$$x > 1 \quad or \quad x < -2$$

Solution

The solution set is the union of the individual sets. The graph of the solution set is as shown in Fig. 3.35. In interval notation the solution set is $(-\infty, -2) \cup (1, \infty)$.

Figure 3.35

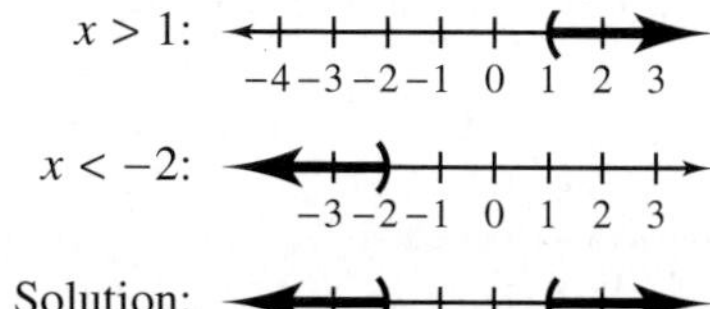

To solve a disjunction, we solve each individual inequality and then form the union of the individual solution sets.

EXAMPLE 7

Solving a Disjunction

Solve the disjunction $3x - 2 < 1$ or $x - 3 \geq 1$.

Solution

Solve each inequality individually.

$$\begin{aligned} 3x - 2 &< 1 & \text{or} \quad x - 3 &\geq 1 \\ 3x - 2 + 2 &< 1 + 2 & \text{or} \quad x - 3 + 3 &\geq 1 + 3 \\ 3x &< 3 & \text{or} \quad x &\geq 4 \\ x &< 1 & \text{or} \quad x &\geq 4 \end{aligned}$$

The connective is *or*, so the solution set for the compound inequality is the union of the two sets. From the graphs in Fig. 3.36, we see the points that belong to at least one of the sets. The solution set is $(-\infty, 1) \cup [4, \infty)$.

Figure 3.36

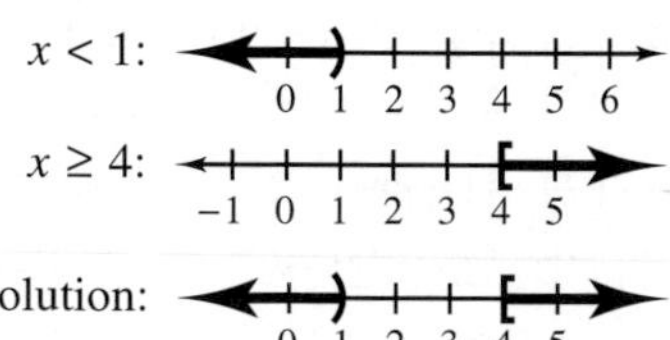

EXAMPLE 8

Solving a Disjunction

Solve the disjunction $x + 4 \geq 4$ or $x - 5 \leq 0$.

Figure 3.37

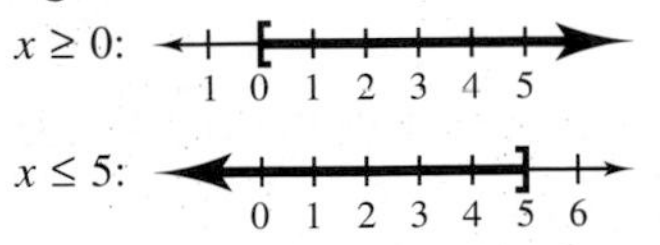

Think About It

What symbol would you use to describe the following sets?
(a) $\mathbf{R} \cap \emptyset$ (b) $\mathbf{R} \cup \emptyset$

Solution

Solve each inequality individually.

$$\begin{aligned} x + 4 &\geq 4 & \text{or} \quad x - 5 &\leq 0 \\ x + 4 - 4 &\geq 4 - 4 & \text{or} \quad x - 5 + 5 &\leq 0 + 5 \\ x &\geq 0 & \text{or} \quad x &\leq 5 \end{aligned}$$

We can see from the graphs in Fig. 3.37 that every real number belongs to at least one of the two sets. The solution set for the compound inequality is **R.**

Real-Life Applications

In Example 9 we consider an application involving a compound inequality.

EXAMPLE 9

Fencing a Pasture

Farm Resources specializes in installing fencing. A customer, who cannot spend more than \$6000, needs a 4-foot-high fence around a rectangular field. The width of the field must be 60% of the length, and the perimeter must be at least 800 feet. If Farm Resources charges \$5.00 per linear foot for this fencing, what are the minimum and maximum lengths that the field can be?

Solution

The minimum amount of fencing is 800 feet, and the maximum is the customer's budget divided by the cost per linear foot: $\frac{6000}{5} = 1200$. Thus the perimeter must be between 800 feet and 1200 feet, inclusive.

Figure 3.38

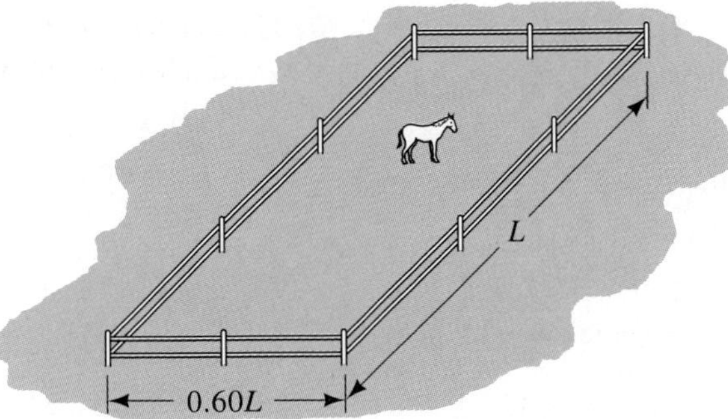

As shown in Fig. 3.38, we let $L =$ the length of the field. Then the width is 60% of L, or $0.60L$.

$$\begin{aligned} &800 \leq 2L + 2W \leq 1200 && \text{The perimeter of a rectangle is } 2L + 2W. \\ &800 \leq 2L + 2(0.60L) \leq 1200 && \text{Replace } W \text{ with } 0.60L. \\ &800 \leq 2L + 1.2L \leq 1200 \\ &800 \leq 3.2L \leq 1200 \\ &\frac{800}{3.2} \leq \frac{3.2L}{3.2} \leq \frac{1200}{3.2} \\ &250 \leq L \leq 375 \end{aligned}$$

Thus the length of the field will be at least 250 feet and at most 375 feet.

3.7 QUICK REFERENCE

Double Inequalities

- A **compound inequality** consists of two inequalities connected with the words *and* or *or*.
- A **double inequality** is a statement that the value of an expression is between two values. It is a special kind of compound inequality with the implied connective *and*.
- To solve a double inequality graphically, graph the three components of the double inequality on the same coordinate axes. Then estimate the values of x for which the graph of the middle expression is between the graphs of the outer expressions.
- To solve a double inequality algebraically, apply the properties of inequalities to all three components at once.

Conjunctions

- A **conjunction** is a compound inequality whose connective is *and*. To solve a conjunction, solve each individual inequality and then form the *intersection* of the individual solution sets.

Disjunctions

- A **disjunction** is a compound inequality whose connective is *or*. To solve a disjunction, solve each individual inequality and then form the *union* of the individual solution sets.

3.7 EXERCISES

Concepts and Skills

1. Explain the use of the word *and* as it is used in compound inequalities.

2. Explain the use of the word *or* as it is used in compound inequalities.

In Exercises 3–10, determine whether the given numbers are solutions of the compound inequality.

3. $5x < 12$ and $3x > 3$; 2, 1

4. $-3x < 15$ or $2x > -5$; $-8, 4$

5. $2x + 7 < -3$ or $3x + 5 > 11$; $5, -7$

6. $4 - 3x < 2$ and $3 + 2x > -1$; $3, -6$

7. $0.5x > 4.5$ or $0.3x > 1.8$; 7, 11

8. $3 - 0.2x > 7$ and $1.2x > 14$; 8, 12

9. $0.5 \leq 2x - 3 < 5$; 3.75, 5

10. $-3 < x + 2 \leq 5$; $-5, 3, 0$

11. What is the difference between a conjunction and a disjunction?

12. For each of the given pairs of compound inequalities, explain the difference between their solution sets.

(a) $x < -7$ or $x \geq 1$ $\quad$ $x < -7$ and $x \geq 1$

(b) $x < 1$ or $x \geq -7$ $\quad$ $x < 1$ and $x \geq -7$

In Exercises 13–18, graph the solution set of each conjunction.

13. $x \geq -5$ and $x < 3$ $\quad$ **14.** $x \geq -6$ and $x \leq 2$

15. $x < 5$ and $x \leq 1$ $\quad$ **16.** $x \geq 0$ and $x > -3$

17. $x \leq 4$ and $x > 6$ $\quad$ **18.** $x > -1$ and $x \leq -7$

In Exercises 19–24, graph the solution set of each disjunction.

19. $x < -5$ or $x > 4$ $\quad$ **20.** $x \leq -7$ or $x \geq 8$

21. $x \geq -3$ or $x > 4$ $\quad$ **22.** $x \leq 2$ or $x < 7$

23. $x \leq 4$ or $x > 0$ $\quad$ **24.** $x \geq -3$ or $x < 5$

In Exercises 25 and 26, explain how to interpret the given graphics display to estimate the solution set of each of the following.

(a) $y_1 \le y_2 \le y_3$ (b) $y_2 \le y_1$ or $y_2 \ge y_3$

25.

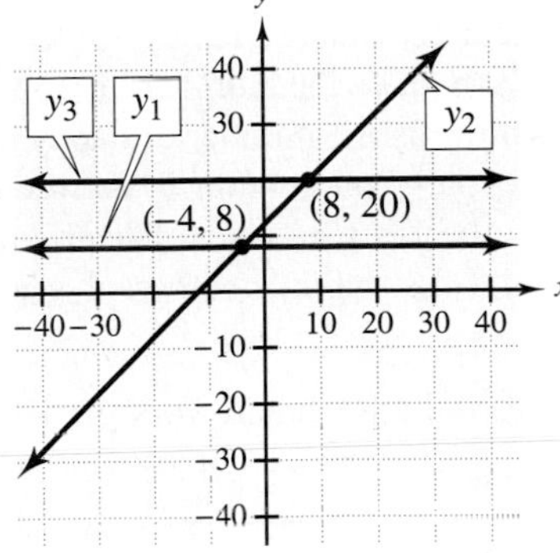

26.

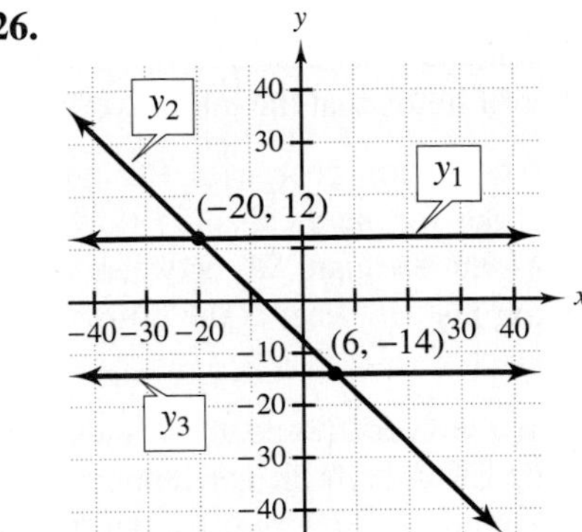

In Exercises 27–30, use the graphing method to estimate the solutions of the given inequality and write the solution set with interval notation.

27. $-12 \le 3x - 2 \le 8$

28. $0 < 3x - 1 < 8$

29. $-2 < \dfrac{x-1}{-3} < 2$

30. $-4 \le \dfrac{3-x}{2} < 4$

31. For any real number c, describe the solution set of each compound inequality.

(a) $x \le c$ and $x \ge c$ (b) $x > c$ and $x < c$

(c) $x \le c$ or $x \ge c$ (d) $x > c$ or $x < c$

32. Explain whether each of the following pairs of compound inequalities are equivalent or not equivalent and why.

(a) $x > -3$ and $x < 7$ $\quad -3 < x < 7$

(b) $3 < x$ or $x < -2$ $\quad 3 < x < -2$

(c) $1 \le x$ and $x \ge 5$ $\quad 1 \le x \ge 5$

In Exercises 33–46, solve the inequalities algebraically.

33. $-4 < x + 3 < 7$

34. $-3 < x - 2 < 5$

35. $2 \le -3t \le 18$

36. $-14 \le -2y < -6$

37. $9 \ge 2x + 1 \ge -2$

38. $7 > 3x - 2 \ge -5$

39. $-5 \le -2x - 7 < 2$

40. $-3 < 5 - x < 4$

41. $-\dfrac{7}{4} \le \dfrac{3}{4}x - 1 < 2$

42. $3 > \dfrac{t}{2} + 1 \ge 0$

43. $-\dfrac{17}{6} < \dfrac{2}{3}x - \dfrac{1}{6} \le \dfrac{11}{3}$

44. $-2 < \dfrac{1}{2}x - \dfrac{1}{4} < \dfrac{7}{4}$

45. $6.5 \ge 0.75x + 0.5 > -2.5$

46. $-3.25 < 0.75 - 0.8x < 4.75$

47. Describe how the solution set of a disjunction can be all real numbers but the solution set of the corresponding conjunction is the empty set.

48. Explain why the solution set of a disjunction of linear inequalities can never be the empty set.

In Exercises 49–54, solve each conjunction algebraically and write the solution set with interval notation.

49. $-x \le 5$ and $-x \ge -7$

50. $-x \ge -12$ and $-x \le 10$

51. $x + 1 \ge 5$ and $x - 2 < 12$

52. $x - 2 > 3$ and $x - 5 < 4$

53. $2x - 5 \le 7$ and $x + 4 > -7$

54. $2x + 3 \ge -1$ and $3x + 6 < 21$

In Exercises 55–60, solve each disjunction algebraically and write the solution set with interval notation.

55. $-x \ge 4$ or $-x \le -3$

56. $-x \ge 8$ or $-x \le -2$

57. $x + 4 < -3$ or $x + 4 > 3$

58. $x - 5 < -4$ or $x - 5 > 4$

59. $-4x - 5 > 0$ or $3x - 2 \ge 1$

60. $-3x - 6 > 3$ or $4x + 3 \ge 5$

In Exercises 61–76, solve each compound inequality and write the solution set with interval notation.

61. $2x - 5 \le -17$ and $3x > -27$

62. $-8x \le 24$ and $-2x \ge -8$

63. $x - 2 > 3$ or $x + 5 < 4$

64. $4 - x \ge 5$ or $2x - 3 > 5$

65. $7 - x \ge 10$ or $3x - 5 > 7$

66. $x + 3 > 5$ or $x + 4 < 0$

67. $-4x - 5 \ge 0$ and $3x - 2 \ge 1$

68. $2x - 3 \geq 5$ and $3x + 4 < -2$

69. $\frac{1}{2}x - 2 \geq 1$ or $\frac{2}{3}x - 1 \geq \frac{5}{3}$

70. $-\frac{1}{3}x + 1 \geq \frac{5}{3}$ or $\frac{4}{9}x - 1 < \frac{1}{3}$

71. $\frac{2}{5}x + \frac{1}{2} \geq \frac{1}{2}$ and $\frac{1}{4}x - \frac{1}{3} \leq \frac{1}{6}$

72. $\frac{1}{3}x - 1 > -\frac{5}{3}$ and $\frac{1}{4}x + \frac{1}{2} < \frac{1}{4}$

73. $0.8 - 0.2x < -0.4$ or $0.75x - 0.25 < -0.25$

74. $-0.2x + 0.6 < 0$ or $4.6x - 11.5 \geq 6.9$

75. $1.8x + 3.6 < 5.4$ and $2.35x + 7.05 > -4.7$

76. $-19.2 < 4.8x - 8$ and $8.1x - 13.5 \leq -10.8$

In Exercises 77–92, solve the compound inequality algebraically.

77. $-2x > x - 6$ and $2x - 4 \geq 1 + x$

78. $2 - x > 2 + 2x$ and $-2x > -3x$

79. $5x - 7 \geq 8$ and $2 - 3x < -4$

80. $-4x > x - 10$ and $x + 7 \leq 4$

81. $5 - 2x < -9$ or $6x - 2 > 5x$

82. $-5x \leq 10$ or $-3x \geq -4x - 2$

83. $x - 3 \leq 1$ and $x - 4 \geq 0$

84. $4x \leq 0$ and $x - 5 \geq -5$

85. $7 - x < 0$ or $x + 2 > 0$

86. $2x \geq 0$ or $2x \leq 12 + 5x$

87. $-2x + 6 \geq -4$ and $2x - 1 > 9$

88. $3x - 5 < -5$ and $2(x - 4) > -8$

89. $-x \leq 20 - 3x$ and $4 - x \leq -6$

90. $3(2x + 1) \geq 5x$ and $2x - 5 \leq -11$

91. $5 - x \leq 5$ or $3(2x + 1) < 3 + 5x$

92. $1 - 2x < 7$ or $2x + 8 < 7 + x$

93. Which of the following is a correct interpretation of $3 < x > 8$? Explain your answer.

(i) x is between 3 and 8.

(ii) x is greater than 3 and x is greater than 8.

94. If x is a real number such that $0.99 < x < 1$, would you describe the number of solutions as none, one, or infinitely many? Why?

Real-Life Applications

95. A doll manufacturer makes happy dolls and grumpy dolls. Each day, the number of grumpy dolls made is two-thirds the number of happy dolls made. If the total daily production is between 75 and 100, what is the least number of happy dolls made? What is the maximum number of grumpy dolls made?

96. A senior, a junior, and a freshman work on the student newspaper. Each month, the junior works two-thirds the number of hours that the senior works, and the freshman works one-fifth the number of hours that the senior does. If their combined hours for the month are at least 56, what is the minimum number of hours that the junior works?

97. An automobile repair shop charges a flat rate of \$50 plus \$32 per hour for the mechanic's time. If a customer receives an estimate of between \$130 and \$170 for fixing his car, what is the estimate of the mechanic's time?

98. The distance from a lawyer's home to the shopping center is twice the distance from her home to her office. The office and the shopping center are 4 miles apart. The lawyer leaves her home, goes to her office and to the shopping center, and returns home. If the total distance traveled was between 11.8 and 12.4 miles, how far is it from the lawyer's home to her office?

99. A drugstore sells toothbrushes for \$1.30 and toothpaste for \$3.00. For an average month the store sells twice as much toothpaste as toothbrushes, and combined sales for these items are between \$211.70 and \$255.50. How many toothbrushes are sold in an average month?

100. A builder constructs rectangular swimming pools. For every design the width is one-half of the length. If the perimeter of the pool can range between 120 feet and 192 feet, what pool widths does the builder offer?

101. In each of a player's ten basketball games, the number of free throws he made was two-thirds the number of two-point field goals he made. The player never scored less than 8 points and never scored more than 16 points. What was the minimum and maximum number of two-point field goals the player made?

102. In a small community the number of registered Republicans is 1.2 times the number of registered Democrats. It is known that 60% of the Democrats and 40% of the Republicans will vote in the special election. If the expected voter turnout is between 216 and 243, how many registered Democrats are there?

103. A packet of marigold seeds has twice as many seeds as a packet of delphinium seeds. The germination rate for marigolds is 90%, and the germination rate for delphiniums is 50%. A gardener plants all the seeds and expects to have a total of between 92 and 115 plants. How many marigold seeds were planted?

104. A light-bulb manufacturer finds that the number of defective 60-watt bulbs averages 1.4 times the number of defective 100-watt bulbs. On a given day the total number of defective bulbs is between 36 and 48. How many 100-watt bulbs are defective?

Modeling with Real Data

105. In 1993 the average annual cost of operating an automobile was $4486 plus 9.3 cents per mile. (Source: American Automobile Manufacturers Association.)

(a) Write an expression that models the average annual cost for an automobile that is driven m miles.

(b) Suppose that a company budgets between $5000 and $7000 for each car in the company pool. Write a double inequality to describe the number of miles that a car can be driven.

(c) Solve your inequality in part (b) and interpret the solution.

106. Using a suitable window setting, produce a graph of the model expression in Exercise 105. Then trace the graph to complete the following table.

Miles Driven	Cost
	$5000–$6000
11,000–12,000	
	$7000–$8000

107. Major league baseball owners lost more than $700 million during the 232-day strike in 1994–1995. However, as the following table shows, their profits were decreasing even before the strike. (Source: Associated Press.)

Year	Revenue (billions)	Expenses (billions)	Profit/Loss (millions)
1990	1.337	1.194	143
1991	1.537	1.438	99
1992	1.663	1.641	22
1993	1.865	1.829	36
1994	1.208	1.584	−376
1995	1.355	1.680	−325

Before the strike (1990–1993), revenue R and expenses E (in billions of dollars) can be modeled as follows.

$$R(t) = 0.17t + 1.35$$

$$E(t) = 0.21t + 1.20$$

In both models, t is the number of years since 1990.

(a) If the prestrike model had remained valid, write and solve an equation that describes the year when the owners would have only broken even.

(b) Write and solve an inequality that describes the years when the owners were projected to lose money.

108. Using the functions in Exercise 107, write and solve a compound inequality that describes the years when both revenue and expenses would exceed $2.5 billion.

Group Project: Debit Cards

In addition to using bank cards at ATM machines, consumers are using them as debit cards to pay for a variety of goods and services. The accompanying bar graph shows the average number of transactions per month by card holders during 1990–1995. (Source: Star System Consumer Survey.)

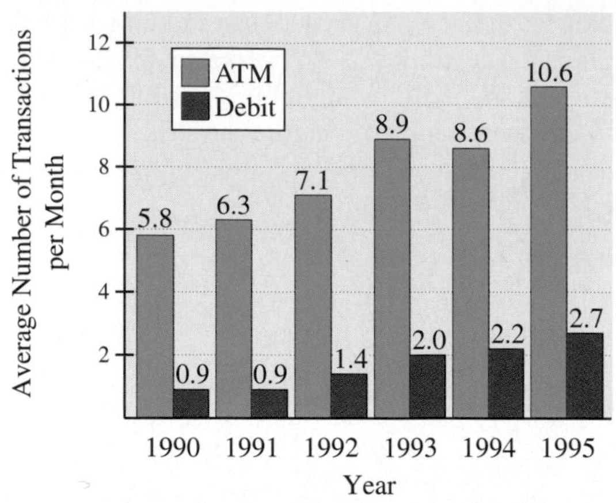

The average number T of ATM transactions and the average number B of debit card transactions can be modeled by the following functions, where x is the number of years since 1990.

$$T(x) = 0.93x + 5.55$$

$$B(x) = 0.40x + 0.72$$

109. Suppose that bank research indicates that there are enough ATM machines to serve customers as long as the average number of transactions does not exceed 16 per month. Write an inequality that models this condition.

110. Suppose that bank research also indicates that debit card usage is most profitable when the average number of transactions is at least 4 per month. Write an inequality that models this condition.

111. Write and solve a compound inequality that describes the years in which the conditions in both Exercises 109 and 110 are met.

112. What are some advantages and risks to you as a consumer in the use of debit cards?

Challenge

113. Given $x + k < 5$ and $k - x < 1$, for what values of k is the solution set empty?

114. Given $x + k > -2$ or $k - x > -1$, for what values of k is the solution set all real numbers?

115. Write a complete explanation of how to solve the following compound inequality graphically.

$$0.5x - 15 < x - 5 < -x + 21$$

Include these items in your explanation.

(a) What functions are entered in the calculator and graphed?

(b) What is the significance of the points of intersection?

(c) What portion of the total graph reveals the solutions? What is the solution set?

116. Use the following guide for algebraically solving the compound inequality in Exercise 115.

(a) Write the double inequality in the form of a conjunction.

(b) Solve each inequality.

(c) Form the intersection of the two solution sets in part (b).

(d) Compare your result with the solution set obtained in Exercise 115.

In Exercises 117–120, solve the compound inequality algebraically and write the solution set with interval notation.

117. $3x + 2 \le x + 8 \le 2x + 15$

118. $2x - 1 < x + 1 < 7 - x$

119. $x < x + 1 < x + 2$

120. $x + 1 < x < x + 2$

3.8 ABSOLUTE VALUE: EQUATIONS AND INEQUALITIES

Graphs and Number Lines • Algebraic Interpretation • Special Cases • Algebraic Methods • Real-Life Applications

Graphs and Number Lines

Equations and inequalities sometimes involve absolute values. The following are examples of **absolute value equations** and **absolute value inequalities.**

$$|2x - 3| = 5 \qquad |1 - x| \geq 9 \qquad |x^2 + 4x - 12| < 10$$

We can use graphing methods to develop generalizations about the solution sets of absolute value equations and inequalities.

EXPLORING THE CONCEPT

Solutions of Absolute Value Equations and Inequalities

Figure 3.39 shows the graphs of $y_1 = |x|$ and $y_2 = 15$.

Figure 3.39

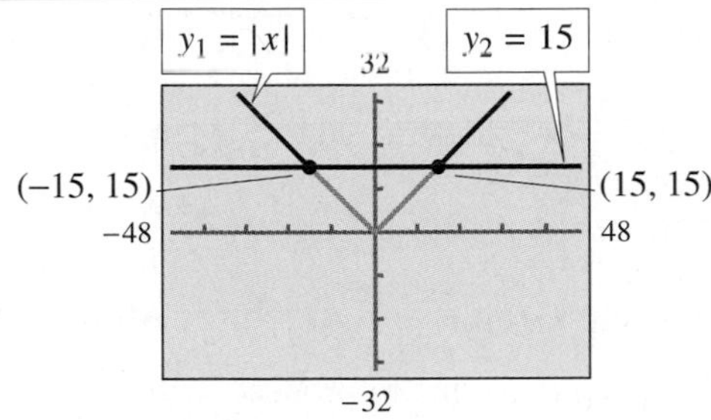

The graphs intersect at $(-15, 15)$ and $(15, 15)$. The x-coordinates of these points represent the solutions of the equation $|x| = 15$, so the solution set is $\{-15, 15\}$. Note that these two solutions are 15 units from the origin on the number line. (See Fig. 3.40.)

Figure 3.40

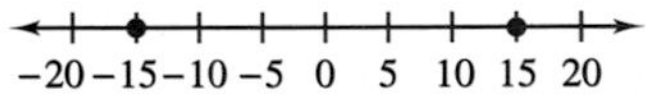

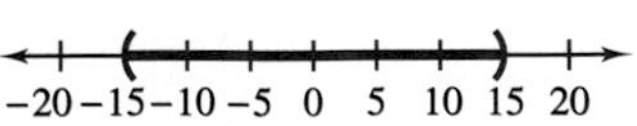

Figure 3.39 also shows that the graph of y_1 is below the graph of y_2 for all x-coordinates between -15 and 15. Therefore, the solution set of the inequality $|x| < 15$ is $\{x \mid -15 < x < 15\}$. Note that these solutions are less than 15 units from the origin on the number line. (See Fig. 3.41.)

Figure 3.41

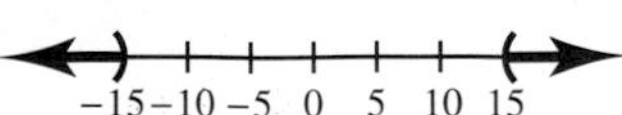

Finally, Fig. 3.39 shows that the graph of y_1 is above the graph of y_2 for all x-coordinates less than -15 and for all x-coordinates greater than 15. Therefore, the solution set of the inequality $|x| > 15$ is $\{x \mid x < -15 \text{ or } x > 15\}$. These solutions are all more than 15 units from the origin on the number line. (See Fig. 3.42.)

Figure 3.42

−15 −10 −5 0 5 10 15

Algebraic Interpretation

As we used graphing methods to estimate solutions of absolute value equations and inequalities, we observed that the solutions can be interpreted in terms of their distance from the origin on the number line. These observations can be generalized. For any expression A and any positive number C, the following are true:

	Graphic Interpretation	*Algebraic Interpretation*
1. $\lvert A\rvert = C$	A represents a number that is C units from the origin.	$A = C$ or $A = -C$

Figure 3.43

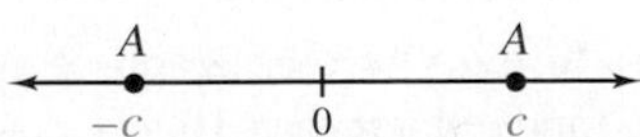

2. $|A| > C$ — A represents a number that is more than C units from the origin. — $A < -C$ or $A > C$

Figure 3.44

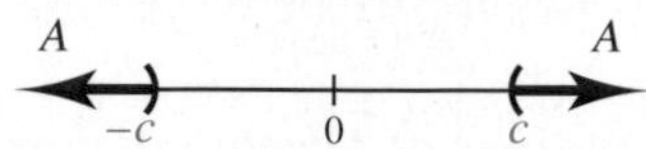

3. $|A| < C$ — A represents a number that is less than C units from the origin. — $A > -C$ and $A < C$

Figure 3.45

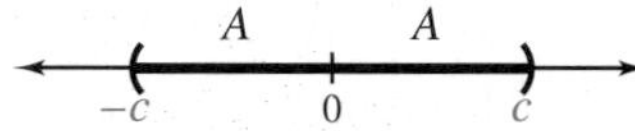

In Example 1 we illustrate these algebraic interpretations with the assistance of number line graphs.

EXAMPLE 1

Combining Graphs and Algebraic Methods

Use the integer setting to produce the graphs of $y_1 = |3 - 2x|$ and $y_2 = 17$ on the same coordinate axes. Use the graphs to estimate the solution set for each of the following. Then determine the solutions algebraically.

(a) $|3 - 2x| = 17$ (b) $|3 - 2x| > 17$ (c) $|3 - 2x| < 17$

Solution

Figure 3.46

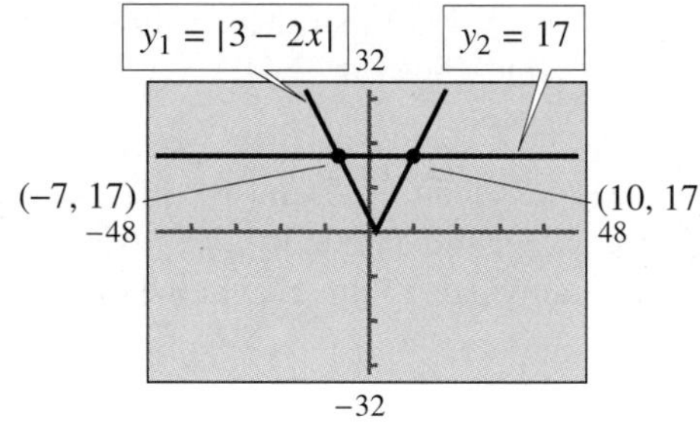

Figure 3.47

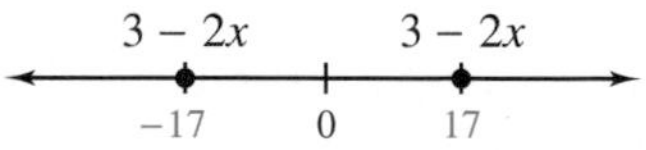

$3 - 2x$ $3 - 2x$

−17 0 17

(a) In Fig. 3.46 there are two points of intersection, and the solutions appear to be -7 and 10.

Because $|3 - 2x| = 17$, the expression $3 - 2x$ represents a number that is 17 units from the origin. (See Fig. 3.47.)

This suggests the following equations. Solve each one.

$$\begin{aligned} 3 - 2x &= 17 \quad \text{or} \quad 3 - 2x = -17 \\ -2x &= 14 \quad \text{or} \quad -2x = -20 \\ x &= -7 \quad \text{or} \quad x = 10 \end{aligned}$$

The solutions are -7 and 10.

(b) In Fig. 3.46 the graph of y_1 is above the graph of y_2 for x-values less than -7 and greater than 10.

Figure 3.48

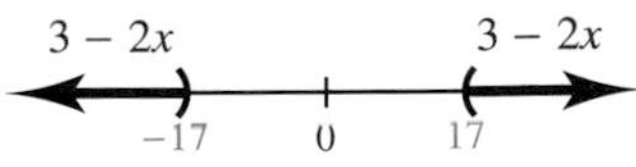

Because $|3 - 2x| > 17$, the expression $3 - 2x$ represents a number that is more than 17 units from the origin. (See Fig. 3.48.)

This suggests the following inequalities. Solve each one.

$$\begin{aligned} 3 - 2x &< -17 \quad \text{or} \quad & 3 - 2x &> 17 \\ -2x &< -20 \quad \text{or} \quad & -2x &> 14 \\ x &> 10 \quad \text{or} \quad & x &< -7 \end{aligned}$$

The solution set is $(-\infty, -7) \cup (10, \infty)$.

(c) In Fig. 3.46 the graph of y_1 is below the graph of y_2 for x-values between -7 and 10.

Figure 3.49

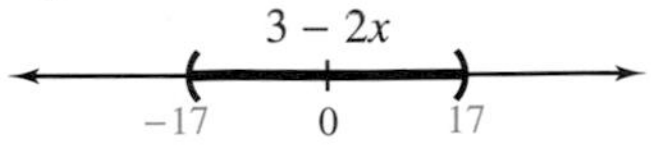

3 − 2x

−17 0 17

Because $|3 - 2x| < 17$, the expression $3 - 2x$ represents a number that is less than 17 units from the origin. (See Fig. 3.49.)

This suggests the following inequalities. Solve each one.

$$\begin{aligned} 3 - 2x &> -17 \quad \text{and} \quad & 3 - 2x &< 17 \\ -2x &> -20 \quad \text{and} \quad & -2x &< 14 \\ x &< 10 \quad \text{and} \quad & x &> -7 \end{aligned}$$

The solution set is $(-7, 10)$.

Special Cases

The absolute value of a number is never negative, and the absolute value of 0 is 0. These limitations can present some interesting special situations when we solve absolute value equations and inequalities.

EXAMPLE 2

Special Cases

Solve the following equations and inequalities. In each case note the value of the constant on the right side. What general conclusions can you draw?

(a) $|x - 6| = -12$ (b) $|x - 6| > -12$

(c) $|2x + 14| = 0$ (d) $|2x + 14| < 0$

Solution

Figure 3.50

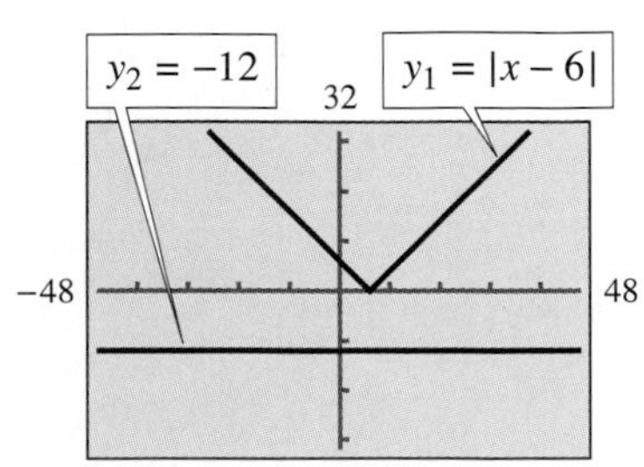

(a) In Fig. 3.50 we produce the graphs of $y_1 = |x - 6|$ and $y_2 = -12$ on the same coordinate axes.

There is no point of intersection, so the equation has no solution.

Note that $|x - 6|$ cannot be negative. Therefore, $|x - 6| = -12$ is false.

Generalization: An equation $|A| = C$, where $C < 0$, has no solution.

(b) In Fig. 3.50, every point of the graph of y_1 is above the graph of y_2, so every real number is a solution.

Because $|x - 6| \geq 0$, $|x - 6| > -12$ is always true.

Generalization: An inequality $|A| > C$, where $C < 0$, is satisfied by all real numbers.

Figure 3.51

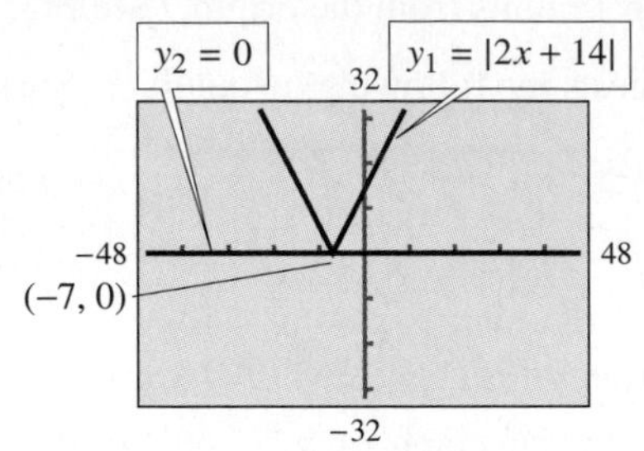

(c) In Fig 3.51 we produce the graphs of $y_1 = |2x + 14|$ and $y_2 = 0$. Note that the graph of y_2 is the x-axis.

There appears to be one point of intersection, $(-7, 0)$, so the equation has one solution, -7.

Observe that $|2x + 14| = 0$ only if $2x + 14 = 0$, which implies that $x = -7$.

Generalization: An equation $|A| = 0$ implies that $A = 0$.

(d) In Fig. 3.51 it appears that no point of the graph of y_1 is below the graph of y_2, so there is no solution.

Again, $|2x + 14|$ cannot be negative. Therefore, $|2x + 14| < 0$ is false.

Generalization: An inequality $|A| < 0$ has no solution.

LEARNING TIP

Rather than memorizing the various possibilities for absolute value inequalities, use the definition of absolute value and graphs to see the associated inequalities.

Algebraic Methods

The graphing method is useful for estimating the number of solutions or for estimating the solutions themselves, but to determine exact solutions, we use algebraic methods.

We can use algebraic methods to solve the compound equations or inequalities arising from absolute value equations or inequalities. We normally begin by isolating the absolute value expression.

EXAMPLE 3

Solving Absolute Value Equations and Inequalities Algebraically

Solve each of the following algebraically.

(a) $|2x + 5| - 4 = -2$ (b) $|1 - 3x| - 6 \geq 1$ (c) $9 - |x - 2| > 1$

Solution

(a)

$$|2x + 5| - 4 = -2$$

$$|2x + 5| - 4 + 4 = -2 + 4 \quad \text{Add 4 to both sides.}$$

$$|2x + 5| = 2 \quad \text{The equation is in the form } |A| = C.$$

Now write $|2x + 5| = 2$ as two linear equations and solve each one.

$$2x + 5 = 2 \quad \text{or} \quad 2x + 5 = -2 \qquad \text{If } |A| = C, \text{ then } A = C \text{ or } A = -C.$$

$$2x = -3 \quad \text{or} \quad 2x = -7$$

$$x = -\frac{3}{2} \quad \text{or} \quad x = -\frac{7}{2}$$

The solutions are $-\frac{3}{2}$ and $-\frac{7}{2}$.

Think About It

Can $(-\infty, -2] \cup (2, \infty)$ represent the solutions of an absolute value inequality of the form $|Ax + B| > C$ or $|Ax + B| \geq C$?

(b)
$$|1 - 3x| - 6 \geq 1$$
$$|1 - 3x| - 6 + 6 \geq 1 + 6 \quad \text{Add 6 to both sides.}$$
$$|1 - 3x| \geq 7 \quad \text{The inequality is in the form } |A| \geq C.$$

The resulting inequality can be written as a disjunction.

$$1 - 3x \leq -7 \quad \text{or} \quad 1 - 3x \geq 7 \quad \text{If } |A| \geq C \text{, then } A \leq -C \text{ or } A \geq C.$$
$$-3x \leq -8 \quad \text{or} \quad -3x \geq 6$$
$$x \geq \frac{8}{3} \quad \text{or} \quad x \leq -2$$

Figure 3.52

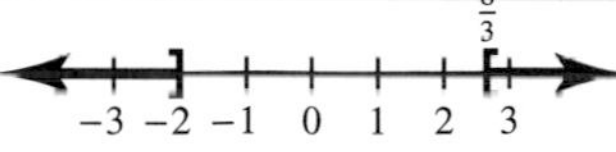

The solution set is $(-\infty, -2] \cup \left[\frac{8}{3}, \infty\right)$. (See Fig. 3.52.)

(c)
$$9 - |x - 2| > 1$$
$$9 - |x - 2| - 9 > 1 - 9 \quad \text{Subtract 9 from both sides.}$$
$$-|x - 2| > -8 \quad \text{Multiply both sides by } -1 \text{ and reverse the inequality symbol.}$$
$$|x - 2| < 8 \quad \text{The inequality is in the form } |A| < C.$$

The resulting inequality can be written as a conjunction.

$$x - 2 > -8 \quad \text{and} \quad x - 2 < 8 \quad \text{If } |A| < C \text{, then } A > -C \text{ and } A < C.$$
$$x > -6 \quad \text{and} \quad x < 10$$

Figure 3.53

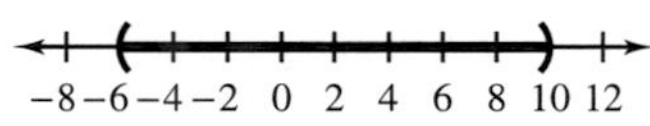

The solution set is $(-6, 10)$. (See Fig. 3.53.)

If two expressions have the same absolute value, then the expressions must be equal or they must be opposites. Thus an equation of the form $|A| = |B|$ is equivalent to the disjunction $A = B$ or $A = -B$.

EXAMPLE 4

Solving Equations of the Form $|A| = |B|$

Solve the following equations.

(a) $|4 - x| = |3 - 2(5 - x)|$ (b) $|4 - x| = |x - 7|$

Solution

(a)
$$|4 - x| = |3 - 2(5 - x)|$$
$$|4 - x| = |3 - 10 + 2x| \quad \text{Distributive Property}$$
$$|4 - x| = |-7 + 2x| \quad \text{The equation is in the form } |A| = |B|.$$
$$4 - x = -7 + 2x \quad \text{or} \quad 4 - x = -(-7 + 2x) \quad \text{If } |A| = |B| \text{ then } A = B \text{ or } A = -B.$$
$$-3x = -11 \quad \text{or} \quad 4 - x = 7 - 2x$$
$$x = \frac{11}{3} \quad \text{or} \quad x = 3$$

The solutions are $\frac{11}{3}$ and 3.

(b) $|4 - x| = |x - 7|$

$$\begin{aligned} 4 - x &= x - 7 \quad \text{or} \quad 4 - x = -(x - 7) \\ 4 - x &= x - 7 \quad \text{or} \quad 4 - x = 7 - x \\ 11 &= 2x \quad \text{or} \quad 4 = 7 \quad \text{False} \\ x &= \frac{11}{2} \end{aligned}$$

The solution is $\frac{11}{2}$.

Real-Life Applications

Absolute value inequalities typically arise in applied situations in which the difference between two quantities is less than or greater than a fixed amount. The following example illustrates how absolute value plays a role in matters of *tolerance* in measurements.

EXAMPLE 5

Temperature Tolerances

A nuclear reactor shuts down automatically if the temperature in the reactor differs from 1200°F by more than 30°. Write and solve an absolute value inequality that describes these conditions.

Solution

Let T = reactor temperature.

$$|T - 1200| \le 30 \quad \text{The difference between } T \text{ and 1200 is at most 30.}$$

$$T - 1200 \ge -30 \quad \text{and} \quad T - 1200 \le 30$$

$$T \ge 1170 \quad \text{and} \quad T \le 1230$$

In interval notation, the solution set is [1170, 1230]. The reactor will operate as long as the temperature is between 1170° and 1230°, inclusive.

3.8 QUICK REFERENCE

Graphs and Number Lines

- To estimate the solution set of an absolute value equation or inequality graphically, graph the left and right sides of the equation or inequality. Determine the point(s) of intersection.
  1. If the symbol is $=$, $\le$, or $\ge$, the x-coordinate of each point of intersection is a solution.
  2. If the symbol is $<$, determine the x-values of all points for which the graph of the left side is *below* the graph of the right side.
  3. If the symbol is $>$, determine the x-values of all points for which the graph of the left side is *above* the graph of the right side.
- We can interpret the solutions of an absolute value equation or inequality in terms of their distance from 0 on the number line.

Algebraic Interpretation

- If $C > 0$, then
  1. $|A| = C$ is equivalent to $A = C$ or $A = -C$.
  2. $|A| > C$ is equivalent to $A > C$ or $A < -C$.
  3. $|A| < C$ is equivalent to $A < C$ and $A > -C$.

Special Cases

- If $C = 0$, then
  1. $|A| = C$ is equivalent to $A = 0$.
  2. $|A| > C$ is true for every real number except for the solution of $A = 0$.
  3. $|A| < C$ has no solution.
- If $C < 0$, then
  1. $|A| = C$ has no solution.
  2. $|A| > C$ is true for every real number.
  3. $|A| < C$ has no solution.

Algebraic Methods

- To solve a compound absolute value equation or inequality algebraically:
  1. Isolate the absolute value expression.
  2. Translate the absolute value equation or inequality into an equivalent conjunction or disjunction.
  3. Solve each component of the compound equation or inequality and form the union or intersection of the solution sets.

3.8 EXERCISES

Concepts and Skills

1. Interpret $|x - 2| = 5$ in terms of distance on a number line.

2. How does the solution set for $|x + 3| = 5$ differ from the solution set for $x + 3 = 5$?

In Exercises 3–8, estimate the solutions of the given equations by graphing.

3. $|x| = 12$	**4.** $|x - 8| = 9$
5. $|2x - 10| + 7 = 7$	**6.** $|6 - x| - 9 = 5$
7. $-|x + 10| = -15$	**8.** $10 - |x + 8| = 17$

In Exercises 9–14, solve the equation algebraically.

9. $|2x - 5| = 10$	**10.** $|3 - 4x| = 12$
11. $|5x - 2| = 0$	**12.** $|5 - x| = 0$
13. $|x + 4| - 3 = 7$	**14.** $6 + |3 - x| = 7$

In Exercises 15 and 16, use the graphics display to determine the solution set of each of the following.

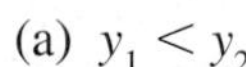

(a) $y_1 < y_2$

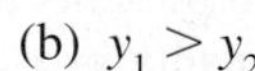

(b) $y_1 > y_2$

(c) $y_1 = y_2$

15.

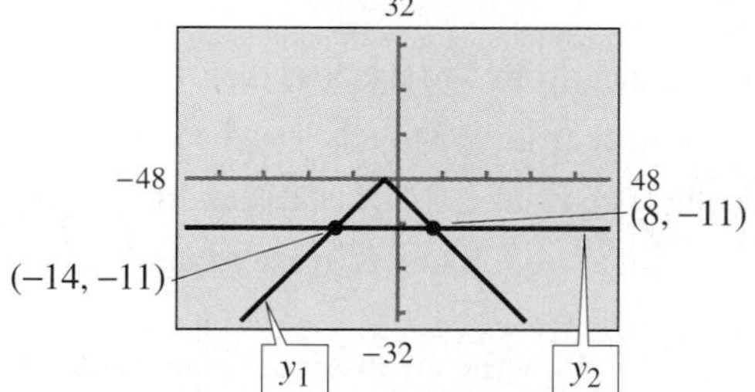

16.

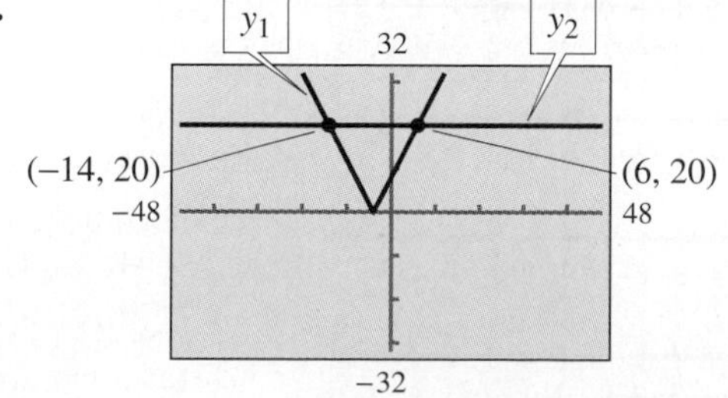

In Exercises 17–22, estimate the solutions of the inequality by graphing. Write the solution set with interval notation.

17. $|x| < 12$

18. $|x - 8| \geq 9$

19. $15 - |x - 6| > 6$

20. $-|x + 10| \leq -15$

21. $|2x - 10| + 7 \geq 7$

22. $10 - |x + 8| \leq 17$

23. What is the difference between the solution sets of $|x| \geq 5$ and $|x| \leq 5$?

24. Explain how the graphing method shows that the inequality $|x| < -5$ has no solutions.

In Exercises 25–32, solve the inequality algebraically. Write the solution set with interval notation.

25. $|1 - 2t| < 9$

26. $|3y + 2| < 14$

27. $|7 + x| \leq 0$

28. $|x - 1| \leq 0$

29. $|4 - x| - 5 \leq 2$

30. $|2x - 1| - 3 \leq 10$

31. $|x + 5| < -3$

32. $|4 + 3x| < -5$

In Exercises 33–40, solve the inequality algebraically. Write the solution set with interval notation.

33. $|3x + 2| \geq 7$

34. $|5 - 2x| \geq 11$

35. $|x - 5| > 0$

36. $|3 + x| > 0$

37. $|3x - 2| \geq -5$

38. $|5 - 7x| \geq -1$

39. $|4t + 3| + 2 > 10$

40. $|2 - 3t| - 9 > 8$

In Exercises 41–44, x represents any real number. Write an absolute value inequality in x to describe each situation below.

41. The distance from 5 to x is less than 7.

42. The distance from -5 to x is less than or equal to 4.

43. The distance from -7 to x is at least 10.

44. The distance from 7 to x is greater than 12.

In Exercises 45–48, write an absolute value inequality in x to describe each graph.

45. −5 −4 −3 −2 −1 0 1 2 3 4 5 6

46. −5 −4 −3 −2 −1 0 1 2 3 4 5 6

47. −4 −3 −2 −1 0 1 2 3 4 5

48. −4 −3 −2 −1 0 1 2 3 4 5

In Exercises 49–64, solve each equation algebraically.

49. $|2k - 3| = 8$

50. $|4 - 3m| = 15$

51. $|y + 1| - 3 = 7$

52. $|2x + 3| - 5 = 12$

53. $|t| + 8 = 5$

54. $|-t| - 7 = -12$

55. $5 - 3|x - 5| = -13$

56. $2 - |x + 1| = -7$

57. $|t + 4| - 3 = -23$

58. $|m - 23| + 9 = 4$

59. $|3x - 4| - 4 = -4$

60. $|5 - 2x| + 5 = 5$

61. $5 + 3|2x + 1| = 20$

62. $7 - 2|1 - 3x| = -13$

63. $\left|\dfrac{5 - 3x}{4}\right| = 3$

64. $\left|x - \dfrac{2}{3}\right| = \dfrac{3}{4}$

In Exercises 65–70, solve the given equation.

65. $|x - 1| = |x - 19|$

66. $|x + 3| = |x - 3|$

67. $|2x - 5| = |x|$

68. $|t| = |3 - 4t|$

69. $|x - 2| = |2 - x|$

70. $|1 - 2x| = |2x - 1|$

71. What is the first step to solve $-3|x - 2| \geq 12$?

72. Explain why each given pair of inequalities is equivalent or not equivalent.

(a) $|x| > 5$ $\quad$ $x > 5$

(b) $|x - 1| \leq 3$ $\quad$ $-3 \leq x - 1 \leq 3$

In Exercises 73–90, solve each inequality algebraically. Write the solution set with interval notation.

73. $-5|2x + 3| < -35$

74. $-2|7 - x| > -16$

75. $5 - |4 - 3x| > 2$

76. $12 - |6 - 5p| < 9$

77. $|4x + 3| + 9 \geq 4$

78. $|3x + 2| \leq -3$

79. $|3y - 5| + 4 \geq 10$

80. $|2x - 7| - 3 \leq 8$

81. $|3y - 5| + 6 \leq 6$

82. $|2k + 7| - 7 \geq 8 - 15$

83. $|5 - 3x| + 7 > 9$

84. $|4 - 3x| - 5 < 2$

85. $5 - |x - 2| \geq 7$

86. $24 - |m + 3| \leq 27$

87. $5 - 2|2x - 1| > -1$

88. $4 - 3|x + 5| < -8$

89. $\left|\dfrac{2x + 1}{4}\right| > 1$

90. $\left|\dfrac{4x - 3}{2}\right| + 5 \leq 7$

In Exercises 91–94, use the graphing method to estimate the solution set of the given inequality.

91. $|x| \geq x + 3$

92. $|x| \geq x$

93. $|x - 3| < 1 - 2x$

94. $2x + 7 < |x + 2|$

In Exercises 95–98, use the graphing method to estimate the solution set of the given inequality.

95. $|x + 6| < |x - 4|$ **96.** $|x + 5| > |x - 5|$

97. $|t| \geq |15 - 2t|$

98. $|2x - 11| \leq |x + 6|$

99. If x_1 and x_2 are two numbers on a number line, the distance between them is $|x_1 - x_2|$. Why is it necessary to take the absolute value of the difference?

100. Solving the inequality $|-x| < 3$, a student states that $|-x| = x$ and therefore $x < 3$. Explain the error in the student's reasoning.

Modeling Conditions

In Exercises 101–104, describe the given conditions with an absolute value inequality and write the corresponding double inequality.

101. A carpenter cuts a board to a length of 12 feet with no more than a 2% error. (Let L = the length of the cut board.)

102. A scale is never off its measurement by more than 3 pounds. The scale indicates that a person weighs 135 pounds. (Let w = the person's weight.)

103. A pipe 3 inches in diameter must have a tolerance of one-sixteenth of an inch so that the pipe can fit into a specific hole in the wall. (Let d = the diameter of the pipe.)

104. A freezer is most efficient when the temperature is 5°F plus or minus 6°. (Let t = temperature of the freezer.)

Real-Life Applications

In Exercises 105–108, write and solve an absolute value inequality that describes the given conditions.

105. On a psychological test the actual score s differs from the expected score of 50 by no more than 6 points. What is the range of actual scores?

106. For submarine duty a sailor's height h can differ from 6 feet by no more than 2 inches. What range of heights is required?

107. A medical clinic considers a patient's body temperature to be normal if it deviates from 98.6°F by no more than 0.5°. What is the normal temperature range?

108. A housing inspector approves residential wiring if the incoming voltage deviates from 220 volts by less than 5 volts. What range of voltages would pass inspection?

Group Project: Health Care Costs

In 1992 the average annual per capita out-of-pocket expenditure for health care was \$1634. However, the per capita expenditure for those under 25 years of age was \$416 compared with \$2474 for people over age 65. (Source: U.S. Bureau of Labor Statistics.) The average annual per capital expenditure c can be modeled by $c(t) = 37.7t - 170$, where t is a person's age.

109. Write an inequality that can be used to predict the age groups for which the annual expenditure differed from the average by at least \$400.

110. Solve the inequality in Exercise 109 and interpret the solution.

111. Write an inequality that can be used to predict the age groups for which the annual expenditure differed from the average by less than 20%. Then solve the inequality and interpret the solution.

112. The data refer to out-of-pocket expenditures. How might the given model differ from a model that reflected total health care costs? Would you expect the solutions to the inequalities in Exercises 109 and 111 to differ significantly from the solutions for this new model?

Challenge

113. For what values are the following true? Explain your reasoning.

(a) $|x - 2| = x - 2$

(b) $|x - 2| = -(x - 2)$

(c) $|x - 2| = |2 - x|$

In Exercises 114–117, solve the given equation.

114. $|x - 4| = x + 2$ **115.** $|t - 3| = t + 3$

116. $|x - 1| = 7 - (x - 1)$

117. $|t + 5| = -t$

In Exercises 118 and 119, determine the values of x for which the given equation is true.

118. $x + |x| = 2x$ **119.** $x + |x| = 0$

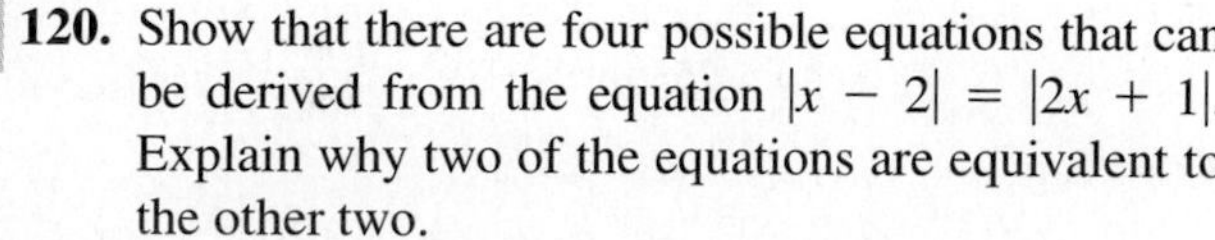

120. Show that there are four possible equations that can be derived from the equation $|x - 2| = |2x + 1|$. Explain why two of the equations are equivalent to the other two.

3. CHAPTER PROJECT Juvenile Crime

Referring to the table in the Group Project for Section 3.5, we see that the greatest percentage increase in the number of state court cases from 1984 to 1994 was in the juvenile category. In 1994, the lowest violent crime rate among juveniles was in Vermont, while the highest was in New York. (Source: Center for the Study of Social Policy.)

In 1991, U.S. juvenile courts handled 260,300 cases. The following table shows the general categories of offenses.

Offense	Percentage of Cases
Assault	75%
Robbery	12%
Delinquency	7%
Sex crimes	5%
Homicide	1%

1. Use the information in the table to produce a bar graph showing the categories of offenses and the actual number of cases in those categories.
2. In 1987 there were 7.1 juvenile cases for every 1000 juveniles in the United States. By 1991 that figure had increased by 41%. How many juvenile cases per 1000 juveniles were there in 1991?
3. Juvenile drug cases decreased by 23% from 1988 to 1991, but there was a 15% increase from 1991 to 1992. What was the overall percentage change in drug cases from 1988 to 1992?
4. The 1996–1997 National Debate topic was juvenile crime prevention. Suggested ideas included the following:
   (a) Increase the number of school hours.
   (b) Increase the number of hours that juveniles are allowed to work.
   (c) Lower the minimum wage for juvenile offenders.
   (d) Confiscate juvenile offenders' automobiles.
   (e) Revoke free lunch programs for offenders.
   (f) Legalize drugs.

   In groups, use these or ideas of your own to prepare a proposal for reducing juvenile crime. Whenever possible, back up your position with data, tables, graphs, or other supporting evidence.
5. Assess each group's report. In doing so, give special weight to the relevance of the data and to the quality of their presentation.

3. CHAPTER REVIEW EXERCISES

Section 3.1

In Exercises 1 and 2, state whether the equation is a linear equation.

1. $3x^2 - 4x = 7$

2. $\frac{2}{3}x - 5 = 9$

In Exercises 3 and 4, verify that the given number is a solution of the equation.

3. $3x - 7 = 2(x - 5)$; -3

4. $\frac{2}{7}(x - 5) + \frac{1}{3}(x + 7) = x + 2$; $-\frac{23}{8}$

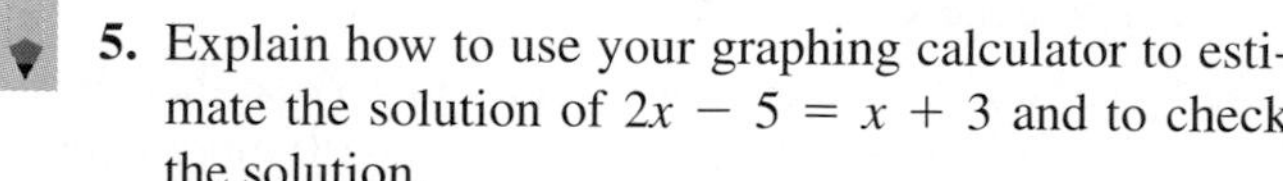

5. Explain how to use your graphing calculator to estimate the solution of $2x - 5 = x + 3$ and to check the solution.

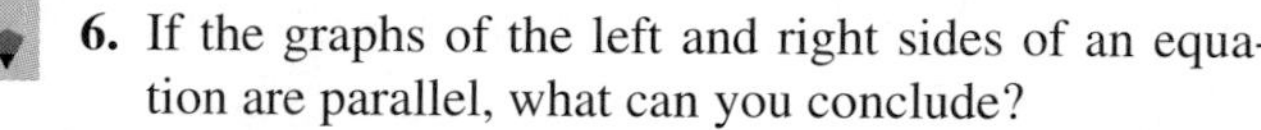

6. If the graphs of the left and right sides of an equation are parallel, what can you conclude?

In Exercises 7–10, use the graphing method to estimate the solution of the given equation.

7. $4 - x = -2$

8. $4(2x + 1) = 8x - 1$

9. $2(x + 3) = 2(x + 1) + 4$

10. $\frac{3}{2}x = \frac{5}{2} + x$

Section 3.2

In Exercises 11–18, solve the equation.

11. $\frac{-y}{3} = 10$

12. $4(5x - 6) - 6(3x - 2) = 4$

13. $0.2n + 0.3 = 0.3n - 5.2$

14. $\frac{3}{2}p + \frac{1}{6} = \frac{5}{3}$

15. $7x - 5 = 12x - 5 - 4x$

16. $\frac{x - 2}{6} = \frac{x}{3}$

17. $4(2s + 1) = s + 3(2s - 1)$

18. $\frac{2}{3}(k + 2) + \frac{1}{4}(k - 4) = k - \frac{1}{6}$

In Exercises 19–22, solve the equation. Then state whether the equation is an identity, an inconsistent equation, or a conditional equation.

19. $3x - 2(1 - 4x) = 6x + 5$

20. $3c - 4(1 + c) = 5 - c$

21. $2 - (2y - 3) = 5 - 2y$

22. $0.4q - 1.4 + 0.6q = 1 - 0.8q + 1.4$

In Exercises 23 and 24, determine k so that the equation meets the given condition.

23. $3 - 2x = kx - 2 + 3x$
The equation is inconsistent.

24. $-2(k - x) = 2x + 6$
The equation is an identity.

Section 3.3

In Exercises 25–30, solve each formula for the given variable.

25. $V = LWH$ for H

26. $R_T = \frac{R_1 + R_2}{2}$ for R_2

27. $\frac{ac}{x} + \frac{cd}{y} = 6$ for d

28. $A = \frac{1}{2}h(a + b)$ for b

29. $E = IR$ for I

30. $h = vt - 16t^2$ for v

In Exercises 31 and 32, solve the equation for y.

31. $3x - 4y - 25 = 0$

32. $\frac{4}{5}x = \frac{2}{3}y - \frac{2}{5}$

33. Solve the formula $A = P + Prt$ for r. Then use the formula to determine the simple interest rate (r) required for an investment of \$2400 (P) to be worth \$2556 (A) at the end of one year (t).

34. To obtain a building permit for a small office building, the developer must have at least 4500 square feet of paved parking area. (See figure on page 202.) If the trapezoidal area in front of the building is paved for parking, will the building permit requirement be met? What is the actual paved area?

Figure for 34

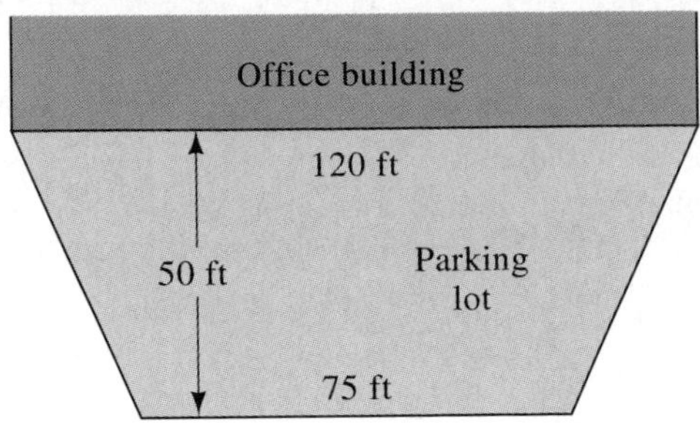

Figure for 38

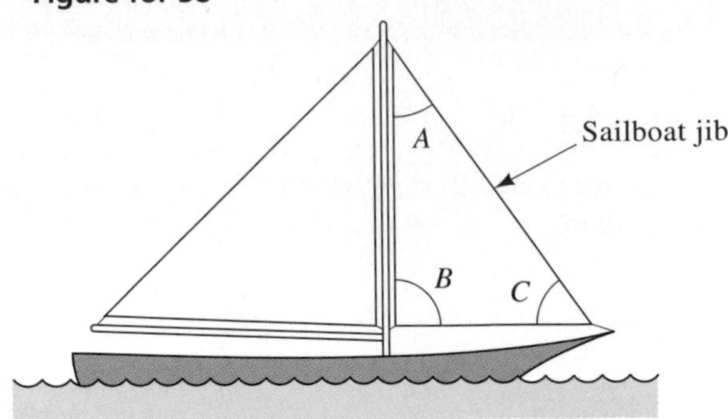

35. A circular goldfish pool is surrounded by a sidewalk. (See figure.) The circumference of the outside of the sidewalk is 43.98 feet, and the circumference of the inside of the sidewalk is 31.42 feet. What is the width of the sidewalk?

Figure for 35

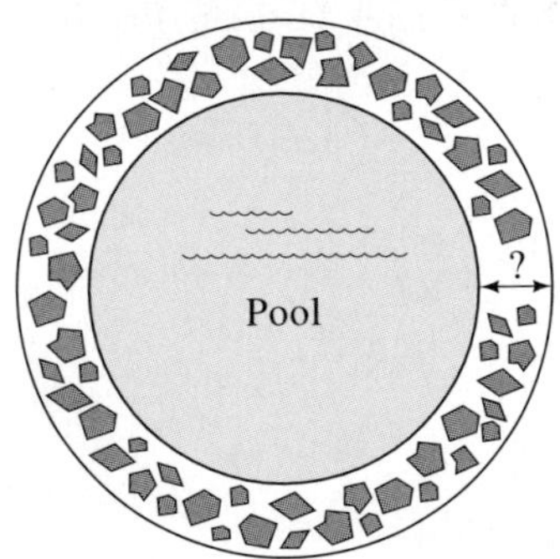

Section 3.4

36. One number is 3 less than another number. If one-fifth of the larger number is added to one-half of the smaller number, the result is 9. What is the smaller number?

37. A 100-foot rope is to be cut into three pieces whose lengths are consecutive integers. One foot of rope will be left over. How long should each piece of rope be?

38. A sailboat jib is not quite a right triangle, with angle A being 10° less than one-half of angle B. Angle C is 15° more than angle B. What are the measures of the three angles? (See figure.)

39. A coin collection jar is full of just nickels and dimes. If there are 355 coins valued at \$24.25, how many nickels are in the jar?

40. Determine two consecutive odd integers so that the difference between twice the first and 20 more than the second is 31.

Section 3.5

41. Menswear Unlimited advertised, "All suits are 40% off." If a suit is on sale for \$93, what was the original price?

42. How much money was invested at 6% simple interest if the value of the investment was \$1653.60 at the end of 1 year?

43. Stewart Construction Company rents a large crane for \$275 per day and provides an operator for \$20 per hour. The crane rental bill for July 29 was \$455. How long did the operator work?

44. A 10% salt solution is to be mixed with a 20% salt solution to obtain 15 gallons of a 12% salt solution. How many gallons of each should be used?

45. Two trucks leave Kansas City at midnight and travel in opposite directions. The truck traveling east is averaging 6 mph less than the truck going west. At 3:30 A.M. the trucks are 469 miles apart. What is the average rate of each truck?

46. Two investments yield 7.5% and 9.3%, respectively. If a total of \$8000 is invested and the income for the year is \$695.40, how much is invested at each rate?

47. Two angles are complementary. The difference between one-third of the larger angle and the smaller angle is 10. What is the measure of each angle?

48. At a chemical plant, a mixing vat receives water from a large pipe and an alkaline solution from a small pipe. The large pipe delivers liquid 1.5 times as fast as the small pipe. The small pipe is opened at noon and the large pipe is opened at 1:45 P.M. At 5:30 P.M., the tank contains 133.5 gallons of liquid. At what rate does each pipe deliver liquid?

Section 3.6

In Exercises 49–51, for the given graph write the set with interval notation.

49. 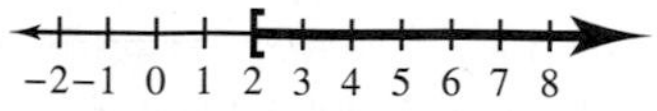

50. 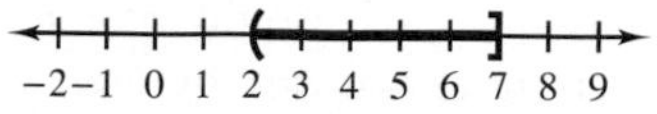

51. 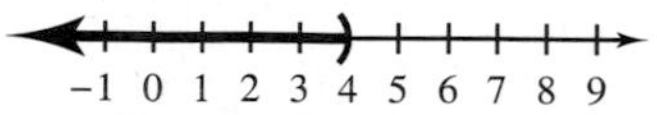

In Exercises 52–54, illustrate the set on a number line.

52. $(-\infty, 7]$ **53.** $[0, 3)$ **54.** $(-2, \infty)$

In Exercises 55–58, use the graphing method to estimate the solutions of the inequality and write the solution set with interval notation.

55. $2x - 1 > 7$ **56.** $-2(x - 2) < 1 - 2x$

57. $5x + 7 - 3x \geq 7 + 2x$

58. $-8 < \frac{1}{2}x - 2 \leq 7$

In Exercises 59–70, solve the inequality.

59. $3x - 4 < 14$ **60.** $7 - 3t \leq -8$

61. $4x - 4 \leq 7x - 13$ **62.** $5x - 3 \geq 6x + 5$

63. $-17 < 4x - 7 \leq -8$

64. $6 + 3(2x - 5) \geq 2x + 3(x - 2) - 7$

65. $-\frac{2}{3}(x - 4) \leq \frac{2x - 3}{-4}$

66. $-3 < \frac{2x - 1}{-1} < 5$

67. $7 > x + 3 > -4$

68. $\frac{1}{6}t + \frac{2}{3} < \frac{1}{3}t - \frac{1}{2}$

69. $-8 \leq -4x < 20$ **70.** $\frac{t}{-3} \geq -9$

71. The width of a rectangular lot is 7 meters less than the length. The perimeter of the lot can be no more than 254 meters. What is the maximum width?

72. A shipping company requires that the total of the length, width, and height of a box not exceed 153 inches. If the height is 75% of the length and the length is twice the width, what are the possible values for the dimensions?

73. Frugal Car Rental charges \$20 per day plus 10 cents per mile. Big C Car Rental charges \$26 per day plus 9 cents per mile. For what range of miles is Frugal Car Rental the best choice for a 1-day rental?

74. If 3 is subtracted from twice a positive integer, the result is negative. Show that there is only one solution.

Section 3.7

In Exercises 75–86, solve the compound inequality and write the solution set with interval notation.

75. $-t \geq -2$ or $\frac{t}{4} > -1$

76. $-5 \leq t - 7 < 5$

77. $\frac{x + 2}{4} \leq 1$ and $\frac{x + 2}{4} \geq -1$

78. $2x - 3 \geq 5$ and $x > 0$

79. $x + 1 > 0$ or $3x - 4 < 0$

80. $t + 3 \leq 5$ and $4t < -4$

81. $-5x + 2 \geq 12$ or $3x + 4 \geq 25$

82. $x - 6 > 4$ or $x + 5 \leq 20$

83. $3x + 4 < x$ or $x + 2 > 8 - x$

84. $4x - 3 \leq 3x$ and $5 - 2x \leq 1$

85. $3x - 4 > 5x - 2$ and $3x - 2 < 2x + 3$

86. $x + \frac{2}{3} < -\frac{1}{3}$ and $\frac{3}{2}x + 1 > \frac{1}{2}$

87. How much money must be invested at 8% simple interest to realize a yearly income between \$150 and \$200?

88. Equal amounts of money are invested in two funds that pay 7% and 8.5% simple interest. The first investment must earn at least \$119 per year and interest from the other investment must not exceed \$238 per year. How much should be invested in each fund?

Section 3.8

In Exercises 89–94, solve the equation.

89. $|-y| = 8$

90. $|3 + 2x| = 18$

91. $|2x - 3| - 7 = 5$

92. $|3x + 5| = -2$

93. $2 + |x - 2| = 3 - |x - 2|$

94. $|-x| + 5 = 1$

In Exercises 95 and 96, use the graphing method to estimate the solutions of the equation.

95. $|x - 3| = \frac{1}{2}x - 1$

96. $|x + 2| = 4 - |2x - 3|$

In Exercises 97–102, solve the absolute value inequality. Write the solution set with interval notation.

97. $|3x + 5| < 14$ **98.** $|4 - x| > 5$

99. $7 - |c - 5| \leq 4$ **100.** $8 - |5 - d| \leq 12$

101. $|-t| \geq 7$ **102.** $|t| \leq -7$

103. Write a double inequality that is equivalent to the inequality $|x - 3| \leq 5$.

104. Write an absolute value inequality whose solution set is $(-\infty, -4) \cup (8, \infty)$.

In Exercises 105–108, use the graphing method to estimate the solutions of the inequality. Write the solution set with interval notation.

105. $|x + 3| > 2x + 1$ **106.** $|x + 2| \leq |3 - x|$

107. $|3 + x| < x + 4$ **108.** $|x + 2| < 3x$

109. Two brothers were born exactly 3 years apart. If one boy is 9 years old, how old is his brother? Write an absolute value equation and solve it.

110. A wrestler can remain in his weight class as long as his present weight does not vary by more than 5 pounds. If his current weight is 136 pounds, what is the range of weights in his weight class?

111. If a number is reduced by 1, the absolute value of the result is less than the original number. Show that the number cannot be negative.

LOOKING AHEAD

The following exercises review concepts and skills that you will need in Chapter 4.

1. Evaluate each expression for the given value(s) of the variable(s).

(a) $5x - 2y$; $x = 2, y = \frac{3}{2}$ (b) $-x + 11$; -7

2. Which pairs of numbers are represented by points of the graph of the expression $1 - 2x$?

(i) $(3, -5)$ (ii) $(-1, 1)$

(iii) $(0, 1)$ (iv) $(0.5, 0)$

3. Use a graph to estimate the x- and y-intercepts of $f(x) = \frac{1}{2}x - 10$.

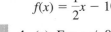

4. (a) For $c \neq 0$, which axis contains the point $(c, 0)$? Why?

(b) Describe the points of the y-axis.

In Exercises 5 and 6, solve for y.

5. $x - y = 5$ **6.** $2x + 3y = 12$

7. Evaluate the expression $\frac{y_2 - y_1}{x_2 - x_1}$ for $x_1 = -2$, $y_1 = 4$, $x_2 = 5$, and $y_2 = 8$.

8. Identify the coefficient of x and the constant term in the expression $5 - \frac{3x}{4}$.

9. Write an expression or equation to model the given information.

(a) The cost in dollars of shipping a package at \$2.00 plus \$1.50 for each pound p.

(b) The length L is 4 meters less than the width W.

10. Solve for x.

(a) $3 - 2x \geq 11$ (b) $\frac{2}{3}x + 8 = 0$

11. Determine whether the given number is a solution of the equation.

(a) $2x - 15 = 7$; 11 (b) $1 - 3x = x - 9$; -2

12. Show that the triangle with the vertices $A(-4, 4)$, $B(6, 9)$, and $C(2, -8)$ is a right triangle.

3. CHAPTER TEST

1. State the property that justifies each step.

$$2x - 3 = 7$$
$$2x - 3 + 3 = 7 + 3 \quad \text{(a)}$$
$$2x = 10 \quad \text{(b)}$$
$$\frac{2x}{2} = \frac{10}{2} \quad \text{(c)}$$
$$x = 5$$

In Questions 2–4, solve the equation. Then state whether the equation is an identity, an inconsistent equation, or a conditional equation.

2. $10 - 3x = 6x - 17$

3. $3x - 4 = 5x - 4 - 2x$

4. $3 - 2(x - 7) = 3(x + 1) - 5x$

5. Solve the formula $A = P + Prt$ for t.

6. Solve the equation $2x + 3y = 12$ for y.

7. What was the original investment at a simple interest rate of 8% if the value of the investment at the end of a year is \$2484?

8. A person has 200 meters of fencing to use to enclose a rectangular garden. If the width is three-fifths of the length, what are the dimensions of the garden?

9. A radiator with a 6-gallon capacity is filled with a 50% antifreeze solution. How much must be drained so that adding pure antifreeze gives a 60% antifreeze solution?

10. Two cars leave Wall, South Dakota, at the same time. The first car travels west, while the other travels east at 12 mph less than the first car. After 1.5 hours they are 144 miles apart. What is the speed of each car?

11. Use interval notation to describe the set whose graph is given.

In Questions 12–15, solve the given inequality.

12. $3x + 1 > 13$

13. $4 - 2(x - 3) < 1 - 5x$

14. $\dfrac{2x + 13}{-2} \geq 1$

15. $x + \dfrac{2}{3}(x - 3) < \dfrac{1}{2}$

In Questions 16–19, solve the compound inequality.

16. $-3 \leq 2x + 1 \leq 5$

17. $3 - x < 5$ and $2x - 1 < 5$

18. $3 - 2x \leq 9$ or $x + 1 \geq 0$

19. $2x \geq 0$ or $x - 3 \leq 1$

20. Solve $|x + 2| - 5 = -3$.

In Questions 21–23, solve the absolute value inequality.

21. $3 - |x - 1| < 1$

22. $|2x - 1| \leq 5$

23. $|x - 3| < -2$

24. The charge for cellular phone service is \$25 per month plus 50 cents per minute that the phone is used. If the phone bill is between \$31 and \$46 per month, what is the most and least number of minutes the phone was used?

25. The average snowfall for the first week of January is 8.7 inches. If the snowfall totals for the first 6 days are 1.2, 0, 4.25, 1.5, 0, and 0.4 inches, how much snow could fall on January 7 without the total exceeding the average?

1–3 CUMULATIVE TEST

1. Give the name of the set of numbers described in each of the following.

(a) Terminating and repeating decimals

(b) Nonnegative integers

(c) Cannot be written in the form p/q, where p and q are integers and $q \neq 0$

(d) Union of the set of rational numbers and the set of irrational numbers

2. Perform the indicated operations.

(a) $\dfrac{5}{16} - \left(-\dfrac{1}{8}\right)$ (b) $-\sqrt{|-9|}$ (c) $(-5)^2 + (-5)$

(d) $(-3)(-2)(-2)(-1)(-1)$

3. Select the property in column B that justifies the statement in column A.

Column A	*Column B*
(a) $3(x + 2) = 3x + 6$	Associative Property of Multiplication
(b) $2\left(\frac{1}{2}x\right) = \left(2 \cdot \frac{1}{2}\right)x$	Property of Additive Inverses
(c) $-(a + b) + (a + b) = 0$	Distributive Property
(d) $3x + 5x = (3 + 5)x$	Multiplication Property of 1

4. Evaluate each expression.

(a) $4 - 4(1 - 2)^2$

(b) $\dfrac{3x + 2y}{x^2(x + y)}$ for $x = 2$, $y = -1$

5. Simplify each expression.

(a) $2 + 3[-(1 - x)] - 4(x + 2)$

(b) $-x^2 + (-x)^2$

6. Simplify each expression.

(a) $-7x^0$ (b) $\dfrac{x^{-2}y^5}{x^3y}$

7. Simplify each expression.

(a) $x^3(x^3)^3$ (b) $\left(\dfrac{-2x^{-1}}{y^3}\right)^{-2}$

8. For $P(-6, 3)$ and $Q(8, -7)$, determine each of the following.

(a) PQ (b) The midpoint of $\overline{PQ}$

9. Explain how to use the Vertical Line Test to determine whether the graph of a relation represents a function.

10. Use a graph of the expression $0.5x^2 + 8x - 7$ to estimate the value(s) of x for which the value of the expression is 11.

11. If you trace the graph of the expression $\dfrac{x - 3}{5}$ to the point whose x-coordinate is -4, what is the y-coordinate?

12. Evaluate each function as indicated.

(a) $g(x) = 4(3x - 1) - (2 - x)$; $g(-1)$

(b) $f(x) = \sqrt{2x + 7} - x^2$; $f(3)$

13. Produce the graph of $y = x^3 - 5x$ on your calculator. Estimate each of the following to the nearest tenth.

(a) All intercepts

(b) Local maximum and local minimum

14. To estimate the solution of $5x - 3 = 2x + 9$, we graph $y_1 = 5x - 3$ and $y_2 = 2x + 9$ and note that the point of intersection of the two graphs is (4, 17). Explain the significance of the 4 and the 17.

15. Solve each equation algebraically.

(a) $2(3x - 1) - 2x = -(x - 4)$

(b) $\frac{1}{3}(x + 1) - \frac{1}{2} = \frac{3}{5}(2x - 3)$

16. After three tests of equal weight, a student has an 81 average. If the first and last test scores were 80 and 91, what was the second test score?

17. The second angle of a triangle is 2° larger than the first angle. If the third angle is 2° more than one-fourth the second angle, what are the measures of the angles?

18. A car and a bicycle are 39 miles apart. The car driver and the cyclist leave at 1:00 P.M. and head directly toward each other. If the driver averages 40 mph and the cyclist averages 12 mph, at what time will they meet?

19. Solve each inequality algebraically.

(a) $x - 2(x - 2) < 5x + 8$ (b) $\dfrac{3 - x}{-4} \geq x$

20. Solve each compound inequality. Write the solution set with interval notation.

(a) $-5x + 1 > 6$ or $2x - 1 \geq 5$

(b) $4 - x < 1$ and $\frac{1}{3}x < 2$

21. Solve each of the following.

(a) $|x - 3| < 8$ (b) $2|x + 1| - 3 = 5$

(c) $-2|3 + x| \leq -4$

22. The length of a rectangle is 3 more than twice the width. If the perimeter of the rectangle does not exceed 42 inches, what is the maximum length of the rectangle?

Chapter 4

Properties of Lines

When Europeans first set foot in North America, there were an estimated half million bald eagles. By 1963, that number had dwindled to a mere 417 pairs. The accompanying graph shows the modest increases that have resulted from conservationists' efforts.

To visualize the trend, we can connect the data points for 1963 and 1995 with a line. This line has a **slope** that represents the average **rate of change** in the bald eagle population. Moreover, the line can be described by a **linear function,** which can be used to model the number of pairs of bald eagles for any given year. (For more on this topic, see Exercises 101–104 at the end of Section 4.3, and see the Chapter Project.)

In Chapter 4 we consider linear equations in two variables and the properties of their graphs. In particular, we introduce the concept of slope and its many applications. We conclude with a discussion of linear inequalities and their graphs.

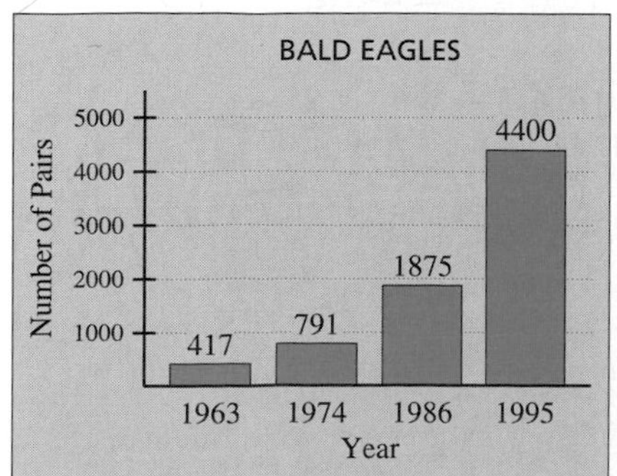

(Source: U.S. Fish and Wildlife Service.)

4.1 LINEAR EQUATIONS IN TWO VARIABLES

The Standard Form • Solutions and Graphs • Intercepts • Special Cases

The Standard Form

Linear equations in one variable typically arise when a problem involves a single unknown quantity or when all unknown quantities can be easily represented with just one variable.

When problems involve two unknown quantities, it is often more convenient to use two different variables and to write equations in those variables. One such type of an equation is a **linear equation in two variables.**

Definition of Linear Equation in Two Variables

A **linear equation in two variables** is an equation that can be written in the *standard form* $Ax + By = C$, where A, B, and C are real numbers and A and B are not both zero.

Each of the following equations is a linear equation in two variables.

Equation	*Standard Form*	
$3x + 2y = 12$	$3x + 2y = 12$	$A = 3, B = 2, C = 12$
$y = 2x + 3$	$-2x + y = 3$	$A = -2, B = 1, C = 3$
$y = 3$	$0x + y = 3$	$A = 0, B = 1, C = 3$
$x = -2$	$x + 0y = -2$	$A = 1, B = 0, C = -2$

An equation may have two variables but not be a linear equation. The following are some examples of equations that *cannot* be written in the standard form $Ax + By = C$.

$x^2 + 3y = 5$ The exponents on both variables must be 1.

$\frac{2}{x} + y = -2$ No variable can appear in the denominator of a fraction.

$xy = 1$ The variables cannot be part of the same term.

Solutions and Graphs

To solve a linear equation in two variables, we must find a replacement for each variable so that the resulting equation is true. We say that such replacements *satisfy* the equation.

Definition of a Solution of a Linear Equation in Two Variables

A **solution** of a linear equation in two variables is a pair of numbers (x, y) that satisfies the equation.

The **graph of an equation** is a visual representation of the solutions of the equation.

EXPLORING THE CONCEPT

Graphs of Linear Equations

Consider the equation $2x + y = 8$. We can determine a solution of this equation by selecting a value for either x or y and calculating the corresponding value of the other variable. This process is easier if we first solve the equation for one of the variables. Because we will use a calculator to assist us, we choose to solve the equation for y: $y = -2x + 8$.

By entering $-2x + 8$ as Y_1 on the Y screen, we can create a calculator table of values that satisfy the equation $y = -2x + 8$. (Of course, you also can perform the arithmetic by hand.) The following table gives a partial list of solutions of $y = -2x + 8$.

x	-4	-1	0	3	4	9
y	16	10	8	2	0	-10
(x, y)	$(-4, 16)$	$(-1, 10)$	$(0, 8)$	$(3, 2)$	$(4, 0)$	$(9, -10)$

By scrolling through the calculator table, you can see that there is no end to the number of solutions; that is, the equation has infinitely many solutions.

Figure 4.1 shows the points associated with the ordered pairs in the preceding table. The points appear to be points of a straight line. Plotting additional points would provide evidence that the graph of $y = -2x + 8$ is indeed a straight line.

Figure 4.1

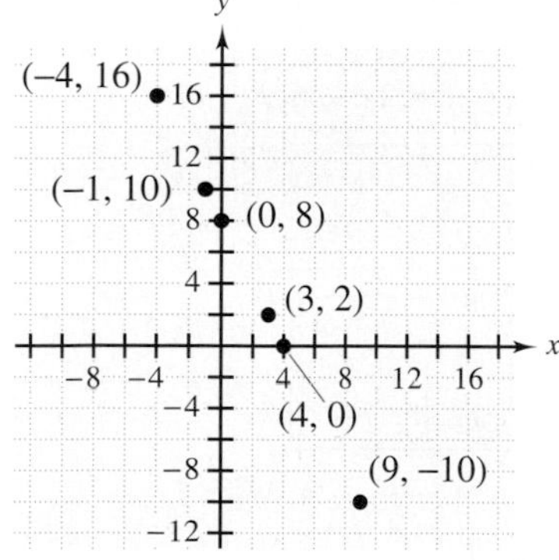

Figure 4.2

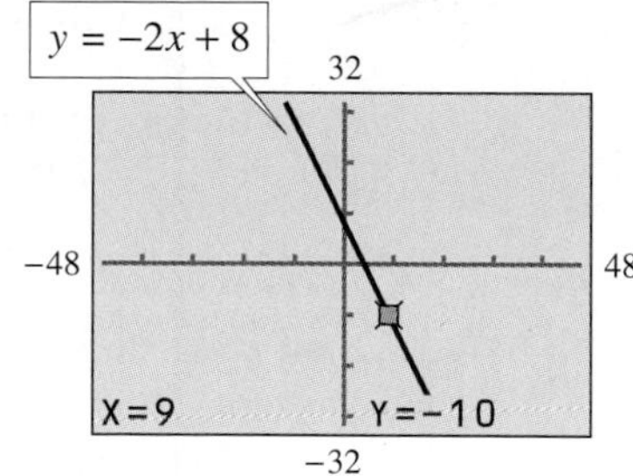

Figure 4.2 is a calculator display of the graph of $y = -2x + 8$. The tracing cursor is on $(9, -10)$, which is one of the solutions in the table. In fact, we can trace to any point in the table as well as to any point in the calculator table.

These results suggest the following generalizations about a linear equation in two variables.

1. The graph is a straight line.
2. There is a **one-to-one correspondence** between the solutions of the equation and the points of the graph. In other words,
   (a) Every solution of the equation is represented by a point of the graph.
   (b) Every point of the graph represents a solution of the equation.

Intercepts

The graph of a linear equation in two variables may intersect the y-axis and/or the x-axis. In Chapter 2 we defined these points as **intercepts.** Because of the importance of these points, we repeat the definitions.

Definitions of Intercepts

A point where a graph intersects the y-axis is called a ***y*-intercept.** A point where a graph intersects the x-axis is called an ***x*-intercept.**

Although we define y-intercept and x-intercept as *points*, the words are often used to refer to the coordinates of the points.

EXAMPLE 1

Estimating Intercepts with a Calculator

Use the integer setting to produce the graph of $y = 2x - 10$. Then trace the graph to estimate the intercepts.

Solution

Figure 4.3

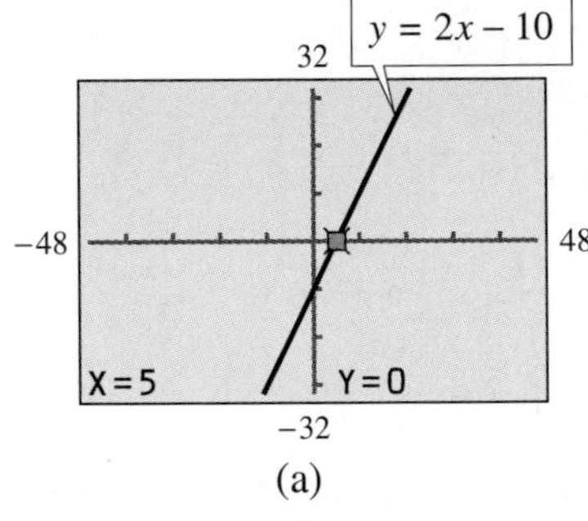

(a)

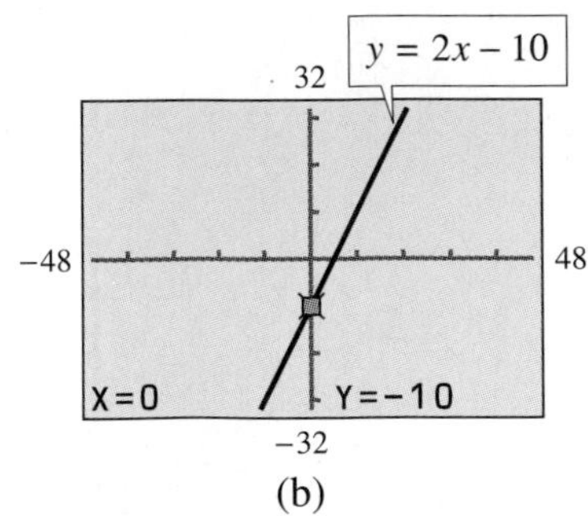

(b)

The x-intercept is $(5, 0)$; the y-intercept is $(0, -10)$. [See Fig. 4.3(a) and (b).]

LEARNING TIP

Visualizing the y-intercept as a point of the y-axis, where the x-coordinate is always 0, provides a way to remember that x is replaced with 0. Similarly, visualize the x-intercept as a point of the x-axis, where the y-coordinate is always 0.

Although we can use graphing methods to provide reasonable estimates of the x- and y-intercepts, we can determine the exact coordinates algebraically.

1. The y-intercept is the point whose x-coordinate is 0. Replace x with 0 and calculate the corresponding value of y.
2. The x-intercept is the point whose y-coordinate is 0. Replace y with 0 and calculate the corresponding value of x.

EXAMPLE 2

Determining Intercepts Algebraically

Use algebraic methods to determine the intercepts for $y = 2x - 3$.

Solution

x-Intercept

$y = 2x - 3$

$0 = 2x - 3$ Replace y with 0 and solve for x.

$3 = 2x$

$x = \frac{3}{2}$

The x-intercept is $\left(\frac{3}{2}, 0\right)$.

y-Intercept

$y = 2x - 3$

$y = 2(0) - 3$ Replace x with 0 and solve for y.

$y = 0 - 3$

$y = -3$

The y-intercept is $(0, -3)$.

By examining the graphs of linear equations in the form $y = ax + b$, we can observe the effect of the constant term b on the y-intercept.

EXPLORING THE CONCEPT

Determining y-Intercepts from Equations

Figure 4.4 is a calculator display of the graphs of the following four equations.

$y = x + 12 \qquad y = x + 5 \qquad y = x \qquad y = x - 8$

Figure 4.4

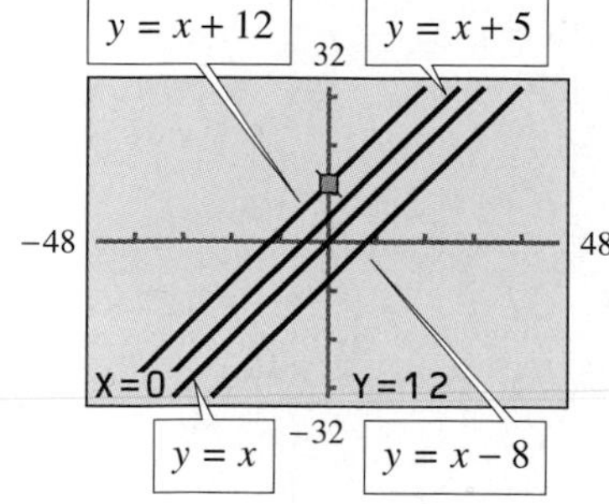

Note that the tracing cursor is on the y-intercept of the graph of $y = x + 12$. By tracing to the y-intercepts of the other three graphs, we obtain the following results.

Equation	*y-Intercept*
$y = x + 12$	$(0, 12)$
$y = x + 5$	$(0, 5)$
$y = x + 0$	$(0, 0)$
$y = x - 8$	$(0, -8)$

In each case, the y-coordinate of the y-intercept is the same as the constant term of the equation.

This conclusion can be supported as follows: When we solve a linear equation in two variables for y, the resulting equation has the form $y = ax + b$, where a and b are real numbers. We determine the y-intercept by replacing x with 0.

$$y = ax + b$$
$$y = a(0) + b$$
$$y = 0 + b$$
$$y = b$$

The y-intercept is $(0, b)$.

In summary, when a linear equation in two variables is written in the form $y = ax + b$, the y-intercept $(0, b)$ can be determined from the constant term of the equation.

Think About It

The y-intercept of the graph of $y = ax + b$ is $(0, b)$. Can you write a general ordered pair for the x-intercept?

EXAMPLE 3

Determining the y-Intercept from the Equation

Write each of the following equations in the form $y = ax + b$. Then determine the y-intercept in each case.

(a) $y = x - 4$ (b) $5x + 7y = 9$

Solution

(a) The equation is already solved for y. The constant term is -4. Therefore, the y-intercept is $(0, -4)$.

(b)
$$5x + 7y = 9 \quad \text{Solve the equation for } y.$$
$$5x + 7y - 5x = -5x + 9 \quad \text{Subtract } 5x \text{ from both sides.}$$
$$7y = -5x + 9 \quad \text{Combine like terms.}$$
$$\frac{7y}{7} = \frac{-5x}{7} + \frac{9}{7} \quad \text{Divide each term on both sides by 7.}$$
$$y = -\frac{5}{7}x + \frac{9}{7}$$

The constant term is $\frac{9}{7}$. Therefore, the y-intercept is $\left(0, \frac{9}{7}\right)$.

Determining the y-intercept of a graph is easy when the equation is written in the form $y = ax + b$. However, the y-intercept can be determined from any equation form by replacing x with 0 and computing the corresponding y-value.

Consider again the equation in part (b) of Example 3.

$$5x + 7y = 9$$
$$5(0) + 7y = 9 \quad \text{Replace } x \text{ with 0.}$$
$$7y = 9 \quad \text{Solve for } y.$$
$$y = \frac{9}{7}$$

The y-intercept is $\left(0, \frac{9}{7}\right)$.

Special Cases

At the beginning of this section we stated that $y = 3$ and $x = -2$ are both examples of linear equations in two variables. In both equations, one of the variables is missing. For this reason, we consider the equations to be special cases.

EXPLORING THE CONCEPT

Special Cases

We interpret the equation $y = 4$ to mean that every point of the graph has a y-coordinate of 4. Figure 4.5 shows this graph along with the graphs of $y = 10$ and $y = -17$

Figure 4.5

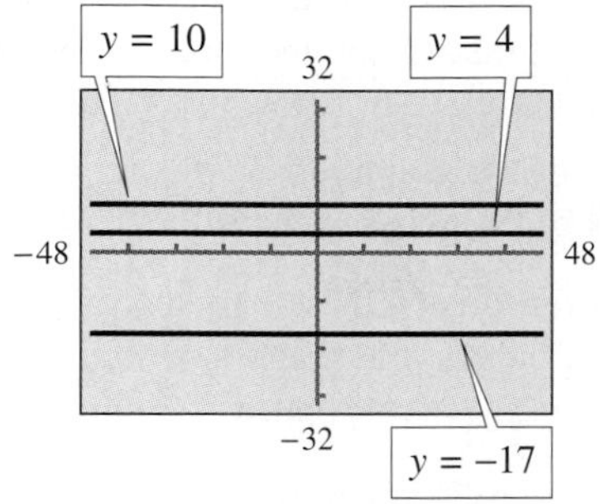

Observe that the lines in Fig. 4.5 are horizontal and have no x-intercept. The constant value of y indicates how far above or below the x-axis the line is drawn. By tracing each graph, you can easily see that every point of the graph has the same y-coordinate. The y-intercepts are $(0, 4)$, $(0, 10)$, and $(0, -17)$.

Similarly, we interpret the equation $x = 7$ to mean that every point of the graph has an x-coordinate of 7. Figure 4.6 shows this graph and the graph of $x = -6$.

Figure 4.6

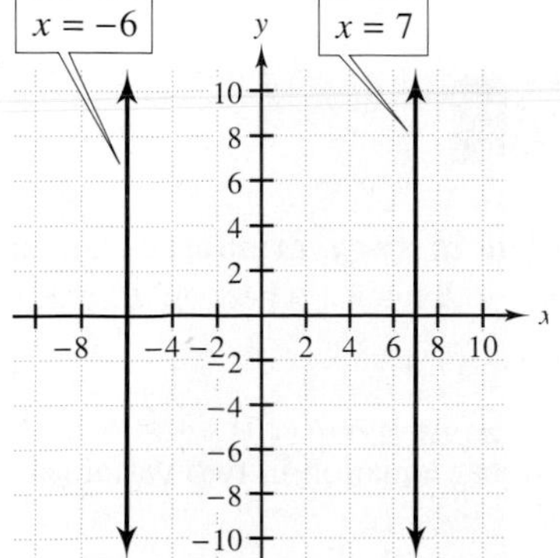

Note: Because an equation of the form $x = c$ cannot be solved for y, no expression can be entered on the Y screen to produce a calculator graph in the usual way.

The lines in Fig. 4.6 are vertical, with every point of each graph having the same

x-coordinate. The constant value of x indicates how far to the left or right of the y-axis the line is drawn. The x-intercepts are (7, 0) and (−6, 0). Neither line has a y-intercept.

Note: Because a vertical line does not pass the Vertical Line Test, the graph of $x = c$ does not represent a function.

In summary, the following characteristics apply to all equations of the form $x = c$ and $y = b$, where c and b are constants.

1. For $x = c$, the first coordinate of all ordered pairs is c; for $y = b$, the second coordinate of all ordered pairs is b.
2. For $x = c$, the graph is a vertical line, where c indicates how far to the right or left of the y-axis to draw the line. For $y = b$, the graph is a horizontal line, where b indicates how far above or below the x-axis to draw the line.
3. The x-intercept of the graph of $x = c$ is $(c, 0)$, and if $c \neq 0$, there is no y-intercept. The y-intercept of the graph of $y = b$ is $(0, b)$, and if $b \neq 0$, there is no x-intercept.

It will be very useful to keep these two special cases in mind. As is true for all linear equations in two variables, the graph is a straight line. Knowing that the line is horizontal when $y = b$ and vertical when $x = c$ simplifies sketching the graph.

Here is a summary of what is known so far about the two special cases.

Summary of Special Cases of Linear Equations

Equation	*Graph*	*Intercepts*	*Function?*
$x = c$	Vertical	x-Intercept: $(c, 0)$ y-Intercept: if $c \neq 0$, none.	No
$y = b$	Horizontal	x-Intercept: if $b \neq 0$, none. y-Intercept: $(0, b)$	Yes

4.1 QUICK REFERENCE

The Standard Form

- A **linear equation in two variables** is an equation that can be written in the *standard form* $Ax + By = C$, where A, B, and C are real numbers and A and B are not both zero.

Solutions and Graphs

- A **solution** of a linear equation in two variables is a pair of numbers (x, y) that satisfies the equation.
- Solutions can be determined by replacing one variable with some value and solving the equation for the other variable.
- There exists a one-to-one correspondence between the solutions of an equation and the points of its graph.
- The graph of a linear equation in two variables is a straight line.

- To produce the graph of a linear equation on a calculator, solve the equation for y and enter the resulting function.

Intercepts

- A point where a graph intersects the y-axis is called a ***y*-intercept.** A point where a graph intersects the x-axis is called an ***x*-intercept.**
- We can estimate the x- and y-intercepts with a calculator by locating the points where the graph intersects the axes.
- We determine the exact x-intercept algebraically by replacing y with 0 and solving for x. Similarly, we determine the exact y-intercept algebraically by replacing x with 0 and solving for y.
- If a linear equation is solved for y, the form of the equation is $y = ax + b$. From this, we can see that the y-intercept is $(0, b)$.

Special Cases

- The graph of the equation $y = b$ is a horizontal line. The constant b indicates how far the line is above or below the x-axis. Excluding the graph of $y = 0$, the graph has the y-intercept $(0, b)$ and no x-intercepts.
- The graph of the equation $x = c$ is a vertical line. The constant c indicates how far the line is to the right or left of the y-axis. Excluding the graph of $x = 0$, the graph has the x-intercept $(c, 0)$ and no y-intercepts.

4.1 EXERCISES

Concepts and Skills

1. Explain why $x = -5$ can be a linear equation in two variables even though there is no y-term.

2. How are the points of the graph of an equation related to the solutions of the equation?

In Exercises 3–8, state whether the equation is a linear equation in two variables.

3. $\frac{x}{2} + \frac{y}{3} - \frac{7}{8} = 0$
4. $y = \frac{2x}{3} + 5$
5. $y = x^2 - 3$
6. $4y = \frac{3}{x} + \frac{4}{5}$
7. $-6 = 2xy$
8. $4y + 5 = -7$

In Exercises 9–14, determine whether the given ordered pair is a solution of the given equation.

9. $(4, 5)$; $y = 2x - 3$
10. $(-2, 2)$; $y = 3x + 4$
11. $\left(2, \frac{3}{2}\right)$; $3x - 4y = 12$
12. $(4, 5)$; $\frac{1}{4}x + \frac{2}{5}y = 3$
13. $(-2, -1)$; $x - 5 = 3$
14. $(-2, -3)$; $7 + y = 4$

In Exercises 15–20, determine k so that the given ordered pair is a solution of the equation.

15. $2x - ky = 7$; $(1, -1)$
16. $y + 3x = k$; $(2, -5)$
17. $y = -\frac{2}{3}x + k$; $(12, -3)$
18. $kx + y = 0$; $(-2, 6)$
19. $y = -2x + k$; $(k, k + 1)$
20. $kx - y = 9$; $(3, k - 5)$

In Exercises 21–26, find the values of a, b, and c so that the ordered pairs are solutions of the given equation.

21. $y = -2x + 5$; $(-3, a)$, $(4, b)$, $(c, -7)$
22. $y = 4x - 3$; $(a, 5)$, $(b, -3)$, $(3, c)$
23. $x - 5 = -2$; $(a, 5)$, $(b, 9)$, $(c, -6)$
24. $5 - 2y = -1$; $(-1, a)$, $(4, b)$, $(-12, c)$
25. $x + 3y - 4 = 0$; $(a, 3)$, $(7, b)$, $(c, -4)$
26. $5y - 3x = -10$; $(a, -2)$, $(-5, b)$, $(10, c)$

27. For each equation or inequality in column A, a solution or solution set is given in column B. Explain how the differences in notation affect the meaning of each entry in column B.

Column A	*Column B*
$\lvert x \rvert = 2$	$\{-2, 2\}$
$\lvert x \rvert < 2$	$(-2, 2)$
$y = x + 4$	$(-2, 2)$

28. Explain the difference between the solution set of $x = 1$, a linear equation in one variable, and $x = 1$, a linear equation in two variables.

In Exercises 29–32, complete the table.

29. $y = 4 - x$

x	10	−6	
y			−3

30. $x = 2y + 3$

x	5	−13	
y			−6

31. $f(x) = 3x + 1$

x	0	3	−5
$f(x)$			
(x, y)			

32. $g(x) = -2x - 5$

x	−2	−7	3
$g(x)$			
(x, y)			

33. (a) Which of the following equations is graphed in the accompanying figure?

$x - 2y - 5 = 0$

$2x - 3y - 6 = 0$

$5x - 6y - 18 = 0$

Figure for 33

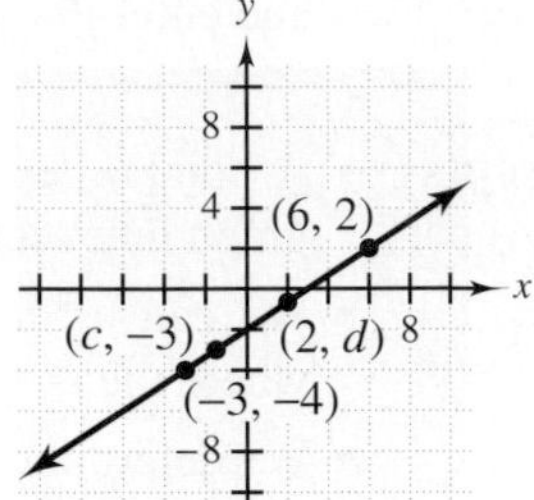

(b) What are the values of c and d in the figure?

34. (a) Which of the following equations is graphed in the accompanying figure?

$3x + 4y + 9 = 0$

$5x + 8y + 11 = 0$

$7x + 8y + 13 = 0$

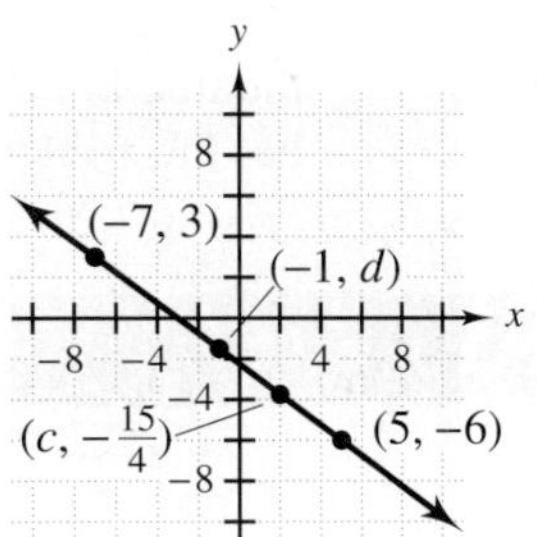

(b) What are the values of c and d in the figure?

In Exercises 35–48, determine three solutions of the given equation and sketch the graph. If possible, produce the graph of the equation on your calculator. Compare your sketch with the calculator's graph.

35. $y = 3x - 4$

36. $y = 4x + 3$

37. $y = \frac{2}{3}x - 2$

38. $y = -\frac{3}{4}x + 5$

39. $3x - 5y = 15$

40. $2x - 3y = 6$

41. $y - 2 = 3$

42. $y + 5 = 7$

43. $x - 9 = -11$

44. $x + 7 = 9$

45. $y = x$

46. $y = -x$

47. $3x = 21 - 7y$

48. $5x + 2y - 10 = 0$

In Exercises 49–54, use the integer setting to produce the graph of the given equation. Then trace the graph to estimate the intercepts.

49. $y = 12 - 3x$

50. $y = -12 + x$

51. $x = 3y + 30$

52. $x = 15 - y$

53. $3x - 4y = 36$

54. $2x + 3y = 30$

In Exercises 55 and 56, select from the given equations the one whose graph has the given y-intercept. (There could be more than one answer.)

55. The y-intercept is $(0, 3)$.

$5x + 3y - 9 = 0$ $\quad$ $6x + 2y - 6 = 0$

$3x + 4y + 12 = 0$

56. The y-intercept is $(0, -2)$.

$5x + 3y - 6 = 0$ $\quad$ $5x + 3y + 6 = 0$

$2x + 4y + 8 = 0$

In Exercises 57–60, write each equation in the form $y = ax + b$. Then determine the y-intercept.

57. $2x + 5y = 14$

58. $4x - 3y = 15$

59. $y - 2 = -4(x + 3)$

60. $\frac{1}{3}x + \frac{3}{4}y - 9 = 0$

In Exercises 61–68, determine the intercepts algebraically.

61. $y = -2x$

62. $5x - 4y = 0$

63. $3x - 4y = 12$

64. $2x - 3y = 12$

65. $\frac{x}{3} + \frac{y}{4} = 1$

66. $\frac{x}{5} - \frac{y}{3} = 1$

67. $y + 2 = 17$

68. $2x + 9 = 7$

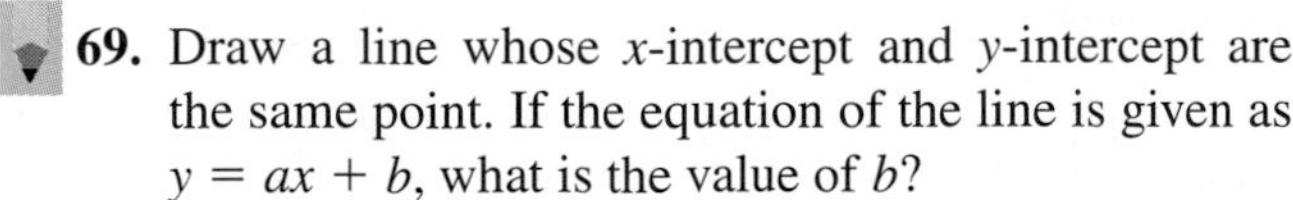

69. Draw a line whose x-intercept and y-intercept are the same point. If the equation of the line is given as $y = ax + b$, what is the value of b?

70. Describe the form of an equation whose graph is a

(a) vertical line.

(b) horizontal line.

In Exercises 71 and 72, produce the graph of the given equation. Then determine the equations of the lines described in parts (a) and (b). To check your equations, produce their graphs on the same axes.

71. $y = x$

(a) A line shifted upward 4 units from the given line

(b) A line shifted downward 3 units from the given line

72. $y = -1$

(a) A line shifted upward 5 units from the given line

(b) A line shifted downward 4 units from the given line

In Exercises 73–76, determine a, b, and c so that the given equation has the specified intercepts.

73. $ax + 4y = c$; $(0, 2), (-4, 0)$

74. $x - 2by = c$; $(0, 5), (-1, 0)$

75. $ax + by = 15$; $(0, -5), (3, 0)$

76. $\frac{x}{a} + \frac{y}{b} = 1$; $(0, 1), (1, 0)$

In Exercises 77–82, determine whether the graph of the given equation is horizontal, vertical, or neither.

77. $3x = 12$

78. $x - 5 = 6$

79. $x - y = 0$

80. $y = x - 7$

81. $-2y = -8$

82. $y + 3 = -2$

In Exercises 83–88, write two ordered pairs whose coordinates satisfy the given conditions. Then translate the conditions into an equation. Produce the graph of the equation and trace it to verify that your two ordered pairs are represented by points of the graph.

83. One number y is the opposite of another number x.

84. The numbers x and y are equal.

85. One number y is half of another number x.

86. One number y is 3 less than twice another number x.

87. The sum of two numbers x and y is 5.

88. The difference of two numbers x and y is 2.

Real-Life Applications

In Exercises 89–92, translate the given information into a function with an appropriate domain. Graph the function and trace the graph to answer the question(s).

89. A student takes a 32-question final exam.

(a) Write a linear function that models the number y of questions remaining as a function of the number x of questions answered.

(b) Interpret the intercepts of the graph of this function.

90. A package delivery service employs 40 drivers, of whom x are men.

(a) Write a linear function that describes the number y of women drivers as a function of x.

(b) Determine and interpret the intercepts of the graph of this function.

91. A taxi fare is \$4.00 plus 15 cents per tenth of a mile. What are all the possible fares for a trip of between 2 and 2.5 miles, inclusive?

92. When the daily high temperature is 75°F, the number of people using the local jogging path is 120 per day. For each 1° temperature increase over 75°F (up to 100°F), the number of joggers decreases by 4. For what temperature range will the number of joggers be between 88 and 100, inclusive?

Modeling with Real Data

93. The accompanying table shows the total purchases of books (in millions) in the United States during the period 1991–1993. (Source: Consumer Research Study on Book Purchasing.)

Year	Books Purchased (millions)
1991	776
1992	897
1993	986

These data can be modeled by the linear function $B(x) = 105x + 676$, where B is the number of books purchased (in millions) and x is the number of years since 1990.

(a) What does the y-intercept of the graph of B represent?

(b) In what year does the model function predict that book sales will be twice the sales in 1993?

94. In Exercise 93, the linear function $B(x) = 105x + 676$ was used to model the number of book purchases.

(a) Produce a calculator table for B and estimate the number by which B is increasing each year.

(b) Compare your result in part (a) with the coefficient of x in the model function.

95. The accompanying bar graph shows the number of 1-pound loaves of bread consumed each year per person from 1988 to 1992. (Source: Annual Survey of Manufacturers.)

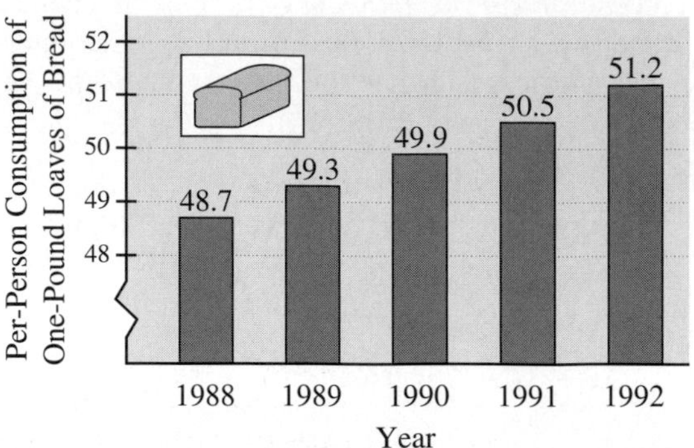

The data can be modeled by the linear function $L(Y) = 0.625Y - 1193.8$, where L is the number of loaves of bread and Y is a year, $1988 \leq Y \leq 1992$.

(a) According to the model function, what is the average annual increase in the per-person consumption of bread during this period?

(b) Determine and interpret the x-intercept of the graph of L. How realistic is this result?

96. For the hundred years prior to 1988, how would you expect the shape of the actual graph of bread consumption to differ from the graph of L?

Group Project: Alternative Fuels

The Clean Air Act of 1992 prompted increased attention to alternative fuels. The bar graph shows the consumption of 85% ethanol (E85) fuel during the period 1992–1995. (Source: Alternative Fuels Data Center.)

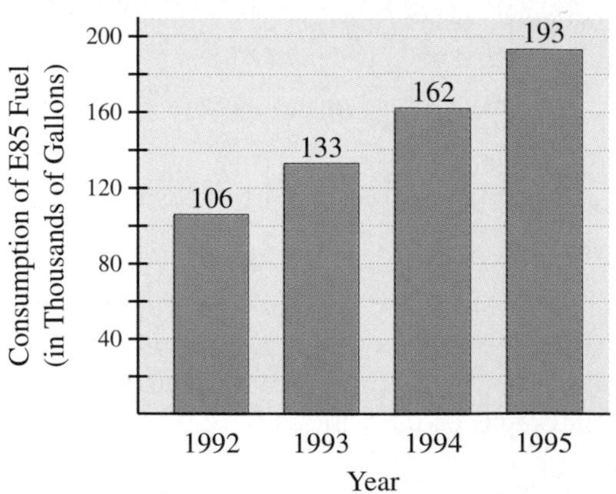

The data can be modeled by the linear equation $C = 29x - 57{,}663$, where C is the consumption of E85 (in thousands of gallons) and x is the year, where $1992 \leq x \leq 1995$.

97. What point of the graph of the model equation would represent the year in which there was no consumption of E85 fuel? Determine this point and judge whether the represented data are reasonable.

98. What is the y-intercept of the graph of the model equation? Is this a meaningful data point? Why?

99. Beginning in 1996, Ford and General Motors were committed to large increases in the production of E85-fueled vehicles. In view of this, would you expect the model to remain linear?

100. In 1995, nearly 100 American cities were involved in various Clean Cities programs. What interest do these cities have in the development and expanded use of alternative fuels?

Challenge

101. Interpret the roles of g and h in the equation

$$\frac{x}{g} + \frac{y}{h} = 1$$

(Hint: Consider where the graph crosses the axes.)

In Exercises 102–104, determine and plot the intercepts of the graph of the given equation. Be sure to take into account the conditions on a and b. Then sketch the graph.

102. $ax + by = ab; \quad a > 0, b < 0$

103. $2x + by = 4b; \quad b < 0$

104. $ax + 2y = -2a; \quad a > 0$

In Exercises 105–108, sketch or produce the graph using the specified domain.

105. $y = 2x + 1, \quad x \geq 0$

106. $y = 1 - x, \quad x \leq 0$

107. $y = x - 5, \quad 10 \leq x \leq 15$

108. $y = 3x, \quad -2 \leq x \leq 3$

4.2 SLOPE OF A LINE

The Slope Formula • The Slope-Intercept Form • Graphing with Slope

The Slope Formula

In the preceding section we learned the significance of the constant term b when a linear equation is written in the form $y = ax + b$. The y-intercept of the graph of the equation is $(0, b)$. We now consider the coefficient a and learn how to interpret this number graphically.

EXPLORING THE CONCEPT

The Effect of a on the Graph of $y = ax + b$

In Figs. 4.7 and 4.8, all the equations have the form $y = ax + 3$, and so all the graphs have the same y-intercept, $(0, 3)$.

Figure 4.7

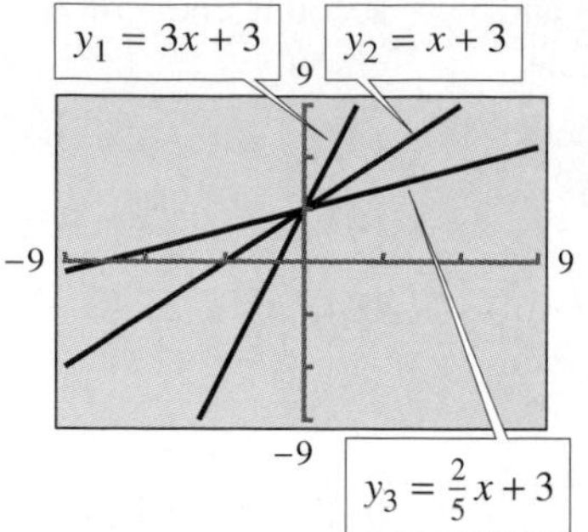

Figure 4.8

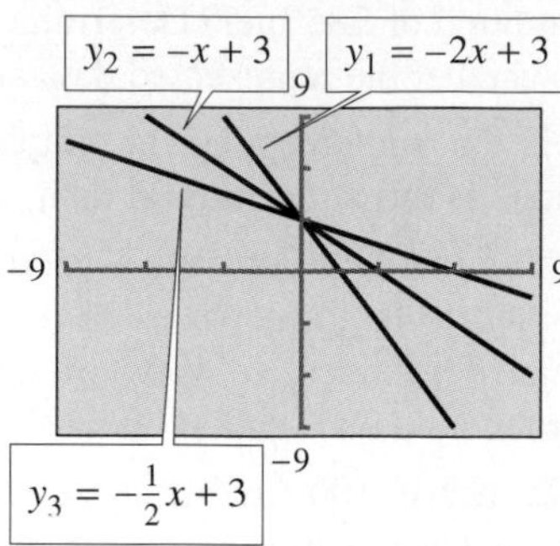

For the equations in Fig. 4.7, the coefficient a is positive, and the lines rise from left to right. For the equations in Fig. 4.8, the coefficient a is negative, and the lines fall from left to right.

We can also observe that the coefficient a influences the steepness of the line. In both figures, larger values of $|a|$ are associated with steeper lines.

Steepness is a relative word, and it is too vague to describe the orientation of a line specifically. What we need is a numerical measure of steepness. That measurement is called the **slope** of the line.

EXPLORING THE CONCEPT

The Slope of a Line

Consider the two lines in Figs. 4.9 and 4.10.

Figure 4.9

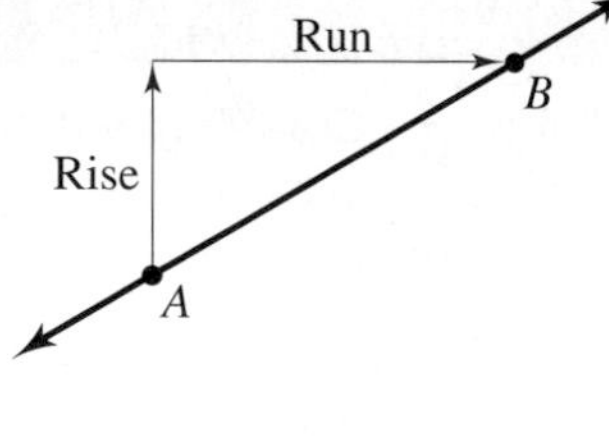

Figure 4.10

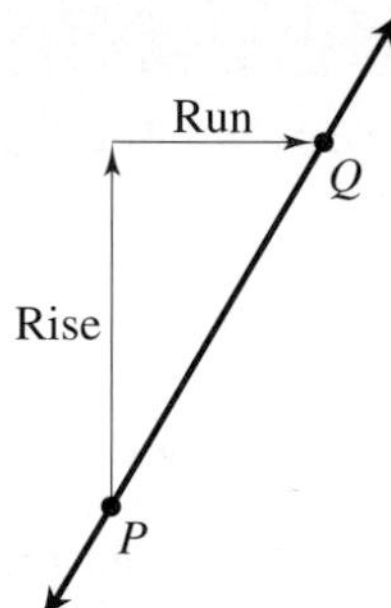

One way to travel from point A to point B in Fig. 4.9 is to start at A, travel vertically (rise) upward to the point directly across from B, and then travel horizontally (run) to the right to reach B. We can travel from point P to point Q in Fig. 4.10 in a similar way.

We can see the difference in the steepness of the two lines by comparing the rise to the run in each case. For the line that is rising more gradually (Fig. 4.9), the rise is relatively small compared with the run. For the steeper line (Fig. 4.10), the rise is relatively large compared with the run.

This suggests a way of describing the steepness, or slope, of a line. We compare the rise to the run with a ratio.

In Fig. 4.11, $P(x_1, y_1)$ and $Q(x_2, y_2)$ are two general points of a line. Point $R(x_1, y_2)$ is on a vertical line with P and a horizontal line with Q. The triangle formed by points P, Q, and R is called a **slope triangle.**

Figure 4.11

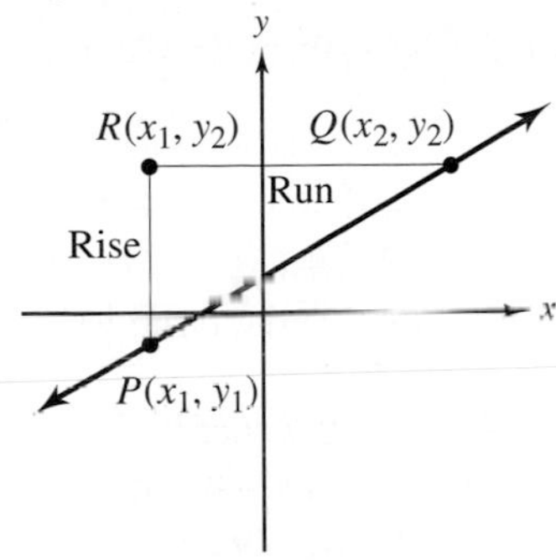

The rise PR is the difference between the y-coordinates.

$$\text{Rise} = y_2 - y_1$$

The run QR is the difference between the x-coordinates.

$$\text{Run} = x_2 - x_1$$

With these formulas for rise and run, we can now define the slope of a line.

Definition of Slope

The **slope** m of a line containing $P(x_1, y_1)$ and $Q(x_2, y_2)$ is defined as follows.

$$m = \frac{\text{rise}}{\text{run}} = \frac{y_2 - y_1}{x_2 - x_1}, \qquad x_2 \neq x_1$$

Note that the conventional symbol for slope is the letter m.

There are several important observations to be made from this definition. First, we can use the formula to calculate the slope of any line except a vertical line. The slope of a vertical line is undefined.

Next, the order in which we subtract the coordinates does not matter as long as we are consistent. We can choose to compute $y_2 - y_1$ for the rise, but then we must compute $x_2 - x_1$ for the run. Or we can choose to compute $y_1 - y_2$ for the rise, but then we must compute $x_1 - x_2$ for the run.

Finally, the definition refers to any two points of the line. No matter which two points we use, the calculated slope will be the same.

EXAMPLE 1

Determining Slope Given Two Points of a Line

Sketch the line determined by the given points and determine the slope of the line.

(a) $A(2, -1)$ and $B(4, 5)$ (b) $C(-2, 5)$ and $D(2, -3)$

Figure 4.12

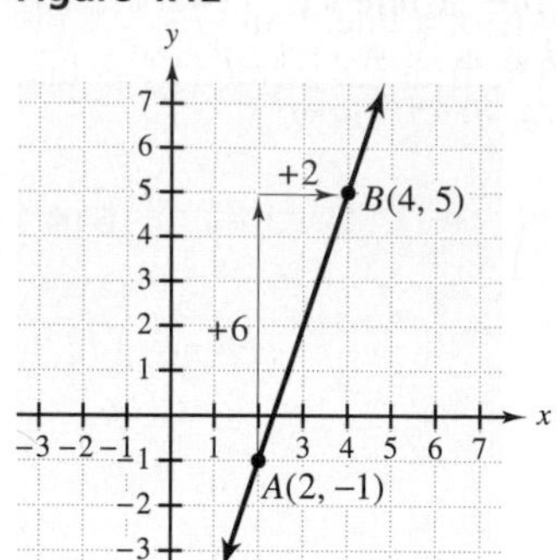

Solution

(a) In Fig. 4.12 the rise is +6 and the run is +2.

$$m = \frac{\text{rise}}{\text{run}} = \frac{6}{2} = 3$$

We can also use the slope formula to determine the slope.

$$m = \frac{y_2 - y_1}{x_2 - x_1} = \frac{5 - (-1)}{4 - 2} = \frac{6}{2} = 3$$

(b) In Fig. 4.13 the rise is −8 and the run is +4.

$$m = \frac{\text{rise}}{\text{run}} = \frac{-8}{4} = -2$$

Using the slope formula, we obtain the same result.

$$m = \frac{-3 - 5}{2 - (-2)} = \frac{-8}{4} = -2$$

Figure 4.13

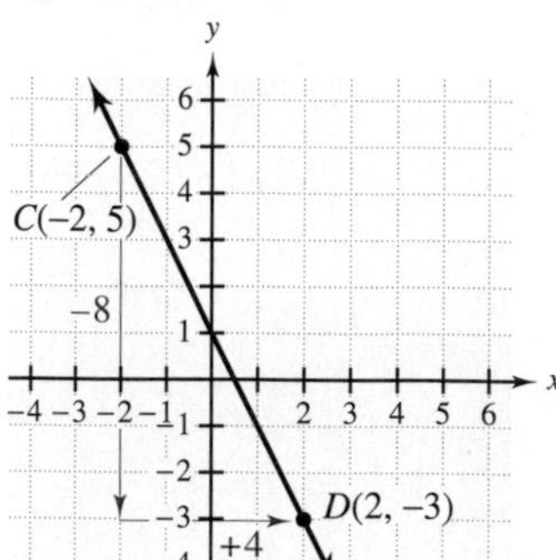

EXAMPLE 2

Special Cases

Determine the slope of the following lines.

(a) The equation of the line is $y = 5$. (b) The line contains $E(1, 3)$ and $F(1, 7)$.

Solution

(a) Two points of the line $y = 5$ are shown in Fig. 4.14. Note that points A and B have the same y-coordinate.

$$m = \frac{y_2 - y_1}{x_2 - x_1} = \frac{5 - 5}{-3 - 2} = \frac{0}{-3 - 2} = \frac{0}{-5} = 0$$

Figure 4.14

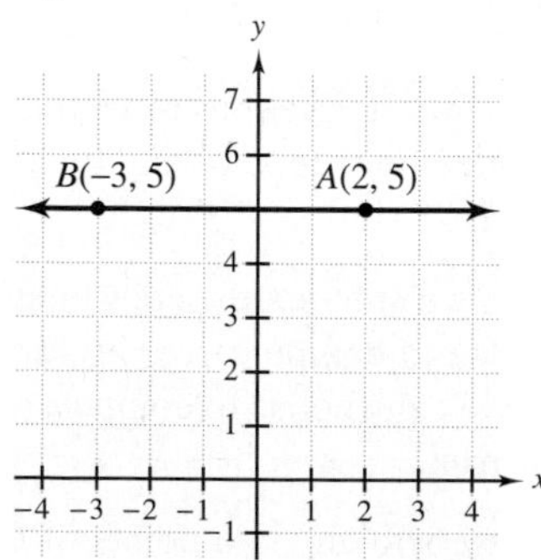

Figure 4.15

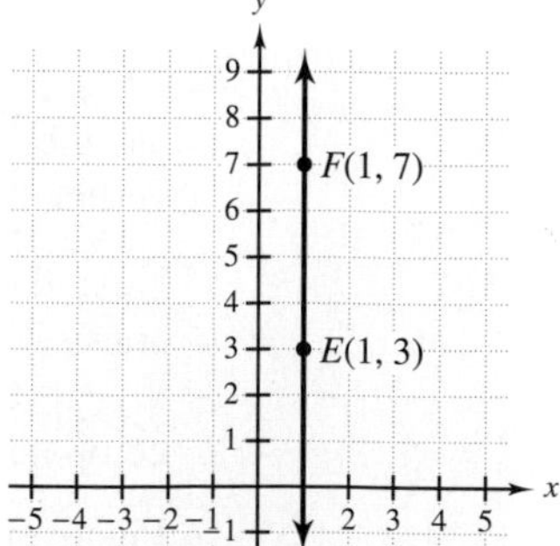

(b) In Fig. 4.15 the line is vertical because points E and F have the same x-coordinate. The slope formula does not apply because the ratio is

$$\frac{y_2 - y_1}{x_2 - x_1} = \frac{7 - 3}{1 - 1} = \frac{4}{0}$$

which is undefined. In general, the slope of any vertical line is undefined.

Note: For these special cases, it is important to know the difference between *zero* and *undefined.* A horizontal line has a defined slope, and it is 0. A vertical line does not have a defined slope. *Zero* does not mean the same thing as *undefined.*

One of the special cases, that of a vertical line, is handled in the definition of slope. The definition states that the slope of a vertical line is undefined.

In Example 2 we found that the slope of a horizontal line is 0. Because each point of a horizontal line has the same y-coordinate, $y_2 - y_1 = 0$ for any two points. Therefore,

$$m = \frac{y_2 - y_1}{x_2 - x_1} = \frac{0}{x_2 - x_1} = 0$$

LEARNING TIP

Compare the slope of a line to the *pitch* of a roof. A roof can have a small or large pitch (slope). A flat roof has no pitch (zero slope), and there is no such thing as a vertical roof (undefined slope).

We can now add this new knowledge about slopes to our summary for special cases.

Summary of Special Cases

Equation	*Graph*	*Intercepts*	*Slope*
$x = c$	Vertical	x-Intercept: $(c, 0)$ y-Intercept: If $c \neq 0$, none.	Undefined
$y = b$	Horizontal	y-Intercept: $(0, b)$ x-Intercept: If $b \neq 0$, none.	0

The Slope–Intercept Form

When a linear equation is written in the form $y = ax + b$, we have seen that the coefficient a affects the slope of the line. We now consider this relationship in more detail.

EXPLORING THE CONCEPT

Determining Slopes from Equations

The following table lists four equations. For each equation, two solutions are given. Note that the y-intercept of each graph is (0, 1), and so (0, 1) is a solution of each equation. We use these solutions in the slope formula to calculate the slope of each line.

Equation	First Solution	Second Solution	Slope
$y = \frac{1}{2}x + 1$	(0, 1)	(2, 2)	$m = \frac{2-1}{2-0} = \frac{1}{2}$
$y = 2x + 1$	(0, 1)	(1, 3)	$m = \frac{3-1}{1-0} = \frac{2}{1} = 2$
$y = -1x + 1$	(0, 1)	(1, 0)	$m = \frac{0-1}{1-0} = \frac{-1}{1} = -1$
$y = -3x + 1$	(0, 1)	(−1, 4)	$m = \frac{4-1}{-1-0} = \frac{3}{-1} = -3$

In each case, we observe that the coefficient of x is the same as the slope of the line. These results suggest the following generalization: If an equation is written in the form $y = ax + b$, the coefficient a is the same as the slope m. (The proof of this generalization is left as an exercise.)

Because $a = m$, we can write $y = ax + b$ in the more meaningful form $y = mx + b$, which is called the **slope–intercept form** of a linear equation.

Definition of Slope–Intercept Form

The **slope–intercept form** of a linear equation is the form $y = mx + b$. The slope of the graph is m and the y-intercept is $(0, b)$.

Graphing with Slope

If we know the slope and one point of a line, we can sketch the line. To do this, we use a starting point and the slope to construct a slope triangle.

EXAMPLE 3

Using Slope to Draw a Line

Draw the line that contains the point $P(-3, 4)$ and whose slope is -2.

Solution

Plot the point $P(-3, 4)$. We can write the slope as $\frac{-2}{1}$, which means that the rise is -2 and the run is 1.

Starting at $P(-3, 4)$, move *down* 2 units and to the *right* 1 unit to locate a second point Q of the line. (See Fig. 4.16.) Points P and Q determine the line.

Figure 4.16

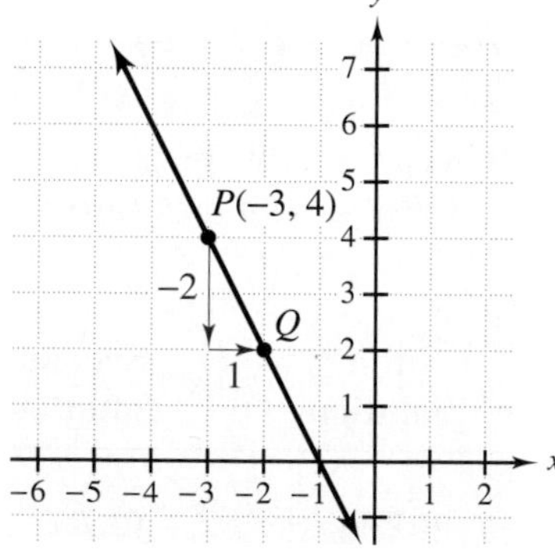

Figure 4.17

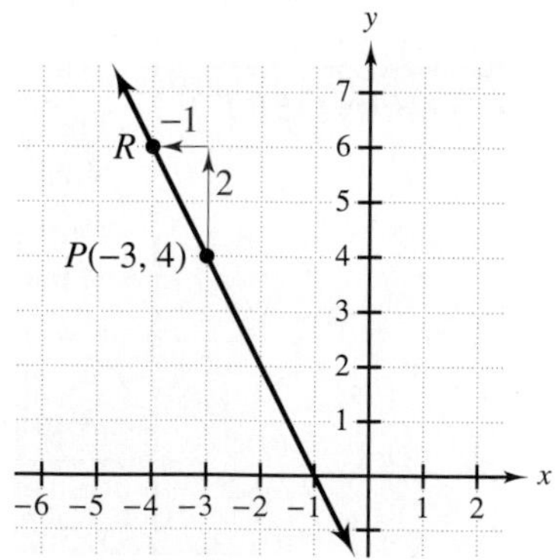

Note that the slope also can be written $\frac{2}{-1}$, which means that the rise is 2 and the run is -1. Starting at $P(-3, 4)$, move *up* 2 units and to the *left* 1 unit to locate a second point R of the line. (See Fig. 4.17.) The line determined by points P and R is the same as the line determined by points P and Q.

Note: When using a point and the slope to graph a line, always start the slope triangle at the plotted point, not at the origin.

Think About It

Why do the lines $y = -\frac{-2}{-3}x$, $y = -\frac{2}{3}x$, $y = \frac{-2}{3}x$, and $y = \frac{2}{-3}x$ all have the same slope?

If we are given just the equation of a line, we can write the equation in slope–intercept form, from which it is easy to determine the slope and y-intercept. Using this information, we can then draw the graph as we did in Example 3.

The following summarizes the approach to graphing a line with information about its slope and y-intercept.

Using a Line's Slope and y-Intercept to Graph the Line

1. Write the equation in the form $y = mx + b$.
2. Plot the y-intercept $(0, b)$.
3. Write the slope m as $\frac{\text{rise}}{\text{run}}$. Starting at the plotted point, move up or down as indicated by the rise. Then move left or right as indicated by the run. Plot the destination point, which is a second point of the line.
4. Draw the line through the two plotted points.

EXAMPLE 4

Using the Slope and y-Intercept to Draw a Line

Sketch the graph of $3x - 2y = 8$.

Solution

First, we write the equation in slope–intercept form.

$$\begin{aligned} 3x - 2y &= 8 \\ -2y &= -3x + 8 \\ y &= \frac{3}{2}x - 4 \end{aligned}$$

Plot the y-intercept, $P(0, -4)$.

The slope is $\frac{3}{2}$, which means the rise is 3 and the run is 2. From the y-intercept, draw the slope triangle by moving up 3 units and right 2 units. The destination is a second point Q of the line. (See Fig. 4.18.) The points P and Q determine the line.

Figure 4.18

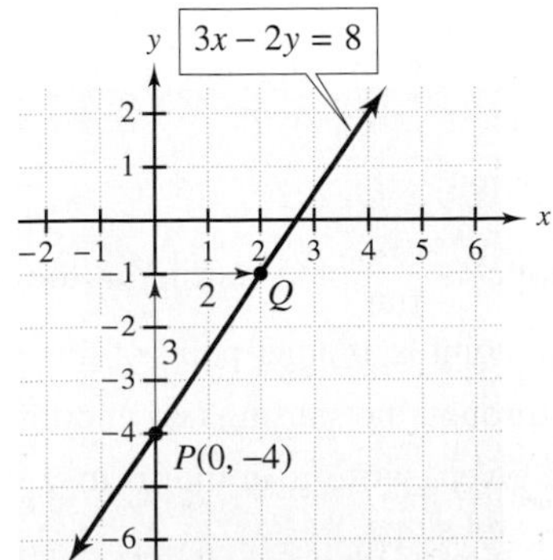

To graph an equation with its slope and y-intercept, we need to write the equation in the form $y = mx + b$. But if we need to solve for y, why not just enter $y = mx + b$ in a calculator and produce the graph electronically? Because this is the quickest of all methods, it is usually what we will want to do.

There will be times, however, when we will have information about slopes and points, but we will not know the equation. Because we will still want to be able to produce the graph, learning pencil and paper methods is important. Although using a graphing calculator has many advantages, you should try to avoid becoming so dependent on a calculator that you fail to understand the underlying mathematical concepts.

4.2 QUICK REFERENCE

The Slope Formula

- Given points A and B of a line, the steepness, or **slope,** of the line is measured by the ratio of the vertical change from A to B (**rise**) and the horizontal change from A to B (**run**).

  $$\text{Slope} = \frac{\text{rise}}{\text{run}}$$

- The slope of a line does not depend on the points of the line used to calculate it or on the points designated as the starting and ending points.
- For two points (x_1, y_1) and (x_2, y_2) of a line, $x_1 \neq x_2$, the slope m is defined as follows:

  $$m = \frac{y_2 - y_1}{x_2 - x_1} = \frac{y_1 - y_2}{x_1 - x_2}$$

- The slope of any horizontal line is 0; the slope of any vertical line is undefined.
- A line with a positive slope rises from left to right; a line with a negative slope falls from left to right.

The Slope-Intercept Form

- The **slope–intercept form** of a linear equation is the form $y = mx + b$.
- The constant term b is the y-coordinate of the y-intercept of the line. The coefficient m of the x-term is the slope of the line.

Graphing with Slope

- If we know the slope of a line and one of its points, we can use the following procedure to draw the line.
  1. Plot the given point.
  2. Interpreting slope as $\frac{\text{rise}}{\text{run}}$, and starting at the plotted point, draw a slope triangle. The destination point is another point of the line.
  3. Draw the line through the starting and ending points.
- If the equation is given, write it in the form $y = mx + b$ to determine the starting point (y-intercept) and slope.

4.2 EXERCISES

Concepts and Skills

1. What is the geometric interpretation of the slope of a line?

2. Describe a line that has a
   (a) positive slope.
   (b) negative slope.

In Exercises 3–10, predict whether the slope is positive, negative, 0, or undefined. Then calculate the slope.

3.

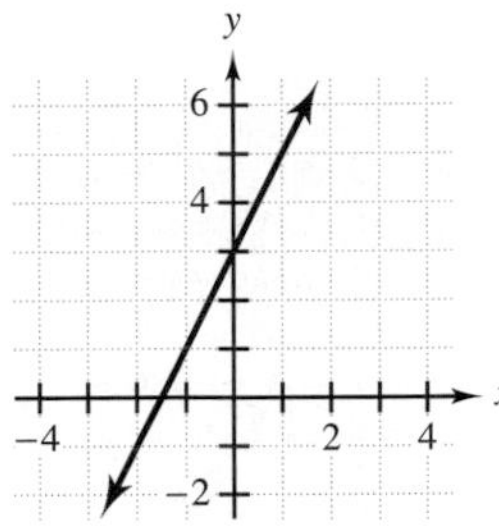

4.

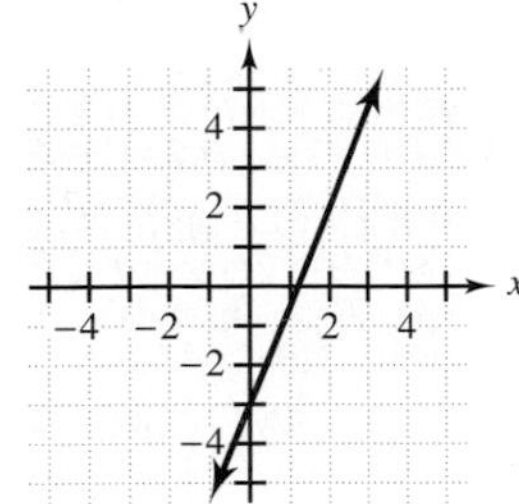

5.

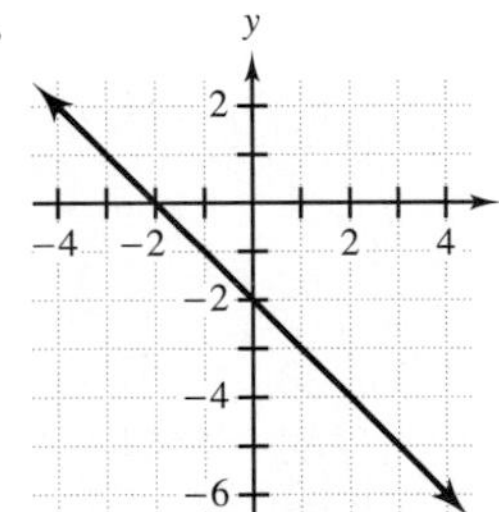

6.

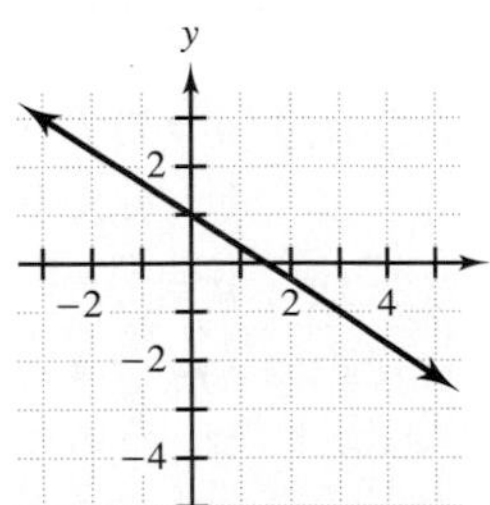

7.

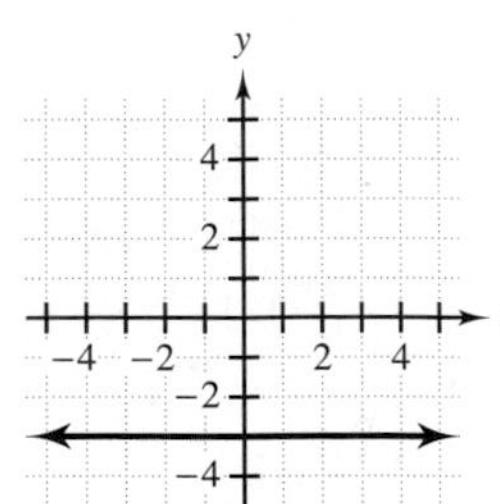

8.

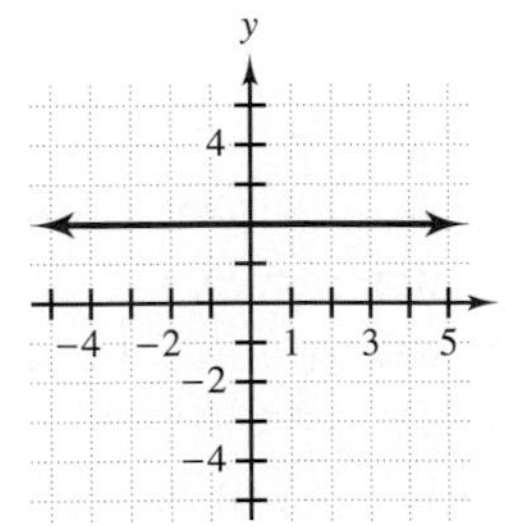

9.

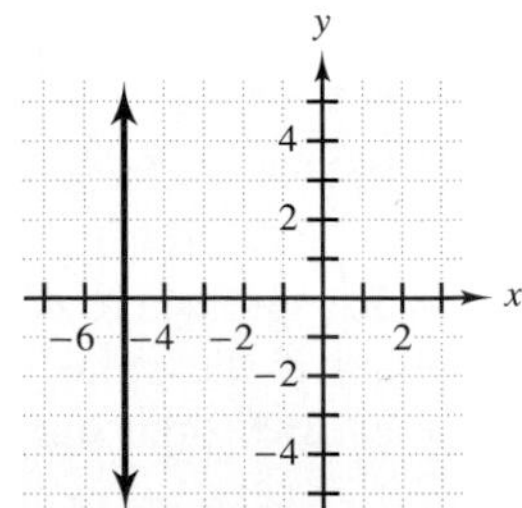

10.

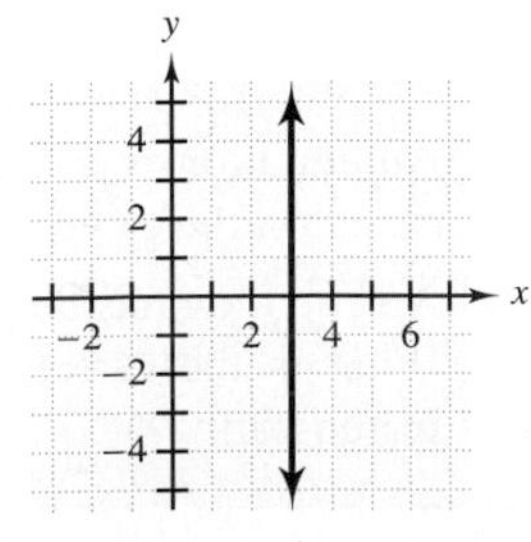

In Exercises 11–24, determine the slope of the line that contains the given points.

11. $(8, -2)$, $(6, -12)$
12. $(-3, -4)$, $(-6, -10)$
13. $(3, 4)$, $(5, 11)$
14. $(-2, 4)$, $(5, 8)$
15. $(5, -6)$, $(5, 7)$
16. $(2, -4)$, $(4, -4)$
17. $(2.3, -5.9)$, $(-1.7, 5.4)$
18. $(0, 6.94)$, $(2.45, -3.5)$
19. $(-2, 5)$, $(8, -3)$
20. $(-1, -6)$, $(-7, -4)$
21. $(-3, 7)$, $(6, 7)$
22. $(-3, 7)$, $(-3, 4)$
23. $\left(\frac{1}{2}, \frac{1}{3}\right)$, $\left(\frac{2}{3}, \frac{3}{4}\right)$
24. $\left(\frac{1}{3}, \frac{2}{3}\right)$, $\left(\frac{3}{4}, \frac{1}{2}\right)$

25. If the slope of a line is -3, two of the following procedures are correct ways to draw a slope triangle from a given point of the line. Identify the two correct procedures and explain why the third procedure is incorrect.
   (i) Move down 3 units and right 1 unit.
   (ii) Move down 3 units and left 1 unit.
   (iii) Move up 3 units and left 1 unit.

26. What is the difference between a zero slope and an undefined slope?

In Exercises 27–32, determine the unknown coordinate so that the line containing the points has the given slope.

27. $(-3, y)$, $(2, -2)$; $m = -1$
28. $(0, 8)$, $(2, y)$; $m = -\frac{5}{4}$
29. $(2, -1)$, $(x, 0)$; slope is undefined
30. $(x, -3)$, $(-5, 7)$; slope is undefined
31. $(-3, 4)$, $(2, y)$; $m = 0$
32. $(0, y)$, $(6, 5)$; $m = 0$

33. A line contains the points $P(3, -2)$ and $Q(-11, 6)$. Which of the following is a correct way to calculate the slope of the line? Why?

(i) $m = \dfrac{6-(-2)}{3-(-11)} = \dfrac{8}{14} = \dfrac{4}{7}$

(ii) $m = \dfrac{-2-6}{3-(-11)} = \dfrac{-8}{14} = -\dfrac{4}{7}$

34. The slope of a line containing the points (x_1, y_1) and (x_2, y_2) is $\frac{4}{5}$. Does this mean that $y_2 - y_1 = 4$ and $x_2 - x_1 = 5$? Explain.

In Exercises 35–40, determine two solutions of the given equation and use the points to determine the slope of the line.

35. $3x + 5y = 15$

36. $x - 2y - 6 = 0$

37. $3 + y = 12$

38. $7 + 3x = 22$

39. $y = \dfrac{3}{4}x - 3$

40. $y = \dfrac{2}{3}x + 2$

In Exercises 41–54, write the equation in slope–intercept form, if possible. Then determine the slope of the line.

41. $y + 2x = 17$

42. $y - 3x = 10$

43. $2x - y = 0$

44. $y + 3x = 0$

45. $y + 2 = 17$

46. $y - 3 = 5$

47. $x + 3y - 6 = 0$

48. $3x + 4y = 10$

49. $4x - 3 = 13$

50. $2x + 9 = 7$

51. $3y = 5x - 7$

52. $7x = 5y + 8$

53. $\dfrac{3}{4}x - \dfrac{2}{3}y = 12$

54. $\dfrac{1}{4}x + \dfrac{2}{5}y = 10$

55. If a linear equation is in the form $Ax + By = C$, $B \neq 0$, show that the slope of the graph is $-\dfrac{A}{B}$.

56. If a linear equation is in the form $Ax + By = C$, $B \neq 0$, show that the y-intercept of the graph is $\dfrac{C}{B}$.

57. A line has a slope of -2. Which of the following could be an equation of the line?

(i) $6x + 3y - 7 = 0$

(ii) $4x - 2y + 3 = 0$

(iii) $x + \dfrac{1}{2}y - 1 = 0$

58. A line has a slope of $\frac{1}{3}$. Which of the following could be an equation of the line?

(i) $\dfrac{1}{9}x = \dfrac{1}{3}y + 3$

(ii) $9x - 3y = 4$

(iii) $12x + 2 - 4y = 0$

59. What is the rise of a line if the line's slope is zero?

60. What is the run of a line if the line's slope is undefined?

In Exercises 61–72, draw a line that contains the given point and has the given slope.

61. $(0, 3)$; $m = \dfrac{3}{4}$

62. $(0, -2)$; $m = -\dfrac{3}{4}$

63. $(0, -1)$; $m = -4$

64. $(0, 3)$; $m = 2$

65. $(-5, 0)$; m is undefined

66. $(0, 2)$; $m = 0$

67. $(-2, -3)$; $m = \dfrac{2}{3}$

68. $(-3, -5)$; $m = -\dfrac{3}{5}$

69. $(3, 7)$; $m = 0$

70. $(-4, -3)$; m is undefined

71. $(2, 6)$; $m = -5$

72. $(-5, 2)$; $m = 4$

In Exercises 73–84, sketch the graph of the given equation. When possible, produce the graph on your calculator and compare it to your sketch.

73. $y = \dfrac{3}{4}x + 5$

74. $y = -\dfrac{2}{3}x - 2$

75. $y = -3x + 4$

76. $y = 2x + 3$

77. $y = 2$

78. $y = -2$

79. $y = x$

80. $y = -x$

81. $y = -\dfrac{3}{4}x$

82. $y = \dfrac{2}{5}x$

83. $x = 3$

84. $x = -4$

85. An equation is in the form $y = 3x + b$. What is the effect on the graph if the value of b increases?

86. An equation has the form $y = mx + b$, where $m > 0$. What is the effect on the graph if the value of m increases?

In Exercises 87–96, if possible, write the given equation in the form $y = mx + b$. Then sketch the graph.

87. $2x + 5y = 10$

88. $4x - 5y = 20$

89. $x - 3 = 4$

90. $x + 6 = 2$

91. $2x - y - 6 = 2$

92. $x + 3y - 6 = 0$

93. $y + 3 = 10$

94. $y - 2 = 4$

95. $y - 2x = 0$

96. $y + 3x = 0$

Real-Life Applications

97. A pediatrician uses the rule of thumb that a boy grows about 2 inches per year between the ages of 6 and 14.

(a) If the average height of a boy age 6 is 3 feet 9 inches, what is his expected height at age 14?

(b) Using the two data points in part (a), and assuming that the growth is linear, what would be the slope of the graph?

(c) How does your answer in part (b) compare with the pediatrician's average growth rate figure?

(d) Starting at the data point for age 6, suppose that you rise 8 units and run 4 units. What information would be represented by the destination point?

98. In recent years, math SAT scores have remained relatively constant. What does this imply about the slope of a line used to model the data?

99. The population of a small rural town was 1200 in 1990, but the population has decreased by an average of 75 people per year since then. Suppose that the slope–intercept form $y = mx + b$ is used to model the population y, where x represents the number of years since 1990.

(a) What is the value of m?

(b) What is the value of b?

(c) If you move from the data point for 1990 to the data point for 1997, what would be the rise and run?

(d) In what year does the model equation predict that the town will be a ghost town?

100. A standard stairway has 9-inch treads. How long should the risers be in order for the stairway to have a slope of 0.8?

Modeling with Real Data

101. By the year 2000, 52% of all automobiles are projected to be equipped with airbags. The accompanying bar graph shows actual and projected data. (Source: National Highway Traffic Safety Administration.)

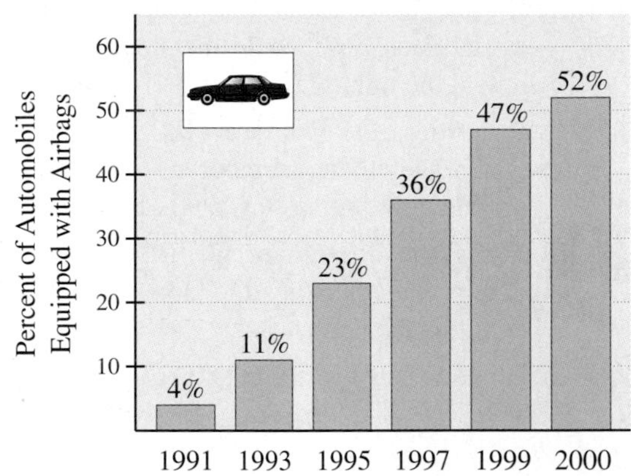

The percentage B of automobiles with airbags can be modeled by the linear function $B(t) = 5.6t - 3.6$, where t is the number of years since 1990.

(a) The model function is selected to fit all the data points as closely as possible. What is the slope of the graph of B?

(b) If a line were drawn through just the data points for 1991 and 2000, what would be the slope of that line?

102. In both parts (a) and (b) of Exercise 101, what does the slope indicate?

103. In Section 2.4 we observed the trends in the number of private and semiprivate golf courses from 1990 to 1995. The accompanying bar graph shows the data for the beginning and end of that period. (Source: National Golf Federation.)

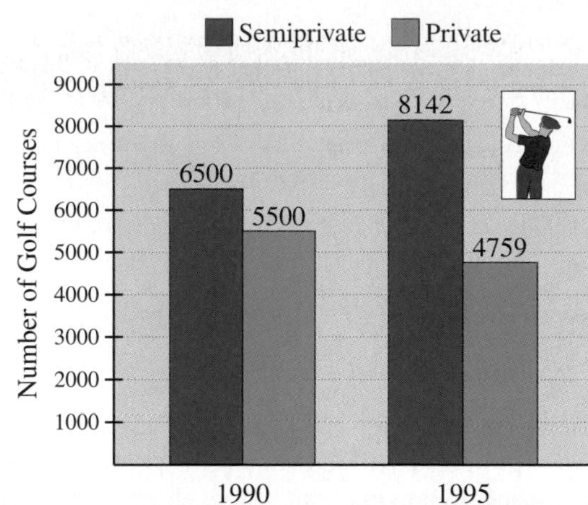

(a) Assuming that the number of courses of each type can be modeled with a linear function, what is the slope of each line?

(b) If you were given line graphs without the actual data, how would the slopes of the lines reveal the trends?

104. (a) Referring to Exercise 103, calculate the difference in the number of semiprivate and private golf courses for 1990 and 1995.

(b) Using the differences for 1990 and 1995 as data points, calculate the slope of the line that contains the points.

(c) What does the result in part (b) indicate?

Group Project: Female State Legislators

The accompanying bar graph shows the percentage of female state legislators during the period 1971–1995. (Source: Center for the American Woman and Politics.)

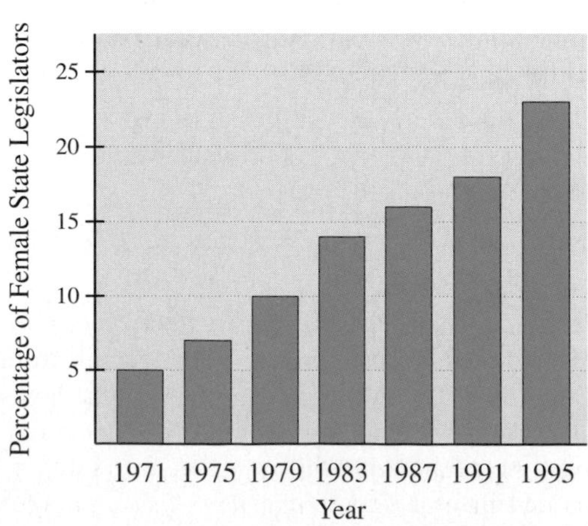

The data can be modeled by the linear equation

$$P = 0.705x - 1385.29, x \geq 1971$$

where P is the percentage of women in state legislatures in the year x.

105. Observe that the coefficient of x is positive. What does this tell you about the graph of the model equation? Is this consistent with the shape of the bar graph?

106. Assuming that the model remains valid, by what year will the percentage of women in state legislatures reach 50%?

107. If a corresponding linear equation were written to model the percentage of male state legislators during the same period, what can you predict about the coefficient of x? Why?

108. Some believe that increasing numbers of women in elected offices will result in greater attention to issues such as child welfare and education. What is your opinion?

Challenge

109. If the slope of a line is 50, the line may appear to be vertical in the default viewing rectangle. Which setting, X or Y, do you need to adjust so that the line does not appear to be vertical and so that the intercepts are more clearly displayed? Why?

In Exercises 110–113, adjust the window setting so that both intercepts of the graph are clearly displayed and the line does not appear vertical or horizontal.

110. $y = 100x - 7$

111. $y = 90x + 200$

112. $y = 0.02x + 0.05$

113. $y = 35 - 0.02x$

114. Suppose $P(x_1, y_1)$ and $Q(x_2, y_2)$ are two points of the graph of $y = ax + b$.

(a) Write an equation expressing the fact that (x_1, y_1) is a solution of $y = ax + b$.

(b) Write an equation expressing the fact that (x_2, y_2) is a solution of $y = ax + b$.

(c) Subtract the equation in part (b) from the equation in part (a) and solve for a.

(d) Explain why a is the slope of the graph of $y = ax + b$.

4.3 APPLICATIONS OF SLOPE

Geometric Applications • Rate of Change • Direct Variation

Geometric Applications

In geometry, if two lines in the same plane do not intersect, then the lines are **parallel.** We begin this section by investigating the relationship between parallel lines and their slopes.

EXPLORING THE CONCEPT

Slopes of Parallel Lines

Figure 4.19 shows a calculator display of the graphs of the following four equations.

$$y_1 = 2x - 9 \qquad y_2 = 2x - 1$$
$$y_3 = 2x + 3 \qquad y_4 = 2x + 8$$

Figure 4.19

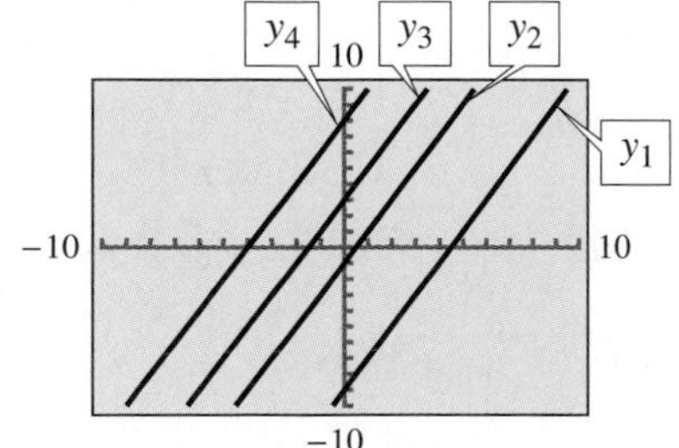

We can see from Fig. 4.19 that the lines appear to be parallel. We also observe that the slopes of all four lines are equal.

These results suggest the generalization that distinct lines with the same slope are parallel.

Note: The word *distinct* means that the lines are different lines. If two lines have the same slope and the same y-intercept, then the lines coincide. Such lines are not parallel because they intersect everywhere.

The following relationship can be proven to be true.

Parallel Lines and Slopes

Two distinct, nonvertical lines are **parallel** if and only if they have the same slope. Also, any two distinct vertical lines are parallel.

Roughly speaking, a *quadrilateral* is a geometric figure with four sides. In Example 1 we consider a *parallelogram*, which is a quadrilateral whose opposite sides are parallel.

EXAMPLE 1

Determining Whether a Quadrilateral Is a Parallelogram

Figure 4.20 shows a quadrilateral drawn in a coordinate system. Show that the quadrilateral is a parallelogram.

Figure 4.20

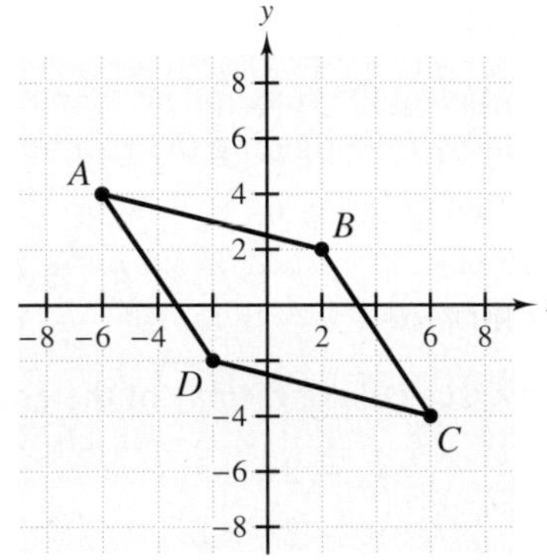

Solution

Use the slope formula to determine the slopes of the sides of the quadrilateral.

Side	*Points*	*Slope*
$\overline{AB}$	$A(-6, 4), B(2, 2)$	$m = \dfrac{4 - 2}{-6 - 2} = \dfrac{2}{-8} = -\dfrac{1}{4}$
$\overline{DC}$	$D(-2, -2), C(6, -4)$	$m = \dfrac{-2 - (-4)}{-2 - 6} = \dfrac{2}{-8} = -\dfrac{1}{4}$
$\overline{AD}$	$A(-6, 4), D(-2, -2)$	$m = \dfrac{4 - (-2)}{-6 - (-2)} = \dfrac{6}{-4} = -\dfrac{3}{2}$
$\overline{BC}$	$B(2, 2), C(6, -4)$	$m = \dfrac{2 - (-4)}{2 - 6} = \dfrac{6}{-4} = -\dfrac{3}{2}$

The opposite sides are parallel because they have the same slopes. Therefore, the quadrilateral is a parallelogram.

Think About It

A rhombus is a parallelogram with sides of equal length. To show that a quadrilateral is a rhombus, you would need to show that the opposite sides are parallel. What other formula would be needed to complete the proof?

Square

Tick marks on the two axes should be equally spaced when the shape of the graph or the relationship between graphs is visually important.

Two lines are **perpendicular** if they intersect to form a right angle. Although the relationship between parallel lines and their slopes is intuitively evident, the relationship between perpendicular lines and their slopes is not obvious.

EXPLORING THE CONCEPT

Slopes of Perpendicular Lines

Figure 4.21 shows the graphs of $y_1 = \frac{1}{2}x + 3$ and $y_2 = -2x - 4$. By using a square setting, we obtain a calculator display in which the lines appear to be perpendicular.

Figure 4.21

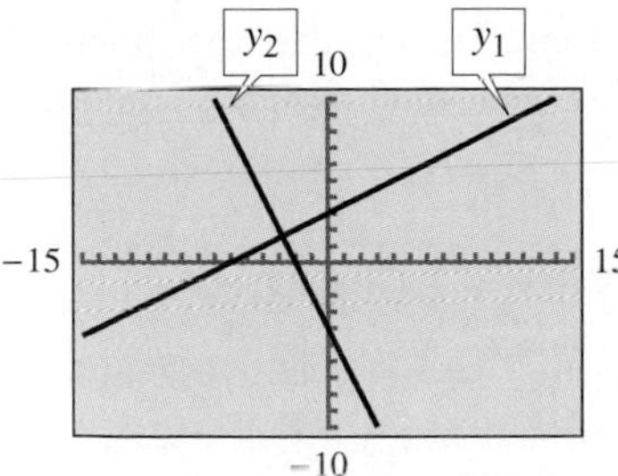

The slopes of the two lines are $m_1 = \frac{1}{2}$ and $m_2 = -2$. There are many relationships between these two numbers that one could imagine. However, one relationship can be proven to be true for any two nonvertical perpendicular lines: The product of the slopes is -1.

$$m_1 \cdot m_2 = \frac{1}{2} \cdot (-2) = -1$$

Note: For this relationship to hold, neither line can be vertical, because the slope of a vertical line is undefined.

LEARNING TIP

If we know the slope of one line, then an easy way to determine the slope of a perpendicular line is to take the opposite of the reciprocal of the known slope. For instance, if the slope of a line is $-\frac{7}{5}$, then $\frac{5}{7}$ is the slope of a perpendicular line.

These conclusions are true in general.

Perpendicular Lines and Slopes

If neither of two lines is vertical, then the lines are **perpendicular** if and only if the product of their slopes is -1. Also, a vertical line and a horizontal line are perpendicular.

Note: Symbolically, if the slopes of two lines are m_1 and m_2, then the lines are perpendicular if and only if $m_1m_2 = -1$. This relationship can also be expressed as $m_1 = -\dfrac{1}{m_2}$ or $m_2 = -\dfrac{1}{m_1}$.

EXAMPLE 2

Determining Whether Lines Are Perpendicular

Figure 4.22 shows the graphs of $y_1 = \frac{5}{3}x - 5$ and $y_2 = -\frac{3}{5}x + 8$. Show that the lines are perpendicular.

Figure 4.22

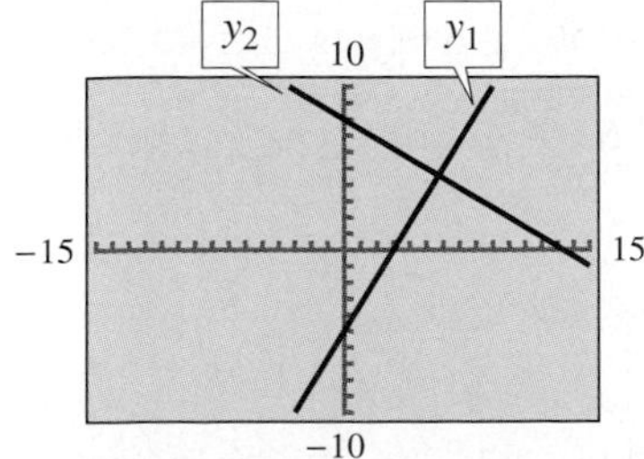

Solution

From the equations, we see that the slopes are $m_1 = \frac{5}{3}$ and $m_2 = -\frac{3}{5}$. Because $m_1m_2 = \frac{5}{3} \cdot \left(-\frac{3}{5}\right) = -1$, the lines are perpendicular.

EXAMPLE 3

Determining Whether Lines Are Perpendicular

Use a square setting (if necessary) to produce the graphs of the linear equations $7x - 12y = -36$ and $11x + 6y = -24$ on your calculator. Do the lines appear to be perpendicular? Verify your conclusion.

Solution

Write the equations in the slope–intercept form and produce their graphs.

$$y = \frac{7}{12}x + 3$$

$$y = -\frac{11}{6}x - 4$$

Figure 4.23

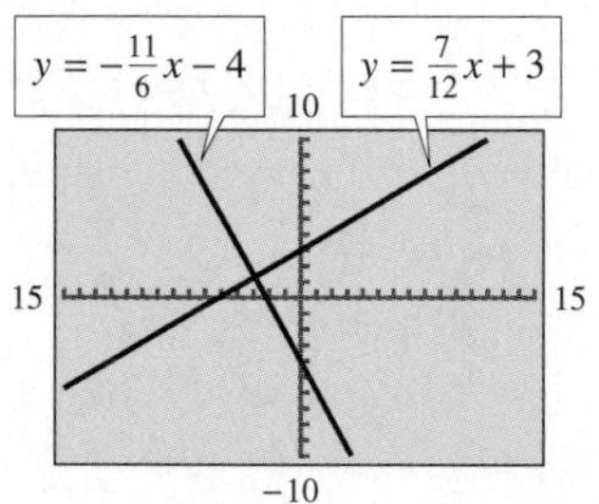

The lines in Fig. 4.23 appear to be perpendicular. However, multiplying the slopes leads to a different conclusion.

$$\frac{7}{12} \cdot \left(-\frac{11}{6}\right) = -\frac{77}{72} \approx -1.07$$

Because the product of the slopes is not -1, the lines are not perpendicular.

EXAMPLE 4

Parallel and Perpendicular Lines

Determine whether the lines in each pair are parallel, perpendicular, or neither.

(a) $y = -x + 3$ and $x - 1 + y = 0$

(b) $3x + 2y = 6$ and $-y = 2 - \frac{2}{3}x$

(c) $3x - y = 5$ and $x - 3y = 0$

Solution

	Equation	*Slope–Intercept Form*	*Slope*
(a)	$y = -x + 3$	$y = -x + 3$	-1
	$x - 1 + y = 0$	$y = -x + 1$	-1
(b)	$3x + 2y = 6$	$y = -\frac{3}{2}x + 3$	$-\frac{3}{2}$
	$-y = 2 - \frac{2}{3}x$	$y = \frac{2}{3}x - 2$	$\frac{2}{3}$
(c)	$3x - y = 5$	$y = 3x - 5$	3
	$x - 3y = 0$	$y = \frac{1}{3}x$	$\frac{1}{3}$

Because the lines in (a) have the same slope and different y-intercepts, they are parallel. The product of the slopes in (b) is -1, so the lines are perpendicular. Because the slopes in (c) are not equal and their product is not -1, the lines are neither parallel nor perpendicular.

EXAMPLE 5

Slopes of Perpendicular Lines

Suppose the equation of line L is $y = \frac{2}{3}x - 5$. If line N is perpendicular to line L, what is the slope of line N?

Solution

From the equation of line L, the slope of line L is $\frac{2}{3}$. Let m represent the slope of line N.

$$\frac{2}{3} \cdot m = -1$$

Because lines L and N are perpendicular, the product of their slopes is -1.

$$\frac{3}{2} \cdot \frac{2}{3} \cdot m = \frac{3}{2} \cdot (-1)$$

Multiply both sides by $\frac{3}{2}$ to solve for m.

$$m = -\frac{3}{2}$$

The slope of line N is $-\frac{3}{2}$.

Previously, we determined whether a triangle was a right triangle by applying the Pythagorean Theorem. Now we can determine if any two sides are perpendicular by examining their slopes.

EXAMPLE 6

Determining Whether a Triangle Is a Right Triangle

The points $A(2, 5)$, $B(3, 7)$, and $C(6, 3)$ are the vertices of a triangle. (See Fig. 4.24.) Is the triangle a right triangle?

Figure 4.24

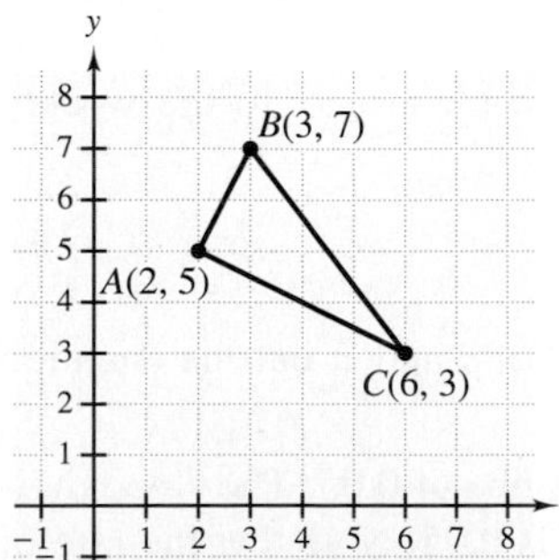

Solution

From Fig. 4.24 it appears that line segments $\overline{AB}$ and $\overline{AC}$ are perpendicular. We use the slope formula to find the slopes of these line segments.

Points	*Slope*
$A(2, 5)$ and $B(3, 7)$	$m = \frac{7-5}{3-2} = \frac{2}{1} = 2$
$A(2, 5)$ and $C(6, 3)$	$m = \frac{3-5}{6-2} = \frac{-2}{4} = -\frac{1}{2}$

Because the product of the slopes is $2 \cdot \left(-\frac{1}{2}\right) = -1$, the line segments are perpendicular. Therefore, the triangle is a right triangle.

Rate of Change

An important concept in mathematics involves the rate at which a function changes. Specifically, for the function $y = mx + b$, the **rate of change** of y with respect to x is the amount by which y changes for a unit change in x.

Note: A *unit change in x* means a 1-unit increase or decrease in the value of x. The corresponding change in y can be positive or negative.

EXPLORING THE CONCEPT

Rate of Change and Slope

Figure 4.25 shows a calculator table of values for the function $y = 2x - 8$.

Figure 4.25

X	Y_1	
2	-4	
3	-2	
4	0	
5	2	
6	4	
7	6	
8	8	
X=2		

By scrolling through the table, we see that y always increases by 2 units for each unit increase in x. We can compare these changes with a ratio, which is called the rate of change of y with respect to x.

$$\frac{\text{Change in } y}{\text{Change in } x} = \frac{2}{1} = 2$$

Noting that the slope of the line $y = 2x - 8$ is 2, we observe that the rate of change of y with respect to x is the same as the slope of the line.

This conclusion is true in general.

Slope as a Rate of Change

For a linear equation in two variables, the **rate of change of y with respect to x** is the slope of the graph of the equation. This rate of change, like the slope, is constant.

Many real-life situations can be modeled by an equation for which the rate of change is constant.

EXAMPLE 7

Hourly Pay as Rate of Change

A salesperson earns a base salary of \$300 per week plus a 5% commission on all sales.

(a) Write an equation for the salesperson's weekly income I as a function of weekly sales s.

(b) What is the rate of change in the weekly income with respect to the weekly sales?

Solution

(a) $I = 0.05s + 300$. The variables are I and s instead of y and x, but the equation is in slope–intercept form.

(b) The graph of the equation is a straight line with a slope of 0.05. The slope indicates that the rate of change of I with respect to s is 0.05. Note that the rate of change is simply the commission.

Direct Variation

Suppose that a neighborhood association collects dues of \$25 per residence. Then the amount y that is collected from x residences is $y = 25x$. In this case we say that y **varies directly** with x.

Definition of Direct Variation

The value of y **varies directly** with the value of x if there is a constant k such that $y = kx$.

Note that $y = kx$ is a special case of the slope–intercept form $y = mx + b$. The constant k is called the **constant of variation**. A direct variation also can be written $\frac{y}{x} = k$, $x \neq 0$. In this case we say that y is **directly proportional** to x, and k is called the **constant of proportionality.**

There are many examples of one variable varying directly with another variable. In each case, the y-intercept of the graph is the origin, and the slope m is the same as the constant k of variation.

EXAMPLE 8

Proportional Distances

The actual distance between two towns varies directly with the distance between them on a map. On a certain map, the distance between Tortilla Flat, Arizona, and Tombstone, Arizona, is $6\frac{3}{4}$ inches. Suppose that on this map, $\frac{1}{3}$ of an inch represents 7 miles. What is the actual distance between these towns?

Solution

Let d represent the actual distance (in miles), and let x represent the map distance (in inches). Because d varies directly with x, the relation is $d = kx$.

A map distance of $\frac{1}{3}$ of an inch corresponds to an actual distance of 7 miles. Replace x with $\frac{1}{3}$ and d with 7 to determine k.

$$d = kx$$

$$7 = k \cdot \frac{1}{3}$$

$$21 = k$$

The constant of variation is 21, so $d = 21x$. Since the map distance between the two towns is 6.75,

$$d = 21(6.75) = 141.75$$

The distance from Tortilla Flat to Tombstone is 141.75 miles.

Note that the slope of the graph of $d = 21x$ is the same as the constant of variation, $k = 21$. There are 21 miles of actual distance for each inch of map distance.

4.3 QUICK REFERENCE

Geometric Applications

- Two distinct, nonvertical lines are **parallel** if and only if they have the same slope. Any two distinct vertical lines are parallel.
- If neither of two lines is vertical, then the lines are **perpendicular** if and only if the product of their slopes is -1. A vertical line and a horizontal line are perpendicular.

Rate of Change

- For a linear function $y = mx + b$, the **rate of change** of y with respect to x is constant, and it is the slope of the graph of the function.

Direct Variation

- We say that y **varies directly** with x if there is a constant k such that $y = kx$. The linear function $y = kx$ is called a **direct variation**, where k is the **constant of variation.**
- If $x \neq 0$, a direct variation can be written $\frac{y}{x} = k$, where k is called the **constant of proportionality.**

4.3 EXERCISES

Parallel and Perpendicular Lines

In all exercises in this section, L_1 refers to line 1 and L_2 refers to line 2. The symbol m_1 is used to represent the slope of line 1, and m_2 is used to represent the slope of line 2.

1. What is the geometric definition of perpendicular lines?

2. What is the geometric definition of parallel lines?

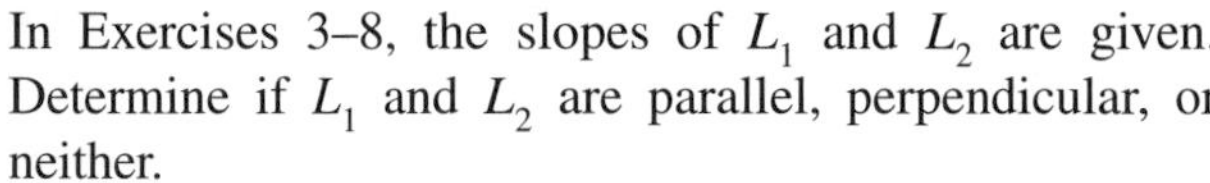

In Exercises 3–8, the slopes of L_1 and L_2 are given. Determine if L_1 and L_2 are parallel, perpendicular, or neither.

3. $m_1 = 2;\ m_2 = -2$
4. $m_1 = \frac{1}{4};\ m_2 = 4$
5. $m_1 = \frac{1}{2};\ m_2 = 0.5$
6. $m_1 = -4;\ m_2 = 0.25$
7. $m_1 = \frac{3}{5};\ m_2 = -\frac{5}{3}$
8. $m_1 = \frac{7}{9};\ m_2 = \frac{9}{7}$

In Exercises 9–14, for the given slope, determine m_2 so that

(a) L_2 is parallel to L_1.
(b) L_2 is perpendicular to L_1.

9. $m_1 = -3$
10. $m_1 = \frac{5}{8}$
11. $m_1 = -\frac{3}{4}$
12. $m_1 = 2.5$
13. $m_1 = 0$
14. m_1 undefined

In Exercises 15–20, two points of L_1 and two points of L_2 are given. Determine if the lines are parallel, perpendicular, or neither.

15. L_1: $(3, 5)$ and $(-2, 1)$
    L_2: $(-6, 9)$ and $(-1, 13)$
16. L_1: $(-6, 4)$ and $(-2, -8)$
    L_2: $(108, 120)$ and $(100, 144)$
17. L_1: $(0, 0)$ and $(8, 8)$
    L_2: $(-1, 4)$ and $(3, -7)$
18. L_1: $(4, 8)$ and $(1, -2)$
    L_2: $(-3, 5)$ and $(2, 2)$
19. L_1: $(-4, 2)$ and $(-4, 7)$
    L_2: $(0, 5)$ and $(3, 5)$

20. L_1: $(-2, 3)$ and $(1, 5)$
L_2: $(0, 4)$ and $(2, 1)$

21. A vertical line and a horizontal line are perpendicular, but the relation $m_1m_2 = -1$ is not valid. Why?

22. What can you conclude about the lines L_1 and L_2 if $m_1 = -(1/m_2)$, where $m_2 \neq 0$? Why?

In Exercises 23–28, the equations of L_1 and L_2 are given. Determine whether the lines are parallel, perpendicular, or neither.

23. L_1: $3y = 5 - 2x$ $\quad L_2$: $y = -\frac{2}{3}x + 9$

24. L_1: $y = \frac{2}{3}x + 1$ $\quad L_2$: $3x + 2y = 0$

25. L_1: $x = -4$ $\quad L_2$: $y = -4$

26. L_1: $y - 3 = 12$ $\quad L_2$: $5 - y = 14$

27. L_1: $y = 5x - 3$ $\quad L_2$: $y = -5x + 3$

28. L_1: $\frac{1}{2}x - \frac{1}{3}y = 0$ $\quad L_2$: $\frac{5}{6}x - \frac{5}{9}y = 1$

In Exercises 29–34, the equation of one line is given and two points of another line are given. Are the lines parallel, perpendicular, or neither?

29. L_1: $x - y = 2$ $\quad L_2$: $(5, 9)$ and $(6, 8)$

30. L_1: $(-3, 1)$ and $(4, 2)$ $\quad L_2$: $y = \frac{1}{7}x - 2$

31. L_1: $(-2, -5)$ and $(0, 4)$ $\quad L_2$: $y = \frac{9}{2}x - 4$

32. L_1: $-2x + 5y = 10$ $\quad L_2$: $(-6, 6)$ and $(-4, 1)$

33. L_1: $y = 5$ $\quad L_2$: $(2, 3)$ and $(-6, 0)$

34. L_1: $(2, -1)$ and $(-7, -1)$ $\quad L_2$: $x = -1$

In Exercises 35–38, what can you conclude about the slope of line L?

35. L is parallel to the y-axis.

36. L is perpendicular to the x-axis.

37. L is perpendicular to the y-axis.

38. L is parallel to the x-axis.

In Exercises 39–44, determine the slope of a line that is

(a) parallel to the given line.

(b) perpendicular to the given line.

39. $2x - y = 6$

40. $3x + y - 7 = 0$

41. $4x + 3y - 6 = 0$

42. $2x = 5y - 5$

43. $3y - 9 = 0$

44. $2(x - 3) = 1$

In Exercises 45–48, determine the slope of the line and sketch the graph.

45. The line contains $(-1, -4)$ and is parallel to the line containing $(-2, 3)$ and $(4, -1)$.

46. The line contains $(-3, 1)$ and is perpendicular to the line $y = 2x - 3$.

47. The x-intercept of the line is $(3, 0)$, and the line is parallel to the y-axis.

48. The y intercept of the line is $(0, -6)$, and the line is parallel to the x-axis.

In Exercises 49 and 50, information is given about two lines L_1 and L_2. Determine the value of k so that

(a) L_1 and L_2 are parallel.

(b) L_1 and L_2 are perpendicular.

49. L_1 contains the points $(3, -4)$ and $(-1, 2)$. L_2 contains the points $(5, 0)$ and $(-2, k)$.

50. The equation of L_1 is $3x + 12y = 15$. The equation of L_2 is $y = kx + 7$.

Geometric Models

In Exercises 51 and 52, show that the triangles in the figures are right triangles.

51.

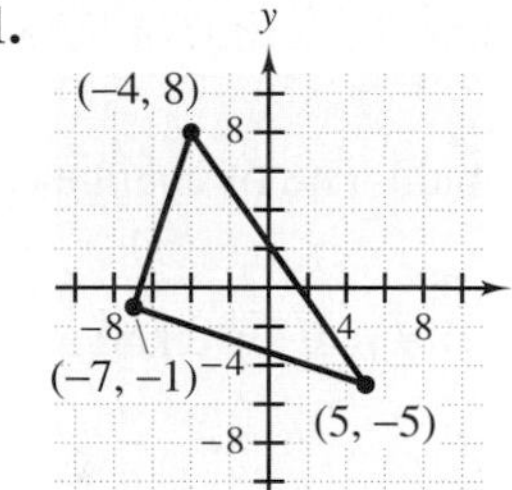

52.

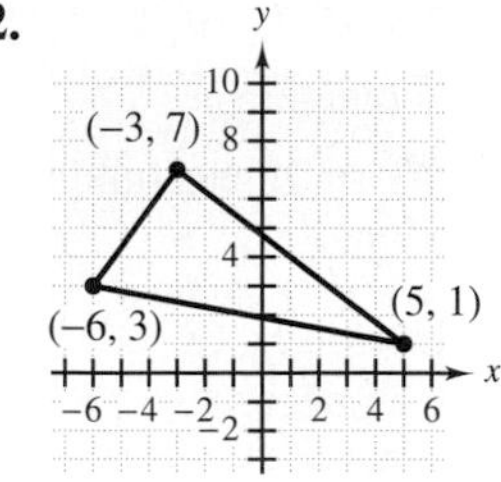

In Exercises 53 and 54, determine whether the three given points are collinear. (Points are **collinear** if they are points of the same straight line.)

53. $P(-5, 8)$
$Q(-1, 3)$
$R(7, -7)$

54. $A(-7, -2)$
$B(0, 0)$
$C(21, 5)$

55. The plan for a retaining wall and a shrub border is laid out on a coordinate system. The retaining wall is drawn from point $A(3, -1)$ to point $B(-7, 9)$ and then to point $C(-2, 14)$. The shrub border connects point $D(8, 4)$ with points A and C. Show that the region enclosed by the retaining wall and the shrub border is a rectangle.

56. A coordinate system design for a park shows a low marble wall connecting points $A(-4, -7)$, $B(-1, -3)$, $C(12, 1)$, and $D(2, -9)$. A stone path is drawn from point A to point C, and a brick walk is drawn from point B to point D. Show that the stone path and brick walk are perpendicular.

57. A plan for a small airport is laid out on a coordinate system. A short runway extends from point $A(-2, 9)$ to point $D(-12, 4)$, and a longer runway extends from point $B(4, -3)$ to point $C(-20, -15)$. A taxiway connects points A and B. Show that the runways are parallel and that the taxiway is perpendicular to each runway.

58. Four aircraft carriers are positioned on a coordinate system. *America* is at $A(4, 11)$, *Eisenhower* is at $B(12, 3)$, *Kennedy* is at $C(-8, -7)$, and *Halsey* is at $D(-16, 1)$. Show that the aircraft carriers are at the vertices of a parallelogram.

Rate of Change

59. Describe the slope of a line in terms of rate of change.

60. For a linear function, why is the rate of change constant?

In Exercises 61–64, determine the rate of change of y with respect to x.

61. $y = \frac{1}{2}x + 3$ **62.** $y = -\frac{2}{3}x + 5$

63. $y = -4x + 9$ **64.** $y = 3x - 2$

In Exercises 65–68, determine the rate of change of y with respect to x for the line containing the given points.

65. $(-1, -2)$ and $(3, 4)$ **66.** $(-2, 3)$ and $(3, 1)$

67. $(2, -2)$ and $(-1, 5)$ **68.** $(-3, 1)$ and $(2, 5)$

In Exercises 69–76, the rate of change of a line is given. For the given increase/decrease of one variable, determine the corresponding increase/decrease of the other variable.

69. $m = 3$; x increased by 2

70. $m = -2$; y increased by 6

71. $m = -\frac{5}{4}$; y increased by 5

72. $m = \frac{1}{3}$; x increased by 9

73. $m = 5$; x decreased by 2

74. $m = -4$; y decreased by 12

75. $m = -\frac{3}{2}$; y decreased by 6

76. $m = \frac{3}{5}$; x decreased by 20

Direct Variation

77. Explain the meaning of the statement, "A varies directly with B."

78. Suppose that y varies directly with x. If $x = a$ when $y = b$ and $x = c$ when $y = d$, explain why $\frac{a}{b} = \frac{c}{d}$, where a, b, c, and d are nonzero numbers.

In Exercises 79–84, determine whether the given equation represents a direct variation.

79. $y = 4x$ **80.** $4 = xy$

81. $\frac{y}{4} = x$ **82.** $y = 2x + 3$

83. $y = 5x - 3$ **84.** $y = 3$

85. Suppose that $y = kx$, $k > 0$. If x increases, then y ______, and if y decreases, then x ______.

86. Suppose y varies directly with x and k is the constant of proportionality. Then $y =$ ______ and $k =$ ______.

Real-Life Applications

Rate of Change

87. Ace Auto Rental rents a car for \$29 a day plus 10 cents per mile. For a 1-day rental, let m represent the number of miles driven and let c represent the total cost.

(a) Write an equation for total rental cost.

(b) What is the rate of change in the total cost with respect to the number of miles driven?

88. A telemarketer earns \$100 per week plus a 30% commission on all sales. Let I represent the weekly income and let s represent the weekly sales.

(a) Write an equation for the total weekly income.

(b) What is the rate of change in income with respect to sales?

In Exercises 89 and 90, assume that the described relationship is linear.

89. In 1950 the cost of a soft drink was 5 cents. In 1990 the cost of the same soft drink was 55 cents. During this period, what was the annual rate of change in the cost of the drink?

90. On a certain January day in Minnesota, the temperature at 6:00 A.M. was $-15°$F. By noon, the temperature was up to 12°F. What was the hourly rate of temperature change?

Direct Variation

91. The weight of a load of bricks varies directly with the number of bricks in the load. If a load of 500 bricks weighs 1175 pounds, what will a load of 1200 bricks weigh?

92. The weight that can be lifted by an automobile jack varies directly with the force exerted downward on the jack handle. If a force of 9 pounds will lift 954 pounds, what weight will be lifted by a force of 15 pounds?

93. If the height of a triangle remains constant, then the area of the triangle varies directly with the length of the base. If a triangular sail with a fixed height has an area of 165 square feet when the base is 12 feet, what would the area be if the base were increased to 18 feet?

94. In physics, Hooke's law states the following: The distance d a hanging spring is stretched varies directly with the weight of an attached object. If a weight of 4.5 pounds stretches a spring a distance of 3 inches, how far will a weight of 9.5 pounds stretch this spring?

95. The amount paid for typing varies directly with the number of pages typed. If \$100 is paid for typing 16 pages, how much will be paid to type 25 pages?

96. The discount on a blouse varies directly with the marked (selling) price. If the discount is \$11.84 on a blouse marked \$37, what is the discount on a blouse marked \$50?

Modeling with Real Data

97. In Section 3.6 we learned that the average daily cost of a car vacation for a family of four was \$233 in 1996. (Source: American Automobile Association.)

(a) Using data points of the form (days, cost), write a linear function to model the cost of a trip.

(b) Explain why your function is a direct variation.

(c) What is the constant of variation?

98. Hubbell, Inc., is a manufacturer of electrical equipment. This company's sales rose from \$786 million in 1992 to \$1.16 billion in 1995. (Source: Hubbell, Inc.) Assuming a constant rate of increase, what was the average rate of change in sales during this period?

99. The pole speeds at the Charlotte Motor Speedway have increased steadily. The accompanying bar graph shows the pole speeds for selected years. (Source: Charlotte Motor Speedway.)

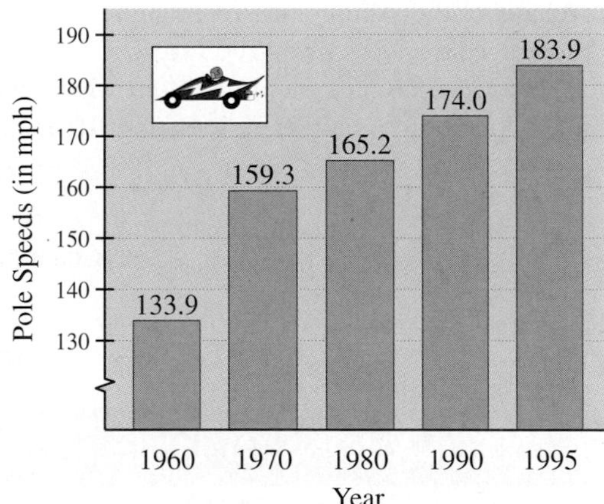

(a) The speed S can be modeled by the linear function $S(t) = 1.3t + 139$, where t is the number of years since 1960. What is the slope of the graph of S, and what does it indicate?

(b) What is the average rate of change in the pole speeds from 1980 to 1995? Compare your result with the slope in part (a).

(c) Does the model function predict a realistic pole speed for the year 2010?

100. The per-share dividends for Wachovia Corporation increased from 92 cents in 1991 to \$1.48 in 1996. (Source: Wachovia Corporation.) Assume that the rate of change was constant.

(a) Using data points of the form $(x, \text{dividend})$, where x is the number of years since 1991, would a linear function describing this information be a direct variation? Why?

(b) If the function describing this information is in the slope–intercept form, what is m?

(c) What is your interpretation of the value of m?

Group Project: Bald Eagles

Although once about to become extinct, the population of bald eagles in the United States increased from 417 pairs in 1963 to 4400 pairs in 1995. (Source: U.S. Fish and Wildlife Service.)

101. Plot the two given data points of the form (year, population) and draw a line graph.

102. Calculate the slope of the line in Exercise 101.

103. What was the average rate of change in the population of bald eagles from 1963 to 1995? If this rate of change continues indefinitely, in what year will the population of bald eagles reach 5646 pairs?

104. Conservationists always try to prevent the extinction of a species. For what special reason might we try to protect the bald eagle?

Challenge

In Exercises 105–110, the equation of a line is given. Determine k so that the given condition is met.

105. $kx + 3y = 10$; the slope of the line is -2.

106. $2x - ky = 2$; the line is parallel to the line $y = \frac{1}{2}x - 5$.

107. $3kx - 3y = 4$; the line is perpendicular to the line $x + y = 0$.

108. $5x + 2ky = 7$; the slope of the line is undefined.

109. $kx - 5y = 20$; the line is parallel to the x-axis.

110. $kx + ky = 4$; the slope of the line is -1.

111. A **rhombus** is a parallelogram whose sides are all the same length. The vertices of a rhombus are $A(0, 0)$, $B(3, 4)$, $C(8, 4)$, and $D(5, 0)$. Show that the diagonals of the rhombus are perpendicular.

112. If L_1, L_2, and L_3 are three lines in the same plane such that L_1 and L_2 are both perpendicular to L_3, show that L_1 and L_2 are parallel.

4.4 EQUATIONS OF LINES

Slope–Intercept Form • Point–Slope Form • Special Cases • Analyzing Conditions • Real-Life Applications • Linear Regression

Slope–Intercept Form

By writing a linear equation in the form $y = mx + b$, we can easily determine the slope and y-intercept and produce the graph of the equation. Conversely, if we know enough about a nonvertical line that we can sketch the line, then we can use the model $y = mx + b$ to write an equation of the line.

EXAMPLE 1

Writing an Equation of a Line Given the Slope and the y-Intercept

Write an equation of a line whose slope is 2 and whose y-intercept is $(0, -3)$.

Solution

We are given that $m = 2$ and $b = -3$

$y = mx + b$	Slope–intercept form of the required equation
$y = 2x - 3$	Replace m with 2 and b with -3.

The equation of the line is $y = 2x - 3$.

Suppose we are given the slope of a line along with some point other than the y-intercept. To determine an equation of the line, we use the key concept that relates equations and graphs: Every point of a line has coordinates that are solutions of the equation of the line.

EXAMPLE 2

Writing an Equation of a Line Given the Slope and One Point

Write an equation of the line that contains the point $P(1, 3)$ and whose slope is 2.

Solution

Because $m = 2$, we can start the equation as follows.

$y = mx + b$	Slope–intercept form of a linear equation
$y = 2x + b$	Replace m with 2.

Because $P(1, 3)$ is a point of the line, the coordinates satisfy the equation.

$3 = 2(1) + b$	Replace x with 1 and y with 3.
$3 = 2 + b$	
$1 = b$	

Now we know that $b = 1$, so the equation can be completed.

$y = 2x + 1$	Replace b with 1.

Point–Slope Form

As we have seen, the form $y = mx + b$ can be used to write an equation of a line even if the given point is not the y-intercept.

Figure 4.26

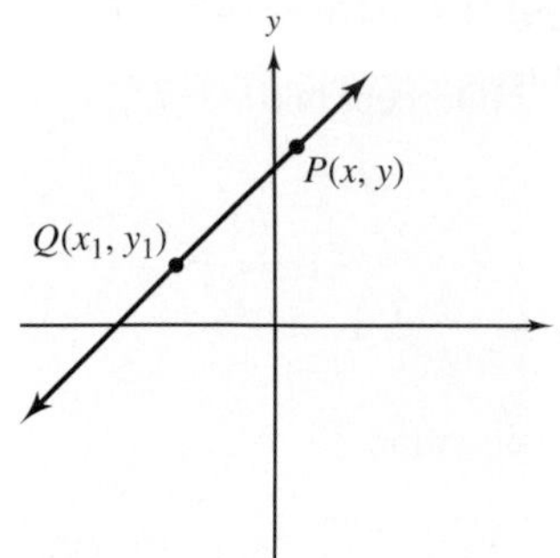

An alternative model can be used in such cases. Figure 4.26 shows a line with a *fixed* point $Q(x_1, y_1)$ and point $P(x, y)$, which represents *any* other point of the line.

The slope of the line is given by

$$m = \frac{y - y_1}{x - x_1}$$

Because this equation is in the form of a direct variation, it is equivalent to

$$y - y_1 = m(x - x_1)$$

This result is known as the **point–slope form** of a linear equation.

The Point–Slope Form of a Linear Equation

The **point–slope form** of a linear equation is

$$y - y_1 = m(x - x_1)$$

where m is the slope of the line and (x_1, y_1) is a fixed point of the line.

The point–slope form is particularly useful for writing an equation of a line when the slope and one point of the line are given.

EXAMPLE 3

Writing an Equation of a Line Given the Slope and One Point

Use the point–slope form to write an equation of a line that contains $P(-5, -3)$ and whose slope is $-\frac{2}{5}$.

Solution

$$y - y_1 = m(x - x_1) \quad \text{Point–slope form of a linear equation}$$

$$y - (-3) = -\frac{2}{5}[x - (-5)] \quad \text{Replace } m \text{ with } -\tfrac{2}{5},\ x_1 \text{ with } -5, \text{ and } y_1 \text{ with } -3.$$

$$y + 3 = -\frac{2}{5}(x + 5)$$

$$y + 3 = -\frac{2}{5}x - 2 \quad \text{Distributive Property.}$$

$$y = -\frac{2}{5}x - 5 \quad \text{Subtract 3 from both sides.}$$

The slope–intercept form and the point–slope form both require that we know the slope of a line in order to write its equation. If we are given just two points of a line, we can calculate the slope and write an equation of the line with either of the forms.

EXAMPLE 4

Writing an Equation of a Line Given Two Points

Write an equation of the line containing the points $P(4, 2)$ and $Q(-3, -5)$.

Solution

We use the slope formula to compute the slope of the line.

$$m = \frac{y_2 - y_1}{x_2 - x_1} = \frac{-5 - 2}{-3 - 4} = \frac{-7}{-7} = 1$$

We use the point–slope form to begin writing the equation.

$$y - y_1 = m(x - x_1)$$

$$y - y_1 = 1(x - x_1) \quad \text{Replace } m \text{ with 1.}$$

Because points P and Q both represent solutions, we can use either one as the fixed point. We use point $P(4, 2)$.

$$y - 2 = 1(x - 4) \quad \text{Replace } x_1 \text{ with 4 and } y_1 \text{ with 2.}$$

$$y - 2 = x - 4$$

$$y = x - 2 \quad \text{Add 2 to both sides.}$$

Special Cases

As we have seen, the special cases of linear equations have graphs with the following characteristics.

Equation	*Graph Characteristics*
$y = b$	Horizontal line; slope is 0.
$x = c$	Vertical line; slope is undefined.

If the slope is 0, we can use either the slope–intercept form or the point–slope form to write an equation of the line. However, it is easier to use the special case form $y = b$. If the slope is undefined, then neither model applies, and we must use the special case form $x = c$.

EXAMPLE 5

Special Cases

(a) Write an equation of a line whose slope is 0 and that contains the point $R(2, 4)$.

(b) Write an equation of the line containing the points $C(4, 2)$ and $D(4, 7)$.

Figure 4.27

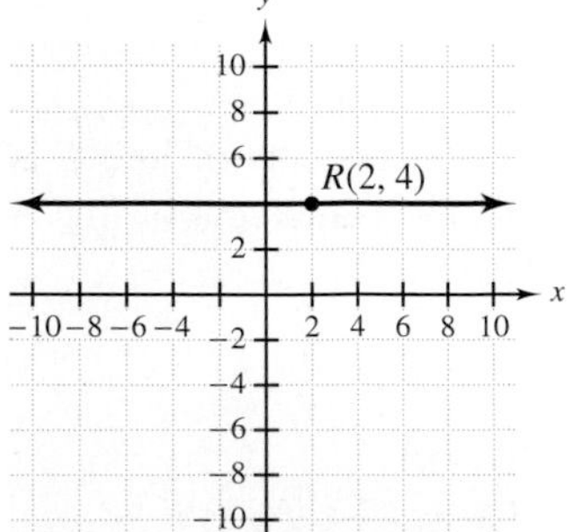

Solution

(a) Because the slope is 0, the line is horizontal. (See Fig. 4.27.) Therefore, the equation is of the form $y = b$, where b is the y-coordinate of every point of the line.

Because $R(2, 4)$ is a point of the line and its y-coordinate is 4, every point of the line has a y-coordinate of 4. The equation is $y = 4$.

(b) Because the x-coordinates of points C and D are the same, the line is vertical. (See Fig. 4.28.) Therefore, the equation is of the form $x = c$. Points C and D and all other points of the line have an x-coordinate of 4. The equation is $x = 4$.

Figure 4.28

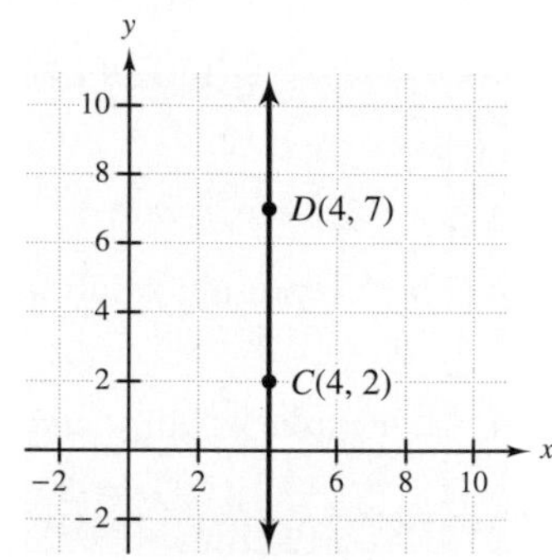

Analyzing Conditions

Sometimes a line is described in reference to one or more other lines. If enough information is furnished about the given line(s), we can deduce the features of the required line and write its equation.

EXAMPLE 6

Figure 4.29

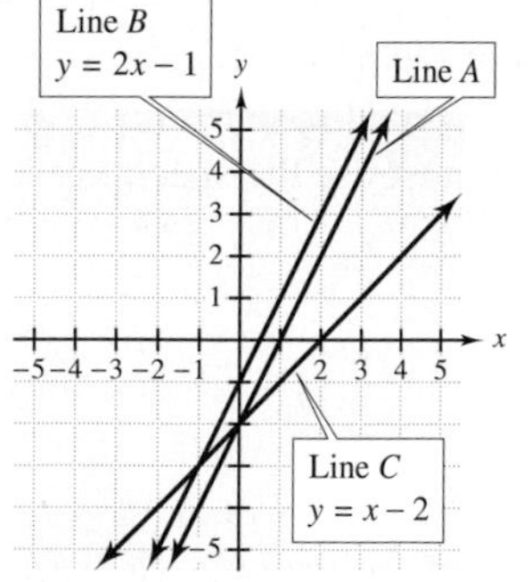

Describing a Line Relative to Other Lines

The equation of line B is $y = 2x - 1$, and the equation of line C is $y = x - 2$. Line A is parallel to line B and has the same y-intercept as line C. (See Fig. 4.29.) Write an equation of line A.

Solution

The equation of line B is $y = 2x - 1$, which means that the slope of line B is 2. Because line A is parallel to line B, the slope of line A is also 2. Thus we can begin writing the equation of line A.

$y = mx + b$

$y = 2x + b$ Replace m with 2.

The equation of line C is $y = x - 2$, which means that the y-intercept of line C is $(0, -2)$. Because line A has the same y-intercept as line C, the y-intercept of line A is also $(0, -2)$. Now the equation of line A can be completed.

$y = 2x - 2$ Replace b with -2.

LEARNING TIP

When you work problems involving two or more lines, sketching the lines and visualizing their relationship will be a great help in understanding what to do.

EXAMPLE 7

Horizontal and Vertical Lines

The equation of line L is $y = -2$. Line M is perpendicular to line L and contains the point (3, 4). (See Fig. 4.30.) Write an equation of line M.

Figure 4.30

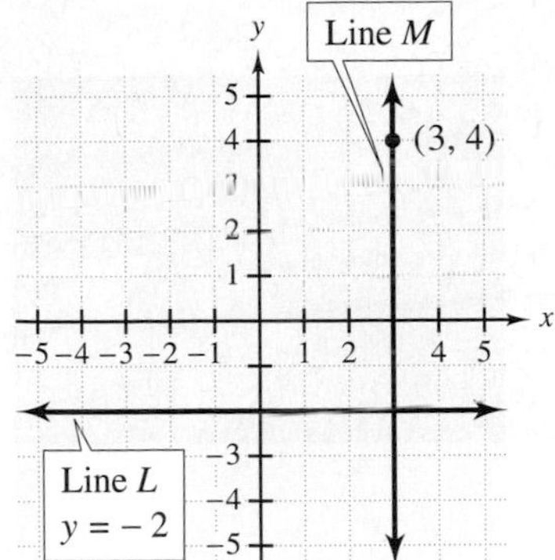

Think About It

Suppose that two lines are perpendicular and you know their point of intersection. Why is this information insufficient for writing the equations of the lines?

Solution

The equation of line L is of the form $y = b$. Therefore, line L is a horizontal line. Because line M is perpendicular to line L, line M must be a vertical line. Therefore, its equation is of the form $x = c$, where the constant is the x-coordinate of every point of the line.

Because (3, 4) is a point of the line and its x-coordinate is 3, every point of the line has an x-coordinate of 3. The equation of line M is $x = 3$.

Real-Life Applications

Sometimes we need to write a linear equation to model all or part of the conditions in an applied problem.

EXAMPLE 8

Modeling with a Linear Equation

At age 8, a boy competed in the high jump and managed a personal best of 3 feet. By age 12, the boy was jumping 5 feet.

His coach decided that the boy's progress in the high jump could be modeled by a linear equation. He let x represent the boy's age and y represent the height of his jump.

(a) What linear equation did the coach write?

(b) If he uses this equation to predict how high the boy will be able to jump at age 20, what will his result be?

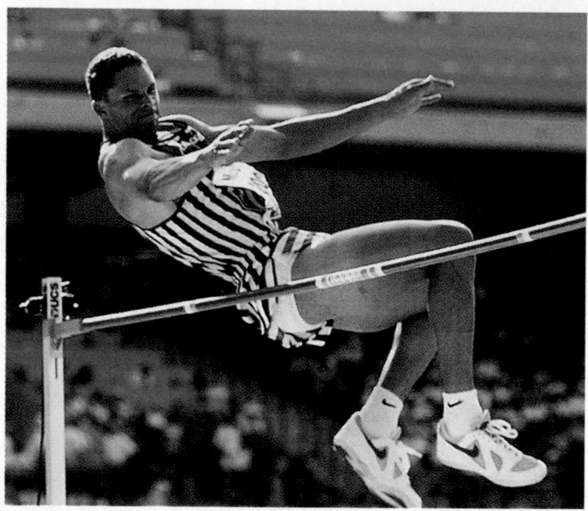

Solution

(a) If we write ordered pairs in the form (age, height) or (x, y), two solutions of the equation are $(8, 3)$ and $(12, 5)$.

The slope is $m = \frac{5-3}{12-8} = \frac{2}{4} = \frac{1}{2}$. Thus we can begin to write the equation.

$$y = \frac{1}{2}x + b$$

Use either ordered pair as replacement for x and y.

$$3 = \frac{1}{2} \cdot 8 + b \qquad \text{Replace } x \text{ with 8 and } y \text{ with 3.}$$

$$3 = 4 + b$$

$$b = -1$$

Now we can write the complete equation.

$$y = \frac{1}{2}x - 1$$

(b) The coach wants to know the height that the boy will be able to jump when his age x is 20.

$$y = \frac{1}{2} \cdot 20 - 1 = 10 - 1 = 9$$

Using the linear equation as a model, the coach predicts that the boy will be able to jump 9 feet at age 20.

Note: Try using the model in Example 8 to predict the height the boy will be able to jump at age 60. Because of the nature of a given problem, a mathematical model may be reasonably accurate only for a limited domain.

Linear Regression

Regression analysis is a process of fitting a graph as closely as possible to two or more given points. The equation of this graph is called a **regression equation.**

Most calculators have the capability of determining certain regression equations. When only two points are given, the calculator generates a linear regression equation, which is simply the equation of the line containing the two points. Typically, the calculator displays the slope and the y-intercept rather than the equation itself.

Erase Data

It is usually best to clear the calculator's memory of any old data before entering new data.

Enter Data

The coordinates of the data points are entered as new data.

Linreg

Usually a menu of different kinds of regression equations is displayed. For an equation of a line, we select a linear regression equation.

EXAMPLE 9

Using Linear Regression to Write an Equation of a Line

Two points of a line are $P(75, 45)$ and $Q(120, 54)$. Use the linear regression capability of your calculator to determine an equation of the line containing P and Q.

Solution

We begin by entering the coordinates of the data points (75, 45) and (120, 54). (See Fig. 4.31.)

Figure 4.31

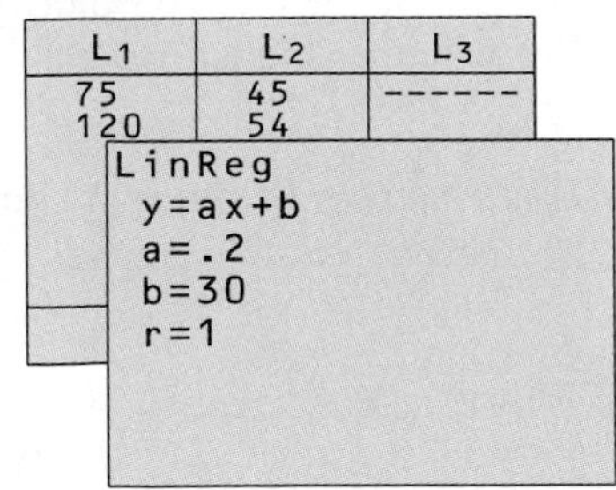

Figure 4.31 also shows a typical display of the reported slope and y-intercept. The linear regression equation is $y = 0.2x + 30$.

Note: In Example 9 the number r in Fig. 4.31 is a statistical measure of *correlation* and can be ignored.

4.4 QUICK REFERENCE

Slope–Intercept Form

- The **slope–intercept form** $y = mx + b$ can be used as a model for writing an equation of a nonvertical line.

- If the slope m and y-intercept b of a line are given, we can write an equation of the line by substituting directly into the slope–intercept form $y = mx + b$.
- We can also use the slope–intercept form if the slope and a point (x_1, y_1) of the line are given. Replace m with the given slope and replace x and y with x_1 and y_1, respectively. Then we can solve for b.

Point–Slope Form

- An equivalent model for writing a linear equation is the **point–slope form** $y - y_1 = m(x - x_1)$. The point–slope form is particularly useful for writing an equation of a line for which the given point is not the y-intercept.
- If two points of a line are given, we can write an equation of the line by using the slope formula to calculate the slope. Then, using the point–slope form, we can replace m with the slope and x_1 and y_1 with the coordinates of either of the given points.

Special Cases

- The equation of a horizontal line can be found with either the slope–intercept form or the point–slope form, but it is much easier to use the model $y = b$. The constant is the y-coordinate of any point of the line.
- The equation of a vertical line cannot be found with the slope–intercept or point–slope models because the slope is undefined. Instead, use the model $x = c$, where the constant is the x-coordinate of any point of the line.

Analyzing Conditions

- If a nonhorizontal or nonvertical line is described in reference to one or more other lines, we use the given information to determine the slope of the required line and one of its points. Then we can write the equation of the line with the slope–intercept or point–slope model.
- If the given conditions indicate that a line is horizontal or vertical, then we can write the equation of the line with the simpler models $y = b$ or $x = c$.

Linear Regression

- **Regression analysis** is a process of fitting a graph as closely as possible to two or more given points. The equation of this graph is called a **regression equation.**

4.4 EXERCISES

Concepts and Skills

1. What information is needed to write an equation of a line with the slope–intercept form?

2. Why can the slope–intercept form be used to write an equation of a horizontal line but not an equation of a vertical line?

In Exercises 3–6, write the equation of the line.

3. $m = -4, b = -3$
4. $m = 3, b = 4$
5. $m = \frac{1}{2}, b = 3$
6. $m = -\frac{2}{3}, b = -2$

In Exercises 7–12, write an equation of the line having the given slope and containing the given y-intercept.

7. $m = -5, (0, 1)$
8. $m = 6, (0, -6)$
9. $m = -\frac{3}{4}, (0, -5)$
10. $m = \frac{5}{8}, (0, 8)$

11. $m = 7, \left(0, -\frac{2}{3}\right)$

12. $m = -4, \left(0, \frac{1}{2}\right)$

In Exercises 13–18, write an equation for the line whose graph is shown in the figure.

13.

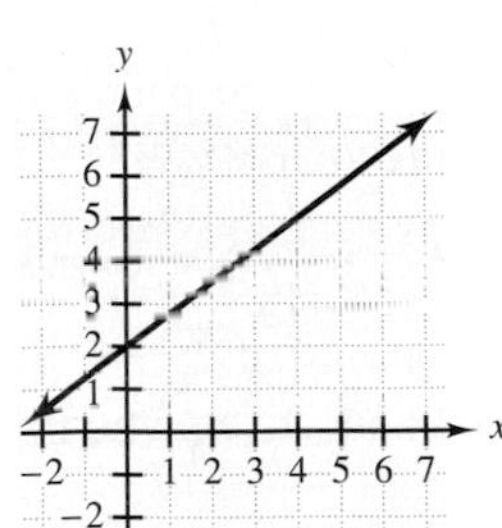

14.

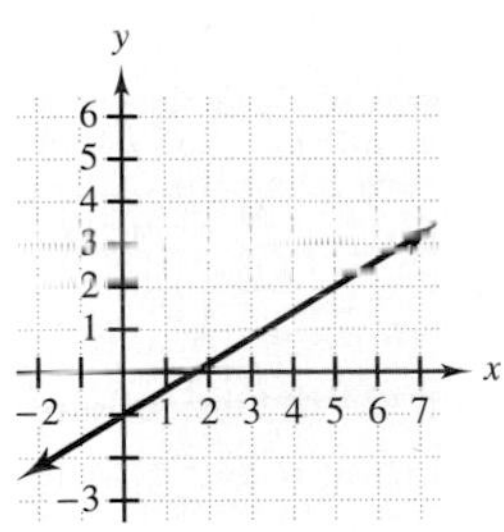

15.

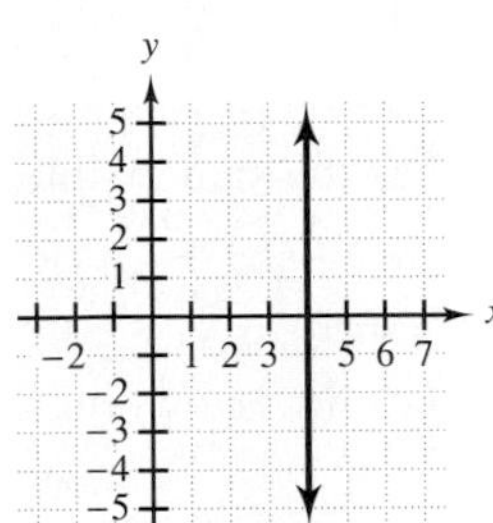

16.

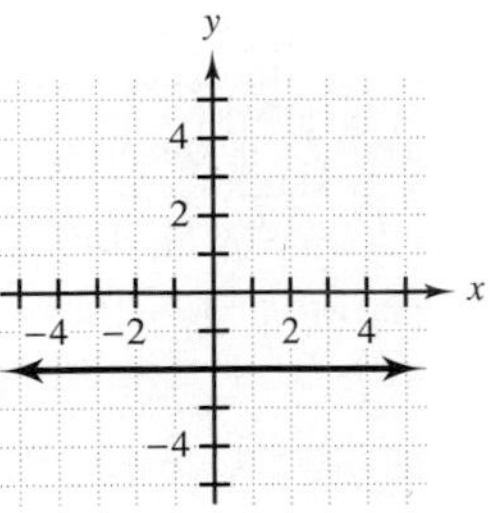

17.

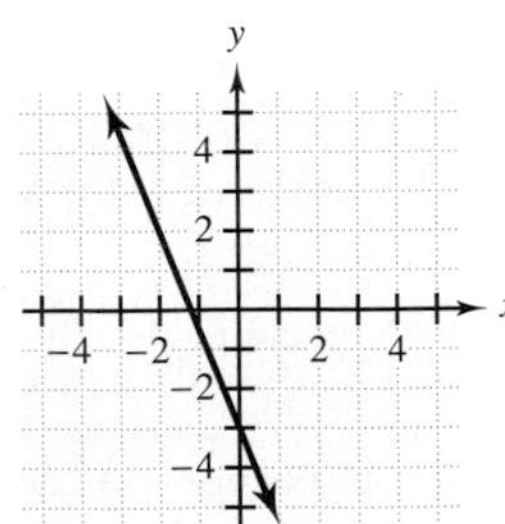

18.

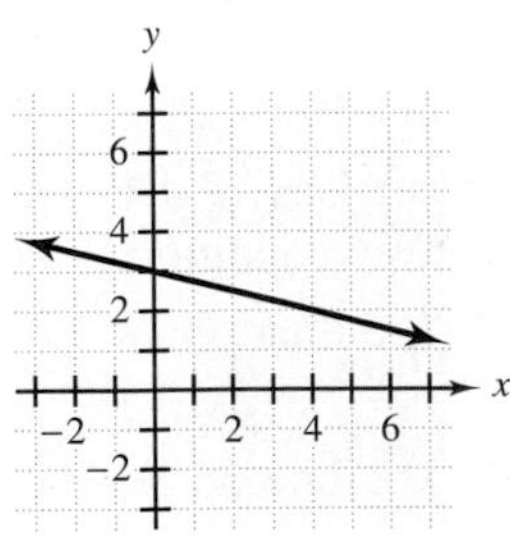

19. Why can the point–slope form be used to write an equation of a horizontal line but not an equation of a vertical line?

20. What information is needed to write an equation of a line with the point–slope form?

In Exercises 21–30, write an equation of the line with the given slope and containing the given point. Express the equation in the form $y = mx + b$ if possible.

21. $m = 2, (1, 3)$

22. $m = -3, (2, 5)$

23. $m = \frac{2}{5}, (10, -4)$

24. $m = -\frac{1}{3}, (-6, 3)$

25. $m = -\frac{2}{3}, (-2, -5)$

26. $m = \frac{4}{5}, (-1, -2)$

27. $m = 0, (2, -5)$

28. $m = 0, (-3, -4)$

29. m is undefined, $(-3, 4)$

30. m is undefined, $(-4, 5)$

In Exercises 31–36, write an equation of the line with the given slope and containing the given point. Express the equation in the form $Ax + By = C$.

31. $m = 1, (7, 4)$

32. $m = 2, \left(2, \frac{1}{2}\right)$

33. $m = \frac{5}{3}, (2, 0)$

34. $m = -2, \left(-1, \frac{2}{3}\right)$

35. $m = -\frac{1}{4}, \left(\frac{1}{2}, -1\right)$

36. $m = -\frac{5}{6}, (-3, 3)$

37. If we are given two points of a line but not the slope of the line, how can we use the point–slope or slope–intercept model to write an equation of the line?

38. If a line contains $A(2, -3)$ and $B(2, 5)$, why is the point–slope form not an appropriate model for writing an equation of the line?

In Exercises 39–52, write an equation of the line that contains the given pair of points.

39. (3, 4) and (1, 8)

40. (4, 7) and (6, 11)

41. (−2, 7) and (6, −9)

42. (5, −9) and (−1, 9)

43. (2, 3) and (−3, −2)

44. (12, 1) and (4, −1)

45. (2, −3) and (−4, 1)

46. (3, −2) and (−5, 4)

47. (−4, 2) and (−4, 5)

48. (2, −3) and (2, 4)

49. (−3, −4) and (1, −2)

50. (5, −4) and (2, −2)

51. (−3, 5) and (4, 5)

52. (2, −3) and (5, −3)

53. Describe the equation of a line that is

(a) perpendicular to a vertical line.

(b) parallel to the y-axis.

54. If the slope of L_1 is m_1, what additional information is needed in order to write an equation of L_1?

In Exercises 55–66, write an equation of the described line.

55. The line is horizontal and contains the point (3, −4).

56. The line is horizontal and contains the point (−5, 6).

57. The line contains the point $(-2, 5)$ and has the same slope as the line whose equation is $y = 3$.

58. The line contains the point $(5, -3)$ and has the same slope as the line whose equation is $y = -5$.

59. The line contains the point $(-2, -3)$ and has the same slope as the line whose equation is $2x + 3y = 1$.

60. The line contains the point $(3, 5)$ and has the same slope as the line whose equation is $3x - 4y = 12$.

61. The line is vertical and contains the point $(-2, 5)$.

62. The line is vertical and contains the point $(7, -4)$.

63. The line contains the point $(1, 5)$ and has the same y-intercept as the line whose equation is $2x + 3y = 6$.

64. The line contains the point $(2, 3)$ and has the same y-intercept as the line whose equation is $3x - 2y = 8$.

65. The line contains the point $(-2, -3)$ and has an undefined slope.

66. The line contains the point $(3, 5)$ and has an undefined slope.

In Exercises 67–70, use the given information to write an equation of L_1.

67. The equation of L_2 is $y = 4x - 7$. L_1 is parallel to L_2, and its y-intercept is $(0, 2)$.

68. The equation of L_2 is $2x - 3y = 1$. L_1 is perpendicular to L_2, and its y-intercept is $(0, -4)$.

69. The equation of L_2 is $y = 4$. L_1 is perpendicular to L_2 and contains the point $(-2, 7)$.

70. The equation of L_2 is $x = -1$. L_1 is parallel to L_2 and contains the point $(3, 5)$.

In Exercises 71 and 72, L_1 is perpendicular to L_2. Refer to the figures and write equations for L_1 and L_2.

71.

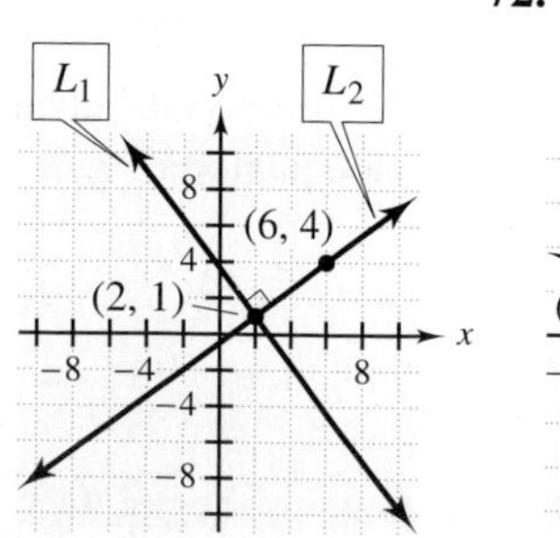

72.

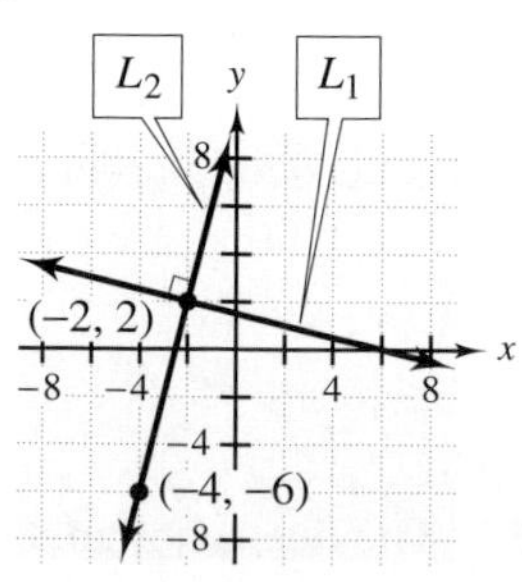

73. Consider the equations of L_1 and L_2.

L_1: $y = 3x - 7$ L_2: $x + y = 3$

Write an equation of L_3 where L_3 is parallel to L_2 and L_3 has the same y-intercept as L_1.

74. Consider the equations of L_1 and L_2.

L_1: $y = 3x - 7$ L_2: $x + y = 3$

Write an equation of L_3 where L_3 is perpendicular to L_1 and L_3 has the same y-intercept as L_2.

In Exercises 75 and 76, the equations of L_1 and L_2 are given. Write an equation of L_3 and an equation of L_4 according to the given conditions.

75. L_1: $3x + 4y = 12$ L_2: $x - y = 3$

(a) L_3 is perpendicular to L_2 and has the same y-intercept as L_1.

(b) L_4 is parallel to L_1 and has the same y-intercept as L_2.

76. L_1: $x - 2y = 6$ L_2: $4 - y = x$

(a) L_3 is parallel to L_1 and has the same y-intercept as L_2.

(b) L_4 is perpendicular to L_1 and has the same x-intercept as L_2.

In Exercises 77–84, determine the value of k for which the given condition is met.

77. The line $kx + 3y = 10$ is parallel to the line $y = 3 - 2x$.

78. The line $3y + 2kx - 8 = 0$ is parallel to the line $y = -2 + x$.

79. The line $2x - ky = 2$ is perpendicular to the line $y = \dfrac{3 - 4x}{2}$.

80. The line $x = 5 - 3ky$ is perpendicular to the line $y = \dfrac{15 - 2x}{-3}$.

81. The line $x + 2y = 6k$ has the same y-intercept as $y = \dfrac{1 - x}{-2}$.

82. The line $3x - 5y - k = 0$ has the same y-intercept as $y = 2 - 4x$.

83. The line $x - y = 2k$ has the same x-intercept as $y = 4x - 8$.

84. The line $kx - 5y - 20 = 0$ has the same x-intercept as $y = 3x + 12$.

In Exercises 85–88, use the linear regression capability of your calculator to determine an equation of the line containing P and Q.

85. $P(75, 30)$ and $Q(-245, -130)$

86. $P(692, -536)$ and $Q(-108, 64)$

87. $P(-173, -212)$ and $Q(-573, 288)$

88. $P(250, 140)$ and $Q(450, -20)$

In Exercises 89 and 90, use the linear regression capability of your calculator to determine an equation of the line containing P and Q and of the line containing P and R. Then state whether P, Q, and R are collinear and explain why.

89. $P(-3, 1)$, $Q(6, 4)$, $R(-9, -1)$

90. $P(-1.9, -4.9)$, $Q(1.1, 4.1)$, $R(2.9, 9.8)$

Real-Life Applications

91. In 1996 a person bought a computer for $1500. She estimated the useful life of this computer to be 7 years, with a remaining value of $100 at the end of this period. She used the straight-line (linear) method of depreciation to estimate the loss in value of this equipment over time.

(a) Write the equation for the value v of the computer t years from 1996. (Let $t = 0$ for year 1996.)

(b) Use the equation in part (a) to determine the value of the computer in the year 2000.

92. The population of a small town grew from 35 in 1985 to 62 in 1995. Assuming the growth is linear, write an equation to describe the population for any year. What is the rate of change in the population?

93. The net loans at a certain bank increased from $92.1 million in 1995 to $105.7 million in 1997. Using a linear equation, predict the net loans in 2004. What is the rate of change of the net loans?

94. A library needs one children's librarian for each 1000 children attending story hour during a calendar quarter. The number of children attending story hour increased from 1153 during the first quarter of 1996 to 1732 during the same quarter of 1997. Using a linear equation, determine the year in which the library can fully support three children's librarians.

Modeling with Real Data

95. The accompanying table shows that the number of students receiving bachelor's degrees increased at a constant rate from 1993 to 1996. (Source: National Center for Educational Statistics.)

Year	Degrees Awarded (millions)
1993	1.17
1994	1.18
1995	1.19
1996	1.20

(a) If the data in the table were modeled with a linear function, what would be the slope of the graph?

(b) Using the data for 1993 and 1996, write a model function in the slope–intercept form, where y is the number of degrees awarded (in millions) and x is the number of years since 1993.

(c) How would your function in part (b) change if you let x represent the number of years since 1995?

96. Leasing rather than purchasing automobiles has become increasingly popular. The percentage of car deals that were leases increased from 10% in 1984 to 32% in 1995. (Source: CNW Marketing.) Write a linear function to model the trend in leasing.

97. Fortune 500 companies show an increase in the number of minorities serving on boards of directors. For example, from 1990 to 1995, the percentage of companies with at least one Hispanic director doubled from 6% to 12%. (Source: SpencerStuart.)

(a) If you assume that the increase in the percentage was linear, what percent of companies would have had at least one Hispanic director in 1993?

(b) Suppose that we model the data with a function in the form $y = mx + b$. How must we define x in order for the y-intercept of the graph to be $(0, 6)$?

(c) Model the data with a function $y = mx + b$, where x is the number of years since 1985.

98. In 1990 the circulation of *TV Guide* was 16 million. In 1994 the circulation was 14 million. (Source: Audit Bureau of Circulations.)

(a) If a linear function were used to model the given data, what can you predict about the value of m?

(b) Suppose that the function $y = -0.5x + 16$ were used to model the data, where y is the circulation (in millions). How is x defined?

Group Project: Library Management

The accompanying bar graph shows the circulation statistics for the Gilmer Library during selected years in the period 1989–1995. (Source: Sequoyah Regional Library Statistical Report.)

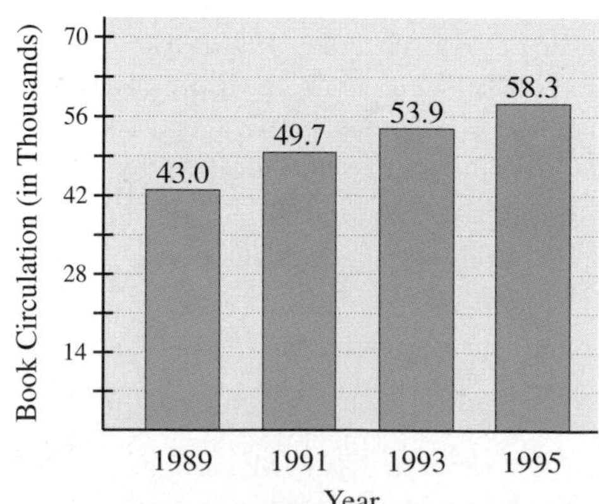

In 1995 the librarian was asked to project circulation for 1997, 1999, and 2001 to plan for increased funding.

99. Let x represent the number of years since 1988, and let C represent the circulation (in thousands) for year x. Use your calculator's linear regression capability to write a linear model of the given data.

100. Suppose that the annual funding for books in 1995 was \$10 per circulated volume and that this amount was to increase by \$0.50 per year for the indefinite future. What was the projected funding for 2001?

101. Suppose that the library is allowed three librarians plus one additional librarian for each 20,000 circulated volumes or fraction thereof. What was the projected staffing requirement for the year 1999?

102. What advice would you give to students of library science who doubt the need for algebra in their program of study?

Challenge

103. If (x_1, y_1) and (x_2, y_2) are points of a line, show that an equation of the line can be written as

$$y - y_1 = \frac{y_2 - y_1}{x_2 - x_1}(x - x_1), \quad x_1 \neq x_2$$

104. Show that the equation of the line with intercepts $(a, 0)$ and $(0, b)$ can be written as

$$\frac{x}{a} + \frac{y}{b} = 1, \quad a \neq 0 \text{ and } b \neq 0$$

In Exercises 105–108, use the results of Exercise 104 to determine the intercepts directly from the given equation.

105. $\frac{x}{3} + \frac{y}{2} = 1$

106. $\frac{x}{-4} - \frac{y}{2} = 1$

107. $\frac{2x}{5} - \frac{3y}{4} = 1$

108. $4x + 3y = 12$

109. Write an equation of the line that contains the point (a, b) and is perpendicular to the line $ax + by = c$.

110. Write an equation of the line that contains the point (b, a) and is parallel to the line whose equation is $ax + by = c$.

4.5 GRAPHS OF LINEAR INEQUALITIES

Linear Inequalities • Graphs of Solution Sets • Graphs of Compound Inequalities

Linear Inequalities

The definition of a linear inequality in two variables is very similar to the definition of a linear equation in two variables.

Definition of Linear Inequality in Two Variables

A **linear inequality in two variables** is an inequality that can be written in the form $Ax + By < C$, where A, B, and C represent any real numbers, and A and B are not both zero. Similar definitions can be stated with $>$, $\geq$, and $\leq$.

The following are some examples of linear inequalities along with their standard forms.

Inequality	*Standard Form*
$2x - 3y < 1$	$2x - 3y < 1$
$y \leq 4x - 1$	$-4x + y \leq -1$
$x > 3$	$1x + 0y > 3$
$y + 3 \geq 5$	$0x + 1y \geq 2$

Definitions of a Solution and a Solution Set

A **solution** of a linear inequality in two variables is a pair of numbers (x, y) that makes the inequality true. The **solution set** is the set of all solutions of the inequality.

As is true with equations, it is a simple matter to determine whether a particular pair of numbers is a solution of a linear inequality.

Test

You may be able to use your calculator to determine whether a given ordered pair is a solution of a linear inequality.

EXAMPLE 1

Verifying Solutions of an Inequality

Given the linear inequality $y \leq 4x - 3$, determine whether the following are solutions.

(a) $(0, -3)$ (b) $(1, 2)$ (c) $(1, -5)$

Solution

For each pair, replace x with the first coordinate and y with the second coordinate.

(a) $(0, -3)$	(b) $(1, 2)$	(c) $(1, -5)$
$y \le 4x - 3$	$y \le 4x - 3$	$y \le 4x - 3$
$-3 \le 4(0) - 3$	$2 \le 4(1) - 3$	$-5 \le 4(1) - 3$
$-3 \le -3$ True	$2 \le 1$ False	$-5 \le 1$ True

Both $(0, -3)$ and $(1, -5)$ are solutions.

Graphs of Solution Sets

The **graph** of the solution set of an inequality is a picture of all the solutions of the inequality. We can use a calculator to make some preliminary judgments about the nature of such a graph.

EXPLORING THE CONCEPT

Graphs of Linear Inequalities

Consider the inequality $y \le x - 9$. Figure 4.32 shows the graph of the companion equation $y = x - 9$. Note that the points of this line represent solutions of $y \le x - 9$ because the inequality symbol is $\le$.

Figure 4.32

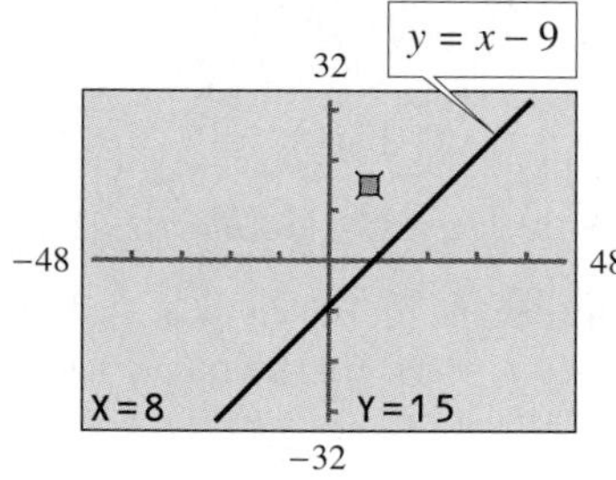

Figure 4.33

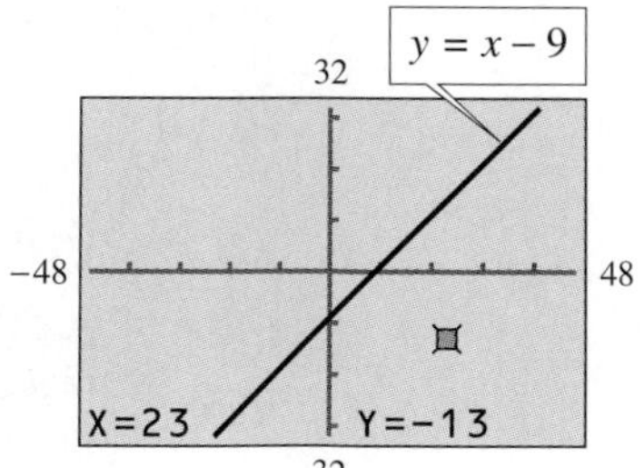

Figure 4.32 shows the general cursor at the point $(8, 15)$. Because $15 \le 8 - 9$ is false, this point does not represent a solution. You can move your cursor to other points above the line $y = x - 9$ and verify that none of those points represent solutions.

In Fig. 4.33 the general cursor is at the point $(23, -13)$. This point does represent a solution because $-13 \le 23 - 9$ is true. If you move your cursor to other points below the line, you will find that these points also represent solutions.

These results suggest that the solution set of $y \le x - 9$ is represented by all points of the line and below the line. To depict these solutions, we use a solid line for the graph of $y = x - 9$, and we shade the region below the line. (See Fig. 4.34.)

Figure 4.34

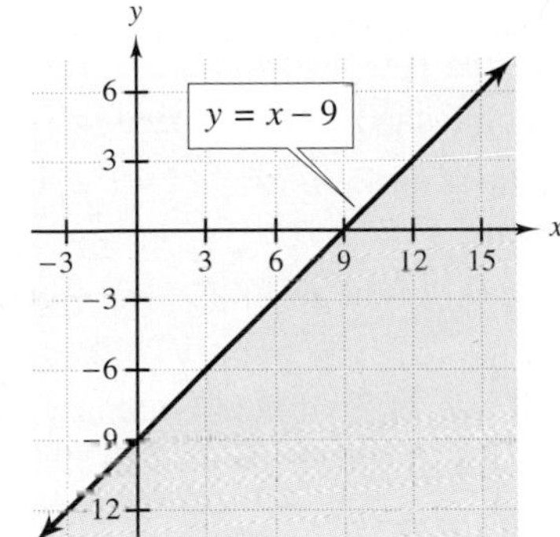

Figure 4.35

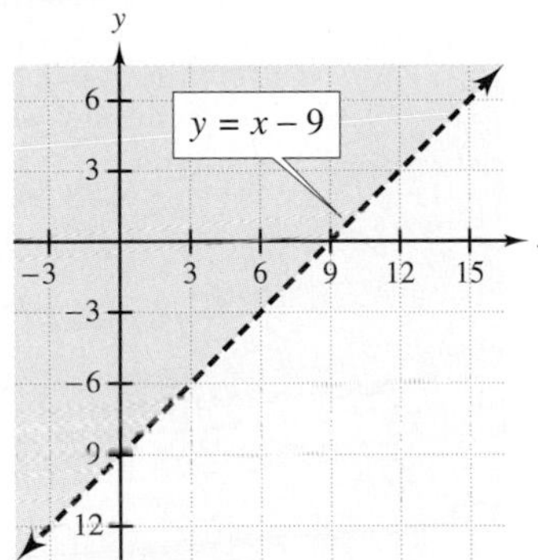

In contrast, Fig. 4.35 shows the graph of the solution set of $y > x - 9$. This time the solutions of the companion equation $y = x - 9$ are *not* solutions of $y > x - 9$, so the line is dashed. The region above the line is shaded to indicate that points above the line represent solutions of the inequality.

The graph of the companion equation is called the **boundary line,** and the regions are called **half-planes.** We indicate that points of the boundary line represent solutions of $\leq$ and $\geq$ inequalities by drawing a solid line. For $<$ or $>$ inequalities, we draw a dashed line to indicate that points of the line do not represent solutions.

To determine which half-plane contains points that represent solutions of an inequality, we can select any **test point** on either side of the boundary line and test the coordinates to see whether they satisfy the inequality.

The following is a summary of this method for sketching the graph of a linear inequality.

Sketching the Graph of a Linear Inequality

1. Either with a calculator or by hand, produce the graph of the companion equation. The boundary line is solid for $\leq$ and $\geq$ inequalities and dashed for $<$ and $>$ inequalities.
2. Select any test point in either half-plane (not on the boundary line). By substitution, determine whether the point represents a solution. If the test point represents a solution, shade the half-plane in which the test point lies. If the test point does not represent a solution, shade the other half-plane.

When a linear inequality is in the form $y < mx + b$, all points *below* the boundary line represent solutions. Similarly, when a linear inequality is in the form $y > mx + b$, all points *above* the boundary line represent solutions. This suggests an alternative graphing method.

Sketching the Graph of a Linear Inequality

1. Solve the inequality for y.
2. Draw the solid or dashed boundary line.
3. (a) For inequalities of the form $y < mx + b$ or $y \leq mx + b$, shade the region below the line.

   (b) For inequalities of the form $y > mx + b$ or $y \geq mx + b$, shade the region above the line.
4. For the special case of a vertical boundary line, shade the half-plane on the left for $x < c$ or $x \leq c$, and shade the half-plane on the right for $x > c$ or $x \geq c$.

EXAMPLE 2

Graphing Linear Inequalities

Sketch the graph of each inequality.

(a) $y > 2x$ (b) $2x - 3y \geq 12$

(c) $y - 5 \leq 2$ (d) $x < -2$

Solution

(a) The boundary line $y = 2x$ is dashed because the inequality symbol is $>$. Because the inequality is in the form $y > mx + b$, we shade above the boundary line. (See Fig. 4.36.)

Figure 4.36 also shows the point $(2, -3)$, which we can test to determine whether it represents a solution.

$$\begin{aligned} y &> 2x \\ -3 &> 2(2) \\ -3 &> 4 \quad \text{False} \end{aligned}$$

Because the test point $(2, -3)$ does not represent a solution, we shade the half-plane above the line.

Figure 4.36

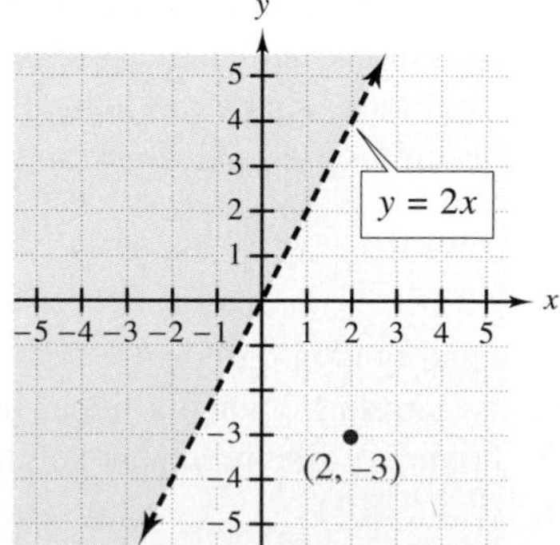

Figure 4.37

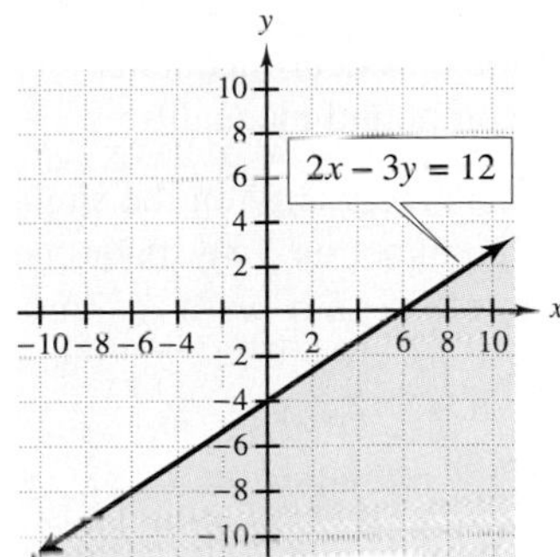

(b) $2x - 3y \geq 12$ — Solve the inequality for y.

$-3y \geq -2x + 12$ — Subtract $2x$ from both sides.

$\frac{-3y}{-3} \leq \frac{-2x}{-3} + \frac{12}{-3}$ — Divide both sides by -3 and reverse the inequality symbol.

$y \leq \frac{2}{3}x - 4$

The inequality symbol is $\leq$, so we draw the solid boundary line $y = \frac{2}{3}x - 4$. The inequality is in the form $y \leq mx + b$, so we shade below the line. (See Fig. 4.37.)

(c) $y - 5 \leq 2$ — Solve the inequality for y.

$y \leq 7$

The boundary line is the solid horizontal line $y = 7$. Because the inequality symbol is $\leq$, we shade below the line. (See Fig. 4.38.)

Figure 4.38

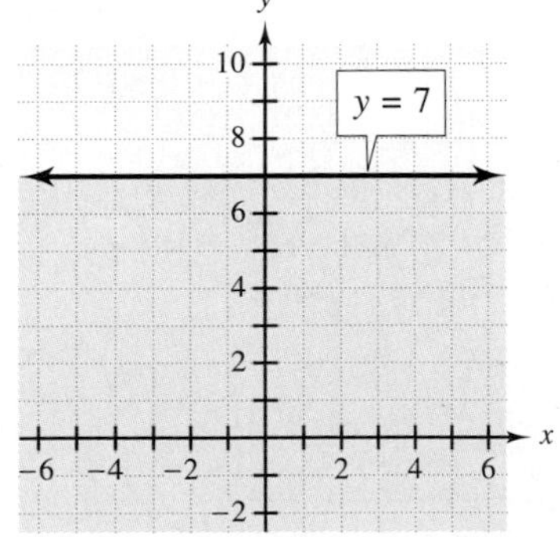

Figure 4.39

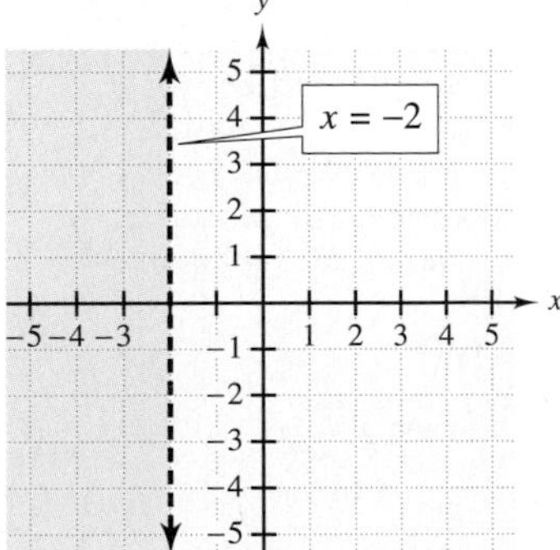

(d) Draw the dashed vertical line $x = -2$. Observe that the inequality symbol is $<$, so we shade to the left of the line. (See Fig. 4.39.)

Shade

We can use a graphing calculator to produce the graph of the solution set of a linear inequality. Figure 4.40 is a typical display of the graph in Example 2(b).

Figure 4.40

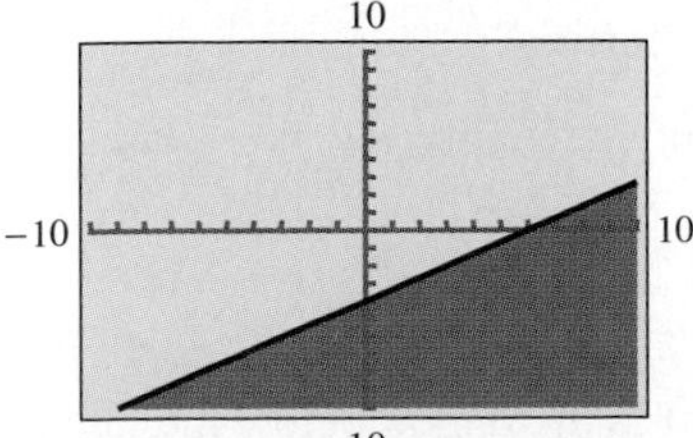

Graphs of Compound Inequalities

In Section 3.7 we solved compound inequalities in one variable. Linear inequalities in two variables also can be connected with *and* or *or* to form compound inequalities.

To produce the graph of a compound inequality, we graph each inequality on the same coordinate system. For an inequality connected with *and* (a conjunction), we shade the intersection of the solution sets, and for an inequality connected with *or* (a disjunction), we shade the union of the solution sets.

EXAMPLE 3

Graphing a Conjunction

Graph the compound inequality $y \geq 2$ and $y \leq 2x + 5$.

LEARNING TIP

Think of the graph of each inequality in a compound inequality as being drawn on a transparent sheet with one sheet placed on top of the other. For a conjunction, the graph is the region where the individual graphs overlap. For a disjunction, the graph includes any region that is shaded.

Solution

The graph of $y \geq 2$ is the line $y = 2$ and the region above the line. The graph of $y \leq 2x + 5$ is the line $y = 2x + 5$ and the region below the line. (See Fig. 4.41.)

Figure 4.41

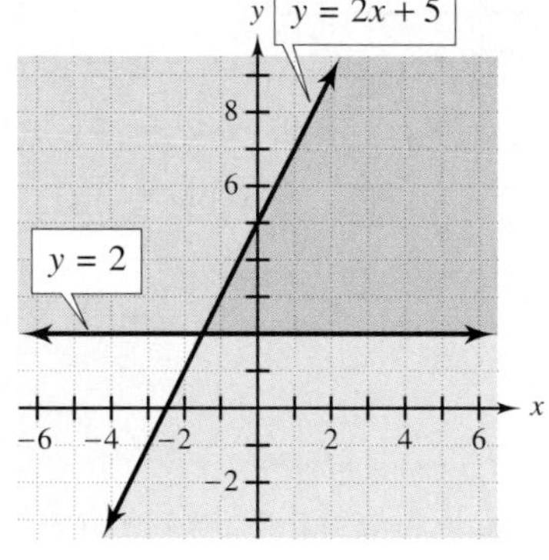

Figure 4.42

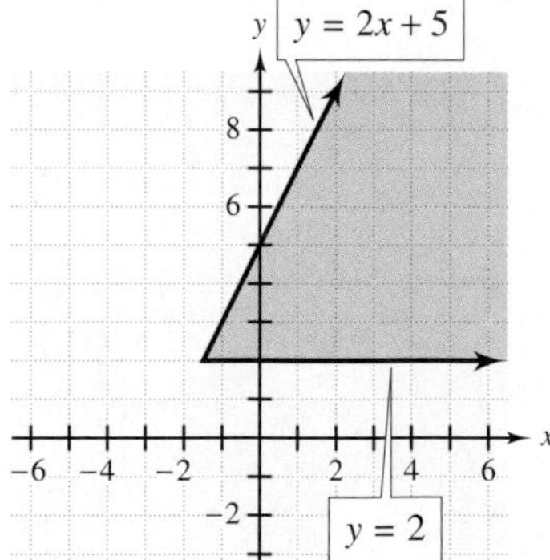

Because the inequalities are connected with *and*, the graph of the compound inequality is the intersection of the two graphs. (See Fig. 4.42.)

EXAMPLE 4

Graphing a Disjunction

Graph the compound inequality $y \geq x - 3$ or $2x + y < 1$.

Solution

The graph of $y \geq x - 3$ is the line $y = x - 3$ and the region above the line. The graph of $2x + y < 1$ is the region below the line $y = -2x + 1$. (See Fig. 4.43.)

Figure 4.43

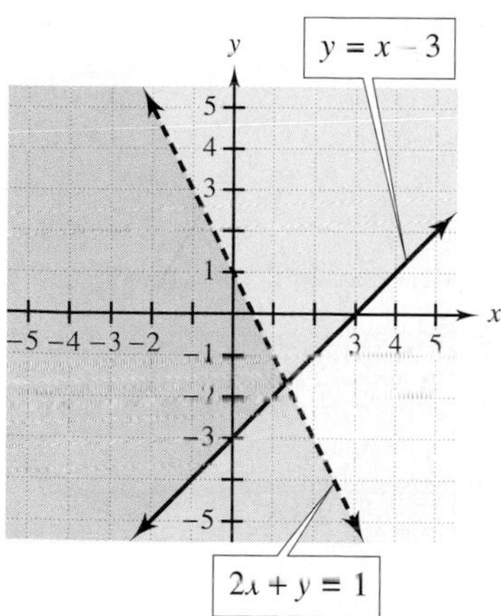

Figure 4.44

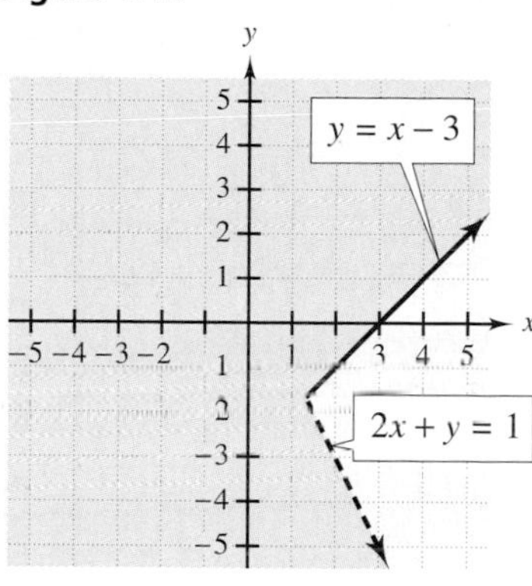

Because the inequalities are connected with *or*, the graph of the compound inequality is the union of the two graphs. (See Fig. 4.44.)

To graph inequalities involving absolute value, we must recall the following for any expression A and any positive number C.

1. If $|A| < C$, then $A < C$ *and* $A > -C$.
2. If $|A| > C$, then $A > C$ *or* $A < -C$.

Similar rules hold for the symbols $\leq$ and $\geq$.

EXAMPLE 5

Graphing an Absolute Value Inequality

Graph the absolute value inequality $|x + y| > 2$.

Solution

The inequality means $x + y > 2$ or $x + y < -2$. We solve both inequalities for y to obtain $y > -x + 2$ or $y < -x - 2$.

The graph of $y > -x + 2$ is the region above the line $y = -x + 2$. The graph of $y < -x - 2$ is the region below the line $y = -x - 2$.

Because the inequalities are connected with *or*, the graph of the absolute value inequality is the union of the two graphs. (See Fig. 4.45.)

Figure 4.45

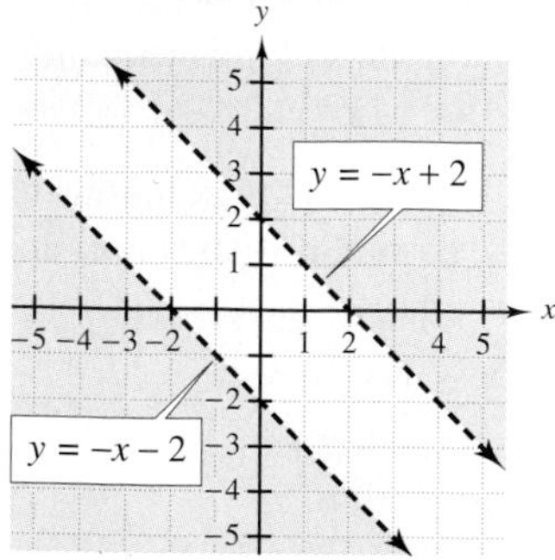

EXAMPLE 6

Graphing an Absolute Value Inequality

Graph the absolute value inequality $|y + x| \leq 4$.

Solution

The absolute value inequality means that $y + x \leq 4$ and $y + x \geq -4$. Solving each inequality for y, we obtain $y \leq -x + 4$ and $y \geq -x - 4$.

For $y \leq -x + 4$ we shade below the line, and for $y \geq -x - 4$ we shade above the line. (See Fig. 4.46.)

Figure 4.46

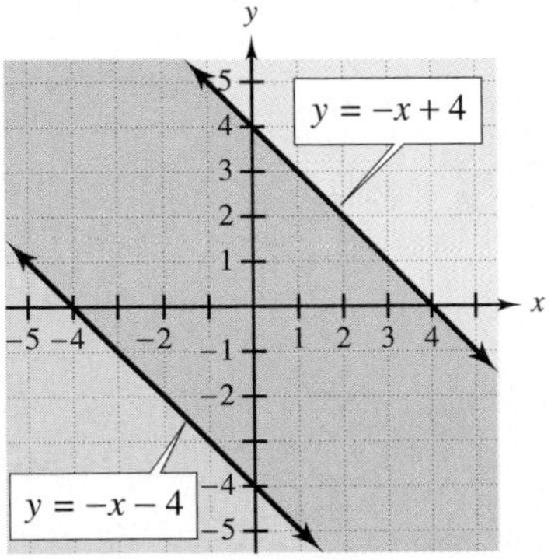

Figure 4.47

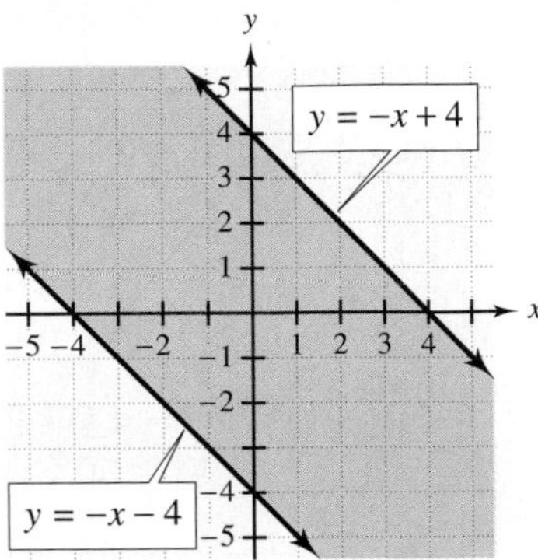

Think About It

Compare the graphs of $|x - y| < 0$ and $|x - y| \geq 0$.

Because the inequalities are connected with *and*, the graph of the absolute value inequality is the intersection of the two graphs. (See Fig. 4.47.)

4.5 QUICK REFERENCE

Linear Inequalities

- A **linear inequality in two variables** is an inequality that can be written in the form $Ax + By < C$, where A, B, and C represent any real numbers, and A and B are not both zero. A similar definition can be stated with $>$, $\geq$, and $\leq$.
- A **solution** of a linear inequality in two variables is a pair of numbers (x, y) that makes the inequality true. The **solution set** is the set of all solutions of the inequality.

Graphs of Solution Sets

- The **graph** of a linear inequality in two variables is a picture of its solutions. It consists of a boundary line (solid if it represents solutions, dashed if it does not) and a shaded region (a **half-plane**) on one side of the boundary line.
- The recommended method for graphing a linear inequality is as follows:
  1. Solve the inequality for y and draw the solid (for $\geq$ and $\leq$) or dashed (for $>$ and $<$) boundary line that represents the companion equation.
  2. Determine the region to shade by one of these methods:
     (a) If the coordinates of a test point satisfy the inequality, shade the half-plane in which the test point lies; otherwise, shade the other half-plane.
     (b) For inequalities of the form $y < mx + b$ or $y \leq mx + b$, shade below the boundary line; for inequalities of the form $y > mx + b$ or $y \geq mx + b$, shade above the boundary line.

Graphs of Compound Inequalities

- Two or more linear inequalities in two variables can be connected by *and* (conjunction) or *or* (disjunction).

- To sketch a graph of a compound inequality, draw the graph of each component of the inequality on the same coordinate system. For a conjunction, shade the intersection of the two graphs; for a disjunction, shade the union of the two graphs.
- To sketch a graph of an absolute value inequality, translate the inequality into a conjunction or disjunction, and graph accordingly.

4.5 EXERCISES

Concepts and Skills

1. How do the solution sets of $x + 2y < 6$ and of $x + 2y \leq 6$ differ?
2. Does the inequality $y > x + 2$ define a function? Explain.

In Exercises 3–8, determine whether each ordered pair is a solution of the given inequality.

3. $y < 3x - 5$; $(1, 0), (0, -6), (2, 1)$
4. $y \geq 1 - 2x$; $(5, -9), (-3, 8), (2, -7)$
5. $2x - 7y \geq 12$; $(12, 1), (6, 0), (-4, -2)$
6. $4 - 3x - 5y < 0$; $(4, 3), (-5, -4), (7, 3.4)$
7. $y \leq 4$; $(1, 10), (0, 4), (4, 5)$
8. $x > -6$; $(-5, -7), (-5.99, 3), (-6.01, 4)$
9. For the inequality $x + 3 > 7$, we might write the solution set as $(4, \infty)$, or we might graph the solution set as a shaded half-plane. What assumptions would we be making in each case?
10. To sketch the graph of $x - 2y < 8$, you must draw the boundary line and then decide which half-plane to shade. A classmate tells you to shade below the line because the inequality symbol is <. Is your classmate correct? Why, or why not?

In Exercises 11–30, graph the inequality.

11. $y < 7 - 3x$
12. $y \geq \frac{1}{2}x - 5$
13. $-y \leq -2x$
14. $y \geq x$
15. $3x + 2y \leq 6$
16. $4x + 5y \geq 20$
17. $-2y \geq 4x - 7$
18. $2x \geq 3 - y$
19. $-6 \leq x - 2y$
20. $x - 3y \leq 0$
21. $2x - 5y \geq 10$
22. $x - 6 < -y$
23. $y - 3 \leq -2(x - 3)$
24. $y + 2 - 4(x - 1) > 0$
25. $\frac{x}{5} + \frac{y}{2} < 1$
26. $\frac{x}{3} - \frac{y}{2} \geq 1$
27. $y \geq 20$
28. $y - 3 \leq 2$
29. $x < 5$
30. $2x + 6 \geq 0$

In Exercises 31–36, write an inequality for the given graph.

31.

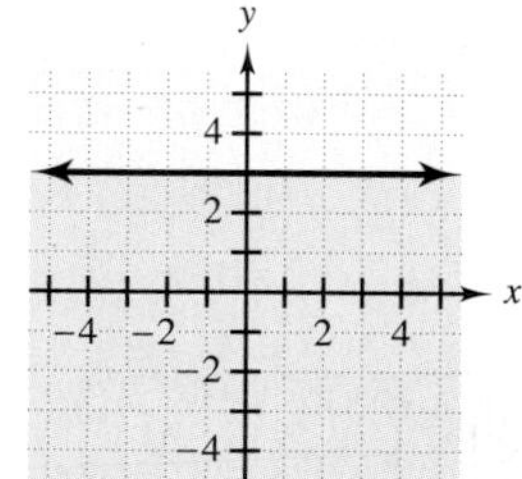

32.

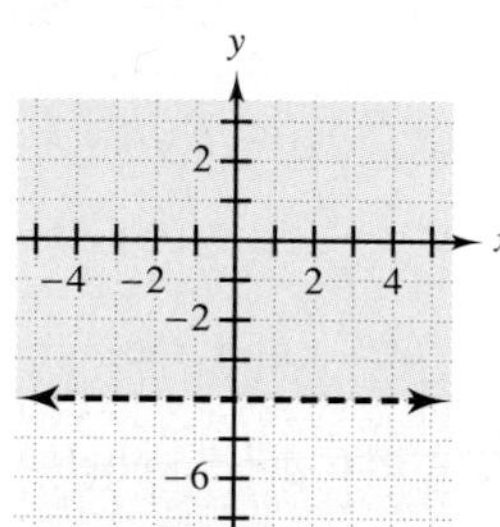

33.

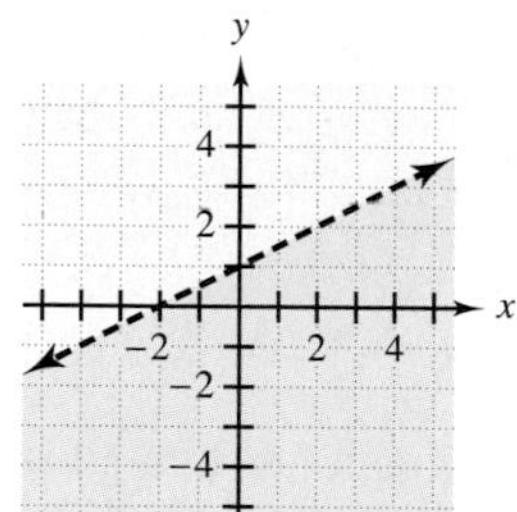

34.

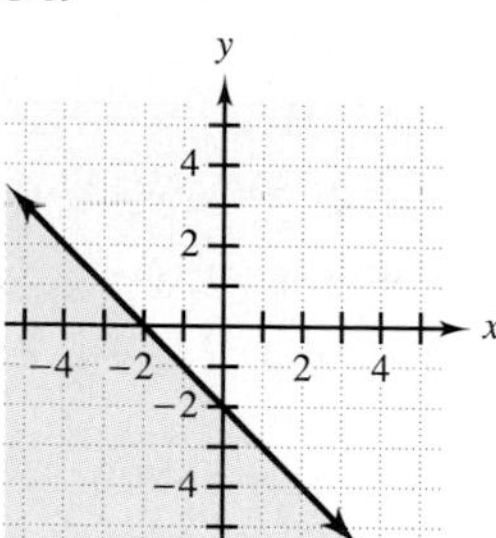

35.

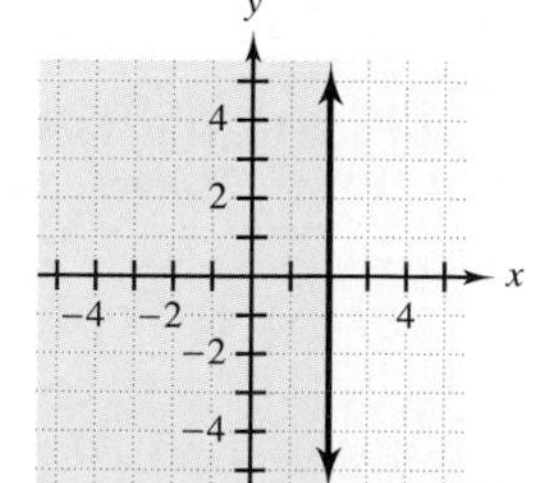

36.

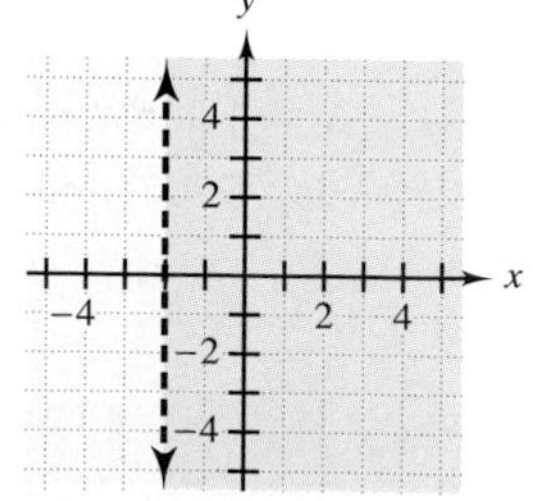

In Exercises 37–42, determine an ordered pair of numbers that satisfies the given condition. Then translate the given information into a linear inequality, produce its graph, and test whether your pair is a solution.

37. One number x is at least 1 more than another number y.

38. One number y is at most twice another number x.

39. One number y is not less than 4 less than 3 times another number x.

40. One number x is no greater than half of another number y.

41. The sum of a number x and twice another number y exceeds 10.

42. The difference of a number x and two-thirds of another number y exceeds 7.

In Exercises 43–48, graph the conjunction.

43. $x \leq 2$ and $x \geq 0$

44. $y > -3$ and $x > 2$

45. $x + y \geq 3$ and $x \geq 0$

46. $y < -3x - 10$ and $y \leq 0$

47. $y < \frac{1}{2}x$ and $y \geq 3x$

48. $x + y \geq -3$ and $y > x - 5$

In Exercises 49–54, graph the disjunction.

49. $y > -1$ or $y \leq -4$

50. $x \geq 2$ or $y > -3$

51. $y > -\frac{2}{3}x$ or $y + 3x \leq 0$

52. $x + 2y > 2$ or $x \geq 0$

53. $y < x$ or $y \geq 0$

54. $3x < y + 2$ or $y > 4 - x$

In Exercises 55–66, sketch the graph of the given compound inequality.

55. $y \leq -x + 3$ and $y \leq 2x + 1$

56. $y \geq 4$ or $y < -2x + 3$

57. $3x + y < 5$ or $x - y < 2$

58. $3x + y < 5$ and $x - y < 2$

59. $2x - y - 5 < 0$ and $4x - 2y > 0$

60. $\frac{y}{2} - \frac{x}{3} \leq 1$ or $\frac{x}{3} + \frac{y}{2} \geq 1$

61. $5y + 3x < 30$ and $5y + 3x < -20$

62. $2x - 3y < 9$ or $2x - 3y < -12$

63. $x \geq y$ and $x \leq y$

64. $x - 3 > y$ or $x - 3 < y$

65. $x \geq 0$, $y \geq 0$, and $y \leq 10 - x$

66. $x \geq 0$, $y \geq 0$, and $y \geq 12 - 3x$

In Exercises 67–72, write a compound inequality that describes the graph.

67.

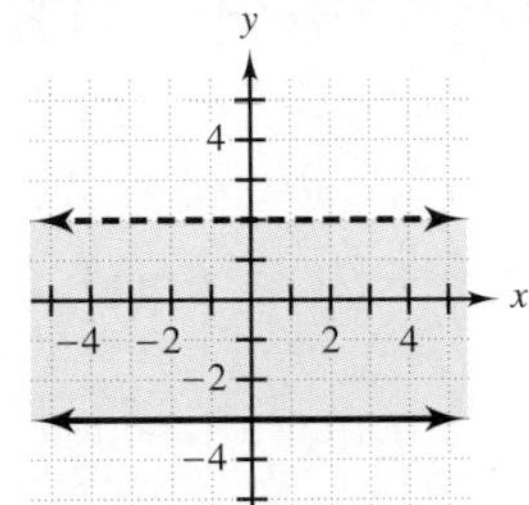

68.

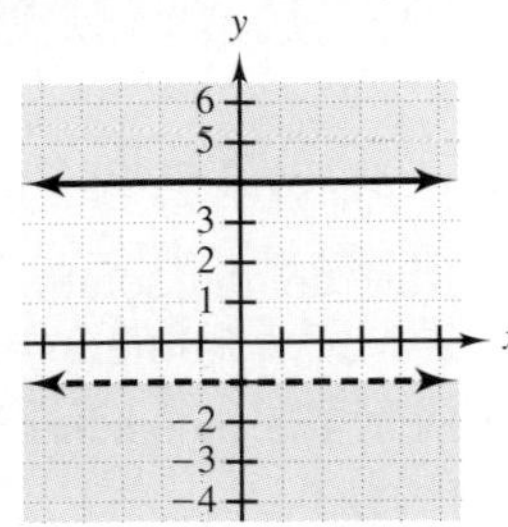

69.

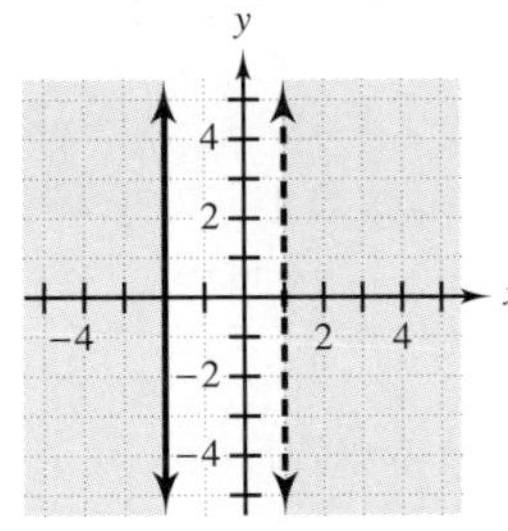

70.

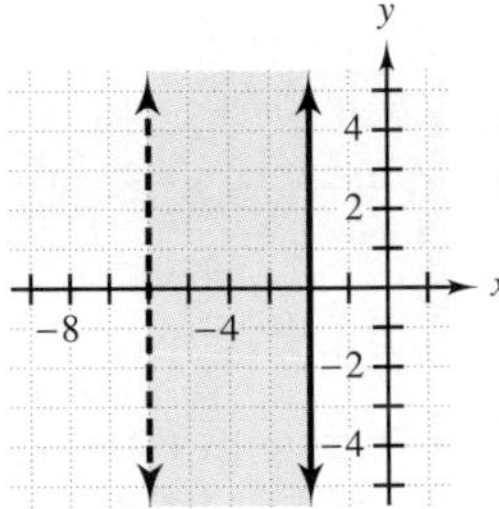

71.

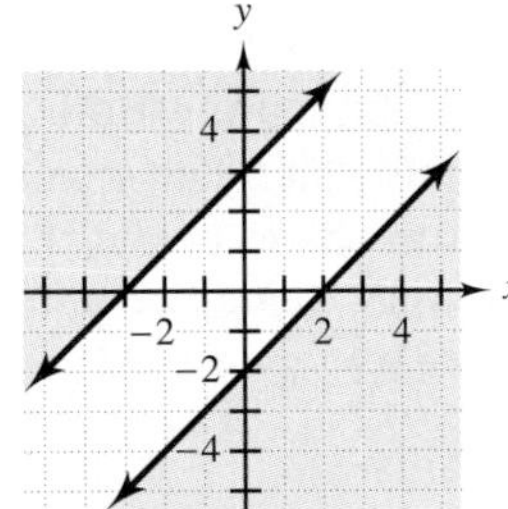

72.

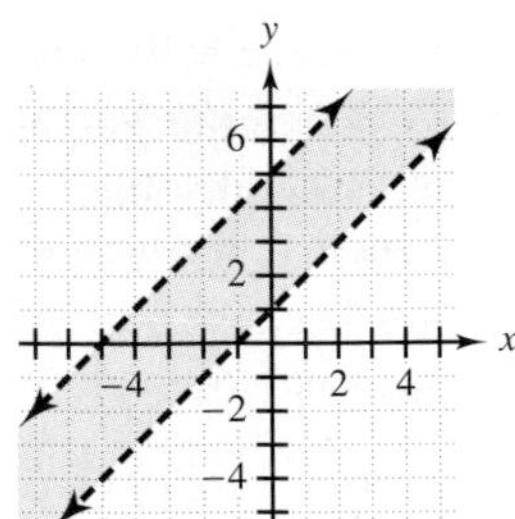

In Exercises 73–84, sketch the graph of the given absolute value inequality.

73. $|y| \leq 3$

74. $|x| \geq 4$

75. $|x - 2y| > 4$

76. $|2x + y| < 5$

77. $|y - x| \geq 3$

78. $|y + x| \leq 6$

79. $|x| \leq -2$

80. $|y| \geq -1$

81. $|x - y| \leq 0$

82. $|2x + 3y| \leq 0$

83. $|x + y| > 0$

84. $|y - 2x| > 0$

Real-Life Applications

85. The main hold of a supply ship has an area of 6000 square feet. Pallets of two sizes are used to store materials in the hold. One pallet is 4 feet by 4 feet, and the other is 3 feet by 5 feet. Write a linear inequality that describes the number of pallets that can be used and produce its graph.

86. A hospital food service can serve at most 1000 meals per day. Patients on a normal diet receive three meals each day, and patients on special diets receive four meals each day. Write a linear inequality that describes the number of patients that can be served and produce its graph.

87. Two basketball teams are considered evenly matched if the difference in their scores does not exceed 5 points. Write an absolute value inequality to describe this condition and produce its graph.

88. A community organization has determined that social harmony among its members is greatest when the difference between the number of male and female members is less than 10. Write an absolute value inequality to describe this condition and produce its graph.

89. A county animal shelter houses only cats and dogs. The maximum number of animals that can be kept is 100, but the county will close the shelter if the population falls below 20. Write a compound inequality that describes these conditions and produce its graph.

90. To be eligible for certain development grants, a local arts council must have at least 50 members, and the difference between the number of male and female members must be at most 10. Write a compound inequality that describes these conditions and produce the graph. Use the graph to show that not all nonnegative pairs of numbers satisfy the given conditions.

Modeling with Real Data

91. In 1996 the estimated number of registered voters in the United States was 190 million, and the estimated maximum voter turnout for the presidential election was 53.6%. (Source: Federal Elections Commission.)

(a) Let R represent the number of votes for the Republican candidate, and let D represent the number of votes for the Democratic candidate. Assuming that there are only two candidates, write an inequality that relates R and D to the total number of votes cast.

(b) Interpret the meaning of the points of the graph of the inequality in part (a).

92. (a) What change would you make to the inequality in Exercise 91 if there were other candidates in the race?

(b) How would the change in part (a) affect the graph of the inequality in Exercise 91?

93. In 1996, taxpayers who were married and filing jointly and who had a taxable income between \$39,000 and \$94,250 paid a tax of \$5850 plus 28% of the taxable income over \$39,000. (Source: Internal Revenue Service.)

(a) Suppose that a couple in this category had an adjusted gross income (before deductions) of x dollars. Write an inequality that describes the maximum tax y that the couple would pay.

(b) If you were to draw the graph of the inequality in part (a), would you shade above or below the boundary line?

94. In Exercise 93, if the base tax for the described tax bracket remained at \$5850 but the tax rate were to increase, what would be the effect on the graph of the inequality?

Group Project: Blood Alcohol Level

Research shows that when a person's blood alcohol level reaches 0.05, that person's chances of having an automobile accident double. The blood alcohol level depends on the person's weight as well as the amount of alcohol consumed. For example, a 100-pound person would reach a blood alcohol level of 0.05 with two drinks in a 2-hour period, while a 240-pound person could have four drinks in the same time period. (Source: Alcohol Safety Action Project.)

The equation $0.0875D - 0.00125W = 0.05$ models combinations of weight W and number of drinks D that will produce a blood alcohol level of 0.05 in 2 hours.

95. To the nearest whole number, how many drinks in 2 hours will produce a blood alcohol level of 0.05 for a 180-pound person?

96. What does $0.0875D - 0.00125W > 0.05$ describe?

97. If you were to graph the inequality in Exercise 96, with weight along the vertical axis and number of drinks along the horizontal axis, would you shade above or below the boundary line? (Hint: To help you decide, solve the inequality for W.)

98. In the 1990s, alcohol became the drug of choice for high school and college students. How has this fact affected automobile insurance rates for drivers between the ages of 16 and 25?

Challenge

In Exercises 99–102, graph the compound inequality.

99. $|x| > 3$ and $|y| \geq 2$

100. $|x| < 4$ or $|y| < 3$

101. $|x + y| > -2$ or $|x| < -5$

102. $|y + 2x| < -1$ and $|y| > -1$

In Exercises 103–106, draw an example of a graph of the given inequality.

103. $y < b, b > 0$

104. $y \geq mx, m < 0$

105. $y < mx + b, m > 0, b < 0$

106. $ax + by \leq ab, a > 0, b < 0$

4. CHAPTER PROJECT Bald Eagle Populations

The Bald Eagle Protection Act was passed in 1940. Alaska was exempted from this act because of pressure from the salmon fishing industry and because the primary problem was in the lower 48 states. The Alaska exemption was removed in 1952.

Bald eagles also were protected by the Endangered Species Act, but in 1995, when the population had reached 4400 pairs, bald eagles were listed as "threatened" rather than "endangered."

1. In the last two decades, the population of bald eagles has been estimated to have doubled every 7 years. Based on the data for 1995, what was the estimated population in 1988?
2. The term *occupied territories* indicates that a pair of bald eagles has established a breeding territory and a nest site. In 1963 there were only 417 active nests with an average of 0.59 young birds produced per active nest. Approximately how many young bald eagles were produced that year?
3. By 1994 there were 4110 occupied territories, and the average number of young bald eagles per nest site had increased by 98%. Approximately how many young bald eagles were produced that year?
4. In 1995 the expenditures by public and private agencies for the recovery and protection of the bald eagle were about \$1 million. Based on the bald eagle population that year, what was the apparent cost per pair of eagles? Given the purposes of the expenditures, do you think that antienvironmentalists are justified in using this statistic?
5. Referring to the accompanying bar graph, do you think that a line connecting the data points for 1963 and 1995 is the best model for the population of bald eagles during that period? For what years does the line overestimate the population? For what years will the line probably underestimate the population?

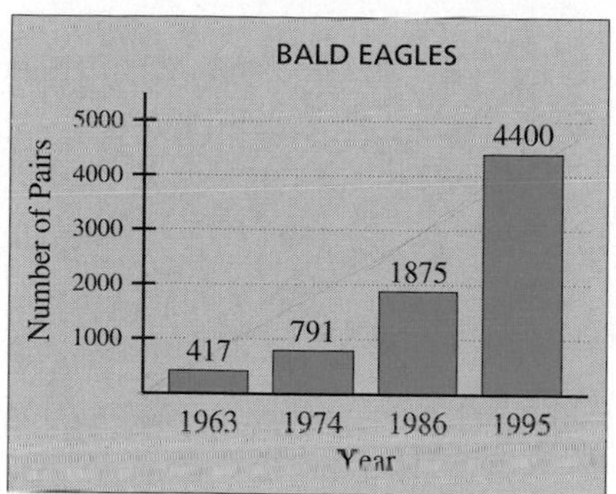

6. For what population do you think that bald eagles can be removed from the "threatened" list? To support your position with research, ask your librarian to help you locate related articles by the National Audubon Society and the Sierra Club.

4. CHAPTER REVIEW EXERCISES

Section 4.1

In Exercises 1–4, state whether the equation is a linear equation in two variables.

1. $3x = \frac{2}{y} + \frac{3}{2}$

2. $3x - 8 = 15$

3. $x = \frac{y}{3} - \frac{4}{5}$

4. $5 = \frac{-3xy}{4}$

In Exercises 5 and 6, verify that the given ordered pair is a solution of the given equation.

5. $(5, -11)$; $y = -3x + 4$

6. $(-14, -6)$; $2x - 7y = 14$

In Exercises 7 and 8, find a, b, and c so that the ordered pairs are solutions of the given equation.

7. $y = 7 - x$; $(-2, a)$, $(b, 14)$, $(c, -5)$

8. $4y - 3x = 7$; $(1, a)$, $(b, 4)$, $(-9, c)$

In Exercises 9 and 10, complete the table of solutions of the given equation.

9. $2y - 3x = 9$

x	3	7	−7
y			

10. $2x - 4 = 3y$

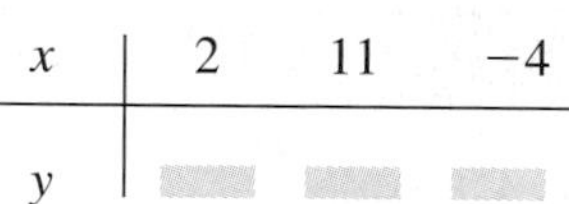

x	2	11	−4
y			

In Exercises 11–14, determine the x- and y-intercepts.

11. $7x - 3y = 21$

12. $y = -2x + 3$

13. $y - 2 = 18$

14. $2x - 5 = 7$

In Exercises 15–18, write each equation in the form $y = mx + b$. Then determine the y-intercept.

15. $15 - 3x + 6y = 0$

16. $4x - 5y = 19$

17. $2x - y = 8$

18. $\frac{1}{2}x - \frac{2}{5}y = 6$

In Exercises 19–22, determine three solutions to the equation and sketch the graph.

19. $5x + 3y = 15$

20. $\frac{1}{4}x - \frac{2}{5}y = 10$

21. $y + 3 = 9$

22. $2x + 1 = 7$

23. Which of the following equations does the graph represent?

(i) $3x + 4y = 12$

(ii) $3x + 4y = -12$

(iii) $3x - 4y = -12$

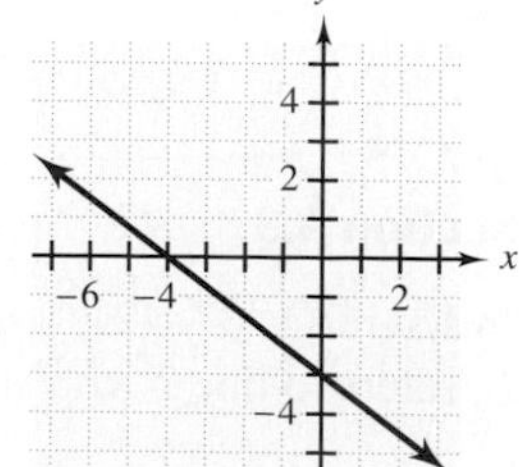

Section 4.2

24. In the accompanying figure, name the line whose slope is
(a) negative.
(b) positive.
(c) undefined.
(d) zero.

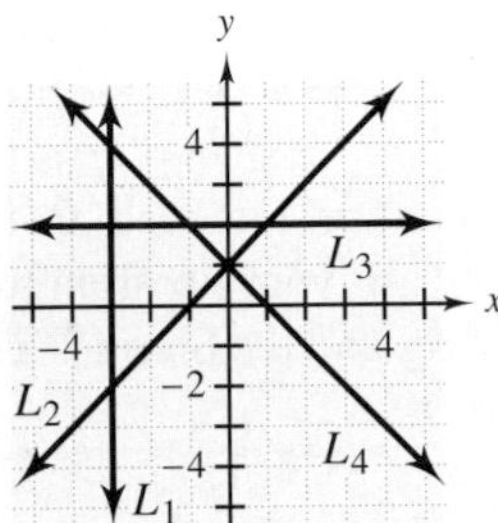

In Exercises 25–28, determine the slope of the line that contains the given points.

25. $(-5, -8)$ and $(2, -3)$

26. $(3, -4)$ and $(3, 5)$

27. $(2, -4)$ and $(-7, 6)$

28. $(-4, 5)$ and $(4, 5)$

In Exercises 29 and 30, write each equation in the form $y = mx + b$. Then determine the slope.

29. $3x - 4y = 15$

30. $x + 2y = 4$

In Exercises 31 and 32, draw a line through the given point with the given slope.

31. $(0, -4)$; $m = -\dfrac{3}{5}$

32. $(2, -3)$; $m = 0.5$

In Exercises 33–36, sketch the graph of the given equation.

33. $y = -x + 3$

34. $3y + 2x = 3$

35. $y = 3$

36. $x - 2 = 0$

Section 4.3

In Exercises 37–42, L_1 refers to line 1 with slope m_1 and L_2 refers to line 2 with slope m_2.

In Exercises 37–40, the slopes of L_1 and L_2 are given. Determine if L_1 and L_2 are parallel, perpendicular, or neither.

37. $m_1 = \dfrac{1}{4}$; $m_2 = 0.25$

38. $m_1 = 5$; $m_2 = -5$

39. $m_1 = \dfrac{1}{3}$; $m_2 = -3$

40. $m_1 = 0$; m_2 is undefined

In Exercises 41 and 42, the equations of L_1 and L_2 are given. Determine if the lines are parallel, perpendicular, or neither.

41. L_1: $y = 0.5x - 3$ L_2: $y = -2x + 4$

42. L_1: $\dfrac{2}{3}x + \dfrac{3}{5}y = 1$ L_2: $15 - 10x = 9y$

43. Determine whether a triangle whose vertices are $A(-3, 6)$, $B(10, 0)$, and $C(2, -4)$ is a right triangle.

44. Determine whether the points $P(-4, 3)$, $Q(2, 1)$, and $R(5, -1)$ are collinear.

In Exercises 45 and 46, determine the rate of change of y with respect to x.

45. $y = -3x + 5$

46. $y = \dfrac{2}{3}x + 4$

In Exercises 47 and 48, the constant rate of change of a line is given along with an increase or decrease in one variable. Determine the increase or decrease in the other variable.

47. $m = -5$; x increased by 3

48. $m = \dfrac{5}{3}$; y increased by 10

In Exercises 49 and 50, assume a constant rate of change.

49. The price of a ticket to a college football game increased from \$5.00 in 1986 to \$20.00 in 1998. What was the annual rate of change of the price of the ticket?

50. If the temperature fell from 20°F at 5:00 P.M. to −6°F at 6:00 A.M., what was the hourly rate of change in temperature?

In Exercises 51–54, determine whether the given equation represents a direct variation.

51. $y = 4x - 7$

52. $y = \dfrac{x}{-3}$

53. $y = 7x$

54. $xy = 4$

55. The number n of toothpicks produced by a machine varies directly with the number of hours h the machine is operating. If it produces 15,000 toothpicks in 6 hours, how many toothpicks are made in 45 hours by this one machine?

56. The per-share dividend that a company declares varies directly with the earnings per share. If a company pays a $1.50 per-share dividend when earnings are $4.00 per share, what would the earnings per share be for a dividend of $2.00 per share?

Section 4.4

In Exercises 57 and 58, write an equation of the line having the given slope and containing the given point.

57. $m = -3$; $(-2, 5)$

58. $m = \frac{4}{3}$; $(1, 2)$

In Exercises 59–62, write an equation of the line containing the given pair of points.

59. $(2, -4)$ and $(-5, 3)$

60. $(2, -5)$ and $(2, 4)$

61. $(-3, -6)$ and $(1, -1)$

62. $(-2, -5)$ and $(4, -5)$

In Exercises 63–68, write an equation of the line with the following conditions.

63. The line is horizontal and contains the point $(-3, 6)$.

64. The line is vertical and contains the point $(6, -3)$.

65. The line contains the point $(1, 5)$ and is parallel to the line whose equation is $2x + 3y = 6$.

66. The line contains the point $(2, -3)$ and is perpendicular to the line $2x - 3y = 7$.

67. The line is graphed in the accompanying figure.

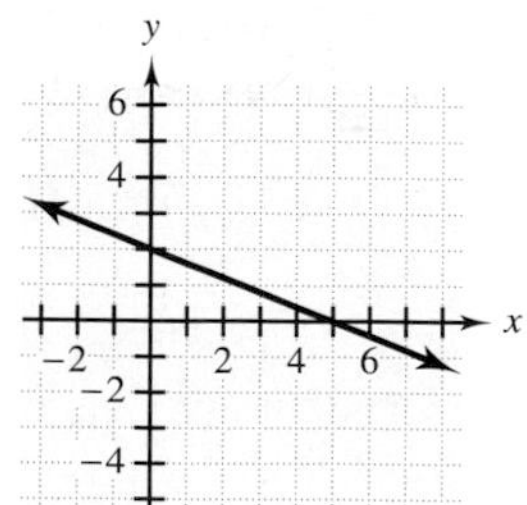

68. The line is perpendicular to the line whose equation is $y = 3x + 4$ and has the same y-intercept as the line whose equation is $y = 3x - 5$.

69. A store owner decides to expand the facilities when annual sales reaches $800,000. Sales were $150,000 in 1994 and $400,000 in 1996. Assuming the increase in sales is linear, predict the year in which the store owner will expand.

70. Suppose a public school system will lose its accreditation if its student–teacher ratio rises above 30 to 1. In 1996, the system had 1000 students and 40 teachers. During the following year, there were 1100 students and 45 teachers. Assuming a linear change in the student–teacher ratio, determine whether the system is headed for a loss of accreditation.

Section 4.5

In Exercises 71–74, determine whether each of the given ordered pairs is a solution of the given inequality.

71. $4x - y \geq -2$; $(3, 14), (-2, 0), (-4, -20)$

72. $2x - 3y < 0$; $(0, 0), (5, 4), (-4, -4)$

73. $y \leq -3$; $(1, 10), (0, -4), (4, 5)$

74. $x > 5$; $(-5, -7), (4.99, 5), (5.01, 8)$

In Exercises 75–78, sketch the graphs of the following inequalities.

75. $3x - 4y \leq 12$

76. $x \leq 2y$

77. $y + 4 \leq 7$

78. $8 - 2x \geq 0$

In Exercises 79–82, sketch the graph of the compound inequality.

79. $5y - 2x > 7$ or $y + 3x < -2$

80. $y \leq 2x - 3$ and $y \geq 4 - x$

81. $y \leq 4$ and $x + 2 \leq 0$

82. $x \leq -3$ or $y < 2x - 1$

In Exercises 83–86, sketch the graph of the absolute value inequality.

83. $|y| \geq 2$

84. $|4x - 3y| \geq 5$

85. $|x - 2| - 2 < 5$

86. $|3x + 2y| \leq -1$

87. At a minor league stadium, grandstand seats cost $6 and bleacher seats cost $4. The total receipts from the ticket sales for a certain game were less than $11,400. Write a linear inequality that describes the number of each kind of ticket sold.

88. An **isosceles triangle** is a triangle with two sides of equal length. The perimeter of a certain isosceles triangle is at most 42 inches. Write a linear inequality that describes the lengths of the sides of the triangle.

89. A community athletics council sponsors coed soccer teams. The number of boys and girls enrolled in the league cannot exceed 400, and there must be at least 100 girls. Write a compound inequality that describes the number of youngsters who can participate in the program.

90. In a survey a certain percentage of those polled were in favor of a local bond issue. The survey is known to be accurate to within 3 percentage points. Write an absolute value inequality that describes the relationship between the percentage of the sample who were in favor and the actual percentage of all voters who were in favor.

LOOKING AHEAD

The following exercises review concepts and skills that you will need in Chapter 5.

1. (a) Graph the equation $2x - 3y = 6$.

(b) Graph the inequality $2x - 3y \geq 6$.

2. Determine whether the given ordered pair is a solution of the equation $2x + y = 1$.

(i) $(-2, 5)$ (ii) $(-1, 1)$ (iii) $\left(\frac{1}{2}, 0\right)$

3. Describe the graphs of $x = 3$ and $y = 3$.

4. Use the graphing method to estimate the solution of the equation $2x - 2 = 13 - 3x$.

5. Use a graph to estimate the point of intersection of the graphs of $f(x) = -5$ and $g(x) = 3x - 2$.

In Exercises 6 and 7, solve the equation.

6. $3x + 2(4x - 13) = 29$

7. $2y - (y - 5) + 17 = 0$

In Exercises 8 and 9, the equations of line L_1 and line L_2 are given. Determine whether the lines are parallel.

8. L_1: $y + 3x = 6$ L_2: $3x - y + 8 = 0$

9. L_1: $y = -x - 3$ L_2: $x + y = 4$

10. (a) Suppose that you solve an equation and obtain $-2 = -2$. What is the equation called, and what is the solution set?

(b) Suppose that you solve an equation and obtain $5 = 2$. What is the equation called, and what is the solution set?

11. Evaluate each expression for $x = 4$, $y = \frac{1}{2}$, and $z = -3$.

(a) $-3x + 4y + z$ (b) $2x - 2y + 3z$

12. Write an expression that models the given information.

(a) The total cost in dollars of mailing x first-class letters at 32¢ each and y post cards at 20¢ each

(b) The total number of gallons of boric acid in x gallons of a 20% solution and y gallons of a 9% solution

(c) The total distance traveled at x mph for 2 hours and y mph for 3 hours

4. CHAPTER TEST

1. Complete the ordered pairs so that they are solutions of $2x - y = -2$.

(a) (, 0) (b) (0,)

(c) (−2,) (d) (, 3)

In Questions 2–5, sketch the graph of each equation. Give the x- and y-intercepts and the slope of each.

2. $y = x$

3. $y = -x - 2$

4. $2x - 3y = 9$

5. $x - 2 = 3$

6. Refer to the graph and fill in the blanks with the symbols $<$ or $>$: m ___ 0 and b ___ 0.

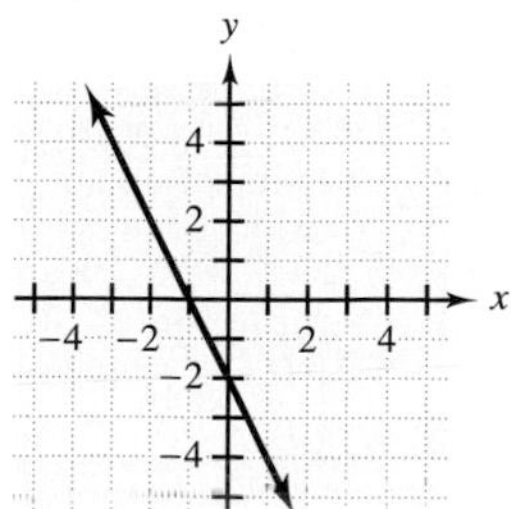

In Questions 7–12, determine an equation of the line that satisfies the given information.

7. The line contains the points $(2, -3)$ and $(-1, -5)$.
8. The line contains the point $(-1, 3)$ and has a slope of 2.
9. The line is vertical and contains the point $(2, 5)$.
10. The line contains the point $(-1, 3)$ and is parallel to the line whose equation is $y = -4$.
11. The line contains the point $(1, 4)$ and is perpendicular to the line whose equation is $y = -\frac{1}{2}x + 3$.
12. The graph of the line is shown in the figure.

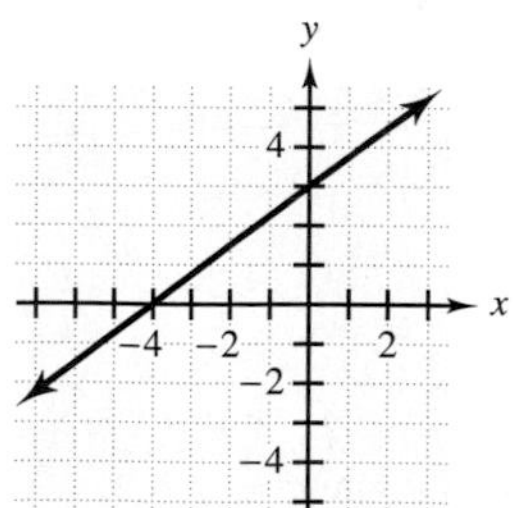

In Questions 13 and 14, graph the given inequalities.

13. $2x - y \leq 5$

14. $y - 3 > 0$

In Questions 15 and 16, graph the given compound inequalities.

15. $y \geq 2x + 1$ and $x + y \geq 3$
16. $x - 2y < 2$ or $2y - x > 6$

In Questions 17 and 18, graph the given absolute value inequalities.

17. $|x + y| \leq 3$

18. $|x + 2y| > 5$

19. Show that the points $(1, -2)$, $(-2, 1)$, and $(5, 2)$ are the vertices of a right triangle.
20. Draw a line through $(-1, 3)$ with a slope of -2.

In Questions 21 and 22, determine whether the lines are perpendicular, parallel, or neither.

21. $2x + 3y = 3$ and $2y - 3x = -4$
22. $y = 3x + 1$ and $x - 3y = -1$
23. The velocity of an object dropped from rest varies directly with the time of the fall. After 0.5 seconds, the speed is 16 feet per second. Write the relation between the velocity and the time. What is the velocity after 2 seconds?
24. The earnings for a company dropped from \$1.35 per share in 1994 to \$0.45 in 1996. Assuming that the relationship is linear, what was the rate of change in per-share earnings per year?
25. At 6:00 A.M., the temperature in Death Valley was 88°F. At 10:00 A.M., the temperature was 98°F. Assuming the rise in temperature is linear, write an equation to describe the rise in temperature. Use the equation to predict the temperature at 3:00 P.M.

Chapter 5

Systems of Linear Equations

The percentage of office communications conducted by E-mail is increasing rapidly, while the percentage conducted by hard copy (paper) is decreasing. (Source: Cognitive Communications Survey.) The trends in the use of E-mail and in the use of paper can each be modeled by a linear equation, and the two equations together form a **system of equations.** The **solution** of this system is an ordered pair of numbers that indicates, for example, the year when the percentages for E-mail and paper are predicted to be the same. (For more on this topic, see Exercises 89–92 at the end of Section 5.2, and see the Chapter Project.)

In this chapter we introduce systems of linear equations in two and three variables. We discuss both graphic and algebraic methods for solving systems. These methods include Cramer's Rule, for which we need to learn about matrices and determinants. After using our new skills to solve application problems, we conclude with the topic of systems of linear inequalities.

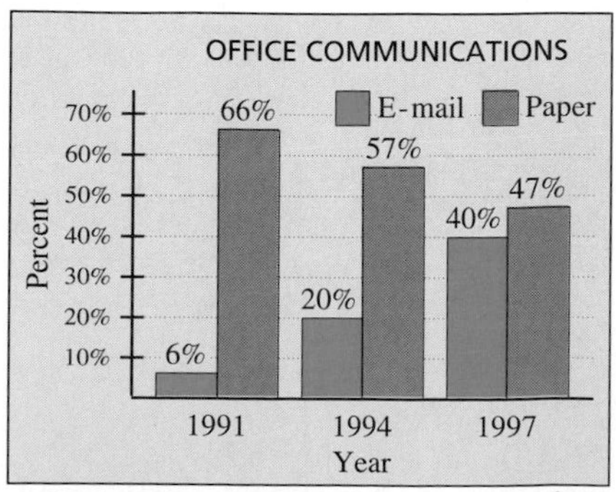

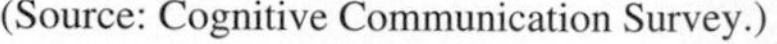
(Source: Cognitive Communication Survey.)

5.1 GRAPHING AND SUBSTITUTION METHODS

Systems of Equations • Estimating Solutions by Graphing • Substitution Method • Special Cases • Applications

Systems of Equations

We write equations to describe the conditions and relationships given in applied problems. To this point, applications have led to a single equation in one or two variables. However, applications often lead to more than one equation.

EXPLORING THE CONCEPT

Systems of Equations

Suppose that x and y are two numbers whose sum is 6. We can model this condition with the equation $x + y = 6$. We write this equation as $y = -x + 6$ and produce its graph. (See Fig. 5.1.) The cursor highlights one pair of numbers whose sum is 6. By tracing the graph, you will find infinitely many other pairs whose sum is 6.

Figure 5.1

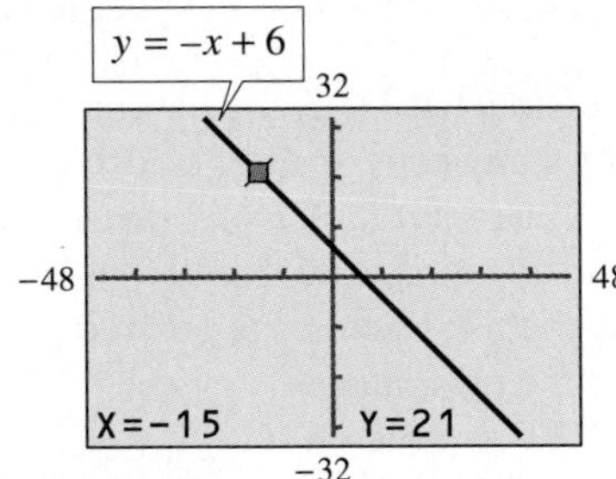

Figure 5.2

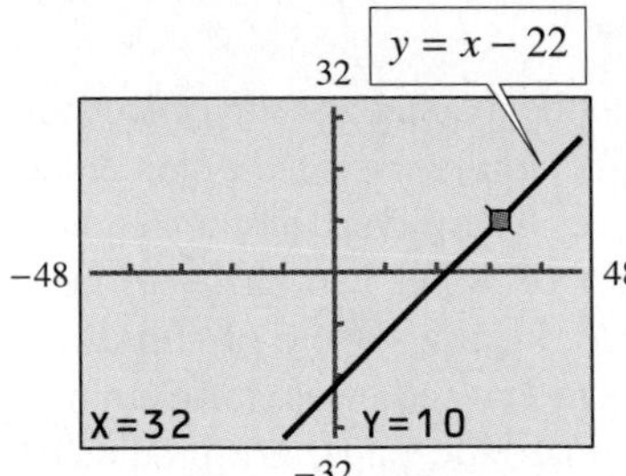

Turning to a second situation, suppose that x and y are two numbers whose difference is 22. If x is the larger number, then a model equation is $x - y = 22$. We write this equation as $y = x - 22$ and produce its graph. (See Fig. 5.2.) The cursor highlights (32, 10), one of the infinitely many pairs of numbers whose difference is 22.

Now suppose that we produce both graphs in the same coordinate system. (See Fig. 5.3.)

Figure 5.3

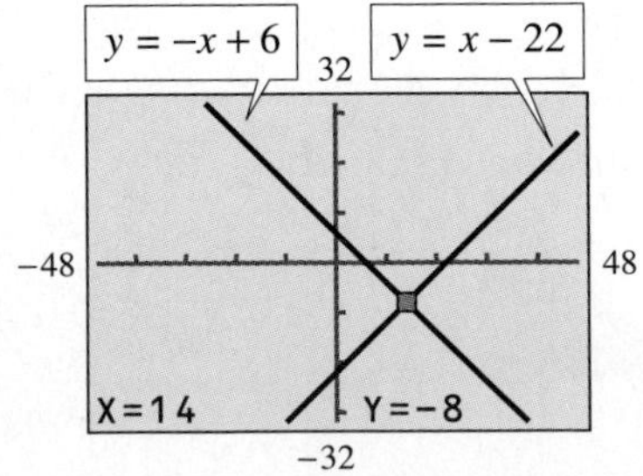

If we trace either graph, we find that the point of intersection is (14, −8). Because this point belongs to both graphs, it represents a pair of numbers whose sum is 6 *and* whose difference is 22.

Considered simultaneously, the pair of equations $x + y = 6$ and $x - y = 22$ is an example of a **system of equations.** The ordered pair (14, −8) is a **solution** of this system because both equations are satisfied when $x = 14$ and $y = -8$.

Definition of a System of Equations and Solution

Two or more equations considered simultaneously form a **system of equations.** A **solution** of a system of equations in two variables is an ordered pair (a, b) that satisfies all equations of the system.

A system of equations is a *conjunction*, but we usually write the system without explicitly including the word *and*.

$$\begin{aligned} x + y &= 6 \\ x - y &= 22 \end{aligned}$$

In certain applications there can be many equations and many variables. In this section we consider only systems of two linear equations in two variables.

EXAMPLE 1

Verifying Solutions of a System of Equations

For the following system of equations, determine whether the given pair of numbers is a solution.

$$\begin{aligned} x - y &= 2 \\ x - 2y &= -2 \end{aligned}$$

(a) (5, 3) (b) (6, 4)

Solution

(a) In each equation replace x with 5 and y with 3.

$$\begin{aligned} x - y &= 2 \\ 5 - 3 &= 2 \\ 2 &= 2 \quad \text{True} \end{aligned} \qquad \begin{aligned} x - 2y &= -2 \\ 5 - 2(3) &= -2 \\ 5 - 6 &= -2 \\ -1 &= -2 \quad \text{False} \end{aligned}$$

Although (5, 3) satisfies the first equation, it does not satisfy the second equation. Therefore, the pair is not a solution of the system of equations.

(b) In each equation replace x with 6 and y with 4.

$$\begin{aligned} x - y &= 2 \\ 6 - 4 &= 2 \\ 2 &= 2 \quad \text{True} \end{aligned} \qquad \begin{aligned} x - 2y &= -2 \\ 6 - 2(4) &= -2 \\ 6 - 8 &= -2 \\ -2 &= -2 \quad \text{True} \end{aligned}$$

Because (6, 4) satisfies both equations, the pair is a solution of the system of equations.

Estimating Solutions by Graphing

At the beginning of this section we used a graphing method for estimating the solution of the system $x + y = 6$ and $x - y = 22$. The general procedure is given in the following summary.

Estimating Solutions of Systems by Graphing

1. Produce the graph of each equation in the same coordinate system.
2. Trace to the point of intersection of the two graphs.
3. This point represents the estimated solution of the system of equations.

EXAMPLE 2

Estimating Solutions by Graphing

Use the graphing method to estimate the solution of the following system of equations.

$$2x - y = 9$$
$$3x - 4y = -24$$

Solution

Solve each equation for y and produce the graphs of both equations. (See Fig. 5.4.)

Figure 5.4

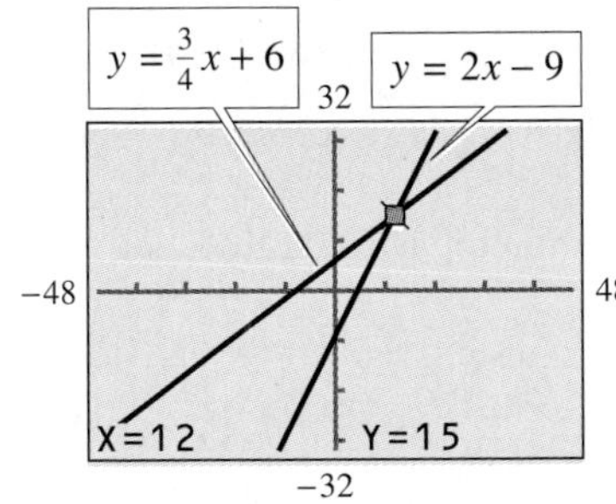

Figure 5.5

The point of intersection (12, 15) is the estimated solution of the system. In Fig. 5.5 we verify that (12, 15) is the solution by showing that when x is 12, y is 15 in both equations of the system.

Substitution Method

When we use the graphing method to estimate the solution of a system of equations, the point of intersection of the two graphs corresponds to the solution of the system.

At that point, one value of x gives the same value for y in each equation. The substitution method is an algebraic method that uses this same idea.

The substitution method is a particularly useful method when it is easy to solve one of the equations for a certain variable. In the following summary of the method, we begin by solving one equation for y. However, either x or y may be chosen.

The Substitution Method

1. Solve one equation for y to obtain $y =$ expression.
2. Replace y in the other equation with the expression. This results in an equation in one variable x.
3. Solve that equation.
4. Substitute the solution for x in either original equation and solve for y.
5. Check the solution.

EXAMPLE 3

Solving a System with the Substitution Method

Solve the following system with the substitution method.

$$\begin{aligned} 2x - y &= 9 \\ 3x - 4y &= -24 \end{aligned}$$

Solution

We choose to solve the first equation for y.

$$\begin{aligned} 2x - y &= 9 && \text{First equation} \\ -y &= -2x + 9 && \text{Subtract } 2x \text{ from both sides.} \\ y &= 2x - 9 && \text{Multiply both sides by } -1. \end{aligned}$$

Substitute $2x - 9$ for y in the second equation.

$$\begin{aligned} 3x - 4y &= -24 && \text{Second equation} \\ 3x - 4(2x - 9) &= -24 && \text{Replace } y \text{ with } 2x - 9. \\ 3x - 8x + 36 &= -24 && \text{Distributive Property} \\ -5x + 36 &= -24 && \text{Combine like terms.} \\ -5x &= -60 && \text{Subtract 36 from both sides.} \\ x &= 12 && \text{Divide both sides by } -5. \end{aligned}$$

Now determine the y-value of the solution by substituting 12 for x in the equation that is already solved for y.

$$\begin{aligned} y &= 2x - 9 \\ y &= 2(12) - 9 && \text{Replace } x \text{ with 12.} \\ y &= 24 - 9 \\ y &= 15 \end{aligned}$$

LEARNING TIP

The solution of a system of equations in two variables is a pair of numbers. After solving for one variable, remember to substitute to determine the value of the other variable.

Think About It

The solution to the following system is easy to determine without the graphing method and without the substitution method.

$y = 2x - 7$
$y = -x - 7$

What is the solution? Why?

The solution is (12, 15). This system was solved graphically in the preceding example. Recall that the coordinates of the point of intersection were (12, 15).

Special Cases

A system of two linear equations in two variables has a unique solution only if the graphs of the equations intersect at exactly one point. However, two lines may not intersect, or they may coincide.

EXPLORING THE CONCEPT

Inconsistent Systems

Consider the following system of equations.

$$x - 2y = 5 \quad \rightarrow \quad y = \frac{1}{2}x - \frac{5}{2}$$

$$2x - 4y = 4 \quad \rightarrow \quad y = \frac{1}{2}x - 1$$

Figure 5.6

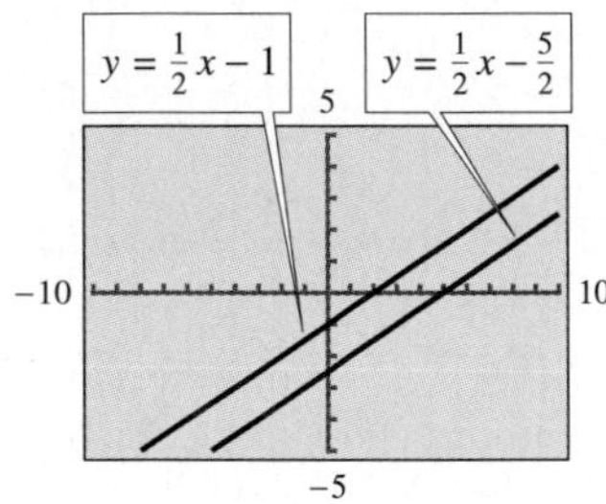

Figure 5.6 shows the graphs of the two equations.

Because the slope of each line is $\frac{1}{2}$, the lines are parallel. Thus there is no point of intersection; the system has no solution.

To solve the system with the substitution method, we choose to solve the first equation for x.

$$x - 2y = 5 \qquad \text{First equation}$$
$$x = 2y + 5 \qquad \text{Add } 2y \text{ to both sides.}$$

Substitute $2y + 5$ for x in the second equation and solve for y.

$$2x - 4y = 4 \qquad \text{Second equation}$$
$$2(2y + 5) - 4y = 4 \qquad \text{Replace } x \text{ with } 2y + 5.$$
$$4y + 10 - 4y = 4 \qquad \text{Distributive Property}$$
$$10 = 4 \qquad \text{False}$$

Because the resulting equation is false, it has no solution. Therefore, the system of equations has no solution.

If a system of equations has at least one solution (the graphs intersect), the system is called **consistent.** A system that has no solution (the graphs do not intersect) is called **inconsistent.**

EXPLORING THE CONCEPT

Dependent Equations

Another special case is illustrated by the following system of equations.

$$2x + y = 3 \quad \rightarrow \quad y = -2x + 3$$
$$4x + 2y = 6 \quad \rightarrow \quad y = -2x + 3$$

Because the two equations are exactly the same, their graphs coincide, and every point of the line is a point of intersection. Thus the system has infinitely many solutions, and they are described by the equation $y = -2x + 3$.

As before, the substitution method will detect this special case.

$$2x + y = 3 \quad \text{First equation}$$
$$2x + (-2x + 3) = 3 \quad \text{Replace } y \text{ with } -2x + 3 \text{ from the second equation.}$$
$$3 = 3 \quad \text{True}$$

Because the resulting equation is true, the system is satisfied by the infinitely many solutions of $y = -2x + 3$.

If a system has infinitely many solutions (the graphs coincide), the equations of the system are called **dependent.** Otherwise, the equations are called **independent.**

Figure 5.7 summarizes the three possible outcomes when we solve a system of two linear equations in two variables.

Figure 5.7

(a)

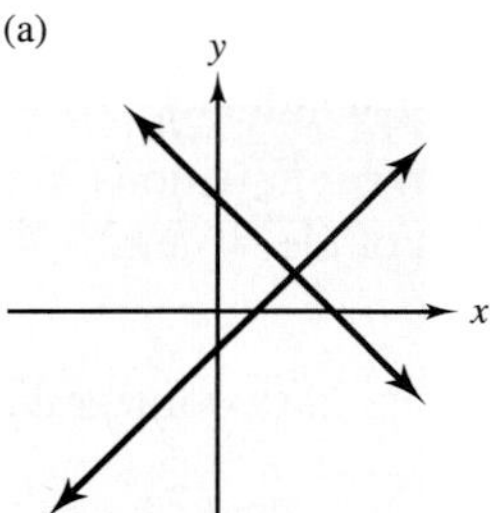

Solution is unique.
Equations are independent.
System is consistent.
The lines intersect.

(b)

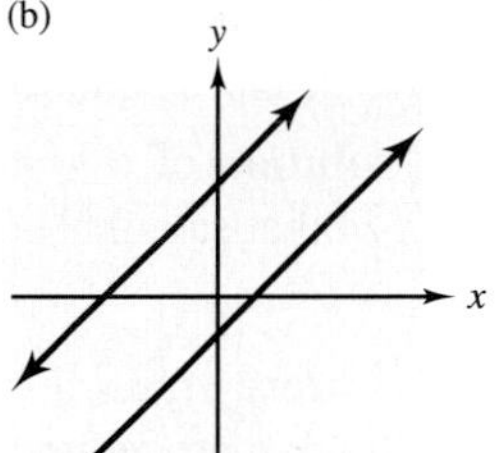

No solution.
Equations are independent.
System is inconsistent.
The lines are parallel.

(c)

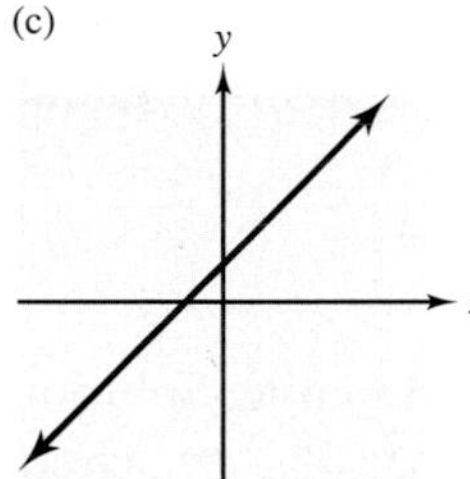

Infinitely many solutions.
Equations are dependent.
System is consistent.
The lines coincide.

Applications

Many applied problems can be modeled and solved with an equation that has a single variable. However, when the problem involves two unknown quantities, it is often more convenient to use two variables.

EXAMPLE 4

Using a System of Equations to Solve an Application Problem

Two angles are supplementary. One angle is 20° less than the other. What are the measures of the two angles?

Solution

Let x = measure of one angle and
y = measure of the other angle.

The following system of equations describes the given conditions.

$$x + y = 180 \quad \text{The angles are supplementary.}$$
$$y = x - 20 \quad \text{One angle is 20° less than the other angle.}$$

Replace y in the first equation with $x - 20$.

$$x + (x - 20) = 180$$

$$2x - 20 = 180 \quad \text{Remove parentheses and combine like terms.}$$

$$2x = 200 \quad \text{Add 20 to both sides.}$$

$$x = 100 \quad \text{Divide both sides by 2.}$$

Now we use the second equation of the system to determine y.

$$y = x - 20$$

$$y = 100 - 20 = 80 \quad \text{Replace } x \text{ with 100.}$$

The measures of the angles are 100° and 80°.

5.1 QUICK REFERENCE

Systems of Equations

- Two or more equations considered simultaneously form a **system of equations.**
- A **solution** of a system of two linear equations in two variables is an ordered pair (a, b) that satisfies both equations of the system.

Estimating Solutions by Graphing

- To estimate the solution of a system of two linear equations in two variables, use the following procedure:
  1. Produce the graph of each equation in the same coordinate system.
  2. Trace to the point of intersection of the two graphs.
  3. This point represents the estimated solution of the system of equations.

Substitution Method

- The substitution method is an algebraic method for solving a system of equations.
- In the following summary of the method, we begin by solving one equation for y. However, either x or y may be chosen.
  1. Solve one equation for y to obtain $y =$ expression.
  2. Replace y in the other equation with the expression. This results in an equation in one variable x.
  3. Solve that equation.
  4. Substitute the solution for x in either original equation.
  5. Check the solution.

Special Cases

- Two special cases may arise when solving a system of two linear equations in two variables.
  1. The system of equations is **inconsistent.** The graphs of the equations are parallel, and the system has no solution.
  2. The equations of the system are **dependent.** The graphs of the equations coincide, and the solution set of the system contains the infinitely many solutions of either equation.

5.1 EXERCISES

Concepts and Skills

1. Explain how to determine whether an ordered pair is a solution of a system of equations in two variables.

2. The pair (1, 2) does not satisfy the first equation of the following system.

$$2x + 3y = 7$$
$$x - 2y = 1$$

Why can you conclude, even without testing the pair in the second equation, that (1, 2) is not a solution of the system?

In Exercises 3–6, determine whether each ordered pair is a solution of the given system of linear equations.

3. $(0, 1), (2, -3), (4, 0)$
$$2x + y = 1$$
$$3x - 2y = 12$$

4. $\left(4, -\frac{5}{3}\right), (-0.5, 0), \left(\frac{1}{2}, \frac{2}{3}\right)$
$$2x + 3y = 3$$
$$8x = 12y - 4$$

5. $(1, 1), \left(-2, \frac{8}{7}\right), (2, 0)$
$$2x + 7y = 4$$
$$14y + 4x = 8$$

6. $(4, -2), (-2, 1), (-6, 13)$
$$3x = 2(4 - y)$$
$$2y + 1 = -3(x - 3)$$

In Exercises 7–10, determine the value of a, b, or c so that the given ordered pair is a solution of the system of equations.

7. $ax - 2y = 5$, $x + by = 1$ $\quad (3, -1)$

8. $ax + y = 5$, $x + by = 7$ $\quad (1, 2)$

9. $ax + 3y = 9$, $2x - y = c$ $\quad (3, -1)$

10. $3x - y = c$, $x + by = 2$ $\quad (8, 2)$

11. Explain why two linear equations in two variables cannot have exactly two solutions.

12. Describe the graph of a system that has

(a) no solution.

(b) a unique solution.

(c) infinitely many solutions.

In Exercises 13–18, determine the number of solutions of the system without graphing or solving.

13. $y = 3x - 4$, $y = 3x + 5$

14. $y = -2x - 3$, $y + 2x = 4$

15. $y + 2x = 6$, $y - 2x = 8$

16. $y = \frac{2}{3}x + 5$, $y = -\frac{3}{2}x - 4$

17. $y + x = 5$, $y = -x + 5$

18. $y + 3x = 4$, $y = -3x + 4$

19. (a) How do the graphs of the following systems of equations differ?

(i) The equations are independent.

(ii) The equations are dependent.

(b) Is it possible for a system of two independent equations to have no solution? Explain.

20. (a) How do the graphs of the following systems of equations differ?

(i) The system is consistent.

(ii) The system is inconsistent.

(b) Is it possible for a consistent system to have infinitely many solutions? Explain.

In Exercises 21–24, estimate the solution of the system of equations.

21.

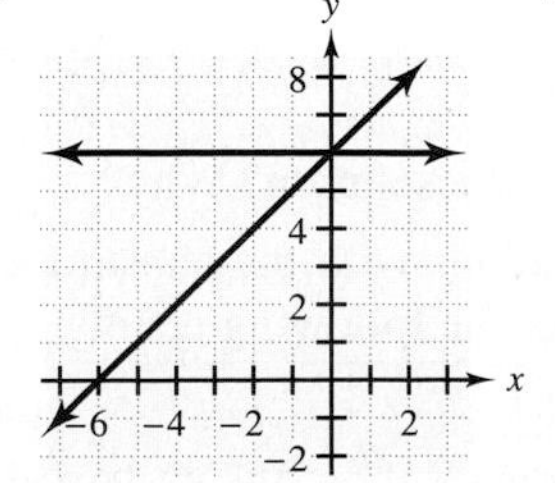

22.

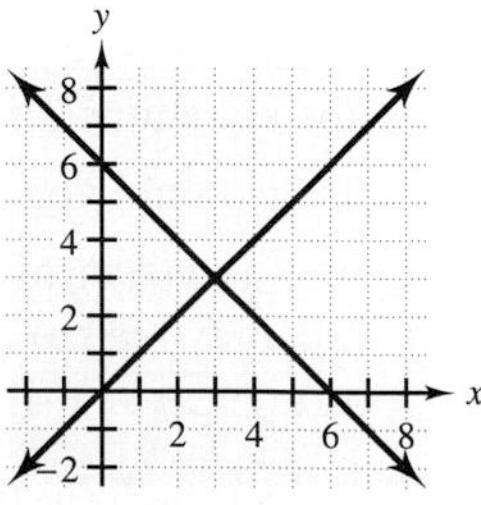

23.

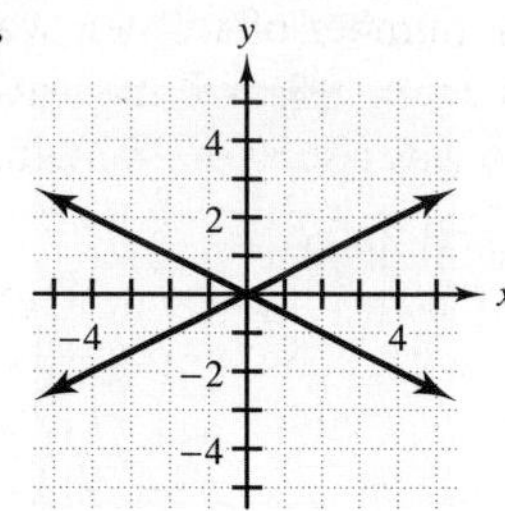

24.

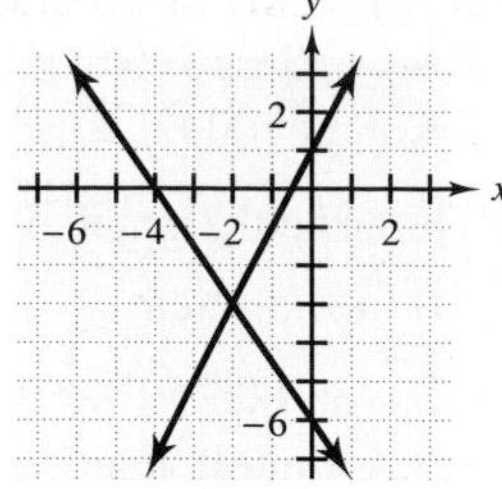

In Exercises 25–36, use the graphing method to estimate the solution of the given system of equations. Then use the substitution method to determine the solution algebraically.

25. $y = 2x - 2$
$y = -3x + 13$

26. $y = 4x - 13$
$3x + 2y = 29$

27. $y = 0.6x - 10$
$5y - 3x = 4$

28. $3x + y = 12$
$2y + 10 = -6x$

29. $2x + y = 8$
$y + 2 = 0$

30. $y = 3x - 2$
$y = 4$

31. $x - y + 6 = 0$
$2x - y - 5 = 0$

32. $3x + 2y + 8 = 0$
$x + 2y - 16 = 0$

33. $y = -2(4 - x)$
$y + 8 = 2x$

34. $2y - 3x = 10$
$-1.5x = 5 - y$

35. $\frac{1}{3}x + \frac{2}{9}y = 5$
$\frac{1}{9}x - \frac{2}{3}y = 15$

36. $\frac{1}{6}x + \frac{1}{3}y = 6$
$\frac{2}{3}x + \frac{5}{6}y = 12$

In Exercises 37–40, sketch the graphs of the equations of the given system and estimate the solution.

37. $x + 3 = 0$
$2x + 3y = 6$

38. $x - 2 = 0$
$x - y = -1$

39. $x = -4$
$y = 7$

40. $x = 3$
$y = -5$

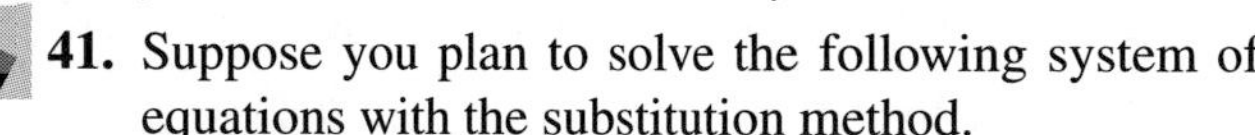

41. Suppose you plan to solve the following system of equations with the substitution method.

$$3x - 4y = 16$$
$$2x + y = 7$$

(a) Would it be correct to begin by solving one of the equations for x?

(b) How would you begin the solution? Why?

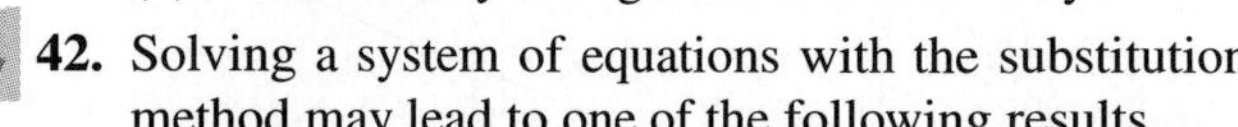

42. Solving a system of equations with the substitution method may lead to one of the following results.

(a) $-3 = -3$ (b) $13 = 7$

What does each result indicate about the system and its solution?

In Exercises 43–54, use the substitution method to solve the system of equations. Identify inconsistent systems and dependent equations.

43. $x - y - 5 = 0$
$2x - y + 17 = 0$

44. $2x + 3y + 15 = 0$
$x + 2y + 18 = 0$

45. $y + 3x - 4 = 0$
$12 = 9x + 3y$

46. $2x = y - 5$
$3y = 6x + 15$

47. $x = 3y + 9$
$3y - x = 6$

48. $x - 2y = 5$
$3x = 6y + 8$

49. $y = 2x - 5$
$4x - 2y - 10 = 0$

50. $x - 2y = 5$
$2x = 4y + 10$

51. $y = 2x - 7$
$3x + 2y = 0$

52. $x = 2y - 5$
$2x - 3y = -6$

53. $6x = 5y$
$12x + 7 = 10y$

54. $2x + 3y + 2 = 0$
$0 = 6x + 9y$

In Exercises 55–58, use your calculator to produce the graph of the given system of equations. From the graph, predict the number of solutions of the system. Then, use the substitution method to solve the system.

55. $y = -2x + 5$
$21x + 10y = 50$

56. $1.5x + y = 9$
$y = -1.52x + 9$

57. $x = y$
$11x - 10y = 80$

58. $y = \frac{7}{20}x$
$3y - x = -36$

In Exercises 59–62, determine the value of k so that the equations are dependent.

59. $y = \frac{2}{3}x + 1$
$kx + 3y = 3$

60. $3x + ky = 15$
$y = 5 - x$

61. $y = 3x + k$
$y + (2 - k)x = 5$

62. $kx - 4y = -2$
$4y - 3x = k - 1$

In Exercises 63–66, determine the value of k so that the system is inconsistent.

63. $kx + y = 2k$
$y = 2x + 5$

64. $2y - x = 6$
$x + ky = 0$

65. $kx + 3y = 12$
$y = \frac{4}{3}x + (k + 1)$

66. $x = ky - 6$
$3x + y = 9k$

In Exercises 67–70, translate the given information into a system of equations. Then solve the system to determine the two numbers.

67. The sum of two numbers is 16 and their difference is 4.

68. The difference of two numbers is one-third of the smaller number. The sum of the numbers is 6 less than twice the larger number.

69. One number exceeds a second number by 8. Three-fourths of the larger number is 3 more than the smaller number.

70. A number is 6 less than a larger number. Two-thirds of the larger number plus one-fourth of the smaller number is 15.

Geometric Models

71. Two angles are complementary. The difference between twice the larger angle and 5 times the smaller angle is 19°. What is the measure of each angle?

72. Two angles are complementary. The smaller angle is one-third of 10° less than the larger angle. What is the measure of each angle?

73. Two angles are supplementary. The larger angle is 24° less than twice the smaller angle. What is the measure of each angle?

74. Two angles are supplementary. Two-thirds of the larger angle is 30° more than the smaller angle. What is the measure of each angle?

75. The perimeter of a rectangle is 74 feet. The length is 7 feet more than twice the width. What are the dimensions of the rectangle?

76. The perimeter of a rectangle is 44 meters. The width is 4 meters less than the length. What are the dimensions of the rectangle?

77. The perimeter of a rectangle is 80 inches. The width is one-third of the length. What are the dimensions of the rectangle?

78. The perimeter of a rectangle is 32 meters. The length is 4 meters less than 3 times the width. What are the dimensions of the rectangle?

Real-Life Applications

79. Mills and Lopez were the only candidates in a county election. Of the 24,000 registered voters, only 30% turned out for the election, and they all voted for one of the candidates. If Lopez won the election by 200 votes, how many voted for Mills?

80. On an AAU swim team, the number of women was 1 more than the number of men. When 2 men and 1 woman left the team, there was a total of 26 swimmers. How many men were originally on the team?

81. A coed Little League team had 18 players. Even though it was raining, half the girls and one-third of the boys showed up for a game, but these 7 players went home when the game was canceled. How many girls were on the team?

82. Perfecto and CTV were cable TV companies competing for a local area franchise. Perfecto offered the same channels as CTV plus two other channels. One-tenth of Perfecto's channels were Pay-Per-View, and one-eighth of CTV's channels were Pay-Per-View. CTV offered one more Pay-Per-View channel than Perfecto. How many channels did CTV offer?

Modeling with Real Data

83. In Exercise 81 of Section 3.1 we saw that the equation $y = 1.16x + 9.27$ models the percentage y of people of age x who pay off all their credit card bills each month. (Source: *USA Today*.)

(a) Use the given equation to write an equation that models the percentage y of people of age x who do *not* pay off all their credit card bills each month. (Note that the two percentages should total 100%.)

(b) Solve the system consisting of the given equation and your equation in part (b).

(c) Interpret the solution.

84. The accompanying table shows that tuitions at both public and private four-year colleges increased dramatically in the period 1986–1995. (Source: The College Board.)

	Tuition	
Year	**Private**	**Public**
1986	\$ 5,641	\$1,157
1995	10,798	2,509

The following equations model the tuition y (in dollars), where x is the number of years since 1980.

Private: $y = 573x + 2203$

Public: $y = 150.2x + 255.7$

(a) Does the system of these model equations have a solution? If so, what does it represent?

(b) Suppose that you are interested only in years beyond 1995. Does the system have a solution for those values of x? How can you know without solving?

Group Project: Office Paper Recycling

The accompanying table shows the amount (in millions of tons) of office paper discarded and recycled for the years 1990 and 1995. (Source: Environmental Protection Agency.)

Year	Discarded	Recycled
1990	10.9	1.7
1995	12.5	5.1

85. Let y represent a quantity of office paper (in tons). Write equations in the form $y = mx + b$ to represent the amount of paper recycled and the amount of paper discarded, where $x = 0$ for 1990 and $x = 5$ for 1995.

86. Using a suitable window setting, produce the graphs of the two equations in Exercise 85. Moving from one graph to the other, trace along the graphs to estimate the year when 50% of all discarded office paper was recycled.

87. Use the substitution method to solve the system in Exercise 85. What does the solution represent?

88. Even though many state governments and many businesses require that office paper be recycled and that recycled paper be purchased, how realistic is the result in Exercise 87?

Challenge

In Exercises 89 and 90, solve the system of equations.

89. $x^2 + y = 9$
$2x^2 - y = 3$

90. $y = |x| - 7$
$x + 2y + 2 = 0$

In Exercises 91 and 92, determine the vertices of the polygon formed by the graphs of the given equations. Then identify the polygon.

91. $x = y + 12$
$y - x = 9$
$2x + y = 12$
$y + 6 = -2x$

92. $y - 20 = 0$
$3y - 4x = 24$
$4x = 24 - 3y$

In Exercises 93 and 94, solve the system of equations.

93. $y - x = 5$
$3y = x + 9$
$x + y + 1 = 0$

94. $3y = x - 9$
$y + 2x = 4$
$-22 = x + 2y$

95. Show whether there is a value of k such that the given system has exactly one solution.

$$x - y = k$$
$$3x - 3y = 2$$

96. Is it possible to determine a and c so that the given system of equations has

(a) no solution?

(b) infinitely many solutions?

$$2ax + ay = 1$$
$$y - x = c$$

5.2 THE ADDITION METHOD

Addition Method • Special Cases • Applications

Addition Method

The substitution method is not always the most convenient algebraic method for solving a system of equations. An alternative method is the **addition method,** sometimes called the **elimination method.**

The Addition Property of Equations allows us to add the same number to both sides of an equation: If $A = B$, then $A + C = B + C$. If $C = D$, then we can write the result as follows:

$$\begin{array}{r} A = B \\ \underline{C = D} \\ A + C = B + D \end{array}$$

We call this process *adding the equations.*

The goal of the addition method is to eliminate one of the variables in a system of equations by adding the two equations, with the result being a linear equation in one variable.

EXAMPLE 1

Solving a System of Equations with the Addition Method

Use the addition method to solve the following system of equations.

$$\begin{aligned} x + y &= 6 \\ x - y &= 2 \end{aligned}$$

Solution

$$\begin{array}{rl} x + y = 6 & \text{Add the two equations.} \\ \underline{x - y = 2} & \text{The coefficients of } y \text{ are opposites.} \\ 2x = 8 & \text{The variable } y \text{ is eliminated.} \\ x = 4 & \text{Solve for } x. \end{array}$$

Replace x with 4 in either of the original equations, and solve for y.

$$\begin{array}{rl} x + y = 6 & \text{First equation} \\ 4 + y = 6 & \text{Replace } x \text{ with 4.} \\ y = 2 & \end{array}$$

The solution of the system is (4, 2).

Adding two equations results in the elimination of a variable only when the variable appears in both equations and has coefficients that are opposites. If this is not the case, we can make them opposites by applying the Multiplication Property of Equations.

EXAMPLE 2

Using the Multiplication Property of Equations with the Addition Method

Solve the following system of equations.

$$\begin{aligned} 2x - y &= 9 \\ 3x - 4y &= -24 \end{aligned}$$

LEARNING TIP

Try to recognize the variable whose elimination requires the least amount of work. For instance, a variable with a coefficient of -1 or 1 is a good choice for a variable to eliminate.

Solution

If we choose to eliminate y, its coefficients must be opposites. Therefore, we multiply both sides of the first equation by -4 and then add the equations.

$$-4(2x - y) = -4(9) \rightarrow -8x + 4y = -36 \quad \text{Multiply first equation by } -4.$$
$$3x - 4y = -24 \rightarrow 3x - 4y = -24 \quad \text{Second equation}$$
$$-5x = -60 \quad \text{Add the equations.}$$
$$x = 12 \quad \text{Solve for } x.$$

Replace x with 12 in either equation to determine y.

$$2x - y = 9 \quad \text{First equation}$$
$$2(12) - y = 9 \quad \text{Replace } x \text{ with 12.}$$
$$24 - y = 9$$
$$-y = -15$$
$$y = 15$$

The solution is (12, 15).

Deciding which variable to eliminate is arbitrary. Usually, we choose the variable whose elimination requires the least amount of work.

Writing each equation in the same form is a good idea when using the addition method because then the like terms are lined up in columns. It is common to write both equations in the standard form $Ax + By = C$.

The Multiplication Property of Equations requires us to multiply both sides of an equation by the same number. However, we can multiply both sides of one equation by one number and both sides of the other equation by a *different* number. It is not necessary to multiply both equations by the same number.

EXAMPLE 3

Using the Multiplication Property of Equations with the Addition Method

Solve the following system.

$$3x = 33 - 5y$$
$$3y = 4x - 15$$

Solution

$$3x + 5y = 33 \quad \text{Write each equation in standard form.}$$
$$-4x + 3y = -15$$

To eliminate x, multiply the first equation by 4 and the second equation by 3. Then add the equations.

$$4(3x + 5y) = 4(33) \rightarrow 12x + 20y = 132$$
$$3(-4x + 3y) = 3(-15) \rightarrow -12x + 9y = -45$$
$$29y = 87$$
$$y = 3$$

Replace y with 3 in either equation.

$$3x = 33 - 5y \quad \text{First equation}$$
$$3x = 33 - 5(3) \quad \text{Replace } y \text{ with 3.}$$
$$3x = 18$$
$$x = 6$$

The solution is (6, 3).

When the equations involve fractions, we can multiply by the LCD to clear them.

EXAMPLE 4

Clearing Fractions

Solve the following system.

$$\frac{3}{2}x - \frac{3}{4}y = \frac{15}{4}$$
$$\frac{4}{3}x - \frac{5}{3}y = 3$$

Solution

Clear the fractions by multiplying each term by the LCD. For the first equation, the LCD is 4; for the second equation, the LCD is 3.

$$4 \cdot \frac{3}{2}x - 4 \cdot \frac{3}{4}y = 4 \cdot \frac{15}{4} \quad \rightarrow \quad 6x - 3y = 15$$
$$3 \cdot \frac{4}{3}x - 3 \cdot \frac{5}{3}y = 3 \cdot 3 \quad \rightarrow \quad 4x - 5y = 9$$

Now use the Multiplication Property of Equations again, this time to adjust the coefficients. To eliminate x, multiply the first equation by 2 and the second equation by -3. Then add the equations.

$$2(6x - 3y) = 2(15) \quad \rightarrow \quad 12x - 6y = 30$$
$$-3(4x - 5y) = -3(9) \quad \rightarrow \quad -12x + 15y = -27$$
$$9y = 3$$
$$y = \frac{3}{9} = \frac{1}{3}$$

There is an alternative to replacing y with $\frac{1}{3}$ in order to solve for x. We can return to the system in which the fractions have been cleared, eliminate y, and solve for x.

$$30x - 15y = 75 \quad \text{Multiply the first equation by 5.}$$
$$-12x + 15y = -27 \quad \text{Multiply the second equation by } -3.$$
$$18x = 48$$
$$x = \frac{48}{18} = \frac{8}{3}$$

The solution is $\left(\frac{8}{3}, \frac{1}{3}\right)$.

The following is a summary of the addition method for solving a system of equations.

The Addition Method

1. Write both equations in the form $Ax + By = C$.
2. If necessary, multiply one or both of the equations by appropriate numbers so that the coefficients of one of the variables are opposites.
3. Add the equations to eliminate a variable.
4. Solve the resulting equation.
5. Substitute that value in either of the original equations and solve for the other variable.

Special Cases

In Section 5.1 we saw how the substitution method detected the two special cases: lines that are parallel and lines that coincide. When both variables were eliminated, the result was either a false equation or an identity.

Example 5 shows that special cases can be identified algebraically with the addition method in exactly the same way that they are identified with the substitution method.

EXAMPLE 5

Special Cases

Use the addition method to solve the following systems of equations.

(a) $6x - 9y = -34$
$2x - 3y = -12$

(b) $x - 0.2y = -0.4$
$5x - y = -2$

Solution

(a)
$$\begin{aligned} 6x - 9y &= -34 && \text{First equation} \\ -6x + 9y &= 36 && \text{Multiply the second equation by } -3. \\ \hline 0 &= 2 && \text{False} \end{aligned}$$

The system is inconsistent and has no solution.

(b)
$$\begin{aligned} -5x + y &= 2 && \text{Multiply the first equation by } -5. \\ 5x - y &= -2 && \text{Second equation} \\ \hline 0 &= 0 && \text{True} \end{aligned}$$

The equations are dependent, and the system has infinitely many solutions.

Think About It

A system of linear equations has infinitely many solutions when an algebraic method results in an equation such as $0 = 0$. However, not *every* pair (x, y) is a solution. Why?

Applications

When solving applied problems involving two variables, the initial work is always the same. We assign variables and write a system of equations. The only decision that remains concerns our choice of method for solving the system.

EXAMPLE 6

A Collection of Coins

A collection jar for the Kidney Foundation contained 18 one-dollar bills and a total of 77 coins consisting only of quarters and dimes. If the value of all the money was $30.95, how many quarters and dimes were there in the jar?

Solution

Let q = the number of quarters and
d = the number of dimes.

The total value of the coins was $30.95 − $18.00 = $12.95.

$q + d = 77$ There were 77 coins in all.

$25q + 10d = 1295$ We choose cents as the monetary unit.

$-10q - 10d = -770$ Multiply the first equation by −10.

$25q + 10d = 1295$ Second equation

$15q = 525$ Add the equations.

$q = 35$ Solve for q.

$q + d = 77$ First equation

$35 + d = 77$ Replace q with 35.

$d = 42$

The jar contained 35 quarters and 42 dimes.

We now have three methods for solving a system of equations. The graphing method has the advantage of being visually helpful, and it is a quick method for obtaining an estimate of the solution.

The substitution method and the addition method are both algebraic methods. Choose the one that is most convenient for the system. Either method will work for all linear systems.

The ideal solving process is a combination of graphing and algebraic methods. This will help you to remain aware of the conceptual nature of the problem and to avoid falling into a mere symbol manipulation routine.

5.2 QUICK REFERENCE

Addition Method

- The addition method is another algebraic method for solving a system of two equations. The following is a summary of the method:
  1. Write both equations in the form $Ax + By = C$.
  2. If necessary, multiply one or both of the equations by appropriate numbers so that the coefficients of one of the variables are opposites.
  3. Add the equations to eliminate a variable.
  4. Solve the resulting equation.

5. Substitute that value in either of the original equations and solve for the other variable.

- It is usually best to clear fractions before using the addition method.

Special Cases

- After you add the equations to eliminate one variable, the resulting equation may be false, or it may be an identity.
  1. If it is false, then the system is inconsistent and has no solution.
  2. If it is an identity, then the two equations of the system are dependent, and the system has infinitely many solutions.

5.2 EXERCISES

Concepts and Skills

1. Suppose you wish to solve the following system.

$$\begin{aligned} 3x - y &= 8 \\ x + 4y &= 7 \end{aligned}$$

(a) Describe the operation you would perform to eliminate x.

(b) Describe the operation you would perform to eliminate y.

2. Suppose you wish to use the addition method to solve the following system. Describe the easiest way to eliminate one of the variables.

$$\begin{aligned} 2x + y &= 3 \\ 3x + 4y &= -2 \end{aligned}$$

In Exercises 3–10, use the addition method to solve each system.

3. $x + y = 10$, $x - y = 2$

4. $2x - y = 7$, $2x + y = 13$

5. $x + 2y = 4$, $-x - y = 3$

6. $3x - 3y = 14$, $3x + 3y = -2$

7. $3x + 2y = 4$, $4y = 3x + 26$

8. $x - 3y = 9$, $5y = x - 17$

9. $4x = 3y + 6$, $12 = 5x + 3y$

10. $2x = y + 9$, $3y = 2x - 19$

11. Suppose you solve a system of equations and the resulting equation is $0 = 0$. Does this mean that the solution of the system is $(0, 0)$? Explain.

12. Suppose that you solve a system of equations in x and y, and you eliminate y to obtain $x = 4$. Is 4 the solution of the system? Explain.

In Exercises 13–20, use the addition method to solve each system.

13. $2x + 3y + 1 = 0$, $5x + 3y = 29$

14. $5x - 2y = 0$, $4y = 3x + 14$

15. $3x - 2y = 21$, $5x + 4y = 13$

16. $2x = 3y - 14$, $4x + 5y = 16$

17. $2x + 3y = 3$, $4x = 6y - 2$

18. $3x + 4y + 1 = 0$, $6x = 8y - 14$

19. $6x = 5y + 10$, $3x + 2y = 23$

20. $4x + 3y = 2$, $5x + 6y = 7$

21. Suppose the solution of a system of two linear equations is $(3, 2)$. Try to visualize the general appearance of the graph of the system. Now explain why there are many other systems whose solutions are also $(3, 2)$.

22. Explain why the addition method cannot be used to solve the following system of equations. What is the solution of the system?

$$\begin{aligned} x &= -3 \\ y &= 7 \end{aligned}$$

In Exercises 23–30, use the addition method to solve each system.

23. $6x + 11y = 17$, $4x - 5y = -1$

24. $5x - 3y = 2$, $3x - 2y = -1$

25. $-6x + 5y = 10$, $5x + 4y = 8$

26. $-3x + 7y = 10$, $5x - 2y = -6$

27. $2x - 7y + 11 = 0$
$7x + 4y + 10 = 0$

28. $3x + 2y + 7 = 0$
$2x - 3y = 4$

29. $5y - 2x = 5$
$5x + 2y = 2$

30. $4x = 3y + 8$
$3x = 4y + 6$

31. When you use the addition method, how can you tell when the equations of a system are dependent?

32. When you use the addition method, how can you tell when a system is inconsistent?

In Exercises 33–44, use the addition method to solve the system.

33. $2x + 3y + 18 = 0$
$5y = 6x - 2$

34. $y = 3x - 5$
$2x + 3y = 6$

35. $3x - 2y = 8$
$6x = 4y + 17$

36. $4x + 12y = 32$
$7 - x = 3y$

37. $3x - y = 7$
$2y = 6x - 14$

38. $4x + 12y = 36$
$9 - x = 3y$

39. $2 + 7y = 5x$
$15x = 21y - 6$

40. $5 = x - 4y$
$12y - 3x = -8$

41. $6y + 7x = 16$
$3x = 2y + 16$

42. $5y = 2(3x - 2)$
$8 + 3y = -2x$

43. $2y = 3x - 5$
$-12x = -20 - 8y$

44. $8x + 9y = -3$
$6 + 18y = -16x$

In Exercises 45–56, use the addition method to solve each system.

45. $4(x - 2) - 5(y - 3) = 14$
$3(x + 4) - 2(y + 5) = 13$

46. $3(x - 5) + (y + 7) = 7$
$2(x + 3) - 5(y + 1) = 11$

47. $\frac{1}{2}x + \frac{3}{10}y = \frac{1}{2}$
$-\frac{5}{3}x - y = \frac{4}{3}$

48. $-0.4x + 1.2y = 1$
$-0.6x + 1.8y = -1$

49. $x - 1.5y = 0.25$
$0.9y - 0.6x = -0.15$

50. $\frac{3}{2}x - y = \frac{3}{4}$
$2x - \frac{4}{3}y = 1$

51. $0.1x - 0.25y = 1.05$
$0.625x + 0.4y = 0.675$

52. $0.4x - 0.3y = -0.1$
$0.5x + 0.2y = 1.6$

53. $\frac{x+2}{9} + \frac{y+2}{6} = 1$
$\frac{x-1}{4} - \frac{y+1}{3} = -1$

54. $\frac{x-2}{12} - \frac{y-2}{4} = 1$
$\frac{x+3}{2} + \frac{y-4}{4} = 1$

55. $\frac{x-y}{6} + \frac{x+y}{3} = 1$
$y = 6 - 3x$

56. $\frac{x+y}{2} = \frac{1}{3} + \frac{x-y}{2}$
$\frac{x+2}{2} - \frac{y+4}{2} = 4$

In Exercises 57–60, determine values of a, b, and c so that the system has the given solution.

57. $x + by = -2a$
$ax - 4y = b$
$(-2, -2)$

58. $5x + 2ay = c$
$ax - 3y = -6c$
$(2, -2)$

59. $ax - by = -17$
$bx + ay = -1$
$(-2, 5)$

60. $ax - y = c + 6$
$x - (3a + 4)y = c$
$(7, 1)$

In Exercises 61–64, determine k so that the system of equations will be inconsistent.

61. $x + 2y = 5$
$2x - ky = 7$

62. $3x - 2y = 10$
$kx + 6y = 5$

63. $y = -2x + 3$
$kx - 3y = 4$

64. $x = 2y - 5$
$4x + ky = 2$

In Exercises 65–68, determine a, b, or c so that the equations will be dependent.

65. $x + 2y = c$
$3x + 6y = 12$

66. $2x + by = 5$
$-6x - 12y = -15$

67. $ax + by = 3$
$x - y = 1$

68. $2x + 3y = -4$
$ax - 6y = c$

In Exercises 69–76, solve the given system and obtain a solution in terms of a, b, and c.

69. $x + y = b$
$x - y = c$

70. $2x + y = 3$
$x + y = c$

71. $ax + y = 3$
$x - y = c$

72. $x + by = c$
$x - by = c$

73. $2ax + y = 3$
$-ax + 3y = 2$

74. $ax + by = c$
$bx + ay = c$

75. $ax + by = a$
$x - y = 1$

76. $2x + y = 4 + c$
$x - 2 = y$

77. If 8 hamburgers and 4 milk shakes cost \$20 and 3 hamburgers and 2 milk shakes cost \$8.10, then what are the prices of a hamburger and a milk shake?

78. If 5 root beers and 7 orders of french fries cost \$7.25 and 4 root beers and 5 orders of french fries cost \$5.50, then what are the prices of one root beer and one order of french fries?

79. A parking meter contains \$3.85 in nickels and dimes. If there is a total of 50 coins, how many dimes are in the meter?

80. A coin jar contains 34 coins. If the coins are all quarters and dimes and their total value is \$6.55, how many quarters are in the jar?

81. There are 64 coins in a cash register and their total value is \$4.25. If the coins are all nickels and dimes, how many of each coin are there?

82. A vending machine attendant fills a coin dispenser with \$6.85 in nickels and quarters. If there is a total of 61 coins in the dispenser, how many of each coin are there?

Real-Life Applications

83. A subcontractor employs carpenters at \$22 per hour and unskilled labor at \$8 per hour. If the hourly cost for the company's total of 16 employees is \$212, how many carpenters are employed?

84. Tees Unlimited sold 35 more tee-shirts than sweatshirts last Monday. Tee-shirts sell for \$11 each, and sweatshirts sell for \$20 each. In all, \$1718 worth of shirts were sold that day. How many of each type of shirt were sold?

85. A person working at home obtains contracts for word processing projects from small businesses. She charges \$2 per text page, but she adds \$3 to that charge for any pages that contain graphics. If her total charge for a 60-page report is \$147, what percent of the pages have graphics?

86. A farmer received 90 cents per bushel more for his wheat than he did for his corn. If he sold 6000 bushels of wheat and 4000 bushels of corn for a total of \$30,400, what was the price per bushel of each?

Modeling with Real Data

87. As a result of nuclear arms agreements and the breakup of the Soviet Union, the number of nuclear warheads (mounted on missiles or carried in bombers) decreased from 1990 to 1996. The accompanying table shows data for five countries. (Source: U.S. Department of Defense.)

	Number of Nuclear Warheads	
Country	**1990**	**1996**
United States	14,000	8,500
Russia	7,500	6,700
France	500	500
China	450	450
United Kingdom	200	200

In the following questions, let y represent the number of nuclear warheads, and let x represent the number of years since 1990.

(a) Write linear equations that model the number of warheads in France, China, and the United Kingdom. If you were to form a system with any two of these equations, why would the system have no solution?

(b) The following equations model the data for the United States and Russia.

United States: $y = -916.7x + 14,000$

Russia: $y = -133.3x + 7500$

Solve the system of these model equations.

(c) What does your solution in part (b) represent?

88. For the fall of 1995, medical schools were able to accept approximately 16,000 students from the

45,591 applicants. (Source: Association of American Medical Colleges.)

Year	Number of Applicants: Female	Number of Applicants: Male
1989	10,546	20,000
1995	19,779	25,812

The following equations model the number y of applicants (in thousands), where x is the number of years since 1985.

Female: $y = 1.54x + 4.39$

Male: $y = 0.97x + 16.12$

(a) In the model equations, what do the coefficients of x indicate?

(b) Solve the system of the model equations. What does the solution indicate?

Group Project: E-mail

The accompanying bar graph shows that the percentage of office communications conducted by E-mail is increasing rapidly, while the percentage conducted by hard copy (paper) is decreasing. (Source: Cognitive Communication Survey.)

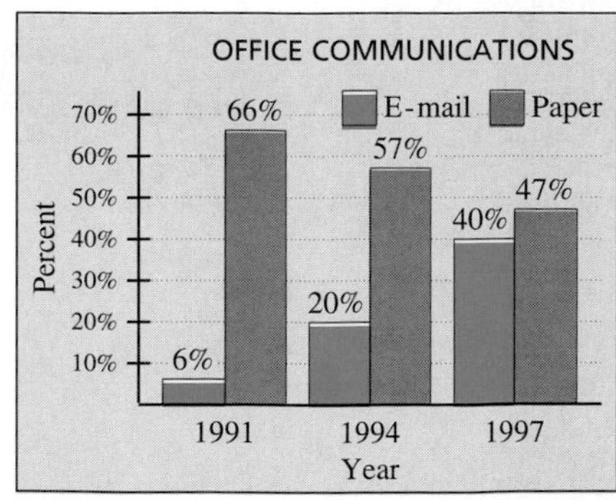

(Source: Cognitive Communication Survey.)

If we let y represent the percentage and x represent the number of years since 1990, the following equations model the use of E-mail and paper.

E-mail: $17x - 3y = 2$

Paper: $19x + 6y - 416 = 0$

89. Solve each of the model equations for y, and use the graphing method to estimate the solution of the system of those two equations.

90. Solve the system algebraically, and compare the result with your estimate in Exercise 89.

91. In the given data, the percentages for E-mail and paper do not total 100%. Why is this so?

92. A commercial claims that you cannot shake a customer's hand with a fax. What is an advantage and a disadvantage of E-mail over other forms of communication?

Challenge

In Exercises 93–96, replace $\frac{1}{x}$ with u and $\frac{1}{y}$ with v. Solve the resulting system for u and v. Then use these values to determine x and y.

93. $\frac{3}{x} + \frac{2}{y} = 2$
$\frac{1}{x} - \frac{6}{y} = -1$

94. $\frac{3}{x} + \frac{2}{y} = 0$
$\frac{9}{x} - \frac{4}{y} = 5$

95. $\frac{5}{x} + \frac{10}{y} = 3$
$\frac{1}{x} + \frac{3}{y} = 1$

96. $\frac{1}{x} + \frac{1}{y} = -1$
$\frac{5}{x} - \frac{2}{y} = -1$

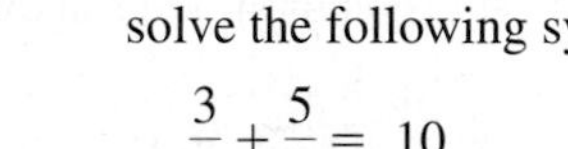

97. Using the technique described in Exercises 93–96, solve the following system. Explain the result.

$$\frac{3}{x} + \frac{5}{y} = 10$$
$$\frac{2}{x} + \frac{1}{y} = 2$$

In Exercises 98–101, determine a, b, or c so that the system will be consistent.

98. $x - 2y = 3$
$ax - 2y = 4$

99. $-3x + 2y = 5$
$6x + by = 7$

100. $3x - 2y = c$
$-6x + 4y = -10$

101. $4x - 8y = 20$
$x - 2y = c$

In Exercises 102–105, determine a, b, or c so that the equations will be independent.

102. $4x - 7y = -1$
$by - 8x = 2$

103. $ax + 3y = 2$
$2x + 6y = 4$

104. $x - 2y = c$
$6y - 3x = 3$

105. $x + y = 5$
$2x + 2y = c$

5.3 SYSTEMS OF EQUATIONS IN THREE VARIABLES

Linear Equations in Three Variables • Graphical Interpretation • Algebraic Methods • Special Cases • Real-Life Applications

Linear Equations in Three Variables

Although we have concentrated on linear equations in two variables, linear equations can be defined for more than two variables.

Definition of a Linear Equation in Three Variables

A **linear equation in three variables** is any equation that can be written in the *standard form* $Ax + By + Cz = D$, where A, B, C, and D are real numbers and A, B, and C are not all zero.

Similar definitions may be stated for linear equations in more than three variables.

The following are some examples of linear equations in three variables.

Equation	*Standard Form*
$3x - y + 2z = 5$	$3x + (-1)y + 2z = 5$
$y = 2x + 1$	$-2x + 1y + 0z = 1$
$x = 1$	$1x + 0y + 0z = 1$

A solution of a linear equation in *two* variables is an ordered *pair* (x, y) that satisfies the equation. We can extend this definition for linear equations in three variables.

Definition of Solution

A **solution** of a linear equation in three variables is an **ordered triple** (x, y, z) of numbers that satisfies the equation.

EXAMPLE 1

Verifying a Solution

Verify that (5, 3, 6) is a solution of the equation $x + y - z = 2$.

Solution

$x + y - z = 2$

$5 + 3 - 6 = 2$ Replace x with 5, y with 3, and z with 6.

$2 = 2$ True

The ordered triple (5, 3, 6) is a solution.

The following is an example of a system of three linear equations in three variables.

$$\begin{aligned} 3x - y + 2z &= 0 \\ 2x + 3y + 8z &= 8 \\ -x + y + 6z &= 0 \end{aligned}$$

A solution of such a system is an ordered triple (x, y, z) that satisfies each equation of the system.

In previous sections we developed graphic and algebraic methods for solving systems of two linear equations in two variables. We can extend those methods to solve systems of three linear equations in three variables.

Graphical Interpretation

We know that the graph of a linear equation in *two* variables is a *line* drawn in a *two*-dimensional (rectangular) coordinate system.

The graph of a linear equation in *three* variables is a *plane* drawn in a *three*-dimensional coordinate system, also called a **coordinate space.** (See Fig. 5.8.)

Think About It

How would you define a linear equation in four variables? Can you imagine the graph of such an equation?

Note that a coordinate space includes three axes that are all perpendicular to each other. We must visualize the x-axis as coming out from the page.

Figure 5.9 shows the point $P(4, 2, 3)$ plotted in a coordinate space.

Figure 5.8 **Figure 5.9**

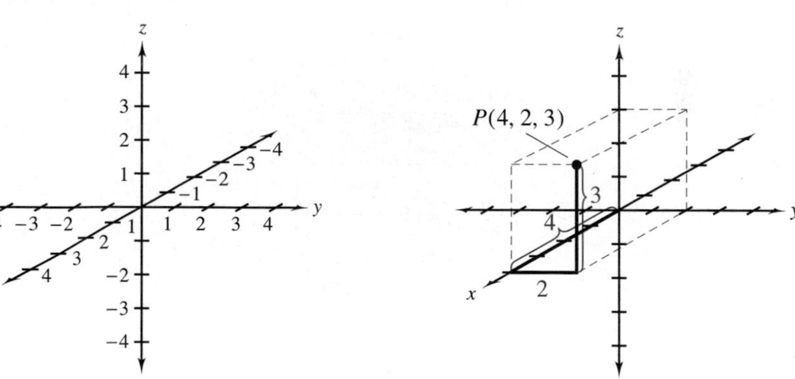

Figure 5.10

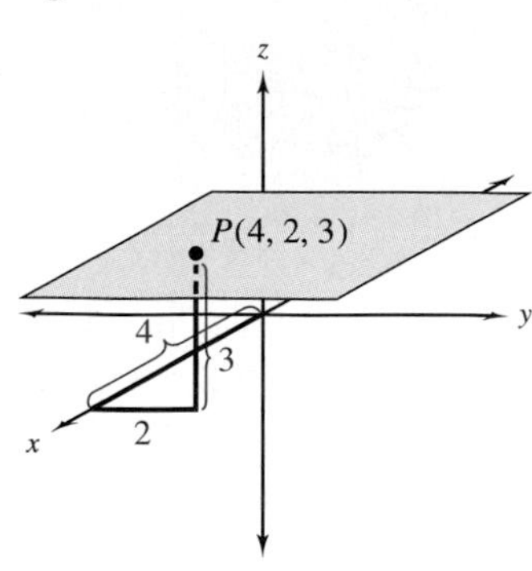

Figure 5.10 shows one of many planes that contains point P. Planes, like lines, extend without end. However, a plane in a coordinate space is conventionally drawn as a bounded figure so that its location can be visualized.

Figure 5.11

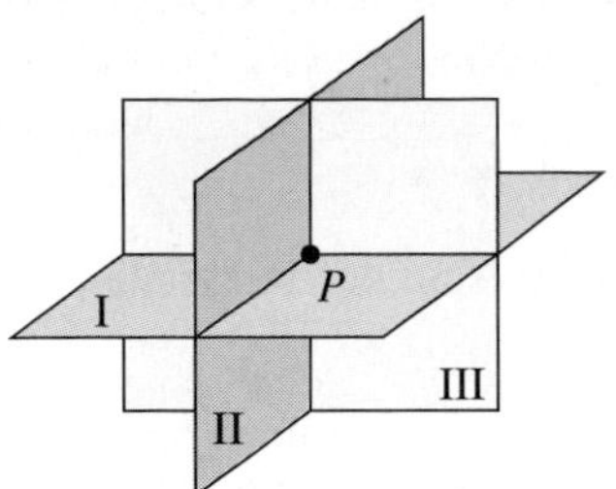

For a system of three linear equations in three variables, the graph of each equation of the system is a plane. The solution set of such a system is represented by the intersection of the graphs (planes) of the three equations. Any point in common to all three planes represents a solution of the system. There are three possible outcomes.

First, the solution could be unique. This means that the three planes intersect at one point P. (See Fig. 5.11.)

Second, there could be no solution. This can occur when at least two of the planes are parallel (see Fig. 5.12) or when one plane is parallel to the line L of intersection of the other two planes (see Fig. 5.13).

Figure 5.12

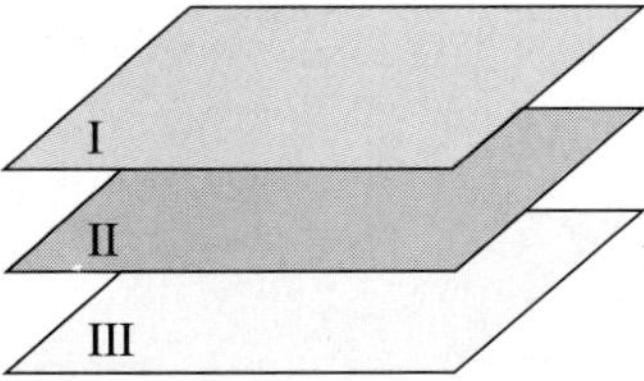

Figure 5.13

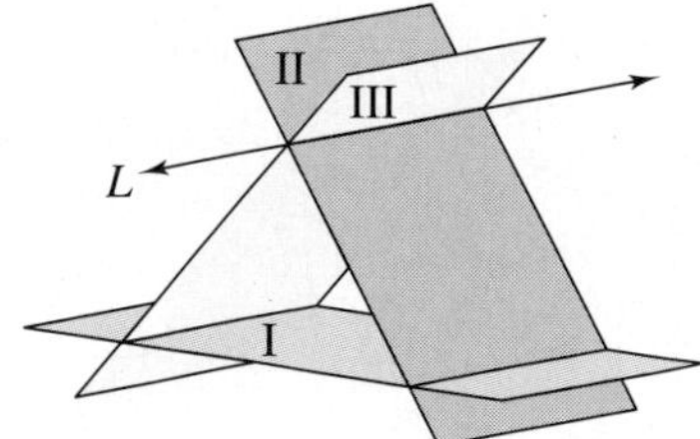

Third, there could be infinitely many solutions. This can occur when the planes coincide (see Fig. 5.14) or when they intersect in a common line L (see Fig. 5.15).

Figure 5.14

Figure 5.15

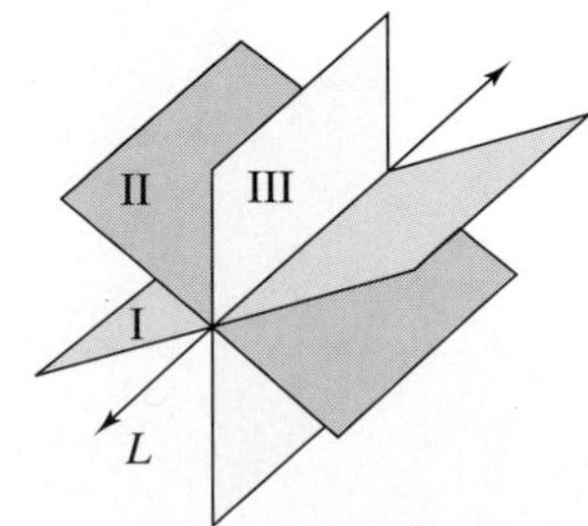

Because of the difficulties inherent in drawing pictures in a coordinate space, we will not use graphing methods for estimating solutions of systems of three or more linear equations in three variables. Nevertheless, your ability to visualize the various ways in which planes can be oriented in space will provide you with a geometric interpretation of the system and its solution.

Algebraic Methods

The substitution method we used for solving systems of two linear equations in two variables can also be used to solve systems of three linear equations in three variables.

Although the substitution method can be used for any system, the addition method is more generally applicable. We will limit our discussion to that one algebraic method.

EXAMPLE 2

Solving a System of Three Equations with the Addition Method

Use the addition method to solve the following system.

$$\begin{aligned} 3x - y + 2z &= 0 \quad (1) \\ 2x + 3y + 8z &= 8 \quad (2) \\ x + y + 6z &= 0 \quad (3) \end{aligned}$$

Solution

LEARNING TIP

In some systems it might be tempting to eliminate a variable from one pair of equations and to eliminate another variable from a second pair of equations. The addition method will not work unless you eliminate the same variable from each pair of equations.

The plan is to eliminate a variable from one pair of equations. Then we eliminate the same variable from another pair of equations. In the following we eliminate y from equations (1) and (3) and from equations (2) and (3).

$$\begin{aligned} 3x - y + 2z &= 0 \quad (1) \\ x + y + 6z &= 0 \quad (3) \end{aligned} \qquad \begin{aligned} 2x + 3y + 8z &= 8 \quad (2) \\ x + y + 6z &= 0 \quad (3) \end{aligned}$$

In the first pair, eliminate y by adding the equations. In the second pair, multiply both sides of equation (3) by -3 and then add the equations.

$$\begin{aligned} 3x - y + 2z &= 0 \\ x + y + 6z &= 0 \\ \hline 4x \quad + 8z &= 0 \end{aligned} \qquad \begin{aligned} 2x + 3y + 8z &= 8 \\ -3x - 3y - 18z &= 0 \\ \hline -x \quad - 10z &= 8 \end{aligned}$$

The result is a system of two equations in two variables.

$$\begin{aligned} 4x + 8z &= 0 \quad (4) \\ -x - 10z &= 8 \quad (5) \end{aligned}$$

Now we apply the addition method again to solve this system.

$$\begin{aligned} 4x + 8z &= 0 && (4) \\ -4x - 40z &= 32 && \text{Multiply equation (5) by 4.} \\ \hline -32z &= 32 && \text{Add the equations.} \\ z &= -1 \end{aligned}$$

Now substitute -1 for z in equation (4) to determine x.

$$\begin{aligned} 4x + 8z &= 0 && \text{(4)} \\ 4x + 8(-1) &= 0 && \text{Replace } z \text{ with } -1. \\ 4x - 8 &= 0 \\ 4x &= 8 \\ x &= 2 \end{aligned}$$

Finally, to determine the value of y, substitute 2 for x and -1 for z in any of the three original equations.

$$\begin{aligned} x + y + 6z &= 0 && \text{(3)} \\ 2 + y + 6(-1) &= 0 && \text{Replace } x \text{ with 2 and } z \text{ with } -1. \\ 2 + y - 6 &= 0 \\ y - 4 &= 0 \\ y &= 4 \end{aligned}$$

The solution is the ordered triple $(2, 4, -1)$.

The solution in Example 2 can be interpreted graphically. The graphs of the three equations of the system are planes. These planes intersect at a point, as shown in Fig. 5.11. In this case, the coordinates of the point of intersection P are $(2, 4, -1)$.

Our work is simplified when one or more variables are missing from the equations of a system.

EXAMPLE 3

Systems with Missing Variables

Solve the following system of equations.

$$\begin{aligned} 3x + 4y \quad\quad &= 4 && \text{(1)} \\ -3x \quad\quad + 2z &= -3 && \text{(2)} \\ y + z &= 1 && \text{(3)} \end{aligned}$$

Solution

Note that x is missing from equation (3). We can easily eliminate x in equations (1) and (2).

$$\begin{aligned} 3x + 4y \quad\quad &= 4 && \text{(1)} \\ -3x \quad\quad + 2z &= -3 && \text{(2)} \\ \hline 4y + 2z &= 1 && \text{(4)} \end{aligned}$$

Now we use equations (4) and (3) to form a system of equations in two variables.

$$\begin{aligned} 4y + 2z &= 1 && \text{(4)} \\ y + z &= 1 && \text{(3)} \end{aligned}$$

$$\begin{aligned} 4y + 2z &= 1 \quad && (4) \\ -2y - 2z &= -2 && \text{Multiply equation (3) by } -2. \\ \hline 2y \qquad &= -1 \\ y &= -\frac{1}{2} \end{aligned}$$

$$-\frac{1}{2} + z = 1 \qquad \text{Substitute } -\tfrac{1}{2} \text{ for } y \text{ in equation (3).}$$

$$z = \frac{3}{2} \qquad \text{Solve for } z.$$

$$3x + 4\left(-\frac{1}{2}\right) = 4 \qquad \text{Substitute } -\tfrac{1}{2} \text{ for } y \text{ in equation (1).}$$

$$x = 2 \qquad \text{Solve for } x.$$

The solution is the ordered triple $\left(2, -\frac{1}{2}, \frac{3}{2}\right)$.

Special Cases

The addition method will detect the special cases in which the system has no solution or has infinitely many solutions.

EXAMPLE 4

An Inconsistent System of Three Linear Equations

Solve the following system.

$$\begin{aligned} 2x - y - 5z &= 3 \quad && (1) \\ 5x - y + 14z &= -11 && (2) \\ 7x - 2y + 9z &= -5 && (3) \end{aligned}$$

Solution

We eliminate y in two pairs of equations.

$$\begin{aligned} 2x - y - 5z &= 3 \quad && (1) \\ 7x - 2y + 9z &= -5 && (3) \end{aligned} \qquad\qquad \begin{aligned} 5x - y + 14z &= -11 \quad && (2) \\ 7x - 2y + 9z &= -5 && (3) \end{aligned}$$

Multiply both sides of equations (1) and (2) by -2, and then add each pair of equations.

$$\begin{aligned} -4x + 2y + 10z &= -6 \\ 7x - 2y + 9z &= -5 \\ \hline 3x \qquad + 19z &= -11 \end{aligned} \qquad\qquad \begin{aligned} -10x + 2y - 28z &= 22 \\ 7x - 2y + 9z &= -5 \\ \hline -3x \qquad - 19z &= 17 \end{aligned}$$

When we add the two resulting equations, both variables are eliminated.

$$\begin{aligned} 3x + 19z &= -11 \\ -3x - 19z &= 17 \\ \hline 0 &= 6 \quad \text{False} \end{aligned}$$

The resulting false statement indicates that there is no solution of the system of equations. The solution set is Ø, and the system is inconsistent.

Determining the orientation of the three planes in Example 4 is a matter that we must leave for more advanced courses. As a point of interest, the planes representing equations (2) and (3) intersect in a line. The plane representing equation (1) is parallel to that line. This situation was illustrated in Fig. 5.13.

EXAMPLE 5

A System of Three Dependent Equations

Solve the following system of equations.

$$\begin{aligned} 6x - 4y + 2z &= 12 \quad (1) \\ 18x - 12y + 6z &= 36 \quad (2) \\ -3x + 2y - z &= -6 \quad (3) \end{aligned}$$

Solution

Notice that multiplying equation (1) by 3 results in equation (2). Multiplying equation (3) by -6 also results in equation (2). In other words, the equations are all equivalent.

Graphically, it can be shown that the three planes coincide, as was illustrated in Fig. 5.14. Therefore, the solution set contains all ordered triples that satisfy any one of the equations. We choose equation (3). The solution set is

$$\{(x, y, z) \mid -3x + 2y - z = -6\}$$

Real-Life Applications

An applied problem involving three unknown quantities may be solved most conveniently with a system of three linear equations in three variables.

EXAMPLE 6

Three Different Exhibit Fees

Each fall the Riverfest Craft Fair rents three types of spaces to exhibitors. The first year there were 40 spaces for food vendors, 50 spaces to sell crafts, and 20 spaces to demonstrate crafts. The total receipts were $3700.

The next year there were 30 food vendors, 60 spaces for selling crafts, and 15 demonstration booths, and the total receipts were $3900. The third year, with 25 food vendors, 80 spaces for selling crafts, and 30 demonstration booths, the total receipts were $4925.

Assuming that the fees did not change over these 3 years, what was the fee for each type of space?

Solution

Let $x =$ the fee for a food booth,

$y =$ the fee for a craft booth, and

$z =$ the fee for a demonstration booth.

The equations are as follows:

$$40x + 50y + 20z = 3700 \quad (1)$$
$$30x + 60y + 15z = 3900 \quad (2)$$
$$25x + 80y + 30z = 4925 \quad (3)$$

Because z has the smallest coefficients, we choose to eliminate z. We select equations (1) and (3) for one pair and equations (2) and (3) for the other pair.

$$40x + 50y + 20z = 3700 \quad (1) \qquad 30x + 60y + 15z = 3900 \quad (2)$$
$$25x + 80y + 30z = 4925 \quad (3) \qquad 25x + 80y + 30z = 4925 \quad (3)$$

In the first pair, multiply equation (1) by -3 and equation (3) by 2. In the second pair multiply equation (2) by -2.

$$\begin{array}{rcl} -120x - 150y - 60z &=& -11100 \\ 50x + 160y + 60z &=& 9850 \\ \hline -70x + 10y &=& -1250 \end{array} \qquad \begin{array}{rcl} -60x - 120y - 30z &=& -7800 \\ 25x + 80y + 30z &=& 4925 \\ \hline -35x - 40y &=& -2875 \end{array}$$

Now we have a system of two equations in two variables.

$$-70x + 10y = -1250 \quad (4)$$
$$-35x - 40y = -2875 \quad (5)$$

To eliminate y, multiply equation (4) by 4.

$$\begin{array}{rcl} -280x + 40y &=& -5000 \\ -35x - 40y &=& -2875 \quad (5) \\ \hline -315x &=& -7875 \\ x &=& 25 \end{array}$$

Verify that the corresponding values for y and z are $y = 50$ and $z = 10$. The food vendors paid \$25, the craft vendors paid \$50, and the demonstrators paid \$10.

5.3 QUICK REFERENCE

Linear Equations in Three Variables

- A **linear equation in three variables** is any equation that can be written in the standard form $Ax + By + Cz = D$, where A, B, C, and D are real numbers and A, B, and C are not all zero.
- A **solution** of a linear equation in three variables is an **ordered triple** (x, y, z) of numbers that satisfies the equation.
- A **solution** of a system of linear equations in three variables is an ordered triple (x, y, z) that satisfies each equation of the system.

Graphical Interpretation

- The graph of a linear equation in three variables is a plane drawn in a three-dimensional **coordinate space.**
- For a system of three linear equations in three variables, any point that is common to all three planes represents a solution.

- When solving a system of three linear equations in three variables, there are three possible outcomes.
  1. *Unique solution*. The three planes intersect at exactly one point.
  2. *No solution*. At least two of the planes are parallel or one plane is parallel to the line of intersection of the other two planes.
  3. *Infinitely many solutions*. The three planes coincide or the three planes intersect in a common line.

Algebraic Methods

- The addition method is the algebraic method discussed in this section. To use this method, we do the following:
  1. Write all equations in the form $Ax + By + Cz = D$.
  2. Add two pairs of equations to eliminate the same variable in each pair. It may be necessary to multiply one or both of the equations in each pair so that the variable will be eliminated when you add the equations.
  3. Solve the resulting system of two equations in two variables.
  4. Substitute to determine the values of the other two variables.

Special Cases

- When we use the addition method, we can detect special cases after the first elimination of a variable.
  1. If the resulting system of two linear equations in two variables is inconsistent, then the original system has no solution.
  2. If the two linear equations of the resulting system are dependent, then the original system has infinitely many solutions.

5.3 EXERCISES

Concepts and Skills

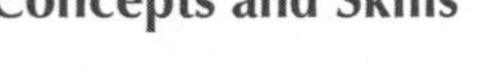

1. The equation $0 = 0$ can be written as follows.

$$0x + 0y + 0z = 0$$

Does this mean that $0 = 0$ is a linear equation in three variables? Explain.

2. Written beside each equation of the following system is a solution of that equation.

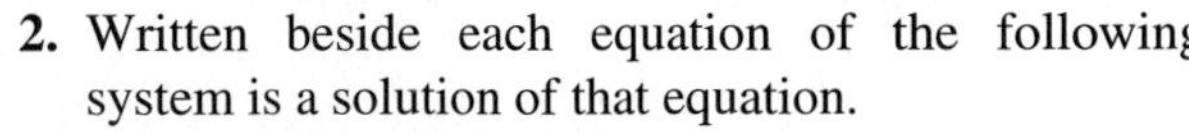

$$\begin{aligned} x + y + z &= 1 \qquad (0, 0, 1) \\ x - y + z &= 1 \qquad (1, 0, 0) \\ x + y - z &= 1 \qquad (1, 1, 1) \end{aligned}$$

Is $\{(0, 0, 1), (1, 0, 0), (1, 1, 1)\}$ the solution set of the system? Explain.

In Exercises 3–6, determine whether each of the given ordered triples is a solution of the system of equations.

3. $$\begin{aligned} 5x + 7y - 2z &= -1 \\ x - 2y + z &= 8 \\ 3x - y + 3z &= 14 \end{aligned}$$
$(3, -2, 1), (2, -1, 2)$

4. $$\begin{aligned} 3x - 2y - z &= 5 \\ x - y + 2z &= -5 \\ 2x + y - z &= 18 \end{aligned}$$
$(0, -1, -3), (5, 6, -2)$

5. $$\begin{aligned} 3x + 4y - z &= -12 \\ -x + 2y + 3z &= 8 \\ 2x + 6y + z &= -8 \end{aligned}$$
$(-3, 1, 7), (0, -2, 4)$

6. $$\begin{aligned} -3x + 4y + z &= -13 \\ 2x - 2y + 3z &= -2 \\ 5x + 6y - 2z &= 29 \end{aligned}$$
$\left(4, \frac{1}{2}, -3\right), (2, -1, -3)$

In Exercises 7–10, determine the value of a, b, c, or d so that the ordered triple is a solution to the system.

7. $$\begin{aligned} ax - 2y + z &= 7 \\ x - y + cz &= -2 \\ 2x + by - z &= 0 \end{aligned}$$ $(1, -1, 2)$

8. $$\begin{aligned} x + y + cz &= 0 \\ 3x + by + z &= 2 \\ ax - y - z &= 6 \end{aligned}$$ $(2, -2, -4)$

9. $$\begin{aligned} 2x + y + z &= d \\ x + by - z &= 7 \\ -3x + 2y + cz &= 4 \end{aligned}$$ $(0, 2, -1)$

10. $$\begin{aligned} ax - y + z &= 0 \\ x - 3y + z &= d \\ x + y + cz &= 8 \end{aligned}$$ $(3, 2, -1)$

11. Compare the graph of a linear equation in two variables with the graph of a linear equation in three variables.

12. If a system of three linear equations in three variables has a unique solution, describe the graph of the system.

13. Describe one way in which the graph of a system of three linear equations in three variables would indicate that the system has

(a) no solution.

(b) infinitely many solutions.

14. Suppose you are solving the following system.

$$\begin{aligned} 2x + y - 3z &= 12 \\ x - y + z &= 0 \\ 3x + 2y - z &= 7 \end{aligned}$$

It is easy to eliminate y in the first two equations and to eliminate z in the last two equations. If you do this, are you making progress toward solving the system? Explain.

In Exercises 15–18, use the addition method to solve the system of equations.

15. $$\begin{aligned} x + y + z &= 0 \\ 3x + y &= 0 \\ y - 2z &= 7 \end{aligned}$$

16. $$\begin{aligned} 2x + 4y + 5z &= -3 \\ y + 4z &= 1 \\ 3x - y &= 9 \end{aligned}$$

17. $$\begin{aligned} x + 2y - 3z &= 5 \\ x + 2z &= 15 \\ 2y - z &= 6 \end{aligned}$$

18. $$\begin{aligned} x - y + z &= 0 \\ 2y + 3z &= 1 \\ x + 2z &= 0 \end{aligned}$$

In Exercises 19–30, use the addition method to solve the system of equations.

19. $$\begin{aligned} 2x - y + z &= 9 \\ 3x + 2y - z &= 4 \\ 4x + 3y + 2z &= 8 \end{aligned}$$

20. $$\begin{aligned} 2x + 3y - 3z &= 1 \\ x - y + 2z &= 7 \\ 3x + 2y + z &= 6 \end{aligned}$$

21. $$\begin{aligned} x + 2y - 3z &= 1 \\ x + y + 2z &= -1 \\ 3x + 3y - z &= 4 \end{aligned}$$

22. $$\begin{aligned} x + 3y - 5z &= -8 \\ -2x + y + 3z &= 9 \\ 3x - 2y + z &= 2 \end{aligned}$$

23. $$\begin{aligned} 4x + 5y - 2z &= 23 \\ -6x + 2y + 7z &= -14 \\ 8x + 3y + 3z &= 11 \end{aligned}$$

24. $$\begin{aligned} -x + y + z &= 0 \\ x - 2y + 3z &= 7 \\ -2x + y - 2z &= 9 \end{aligned}$$

25. $$\begin{aligned} 7x - 2y + 3z &= 19 \\ x + 8y - 6z &= 9 \\ 2x + 4y + 9z &= 5 \end{aligned}$$

26. $$\begin{aligned} x + y - 4z &= -3 \\ 2x - 3y - 2z &= 5 \\ x + y + z &= 2 \end{aligned}$$

27. $$\begin{aligned} 2x + 3y + 2z &= 2 \\ -3x - 6y - 4z &= -3 \\ -\frac{1}{6}x + y + z &= 0 \end{aligned}$$

28. $$\begin{aligned} 3x + 4y - 2z &= -4 \\ -5x + 3y + 4z &= 6 \\ -2x - 2y + 7z &= -3 \end{aligned}$$

29. $$\begin{aligned} 4x + 3y - 3z &= 5 \\ 2x - 6y + 9z &= -7 \\ 6x + 6y - 3z &= 7 \end{aligned}$$

30. $$\begin{aligned} 2x + 4y + 3z &= 1 \\ 2x - 8y - 9z &= 6 \\ 4x + 12y + 9z &= 0 \end{aligned}$$

31. How do we determine algebraically whether a system of equations has

(a) no solution?

(b) infinitely many solutions?

32. Describe a method other than the addition method to solve the following system of equations.

$$\begin{aligned} x + y - z &= 8 \\ 2y + z &= 1 \\ z &= 3 \end{aligned}$$

In Exercises 33–40, determine the solution of the given system. Describe the solution set for special cases.

33. $$\begin{aligned} x + 2y - z &= 5 \\ x - 2y + z &= 2 \\ 2x + 4y - 2z &= 7 \end{aligned}$$

34. $$\begin{aligned} 3x - 3y - 4z &= -7 \\ 7x - 6y + 6z &= 4 \\ 4x - 3y + 10z &= -5 \end{aligned}$$

35. $$\begin{aligned} x + 2y - z &= 3 \\ -4x - 8y + 4z &= -12 \\ 3x + 6y - 3z &= 9 \end{aligned}$$

36. $$\begin{aligned} 2x - y + 4z &= 6 \\ 6x - 3y + 12z &= 18 \\ -4x + 2y - 8z &= -12 \end{aligned}$$

37. $$\begin{aligned} x + y &= 1 \\ 2x + 3y - z &= -1 \\ 3x - y - 4z &= 7 \end{aligned}$$

38. $$\begin{aligned} x - 2z &= 8 \\ 2x + y - z &= 5 \\ 3x - 2y - 4z &= 6 \end{aligned}$$

39. $x - y + z = 2$
$2x + y - 2z = -2$
$3x - 2y + z = 2$

40. $x + y - z = 2$
$2x - 3y + z = 5$
$3x + 2y - 4z = 3$

In Exercises 41–52, describe the solution set of each system of equations.

41. $x + y + z = 0$
$2x + 3y + 2 = 0$
$y - 4z = 0$

42. $3x + y + z = 2$
$x - 2z = 7$
$y + z + 1 = 0$

43. $x + y = 1$
$y - 2z = 2$
$x - 3z = 14$

44. $x + y = 1$
$y - 2z = 5$
$x - 3z = 1$

45. $x + z = y$
$\frac{1}{2}y + \frac{3}{4}z = \frac{1}{4}$
$\frac{1}{2}x + z + \frac{1}{2} = 0$

46. $x + y = 3 + z$
$\frac{1}{4}y - \frac{3}{4}z = 2$
$\frac{1}{2}x - z = \frac{3}{2}$

47. $3x + 2z = y + 3$
$2(x + y) = 2 - z$
$4 + 3y = x + z$

48. $6x + 3z = 1 + 3y$
$4 + y = 2x + z$
$4x - 2y = 7 - 2z$

49. $4x + 6z = 12 + 3y$
$6 + 1.5y = 2x + 3z$
$0.5x - 0.375y + 0.75z = 1.5$

50. $5x - 2y + 15z = 20$
$x - 0.4y + 3z = 4$
$2.5x - y + 7.5z = 10$

51. $\frac{2}{3}x + \frac{5}{2}y - z = -21$
$x - \frac{1}{2}y + \frac{1}{8}z = \frac{1}{2}$
$y + 2z = 8 + 2x$

52. $\frac{1}{4}x + \frac{1}{3}y - \frac{1}{6}z = \frac{1}{3}$
$x + y + \frac{1}{2}z = 1$
$\frac{1}{3}x - \frac{1}{2}y + \frac{1}{12}z = \frac{1}{4}$

In Exercises 53–56, extend the substitution method to solve the given system of three equations in three variables.

53. $2x + y = -4$
$x + z = 2$
$2y - z = -1$

54. $x - y = 0$
$3x + z = 11$
$y - z = 1$

55. $x + y + z = -2$
$x - 3y = 8$
$z - 2y = 8$

56. $2x - 3y = 0$
$2z + y = -2$
$2x - 2y + 3z = -4$

In Exercises 57–62, translate the given information into a system of equations, solve the system, and answer the question.

57. In a triangle, angle B is 5° less than 3 times angle A. Angle C is 6° more than the sum of the other two angles. Determine the sizes of the three angles.

58. In a triangle, angle B is 8° more than twice angle A. Angle C is 11° less than angle B. Determine the sizes of the three angles.

59. A 100-yard rope is cut into three pieces. The second piece is 12 feet longer than the sum of the lengths of the other two pieces. The third piece is 6 yards less than the length of the first piece. How long is each piece?

60. A 14-foot board is cut into three pieces. The second piece is 2 feet less than twice the length of the first piece. The third piece is 36 inches less than the sum of the lengths of the other two pieces. How long is each piece?

61. A collection of 155 coins consists of nickels, dimes, and quarters. The total value of the coins is $24.95. If the number of dimes is two more than twice the number of nickels, how many of each coin is in the collection?

62. A cash register contains a total of 90 coins consisting of pennies, nickels, dimes, and quarters. There are only three pennies and the total value of the coins is $9.28. If there are three times as many dimes as quarters, how many of each coin are in the cash register?

Real-Life Applications

63. At a small college of 1200 students, only 8% belonged to either the Republican Club or the Democratic Club, with all the remaining students calling themselves Independents. There were 4 more Republican Club members than Democratic Club members.

(a) Using the variables R, D, and I, write a system of equations in three variables that models the given information.

(b) Solve the system to determine the number of students in each category.

64. A store has a sale on tee-shirts, sweatshirts, and tank tops. A purchase of 3 tee-shirts, 2 sweatshirts, and 5 tank tops costs $77. It costs $60 for 5 tee-shirts and 1 sweatshirt or for 4 tee-shirts and 6 tank tops. What is the sale price for each shirt?

65. Suppose that a marching band has 100 instruments consisting only of brass, woodwind, and percussion instruments. There are twice as many brass as woodwind instruments, and there are three times as many woodwind as percussion instruments.

(a) Using the variables b, w, and p, write a system of equations in three variables that models the given information.

(b) Solve the system to determine the number of instruments in each category.

66. A total of $10,000 is to be invested in three funds. The following table shows three possible combinations of investments along with the total return for each combination.

Growth Fund	Income Fund	Money Market	Total Return
$2000	$5000	$3000	$460
5000	3000	2000	510
7000	2000	1000	550

What is the percentage rate of return of each fund?

Group Project: Foreign-Born Population

The accompanying bar graph shows that the percentage of the U.S. population that was foreign-born decreased from 11.6% in 1930 to 4.8% in 1970. Then the percentage began to increase and reached 8.7% by 1994. (Source: Bureau of Census.)

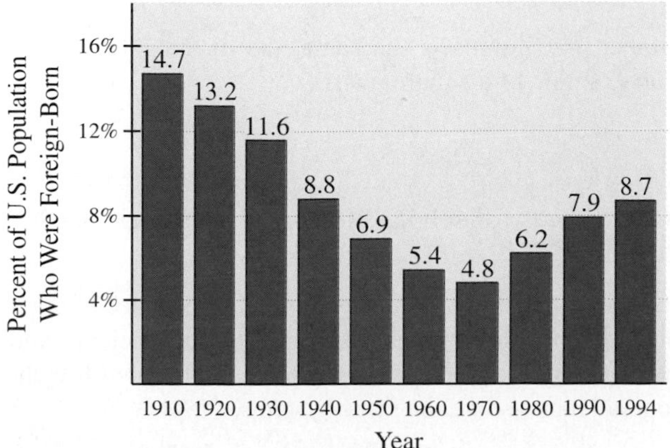

67. Write the data for 1930, 1970, and 1994 as ordered pairs of the form (t, p), where t is the number of years since 1900 and p is that year's percentage of the U.S. population that was foreign-born.

68. Assume that the data can be modeled with a function of the form $p(t) = at^2 + bt + c$. By substituting your three ordered pairs from Exercise 67, create a system of three linear equations in the variables a, b, and c.

69. Solve the system in Exercise 68. Then write the model function.

70. Produce the graph of the model function in Exercise 69. How do you think the shape of this graph would change if illegal aliens were included in the percentages?

Challenge

In Exercises 71 and 72, extend the addition method to solve a system of four linear equations in four variables.

71.
$$\begin{aligned} x + y + z - w &= -1 \\ 2x - y + 2z + w &= 1 \\ x - y - z + 2w &= 6 \\ x + 2y + 3z - 2w &= -4 \end{aligned}$$

72.
$$\begin{aligned} x + y + z &= 2 \\ y + z + w &= 1 \\ x + z + w &= -1 \\ x + y + w &= -2 \end{aligned}$$

In Exercises 73 and 74, make an appropriate substitution to solve the systems of equations. (See Section 5.2 Exercises 95–98.)

73.
$$\begin{aligned} \frac{3}{x} + \frac{2}{y} - \frac{1}{z} &= -1 \\ \frac{1}{x} + \frac{1}{y} + \frac{1}{z} &= 2 \\ \frac{1}{x} - \frac{1}{y} + \frac{2}{z} &= 6 \end{aligned}$$

74.
$$\begin{aligned} \frac{2}{x} + \frac{3}{y} &= 2 \\ \frac{3}{y} - \frac{1}{z} &= 0 \\ \frac{4}{x} - \frac{2}{z} &= 0 \end{aligned}$$

75. Solve the given system for x, y, and z. Express your solutions in terms of a, b, and c.

$$\begin{aligned} x + y \quad &= a \\ y + z &= b \\ x \quad + z &= c \end{aligned}$$

76. For what value(s) of a is the solution of the system unique?

$$\begin{aligned} ax + 2y + z &= 0 \\ 2x - y + 2z &= 0 \\ x + y + 3z &= 0 \end{aligned}$$

5.4 MATRICES AND SYSTEMS OF EQUATIONS

Matrices • Row Operations • Special Cases • Automatic Solving with a Calculator

Matrices

A **matrix** is an array of numbers. As is true with ordered pairs, the position of each of the numbers is meaningful. The following are examples of matrices.

$$A = \begin{bmatrix} 3 & -5 & 0 \\ 2 & 4 & 1 \\ 2 & 1 & 6 \end{bmatrix} \qquad B = \begin{bmatrix} 2 & 4 & -5 & -1 \\ 0 & 2 & 7 & -2 \end{bmatrix}$$

The matrices are given letter names for easy reference. The numbers in the array are called the **elements** of the matrix. The **rows** of the matrix are read horizontally, and the **columns** are read vertically. In matrix B the elements in the first row are 2, 4, -5, and -1. The elements in the first column are 2 and 0.

We refer to the rows and columns of a matrix with the symbols R and C. For example, R_2 refers to row 2 and C_1 refers to column 1.

The **dimensions** of a matrix are the number of rows and the number of columns. Matrix A has 3 rows and 3 columns, and we refer to it as a 3×3 (read "3 by 3") matrix. Matrix B has 2 rows and 4 columns, and we refer to it as a 2×4 (read "2 by 4") matrix. In general, an $m \times n$ matrix has m rows and n columns.

If the number of rows and number of columns is the same, the matrix is a **square matrix.** The 3×3 matrix A is an example of a square matrix.

Row Operations

We have discussed two algebraic methods for solving systems of equations. Another solving method involves the use of matrices.

Recall that we solve systems by using the properties of equations to produce equivalent equations. However, when we apply those properties, only the coefficients and constants change. Therefore, we can use matrices of these numbers along with techniques analogous to algebraic operations to solve the same systems.

EXPLORING THE CONCEPT

Matrices and Systems of Equations

Consider the following system of equations.

$$\begin{aligned} 2x + 3y &= -5 \\ 5x - 2y &= 16 \end{aligned}$$

We define the **coefficient matrix** to be a matrix whose elements are the coefficients of the variables in the equations. For this system, the coefficient matrix is

$$A = \begin{bmatrix} 2 & 3 \\ 5 & -2 \end{bmatrix}$$

The **constant matrix** is a matrix whose elements are the constants on the right side of the system of equations. For the given system, the constant matrix is

$$B = \begin{bmatrix} -5 \\ 16 \end{bmatrix}$$

Think About It

If the dimensions of a coefficient matrix are $n \times n$, what are the dimensions of the corresponding augmented matrix?

The **augmented matrix** is the combination of the coefficient matrix and the constant matrix. For the given system, the augmented matrix is

$$M = \left[\begin{array}{cc|c} 2 & 3 & -5 \\ 5 & -2 & 16 \end{array}\right]$$

To solve the original system of equations, our strategy is to try to transform the system's augmented matrix into a matrix of the form

$$\left[\begin{array}{cc|c} 1 & 0 & n_1 \\ 0 & 1 & n_2 \end{array}\right]$$

where n_1 and n_2 are numbers. If we can do this, then when we translate this matrix back into a system of equations, we obtain

$$\begin{aligned} 1x + 0y &= n_1 \\ 0x + 1y &= n_2. \end{aligned}$$

With the system written in this way, the solutions are obvious: $x = n_1$ and $y = n_2$.

Note: The augmented matrix $\left[\begin{array}{cc|c} 1 & 0 & n_1 \\ 0 & 1 & n_2 \end{array}\right]$ has 1's along the diagonal of the coefficient matrix and 0's elsewhere. We call this the **reduced row echelon form (RREF).**

In order to transform a matrix in this way, we use **row operations.** Row operations correspond to the algebraic operations we perform on the equations of a system. Just as the properties of equations allow us to produce equivalent systems, row operations can be used to produce **row-equivalent matrices.**

The following is a summary of permissible row operations. Included in this summary are examples of the notation we use to indicate each operation.

The Elementary Row Operations on Matrices

1. Two rows may be interchanged: $R_2 \leftrightarrow R_1$.
2. A row may be multiplied by a nonzero number n, and the original row may be replaced by the result: $n \cdot R_1 \to R_1$.
3. A row may be replaced by the sum of itself and another row: $R_1 + R_2 \to R_1$.
4. A row may be replaced by the sum of itself and a multiple of another row: $R_1 + n \cdot R_2 \to R_1$.

EXAMPLE 1

Using Matrices to Solve a System of Two Equations in Two Variables

Solve the following system of equations.

$$2x + y = 1$$
$$x - y = 5$$

Solution

Row Operation	*Row-Equivalent Matrix*		*System of Equations*
	$\left[\begin{array}{cc\|c} 2 & 1 & 1 \\ 1 & -1 & 5 \end{array}\right]$	Augmented matrix	$2x + 1y = 1$ $1x - 1y = 5$
$R_2 \leftrightarrow R_1$	$\left[\begin{array}{cc\|c} 1 & -1 & 5 \\ 2 & 1 & 1 \end{array}\right]$	Interchange rows to obtain a 1 in row 1, column 1.	$1x - 1y = 5$ $2x + 1y = 1$
$-2 \cdot R_1 + R_2 \to$	$\left[\begin{array}{cc\|c} 1 & -1 & 5 \\ 0 & 3 & -9 \end{array}\right]$	Multiply row 1 by −2 and add the result to row 2. Now we have a 0 in row 2, column 1.	$1x - 1y = 5$ $0x + 3y = -9$
$R_2 \div 3 \to$	$\left[\begin{array}{cc\|c} 1 & -1 & 5 \\ 0 & 1 & -3 \end{array}\right]$	Divide row 2 by 3 to obtain a 1 in row 2, column 2.	$1x - 1y = 5$ $0x + 1y = -3$
$R_1 + R_2 \to$	$\left[\begin{array}{cc\|c} 1 & 0 & 2 \\ 0 & 1 & -3 \end{array}\right]$	Replace row 1 with the sum of rows 1 and 2. Now we have a 0 in row 1, column 2.	$1x - 0y = 2$ $0x + 1y = -3$

The system of equations represented by this matrix is equivalent to the original system. The solution of the system is the ordered pair (2, −3).

The matrix method for solving a system of three equations in three variables involves the same strategy as was illustrated in Example 1, but because the augmented matrix is bigger (3×4), more row operations are usually needed to achieve the result.

The goal remains the same. The coefficient matrix must have 1's down its diagonal and 0's elsewhere. Then the solution (ordered triple) is determined from the numbers in the right-hand column.

EXAMPLE 2

Using Matrices to Solve a System of Three Equations in Three Variables

Solve the following system of equations.

$$\begin{aligned} x + y + 2z &= -1 \\ 3x + y - 6z &= 7 \\ -x + 2y + 2z &= 0 \end{aligned}$$

Solution

$$\left[\begin{array}{ccc|c} 1 & 1 & 2 & -1 \\ 3 & 1 & -6 & 7 \\ -1 & 2 & 2 & 0 \end{array}\right]$$ Augmented matrix

$$-3 \cdot R_1 + R_2 \rightarrow \left[\begin{array}{ccc|c} 1 & 1 & 2 & -1 \\ 0 & -2 & -12 & 10 \\ -1 & 2 & 2 & 0 \end{array}\right]$$ Multiply row 1 by -3, and add the result to row 2.

$$R_1 + R_3 \rightarrow \left[\begin{array}{ccc|c} 1 & 1 & 2 & -1 \\ 0 & -2 & -12 & 10 \\ 0 & 3 & 4 & -1 \end{array}\right]$$ Replace row 3 with the sum of rows 1 and 3.

$$R_2 \div (-2) \rightarrow \left[\begin{array}{ccc|c} 1 & 1 & 2 & -1 \\ 0 & 1 & 6 & -5 \\ 0 & 3 & 4 & -1 \end{array}\right]$$ Divide row 2 by -2.

$$-1 \cdot R_2 + R_1 \rightarrow \left[\begin{array}{ccc|c} 1 & 0 & -4 & 4 \\ 0 & 1 & 6 & -5 \\ 0 & 3 & 4 & -1 \end{array}\right]$$ Multiply row 2 by -1, and add the result to row 1.

$$-3 \cdot R_2 + R_3 \rightarrow \left[\begin{array}{ccc|c} 1 & 0 & -4 & 4 \\ 0 & 1 & 6 & -5 \\ 0 & 0 & -14 & 14 \end{array}\right]$$ Multiply row 2 by -3, and add the result to row 3.

$$R_3 \div (-14) \rightarrow \left[\begin{array}{ccc|c} 1 & 0 & -4 & 4 \\ 0 & 1 & 6 & -5 \\ 0 & 0 & 1 & -1 \end{array}\right]$$ Divide row 3 by -14.

$$4 \cdot R_3 + R_1 \rightarrow \left[\begin{array}{ccc|c} 1 & 0 & 0 & 0 \\ 0 & 1 & 6 & -5 \\ 0 & 0 & 1 & -1 \end{array}\right]$$ Multiply row 3 by 4, and add the result to row 1.

$$-6 \cdot R_3 + R_2 \rightarrow \left[\begin{array}{ccc|c} 1 & 0 & 0 & 0 \\ 0 & 1 & 0 & 1 \\ 0 & 0 & 1 & -1 \end{array}\right]$$ Multiply row 3 by -6, and add the result to row 2.

The solution is $(0, 1, -1)$.

LEARNING TIP

Unless you have a plan, it is easy to become lost when solving a system with the matrix method. Begin by obtaining a 1 in the first row, first column. Then perform row operations to obtain a 0 in each of the other positions in the first column. Similarly, next obtain a 1 in the second row, second column. Then obtain a 0 in each of the other positions in the second column. Continue this process until the matrix is in the proper form.

Special Cases

Matrix solution methods reveal inconsistent systems and dependent equations in the same way as algebraic methods.

EXAMPLE 3

An Inconsistent System

Solve the following system of equations.

$$x - 2y = 5$$
$$2x - 4y = 4$$

Solution

$$\left[\begin{array}{cc|c} 1 & -2 & 5 \\ 2 & -4 & 4 \end{array}\right]$$ Augmented matrix

$$-2R_1 + R_2 \rightarrow \left[\begin{array}{cc|c} 1 & -2 & 5 \\ 0 & 0 & -6 \end{array}\right]$$ Multiply row 1 by −2, and add the result to row 2.

The second row corresponds to the equation $0x + 0y = -6$ or $0 = -6$. This false equation indicates that the system has no solution.

EXAMPLE 4

Dependent Equations

Solve the following system of equations.

$$2x + y = 3$$
$$4x + 2y = 6$$

Solution

$$\left[\begin{array}{cc|c} 2 & 1 & 3 \\ 4 & 2 & 6 \end{array}\right]$$ Augmented matrix

$$-2R_1 + R_2 \rightarrow \left[\begin{array}{cc|c} 2 & 1 & 3 \\ 0 & 0 & 0 \end{array}\right]$$ Multiply row 1 by −2, and add the result to row 2.

The second row corresponds to the equation $0x + 0y = 0$ or $0 = 0$, which is true. The system has infinitely many solutions.

Automatic Solving with a Calculator

These same matrix methods can be used to solve larger systems of equations. Of course, larger systems require more row operations.

 Solve

A calculator can perform row operations and display the resulting matrix in reduced row echelon form.

EXAMPLE 5

Automatic Solving with a Calculator

Use a calculator to solve the system of equations in Example 2.

$$x + y + 2z = -1$$
$$3x + y - 6z = 7$$
$$-x + 2y + 2z = 0$$

Solution

Figure 5.16

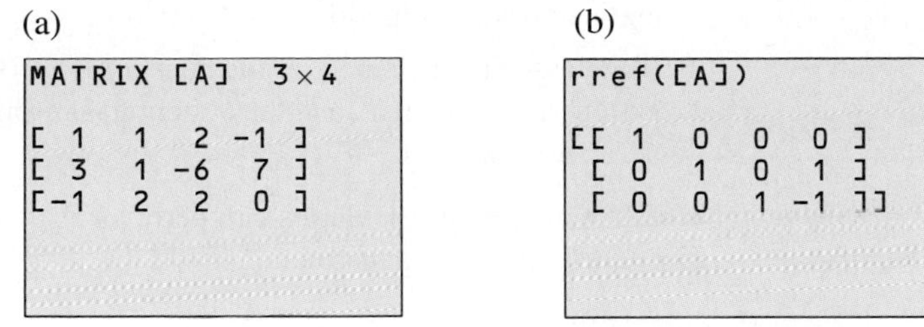

Figure 5.16(a) shows a screen display of the augmented matrix entered as matrix A. The equivalent matrix in reduced row echelon form is shown in Fig. 5.16(b). The solution of the system is read from the last column: $(0, 1, -1)$.

Note: On some calculator models the method is to enter the coefficient matrix as A and the constant matrix as B. Then calculate $A^{-1}B$ on the home screen. The result is the solution matrix.

5.4 QUICK REFERENCE

Matrices

- A **matrix** is an array of numbers called **elements.**
- The **dimensions** of a matrix are the number of rows and the number of columns. (An $m \times n$ matrix has m rows and n columns.) If the dimensions are the same, the matrix is a **square matrix.**

Row Operations

- For a system of equations written in standard form,
  1. the **coefficient matrix** is the matrix of coefficients of the variables.
  2. the **constant matrix** is the matrix of constants on the right sides of the equations.
  3. the **augmented matrix** is the combination of the coefficient and constant matrices.
- For any matrix, the permissible row operations are as follows:
  1. Rows may be interchanged.
  2. A row may be multiplied by a number.
  3. A row may be replaced by the sum of itself and another row.
  4. A row may be replaced by the sum of itself and a multiple of another row.
- When solving systems with matrix methods, the goal is to write the matrix in the **reduced row echelon form,** that is, with 1's along the diagonal of the coefficient matrix and 0's elsewhere. The solution is then determined from the numbers in the last column.

Special Cases
- When solving a system of equations with matrix methods, a special case exists if at any time a row of the coefficient matrix is all 0's.
  1. If the associated element of the constant matrix is not 0, the system is inconsistent and has no solution.
  2. If the associated element of the constant matrix is 0, at least two equations of the system are dependent, and the system has infinitely many solutions.

Automatic Solving with a Calculator
- Most sophisticated calculators can perform row operations and display the reduced row echelon form of a matrix.

5.4 EXERCISES

Concepts and Skills

1. What is a 2×4 matrix?

2. For the matrix $\begin{bmatrix} 0 & 5 \\ 1 & 9 \end{bmatrix}$, are the elements of the second column 1 and 9 or 5 and 9? Why?

In Exercises 3–8, state the dimensions of the matrix.

3. $\begin{bmatrix} 2 & -3 & -5 & 4 \\ 1 & 0 & -6 & 7 \end{bmatrix}$

4. $\begin{bmatrix} 2 & -1 \\ -1 & 4 \\ 5 & -7 \end{bmatrix}$

5. $\begin{bmatrix} 2 & 4 \\ -5 & 0 \end{bmatrix}$

6. $\begin{bmatrix} 1 & 2 & -3 \\ 0 & -4 & 5 \\ 6 & 0 & -7 \end{bmatrix}$

7. $\begin{bmatrix} 2 & -3 & 4 & -5 \\ 0 & 1 & -7 & 6 \\ 5 & 8 & 0 & -9 \end{bmatrix}$

8. $\begin{bmatrix} 1 \\ 2 \\ -3 \\ 0 \end{bmatrix}$

9. Describe each of the following.
   (a) A coefficient matrix
   (b) A constant matrix
   (c) An augmented matrix

10. For a system of n linear equations in n variables, explain why the coefficient matrix must be a square matrix.

In Exercises 11–14, write the augmented matrix for the given system of linear equations.

11. $\begin{aligned} 3x + 2y &= 6 \\ x - 4y &= 9 \end{aligned}$

12. $\begin{aligned} 3x - 2y &= 8 \\ 6x &= 4y + 17 \end{aligned}$

13. $\begin{aligned} x + 2y - 3z &= 5 \\ x + 2z &= 15 \\ 2y - z &= 6 \end{aligned}$

14. $\begin{aligned} x + y - z &= 0 \\ 2y + 3z &= 1 \\ x + 2z &= 0 \end{aligned}$

In Exercises 15–18, write a system of linear equations associated with the augmented matrix.

15. $\left[\begin{array}{cc|c} 2 & 1 & 1 \\ 3 & -2 & 12 \end{array}\right]$

16. $\left[\begin{array}{cc|c} 2 & 1 & 8 \\ 0 & 1 & 2 \end{array}\right]$

17. $\left[\begin{array}{ccc|c} 1 & 1 & -1 & 2 \\ 2 & -3 & 1 & 5 \\ 3 & 2 & -4 & 3 \end{array}\right]$

18. $\left[\begin{array}{ccc|c} 1 & 1 & -1 & 3 \\ 0 & 1 & -3 & 8 \\ 1 & 0 & -2 & 3 \end{array}\right]$

19. When using matrices to solve a system of linear equations, what is your goal with respect to the coefficient matrix?

20. For a system of n linear equations in n variables, what are the dimensions of the constant matrix? Why?

In Exercises 21–24, state the row operation that has been performed in each step of the matrix solution of the given system of linear equations. Then, state the solution.

21. $\begin{aligned} 3x + 6y &= 12 \\ 2x - 3y &= 1 \end{aligned}$

Matrix	*Row Operation*
$\left[\begin{array}{cc\|c} 3 & 6 & 12 \\ 2 & -3 & 1 \end{array}\right]$	Augmented matrix
$\left[\begin{array}{cc\|c} 1 & 2 & 4 \\ 2 & -3 & 1 \end{array}\right]$	(a) ______
$\left[\begin{array}{cc\|c} 1 & 2 & 4 \\ 0 & -7 & -7 \end{array}\right]$	(b) ______

$\left[\begin{array}{cc|c} 1 & 2 & 4 \\ 0 & 1 & 1 \end{array}\right]$ (c) ______

$\left[\begin{array}{cc|c} 1 & 0 & 2 \\ 0 & 1 & 1 \end{array}\right]$ (d) ______

(e) The solution is ______.

22. $2x - 6y = -18$
$3x + 4y = -1$

Matrix	*Row Operation*
$\left[\begin{array}{cc|c} 2 & -6 & -18 \\ 3 & 4 & -1 \end{array}\right]$	Augmented matrix
$\left[\begin{array}{cc|c} 1 & -3 & -9 \\ 3 & 4 & -1 \end{array}\right]$	(a) ______
$\left[\begin{array}{cc|c} 1 & -3 & -9 \\ 0 & 13 & 26 \end{array}\right]$	(b) ______
$\left[\begin{array}{cc|c} 1 & -3 & -9 \\ 0 & 1 & 2 \end{array}\right]$	(c) ______
$\left[\begin{array}{cc|c} 1 & 0 & -3 \\ 0 & 1 & 2 \end{array}\right]$	(d) ______

(e) The solution is ______.

23. $x + y - z = 4$
$2x - y + z = -1$
$x + y - 2z = 5$

Matrix	*Row Operation*
$\left[\begin{array}{ccc|c} 1 & 1 & -1 & 4 \\ 2 & -1 & 1 & -1 \\ 1 & 1 & -2 & 5 \end{array}\right]$	Augmented matrix
$\left[\begin{array}{ccc|c} 1 & 1 & -1 & 4 \\ 0 & -3 & 3 & -9 \\ 0 & 0 & -1 & 1 \end{array}\right]$	(a) ______ (b) ______
$\left[\begin{array}{ccc|c} 1 & 1 & -1 & 4 \\ 0 & 1 & -1 & 3 \\ 0 & 0 & 1 & -1 \end{array}\right]$	(c) ______ (d) ______
$\left[\begin{array}{ccc|c} 1 & 1 & 0 & 3 \\ 0 & 1 & 0 & 2 \\ 0 & 0 & 1 & -1 \end{array}\right]$	(e) ______ (f) ______
$\left[\begin{array}{ccc|c} 1 & 0 & 0 & 1 \\ 0 & 1 & 0 & 2 \\ 0 & 0 & 1 & -1 \end{array}\right]$	(g) ______

(h) The solution is ______.

24. $x + y - z = 4$
$2y - 5z = 2$
$7z = 0$

Matrix	*Row Operation*
$\left[\begin{array}{ccc|c} 1 & 1 & -1 & 4 \\ 0 & 2 & -5 & 2 \\ 0 & 0 & 7 & 0 \end{array}\right]$	Augmented matrix
$\left[\begin{array}{ccc|c} 1 & 1 & -1 & 4 \\ 0 & 2 & -5 & 2 \\ 0 & 0 & 1 & 0 \end{array}\right]$	(a) ______
$\left[\begin{array}{ccc|c} 1 & 1 & 0 & 4 \\ 0 & 2 & 0 & 2 \\ 0 & 0 & 1 & 0 \end{array}\right]$	(b) ______ (c) ______
$\left[\begin{array}{ccc|c} 1 & 1 & 0 & 4 \\ 0 & 1 & 0 & 1 \\ 0 & 0 & 1 & 0 \end{array}\right]$	(d) ______
$\left[\begin{array}{ccc|c} 1 & 0 & 0 & 3 \\ 0 & 1 & 0 & 1 \\ 0 & 0 & 1 & 0 \end{array}\right]$	(e) ______

(f) The solution is ______.

In Exercises 25–28, fill in the blanks to indicate the results of performing row operations. State the solution.

25. $\left[\begin{array}{cc|c} 4 & 8 & 52 \\ 3 & -1 & -17 \end{array}\right]$

$\left[\begin{array}{cc|c} 1 & \text{(a)} & 13 \\ 3 & -1 & -17 \end{array}\right]$

$\left[\begin{array}{cc|c} 1 & \text{(b)} & 13 \\ 0 & -7 & \text{(c)} \end{array}\right]$

$\left[\begin{array}{cc|c} 1 & \text{(d)} & 13 \\ 0 & \text{(e)} & 8 \end{array}\right]$

$\left[\begin{array}{cc|c} 1 & 0 & \text{(f)} \\ 0 & \text{(g)} & 8 \end{array}\right]$

(h) Solution: ______

26. $\left[\begin{array}{cc|c} 6 & 2 & -2 \\ 1 & -1 & 1 \end{array}\right]$

$\left[\begin{array}{cc|c} \text{(a)} & -1 & \text{(b)} \\ 6 & 2 & -2 \end{array}\right]$

$\left[\begin{array}{cc|c} \text{(c)} & -1 & \text{(d)} \\ 0 & \text{(e)} & -8 \end{array}\right]$

$\left[\begin{array}{cc|c} \text{(f)} & -1 & \text{(g)} \\ 0 & 1 & \text{(h)} \end{array}\right]$

$$\left[\begin{array}{cc|c} \text{(i)} & 0 & \text{(j)} \\ 0 & 1 & \text{(k)} \end{array}\right]$$

(l) Solution: ______

27. $\left[\begin{array}{ccc|c} 2 & -3 & 1 & 0 \\ 1 & 1 & -1 & 3 \\ 3 & -2 & -2 & 7 \end{array}\right]$

$$\left[\begin{array}{ccc|c} 1 & 1 & -1 & \text{(a)} \\ 2 & \text{(b)} & 1 & 0 \\ 3 & -2 & -2 & 7 \end{array}\right]$$

$$\left[\begin{array}{ccc|c} 1 & 1 & -1 & 3 \\ 0 & -5 & \text{(c)} & -6 \\ 0 & -5 & 1 & \text{(d)} \end{array}\right]$$

$$\left[\begin{array}{ccc|c} 1 & \text{(e)} & -1 & 3 \\ 0 & -5 & 3 & -6 \\ 0 & 0 & \text{(f)} & 4 \end{array}\right]$$

$$\left[\begin{array}{ccc|c} 1 & 1 & -1 & 3 \\ 0 & -5 & 3 & \text{(g)} \\ 0 & 0 & 1 & \text{(h)} \end{array}\right]$$

$$\left[\begin{array}{ccc|c} 1 & 1 & 0 & \text{(i)} \\ 0 & -5 & \text{(j)} & 0 \\ 0 & 0 & 1 & -2 \end{array}\right]$$

$$\left[\begin{array}{ccc|c} 1 & \text{(k)} & 0 & 1 \\ 0 & 1 & 0 & \text{(l)} \\ 0 & 0 & 1 & -2 \end{array}\right]$$

$$\left[\begin{array}{ccc|c} 1 & 0 & \text{(m)} & 1 \\ \text{(n)} & 1 & 0 & 0 \\ 0 & 0 & 1 & -2 \end{array}\right]$$

(o) Solution: ______

28. $\left[\begin{array}{ccc|c} 3 & 6 & -3 & 9 \\ 1 & -2 & 1 & 3 \\ 2 & -1 & -1 & 0 \end{array}\right]$

$$\left[\begin{array}{ccc|c} \text{(a)} & 2 & -1 & 3 \\ \text{(b)} & -2 & 1 & 3 \\ 2 & -1 & -1 & 0 \end{array}\right]$$

$$\left[\begin{array}{ccc|c} 1 & 2 & -1 & 3 \\ 0 & \text{(c)} & 2 & 0 \\ 0 & \text{(d)} & 1 & -6 \end{array}\right]$$

$$\left[\begin{array}{ccc|c} 1 & -3 & \text{(e)} & -3 \\ 0 & 6 & 0 & \text{(f)} \\ 0 & -5 & 1 & -6 \end{array}\right]$$

$$\left[\begin{array}{ccc|c} 1 & -3 & 0 & -3 \\ 0 & \text{(g)} & 0 & 2 \\ 0 & -5 & 1 & \text{(h)} \end{array}\right]$$

$$\left[\begin{array}{ccc|c} 1 & 0 & 0 & \text{(i)} \\ 0 & 1 & 0 & 2 \\ 0 & 0 & 1 & \text{(j)} \end{array}\right]$$

(k) Solution: ______

29. For the system

$$2x = 3y + 2$$
$$y + 1 = x$$

what must be done before the coefficient matrix and constant matrix can be written?

30. Each of the following shows the last row of an augmented matrix after row operations are performed. In each part, describe the solution set of the associated system of equations.

(a) $0 \quad 0 \mid 6$

(b) $0 \quad 0 \mid 0$

(c) $0 \quad 1 \mid 0$

In Exercises 31–42, use matrix methods to solve the given system of equations. Identify inconsistent systems and dependent equations.

31. $2x - 3y = 18$
$x + 2y = -5$

32. $3x + 4y = 0$
$x + 2y = 2$

33. $4x + 8y = 0$
$x - 2y = 1$

34. $y - x = 1$
$2x + y = 0$

35. $2x - 4y = 7$
$2y - x = 4$

36. $2x - 3y = 7$
$4x = 6y + 7$

37. $3x - 2y = 6$
$6y - 9x = -18$

38. $4x - 5y = 7$
$10y = 8x - 14$

39. $x + z = 1$
$y - z = 5$
$x - y = 2$

40. $x - y = 8$
$x + z = 7$
$y + 2z = 1$

41. $x + y - z = -4$
$-x - 2y + z = 7$
$2x + 2y + z = -2$

42. $-x + 2y - z = -7$
$3x - 7y - 2z = 23$
$-2x + 2y + 3z = -10$

In Exercises 43–46, write the augmented matrix that is associated with the given system of equations.

43. $2x - 5y = -5$
$2y + 5x = 2$

44. $x - 3y = 9$
$5y = x - 17$

45. $z + 3x = 4y + 2$
$1 - y = x + 3z$
$z = 2x + 3y$

46. $z = x + y$
$y - 2 = 2x + 3z$
$z - 1 = 3x + 2y$

In Exercises 47–54, use your calculator to solve the given system of equations.

47. $3x + 4y = -5$
$x + 3 = 0$

48. $y - 2 = 4$
$2x - y = -2$

49. $3x + y + z = 5$
$2x - y + z = 6$
$y = 2z + 2$

50. $x + 3y + z = 2$
$2y + z = 1$
$x - 2y - 3z = 4$

51. $x - 2y - 2z = -2$
$2x + 3y + z = -1$
$3x + y + z = 1$

52. $x + y + z = 0$
$2x - 2y + 3z = 6$
$5x - y + 2z = 0$

53. $x + y - z = 0$
$2x - 3y + 4z = 4$
$3x - 2y + z = 6$

54. $3x - 2y + 5z = -16$
$2x + 5y + 4z = 0$
$4x - 6y - 7z = 13$

Modeling with Real Data

55. The accompanying table shows the average number of minutes in the work day of employed men and women in the United States, in other industrialized countries, and in developing countries. (Source: *Human Development Report*, 1995.)

	Time (in minutes)	
	Men	**Women**
United States	428	453
Other industrialized countries	408	430
Developing countries	483	544

(a) Write the data as a 3×2 matrix. What does the entry in row 3, column 2 represent?

(b) Write the data as a 2×3 matrix. Having done so, in what row or column would you look to determine the difference between men's and women's times in the United States?

56. In part (a) of Exercise 55, suppose that you perform the following operations.

$$\frac{1}{60} \cdot R_i \to R_i \qquad \text{where } i = 1, 2, 3$$

What would the entries in the resulting matrix represent?

Challenge

57. (a) Suppose you use your calculator to solve the following system.

$$\begin{aligned} z + x - 2y &= 11 \\ 3z - 2x + y &= -22 \\ -z + 4x - y &= 24 \end{aligned}$$

You enter the augmented matrix

$$\left[\begin{array}{rrr|r} 1 & 1 & -2 & 11 \\ 3 & -2 & 1 & -22 \\ -1 & 4 & -1 & 24 \end{array}\right]$$

and the calculator result is

$$\left[\begin{array}{rrr|r} 1 & 0 & 0 & -3 \\ 0 & 1 & 0 & 4 \\ 0 & 0 & 1 & -5 \end{array}\right]$$

What is the solution?

(b) Now write each equation of the system in the form $Ax + By + Cz = D$ and solve the resulting system. Is the solution the same as the one you obtained in part (a)?

In Exercises 58–61, use a calculator to solve the system.

58. $2x + y - z + w = -7$
$x - 2y + z - 2w = 16$
$-x + 2y + 3z + 4w = -15$
$3x - y - 2z - w = 3$

59. $x + y + z + w = -2$
$5x - 2y + 3z + w = 1$
$-x + 2y - z - 2w = -9$
$3x + 4y + 2z + 3w = -10$

60. $x - z = 6$
$y + w + 2 = 0$
$x - y + w + 6 = 0$
$y - x = 1$

61. $x + y = 0$
$y + 2z + 3 = 0$
$2z + 3w - 4 = 0$
$3w + 4x = 10$

62. Use matrix methods to solve the following system.

$$x + y = 1$$
$$x + cy = 5$$

For what value(s) of c is the solution unique? For what value(s) of c is there no solution?

63. Use matrix methods to solve the following system.

$$x + 3y = 2$$
$$ax + y = 1$$

For what value(s) of a is the solution unique? For what value(s) of a is there no solution?

64. Use matrix methods to determine p, q, and r so that the given system of equations has

(a) a unique solution.

(b) no solution.

(c) infinitely many solutions.

$$x - y = p$$
$$y - z = q$$
$$z - x = r$$

5.5 DETERMINANTS AND CRAMER'S RULE

Determinant of a 2 × 2 Matrix • Determinant of a 3 × 3 Matrix • Cramer's Rule • Real-Life Applications

Determinant of a 2 × 2 Matrix

We can write the general elements of a matrix by using subscripts. For example, we can write a 2 × 2 matrix as follows.

$$\begin{bmatrix} a_1 & b_1 \\ a_2 & b_2 \end{bmatrix}$$

Note that the subscript on each element indicates the row of the matrix in which the element is located.

Associated with each square matrix is a real number called the **determinant** of the matrix.

LEARNING TIP

Be sure that you understand the difference between a matrix and a determinant. A matrix is an array or table of numbers, whereas a determinant is a *number*.

Determinant of a 2 × 2 Matrix

The **determinant** of a 2 × 2 matrix is defined by

$$\begin{vmatrix} a_1 & b_1 \\ a_2 & b_2 \end{vmatrix} = a_1b_2 - a_2b_1$$

Note: A matrix is enclosed by square brackets, whereas a determinant is enclosed by vertical lines.

As indicated in the diagram, an easy way to evaluate the determinant of a 2 × 2 matrix is to multiply the numbers on the diagonals and subtract.

$$\begin{vmatrix} a_1 & b_1 \\ a_2 & b_2 \end{vmatrix} = a_1b_2 - a_2b_1$$

EXAMPLE 1

Evaluating the Determinant of a 2 × 2 Matrix

Evaluate the determinant of the matrix $\begin{bmatrix} 2 & 1 \\ 1 & -1 \end{bmatrix}$.

Solution

$$\begin{vmatrix} 2 & 1 \\ 1 & -1 \end{vmatrix} = 2(-1) - 1(1) = -2 - 1 = -3$$

Think About It

Consider the following matrices.

$$\begin{bmatrix} a_1 & b_1 \\ a_2 & b_2 \end{bmatrix} \quad \begin{bmatrix} a_2 & b_2 \\ a_1 & b_1 \end{bmatrix}$$

Are the matrices equivalent? Describe the relationship between the associated determinants of these matrices.

 Determinant

You can use a calculator to evaluate the determinant of a matrix.

Determinant of a 3 × 3 Matrix

Defining the determinant for matrices larger than 2 × 2 requires the definition of a **minor** of a matrix.

Definition of a Minor of a Matrix

For each element of a matrix, the **minor** of the matrix corresponding to that element is the determinant of the matrix obtained by eliminating the row and column containing that element.

EXAMPLE 2

Determining Minors of a Matrix

For the following matrix, evaluate the minor corresponding to each of the given elements.

$$\begin{bmatrix} 5 & 2 & 9 \\ 3 & 7 & 4 \\ 6 & 1 & 8 \end{bmatrix}$$

(a) 2 (b) 8 (c) 3

Element	*Matrix*	*Minor*	*Value*
(a) 2	$\begin{bmatrix} 5 & 2 & 9 \\ 3 & 7 & 4 \\ 6 & 1 & 8 \end{bmatrix}$	$\begin{vmatrix} 3 & 4 \\ 6 & 8 \end{vmatrix}$	$= 3 \cdot 8 - 4 \cdot 6 = 24 - 24 = 0$
(b) 8	$\begin{bmatrix} 5 & 2 & 9 \\ 3 & 7 & 4 \\ 6 & 1 & 8 \end{bmatrix}$	$\begin{vmatrix} 5 & 2 \\ 3 & 7 \end{vmatrix}$	$= 5 \cdot 7 - 2 \cdot 3 = 35 - 6 = 29$
(c) 3	$\begin{bmatrix} 5 & 2 & 9 \\ 3 & 7 & 4 \\ 6 & 1 & 8 \end{bmatrix}$	$\begin{vmatrix} 2 & 9 \\ 1 & 8 \end{vmatrix}$	$= 2 \cdot 8 - 9 \cdot 1 = 16 - 9 = 7$

We define the determinant of a 3×3 matrix in terms of the minors of the elements of a row or a column. The following method for evaluating a determinant is called **expanding by minors along the first row.**

$$\begin{vmatrix} a_1 & b_1 & c_1 \\ a_2 & b_2 & c_2 \\ a_3 & b_3 & c_3 \end{vmatrix} = a_1 \begin{vmatrix} b_2 & c_2 \\ b_3 & c_3 \end{vmatrix} - b_1 \begin{vmatrix} a_2 & c_2 \\ a_3 & c_3 \end{vmatrix} + c_1 \begin{vmatrix} a_2 & b_2 \\ a_3 & b_3 \end{vmatrix}$$

We can evaluate a determinant by expanding along any row or any column. To determine whether to add or subtract the product of an element and its minor, we use the following sign pattern.

$$\begin{vmatrix} + & - & + \\ - & + & - \\ + & - & + \end{vmatrix}$$

EXAMPLE 3

Evaluating the Determinant of a 3 × 3 Matrix

Evaluate the determinant of the matrix $\begin{bmatrix} 3 & -5 & 0 \\ 2 & 4 & 1 \\ 2 & 1 & 6 \end{bmatrix}$.

Solution

We expand along the first row.

$$\begin{vmatrix} 3 & -5 & 0 \\ 2 & 4 & 1 \\ 2 & 1 & 6 \end{vmatrix} = 3 \cdot \begin{vmatrix} 4 & 1 \\ 1 & 6 \end{vmatrix} - (-5) \cdot \begin{vmatrix} 2 & 1 \\ 2 & 6 \end{vmatrix} + 0 \cdot \begin{vmatrix} 2 & 4 \\ 2 & 1 \end{vmatrix}$$
$$= 3(23) + 5(10) + 0(-6) = 119$$

Expanding along another row or column gives the same result. For instance, we can expand along the second column. Note the sign pattern.

$$\begin{vmatrix} 3 & -5 & 0 \\ 2 & 4 & 1 \\ 2 & 1 & 6 \end{vmatrix} = -(-5) \cdot \begin{vmatrix} 2 & 1 \\ 2 & 6 \end{vmatrix} + 4 \cdot \begin{vmatrix} 3 & 0 \\ 2 & 6 \end{vmatrix} - 1 \cdot \begin{vmatrix} 3 & 0 \\ 2 & 1 \end{vmatrix}$$
$$= 5(10) + 4(18) - 1(3) = 119$$

Using a calculator to evaluate a determinant eliminates the need to memorize sign patterns and perform the tedious arithmetic.

Cramer's Rule

When we solve a system of equations with the addition method, we perform the same operations repeatedly.

EXPLORING THE CONCEPT

Cramer's Rule

Consider the following general system of equations.

$$a_1x + b_1y = k_1$$
$$a_2x + b_2y = k_2$$

If we apply the addition method to this system of equations, we obtain the following formulas for the solution of the system.

$$x = \frac{k_1b_2 - k_2b_1}{a_1b_2 - a_2b_1} \quad \text{and} \quad y = \frac{a_1k_2 - a_2k_1}{a_1b_2 - a_2b_1}$$

(The derivation of these formulas is left as an exercise.)

Note that the denominator of each formula is the determinant of the coefficient matrix.

$$D = \begin{vmatrix} a_1 & b_1 \\ a_2 & b_2 \end{vmatrix} = a_1b_2 - a_2b_1$$

The numerator of the formula for x is the determinant D_X of the matrix obtained by replacing a_1 and a_2 (the coefficients of x in the coefficient matrix) with the constants k_1 and k_2. Similarly, the numerator of the formula for y is the determinant D_Y of the matrix obtained by replacing b_1 and b_2 (the coefficients of y in the coefficient matrix) with the constants k_1 and k_2.

$$D_X = \begin{vmatrix} k_1 & b_1 \\ k_2 & b_2 \end{vmatrix} = k_1b_2 - k_2b_1 \qquad D_Y = \begin{vmatrix} a_1 & k_1 \\ a_2 & k_2 \end{vmatrix} = a_1k_2 - a_2k_1$$

We can write the formulas for x and y with D, D_X, and D_Y.

$$x = \frac{D_X}{D} \quad \text{and} \quad y = \frac{D_Y}{D}$$

These conclusions are stated in Cramer's Rule.

Cramer's Rule for a 2 × 2 System

The solution of the system

$$a_1x + b_1y = k_1$$
$$a_2x + b_2y = k_2$$

is given by $x = \dfrac{D_X}{D}$ and $y = \dfrac{D_Y}{D}$,

where $D = \begin{vmatrix} a_1 & b_1 \\ a_2 & b_2 \end{vmatrix} \neq 0$, $D_X = \begin{vmatrix} k_1 & b_1 \\ k_2 & b_2 \end{vmatrix}$, and $D_Y = \begin{vmatrix} a_1 & k_1 \\ a_2 & k_2 \end{vmatrix}$

EXAMPLE 4

Using Cramer's Rule to Solve a 2 × 2 System

Use Cramer's Rule to solve the following system of equations.

$$2x + y = 1$$
$$x - y = 5$$

Solution

Evaluate D, D_X, and D_Y.

$$D = \begin{vmatrix} 2 & 1 \\ 1 & -1 \end{vmatrix} = 2(-1) - 1(1) = -3$$

$$D_X = \begin{vmatrix} 1 & 1 \\ 5 & -1 \end{vmatrix} = 1(-1) - 1(5) = -6$$

$$D_Y = \begin{vmatrix} 2 & 1 \\ 1 & 5 \end{vmatrix} = 2(5) - 1(1) = 9$$

Now find x and y.

$$x = \frac{D_X}{D} = \frac{-6}{-3} = 2 \quad \text{and} \quad y = \frac{D_Y}{D} = \frac{9}{-3} = -3$$

The solution is $(2, -3)$.

Note: If the determinant of the coefficient matrix is zero, then the system either has no solution or has infinitely many solutions. Cramer's Rule cannot be applied.

Cramer's Rule can be extended to larger systems of equations.

Cramer's Rule for a 3 × 3 System

The solution of the system of equations

$$\begin{aligned} a_1x + b_1y + c_1z &= k_1 \\ a_2x + b_2y + c_2z &= k_2 \\ a_3x + b_3y + c_3z &= k_3 \end{aligned}$$

is given by $x = \dfrac{D_X}{D}$, $y = \dfrac{D_Y}{D}$, and $z = \dfrac{D_Z}{D}$,

where $D = \begin{vmatrix} a_1 & b_1 & c_1 \\ a_2 & b_2 & c_2 \\ a_3 & b_3 & c_3 \end{vmatrix} \neq 0$, $D_X = \begin{vmatrix} k_1 & b_1 & c_1 \\ k_2 & b_2 & c_2 \\ k_3 & b_3 & c_3 \end{vmatrix}$, $D_Y = \begin{vmatrix} a_1 & k_1 & c_1 \\ a_2 & k_2 & c_2 \\ a_3 & k_3 & c_3 \end{vmatrix}$,

and $D_Z = \begin{vmatrix} a_1 & b_1 & k_1 \\ a_2 & b_2 & k_2 \\ a_3 & b_3 & k_3 \end{vmatrix}$

EXAMPLE 5

Using Cramer's Rule to Solve a 3 × 3 System

Use Cramer's Rule to solve the following system of equations.

$$\begin{aligned} x - y - z &= 1 \\ x + 2y - z &= 1 \\ x + y + z &= 3 \end{aligned}$$

Solution

Using a calculator to calculate D, D_X, D_Y, and D_Z, we obtain the following results.

$$D = \begin{vmatrix} 1 & -1 & -1 \\ 1 & 2 & -1 \\ 1 & 1 & 1 \end{vmatrix} = 6 \qquad D_X = \begin{vmatrix} 1 & -1 & -1 \\ 1 & 2 & -1 \\ 3 & 1 & 1 \end{vmatrix} = 12$$

$$D_Y = \begin{vmatrix} 1 & 1 & -1 \\ 1 & 1 & -1 \\ 1 & 3 & 1 \end{vmatrix} = 0 \qquad D_Z = \begin{vmatrix} 1 & -1 & 1 \\ 1 & 2 & 1 \\ 1 & 1 & 3 \end{vmatrix} = 6$$

Now we use the formulas from Cramer's Rule to calculate x, y, and z.

$$x = \frac{D_X}{D} = \frac{12}{6} = 2 \quad y = \frac{D_Y}{D} = \frac{0}{6} = 0 \quad z = \frac{D_Z}{D} = \frac{6}{6} = 1$$

The solution is (2, 0, 1).

Real-Life Applications

In the following application, we use Cramer's Rule to solve the system of equations.

EXAMPLE 6

Frozen Yogurt Sales

In one day a frozen yogurt shop sold 120 waffle cones. There were 10 fewer strawberry cones sold than chocolate cones. The number of vanilla cones sold was 10 more than twice the number of chocolate cones sold. How many cones of each flavor were sold?

Solution

Let x = number of vanilla cones sold,
y = number of chocolate cones sold, and
z = number of strawberry cones sold.

The following system describes the conditions of the problem.

$x + y + z = 120$ A total of 120 cones were sold.

$z = y - 10$ There were 10 fewer strawberry cones sold than chocolate cones.

$x = 2y + 10$ The number of vanilla cones sold was 10 more than twice the number of chocolate cones sold.

This system can be written in the following form.

$$\begin{aligned} x + y + z &= 120 \\ y - z &= 10 \\ x - 2y &= 10 \end{aligned}$$

The associated determinants are as follows.

$$D = \begin{vmatrix} 1 & 1 & 1 \\ 0 & 1 & -1 \\ 1 & -2 & 0 \end{vmatrix} \qquad D_X = \begin{vmatrix} 120 & 1 & 1 \\ 10 & 1 & -1 \\ 10 & -2 & 0 \end{vmatrix}$$

$$D_Y = \begin{vmatrix} 1 & 120 & 1 \\ 0 & 10 & -1 \\ 1 & 10 & 0 \end{vmatrix} \qquad D_Z = \begin{vmatrix} 1 & 1 & 120 \\ 0 & 1 & 10 \\ 1 & -2 & 10 \end{vmatrix}$$

We use a calculator to evaluate the four determinants.

$$D = -4 \qquad D_X = -280 \qquad D_Y = -120 \qquad D_Z = -80$$

Finally, we use Cramer's Rule to determine the values of x, y, and z.

$$x = \frac{D_X}{D} = \frac{-280}{-4} = 70 \qquad y = \frac{D_Y}{D} = \frac{-120}{-4} = 30$$

$$z = \frac{D_Z}{D} = \frac{-80}{-4} = 20$$

There were 70 vanilla cones, 30 chocolate cones, and 20 strawberry cones sold.

5.5 QUICK REFERENCE

Determinant of a 2 × 2 Matrix

- Associated with each square matrix is a real number called a **determinant.**
- For the matrix $\begin{bmatrix} a_1 & b_1 \\ a_2 & b_2 \end{bmatrix}$, we define the determinant by $\begin{vmatrix} a_1 & b_1 \\ a_2 & b_2 \end{vmatrix} = a_1b_2 - a_2b_1$.
- A calculator can be used to evaluate a determinant.

Determinant of a 3 × 3 Matrix

- The **minor** corresponding to an element of a matrix is the determinant of the matrix obtained by deleting the row and column containing that element.
- We can evaluate a determinant by expanding by minors along any row or any column.

Cramer's Rule

- For a system of equations written in standard form, the following notation is used in connection with Cramer's Rule:
  1. The determinant D is the determinant of the coefficient matrix.
  2. The determinants D_X, D_Y, and D_Z are the determinants of the coefficient matrix after the x-coefficients, y-coefficients, and z-coefficients, respectively, have been replaced by the constants.
- Cramer's Rule states that if $D \neq 0$,
  1. the solution of a 2×2 system is given by $x = \frac{D_X}{D}$ and $y = \frac{D_Y}{D}$.
  2. the solution of a 3×3 system is given by $x = \frac{D_X}{D}$, $y = \frac{D_Y}{D}$, and $z = \frac{D_Z}{D}$.

5.5 EXERCISES

Concepts and Skills

1. What is the difference between a matrix and a determinant?

2. Note that matrix B is obtained by exchanging the rows of A. Are the determinants of the two matrices the same? Explain.

$$A = \begin{bmatrix} p & q \\ r & s \end{bmatrix} \qquad B = \begin{bmatrix} r & s \\ p & q \end{bmatrix}$$

In Exercises 3–6, evaluate the determinant of the given matrix.

3. $\begin{bmatrix} 0 & 1 \\ -7 & 5 \end{bmatrix}$ **4.** $\begin{bmatrix} 3 & 1 \\ -2 & 4 \end{bmatrix}$

5. $\begin{bmatrix} 3 & 2 \\ 9 & 6 \end{bmatrix}$ **6.** $\begin{bmatrix} 5 & -2 \\ -1 & -3 \end{bmatrix}$

In Exercises 7–10, evaluate the given determinant.

7. $\begin{vmatrix} 4 & 2 \\ 2 & 1 \end{vmatrix}$ **8.** $\begin{vmatrix} 0 & -1 \\ 1 & 1 \end{vmatrix}$

9. $\begin{vmatrix} \frac{3}{4} & \frac{2}{3} \\ 3 & 4 \end{vmatrix}$ **10.** $\begin{vmatrix} \sqrt{18} & 2 \\ 3 & \sqrt{2} \end{vmatrix}$

In Exercises 11–16, evaluate the minor corresponding to the given element in matrix A.

$$A = \begin{bmatrix} 3 & -2 & 5 \\ 1 & 2 & -1 \\ 6 & -4 & -7 \end{bmatrix}$$

11. -2 **12.** 1 **13.** -1

14. -4 **15.** 2 **16.** -7

In Exercises 17–24, use your calculator to evaluate the determinant.

17. $\begin{vmatrix} 2 & 0 & 0 \\ 0 & 1 & 0 \\ 0 & 0 & -4 \end{vmatrix}$ **18.** $\begin{vmatrix} 1 & 3 & -2 \\ 0 & 1 & 5 \\ 0 & 0 & 2 \end{vmatrix}$

19. $\begin{vmatrix} 2 & 0 & 3 \\ -5 & 0 & 1 \\ 2 & 0 & -5 \end{vmatrix}$ **20.** $\begin{vmatrix} 3 & -1 & 2 \\ 5 & 2 & -4 \\ 0 & 0 & 0 \end{vmatrix}$

21. $\begin{vmatrix} 1 & 3 & 1 \\ 1 & 2 & 4 \\ 1 & -1 & -3 \end{vmatrix}$ **22.** $\begin{vmatrix} 2 & 4 & -1 \\ 1 & 1 & 1 \\ 3 & -2 & 7 \end{vmatrix}$

23. $\begin{vmatrix} -1 & -3 & 2 \\ 2 & 6 & -4 \\ 1 & 0 & -2 \end{vmatrix}$ **24.** $\begin{vmatrix} -3 & 1 & 0 \\ 6 & -2 & 2 \\ 3 & -1 & 1 \end{vmatrix}$

In Exercises 25–28, use the method for evaluating a determinant to write an equivalent algebraic expression.

25. $\begin{vmatrix} x & 9 \\ 1 & x \end{vmatrix}$ **26.** $\begin{vmatrix} 2 & y \\ y & 8 \end{vmatrix}$

27. $\begin{vmatrix} x-y & 2 \\ x+y & 3 \end{vmatrix}$ **28.** $\begin{vmatrix} x+2 & 1 \\ 3-x & -1 \end{vmatrix}$

In Exercises 29–32, solve for a.

29. $\begin{vmatrix} 4 & a \\ 5 & 7 \end{vmatrix} = -7$ **30.** $\begin{vmatrix} 2 & 5 \\ 3 & a \end{vmatrix} = 3$

31. $\begin{vmatrix} a & 4 \\ -1 & 3 \end{vmatrix} = 7$ **32.** $\begin{vmatrix} 2 & -2 \\ a & 7 \end{vmatrix} = -2$

In Exercises 33 and 34, use the definition of the determinant of a 3 × 3 matrix to solve for a.

33. $\begin{vmatrix} 1 & -2 & 0 \\ 0 & 2 & -1 \\ 3 & 0 & a \end{vmatrix} = 14$ **34.** $\begin{vmatrix} 1 & 0 & 4 \\ -1 & a & 0 \\ 0 & 3 & -2 \end{vmatrix} = 10$

35. Explain why we cannot use Cramer's Rule if the determinant of the coefficient matrix is zero.

36. When solving a 3 × 3 system with Cramer's Rule, why is it wise to calculate D before calculating D_x, D_y, and D_z?

In Exercises 37–42, determine whether Cramer's Rule can be used to solve the system of equations.

37. $x + 2y = 7$
$6x = -12y + 28$

38. $x - 2y = 8$
$3x - 5y = 10$

39. $x = 3y - 4$
$4x - 12y = 28$

40. $y = 4x + 5$
$12x - 3y = 11$

41. $y = 2x - 9$
$3x - 2y = 8$

42. $x = 4y + 3$
$x - 8y = 4$

In Exercises 43–50, use Cramer's Rule to solve the system of equations.

43. $3x + 2y = 16$
$5x + 4y = 30$

44. $x - 3y = 7$
$3x + y = -9$

45. $y = 3x + 5$
$3x + 2y = 1$

46. $x = y + 12$
$2x + y = 3$

47. $2x - \frac{4}{3}y = 6$
$0.3x - y = -1.5$

48. $3x + 1.5y = 0.5$
$3x + 2y = \frac{11}{6}$

49. $1.2x + 2.7y = 9$
$0.4x + 3y = 7.2$

50. $4.5x - 1.2y = -12$
$3.7x + 5.4y = 54$

In Exercises 51–58, solve the given system only if Cramer's Rule applies. If Cramer's Rule does not apply, so indicate.

51. $2x + 3y = 16$
$4x + 5y = 30$

52. $x - 2y = 7$
$3x + y = 7$

53. $2y - \frac{x}{3} = 0$
$2x + 6y = 9$

54. $x + \frac{2y}{3} = 1$
$4x = y - 7$

55. $y = 3x + 4$
$1.5x - 0.5y = 2$

56. $x = 0.2y + 1.4$
$0.5y = 2.5x - 3.5$

57. $\frac{y}{2} - \frac{2x}{3} = \frac{1}{2}$
$\frac{4x}{3} - 2y = -\frac{5}{3}$

58. $\frac{x}{5} - \frac{y}{6} = 1$
$x + \frac{2y}{3} = -1$

In Exercises 59–70, solve the given system only if Cramer's Rule applies. If Cramer's Rule does not apply, so indicate.

59. $x + y + z = 6$
$3x - 2y + 4z = 5$
$2x - y + z = 1$

60. $x - 2y + z = 15$
$2x + 3y + z = -3$
$3x - 2y - 3z = -1$

61. $x = y + 2z - 7$
$y = z - 7$
$3x - 6y - z = 8$

62. $x + y = z + 6$
$y + 2z = x - 3$
$2x + 3z = 3$

63. $x - 2y + z = 7$
$x + 2y - 3z = 5$
$4x - 8y + 4z = 14$

64. $x = 2y + z + 5$
$y = 3z + 7$
$3x - 6y - 3z = 15$

65. $x - y = 3$
$x - 2z = 6$
$z - 3y = 1$

66. $2x + y = 1$
$-3z - y = 0$
$z + x = -3$

67. $0.6x - 2y - 1.2z = 3.8$
$1.5x - 0.5y + z = 5$
$x - y + 0.3z = 4$

68. $1.5x - y + 2z = 10$
$x - y + 2.4z = 8.8$
$2.5x + y + 1.6z = 13.2$

69. $\frac{2}{3}x - \frac{1}{6}y - \frac{1}{2}z = -2$
$\frac{5}{4}x + \frac{1}{4}y + \frac{3}{2}z = 1$
$\frac{3}{2}x - \frac{1}{4}y - \frac{3}{4}z = -\frac{7}{2}$

70. $\frac{1}{5}x - \frac{1}{2}y + \frac{3}{4}z = -7$
$\frac{1}{3}x + 2y - \frac{1}{2}z = 25$
$x - \frac{1}{4}y + \frac{1}{2}z = 9$

71. In the first determinant, the number n is a common factor in the first row. In the second determinant, n is a common factor in the first column.

$$\begin{vmatrix} na & nb \\ c & d \end{vmatrix} \qquad \begin{vmatrix} na & b \\ nc & d \end{vmatrix}$$

Show that the number n can be factored out of either determinant; that is, show that

$$n\begin{vmatrix} a & b \\ c & d \end{vmatrix} = \begin{vmatrix} na & nb \\ c & d \end{vmatrix} = \begin{vmatrix} na & b \\ nc & d \end{vmatrix}$$

72. Show that the slope–intercept form of the equation of a line can be written

$$y = \begin{vmatrix} m & b \\ -1 & x \end{vmatrix}$$

In Exercises 73–76, determine the value of c so that Cramer's Rule will not apply.

73. $2x + 3y = 6$
$6x - cy = 18$

74. $3x - 4y = 12$
$cx + 12y = 36$

75. $5x - 4y = 7$
$20x - cy = 8$

76. $4x - 7y = 3$
$cx - 21y = 4$

In Exercises 77–80, use the definition of the determinant of a 3×3 matrix to determine the value of c so that Cramer's Rule does not apply.

77. $2x - y + z = 2$
$y - 2z = -5$
$x + y + cz = 3$

78. $cx + y = 0$
$-2x + 4y + z = 4$
$3y + z = -7$

79. $2x + y = 1$
$x + cy - 3z = 2$
$cx + y + 2z = 6$

80. $2x + cy + z = -2$
$x - 2y + cz = 3$
$y - z = -1$

Real-Life Applications

81. A local grocery store has been in business 4 years longer than the pharmacy. Six years ago, the grocery store had been in business twice as long as the pharmacy had been. How long has each store been in business?

82. A home was built 9 years before the home next door was built. Five years ago, the home was 4 times as old as the home next door was. How old is each home?

83. A checking account currently has twice as much money in it as a savings account. If \$800 is deposited into each account next month, the checking account will have 1.5 times as much money in it as the savings account. How much money is currently in each account?

84. One spring is 3 centimeters shorter than a second spring. If each spring is stretched by 4 centimeters, the shorter spring will be $\frac{4}{5}$ as long as the second spring. How long is each spring prior to being stretched?

85. A collection of 41 coins consisting of nickels, dimes, and quarters has a value of \$3.50. If there are three times as many nickels as dimes in the collection, how many of each coin is in the collection?

86. In a triangle the measure of angle B is twice the measure of angle A. The measure of angle C is 20° more than the measure of angle B. Determine the measures of the three angles.

Challenge

In Exercises 87 and 88, solve the system of equations.

87. $\sqrt{3}x - 4\sqrt{2}y = 14$
$3\sqrt{3}x + \sqrt{2}y = 16$

88. $2\sqrt{5}x + \sqrt{6}y = -18$
$3\sqrt{5}x + 4\sqrt{6}y = 3$

In Exercises 89–92, write the given expression as a determinant.

89. $2x - 3y$ **90.** $x + 4y$

91. $ax + by$ **92.** $mx + b$

93. Consider the following system of equations.

$$a_1x + b_1y = k_1$$
$$a_2x + b_2y = k_2$$

Use the addition method to derive the following formulas for the solution of the system.

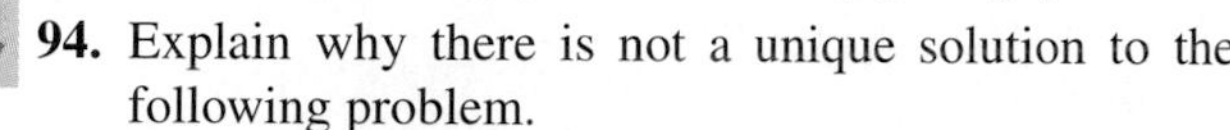

$$x = \frac{k_1b_2 - k_2b_1}{a_1b_2 - a_2b_1} \quad \text{and} \quad y = \frac{a_1k_2 - a_2k_1}{a_1b_2 - a_2b_1}$$

94. Explain why there is not a unique solution to the following problem.

The total value of 82 coins consisting of nickels, dimes, and quarters is \$8.20. There are three times as many nickels as quarters. Determine the number of each type of coin in the collection.

95. In each of the following matrices, either two rows or two columns are the same. Show that the determinant of each is 0.

$$\begin{bmatrix} a & b & c \\ a & b & c \\ x & y & z \end{bmatrix} \qquad \begin{bmatrix} a & a & x \\ b & b & y \\ c & c & z \end{bmatrix}$$

96. In each of the following matrices, either one row or one column consists of zeros. Show that the determinant of each is 0.

$$\begin{bmatrix} a & b & c \\ x & y & z \\ 0 & 0 & 0 \end{bmatrix} \qquad \begin{bmatrix} 0 & a & x \\ 0 & b & y \\ 0 & c & z \end{bmatrix}$$

5.6 APPLICATIONS

2 × 2 Systems • 3 × 3 Systems • Curve Fitting

2 × 2 Systems

In this section we use our ability to solve systems of equations for solving a variety of application problems. While our discussion focuses on specific examples, our goal remains to develop a general understanding of problem-solving strategies. We begin with examples of applications that lead to systems of two equations in two variables.

Mixture problems typically involve two or more items, each with a different cost. The items are combined to create a mixture whose cost is between the costs of the individual items.

EXAMPLE 1

A Mixture of Cashews and Peanuts

A snack food distributor provides cashews for \$6 per pound and peanuts for \$2 per pound. The distributor determines that the total sales would increase if a mixture of the two nuts for \$3 per pound were offered. How many pounds of each nut should be used for a 10-pound mixture?

Solution

Let c = the number of pounds of cashews and

p = the number of pounds of peanuts.

We can use a table to organize the information.

	Weight in Pounds	×	Cost per Pound	=	Total Cost
Cashews	c		6		$6c$
Peanuts	p		2		$2p$
Mixture	10		3		30

From this table we write the following system of equations.

$c + p = 10$ The total weight of the mixture is 10 pounds.

$6c + 2p = 30$ The total cost of the mixture is \$30.00.

LEARNING TIP

Beside each equation in the examples, we indicate the information that the equation models. As you write equations, consider making such notes to yourself as reminders of what the equations represent.

$$\begin{aligned} -2c - 2p &= -20 \\ 6c + 2p &= 30 \\ \hline 4c &= 10 \\ c &= \frac{10}{4} = 2.5 \end{aligned}$$

Multiply the first equation by −2, and add the equations.

$c + p = 10$ First equation

$2.5 + p = 10$ Replace c with 2.5.

$p = 7.5$

The mixture consists of 2.5 pounds of cashews and 7.5 pounds of peanuts.

Dual investment problems involve different amounts of money in two or more investments, each paying a different simple interest rate.

EXAMPLE 2

Dual Investments with Simple Interest

A bank customer invested in two certificates of deposit, one paying 8% and the other paying 3.8%. The total of her two investments was \$10,000. If the certificates paid yearly simple interest and the total interest earned at the end of the year was \$674, how much did this customer invest in each certificate?

Solution

Let $x =$ the amount invested in the 8% certificate and
$y =$ the amount invested in the 3.8% certificate.

We use a table to organize the information.

	Amount Invested	×	Interest Rate	=	Interest Earned
CD paying 8%	x		0.08		$0.08x$
CD paying 3.8%	y		0.038		$0.038y$
Totals	10,000				674

From the table, we can write a system of equations.

$$x + y = 10{,}000 \quad \text{The total amount invested was \$10,000.}$$
$$0.08x + 0.038y = 674 \quad \text{The total interest earned was \$674.}$$

$$\begin{aligned} -0.08x - 0.080y &= -800 \\ 0.08x + 0.038y &= 674 \end{aligned} \quad \text{Multiply the first equation by } -0.08\text{, and add the equations.}$$

$$-0.042y = -126$$
$$y = 3000$$

$$x + y = 10{,}000 \quad \text{First equation}$$
$$x + 3000 = 10{,}000 \quad \text{Replace } y \text{ with 3000.}$$
$$x = 7000$$

The bank customer invested \$7000 in the 8% certificate and \$3000 in the 3.8% certificate.

Recall that a liquid solution is some substance, such as acid, mixed with water. The *concentration* of the substance is the amount of the substance compared with the total amount of solution. Refer to Section 3.5 to review the details concerning this type of problem.

EXAMPLE 3

Creating a Liquid Solution with a Specified Concentration

A nurse has a 20% alcohol solution and a 45% alcohol solution. How many liters of each should be used to make 5 liters of a 30% alcohol solution?

Solution

Let $x =$ the number of liters of the 20% (weak) solution and
$y =$ the number of liters of the 45% (strong) solution.

	Liters of Solution	×	Concentration of Alcohol	=	Liters of Alcohol
Weak	x		0.20		$0.20x$
Strong	y		0.45		$0.45y$
Mix	5		0.30		1.5

From the table we write the following system of equations.

$x + y = 5$ The total amount of solution is 5 liters.

$0.20x + 0.45y = 1.5$ The total amount of alcohol in the solution is 1.5 liters.

This time we use the substitution method to solve the system.

$x = 5 - y$ Solve the first equation for x.

$0.20(5 - y) + 0.45y = 1.5$ Replace x in the second equation with $5 - y$.

$1 - 0.20y + 0.45y = 1.5$ Distributive Property

$1 + 0.25y = 1.5$ Combine like terms.

$0.25y = 0.5$

$$y = \frac{0.5}{0.25} = 2$$

$x = 5 - y$ Replace y with 2 in the equation solved for x.

$x = 5 - 2$

$x = 3$

The nurse should use 3 liters of the 20% solution and 2 liters of the 45% solution.

Distance d, rate r, and time t are related by the formula $d = rt$. Water current and wind speed have the effect of decreasing or increasing a traveler's rate.

EXAMPLE 4

Determining Boat Speed and Rate of Current

A river boat takes tourists from St. Louis up the Mississippi River 72 miles to Clarksville. The trip up the river takes 6 hours, and the return downstream takes 2 hours less time. What is the speed of the current, and what is the speed of the river boat?

Solution

Let b = speed of the boat in still water and

c = rate of the current.

As it travels upstream, the boat is slowed down by the current. The net rate is the boat's rate minus the rate of the current and is given by the expression $b - c$.

The boat is helped by the current as it returns downstream. The net rate is the boat's rate plus the rate of the current and is given by the expression $b + c$.

	Rate (mph)	×	Time (hours)	=	Distance (miles)
Upstream	$b - c$		6		$6(b - c)$
Downstream	$b + c$		4		$6(b + c)$

Because the boat traveled 72 miles upstream and 72 miles downstream, we write the following system of equations.

$$6(b - c) = 72$$

$$4(b + c) = 72$$

$$b - c = 12 \qquad \text{Divide both sides of the first equation by 6.}$$

$$b + c = 18 \qquad \text{Divide both sides of the second equation by 4.}$$

Figure 5.17

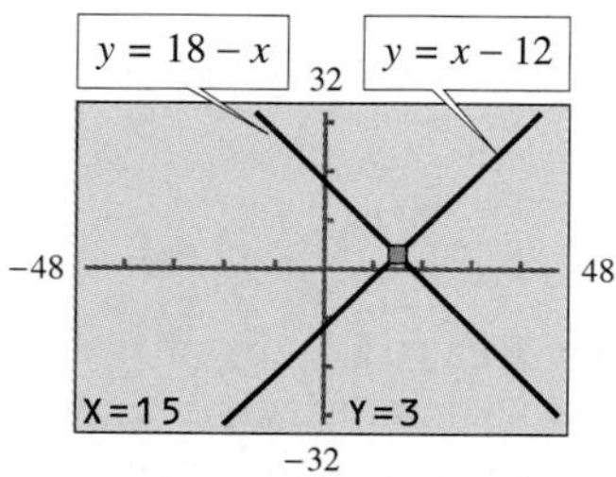

We can estimate the solution of the system by solving the two equations for c and producing their graphs.

$$c = b - 12$$

$$c = 18 - b$$

Because we solved the equations for c, points of the lines have coordinates of the form (b, c). From Fig. 5.17 we estimate the solution to be $(15, 3)$. This solution can easily be verified by substitution.

The speed of the boat in still water is 15 mph, and the rate of the current is 3 mph.

When we model the conditions of a problem with a system of equations, we may learn that the conditions cannot be met.

EXAMPLE 5

Inventory of New and Used Vehicles

An auto dealer sells new and used cars and trucks. He is planning his inventory for next month. His plan is to double the total number of cars and have a total of 190 cars and trucks. He also wants 25% of his cars and 25% of his trucks to be used and the total of used cars and trucks is to be 42. How many cars and trucks will he have in his inventory?

Think About It

In Example 5, suppose that the dealer has a total of 168 cars and trucks. How many cars and how many trucks will he have in his inventory?

Solution

Let x = the current total number of new and used cars and

y = the current total number of new and used trucks.

If the dealer doubles the number of cars, his total number of vehicles can be described by

$$2x + y = 190$$

The number of used cars and trucks can be described by

$$0.25(2x) + 0.25y = 42$$

We solve each equation for y.

$$2x + y = 190 \qquad\qquad 0.25(2x) + 0.25y = 42$$
$$y = -2x + 190 \qquad\qquad 0.50x + 0.25y = 42$$
$$0.25y = -0.50x + 42$$
$$y = -2x + 168$$

From the equations in the form $y = mx + b$, we can see that the slopes of their graphs are the same and the y-intercepts are different. Therefore, the graphs will be parallel. This system of equations has no solution, and the auto dealer cannot carry out his plan.

3 × 3 Systems

Example 6 shows how an application involving three unknowns can be modeled with a system of three linear equations in three variables.

EXAMPLE 6

A Furniture Production Schedule

The Wood Shop makes wooden porch furniture. Rockers require 5 hours of sawing time, 3 hours of assembly time, and 3 hours of finishing time. Chairs require 4 hours of sawing time, 2.5 hours of assembly time, and 2 hours of finishing time. Tables require 3 hours of sawing time, 2 hours of assembly time, and 2 hours of finishing time. The shop employs four full-time people to saw, two full-time and one half-time for assembly, and two full-time and one part-time (10 hours per week) for finishing. Assuming full time is 40 hours per week, how many rockers, chairs, and tables can the shop produce per week?

Solution

Let r = the number of rockers produced per week,
c = the number of chairs produced per week, and
t = the number of tables produced per week.

Four full-time workers provide $4(40) = 160$ available hours for sawing. Two full-time workers and one half-time worker provide $2.5(40) = 100$ available hours for assembly. Two full-time workers and one part-time (one-fourth of full time) worker provide $2.25(40) = 90$ hours for finishing.

	Hours Required			
	Rocker	**Chair**	**Table**	**Available Hours**
Sawing	5	4	3	160
Assembly	3	2.5	2	100
Finishing	3	2	2	90

The system of equations is as follows.

$$\begin{aligned} 5r + 4c + 3t &= 160 \\ 3r + 2.5c + 2t &= 100 \\ 3r + 2c + 2t &= 90 \end{aligned}$$

To use the solving procedure on a calculator, we need the augmented matrix A.

$$A = \left[\begin{array}{ccc|c} 5 & 4 & 3 & 160 \\ 3 & 2.5 & 2 & 100 \\ 3 & 2 & 2 & 90 \end{array}\right]$$

The reported solution is (10, 20, 10). The shop can produce 10 rockers, 20 chairs, and 10 tables each week.

Curve Fitting

In Chapter 4 we saw that two given points uniquely determine a straight line, and we learned how to find an equation of the line.

Similarly, three given noncollinear points uniquely determine the graph of a function whose form is $y = ax^2 + bx + c$. We can use the three given points to create a system of three equations. Then we can solve the system for a, b, and c and thereby learn what function the graph represents.

This process of determining a function whose graph contains given points is known as **curve fitting**.

EXAMPLE 7

Curve Fitting

Determine the values of a, b, and c so that the graph of $y = ax^2 + bx + c$ contains the points $(-2, 4)$, $(1, -5)$, and $(4, 4)$. Use your calculator to produce the graph of the equation, and trace to verify that the points are on the graph.

Solution

Substitute each ordered pair into $y = ax^2 + bx + c$.

$$\begin{aligned} (-2, 4)&: & 4 &= a(-2)^2 + b(-2) + c \\ (1, -5)&: & -5 &= a(1)^2 + b(1) + c \\ (4, 4)&: & 4 &= a(4)^2 + b(4) + c \end{aligned}$$

After simplifying each equation, the system of equations is as follows.

$$\begin{aligned} 4a - 2b + c &= 4 \\ a + b + c &= -5 \\ 16a + 4b + c &= 4 \end{aligned}$$

To solve this system with a calculator, enter the augmented matrix [see Fig. 5.18(a)] and obtain the reduced row echelon form [see Fig. 5.18(b)].

Figure 5.18

(a)

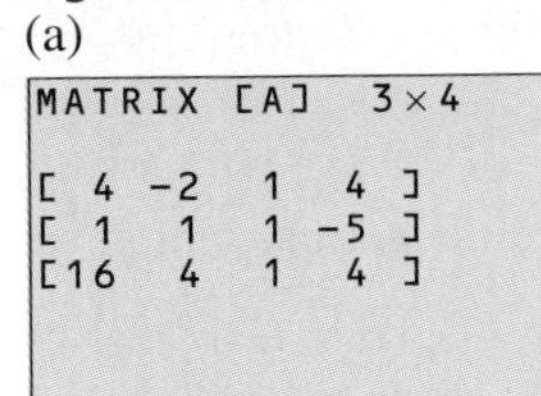

(b)

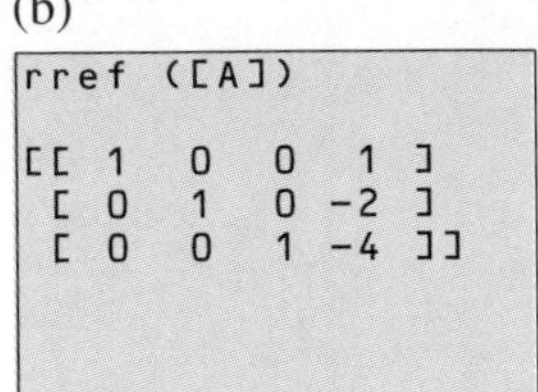

Figure 5.19

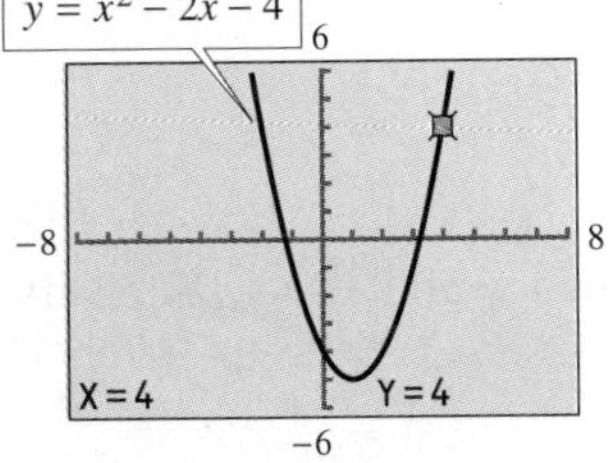

The matrix in Fig. 5.18(b) shows that $a = 1$, $b = -2$, and $c = -4$. Substituting these values into $y = ax^2 + bx + c$, we have the desired equation: $y = x^2 - 2x - 4$.

Figure 5.19 shows the graph of $y = x^2 - 2x - 4$ and the point (4, 4). The other two points, (−2, 4) and (1, −5), also can be traced.

Quadreg

We can use a calculator's regression equation capability to create an equation of the form $y = ax^2 + bx + c$ when three noncollinear points are given.

In Example 7, after entering the data points (−2, 4), (1, −5), and (4, 4), we select the quadratic regression option from the menu of regression equations. Figure 5.20 shows the reported values of a, b, and c.

Figure 5.20

L1	L2	L3
-2	4	------
1	-5	
4	4	

QuadReg
y=ax²+bx+c
a=1
b=-2
c=-4

5.6 EXERCISES

Real-Life Applications

Mixtures

1. A candy store offers a mixture of candies for \$3.25 per pound. A 20-pound mixture is produced by mixing assorted creams worth \$2.95 per pound with chocolate-covered almonds worth \$3.70 per pound. How many pounds of assorted creams are in the 20-pound mixture?

2. A caterer is able to purchase nut mixtures from a wholesale company. Preparing for a convention, the caterer bought a 45-pound nut mixture for \$5.40 per pound. The mixture was produced by mixing cashews worth \$6.00 per pound with pecans worth \$5.00 per pound. How many pounds of each kind of nut were used to make the mixture?

3. A Kroger's shopper bought a total of 8 pounds of fruit for \$6.12. The apples cost \$0.45 per pound, and the grapes cost \$1.29 per pound. How many pounds of apples did the shopper buy?

4. A landscaper specializes in azaleas and rhododendrons. A commercial customer has a budget of \$2250.00 and wants the landscaper to plant a mixture of 100 azaleas and rhododendrons around a new office building. Including planting costs, the landscaper charges \$40.00 per rhododendron and

$15.00 per azalea. How many of each shrub can the landscaper plant?

5. During a bake sale, an organization sold all its cakes and cookies and three-fourths of its pies. The cakes were priced at $3.50 each, the pies at $4.00 each, and the cookies at $1.40 per dozen. A sum of $361.00 was earned at the sale. How many baked items were donated if there were six more pies than cakes and if 128 items were sold? (A dozen cookies represents one baked item.)

6. A caterer made 30 pounds of a party mix to sell for $4.05 a pound. Peanuts selling for $2.95 a pound, cashews selling for $6.25 a pound, and coated chocolates selling for $3.50 a pound were all used in the party mix. If two more pounds of peanuts than cashews were used, how many pounds of each item were in the party mix?

Tickets

7. Arthur Miller's *Death of a Salesman* was the spring production at a certain college. The opening of the play was attended by 456 people. Patron's tickets cost $2, and all other tickets cost $3. If the total box office receipts were $1131, how many of each kind of ticket were sold?

8. A college basketball game was attended by 1243 people. Student tickets cost $2, and all other tickets cost $8. If the total receipts were $5018, how many students and how many nonstudents attended the game?

9. The attendance at an Atlanta Braves and Los Angeles Dodgers game was 42,245. All seats were either box seats costing $30.00 or reserved seats costing $20.00. The total receipts for the game were $987,510. How many box seats were sold?

10. Tickets for a charity ball were $10 apiece or $15 per couple. If 802 people attended the dance and a total of $6530 was collected from ticket sales, how many couples attended?

11. The attendance at an opera was 5500. Orchestra seats were $65 each, loge seats were $50 each, and balcony seats were $30 each. The total receipts for the performance were $246,750. If there were 100 more loge seats than orchestra seats sold, how many of each type of seat were sold?

12. At a state fair the total receipts for one evening were $17,000 for the 9000 people who attended. The entrance tickets were $3.00 for each adult, $1.50 for each student, and $1.00 for each child under 10 years old. If there were twice as many students as children under 10, how many adults attended the state fair that evening?

Investments

13. A financial adviser recommended that her client's $10,000 investment be diversified, part at 8.5% simple interest and the remainder at 6.9% simple interest. How much did her client invest at each rate if the total interest income at the end of 1 year was $745.84?

14. A lottery winner invested $11,000, part at 6.5% simple interest and the remainder at 9% simple interest. How much was invested at each rate if the total interest income at the end of 1 year was $830.85?

15. A used car buyer borrowed $7000 from two sources. One loan was to be repaid at 6% simple interest and the other loan at 7% simple interest. If the borrower paid a total of $452 in interest for the year, how much was borrowed at each rate?

16. Newlyweds purchased kitchen appliances for $8500. They borrowed the money, part at 9.5% and the remainder at 8% simple interest. If the couple owed a total of $741.50 in interest at the end of the year, how much did they borrow at each rate?

17. On behalf of a client, a stockbroker invested a total of $16,500 in a stock fund, a bond fund, and a mutual fund with yields of 6%, 7.5%, and 8% simple interest, respectively. At the end of 1 year, the total income from all three funds was $1191.50. If the broker invested $1200 less in the mutual fund than twice the amount that was invested in the bond fund, how much was invested in each fund?

18. A retiree placed his $60,000 lump-sum pension distribution into three investments, each paying simple interest. He placed some of the money in a high-risk growth fund that paid 12%, some in a GNMA income fund that paid 6.4%, and the rest in a conservative money-market account that paid 3.7%. The high-risk investment was $1000 less than half the money-market investment. If the total interest earned at the end of 1 year was $3806, how much money was invested in each account?

Liquid Solutions

19. Suppose that a 20-ounce acid solution has a concentration of 10% acid. Explain how to determine the number of ounces of acid in the solution.

20. Suppose that three beakers each contain 10 ounces of an alcohol solution. The concentrations are 40%, 1%, and 2%. If the three solutions are mixed together, would you expect the resulting alcohol concentration to be less than or greater than 20%? Why?

21. A brick mason has a 34% solution of muriatic acid and a 55% solution of muriatic acid. How much of each solution should be mixed together to produce 70 liters of a 40% acid solution?

22. An exterminator needed 26 liters of a 49% insecticide solution. To obtain this, he mixed solution A, which had a 34% concentration, with solution B, which had a 60% concentration. How much of each solution was required?

23. At a water purification research facility, 600 gallons of a 2.5% chlorine solution were needed. How much 3.5% chlorine solution had to be added to a 2% chlorine solution to obtain the required result?

24. A nurse needs 15 liters of a 40% alcohol solution. A solution of 20% alcohol and another solution of 70% alcohol are available. How much of each solution should be mixed?

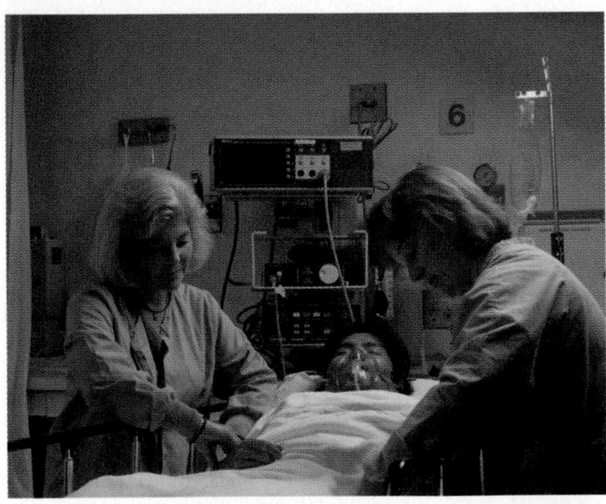

25. An infirmary has solutions of 10%, 20%, and 50% alcohol content. A nurse needs 18 liters of a 30% alcohol solution. How much of each solution is needed if twice as much of the 50% alcohol solution as the 20% alcohol solution is to be used?

26. A chemist is analyzing three acid solutions labeled A, B, and C. Each solution was formed by mixing together two of three other acid solutions labeled 1, 2, and 3. The following table summarizes the amounts of solutions 1, 2, and 3 that were used to make solutions A, B, and C. All amounts are measured in cubic centimeters. The numbers in parentheses indicate the acid concentrations in solutions A, B, and C.

	Amount of Solution 1	Amount of Solution 2	Amount of Solution 3
A (28%)	2	8	0
B (31%)	0	12	3
C (46%)	9	0	6

What are the acid concentrations in solutions 1, 2, and 3?

Distance, Rate, and Time

27. If the still-water rate of a boat is r and the rate of the current is c, then the downstream rate of the boat is $r + c$. Explain why this is so.

28. Suppose that a boater travels upstream 5 miles and then returns downstream to the starting point. If the river has a current and the time was the same for both trips, compare the rates for the two trips.

29. A man paddles his canoe 21 miles upstream in 7 hours. Then he paddles downstream to his starting point in 3 hours. What is the speed of the canoe in still water, and what is the speed of the current?

30. A motor boat travels 137.5 miles upstream in 12.5 hours. The next day it travels downstream to its starting point in 5.5 hours. What is the speed of the boat in still water, and what is the speed of the current?

31. The distance from Cincinnati to Dublin is 3360 miles. Suppose that an airplane made a flight with the wind in 7 hours and returned against the wind in 8 hours. What was the speed of the airplane in calm air, and what was the speed of the wind?

32. An airplane flew from Minneapolis to San Juan, a distance of 2860 miles. Because there was a tail wind, the flight took 5.5 hours. There was a head

wind on the return flight, so the trip took 6.5 hours. What was the speed of the airplane in calm air, and what was the speed of the wind?

33. A vacationer paddled his canoe downstream in 2 hours and paddled upstream back to his starting point in 6 hours. At noon the next day, he decided to drift downstream. At 5:00 P.M. he fell asleep. He woke up at 7:00 P.M. to discover that he had drifted 2 miles past his turn-around point of the previous day. Determine the rate of the current and the rate of the canoe in still water.

34. An athlete is training for the triathlon and has discovered that her rates seem to be staying the same. On Monday she jogged for 30 minutes and rode the bike for 30 minutes and covered 21 miles. On Tuesday she rode the bike for one-half hour and swam for 1 hour and covered 18.4 miles. On Wednesday she jogged for 15 minutes and biked for 45 minutes and covered 25.5 miles. What was her rate for each sport?

Curve Fitting

In Exercises 37 and 40, determine the values of a and b so that the graph of $y = ax + b$ contains the given points. Then write the equation.

35. $(3, -2)$ and $(-1, 6)$ **36.** $(-4, 5)$ and $(-6, -1)$

In Exercises 37–40, determine the values of a, b, and c so that the graph of $y = ax^2 + bx + c$ contains the given points. Then write the equation.

37. $(-3, 22)$, $(4, 15)$, and $(1, -6)$

38. $(-2, -11)$, $(3, -6)$, and $(1, -2)$

39. $(-1, 2)$, $(2, -7)$, and $(3, -26)$

40. $(-1, 2)$, $(3, 10)$, and $(-2, 15)$

Miscellaneous Applications

41. When a father and his son play golf, the father agrees to pay twice as much as his son pays. The total cost of one round of golf is \$51. How much does each pay for one round of golf?

42. In *The Joy of Cooking*, a torte recipe calls for three times as much flour as sugar. If 6 cups of the mixture are to be used, how many cups of flour are needed?

43. At the close of business, a store owner had \$505 in five- and ten-dollar bills in the cash register. If there was a total of 70 of these bills, how many of each denomination were there?

44. A biology class had five more students than a French class. After four more students enrolled in each class, the French class had three-fourths as many students as the biology class. Determine the original enrollment in each class.

45. Your building-supply store stocks two brands of electric drills. The retail price of the heavy-duty drill is \$46.50, and the price of the homeowner's drill is \$18.95. You have hired a part-time employee to help with your annual inventory. You neglect to tell the employee that you need to know how many of each kind of drill you have in your inventory. Therefore, one line of your inventory report is this.

Item	*Quantity*	*Retail Value*
Drills	31	\$918.05

From this report, determine how many of each kind of drill you have.

46. A city school system wants to ensure a balance of minority students and teachers in the schools. At Hamilton Middle School, there is a combined total of 435 students and faculty, of whom 167 are minorities. If 40% of the students and 20% of the faculty are minorities, how many students attend the school?

47. A paint store has a total of 75 one-gallon cans of paint, with high-gloss enamel selling at \$19.95 per gallon and flat latex selling at \$10.95 per gallon. If the total value of the one-gallon cans is \$1109.25, how many of each type are in stock?

48. The owner of a precious metals shop wishes to produce 69.8 ounces of an alloy that is 25% gold. Some 20% gold alloy and some 28% gold alloy are available. How many ounces of each alloy should be melted and mixed to produce the desired alloy?

49. A florist sells orchid corsages for \$23 each and carnation corsages for \$14 each. If the total receipts for 95 corsages were \$1663, how many of each type of corsage were sold?

50. At a concession stand, students sell candy bars for 75 cents each and popcorn for \$1.25 a box. If the total receipts for 365 items were \$352.75, how many of each item did they sell?

51. At a local election, 536 of the 700 voters voted in favor of a bond referendum. If 88% of the Democrats and 72% of the Republicans supported the referendum, how many Republicans voted against it?

52. A computer programmer has produced two programs, one in the BASIC language and one in the Pascal language. Together the programs total 950 lines of code, of which a total of 15 lines contain errors. If 1% of the BASIC program lines contain errors and 2% of the Pascal program lines contain errors, how many lines of code were written in Pascal?

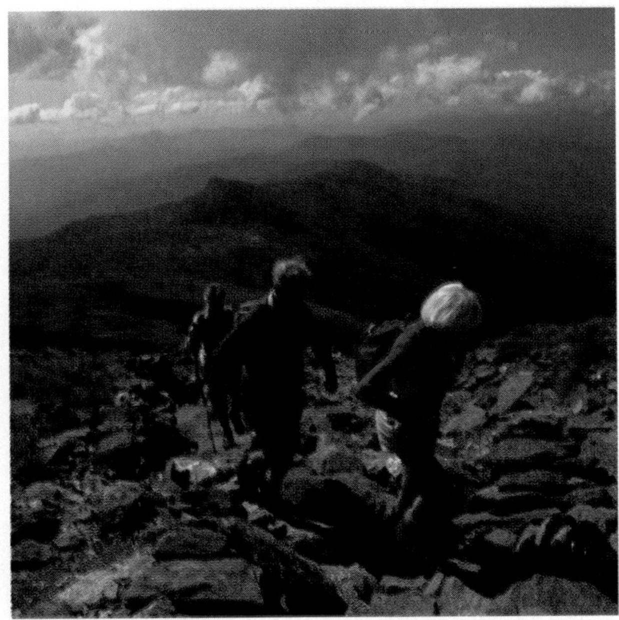

53. A hiking club of 40 members climbed a mountain on three different trails. Part of the group climbed an 8-mile trail, another part climbed a rougher 6-mile trail, and the third group climbed the steepest trail of 5.5 miles. The second group had two more people than the third group. If the members of the club walked a total of 280 person-miles, how many hikers climbed each trail? (Note: If a group of 25 people walks 10 miles, the group has walked a total of $10 \cdot 25$ or 250 person-miles.)

54. The church secretary, the organist, and a volunteer addressed envelopes for a bulk mailing. The organist can address 120 envelopes per hour, the volunteer 116 per hour, and the secretary 140 per hour. They worked a total of 6 person-hours and addressed a total of 740 envelopes. The volunteer worked 1 hour longer than the secretary. How long did each work?

55. At a certain college, A's are worth 4 points, B's are worth 3 points, and C's are worth 2 points. A student's *quality points* are determined by multiplying the number of credit hours for each course times the point value of the grade earned in the course. One semester a student earned 52 quality points for 17 credit hours of course work. The number of credit hours in which the student received B's was twice the number of credit hours in which the student received C's. If none of the student's grades was below C, what was the number of credit hours in which the student earned each grade?

56. The sum of four digits on a license plate is 18. The first and last digits are the same, and the sum of the first and the second digits is 10. The second digit is twice the third. What is the number on the license plate?

57. A bookshelf 4 feet in length holds exactly three sets of books totaling 35 volumes. The books in the first set are 1 inch thick, those in the second set are 1.5 inches thick, and those in the third set are 2 inches thick. If the number of books in the first set is one more than the number in the second set, how many volumes are in each set?

58. The perimeter of a triangle is 40 inches. The length of the shortest side is 4 inches more than the positive difference between the lengths of the other two sides. The length of the longest side is 4 inches less than the sum of the lengths of the other two sides. Which of the three sides can you determine from this information? What can you say about the other two sides?

59. The following shows three orders placed at a fast-food restaurant.

Order	*Cost*
2 hamburgers, 3 fries, 2 shakes	\$ 9.75
4 hamburgers, 2 shakes	\$10.80
5 fries, 3 shakes	\$ 9.25

What is the price of each item?

60. Suppose that your hardware store carries 4-penny, 6-penny, and 8-penny nails and you have a total of 85 pounds in your inventory. The nails sell for \$3.50 per pound, \$4.00 per pound, and \$4.50 per pound, respectively, and the total retail value of your inventory is \$350. How many more pounds of 8-penny nails than 4-penny nails do you have in stock?

Modeling with Real Data

61. The accompanying bar graph shows how much longer people were keeping their cars and trucks in 1994 compared with 1970. (Source: The Polk Company.)

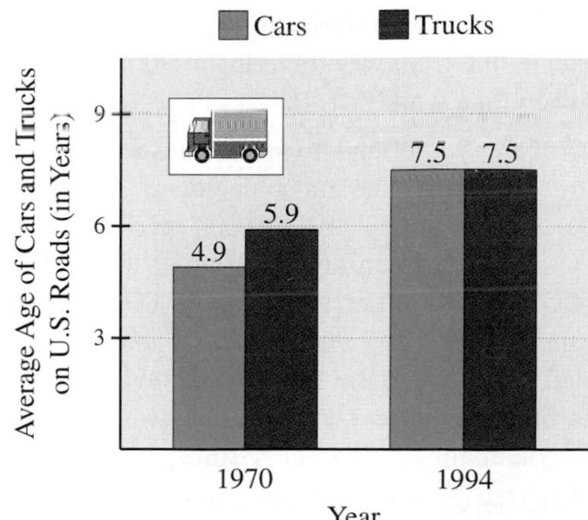

(a) Write a linear equation that models the average age of cars as a function of the number of years since 1970.

(b) Write a linear equation that models the average age of trucks as a function of the number of years since 1970.

(c) Referring to the bar graph, what is your estimate of the solution of the system of these two equations?

62. (a) Verify your estimate in part (c) of Exercise 61 by solving the system algebraically.

(b) Interpret the solution of the system.

63. The accompanying table shows the cost of college as a percentage of the median family income for the given year. For example, the median income in 1994 was \$42,000, and the cost of attending a public college was 14.4% of that income. (Source: U.S. Department of Education.)

	Percentage of Median Income	
Year	**Public**	**Private**
1979	9.1%	20.7%
1994	14.4%	38.7%

The percentage y can be modeled with the following equations, where x is the number of years since 1974.

Public: $y = 0.35x + 7.3$

Private: $y = 1.20x + 14.7$

Does the system of these two equations have a solution? If so, what does it represent?

64. (a) In Exercise 63, how can you tell just by looking at the equations that the system has no solution for years beyond 1974?

(b) If the model equations were to remain valid indefinitely, by what year would attending a private college cost half of a median family income?

Group Project: Small Business Loans

One function of the Small Business Administration (SBA) is to make or guarantee loans to minority business owners. The accompanying bar graph shows the trend in such loans during the period 1988–1994. (Source: Small Business Administration.)

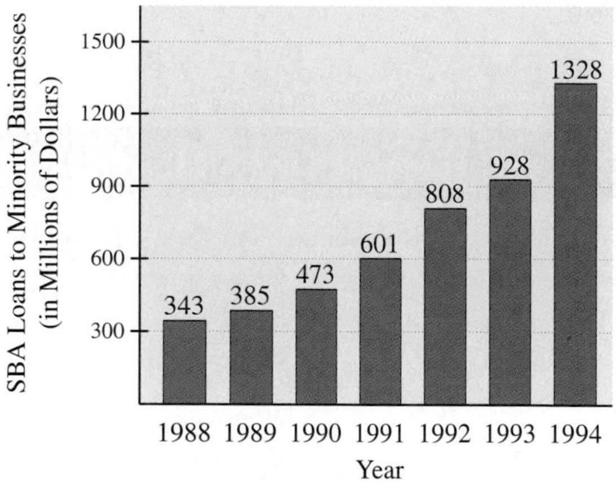

We can model the data in the bar graph with an equation of the form $y = ax^2 + bx + c$.

65. For the data given for 1990 ($x = 2$), 1992 ($x = 4$), and 1994 ($x = 6$), replace y with the amount of the loan and x with the year number. Write the result as a system of three equations in three variables a, b, and c.

66. Use the matrix method to solve the system in Exercise 65. Then write the equation that models the data points by replacing a, b, and c in the equation $y = ax^2 + bx + c$ with their respective values.

67. Use your calculator's quadratic regression equation capability to create a model equation for the same three data points described in Exercise 65. Compare this model to the one that you obtained in Exercise 66.

68. How might the trend indicated in the bar graph be changed if political efforts are made to curtail Affirmative Action programs?

Challenge

69. A two-cycle engine requires an oil and gas mixture that is 2.5% oil. A man mixed 6 gallons of gas and oil but then discovered he had mistakenly made a 2% mixture. How many *ounces* of oil does he need to add to the mixture to correct the problem? (There are 128 ounces in a gallon.)

70. On a pan balance, two baseballs and three softballs balance with one softball and five baseballs. In another experiment, two softballs and one baseball balance with four baseballs. How many softballs are needed to balance with three baseballs?

71. A track inspector and a crew supervisor were standing in a tunnel whose length is 0.6 miles. They departed in opposite directions to leave the tunnel before the next train came through. Both men exited their ends of the tunnel in 3 minutes, and the inspector reached the end of the tunnel just as the train arrived. The train passed the supervisor, who had walked 0.12 miles beyond the other end of the tunnel. If the train was traveling at 30 mph, how fast did each man walk?

72. A woman is applying to be a salesperson for Ace Auto Company. She must decide if she wants to have a base salary of $20,000 per year with a 12% commission on annual sales or to have a base salary of $30,000 per year with an 8% commission on annual sales. If x represents the annual sales and y_1 and y_2 represent the first and second annual earnings options, write a system of equations and use your calculator to solve the system graphically.

    At what level of annual sales will the annual earnings be the same? For what range of sales should the salesperson choose the first option? For what range of annual sales should she choose the second option?

5.7 SYSTEMS OF LINEAR INEQUALITIES

Systems of Inequalities • Real-Life Applications

Systems of Inequalities

When a problem involving inequalities requires two or more conditions to be satisfied, a system of inequalities is needed.

Definition of a System of Inequalities, Solution, and Solution Set

Two or more inequalities considered simultaneously form a **system of inequalities.** A **solution** of a system of inequalities in two variables is an ordered pair that satisfies each inequality. The **solution set** is the set of all solutions.

In this section we consider only systems of linear inequalities in two variables. For brevity, we refer to such systems simply as *systems of inequalities.*

The inequalities of a system are implicitly joined by the word *and.* Therefore, a system of inequalities is an instance of a compound inequality, a concept that was introduced in Section 4.5.

EXAMPLE 1

Verifying Solutions of a System of Inequalities

Determine whether each ordered pair is a solution of the given system of inequalities.

$$x - 2y < 5$$
$$y \geq 3 - x$$

(a) $(6, 11)$ (b) $(-5, 3)$

Solution

(a) Replace x with 6 and y with 11.

$x - 2y < 5$		$y \geq 3 - x$	
$6 - 2(11) < 5$		$11 \geq 3 - 6$	
$-16 < 5$	True	$11 \geq -3$	True

Because $(6, 11)$ satisfies both inequalities, it is a solution of the system.

(b) Replace x with -5 and y with 3.

$x - 2y < 5$		$y \geq 3 - x$	
$-5 - 2(3) < 5$		$3 \geq 3 - (-5)$	
$-11 < 5$	True	$3 \geq 8$	False

Although $(-5, 3)$ satisfies the first inequality, it does not satisfy the second inequality. Therefore, $(-5, 3)$ is not a solution of the system.

A graph of a system of inequalities is a picture of all the solutions of the system. Because a solution of a system of inequalities satisfies each inequality, the graph of the solution set of the system is the intersection of the graphs of the inequalities of the system.

Graphing the Solution Set of a System of Inequalities

1. Graph each inequality.
2. Determine the *intersection* of the solution sets.

EXAMPLE 2

The Graph of a System of Inequalities

Graph the following system of inequalities.

$$3x - y \leq 9$$
$$y + 6 > -x$$

Solution

To produce the graphs, we solve each inequality for y.

$$y \geq 3x - 9$$
$$y > -x - 6$$

The graphs of these inequalities are the half-planes shown in Figs. 5.21 and 5.22.

Figure 5.21

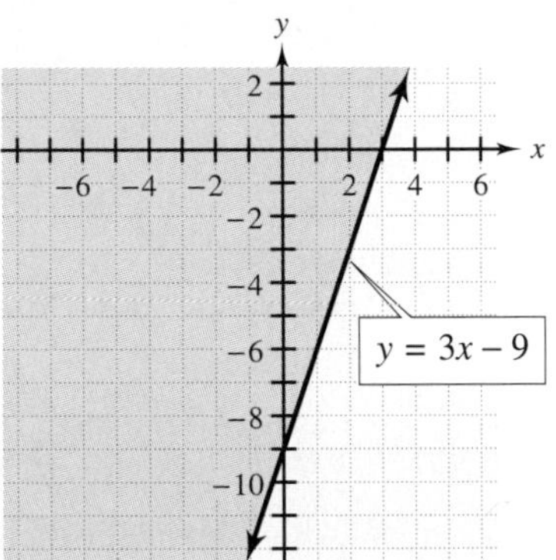

Figure 5.22

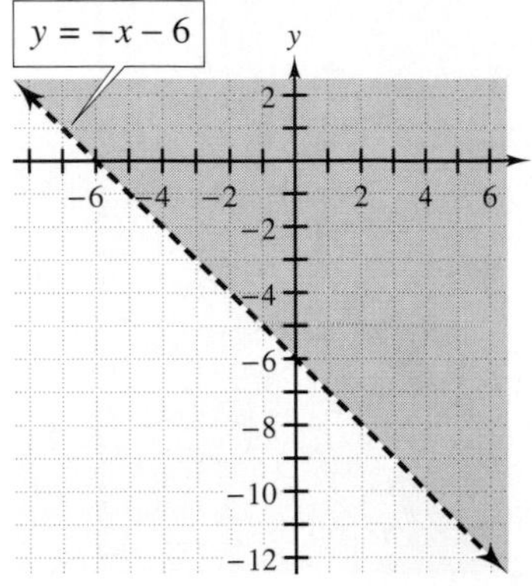

Figure 5.23 shows the two half-planes in the same coordinate system. Their intersection is shown in Fig. 5.24.

LEARNING TIP

After you have graphed the solution set, test a couple of points inside and outside the shaded region. It is impossible to check every solution, but checking a few may catch errors.

Figure 5.23

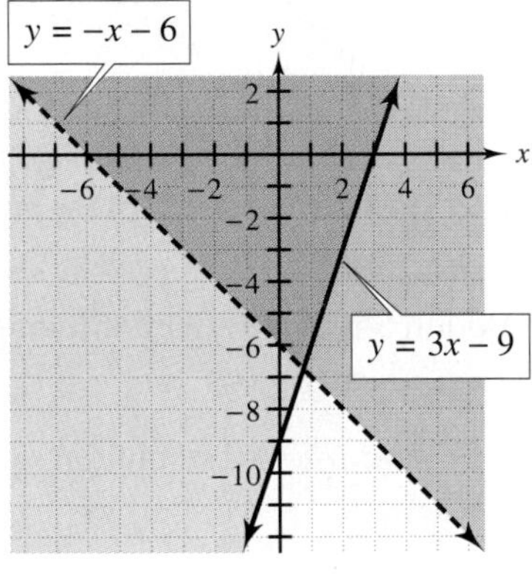

Figure 5.24

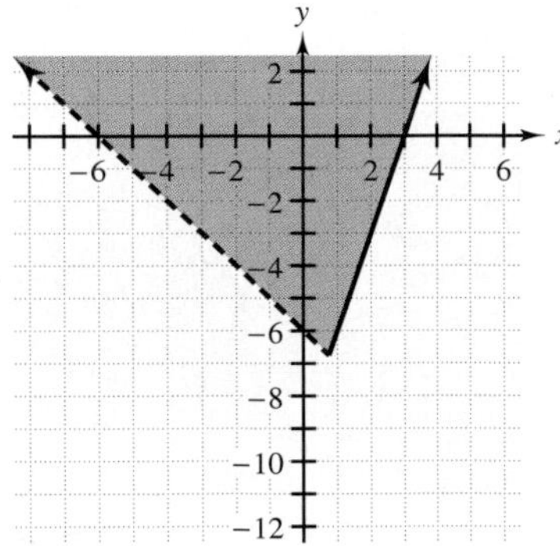

EXAMPLE 3

The Graph of a System of Inequalities

Graph the following system of inequalities.

$$y \leq x - 7$$
$$x \leq y + 3$$

Solution

To produce the graphs, we solve each inequality for y.

$$y \leq x - 7$$
$$y \geq x - 3$$

Think About It

Can the entire coordinate plane represent the solution set of a system of linear inequalities?

The graph of each inequality is shown in Fig. 5.25. Because the graphs do not intersect, the system of inequalities has no solution.

Figure 5.25

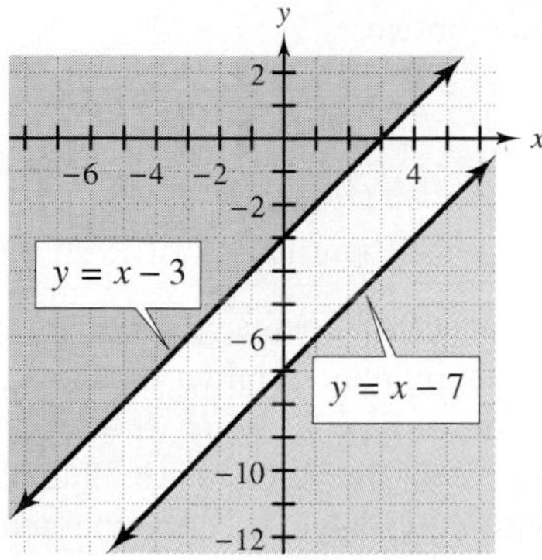

EXAMPLE 4

A Graph of a System of Three Inequalities

Graph the following system of inequalities.

$$x - y < 3$$
$$2x + y > 1$$
$$x - 4 \leq 0$$

Solution

To produce the graphs, we solve the first two inequalities for y and the third inequality for x.

$x - y < 3$	$2x + y > 1$	$x - 4 \leq 0$
$y > x - 3$	$y > -2x + 1$	$x \leq 4$

The graph of each inequality is a half-plane. The solution set of the system of inequalities is the intersection of the three half-planes. (See Fig. 5.26.)

Figure 5.26

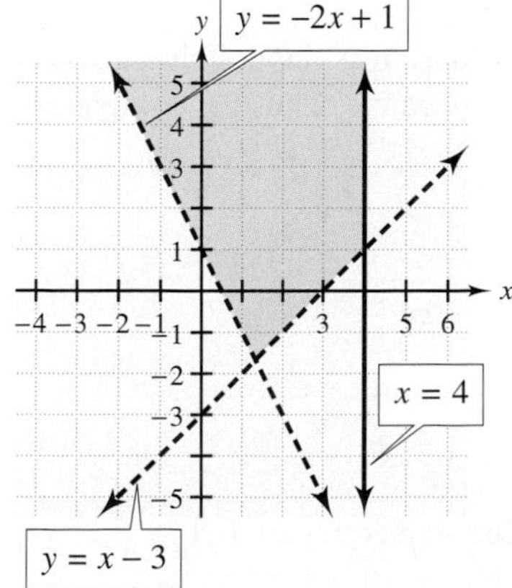

Real-Life Applications

Application problems may require two or more inequalities to describe the conditions of the problem.

EXAMPLE 5

Feasible Production of Recycled Products

A recycling company collects type 1 and type 2 plastics for the manufacture of recycling bins and litter boxes. A recycling bin requires 5 units of type 1 plastic and 1 unit of type 2 plastic. A litter box requires 3 units of type 1 plastic and 2 units of type 2 plastic. If there are 75 units of type 1 plastic available and 36 units of type 2 plastic available, describe the number of recycling bins and litter boxes that can be manufactured.

Solution

	Units Required	
	Type 1	**Type 2**
Recycling bins	5	1
Litter boxes	3	2
Available	75	36

Let $x =$ the number of recycling bins manufactured and
$y =$ the number of litter boxes manufactured.

The amount of plastic available gives two inequalities.

$$5x + 3y \leq 75 \quad \text{or} \quad y \leq -\frac{5}{3}x + 25$$

$$x + 2y \leq 36 \quad \text{or} \quad y \leq -\frac{1}{2}x + 18$$

Because it is impossible to make a negative number of items, each variable must be nonnegative. Thus the complete system of inequalities is as follows.

$$y \leq -\frac{5}{3}x + 25$$

$$y \leq -\frac{1}{2}x + 18$$

$$x \geq 0$$

$$y \geq 0$$

The shaded region in Fig. 5.27 represents the solution set of the system.

Figure 5.27

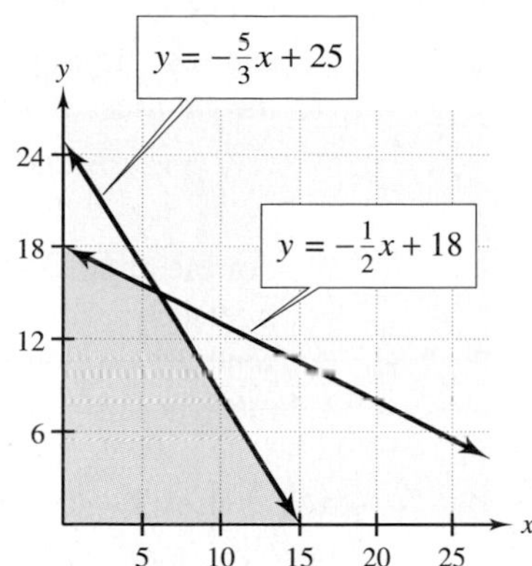

Note that every point in the shaded area, as well as every point of the line segments that bound the shaded area, represents a solution of the system. However, because of the nature of the problem, we can consider only those points whose coordinates are integers.

5.7 QUICK REFERENCE

Systems of Inequalities

- A **system of inequalities** consists of two or more inequalities considered simultaneously. In this section we consider systems of linear inequalities in two variables.
- A **solution** of a system of inequalities is a pair of numbers that satisfies each inequality of the system.
- To graph the solution set of a system of inequalities,
  1. graph each inequality and
  2. determine the intersection of the solution sets.

5.7 EXERCISES

Concepts and Skills

1. Use the word *half-plane* to describe the graph of the solution set of a system of linear inequalities.

2. Explain why a system of inequalities is a conjunction.

In Exercises 3–14, graph the solution set of the given system of linear inequalities.

3. $x \geq 3$
$y \geq 2$

4. $y > 1$
$x < -2$

5. $16 \geq 3x - 4y$
$y < 6$

6. $5y + 3x > -20$
$x \geq -2$

7. $3y - 4x < 15$
$y > 0$

8. $x \geq 0$
$x + 3y \leq 12$

9. $y \geq \frac{1}{2}x$
$y \leq 3x$

10. $y < 2x - 5$
$y > -\frac{2}{3}x - 5$

11. $6 \leq 2x + 3y$
$2y - 5x \geq 14$

12. $x - y > 3$
$2x + y \leq 3$

13. $y - 2x > 7$
$y + 3 < 2x$

14. $2x + 3y \leq -15$
$y \geq -\frac{2}{3}x - 1$

In Exercises 15–18, write the system of linear inequalities whose solution set is represented by the shaded region of the graph.

15.

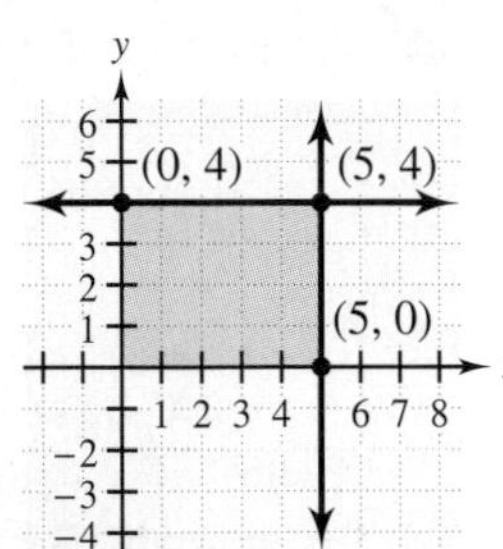

16.

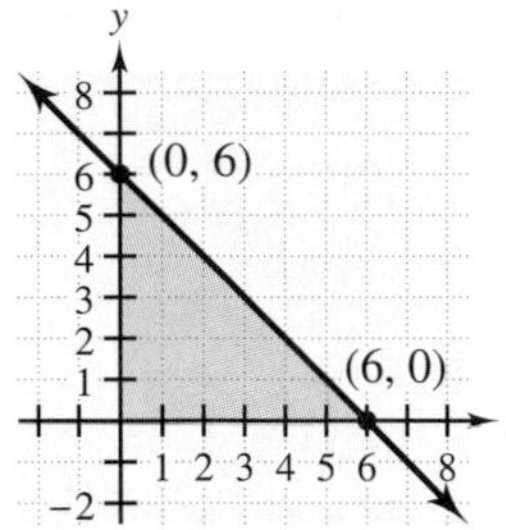

17.

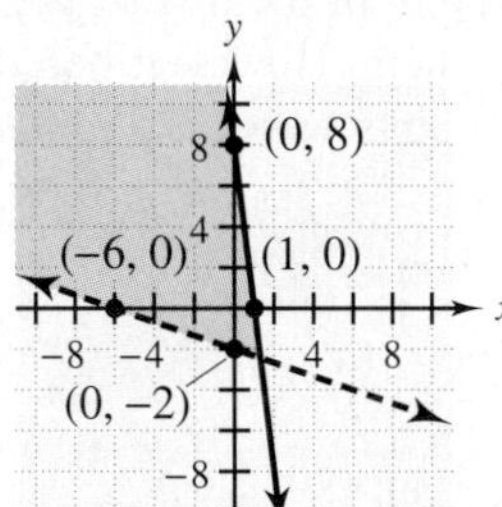

18.

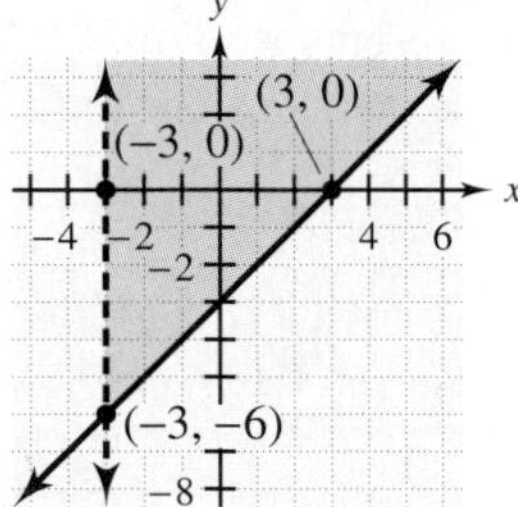

19. Describe the graph of the solution set of the following system of linear inequalities.

$$y \le 5$$
$$y \ge -5$$
$$x \le 5$$
$$x \ge -5$$

20. Suppose you want to determine whether (a, b) is a solution of a system of three inequalities, and you find that the pair does not satisfy the first inequality. Why is it not necessary to test the pair in the other two inequalities?

In Exercises 21–26, graph the system of linear inequalities.

21. $y + 3x \le 3$, $y + x + 1 \ge 0$, $y < x + 7$

22. $y \le x + 5$, $y - 3x - 2 > 0$, $y + x + 2 \ge 0$

23. $x + 5 \ge 0$, $y - x > 1$, $2x + y < 6$

24. $x + y \le 7$, $2y > x$, $y < 5x$

25. $y - 2x < 1$, $y + x + 2 \ge 0$, $y + 7 \ge 0$

26. $y \ge 0$, $x + 2y \le 8$, $3x - 2y < 4$

27. For an application involving a number x of men and a number y of women, suppose the given conditions can be translated into the following two linear inequalities. State two additional implied inequalities.

$$2x + y \le 30$$
$$x + y \ge 10$$

28. For the application in Exercise 27, every point in the intersection of the graphs of the inequalities represents a solution of the system. However, not all of these points can be used to solve the problem. Why?

In Exercises 29–40, graph the system of linear inequalities.

29. $2x + 3y \le 18$, $x \ge 0$, $y \ge 0$

30. $3x + y \le 10$, $x \ge 0$, $y \ge 0$

31. $y \ge x$, $y \le 8$, $x \ge 0$, $y \ge 0$

32. $x \le 6$, $y \le 4$, $x \ge 0$, $y \ge 0$

33. $-2x + y \le 7$, $x \le 5$, $x \ge 0$, $y \ge 0$

34. $2x - y \le 5$, $y \le 7$, $x \ge 0$, $y \ge 0$

35. $-x + y \le 1$, $3x + 4y \ge 32$, $x \ge 0$, $y \ge 0$

36. $2x - y \le 7$, $x + y \ge 2$, $x \ge 0$, $y \ge 0$

37. $3x + y \ge 5$, $x + y \ge 3$, $y \ge \frac{1}{2}x$, $x \ge 0$, $y \ge 0$

38. $4x + 3y \ge 18$, $x - y \le 8$, $2x + 3y \ge 12$, $x \ge 0$, $y \ge 0$

39. $-x + y \ge 3$, $x + y \le -2$, $x \ge 0$, $y \ge 0$

40. $x - y \ge 5$, $x + y \le 0$, $x \ge 0$, $y \ge 0$

Real-Life Applications

41. A bank invests at most \$10 million either in short-term notes or in mortgages. If the amount in mortgages cannot exceed \$7 million, describe the distribution of funds with a system of inequalities and graph the system. If \$2 million is invested in short-term notes, what is the most that can be invested in mortgages? If \$6 million is invested in short-term notes, what is the most that can be invested in mortgages?

42. A farmer allocates at most 2000 acres of land for either corn or soybeans. At most 40% of the total available acres can be used for soybeans. The number of acres for corn cannot exceed the number of acres of soybeans by more than 700 acres. Write and graph the system of inequalities that describes these conditions. If 1100 acres are in corn, what are the greatest and fewest number of acres that can be planted in soybeans?

43. A restaurant prepares tuna fruit salads and chicken Caesar salads for luncheon parties. The number of people who request tuna fruit salad never exceeds two-thirds of the number who order chicken Caesar salads. The smallest group that the restaurant books is 50. Write a system of inequalities that describes the number of salads to be made and graph the system. If 60 chicken Caesar salads are prepared, what is the possible number of tuna fruit salads prepared?

44. A coffee company produces at least 100 pounds of Irish cream coffee and Hawaiian macadamia coffee. The number of pounds of Irish cream cannot exceed twice the number of pounds of Hawaiian macadamia. Describe and graph the number of pounds of each that can be produced. What is the fewest number of pounds of Hawaiian macadamia that can be produced?

45. A builder specializes in the construction of small barns and tool buildings. The following timetable shows the number of hours needed to frame and roof the buildings and the maximum number of hours available for each kind of construction.

	Hours to Frame	Hours to Roof
Barn	15	7
Tool shed	5	6
Maximum hours available	90	84

(a) Write a system of linear inequalities and draw the graph of the system. Shade the region that represents the possible number of barns and tool buildings that can be built under these conditions.

(b) Can the builder construct 4 barns and 5 tool buildings?

(c) Can the builder construct 5 barns and 4 tool buildings?

46. A businesswoman runs a full-service catering business. She prepares food ahead of time and, on the day of the event, she sets up the tables and provides decorations. Her time requirements and constraints are summarized in the following table.

	Food Preparation Hours	Set-Up Hours
Banquet	13	8
Reception	3	5
Maximum hours available	39	40

(a) Write a system of linear inequalities and draw the graph of the system. Shade the region showing the possible number of banquets and receptions that can be catered.

(b) Can she cater 3 banquets and 2 receptions?

(c) Can she cater 2 banquets and 3 receptions?

Challenge

47. Refer to Exercise 41. If mortgages yield 7% and short-term notes yield 5%, how should the funds be invested to maximize the return to the bank?

48. Refer to Exercise 42. Suppose the profit on soybeans is 70 cents per bushel and the profit on corn is 40 cents per bushel. If the farmer can produce 35 bushels of soybeans per acre and 150 bushels of corn per acre, what land allocation would produce a maximum profit?

5. CHAPTER PROJECT

Office Communications

The table shows the percentage of office communications conducted by E-mail and by paper for three selected years. (Source: Cognitive Communication Survey.)

	Percent of Office Communications	
Year	**E-mail**	**Paper**
1991	6%	66%
1994	20%	57%
1997	40%	47%

1. For 1994 and 1997, what was the percentage of office communications conducted *either* by E-mail or by paper?
2. Assume that all other office communications are conducted orally (telephone, personal contact, meetings). What trend in oral communication can be deduced from your data in Question 1?
3. In Section 5.2 we modeled the percentage of office communications conducted by E-mail with the equation $17x - 3y = 2$, where y represents the percentage and x represents the number of years since 1990. Use this model to predict the year when half of all office communications will be conducted by E-mail.
4. Suppose that an office has 1 million communications events in the year that you estimated in Question 3. Suppose further that 10% of all E-mail is printed, with each E-mail message occupying one sheet of paper. How many reams of paper (500 sheets per ream) would be used for printed E-mail?
5. Given your results in Question 4, do you think the percentages in the table for paper are an accurate reflection of paper consumption in office communications? Why?
6. The E-mail address of the President of the United States is president@whitehouse.gov. Compose a message that expresses your views on some political subject. Then ask your instructor to help you send your message to the President via E-mail. During the next few days, look for a reply.

5. CHAPTER REVIEW EXERCISES

Section 5.1

1. Suppose you draw the graphs of $y_1 = ax + b$ and $y_2 = cx + d$ on the same coordinate axes. If the graphs intersect only at point Q, what is the significance of that point?

2. Suppose you use the substitution method to solve the following system.

$$2x + 3y = 3$$
$$y = 7x + 1$$

What is the first equation in one variable that you would write? Why?

In Exercises 3 and 4, determine whether the given ordered pair is a solution of the given system of equations.

3. (2, 1); $3x - 5y = 1$
$y = 2x - 3$

4. (−4, 5); $3x + y = -7$
$2x - 3y = 12$

In Exercises 5–8, use the graphing method to estimate the solution of the given systems of equations.

5. $y = 2x - 5$
$y = -x + 4$

6. $x - 3y = 7$
$3x = 9y + 2$

7. $2x - 5y = 10$
$10y - 4x + 20 = 0$

8. $2x + y + 2 = 0$
$3x + 2y - 1 = 0$

In Exercises 9–12, determine the number of solutions of the system without graphing or solving algebraically.

9. $y = -x + 4$
$y = x - 3$

10. $7x - 6y = 5$
$7x - y = 5(y + 1)$

11. $y = 5$
$x - 2 = 4$

12. $3x = y + 2$
$3y = 9x + 21$

In Exercises 13–18, use the substitution method to solve the given system of equations.

13. $y = x + 1$
$2x + 3y = 6$

14. $x = -2y + 3$
$2y + x = 4$

15. $y = x + 3$
$2x - 2y + 6 = 0$

16. $x - 2y = 6$
$2x + y = 4$

17. $3x + y = 2$
$4x + 8y = 1$

18. $\frac{5}{3}x + \frac{3}{2}y = 1$
$15x + 18y = 6$

Section 5.2

19. Suppose that as you solve a system of equations with the addition method, you obtain $x = \frac{17}{31}$. Explain how you can determine the y-value without using substitution.

20. Suppose that as you solve a system of equations with the addition method, you obtain an equation with no variable. Explain how to interpret this result.

In Exercises 21–28, use the addition method to solve the given system of equations.

21. $3x - 4y = 22$
$2x + 5y = -16$

22. $x - 2y + 4 = 0$
$3x + y + 5 = 0$

23. $y - 3x = 4$
$6x - 2y + 8 = 0$

24. $y = 2x + 3$
$4x - 2y - 6 = 0$

25. $\frac{3}{4}x + \frac{1}{3}y - 4 = 0$
$\frac{1}{2}x = \frac{2}{3}y + 8$

26. $\frac{3}{4}x + \frac{1}{3}y - 3 = 0$
$9x = 36 - 4y$

27. $\frac{1}{5}x - \frac{2}{3}y = 2$
$3x = 10y + 19$

28. $\frac{2}{5}x + \frac{1}{2}y = 2$
$\frac{3}{10}x + \frac{1}{4}y = 0$

In Exercises 29 and 30, determine c so that the given system of equations will be inconsistent.

29. $x + 3y = 6$
$4x - cy = 8$

30. $y - 4x = c$
$y - (c + 1)x = -7$

In Exercises 31 and 32, determine c so that the equations of the given system will be dependent.

31. $x - 2y = 5$
$cx + 6y + 15 = 0$

32. $\frac{1}{2}x + y - 3 = 0$
$cx + 4y - 12 = 0$

Section 5.3

33. What geometric figure represents the solutions of a linear equation in three variables? What do we call the coordinate system in which such a figure is drawn?

34. Suppose you want to solve the following system of equations.

$$2x + y + 3z = 7$$
$$3x - y + z = 1$$
$$x + 4y - z = 2$$

You can easily eliminate y from the first two equations and z from the last two equations. What is your opinion of this initial approach?

In Exercises 35 and 36, determine whether the given ordered triple is a solution of the given system of equations.

35. $x + y + z = 2$
$2x - 3y + 3z = 2$
$x - 2y + z = 1$
$(-2, 1, 3)$

36. $3x - z = 3$
$y + 2x - 2 = 0$
$x + y + z = 5$
$(2, 0, 3)$

In Exercises 37–46, use the addition method to solve the given system of equations.

37. $x + 2y + z = 5$
$2x + y - 2z = -3$
$4x + 3y - z = 0$

38. $y = x + 2$
$x + 2y + 4z = 9$
$x + y - z = 9$

39. $5x - 2y + 7z = -19$
$x + 3y - 2z = 3$
$4x - y - z = 16$

40. $x + 2y + 3z = -10$
$3x + 2y + z = -2$
$x + 3y + 2z = -8$

41. $2y = x + z - 3$
$7x + 5z = 3y + 1$
$2x + 2z + 5 = 4y$

42. $2x - 5y + 7z = -3$
$x + 2y - 4z = 0$
$2x - 5y + 3z = -7$

43. $2x - 2y + 10z = 12$
$x - 2y + 3z = -1$
$4x - 3y + z = -32$

44. $3x + 4z = 2y - 6$
$2(y - x) = 4z + 1$
$2(z - 2) = y - x$

45. $x + y - z = 8$
$y + 2z + 8 = 0$
$x - 2y + 3 = 0$

46. $x - y + z = 9$
$2x + y - z = 6$
$5x + 2y - 2z = -4$

Section 5.4

47. What is the difference between a coefficient matrix and an augmented matrix?

48. Describe the row operation used to transform

$$\begin{bmatrix} 2 & 5 \\ -1 & 3 \end{bmatrix} \text{ into } \begin{bmatrix} 0 & 11 \\ -1 & 3 \end{bmatrix}$$

In Exercises 49 and 50, write the augmented matrix for the given system of equations.

49. $4x - 3y = 12$
$x - 2y + 3 = 0$

50. $y - 2z - 7 = -11$
$2z - 2x = 5$
$2y - 6x = 1$

In Exercises 51 and 52, write a system of linear equations associated with the given augmented matrix.

51. $\left[\begin{array}{ccc|c} 1 & 2 & -3 & 4 \\ 3 & 1 & -2 & 5 \\ 1 & 1 & 0 & 0 \end{array}\right]$

52. $\left[\begin{array}{cc|c} 5 & -2 & -2 \\ 6 & 0 & 7 \end{array}\right]$

In Exercises 53–58, use the matrix method to solve the given system of equations.

53. $2x - y = 5$
$x - 2y = 4$

54. $y + 9 = 1.2x$
$15 - x = 0$

55. $x - y - 5 = 0$
$2x - y + 17 = 0$

56. $4x - 3y - 6z = 6$
$x - 3y + 6z = -1$
$x - 6y + 12z = -1$

57. $x - y - 2z = -2$
$x + y + 6z = 6$
$y - z = -1$

58. $3x - 4y + 2z = 1$
$-3x + 3y - z = 0$
$-6x + 9y - 3z = 3$

In Exercises 59–62, use your calculator to solve the given system of equations.

59. $x + 10 = 0$
$y + 0.3x = 6$

60. $4x - 3y = -12$
$5x + 2y = 1$

61. $1.6x + 1.2y - z = 3.2$
$x - 3y - 0.75z = -13.0$
$0.5x + y + z = 3.5$

62. $x - y = 1$
$x + 3z = 6$
$y - 2z = 0$

Section 5.5

63. Suppose that you apply Cramer's Rule to solve a system of equations and obtain the following expression for x. What can you conclude?

$$x = \frac{1}{\begin{vmatrix} 2 & 4 \\ 3 & 6 \end{vmatrix}}$$

64. The number 13 is associated with the following matrix. What do we call this number, and how is it calculated?

$$\begin{bmatrix} 2 & -1 \\ 3 & 5 \end{bmatrix}$$

In Exercises 65 and 66, evaluate the determinant.

65. $\begin{vmatrix} 7 & 20 \\ 3 & 10 \end{vmatrix}$ **66.** $\begin{vmatrix} 0 & 0 \\ 0 & 1 \end{vmatrix}$

67. For the following matrix, evaluate the minor for the element 6.

$$\begin{bmatrix} 1 & 2 & 3 \\ 4 & 5 & 6 \\ 7 & 8 & 9 \end{bmatrix}$$

68. Use your calculator to evaluate the determinant associated with the matrix in Exercise 67.

In Exercises 69 and 70, use Cramer's Rule to solve the system of equations.

69. $\begin{aligned} 7x + 3y &= -4 \\ -11x - 13y &= -2 \end{aligned}$

70. $\begin{aligned} 2x - 5y + z &= -1 \\ x + y - 4z &= 13 \\ 3x - 2y + 5z &= -12 \end{aligned}$

Section 5.6

In Exercises 71–76, assign variables to the unknown quantities and write a system of equations to describe the situation. Use any method to solve the system.

71. A cash register drawer contains \$705 in five- and twenty-dollar bills. If there is a total of 60 bills, how many bills of each denomination are in the drawer?

72. The difference between one number and twice a smaller number is 20. The smaller number plus 10 is one-half the larger number. What are the numbers?

73. A hardware store stocks two types of hammers. The store has a total of 42 hammers, with sledge hammers selling at \$22.95 each and claw hammers selling at \$10.95 each. If the total value of the hammers is \$639.90, how many of each type are in stock?

74. A forestry ranger had to hike from the park office up into the mountains to reach the newly planted seedlings that she had to inspect. She hiked at an average rate of 3.5 mph to reach this area. The ranger then averaged 2.5 mph through the inspection area. On her return to the park office, she again averaged 3.5 mph. The total length of her hike was 17 miles, which she covered in 6 hours. How long was she in the planted area? How far from the park office is the inspection site?

75. The sum of the three digits on a license plate is 12. The sum of the first two digits is 7, and the sum of the last two digits is 8. What is the three-digit number on the license plate?

76. Determine the values of a, b, and c so that the graph of the equation $y = ax^2 + bx + c$ contains the points (4, 8), (2, 2.5), and (−8, 35). Then write the specific equation.

Section 5.7

In Exercises 77–86, graph the given system of inequalities.

77. $\begin{aligned} y &< 4 \\ 2x - 5y &< 10 \end{aligned}$

78. $\begin{aligned} x &\geq -3 \\ 2x + 3y &\leq 5 \end{aligned}$

79. $\begin{aligned} y &< 3x + 1 \\ 2y + 3x &\leq 10 \end{aligned}$

80. $\begin{aligned} y + x &< -5 \\ y - x &\geq -5 \end{aligned}$

81. $\begin{aligned} y &\leq 3x \\ 3y &\geq x \\ x + y &\leq 9 \end{aligned}$

82. $\begin{aligned} 4y - 3x &< 16 \\ y &> -5 \\ 6y &< 5x + 36 \end{aligned}$

83. $\begin{aligned} y &\geq 2 \\ y &\geq -3 \\ x &\geq -5 \end{aligned}$

84. $\begin{aligned} y - x &> 3 \\ y &< x - 5 \\ 3x + 7y &< 14 \end{aligned}$

85. $\begin{aligned} x + y &\leq 10 \\ x + y &\geq 2 \\ x &\geq 0 \\ y &\geq 0 \end{aligned}$

86. $\begin{aligned} 2x - 3y &\leq 6 \\ y &\leq 7 \\ x &\leq 9 \\ x &\geq 0 \\ y &\geq 0 \end{aligned}$

87. A certain pickup truck can carry no more than 1000 pounds and no more than 20 bags of mortar mix and/or lime. A bag of mortar mix weighs 80 pounds and a bag of lime weighs 40 pounds. Write a system of inequalities that describes the given conditions and graph the solution set.

LOOKING AHEAD

The following exercises review concepts and skills that you will need in Chapter 6.

1. Evaluate the expression $3x^2 - 2x - 7$ for $x = 2$.

2. Let $x = -2$ and $y = 3$. Evaluate $4y^2 + \frac{1}{2}x^3$.

3. Use the commutative and associative properties to rewrite the expression $(3x)(-5y)$ in simpler form.

4. Use the Distributive Property to rewrite each expression.

(a) $5(2 - 3x)$ (b) $-1(4 + y)$

(c) $3x + 3y$ (d) $2ab - 7b$

5. Simplify the given expression.
   (a) $3x + 1 - x + 4$
   (b) $12x^3 - x^2 + x^3 - 6x^2$
6. Simplify the given expression.
   (a) $3(t - 2) - t$ (b) $(a + 3) - (2a - 1) + 6$
7. Write each product in exponential form and each exponential form in factored form.
   (a) $x \cdot x \cdot x \cdot x \cdot x$ (b) $(a - 1)(a - 1)$
   (c) $3y^2$ (d) $(t + 2)^2$
8. Evaluate each function as indicated.
   (a) $f(x) = 2x - 15 - x^2;\quad f(3)$
   (b) $g(x) = 7 - 15x^2 - 8x;\quad g(-1)$
9. Find two numbers whose sum is the first number and product is the second number.
   (a) 9, 20   (b) −5, −6
   (c) −7, 12   (d) 6, −16

In Exercises 10 and 11, solve the given equation.

10. (a) $2x + 1 = 0$ (b) $-5x = 0$
11. $|x - 7| = 8$
12. Explain how to use a calculator to estimate an x-intercept of the graph of a function.

5. CHAPTER TEST

1. Use the graphing method to estimate the solution of the following system of equations. Show your verification of the solution by substitution.
$$y = -2x + 3$$
$$x - y = 3$$

In Questions 2 and 3, use the substitution method to solve the system of equations.

2. $x + 2y = 10$; $3x + 4y = 8$
3. $2x + y = 7$; $x - y = 8$

In Questions 4 and 5, use the addition method to solve the system of equations.

4. $x - 2y = 3$; $3x = 6y + 9$
5. $x - \frac{3}{2}y = 13$; $\frac{3}{2}x - y = 17$

6. Use the addition method to solve the system of equations.
$$x - 2y + 3z = 7$$
$$2x + y + z = 4$$
$$-3x + 2y - 2z = -10$$

In Questions 7 and 8, use matrix methods to solve the system of equations.

7. $3x - 6y = 24$; $5x + 4y = 12$
8. $x + y - z = 6$; $3x - 2y + z = -5$; $x + 3y - 2z = 14$

9. Use your calculator to solve the system of equations.
$$x + 2y - z = -3$$
$$2x - 4y + z = -7$$
$$-2x + 2y - 3z = 4$$

In Questions 10–12, classify the given system of equations as a consistent system with a unique solution, an inconsistent system, or a system of dependent equations.

10. $2x + 3y = 5$; $4x + 6y = 10$
11. $2x + y = 2$; $8x + 4y = 7$
12. $x + y = 6$; $x - y = 4$

13. Describe the graph of a system of two linear equations in two variables for each of the following.
   (a) The equations are dependent.
   (b) The system has a unique solution.
   (c) The system is inconsistent.
14. Determine k so that the following equation is true.
$$\begin{vmatrix} 2 & -1 \\ 4 & 6 \end{vmatrix} = \begin{vmatrix} -3 & k \\ -1 & -5 \end{vmatrix}$$
15. Use Cramer's Rule to solve the system of equations.
$$x - 3y = -8$$
$$2x + 7y = -3$$

In Questions 16 and 17, graph the solution set of the system of inequalities.

16. $y < -2x$
$y > -\frac{2}{3}x$

17. $2x + y \le 12$
$y \le 8$
$x \ge 0$
$y \ge 0$

In Questions 18–21, define variables, translate the problem into a system of equations, solve the system of equations, and state the answer to the question.

18. With a tail wind an airplane flies 600 miles in 3 hours, and against a head wind the plane flies the same distance in 4 hours. What is the wind speed and the airspeed of the plane?

19. Two angles are complementary. The measure of one angle is 5° less than twice the measure of the other. What is the measure of each angle?

20. A chemistry lab has two containers of hydrochloric acid. One is a 15% solution and the other a 40% solution. How many liters of each is required to make 100 liters of a 30% solution?

21. For a seafood party the caterer estimated that the number of pounds of shrimp must be half the combined weight of the crabs and lobster. The number of pounds of crabs should be 5 less than the number of pounds of lobster. The cost of shrimp is \$5 per pound, crab is \$7 per pound, and lobster is \$9 per pound. How many pounds of each can the caterer purchase for \$960.50?

In Question 22, write a system of inequalities that describes the given conditions and draw a graph of the solution set.

22. On a vacant lot the total number of pines and dogwoods does not exceed 20. There are at least twice as many pines as dogwoods. What are the possible numbers of pines and dogwoods?

4–5 CUMULATIVE TEST

1. Find the values of a and b so that the ordered pairs are solutions of the given equation.

   $2x + 3y = 7$ $(3, a)$ $(b, 1)$

2. Determine the intercepts of the graph of the equation $5x + y = -10$.

3. For each of the given equations, state whether the slope of the graph is positive, negative, zero, or undefined.

   (a) $x + 2y = 3$ (b) $x + 2 = 3$
   (c) $0 \cdot x + 2y = 3$ (d) $x = 2y + 3$

4. Write $x - 2y + 4 = 0$ in slope–intercept form. Use the slope and y-intercept to draw a graph of the equation.

5. Consider the equation $y = \frac{2}{3}x - 1$. Describe the rate of change of y with respect to x.

6. Line L_1 contains point $P(-4, 3)$. Line L_2 is perpendicular to L_1. If L_1 and L_2 intersect at $Q(3, -2)$, what is the slope of L_2?

7. Write an equation of a line that contains points $A(-2, -5)$ and $B(6, 3)$. Express the equation in the form $Ax + By = C$.

8. The equation of line L_1 is $2x + 3y = 6$. Line L_2 is parallel to L_1 and contains the point $P(1, -4)$. Write the equation of L_2 in slope–intercept form.

9. The graph of $x - y < 7$ is a shaded half-plane.

   (a) Describe the boundary line.

   (b) Describe two methods for determining which half-plane to shade.

10. Draw the graph of $|x - y| \le 5$.

11. Each of the following is a graph of a system of two linear equations in two variables. From the graph, describe the solution set of the system.

   (a)

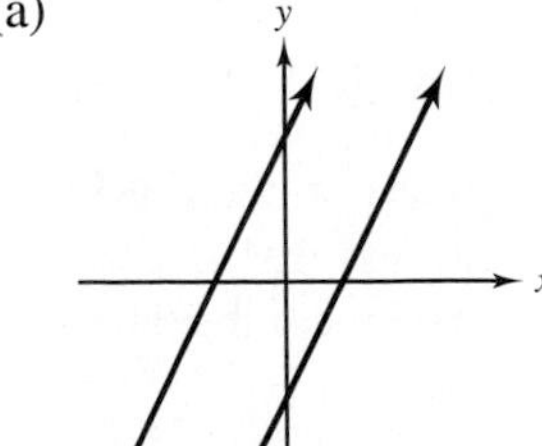

   (b)

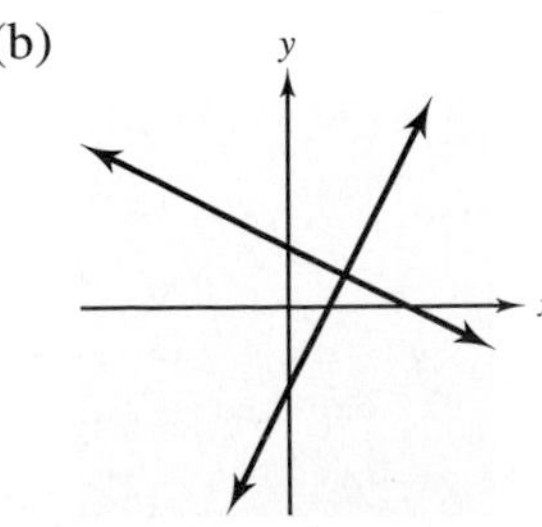

(c)

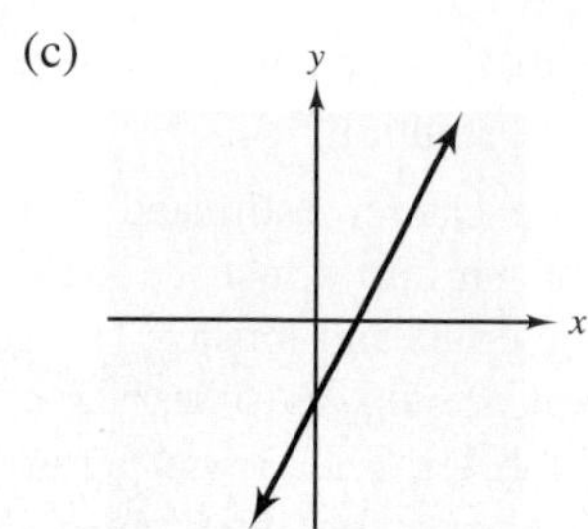

12. Use the substitution method to solve the following system of equations.

$$x + 7y = 40$$
$$3x - 2y = -18$$

13. Use the addition method to solve each of the following systems. Indicate if a system is inconsistent or if the equations are dependent.

(a) $2x - 3y = -21$
$y = \frac{2}{3}x + 7$

(b) $4x - y = -9$
$2y - x = 11$

(c) $y = 5x + 1$
$x - \frac{y}{5} = 0.2$

14. A college bookstore placed a textbook order for Mathematics 101 and English 122. The mathematics textbooks cost \$35 each and the English textbooks cost \$28 each. If a total of 420 books were ordered and the total cost was \$13,160, how many of each book were ordered?

15. Use the addition method to solve the following system of equations.

$$2x + 3y + 5z = 18$$
$$3x - 2y + 4z = 13$$
$$4x - 3y + 3z = 8$$

16. Two angles are supplementary. A third angle is half as large as the first angle and 30° larger than the second angle. Find the measures of the three angles.

17. Use matrix methods to solve the following system of equations. Then verify the solution with your calculator's automatic solve procedure.

$$x + y + z = 2$$
$$2x - y + z = 9$$
$$-x + 2y + 2z = -5$$

18. For what value of k is the following equation true?

$$\begin{vmatrix} 2 & -7 \\ k & -2 \end{vmatrix} = -25$$

19. Show why Cramer's Rule *cannot* be used to solve the following system.

$$-2x + y + 3z = 1$$
$$x - 2y + z = 2$$
$$3x - 3y - 2z = -2$$

20. A cyclist rides 32 miles against the wind in 2 hours. The return trip takes 40 minutes less time. What is the wind speed?

21. A cruise ship has a total of 500 living accommodations on decks A, B, and C. For a 4-day cruise, A-deck spaces cost \$1200, B-deck spaces cost \$900, and C-deck spaces cost \$500. The combined number of A- and B-deck spaces equals the number of C-deck spaces. If the total receipts from a sold-out cruise are \$380,000, how many B-deck spaces are there?

22. Graph the solution set of the following system of inequalities.

$$x + y \leq 9$$
$$2x - y \leq 0$$
$$y \geq 2$$

23. The perimeter of a rectangle cannot exceed 40 inches. The width of the rectangle can be no longer than half the length. Write a system of *all* inequalities that describe these conditions.

Chapter 6

Polynomials

The accompanying table shows the number (in thousands) of Americans who traveled to Europe during the period 1985–1992. The table also shows the average amount of money spent per trip.

These data can be modeled with polynomial expressions. Various techniques for **factoring** polynomials can be used to solve **quadratic equations** and to learn about the features of the graphs of polynomials. (For more on this topic, see Exercises 113–116 at the end of Section 6.4, and see the Chapter Project.)

In this chapter we learn how to perform the four basic operations with polynomials. We also learn numerous techniques for factoring a polynomial, and we apply this skill to equation solving. We conclude with division, including synthetic division, and two important theorems about polynomials.

Year	Number of Travelers (in thousands)	Average Expenditures (in dollars)
1985	6780	1391
1988	7438	1526
1989	7233	1655
1990	8043	1740
1991	6316	1825
1992	7136	1918

(Source: U.S Travel and Tourism Administration.)

6.1 ADDITION AND SUBTRACTION

Definitions • Addition of Polynomials • Subtraction of Polynomials • Polynomial Functions

Definitions

Recall that **terms** are parts of an expression that are separated by plus signs. In this section we consider one kind of term called a **monomial,** and we consider expressions called **polynomials** whose terms are monomials.

> **Definition of Monomial and Polynomial**
>
> A **monomial** is a number or a product of a number and one or more variables with nonnegative integer exponents. A **polynomial** is a finite sum of monomials.

An example of a polynomial is $4x^5 - 5x^3 + 7x + 2$. We can write this expression as a *sum* of monomials: $4x^5 + (-5x^3) + 7x + 2$. The terms are $4x^5$, $-5x^3$, $7x$, and 2 (or $2x^0$).

The **numerical coefficient** (usually just called the **coefficient**) of a term is the numerical factor of the term. In our example, the coefficients are 4, -5, 7, and 2. We also say that 2 is the **constant term.**

The following are some examples of algebraic expressions that are not polynomials.

$5x^2 + 3x^{-5}$ Exponents must be nonnegative integers.

$\dfrac{x - 4}{x^2 + 1}$ Variables cannot be in the denominator of a fraction.

$\sqrt{x} + 3$ Variables cannot appear in a radicand.

A polynomial consists of one or more terms (monomials). When a polynomial has only one term, we usually refer to the expression simply as a monomial. A polynomial with two nonzero terms is called a **binomial,** and a polynomial with three nonzero terms is called a **trinomial.**

Polynomial	*Number of Terms*	*Name*
$t^2 - 6 + 3t$	3	trinomial
$16 - x^2$	2	binomial
$5x^4$	1	monomial
-15	1	monomial

Except for 0, the **degree of a monomial** in one variable is the value of the exponent on the variable. The degree of a monomial in more than one variable is the sum of the exponents on the variables. For instance, the degree of $5x^4$ is 4, and the degree of a^2b^5 is $2 + 5$, or 7. Because a constant term such as 2 can be written $2x^0$, the degree of a nonzero constant is 0. The degree of the monomial 0 is not defined because 0 can be written $0x^n$ for any nonnegative integer n.

The **degree of a polynomial** is the degree of the term with the highest degree.

EXAMPLE 1

Determining the Degree of a Polynomial

Polynomial	*Degree*	
$2y$	1	$2y = 2y^1$
$7 - 3x^4$	4	The degree of $-3x^4$ is 4.
$4t^5 - 5t^3 + 7t + 2$	5	The largest exponent is 5.
$3x^2y^4 - 4xy^3 + x^3y^5$	8	The degree of x^3y^5 is $3 + 5$, or 8.

The conventional way to write a polynomial in one variable is to write the terms in order of descending (or decreasing) degrees. We say that the polynomial is written in **descending order.**

In some instances, it is convenient to write the polynomial in **ascending** (or increasing) order. In such cases, we begin with the term with the lowest exponent. Then the terms that follow have increasing degrees.

Polynomial	*Descending Order*	*Ascending Order*
$4x^5 - 5x^3 + 7x + 2$	$4x^5 - 5x^3 + 7x + 2$	$2 + 7x - 5x^3 + 4x^5$
$-2x - x^5 + 7x^3$	$-x^5 + 7x^3 - 2x$	$-2x + 7x^3 - x^5$

In the first polynomial, there is no term with x^4 or x^2. We say that these are *missing terms.* The polynomial could have been written with the missing terms:

$$4x^5 + 0x^4 - 5x^3 + 0x^2 + 7x + 2$$

We evaluate polynomials in the same way as we evaluate any expression.

EXAMPLE 2

Evaluating a Polynomial in Two Variables

Evaluate the polynomial $3x^2y^4 - 4xy^3 + x^3y^5$ for $x = 2$ and $y = -1$.

Figure 6.1

```
2→X
                    2
-1→Y
                   -1
3X^2Y^4-4XY^3+X^
3Y^5
                   12
```

Solution

$$\begin{aligned} 3x^2y^4 - 4xy^3 + x^3y^5 &= 3(2)^2(-1)^4 - 4(2)(-1)^3 + (2)^3(-1)^5 \\ &= 3(4)(1) - 4(2)(-1) + 8(-1) \\ &= 12 + 8 - 8 \\ &= 12 \end{aligned}$$

As before, we can use a calculator to evaluate the polynomial. Figure 6.1 shows a typical display.

Addition of Polynomials

Terms of a polynomial are **like terms** if they are both constant terms of if they have the same variables with the same exponents. To add polynomials, we simply combine like terms.

EXAMPLE 3

Adding Polynomials

Add the polynomials $5x^3 - 4x + 2$ and $-x^3 + 5x^2 - 2x + 5$.

Solution

$$\begin{aligned}&(5x^3 - 4x + 2) + (-x^3 + 5x^2 - 2x + 5)\\ &\quad = 5x^3 - 4x + 2 - x^3 + 5x^2 - 2x + 5 && \text{Remove parentheses.}\\ &\quad = 4x^3 + 5x^2 - 6x + 7 && \text{Combine like terms.}\end{aligned}$$

Sometimes we write a polynomial addition problem as column addition. This approach has the advantage that like terms are aligned in columns. When you use the column approach, it is helpful to write any missing terms with a coefficient of 0.

$$\begin{array}{rl} 5x^3 + 0x^2 - 4x + 2 & \text{First polynomial} \\ -1x^3 + 5x^2 - 2x + 5 & \text{Second polynomial} \\ \hline 4x^3 + 5x^2 - 6x + 7 & \text{Add the polynomials.} \end{array}$$

Subtraction of Polynomials

The definition of subtraction of real numbers applies to the subtraction of polynomials. If P and Q represent polynomials, then $P - Q = P + (-Q)$.

EXAMPLE 4

Subtracting Polynomials

Subtract the polynomial $2x^3 - x + 4$ from $-2x^3 + x^2 - 4x + 1$.

LEARNING TIP

When you subtract polynomials, write the intermediate step that shows the result of removing grouping symbols. This will help you to avoid the sign errors that commonly occur in subtraction.

Solution

$$\begin{aligned}&(-2x^3 + x^2 - 4x + 1) - (2x^3 - x + 4)\\ &\quad = -2x^3 + x^2 - 4x + 1 - 2x^3 + x - 4 && \text{Remove parentheses.}\\ &\quad = -4x^3 + x^2 - 3x - 3 && \text{Combine like terms.}\end{aligned}$$

Polynomial subtraction problems can also be done with columns. Again, it is helpful to write any missing terms with the coefficient of 0.

$$\begin{array}{rcr} -2x^3 + 1x^2 - 4x + 1 & & -2x^3 + 1x^2 - 4x + 1 \\ 2x^3 + 0x^2 - 1x + 4 & \text{Change signs.} & -2x^3 - 0x^2 + 1x - 4 \\ \hline & \text{Add the polynomials.} & -4x^3 + 1x^2 - 3x - 3 \end{array}$$

Examples 3 and 4 show that we add and subtract polynomials simply by removing parentheses and combining like terms. We can simplify any polynomial expression in the same way.

EXAMPLE 5

Simplifying Polynomial Expressions

(a) $2(x^3 - 2x^2 - 6x - 1) - 3(-2x^3 + x^2 - 4x - 2)$

$= 2x^3 - 4x^2 - 12x - 2 + 6x^3 - 3x^2 + 12x + 6$ Distributive Property

$= 8x^3 - 7x^2 + 4$ Combine like terms.

(b) $(2x^3y^4 - x^2y^3 + 3xy - 7) + (x^3y^4 + 3x^2y^3 + 5x - 2)$

$= 2x^3y^4 - x^2y^3 + 3xy - 7 + x^3y^4 + 3x^2y^3 + 5x - 2$ Remove parentheses.

$= 3x^3y^4 + 2x^2y^3 + 3xy - 9 + 5x$ Combine like terms.

(c) $(x^2y^3 - 2xy^2 - 5xy - 7) - (x^2y^3 - xy^2 - 5xy + 3)$

$= x^2y^3 - 2xy^2 - 5xy - 7 - x^2y^3 + xy^2 + 5xy - 3$ Remove parentheses.

$= -xy^2 - 10$ Combine like terms.

Polynomial Functions

We can use function notation to give a polynomial a name and to indicate the variable used in the polynomial. For example, to write a polynomial in t, we give the polynomial a name such as p and use the function notation $p(t)$. (We can also use function notation for polynomials in more than one variable, but that topic is beyond the scope of this book.)

Function notation is especially useful when we evaluate a polynomial. We can write "the value of the polynomial p when t has a value of 3" as $p(3)$.

EXAMPLE 6

Evaluating a Polynomial

For $p(t) = -t^3 + 3t - 6$, calculate $p(-2)$.

Solution

$$\begin{aligned} p(t) &= -t^3 + 3t - 6 \\ p(-2) &= -1(-2)^3 + 3(-2) - 6 \\ &= -1(-8) - 6 - 6 \\ &= 8 - 6 - 6 \\ &= -4 \end{aligned}$$

Figure 6.2

```
-2→T
                -2
-T^3+3T-6
                -4
```

You can also use your calculator to evaluate the polynomial. Figure 6.2 shows a typical screen display.

When two polynomial *expressions* are added or subtracted, the result is another polynomial expression. If $p(x)$ and $q(x)$ are two polynomial *functions,* then $p(x) + q(x)$, the sum of the two functions, is also a function. The functional notation used for $p(x) + q(x)$ is $(p + q)(x)$. Similarly, we write $p(x) - q(x) = (p - q)(x)$.

EXAMPLE 7

Evaluating Sums of Polynomial Functions

Given $p(x) = x^2 - 5$ and $q(x) = 3x + 2$, write $(p + q)(x)$. Using your calculator to evaluate the expressions, verify that $(p + q)(3) = p(3) + q(3)$.

Solution

$$\begin{aligned}(p + q)(x) &= p(x) + q(x)\\ &= (x^2 - 5) + (3x + 2)\\ &= x^2 + 3x - 3\end{aligned}$$

To evaluate the expressions with a calculator, we enter $x^2 - 5$, $3x + 2$, and $x^2 + 3x - 3$ as Y_1, Y_2, and Y_3, respectively. Then we let $x = 3$ and show that $Y_1 + Y_2 = Y_3$. Figure 6.3 shows a typical display.

Figure 6.3

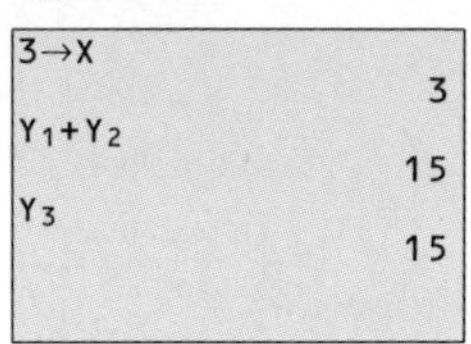

Think About It

We used $(p + q)(x)$ and $(p - q)(x)$ to represent the sum and difference of two polynomial functions. How could we represent the product of two polynomial functions? the quotient?

6.1 QUICK REFERENCE

Definitions

- A **monomial** is a number or the product of a number and one or more variables raised to nonnegative integer powers. A **polynomial** is a finite sum of monomials.
- The **numerical coefficient** of a term is the numerical factor of the term. A **constant term** of a polynomial is a term that is just a number.
- Polynomials with one, two, and three terms are called **monomials, binomials,** and **trinomials,** respectively.
- Except for 0, the **degree of a monomial** in one variable is the value of the exponent on the variable. The degree of a monomial in more than one variable is the sum of the exponents on the variables. The **degree of a polynomial** is the degree of the term with the highest degree.

Addition of Polynomials

- Terms of a polynomial are **like terms** if they are both constant terms or if they have the same variables with the same exponents.
- To add polynomials, we remove parentheses and combine the like terms. Column addition has the advantage that like terms are aligned in columns.

Subtraction of Polynomials

- To subtract polynomials, remove parentheses by changing the sign of each term of the polynomial that is being subtracted. Then combine like terms.
- Column subtraction has the advantage that like terms are aligned in columns. However, remember to change the sign of each term of the polynomial being subtracted. Then add the like terms.
- All polynomial expressions are simplified by removing parentheses and combining like terms.

Polynomial Functions

- We can use function notation to give a polynomial a name and to indicate the variable used in the polynomial. Function notation is especially useful when we evaluate a polynomial.

- When two polynomial functions $p(x)$ and $q(x)$ are added or subtracted, the result is another polynomial function, which can be written as $(p + q)(x)$ for addition or $(p - q)(x)$ for subtraction.

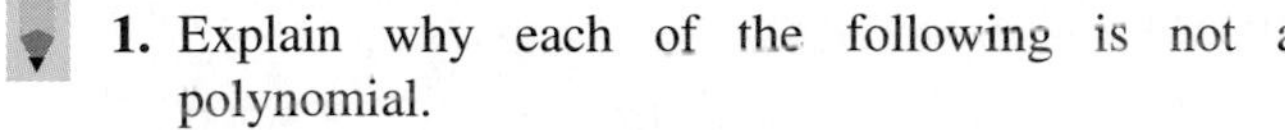

Concepts and Skills

1. Explain why each of the following is not a polynomial.

   (a) $\frac{5}{x+2}$ (b) $\frac{1}{3}x^{-4}$

2. Two of the following statements are true. Explain why the other two statements are not true.

   (i) The term $5x^2$ is both a monomial and a polynomial.

   (ii) The coefficients of $-2x^3 + 5x - 8$ are -2 and 5.

   (iii) An example of a third-degree term is 2^3.

   (iv) The expression $5 + 4x + 3x^2 + 2x^3$ is in ascending order.

In Exercises 3–6, state whether the expression is a polynomial.

3. $5x^2 + 4x - 6$
4. $\frac{5}{x^2} + 4x - 6$
5. $\sqrt{7} - 4x$
6. $4x^3 - 8x + 5x^2$

In Exercises 7–12, give the coefficient and degree of each term.

7. 5
8. $-y$
9. $-4x^3$
10. $2\pi z^2$
11. xy
12. $4ab^2$

13. What is the difference between the degree of a term and the degree of a polynomial with more than one term?

14. The expression $x^2 + 5$ can be written in the form $x^2 + 0x + 5$. Does this mean that $x^2 + 5$ is a trinomial? Why?

In Exercises 15–18, determine the degree of the polynomial.

15. $2x + 3y$
16. $4xy - 3y$
17. $2x^4 + xy^2 - x^2y^3$
18. $7a^4 - 3a^2b$

In Exercises 19–24, simplify each polynomial and determine its degree. Classify the simplified polynomial as a monomial, binomial, trinomial, or none of these.

19. $7x + 3x^5 - 9$
20. $5x^6$
21. $5x^3 + 7x - 2x^2 - 3$
22. $2x^2 + 3x - 1 - 2x^2$
23. $x^2y^3 + xy^2$
24. $3xy^2 + 2^3 - 3xy^2$

In Exercises 25–30, determine the degree of the polynomial. Then write the polynomial in descending order and in ascending order.

25. $6 - 3y^2 + 6y$
26. $7 - 2y^3$
27. $x^3 - 5x^6 + 4x^9 - 7x^8$
28. $3x^4 + 5x^7 - 6x^2 + 4x - 7x^{11}$
29. $4n^8 - 3n^4 + n^{12} - 2n^6$
30. $8p^5 - p^9 + 22p^4 - 5^7$

In Exercises 31–36, evaluate the polynomial for the given values of the variables.

31. $2x^2 + 3xy$

    $x = -1, y = 2$

32. $x^2y - 3y^2$

    $x = 2, y = -1$

33. $ab^2 - 2ab$

    $a = 1, b = -3$

34. $ab^2 + a^2b$

    $a = -2, b = 2$

35. $x^2y + 3xy - 2y^2$

    $x = -2, y = 3$

36. $x^3y^2 - 4x^2y + 5y^3$

    $x = 1, y = -1$

In Exercises 37–40, add the polynomials.

37. $(-a^2 + 2a - 1) + (a^2 + 2a + 1)$
38. $(x^2 - 5x) + (3x^2 - 4x - 2)$
39. $(2xy^2 - y^3) + (y^3 + xy^2)$
40. $(m^2n + mn) + (mn + 1)$

In Exercises 41–46, subtract the polynomials.

41. $(-5x - 3) - (2x - 9)$
42. $(5x - 6) - (-4x + 8)$
43. $(2x^2 - 4x + 7) - (-2x^2 - 3x - 7)$
44. $(x^2 + 6x + 3) - (4x - x^2)$
45. $(-3t + 7) - (2t^2 + 6)$
46. $(m^2 + 3) - (4 - 3m)$

In Exercises 47–52, perform the indicated operations.

47. $(4mn + n^2) - (m^2 - 3mn)$

48. $(x^2y + 3) - (2 - x^2y)$

49. $(x^2y + 2xy + 3y^2) + (-x^2y - 4xy + 2y^2)$

50. $(ab^2 + a^2b - 3ab) + (3ab^2 - 2ab - a^2b)$

51. $(9a^2b + 2ab - 5a^2) - (2ab - 3a^2 + 4a^2b)$

52. $(4y^3 - 3xy^2 + 4) - (2xy^2 - 3y^3 - 2)$

In Exercises 53–60, add or subtract as indicated.

53. Subtract

$$\begin{array}{l} 4x^3 - 10x \\ \underline{6x^3 - 2x} \end{array}$$

54. Subtract

$$\begin{array}{r} 3y - 6 \\ \underline{-5y + 2} \end{array}$$

55. Add

$$\begin{array}{r} x^2 + 3x - 7 \\ \underline{-5x^2 + 2x - 1} \end{array}$$

56. Add

$$\begin{array}{rrrr} 4x^3 & - 5x^2 & & + 6 \\ \hline & 2x^2 & - 5x & + 7 \end{array}$$

57. Subtract

$$\begin{array}{l} 4x^2 - x - 6 \\ \underline{3x^2 - x + 6} \end{array}$$

58. Subtract

$$\begin{array}{llll} 2x^4 & + 3x^2 & & - 6 \\ 3x^4 & & - 2x & + 5 \\ \hline \end{array}$$

59. Add

$$\begin{array}{rrr} x^2 & + 3x & - 2 \\ -2x^2 & - x & \\ 3x^2 & & + 4 \\ \hline \end{array}$$

60. Add

$$\begin{array}{rrrr} y^3 & + 2y^2 & & - 5 \\ -4y^3 & & - 3y & + 2 \\ & 6y^2 & + 2y & - 5 \\ \hline \end{array}$$

61. For the problem "Subtract $x - 2$ from $3x + 4$," which of the following is correct? Why?

(i) $(x - 2) - (3x + 4)$

(ii) $(3x + 4) - (x - 2)$

(iii) $3x + 4 - x - 2$

62. To solve the equation $\frac{1}{2}x - \frac{1}{3} = \frac{3}{4}$, we clear fractions by multiplying by the LCD. Can we do the same for the polynomial $\frac{1}{2}x - \frac{1}{3} - \frac{3}{4}$? Why?

63. Find the sum of $3x^2 - 6x$ and $4x + 7$.

64. Find the sum of $y^3 - 2y + 4$ and $y^2 + 2y + 1$.

65. Subtract $7x^2 - 2x + 9$ from $-5x - 2x^2$.

66. Subtract $2x^3 + 3x - 1$ from $x^3 + 3x - 1$.

67. What must be added to $x - 7$ to obtain 0?

68. What must be added to $3x + 5$ to obtain $x + 9$?

69. What must be subtracted from $x - 3$ to obtain 0?

70. What must be subtracted from $5x - 9$ to obtain $3x + 2$?

In Exercises 71–80, simplify.

71. $(2x - 3) - (3x^2 + 4x - 5) - (3x - 4x^2)$

72. $(-x^2 - x - 1) + (2x^2 + 3x + 4) - (2x - 3x^2)$

73. $-[3x - (2x - 1)]$

74. $-(2y + 4) - (2y - 4)$

75. $(x - 3) + [2x - (3x - 1)]$

76. $(2 - 3x) - [(3x + 1) - (1 - 2x)]$

77. $[3 - (y - 1)] - (2y + 1)$

78. $[(a^2 - 2) + (a + 2)] - (a^2 - 3a)$

79. $(3x^2 - 2xy + y^2) - (4xy + 3x^2 - y^2) + (4xy^2 - 2x^2)$

80. $(4a^2 + 5ab - 7b^2) + (5ab - a^2 + 3b^2) - (2a^2 - 7ab + 3a^2b)$

In Exercises 81–86, evaluate the function as indicated.

81. $P(x) = 3 - x$

$P(-2), P(3)$

82. $P(x) = -2x + 3$

$P(0), P(-1)$

83. $Q(x) = -x^2 + 3x - 1$

$Q(0), Q(-2)$

84. $P(x) = 3x^3 - 2x^2 + 6x + 4$

$P(-3), P(1)$

85. $R(x) = x^3 - 4x^5 + 5x^2 + 2$

$R(2.1), R(-1.7)$

86. $R(x) = x^4 + 5x^3 - 2x^2 - 5x$

$R(-3.1), R(1.2)$

In Exercises 87–90, let $P(x) = 4x + 3$, $Q(x) = x^2 - 5$, and $R(x) = 2x^2 - 7x - 8$. Perform the indicated operations.

87. $P(x) + Q(x)$

88. $Q(x) - R(x)$

89. $R(x) - [P(x) - Q(x)]$

90. $[R(x) - P(x)] - Q(x)$

Real-Life Applications

91. The attendance for a game at a minor league baseball stadium is approximated by the function $A(t) = -2t^2 + 280t - 8000$, where t is the noonday temperature (Fahrenheit). For the following noonday temperatures, what would be the expected attendance?

(a) 80°F (b) 50°F (c) 98°F
(d) 40°F (e) 100°F

(f) Using the results in parts (d) and (e), and taking into account what the function represents, what is the domain of $A(t)$?

92. The number of people swimming at a certain Florida beach is approximated by the function $N(t) = -t^2 + 140t - 4500$, where t is the Fahrenheit temperature of the water. For the following water temperatures, what would be the expected number of people swimming?

(a) 60°F (b) 75°F (c) 86°F
(d) 50°F (e) 90°F

(f) Using the results in parts (d) and (e), and taking into account what the function represents, what is the domain of $N(t)$?

93. The following functions describe costs and revenue (in dollars) from the production and sale of x units of light bulbs.

Cost C of Manufacturing	$C(x) = 20x + 12$
Cost M of Marketing	$M(x) = x^2 - 3x + 10$
Revenue R from Sales	$R(x) = 2x^2 + 8x$

(a) What is the cost of manufacturing 100 units of bulbs?

(b) What is the cost of marketing 100 units of bulbs?

(c) Write a function $T(x) = C(x) + M(x)$ that represents the total cost of manufacturing and marketing x units of bulbs.

(d) What is the total cost of manufacturing and marketing 100 units of bulbs?

(e) Show that $C(100) + M(100) = T(100)$.

(f) What is the revenue from selling 100 units of bulbs?

(g) Assuming profit is revenue minus total cost, write a function $P(x) = R(x) - T(x)$ that represents the profit from selling x units of bulbs.

(h) What is the profit from selling 100 units of bulbs?

(i) Show that $P(100) = R(100) - T(100)$.

94. The following functions describe costs and revenue (in dollars) from the production and sale of x units of radios.

Cost C of Manufacturing	$C(x) = 25x + 700$
Cost M of Marketing	$M(x) = x^2 - 2x + 15$
Revenue R from Sales	$R(x) = 4x^2 + 3x$

(a) What is the cost of manufacturing 70 units of radios?

(b) What is the cost of marketing 70 units of radios?

(c) Write a function $T(x) = C(x) + M(x)$ that represents the total cost of manufacturing and marketing x units of radios.

(d) What is the total cost of manufacturing and marketing 70 units of radios?

(e) Show that $C(70) + M(70) = T(70)$.

(f) What is the revenue from selling 70 units of radios?

(g) Assuming profit is revenue minus total cost, write a function $P(x) = R(x) - T(x)$ that represents the profit from selling x units of radios.

(h) What is the profit from selling 70 units of radios?

(i) Show that $P(70) = R(70) - T(70)$.

Modeling with Real Data

95. In 1995, more than one golf course per day was constructed in the United States. The accompanying table shows the number of courses built in selected years from 1980 to 1995. (Source: National Golf Foundation.)

Year	Courses Built
1980	132
1985	109
1990	289
1995	468

The number G of golf courses built can be modeled by $G(x) = -0.27x^3 + 8.14x^2 - 38.5x + 132$, where x is the number of years since 1980.

(a) Produce the graph of G, and trace it to estimate the year after 1980 when golf course construction was lowest and highest.

(b) What trend does the model predict by the year 2000?

96. What do we call the y-values of the points to which you traced in part (a) of Exercise 95?

Challenge

In Exercises 97–100, determine a, b, and c so that the sum or difference on the left is equal to the polynomial on the right.

97. $(ax^2 + bx - 3) + (x^2 + 2x + c) = 3x + 1$

98. $(by + c) - (ay^3 - 2y + 5) = y^3 - 1$

99. $(x^2 + bx - c) + ax^2 = x^2 + 6x + 9$

100. $(2x^2 + bxy + 3y^2) - (ax^2 - xy - cy^2) = x^2 + xy - 4y^2$

101. If $(ax^2 + 5) + (bx^2 - 7) = 4x^2 - 2$, explain why it is not possible to determine unique values for a and b.

102. If the degree of $p(x)$ is $n > 1$, what is the degree of $p(3)$?

103. Suppose that $P(x) = x^3 - 5x^2 + 6x + 10$ and $P(0) = a + b + c$, where a, b, and c are constants. If c is the sum of a and b, and if b is one less than twice a, what are a, b, and c?

6.2 MULTIPLICATION AND SPECIAL PRODUCTS

Multiplying by a Monomial • Multiplying Polynomials • Multiplying Binomials • Special Products • Simplifying Expressions

Multiplying by a Monomial

In Section 1.6 we learned the Product Rule for Exponents.

$$b^m \cdot b^n = b^{m+n}$$

In words, if the bases are the same, then we add the exponents. To multiply polynomials, we use this rule, often repeatedly.

To multiply two monomials, we use the associative and commutative properties of multiplication with the Product Rule for Exponents.

$$(4x^4)(5x^2) = (4 \cdot 5)(x^4 \cdot x^2) = 20(x^{4+2}) = 20x^6$$

To multiply a monomial times a polynomial we use the Distributive Property.

EXAMPLE 1

The Product of a Monomial and a Polynomial

Multiply $3x^2(5x^3 - x^2 - 4x + 2)$.

Solution

We use the Distributive Property to multiply the monomial times each term of the polynomial.

$$\begin{aligned}&3x^2(5x^3 - x^2 - 4x + 2)\\&\quad = 3x^2(5x^3) - 3x^2(x^2) - 3x^2(4x) + 3x^2(2)\\&\quad = 15x^5 - 3x^4 - 12x^3 + 6x^2\end{aligned}$$

Multiplying Polynomials

Multiplying a polynomial times another polynomial requires repeated use of the Distributive Property.

EXAMPLE 2

The Product of Two Polynomials

Multiply $(2x - 3)(x^2 + 3x - 5)$.

Solution

Multiply the second polynomial by $2x$ and then by -3.

$$\begin{aligned}&(2x - 3)(x^2 + 3x - 5)\\&\quad = 2x(x^2 + 3x - 5) - 3(x^2 + 3x - 5)\\&\quad = 2x(x^2) + 2x(3x) - 2x(5) - 3(x^2) - 3(3x) + 3(5)\\&\quad = 2x^3 + 6x^2 - 10x - 3x^2 - 9x + 15\\&\quad = 2x^3 + 3x^2 - 19x + 15\end{aligned}$$

Another approach to the problem in Example 2 is column multiplication. This method makes it easier to align and combine like terms.

$$\begin{array}{rrrrl} & x^2 + & 3x - & 5 & \\ & & 2x - & 3 & \\ \hline & -\ 3x^2 - & 9x + & 15 & -3(x^2 + 3x - 5) \\ 2x^3 + & 6x^2 - & 10x & & 2x(x^2 + 3x - 5) \\ \hline 2x^3 + & 3x^2 - & 19x + & 15 & \text{Add the results.} \end{array}$$

EXAMPLE 3

The Product of Two Trinomials

Multiply $(x^2 + 3x - 1)(2x^3 - x + 1)$.

Solution

Use the Distributive Property to remove the parentheses. Then combine the like terms.

$$\begin{aligned}&(x^2 + 3x - 1)(2x^3 - x + 1)\\&\quad = x^2(2x^3 - x + 1) + 3x(2x^3 - x + 1) - 1(2x^3 - x + 1)\\&\quad = 2x^5 - x^3 + x^2 + 6x^4 - 3x^2 + 3x - 2x^3 + x - 1\\&\quad = 2x^5 + 6x^4 - 3x^3 - 2x^2 + 4x - 1\end{aligned}$$

With practice, you will soon find that you do not need to write the first step. Also, you may find that the column method is especially appealing when the polynomials are large.

EXAMPLE 4

The Product of Two Binomials

Multiply $(2x - 1)(3x + 5)$.

Solution

Again, we use the Distributive Property.

$$\begin{aligned}(2x - 1)(3x + 5) &= 2x(3x + 5) - 1(3x + 5) \\ &= 6x^2 + 10x - 3x - 5 \\ &= 6x^2 + 7x - 5\end{aligned}$$

Multiplying Binomials

To multiply binomials efficiently, shortcuts or mnemonic devices are often used.

EXPLORING THE CONCEPT

The FOIL Method

Look carefully at the result in Example 4.

$$(2x - 1)(3x + 5) = 6x^2 + 10x - 3x - 5$$

1. The product of the *first terms* ($2x$ and $3x$) of the binomials is the first term of the result ($2x \cdot 3x = 6x^2$).
2. The products of the *outer terms* ($2x$ and 5) and the *inner terms* (-1 and $3x$) are the next two terms of the result $[2x \cdot 5 + (-1)(3x) = 10x - 3x]$.
3. The product of the *last terms* (-1 and 5) of the binomials is the last term of the result $[(-1)(5) = -5]$.

Using the letters in the word FOIL might help you remember which terms are multiplied and allow you to quickly compute the product.

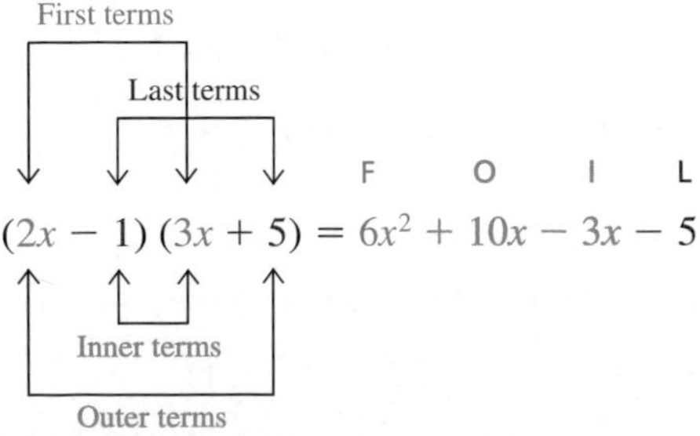

With practice, you should be able to combine like terms mentally.

EXAMPLE 5

Using the FOIL Pattern to Multiply Binomials

$$\text{(a)}\quad (x-3)(4x-1) = \overset{F}{4x^2} - \overset{O}{x} - \overset{I}{12x} + \overset{L}{3}$$
$$= 4x^2 - 13x + 3$$

$$\text{(b)}\quad (x+4)^2 = (x+4)(x+4)$$
$$= \overset{F}{x^2} + \overset{O}{4x} + \overset{I}{4x} + \overset{L}{16}$$
$$= x^2 + 8x + 16$$

Note that the product $(x + 4)^2$ is the product of the same two binomials.

$$\text{(c)}\quad (3x+5)(3x-5) = \overset{F}{9x^2} - \overset{O}{15x} + \overset{I}{15x} - \overset{L}{25}$$
$$= 9x^2 - 25$$

Note that the product $(3x + 5)(3x - 5)$ is the product of the sum of two terms and the difference of the same two terms.

Special Products

Parts (b) and (c) of Example 5 illustrate two special products of binomials.

If A and B represent terms (monomials), the special product of the sum and difference of those two terms can be written as $(A + B)(A - B)$. We can use the FOIL method to find the product.

$$(A + B)(A - B) = A^2 - AB + AB - B^2$$
$$= A^2 - B^2$$

In words, the product of the sum and difference of the same two terms is the difference of their squares.

Product of the Sum and Difference of the Same Two Terms

$$(A + B)(A - B) = A^2 - B^2$$

EXAMPLE 6

The Product of a Sum and Difference of the Same Two Terms

Multiply.

(a) $(3x + 5)(3x - 5)$ (b) $(5 + x^2)(5 - x^2)$

Solution

(a) $(A + B)(A - B) = A^2 - B^2$

$\downarrow \quad \downarrow \; \downarrow \quad \downarrow \quad \downarrow \quad \downarrow$

$(3x + 5)(3x - 5) = (3x)^2 - 5^2 = 9x^2 - 25$

(b) $(5 + x^2)(5 - x^2) = 5^2 - (x^2)^2 = 25 - x^4$

The square of a binomial is another special product. Letting A and B represent terms and using the FOIL method, we find that the square of a binomial is a trinomial of a specific form.

$$(A + B)^2 = (A + B)(A + B) = A^2 + AB + AB + B^2 = A^2 + 2AB + B^2$$
$$(A - B)^2 = (A - B)(A - B) = A^2 - AB - AB + B^2 = A^2 - 2AB + B^2$$

The first term of the result is the square of the first term of the binomial. The last term of the result is the square of the last term of the binomial. The middle term of the result is double the product of the two terms of the binomial.

The Square of a Binomial

$$(A + B)^2 = A^2 + 2AB + B^2$$
$$(A - B)^2 = A^2 - 2AB + B^2$$

EXAMPLE 7

The Square of a Binomial

Determine the following products.

(a) $(x + 4)^2$ (b) $(2x^3 + 5)^2$ (c) $(3 - 2y)^2$

Solution

Think About It

Suppose you display the table for $Y_1 = (x + 3)^2$, $Y_2 = x^2 + 9$, and $Y_3 = x^2 + 6x + 9$. Which columns should be the same? Why?

(a) $(A + B)^2 = A^2 + 2 \cdot A \cdot B + B^2$

$\quad\downarrow\ \downarrow\quad \downarrow\quad \downarrow\ \downarrow\ \downarrow\ \downarrow$

$(x + 4)^2 = x^2 + 2 \cdot (x) \cdot (4) + 4^2 = x^2 + 8x + 16$

(b) $(2x^3 + 5)^2 = (2x^3)^2 + 2(2x^3)(5) + 5^2 = 4x^6 + 20x^3 + 25$

(c) $(3 - 2y)^2 = 9 - 12y + 4y^2$

Simplifying Expressions

Sometimes we must multiply polynomials as part of a larger simplification problem. In the following examples, we use the Distributive Property, we use the FOIL pattern, and we recognize special products in simplifying the expressions.

EXAMPLE 8

Simplifying Expressions

Simplify the following expressions.

(a) $(x + 3)(2x + 1) - (x + 3)(x - 3)$

(b) $2x(x - 5)(x + 2)$

(c) $(x + 5)(x - 3)(x - 5)$

LEARNING TIP

For an expression such as $x(x + 1)(x + 2)$, avoid the error of multiplying x times both $x + 1$ and $x + 2$. Multiplying the binomials first and then the result times x is generally the easiest approach.

Solution

(a) The first product can be found with the FOIL method. The second product is the sum of two terms times the difference of the same two terms, and the result is the difference of their squares.

$$(x + 3)(2x + 1) - (x + 3)(x - 3)$$
$$= (2x^2 + 7x + 3) - (x^2 - 9)$$
$$= 2x^2 + 7x + 3 - x^2 + 9$$
$$= x^2 + 7x + 12$$

In the first step, the product $(x + 3)(x - 3)$ is the *quantity* $(x^2 - 9)$. The quantity must remain in parentheses until the subtraction is performed.

(b) We can multiply $2x$ and $x - 5$ and then multiply the result by $x + 2$.

$$2x(x - 5)(x + 2) = (2x^2 - 10x)(x + 2) = 2x^3 - 6x^2 - 20x$$

Alternatively, we can multiply the binomials and then multiply the result by $2x$.

$$2x(x - 5)(x + 2) = 2x(x^2 - 3x - 10) = 2x^3 - 6x^2 - 20x$$

(c) Look for ways to rearrange the factors so that special product patterns can be used.

$$(x + 5)(x - 3)(x - 5)$$
$$= (x + 5)(x - 5)(x - 3) \quad \text{Commutative Property of Multiplication}$$
$$= (x^2 - 25)(x - 3) \quad (A + B)(A - B) = A^2 - B^2$$
$$= x^3 - 3x^2 - 25x + 75 \quad \text{FOIL}$$

6.2 QUICK REFERENCE

Multiplying by a Monomial

- To multiply a monomial by a monomial, multiply the coefficients and the bases.
- To multiply a monomial by a polynomial, use the Distributive Property to multiply each term of the polynomial by the monomial.

Multiplying Polynomials

- To multiply two polynomials, multiply each term of one polynomial by each term of the other polynomial, and then combine like terms.
- For larger polynomials, multiplying in columns helps to align like terms for easier addition.

Multiplying Binomials

- The letters in the word FOIL (First–Outside–Inside–Last) give us a pattern for remembering how to multiply two binomials quickly.

Special Products

- The product of the sum and difference of the same two terms has the form $(A + B)(A - B)$.

  $$(A + B)(A - B) = A^2 - B^2$$

- The square of a binomial has the form $(A + B)^2$ or $(A - B)^2$. The products have the following forms:

  $$(A + B)^2 = A^2 + 2AB + B^2$$
  $$(A - B)^2 = A^2 - 2AB + B^2$$

Simplifying Expressions

- The Distributive Property, the FOIL pattern, and the special product forms are useful in simplifying algebraic expressions that involve products of polynomials.

6.2 EXERCISES

Concepts and Skills

1. Which of the following statements is false? Why?
   (i) The simplified product of a nonzero monomial and a binomial is a binomial.
   (ii) The simplified product of a binomial and a binomial is a trinomial.

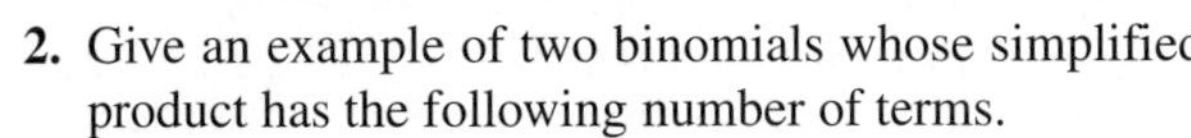

2. Give an example of two binomials whose simplified product has the following number of terms.
   (a) 4 (b) 3 (c) 2

In Exercises 3–8, multiply the monomial times the polynomial.

3. $3x^2(2x - 3)$ **4.** $2x^3(3x + 4)$

5. $-4x^3(2x^2 + 5)$ **6.** $a^4(a^2 - 4a + 3)$

7. $-xy^2(x^2 - xy - 2)$

8. $-6x^2y(x^2 - 2xy + 7xy^2)$

In Exercises 9–18, multiply the binomials.

9. $(x - 6)(x + 4)$ **10.** $(y - 7)(y - 2)$

11. $(y + 5)(3 - y)$ **12.** $(5x - 6)(4x + 3)$

13. $(9x + 2)(x + 3)$ **14.** $(4 - 5y)(2y + 1)$

15. $(x - 9y)(x - 4y)$ **16.** $(a + 5b)(a - 2b)$

17. $(6x - 5y)(2x - 3y)$ **18.** $(2x^2 - 3)(3x^2 - 2)$

In Exercises 19–28, multiply the polynomials.

19. $(x + 2)(x^2 - 4x + 5)$

20. $(x - 3)(x^2 + 5x - 3)$

21. $(2x - 3)(3x^2 + x - 1)$

22. $(3x + 2)(2x^2 - 2x + 1)$

23. $(x + y)(x^2 - xy + y^2)$

24. $(x - 2y)(x^2 + 2xy + 4y^2)$

25. $(2x - 1)(2x^3 - x^2 + 3x - 4)$

26. $(3x + 2)(3x^3 - 2x^2 + 5x - 2)$

27. $(x^2 - 2x + 1)(x^2 + 2x + 1)$

28. $(x^2 + 4x + 4)(x^2 - 4x + 4)$

In Exercises 29–32, multiply the polynomials.

29. $\begin{array}{r} x^2 + 2x + 4 \\ \underline{x - 2} \end{array}$ **30.** $\begin{array}{r} x^2 - 3x + 9 \\ \underline{x + 3} \end{array}$

31. $\begin{array}{r} 2x^2 - 4x + 5 \\ \underline{x^2 - 2x + 3} \end{array}$ **32.** $\begin{array}{r} 2x^2 + 3x - 2 \\ \underline{x^2 - 2x + 1} \end{array}$

33. Can the FOIL method be used to determine the product $(x + 3)(x - 3)$? What is the advantage of recognizing this as a special product instead?

34. Can the FOIL method be applied to multiply $x - 3(x - 3)$? Why?

In Exercises 35–40, square the binomial.

35. $(x + 3)^2$ **36.** $(2 - y)^2$

37. $(7x - 3)^2$ **38.** $(3a - b)^2$

39. $(2x + 5y)^2$ **40.** $(3x - 4y)^2$

In Exercises 41–46, multiply the binomials.

41. $(x - 4)(x + 4)$ **42.** $(7 - y)(7 + y)$

43. $(2x - 3)(2x + 3)$ **44.** $(4x + 5y)(4x - 5y)$

45. $(9n - 4m)(9n + 4m)$ **46.** $(a - 2b)(a + 2b)$

47. For $x^2(x + 1)(x - 3)$, two approaches are
   (i) multiply the two binomials and then multiply the result by x^2.
   (ii) multiply $x + 1$ by x^2 and then multiply the result times $x - 3$.

   Explain why both approaches are correct. Which approach is easier?

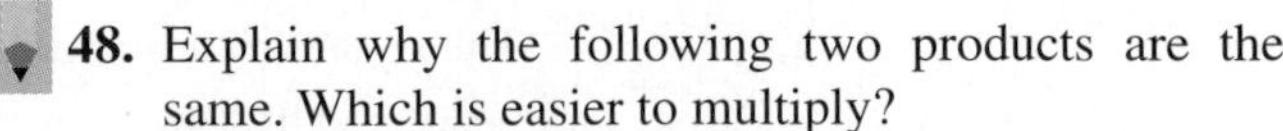

48. Explain why the following two products are the same. Which is easier to multiply?
   (i) $(x - a)[(x + a)(x + a)]$
   (ii) $[(x - a)(x + a)](x + a)$

In Exercises 49–68, multiply and simplify.

49. $(a^3 - b^3)(a^3 + b^3)$

50. $(a^2 + b^2)(a^2 - b^2)$

51. $(4 - pq^2)(4 + p^2q)$

52. $(a^2b^2 - 1)(a^2b^2 - 3)$

53. $(x^2y^2 + 2)^2$

54. $(x^3 + 4yz^2)^2$

55. $2x^3(2x - 5)(3x + 1)$

56. $-3y(y^2 + 1)(y^2 - 1)$

57. $5x - 6(4x + 3)$

58. $2x + 3(4x - 5)$

59. $(b - 3)(2b + 5)(b - 2)$

60. $(a - 5)(3a + 7)(a + 2)$

61. $(2x + 1)(4x^2 + 1)(2x - 1)$

62. $(1 + 3a)(9a^2 - 1)(3a - 1)$

63. $(2x + 5)(2x - 5) - (2x + 5)^2$

64. $(x + 7)(x - 7) - (x + 3)(x - 3)$

65. $(x + 6)(x - 4) - (2x - 3)(x + 2)$

66. $(3x + 5)(2x + 4) - (x - 6)(4x - 5)$

67. $(x + 3)^2 - (x - 3)^2$

68. $(x - 2)^2 + (5 - 2x)(1 + x)$

69. Multiply the square of the quantity $4x - 3$ by $2x^3$.

70. Multiply the square of the quantity $2 + 3y$ by $y + 1$.

71. Subtract the square of the quantity $5 - 2y$ from the product of the sum and difference of $3y$ and 2.

72. Subtract the product of $3 - 5x$ and $x + 4$ from $7x^2 + 5x + 6$.

In Exercises 73–76, determine a and b so that the product is equal to the polynomial.

73. $(ax + 3)(2x + b) = 8x^2 - 2x - 6$

74. $(ax + b)(3x + 1) = 6x^2 - 7x + b$

75. $(ax - 2)(3x + b) = 9x^2 - 4$

76. $(ax - b)^2 = 49 + 14x + x^2$

77. At noon, a car left a gas station and traveled north at an average of r miles per hour. The car reached its destination t hours later. At 1:00 P.M., a van left the same gas station and traveled south. Its average speed was 5 mph greater than the car's average speed. Write a polynomial expression (in two variables) that represents the total distance between the car and the van at the time that the car reached its destination.

78. Write a polynomial expression that represents the product of three consecutive integers.

79. A gardener uses 300 feet of fencing to enclose three sides of a rectangular garden. With w representing the width of the garden, write a polynomial expression that represents the area of the garden.

80. The lengths of the two legs of a right triangle are $(3x + 2)$ and $(x - 5)$. Use the Pythagorean Theorem to derive a polynomial expression for the square of the hypotenuse.

Group Project: College Textbook Sales

The following table shows the number (in millions) of college textbooks sold during the period 1985–1993. Also shown are the average wholesale prices (in dollars) for college textbooks during the same period. (Source: Book Industry Study Group, Inc.)

	1985	1987	1989	1991	1993
Number sold	110	119	136	137	136
Average price	14.31	15.15	15.76	17.33	18.37

81. Use the concept of rate of change to explain whether you think a linear model would be appropriate for the number of books sold.

82. Letting x represent the number of years since 1985, use your calculator's regression capability to create a linear model of the average wholesale price data. Write the result in the form $P(x) = mx + b$.

83. Suppose that we model the number of books sold with $B(x) = -0.64x^2 + 8.64x + 108.46$. Write a simplified expression for $E(x) = B(x) \cdot P(x)$, and interpret what the resulting function represents. What does the graph of $E(x)$ imply?

84. What do you think should be the guiding principle in determining the amount of markup on textbooks at a college bookstore?

Challenge

In Exercises 85–88, multiply.

85. $(x^n + 2)(x^n + 4)$

86. $(x^n - 2)^2$

87. $(x^n + 4)(x^n - 4)$

88. $(x + 1)(x^{2n} - x^n + 1)$

89. In the accompanying figure, the total area of the largest rectangle is the sum of the areas of the four inner rectangles. Use this fact to show the following.

$$(a + b)^2 = a^2 + 2ab + b^2$$

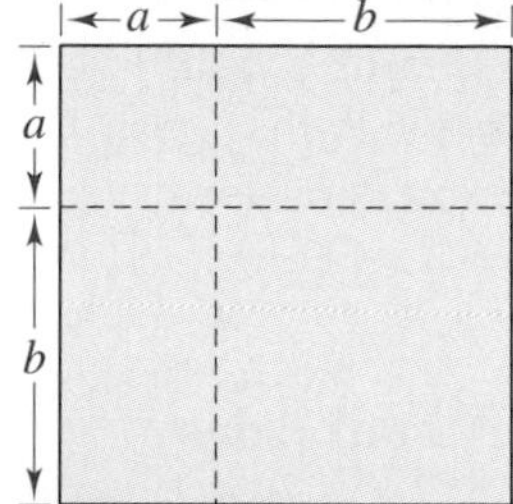

90. In the accompanying figure, the total area of the largest rectangle is the sum of the areas of the three inner rectangles. Use this fact to show the following.

$$(a - b)(a + b) = a^2 - b^2$$

Figure for 90

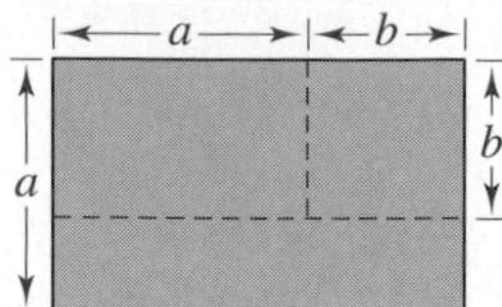

91. Explain how to use a special product to find the product of the trinomials

$$(x + 3y - 4)(x + 3y + 4)$$

In Exercises 92 and 93, decide how to rearrange terms and to group within the factors to obtain a special product. Then perform the multiplication.

92. $(x + y + 2)(2 + x + y)$

93. $(x^2 + 5x + 3)(x^2 + 5x - 3)$

In Exercises 94 and 95, multiply.

94. $[2y - (x - 3)]^2$ **95.** $[(x - 3) + 5y][(x - 3) - 5y]$

6.3 COMMON FACTORS AND GROUPING

Greatest Common Factor • Common Factors in Polynomials • Factoring by Grouping

Greatest Common Factor

To factor a number, we write the number as a product of numbers. Sometimes there is more than one way to factor a number.

$$24 = 1 \cdot 24 \qquad 24 = 2 \cdot 12 \qquad 24 = 3 \cdot 8 \qquad 24 = 4 \cdot 6$$

The numbers on the right are numbers that can be divided into 24 with a remainder of 0. These numbers are called **factors** of 24.

A **prime number** is a natural number greater than 1 that has exactly two different natural number factors, 1 and itself. To factor a number *completely* means to write the number as the product of prime numbers. For example, to factor 24 completely, we write

$$24 = 3 \cdot 2 \cdot 2 \cdot 2$$

The prime factors of 24 are 3 and 2.

The **greatest common factor (GCF)** of two numbers a and b is the largest integer that is a factor of both a and b.

EXAMPLE 1

Finding the Greatest Common Factor

Find the GCF for 36, 48, and 120.

Solution

$$36 = 3 \cdot 3 \cdot 2 \cdot 2 = 12 \cdot 3$$ Factor each number completely.

$$48 = 3 \cdot 2 \cdot 2 \cdot 2 \cdot 2 = 12 \cdot 4$$

$$120 = 3 \cdot 2 \cdot 2 \cdot 2 \cdot 5 = 12 \cdot 10$$

The GCF is 12.

We use a similar method to determine the GCF of two or more monomials.

$$10x^3 = 2 \cdot 5 \cdot x^3$$

$$15x^5 = 3 \cdot 5 \cdot x^3 \cdot x^2$$

The factors 5 and x^3 are the greatest factors that are common to both monomials. Thus the GCF is $5x^3$. Observe that for exponential factors whose base is common to each monomial, we can simply use the smallest exponent.

Determining the GCF of Two or More Monomials

1. Determine the GCF of the numerical coefficients.
2. Determine the smallest exponent on each exponential factor whose base is common to the monomials. Write the base with that exponent.
3. The product of the results of steps 1 and 2 is the GCF of the monomials.

Note: The numbers 1 and -1 are factors of every monomial. When we determine a GCF, we usually ignore these factors.

EXAMPLE 2

Determining the GCF in Monomials

Determine the GCF of the given monomials.

(a) $6y^4, 9y^6$ (b) $75x^3y^4, 45x^2y^5, 225x^4y^7$

Solution

(a) $6y^4 = 2 \cdot 3 \cdot y^4$ Factor the monomials completely.

$9y^6 = 3 \cdot 3 \cdot y^6$

The GCF of 6 and 9 is 3, and the smallest exponent on y is 4. Thus the GCF of the monomials is $3y^4$.

The monomials can be written with the GCF as a factor.

$$6y^4 = 3y^4 \cdot 2 \qquad 9y^6 = 3y^4 \cdot 3y^2$$

(b) $75x^3y^4 = 3 \cdot 5 \cdot 5 \cdot x^3 \cdot y^4$

$45x^2y^5 = 3 \cdot 3 \cdot 5 \cdot x^2 \cdot y^5$

$225x^4y^7 = 3 \cdot 3 \cdot 5 \cdot 5 \cdot x^4 \cdot y^7$

The GCF of the numerical coefficients is $3 \cdot 5 = 15$. The smallest exponents on x and y are 2 and 4, respectively. Thus the GCF of the three monomials is $15x^2y^4$.

The monomials can be written with the GCF as one factor.

$$75x^3y^4 = 15x^2y^4 \cdot 5x$$
$$45x^2y^5 = 15x^2y^4 \cdot 3y$$
$$225x^4y^7 = 15x^2y^4 \cdot 15x^2y^3$$

Common Factors in Polynomials

To **factor a polynomial,** we write the polynomial as a product of two or more polynomials. We begin by determining whether the terms of the polynomial have a GCF other than 1 or -1.

We use the Distributive Property to multiply a polynomial by a monomial. To factor a polynomial, we reverse the process. For example, to factor $12x^3 - 20x^2$, we begin by observing that the GCF of the two terms is $4x^2$.

$$12x^3 - 20x^2 = 4x^2 \cdot 3x - 4x^2 \cdot 5$$ Factor each term with the GCF as one factor.

$$= 4x^2(3x - 5)$$ Apply the Distributive Property.

If the polynomial to be factored has integer coefficients, then we try to factor so that the resulting expression has integer coefficients. If this is not possible, we consider the expression to be a **prime polynomial**.

A polynomial is **factored completely** if each factor other than a monomial factor is prime. For example, although $12x^3 - 20x^2$ can be written as $2x^2(6x - 10)$, we usually want to factor completely and write $4x^2(3x - 5)$.

Factoring Out a GCF from a Polynomial

1. Find the GCF of the terms.
2. Factor each term with the GCF as one factor.
3. Apply the Distributive Property to factor the polynomial with the GCF as a factor.

EXAMPLE 3

Factoring Out the Greatest Common Monomial Factor

Factor completely.

(a) $10x^3 - 5x^2$ (b) $-4x^3 - 8y^2$ (c) $-12x^2y^3 + 10xy^3$

Solution

(a) $10x^3 - 5x^2 = (5x^2)(2x) - (5x^2)(1)$ Determine the GCF of each term.

$= 5x^2(2x - 1)$ Distributive Property

(b) $-4x^3 - 8y^2 = (-4)x^3 + (-4)(2y^2)$ — We regard the common numerical factor as negative.

$= -4(x^3 + 2y^2)$ — Distributive Property

(c) $-12x^2y^3 + 10xy^3 = (-2xy^3)(6x) - (-2xy^3)(5)$ — GCF $= -2xy^3$

$= -2xy^3(6x - 5)$ — Distributive Property

EXAMPLE 4

Factoring Out the Greatest Common Monomial Factor

Factor the trinomial $54x^3y^5 + 36x^4y^2 - 90x^6y^3$.

Solution

$54x^3y^5 = 18 \cdot 3 \cdot x^3 \cdot y^5$ — Factor the monomials.

$36x^4y^2 = 18 \cdot 2 \cdot x^4 \cdot y^2$

$90x^6y^3 = 18 \cdot 5 \cdot x^6 \cdot y^3$

$\text{GCF} = 18x^3y^2$ — Use the smallest exponent on each common factor.

In factored form, $54x^3y^5 + 36x^4y^2 - 90x^6y^3 = 18x^3y^2(3y^3 + 2x - 5x^3y)$.

EXAMPLE 5

Factoring Out the Greatest Common Binomial Factor

Factor each of the following.

(a) $3x(x + 4) - y(x + 4)$

(b) $5(2x - 1) - 2y(1 - 2x) + z(2x - 1)$

(c) $x^2(x + 5) + 3x^4(x + 5) - 6x^3(x + 5)$

(d) $6x^3(y - 1)^2 + 9x(y - 1)^5$

Solution

(a) $3x(x + 4) - y(x + 4)$ — The binomial $(x + 4)$ is common to both terms.

$= (x + 4)(3x - y)$

(b) $5(2x - 1) - 2y(1 - 2x) + z(2x - 1)$ — Use the rule $a - b = -(b - a)$ to rewrite $1 - 2x$.

$= 5(2x - 1) + 2y(2x - 1) + z(2x - 1)$

$= (2x - 1)(5 + 2y + z)$ — Now $(2x - 1)$ is common to all three terms.

(c) $x^2(x + 5) + 3x^4(x + 5) - 6x^3(x + 5)$ — Note that x^2 and $(x + 5)$ are common to all three terms.

$= x^2(x + 5)(1 + 3x^2 - 6x)$

(d) $6x^3(y - 1)^2 + 9x(y - 1)^5$ — Note that 3, x, and $(y - 1)^2$ are common to both terms.

$= 3x(y - 1)^2[2x^2 + 3(y - 1)^3]$

Factoring by Grouping

In some cases we must be creative to find the GCF.

EXPLORING THE CONCEPT

Factoring by Grouping

At first glance, the polynomial $x^3 - 3x^2 + 2x - 6$ does not appear to have a common factor. A factor of x appears in three of the terms but not in the last term.

However, suppose the first two terms and last two terms are grouped together.

$$x^3 - 3x^2 + 2x - 6 = (x^3 - 3x^2) + (2x - 6)$$

The first group has a GCF of x^2; the second group has a GCF of 2.

$$x^3 - 3x^2 + 2x - 6 = x^2(x - 3) + 2(x - 3)$$

The polynomial is still not factored because it is a sum of two terms, not a product. Observe that the two terms have a common binomial factor $(x - 3)$.

$$\begin{aligned} x^3 - 3x^2 + 2x - 6 &= x^2(x - 3) + 2(x - 3) \\ &= (x - 3)(x^2 + 2) \end{aligned}$$

In factored form, $x^3 - 3x^2 + 2x - 6 = (x - 3)(x^2 + 2)$.

This method of factoring is called the **grouping method**.

Factoring by the Grouping Method

1. Group the terms so that each group has a common factor.
2. Factor out the GCF in each group.
3. If there is a common factor in the resulting terms, factor it out. If there is no common factor, a different grouping may produce the desired result.

EXAMPLE 6

Factoring by the Grouping Method

Factor completely.

(a) $ax + bx + ay + by$

(b) $6x - 3y - 2ax + ay$

(c) $x^3 + 2x^2 - 5x - 10$

Solution

Think About It

Use the grouping method to factor the following expression $xy(x + 1) + x(x + 1) + y^2(x + 1) + y(x + 1)$

(a) $ax + bx + ay + by$

$= (ax + bx) + (ay + by)$	Group the expression in pairs.
$= x(a + b) + y(a + b)$	Factor out the GCF in each pair.
$= (a + b)(x + y)$	In the resulting two terms, the GCF is $(a + b)$.

(b) When the third term is negative, care must be taken when we group the terms. Just as we change signs when we *remove* parentheses from $-(m - n) = -m + n$, we must change signs when we *insert* parentheses in $-m + n = -(m - n)$.

$6x - 3y - 2ax + ay$

$= (6x - 3y) - (2ax - ay)$	Change signs in the second group.
$= 3(2x - y) - a(2x - y)$	Factor out the GCF in each pair.
$= (2x - y)(3 - a)$	In the resulting two terms, the GCF is $(2x - y)$.

LEARNING TIP

In Example 6(c), we insert parentheses and then factor. In such cases, you may find it easier just to factor out -5 from the last two terms. Either way, be careful to make the necessary sign changes.

(c) $x^3 + 2x^2 - 5x - 10$

$= (x^3 + 2x^2) - (5x + 10)$ Change signs in the second group.

$= x^2(x + 2) - 5(x + 2)$ Factor out the GCF in each pair.

$= (x + 2)(x^2 - 5)$ In the resulting two terms, the GCF is $(x + 2)$.

Note: In these examples, we grouped by pairs. Different grouping arrangements may be needed in other cases.

6.3 QUICK REFERENCE

Greatest Common Factor

- A **prime number** is a natural number greater than 1 that has exactly two different natural number factors, 1 and itself.
- To factor a number *completely* means to express the number as a product of prime numbers.
- The **greatest common factor (GCF)** of two numbers a and b is the largest integer that is a factor of both a and b.
- To determine the GCF of two or more monomials, perform the following steps:
  1. Determine the GCF of the numerical coefficients.
  2. Determine the smallest exponent on each exponential factor whose base is common to the monomials. Write the base with that exponent.
  3. The product of the results of steps 1 and 2 is the GCF of the monomials.

Common Factors in Polynomials

- **Factoring a polynomial** means expressing the polynomial as a product of two or more polynomials.
- If the polynomial to be factored has integer coefficients, then we try to factor so that the resulting expression has integer coefficients. If this is not possible, we consider the expression to be a **prime polynomial**.
- A polynomial is **factored completely** if each factor other than a monomial factor is prime.
- To factor out a GCF from a polynomial, perform the following steps:
  1. Find the GCF of the terms.
  2. Factor each term with the GCF as one factor.
  3. Apply the Distributive Property to factor the polynomial with the GCF as a factor.

Factoring by Grouping

- To factor with the grouping method, perform the following steps:
  1. Group the terms so that each group has a common factor.
  2. Factor out the GCF in each group.
  3. If there is a common factor in the resulting terms, factor it out. If there is no common factor, a different grouping may produce the desired result.

6.3 EXERCISES

Concepts and Skills

1. For the monomials x^5y^3, x^7y^2, and x^4y^6, explain how you can determine the GCF by inspecting exponents.

2. Is the GCF of an expression always a monomial? Give an example that supports your answer.

In Exercises 3–20, determine the GCF and factor the given expression completely.

3. $3x + 12$
4. $2x + 6$
5. $5x + 10y - 30$
6. $6 - 24a - 18b$
7. $12x^2 - 4x$
8. $6x^2 + 2x$
9. $7x^2 - 10y$
10. $5a^3 - 8b^2$
11. $x^3 - x^2$
12. $c^4 + 3c^3$
13. $3x^3 - 9x^2 + 12x$
14. $15b^5 + 5b^7 - 10b^4$
15. $x^5y^4 - x^4y^3 + xy^2$
16. $x^7y^8 + xy^5 - x^5y$
17. $3x(a + by) - 2y(a + by)$
18. $2x(x - 3) - 5y(x - 3)$
19. $4x^2(2x + 3) - 7y^2(2x + 3)$
20. $5x^5(x - 2) - 9y^3(x - 2)$

21. Determine whether each of the following is a correct factorization of $4x - 12$. If it is incorrect, explain why.
(a) $4(x - 3)$
(b) $-4(-x + 3)$
(c) $4(-3 + x)$
(d) $-4(3 - x)$

22. Explain how the Multiplication Property of -1 can be used to factor $x - 2y$ so that one of the factors is $2y - x$.

In Exercises 23–26, factor out the GCF and the opposite of the GCF.

23. $4x - 12$
24. $10 - 4x$
25. $6y + 3xy - x^2y$
26. $-a^2b + 5b$

In Exercises 27–32, factor the given expression.

27. $y(y - 3) + 2(3 - y)$
28. $x(x - 2) - 3(2 - x)$
29. $5(3 - z) - z(z - 3)$
30. $4(6 - a) + a(a - 6)$
31. $(x + 3)(x - 4) - (x + 2)(4 - x)$
32. $(3 - a)(2a - 1) + (a - 2)(a - 3)$

33. If $(2x - 5)(x + 3)$ is the result of factoring an expression by grouping, what were the four terms in the original expression?

34. Which of the following is the correct way to group $x^3 + x^2 - x - 1$ in pairs? Explain.
(i) $(x^3 + x^2) - (x - 1)$
(ii) $(x^3 + x^2) - (x + 1)$

In Exercises 35–52, use the grouping method to factor the given expression completely.

35. $ab + bx + ay + xy$
36. $ax + 3a + 2x + 6$
37. $cd + 3d - 4c - 12$
38. $ac + 5a - 2c - 10$
39. $ab - ay - bx - xy$
40. $ab - by + ax + xy$
41. $ax^2 + 3x^2 + ay + 3y$
42. $ax^2 + 2a - x^2y - 2y$
43. $x^3 + 2x^2 + 3x + 6$
44. $x^3 + 4x^2 - 3x - 12$
45. $3y^3 + 9y^2 + 12y + 36$
46. $4y^3 - 12y^2 + 20y - 60$
47. $a^2b + a^2y + b + y$
48. $ab^2 - abx - b + x$
49. $x^3 - 15 + 3x - 5x^2$
50. $2x^3 + 12 - 3x - 8x^2$
51. $6x^3 - 12 + 8x - 9x^2$
52. $6x^3 + 6 + 9x + 4x^2$

In Exercises 53–56, factor completely.

53. $18x^3 + 6x^2 - 45x - 15$
54. $2a^3 + 2a^2b + 6ab + 6b^2$
55. $3xy^3 - 6xy^2 + 7xy - 14x$
56. $r^2t^2 + r^2t + rt^2 + rt$

In Exercises 57–60, complete the factorization of each expression.

57. $\frac{2}{3}y - \frac{5}{6} = \frac{1}{6}(\quad\quad)$
58. $1 - \frac{3}{4}x = \frac{1}{4}(\quad\quad)$

59. $\frac{1}{2}x^2 + \frac{1}{3}x + 1 = \frac{1}{6}(\quad)$

60. $x^2 + \frac{1}{4}x - \frac{1}{2} = \frac{1}{4}(\quad)$

In Exercises 61–64, factor the given expression as in Exercises 57–60.

61. $\frac{1}{3}x + 5$

62. $\frac{1}{4}a - \frac{3}{2}b$

63. $\frac{1}{6}x^2 + \frac{1}{4}x + 3$

64. $x^2 - \frac{1}{2}x + \frac{1}{5}$

In Exercises 65–72, factor out the GCF in the expression.

65. $d^{20} - d^{15}$

66. $y^{12} + y^{20}$

67. $88x^{48} + 11x^{51}$

68. $100a^{50} + 10a^{25}$

69. $24a^5b^3 - 6ab^4 + 12a^3b^2$

70. $16c^6d^4 + 8c^2d^3 - 12c^2d$

71. $4a^2b^3(x - 4)^2 + 12a^3b^2(x - 4)^3$

72. $15c^4d^5(x + 3)^5 + 5c^5d^4(x + 3)^3$

In Exercises 73–76, write the given formula in completely factored form.

73. If P dollars are invested at a simple interest rate r for one year, the account value A is given by $A = P + Pr$.

74. If an object is propelled upward at an initial velocity of 560 feet per second, the height h of the object after t seconds is given by $h = 560t - 16t^2$.

75. The total surface area A of a right circular cylinder is given by $A = 2\pi rh + 2\pi r^2$, where h is the height of the cylinder and r is its radius.

76. The area A of a trapezoid is given by the formula $A = \frac{1}{2}b_1h + \frac{1}{2}b_2h$, where h is the altitude of the trapezoid and b_1 and b_2 are the lengths of the bases.

Challenge

77. Explain how to verify the following.

$$3x^{-4} + 2x^{-6} = x^{-6}(3x^2 + 2)$$

78. Although the expression $2x^{-3} + 5x^{-4}$ is not a polynomial, can we still determine the GCF by using the smaller exponent on x? Is the GCF x^{-4} in this case? If so, what is the other factor?

In Exercises 79 and 80, complete the factorization of each expression.

79. $2x^{-3} - 5x^{-4} = x^{-4}(\quad)$

80. $x^{-5} + x^5 = x^{-5}(\quad)$

In Exercises 81–84, factor completely as in Exercises 79 and 80.

81. $2x^3 - 10x^{-4}$

82. $y^{-2} + 3y^{-4}$

83. $5x^{-4}y^7 - 15x^3y^{-2}$

84. $3x^{-1}y^{-2} + 9xy^{-3}$

85. Use the grouping method to factor the following expression.

$$a^2x + 3a^2 + bx + 3b - 5x - 15$$

86. If n is an integer, show that when $4n^2 + 2n$ is factored completely, one factor is an even number and the other factor is an odd number.

6.4 SPECIAL FACTORING

Difference of Two Squares • Perfect Square Trinomials • Sum and Difference of Two Cubes

Difference of Two Squares

We can reverse the special product $(A + B)(A - B) = A^2 - B^2$ and write

$$A^2 - B^2 = (A + B)(A - B)$$

We recognize a **difference of two squares** as two perfect square terms separated by a minus sign. If the binomial we want to factor matches the pattern $A^2 - B^2$, then the factors are $A + B$ and $A - B$.

EXAMPLE 1

Factoring a Difference of Two Squares

Factor completely.

(a) $81 - 4y^2$ (b) $25x^2 - 49y^2$ (c) $(x - 1)^2 - 25$

Solution

The binomial in each part is in the form $A^2 - B^2$.

(a) $81 - 4y^2 = 9^2 - (2y)^2 = (9 + 2y)(9 - 2y)$

(b) $25x^2 - 49y^2 = (5x)^2 - (7y)^2 = (5x + 7y)(5x - 7y)$

(c) $(x - 1)^2 - 25 = (x - 1)^2 - 5^2$
$= [(x - 1) + 5][(x - 1) - 5]$
$= (x + 4)(x - 6)$

EXAMPLE 2

Combining Methods

Factor completely.

(a) $5x^4 - 80$ (b) $8x^3 - 2x$

Solution

(a) $5x^4 - 80 = 5(x^4 - 16)$ Factor out the GCF 5.
$= 5[(x^2)^2 - 4^2]$ The other factor is a difference of two squares.
$= 5(x^2 + 4)(x^2 - 4)$ $A^2 - B^2 = (A + B)(A - B)$
$= 5(x^2 + 4)(x^2 - 2^2)$ The last factor is a difference of two squares.
$= 5(x^2 + 4)(x + 2)(x - 2)$

Note that the factor $(x^2 + 4)$ cannot be factored further.

(b) $8x^3 - 2x = 2x(4x^2 - 1) = 2x(2x + 1)(2x - 1)$

Perfect Square Trinomials

Previously, we have seen the results of squaring binomials.

$$(A + B)^2 = A^2 + 2AB + B^2$$
$$(A - B)^2 = A^2 - 2AB + B^2$$

The trinomials on the right are called **perfect square trinomials** because they are trinomials that are the squares of binomials. By reversing these special products, we can establish two factoring patterns.

$$A^2 + 2AB + B^2 = (A + B)^2$$
$$A^2 - 2AB + B^2 = (A - B)^2$$

EXAMPLE 3

Factoring Perfect Square Trinomials

Factor completely.

(a) $x^2 - 12x + 36$ (b) $25x^2 + 70xy + 49y^2$ (c) $x^6 + 10x^3y^2 + 25y^4$

Solution

$$\begin{array}{l} A^2 - 2\,A\,B + B^2 = (A - B)^2 \\ \downarrow \quad \downarrow\downarrow\downarrow \quad \downarrow \quad \downarrow\ \downarrow \end{array}$$

(a) $x^2 - 12x + 36 = (x)^2 - 2(x)(6) + (6)^2 = (x - 6)^2$

(b) $25x^2 + 70xy + 49y^2 = (5x)^2 + 2(5x)(7y) + (7y)^2 = (5x + 7y)^2$

(c) $x^6 + 10x^3y^2 + 25y^4 = (x^3)^2 + 2(x^3)(5y^2) + (5y^2)^2 = (x^3 + 5y^2)^2$

Sum and Difference of Two Cubes

Sometimes a factoring pattern can be discovered by experimenting with multiplication. Consider the following two products.

$$\begin{aligned}(A + B)(A^2 - AB + B^2) &= A^3 - A^2B + AB^2 + A^2B - AB^2 + B^3 \\ &= A^3 + B^3 \\ (A - B)(A^2 + AB + B^2) &= A^3 + A^2B + AB^2 - A^2B - AB^2 - B^3 \\ &= A^3 - B^3\end{aligned}$$

We call the first result a **sum of two cubes,** and we call the second result a **difference of two cubes**. By reversing the order of the results, we have the following factoring patterns.

$$A^3 + B^3 = (A + B)(A^2 - AB + B^2)$$
$$A^3 - B^3 = (A - B)(A^2 + AB + B^2)$$

EXAMPLE 4

LEARNING TIP

Be aware of the difference between $A^2 + 2AB + B^2$, which is a perfect square trinomial pattern, and $A^2 + AB + B^2$, which is the second factor for a difference of two cubes and is prime.

Think About It

By combining like terms, $x^3 + x^3 = 2x^3$. Use the factoring pattern for a sum of two cubes on $x^3 + x^3$ to see whether you obtain the same result.

Factoring Sums and Differences of Two Cubes

Factor completely.

(a) $y^3 + 64$ (b) $54x^3 + 2y^3$ (c) $t^6 - 125$

Solution

$$\begin{array}{l} A^3 + B^3 = (A + B)(A^2 - A \cdot B + B)^2 \\ \downarrow \quad \downarrow \quad \downarrow\ \downarrow\ \downarrow \quad \downarrow\ \downarrow\ \downarrow \end{array}$$

(a) $y^3 + 64 = (y)^3 + (4)^3 = (y + 4)(y^2 - y \cdot 4 + 4^2)$
$= (y + 4)(y^2 - 4y + 16)$

(b) $54x^3 + 2y^3$

$= 2(27x^3 + y^3)$ Note the GCF 2.

$= 2[(3x)^3 + y^3]$ The other factor is a sum of two cubes.

$= 2(3x + y)[(3x)^2 - (3x)(y) + y^2]$ $A^3 + B^3 = (A + B)(A^2 - AB + B^2)$

$= 2(3x + y)(9x^2 - 3xy + y^2)$

(c) $t^6 - 125 = (t^2)^3 - 5^3$ The expression is a difference of two cubes.

$= (t^2 - 5)[(t^2)^2 + t^2 \cdot 5 + 5^2]$ $A^3 - B^3 = (A - B)(A^2 + AB + B^2)$

$= (t^2 - 5)(t^4 + 5t^2 + 25)$

6.4 QUICK REFERENCE

Difference of Two Squares

- An expression consisting of two perfect square terms separated by a minus sign is called a **difference of two squares.**
- The special factoring pattern for a difference of two squares is

$$A^2 - B^2 = (A + B)(A - B)$$

Perfect Square Trinomials

- A **perfect square trinomial** is a trinomial of the form

$$A^2 + 2AB + B^2 \quad \text{or} \quad A^2 - 2AB + B^2$$

- The special factoring patterns for perfect square trinomials are as follows:

$$A^2 + 2AB + B^2 = (A + B)^2$$
$$A^2 - 2AB + B^2 = (A - B)^2$$

Sum and Difference of Two Cubes

- A **sum of two cubes** is a binomial of the form $A^3 + B^3$; a **difference of two cubes** is a binomial of the form $A^3 - B^3$.
- The special factoring patterns for the sum of two cubes and the difference of two cubes are as follows:

$$A^3 + B^3 = (A + B)(A^2 - AB + B^2)$$
$$A^3 - B^3 = (A - B)(A^2 + AB + B^2)$$

6.4 EXERCISES

Concepts and Skills

1. In words, how do we recognize a difference of two squares?
2. Why can we factor $9x^2 + 36$ but not $9x^2 + 16$?

In Exercises 3–16, factor completely.

3. $x^2 - 9$
4. $9x^2 - 16$
5. $1 - 4x^2$
6. $25 - a^2$
7. $16x^2 - 25y^2$
8. $49x^2 - 64y^2$
9. $\frac{1}{4}x^2 - 1$
10. $\frac{1}{9}y^2 - \frac{1}{16}$
11. $a^6 - b^{16}$
12. $x^4 - y^2$
13. $(a + b)^2 - 4$
14. $(x - 3)^2 - y^2$
15. $16 - (x - y)^2$
16. $25 - (x + 4)^2$

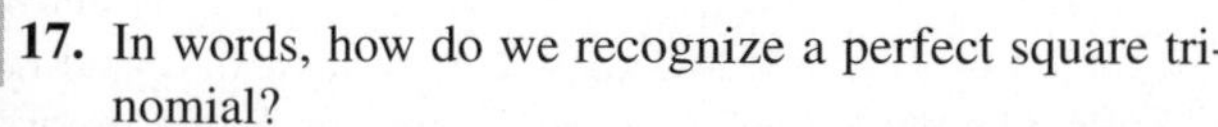

17. In words, how do we recognize a perfect square trinomial?
18. For $x^2 + 6xy + 9y^2$, how do we check the suspected factorization $(x + 3y)^2$?

In Exercises 19–28, factor completely.

19. $a^2 - 4a + 4$
20. $x^2 + 8x + 16$
21. $4x^2 + 12x + 9$
22. $9x^2 - 12x + 4$
23. $x^2 - 14xy + 49y^2$
24. $x^2 + 12xy + 36y^2$
25. $x^2 + x + \frac{1}{4}$
26. $y^2 - \frac{4}{3}y + \frac{4}{9}$
27. $x^4 - 10x^2 + 25$
28. $y^4 + 6y^2 + 9$

In Exercises 29–38, factor completely.

29. $x^3 + 1$
30. $8 - a^3$
31. $27y^3 - 8$
32. $64y^3 + 27$
33. $125 - 8x^3$
34. $64 - 27z^3$
35. $27x^3 + y^3$
36. $8a^3 - b^3c^3$
37. $x^6 + 1$
38. $b^6 - 125$

In Exercises 39–64, factor completely.

39. $x^2 + 2x + 1$
40. $x^2 - 10x + 25$

41. $0.25x^2 - 0.49$ **42.** $1.44a^2 - 0.0016$

43. $x^3 - 1$ **44.** $27 + b^3$

45. $c^2 + 100 - 20c$ **46.** $d^2 + 81 + 18d$

47. $2x^3 + 54$ **48.** $3x^3 - 24$

49. $81 + y^2$ **50.** $32a^2 + 50$

51. $36 - 84y + 49y^2$ **52.** $25 + 60p + 36p^2$

53. $4x^2 - 64$ **54.** $16 - 4y^2$

55. $25x^2 - 40xy + 16y^2$

56. $16x^2 + 40xy + 25y^2$

57. $x^4 - 16$ **58.** $81a^4 - b^4$

59. $8 + 125x^6$ **60.** $125 - 27x^6$

61. $y^8 - 6y^4 + 9$ **62.** $x^8 + 4x^4 + 4$

63. $216x^3 - y^3$ **64.** $a^3b^3 + 8$

In Exercises 65–92, factor completely.

65. $x^6 - 16x^3 + 64$ **66.** $x^6 + 64 + 16x^3$

67. $x^{12} - 4y^{20}$ **68.** $49 - 81x^4$

69. $x^6 + 4 - 4x^3$ **70.** $y^6 + 9 + 6y^3$

71. $98 - 2y^6$ **72.** $125x^2 - 5x^8$

73. $125x^5y^3 - x^2$ **74.** $27x^9y^7 - 8y^4$

75. $49x^7 - 70x^5 + 25x^3$

76. $9x^6 - 42x^4 + 49x^2$

77. $x^4 + 36y^4$ **78.** $16a^4 + 25b^6$

79. $64 + 27x^3y^9$ **80.** $1 - 27x^6y^3$

81. $x^4 - 8x^2 + 16$ **82.** $x^4 - 18x^2 + 81$

83. $16x^4 - 25y^2$ **84.** $c^8 - 1$

85. $64x^4 + 80x^2y + 25y^2$

86. $49x^6 + 126x^3y^2 + 81y^4$

87. $16a^4 - 81d^4$ **88.** $x^4 - c^4$

89. $2x^4 + 16x$ **90.** $3y^5 - 81y^2$

91. $125x^6 - 27y^{12}$ **92.** $27x^9 + 8y^6$

In Exercises 93–96, factor completely.

93. $4a(x^2 - y^2) + 8b(x^2 - y^2)$

94. $10x(a^2 - b^2) + 5y(a^2 - b^2)$

95. $x^2(a + b) - y^2(a + b)$

96. $4a^2(x - y) - b^2(x - y)$

In Exercises 97–100, factor the expression completely. (Hint: Replace the binomial with u and factor the resulting expression. Then replace u with the original binomial and simplify.)

97. $(2x + 1)^2 - 10(2x + 1) + 25$

98. $36(x - 1)^2 + 36(x - 1) + 9$

99. $(x^2 - 2)^2 - 4(x^2 - 2) + 4$

100. $(x^2 - 9)^2 + 16(x^2 - 9) + 64$

101. Match the expression $x^6 - y^6$ with the appropriate special factoring expression to show that the expression is a difference of two squares and is also a difference of two cubes.

102. The expression $x^6 + y^6$ is both a sum of two squares and a sum of two cubes. Which way should you regard the expression if you want to factor it?

In Exercises 103–108, determine a, b, or c so that the trinomial is a perfect square.

103. $9x^2 + bx + 1$ **104.** $x^2 + bx + 100$

105. $ax^2 - 10x + 25$ **106.** $ax^2 + 12x + 36$

107. $9x^2 - 24x + c$ **108.** $4x^2 - 20x + c$

Real-Life Applications

109. A circular swimming pool of radius r is surrounded by a circular deck of radius R. (See figure.) Write an expression for the area of the deck. Then write the expression in factored form.

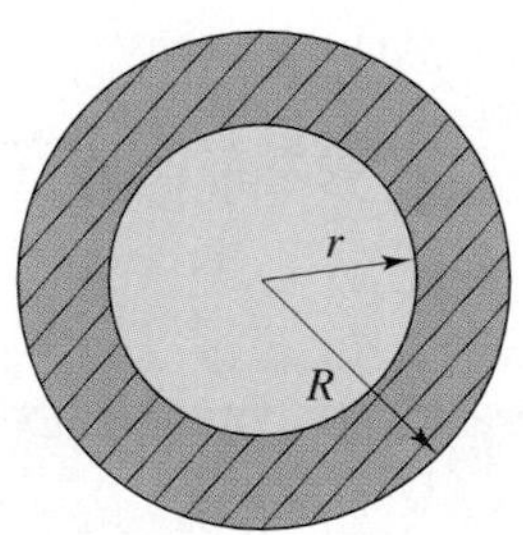

110. A Sea-Land shipping container for machine parts is a rectangular box with a volume that is represented by $x^3 + 10x^2 + 25x$. The length and width of the container are represented by binomials with integer coefficients. By factoring the expression for the volume, show that the height of the container is 5 units less than the length and width.

111. In physics, kinetic energy is the energy an object has by virtue of its motion. The formula for kinetic energy K is $K = \frac{1}{2}mv^2$, where m is the mass of the object and v is the velocity of the object. Suppose two objects of equal mass have velocities v_1 and v_2, respectively. Write an expression for the difference in their kinetic energies. Then write the expression in factored form.

112. A pharmaceutical company determines that a drug costing x dollars to develop and market will yield x^4 dollars in revenue. Write an expression for the net earnings (revenue minus cost) of the drug. Then write the expression in factored form.

Group Project: American Tourists in Europe

The accompanying table shows the number (in thousands) of Americans who traveled to Europe during the period 1985–1992. The table also shows the average amount of money spent per trip. (Source: U.S. Travel and Tourism Administration.)

Year	Number of Travelers (in thousands)	Average Expenditures (in dollars)
1985	6780	1391
1988	7438	1526
1989	7233	1655
1990	8043	1740
1991	6316	1825
1992	7136	1918

Let x represent the number of years before (negative) or after (positive) 1990. Then $T(x) = -51x^2 + 7344$ models the number of tourists (in thousands), and $E(x) = 80x + 1740$ models the average expenditure (in dollars) per trip.

113. Write function T in factored form.

114. Write function E in factored form.

115. Write a model function $C(x)$ (in factored form) that models the total expenditures.

116. What are two factors that might influence the number of American tourists who travel to Europe?

Challenge

In Exercises 117–120, factor completely. (Hint: In each case, begin by factoring the expression as a difference of two squares.)

117. $x^6 - 64$

118. $c^6 - 1$

119. $a^6 - b^6$

120. $x^6 - 64y^6$

In Exercises 121–124, factor completely. (Hint: Start with the grouping method.)

121. $4x^5 - x^3 - 32x^2 + 8$

122. $x^5 - 32 + 8x^2 - 4x^3$

123. $x^2y^3 - x^2z^3 - y^3 + z^3$

124. $y^2 - 9x^2 + 12x - 4$

6.5 FACTORING TRINOMIALS OF THE FORM $ax^2 + bx + c$

Leading Coefficient $a = 1$ • Leading Coefficient $a \neq 1$ • General Strategy

Leading Coefficient $a = 1$

The product of two binomials, such as $(x + 4)(x - 7)$, is a trinomial called a **quadratic polynomial.**

$$(x - 3)(x + 5) = x^2 + 2x - 15$$

In a quadratic polynomial $ax^2 + bx + c$, where $a \neq 0$, we call ax^2 the **quadratic term,** bx the **linear term,** and c the **constant term.** We call the coefficient a of the quadratic term the **leading coefficient.**

Note: Although the leading coefficient a of a quadratic polynomial cannot be 0, b or c can be 0. Thus there must be a quadratic term, but the linear term or constant term may be missing.

We can factor certain quadratic polynomials by factoring out a GCF or by using special factoring patterns.

$$3x^2 + 8x = x(3x + 8) \qquad \text{Factor out the GCF } x.$$

$$4y^2 - 25 = (2y + 5)(2y - 5) \qquad \text{Factor the difference of two squares.}$$

In this section we factor quadratic polynomials that have all three terms.

EXPLORING THE CONCEPT

Factoring a Quadratic Trinomial

When the leading coefficient a is 1, the trinomial has the form $x^2 + bx + c$. From our experience with multiplication, we may recall that $x^2 + bx + c$ was often the result of multiplying two binomials of the form $(x + m)(x + n)$, where m and n are constants.

Factors	F O I L	$x^2 + bx + c$	
$(x + 4)(x - 2)$	$= x^2 - 2x + 4x - 8$	$= x^2 + (-2 + 4)x - 8$	$b = -2 + 4, c = -8$
$(x - 5)(x - 3)$	$= x^2 - 3x - 5x + 15$	$= x^2 + (-3 - 5)x + 15$	$b = -3 - 5, c = 15$
$(x + 6)(x + 1)$	$= x^2 + x + 6x + 6$	$= x^2 + (1 + 6)x + 6$	$b = 1 + 6, c = 6$
$(x + m)(x + n)$	$= x^2 + nx + mx + mn$	$= x^2 + (m + n)x + mn$	$b = m + n, c = mn$

These patterns suggest that we can factor the trinomial $x^2 + bx + c$ if we can find numbers m and n such that $mn = c$ and $m + n = b$.

EXAMPLE 1

Factoring a Quadratic Trinomial $x^2 + bx + c$

Factor each trinomial.

(a) $x^2 + 8x + 12$ (b) $x^2 - 5x + 6$

Solution

(a) Because $c = 12$ and $b = 8$, we find pairs of numbers m and n whose product is 12 and whose sum is 8.

Factors of 12	*Sums*
1, 12 and −1, −12	13 and −13
2, 6 and −2, −6	8 and −8
3, 4 and −3, −4	7 and −7

The required values for m and n are 2 and 6. Thus

$$x^2 + 8x + 12 = (x + 2)(x + 6)$$

We can (and should) verify that the factorization is correct by confirming that $(x + 2)(x + 6) = x^2 + 8x + 12$.

(b) We look for a pair of numbers m and n whose product is 6 and whose sum is −5. The required numbers are −2 and −3.

$$x^2 - 5x + 6 = (x - 2)(x - 3)$$

In both parts of Example 1, $c > 0$. In Example 2 we consider quadratic trinomials in which $c < 0$.

EXAMPLE 2

Factoring a Quadratic Trinomial $x^2 + bx + c$

Factor each trinomial.

(a) $x^2 - x - 20$ (b) $x^2 + 5x - 14$

Solution

(a) Because $c = -20$ and $b = -1$, we find pairs of numbers m and n whose product is -20 and whose sum is -1.

Factors of -20	*Sums*
$1, -20$ and $-1, 20$	-19 and 19
$2, -10$ and $-2, 10$	-8 and 8
$4, -5$ and $-4, 5$	-1 and 1

The required values for m and n are 4 and -5. Thus

$$x^2 - x - 20 = (x + 4)(x - 5)$$

(b) We look for a pair of numbers m and n whose product is -14 and whose sum is 5. The required numbers are 7 and -2.

$$x^2 + 5x - 14 = (x + 7)(x - 2)$$

With practice, you can determine the pair of numbers m and n mentally without making the list of factors and sums.

EXAMPLE 3

Extending the Methods

Factor each trinomial.

(a) $12 - x - x^2$ (b) $3x^2 + 18x - 21$

(c) $t^2 + 11st + 18s^2$ (d) $y^4 - 5y^2 - 14$

Solution

(a) The trinomial $12 - x - x^2$ is in ascending order, and $a = -1$. Verify that $12 - x - x^2 = (4 + x)(3 - x)$.

An alternative approach is to write the trinomial in descending order and factor out -1.

$$12 - x - x^2 = -x^2 - x + 12 = -1(x^2 + x - 12)$$

To factor $x^2 + x - 12$, we determine numbers whose product is -12 and whose sum is 1. The required numbers are 4 and -3.

$$12 - x - x^2 = -1(x^2 + x - 12) = -1(x + 4)(x - 3)$$

Note that the two approaches lead to equivalent factorizations.

(b) $3x^2 + 18x - 21 = 3(x^2 + 6x - 7)$ Factor out the GCF 3.

$= 3(x + 7)(x - 1)$ Choose 7 and -1 because their product is -7 and their sum is 6.

(c) $t^2 + 11st + 18s^2 = (t + 2s)(t + 9s)$ Choose 2 and 9 because their product is 18 and their sum is 11.

(d) Although $y^4 - 5y^2 - 14$ is a fourth-degree polynomial, not a quadratic polynomial, the method still applies.

$$y^4 - 5y^2 - 14 = (y^2 - 7)(y^2 + 2)$$ Choose -7 and 2 because their product is -14 and their sum is -5.

Both factors have a degree greater than 1, so they should be inspected to see if they can be factored further. In this case, neither of them can be.

Leading Coefficient $a \neq 1$

When the leading coefficient of a quadratic trinomial is a number other than 1, an approach to factoring the trinomial is the *trial-and-check method.*

EXPLORING THE CONCEPT

The Trial-and-Check Method

Suppose that we want to factor $6x^2 + 7x - 3$. To write the trinomial as the product of binomials, recall that the product of the first terms must be $6x^2$, and the product of the last terms must be -3.

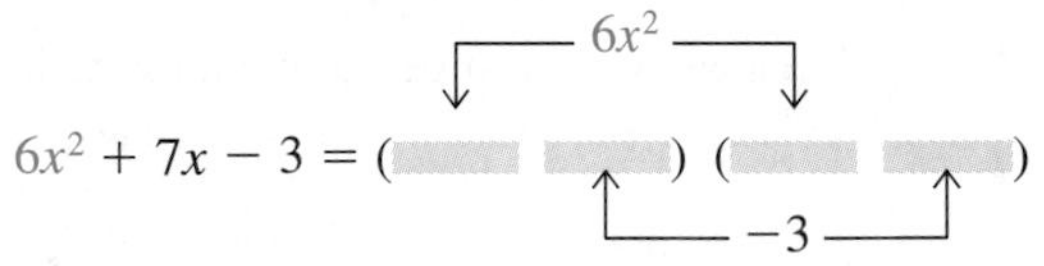

$$6x^2 + 7x - 3 = (\square \ \square)(\square \ \square)$$

(first terms: $6x^2$; last terms: -3)

Here are the possibilities.

$$6x^2 = 6x \cdot x \quad \text{or} \quad 2x \cdot 3x \qquad -3 = 1(-3) \quad \text{or} \quad -1 \cdot 3$$

Using all the combinations of these possibilities, we can list the possible factors of $6x^2 + 7x - 3$. Then we check the middle term resulting from multiplying the factors.

Possible Factors

$(6x + 1)(x - 3)$	The middle term is $-17x$.
$(6x - 3)(x + 1)$	The factor $6x - 3$ has a common factor 3.
$(6x - 1)(x + 3)$	The middle term is $17x$.
$(6x + 3)(x - 1)$	The factor $6x + 3$ has a common factor 3.
$(2x + 1)(3x - 3)$	The factor $3x - 3$ has a common factor 3.
$(2x - 3)(3x + 1)$	The middle term is $-7x$.
$(2x - 1)(3x + 3)$	The factor $3x + 3$ has a common factor 3.
$(2x + 3)(3x - 1)$	The middle term is $7x$.

Because $6x^2 + 7x - 3$ does not have a common factor, none of the factors of $6x^2 + 7x - 3$ has a common factor. This means that we can eliminate any factorization in which a binomial factor has a common factor.

The last possible factorization in the list produces the correct middle term, $7x$.

$$6x^2 + 7x - 3 = (2x + 3)(3x - 1)$$

As always, factoring out the GCF, if any, is the first step in factoring. Once you have done so, it is not necessary to list or try any combination in which a binomial factor has a common factor.

EXAMPLE 4

Factoring Quadratic Trinomials by Trial and Check

Factor each trinomial.

(a) $4x^2 + 11x + 6$ (b) $12x^2 + 7x - 12$ (c) $2t^2 - 3t - 1$

LEARNING TIP

If a is positive, then the signs in a trinomial provide valuable clues to factoring. For example, if $c > 0$, then the binomial factors have the same sign as b. If $c < 0$, then the binomial factors have opposite signs.

Solution

(a) $4x^2 + 11x + 6$

Because $4x^2 = 4x \cdot x$ or $2x \cdot 2x$, the possible factorizations have the following forms:

$$(4x + \square)(x + \square) \quad \text{or} \quad (2x + \square)(2x + \square)$$

Because b and c are both positive, the factors of c must be positive: $6 \cdot 1$ or $3 \cdot 2$. The only factorizations that do not have a common factor are as follows.

$(4x + 1)(x + 6)$ The middle term is $25x$.

$(4x + 3)(x + 2)$ The middle term is $11x$.

Because $b = 11$, the correct factorization is $(4x + 3)(x + 2)$.

(b) $12x^2 + 7x - 12$

The factors of 12 are $12 \cdot 1$, $6 \cdot 2$, and $4 \cdot 3$. Taking signs into account and eliminating combinations that have a common factor, we have the following possible factorizations.

$(12x + 1)(x - 12)$ The middle term is $-143x$.

$(12x - 1)(x + 12)$ The middle term is $143x$.

$(4x + 3)(3x - 4)$ The middle term is $-7x$.

$(4x - 3)(3x + 4)$ The middle term is $7x$.

Because $b = 7$, the correct factorization is $(4x - 3)(3x + 4)$.

(c) $2t^2 - 3t - 1$

The only possibilities are as follows:

$(2t + 1)(t - 1)$ The middle term is $-t$.

$(2t - 1)(t + 1)$ The middle term is t.

Because neither possibility gives us the correct middle term, $-3t$, we conclude that the trinomial is prime.

EXAMPLE 5

Extending the Methods

Factor each trinomial

(a) $30x^2 - 65x + 10$ (b) $3y^2 + yz - 10z^2$

Solution

(a) First, we factor out the GCF 5.

$$30x^2 - 65x + 10 = 5(6x^2 - 13x + 2)$$

The quadratic trinomial factor may be factorable. Because c is positive and b negative, the factors of the constant term must be negative. The only possibilities are as follows:

$(x - 2)(6x - 1)$ The middle term is $-13x$.

$(3x - 2)(2x - 1)$ The middle term is $-7x$.

Because $b = -13$, the correct factorization of the trinomial $6x^2 - 13x + 2$ is $(x - 2)(6x - 1)$. Thus the complete factorization of $30x^2 - 65x + 10$ is $5(x - 2)(6x - 1)$.

(b) $3y^2 + yz - 10z^2$

The only possible factorizations are as follows:

Possible Factors	*Middle Term*	*Possible Factors*	*Middle Term*
$(3y - 10z)(y + z)$	$-7yz$	$(3y + 10z)(y - z)$	$7yz$
$(3y - 5z)(y + 2z)$	yz	$(3y + 5z)(y - 2z)$	$-yz$
$(3y - z)(y + 10z)$	$29yz$	$(3y + z)(y - 10z)$	$-29yz$
$(3y - 2z)(y + 5z)$	$13yz$	$(3y + 2z)(y - 5z)$	$-13yz$

Because $b = 1$, the correct factorization is $(3y - 5z)(y + 2z)$.

Think About It

If you forget the GCF, you might factor $6x^2 + 6x - 36$ as $(3x + 9)(2x - 4)$. What can you do now to factor completely?

An alternative to the trial-and-check method is the grouping method.

Factoring $ax^2 + bx + c$ with the Grouping Method

1. Find the product of the leading coefficient a and the constant term c.
2. List the pairs of numbers whose product is ac.
3. Choose the pair whose sum is b, the coefficient of the middle term.
4. Use these numbers to write the middle term as a sum of two terms.
5. Factor by grouping.

EXAMPLE 6

Factoring Trinomials with the Grouping Method

Factor $6x^2 - x - 2$.

Solution

Because $ac = -12$ and $b = -1$, we look for pairs of numbers whose product is -12 and whose sum is -1.

Factors of -12	*Sums*
$1, -12$ and $-1, 12$	-11 and 11
$2, -6$ and $-2, 6$	-4 and 4
$3, -4$ and $-3, 4$	-1 and 1

We use 3 and -4 to rewrite the middle term as $3x - 4x$.

$$
\begin{aligned}
6x^2 - x - 2 &= 6x^2 + 3x - 4x - 2 && \text{Write } -x \text{ as } 3x - 4x.\\
&= (6x^2 + 3x) - (4x + 2) && \text{Group in pairs.}\\
&= 3x(2x + 1) - 2(2x + 1) && \text{Factor each group.}\\
&= (2x + 1)(3x - 2) && \text{Factor out } 2x + 1.
\end{aligned}
$$

The factorization of $6x^2 - x - 2$ is $(2x + 1)(3x - 2)$.

EXAMPLE 7

Using the Grouping Method

Factor $10x^2 - 21x + 9$.

Solution

We see that $ac = 90$ and $b = -21$. Because the numbers we are seeking must have a positive product and a negative sum, the numbers must be negative. The required numbers are -15 and -6.

$$
\begin{aligned}
10x^2 - 21x + 9 &= 10x^2 - 15x - 6x + 9 && \text{Write } -21x \text{ as } -15x - 6x.\\
&= (10x^2 - 15x) - (6x - 9) && \text{Group in pairs.}\\
&= 5x(2x - 3) - 3(2x - 3) && \text{Factor each group.}\\
&= (2x - 3)(5x - 3) && \text{Factor out } 2x - 3.
\end{aligned}
$$

Remember that you should always verify your factorization.

$$(2x - 3)(5x - 3) = 10x^2 - 21x + 9$$

General Strategy

As we solve problems in algebra, we sometimes must factor a polynomial and need to decide which of the methods of factoring we have studied is appropriate. It is helpful to have a general strategy to guide us in making the decision.

A polynomial is factored when it is written as the product of polynomials. A polynomial is factored completely if each factor other than a monomial factor is prime. Sometimes more than one of these methods is needed to factor the polynomial completely:

A General Strategy for Factoring

1. When factoring a polynomial, always look for a greatest common factor first. Even if other methods of factoring are required, factoring out the GCF will make other factoring methods easier to use.
2. When factoring a polynomial with just two terms, determine whether it matches one of the special product patterns.
3. When factoring a polynomial that has three terms, check whether it is a perfect square trinomial, and if so, use the special product for the square of a binomial.
4. When factoring a trinomial that is not a perfect square trinomial, use the grouping method or the trial-and-check method.
5. When factoring a polynomial that has more than three terms, try the grouping method.

6.5 QUICK REFERENCE

Leading Coefficient $a = 1$

- A trinomial of the form $ax^2 + bx + c$ is called a **quadratic trinomial**. The number a is called the **leading coefficient**.
- To factor a quadratic trinomial $x^2 + bx + c$ (the leading coefficient is 1), one approach is as follows:
  1. List the pairs of numbers whose product is the constant term c.
  2. Choose the pair whose sum is b, the coefficient of the middle term.
  3. These numbers are the constant terms of the binomial factors.

Leading Coefficient $a \neq 1$

- To factor a quadratic trinomial $ax^2 + bx + c$ by the trial-and-check method, follow these steps:
  1. List all the pairs of numbers whose product is the leading coefficient a.
  2. List all the pairs of numbers whose product is the constant term c.
  3. Use these two sets of numbers to form possible binomial factors of the trinomial. (Eliminate all possibilities that would produce a common factor in the terms of either of the binomial factors.)
  4. Multiply the remaining possibilities to determine the correct factorization.
- An alternative approach is the grouping method:
  1. Find the product of the leading coefficient a and the constant term c.
  2. List the pairs of numbers whose product is ac.
  3. Choose the pair whose sum is b, the coefficient of the middle term.
  4. Use these numbers to write the middle term as a sum of two terms.
  5. Factor by grouping.

General Strategy

- The first step in factoring is to factor out the greatest common factor.
- If the polynomial has just two terms, determine whether it matches the pattern of one of the special products.
  1. Difference of two squares
     $$A^2 - B^2 = (A - B)(A + B)$$
  2. Difference of two cubes
     $$A^3 - B^3 = (A - B)(A^2 + AB + B^2)$$
  3. Sum of two cubes
     $$A^3 + B^3 = (A + B)(A^2 - AB + B^2)$$
- If the polynomial has three terms, determine whether it is a perfect square trinomial.
  1. $A^2 + 2AB + B^2 = (A + B)^2$
  2. $A^2 - 2AB + B^2 = (A - B)^2$
- If the trinomial to be factored is not a perfect square trinomial, use the grouping method or the trial-and-check method.
- If the polynomial has more than three terms, try the grouping method. It might be necessary to rearrange terms or to group in ways other than in pairs.

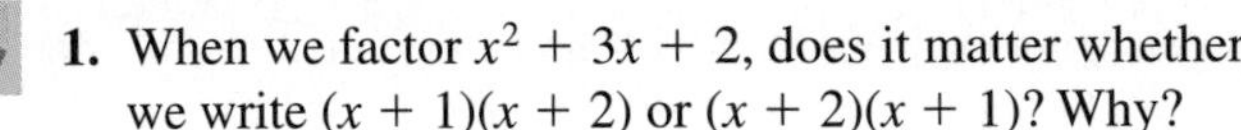

6.5 EXERCISES

Concepts and Skills

1. When we factor $x^2 + 3x + 2$, does it matter whether we write $(x + 1)(x + 2)$ or $(x + 2)(x + 1)$? Why?

2. After you factor an expression, how can you check the accuracy of your work?

In Exercises 3–12, factor the quadratic trinomial completely.

3. $x^2 + 3x + 2$	**4.** $x^2 - 4x + 3$
5. $x^2 + 18 - 9x$	**6.** $x^2 - 18 + 3x$
7. $14 + 9x + x^2$	**8.** $21 + 10x + x^2$
9. $x^2 + 2x + 2$	**10.** $x^2 + 3x + 5$
11. $x^2 - 6xy - 16y^2$	**12.** $x^2 - 4xy - 45y^2$

In Exercises 13–22, factor the trinomial completely.

13. $2x^2 + 5x + 2$	**14.** $2x^2 + 7x + 6$
15. $12 + 4x - x^2$	**16.** $4 + 3x - x^2$
17. $3x^2 - 8x - 3$	**18.** $3x^2 - 4x - 15$
19. $4x^2 - 17x - 15$	**20.** $12 + 7x - 10x^2$
21. $12x^2 + 19x + 21$	**22.** $12x^2 + 13x + 14$

In Exercises 23–58, factor the trinomial completely.

23. $c^2 + c - 2$	**24.** $x^2 - 2x - 3$
25. $6x^2 + 13x - 15$	**26.** $15x^2 + 8x - 7$
27. $x^2 + 5x - 24$	**28.** $3x + x^2 - 40$
29. $24 + 2x - x^2$	**30.** $15 + 2x - x^2$
31. $b^2 - 2b - 15$	**32.** $c^2 - 4c - 12$
33. $3x^2 + 5x + 2$	**34.** $3x^2 - 7x - 6$
35. $15 + 6x^2 - 19x$	**36.** $6 + x - 12x^2$
37. $x^2 + 25 + 10x$	**38.** $16 + y^2 + 8y$
39. $2 + x - 15x^2$	**40.** $2 - 11x + 15x^2$
41. $18 - 11x - 10x^2$	**42.** $3x^2 + 10x - 8$
43. $24 + a^2 - 10a$	**44.** $b^2 - 2b - 24$
45. $9x^2 - 9x - 10$	**46.** $6x^2 + x - 15$
47. $x^2 - 16x + 63$	**48.** $x^2 - 2x - 63$
49. $9x^2 - 9x - 18$	**50.** $9x^2 - 9x - 54$
51. $12x^2 + 16x - 3$	**52.** $13x + 35x^2 - 12$
53. $8x^2 - 2xy - 21y^2$	**54.** $8x^2 - 2xy - 15y^2$
55. $2x^2 - 12 + 2x$	**56.** $3x^2 - 12 + 9x$
57. $24x + 36 + 4x^2$	**58.** $48x + 40 + 8x^2$

59. Explain why it is easier to factor $17x^2 + 392x + 23$ than $24x^2 + 11x - 18$.

60. Which of the following is a correct factorization of $6 - 5x + x^2$? Explain.
(i) $(2 - x)(3 - x)$ (ii) $(x - 2)(x - 3)$

In Exercises 61–66, factor completely.

61. $8x^3y - 24x^2y^2 - 80xy^3$	**62.** $6x^6 - 60x^5 + 96x^4$
63. $45x + 12x^2 - 12x^3$	**64.** $10y^6 + 55y^5 + 75y^4$
65. $28x^2y - 30xy - 18y$	**66.** $15y^3 - 84y^2 - 36y$

In Exercises 67–78, factor the expression completely.

67. $a^2b^2 - 3ab - 10$	**68.** $c^2d^2 - 4cd - 21$
69. $12x^2y + 36xy + x^3y$	**70.** $28x^2 + 98x + 2x^3$
71. $x^2y^2 - 2xy - 48$	**72.** $36 - 9ab - a^2b^2$
73. $18x^2 - 12xy + 2y^2$	**74.** $8x^2y^2 + 18xy - 5$

75. $6m^2 + 5mn - n^2$

76. $16m^2n^4 + 12mn^5 + 2n^6$

77. $16x^3y^2 - 12x^2y^3 + 2xy^4$

78. $30ab^4 + 24a^2b^3 - 6a^3b^2$

In Exercises 79–84, factor the expression completely.

79. $x^3(x - 1) + 11x^2(x - 1) - 42x(x - 1)$

80. $x^3(2x - 1) + 4x^2(2x - 1) - 21x(2x - 1)$

81. $x^4(x^2 - 25) - 10x^2(x^2 - 25) + 9(x^2 - 25)$

82. $3x^4(y^2 - 4) + 7x^2(y^2 - 4) - 6(y^2 - 4)$

83. $2x^2y^2(xy + 1) + 3xy(xy + 1) - 9(xy + 1)$

84. $2x^2(2 - y) + xy(2 - y) - 10y^2(2 - y)$

85. If we write $x^4 - x^2 - 12 = (x^2 + 3)(x^2 - 4)$, have we factored completely? Explain.

86. How can you tell at a glance that $x^4 + 7x + 10$ cannot be factored with our current methods for factoring trinomials?

In Exercises 87–98, factor completely.

87. $x^4 + 28x^2 - 60$	**88.** $x^4 - 17x^2 - 60$
89. $x^6 + 17x^3 + 60$	**90.** $x^6 - 16x^3 + 60$

91. $x^4 - 13x^2 + 36$

92. $x^4 - 5x^2 + 4$

93. $5x^8 + 3x^4 - 14$

94. $9x^8 + 77x^4 - 36$

95. $4x^4 - 37x^2 + 9$

96. $4x^4 + 11x^2 - 3$

97. $9x^4 - 13x^2 + 4$

98. $9x^4 - 31x^2 - 20$

In Exercises 99–104, factor completely. (Hint: Replace the binomial with u and factor the resulting expression. Then replace u with the binomial and simplify.)

99. $(x - 1)^2 + 11(x - 1) + 28$

100. $(x + 2)^2 + 11(x + 2) + 24$

101. $(2x + 1)^2 - 15(2x + 1) + 56$

102. $12(2 - 3y)^2 - 29(2 - 3y) + 14$

103. $2x^2(x + 2)^2 + x(x + 2) - 6$

104. $2x^2(2 + y)^2 - x(2 + y) - 15$

Geometric Models

105. Real data applications can rarely be modeled with polynomials whose coefficients are integers. For example, an architect's design shows a building with a rectangular floor plan for which the area is modeled by $1.01x^2 + 4.99x - 24.05$, a polynomial that cannot be factored with our present methods. Round off the coefficients of this expression, and approximate a factorization.

106. A sheet metal fabricator produces a triangular piece of aluminum with area $\frac{1}{2}x^2 + 7x + 20$, where the base and height are polynomials with integer coefficients. Factor the expression to determine the difference between the base and the height of the piece of aluminum.

Group Project: Food Costs

The accompanying bar graph shows the average annual food costs (in dollars) for couples 20–50 years old during the period 1986–1994. (Source: U.S. Department of Agriculture.)

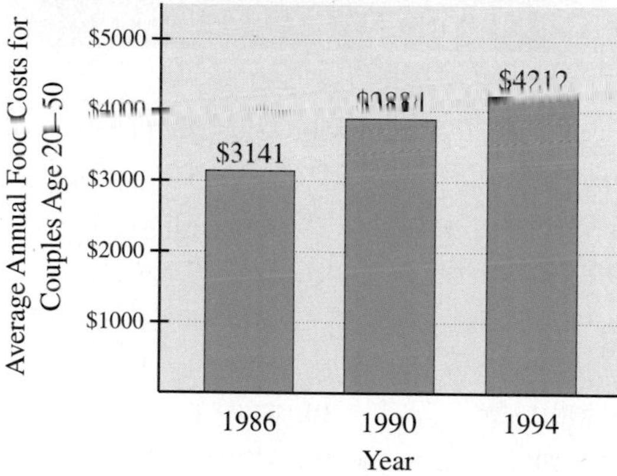

A model function for the average annual food cost is $f(t) = 130t + 3120$, where t is the number of years since 1985.

107. Write $f(t)$ in factored form with 52 as a common factor.

108. What does the other factor represent?

109. What does the coefficient of t in the original function indicate about the rate of increase in annual food costs? How can you interpret the coefficient of t in the factored form of the function?

110. Would you expect the average annual food costs to increase or decrease after age 50?

Challenge

In Exercises 111–114, determine all the integers k such that the trinomial can be factored.

111. $x^2 + kx + 10$

112. $x^2 + kx - 12$

113. $3x^2 + kx + 5$

114. $7x^2 + kx - 2$

In Exercises 115 and 116, determine a positive number k so that the polynomial can be factored.

115. $x^2 + 3x + k$

116. $x^2 + 2x + k$

In Exercises 117–123, factor completely.

117. $x^{2n} + 11x^n + 24$

118. $y^{2n} + 11y^n + 30$

119. $y^{4n} - 20y^{2n} + 99$

120. $x^{4n} - 17x^{2n} + 72$

121. $5x^{2n} + 2x^n - 3$

122. $7x^{2n} + 16x^n - 15$

123. $3x^{4n} + 5x^{3n} + 2x^{2n}$

124. If x is an odd integer, show that the expression $3x^2 + 12x + 20$ is the sum of the squares of three consecutive odd integers. (Hint: Write the term $3x^2$ as $x^2 + x^2 + x^2$, write the term $12x$ as $4x + 8x$, and write the term 20 as $4 + 16$.)

6.5 SUPPLEMENTARY EXERCISES

In Exercises 1–56, factor the expression completely.

1. $5x^2 - 7x - 6$

2. $5x^2 - 31x + 6$

3. $25 - 81y^2$

4. $9x^2 - 81$

5. $2x^3 + x^2 - 6x - 3$

6. $3x^3 + 9x^2 + x + 3$

7. $3x(x - 2) + 5y(x - 2)$

8. $y^2(y - 3) - 5(y - 3)$

9. $125x^3 + 1$

10. $64 - 27a^3$

11. $6x + x^2 + 8$

12. $x^2 - x - 12$

13. $x^2 + 9$

14. $x^2 - 5x + 36$

15. $4x^2 + 20x + 25$

16. $45a^2 - 30a + 5$

17. $ax - 3a - bx + 3b$

18. $xy + 3y - 5x - 15$

19. $25y^3 - 25$

20. $a^3b^3 + 8$

21. $y^4 - 16x^4$

22. $5x^2 - 20$

23. $a^2b^7 + 3a^3b^5 - a^2b^4$

24. $3x^4 + 15x^3 - 3x^2$

25. $x + 21x^2 - 2$

26. $3 - 22y + 7y^2$

27. $16 - 8a + a^2$

28. $x^2 + 6xy + 9y^2$

29. $50 - 18x^4$

30. $3x^4 + 14x^2 + 8$

31. $2y + 10 - 3xy - 15x$

32. $2x^3 - 3x^2 - 18x + 27$

33. $24x^4 - 24x^2 + 20x^3$

34. $8a^2b - 27ab + 9b$

35. $c^{20} - d^8$

36. $81 - t^4$

37. $x^2 + xy + y^2$

38. $x^2 + 4x + 16$

39. $a^2x^3 - a^2y^3 - 4x^3 + 4y^3$

40. $18a^3 - 8a + 45a^2 - 20$

41. $a^5b^4 - a^7b^8 + a^6b^3$

42. $(a + b)(c + d) - 2(a + b)(c - d)$

43. $m^2 - 10m + 16$

44. $8 + 7y - y^2$

45. $4x^4 - 64y^4$

46. $8x^2 - 98y^2$

47. $(a + b)^3 - 8$

48. $(c + d)^2 - 16$

49. $20x^2 - 60x + 45$

50. $x^2y^2 + 3x^2y - 88x^2$

51. $30a^6 - 104a^5 + 90a^4$

52. $12a^2 - 22a - 42$

53. $c^5 - 4c^3 - c^2 + 4$

54. $16x^5 - 54x^2$

55. $x^4 + 2x^2y^2 - 99y^4$

56. $a^2 - 13a + 40$

6.6 SOLVING EQUATIONS BY FACTORING

Quadratic Equations • Solving by Factoring

Quadratic Equations

In Chapter 3 we developed a routine for solving linear equations in one variable. Equations involving polynomials of degree two or higher require different solution techniques.

An equation involving a polynomial of degree 2 is called a **second-degree** or **quadratic equation.**

Definition of a Quadratic Equation

A **quadratic equation** is an equation that can be written in the *standard form*

$$ax^2 + bx + c = 0$$

where a, b, and c are real numbers and $a \neq 0$.

When a quadratic equation is written in standard form, the term ax^2 is called the **quadratic term,** the term bx is called the **linear term,** and the term c is called the **constant term.**

In Chapter 3 we used the graphing method to estimate the solutions of first-degree equations. We can also use the graphing method to estimate the solutions of a quadratic equation.

EXPLORING THE CONCEPT

Solutions of a Quadratic Equation

Consider the quadratic equation $x^2 - 2x - 24 = 11$. Figure 6.4 shows the graph of $y = x^2 - 2x - 24$. Tracing the graph, we find that there are two values of x for which the value of $x^2 - 2x - 24$ is 11. Those x-values, -5 and 7, are the estimated solutions of the equation.

Figure 6.4

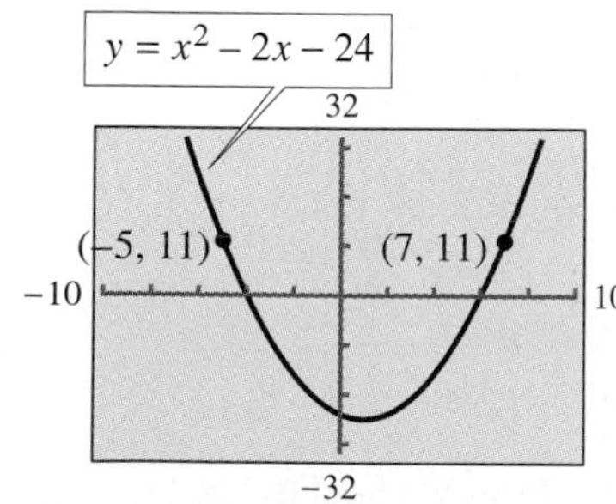

Figure 6.5

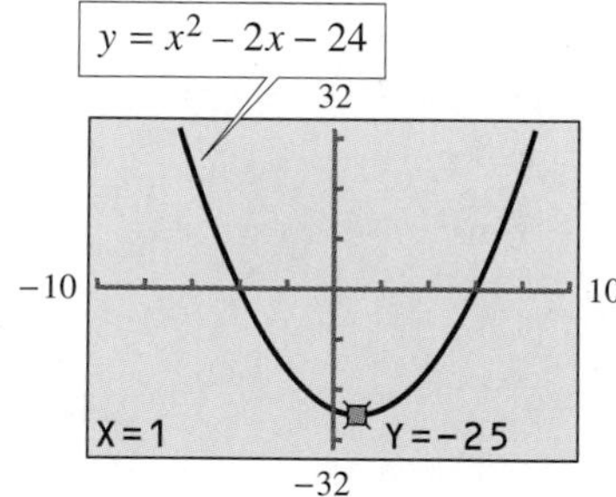

For the equation $x^2 - 2x - 24 = -25$, we trace the graph of $y = x^2 - 2x - 24$ to the point whose y-coordinate is -25. In Fig. 6.5, we see that there appears to be only one such point, $(1, -25)$. Thus the estimated solution of the equation is 1.

Finally, the point $(1, -25)$ appears to be the lowest point of the graph, and so $x^2 - 2x - 24$ cannot have a value less than -25. For example, the quadratic equation $x^2 - 2x - 24 = -30$ has no solution.

Think About It

Produce the graph of $f(x) = 0.43x^2 - 2.6x + 4$ and estimate the number of solutions of $f(x) = 0$. Now zoom and trace several times at the point that appears to represent a solution. What do you conclude?

These results suggest that a quadratic equation can have zero, one, or two real-number solutions.

Solving by Factoring

Even though we can use a graph to estimate solutions of a quadratic equation, we want to develop an algebraic method to determine the solutions exactly, just as we did for linear equations.

LEARNING TIP Although the algebraic methods for solving quadratic and linear equations differ, the graphing techniques for estimating solutions remain the same. As with all equation solving, a combination of graphing and algebraic methods is the ideal way to proceed.

A simple but important property of real numbers is central to the method we will develop. The Zero Factor Property states that if the product of two numbers is zero, then one of the numbers must be zero.

Zero Factor Property

For real numbers a and b, if $ab = 0$, then $a = 0$ or $b = 0$.

Note: To apply the Zero Factor Property, one side of the equation must be expressed as a product, and the other side must be zero. The property can be extended to products with any number of factors.

EXAMPLE 1

Using the Zero Factor Property to Solve an Equation

Solve the equation $(x - 5)(x + 3) = 0$.

Solution

$(x - 5)(x + 3) = 0$	The equation meets the conditions of the Zero Factor Property.
$x - 5 = 0$ or $x + 3 = 0$	One of the factors must be 0.
$x = 5$ or $x = -3$	Solve each first-degree equation.

The solutions are 5 and -3.

If one side of an equation is not a product, we can create a product by factoring the expression.

$x^2 - 2x - 15 = 0$	The Zero Factor Property does not apply.
$(x - 5)(x + 3) = 0$	Factor the expression.

Now the Zero Factor Property applies, and we can proceed as in Example 1.

The factoring method for solving a polynomial equation is given in the following summary.

Solving a Polynomial Equation by Factoring

1. Write the equation in standard form.
2. Factor the polynomial completely.
3. Set each factor equal to 0.
4. Solve each of the resulting equations.

EXAMPLE 2

Solving Quadratic Equations by Factoring

Solve each equation.

(a) $2x^2 - 9x - 5 = 0$ (b) $9x^2 + 4 = 12x$ (c) $x^2 + 4x - 3 = 4$

Figure 6.6

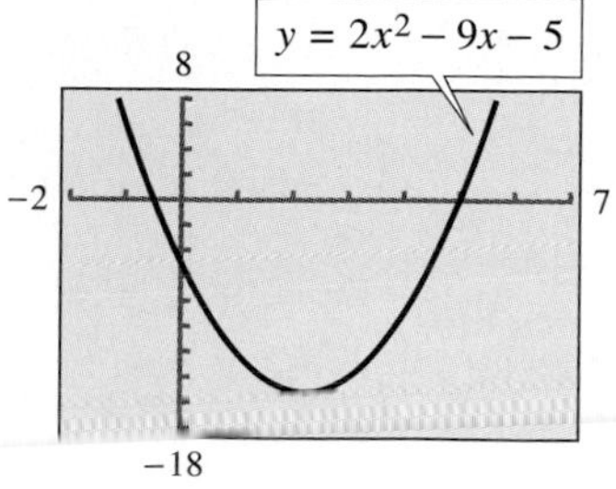

Solution

(a) In Fig. 6.6, the graph of $y = 2x^2 - 9x - 5$ has two x-intercepts, that is, points for which $y = 0$. This indicates that the equation has two solutions.

$$2x^2 - 9x - 5 = 0 \qquad \text{The equation is in standard form.}$$

$$(2x + 1)(x - 5) = 0 \qquad \text{Factor the expression on the left side.}$$

$$2x + 1 = 0 \quad \text{or} \quad x - 5 = 0 \qquad \text{Zero Factor Property}$$

$$x = -\frac{1}{2} \quad \text{or} \quad x = 5 \qquad \text{Solve each case for } x.$$

These exact solutions are consistent with the solutions approximated by the graph.

(b)

$$9x^2 + 4 = 12x$$

$$9x^2 - 12x + 4 = 0 \qquad \text{Write the equation in standard form.}$$

Figure 6.7

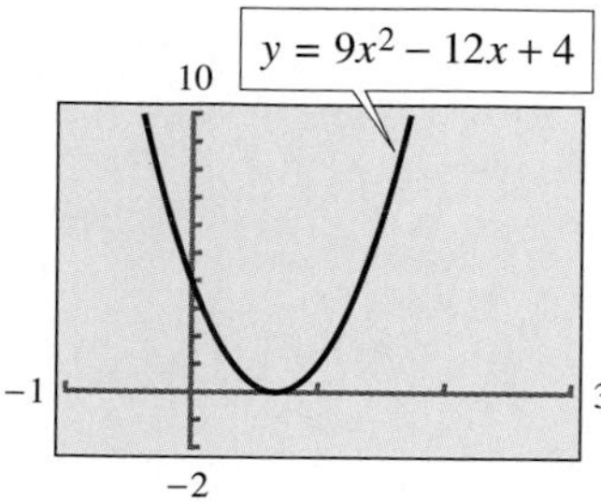

The graph of $y = 9x^2 - 12x + 4$ is shown in Fig. 6.7. There appears to be only one x-intercept, which suggests that the equation has only one solution. To determine the solution algebraically, we solve by factoring.

$$9x^2 - 12x + 4 = 0 \qquad \text{The trinomial is a perfect square.}$$

$$(3x - 2)(3x - 2) = 0 \qquad \text{The factors are identical.}$$

$$3x - 2 = 0 \qquad \text{We obtain only one first-degree equation.}$$

$$x = \frac{2}{3} \qquad \text{Solve for } x.$$

Because the two factors are the same, we obtain just one distinct solution (or root). We say that $\frac{2}{3}$ is a **double root.**

(c)

$$x^2 + 4x - 3 = 4$$

$$x^2 + 4x - 7 = 0 \qquad \text{Write the equation in standard form.}$$

Because we cannot factor $x^2 + 4x - 7$, we are unable to use our current algebraic method for solving the equation. Note, however, that this does not necessarily mean that the equation has no solution. In fact, the graph of $y = x^2 + 4x - 7$ clearly shows that there are two solutions, one between -6 and -5 and another between 1 and 2. (See Fig. 6.8.)

Figure 6.8

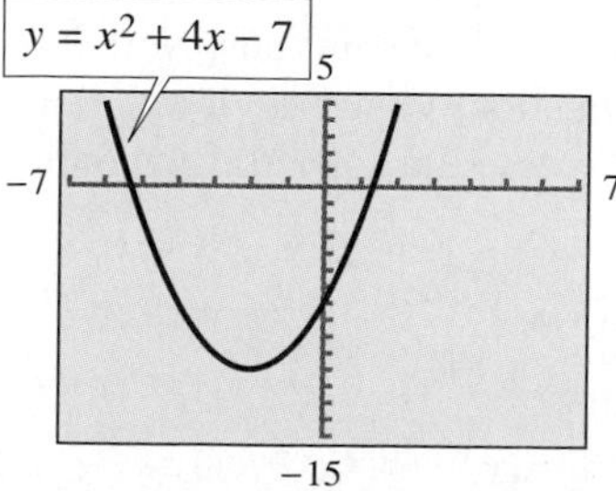

EXAMPLE 3

Solving Quadratic Equations Algebraically

Solve each equation.

(a) $12x^2 = 3x$ (b) $x(6x + 1) = 12$

Solution

(a)

$$12x^2 = 3x$$

$$12x^2 - 3x = 0 \quad \text{Write the equation in standard form.}$$

$$3x(4x - 1) = 0 \quad \text{Factor the left side.}$$

$$3x = 0 \quad \text{or} \quad 4x - 1 = 0 \quad \text{Zero Factor Property}$$

$$x = 0 \quad \text{or} \quad x = \frac{1}{4} \quad \text{Solve each case for } x.$$

The two solutions are 0 and $\frac{1}{4}$.

(b)

$$x(6x + 1) = 12$$

$$6x^2 + x = 12 \quad \text{Distributive Property}$$

$$6x^2 + x - 12 = 0 \quad \text{Write the equation in standard form.}$$

$$(2x + 3)(3x - 4) = 0 \quad \text{Factor the left side.}$$

$$2x + 3 = 0 \quad \text{or} \quad 3x - 4 = 0 \quad \text{Zero Factor Property}$$

$$x = -\frac{3}{2} \quad \text{or} \quad x = \frac{4}{3} \quad \text{Solve each case for } x.$$

The two solutions are $-\frac{3}{2}$ and $\frac{4}{3}$.

While the factoring method is a classic technique for solving quadratic equations, it can also be used for other types of equations.

EXAMPLE 4

Solving Other Types of Equations by Factoring

Solve the following equations.

(a) $4x^3 + 12x^2 - x - 3 = 0$ (b) $|x^2 - 13| = 12$

Solution

Figure 6.9

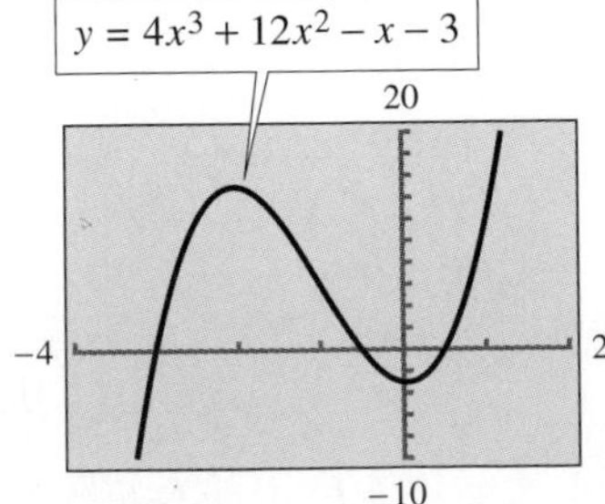

(a) From the graph of $y = 4x^3 + 12x^2 - x - 3$ in Fig. 6.9, we anticipate that the equation has three solutions. Note that this is a third-degree equation, not a quadratic equation. However, if we can factor the left side, the factoring method can still be used to solve the equation.

$$4x^3 + 12x^2 - x - 3 = 0$$

$$(4x^3 + 12x^2) - (x + 3) = 0 \quad \text{Group the left side in pairs.}$$

$$4x^2(x + 3) - (x + 3) = 0 \quad \text{Factor out the GCF in the first pair.}$$

$$(x + 3)(4x^2 - 1) = 0 \quad \text{Factor out the GCF } x + 3.$$

$$(x + 3)(2x + 1)(2x - 1) = 0 \quad \text{Factor the difference of two squares.}$$

$$x + 3 = 0 \quad \text{or} \quad 2x + 1 = 0 \quad \text{or} \quad 2x - 1 = 0 \qquad \text{Zero Factor Property}$$

$$x = -3 \quad \text{or} \quad x = -\frac{1}{2} \quad \text{or} \quad x = \frac{1}{2} \qquad \text{Solve each case for } x.$$

The solutions are -3, $-\frac{1}{2}$, and $\frac{1}{2}$.

Figure 6.10

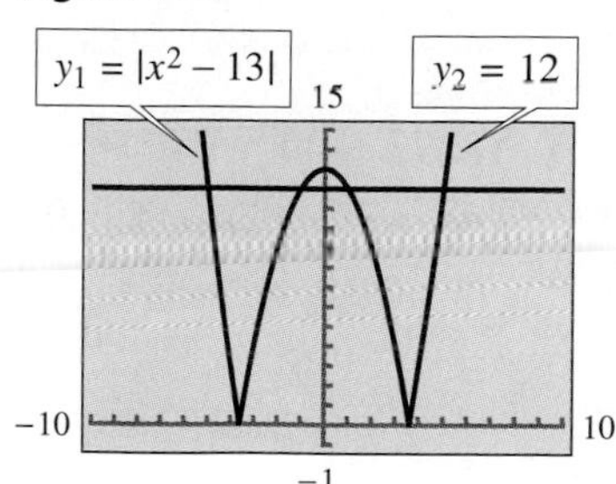

(b) To use the graphing method, we let $y_1 = |x^2 - 13|$ and $y_2 = 12$. In Fig. 6.10 we see that the two graphs intersect in four points. Thus we expect the equation to have four solutions.

The equation $|x^2 - 13| = 12$ states that $x^2 - 13$ represents a number whose absolute value is 12. Therefore, $x^2 - 13$ represents either 12 or -12.

$$x^2 - 13 = 12 \quad \text{or} \quad x^2 - 13 = -12$$

Each possibility is a quadratic equation that we will try to solve by the factoring method.

$$x^2 - 13 = 12 \qquad\qquad x^2 - 13 = -12$$
$$x^2 - 25 = 0 \qquad\qquad x^2 - 1 = 0$$
$$(x + 5)(x - 5) = 0 \qquad\qquad (x + 1)(x - 1) = 0$$
$$x + 5 = 0 \quad \text{or} \quad x - 5 = 0 \qquad\qquad x + 1 = 0 \quad \text{or} \quad x - 1 = 0$$
$$x = -5 \quad \text{or} \quad x = 5 \qquad\qquad x = -1 \quad \text{or} \quad x = 1$$

The four solutions are 5, -5, 1, and -1.

6.6 QUICK REFERENCE

Quadratic Equations

- A **quadratic equation** is an equation that can be written in the *standard form*

  $$ax^2 + bx + c = 0$$

  where a, b, and c are real numbers and $a \neq 0$.
- The term ax^2 is the **quadratic term,** the term bx is the **linear term,** and the term c is the **constant term.**

Solving by Factoring

- The **Zero Factor Property** is as follows:

  For real numbers a and b, if $ab = 0$, then $a = 0$ or $b = 0$.

  In words, if a product of two or more factors is equal to zero, then one of the factors must equal zero.
- For the Zero Factor Property to apply,
  1. one side of the equation must be a product, and
  2. the other side of the equation must be 0.
- To solve a quadratic equation by factoring:
  1. Write the equation in standard form.
  2. Factor the polynomial completely.
  3. Set each factor equal to zero.
  4. Solve each resulting equation.
  5. Check the solution(s).

- In the solving of a quadratic equation, if two factors are the same, each results in the same solution. The solution is called a **double root.**

6.6 EXERCISES

Concepts and Skills

1. The equation $x^2 + 3x - 1 = 0$ has two solutions, but they cannot be determined by factoring. Describe a method that can be used to estimate the solutions.
2. Describe the graphs of second-degree equations that have 0, 1, and 2 solutions.
3. Produce the graph of $y = x^2 - 4x - 12$, and trace the graph to estimate the solution(s) of the following equations.
   (a) $x^2 - 4x - 12 = 0$
   (b) $x^2 - 4x - 12 = -18$
   (c) $x^2 - 4x - 12 = -16$
   (d) $x^2 - 4x - 12 = 20$
4. Produce the graph of $y = x^2 - 2x - 8$, and trace the graph to estimate the solution(s) of the following equations.
   (a) $x^2 - 2x - 8 = -9$
   (b) $x^2 - 2x - 8 = -16$
   (c) $x^2 - 2x - 8 = 0$
   (d) $x^2 - 2x - 8 = 16$

In Exercises 5–12, use the Zero Factor Property to solve the equation.

5. $(x - 4)(x + 3) = 0$
6. $3(2x + 5)(x + 1) = 0$
7. $5x(6 - x) = 0$
8. $x(1 - 2x) = 0$
9. $(x - 5)^2 = 0$
10. $4(x + 4)^2 = 0$
11. $(x + 5)(2x - 6)(3x + 3) = 0$
12. $x(3 + 2x)(1 - x) = 0$
13. How do we distinguish between a first-degree equation and a second-degree equation?
14. Explain why the Zero Factor Property does not apply to the equation $(x + 2)(x - 3) = -6$.

In Exercises 15–22, solve algebraically.

15. $x^2 + 7x = 0$
16. $2c^2 - 5c = 0$
17. $2g^2 - 18 = 0$
18. $h^2 - 49 = 0$
19. $x^2 - 5x + 6 = 0$
20. $-3b^2 - 7b + 6 = 0$
21. $\frac{2}{5}x^2 - \frac{3}{5}x - 1 = 0$
22. $x^2 - \frac{7}{6}x - \frac{1}{2} = 0$

In Exercises 23–38, solve algebraically.

23. $a^2 = 5a$
24. $x^2 = 25$
25. $(x - 2)^2 = 25$
26. $(a + 3)^2 - 64 = 0$
27. $x^2 + 5x = 6$
28. $y^2 - 2y = 8$
29. $x^2 + 35 = 12x$
30. $x^2 + 12 = 7x$
31. $12 = 11d - 2d^2$
32. $15 = 26g - 7g^2$
33. $-8 = -9y^2 - 14y$
34. $-x - 6 = -12x^2$
35. $x(x + 1) = 30$
36. $x(x - 1) = 42$
37. $a(a - 32) + 60 = 0$
38. $c(c + 23) + 60 = 0$

In Exercises 39–48, solve algebraically.

39. $3x(2 + x) = 24$
40. $x(2 - 3x) = \frac{1}{3}$
41. $\frac{2}{5}x(x - 2) = x - 2$
42. $3(x + 1)^2 = 9 - x$
43. $(x - 3)(x + 2) = 6$
44. $(x - 7)(x - 1) = -8$
45. $(3x - 2)(x + 2) = (x - 2)(x + 1) + 10$
46. $3(x - 3)(x + 1) = (2x - 1)(x - 1) + 8$
47. $(4x - 3)^2 + (4x - 3) = 6$
48. $5(x + 2)^2 - 17(x + 2) - 12 = 0$

In Exercises 49–64, solve algebraically.

49. $(x - 5)(x^2 + 3x - 18) = 0$
50. $(2x - 5)(8x^2 + 14x - 15) = 0$
51. $3x^3 - 27x = 0$
52. $8x^2 - 50x^4 = 0$
53. $6x^3 + 2x^2 = 4x$
54. $6y^2 = 3y^3 + 9y^4$

55. $x^3 - 5x^2 - x + 5 = 0$

56. $x^3 + 3x^2 - x - 3 = 0$

57. $2x^3 - 3x^2 - 8x + 12 = 0$

58. $3x^3 + 4x^2 - 27x - 36 = 0$

59. $x^4 + 36 = 13x^2$

60. $x^4 - 26x^2 + 25 = 0$

61. $6(y - 2)^3 - (y - 2)^2 - (y - 2) = 0$

62. $y^2(4y^2 - 9) - 16y(4y^2 - 9) + 48(4y^2 - 9) = 0$

63. $3x^2(3 - x) + 3x(3 - x) + 18(x - 3) = 0$

64. $10(3 - 2x)x^2 - 27(3 - 2x)x - 5(2x - 3) = 0$

In Exercises 65–68, one solution of the equation is given. Determine a, b, or c and another solution of the equation.

65. $2x^2 + bx + 2 = 0; \quad x = -2$

66. $(2a + 1)x^2 + x + 10 = 0; \quad x = 2$

67. $2x^2 + x + (3 - 4c) = 0; \quad x = 3$

68. $-x^2 + x + c = 0; \quad x = -1$

In Exercises 69–74, write an equation with the given solutions.

69. $3, -4$

70. $-2, 5$

71. 4 (double root)

72. -2 (double root)

73. $0, \frac{1}{2}$

74. $\frac{2}{3}, -2$

In Exercises 75–80, solve for x.

75. $x^2 - cx - 2c^2 = 0$

76. $x^2 + xy - 2y^2 = 0$

77. $3y^2 - 4xy = 4x^2$

78. $cx + 2c^2 = 6x^2$

79. $x^2 + 2ax = 3a^2$

80. $4a^2 = 10x^2 + 3ax$

In Exercises 81–86, use the graphing method to estimate the solution(s) of the given equation.

81. $|x^2 - 4x| = 3$

82. $|2x^2 - 5x| = 2$

83. $|2x^2 - 3x| = 2$

84. $|x^2 - x| = 6$

85. $|6x^2 - 5x - 2| = 2$

86. $|2x^2 - 9x - 2| = 3$

In Exercises 87–92, solve algebraically.

87. $|x^2 + 5x| = 6$

88. $|3x^2 - 5x| = 2$

89. $|x^2 + 10x + 15| = 6$

90. $|x^2 + 7x + 8| = 2$

91. $|x^2 - 3x - 1| = 3$

92. $|x^2 - 5x + 5| = 1$

In Exercises 93–100, write an equation describing the conditions of the problem. Then solve the equation to determine the number(s).

93. The sum of a number and its square is 72.

94. The sum of twice a number and its square is 63.

95. The sum of the squares of three consecutive integers is 50.

96. The sum of the squares of three consecutive integers is 77.

97. The sum of the cube of a number and 6 times the number is equal to 5 times the square of the number.

98. The sum of the cube of a number and 12 times the number is equal to 8 times the square of the number.

99. Two numbers have a sum of 3 and a product of -10.

100. The sum of two numbers is 12 and the sum of their squares is 74.

Geometric Models

101. The length of a rectangle is 5 meters more than the width. The area of the rectangle is 594 square meters. What are the dimensions of the rectangle?

102. The width of a rectangular living room is 7 feet less than the length. The area of the room is 294 square feet. What are the dimensions of the rectangle?

103. One leg of a right triangle is 7 inches more than the other leg. The area of the triangle is 60 square inches. What are the lengths of the three sides?

104. One leg of a right triangle is 5 feet more than the other leg. The area of the triangle is 150 square feet. What are the lengths of the three sides?

Real-Life Applications

105. A landscape designer decided to surround a rectangular garden with a path of uniform width. The garden was 14 feet wide and 20 feet long. If the total area of the path was 152 square feet, how wide was the path?

106. A farmer wants to use 500 feet of fencing to enclose a rectangular area of 15,000 square feet. To meet these conditions, what should be the dimensions of the fenced area?

107. A product packaging firm manufactures cardboard trays for holding 24 twelve-ounce cans of soft drinks. The process begins with a rectangular piece of cardboard that is 5 inches longer than it is wide. Two-inch squares are cut from each corner, and the flaps are folded up to form the tray.

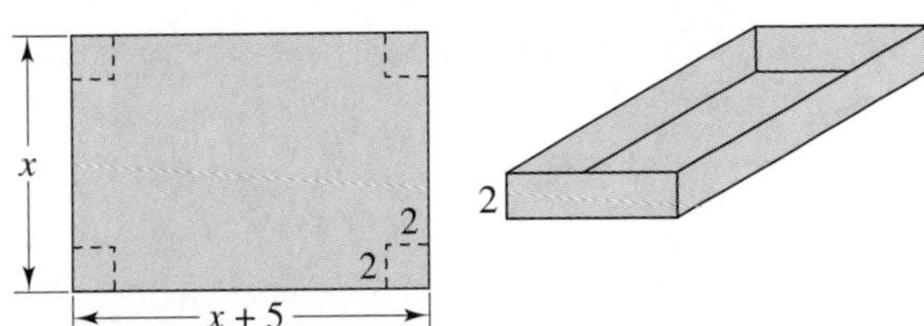

If the area of the bottom of the tray must be 176 square inches, what are the dimensions of the original cardboard rectangle?

108. A small software development firm finds that its profits (in hundreds of thousands of dollars) can be modeled by the expression $-0.4x^2 + 4x$, where x is the number of sales representatives. For how many sales representatives (other than none) will the company earn no profit?

Modeling with Real Data

109. A survey of coffee drinkers revealed that 62% of those in the 35–49 age group become upset if they do not have coffee at a regular time. The accompanying table gives the results for the three age groups that were surveyed. (Source: LMK Associates.)

Age Group	Percent
18–34	50%
25–49	62%
50 and over	56%

The function $C(x) = -0.05x^2 + 4.4x - 32$ models the percents as a function of age x.

(a) Write $C(x)$ with a common factor of -0.05.

(b) Write the function $C(x)$ as a completely factored polynomial.

(c) Solve the equation $C(x) = 0$.

(d) What do the solutions represent?

110. In 1996 the average annual cost of utilities was \$1984. However, the accompanying table shows that the average annual cost varies with age. (Source: Bureau of Labor Statistics.)

Age Group	Average Cost
Under 25	\$1024
25–34	1797
35–44	2232
45–54	2375
55–64	2255
65 and over	1816

The function $f(x) = -1.4x^2 + 142.8x - 1288$ models the average costs as a function of age x.

(a) Trace the graph of the function $f(x)$ to estimate the x-intercepts.

(b) Write an equation that can be used to determine the x-intercepts algebraically.

(c) Use the factoring method to solve your equation. (Hint: Begin by factoring out the common factor -1.4.)

(d) Interpret the x-intercepts. Are they meaningful?

Challenge

111. Produce the graph of $p(x) = x^3 + 1$. The graph suggests only one value of c for which $p(c) = 0$. Determine that number c by solving $x^3 + 1 = 0$ by factoring. Can the quadratic factor be factored more? Produce the graph of the quadratic factor to support your conclusion.

112. Produce the graph of $p(x) = 2x^4 - 3x^2 + 2$. Explain how the graph tells you that $p(x)$ cannot be factored.

113. Solve the following equation.

$$(x + y + 1)(1 - x) + (x + y - 1)(x - 1) = 0$$

In Exercises 114 and 115, solve the equation. (Hint: Recall the substitution technique: $u = 1/x$.)

114. $3\left(\frac{1}{x}\right)^2 + 14\left(\frac{1}{x}\right) - 24 = 0$

115. $\frac{2}{x^2} - \frac{1}{x} - 1 = 0$

116. A rectangular swimming pool is surrounded by a concrete deck. The diagram shows the relative dimensions of the pool and deck.

The combined area of the pool and deck is 2116 square feet. The owner plans to cover the concrete deck with tile that costs $80 per hundred square feet. Determine the total cost of the job.

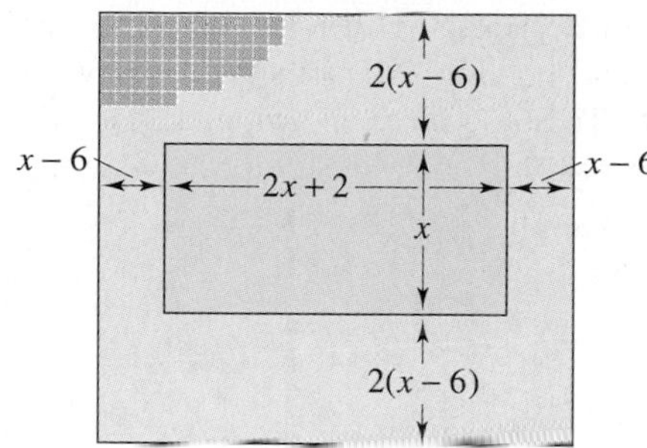

6.7 DIVISION

Quotient of a Polynomial and a Monomial • Long Division • Synthetic Division

Quotient of a Polynomial and a Monomial

To simplify the quotient of two monomials, we use the Quotient Rule for Exponents.

$$\frac{15x^5}{9x^3} = \frac{15}{9} \cdot \frac{x^5}{x^3} = \frac{5}{3} \cdot \frac{x^{5-3}}{1} = \frac{5}{3} \cdot \frac{x^2}{1} = \frac{5x^2}{3}$$

$$\frac{4y^3}{20y^5} = \frac{4}{20} \cdot \frac{y^3}{y^5} = \frac{1}{5} \cdot \frac{y^{3-5}}{1} = \frac{1}{5} \cdot \frac{y^{-2}}{1} = \frac{1}{5y^2}$$

To divide a polynomial by a monomial, we reverse the rule for adding fractions with a common denominator to write the quotient as a sum of quotients of monomials.

Dividing a Polynomial by a Monomial

For monomials a, b, and c, $c \neq 0$,

$$\frac{a+b}{c} = \frac{a}{c} + \frac{b}{c}$$

EXAMPLE 1

Dividing a Polynomial by a Monomial

Find the following quotients.

(a) $\dfrac{12x^3 - 8x^2 + 6x}{2x}$ (b) $\dfrac{15x^6 + 21x^4 - 2x^2 + 3x}{3x^2}$

Solution

(a) $\dfrac{12x^3 - 8x^2 + 6x}{2x} = \dfrac{12x^3}{2x} - \dfrac{8x^2}{2x} + \dfrac{6x}{2x}$ Write the quotient as a sum of quotients.

$= 6x^2 - 4x + 3$ Simplify each quotient.

(b) $\dfrac{15x^6 + 21x^4 - 2x^2 + 3x}{3x^2} = \dfrac{15x^6}{3x^2} + \dfrac{21x^4}{3x^2} - \dfrac{2x^2}{3x^2} + \dfrac{3x}{3x^2}$

$= 5x^4 + 7x^2 - \dfrac{2}{3} + \dfrac{1}{x}$

Long Division

To divide a polynomial by a polynomial with more than one term, we use a procedure similar to long division in arithmetic.

$$\begin{array}{rrl} & 5 & \leftarrow \text{Quotient} \\ \text{Divisor} \rightarrow & 41\overline{)211} & \leftarrow \text{Dividend} \\ & \underline{205} & \\ & 6 & \leftarrow \text{Remainder} \end{array}$$

The result can be written in the following form:

$$\frac{\text{Dividend}}{\text{Divisor}} = \text{Quotient} + \frac{\text{Remainder}}{\text{Divisor}}$$

$$\frac{211}{41} = 5 + \frac{6}{41} = 5\frac{6}{41}$$

To check the result, we use the following relation:

$$\begin{array}{ccccccc} \text{Quotient} & \cdot \ \text{Divisor} & + & \text{Remainder} & = & \text{Dividend} \\ \downarrow & \downarrow & & \downarrow & & \downarrow \\ 5 & \cdot \ 41 & + & 6 & = & 211 \end{array}$$

EXAMPLE 2

Dividing a Polynomial by a Binomial

Divide $(2x^3 - 3x^2 - x + 2)$ by $(x - 2)$.

Solution

$x - 2\overline{)2x^3 - 3x^2 - x + 2}$ Write the problem in long division format.

$$\begin{array}{r} 2x^2 \\ x - 2\overline{)2x^3 - 3x^2 - x + 2} \end{array}$$

Divide the first term of the dividend $2x^3$ by the first term of the divisor x.

$$\begin{array}{r} 2x^2 \\ x - 2\overline{)2x^3 - 3x^2 - x + 2} \\ 2x^3 - 4x^2 \qquad\quad \end{array}$$

Multiply the result $2x^2$ times the divisor $x - 2$. Write the result under the dividend.

$$\begin{array}{r} 2x^2 \\ x - 2\overline{)2x^3 - 3x^2 - x + 2} \\ \underline{2x^3 - 4x^2} \qquad\quad \\ x^2 - x \qquad \end{array}$$

Subtract (by changing signs and adding) and then bring down the next term of the dividend $-x$.

LEARNING TIP

In long division of polynomials, the most frequent error is in the subtraction step. Rather than scribbling out signs, you might want to circle sign changes to make them more prominent.

$$\begin{array}{r}
2x^2 + x \\
x-2\overline{)\,2x^3 - 3x^2 - x + 2} \\
\underline{2x^3 - 4x^2} \\
x^2 - x
\end{array}$$

Divide the first term of the result x^2 by the first term of the divisor x.

$$\begin{array}{r}
2x^2 + x \\
x-2\overline{)\,2x^3 - 3x^2 - x + 2} \\
\underline{2x^3 - 4x^2} \\
x^2 - x
\end{array}$$

Multiply the result x times the divisor $x - 2$.

$$\begin{array}{r}
x^2 - 2x
\end{array}$$

Write the result under the dividend.

$$\begin{array}{r}
2x^2 + x \\
x-2\overline{)\,2x^3 - 3x^2 - x + 2} \\
\underline{2x^3 - 4x^2} \\
x^2 - x \\
\underline{x^2 - 2x} \\
x + 2
\end{array}$$

Subtract (by changing signs and adding) and then bring down the next term of the dividend 2.

$$\begin{array}{r}
2x^2 + x + 1 \\
x-2\overline{)\,2x^3 - 3x^2 - x + 2} \\
\underline{2x^3 - 4x^2} \\
x^2 - x \\
\underline{x^2 - 2x} \\
x + 2 \\
\underline{x - 2} \\
4
\end{array}$$

Divide the first term of the result x by the first term of the divisor x. Then multiply the result 1 by the divisor $x - 2$.

Then subtract.

The quotient is $2x^2 + x + 1$, and the remainder is 4. We can write the result in the following form:

$$\frac{2x^3 - 3x^2 - x + 2}{x - 2} = 2x^2 + x + 1 + \frac{4}{x - 2}$$

We check the result with the following relation:

Quotient	·	Divisor	+	Remainder	=	Dividend
$(2x^2 + x + 1)$	·	$(x - 2)$	+	4	=	$2x^3 - 3x^2 - x + 2$

Note: When using long division to divide polynomials, the dividend and the divisor should be in descending order. We continue the division operation until the remainder is 0 or the degree of the remainder is less than the degree of the divisor.

EXAMPLE 3

Missing Terms in the Dividend

Divide $x^3 + 27$ by $x + 3$.

Solution

As we write the problem in the long division format, it is helpful to include the missing terms in the dividend with zero coefficients.

$$\begin{array}{r l}
\begin{array}{r}
x^2 - 3x + 9 \\
x + 3 \overline{)\, x^3 + 0x^2 + 0x + 27} \\
\underline{x^3 + 3x^2 \qquad\qquad} \\
-3x^2 + 0x \qquad \\
\underline{-3x^2 - 9x \qquad} \\
9x + 27 \\
\underline{9x + 27} \\
0
\end{array}
&
\begin{array}{l}
\\
\text{Include the missing } x^2 \text{ and } x \text{ terms.} \\
x^2(x + 3) = x^3 + 3x^2 \\
\text{Subtract and bring down } 0x. \\
-3x(x + 3) = -3x^2 - 9x \\
\text{Subtract and bring down 27.} \\
9(x + 3) = 9x + 27 \\
\text{Subtract. The remainder is 0.}
\end{array}
\end{array}$$

The quotient is $x^2 - 3x + 9$, and the remainder is 0.

$$\frac{x^3 + 27}{x + 3} = x^2 - 3x + 9$$

To check, multiply $(x^2 - 3x + 9)$ by $(x + 3)$. The result is $x^3 + 27$.

When we divide 20 by 4, we obtain a quotient of 5. Because the remainder is 0, we say that 20 is *divisible* by 4, and we can write the factorization $20 = 5 \cdot 4$.

Similarly, in Example 3 the remainder is 0, and so we say that $x^3 + 27$ is divisible by $x + 3$, and we can write the dividend in factored form.

$$\begin{array}{ccccc}
\text{Dividend} & = & \text{Quotient} & \cdot & \text{Divisor} \\
x^3 + 27 & = & (x^2 - 3x + 9) & \cdot & (x + 3)
\end{array}$$

EXAMPLE 4

Missing Terms in the Dividend and the Divisor

Divide $x^3 + 2x - 1$ by $2x^2 + 4$.

Solution

Write the problem in the long division format with zero for the coefficients of the missing terms in the dividend and the divisor.

$$2x^2 + 0x + 4 \overline{)\, x^3 + 0x^2 + 2x - 1}$$

$$\begin{array}{r l}
\begin{array}{r}
\tfrac{1}{2}x \qquad \\
2x^2 + 0x + 4 \overline{)\, x^3 + 0x^2 + 2x - 1} \\
\underline{x^3 + 0x^2 + 2x \qquad} \\
-1
\end{array}
&
\begin{array}{l}
\text{Divide } x^3 \text{ by } 2x^2. \\
\\
\tfrac{1}{2}x(2x^2 + 0x + 4) = x^3 + 0x^2 + 2x \\
\text{Subtract and bring down } -1.
\end{array}
\end{array}$$

We stop dividing because the degree of the remainder is less than the degree of the divisor. The quotient is $\frac{1}{2}x$; the remainder is -1.

Synthetic Division

Synthetic division is a numerical method for dividing a polynomial by a binomial of the form $x - c$. The technique involves writing only the essential parts of a long division problem.

EXPLORING THE CONCEPT

Synthetic Division

Consider the following long division problem on the left and compare it with the numerical arrangement on the right.

$$
\begin{array}{r}
2x^2 + 3x + 8 \\
x - 3 \overline{)\, 2x^3 - 3x^2 - 1x + 2} \\
\underline{2x^3 - 6x^2} \phantom{{}- 1x + 2} \\
3x^2 - x \phantom{{}+ 2} \\
\underline{3x^2 - 9x} \phantom{{}+ 2} \\
8x + 2 \\
\underline{8x - 24} \\
26
\end{array}
\qquad\qquad
\begin{array}{c|rrrr}
\textcircled{3} & 2 & -3 & -1 & 2 \\
 & & 6 & 9 & 24 \\
\hline
 & 2 & 3 & 8 & \boxed{26}
\end{array}
$$

We can make the following observations about the numbers on the right.

1. The circled number 3 is the same as the 3 in the divisor $x - 3$. We refer to this number as the *divider*.
2. The numbers 2, -3, -1, and 2 along the first row are the coefficients of the dividend $2x^3 - 3x^2 - x + 2$.
3. The numbers 2, 3, and 8 along the bottom row are the coefficients of the quotient $2x^2 + 3x + 8$.
4. The boxed number 26 is the remainder.
5. Each number in the bottom row is the sum of the two numbers above it.
6. Each number in the middle row is the product of the divider 3 and the number in the bottom row of the preceding column.

The following illustrates the use of synthetic division to divide $2x^3 - 3x^2 - x + 2$ by $x - 3$.

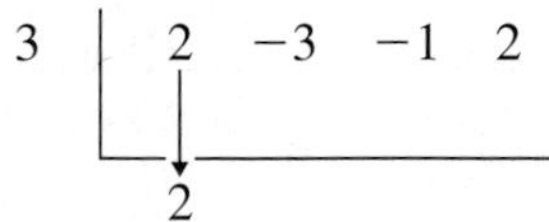

Write the divider 3 and the coefficients of the dividend.

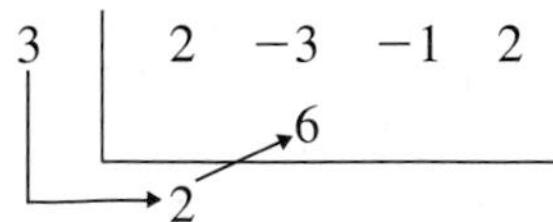

Bring the first number of the top row down to the bottom row.

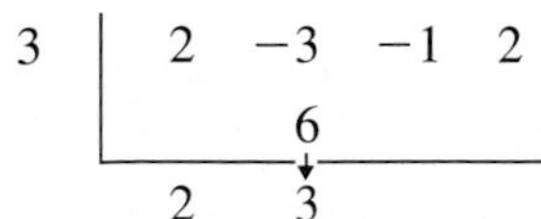

Multiply the first number 2 in the bottom row by the divider 3 to obtain 6, the first number in the middle row.

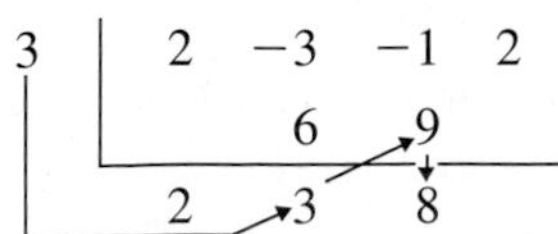

Add the -3 and 6 to obtain the second number 3 in the bottom row.

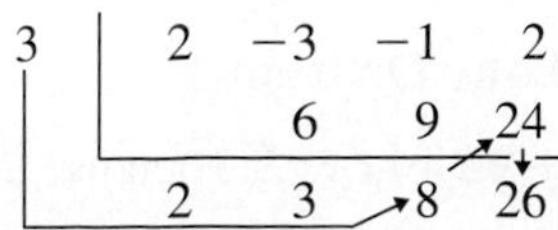

Similarly, multiply the 3 in the bottom row by the divider 3 to obtain 9, the second number in the middle row. Then add the column to obtain 8.

$$
\begin{array}{c|rrrr}
3 & 2 & -3 & -1 & 2 \\
 & & 6 & 9 & 24 \\
\hline
 & 2 & 3 & 8 & 26
\end{array}
$$

Finally, multiply this 8 by the divider 3 to obtain 24, the last number in the middle row. Then add the column to obtain 26.

The entries in the bottom row are the coefficients of the quotient and the remainder.

Quotient: $2x^2 + 3x + 8$ Remainder: 26

Note that the degree of the quotient is one less than the degree of the dividend. Because the degree of the dividend is 3, the degree of the quotient is 2. Thus we begin the quotient with $2x^2$.

Synthetic division is a valid shortcut to long division provided certain requirements are met.

1. The divisor must be of the form $x - c$. For a divisor such as $x - 7$, the divider is 7. For a divisor such as $x + 5$, we write $x - (-5)$, and the divider is -5.
2. The dividend must be written in descending order.
3. We write 0 as the coefficient of any missing term.

The advantages of synthetic division are the ease and speed with which it can be performed. The method eliminates much of the repetition involved in long division. For example, it is unnecessary to keep writing the variable factors of the terms.

EXAMPLE 5

Dividing with Synthetic Division

Use synthetic division to divide $x^4 - 3x^3 + 2x - 4$ by $x - 3$.

Solution

$$\begin{array}{r|rrrrr} 3 & 1 & -3 & 0 & 2 & -4 \\ & & 3 & 0 & 0 & 6 \\ \hline & 1 & 0 & 0 & 2 & 2 \\ & \downarrow & \downarrow & \downarrow & \downarrow & \\ & 1x^3 + & 0x^2 + & 0x + & 2 & \end{array}$$

Include a 0 as the coefficient of the missing x^2 term.

The remainder is 2.

Quotient

The degree of the dividend is 4, so the degree of the quotient is 3. The quotient is $x^3 + 2$, and the remainder is 2.

We can also write the result in the following form:

$$\frac{x^4 - 3x^3 + 2x - 4}{x - 3} = x^3 + 2 + \frac{2}{x - 3}$$

To check, multiply the quotient by the divisor and add the remainder. The result is the dividend.

$$(x^3 + 2)(x - 3) + 2 = x^4 - 3x^3 + 2x - 4$$

EXAMPLE 6

Dividing with Synthetic Division

Divide $x^3 - 6x^2 - x + 30$ by $x + 2$. Then use the result to factor $x^3 - 6x^2 - x + 30$ completely.

Solution

The divisor must be written as $x - (-2)$ with $c = -2$.

$$\begin{array}{r|rrrr} -2 & 1 & -6 & -1 & 30 \\ & & -2 & 16 & -30 \\ \hline & 1 & -8 & 15 & 0 \\ & \downarrow & \downarrow & \downarrow & \\ & 1x^2 - & 8x + & 15 & \end{array}$$

The remainder is 0.

Quotient

Think About It

Suppose you use synthetic division to divide $x^{37} - 1$ by $x - 1$. Describe the bottom row of numbers and the remainder. What can you conclude?

Because the remainder is 0, we can write the dividend in factored form.

$$x^3 - 6x^2 - x + 30 = (x + 2)(x^2 - 8x + 15)$$ Complete the factorization by factoring the trinomial.

$$= (x + 2)(x - 5)(x - 3)$$

6.7 QUICK REFERENCE

Quotient of a Polynomial and a Monomial

- To divide a monomial by a monomial, use the Quotient Rule for Exponents.
- To divide a polynomial by a monomial, divide each term of the polynomial by the monomial.

Long Division

- To divide a polynomial by a polynomial, we can use a method similar to long division in arithmetic.
- To keep like terms in columns when dividing polynomials, use coefficients of 0 for missing terms.
- When dividing polynomials, continue the division operation until the remainder is 0 or the degree of the remainder is less than the degree of the divisor.
- To check the answer, multiply the quotient times the divisor and add the remainder. The result should be the dividend.

Synthetic Division

- To divide a polynomial (written in descending order) by $x - c$, where c is a constant called the *divider*, follow these steps:
  1. In the first row, write the coefficients of the dividend with zero as the coefficient of any missing term.
  2. Write the divider c to the left of the first row in step 1.
  3. Bring the first coefficient in the top row down to the bottom row.
  4. Multiply the divider c times the new entry in the bottom row, and place the result in the middle row of the next column.
  5. Add the column and write the result underneath in the bottom row.
  6. Repeat steps 4 and 5 until the bottom row is completely filled.
  7. The bottom row then gives the coefficients of the quotient. The last number in the bottom row is the remainder. The degree of the quotient will be 1 less than the degree of the dividend.

6.7 EXERCISES

Concepts and Skills

1. Identify the dividend, divisor, quotient, and remainder in the following long division problem.

$$\begin{array}{r} 1 \\ x\overline{)x+7} \\ \underline{x} \\ 7 \end{array}$$

2. Suppose that you divide $x^2 + 5x + 9$ by $x + 1$ and obtain the quotient $x + 4$ and a remainder of 5. Explain how you can check your work. Then perform the check to determine whether this result is correct.

In Exercises 3–12, determine the quotient.

3. $\dfrac{10x^4 + 20x^3}{5x^2}$

4. $\dfrac{4x^3 - 2x^2}{2x^2}$

5. $\dfrac{4x^5 + 6x^3 + x^2}{2x^2}$

6. $\dfrac{6x^6 - 3x^5 + 2x^3}{3x^3}$

7. $\dfrac{16a^4 + 12a^3 - 6a^2 + 8a}{2a}$

8. $\dfrac{18a^4 - 12a^3 - 15a^2 + 9a}{3a}$

9. $\dfrac{20x^5 - 12x^4 + 6x^2 - 8x}{4x^2}$

10. $\dfrac{20x^5 + 10x^4 - 3x^2 + 15x}{5x^2}$

11. $\dfrac{15c^5d^7 + 6c^4d^5 - 9cd^3}{3c^2d^2}$

12. $\dfrac{18c^4d^7 - 10c^2d^5 + 8cd^2}{2cd^3}$

In Exercises 13–16, determine the quotient. (Hint: In Exercises 15 and 16, begin by simplifying the numerator.)

13. $\dfrac{6x^3(3x + 1) + 4x^2(2x - 3)}{2x^2}$

14. $\dfrac{9x^2(2x - 1) - 3x(x - 2)}{3x}$

15. $\dfrac{(x + 4)^2 - (x - 4)^2}{2x}$

16. $\dfrac{(3 - x)^2 - (3 + x)^2}{3x}$

17. When we perform long division, when do we stop dividing?

18. If a polynomial of degree $n > 2$ is divided by a polynomial of degree 2, what are the possible degrees of the remainder? Explain.

In Exercises 19–28, use long division to determine the quotient and remainder.

19. $\dfrac{2x + 1}{x - 3}$

20. $\dfrac{4x + 3}{x + 2}$

21. $(x^2 + 4x + 3) \div (x + 2)$

22. $(x^2 - x - 6) \div (x + 3)$

23. $\dfrac{3x^2 + 2}{x - 1}$

24. $\dfrac{x^2 - 2}{x + 1}$

25. $(4a^3 - a^2 + 5a + 4) \div (4a + 3)$

26. $(3x^4 + 2x^3 - 4x^2 + 7x - 5) \div (3x - 1)$

27. $\dfrac{12x^3 + 13x^2 - 1}{4x - 1}$

28. $\dfrac{10x^3 + x^2 + 12x + 9}{5x + 3}$

In Exercises 29–34, use long division to determine the quotient and remainder.

29. $\dfrac{x^2 + 3}{x^2 - 1}$

30. $\dfrac{x^2 - 4}{x^2 + 1}$

31. $\dfrac{x^2 + 3x - 4}{x^2 + 2x}$

32. $\dfrac{3x^3 + 2x - 1}{x^2 + 2}$

33. $(a^4 + 7a - a^2 - 3) \div (a^2 - 2a + 3)$

34. $(2c^4 + 3 - 8c) \div (c^2 - 3c + 1)$

In Exercises 35–38, use long division to determine the quotient and remainder.

35. $\dfrac{x^2 + 4xy + 3y^2}{x - 2y}$

36. $\dfrac{4x^2 - 2xy - y^2}{2x + y}$

37. $\dfrac{4x^3 - 3xy^2 - 2y^3}{x + y}$

38. $\dfrac{x^3 - x^2y - y^3}{x - 2y}$

39. To use synthetic division to divide a polynomial by $x + 3$, show how to write the divisor in order to determine the divider and state what the divider is.

40. Suppose that you want to use synthetic division to divide $\dfrac{x - x^5}{x - 2}$. How should you write the row of dividend coefficients?

In Exercises 41–50, use synthetic division to determine the quotient and remainder.

41. $\dfrac{3x^2 - 10x - 8}{x - 4}$ **42.** $\dfrac{x^2 - 2x - 15}{x + 3}$

43. $(6x^2 + x + 1) \div \left(x - \dfrac{1}{2}\right)$

44. $(2x^2 - 3x - 10) \div \left(x + \dfrac{5}{2}\right)$

45. $(2x^4 - 3x^3 + x^2 - 3x + 4) \div (x - 2)$

46. $(3x^3 - 4x^2 - 2x - 5) \div (x - 3)$

47. $(x^2 + x^4 - 14) \div (x + 2)$

48. $(x^2 - 18x + 2x^3 + 42) \div (x + 4)$

49. $\dfrac{2x^2 - 6 + x^3 - 3x}{x - 2}$ **50.** $\dfrac{4x + x^3 - 13}{x - 3}$

51. Can you use synthetic division to divide the polynomial $x^4 + x - 2$ by $x^2 - 1$? Why?

52. When we say, "Find the quotient of $x + 2$ and $x^2 + 5x - 3$," do we divide $x + 2$ by $x^2 + 5x - 3$ or do we divide $x^2 + 5x - 3$ by $x + 2$?

In Exercises 53 and 54, by inspecting the given synthetic division problem, determine a, b, c, d, and e.

53.
$$\begin{array}{r|rrrrr} 2 & a & b & c & d & e \\ & & 2 & 0 & 0 & 2 \\ \hline & 1 & 0 & 0 & 1 & 0 \end{array}$$

54.
$$\begin{array}{r|rrrrr} -3 & a & b & c & d & e \\ & & -6 & 15 & 3 & -12 \\ \hline & 2 & -5 & -1 & 4 & -6 \end{array}$$

In Exercises 55 and 56, the given synthetic division problem represents the division of a third-degree polynomial by a divisor of the form $x - c$. State the dividend and the divisor.

55.
$$\begin{array}{|rrrr} 1 & 3 & -2 & -4 \\ & -1 & -2 & 4 \\ \hline 1 & 2 & -4 & 0 \end{array}$$

56.
$$\begin{array}{|rrrr} 1 & -4 & 3 & -11 \\ & 4 & 0 & 12 \\ \hline 1 & 0 & 3 & 1 \end{array}$$

In Exercises 57–68, divide.

57. $\dfrac{x^2 - 2x - 9}{x - 6}$ **58.** $\dfrac{3x^2 + 14x + 5}{x + 3}$

59. $(3x^2 + 5x + 2) \div (x - 3)$

60. $(2x^2 - 4x - 1) \div (x + 2)$

61. $\dfrac{x^2 + 9x + 20}{x + 4}$

62. $\dfrac{x^2 - 8x + 16}{x - 4}$

63. $(6a^3 + 5a^2 + 3) \div (3a - 2)$

64. $(4x^4 + 5x^2 - 7x) \div (2x + 1)$

65. $(2x^3 + 5x^2 - 3x + 2) \div (x + 2)$

66. $(3x^4 + x^3 - 5x^2 + 2x - 3) \div (x + 1)$

67. $\dfrac{x + 5 - 2x^3 + x^5}{x + 2}$

68. $\dfrac{2x^4 - 25x + x^5 - 5}{x + 3}$

In Exercises 69–74, divide the first polynomial by the given binomial. Then use the result to factor the polynomial completely.

69. $x^3 - 13x + 12;\quad x - 1$

70. $x^3 - 49x + 120;\quad x + 8$

71. $x^3 - 3x^2 - 10x + 24;\quad x + 3$

72. $x^3 - 6x^2 + 11x - 6;\quad x - 2$

73. $2x^3 - 3x^2 + 1;\quad 2x + 1$

74. $3x^3 - x^2 - 8x - 4;\quad 3x + 2$

In Exercises 75–78, perform the indicated operations.

75. $(2x + 1)(x - 5) \div (x + 2)$

76. $(x^2 + 7x + 12) \div (x + 3) \div (x - 2)$

77. $(3x + 1)^2 \div (x - 1)$

78. $(x^2 + 2)^2 \div (x^2 + 1)$

In Exercises 79–82, determine the missing quantity.

	Quotient	*Divisor*	*Remainder*	*Dividend*
79.	$x + 3$	$2x - 1$	2	
80.	$x^2 + 1$	$x + 1$	-3	
81.	$x - 3$		-1	$x^2 - x - 7$
82.	$x - 2$		1	$x^3 - 2x^2 + x - 1$

83. Divide $x - 3$ into $x - 12 - 17x^2 + x^5 - 3x^3$.

84. Divide $x^4 + x^2 + 2$ by $x + 1$.

85. Find the quotient of $x^3 + 2x - 3$ and the product of x and $x - 3$.

86. Subtract $4x - 1$ from the quotient of $6x^2 + 11x - 7$ and $2x - 1$.

87. What polynomial divided by $3x^2$ results in $3x^3 + 4x^2 + x - 4$?

88. When the trinomial $x^2 - 2x - 3$ is divided by $P(x)$, the quotient is $x + 2$ and the remainder is 5. What is $P(x)$?

Real-Life Applications

89. A physicist found that a particle traveled a distance of $x^2 + 5x - 24$ meters in $x - 3$ seconds. What expression represents the average speed of the particle in meters per second?

90. A moving company loads cubic boxes into a moving van. Suppose that $w + 8$ boxes can be placed from the front to the back of the van and w boxes can be stacked on each other. If the capacity of the truck is $w^3 + 10w^2 + 16w$ boxes, what expression represents the number of boxes that can be placed along one end of the truck?

91. The length in feet of a roll of fencing is represented by $x^3 - 65x - 6$. The fencing is to be stretched across $x + 9$ posts, which are evenly spaced.
    (a) Write an expression that represents the distance between the posts. (Hint: The number of spans between the posts is 1 less than the number of posts.)
    (b) How much fencing will be left over?

92. A production engineer determined that the cost of producing n rolls of fiberoptic cable is $n^2 + 52n$ dollars. If 50 additional rolls are produced, the total cost is increased by \$100.
    (a) What is the unit cost of the original n rolls?
    (b) Does producing 50 additional rolls increase or decrease the unit cost? By how much?

Challenge

In Exercises 93 and 94, determine k so that the quotient has the specified remainder.

93. $(x^2 + 4x + k) \div (x - 2), \quad R = 4$

94. $(3x^2 + 7x + k) \div (3x + 1), \quad R = 0$

In Exercises 95 and 96, determine the value of k so that the remainder will be zero.

95. $(x^3 + kx^2 - 2k + 3) \div (x - 1)$

96. $(4x^3 + kx^2 + 7) \div (x + 1)$

In Exercises 97 and 98, determine the quotient.

97. $\dfrac{16x^{4n} - 64x^{2n}}{4x^{2n}}$

98. $\dfrac{6x^{6n} + 12x^{5n} - 9x^{4n}}{3x^{2n}}$

In Exercises 99 and 100, divide.

99. $(x^{4n} + 3x^{3n} - x^{2n} + 2x^n + 1) \div (x^n + 3)$

100. $(x^{2n} + 2x^n - 4) \div (x^n + 1)$

101. For $(x^2 + bx + b^2) \div (x - b)$, determine the positive value of b for which the remainder is 12.

102. Suppose that when $ax^2 + 3x + c$ is divided by $x + 1$, the remainder is 0. If a is twice c, what are the values of a and c?

6.8 THE REMAINDER AND FACTOR THEOREMS

The Remainder Theorem • The Factor Theorem

The Remainder Theorem

In Section 6.7 we discussed methods for dividing a polynomial $P(x)$ by a binomial in the form $x - c$. We now examine the relationship between the remainder and the value of $P(c)$.

EXPLORING THE CONCEPT

The Remainder Theorem

Consider the following quotient:

$$\frac{2x^3 - 3x^2 - x + 2}{x - 2}$$

In this instance, $P(x) = 2x^3 - 3x^2 - x + 2$, and c is 2. In the following, we use synthetic division to determine the remainder, and we evaluate $P(c)$.

$$\begin{array}{r|rrrr} 2 & 2 & -3 & -1 & 2 \\ & & 4 & 2 & 2 \\ \hline & 2 & 1 & 1 & 4 \end{array} \leftarrow \text{Remainder} \rightarrow \qquad \begin{aligned} P(2) &= 2(2)^3 - 3(2)^2 - 2 + 2 \\ &= 16 - 12 - 2 + 2 \\ &= 4 \end{aligned}$$

Observe that when we divide $P(x)$ by $x - 2$, we obtain a remainder that is equal to $P(2)$.

This result can be generalized, and it is stated in the following theorem.

The Remainder Theorem

If a polynomial $P(x)$ is divided by $x - c$, where c is a constant, then the remainder is $P(c)$.

We can justify the Remainder Theorem with the following reasoning. Suppose that when $P(x)$ is divided by $x - c$, the quotient is $Q(x)$, and the remainder is R. We know that the relationship among these quantities can be expressed in the following way:

$$P(x) = Q(x) \cdot (x - c) + R$$

When we evaluate $P(x)$ for $x = c$, we obtain the following:

$$\begin{aligned} P(c) &= Q(c) \cdot (c - c) + R \\ &= 0 + R \\ &= R \end{aligned}$$

Therefore, the value of $P(c)$ is the same as the remainder when $P(x)$ is divided by $x - c$.

Note: The Remainder Theorem applies only if the divisor is of the form $x - c$. Divisors of the form $x + c$ must first be changed to $x - (-c)$.

The Remainder Theorem gives us a second way to evaluate a polynomial for $x = c$ and to determine the remainder when $P(x)$ is divided by $x - c$. We can evaluate $P(c)$ directly, or we can use synthetic division.

EXAMPLE 1

Using the Remainder Theorem

Suppose that $p(x) = x^4 - 2x^3 + x^2 - 3x - 2$. Use the Remainder Theorem to evaluate $p(-1)$.

Solution

The value of $p(-1)$ is the remainder when $p(x)$ is divided by $x - (-1)$. Use synthetic division to divide $p(x)$ by $x + 1$.

$$\begin{array}{r|rrrrr} -1 & 1 & -2 & 1 & -3 & -2 \\ & & -1 & 3 & -4 & 7 \\ \hline & 1 & -3 & 4 & -7 & 5 \end{array} \leftarrow \text{Remainder}$$

Because the remainder is 5, $p(-1) = 5$. This result can be verified by entering $x^4 - 2x^3 + x^2 - 3x - 2$ in a calculator and evaluating the expression for $x = -1$.

When you divide $P(x)$ by $x - c$, determining the remainder by evaluating $P(c)$ is a good use of a calculator. However, if you also want to know the quotient, then synthetic division is the best approach.

The Factor Theorem

In Section 6.7 we observed that when a polynomial $P(x)$ is divided by $x - c$ and the remainder is 0, we can write $P(x) = Q(x)(x - c)$, and thus $x - c$ is a factor of $P(x)$. Combining this result with the Remainder Theorem, we obtain the Factor Theorem.

The Factor Theorem

For a polynomial $P(x)$ and a constant c, $P(c) = 0$ if and only if $x - c$ is a factor of $P(x)$.

The Factor Theorem implies that if we find a value of c such that $P(c) = 0$, then $x - c$ is a factor of $P(x)$. Conversely, if $x - c$ is a factor of $P(x)$, then $P(c) = 0$.

EXAMPLE 2

Using the Factor Theorem

(a) Given that $x + 3$ is a factor of $P(x) = x^3 - 7x + 6$, what is $P(-3)$?

(b) If $Q(x)$ is a polynomial such that $Q(4) = 0$, what is one factor of $Q(x)$?

Solution

The Factor Theorem applies to both parts.

(a) Because $x + 3 = x - (-3)$ is a factor of $P(x)$, $P(-3) = 0$.

(b) Because $Q(4) = 0$, a factor of $Q(x)$ is $x - 4$.

The Factor Theorem also has implications for equations. If $x - c$ is a factor of $P(x)$, then $P(c) = 0$. In other words, c is a **solution** (or **root**) of the equation $P(x) = 0$. We also call c a **zero** of function P.

We can also use the Factor Theorem for factoring a polynomial.

EXPLORING THE CONCEPT

Factoring a Polynomial

Consider the polynomial $p(x) = x^3 + x^2 - 17x + 15$, whose graph is shown in Fig. 6.11.

Figure 6.11

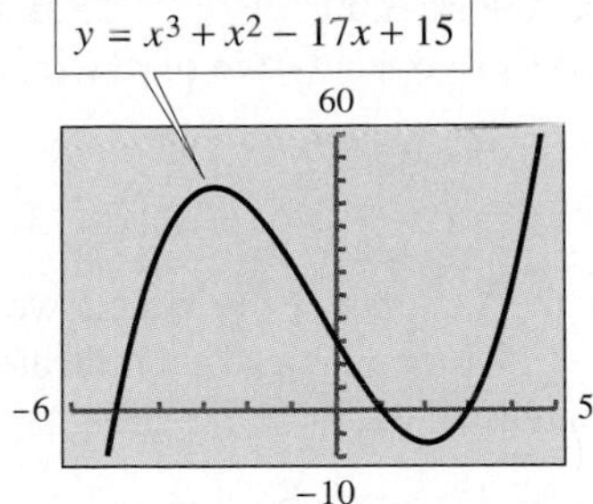

From the graph we see that the x-intercepts appear to be $(-5, 0)$, $(1, 0)$, and $(3, 0)$. We can verify these estimates by showing that $p(-5) = p(1) = p(3) = 0$. Thus, according to the Factor Theorem, $x + 5$, $x - 1$, and $x - 3$ are factors of the polynomial.

The following summary outlines this procedure for factoring a polynomial.

LEARNING TIP

To use synthetic division as much as possible, begin by estimating the intercepts with integer x-coordinates. Then use synthetic division to divide by the factors associated with those intercepts.

Factoring with the Factor Theorem

To factor a polynomial $p(x)$, perform the following steps:

1. Produce the graph of $p(x)$.
2. Estimate each x-intercept $(c, 0)$.
3. Verify that $p(c) = 0$ for each x-intercept.
4. For each x-intercept $(c, 0)$, $x - c$ is a factor of $p(x)$.

The factoring methods discussed earlier in this chapter are generally the best methods to use for first- and second-degree polynomials. For third- and higher-degree polynomials, the use of a graph in conjunction with the Factor Theorem is a good approach.

EXAMPLE 3

Using a Graph and the Factor Theorem to Factor a Polynomial

Factor $p(x) = 2x^3 - 5x^2 - 4x + 3$ completely.

Solution

We produce the graph of $p(x) = 2x^3 - 5x^2 - 4x + 3$.

Figure 6.12

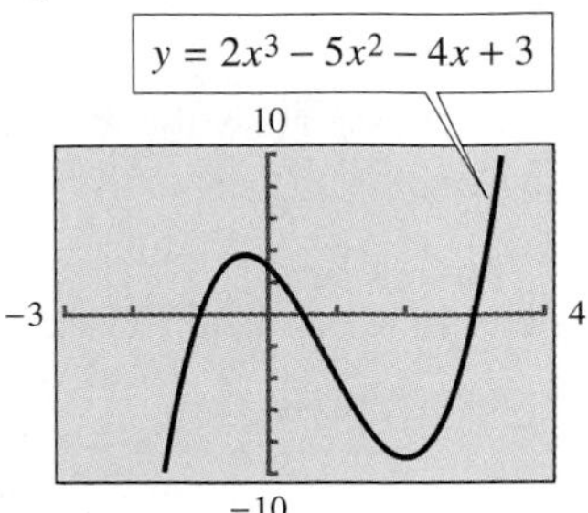

From the graph in Fig. 6.12, two of the x-intercepts appear to be $(-1, 0)$ and $(3, 0)$. We can easily verify that $p(-1) = 0$ and $p(3) = 0$. Therefore, according to the Factor Theorem, two factors of $p(x)$ are $x + 1$ and $x - 3$.

We could also use the graph to estimate the third x-intercept, but in the following, we use an alternative approach.

Synthetically divide $p(x)$ by the factor $x - 3$. (Dividing by $x - 3$ is an arbitrary choice. The same factorization would be obtained by dividing by $x + 1$.)

$$\begin{array}{r|rrrr} 3 & 2 & -5 & -4 & 3 \\ & & 6 & 3 & -3 \\ \hline & 2 & 1 & -1 & 0 \end{array}$$

The quotient is $2x^2 + x - 1$, and the remainder is 0 [as we knew it should be because $x - 3$ is a factor of $p(x)$]. Therefore, we can write $p(x)$ as follows:

$$\begin{aligned} p(x) &= (2x^2 + x - 1)(x - 3) \\ &= (2x - 1)(x + 1)(x - 3) \end{aligned}$$

Note the factor $x + 1$, which we had previously anticipated. Also, $p(x) = 0$ if $2x - 1 = 0$ or $x = \frac{1}{2}$. This indicates that the x-intercept that we did not attempt to estimate is $\left(\frac{1}{2}, 0\right)$.

In the following summary we list equivalent statements that are implied by the Factor Theorem. If any of these statements is true, then all the other statements are also true.

Think About It

Suppose that 4 is a solution of $P(x) = 0$, that an x-intercept of the graph of $P(x)$ is $(-3, 0)$, and that $P(2) = 0$. If P is a third-degree polynomial, write $P(x)$ in factored form.

Equivalent Statements Implied by the Factor Theorem

1. For a polynomial $P(x)$ and constant c, $x - c$ is a factor of $P(x)$.
2. $P(c) = 0$.
3. The remainder $R = 0$ when $P(x)$ is divided by $x - c$.
4. The number c is a solution (or root) of $P(x) = 0$.
5. An x-intercept of the graph of $y = P(x)$ is $(c, 0)$.

6.8 QUICK REFERENCE

The Remainder Theorem

- The Remainder Theorem states that if a polynomial $P(x)$ is divided by $x - c$, where c is a constant, then the remainder is $P(c)$. The divisor must be in the form $x - c$.
- The Remainder Theorem is useful if only the remainder from a division problem is sought. If the quotient is also needed, then synthetic division is the best method to use.

The Factor Theorem

- The Factor Theorem states that for a polynomial $P(x)$ and a constant c, $P(c) = 0$ if and only if $x - c$ is a factor of $P(x)$.
- The following procedure can be used to factor a polynomial $P(x)$:
  1. Produce the graph of $y = P(x)$ and estimate the x-intercepts.
  2. Each x-intercept has the form $(c, 0)$. Verify that $P(c) = 0$. Then the Factor Theorem states that $x - c$ is a factor of $P(x)$.
- A variation of this procedure is as follows:
  1. Estimate and verify one x-intercept $(c, 0)$ in order to determine one factor $x - c$.
  2. Divide $P(x)$ synthetically by $x - c$ to determine the quotient $Q(x)$.
  3. Write $P(x) = Q(x) \cdot (x - c)$ and then factor $Q(x)$.
  4. Repeat this process until $P(x)$ is factored completely.
- The following are all equivalent statements:
  1. For a polynomial $P(x)$ and constant c, $x - c$ is a factor of $P(x)$.
  2. $P(c) = 0$.

3. The remainder $R = 0$ when $P(x)$ is divided by $x - c$.
4. The number c is a solution (or root) of $P(x) = 0$.
5. An x-intercept of the graph of $y = P(x)$ is $(c, 0)$.

6.8 EXERCISES

Concepts and Skills

1. If $P(x) = x^4 + 3x^3 + 4x - 1$ is divided by $x - c$, what is the remainder if $P(c) = 5$? Why?

2. If $x - c$ is a factor of $P(x) = x^3 + 6x^2 - 5$, explain why $c^3 + 6c^2 - 5 = 0$.

In Exercises 3–12, use the Remainder Theorem to determine the remainder when $P(x)$ is divided by $x - c$. Verify your result with synthetic division.

3. $P(x) = 3x^3 - 4x^2 + 5x - 6;\quad c = 1$
4. $P(x) = 2x^3 - 3x^2 - 5x + 7;\quad c = -1$
5. $P(x) = 4x^3 - 5x + 7;\quad c = -2$
6. $P(x) = 5x^3 - 3x^2 + x;\quad c = 2$
7. $P(x) = x^4 - 3x^2 + 5;\quad c = 2$
8. $P(x) = 2x^4 - 3x^3 + 2x;\quad c = 3$
9. $P(x) = x^5 + 3x^3 - 7;\quad c = -3$
10. $P(x) = x^6 - 3x;\quad c = 2$
11. $P(x) = -8x^4 - x^3 - x;\quad c = \frac{1}{2}$
12. $P(x) = 24x^6 - 3x^2;\quad c = -\frac{1}{2}$

13. If $P(x) = 2x^4 + x^3 - x^2 + 5x - 6$ and $P(c) = 0$, what do we call c? Why?

14. If $P(x) = 2x^4 + x^3 - x^2 + 5x - 6$ and $P(c) = 0$, what do we call $x - c$? Why?

In Exercises 15–18, produce the graph of the polynomial and verify that the ordered pair $(c, 0)$ is an x-intercept. Then verify algebraically that c is a solution of the equation.

15. $x^4 + 3x^3 - 2x^2 + 5x - 7 = 0;\quad c = 1$
16. $x^5 - 3x^4 + 2x^3 + 2x - 4 = 0;\quad c = 2$
17. $x^4 - 16x^2 - 3x - 12 = 0;\quad c = -4$
18. $3x^3 - 7x^2 + 7x - 10 = 0;\quad c = 2$

In Exercises 19–26, use the Factor Theorem to show that the given binomial is a factor of the given polynomial.

19. $x + 2;\quad 2x^3 + 7x^2 + 14x + 16$
20. $x - 3;\quad x^3 - x^2 - x - 15$
21. $x - 1;\quad 2x^3 - 7x^2 + 11x - 6$
22. $x + 3;\quad x^3 + x^2 - 4x + 6$
23. $x + 1;\quad x^3 - x^2 + 3x + 5$
24. $x + 4;\quad x^3 + 3x^2 - 2x + 8$
25. $x - \frac{3}{2};\quad x^3 - \frac{3}{2}x^2 - 16x + 24$
26. $x + \frac{3}{4};\quad 16x^3 + 12x^2 + x + \frac{3}{4}$

In Exercises 27–34, use the Factor Theorem to show that the given binomial $D(x)$ is a factor of the given polynomial $P(x)$. Write the polynomial in the form $P(x) = Q(x)D(x)$, where $Q(x)$ is the quotient.

27. $x + 3;\quad x^3 - 5x + 12$
28. $x + 2;\quad x^3 - 2x^2 - 7x + 2$
29. $x + 2;\quad 2x^3 + 7x^2 + 12x + 12$
30. $x - 2;\quad 3x^3 - 5x^2 + 3x - 10$
31. $x - \frac{1}{2};\quad 2x^3 - 5x^2 - 4x + 3$
32. $x + \frac{2}{3};\quad 3x^3 - 4x^2 - x + 2$
33. $x - 4;\quad x^4 - 4x^3 - x^2 + 6x - 8$
34. $x + 3;\quad x^5 + 3x^4 - x^3 - 3x^2 - 3x - 9$

35. One factor of $x^3 - 3x^2 - 13x + 15$ is $x - 1$. Explain how you can use this information to factor the polynomial completely.

36. Explain how to use the graph of $y = P(x)$ as a guide to factoring $P(x)$.

In Exercises 37–48, use synthetic division, the given factor, and the Factor Theorem to factor the polynomial completely.

37. $x + 3;\quad x^3 + 8x^2 + 21x + 18$
38. $x + 2;\quad x^3 - 8x^2 + 5x + 50$

39. $x + 5$; $x^3 - 21x + 20$

40. $x + 4$; $x^3 - 13x + 12$

41. $x - 1$; $2x^3 + x^2 - 5x + 2$

42. $x - 2$; $3x^3 - 7x^2 - 2x + 8$

43. $x - 3$; $4x^3 - 7x^2 - 21x + 18$

44. $x + 2$; $6x^3 + x^2 - 19x + 6$

45. $x + 2$; $8x^3 + 6x^2 - 17x + 6$

46. $x - 2$; $10x^3 - 21x^2 - x + 6$

47. $x - \frac{2}{3}$; $3x^3 - 8x^2 - 5x + 6$

48. $x - \frac{5}{6}$; $6x^3 - 11x^2 - 31x + 30$

In Exercises 49–62, factor the polynomial completely.

49. $x^3 - 4x^2 + x + 6$

50. $x^3 + 5x^2 + 2x - 8$

51. $x^3 - 4x^2 + 5x - 6$

52. $x^3 - x + 6$

53. $2x^3 - 5x^2 - x + 6$

54. $2x^3 + x^2 - 13x + 6$

55. $6x^3 - 7x^2 - 7x + 6$

56. $6x^3 + 7x^2 - 7x - 6$

57. $3x^3 + 8x^2 - 5x - 6$

58. $3x^3 + 7x^2 - 2x - 8$

59. $x^4 + 3x^3 - 7x^2 - 27x - 18$

60. $x^4 + 4x^3 - x^2 - 16x - 12$

61. $x^5 - x^4 - 3x^3 + 3x^2 - 4x + 4$

62. $x^5 - 2x^4 + x^3 - 2x^2 - 2x + 4$

63. Suppose that the graph of $y = P(x)$ has three x-intercepts: $(a, 0)$, $(b, 0)$, and $(c, 0)$. What are three factors of $P(x)$? Can you be certain that $P(x)$ is of degree 3? Why?

64. Suppose $P(x)$ is of degree 2 and two factors of $P(x)$ are $x + 2$ and $x - 3$. Can you be certain that $P(x) = (x + 2)(x - 3) = x^2 - x - 6$? Why?

In Exercises 65–68, write a polynomial with the given degree and x-intercepts.

65. degree 2; $(3, 0)$ and $(-2, 0)$

66. degree 2; $(2, 0)$ and $(-1, 0)$

67. degree 3; $(-4, 0)$, $(0, 0)$, and $(3, 0)$

68. degree 3; $(-2, 0)$, $(1, 0)$, and $(4, 0)$

69. In a division problem, the dividend is x^2, the quotient is $x + 3$, and the remainder is 9. What is the divisor?

70. If the quotient is $x^2 + 2x + 5$, the divisor is $x - 2$, and $P(2) = 5$, what is the dividend $P(x)$?

71. Suppose that $x - c$ is a factor of $P(x)$. What is a solution of the equation $P(x) = 0$? Why?

72. Explain how to determine the remainder when $x^{100} - x^{37} - 1$ is divided by $x + 1$ without using synthetic division. Then, determine the remainder.

In Exercises 73–76, one root of the equation is given. Determine the other solutions.

73. $x^4 + x^3 - 11x^2 - 9x + 18 = 0$; $x = 1$

74. $x^4 + x^3 - 7x^2 - x + 6 = 0$; $x = 2$

75. $x^5 + x^4 - 10x^3 - 10x^2 + 9x + 9 = 0$;
$x = -1$ (double root)

76. $x^5 + 2x^4 - 13x^3 - 26x^2 + 36x + 72 = 0$;
$x = -2$ (double root)

77. Determine whether there is a number c such that $x + c$ is a factor of $x^4 + x^2 + 1$.

78. Suppose $P(x) = p_1(x) \cdot p_2(x)$, where p_1 and p_2 are polynomials. If $x + 2$ is a factor of both p_1 and p_2, and if $P(x) = x^5 + 2x^4 - 5x^3 - 10x^2 + 4x + 8$, what is the complete factorization of $P(x)$?

Group Project: Popularity of Fishing

In 1996, freshwater fishing attracted over 45 million people. The accompanying table shows the percentage of people in various age groups who fished at least once per year. (Source: National Sporting Goods Association.)

Age Group	Percent
7–11	23
12–17	22
18–24	16
25–34	23
45–54	17
55–64	13
65 and over	8

The percentages $F(x)$ of the population who fished can be modeled as a function of age x:

$$F(x) = -0.0075x^2 + 0.375x + 18$$

79. Write the polynomial with a common factor of -0.0075. Then factor the polynomial completely.

80. Solve the quadratic equation $F(x) = 0$.

81. Which solution, if any, is meaningful? Evaluate and interpret $F(0)$. For what values of x does F model the data reasonably well?

82. Fishing is often regarded as a popular pastime for retirees. Do the data in the table support this notion?

Challenge

In Exercises 83–88, determine k so that the polynomial is divisible by the binomial.

83. $x^4 - 3x^3 + kx + 3$; $x - 3$

84. $x^3 + kx^2 + 2x - 4$; $x + 2$

85. $k^2x^3 - 7kx + 6$; $x - 1$

86. $2k^2x^4 - 5kx^3 + 3x$; $x + 1$

87. $x^4 - 16$; $x + k$

88. $x^3 - 4x^2 + 3x - 12$; $x + k$

89. For what integers n is $x + 1$ a factor of $x^n + 1$?

90. For what integers n is $x - c$ a factor of $x^n - c^n$?

6. CHAPTER PROJECT Foreign Currency Exchange

In Section 6.4 we saw that 7,136,000 Americans traveled to Europe in 1992 and spent an average of $1918 per trip. In most cases, these dollars had to be converted to the monetary units of the host country. The following table lists five countries along with the number of monetary units that could be purchased with one dollar at a certain time in 1996.

Country	One Dollar Exchange
England	0.65 pound
Italy	1517.00 lire
Germany	1.48 marks
Norway	6.41 crowns
Spain	125.00 pesetas

1. For each country in the table, convert the average of $1918 to the monetary units of that country.
2. If c represents the number of crowns, d represents the number of dollars, and p represents the number of pounds, write equations of the form $c = k_1d$ and $p = k_2d$, where k_1 and k_2 are determined from the table.
3. Solve the two equations in Exercise 2 for d. Then, using substitution, write an equation of the form $c = k_3p$. What does this formula model?

4. Suppose that you leave for England with \$2000 and you exchange this money for pounds. While in England, you spend 400 pounds. You then travel to Norway, where you exchange your pounds for crowns. How many crowns will you have?
5. Continuing with Exercise 4, if you spend 6000 crowns while in Norway, how many dollars will you receive when you return to the United States and exchange crowns for dollars?
6. Arrange a visit with an economics professor and learn the meanings of a "strong dollar" and a "weak dollar." Based on what you learn, prepare an argument for or against traveling to another country when the dollar is "strong."

6. CHAPTER REVIEW EXERCISES

Section 6.1

In Exercises 1–4, answer each of the following.

(a) State whether the given polynomial is a monomial, binomial, trinomial, or none of these.

(b) State the degree of the polynomial.

(c) Write the polynomial in descending order.

(d) State the coefficient of the x^2 term.

1. $3x - 2x^2$ **2.** x^4

3. $x^2 + 9x^5 - 3$

4. $x^6 - 7x^2 + 5x^5 - 4x$

In Exercises 5 and 6, determine the degree of the given polynomial.

5. $4x^2 + y^2 + x^2y^4$ **6.** $a^3b - b^3$

In Exercises 7 and 8, evaluate the given function as indicated.

7. $P(x) = x^5 - x^3 + 2x^2$; $P(1)$

8. $P(x) = -x^2 - 3x - 5$; $P(-2)$

In Exercises 9 and 10, evaluate the given polynomial as indicated.

9. $x^5y^3 - 4x^3y^4 + x^2$ for $x = 1$ and $y = 2$

10. $a^3 - a^2b + ab^2 - b^3$ for $a = 1$ and $b = -1$

In Exercises 11–14, perform the indicated operations.

11. $(5x^2 - 3x + 4) + (-5x - 3x^2 + 9)$

12. $(3x - 2) - (x^2 - 3x + 2) + (x^2 - 4x - 5)$

13. Find the sum of $3x^2 - 6x$ and $4x + 7$.

14. Subtract $-2x^2 + 3x - 7$ from $5x^2 - x$.

Section 6.2

In Exercises 15–22, multiply.

15. $4x^5(-2x^3)$

16. $2x^3(3x^2 - 4)$

17. $2a^2b(5 - b^2 + 3b)$

18. $(x - 4)(x + 3)$

19. $(2x + 3)(8 - 3x)$ **20.** $(x + 2)(2x^2 - 3x + 1)$

21. $(x + 3)(2x - 5) - (3x + 1)(x + 4)$

22. $3x(x - 1)(x + 2)$

23. Write a polynomial expression for the area of a rectangle with a length of $4x - 5$ and a width of $2x - 3$.

24. If three numbers are represented by $x + y$, $y - 3$, and $4 - x$, write a polynomial expression for the product of the numbers.

25. In words, describe how to recognize a difference of two squares.

26. For what special product is the following template used for multiplying?

$$(\quad)^2 + 2(\quad)(\quad) + (\quad)^2$$

In Exercises 27–32, perform the indicated operations.

27. $(3x - 2)(3x + 2)$ **28.** $(1 - 3y)(1 + 3y)$

29. $(3y - 2)^2$ **30.** $(ab + 4)^2$

31. $(2x - 1)(x - 3) - (x - 3)^2$

32. $[(x + 2y) + 5][(x + 2y) - 5]$

33. Write a polynomial expression for the product of three consecutive integers where x represents the second integer.

34. The width and height of a rectangular carton are the same. The length is 1 foot longer than the height. This carton is then placed inside a second carton in the shape of a cube with the same length as the first carton. (See figure.) Write a polynomial expression for the space that remains inside the second carton.

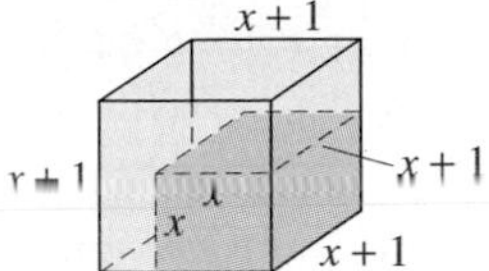

Section 6.3

35. Suppose that you want to factor $-x^3y^2 - 2x^2y^3$ so that one of the factors is a binomial with a positive leading coefficient. What would you select for the GCF?

36. Explain why $2x + 4y + 3x + 9y$ cannot be factored with the grouping method.

In Exercises 37–44, factor completely.

37. $x^{25} + x^{35}$

38. $6x^3 - 36x^2$

39. $x^6y^3 - x^7y^4 - x^3y^5$

40. $3x(x - a) - 4y(x - a)$

41. $x^3 + 3x^2 - 2x - 6$

42. $ac - af + bf - bc$

43. $14x^7 - 35x^5 + 42x^3$

44. $(3x - 2)^4 + 2(3x - 2)^3$

45. An odd number can be represented by $2n - 1$, where n is an integer. Using this representation, show that the difference between the square of an odd number and the odd number itself can be written $2(2n - 1)(n - 1)$.

46. The total cost of z items is given by $xy^2z - 4z$. What binomial represents the unit cost of the items?

Section 6.4

47. How can you quickly tell that $9x^2 - 30x + 24$ is not a perfect square trinomial?

48. Suppose that you factor an expression and obtain the following: $(x + 1)(x^2 - x + 1)$. What expression did you factor?

In Exercises 49–56, factor completely.

49. $49x^2 - 16y^2$

50. $a^{20} - b^{10}$

51. $16 - a^4$

52. $8a^3 - 1$

53. $27b^6 + 8a^3$

54. $4x^2 + 12xy + 9y^2$

55. $c^2 + 64 + 16c$

56. $c^4 + 6c^2 + 9$

57. The base and height of a parallelogram are represented by completely factored binomials. If the area of the parallelogram is given by $a^2 - 16$ square feet, show that the base and height differ by 8 feet.

58. The total revenue from selling n items is given by $4x^2 + 9y^2$. If the total cost of the n items is $12xy$, write a completely factored expression for the profit.

Section 6.5

In Exercises 59–62, factor completely.

59. $x^2 - 5x - 6$

60. $x^2 + 5x + 4$

61. $2x^2 + 12x - 14$

62. $x^4 + 7x^2 - 60$

63. What is a disadvantage of using the trial-and-check method to factor $6x^2 + 25x + 24$?

64. One way to factor $x(2x - 1) + 3(2x - 1)$ is to simplify the expression to obtain $2x^2 + 5x - 3$. Then, using the trial-and-check method, we can write $2x^2 + 5x - 3 = (2x - 1)(x + 3)$. Is this approach correct? What is a more efficient approach?

In Exercises 65–72, factor completely.

65. $3x^2 + 14x + 15$

66. $8x^2 - 3x - 5$

67. $12x^2 + 8x - 15$

68. $12x^2 - 12x - 9$

69. $2x - 8 + x^2$

70. $125x^2 - 80y^2$

71. $125x^3 - 64$

72. $3x^2 - 3x - 60$

73. While factoring $4x^2 + 16xy + 16y^2$, suppose you overlook the GCF of 4 and write $(2x + 4y)^2$. Can you now factor out 2 and write $2(x + 2y)^2$? Why?

74. Last year a billboard designer created a sign L feet long and W feet wide. The area of the sign was A square feet. This year the designer changed the dimensions of the sign, and the new area is given by $A - 5W + 2L - 10$ square feet. If the dimensions of the new sign can be represented by binomials, what changes did the designer make to the sign?

In Exercises 75–82, factor completely.

75. $3x^2 + 4x - 15$

76. $2x^3 + 5x^2 - 18x - 45$

77. $x^9 + 27y^6$

78. $y^4 - x^4$

79. $x^3y^4 - x^4y^3$

80. $50x^3 - 25x^4$

81. $6x^3 + 25x^2 + 21x$

82. $12(2x - 3)^2 + 15(3 - 2x) - 18$

Section 6.6

83. In which of the following equations is the value of a or b definitely known?

$a + b = 0 \qquad ab = 0$

84. The left side of $5(x - 2)(x + 4) = 0$ has three factors, but the equation has only two solutions. Why?

In Exercises 85–94, solve the equation algebraically.

85. $x^2 = 8x$ **86.** $x^2 - 81 = 0$

87. $x^2 + x = 42$

88. $4x^2 - 23x + 15 = 0$

89. $x(x + 2) = 24$

90. $(x + 3)(x - 2) = 6$

91. $x^3 + 3x^2 - 4x = 12$

92. $(x - 1)^2 - 2(x - 1) = 15$

93. $|3x^2 + 7x - 2| = 4$

94. $|5x^2 + 2x + 2| = 5$

95. The sum of twice a number and its square is 63. Determine the number.

96. Standing on the roof of an eight-story building, a person throws a ball upward with an initial velocity of 64 feet per second. If the release point is 80 feet above the ground, then the height h of the ball after t seconds is determined by $h(t) = -16t^2 + 64t + 80$. Use this equation to find the time it takes for the ball to reach the ground.

Section 6.7

97. Explain how to check the result of a long division problem.

98. Interpret the bottom row of numbers in the following synthetic division problem.

$$\begin{array}{r|rrrr} -1 & 2 & 1 & 0 & 4 \\ & & -2 & 1 & -1 \\ \hline & 2 & -1 & 1 & 3 \end{array}$$

In Exercises 99 and 100, determine the quotient.

99. $\dfrac{9x^5 - 12x^3}{3x^2}$

100. $\dfrac{6a^5b^4 - 4a^3b^3 + 8a^2b}{2a^2b^3}$

In Exercises 101–104, use long division to find the quotient and remainder.

101. $(9x^3 - 18x^2 - 7x + 5) \div (3x - 2)$

102. $(12x^4 - 4x^3 + 2x - 17) \div (2x^2 + 3)$

103. $(x^3 - 3x^2y + 3xy^2 - y^3) \div (x - 3y)$

104. $(x^4 - 1) \div (x - 1)^2$

105. If $P(x)$ is divided by $x + 1$, the resulting quotient is $2x^2 - x + 1$ and the remainder is 3. Write a polynomial expression for $P(x)$.

106. For the following synthetic division problem, determine a, b, c, and d.

$$\begin{array}{r|rrrr} 3 & a & b & c & d \\ & & 3 & 3 & 6 \\ \hline & 1 & 1 & 2 & 13 \end{array}$$

In Exercises 107–110, use synthetic division to find the quotient and remainder.

107. $(2x^3 - 3x^2 + 5) \div (x - 1)$

108. $(3x - 4x^3 + x^5 + 2) \div (x + 2)$

109. $(x^5 - 32) \div (x - 2)$

110. $(2x^3 + 3x^2 - 5) \div (x - 0.5)$

Section 6.8

111. Suppose that when $P(x)$ is divided by $x - 2$, the remainder is 3. Explain how you know the value of $P(2)$.

112. Suppose that when $P(x)$ is divided by $x + 3$, the remainder is 0. What does this imply about $x + 3$?

113. Let $P(x) = x^5 - 2x^4 + 3x^3 - 4x^2 + 5$. Use synthetic division and the Remainder Theorem to determine the value of $P(1)$.

114. Use synthetic division and the Factor Theorem to show that $x + 2$ is a factor of $2x^3 + x^2 - 4x + 4$.

115. Suppose you divide $P(x)$ by $x - 5$, by $x - 2$, and by $x + 4$, and each time the remainder is 0. What can you conclude about the graph of $y = P(x)$? Why?

116. Suppose $P(x) = x^3 - 4x + 1$ and c is a number such that $c^3 - 4c + 1 = 0$. What does this imply about $x - c$? Why?

In Exercises 117 and 118, use any method to factor the given polynomial completely.

117. $x^3 - 4x^2 + x + 6$

118. $2x^3 - 5x^2 - 14x + 8$

LOOKING AHEAD

The following exercises review concepts and skills that you will need in Chapter 7.

1. Perform the indicated operations.

   (a) $\frac{-2}{3} + \frac{1}{6}$ (b) $\frac{5}{8} - \frac{3}{10}$

   (c) $\frac{-3}{14} \cdot \frac{4}{5}$ (d) $\frac{9}{5} \div 6$

2. Evaluate the numerical expression.

   $$\left(\frac{1}{2} + \frac{3}{4}\right) \div \left(\frac{1}{3} - 2\right)$$

3. Evaluate the expression $\frac{x-3}{x+6}$ for the given value of x.

   (a) 3 (b) -6 (c) -5

4. Simplify and express the result with positive exponents. Assume that all expressions are defined.

   (a) $3x^{-1}$ (b) $\frac{8x^3y^4}{12x^5y}$

In Exercises 5–7, factor completely.

5. $2x^2 + 6x$ 6. $x^2 - 81$ 7. $6 - x - 2x^2$

8. Explain how to factor $2 - 2x$ so that one factor is

   (a) $1 - x$ (b) $x - 1$

In Exercises 9–11, solve the given equation.

9. $x^2 + 2x - 8 = 0$ 10. $2 - \frac{2}{15}t = \frac{3}{5} + \frac{2t}{3}$

11. $\frac{x}{2} + \frac{5}{6}(x - 1) = \frac{3x - 1}{3}$

12. Write an expression to model the given information.

   (a) The time required to travel 250 miles at r mph

   (b) The price per foot of molding that costs \$84 dollars for f feet

6. CHAPTER TEST

1. Evaluate $-x^2y^3 + 2xy^2$ for $x = -1$ and $y = 2$.

2. Suppose that $p(x) = 3x^4 - 4x^3 - 5x^2 + 6x - 7$ and $q(x) = 5x^3 - 8x + 2$. Show two ways to evaluate each of the following.

   (a) $(p + q)(1)$

   (b) $(p - q)(-1)$

In Questions 3–8, determine the product.

3. $-2x^3(x^2 - 5x + 3)$ 4. $(2t + s)(3t + 4s)$

5. $(x - 3)(x^2 + 3x - 2)$ 6. $(x^2 + 3y)^2$

7. $(5x - 7z)(5x + 7z)$ 8. $(-3a^2b^5)(2a^3b)$

9. Suppose $p(x) = x^2 + 1$ and $q(x) = x^3 - 1$. Write the polynomial $P(x)$, where $P(x) = p(x) \cdot q(x)$. Produce the graph of $y = P(x)$ and determine the quadrant(s), if any, in which the graph is not located.

10. If $P(x)$ is divided by $D(x) = x + 3$, the quotient $Q(x)$ is $2x^2 - x + 3$, and the remainder R is -10. Determine $P(x)$. Then write an equation (with no fractions) that relates $P(x)$, $Q(x)$, $D(x)$, and R.

In Questions 11–14, divide by any method to determine the quotient and remainder.

11. $\frac{3x^3 - 2x^2 + 5x - 7}{x + 2}$ 12. $\frac{15x^6y^4 - 21x^5y^8}{-3xy^2}$

13. $\frac{x^3 + 3x^2 - 3x + 1}{x^2 - 1}$ 14. $\frac{x^5 - 1}{x + 2}$

15. For the following synthetic division calculation, what is the significance of the number k?

$$\begin{array}{r|rrrr} -2 & a & b & c & d \\ & & e & f & g \\ \hline & a & h & j & k \end{array}$$

16. Show that when the sum of the squares of three consecutive even integers is divided by the second integer, the remainder is 8.

In Questions 17–22, factor completely.

17. $18r^3s^5 - 12r^5s^4$ 18. $xy + 3y - x - 3$

19. $27 - 12x^2$ 20. $4y^2 - 20xy + 25x^2$

21. $1 - 8x^3$ 22. $6x^2 + 11x + 4$

23. Use a graph and the Factor Theorem to factor the polynomial.

$x^3 - 3x^2 - 10x + 24$

24. The figure shows the graph of $y = p(x)$. Estimate three factors of $p(x)$.

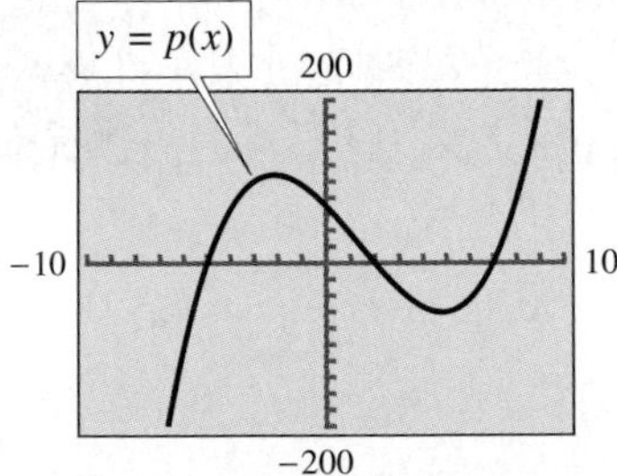

In Questions 25–28, solve.

25. $3x^2 - 4x = 0$

26. $5x^2 + 13x = 6$

27. $2x^3 - x^2 - 8x + 4 = 0$

28. $|2x^2 - 3x - 2| = 3$

29. Verify that $x - 5$ is a factor of $x^3 - 2x^2 - 13x - 10$.

30. A rectangular garden is 3 feet longer than it is wide. If the garden were widened by 3 feet and lengthened by 5 feet, its area would be 10 square feet more than twice the area of the original garden. Find the dimensions of the original garden.

31. A city lot is in the shape of a trapezoid with two right angles. The figure shows the relative lengths of three sides of the lot. If the area of the lot is 5500 square feet, what are the lengths of the parallel sides? (The area A of a trapezoid is $A = \frac{1}{2}h(b_1 + b_2)$, where h is the height and b_1 and b_2 are the lengths of the parallel sides.)

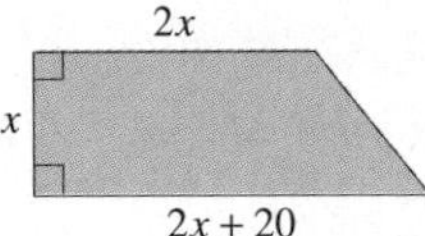

Chapter 7

Rational Expressions

The accompanying bar graph shows the dramatic increase in the average salaries of major league baseball players from 1990 to 1993. Also shown are attendance figures for the same period. Both sets of data can be modeled with a **rational expression,** which is a special kind of fraction.

The analysis of such data is central to the resolution of salary disputes among owners, players, and their agents. For example, does the trend in attendance justify the trend in salaries? What effect do salaries have on ticket prices? (For more on this topic, see Exercises 75–78 at the end of Section 7.4, and see the Chapter Project.)

In this chapter we learn how to perform the basic operations with rational expressions, and we develop methods for solving equations and application problems that involve rational expressions.

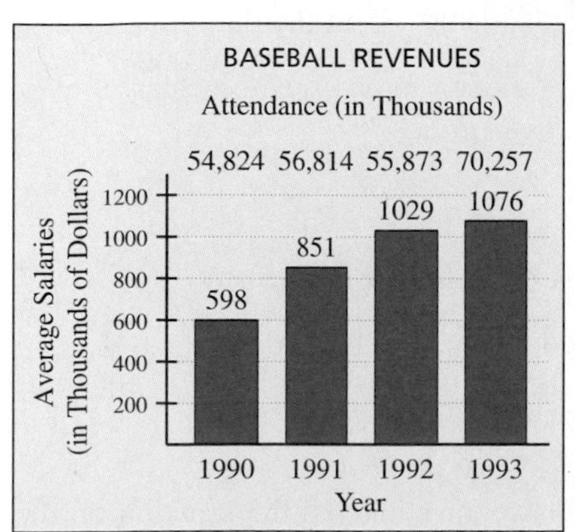

(Source: Major League Baseball Players Association.)

7.1 INTRODUCTION TO RATIONAL EXPRESSIONS

Definitions • Restricted Values • Simplifying Rational Expressions • Equivalent Rational Expressions

LEARNING TIP

Operations with rational expressions are similar to operations with fractions in arithmetic. A review of fractions before beginning this chapter may be beneficial.

Definitions

Recall that a **rational number** is a number that can be written as $\frac{p}{q}$, where p and q are integers and $q \neq 0$. In algebra, **rational expressions** play the same role as rational numbers or fractions in arithmetic. In fact, we sometimes refer to rational expressions as *algebraic fractions*.

Definition of a Rational Expression

A **rational expression** can be written as $\frac{P(x)}{Q(x)}$, where P and Q are polynomials and $Q(x) \neq 0$.

The following are examples of rational expressions.

$$\frac{3 + x}{x - 5} \qquad \frac{x^2 - 2x - 3}{3x^3 - 6x^2} \qquad x^3$$

Note that x^3 is a rational expression because it can be written $\frac{x^3}{1}$.

Restricted Values

A **rational function** is a function of the form $y = R(x)$, where $R(x)$ is a rational expression. Recall that the domain of a function is the set of all permissible replacements for the variable. From the definition of a rational expression $\frac{P(x)}{Q(x)}$, the domain of a rational function must *exclude* values of x for which $Q(x) = 0$. Such values are called *restricted values*.

EXPLORING THE CONCEPT

Domains of Rational Functions

Consider the rational function

$$f(x) = \frac{3 + x}{x - 5}$$

In Fig. 7.1 we have entered the function on the Y screen, and we have produced a table of values.

Figure 7.1

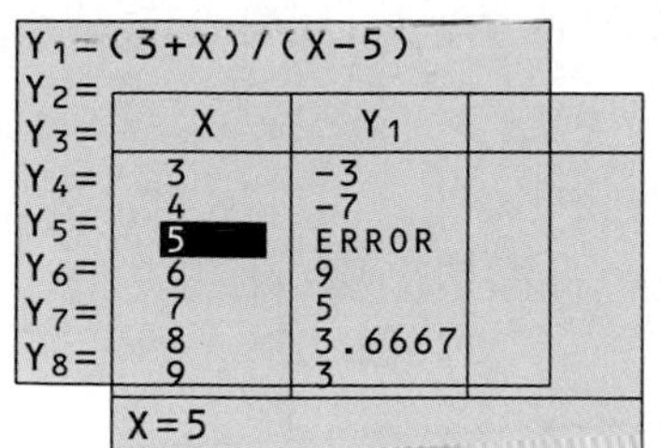

Figure 7.2

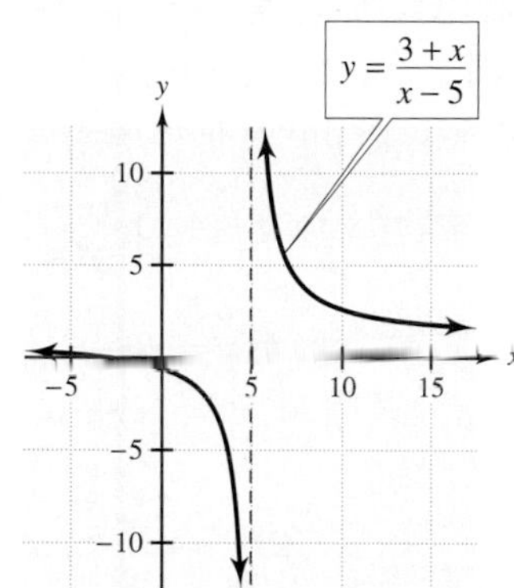

Note that an error is reported when $x = 5$ because replacing x with that value causes the denominator of the rational expression to be 0. Thus 5 is a restricted value. However, the function is defined for all other values of x, so the domain of f is $\{x \mid x \neq 5\}$.

In Fig. 7.2 we see that the graph of f consists of two branches. Because no point of the graph has an x-coordinate of 5, there is a break at $x = 5$, and the branches are not connected. The vertical line shown in Fig. 7.2 is not part of the graph of f. This line is called a **vertical asymptote,** and it signifies the restricted value. Observe that the two branches of the graph approach, but never reach, the vertical asymptote.

In general, to determine the domain of a rational function, we only need to determine the values of the variable for which the denominator is 0.

Determining the Domain of a Rational Function $R(x) = \frac{P(x)}{Q(x)}$

1. Solve the equation $Q(x) = 0$.
2. Any solution of that equation is a restricted value and must be excluded from the domain of the function.

Unlike the graphs of linear functions, which are always straight lines, graphs of rational functions vary in appearance. In Example 1 we produce the graph of each function and observe the relationship between the graph and the domain of the function.

EXAMPLE 1

Determining the Domain of a Rational Function

Determine the domain of each rational function.

(a) $f(x) = \frac{2x}{x + 3}$ (b) $g(x) = \frac{x^2 - 5}{x^2 + 4x}$ (c) $h(x) = \frac{x}{x^2 + 1}$

Solution

(a) By tracing the graph in Fig. 7.3, we can anticipate that -3 is a restricted value of function f. This can be verified algebraically.

$x + 3 = 0$ — Set the denominator equal to 0.

$x = -3$ — Solve for x, and exclude the result from the domain.

The domain of function f is $\{x \mid x \neq -3\}$.

Figure 7.3

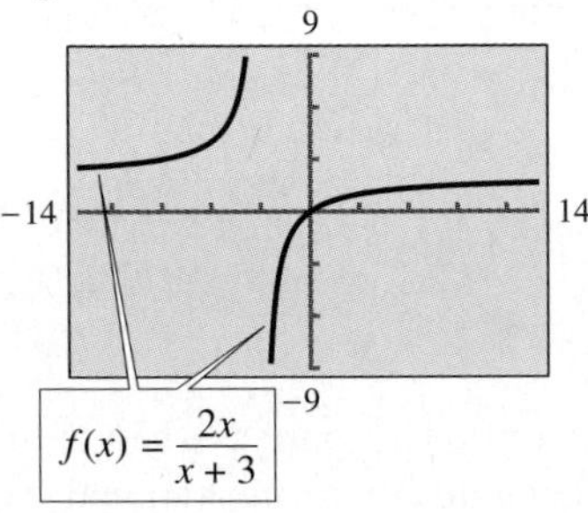

Figure 7.4

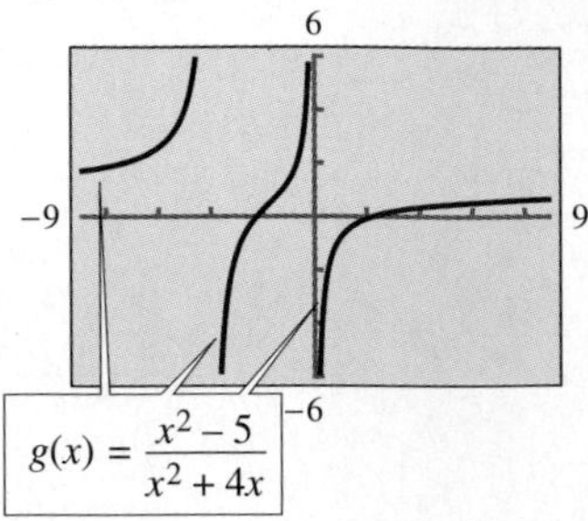

(b) Tracing the graph in Fig. 7.4 suggests that -4 and 0 are restricted values of function g.

$x^2 + 4x = 0$ — Set the denominator equal to 0.

$x(x + 4) = 0$ — Solve with the factoring method.

$x = 0$ or $x + 4 = 0$ — Zero Factor Property

$x = 0$ or $x = -4$ — Solve each equation for x, and exclude the results from the domain.

The domain of g is $\{x \mid x \neq 0 \text{ and } x \neq -4\}$.

(c) There are no breaks in the graph in Fig. 7.5, which suggests that there are no restricted values. Because $x^2 + 1$ is never 0, the domain of h is **R**.

Figure 7.5

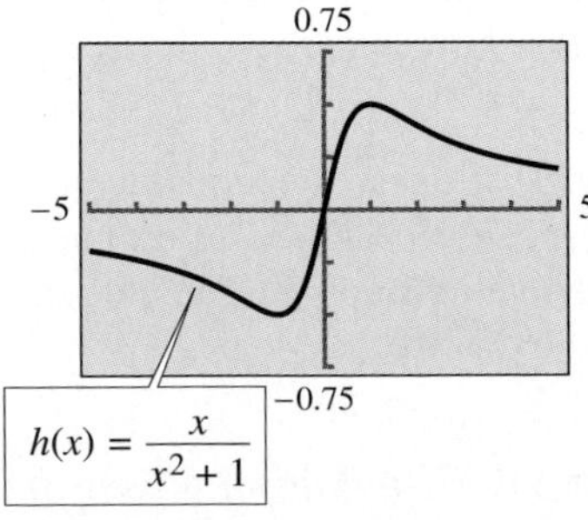

Simplifying Rational Expressions

The process of simplifying a rational expression is similar to that of reducing a fraction to lowest terms.

Property of Equivalent Fractions

If a, b, and c are real numbers, where $b \neq 0$ and $c \neq 0$, then

$$\frac{ac}{bc} = \frac{a}{b}$$

If we let a, b, and c represent algebraic expressions (with the same restrictions), then the Property of Equivalent Fractions is also valid for rational expressions.

To apply the Property of Equivalent Fractions, the numerator and denominator must be products; that is, they must be in factored form. Then, if the numerator and the denominator have a common factor, we can divide the numerator and denominator by that factor. We will refer to this process as "dividing out the common factor."

$$\frac{10}{16} = \frac{2 \cdot 5}{2 \cdot 8} \quad \text{Factor.} \qquad \frac{9x^2y}{15xy^3} = \frac{3xy \cdot 3x}{3xy \cdot 5y^2}$$

$$= \frac{5}{8} \quad \text{Divide out common factors.} \qquad = \frac{3x}{5y^2}$$

We often need to factor both the numerator and the denominator in order to divide out common factors.

$$\frac{5x^2 - 10x}{3x - 6} = \frac{5x(x-2)}{3(x-2)} = \frac{5x}{3}$$

Note: Because 2 is a restricted value, dividing out the common factor $x - 2$ is valid only if $x \neq 2$. Thus $\frac{5x^2 - 10x}{3x - 6} = \frac{5x}{3}$ only if $x \neq 2$. In our discussion of rational expressions, rather than state the restricted values for each expression, we agree that the equalities hold only for values for which the expression is defined.

A rational expression is considered simplified if there is no factor other than 1 or -1 that is common to the numerator and the denominator. The following summarizes the procedure for simplifying a rational expression.

Simplifying Rational Expressions

1. Factor the numerator and the denominator completely.
2. Divide out any factor that is common to both.

EXAMPLE 2

Simplifying a Rational Expression

(a) $$\frac{5x^4 + 10x^3}{15x^2} = \frac{5x^3(x+2)}{15x^2} \quad \text{Factor.}$$

$$= \frac{x(x+2)}{3} \quad \text{Divide out the common factors.}$$

(b) $$\frac{x^2 - 9}{x^2 - 8x + 15} = \frac{(x+3)(x-3)}{(x-5)(x-3)} \quad \text{Factor.}$$

$$= \frac{x+3}{x-5} \quad \text{Divide out the common factor.}$$

Think About It

Find two values of k such that $\frac{x^2 + 8x + 15}{x^2 + 2kx + k^2}$ can be simplified.

If we wish to list restricted values, the best time to do that is when the rational expression is in factored form. If we look only at the simplified result in part (b) of Example 2, we might mistakenly identify 5 as the only restricted value. From the factored form of the expression, we can see that 5 and 3 are restricted values.

When simplifying rational expressions, we sometimes need to use the fact that $b - a = -1(a - b)$.

EXAMPLE 3

Simplifying a Rational Expression by Factoring Out −1

(a) $\dfrac{y-4}{4-y} = \dfrac{1(y-4)}{-1(y-4)}$ Factor.

$= \dfrac{1}{-1}$ Divide out the common factor.

$= -1$

(b) $\dfrac{t^2-25}{5-t} = \dfrac{(t+5)(t-5)}{-1(t-5)}$ Factor.

$= \dfrac{t+5}{-1}$ Divide out the common factor.

$= -(t+5) \quad \text{or} \quad -t-5$

Because the main step in simplifying a rational expression is the factoring step, it is necessary to remember the many techniques we have developed for factoring.

EXAMPLE 4

Simplifying Rational Expressions

(a) $\dfrac{4x^4+6x^3-10x^2}{6x^4-6x^2} = \dfrac{2x^2(2x^2+3x-5)}{6x^2(x^2-1)}$ Factor out the GCF.

$= \dfrac{2x^2(2x+5)(x-1)}{6x^2(x+1)(x-1)}$ Factor the polynomials.

$= \dfrac{2x+5}{3(x+1)}$ Divide out the common factors.

(b) $\dfrac{x^3-1}{x^2-1} = \dfrac{(x-1)(x^2+x+1)}{(x+1)(x-1)}$ Difference of two cubes and difference of two squares

$= \dfrac{x^2+x+1}{x+1}$ Divide out the common factors.

(c) $\dfrac{16-8x+x^2}{(x+y)(x-4)-(x-4)} = \dfrac{x^2-8x+16}{(x-4)[(x+y)-1]}$ GCF is $(x-4)$.

$= \dfrac{(x-4)(x-4)}{(x-4)(x+y-1)}$ Perfect square trinomial

$= \dfrac{x-4}{x+y-1}$ Divide out the common factor.

Equivalent Rational Expressions

We use the rule $\frac{ac}{bc} = \frac{a}{b}$ to simplify a fraction. In reverse, the rule is $\frac{a}{b} = \frac{ac}{bc}$. This means that we can multiply the numerator and the denominator of a fraction by any nonzero number c. For example, to write $\frac{2}{5}$ with a denominator of 15, we multiply the numerator and the denominator by 3.

$$\frac{2}{5} = \frac{2 \cdot 3}{5 \cdot 3} = \frac{6}{15}$$

Note that the appearance of the fraction has changed, but its value has not; that is, the fractions are **equivalent.** This process is sometimes called *renaming the fraction.*

We use a similar method to write an equivalent rational expression with a specified denominator. Determine the factors in the specified denominator that are not in the original denominator. Then multiply the numerator and the denominator of the original fraction by those factors.

EXAMPLE 5

Writing Equivalent Rational Expressions

Rewrite the given rational expression with the specified denominator.

(a) $\frac{2}{3x} = \frac{\quad}{12x^2y^4}$ (b) $\frac{a+2}{a^2 - 2a - 3} = \frac{\quad}{(a+1)^2(a-3)}$

Solution

(a) Factor the specified denominator $12x^2y^4$ so that the original denominator $3x$ is one of the factors.

$$12x^2y^4 = 3x \cdot 4xy^4$$

Because $4xy^4$ is a factor of the specified denominator but not of the original denominator, we multiply the numerator and the denominator of the original fraction by $4xy^4$.

$$\frac{2}{3x} = \frac{2(4xy^4)}{3x(4xy^4)} = \frac{8xy^4}{12x^2y^4}$$

(b) Original denominator: $a^2 - 2a - 3 = (a+1)(a-3)$

Specified denominator: $(a+1)(a+1)(a-3)$

There is one factor of $(a + 1)$ in the specified denominator that is not in the original denominator. Therefore, we multiply the numerator and the denominator of the original fraction by $(a + 1)$.

$$\frac{a+2}{a^2 - 2a - 3} = \frac{a+2}{(a-3)(a+1)} = \frac{(a+1)(a+2)}{(a+1)(a+1)(a-3)}$$

$$= \frac{a^2 + 3a + 2}{(a+1)^2(a-3)}$$

7.1 QUICK REFERENCE

Definitions

- A **rational expression** can be written as $\frac{P(x)}{Q(x)}$, where P and Q are polynomials and $Q(x) \neq 0$.

Restricted Values

- A **rational function** is a function of the form $y = R(x)$, where $R(x)$ is a rational expression.
- A **restricted value** is a number that is not a permissible replacement for a variable. Use the following steps to determine the restricted value(s), if any, of a rational function $R(x) = \frac{P(x)}{Q(x)}$:
  1. Solve the equation $Q(x) = 0$.
  2. Any solution of the equation is a restricted value and must be excluded from the domain.

Simplifying Rational Expressions

- The Property of Equivalent Fractions is used to simplify a rational expression.

  If a, b, and c are real numbers, where $b \neq 0$ and $c \neq 0$, then $\frac{ac}{bc} = \frac{a}{b}$.
- To simplify a rational expression,
  1. factor the numerator and the denominator completely, and
  2. divide out any factor that is common to both.

Equivalent Rational Expressions

- Multiplying the numerator and the denominator of a rational expression by a nonzero number produces an **equivalent** expression.
- To write a rational expression with a specified denominator,
  1. determine the factors in the specified denominator that are not in the original denominator, and
  2. multiply the numerator and the denominator of the original fraction by those factors.

7.1 EXERCISES

Concepts and Skills

1. In general, what numbers are excluded from the domain of a rational function?

2. What is the first step in determining the domain of $f(x) = \frac{x-6}{x^2-2x-15}$?

In Exercises 3 and 4, determine the domain of the function. Then produce the graph of the function on your calculator. What feature of the graph indicates the domain?

3. $y = \frac{10}{x-5}$

4. $y = \frac{x^2-4}{x-2}$

In Exercises 5–8, evaluate the rational expression, if possible, for each of the given values of the variable.

5. $\frac{x}{x-3}$; $-3, 0, 3$

6. $\frac{x-2}{2x+4}$; $-2, 0, 2$

7. $\frac{t-3}{4-t}$; $-4, 3, 4$

8. $\frac{t-6}{6-2t}$; $-3, 3, 6$

In Exercises 9–18, determine the domain of the given rational function.

9. $y = \dfrac{2x+3}{x-4}$ **10.** $y = \dfrac{x+3}{x+3}$

11. $y = \dfrac{x-4}{(x+2)^2}$ **12.** $y = \dfrac{3x-1}{x^2}$

13. $y = \dfrac{2x-6}{x^2-9}$ **14.** $y = \dfrac{x+1}{2x^2+6x}$

15. $y = \dfrac{3}{x^3+2x-8}$ **16.** $y = \dfrac{4x-3}{5x^2-9x-2}$

17. $y = \dfrac{2x-1}{x^2+4}$ **18.** $y = \dfrac{1-2x}{1+2x^2}$

In Exercises 19–22, fill in the blank so that the given number(s) is/are not in the domain of the function.

19. $y = \dfrac{5}{x \underline{\hspace{1cm}} 2}$; -2 **20.** $y = \dfrac{x}{x \underline{\hspace{1cm}} 3}$; 3

21. $y = \dfrac{x}{x^2 \underline{\hspace{1cm}}}$; $5, -5$ **22.** $y = \dfrac{2x}{x^2 \underline{\hspace{1cm}}}$; $3, -3$

23. Explain why the first of the following expressions can be reduced but the second cannot.

(i) $\dfrac{2(x+3)}{(x+3)}$ (ii) $\dfrac{2x+3}{x+3}$

24. If $x \neq 3$, then we know that $\dfrac{3(x-3)}{x-3} = 3$ is true.

Is $\dfrac{3x-3}{x-3} = 3$ also true? Why?

In Exercises 25–32, simplify the rational expression.

25. $\dfrac{6x^3-3x^2}{9x^2}$ **26.** $\dfrac{4a^5+8a^7}{12a^4}$

27. $\dfrac{2(3x+1)}{2+6x}$ **28.** $\dfrac{-3(2x-1)}{3-6x}$

29. $\dfrac{x^2-81}{x+9}$ **30.** $\dfrac{x+8}{x^2-64}$

31. $\dfrac{a^2+5a}{4a-20}$ **32.** $\dfrac{3b-9}{b^2+3b}$

In Exercises 33–44, simplify the rational expression.

33. $\dfrac{2-x}{2+9x-5x^2}$ **34.** $\dfrac{x+3}{12+x-x^2}$

35. $\dfrac{x^2-16}{x^2-x-12}$ **36.** $\dfrac{10x^2-x-2}{6x^2+x-2}$

37. $\dfrac{a^3-a}{6a-6}$ **38.** $\dfrac{5a+10}{a^3-4a}$

39. $\dfrac{x-3}{2x^2-5x-3}$ **40.** $\dfrac{2+x}{6-x-2x^2}$

41. $\dfrac{x^2-14x+45}{x^2-17x+72}$ **42.** $\dfrac{4a^2-13a-12}{a^2-a-12}$

43. $\dfrac{2x^2+6x-1}{2x(x+3)-1}$ **44.** $\dfrac{a+3}{a(a+6)+9}$

In Exercises 45–50, simplify the rational expression.

45. $\dfrac{ab-3a}{b^2-9}$ **46.** $\dfrac{5ab+25b}{a^2-25}$

47. $\dfrac{x+3y}{x^2+2xy-3y^2}$ **48.** $\dfrac{a+4b}{a^2+3ab-4b^2}$

49. $\dfrac{c^2-5cd+6d^2}{2c^2-5cd-3d^2}$ **50.** $\dfrac{2x^2+13xy+20y^2}{2x^2+17xy+30y^2}$

In Exercises 51–54, determine the unknown numerator or denominator.

51. $\dfrac{2}{a-b} = \dfrac{\underline{\hspace{1cm}}}{b-a}$ **52.** $\dfrac{2x-3}{3x-2} = \dfrac{\underline{\hspace{1cm}}}{2-3x}$

53. $\dfrac{x}{x-3} = -\dfrac{x}{\underline{\hspace{1cm}}}$ **54.** $\dfrac{x+2}{2-x} = -\dfrac{x+2}{\underline{\hspace{1cm}}}$

In Exercises 55–58, determine whether the given expressions have a value of 1, -1, or neither.

55. (a) $\dfrac{x+3}{x-3}$ (b) $\dfrac{3+x}{x+3}$ (c) $\dfrac{x-3}{3-x}$

56. (a) $\dfrac{x-y}{-x+y}$ (b) $\dfrac{x+y}{-y+x}$ (c) $-\dfrac{y-x}{x-y}$

57. (a) $\dfrac{-2x+5}{-(5-2x)}$ (b) $\dfrac{2x-5}{-5+2x}$ (c) $\dfrac{2x-5}{2x+5}$

58. (a) $\dfrac{x+y}{y+x}$ (b) $\dfrac{x-y}{y-x}$ (c) $\dfrac{x-y}{x+y}$

In Exercises 59–68, simplify the rational expression.

59. $\dfrac{5x-2}{2-5x}$ **60.** $\dfrac{x-y}{y-x}$

61. $\dfrac{5x-3}{-15x+9}$ **62.** $\dfrac{-3x+6}{4x-8}$

63. $\dfrac{a^2-49}{7-a}$ **64.** $\dfrac{x^2-2x-8}{4-x}$

65. $\dfrac{z-5w}{25w^2-z^2}$ **66.** $\dfrac{x^2-4}{(2-x)^2}$

67. $\dfrac{-2x^2-3x+20}{2x^2+x-15}$ **68.** $\dfrac{16-x^2}{2x^2-9x+4}$

In Exercises 69–74, simplify the rational expression.

69. $\dfrac{x^3 + 3x^2 - 9x - 27}{x^2 + 6x + 9}$

70. $\dfrac{8 - 27y^3}{2 - 3y + 2x - 3xy}$

71. $\dfrac{x^3 + 8}{x^2 - 4}$ **72.** $\dfrac{1 + x + x^2}{1 - x^3}$

73. $\dfrac{18x^5 - 39x^4 + 18x^3}{8x^4 - 18x^2}$

74. $\dfrac{16 - x^4}{x^3 + 2x^2 + 4x + 8}$

In Exercises 75–86, supply the missing numerator.

75. $\dfrac{3}{2xy^2} = \dfrac{\square}{8x^3y^3}$ **76.** $\dfrac{-3}{5ab^2} = \dfrac{\square}{15a^2b^5}$

77. $\dfrac{5}{x - 5} = \dfrac{\square}{5x - 25}$ **78.** $\dfrac{-5}{x - 6} = \dfrac{\square}{7x - 42}$

79. $x + 5 = \dfrac{\square}{x + 5}$ **80.** $3x = \dfrac{\square}{2x^2}$

81. $\dfrac{5}{4x + 12} = \dfrac{\square}{4x^2 - 36}$ **82.** $\dfrac{9}{5x + 5} = \dfrac{\square}{5x^2 - 5}$

83. $\dfrac{-3}{x + 1} = \dfrac{\square}{x^2 + 5x + 4}$ **84.** $\dfrac{5}{x + 1} = \dfrac{\square}{x^2 - 3x - 4}$

85. $\dfrac{9}{x^2 - 5x - 6} = \dfrac{\square}{(x - 6)(x - 5)(x + 1)}$

86. $\dfrac{8}{x^2 - 7x + 6} = \dfrac{\square}{(x - 6)(x^2 + 3x - 4)}$

In Exercises 87–90, supply the denominator.

87. $\dfrac{5t}{6} = \dfrac{-10t^2}{\square}$ **88.** $\dfrac{4}{5x^2} = \dfrac{12x^3y}{\square}$

89. $\dfrac{2x - 1}{1 - x} = \dfrac{1 - x - 2x^2}{\square}$

90. $\dfrac{x + 2}{2 - 3x} = \dfrac{3x^2 + 4x - 4}{\square}$

Real-Life Applications

91. Suppose that you are in charge of cleaning up an oil spill from a tanker and that one of your jobs is to estimate the total cost. The model for estimating the total cost C (in millions of dollars) is

$$C(x) = \frac{100}{100 - x} \qquad 0 \le x < 100$$

where x is the percentage of the oil that is cleaned up.

(a) What is the cost of cleaning up 50% of the oil spill?

(b) According to the given model, can you remove 100% of the oil? Why?

(c) Use a suitable window setting to produce the graph of $C(x)$. Trace to a point where the cost begins to rise rapidly. At that point, what percentage of the oil would be cleaned up and at what cost?

92. In diving board competitions, scoring is on a scale from 1 to 10, with 10 being perfect. A certain diver uses the function $s(x) = \dfrac{10x}{x + 5}$ to predict the score on a particular dive, where x is the number of training dives per day.

(a) Suppose that the diver performs 20 training dives per day. What is the predicted score on the competition dive?

(b) Use the graph of s to estimate the number of training dives per day to achieve a predicted score of 8.5.

(c) What is the predicted increase in score if a diver who currently does 30 dives per day doubles the number of dives per day?

Group Project: Net Worth of Households

The accompanying table shows the percentages of heads of households who are in various age groups along with their average net worth. (Source: *Federal Reserve Bulletin.*)

Age Group	Percent	Net Worth
34 and under	25.9	\$ 60,200
35–44	22.7	157,000
45–54	16.2	304,500
55–64	13.1	371,000
65–74	12.7	369,800
75 and over	9.4	257,600

The function $N(x) = -0.293x^2 + 37.077x - 808.444$ models the net worth $N(x)$, where x is the age of the head of household.

93. Use long (or synthetic) division to simplify the rational expression $\dfrac{N(x)}{x - 28}$.

94. You should have obtained a remainder of 0 in Exercise 93. What does this imply about the expression $x - 28$?

95. Write $N(x)$ in factored form, and solve the equation $N(x) = 0$. Interpret the result.

96. What are some factors that contribute to a low net worth for younger heads of households?

Challenge

In Exercises 97–100, simplify the given rational expression.

97. $\dfrac{2x^{-2} - 3x^{-1}}{3x^2 - 2x}$ **98.** $\dfrac{3x^{-2} + 5x^{-4}}{3x + 5x^{-1}}$

99. $\dfrac{x^2 + 3x}{3x^{-4} + x^{-3}}$ **100.** $\dfrac{2x^5 - x^2}{2x - x^{-2}}$

In Exercises 101–104, determine the domain of the rational function where n is a positive integer.

101. $\dfrac{3}{x^n - 1}$ **102.** $\dfrac{4}{x^n + 1}$

103. $\dfrac{x}{x^{2n} + 1}$ **104.** $\dfrac{x + 3}{x^{2n} - 1}$

105. Evaluate the expression without doing any significant arithmetic. (Hint: Let $x = 5250$.)

$$\frac{5251^2 \quad 5240^2}{5250(5254) - 5251^2}$$

106. Use the x^{-1} key on your calculator to produce the graph of $y = x^{-1}$.

(a) What does the graph suggest about the domain of $y = x^{-1}$?

(b) Use the word *reciprocal* to interpret your answer in part (a).

7.2 MULTIPLICATION AND DIVISION

Multiplication • Division

Multiplication

To multiply fractions, recall that we multiply the numerators, multiply the denominators, and then reduce the answer.

$$\frac{10}{21} \cdot \frac{15}{8} = \frac{150}{168} = \frac{25 \cdot 6}{28 \cdot 6} = \frac{25}{28}$$

An easier method is to reduce first and then multiply the numerators and the denominators.

$$\frac{10}{21} \cdot \frac{15}{8} = \frac{5 \cdot 2}{7 \cdot 3} \cdot \frac{5 \cdot 3}{4 \cdot 2} = \frac{25}{28}$$

In arithmetic, reducing before multiplying is a convenience. For rational expressions, it is nearly a necessity.

The definition of multiplication of rational expressions is similar to the definition of multiplication of arithmetic fractions.

Definition of Multiplication of Rational Expressions

If a, b, c, and d represent algebraic expressions, where b and d are not 0, then

$$\frac{a}{b} \cdot \frac{c}{d} = \frac{ac}{bd}$$

We are guided by our experience with arithmetic in stating the routine for multiplying rational expressions.

To Multiply Rational Expressions

1. Factor each numerator and each denominator completely.
2. Divide out any factors that are common to both a numerator and a denominator.
3. Multiply the numerators and multiply the denominators.

As in arithmetic, we can divide out common factors from the numerator of one fraction and the denominator of another fraction.

$$\frac{xy}{y-1} \cdot \frac{y-1}{y} = \frac{x}{1} = x$$

EXAMPLE 1

LEARNING TIP

Rewrite the problem with the numerators and denominators in factored form. Attempting to skip steps by writing the factors above the polynomial invites errors.

Multiplying Rational Expressions

Determine the product of the rational expressions.

(a) $\dfrac{8x^2 - 4x}{2x^2 + 5x - 3} \cdot \dfrac{x^2 - 9}{4x^2}$

$= \dfrac{4x(2x-1)}{(2x-1)(x+3)} \cdot \dfrac{(x+3)(x-3)}{4 \cdot x \cdot x}$ Factor numerator and denominator.

$= \dfrac{x-3}{x}$ Divide out common factors.

(b) $(x^2 - 4xy + 4y^2) \cdot \dfrac{6}{2x^2 - 8y^2}$

$= \dfrac{(x-2y)(x-2y)}{1} \cdot \dfrac{2 \cdot 3}{2(x+2y)(x-2y)}$ Factor.

$= \dfrac{3(x-2y)}{x+2y}$ Divide out common factors.

(c) $\dfrac{x^2 - y^2}{y - x} \cdot \dfrac{1}{x^3 + y^3}$

$= \dfrac{(x+y)(x-y)}{-1(x-y)} \cdot \dfrac{1}{(x+y)(x^2 - xy + y^2)}$ Factor.

$= \dfrac{1}{-(x^2 - xy + y^2)}$ or $-\dfrac{1}{x^2 - xy + y^2}$ Divide out common factors.

Division

Two rational expressions are **reciprocals** of each other if their product is 1. The following are some examples of rational expressions and their reciprocals.

Rational Expression	*Reciprocal*
$\dfrac{x}{2x+3}$	$\dfrac{2x+3}{x}$
x^2	$\dfrac{1}{x^2}$
$\dfrac{1}{x-1}$	$x-1$

From arithmetic we know that to divide by a fraction, we multiply by the reciprocal of the divisor.

$$10 \div \frac{2}{3} = \frac{10}{1} \cdot \frac{3}{2} = \frac{5 \cdot 2}{1} \cdot \frac{3}{2} = \frac{15}{1} = 15$$

Division of rational expressions is similar to division of fractions in arithmetic.

Definition of Division of Rational Expressions

If a, b, c, and d represent algebraic expressions, where b, c, and d are not 0, then

$$\frac{a}{b} \div \frac{c}{d} = \frac{a}{b} \cdot \frac{d}{c}$$

EXAMPLE 2

Division of Rational Expressions

(a) $\dfrac{3x^2-12}{x^2} \div \dfrac{3x+6}{x} = \dfrac{3x^2-12}{x^2} \cdot \dfrac{x}{3x+6}$ — Definition of division

$= \dfrac{3(x+2)(x-2)}{x \cdot x} \cdot \dfrac{x}{3(x+2)}$ — Factor.

$= \dfrac{x-2}{x}$ — Divide out common factors.

(b) $\dfrac{6x^2+3x}{3x-4} \div (6x^2-5x-4)$

$= \dfrac{6x^2+3x}{3x-4} \cdot \dfrac{1}{6x^2-5x-4}$ — Definition of division

$= \dfrac{3x(2x+1)}{3x-4} \cdot \dfrac{1}{(3x-4)(2x+1)}$ — Factor.

$= \dfrac{3x}{(3x-4)^2}$ — Divide out common factors.

(c) $\dfrac{x^3-x^2+2x-2}{2x^2-5x+3} \div \dfrac{x^4-4}{2x^3+3x^2-4x-6}$

$= \dfrac{x^3-x^2+2x-2}{2x^2-5x+3} \cdot \dfrac{2x^3+3x^2-4x-6}{x^4-4}$ — Definition of division

$= \dfrac{x^2(x-1)+2(x-1)}{(2x-3)(x-1)} \cdot \dfrac{x^2(2x+3)-2(2x+3)}{(x^2+2)(x^2-2)}$ — Factor.

Think About It

Produce the graph of $\frac{x-5}{x+5} \div \frac{x-5}{x+5}$ and compare it with the graph of $y = 1$. Explain the differences.

$$= \frac{(x-1)(x^2+2)}{(2x-3)(x-1)} \cdot \frac{(2x+3)(x^2-2)}{(x^2+2)(x^2-2)}$$

$$= \frac{2x+3}{2x-3}$$ Divide out common factors.

Multiplication and division operations can appear in the same expression.

EXAMPLE 3

Combined Operations

$$\frac{y-3}{y^2+y-2} \div (y+2) \cdot \frac{y^2+2y-3}{9-y^2}$$

$$= \frac{y-3}{y^2+y-2} \cdot \frac{1}{y+2} \cdot \frac{y^2+2y-3}{9-y^2}$$ Definition of division

$$= \frac{-1(3-y)}{(y+2)(y-1)} \cdot \frac{1}{y+2} \cdot \frac{(y+3)(y-1)}{(3+y)(3-y)}$$ Factor.

$$= \frac{-1}{(y+2)^2}$$ Divide out common factors.

7.2 QUICK REFERENCE

Multiplication

- If a, b, c, and d represent algebraic expressions, where b and d are not 0, then

$$\frac{a}{b} \cdot \frac{c}{d} = \frac{ac}{bd}$$

- Follow these steps to multiply rational expressions:
  1. Factor each numerator and denominator completely.
  2. Divide out any factors that are common to both a numerator and a denominator.
  3. Multiply the numerators and multiply the denominators.

Division

- Two rational expressions are **reciprocals** of each other if their product is 1.
- If a, b, c, and d represent algebraic expressions, where b, c, and d are not 0, then

$$\frac{a}{b} \div \frac{c}{d} = \frac{a}{b} \cdot \frac{d}{c}$$

- Follow these steps to divide rational expressions:
  1. Determine the reciprocal of the divisor.
  2. Change the operation from division to multiplication by the reciprocal.
  3. Follow the steps for multiplication.
- In expressions involving both multiplication and division of rational expressions, follow the Order of Operations.

7.2 EXERCISES

Concepts and Skills

1. Consider the following multiplication problem.

$$\frac{x+1}{x^2-2x-15}\cdot\frac{x^2-5x}{x^2-1}$$
$$=\frac{(x+1)(x^2-5x)}{(x^2-2x-15)(x^2-1)}$$
$$=\frac{x^2-4x^2-5x}{x^4-2x^3-16x^2+2x+15}$$

The next step is to simplify the result, but the numerator and denominator will be difficult to factor. How could the problem have been done more easily?

2. Suppose $R(x) = x^{-1} \cdot x$. Write $R(x)$ as a product of rational expressions. Explain why the graph of $y = R(x)$ is not the same as the graph of $y = 1$.

In Exercises 3–14, multiply and simplify.

3. $\dfrac{5a^2b}{b^3c^4}\cdot\dfrac{b^2c^4}{25a^3}$

4. $\dfrac{4x^3y^2}{z^3}\cdot\dfrac{y^3z^4}{2x^5}$

5. $(a+4)\cdot\dfrac{a-4}{3a+12}$

6. $\dfrac{2x+1}{x-5}\cdot(15-3x)$

7. $\dfrac{5x-7}{12-3x}\cdot\dfrac{3x-12}{7-5x}$

8. $\dfrac{5p-25}{2p^2}\cdot\dfrac{6p}{p-5}$

9. $\dfrac{4c^2}{5c+30}\cdot\dfrac{c^2-36}{16c^3}$

10. $\dfrac{2v}{v+7}\cdot\dfrac{v^2+11v+28}{8v^2}$

11. $\dfrac{7w^2-14w}{w^2+3w-10}\cdot\dfrac{w+5}{21w}$

12. $\dfrac{z^2-11z+18}{6z^3}\cdot\dfrac{z^2+2z}{z^2-4}$

13. $\dfrac{y^2-1}{y^2+14y+48}\cdot\dfrac{y^2+5y-6}{(y-1)^2}$

14. $\dfrac{z^2+5z-14}{z^2-8z+12}\cdot\dfrac{z^2+z-42}{z^2+9z+14}$

In Exercises 15–18, fill in the blank to make the resulting equation true.

15. $\dfrac{x-3}{3x-21}\cdot\dfrac{7-x}{\rule{2em}{0.5em}}=\dfrac{1}{3}$

16. $\dfrac{\rule{2em}{0.5em}}{x+1}\cdot\dfrac{x+1}{x-2}=x+2$

17. $\dfrac{3x^2+8x-3}{2x+1}\cdot\dfrac{\rule{2em}{0.5em}}{3x^2-4x+1}=x+3$

18. $\dfrac{x^2-16}{\rule{2em}{0.5em}}\cdot\dfrac{x^2+3x-10}{x^2-6x+8}=\dfrac{x+4}{x+2}$

19. What is the difference between the *opposite* of the expression $x + 3$ and the *reciprocal* of $x + 3$?

20. For

$$R(x)=\frac{5}{x-1}\div\frac{x-2}{3},$$

why is 2 excluded from the domain of R?

In Exercises 21–32, divide and simplify.

21. $\dfrac{24x^3y^7}{16x^4y^3}\div\dfrac{9y^6x}{x^5}$

22. $\dfrac{25x^5y^4}{15y}\div\dfrac{x^3y^3}{3x}$

23. $\dfrac{2x+6}{5x^2}\div\dfrac{x+3}{15x}$

24. $\dfrac{(x+1)^2}{8x+8}\div\dfrac{7x^2+7x}{8x^2}$

25. $\dfrac{5x-x^2}{x+2}\div\dfrac{x^2-6x+5}{x^2-3x+2}$

26. $\dfrac{x^2-64}{x^2-9}\div\dfrac{6x^2+48x}{2x-6}$

27. $\dfrac{x^2+3x-4}{8x}\div\dfrac{x^2+10x+24}{x+6}$

28. $\dfrac{3x-12}{x^2-9x+20}\div\dfrac{9x+54}{x^2+x-30}$

29. $\dfrac{x^2-3x-10}{x^2-25}\div\dfrac{x^2-4}{x^2-x-30}$

30. $\dfrac{x^2-7x-8}{x^2-2x-48}\div\dfrac{6x^2+6x}{x^2+7x+6}$

31. $\dfrac{x^2-16}{2x^2-8x}\div(3x^2+10x-8)$

32. $\dfrac{x^2+12x+36}{x^2+3x-18}\div(x^2+9x+18)$

In Exercises 33–36, fill in the blank to make the resulting equation true.

33. $\dfrac{\underline{\qquad}}{5} \div \dfrac{3x-4}{10} = 4x-2$

34. $\dfrac{x^2+7x}{x-x^2} \div \dfrac{\underline{\qquad}}{7x-7} = -1$

35. $\dfrac{2x^2+4x}{x^2+2x+4} \div \dfrac{x^2-4}{\underline{\qquad}} = 2x$

36. $\dfrac{x^3-5x^2}{\underline{\qquad}} \div \dfrac{x^2-4x-5}{x^2+2x} = \dfrac{x+2}{x^2+x}$

In Exercises 37–54, perform the indicated operation and simplify.

37. $\dfrac{32x^5}{3y^4} \cdot \dfrac{5y^3}{-8x^2}$

38. $\dfrac{6x^3}{7y^4} \div \dfrac{36x^7}{49y^5}$

39. $-2x \cdot \dfrac{x+3}{4x-x^2}$

40. $\dfrac{x-3x^2}{3-x} \div -3x^2$

41. $\dfrac{x^2-4}{x^2-1} \div \dfrac{12x^2-24x}{4x+4}$

42. $\dfrac{b^2-16}{b^2-9} \cdot \dfrac{4b+12}{8b^2+32b}$

43. $\dfrac{4x-8}{x^2-5x+6} \div \dfrac{8x+48}{x^2+3x-18}$

44. $\dfrac{b^2-7b+10}{2b-10} \cdot \dfrac{4b^2+8b}{b^2-4}$

45. $\dfrac{a^2-81}{a^2-5a-36} \cdot \dfrac{a^2+10a+24}{a^2-3a-54}$

46. $\dfrac{x^2-5x-24}{x^2-3x-18} \div \dfrac{x^2-64}{3x-18}$

47. $\dfrac{x^2-3x-4}{x^2-10x+24} \div \dfrac{x^2+5x+4}{x^2-5x-6}$

48. $\dfrac{t^2+t-20}{t^2+2t-24} \cdot \dfrac{t^2+11t+30}{t^2+9t+20}$

49. $\dfrac{3x^3+6x^2}{2x^2+x-6} \div (6x^2-15x)$

50. $(w^2-w-12) \cdot \dfrac{9w^2}{3w^4-27w^2}$

51. $\dfrac{xy+3x-2y-6}{x^2-6x+8} \cdot \dfrac{xy-4y+2x-8}{y^2+5y+6}$

52. $\dfrac{ab-ad+bc-cd}{b^2+bd-2d^2} \div \dfrac{a^2-ac-2c^2}{ab+2ad-bc-2cd}$

53. $\dfrac{x^3-27}{x^2-9} \div \dfrac{x^2+3x+9}{x^2+8x+15}$

54. $\dfrac{x^3+y^3}{x^2-y^2} \cdot \dfrac{x^2-3xy+2y^2}{x-2y}$

In Exercises 55–62, perform the indicated operations.

55. $\dfrac{6x-18}{3x-6} \cdot \dfrac{5+5x}{25x+25} \div \dfrac{12-4x}{9x-18}$

56. $\dfrac{x^2-2x}{4y} \cdot \dfrac{28xy^2}{3x-6} \div \dfrac{2x^2+2x}{21x^2y^2}$

57. $\dfrac{2x+6}{x-5} \div \left[\dfrac{x}{4x+12} \cdot \dfrac{2(x+3)^2}{x^2-5x}\right]$

58. $(2x^2+3x-5) \div \left(\dfrac{6x+15}{x-2} \div \dfrac{3}{x-2}\right)$

59. $\dfrac{x^3-1}{x+1} \cdot \dfrac{x^2+2x+1}{x^2+x+1} \div \dfrac{x+1}{3x^2}$

60. $\dfrac{x^3-8}{x^2-4} \div \dfrac{x^2+2x+4}{x^3-3x^2} \cdot \dfrac{x^2+6x+8}{x^2+x-12}$

61. $\dfrac{x^2+x-2}{x^2-1} \div (x+2) \cdot \dfrac{x^2-9}{x^2+3x}$

62. $\dfrac{x^2+x-6}{x^2-16} \cdot \dfrac{x^2-4x}{x^2-4} \div (x^2+3x)$

Modeling with Real Data

63. British psychologists studied children's preferences for teddy bears with adult features and with baby features. Older children were more likely to prefer baby features than younger children. (Source: *Discover.*)

The following table shows the number of boys and girls, by age, who chose baby-featured bears. The combined data are the averages for the two genders.

Age	Boys	Girls	Combined
4	6	8	7
6	8	10	9
8	10	14	12

The following functions model the data, where x is the age.

Boys: $B(x) = x + 2$

Girls: $G(x) = \frac{3}{2}x + \frac{5}{3}$

(a) Write a rational expression that models the combined data. Then simplify the expression.

(b) Produce the graphs of the three model expressions, and describe the graph of the expression in part (a) relative to the other two graphs.

64. Trace the three graphs in part (b) to estimate the number of children age 5 who preferred baby-featured bears.

Group Project: Food Stamp Program

The following table shows the number (in millions) of Americans receiving food stamps during the period 1990–1994. Also shown are the costs (in billions) of the food stamp program during the same period. (Source: U.S. Department of Agriculture.)

	1990	1991	1992	1993	1994
Number of people	20.1	22.6	25.4	27.0	27.5
Total cost	14.19	17.34	20.91	22.01	22.75

With $P(t)$ representing the number of people receiving food stamps and $C(t)$ representing the total cost, the following functions model the data in the table:

$$P(t) = \frac{32t + 52}{t + 2.6} \qquad C(t) = \frac{26t + 20}{t + 1.5}$$

For both functions, t is the number of years since 1990.

65. Use $P(t)$ and $C(t)$ to write a function $A(t)$ that represents the average cost per person during the given period. [Remember that $P(t)$ is in millions and $C(t)$ is in billions.]

66. Perform the indicated division in Exercise 65 and simplify.

67. Enter $A(t)$ on the Y screen and produce a table for $0 \le t \le 4$. By how much did $A(t)$ change from 1990 to 1994? Describe the predicted trend in $A(t)$ after 1994.

68. Because $C(t) = A(t) \cdot P(t)$, the total cost of the food stamp program could be lowered by reducing either $A(t)$ or $P(t)$. What steps would you propose for reducing $P(t)$?

Challenge

In Exercises 69–72, answer the question by determining the unknown quantity.

69. What rational expression multiplied by $\frac{x + 3}{x - 5}$ is $\frac{x^2 + x - 6}{x^2 - 2x - 15}$?

70. The product of $x^2 + x - 1$ and what rational expression is 1?

71. What rational expression divided by $\frac{2x^2 + x - 15}{2x^2 - 3x - 5}$ is $\frac{2x - 3}{x + 3}$?

72. What rational expression divided into $\frac{x + 4}{x^2 - 1}$ is $\frac{1}{x^2 - 5x + 4}$?

For Exercises 73 and 74, produce the graph of $y = 8/x$ and observe the behavior of the function in Quadrant I.

73. For x-values closer and closer to 0, what is true about the corresponding y-values? Describe how you can select x-values to make the y-values as large as you want.

74. For larger and larger x-values, what is true about the corresponding y-values? Describe how you can select x-values to make the y-values as close to 0 as you want.

7.3 ADDITION AND SUBTRACTION

Like Denominators • Least Common Multiples • Unlike Denominators • Combined Operations

Like Denominators

Unlike multiplication and division, addition and subtraction of fractions requires that the fractions have a common denominator. Recall that to add fractions with the same denominator, we simply add the numerators, retain the common denominator, and reduce the result to lowest terms.

$$\frac{2}{9} + \frac{4}{9} = \frac{6}{9} = \frac{2 \cdot 3}{3 \cdot 3} = \frac{2}{3}$$

The procedure for subtracting fractions with like denominators is similar. We use the same principle for addition and subtraction of rational expressions with the same denominator.

Definition of Addition and Subtraction of Rational Expressions

If a, b, and c are algebraic expressions, where $c \neq 0$, then

$$\frac{a}{c} + \frac{b}{c} = \frac{a+b}{c} \quad \text{and} \quad \frac{a}{c} - \frac{b}{c} = \frac{a-b}{c}$$

The following are the steps for adding or subtracting rational expressions with like denominators.

To Add or Subtract Rational Expressions with Like Denominators

1. Add (or subtract) the numerators.
2. Retain the common denominator.
3. Simplify the result.

When we add or subtract rational expressions, each numerator must be treated as if it were enclosed in parentheses. For a subtraction problem, in particular, we will insert parentheses around each numerator. Then, when we subtract the numerators, we will remove the parentheses to simplify.

EXAMPLE 1

Combining Rational Expressions with Like Denominators

Perform the indicated operations.

(a) $\dfrac{3r-2}{r^2s} + \dfrac{2}{r^2s} = \dfrac{3r-2+2}{r^2s} = \dfrac{3r}{r^2s} = \dfrac{3 \cdot r}{rs \cdot r} = \dfrac{3}{rs}$

(b) $\dfrac{3x+1}{x+3} - \dfrac{3x+2}{x+3}$

$= \dfrac{(3x+1)-(3x+2)}{x+3}$ Insert parentheses.

$= \dfrac{3x+1-3x-2}{x+3}$ Subtract by removing parentheses.

$= \dfrac{-1}{x+3}$ or $-\dfrac{1}{x+3}$ Combine like terms.

(c) $\dfrac{x}{x^2-1} - \dfrac{4x-3}{x^2-1}$

$= \dfrac{x-(4x-3)}{x^2-1}$ Insert parentheses.

$= \dfrac{x-4x+3}{x^2-1}$ Subtract by removing parentheses.

$= \dfrac{-3x+3}{x^2-1}$ Combine like terms.

$= \dfrac{-3(x-1)}{(x+1)(x-1)}$ Factor the numerator and the denominator.

$= \dfrac{-3}{x+1}$ Divide out common factors.

(d) $\dfrac{6x}{x^2-x-12} - \dfrac{4x+7}{x^2-x-12} + \dfrac{x-5}{x^2-x-12}$

$= \dfrac{6x-(4x+7)+(x-5)}{x^2-x-12}$ Insert parentheses.

$= \dfrac{6x-4x-7+x-5}{x^2-x-12}$ Simplify by removing parentheses.

$= \dfrac{3x-12}{x^2-x-12}$ Combine like terms.

$= \dfrac{3(x-4)}{(x-4)(x+3)}$ Factor.

$= \dfrac{3}{x+3}$ Divide out the common factor.

Least Common Multiples

The **least common multiple (LCM)** of two numbers is the smallest number that is a multiple of both numbers. To determine the LCM of two numbers, we begin by factoring the numbers.

$72 = 2^3 \cdot 3^2$

$48 = 2^4 \cdot 3$

LCM: $2^4 \cdot 3^2 = 144$

Note that the LCM is the product of each factor that appears in either factorization, and the exponent on each base is the largest exponent that appears on that base in either factorization. We determine the LCM of polynomials in a similar way.

Determining the LCM of Polynomials

1. Factor each polynomial completely, and write the result in exponential form.
2. Include in the LCM each factor that appears in at least one polynomial.
3. For each factor, use the largest exponent that appears on that factor in any polynomial.

EXAMPLE 2

Determining the LCM of Polynomials

Determine the LCM of the given polynomials.

(a) $3xy^4, 4x^3y, 6x^5y^2$ (b) $y, y - 2$

(c) $3x^2 - 3x, 6x - 6$ (d) $2x^2 - x - 3, 4x^2 - 4x - 3$

(e) $t^2 + 4t + 4, t^2 - t - 6$

Solution

(a) $3xy^4 = 3xy^4$ Factor completely.

$4x^3y = 2^2x^3y$

$6x^5y^2 = 2 \cdot 3x^5y^2$ The bases that appear in at least one expression are 2, 3, x, and y.

LCM $= 2^2 \cdot 3x^5y^4 = 12x^5y^4$ For each base, use the largest exponent that appears in any expression.

(b) LCM $= y(y - 2)$

(c) $3x^2 - 3x = 3x(x - 1)$ Factor completely.

$6x - 6 = 6(x - 1) = 2 \cdot 3(x - 1)$ The bases that appear in at least one expression are 2, 3, x, and $x - 1$.

LCM $= 2 \cdot 3x(x - 1) = 6x(x - 1)$

(d) $2x^2 - x - 3 = (2x - 3)(x + 1)$ Factor completely.

$4x^2 - 4x - 3 = (2x + 1)(2x - 3)$

LCM $= (2x - 3)(x + 1)(2x + 1)$

(e) $t^2 + 4t + 4 = (t + 2)^2$ Factor completely.

$t^2 - t - 6 = (t - 3)(t + 2)$ The bases that appear in at least one expression are $t + 2$ and $t - 3$.

LCM $= (t + 2)^2(t - 3)$ For the base $t + 2$, the largest exponent that appears in either expression is 2.

Think About It

Suppose that $a = 2^m \cdot 3^n$ and $b = 2^n \cdot 3^m$, where $m < n$. What is the GCF of a and b? What is the LCM of a and b?

Unlike Denominators

If the rational expressions to be added or subtracted have unlike denominators, then the first task is to write equivalent rational expressions with the same denominator.

The routine begins with determining the **least common denominator (LCD),** that is, the LCM of the denominators. Then we rename the fractions so that each one has the LCD as its denominator. Finally, we perform the addition or subtraction.

EXAMPLE 3

Combining Rational Expressions with Unlike Denominators

Perform the indicated operations.

(a) $\dfrac{2}{r^2s} - \dfrac{3}{rs^2}$ The LCD is r^2s^2.

$= \dfrac{2 \cdot s}{r^2s \cdot s} - \dfrac{3 \cdot r}{rs^2 \cdot r}$ Write equivalent fractions.

$= \dfrac{2s}{r^2s^2} - \dfrac{3r}{r^2s^2}$

$= \dfrac{2s - 3r}{r^2s^2}$ Combine the numerators.

(b) $\dfrac{x-1}{6} + \dfrac{2x+1}{9}$ The LCD is 18.

$= \dfrac{(x-1) \cdot 3}{6 \cdot 3} + \dfrac{(2x+1) \cdot 2}{9 \cdot 2}$ Write equivalent fractions.

$= \dfrac{3x-3}{18} + \dfrac{4x+2}{18}$

$= \dfrac{3x - 3 + 4x + 2}{18}$ Combine the numerators.

$= \dfrac{7x-1}{18}$ Combine like terms.

(c) $\dfrac{1}{x-3} - \dfrac{2}{3-x}$

$= \dfrac{1}{x-3} - \dfrac{(-1) \cdot 2}{(-1) \cdot (3-x)}$ $-1(a - b) = b - a$

$= \dfrac{1}{x-3} - \dfrac{-2}{x-3}$ Now the denominators are the same.

$= \dfrac{1-(-2)}{x-3}$ Combine the numerators.

$= \dfrac{3}{x-3}$

(d) $x + \dfrac{3}{x+2} = \dfrac{x}{1} + \dfrac{3}{x+2}$ The LCD is $x + 2$.

$= \dfrac{x(x+2)}{1(x+2)} + \dfrac{3}{x+2}$ Write equivalent fractions.

$= \dfrac{x^2+2x}{x+2} + \dfrac{3}{x+2}$

$= \dfrac{x^2+2x+3}{x+2}$ Combine the numerators.

In Example 3 it was not necessary to factor any of the denominators in order to determine the LCD. In the next example, both problems require factoring.

EXAMPLE 4

Factoring to Determine the LCD

(a) $\dfrac{3x-2}{3x^2+3x} - \dfrac{x}{x^2+2x+1}$ Factor the denominators.

$= \dfrac{3x-2}{3x(x+1)} - \dfrac{x}{(x+1)^2}$ The LCD is $3x(x+1)^2$.

$= \dfrac{(3x-2)(x+1)}{3x(x+1)(x+1)} - \dfrac{x \cdot 3x}{3x(x+1)^2}$ Write equivalent fractions.

$= \dfrac{3x^2+x-2}{3x(x+1)^2} - \dfrac{3x^2}{3x(x+1)^2}$ Simplify the numerators.

$= \dfrac{3x^2+x-2-3x^2}{3x(x+1)^2}$ Combine the numerators.

$= \dfrac{x-2}{3x(x+1)^2}$ Combine like terms.

LEARNING TIP

After you rename rational expressions so that they have a common denominator, avoid the temptation to divide out common factors. The simplifying step occurs after the addition or subtraction has been performed.

(b) $\dfrac{x+4}{x^2+x-6} - \dfrac{2x+1}{x^2-6x+8}$ Factor the denominators.

$= \dfrac{x+4}{(x+3)(x-2)} - \dfrac{2x+1}{(x-4)(x-2)}$ The LCD is $(x+3)(x-2)(x-4)$.

$= \dfrac{(x+4)(x-4)}{(x+3)(x-2)(x-4)} - \dfrac{(2x+1)(x+3)}{(x-4)(x-2)(x+3)}$ Write equivalent fractions.

$= \dfrac{x^2-16}{(x-4)(x-2)(x+3)} - \dfrac{2x^2+7x+3}{(x-4)(x-2)(x+3)}$ Simplify the numerators.

$= \dfrac{(x^2-16)-(2x^2+7x+3)}{(x-4)(x-2)(x+3)}$ Combine the numerators.

$= \dfrac{x^2-16-2x^2-7x-3}{(x-4)(x-2)(x+3)}$ Remove parentheses.

$= \dfrac{-x^2-7x-19}{(x-4)(x-2)(x+3)}$ Combine like terms.

Combined Operations

We may find it necessary to perform both addition and subtraction in the same problem.

EXAMPLE 5

Combined Operations with Rational Expressions

$\dfrac{-2x}{x+3} + \dfrac{3}{3-x} - \dfrac{8x-12}{x^2-9}$

$= \dfrac{-2x}{x+3} + \dfrac{-1(3)}{-1(3-x)} - \dfrac{8x-12}{(x+3)(x-3)}$ $-1(a-b) = b-a$

$= \dfrac{-2x}{x+3} + \dfrac{-3}{x-3} - \dfrac{8x-12}{(x+3)(x-3)}$ The LCD is $(x+3)(x-3)$.

$= \dfrac{-2x(x-3)}{(x+3)(x-3)} + \dfrac{-3(x+3)}{(x+3)(x-3)} - \dfrac{8x-12}{(x+3)(x-3)}$ Write equivalent fractions.

$$= \frac{-2x^2 + 6x}{(x+3)(x-3)} + \frac{-3x - 9}{(x+3)(x-3)} - \frac{8x - 12}{(x+3)(x-3)}$$ Simplify the numerators.

$$= \frac{(-2x^2 + 6x) + (-3x - 9) - (8x - 12)}{(x+3)(x-3)}$$ Combine the numerators.

$$= \frac{-2x^2 + 6x - 3x - 9 - 8x + 12}{(x+3)(x-3)}$$ Remove parentheses.

$$= \frac{-2x^2 - 5x + 3}{(x+3)(x-3)}$$ Combine like terms.

$$= \frac{-1(2x^2 + 5x - 3)}{(x+3)(x-3)}$$ Factor the numerator.

$$= \frac{-1(2x - 1)(x + 3)}{(x+3)(x-3)}$$

$$= \frac{-(2x - 1)}{x - 3} \quad \text{or} \quad -\frac{2x - 1}{x - 3}$$ Divide out the common factor.

Expressions with negative exponents sometimes lead to addition or subtraction of rational expressions.

EXAMPLE 6

Expressions with Negative Exponents

Add $x^{-1} + y^{-1}$. Express the result with positive exponents.

Solution

$$x^{-1} + y^{-1} = \frac{1}{x} + \frac{1}{y}$$ Definition of negative exponent

$$= \frac{y}{xy} + \frac{x}{xy}$$ The LCD is xy.

$$= \frac{x + y}{xy}$$ Combine the numerators.

7.3 QUICK REFERENCE

Like Denominators

- If a, b, and c are algebraic expressions, where $c \neq 0$, then

$$\frac{a}{c} + \frac{b}{c} = \frac{a + b}{c} \quad \text{and} \quad \frac{a}{c} - \frac{b}{c} = \frac{a - b}{c}$$

- To add or subtract rational expressions with like denominators, follow these steps:
  1. Add or subtract the numerators.
  2. Retain the common denominator.
  3. Simplify the result.
- When subtracting rational expressions, be sure to treat each numerator as a quantity in parentheses.

Least Common Multiples

- To determine the **least common multiple (LCM)** of two or more polynomials, follow these steps:
  1. Factor each polynomial completely, and write the result in exponential form.
  2. Include in the LCM each factor that appears in at least one polynomial.
  3. For each factor, use the largest exponent that appears on that factor in any polynomial.

Unlike Denominators

- Follow these steps when adding or subtracting rational expressions with unlike denominators:
  1. Determine the **least common denominator (LCD)**, which is the LCM of the denominators.
  2. Rewrite each rational expression as an equivalent expression with the LCD as its denominator.
  3. Combine and simplify the numerators.
  4. Simplify the result.

Combined Operations

- The procedure for adding and subtracting rational expressions can be extended to addition and subtraction involving three or more rational expressions.
- Using the definition of a negative exponent, we can write expressions involving negative exponents as rational expressions.

7.3 EXERCISES

Concepts and Skills

1. Suppose that two rational expressions have the same denominator. When we add the expressions, how do we determine the numerator and denominator of the sum?

2. So far the following work is correct.

$$\frac{x^2 - 15}{x - 3} + \frac{2x}{x - 3} = \frac{x^2 - 15 + 2x}{x - 3}$$

Explain why the problem is not yet done.

In Exercises 3–14, perform the indicated operations.

3. $\frac{5}{a} + \frac{3}{a}$

4. $\frac{5}{t} - \frac{1}{t}$

5. $\frac{5x - 4}{x} - \frac{8x + 5}{x}$

6. $\frac{8a - 2b}{b} + \frac{4a - 5b}{b}$

7. $\frac{4x - 45}{x - 9} + \frac{2x - 9}{x - 9}$

8. $\frac{3x - 63}{x - 6} + \frac{8x - 3}{x - 6}$

9. $\frac{3c + 51d}{c - 8d} - \frac{9c + 3d}{c - 8d}$

10. $\frac{3x - 24y}{x + 7y} - \frac{6x - 3y}{x + 7y}$

11. $\frac{2x + 5}{x^2 - 2x - 8} + \frac{x + 1}{x^2 - 2x - 8}$

12. $\frac{3t + 4}{t^2 + 10t + 25} - \frac{2t - 1}{t^2 + 10t + 25}$

13. $\frac{x^3 + 6x^2 - 20x}{x(x - 2)} - \frac{x^2 - 6x}{x(x - 2)}$

14. $\frac{x(x - 1)}{(2x + 3)(x + 1)} - \frac{2(2x + 3)}{(2x + 3)(x + 1)}$

In Exercises 15–20, perform the indicated operations.

15. $\dfrac{x}{x^2 - x - 6} - \dfrac{x + 1}{x^2 - x - 6} + \dfrac{x - 2}{x^2 - x - 6}$

16. $\dfrac{2x + 1}{x^2 + x - 2} + \dfrac{3x + 4}{x^2 + x - 2} - \dfrac{4x + 3}{x^2 + x - 2}$

17. $\dfrac{6}{2x + 3} - \left[\dfrac{x}{2x + 3} - \dfrac{3 - x}{2x + 3}\right]$

18. $\dfrac{x}{4 - x} - \left[\dfrac{x + 4}{4 - x} + \dfrac{x}{4 - x}\right]$

19. $\left[\dfrac{3x + 1}{x - 2} - \dfrac{2x - 2}{x - 2}\right] \div (x + 3)$

20. $\left[\dfrac{3x + 1}{5 - x} + \dfrac{1 - x}{5 - x}\right] \cdot \dfrac{x - 5}{x + 1}$

In Exercises 21–26, fill in the blank to make the resulting equation true.

21. $\dfrac{2x - 5}{x + 4} + \dfrac{\underline{\qquad}}{x + 4} = \dfrac{3x + 1}{x + 4}$

22. $\dfrac{x - 4}{2x - 3} - \dfrac{\underline{\qquad}}{2x - 3} = \dfrac{2}{2x - 3}$

23. $\dfrac{\underline{\qquad}}{t^2 + 3} - \dfrac{3t - t^2}{t^2 + 3} = \dfrac{t^2 - 3}{t^2 + 3}$

24. $\dfrac{x^2 - 5x + 2}{x^2 + x - 1} + \dfrac{\underline{\qquad}}{x^2 + x - 1} = \dfrac{3x^2 + x - 4}{x^2 + x - 1}$

25. $\dfrac{\underline{\qquad}}{(x + 2)(x - 5)} - \dfrac{x - x^2}{(x + 2)(x - 5)} = \dfrac{3x^2 - 2x}{(x + 2)(x - 5)}$

26. $\dfrac{\underline{\qquad}}{a - 2b} + \dfrac{a^2 + b}{a - 2b} = \dfrac{a^2 + ab}{a - 2b}$

27. Explain how to rewrite the following expression so that the fractions have the same denominator.

$$\frac{15}{x} + \frac{7}{-x}$$

28. Explain how to rewrite the following expression so that the fractions have the same denominator.

$$\frac{x}{2x - 1} - \frac{1}{1 - 2x}$$

In Exercises 29–32, fill in the blanks so that the resulting equation is true.

29. $\dfrac{a}{3} - \dfrac{5}{-3} = \dfrac{a}{3} + \dfrac{\underline{\qquad}}{3} = \dfrac{\underline{\qquad}}{3}$

30. $\dfrac{7}{c} + \dfrac{2}{-c} = \dfrac{7}{c} + \dfrac{\underline{\qquad}}{c} = \dfrac{\underline{\qquad}}{c}$

31. $\dfrac{3x + 9}{x - 2} - \dfrac{8x}{2 - x} = \dfrac{3x + 9}{x - 2} + \dfrac{\underline{\qquad}}{x - 2} = \dfrac{\underline{\qquad}}{x - 2}$

32. $\dfrac{2x - 1}{x - 2} + \dfrac{7x}{2 - x} = \dfrac{2x - 1}{x - 2} + \dfrac{\underline{\qquad}}{x - 2} = \dfrac{\underline{\qquad}}{x - 2}$

In Exercises 33–38, perform the indicated operations.

33. $\dfrac{2x + 9}{7x - 8} - \dfrac{7x - 6}{8 - 7x}$

34. $\dfrac{5x + 3}{9x - 2} + \dfrac{6x - 6}{2 - 9x}$

35. $\dfrac{x^2}{x - 4} + \dfrac{16}{4 - x}$

36. $\dfrac{x^2}{x - 2y} + \dfrac{4y^2}{2y - x}$

37. $\dfrac{a^2}{a - b} - \dfrac{b^2}{b - a}$

38. $\dfrac{x - 4}{x^2 - 25} - \dfrac{x - 6}{25 - x^2}$

In Exercises 39–48, determine the LCM of the given expressions.

39. $5x^2y$, $10xy^2$, $15x^3y$

40. $3x^5y$, $6xy^4$, $9x^3y^7$

41. $6x$, $3x - 12$

42. $2t$, $4t + 8$

43. $t + 5$, $-3t - 15$, $10t$

44. $x - 7$, $4x - 28$, $6x$

45. $3x^2 + x - 2$, $4x^2 + 5x + 1$

46. $3x^2 + 4x - 15$, $x^2 + 5x + 6$

47. $x^2 + 3x + 2$, $x^2 - 4$, $3x + 6$

48. $y^2 + 2y - 3$, $y^2 + y - 2$, $3y - 3y^2$

49. Suppose the LCD of two fractions is $x(x - 4)$. If one of the fractions is $\dfrac{x}{x - 4}$, explain how to rewrite the fraction so that its denominator is the LCD.

50. Suppose the GCF of two denominators is 1. Explain how to determine the LCD for the two fractions.

In Exercises 51–64, perform the indicated operations.

51. $\dfrac{5}{a^3b} + \dfrac{7}{ab^2}$

52. $\dfrac{3}{5c^3d^2} - \dfrac{4}{15c^2d^4}$

53. $x - \dfrac{3}{x}$

54. $3 + \dfrac{5x}{x - 2}$

55. $\dfrac{3}{x} + \dfrac{8}{x - 4}$

56. $\dfrac{3}{t + 3} - \dfrac{2}{t}$

57. $\dfrac{1}{x - 5} - \dfrac{1}{x + 7}$

58. $\dfrac{3}{x - 8} + \dfrac{x}{x - 3}$

59. $\dfrac{2}{t+1} + t + 1$

60. $1 - 2t - \dfrac{2t}{t+3}$

61. $\dfrac{3}{6(t-5)} - \dfrac{2-t}{9(t-5)}$

62. $\dfrac{2x+1}{3(2x-1)} + \dfrac{2-3x}{2(1-2x)}$

63. $\dfrac{5t+2}{t(t+1)^2} - \dfrac{3}{t(t+1)}$

64. $\dfrac{3}{y(y-2)} + \dfrac{2y-1}{y^2(y-2)}$

In Exercises 65–78, perform the indicated operations.

65. $\dfrac{2}{3t^2} + \dfrac{3}{2t^2+t-1}$

66. $\dfrac{3}{2x} - \dfrac{x-3}{2x^2-x-6}$

67. $\dfrac{2x^2-7}{10x^2-3x-4} - \dfrac{x-3}{5x-4}$

68. $\dfrac{9}{x-3} + \dfrac{x}{x^2-6x+9}$

69. $\dfrac{x+9}{4x-36} - \dfrac{x-9}{x^2-18x+81}$

70. $\dfrac{x-3}{2x+6} - \dfrac{x+3}{x^2+6x+9}$

71. $\dfrac{7}{x^2+15x+56} + \dfrac{1}{x^2-64}$

72. $\dfrac{5}{x^2-4x-12} + \dfrac{9}{x^2-15x+54}$

73. $\dfrac{x}{x+6} - \dfrac{72}{x^2-36}$

74. $\dfrac{x}{x+3} - \dfrac{18}{x^2-9}$

75. $\dfrac{x}{x^2+4x+3} - \dfrac{2}{x^2-2x-3}$

76. $\dfrac{x}{x^2+5x+4} + \dfrac{3}{x^2+11x+10}$

77. $\dfrac{x}{x^2+8x+15} + \dfrac{15}{x^2+4x-5}$

78. $\dfrac{x}{x^2+4x+3} - \dfrac{3}{x^2-4x-5}$

79. Write each of the following as a single rational expression with no negative exponents. Are the two expressions equivalent?

$(x+y)^{-1}$ $\qquad x^{-1}+y^{-1}$

80. Write each of the following as a single rational expression with no negative exponents. Are the two expressions equivalent?

$(xy)^{-1}$ $\qquad x^{-1}y^{-1}$

In Exercises 81–84, simplify and express the result as a single rational expression with no negative exponents.

81. $x^{-1} - y^{-1}$

82. $3x^{-1} + 4y^{-1}$

83. $x(x+3)^{-1} + 3(x-2)^{-1}$

84. $4(x-3)^{-1} + x(x-3)^{-2}$

Real-Life Applications

85. A quality control manager has measured the average number of errors that employees make while assembling electronic circuit boards. She has modeled her data with $E_0(x) = \dfrac{5}{x}$, where $E_0(x)$ is the average number of errors made after x hours of training ($x \geq 1$).

(a) Using a window setting to display x-values between 1 and 10 and y-values between 0 and 5, produce the graph of $E_0(x)$. Does the graph suggest that increasing the hours of training reduces errors? Can the average number of errors be reduced to zero if enough training is provided?

(b) Using certain incentives along with training, the manager believes she can reduce the average number of errors by $\dfrac{5}{x+1}$; that is, the average number of errors would be $E_1(x) = \dfrac{5}{x} - \dfrac{5}{x+1}$. Write this new function as a single rational expression, and produce its graph along with the graph of $E_0(x)$. Does the graph suggest that the average number of errors would be reduced even further with the incentive plan?

86. Suppose that one *cycle* of a sound wave can be modeled with the function $y_1 = \dfrac{8x}{x^2+1}$, where where $-3 \leq x \leq 3$. Roughly speaking, the *amplitude* is the vertical distance between the highest and lowest points of the wave.

(a) Using a window setting to display x-values between -3 and 3 and y-values between -5 and 5, produce the graph of one cycle of this sound wave. Then trace the graph to estimate the amplitude of the wave.

(b) Suppose that a second sound wave modeled by the function $y_2 = \dfrac{2x}{x^2+1}$ is mixed with the first sound wave so that the resulting wave is modeled by the sum of y_1 and y_2. Produce and trace the graph of the sum function to estimate the amount by which the amplitude is increased.

Group Project: Automobile Operating Costs

The following table summarizes the costs associated with operating an automobile. Variable costs (in cents per mile) refer to such items as fuel, oil, and maintenance. Fixed costs (in dollars per year) refer to such items as insurance, licenses, finance charges, and depreciation. (Source: American Automobile Manufacturers Association.)

Year	Variable Cost (cents/mile)	Fixed Cost (dollars/year)
1988	7.6	3061
1989	7.9	3534
1990	8.4	3877
1991	9.8	4217
1992	9.1	4538
1993	9.3	4486

The costs can be modeled by the following functions, where t is the number of years since 1985.

Variable costs: $V(t) = \dfrac{20t + 154}{t + 25}$

Fixed costs: $F(t) = \dfrac{1000(11t + 20)}{t + 14}$

87. Write a new function $V_1(t)$ that models the annual variable costs (in dollars) associated with driving 10,000 miles per year.

88. Write the sum of $V_1(t)$ and $F(t)$. Then rewrite the expression with 200 as a common factor.

89. Add the rational expressions in Exercise 88. Use your calculator to perform the necessary arithmetic. What does the resulting expression (including the common factor) represent?

90. In 1993, the Internal Revenue Service allowed a deduction of 22 cents per mile for business-related travel by automobile. Referring to the table, did this deduction fully cover the total costs associated with operating the automobile for 10,000 miles?

Challenge

In Exercises 91 and 92, perform the indicated operations.

91. $\left(\dfrac{1}{x+h} - \dfrac{1}{x}\right) \div h$

92. $\left[\dfrac{1}{(x+h)^2} - \dfrac{1}{x^2}\right] \cdot \dfrac{1}{h}$

In Exercises 93 and 94, perform the indicated operations.

93. $\dfrac{\dfrac{1}{x} - \dfrac{2x}{x^2+1}}{\dfrac{1}{x}}$

94. $\dfrac{\dfrac{x}{x+1} + \dfrac{1}{x}}{\dfrac{1}{x} - 1}$

7.4 COMPLEX FRACTIONS

Order of Operations • Multiplying by the LCD • Expressions with Negative Exponents

Order of Operations

LEARNING TIP

When you write a complex fraction, make sure that you draw the main fraction bar longer than the other fraction bars in order to separate the numerator and the denominator clearly.

A **complex fraction** is a fraction with a rational expression in the numerator, in the denominator, or in both.

$$\frac{\dfrac{x+5}{5}}{\dfrac{x^2-25}{10}} \qquad \frac{y + \dfrac{3}{y-2}}{y+1} \qquad \frac{\dfrac{3}{y} + 2}{\dfrac{5}{y} - 3} \qquad \frac{\dfrac{3}{x+3}}{\dfrac{1}{x+3} - \dfrac{1}{x}}$$

We simplify a complex fraction by eliminating all fractions in the numerator and in the denominator.

A complex fraction can be simplified in two ways. Because the fraction bar indicates division, we can simply rewrite the complex fraction with a division symbol.

EXAMPLE 1

Writing a Complex Fraction with a Division Symbol

$$\frac{\frac{x+5}{5}}{\frac{x^2-25}{10}} = \frac{x+5}{5} \div \frac{x^2-25}{10}$$ The fraction indicates division.

$$= \frac{x+5}{5} \cdot \frac{10}{x^2-25}$$ Definition of division

$$= \frac{x+5}{5} \cdot \frac{5 \cdot 2}{(x+5)(x-5)}$$ Factor.

$$= \frac{2}{x-5}$$ Divide out common factors.

Recall that a fraction bar is also a grouping symbol. When the numerator or the denominator is a sum or difference, we combine the terms and then carry out the division.

EXAMPLE 2

Using the Order of Operations to Simplify a Complex Fraction

$$\frac{\frac{3}{y}+2}{\frac{5}{y^2}-\frac{1}{y}} = \frac{\frac{3}{y}+\frac{2y}{y}}{\frac{5}{y^2}-\frac{y}{y^2}}$$ The LCD of the numerator is y; the LCD of the denominator is y^2.

$$= \frac{\frac{3+2y}{y}}{\frac{5-y}{y^2}}$$ Combine the fractions.

$$= \frac{3+2y}{y} \cdot \frac{y^2}{5-y}$$ Multiply by the reciprocal of the denominator.

$$= \frac{y(3+2y)}{5-y}$$ Divide out the common factor y.

The following summarizes the Order of Operations method for simplifying a complex fraction.

Using the Order of Operations to Simplify a Complex Fraction

1. Perform all indicated operations in the numerator and in the denominator.
2. Perform the division.

Multiplying by the LCD

Another way to simplify a complex fraction is to multiply the numerator and denominator by the LCD of *all the fractions*. Since multiplying the numerator and denominator of the complex fraction by the same number is equivalent to multiplying by 1, the resulting expression will be equivalent to the original complex fraction.

To compare the two methods, we use this approach on the same complex fraction that we simplified in Example 2.

EXAMPLE 3

Using the LCD to Simplify a Complex Fraction

$$\frac{\frac{3}{y}+2}{\frac{5}{y^2}-\frac{1}{y}} = \frac{y^2\cdot\left(\frac{3}{y}+2\right)}{y^2\cdot\left(\frac{5}{y^2}-\frac{1}{y}\right)}$$

Multiply the numerator and the denominator by the LCD y^2.

$$= \frac{y^2\cdot\frac{3}{y}+y^2\cdot 2}{y^2\cdot\frac{5}{y^2}-y^2\cdot\frac{1}{y}}$$

Applying the Distributive Property results in multiplying each term by the LCD.

$$= \frac{3y+2y^2}{5-y}$$

Simplify each product.

LEARNING TIP

Write the LCD beside each term of the complex fraction. Attempting to multiply mentally often leads to errors.

Note: The Order of Operations method may use one LCD for the numerator and another LCD for the denominator. However, for the second method, we must multiply by the *same* LCD, which is the LCD of every fraction in the numerator and in the denominator.

Generally, multiplying both the numerator and the denominator by the LCD is the easier of the two methods. It is the method that we will use in the remaining examples.

Simplifying a Complex Fraction by Multiplying by the LCD

1. Determine the LCD of every fraction in the complex fraction.
2. Multiply the numerator and the denominator of the complex fraction by the LCD.
3. Simplify the result, if possible.

EXAMPLE 4

Simplifying a Complex Fraction

$$\frac{\frac{1}{p-1}+p}{\frac{1}{p-1}-p} = \frac{(p-1)\cdot\left(\frac{1}{p-1}+p\right)}{(p-1)\cdot\left(\frac{1}{p-1}-p\right)}$$

Multiply the numerator and denominator by the LCD, $p-1$.

$$= \frac{(p-1)\cdot\dfrac{1}{p-1} + p(p-1)}{(p-1)\cdot\dfrac{1}{p-1} - p(p-1)}$$ Distributive Property

$$= \frac{1+p^2-p}{1-p^2+p}$$ Simplify the numerator and denominator.

$$\text{or } \frac{1-p+p^2}{1+p-p^2}$$

EXAMPLE 5

Simplifying a Complex Fraction

$$\frac{\dfrac{3}{x+3}}{\dfrac{1}{x+3}-\dfrac{1}{x}} = \frac{x(x+3)\cdot\dfrac{3}{x+3}}{x(x+3)\cdot\left(\dfrac{1}{x+3}-\dfrac{1}{x}\right)}$$ Multiply the numerator and denominator by the LCD, $x(x+3)$.

$$= \frac{x(x+3)\cdot\dfrac{3}{x+3}}{x(x+3)\cdot\dfrac{1}{x+3} - x(x+3)\cdot\dfrac{1}{x}}$$ Distributive Property

$$= \frac{3x}{x-(x+3)}$$ Simplify.

$$= \frac{3x}{x-x-3}$$ Remove parentheses.

$$= \frac{3x}{-3}$$ Combine like terms.

$$= -x$$ Divide out the common factor 3.

Note: The complex fraction in Example 5 has two restricted values, 0 and -3. The very simple expression $-x$ is equivalent to the original complex fraction only if x is not replaced with 0 or -3.

Think About It

Simplify $\dfrac{\frac{bx+a}{b}}{\frac{a}{ax+b}}$. Then show that the result simplifies to a perfect square trinomial if $a = b$.

Expressions with Negative Exponents

When an expression with negative exponents is rewritten with positive exponents, the result is sometimes a complex fraction.

EXAMPLE 6

Simplifying Expressions with Negative Exponents

$$\frac{2r^{-2}+c^{-1}}{r^{-1}+3c^{-2}} = \frac{\dfrac{2}{r^2}+\dfrac{1}{c}}{\dfrac{1}{r}+\dfrac{3}{c^2}}$$ Definition of negative exponent

$$= \frac{r^2c^2 \cdot \left(\frac{2}{r^2} + \frac{1}{c}\right)}{r^2c^2 \cdot \left(\frac{1}{r} + \frac{3}{c^2}\right)}$$ The LCD is r^2c^2.

$$= \frac{r^2c^2 \cdot \frac{2}{r^2} + r^2c^2 \cdot \frac{1}{c}}{r^2c^2 \cdot \frac{1}{r} + r^2c^2 \cdot \frac{3}{c^2}}$$ Distributive Property

$$= \frac{2c^2 + r^2c}{rc^2 + 3r^2}$$ Simplify.

7.4 QUICK REFERENCE

Order of Operations

- A **complex fraction** is a fraction with a rational expression in the numerator, in the denominator, or in both.
- Use the following steps to simplify a complex fraction with the Order of Operations method.
  1. Perform all indicated operations that are in the numerator and in the denominator.
  2. Perform the division.

Multiplying by the LCD

- Use the following steps to simplify a complex fraction by multiplying by the LCD.
  1. Determine the LCD of every fraction in the complex fraction.
  2. Multiply the numerator and the denominator of the complex fraction by the LCD.
  3. Simplify the result, if possible.

Expressions with Negative Exponents

- When expressions with negative exponents are rewritten with positive exponents, the result is sometimes a complex fraction.

7.4 EXERCISES

Concepts and Skills

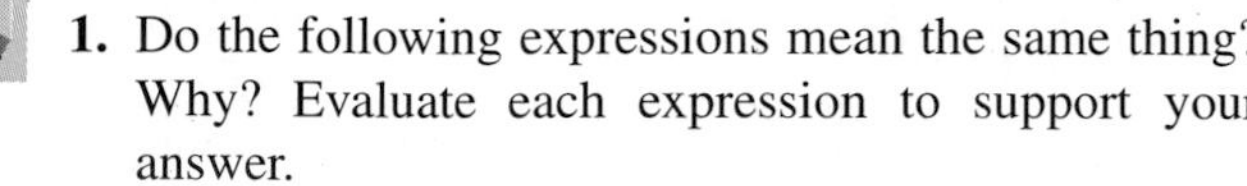

1. Do the following expressions mean the same thing? Why? Evaluate each expression to support your answer.

$$\frac{\frac{2}{3}}{4} \quad \text{and} \quad \frac{2}{\frac{3}{4}}$$

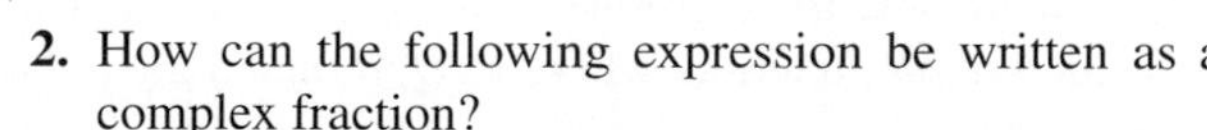

2. How can the following expression be written as a complex fraction?

$$\frac{a+2}{a-1} \div \frac{a-3}{a+5}$$

In Exercises 3–16, simplify the complex fraction.

3. $\dfrac{\frac{3}{x}}{\frac{9}{x^3}}$

4. $\dfrac{-\frac{42}{y^5}}{\frac{35}{y^2}}$

5. $\dfrac{\frac{2x}{x+3}}{\frac{6x^2}{x+3}}$

6. $\dfrac{\frac{c+d}{4}}{\frac{c-d}{3}}$

7. $\dfrac{\frac{5}{x+5}}{\frac{10}{x^2-25}}$

8. $\dfrac{-\frac{2}{x^2-1}}{\frac{2}{1-x}}$

9. $\dfrac{\frac{4x^2}{6x+54}}{\frac{32x^3}{x^2-81}}$

10. $\dfrac{\frac{6x}{x-5}}{\frac{12x^2}{x^2+3x-40}}$

11. $\dfrac{\frac{8x}{x+5}}{\frac{16x^2}{x^2+x-20}}$

12. $\dfrac{\frac{7x-35}{4x^3}}{\frac{x^2-25}{4x^2+20x}}$

13. $\dfrac{\frac{5x}{x^2-4}}{\frac{35x^2}{x^2-x-6}}$

14. $\dfrac{\frac{2x}{x^2-4}}{\frac{8x^2}{x^2-7x-18}}$

15. $\dfrac{\frac{2x^2+3x-2}{x^2-x-6}}{\frac{1-x-2x^2}{x^2+3x+2}}$

16. $\dfrac{\frac{3x^2-13x-30}{6x^2+13x+5}}{\frac{x^2-4x-12}{2x^2+x}}$

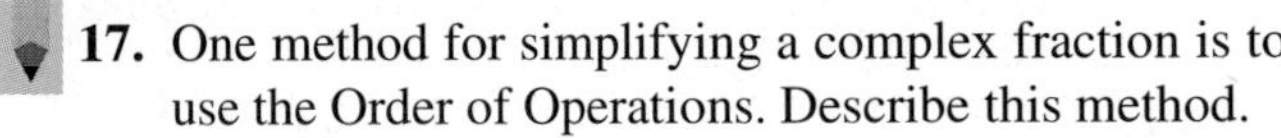

17. One method for simplifying a complex fraction is to use the Order of Operations. Describe this method.

18. We can write the expression $\left(\frac{3}{4}+\frac{2}{3}\right) \div \frac{5}{8}$ as a complex fraction.

$$\dfrac{\frac{3}{4}+\frac{2}{3}}{\frac{5}{8}}$$

Explain why there is no need for parentheses in the complex fraction.

In Exercises 19–30, simplify the complex fraction.

19. $\dfrac{\frac{2}{3}+\frac{4}{t}}{\frac{1}{t}+\frac{1}{2}}$

20. $\dfrac{\frac{3}{x}-\frac{5}{9}}{\frac{4}{3}-\frac{4}{x}}$

21. $\dfrac{\frac{2}{y^2}+\frac{1}{y}}{\frac{2}{y}+1}$

22. $\dfrac{4-\frac{1}{x^2}}{\frac{2}{x}-\frac{1}{x^2}}$

23. $\dfrac{4-\frac{1}{x^2}}{2+\frac{1}{x}}$

24. $\dfrac{1+\frac{3}{x}}{x-\frac{9}{x}}$

25. $\dfrac{\frac{1}{c}-\frac{1}{cd}}{\frac{1}{cd}-\frac{1}{c}}$

26. $\dfrac{\frac{1}{x}+\frac{1}{y}}{\frac{1}{x}+\frac{1}{xy}}$

27. $\dfrac{\frac{5}{x^4}-\frac{2}{x^3}}{\frac{5}{x^3}-\frac{2}{x^2}}$

28. $\dfrac{4-\frac{5}{x}}{\frac{4}{x}-\frac{5}{x^2}}$

29. $\dfrac{x+y}{\frac{1}{x}+\frac{1}{y}}$

30. $\dfrac{\frac{3}{a^2}-\frac{2}{b}}{a+2b}$

In Exercises 31–42, simplify the complex fraction.

31. $\dfrac{\frac{2}{x+3}-\frac{3}{x-3}}{2+\frac{1}{x^2-9}}$

32. $\dfrac{\frac{4}{4-x^2}-1}{\frac{1}{x+2}+\frac{1}{x-2}}$

33. $\dfrac{\frac{3}{2x+1}-(x+1)}{(x-3)-\frac{2}{2x+1}}$

34. $\dfrac{(2x-1)+\frac{1}{x-2}}{x+\frac{1}{x-2}}$

35. $\dfrac{\frac{1}{1-y}-\frac{1}{1+y}}{\frac{1+y}{1-y}-\frac{1-y}{1+y}}$

36. $\dfrac{\frac{x+4}{x-4}+\frac{x-4}{x+4}}{\frac{1}{x-4}+\frac{1}{x+4}}$

37. $\dfrac{3+\frac{1}{x}-\frac{14}{x^2}}{6+\frac{11}{x}-\frac{7}{x^2}}$

38. $\dfrac{6-\frac{1}{x}-\frac{35}{x^2}}{2-\frac{11}{x}+\frac{15}{x^2}}$

39. $\dfrac{5+\frac{2x}{y}+\frac{2y}{x}}{\frac{2x}{y}-\frac{2y}{x}-3}$

40. $\dfrac{\frac{12x}{y^2}+\frac{13}{xy}-\frac{35}{x^3}}{\frac{10}{x^3}-\frac{3}{xy}-\frac{4x}{y^2}}$

41. $\dfrac{x+\frac{8}{x^2}}{1-\frac{2}{x}+\frac{4}{x^2}}$

42. $\dfrac{\frac{8}{y^3}-27}{\frac{4}{y^3}-\frac{12}{y^2}+\frac{9}{y}}$

43. Explain how to determine the domain of the following rational function, where P, Q, R, and S are polynomials.

$$f(x) = \frac{P/Q}{R/S}$$

44. Explain why $R(3)$ is undefined for the following function.

$$R(x) = \frac{\dfrac{x^2}{x+1}}{\dfrac{x-3}{2-x}}$$

In Exercises 45–54, simplify the complex fraction.

45. $\dfrac{\dfrac{4y^2}{2y+x} + (3x - y)}{\dfrac{5x^2}{2y+x} - (3y + 2x)}$

46. $\dfrac{x + 1 + \dfrac{2}{2x-3}}{x + 2 + \dfrac{3}{2x-3}}$

47. $\dfrac{\dfrac{1}{x+h} - \dfrac{1}{x}}{h}$

48. $\dfrac{\dfrac{1}{x+3} - \dfrac{1}{3}}{x}$

49. $\dfrac{a+b}{1 + \dfrac{2b}{a-b}}$

50. $\dfrac{1 + \dfrac{a+6}{a}}{\dfrac{a}{9} - \dfrac{1}{a}}$

51. $\dfrac{x + 3 + \dfrac{4}{x-2}}{x^2 + 3x + 2}$

52. $\dfrac{1 + \dfrac{c}{a+c}}{1 + \dfrac{3c}{a-c}}$

53. $\dfrac{\dfrac{1}{a^3} + \dfrac{1}{b^3}}{b^2 - ab + a^2}$

54. $\dfrac{\dfrac{1}{27} - \dfrac{1}{a^3}}{a^2 + 3a + 9}$

55. Why is it incorrect to multiply the following expression by the LCD as we do when we simplify a complex fraction?

$$\frac{x}{x+3} + \frac{5}{x-3}$$

56. Simplify the following expression with the methods described in parts (a) and (b). Then answer part (c).

$$\frac{x^{-3} - x^{-2}}{2x^{-3} + x^{-2}}$$

(a) Multiply the numerator and denominator by x^3.

(b) Write the expression with positive exponents and then simplify the result.

(c) Which of the two methods did you find easier?

In Exercises 57–64, rewrite the expression with positive exponents and simplify.

57. $\dfrac{a^{-1}b^{-1}}{a^{-1} + b^{-1}}$

58. $\dfrac{a^{-1} - b^{-1}}{a^{-2} - b^{-2}}$

59. $\dfrac{x^{-1} + x}{x^{-2} + x^2}$

60. $\dfrac{a^{-1} + b}{a + b^{-1}}$

61. $\dfrac{1 - x^{-2}}{3 - x^{-1} - 4x^{-2}}$

62. $\dfrac{1 - 4x^{-2}}{2 + 5x^{-1} + 2x^{-2}}$

63. $\dfrac{xy^{-3} + x^{-2}y}{x - y}$

64. $\dfrac{-2a}{a^{-2} - a^{-1}}$

65. Write and simplify a complex fraction that is the reciprocal of $x + \dfrac{5}{x+1}$.

66. Write and simplify a complex fraction that is the average of $\dfrac{1}{a}$ and $\dfrac{1}{a-1}$.

67. Consider any three consecutive integers.

(a) Write a complex fraction whose numerator is the ratio of the first integer to the second integer and whose denominator is the ratio of the second integer to the third integer.

(b) Simplify the complex fraction and show that the denominator of the result is always one more than the numerator.

68. One number is 5 greater than another. If the reciprocal of the sum of the numbers is divided by the reciprocal of the difference of the numbers, the result is $\frac{5}{13}$. What is the larger number?

Real-Life Applications

69. For the fall quarter, x male students and y female students were accepted at a certain college. Of these students, m male students and n female students decided not to attend.

(a) Write a fraction that represents the ratio of accepted male students who actually enrolled. Then do the same for the ratio of accepted female students who actually enrolled.

(b) Write a fraction that represents the ratio of all accepted students who actually enrolled. Is this fraction the sum of the fractions in part (a)?

(c) Write and simplify a complex fraction that is the ratio (male to female) of the two fractions in part (a). Interpret the meaning of this complex fraction.

70. In baseball, a pitcher's earned run average (ERA) is determined by the formula ERA $= R/(I/9)$, where R is the total number of earned runs surrendered and I is the total number of innings pitched. Write an alternative formula for ERA that does not contain a fraction in the numerator or denominator.

71. The length of a rectangular table is $x + 3$, and the width is two units less than the length. A rectangular piece of red paper is placed at the center of the table. The length of the red rectangle is $x + 1$, and the width is $\frac{1}{x^2 + 1}$. Suppose that a dime is tossed onto the table, and assume that the probability of its landing on the red rectangle is the ratio of the area of the red rectangle to the area of the table. Write and simplify a complex fraction to model the probability of the dime's landing in the red rectangle.

72. A small blue circle of radius $\frac{x}{x^2 + 1}$ is drawn in the center of a larger circle of radius $\frac{x^2}{x + 1}$. Suppose that a dart is thrown at the large circle, and assume that the probability of its landing in the blue circle is the ratio of the area of the blue circle to the area of the large circle. Write and simplify a complex fraction to model the probability of the dart's landing in the blue circle.

Modeling with Real Data

73. The accompanying table gives voter information for the 1992 presidential election. (Source: U.S. Bureau of Census.)

Age Group	Voting Age Population (millions)	Percent Registered	Percent Who Voted
18–20	9.7	48.3	38.5
21–24	14.6	55.3	45.7
25–34	41.6	60.6	53.2
35–44	39.7	69.2	63.6
45–64	49.1	75.6	70.0
65 and over	30.8	75.2	70.1

The following functions model the data, where t represents age.

Population (in millions):

$$p(t) = -0.004t^2 + 0.27t - 0.48$$

Percent registered:

$$r(t) = \frac{65t}{0.7t + 10}$$

Percent who voted:

$$v(t) = \frac{58t}{0.7t + 10}$$

Write a model function for the number of registered voters by age.

74. Using the functions in Exercise 73, write a model function for the number of people who voted by age.

Group Project: Baseball Attendance and Salaries

In professional sports, revenue comes from television rights, promotions, and other marketing ventures as well as from ticket sales. The accompanying bar chart shows the average salaries for major league baseball players along with the total attendance figures for the 1990–1993 seasons.

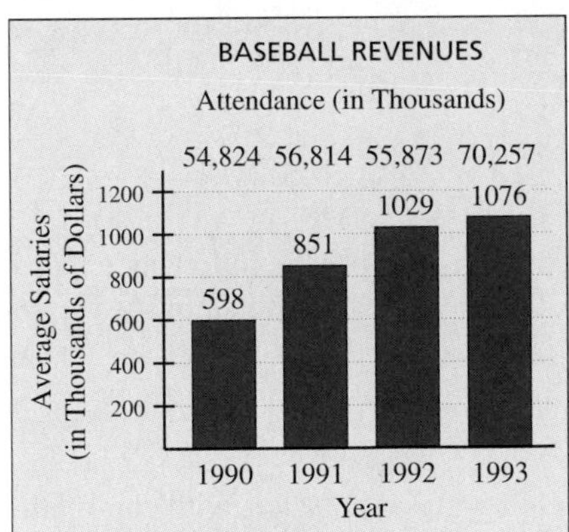

(Source: Major League Baseball Players Association.)

The attendance $A(t)$ and the average salary $S(t)$, both in thousands, are modeled by the following functions, where t is the number of years since 1985.

$$A(t) = \frac{1000(2.8t^3 - 34t^2 + 189t - 5)}{t + 3}$$

$$S(t) = \frac{2049t - 4890}{t + 3}$$

Assume that there were 650 major league players in 1992.

75. Write and simplify the complex fraction $\frac{650 \cdot S(t)}{A(t)}$.

76. If ticket sales were the only source of funds for players' salaries, and if all revenue from ticket sales were used to pay the players, use the model expression in Exercise 75 to estimate the cost of a ticket in 1992.

77. Calculate the percentage increases in attendance and in average salaries from 1992 to 1993. As a player's agent, how would you use this information to argue for higher salaries?

78. As a team owner, how would you counter the argument in Exercise 77? (Hint: Two major league teams were added in 1993.)

Challenge

In Exercises 79–82, use the given function to write

$$\frac{f(x+h)-f(x)}{h}$$

Then simplify the expression.

79. $f(x)=\dfrac{1}{2x-1}$

80. $f(x)=\dfrac{x}{x+2}$

81. $f(x)=\dfrac{x+1}{x-2}$

82. $f(x)=\dfrac{2-3x}{x}$

In Exercises 83–86, simplify the complex fraction.

83. $2x-\dfrac{5+\frac{3}{x}}{7-\frac{2}{x}}$

84. $3x-\dfrac{4+\frac{2}{x}}{\frac{3}{x}-5}$

85. $1+\dfrac{1}{1+\dfrac{1}{1+\frac{1}{2}}}$

86. $\dfrac{1-\dfrac{1}{1-\frac{1}{3}}}{1+\dfrac{1}{1+\frac{1}{3}}}$

7.5 EQUATIONS WITH RATIONAL EXPRESSIONS

Graphing and Algebraic Methods • Solving Formulas

Graphing and Algebraic Methods

In Chapter 3 we solved linear equations containing one or more fractions with constant denominators. The solving technique includes multiplying both sides of the equation by the LCD to eliminate fractions. Example 1 is a review of that method.

EXAMPLE 1

Using the Technique of Clearing Fractions to Solve an Equation

Solve the equation $\dfrac{t}{8}+\dfrac{t+1}{2}=3$.

Solution

$$\frac{t}{8}+\frac{t+1}{2}=3$$

$$8\cdot\frac{t}{8}+8\cdot\frac{t+1}{2}=8\cdot 3$$ Multiply both sides by the LCD 8.

$$t+4(t+1)=24$$ Simplify each term.

$$t+4t+4=24$$ Distributive Property

$$5t+4=24$$ Combine like terms.

$$5t=20$$ Solve for t.

$$t=4$$

We will proceed in a similar way with equations that contain rational expressions with variables in the denominators. In addition, we can use the graphing method to anticipate the number of solutions and to estimate what the solutions are.

EXAMPLE 2

Solving an Equation That Contains a Rational Expression

Solve the equation $\frac{15}{x} + 5 = 2$.

Solution

As always, a graph can help us to estimate the number of solutions of the equation and to approximate the solutions. Figure 7.6 shows the graphs of

$$y_1 = \frac{15}{x} + 5 \quad \text{and} \quad y_2 = 2$$

Figure 7.6

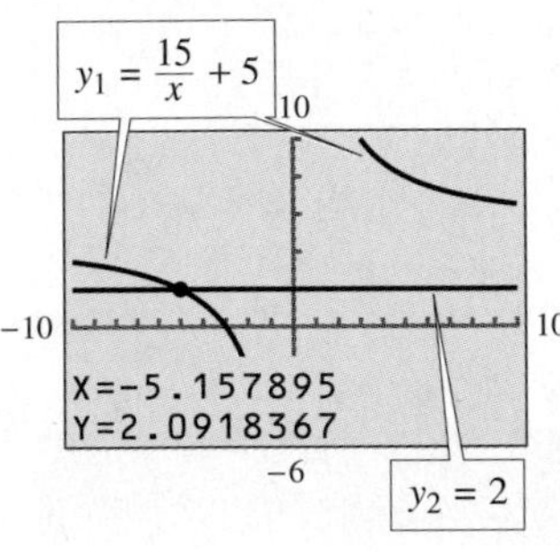

From the graph we estimate that there is one solution, and it appears to be approximately -5.

Now we solve the equation algebraically.

$$\frac{15}{x} + 5 = 2 \qquad \text{Note that 0 is a restricted value.}$$

$$x \cdot \frac{15}{x} + 5 \cdot x = 2 \cdot x \qquad \text{Multiply both sides by the LCD } x.$$

$$15 + 5x = 2x \qquad \text{Simplify each term.}$$

$$15 = -3x \qquad \text{Solve for } x.$$

$$x = -5$$

Using substitution, we can verify that the solution of the equation is -5, as we predicted from the graph.

EXAMPLE 3

Solving an Equation That Contains a Rational Expression

Solve the equation $\frac{2}{x} = x + 1$.

Solution

Figure 7.7 shows the graphs of $y_1 = \frac{2}{x}$ and $y_2 = x + 1$.

Figure 7.7

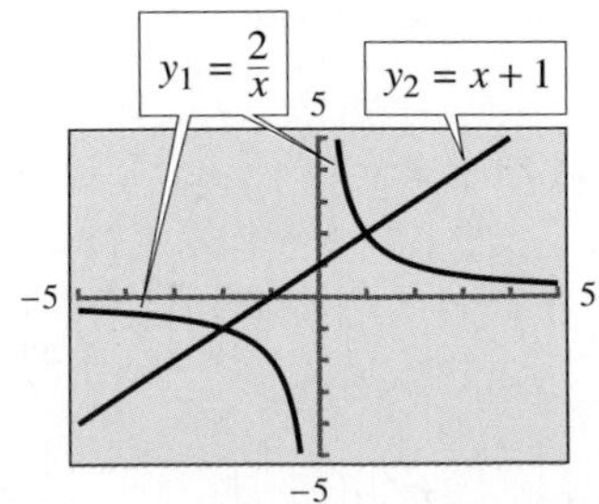

The graphs appear to intersect at two points, $(-2, -1)$ and $(1, 2)$. Thus we estimate that the equation has two solutions, -2 and 1.

The following is an algebraic solution.

$$\frac{2}{x} = x + 1 \qquad \text{Note that 0 is a restricted value.}$$

$$x \cdot \frac{2}{x} = x \cdot (x + 1) \qquad \text{Multiply both sides by the LCD } x.$$

$$2 = x^2 + x \qquad \text{The result is a quadratic equation.}$$

$$0 = x^2 + x - 2 \quad \text{Write the equation in standard form.}$$
$$0 = (x + 2)(x - 1) \quad \text{Factor the trinomial.}$$
$$x + 2 = 0 \quad \text{or} \quad x - 1 = 0 \quad \text{Zero Factor Property}$$
$$x = -2 \quad \text{or} \quad x = 1 \quad \text{Solve each case for } x.$$

The solutions are -2 and 1, which we can verify by substitution. This confirms our estimates from the graph.

Sometimes the graphing method works best if the equation is written with 0 on one side. Then we can estimate the solution(s) of the equation simply by estimating the x-intercept(s) of the graph.

EXPLORING THE CONCEPT

Estimating Solutions of Equations

Consider the equation $\frac{x^2 + 3x - 1}{x - 2} = 2x + \frac{11 - x}{x - 2}$. We write the equation as $\frac{x^2 + 3x - 1}{x - 2} - 2x - \frac{11 - x}{x - 2} = 0$, and we graph the expression on the left side. In Fig. 7.8, the cursor is on the x-intercept (6, 0), and we conclude that 6 is the only solution of the equation.

Figure 7.8

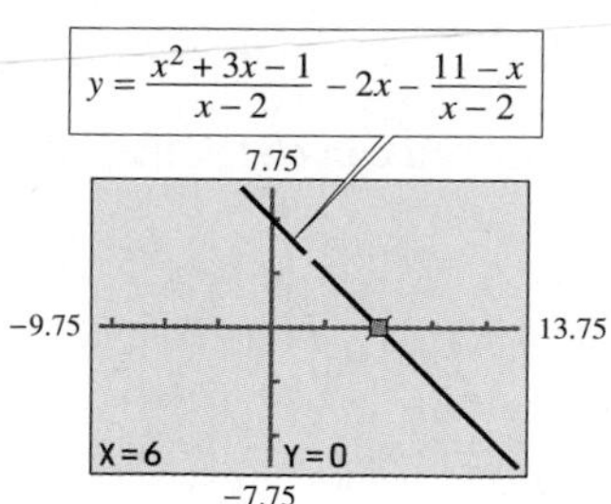

Observe the outcome of solving the same equation algebraically.

$$\frac{x^2 + 3x - 1}{x - 2} = 2x + \frac{11 - x}{x - 2} \quad \text{Multiply both sides by the LCD, } x - 2.$$
$$(x - 2) \cdot \frac{x^2 + 3x - 1}{x - 2} = 2x(x - 2) + (x - 2) \cdot \frac{11 - x}{x - 2}$$
$$x^2 + 3x - 1 = 2x^2 - 4x + 11 - x$$
$$x^2 - 8x + 12 = 0 \quad \text{Write the quadratic equation in standard form.}$$
$$(x - 2)(x - 6) = 0 \quad \text{Factor.}$$
$$x - 2 = 0 \quad \text{or} \quad x - 6 = 0 \quad \text{Zero Factor Property}$$
$$x = 2 \quad \text{or} \quad x = 6$$

When we solve algebraically, the equation appears to have *two* solutions, 2 and 6, whereas the graphing method indicates only *one* solution, 6. However, this discrepancy is easily resolved when we realize that 2 is a restricted value and cannot be a solution. In fact, if you look carefully at the graph in Fig. 7.8, you will see a "hole" in the line where $x = 2$.

An apparent solution that is a restricted value is called an **extraneous solution.**

Note: Using a combined graphing and algebraic approach will help you to identify extraneous solutions. If you do not use a graph, at least make sure that your apparent solution is not a restricted value.

To clear fractions, it may be necessary to factor the denominators in order to determine the LCD.

EXAMPLE 4

Solving Equations That Contain Rational Expressions

Solve the following equations.

(a) $\frac{4}{x+1} - \frac{5}{x} = \frac{20}{x^2+x}$ (b) $\frac{2}{x+3} - \frac{1}{x} = \frac{-6}{x^2+3x}$

Solution

LEARNING TIP

Listing restricted values before solving the equation may help identify an extraneous solution.

(a) $\frac{4}{x+1} - \frac{5}{x} = \frac{20}{x^2+x}$ Begin by factoring the denominators.

$\frac{4}{x+1} - \frac{5}{x} = \frac{20}{x(x+1)}, \quad x \neq 0, -1$ Note that 0 and -1 are restricted values.

Now we see that the LCD is $x(x+1)$. Multiply both sides by the LCD.

$$x(x+1) \cdot \frac{4}{x+1} - x(x+1) \cdot \frac{5}{x} = x(x+1) \cdot \frac{20}{x(x+1)}$$

$4x - 5(x+1) = 20$ Simplify each term.

$4x - 5x - 5 = 20$ Distributive Property

$-x - 5 = 20$ Combine like terms.

$-x = 25$ Solve for x.

$x = -25$

Because it is not a restricted value, -25 is the solution. You can check the solution by hand or with your calculator.

Think About It

Show that solving the equation $\frac{kx}{x-1} = \frac{x+1}{x-1}$ by cross-multiplying always leads to an extraneous solution of 1, whereas solving the equation by equating the numerators leads to an extraneous solution of 2 only if $k = 2$.

(b) $\frac{2}{x+3} - \frac{1}{x} = \frac{-6}{x^2+3x}$ Begin by factoring the denominators.

$\frac{2}{x+3} - \frac{1}{x} = \frac{-6}{x(x+3)}, \quad x \neq 0, -3$ Note that 0 and -3 are restricted values.

The LCD is $x(x+3)$. Multiply both sides of the equation by the LCD to clear the fractions.

$$x(x+3) \cdot \frac{2}{x+3} - x(x+3) \cdot \frac{1}{x} = x(x+3) \cdot \frac{-6}{x(x+3)}$$

$2x - (x+3) = -6$ Simplify each term.

$2x - x - 3 = -6$ Distributive Property

$x - 3 = -6$ Combine like terms.

$x = -3$ Solve for x.

Our solution process leads us to -3 as the apparent solution, but -3 *is a restricted value*. Therefore, -3 cannot be a solution, and the equation has no solution.

EXAMPLE 5

Solving an Equation That Contains Rational Expressions

Solve the equation $\frac{1}{x} - \frac{8}{2x - x^2} = \frac{x+2}{x-2}$.

Solution

$$\frac{1}{x} - \frac{(-1)8}{(-1)(2x - x^2)} = \frac{x+2}{x-2} \qquad -1(a - b) = b - a$$

$$\frac{1}{x} - \frac{-8}{x^2 - 2x} = \frac{x+2}{x-2} \qquad \text{Now the denominators are in descending order.}$$

$$\frac{1}{x} - \frac{-8}{x(x-2)} = \frac{x+2}{x-2}, \quad x \neq 0, 2 \qquad \text{Factor the denominators.}$$

Note that the LCD is $x(x - 2)$ and the restricted values are 0 and 2. Multiply both sides of the equation by the LCD.

$$\frac{1x(x-2)}{x} - \frac{-8x(x-2)}{x(x-2)} = \frac{x(x-2)(x+2)}{x-2}$$

$$(x - 2) - (-8) = x(x + 2) \qquad \text{Simplify each term.}$$

$$x - 2 + 8 = x^2 + 2x \qquad \text{Distributive Property}$$

$$x + 6 = x^2 + 2x \qquad \text{The result is a quadratic equation.}$$

$$0 = x^2 + x - 6 \qquad \text{Write the equation in standard form.}$$

$$0 = (x + 3)(x - 2) \qquad \text{Factor the trinomial.}$$

$$x + 3 = 0 \quad \text{or} \quad x - 2 = 0 \qquad \text{Zero Factor Property}$$

$$x = -3 \quad \text{or} \quad x = 2 \qquad \text{Solve each case for } x.$$

The apparent solutions are -3 and 2. However, because 2 is a restricted value, it is an extraneous solution. Therefore, the only solution is -3.

Solving Formulas

When solving a formula for a specified variable, we regard the specified variable as the only variable and treat all other variables as if they were constants.

Sometimes, formulas contain rational expressions. The method for solving such formulas for specified variables is the same as the method for solving equations containing rational expressions.

EXAMPLE 6

Solving a Formula with a Rational Expression

In electronics, a *resistor* is any device that offers resistance to the flow of electric current. When a light bulb is on, it acts as a resistor. Suppose two light bulbs are placed in a parallel circuit. (See Fig. 7.9.) Let r_1 and r_2 represent their resistances.

Figure 7.9

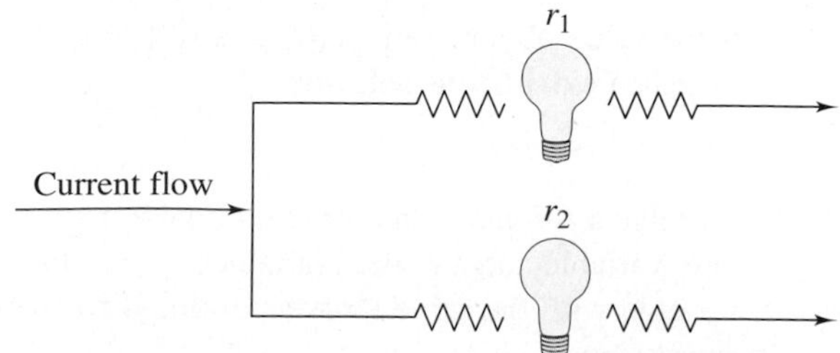

The formula for the total resistance R in this parallel circuit is

$$\frac{1}{R} = \frac{1}{r_1} + \frac{1}{r_2}$$

Solve this formula for r_1.

Solution

$$\frac{1}{R} = \frac{1}{r_1} + \frac{1}{r_2}$$ The LCD is Rr_1r_2.

$$Rr_1r_2 \cdot \frac{1}{R} = Rr_1r_2 \cdot \frac{1}{r_1} + Rr_1r_2 \cdot \frac{1}{r_2}$$ Multiply both sides by the LCD.

$$r_1r_2 = Rr_2 + Rr_1$$ Simplify each term.

$$r_1r_2 - Rr_1 = Rr_2$$ Place the terms containing r_1 on the same side.

$$(r_2 - R)r_1 = Rr_2$$ Factor out the common factor r_1.

$$r_1 = \frac{Rr_2}{r_2 - R}$$ Divide both sides by $r_2 - R$.

7.5 QUICK REFERENCE

Graphing and Algebraic Methods

- By using graphs, we can estimate the number of solutions of an equation with rational expressions, and we can approximate the solutions.
- If a replacement for a variable in a rational expression causes the denominator of the expression to be zero, the replacement is a *restricted value*.
- To solve an equation containing rational expressions algebraically, follow these steps:
  1. Factor each denominator completely and note the restricted values, if any.
  2. Determine the LCD for the fractions.
  3. Clear the fractions by multiplying each term of both sides of the equation by the LCD.
  4. Simplify each side of the equation.
  5. Use previously discussed techniques to solve the resulting equation.
  6. Verify the solution in the original equation.
- If the value obtained in step 5 is a restricted value, it cannot be a solution, and it is called an **extraneous solution.**

Solving Formulas

- To solve a *formula* containing rational expressions, treat the specified variable as the only variable and all other variables as constants. Solve the formula for the specified variable with the same steps as outlined for solving an *equation* containing rational expressions.

7.5 EXERCISES

Concepts and Skills

1. Why are -3 and 0 not permissible replacements for x in the following equation? What do we call -3 and 0 in this case?

$$\frac{1}{x+3} - \frac{1}{x} = 5$$

2. Suppose that before solving the following equation algebraically, you produce the graph of the left side.

$$\frac{x}{3} + \frac{3}{x} = 0$$

(a) Explain how this would allow you to write the solution set immediately.

(b) If you begin the algebraic method by clearing the fractions, how would the resulting equation confirm what the graph suggests?

In Exercises 3–6, state the restricted value(s) for the given equation.

3. $\dfrac{x}{x+3} - \dfrac{2}{x-4} = \dfrac{1}{x}$

4. $\dfrac{3}{a^2} + \dfrac{a}{a-2} = 4$

5. $\dfrac{y}{y^2-9} + \dfrac{7}{y} = \dfrac{3y+2}{y+3}$

6. $\dfrac{2x+1}{x^2+2x-15} - \dfrac{x-3}{x^2-25} = \dfrac{x+5}{2}$

7. To solve $\dfrac{x^2-4}{x-2} = 12$, we clear the fractions and obtain $x^2 - 12x + 20 = 0$, which has the solutions 2 and 10. If we graph the two sides of the original equation, will the graphs intersect at two points? Why?

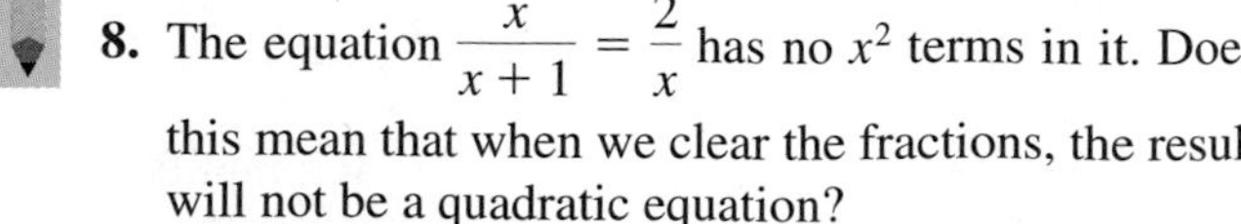

8. The equation $\dfrac{x}{x+1} = \dfrac{2}{x}$ has no x^2 terms in it. Does this mean that when we clear the fractions, the result will not be a quadratic equation?

In Exercises 9–14, estimate the solution(s) of the equation by graphing and verify the solution(s) by substitution.

9. $\dfrac{3}{x} = x + 2$

10. $\dfrac{5}{x+4} = x$

11. $\dfrac{x+3}{x+2} = x + 3$

12. $\dfrac{x-3}{x+5} = x$

13. $\dfrac{12}{x} = x^2 + 3x - 4$

14. $\dfrac{16}{x} = 4 + 4x - x^2$

In Exercises 15–26, solve the equation.

15. $\dfrac{1}{8} = \dfrac{1}{2t} + \dfrac{2}{t}$

16. $\dfrac{5}{4} + \dfrac{4}{3} = \dfrac{7}{x}$

17. $\dfrac{5}{x-2} = 3 - \dfrac{1}{x-2}$

18. $4 + \dfrac{6}{5-x} = \dfrac{2x}{5-x}$

19. $\dfrac{7}{12} + \dfrac{1}{2y-10} = \dfrac{5}{3y-15}$

20. $\dfrac{4}{3x+2} = 1 - \dfrac{2}{9x+6}$

21. $\dfrac{5}{x-2} = \dfrac{1}{x+2} + \dfrac{6}{x^2-4}$

22. $\dfrac{2}{t-5} + \dfrac{3}{t+5} = \dfrac{10}{t^2-25}$

23. $\dfrac{t}{t+4} - \dfrac{2}{t-3} = 1$

24. $\dfrac{1}{x} + \dfrac{1}{x-1} = \dfrac{2}{x+2}$

25. $\dfrac{3}{x+3} = \dfrac{2}{x-4} - \dfrac{10}{x^2-x-12}$

26. $\dfrac{3}{x+1} - \dfrac{4}{2x-1} = \dfrac{5}{2x^2+x-1}$

In Exercises 27–34, solve the equation.

27. $2x - \dfrac{15}{x} = 7$

28. $12 - \dfrac{72}{x^2} = \dfrac{5}{x}$

29. $\dfrac{x}{4} - \dfrac{5}{x} = \dfrac{1}{4}$

30. $\dfrac{x}{2} - \dfrac{2}{x} = 0$

31. $\dfrac{t}{t-2} + 1 = \dfrac{8}{t-1}$

32. $\dfrac{8}{t} + \dfrac{3}{t-1} = 3$

33. $\dfrac{3}{x+2} + 1 = \dfrac{6}{4-x^2}$

34. $\dfrac{x+2}{x-12} = \dfrac{1}{x} + \dfrac{14}{x^2-12x}$

35. Neither of the following equations has a solution, but the reason there is no solution is different in each case. Explain why each equation has no solution.

(i) $\frac{x-4}{x-9} = \frac{5}{x-9}$ (ii) $\frac{4}{x+1} = \frac{4}{x-1}$

36. Verify that multiplying both sides of the following equation by the LCD $x + 1$ results in the equation $1 = 1$.

$$\frac{x}{x+1} - \frac{x-1}{x+1} = \frac{1}{x+1}$$

Because $1 = 1$ is true, should we conclude that all real numbers are solutions of the original equation? Why?

In Exercises 37–60, solve the equation.

37. $\frac{1}{x+1} + \frac{2}{3x+3} = \frac{1}{3}$

38. $\frac{4}{x-2} + \frac{3}{x} = \frac{15}{x^2-2x}$

39. $\frac{x+2}{x+3} = \frac{x-1}{x+1}$

40. $\frac{-2}{3x+2} = \frac{3}{1-2x}$

41. $\frac{5}{x+5} = 4 - \frac{x}{x+5}$

42. $\frac{4}{5x} = \frac{1}{3x}$

43. $\frac{10}{(2x-1)^2} = 4 + \frac{3}{2x-1}$

44. $\frac{1}{x+3} + \frac{21}{(x+3)^2} = 2$

45. $\frac{2}{x^2+x-2} + \frac{3x}{x^2+5x+6} = \frac{5x}{x^2+2x-3}$

46. $\frac{4}{2x^2-7x-15} - \frac{2}{2x^2+13x+15} = \frac{-2}{x^2-25}$

47. $2x + 15 = \frac{8}{x}$

48. $x = \frac{4-11x}{3x}$

49. $\frac{x}{x-3} + \frac{2}{x} = \frac{3}{x-3}$

50. $\frac{x+8}{x+2} + \frac{12}{x^2+2x} = \frac{2}{x}$

51. $\frac{8r}{r-4} - \frac{7}{r^2-16} = 8$

52. $\frac{8}{q-3} + \frac{9}{q^2-11q+24} = \frac{2}{q-8}$

53. $\frac{5}{x-7} + \frac{2}{x+5} = \frac{1}{x^2-2x-35}$

54. $\frac{3}{x+3} = \frac{4}{x+1} - \frac{5}{x^2+4x+3}$

55. $\frac{x+4}{6x^2+5x-6} + \frac{x}{2x+3} = \frac{x}{3x-2}$

56. $\frac{2-6x}{x^2-x-6} = \frac{x-1}{x+2} - \frac{x+1}{3-x}$

57. $\frac{2}{x+1} - \frac{1}{x-3} = \frac{-8}{x^2-2x-3}$

58. $\frac{6}{z-5} + \frac{6}{z^2-11z+30} = \frac{2}{z-6}$

59. $\frac{x}{x+2} + \frac{2}{x+3} + \frac{2}{x^2+5x+6} = 0$

60. $\frac{1}{9x^2+3x-2} = \frac{x}{3x-1} - \frac{1}{9x+6}$

61. To solve the equation

$$\frac{x}{x-1} + \frac{x+3}{x+2} = 3$$

and to perform the addition

$$\frac{x}{x-1} + \frac{x+3}{x+2} + 3$$

we begin in the same way: Determine the LCD. However, what we do with the LCD is completely different for the two problems. Without actually doing the problems, explain what to do with the LCD in each case.

62. In the following problems, identify the one for which an LCD is not needed and explain why.

(i) $\frac{2}{x} - \frac{x^2}{7-x}$ (ii) $\frac{2}{x} = \frac{x^2}{7-x}$

(iii) $\frac{2}{x} \div \frac{x^2}{7-x}$ (iv) $\frac{2}{x} + \frac{x^2}{7-x}$

In Exercises 63–74, solve the equation or simplify the expression.

63. $\frac{3}{x-8} + \frac{x}{x-3}$

64. $\frac{3}{x+4} + \frac{6}{x} + \frac{12}{x^2+4x}$

65. $4 + \frac{19}{2t-3} = \frac{t+2}{3-2t}$

66. $\dfrac{2x+1}{x-2} = \dfrac{2x-3}{x-3}$

67. $\dfrac{y^2-49}{8y+56} \cdot \dfrac{y^2-7y}{(y-7)^2}$

68. $\dfrac{t^2-4}{t^2-25} \cdot \dfrac{3t-15}{6t^2-12t}$

69. $1 - 2t - \dfrac{2t}{t+3}$

70. $\dfrac{3}{6(t-5)} - \dfrac{2-t}{9(t-5)}$

71. $\dfrac{x+5}{x-1} = \dfrac{8}{x-3} - \dfrac{7}{x^2-4x+3}$

72. $9 + \dfrac{3}{x+1} = \dfrac{-4}{x^2+x}$

73. $\dfrac{x^2+x-6}{x^2+9x+18} \div \dfrac{x^2-4}{4x+24}$

74. $\dfrac{3x^3+6x^2}{2x^2+x-6} \div (6x^2-15x)$

75. The current I flowing in an electrical circuit is given by $I = E/R$, where E is the voltage and R is the resistance. Solve this formula for R.

76. The electrostatic force F between two charged bodies is given by $F = \dfrac{q_1q_2}{kr^2}$, where q_1 and q_2 are the electric charges on the two bodies, r is the distance between the two bodies, and k is a constant. Solve this formula for q_1.

77. In chemistry, Charles's law is $\dfrac{P_1V_1}{T_1} = \dfrac{P_2V_2}{T_2}$, where P is pressure, T is absolute temperature, and V is volume. The subscript 1 might be used for initial conditions, and the subscript 2 might be used for final conditions. Solve this formula for T_1.

78. When a metal bar is heated, its length increases according to the formula $L_1 = L_0(kt + 1)$, where L_0 is the length of the bar at 0°C, L_1 is the length of the bar at t°C, and k is the coefficient of linear expansion. Solve this formula for k.

79. An invested amount P can be calculated with the formula $P = \dfrac{I}{rt}$, where I is interest earned, r is the simple interest rate, and t is time. Solve this formula for r.

80. In the accompanying figure, triangle ABC is a right triangle with altitude $\overline{AP}$. In geometry, it can be shown that $\dfrac{a}{h} = \dfrac{h}{b}$. Solve this equation for b.

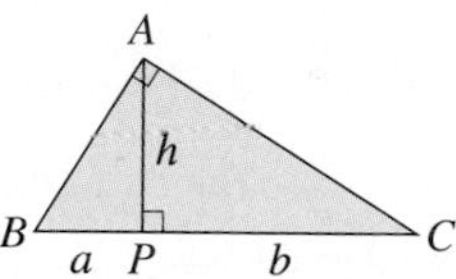

In Exercises 81–86, solve the formula for the specified variable.

81. $F = \dfrac{m_1m_2}{\mu r^2}$; m_2

82. $p = \dfrac{2T}{r}$; r

83. $S = \dfrac{a(1-r^n)}{1-r}$; a

84. $\dfrac{1}{f} = (k-1)\left(\dfrac{1}{p} + \dfrac{1}{q}\right)$; f

85. $\dfrac{x}{a} + \dfrac{y}{b} = 1$; a

86. $\dfrac{x}{a} + \dfrac{y}{b} + \dfrac{z}{c} = 1$; c

In Exercises 87–90, solve for y.

87. $\dfrac{y-4}{x+3} = -2$

88. $\dfrac{y+2}{x-5} = \dfrac{2}{3}$

89. $\dfrac{y}{x+4} = -\dfrac{4}{5}$

90. $\dfrac{y-1}{x} = -\dfrac{3}{4}$

Group Project: DUI Arrests

Statistics show that people in the 25–34 age group account for about half the arrests made for driving under the influence (DUI) of alcohol. (Source: U.S. Bureau of Justice Statistics.)

The number $A(x)$ of arrests per 100,000 drivers can be modeled by the following function, where x represents a driver's age.

$$A(x) = \frac{27{,}720(x-14)}{x^2+9} - 5x$$

91. Using window settings of x from 0 to 70 and y from 0 to 500, produce the graph of the model function. The graph is said to be *skewed* to the right. How can this property of the graph be interpreted?

92. Use the graphing method to estimate the solution of the equation $A(x) = 356$.

93. Interpret the solution in Exercise 92.

94. Some states allow the sale of cold beer at gas stations and convenience stores. To what extent, if any, might this policy contribute to drunk driving?

Challenge

In Exercises 95–98, solve the equation.

95. $\dfrac{\dfrac{x+3}{x-2}}{\dfrac{2}{x-2}} = 5$

96. $\dfrac{1+\dfrac{6}{x}}{1+\dfrac{1}{x}} = 2x$

97. $\dfrac{1}{18} + \dfrac{1}{9x} = \dfrac{1}{2x^2} + \dfrac{1}{x^3}$

98. $1 + \dfrac{2}{x} = \dfrac{1}{x^2} + \dfrac{2}{x^3}$

In Exercises 99 and 100, solve for y.

99. $\dfrac{y - y_1}{x - x_1} = m$

100. $\dfrac{y-b}{x} = m$

In Exercises 101 and 102, use substitution to solve the given equation.

101. $\left(\dfrac{1}{x+1}\right)^2 + 3\left(\dfrac{1}{x+1}\right) - 4 = 0$

102. $\left(\dfrac{x-1}{x+6}\right)^2 + 4\left(\dfrac{x-1}{x+6}\right) - 12 = 0$

7.6 APPLICATIONS

Ratio and Proportion • Percentages • Inverse Variation • Work Rate • Motion

In Chapter 3 we developed a general approach to application problem solving. This approach is appropriate regardless of the type of equation used to model the problem. In this section we present a variety of examples of applications that lead to equations with rational expressions.

Ratio and Proportion

The value of x can be compared with the value of y by way of a ratio, which is written as a quotient $\dfrac{x}{y}$ or in the form $x{:}y$.

A **proportion** is an equation stating that two ratios are equal. If the ratio of x to y is equal to the ratio of a to b, we can write $\dfrac{x}{y} = \dfrac{a}{b}$.

One method for solving a proportion, or any other equation of the form $\dfrac{A}{B} = \dfrac{C}{D}$, is to **cross-multiply.**

If $\dfrac{A}{B} = \dfrac{C}{D}$, then $AD = BC$.

Multiplying both sides of the equation by the LCD produces the same result, but clearing fractions by cross-multiplying is usually a more efficient method. We will illustrate this method in Example 1.

In plane geometry, **similar triangles** are triangles whose corresponding angles have the same measure. In Example 1 we use the theorem from plane geometry that states that the corresponding sides of similar triangles are proportional.

EXAMPLE 1

Similar Triangles

To construct a scale model for wind tunnel testing, an engineer makes a plastic triangle DEF similar to the full-size component ABC. Determine the lengths (in feet) of sides $\overline{AB}$ and $\overline{DE}$.

LEARNING TIP

Sketching a figure allows you to translate the given information into a visual model and to assign variables and expressions to unknown quantities.

Figure 7.10

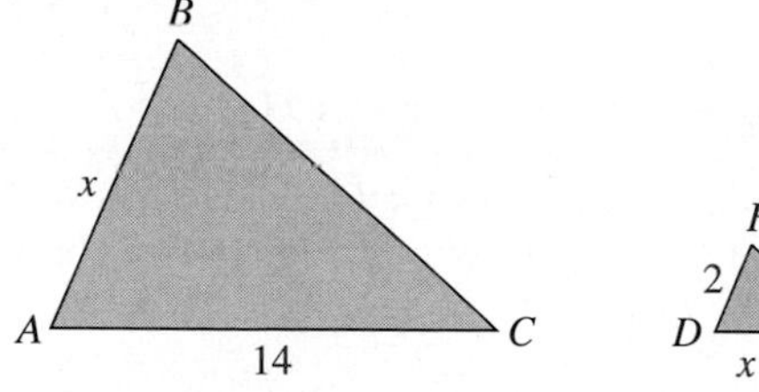

Solution

Because the triangles are similar, the lengths of the corresponding sides are proportional. In Fig. 7.10, $\overline{AB}$ corresponds to $\overline{DF}$, and $\overline{AC}$ corresponds to $\overline{DE}$.

$$\frac{x}{2} = \frac{14}{x-3} \qquad \frac{AB}{DF} = \frac{AC}{DE}$$

$$x(x-3) = 28 \qquad \text{Cross-multiply.}$$

$$x^2 - 3x = 28 \qquad \text{Distributive Property}$$

$$x^2 - 3x - 28 = 0 \qquad \text{Write the quadratic equation in standard form.}$$

$$(x-7)(x+4) = 0 \qquad \text{Factor the trinomial.}$$

$$x - 7 = 0 \quad \text{or} \quad x + 4 = 0 \qquad \text{Zero Factor Property}$$

$$x = 7 \quad \text{or} \quad x = -4 \qquad \text{Solve each case for } x.$$

Because length is not negative, the only solution is 7. Thus the side $\overline{AB}$ of the full-size component is 7 feet and the side $\overline{DE}$ of the model component is 4 feet.

EXAMPLE 2

Win-to-Loss Ratio

At the All-Star break, the ratio of wins to losses for the Pittsburgh Pirates was 3 to 2. If the same ratio holds for the entire 162-game season, how many games will the Pirates win?

Solution

Let x = the number of games that will be won. Then $162 - x$ = the number of games that will be lost.

$$\frac{3}{2} = \frac{x}{162-x} \qquad \text{The ratio of wins to losses is 3 to 2.}$$

$$3(162 - x) = 2x \qquad \text{Cross-multiply.}$$

$$486 - 3x = 2x \qquad \text{Distributive Property}$$

$$486 = 5x \qquad \text{Add } 3x \text{ to both sides.}$$

$$97.2 = x \qquad \text{Solve for } x.$$

If the 3 to 2 win–loss ratio holds, the Pirates' season record will be about 97 wins and 65 losses, probably good enough to win the division championship.

Percentages

Because percentages are actually ratios (multiplied by 100), applications involving percentages sometimes lead to equations with rational expressions.

EXAMPLE 3

Basketball Free Throw Percentage

A basketball player has made 32 free throws in 45 attempts for a current success rate of 71%. How many consecutive free throws must the player make to raise the average to 75%?

Solution

Let $x =$ the required number of consecutive, successful free throws. Then $32 + x =$ the total number of free throws made, and $45 + x =$ the total number of free throws attempted.

The success rate is the ratio of the number of free throws made to the number of free throws attempted.

$$\frac{\text{Free throws made}}{\text{Free throws attempted}} = \frac{32 + x}{45 + x} = 0.75 \qquad \text{The goal is a 75\% success rate.}$$

$$32 + x = 0.75(45 + x) \qquad \text{Cross-multiply.}$$

$$32 + 1.00x = 33.75 + 0.75x \qquad \text{Distributive Property}$$

$$0.25x = 1.75 \qquad \text{Solve for } x.$$

$$x = 7$$

Seven consecutive, successful free throws will raise the player's average to 75%.

Inverse Variation

In Chapter 4 we studied direct variation. Another type of variation is **inverse variation.**

Definition of Inverse Variation

A quantity y **varies inversely** with x if there is a constant k so that $y = \frac{k}{x}$. We sometimes say that y is **inversely proportional** to x. The constant k is called the **constant of variation.**

EXAMPLE 4

Inverse Variation

If the temperature is constant, the volume of a gas varies inversely with the pressure. If the pressure of 2 liters of a gas is 12 newtons per square centimeter, what is the pressure when the volume is 1.6 liters?

Solution

Let $P =$ the pressure and $V =$ the volume.

$$V = \frac{k}{P}$$ Volume varies inversely with pressure.

$$2 = \frac{k}{12}$$ Substitute the intitial conditions $V = 2$ and $P = 12$.

$$k = 24$$ Solve for k.

$$V - \frac{24}{P}$$ Replace k with 24.

$$1.6 = \frac{24}{P}$$ Substitute 1.6 for V.

$$1.6P = 24$$ Cross-multiply.

$$P = 15$$ Solve for P.

The pressure is 15 newtons per square centimeter.

Work Rate

Work rate problems nearly always involve two or more people or machines performing tasks at different rates.

If a machine can perform a task in 5 hours, then it completes $\frac{1}{5}$ of the task each hour. This is the machine's *work rate*. The combined work rate of two or more machines is the sum of their individual work rates.

EXAMPLE 5

Work Rate

At a juice-processing plant, one pipe can fill a storage tank with cranberry juice in 2 hours. Another pipe can fill the same tank with apple juice in 3 hours. If both pipes are open, how long does filling the tank with cran-apple juice take?

Solution

Let $x =$ the number of hours to fill the tank with both pipes open.

	Hours to Fill Tank	Work Rate
Cranberry juice	2	$\frac{1}{2}$
Apple juice	3	$\frac{1}{3}$
Combined	x	$\frac{1}{x}$

The combined work rate is the sum of the individual work rates.

$$\frac{1}{2} + \frac{1}{3} = \frac{1}{x}$$

$$6x \cdot \frac{1}{2} + 6x \cdot \frac{1}{3} = 6x \cdot \frac{1}{x}$$ Multiply by the LCD $6x$.

Think About It

Suppose one person requires k times as long as another person to perform a task. How are the work rates related?

$$3x + 2x = 6 \quad \text{Simplify each term.}$$
$$5x = 6 \quad \text{Combine like terms.}$$
$$x = \frac{6}{5} \quad \text{or} \quad 1.2$$

The time needed to fill the tank is 1.2 hours.

Motion

In our previous work with motion problems we used the formula $d = rt$, where d, r, and t represent distance, rate, and time, respectively. If the focus of a motion problem is on time, then it is convenient to represent time in terms of distance and rate. We can do this by solving the formula $d = rt$ for t to obtain $t = \frac{d}{r}$.

EXAMPLE 6

Distance, Rate, and Time

From Front Royal, Virginia, a couple drove 210 miles on the scenic route to Roanoke, Virginia. From Roanoke they traveled Interstate 81 for 120 miles to Wytheville, Virginia. The average speed on the interstate was 25 mph faster than along the scenic route. The entire trip took 8 hours. How long did they drive on the scenic route and how long on the interstate? What was their average speed on each?

Solution

Let $s =$ the average speed on the scenic route. Then $s + 25$ is the average speed on the interstate.

	d	r	$t = \frac{d}{r}$
Scenic	210	s	$\frac{210}{s}$
Interstate	120	$s + 25$	$\frac{120}{s + 25}$

$$\frac{210}{s} + \frac{120}{s + 25} = 8 \quad \text{The total time is 8 hours.}$$

$$s(s + 25) \cdot \frac{210}{s} + s(s + 25) \cdot \frac{120}{s + 25} = 8 \cdot s(s + 25) \quad \text{Multiply by the LCD } s(s + 25).$$

$$210(s + 25) + 120s = 8s(s + 25) \quad \text{Simplify each term.}$$

$$210s + 5250 + 120s = 8s^2 + 200s \quad \text{Distributive Property}$$

$$330s + 5250 = 8s^2 + 200s \quad \text{Combine like terms.}$$

$$8s^2 - 130s - 5250 = 0 \quad \text{Write the quadratic equation in standard form.}$$

$$2(4s^2 - 65s - 2625) = 0 \quad \text{GCF is 2.}$$

$$2(4s + 75)(s - 35) = 0 \quad \text{Factor the trinomial.}$$

$$
\begin{aligned}
4s + 75 &= 0 \quad \text{or} \quad s - 35 = 0 && \text{Zero Factor Property}\\
4s &= -75 \quad \text{or} \quad s - 35 = 0 && \text{Solve each case for } s.\\
s &= -18.75 \quad \text{or} \quad s = 35
\end{aligned}
$$

Because speed is not negative, the only solution is $s = 35$. The average speed on the scenic route was 35 mph, and the average speed on the interstate was 60 mph.

The driving time on the scenic route was $\frac{210}{35}$ or 6 hours, and the driving time on the interstate was $\frac{120}{60}$ or 2 hours.

EXAMPLE 7

Distance, Rate, and Time

A canoeist can paddle 2 miles up the Olentangy River in the same time she can paddle 6 miles downstream. The speed of the current is 2 mph. What is her speed in still water?

Solution

Let r = rate of canoe in still water.

	d	r	$t = \dfrac{d}{r}$
Upstream	2	$r - 2$	$\dfrac{2}{r-2}$
Downstream	6	$r + 2$	$\dfrac{6}{r+2}$

$$
\begin{aligned}
\frac{2}{r-2} &= \frac{6}{r+2} && \text{The upstream and downstream times are equal.}\\
2(r+2) &= 6(r-2) && \text{Cross-multiply.}\\
2r + 4 &= 6r - 12 && \text{Distributive Property}\\
-4r &= -16 && \text{Solve for } r.\\
r &= 4
\end{aligned}
$$

The canoeist paddles at a rate of 4 mph in still water.

7.6 EXERCISES

Real-Life Applications

Ratio and Proportion

1. Consider the following two problems.

(i) Find the sum of $\frac{1}{3}$ and the reciprocal of a number.

(ii) If the sum of $\frac{1}{3}$ and the reciprocal of a number is 2, what is the number?

For both problems, the LCD is $3x$. Compare the two different ways in which the LCD is used to solve the problems.

2. An arithmetic rule states that if $\dfrac{a}{b} = \dfrac{c}{d}$, then the **cross-products** ad and bc are equal. Explain why this is so.

3. The center of a stained glass window in the shape of an isosceles triangle contains a blue glass piece that is also an isosceles triangle and similar to the window. If the ratio of a side of the window to the base is 3 to 2 and the side of the blue piece is 2 inches longer than the base of the blue piece, what is the length of the base of the blue triangle?

4. A large, modern sculpture includes a bronze plate component in the form of an obtuse triangle. At a certain time of the day, the sun casts a shadow of this component such that the plate and the shadow are similar triangles. (See figure.) Determine the lengths of the sides $\overline{AC}$ and $\overline{DE}$.

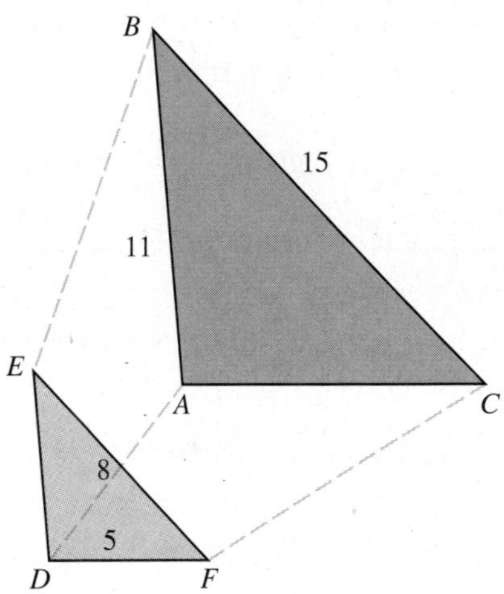

5. At a certain small college, the ratio of traditional age students to nontraditional age students is 4 to 3. If the college enrollment is 1260, how many nontraditional age students attend the college?

6. Following a bombing, 7 more state than federal agents were assigned to investigate the attack. If the ratio of federal agents to the total number of agents was 2 to 5, how many state agents were assigned to the case?

Percentages

7. A softball player has a batting average of 0.252 with 112 hits out of 444 times at bat. If the batter goes on a hitting streak, how many hits in a row would raise the batting average to 0.267?

8. A baseball player has a batting average of 0.295 with 105 hits out of 356 times at bat. The batter then goes into a hitless streak, and the batting average drops to 0.269. How many consecutive at bats without a hit did the batter have?

9. In 1980, 360 male students attended a certain college. By 1990, the number of male students had increased by 300, and the number of female students had increased by 700. If the percentage of male students was the same in 1990 as it was in 1980, what was the total enrollment in 1980?

10. A group of people suffering from a particular ailment agreed to participate in an experimental testing of a new medication. Some of the people received the medication, whereas the others received a placebo (a preparation containing no medicine and given for its psychological effect). Of the group receiving the medication, 35 claimed it helped. Ten people in the group that received the placebo also said they felt much better. For both groups combined, 50% reported no beneficial effect. How many people were tested?

Inverse Variation

11. If y varies inversely as x, and if $y = 9$ when $x = 5$, what is y when $x = 3$?

12. If a varies inversely as c, and if $a = 12$ when $c = 2$, what is a when $c = 10$?

13. If w varies inversely as z^2, and if $z = 2$ when $w = 24$, what is w when $z = 4$?

14. If y varies inversely as x^2, and if $x = 2$ when $y = 150$, what is y when $x = 10$?

15. Suppose that four workers take 9 hours to roof a house. If the time required to do the job varies inversely as the number of people working on it, how long do five workers take to finish a roof of the same size?

16. Suppose that 5 workers take three 8-hour days to landscape a yard completely. If the time required to do the job varies inversely as the number of people working on it, how long would 16 workers take to landscape the same yard?

17. Suppose the price of oil varies inversely with the supply. If an OPEC nation can sell oil for $26.00 per barrel when daily production is 3 million barrels, what will the price of oil be if the daily production is increased to 4 million barrels?

18. The average number of words that can be printed on standard paper varies inversely with the font size. If 320 words can be printed with a 10-point font, how many words can be printed with a 12-point font?

Work Rate

19. Working together, two people can wash their car in 10 minutes. One person, working alone, can wash the car in a half hour. How long does the other person take to do the job working alone?

20. Using a 21-inch mower, a person can mow the front lawn in 45 minutes. Mowing takes only a half hour with a 36-inch mower. If two people work together, one with the smaller mower and the other with the larger mower, how long will they take to mow the lawn?

21. A professional painter can paint the outside trim on a certain house as fast as two apprentices working together. One apprentice can do the painting alone in 15 hours; the other apprentice can finish the job in 10 hours. How long would the painter take to paint the trim without help?

22. Working together, two front-end loaders can fill a railroad car with gravel in 20 minutes. Working alone, the smaller machine would take 9 minutes longer than the larger machine would take working alone. How long would each of the loaders take working alone?

Motion

23. A small plane travels 90 mph faster than a train. The plane travels 525 miles in the same time it takes the train to travel 210 miles. Determine the rate of each.

24. A private plane takes 4 hours less time than a car to make a trip of 495 miles. If the plane's rate is $\frac{9}{5}$ of the car's rate, what is the rate of each?

25. Two cyclists competed in a 5-mile bicycle race. The more experienced cyclist gave the novice cyclist a head start of 0.6 miles and still won the race by 2 minutes. If the experienced cyclist's average speed was 3 mph faster than the novice's average speed, what was the novice's average speed?

26. A pitcher throws a fastball to a catcher standing 66 feet away. At the same instant, the shortstop throws a ball to the first baseman standing 124 feet away. The first baseman catches the throw from the shortstop a half second after the catcher catches the pitch. If the pitcher can pitch 8 feet per second faster than the shortstop can throw, how long does a batter have to react to the pitcher's fastball?

Miscellaneous

27. Consider two data items A and B. If an increase in A causes a corresponding decrease in B, or if a decrease in A causes a corresponding increase in B, we say that A and B are inversely related. Which of the following would you suspect are inversely related?

(a) Humidity (A) and drying time of varnish (B)

(b) Number of amateur skiers (A) and the price of ski equipment (B)

(c) Illiteracy rate (A) and library circulation (B)

(d) Number of auto accidents (A) and insurance premiums (B)

(e) Unemployment (A) and home ownership (B)

28. A soft-drink bottling company uses a vat that can be filled with carbonated water in 1 hour. The vat can be filled with syrup in 9 hours. The vat of soft-drink mixture can be emptied to the bottling apparatus in 3 hours. If the operator opens all three valves and then takes a 90-minute lunch break, describe the operator's state of mind when he returns from lunch and show why.

29. In the figure on the following page, a yardstick is placed perpendicular to the ground, and it casts a shadow 15 inches long. If the nearby oak tree casts a 25-foot shadow, how tall is the oak tree?

Figure for #29

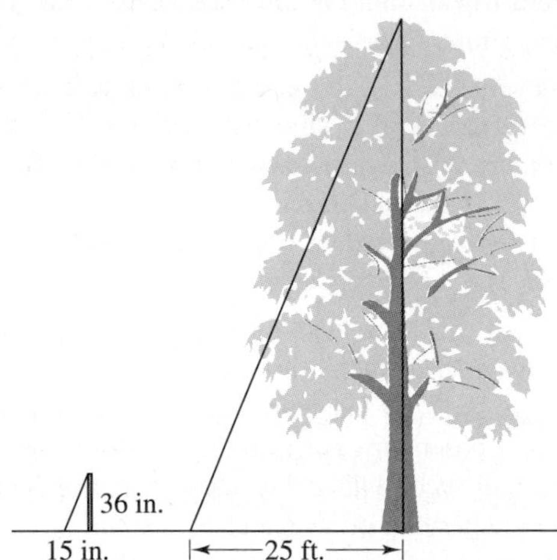

30. To estimate the bass population of a pond, a fish biologist used a net to capture 20 bass. She tagged these fish and released them back into the pond. A few days later, the biologist netted 21 bass and found that 3 of the fish had tags. What was the estimated bass population of the pond?

31. One person takes twice as long to shovel snow from the driveway as another person takes using a snow blower. If the two of them together can clear the driveway in 8 minutes, how long does the person shoveling alone take?

32. Two self-employed house cleaners can wash the windows of a particular client in a half hour. Working alone, one cleaner could wash all the windows in 11 minutes less time than the other would take, but the faster cleaner refuses to do windows. Working alone, how long will the other cleaner take to wash the windows?

33. The number of hairs in an animal's coat varies inversely with the average winter temperature. One year, the average winter temperature was 10° lower than the previous year, and an animal was found to have 1.5 times as many hairs as it had the previous year. What was the average winter temperature during the previous year?

34. A motorist took the interstate from Cincinnati to Columbus, a distance of 80 miles. On the return trip the motorist took some back roads, which reduced the average speed by 15 mph. If the total driving time for the round trip of 160 miles was $3\frac{1}{9}$ hours, what was the average speed on the interstate?

35. A sprinter ran the 100-yard dash for the college track team. It took 1 second longer to run the final 50 yards than it took to run the first 50 yards. If the sprinter's average speed for the total distance was $9\frac{1}{11}$ yards per second, what was the time for the final 50 yards?

36. Two people inherited a total of $76,000. The will specified that the inheritance was to be divided in a 3 to 2 ratio. How much did each receive?

37. When the wind speed is 40 mph, a plane can fly 200 miles against the wind in the same time it can fly 300 miles with the wind. What is the speed of the plane?

38. A 1200-bushel grain bin can be filled in 3 hours and emptied in 2 hours. If grain is added to a full bin at the same time that the bin is being emptied, how long will it be before there is no grain in the bin?

39. An elevated water tank serves both commercial and residential customers. The tank can be filled in 14 hours. Commercial usage alone can empty the tank in 7 hours and residential usage alone can empty the tank in 9 hours. If the intake valve is open at the same time both outtake valves are open, how long will exhausting the water supply take if the tank is initially full?

40. A boat can travel 15 miles upstream and 15 miles back downstream in a total of 4 hours. If the speed of the current is 5 mph, what is the speed of the boat in still water?

41. Two cats can eat a 7-pound bag of cat food in 24 days. One cat can eat a 3.5-pound bag of food in 3 weeks. How long would the other cat take to eat a 3.5-pound bag?

42. A clothing chain has noticed that the number of down parkas sold by any one of their stores is inversely proportional to the average January temperature in the city where the store is located. If the store in Houston, Texas, where the average January temperature is 50°F, sells 500 parkas, how many parkas are sold in Great Falls, Montana, where the average January temperature is 21°F? What would the average January temperature be in a city that sold 4000 parkas?

Modeling with Real Data

43. The accompanying table shows that some progress was made in the recycling of aluminum and glass

from 1970 to 1990. All data are in millions of tons. (Source: Franklin Associates Ltd.)

	Waste Generated		Waste Recovered	
Year	Aluminum	Glass	Aluminum	Glass
1970	0.8	12.7	0.0	0.2
1980	1.8	15.0	0.3	0.8
1990	2.7	13.2	1.0	2.6

The following functions model these data (in millions of tons), where x is the number of years since 1970.

Aluminum generated:
$A_1(x) = 0.095x + 0.80$

Aluminum recovered:
$A_2(x) = 0.050x - 0.06$

Glass generated:
$G_1(x) = -0.020x^2 + 0.43x + 12.7$

Glass recovered:
$G_2(x) = 0.006x^2 + 0.2$

(a) Write a rational expression to model the percentage of aluminum that was recovered.

(b) Write a rational expression to model the percentage of glass that was recovered.

44. (a) Write an equation to determine the year in which the percentages in Exercise 43 are the same.

(b) Use the graphing method to estimate the solution of the equation.

Challenge

45. In Woodville Notch, all registered voters are either Democrats, Republicans, or Independents. The number of Democrats and Republicans are equal. There are 11 Independents, of which 9 are women. In a recent election, five-sixths of the Democrats voted, seven-ninths of the Republicans voted, and all Independents voted. Half the Democrats are women, and two-thirds of the Republicans are women. All the women voted, which means that 75% of those who voted were women. How many registered voters are in Woodville Notch?

46. Pecans worth \$30.00 are mixed with cashews worth \$45.00. There are 3 pounds more of cashews than of pecans. If the unit price of the cashews is 50 cents greater than the unit price of pecans, how many pounds of each kind of nut were used?

47. A builder bought three sizes of lag bolts to build a gazebo. The builder needed 4 pounds more of medium bolts than of small bolts and 3 pounds fewer of large bolts than of small bolts. The sales invoice showed the following:

Lag bolts, small	\$3.20
Lag bolts, medium	8.40
Lag bolts, large	5.50

If the sum of the unit prices for the small and medium bolts is equal to the unit price for the large bolts, how many pounds of each size did the builder buy?

48. The two legs of a right triangle are represented by $\frac{3}{x}$ and $\frac{12}{x+1}$. If the hypotenuse is represented by $\frac{5}{x}$, determine the lengths of the sides of the triangle.

49. For the right triangle ABC in the figure, a geometry theorem states that $(AQ)^2 = BQ \cdot QC$. Use the given representations of lengths to determine AQ.

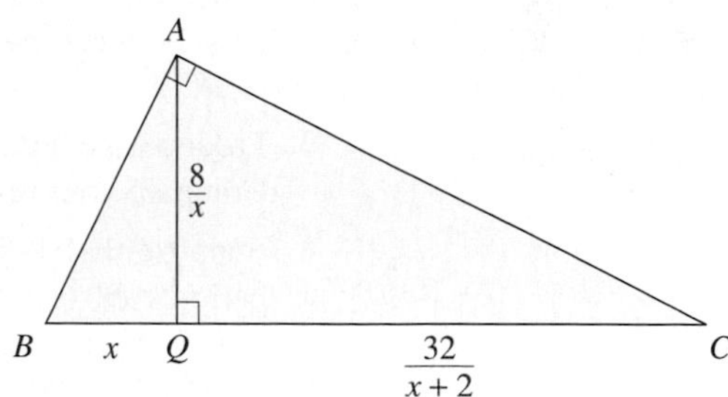

50. In the accompanying figure the square on the left and the rectangle on the right have dimensions as shown. If the area of the rectangle is 3 square inches greater than the area of the square, find the dimensions of each figure.

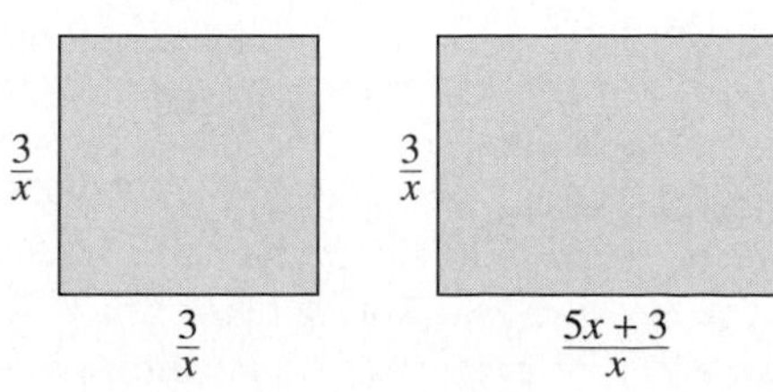

7. CHAPTER PROJECT

Major League Salaries

In Section 7.4 we observed that the average salaries of major league baseball players nearly doubled from 1990 to 1993. By 1996, the average salary was \$1,173,000, but the median salary was \$350,000.

The following table shows the number of players who made at least \$1 million per year in selected years.

Year	Number of Players
1990	138
1994	265
1995	215
1996	241

The function

$$N(x) = \frac{9x^4 - 87x^3 + 175x^2 + 406x + 135}{x + 1}$$

models the number of million-dollar players, where x is the number of years since 1990.

1. Evaluate $N(x)$ for $x = 0$, 4, 5, and 6, and show that the model's values differ from the actual data in the table by less than 4%.
2. Use long or synthetic division to simplify the expression for $N(x)$. Let $P(x)$ represent the resulting expression. How does the domain of $P(x)$ differ from the domain of $N(x)$?
3. Produce the graph of $P(x)$, and trace it to estimate $P(0)$, $P(4)$, $P(5)$, and $P(6)$. Compare your results with those obtained in Exercise 1.
4. Suppose that 1994 is known to be the peak year for million-dollar players in the period 1990–1996. Although $P(x)$ models the data in the table quite well, the model fails badly for the period 1991–1993. Explain how you can tell that this is so from the graph of $P(x)$.
5. In 1996, one-hundred players (13% of all players) made a total of \$480.9 million, which was 53% of the total major league players payroll. Use this information to calculate the total number of major league players and the total payroll in 1996.
6. Form groups that favor either the players or the owners in salary issues. Use the information given in this Chapter Project and in the Group Project for Section 7.4 as well as information that you are able to research on your own. Call the sports editor of your local newspaper to obtain another perspective. Then, using graphs, tables, and other visual aids that support your position, present the opposing views to your class.

7. CHAPTER REVIEW EXERCISES

Section 7.1

1. For $f(x) = \frac{x - 7}{x^2 - 2x - 15}$, find two numbers a_1 and a_2 for which $f(a_1)$ and $f(a_2)$ are undefined. What do we call a_1 and a_2 in this case?

2. Consider the function $g(x) = \frac{1}{x - 5}$.
   (a) Use your calculator to evaluate each of the following.
   (i) $g(6)$ (ii) $g(5.5)$ (iii) $g(5.01)$
   (iv) $g(5.0001)$ (v) $g(5)$
   (b) Describe the values of $g(x)$ as x becomes closer to 5.
   (c) Explain the result in part (v).

In Exercises 3 and 4, determine the domain of the rational function.

3. $y = \frac{5x - 2}{2x - 5}$

4. $y = \frac{3x - 4}{3x^2 + 11x - 20}$

5. Explain why 3 can be divided out of the numerator and denominator of $\frac{3(x + 1)}{3(x + 5)}$ but not of $\frac{3x + 1}{3x + 5}$.

In Exercises 6–9, simplify the rational expression.

6. $\frac{2x - 3}{4x^2 - 9}$

7. $\frac{x^2 + x - 12}{2x^2 - 9x + 9}$

8. $\frac{9 - x^2}{x^2 + x - 12}$

9. $\frac{ab + 4a - 3b - 12}{ab + 2a - 3b - 6}$

10. General Appliance Company has x refrigerators in its inventory. After receiving a shipment of five refrigerators, the total value of the inventory (in thousands of dollars) is $x^2 - 2x - 35$. Write a simplified expression for the average unit value of the refrigerators in the inventory.

In Exercises 11 and 12, supply the missing numerator.

11. $\frac{5}{3x + 12} = \frac{\quad}{3x^2 - 48}$

12. $\frac{3}{x^2 + 2x - 8} = \frac{\quad}{(x - 2)(x^2 - x - 20)}$

Section 7.2

In Exercises 13 and 14, multiply and simplify.

13. $\frac{8x - 40}{4x^3} \cdot \frac{4x^2 + 20x}{x^2 - 25}$

14. $\frac{x^2 - x - 56}{x^2 - 49} \cdot \frac{8x + 64}{x^2 - 64}$

15. What is the product of the rational expressions $\frac{x^2 + 3x + 2}{x^2 + 8x + 12}$ and $\frac{x^2 + 7x + 6}{x^2 - x - 2}$?

16. Multiply $\frac{2x^3}{4x^4 - 36x^2}$ by $x^2 - 2x - 15$.

17. Consider two nonzero numbers x and y.
   (a) What is the reciprocal of their product?
   (b) What is the product of their reciprocals?
   (c) Are the results found in parts (a) and (b) equivalent?

In Exercises 18 and 19, divide and simplify.

18. $\frac{3x}{x + 9} \div \frac{15x^2}{x^2 + 7x - 18}$

19. $\frac{x^2 - 5x + 6}{x^2 + 3x - 18} \div \frac{x^2 - 4}{6x + 36}$

20. Divide $\frac{3x - 3}{8x^3}$ by $\frac{x^2 - 1}{8x^2 + 8x}$.

21. What is the quotient of the rational expressions $\frac{x^2 + 3x - 4}{x^2 - 2x - 24}$ and $\frac{x^2 - 5x + 4}{x^2 - 7x + 6}$?

22. List all the restricted values in Exercise 21.

23. The length and width of a rectangle are, respectively,
$$\frac{x^2 + 5x + 6}{x^2 - 3x - 18} \text{ and } \frac{2x - 12}{x^2 - 4}$$
Write a simplified expression for the area of the rectangle.

24. If a and b are nonzero numbers, explain why dividing by $\frac{a^{-1}}{b^{-1}}$ is the same as multiplying by $\frac{a}{b}$.

Section 7.3

In Exercises 25 and 26, determine the LCD.

25. $\dfrac{1}{x^2 - 49}, \quad \dfrac{1}{9x - 63}$

26. $\dfrac{1}{x^2 + 3x + 2}, \quad \dfrac{1}{x^2 - 4}, \quad \dfrac{1}{x^2 - x - 2}$

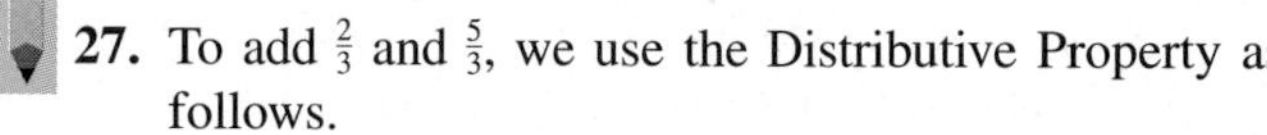

27. To add $\frac{2}{3}$ and $\frac{5}{3}$, we use the Distributive Property as follows.

$$\frac{2}{3} + \frac{5}{3} = \frac{1}{3} \cdot 2 + \frac{1}{3} \cdot 5 = \frac{1}{3}(2 + 5) = \frac{1}{3} \cdot 7 = \frac{7}{3}$$

With this as background, explain why it is not possible to add fractions that have unlike denominators.

In Exercises 28–31, perform the indicated operations.

28. $\dfrac{5x + 3}{x - 1} - \dfrac{9x - 1}{x - 1}$

29. $\dfrac{2x - 1}{5x - 4} + \dfrac{6x - 6}{4 - 5x}$

30. $\dfrac{3}{10x} + \dfrac{8}{5x^2}$

31. $\dfrac{7}{x^2 + x - 12} + \dfrac{3}{x^2 - 16}$

32. Determine the unknown numerator.

$$\frac{x}{5x + 10} = \frac{}{5x^2 - 25x - 70}$$

33. Write $2x^{-1} + 3(x + 2)^{-1}$ as a single rational expression with no negative exponents.

34. What is the sum of the following expressions?

$$\frac{x}{x^2 + 8x + 15} \quad \text{and} \quad \frac{15}{x^2 + 4x - 5}$$

35. Subtract $\dfrac{1}{x^2 - x - 30}$ from $\dfrac{7}{x^2 + 2x - 48}$.

36. Consider any two consecutive odd integers. Find the sum of their reciprocals and show that the numerator of the sum is always an even number.

Section 7.4

37. Describe two different methods for simplifying a complex fraction.

In Exercises 38 and 39, simplify the complex fraction.

38. $\dfrac{9 - \dfrac{1}{x^2}}{3 + \dfrac{1}{x}}$

39. $\dfrac{\dfrac{a}{9} - \dfrac{1}{a}}{\dfrac{1}{3} + \dfrac{a + 4}{a}}$

40. The area of a rectangle is $1 + \dfrac{2}{x} - \dfrac{3}{x^2}$. If the length of the rectangle is $1 - \dfrac{9}{x^2}$, write a simplified expression for the width.

41. For nonzero numbers m and n, show that

$$\frac{\dfrac{m^{-1}}{n^{-1}}}{\dfrac{n}{m}} = 1$$

42. Rewrite $\dfrac{1 - 9x^{-2}}{2 + 3x^{-1} - 9x^{-2}}$ with positive exponents and simplify.

In Exercises 43 and 44, simplify the complex fraction.

43. $\dfrac{\dfrac{x + 4}{x - 4} + \dfrac{x - 4}{x + 4}}{\dfrac{1}{x + 4} + \dfrac{1}{x - 4}}$

44. $\dfrac{\dfrac{12x}{y^2} + \dfrac{13}{xy} - \dfrac{35}{x^3}}{\dfrac{10}{x^3} - \dfrac{3}{xy} - \dfrac{4x}{y^2}}$

Section 7.5

45. Suppose that you make no errors in solving an equation containing rational expressions. How is it possible that the apparent solution does not check? What do we call the apparent solution in this case?

46. Suppose that you are solving an equation containing rational expressions. After you clear the fractions and simplify, you obtain an equation of the form $a = b$, where a and b are constants. How do you interpret this result?

In Exercises 47–54, solve.

47. $10x - \dfrac{2}{x} = 1$

48. $\dfrac{x}{5} + \dfrac{5}{4x} = \dfrac{5}{4}$

49. $\dfrac{x + 3}{x - 5} = \dfrac{8}{x - 5}$

50. $\dfrac{5}{x + 3} + \dfrac{8}{x - 3} = \dfrac{5}{x^2 - 9}$

51. $\dfrac{7}{x + 1} - \dfrac{2}{x - 1} = \dfrac{1}{x^2 - 1}$

52. $\dfrac{9}{3 - x} + \dfrac{9}{x + 3} = \dfrac{-54}{x^2 - 9}$

53. $\dfrac{2x}{x - 2} - \dfrac{16}{x^2 - 4} = 2$

54. $\frac{6}{x-1} + \frac{7}{x^2 - 5x + 4} = \frac{2}{x-4}$

55. Use a calculator graph to estimate the number of solutions of $x^2 + 1 = \frac{3}{x-2}$.

56. Solve $\frac{1}{x} + \frac{1}{y} = \frac{1}{3}$ for x.

Section 7.6

57. If each is working alone, a helper takes twice as long to do a welding job as an experienced welder would take. If the two take 8 hours to complete the job working together, how long does the helper take working alone?

58. Seven members of the Future Farmers of America painted a small barn in 6 hours. If two additional members had helped, how long would it have taken?

59. A chemical tank truck can be filled through an intake pipe in 2.5 hours. With the outtake pipe open, the tank can be emptied in 3 hours. If the tank is initially empty, how long does the tank take to fill with both pipes open?

60. A baseball player has a batting average of 0.244 with 42 hits out of 172 times at bat. If he goes on a hitting streak, how many hits in a row would raise his batting average to 0.278?

61. The total travel time for a family to reach a campsite is 2 hours. They drive 81 miles and hike 2 miles. If they drive 50 mph faster than they hike, what is their average driving speed?

62. During the first half of the season, a basketball player missed 7 free throw attempts. During the second half, she made twice as many free throws as she made during the first half, but she had 16 more attempts than she had during the first half. If her free throw percentage at the end of the season was 75%, how many free throws did she make for the season?

LOOKING AHEAD

The following exercises review concepts and skills that you will need in Chapter 8.

In Exercises 1–4, simplify and express the result with positive exponents. Assume that all expressions are defined.

1. (a) $(x^2)^4$ (b) $(a^{-3})^5$
2. (a) $a^3 \cdot a^5$ (b) $x^{-5} \cdot x^3$
3. (a) $(a^2b^3)^4$ (b) $\left(\frac{3}{x^5}\right)^2$
4. (a) $\frac{3x^{-2}}{y^{-4}}$ (b) $\frac{40a^4b^7}{5a^9b^4}$

5. Evaluate each expression.
   (a) 5^{-2} (b) $-3 \cdot 6^0$ (c) $|-8|$

In Exercises 6 and 7, factor completely.

6. $y^2 + 6y + 9$
7. $a^4b^6 + a^2b^8$
8. Use the Distributive Property to show how to combine like terms in the expression $2n + 5n$.

In Exercises 9–11, solve the given equation.

9. $x^2 = 16$
10. $3x - 1 = 2^3$
11. $x^2 - 3x = 2^2$
12. Show that a triangle with sides 9, 12, and 15 is a right triangle.

7. CHAPTER TEST

1. For $\frac{x-3}{x^2 + 5x - 24}$, explain why 3 is a restricted value.
2. Determine the domain of the function $y = \frac{2x-1}{x^2 + 3x}$.

For Questions 3–5, simplify.

3. $\frac{6x^2 + 12x}{3x^3 - 6x^2}$
4. $\frac{(x+5)^2}{x^2 - 4x - 45}$

5. $\dfrac{x^2 + xy - 2y^2}{y^2 - x^2}$

For Questions 6 and 7, perform the indicated multiplication or division.

6. $\dfrac{t^2 + 3t - 4}{t^2 + 2t - 3} \cdot \dfrac{t^2 - t - 6}{t^2 + t - 12}$

7. $\dfrac{y^2 - 25}{2y + 10} \div (y^2 - 10y + 25)$

8. Determine the missing numerator.

$$\frac{3x}{x^2 - x - 2} = \frac{\qquad}{(x - 2)^2(x + 1)}$$

9. Explain why the following subtraction is easier to perform than it first appears, and show that the result does not involve r.

$$\frac{(m + n)^{-1}}{r^{-1}} - \frac{r - 1}{m + n}$$

For Questions 10–12, perform the indicated addition or subtraction.

10. $\dfrac{7x - 1}{3x + 4} - \dfrac{4x - 5}{3x + 4}$ **11.** $\dfrac{3}{t^2} - \dfrac{5}{t}$

12. $\dfrac{x - 7}{2x^2 + 9x - 5} + \dfrac{4 - x}{4x^2 + 23x + 15}$

13. Which of the following is a complex fraction? Show that two of the expressions are equivalent.

(i) $\dfrac{\frac{m}{n}}{p}$ (ii) $\dfrac{m}{\frac{n}{p}}$ (iii) $\dfrac{\frac{m}{n}}{\frac{1}{p}}$

For Questions 14 and 15, simplify the complex fraction.

14. $\dfrac{\dfrac{x + 3y}{y}}{\dfrac{x^2 - 9y^2}{6y}}$ **15.** $\dfrac{x - 3 + \dfrac{x - 3}{x^2}}{x + 2 + \dfrac{1}{x} + \dfrac{2}{x^2}}$

16. Write $\dfrac{a^{-2}}{a^{-2} - b^{-2}}$ as a single rational expression with positive exponents.

For Questions 17–20, solve the equation.

17. $\dfrac{3}{a + 3} = \dfrac{2}{a - 2}$ **18.** $1 - \dfrac{3}{x} = \dfrac{10}{x^2}$

19. $\dfrac{x}{x - 2} + \dfrac{2}{3} = \dfrac{2}{x - 2}$

20. $\dfrac{8t}{t^2 - 16} = \dfrac{3}{t + 4} + \dfrac{5}{4 - t}$

21. For different reasons, the solution sets of the following equations are empty. Explain why this is so for each equation.

(a) $\dfrac{x - 2}{x - 1} = \dfrac{1}{1 - x}$ (b) $\dfrac{x - 2}{x - 1} = \dfrac{x + 3}{x - 1}$

22. Solve the formula $\dfrac{1}{A} + \dfrac{1}{B} = 2$ for B.

23. In a target-shooting competition, scores are found to vary inversely with the wind speed. If the success rate is 60% when the wind speed is 20 mph, what is the success rate when the wind speed is 30 mph?

24. In the spring a cat sheds so fast that the carpet is covered with hair in 4 hours. If the cat is out of the house, cleaning the carpet takes 1.5 hours. How long does cleaning the carpet take with the cat in the house?

25. The time a person takes to paddle a kayak 2 miles downstream is the same as the time to paddle a half mile upstream. If the rate of the current is 3 mph, what is this person's paddling rate in still water?

6–7 CUMULATIVE TEST

1. Perform the indicated operations.

$(3x^2 - 6x + 8) - (x^3 - 2x^2 - 7) +$
$(2x^3 - 4x^2 + 7x - 16)$

2. Multiply and simplify.

$(x - 3)^2 - (x + 3)(x - 3)$

3. Factor completely.

(a) $x^4y^3 - x^3y^4$

(b) $ax + 3ay - 7bx - 21by$

4. Factor completely.

(a) $(2x + 1)^2 - 16$ (b) $2x^2 - 12x + 18$

(c) $y^3 + 8$

5. Factor completely.

 (a) $z^2 + 2z - 63$ (b) $a^2 - 4ab - 45b^2$

 (c) $12 + 9x - 3x^2$

6. Explain how the Zero Factor Property can be used to solve $x^2 - 3x = 0$.

7. Solve.

 (a) $(x - 4)(x - 3) = 12$ (b) $a^4 + 25 = 26a^2$

8. If the square of a number is decreased by 24, the result is 5 times the number. What is the number?

9. Find the quotient and remainder for the following.

 $(4c^4 + 5c^2 - 7c) \div (2c - 1)$

10. Divide synthetically.

 $(x^5 - 3x^3 - 17x^2 + x - 13) \div (x - 3)$

11. Explain how the Factor Theorem can be used to determine whether $x - c$ (where c is a constant) is a factor of a polynomial $P(x)$.

12. If $P(x) = 2x^3 - 5x^2 + x - 1$, use the Remainder Theorem to evaluate $P(2)$.

13. Simplify $\dfrac{x^2 - x - 12}{4x^2 - 13x - 12}$.

14. Multiply $\dfrac{a^2 - 16}{a^2 - 3a - 4} \cdot \dfrac{a^2 + 3a + 2}{a^2 - 2a - 8}$.

15. Add $\dfrac{3}{y^2 + 11y + 10} + \dfrac{y}{y^2 + 5y + 4}$.

16. Simplify $\dfrac{\dfrac{1}{x+3}}{\dfrac{1}{x} - \dfrac{1}{x+3}}$.

17. Solve $\dfrac{2}{x} - \dfrac{x+8}{x+2} = \dfrac{12}{x^2 + 2x}$.

18. A landscape worker can lay sod in a yard in 6 hours. A helper would take 8 hours to do the same job. How long would sodding the yard take if both people work together?

19. A driver of a car averaged 5 mph faster during the final 120 miles of a 340-mile trip. If the entire trip took 6 hours, at what average speed did the driver travel during the first 220 miles?

Chapter 8

Radical Expressions

Suppose that you are a staff member with the National Park Service. Faced with budgetary cuts, you assemble data on the number of visitors to all areas that are managed by the service (see graph) and on the cost of administering those areas.

The data in the graph can be modeled by an expression with a **rational exponent.** This model, along with the cost data, can be used to show the cost per visitor and to project budgetary requirements for future years. (For more on this topic, see Exercises 99–102 at the end of Section 8.3, and see the Chapter Project.)

In Chapter 8 we discuss expressions with radicals and rational exponents. We learn about the properties of such expressions and about how to simplify them and perform operations with them. Next, we develop methods for solving equations containing radicals and exponents. We conclude by extending the number system to the set of complex numbers.

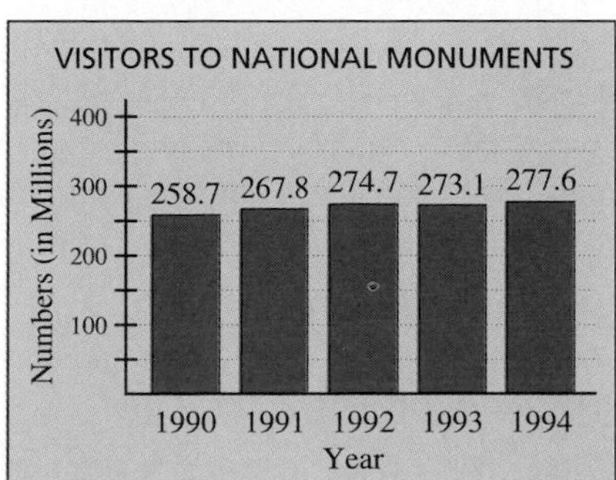

(Source: National Park Service.)

8.1 RADICALS

Higher-Order Roots • Simplifying $(\sqrt[n]{b})^n$ and $\sqrt[n]{b^n}$

Higher-Order Roots

In Section 1.6 we defined a to be a square root of b if $a^2 = b$. For example, 5 and -5 are both square roots of 25 because $5^2 = 25$ and $(-5)^2 = 25$.

The concept of square root can be generalized to higher-order roots. For example, 3 is the **third root** (or **cube root**) of 27 because $3^3 = 27$; 2 and -2 are **fourth roots** of 16 because $2^4 = 16$ and $(-2)^4 = 16$.

Definition of *n*th Root

For any real numbers a and b and any integer $n > 1$, if $a^n = b$, then a is an nth root of b.

EXAMPLE 1

Determining *n*th Roots

(a) The third root of 8 is 2. $2^3 = 8$

(b) The fifth root of -32 is -2. $(-2)^5 = -32$

(c) The fourth roots of 81 are 3 and -3. $3^4 = 81$ and $(-3)^4 = 81$

(d) The fourth root of -81 is undefined. There is no real number a such that $a^4 = -81$.

Example 1 leads to the following observations:

1. Any real number has one odd root.
2. A positive number has two even roots that are opposites, but a negative number has no real number even root.
3. Because $0^n = 0$ for all integers $n > 1$, 0 has one nth root, 0.

In Section 1.6 we defined the **principal square root** of b to be the *nonnegative* square root, and we represented this number with $\sqrt{b}$. We can extend this definition to higher-order even roots by defining the **principal *n*th root** of b to be the nonnegative nth root of b, and we represent this number with $\sqrt[n]{b}$. For odd roots, there is only one nth root, and we represent this number with $\sqrt[n]{b}$.

EXPLORING THE CONCEPT

Principal nth Roots

Consider the graphs of functions of the form $f(x) = \sqrt[n]{x}$, as shown in Figs. 8.1 and 8.2.

Figure 8.1

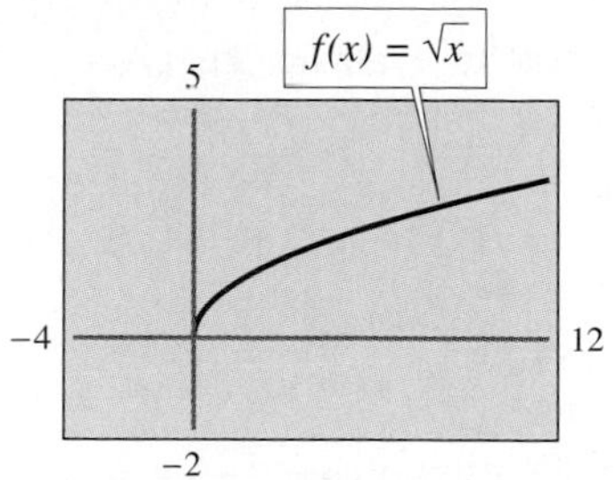

Figure 8.2

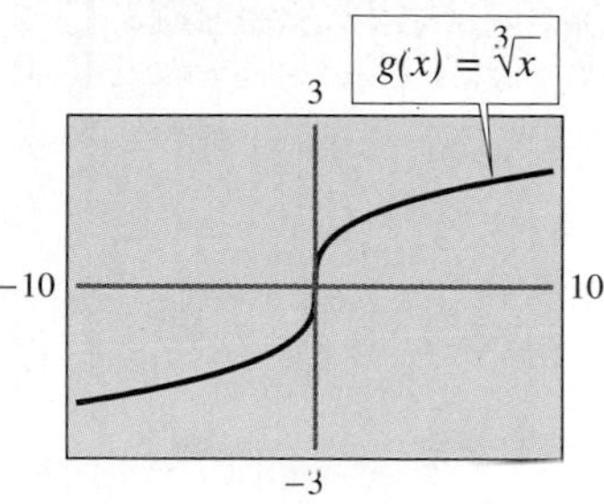

Figure 8.1 shows the graph of $f(x) = \sqrt{x}$. The graph lies on and to the right of the y-axis because $\sqrt{x}$ is a real number only if $x \geq 0$. Also, the graph lies on and above the x-axis because the symbol $\sqrt{x}$ means the nonnegative square root of x.

Figure 8.2 shows the graph of $g(x) = \sqrt[3]{x}$. The graph extends forever to the right and left because $\sqrt[3]{x}$ is defined for all real numbers. Also, the graph extends upward and downward forever because $\sqrt[3]{x}$ can be positive or negative.

The graphs of $y = \sqrt[4]{x}$, $y = \sqrt[6]{x}$, and so on are similar to the graph in Fig. 8.1. Also, the graphs of $y = \sqrt[5]{x}$, $y = \sqrt[7]{x}$, and so on are similar to the graph in Fig. 8.2. Therefore, we can generalize our observations as follows.

LEARNING TIP

For an even index, the radicand must be positive or zero. For an odd index, the radicand can be negative, positive, or zero.

Definition of $\sqrt[n]{b}$

For any even integer $n > 1$ and any nonnegative real number b, $\sqrt[n]{b}$ represents the nonnegative nth root of b.

For any odd integer $n > 1$ and any real number b, $\sqrt[n]{b}$ represents the nth root of b.

In the expression $\sqrt[n]{b}$, the symbol $\sqrt{}$ is called a **radical symbol,** b is called the **radicand,** n is called the **index,** and the entire symbol $\sqrt[n]{b}$ is called a **radical.**

EXAMPLE 2

Evaluating *n*th Roots

(a) $\sqrt[3]{-64} = -4$ — $(-4)^3 = -64$

(b) $\sqrt{-64}$ is not a real number. — Even root of a negative number

(c) $-\sqrt{64} = -8$ — $-1 \cdot \sqrt{64} = -1 \cdot 8 = -8$

(d) $\sqrt[4]{81} = 3$ — $3^4 = 81$

(e) $\sqrt[5]{32} = 2$ — $2^5 = 32$

(f) $\sqrt[6]{-1}$ is not a real number. — Even root of a negative number

(g) $\sqrt[9]{-1} = -1$ — $(-1)^9 = -1$

*n*th Root

Most calculators have square root and cube root keys. Keys for higher-order roots can also be found on some calculators. A typical selection is $\sqrt[x]{}$.

EXAMPLE 3

Calculating Higher-Order Roots

Use your calculator to evaluate the given radical. Round results to the nearest hundredth.

(a) $\sqrt[3]{96} \approx 4.58$

(b) $\sqrt[5]{-6397} \approx -5.77$

(c) $\sqrt[4]{0.012} \approx 0.33$

Simplifying $\left(\sqrt[n]{b}\right)^n$ and $\sqrt[n]{b^n}$

We now consider methods for simplifying expressions of the form $\left(\sqrt[n]{b}\right)^n$ and $\sqrt[n]{b^n}$.

EXPLORING THE CONCEPT

Simplifying $\left(\sqrt[n]{b}\right)^n$ and $\sqrt[n]{b^n}$

To evaluate $\left(\sqrt{25}\right)^2$, we take the square root of 25 and then square the result: $\left(\sqrt{25}\right)^2 = (5)^2 = 25$. Similarly, $\left(\sqrt[3]{8}\right)^3 = (2)^3 = 8$. However, before we conclude that $\left(\sqrt{b}\right)^2$ is always b, we must recall that $\sqrt{b}$ is a real number only if $b \geq 0$. In fact, this provision applies to all even roots. With this limitation in mind, we state our conclusion as follows:

If n is odd, then $\left(\sqrt[n]{b}\right)^n = b$.
If n is even, then $\left(\sqrt[n]{b}\right)^n = b$ only if $b \geq 0$.

Now consider the quantities $\sqrt{7^2}$ and $\sqrt[3]{4^3}$. To evaluate these quantities, we begin with the exponent and then take the root.

$$\sqrt{7^2} = \sqrt{49} = 7 \qquad \sqrt[3]{4^3} = \sqrt[3]{64} = 4$$

These results might suggest that $\sqrt[n]{b^n}$ is always b, and this conclusion is indeed true if n is odd. However, a different result is obtained when n is even and b is negative.

$$\sqrt{(-10)^2} = \sqrt{100} = 10 \qquad \sqrt[4]{(-3)^4} = \sqrt[4]{81} = 3$$

In each of these instances, b is a negative number, but the result is positive because $\sqrt[n]{}$ indicates the *principal* (nonnegative) nth root when n is even. Regardless of the sign of b, we can conclude that $\sqrt[n]{b^n} = |b|$ when n is even.

The following summary gives the rules for simplifying expressions of the form $\left(\sqrt[n]{b}\right)^n$ and $\sqrt[n]{b^n}$.

Simplifying $\left(\sqrt[n]{b}\right)^n$ and $\sqrt[n]{b^n}$

Let b represent any real number.

1. If $\sqrt[n]{b}$ is defined, then $\left(\sqrt[n]{b}\right)^n = b$.
2. For even integers n, $\sqrt[n]{b^n} = |b|$.

   For odd integers n, $\sqrt[n]{b^n} = b$.

Think About It

Raising $\sqrt[n]{b}$ to the nth power reverses or undoes the operation of taking the nth root. Give another example of an operation that reverses an original operation.

Note that for $b < 0$ and even integers n, $\left(\sqrt[n]{b}\right)^n$ is undefined.

EXAMPLE 4

Evaluating $\sqrt[n]{b^n}$ with Numerical Radicands

(a) $\sqrt{(-5)^2} = |-5| = 5$ Because n is even, absolute value symbols are needed.

(b) $\sqrt[5]{(-4)^5} = -4$ Because n is odd, absolute value symbols are not used.

(c) $\sqrt[4]{3^4} = |3| = 3$ Because n is even, absolute value symbols are needed.

When the radicand is a variable expression and the index is even, we must use absolute value symbols around the result if there is a possibility that the result could be negative.

EXAMPLE 5

Simplifying $\sqrt[n]{b^n}$ with Variable Radicands

(a) $\sqrt[4]{a^4} = |a|$ The index is even, and a could be negative. Therefore, absolute value symbols are needed.

(b) $\sqrt[3]{t^3} = t$ Because the index is odd, absolute value symbols are not used.

(c) $\sqrt{(x-3)^2} = |x-3|$ Absolute value symbols are needed because the index is even, and $x - 3$ could be negative.

The rules for simplifying $\sqrt[n]{b^n}$ can be useful for simplifying expressions in which the exponent on b is a multiple of the index.

EXAMPLE 6

Simplifying Radicals

(a) $\sqrt{c^6} = \sqrt{(c^3)^2}$ Write the expression in the form $\sqrt[2]{b^2}$.

$= |c^3|$ The index is even, and c^3 could be negative.

(b) $\sqrt[3]{y^{15}} = \sqrt[3]{(y^5)^3}$ Write the expression in the form $\sqrt[3]{b^3}$.

$= y^5$ The index is odd, so absolute value symbols are not used.

(c) $\sqrt[4]{x^8} = \sqrt[4]{(x^2)^4}$ Write the expression in the form $\sqrt[4]{b^4}$.

$= |x^2|$ The index is even.

$= x^2$ x^2 is not negative.

Note: The techniques used in Example 6 are applicable only if the exponent on the radicand is a multiple of the index. For expressions such as $\sqrt{x^5}$, other methods are needed.

8.1 QUICK REFERENCE

Higher-Order Roots

- For any real numbers a and b and any integer $n > 1$, if $a^n = b$, then a is an ***n*th root** of b.
- The following generalizations apply to nth roots:
  1. Any real number has one odd root.
  2. A positive number has two even roots that are opposites, but a negative number has no real number even root.
  3. Because $0^n = 0$ for all integers $n > 1$, 0 has one nth root, 0.

- For any even integer $n > 1$ and any nonnegative real number b, $\sqrt[n]{b}$ represents the nonnegative nth root of b.

  For any odd integer $n > 1$ and any real number b, $\sqrt[n]{b}$ represents the nth root of b.
- In the expression $\sqrt[n]{b}$, the symbol $\sqrt{\ }$ is called the **radical symbol,** b is called the **radicand,** n is called the **index,** and the entire symbol $\sqrt[n]{b}$ is called a **radical.**

Simplifying $(\sqrt[n]{b})^n$ and $\sqrt[n]{b^n}$

- Let b represent any real number.
  1. If $\sqrt[n]{b}$ is defined, then $(\sqrt[n]{b})^n = b$.
  2. For even integers n, $\sqrt[n]{b^n} = |b|$.

     For odd integers n, $\sqrt[n]{b^n} = b$.

8.1 EXERCISES

Concepts and Skills

1. Why do the following two questions have different answers?

   (a) What are the square roots of 9?

   (b) What is $\sqrt{9}$?

2. In the expression $\sqrt[n]{b}$, what do we call n and b?

In Exercises 3–14, determine the specified root(s), if any, of the given number.

3. Square root of 36
4. Square root of $\frac{16}{25}$
5. Square root of -4
6. Square root of -25
7. Third root of 27
8. Third root of -64
9. Fourth root of 625
10. Fourth root of 16
11. Fifth root of -243
12. Fifth root of -32
13. Sixth root of -64
14. Sixth root of 64

In Exercises 15–30, evaluate.

15. $\sqrt{16}$
16. $\sqrt{64}$
17. $\sqrt[3]{-64}$
18. $\sqrt[5]{32}$
19. $\sqrt{\frac{4}{9}}$
20. $-\sqrt{\frac{25}{144}}$
21. $\sqrt[4]{-16}$
22. $\sqrt[6]{-64}$
23. $-\sqrt{16}$
24. $-\sqrt{36}$
25. $4\sqrt{25}$
26. $-2\sqrt{9}$
27. $\sqrt[3]{64}$
28. $\sqrt[4]{16}$
29. $-\sqrt[5]{-32}$
30. $-\sqrt[3]{-125}$

In Exercises 31–36, use your calculator to evaluate the given radical. Round results to the nearest hundredth.

31. $\sqrt[3]{343}$
32. $\sqrt[3]{-3.375}$
33. $\sqrt[4]{52}$
34. $\sqrt[6]{5600}$
35. $\sqrt[5]{-0.65}$
36. $\sqrt[3]{15}$

37. Why are absolute value symbols required for one of the following results but not for the other?

$$\sqrt{x^6} = |x^3| \qquad \sqrt{x^4} = x^2$$

38. Describe the conditions under which

$$\left(\sqrt[n]{x}\right)^n = \sqrt[n]{x^n} = x$$

In Exercises 39–44, evaluate the radical expression.

39. $\sqrt{(-4)^2}$
40. $\sqrt[4]{(-3)^4}$
41. $\left(\sqrt[4]{7}\right)^4$
42. $\left(\sqrt[5]{-3}\right)^5$
43. $\sqrt[3]{10^3}$
44. $\sqrt[3]{(-5)^3}$

In Exercises 45–48, evaluate each expression.

45. $\sqrt{4^2 - 4(3)(1)}$
46. $\sqrt{9^2 - 4(2)(9)}$
47. $\sqrt{9^2 - 4(10)(-9)}$
48. $\sqrt{19^2 - 4(6)(-7)}$

In Exercises 49–54, translate the given phrase into a radical expression. Assume that all variables represent positive numbers.

49. Twice the square root of x
50. Five less than the cube root of y
51. The difference of the fifth root of a and 9
52. The square root of c plus b
53. The square root of the quantity $c + b$
54. The fourth root of the product of x and y

In Exercises 55–70, determine the indicated root. Assume that all variables represent positive numbers.

55. $\sqrt{x^{10}}$ **56.** $\sqrt{x^{16}}$ **57.** $4\sqrt{y^6}$

58. $-3\sqrt{z^8}$ **59.** $\sqrt[3]{x^{21}}$ **60.** $\sqrt[3]{y^{12}}$

61. $-\sqrt[3]{y^{15}}$ **62.** $4\sqrt[3]{a^6}$ **63.** $\sqrt[5]{x^{35}}$

64. $\sqrt[7]{x^{35}}$ **65.** $-\sqrt[4]{x^8}$ **66.** $5\sqrt[6]{y^{18}}$

67. $\sqrt{(3x)^6}$

68. $\sqrt{(5ac)^4}$

69. $\sqrt{(x+3)^{10}}$

70. $\sqrt{(x-2)^{16}}$

71. In each part determine whether the graphs are the same, and explain why or why not.

(a) $y_1 = \sqrt{x^2}$, $y_2 = x$ (b) $y_1 = \sqrt[3]{x^3}$, $y_2 = x$

72. Which of the following radicals cannot be simplified with our present methods? Why?

(i) $\sqrt{x^{12}}$ (ii) $\sqrt[3]{x^{12}}$ (iii) $\sqrt[4]{x^{12}}$ (iv) $\sqrt[5]{x^{12}}$

In Exercises 73–86, determine the indicated root. Assume that all variables represent any real number.

73. $\sqrt{x^6}$ **74.** $\sqrt{t^{10}}$

75. $-\sqrt{x^8}$ **76.** $5\sqrt{y^{14}}$

77. $\sqrt[5]{x^{15}}$ **78.** $\sqrt[7]{y^{21}}$

79. $\sqrt[4]{a^4}$ **80.** $\sqrt[6]{y^6}$

81. $\sqrt{(x+2)^2}$ **82.** $\sqrt{(x^2-9)^2}$

83. $-2\sqrt{(x-1)^2}$

84. $\sqrt{(3x^3-5x)^2}$

85. $\sqrt[3]{(1-x)^3}$

86. $\sqrt[8]{(x-7)^8}$

In Exercises 87–90, insert =, <, or > to describe the correct relationship between the two given quantities. Use your calculator to evaluate the quantities.

87. $\sqrt{7-5}$ ▭ $\sqrt{7}-\sqrt{5}$

88. $\sqrt{10+3}$ ▭ $\sqrt{10}+\sqrt{3}$

89. $\sqrt{25-9}$ ▭ $\sqrt{25}-\sqrt{9}$

90. $\sqrt{9}+\sqrt{16}$ ▭ $\sqrt{25}$

In Exercises 91–96, supply the missing number to make each statement true.

91. $\sqrt{x^{\square}} = x^2$ **92.** $\sqrt[3]{y^{\square}} = y^4$

93. $\sqrt[\square]{y^{24}} = |y^3|$ **94.** $\sqrt[\square]{y^4} = |y|$

95. $\square\sqrt[4]{16} = -6$ **96.** $\square\sqrt[3]{-8} = 2$

Modeling with Real Data

97. Political action committee (PAC) contributions rose from \$42.3 million in 1986 to \$78.0 million in 1996. (Source: Federal Election Commission.) The function $P(x) = 15.3\sqrt[5]{x^3}$ models PAC contributions (in millions of dollars), where x is the number of years since 1980.

(a) What are the predicted PAC contributions for 2004?

(b) Use a graph to estimate the year in which PAC contributions will reach \$100 million.

98. The senior citizen population in Arizona is expected to increase dramatically. From 1993 to 2020, the over-65 age group is predicted to rise from 529,000 to 1,121,000, and the over-85 age group is predicted to rise from 46,000 to 146,000. The functions $S(x) = 368\sqrt[3]{x}$ (over 65) and $E(x) = 26.5\sqrt{x}$ (over 85) model the age-group populations (in thousands) as functions of the number of years since 1990.

(a) What is the projected population of each age group in 2010? (Round results to the nearest thousand.)

(b) For each age group, what is the projected percentage increase from 2000 to 2010?

Group Project: Windchill Temperatures

The following table lists the average January temperatures (in Fahrenheit degrees) and wind speeds (in miles per hour) for four cities. (Source: U.S. National Oceanic and Atmospheric Administration.)

City	Temperature	Wind Speed
Cheyenne, WY	26.5	15.3
Boston, MA	28.6	13.8
Seattle, WA	40.1	9.7
Norfolk, VA	39.1	8.1

The U.S. Meteorological Service uses the following formula to calculate windchill temperatures.

$$Y = 91.4 - (91.4 - T)\left[0.478 + 0.301\left(\sqrt{x} - 0.02x\right)\right]$$

In this formula, T is the actual air temperature in Fahrenheit degrees, x is the wind speed in miles per hour, and Y is the windchill temperature. Enter this formula as Y_1 in your calculator.

99. To compute the average January windchill temperature for Cheyenne, store 26.5 in T, and calculate $\text{Y}_1(15.3)$. Use this method to compute the average January windchill temperatures for each of the four cities in the table.

100. Suppose that the temperature in Boston is 18°F. Store 18 in T, and produce the graph of the windchill expression. Trace the graph to estimate the wind speed needed for a windchill temperature of 0°F.

101. Suppose that a wind speed of 9 miles per hour in Norfolk creates a windchill temperature of 20°F. Substitute these values into the windchill formula, and solve the equation for T to determine the actual air temperature.

102. When the wind speed is 3 miles per hour, Y and T are approximately equal. Does this mean that when the wind speed is less than 3 miles per hour, it will feel warmer than it actually is? What restrictions would you place on x to make the windchill formula valid?

Challenge

In Exercises 103–108, evaluate the given expression, if possible.

103. $\sqrt[3]{-\sqrt[4]{1}}$

104. $\sqrt[4]{-\sqrt[3]{1}}$

105. $\sqrt[3]{\sqrt{64}}$

106. $\sqrt[3]{-\sqrt{64}}$

107. $\sqrt{\sqrt[3]{64}}$

108. $\sqrt{-\sqrt[3]{64}}$

109. We define $\sqrt[n]{b}$ for $n > 1$.

(a) Why do we exclude $n = 1$? What meaning would $\sqrt[1]{b}$ have?

(b) Why do we exclude $n = 0$? What meaning would $\sqrt[0]{b}$ have? (Hint: Consider the two cases $b = 1$ and $b \neq 1$.)

8.2 RATIONAL EXPONENTS

Definition of $b^{1/n}$ • Definition of $b^{m/n}$ • Negative Rational Exponents

Definition of $b^{1/n}$

Initially we used an exponent as a convenient way to write repeated multiplication: $5 \cdot 5 \cdot 5 \cdot 5 = 5^4$. However, to perform operations with exponents, we had to define a zero exponent and a negative exponent.

In this section we extend the definition of exponent to include all rational numbers.

EXPLORING THE CONCEPT

Interpreting $b^{1/n}$

Figure 8.3 shows a calculator's reported values of various exponential expressions in which the exponent is of the form $1/n$.

Figure 8.3

```
9^(1/2)
                  3
64^(1/3)
                  4
81^(1/4)
                  3
32^(1/5)
                  2
```

(i) $\sqrt{9} = 3$

(ii) $\sqrt[3]{64} = 4$

(iii) $\sqrt[4]{81} = 3$

(iv) $\sqrt[5]{32} = 2$

Aligned with each entry in the calculator display is a radical expression that has the same value as the corresponding exponential expression. Observe that the denominator of each exponent is the same as the index of the corresponding radical.

The results shown in Fig. 8.3 suggest that $b^{1/n} = \sqrt[n]{b}$. This conclusion is supported by an application of the Power to a Power Rule extended to rational exponents. For example,

$$(9^{1/2})^2 = 9^{2\cdot(1/2)} = 9^1 = 9$$

Because $(9^{1/2})^2 = 9$, the definition of square root implies that $9^{1/2}$ is a square root of 9.

In general, $(b^{1/n})^n = b^{n\cdot(1/n)} = b^1 = b$. Thus, by the definition of nth root, $b^{1/n}$ is an nth root of b.

Definition of $b^{1/n}$

For any integer $n > 1$ and for any real number b for which $\sqrt[n]{b}$ is defined, $b^{1/n}$ is defined as $\sqrt[n]{b}$.

Note: This definition implies that b cannot be negative when n is even.

EXAMPLE 1

Evaluating Expressions with Rational Exponents

(a) $25^{1/2} = \sqrt{25} = 5$ $\quad 5^2 = 25$

(b) $(-25)^{1/2}$ is not a real number $\quad n$ is even and $b < 0$.

(c) $-25^{1/2} = -\sqrt{25} = -5$ $\quad -1 \cdot \sqrt{25} = -1 \cdot 5 = -5$

(d) $\left(\frac{4}{9}\right)^{1/2} = \sqrt{\frac{4}{9}} = \frac{2}{3}$ $\quad \left(\frac{2}{3}\right)^2 = \frac{4}{9}$

(e) $8^{1/3} = \sqrt[3]{8} = 2$ $\quad 2^3 = 8$

(f) $(-8)^{1/3} = \sqrt[3]{-8} = -2$ $\quad (-2)^3 = -8$

(g) $256^{1/4} = \sqrt[4]{256} = 4$ $\quad 4^4 = 256$

(h) $32^{1/5} = \sqrt[5]{32} = 2$ $\quad 2^5 = 32$

In Section 8.1 we calculated higher-order roots with the $\sqrt[x]{}$ key. If your calculator does not have such a key, you can still calculate a higher-order root by writing the radical as an expression with a rational exponent.

EXAMPLE 2

Finding Higher-Order Roots with a Calculator

Write each radical in exponential form, and use your calculator to evaluate the resulting expression. Round the results to two decimal places.

(a) $\sqrt[5]{16{,}807}$ (b) $\sqrt[4]{10}$ (c) $\sqrt[7]{-17}$

Figure 8.4

```
16807^(1/5)
                7
10^(1/4)
             1.78
(-17)^(1/7)
            -1.50
```

Solution

(a) $\sqrt[5]{16{,}807} = 16{,}807^{1/5}$

(b) $\sqrt[4]{10} = 10^{1/4}$

(c) $\sqrt[7]{-17} = (-17)^{1/7}$

Figure 8.4 is a typical screen display showing the results for all three parts.

Note that the exponents are enclosed in parentheses. Also, if the base is negative, as in part (c), then the base must be enclosed in parentheses.

Definition of $b^{m/n}$

So far the rational exponents we have used have had 1 as the numerator. However, extending the Power to a Power Rule to rational exponents suggests a logical definition of *any* rational exponent.

Suppose that m and n are positive integers with no common factor except 1. Then for all real numbers b for which $b^{1/n}$ is defined,

$$b^{m/n} = (b^{1/n})^m \qquad \text{Power to a Power Rule}$$

$$= \left(\sqrt[n]{b}\right)^m \qquad \text{Definition of } b^{1/n}$$

Definition of $b^{m/n}$

If m and n are any positive integers with no common factor other than 1, then $b^{m/n} = \left(\sqrt[n]{b}\right)^m$ and $b^{m/n} = \sqrt[n]{b^m}$ for all real numbers b for which $b^{1/n}$ is defined.

Note that when we write $b^{m/n}$ as a radical, n indicates the index of the radical and m indicates the power to which the radical is raised.

EXAMPLE 3

LEARNING TIP

Often the best approach to evaluating $b^{m/n}$ is to find the nth root of b first. Then raise the result to the mth power.

Think About It

Consider $(-16)^{2/4}$ and explain why $b^{m/n} = \left(\sqrt[n]{b}\right)^m$ only if m and n have no common factors.

Evaluating Expressions with Rational Exponents

(a) $8^{2/3} = \left(\sqrt[3]{8}\right)^2 = (2)^2 = 4$

(b) $(-32)^{4/5} = \left(\sqrt[5]{-32}\right)^4 = (-2)^4 = 16$ With parentheses, the base is -32.

(c) $-32^{4/5} = -\left(\sqrt[5]{32}\right)^4 = -1 \cdot (2)^4 = -16$ Without parentheses, the base is 32.

(d) $36^{3/2} = \left(\sqrt{36}\right)^3 = (6)^3 = 216$

Exponential expressions with numerical bases can be evaluated without a calculator as long as the roots and powers are familiar. In Example 4 we do need a calculator.

EXAMPLE 4

Using a Calculator to Evaluate Exponential Expressions

Evaluate the following expressions to two decimal places.

(a) $6^{2/3}$ (b) $(-6)^{2/3}$ (c) $6^{3/2}$ (d) $(-6)^{3/2}$

Solution

Figure 8.5 is a typical screen display for parts (a)–(c).

Figure 8.5

```
6^(2/3)
                3.30
(-6)^(2/3)
                3.30
6^(3/2)
               14.70
```

The expression in part (d) is not a real number because $(-6)^{1/2}$ is not defined.

Even if we cannot evaluate an expression, we can use the definition of $b^{m/n}$ to write the expression in the form of a radical.

EXAMPLE 5

Writing Exponential Expressions as Radicals

Write the following exponential expressions as radicals.

(a) $x^{3/5}$ (b) $(5x)^{3/4}$ (c) $5x^{3/4}$

Solution

(a) $x^{3/5} = \left(\sqrt[5]{x}\right)^3$ or $\sqrt[5]{x^3}$

(b) $(5x)^{3/4} = \sqrt[4]{(5x)^3} = \sqrt[4]{125x^3}$ The base of the exponent is $5x$.

(c) $5x^{3/4} = 5\sqrt[4]{x^3}$ The base of the exponent is x.

Negative Rational Exponents

We can extend the definition of a negative integer exponent to include negative rational exponents.

Definition of a Negative Rational Exponent

If m and n are positive integers with no common factor other than 1, and if $b \neq 0$, then $b^{-m/n} = \dfrac{1}{b^{m/n}}$ for all real numbers b for which $b^{1/n}$ is defined.

EXAMPLE 6

Evaluating Exponential Expressions with Negative Exponents

Evaluate each of the following. Use a calculator for parts (d) and (e) and for verifying parts (a)–(c).

(a) $16^{-3/4}$ (b) $(-125)^{-2/3}$ (c) $\left(\dfrac{16}{25}\right)^{-1/2}$

(d) $150^{-3/4}$ (e) $(-7)^{-3/5}$

Figure 8.6

```
150^(-3/4)
                .023
(-7)^(-3/5)
               -.311
```

Solution

(a) $16^{-3/4} = \dfrac{1}{16^{3/4}} = \dfrac{1}{(\sqrt[4]{16})^3} = \dfrac{1}{2^3} = \dfrac{1}{8}$

(b) $(-125)^{-2/3} = \dfrac{1}{(-125)^{2/3}} = \dfrac{1}{(\sqrt[3]{-125})^2} = \dfrac{1}{(-5)^2} = \dfrac{1}{25}$

(c) $\left(\dfrac{16}{25}\right)^{-1/2} = \left(\dfrac{25}{16}\right)^{1/2} = \sqrt{\dfrac{25}{16}} = \dfrac{5}{4}$ $\qquad \left(\dfrac{a}{b}\right)^{-m/n} = \left(\dfrac{b}{a}\right)^{m/n}$

(d)–(e) Figure 8.6 is a typical screen display for parts (d) and (e). The results are rounded to three decimal places.

8.2 QUICK REFERENCE

Definition of $b^{1/n}$

- For any integer $n > 1$ and for any real number b for which $\sqrt[n]{b}$ is defined, $b^{1/n}$ is defined as $\sqrt[n]{b}$.
- If n is an odd integer, then b can be any real number. If n is an even integer, then b must be nonnegative.
- By writing a radical as an expression with a rational exponent, we can use a calculator to find higher-order roots.

Definition of $b^{m/n}$

- If m and n are any positive integers with no common factor other than 1, then $b^{m/n} = \left(\sqrt[n]{b}\right)^m$ and $b^{m/n} = \sqrt[n]{b^m}$ for all real numbers b for which $b^{1/n}$ is defined.
- The denominator n indicates the index of the radical, and the numerator m indicates the power to which the radical is raised.

Negative Rational Exponents

- If m and n are positive integers with no common factor other than 1, and if $b \neq 0$, then $b^{-m/n} = \dfrac{1}{b^{m/n}}$ for all real numbers b for which $b^{1/n}$ is defined.

8.2 EXERCISES

Concepts and Skills

1. One of the following is false. Identify the false statement and explain why it is not true.
   (i) The square roots of 16 are 4 and -4.
   (ii) $\sqrt{16} = 4$
   (iii) $16^{1/2} = 4$ or -4
   (iv) $64^{1/3} = 4$

2. Only one of the following expressions represents a real number. Identify and evaluate this expression. Then explain why the other expressions do not represent real numbers.
   (i) $(-4)^{1/2}$ (ii) $(-4)^{-1/2}$
   (iii) $4^{-1/2}$ (iv) $\left(-\dfrac{1}{4}\right)^{-1/2}$

In Exercises 3–18, evaluate the given expression.

3. $49^{1/2}$ **4.** $(-32)^{1/5}$

5. $(-64)^{1/3}$ **6.** $81^{1/4}$

7. $(-16)^{1/4}$ **8.** $(-36)^{1/2}$

9. $\left(\frac{8}{27}\right)^{1/3}$ **10.** $\left(\frac{1}{32}\right)^{1/5}$

11. $(-27)^{1/3}$ **12.** $16^{1/4}$

13. $-32^{1/5}$ **14.** $-64^{1/3}$

15. $-(-625)^{1/4}$ **16.** $(-64)^{1/2}$

17. $-625^{1/4}$ **18.** $-36^{1/2}$

In Exercises 19–24, use a calculator to evaluate the given expression. Round results to two decimal places.

19. $15^{1/4}$ **20.** $23^{1/3}$ **21.** $(-12)^{1/2}$

22. $(-7)^{1/4}$ **23.** $-4^{1/4}$ **24.** $-20^{1/2}$

25. When you convert $9^{2/5}$ to a radical, how do you interpret the numerator and denominator of the exponent?

26. Which of the following is the correct way to write $(\sqrt[3]{2})^5$ in exponential form? Use the words *index*, *base*, and *power* to explain how you know.

(i) $2^{5/3}$ (ii) $5^{2/3}$ (iii) $3^{5/2}$

In Exercises 27–34, write the given exponential expression as a radical. Assume all variables represent positive numbers.

27. $3x^{2/3}$ **28.** $-4y^{5/6}$

29. $(3x)^{2/3}$ **30.** $(2ab)^{3/4}$

31. $(2x + 3)^{2/3}$ **32.** $(t + 2)^{5/4}$

33. $3x^{1/2} + (2y)^{1/2}$ **34.** $(-2a)^{1/3} - 2b^{1/3}$

In Exercises 35–44, write the given radical expression in exponential form. Assume all variables represent positive numbers.

35. $\sqrt[3]{y^2}$ **36.** $\sqrt[5]{z^3}$

37. $\sqrt[4]{3x^3}$ **38.** $\sqrt[5]{7x^2y}$

39. $\sqrt{x^2 + 4}$ **40.** $\sqrt{x + 3}$

41. $\sqrt{t} - \sqrt{5}$ **42.** $\sqrt{2y} + 1$

43. $\sqrt[3]{(2x - 1)^2}$ **44.** $\sqrt[4]{(x + 1)^3}$

In Exercises 45–50, use your calculator to evaluate each quantity. Round your results to two decimal places.

45. $\sqrt[5]{-90}$ **46.** $\sqrt[4]{20}$ **47.** $-\sqrt[6]{740}$

48. $4\sqrt[7]{4020}$ **49.** $\sqrt[8]{230^5}$ **50.** $\sqrt[7]{49^3}$

51. Describe two ways to use a calculator to evaluate $\sqrt[5]{470}$.

52. Explain why $8^{2/(-3)}$ can be written as $(\sqrt[3]{8})^{-2}$.

In Exercises 53–74, evaluate the given expression without using a calculator.

53. $9^{-1/2}$ **54.** $(-27)^{-1/3}$

55. $\left(\frac{1}{9}\right)^{-1/2}$ **56.** $\left(\frac{1}{27}\right)^{-1/3}$

57. $27^{2/3}$ **58.** $36^{3/2}$

59. $(-27)^{-2/3}$ **60.** $-16^{-3/2}$

61. $32^{3/5}$ **62.** $25^{3/2}$

63. $(-16)^{-1/2}$ **64.** $(-4)^{3/2}$

65. $8^{-2/3}$ **66.** $8^{-4/3}$

67. $\left(\frac{8}{27}\right)^{-2/3}$ **68.** $\left(\frac{32}{243}\right)^{-4/5}$

69. $-(-8)^{-4/3}$ **70.** $(-8)^{5/3}$

71. $(-36)^{5/2}$ **72.** $-64^{5/6}$

73. $-4^{-5/2}$ **74.** $(-8)^{-2/3}$

In Exercises 75–80, evaluate the given expression.

75. $16^{1/2} + 49^{1/2}$ **76.** $16^{3/2} - 16^{3/4}$

77. $27^{1/3} + 4^{-1/2}$ **78.** $81^{-3/4} + 81^{-3/2}$

79. $(-8)^{2/3}(16^{1/4})$ **80.** $(9^{-3/2})(-27^{1/3})$

In Exercises 81–84, evaluate the given expression without using your calculator.

81. $16^{0.25}$ **82.** $(-27)^{0.\bar{3}}$

83. $9^{-0.5}$ **84.** $49^{1.5}$

In Exercises 85–90, use a calculator to evaluate the given expression. Round results to two decimal places.

85. $16^{2/3}$ **86.** $16^{-2/3}$

87. $\left(\frac{5}{9}\right)^{5/9}$ **88.** $10^{4/5}$

89. $(-5)^{-2/5}$ **90.** $(-9)^{2/3}$

In Exercises 91–96, supply the missing number to make each statement true.

91. $(\square)^{-1/2} = \frac{1}{4}$ **92.** $(\square)^{-1/4} = \frac{1}{2}$

93. $(\square)^{5/3} = -32$ **94.** $(\square)^{2/3} = 9$

95. $(-27)^{(\square)} = 9$ **96.** $(-8)^{(\square)} = \frac{1}{16}$

97. Use a window setting that displays x-values and y-values from 0 to 2 to produce the graphs of $y_1 = x^{1/2}$, $y_2 = x^{1/4}$, and $y_3 = x^{1/8}$.

(a) Where do the three graphs intersect? Why would *all* graphs of $y = x^{1/n}$, $n > 1$, intersect at these points?

(b) For what values of x is $x^{1/2} < x^{1/4} < x^{1/8}$?

(c) For what values of x is $x^{1/2} > x^{1/4} > x^{1/8}$?

98. Show that $16^{(1/2)/(2/3)} = 8$.

Real-Life Applications

99. A researcher has developed a mathematical model for predicting a final exam grade: $G = 60T^{2/9}$. In this model, G is the predicted grade, and T is the number of hours spent studying for the exam. Use a window setting that displays x-values from 1 to 10 and y-values from 0 to 100 to produce the graph of the model function.

(a) If the passing grade is 70, how many hours of study are required to pass the exam?

(b) How many additional hours of study would be needed to earn a 90 on the exam?

(c) What grade is predicted if a student does not study at all? According to the model, can a student earn a 100? How realistic is the model at these extreme points of the graph?

100. Suppose that the Department of Labor uses a mathematical model to relate annual federal funding F (in billions of dollars) for jobs creation to the unemployment rate R: $F = 10R^{-1/3}$. (If the current unemployment rate is 6%, $R = 6$.) Use a window setting that displays x-values from 1 to 10 and y-values from 0 to 10 to produce the graph of the model function.

(a) Approximately what amount of annual funding would be required to maintain an unemployment rate of 5%?

(b) If \$6.8 billion were spent annually for jobs creation, what would be the expected unemployment rate?

(c) Why is the model not valid for zero funding?

Modeling with Real Data

101. Over 70% of taxpayers file their tax returns early, that is, at least 1 month before the tax deadline. The accompanying bar graph shows the percentages of various age groups who file early. (Source: Bruskin/Goldring Research.)

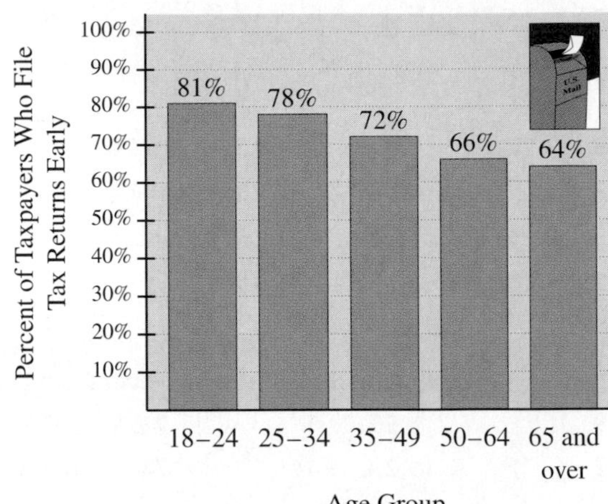

The function $f(x) = 154x^{-1/5}$ models the percentage of those of age x who expect to file early.

(a) What percentage of those of age 25 are expected to file early?

(b) Use a graph of the model function to estimate the age at which 75% are expected to file early.

102. The average age of U.S. commercial jetliners has increased from 1.1 years in 1961 to 14 years in 1995. (Source: Back Associates.) The function $A(x) = x^{3/4}$ models the average age as a function of the number of years since 1960.

(a) According to the model, what is the projected average age of the jetliner fleet in 2002?

(b) What is the predicted increase in the average age from 1997 to 2000?

Challenge

In Exercises 103–106, evaluate.

103. $\sqrt{16^{-1/2}}$

104. $\left(\sqrt{81}\right)^{-1/2}$

105. $\left(\sqrt[3]{64}\right)^{-1/2}$

106. $\sqrt[3]{729^{-1/2}}$

107. Use the integer setting on your calculator to produce the graph of $y = (-2)^{1/x}$. Explain why the graph is displayed as a set of distinguishable points rather than as a smooth curve.

108. Store 2 in N and produce the graph of $y = (x^N)^{1/N}$. Now store 3 in N and produce the graph again. Explain the difference in the graphs.

8.3 PROPERTIES OF RATIONAL EXPONENTS

Simplifying Exponential Expressions • Simplifying Radical Expressions

Simplifying Exponential Expressions

The properties of exponents that we stated for integer exponents in Sections 1.6 and 1.7 are valid for rational exponents, provided the expressions are defined.

Properties of Rational Exponents

Let m and n be rational numbers. For nonzero real numbers a and b for which the expressions are defined, the following are properties of exponents.

$a^m a^n = a^{m+n}$	Product Rule for Exponents
$\dfrac{a^m}{a^n} = a^{m-n}$	Quotient Rule for Exponents
$a^{-n} = \dfrac{1}{a^n}$	Definition of Negative Exponent
$(a^m)^n = a^{mn}$	Power to a Power Rule
$(ab)^n = a^n b^n$	Power of a Product Rule
$\left(\dfrac{a}{b}\right)^n = \dfrac{a^n}{b^n}$	Power of a Quotient Rule
$\left(\dfrac{a}{b}\right)^{-n} = \left(\dfrac{b}{a}\right)^n$	Negative Power of a Quotient Rule

EXAMPLE 1

Evaluating Exponential Expressions

Use the properties of exponents to evaluate each of the following.

(a) $(5^6)^{1/2}$ (b) $2^{1/2} \cdot 8^{1/2}$

(c) $\left(\dfrac{3^6}{5^3}\right)^{-2/3}$ (d) $16^{1/3} \cdot 16^{5/12}$

Solution

(a) $(5^6)^{1/2} = 5^{6\cdot(1/2)}$ Power to a Power Rule

$= 5^3 = 125$

(b) $2^{1/2} \cdot 8^{1/2} = [2\cdot 8]^{1/2}$ Power of a Product Rule

$= 16^{1/2} = \sqrt{16} = 4$

(c) $\left(\dfrac{3^6}{5^3}\right)^{-2/3} = \dfrac{3^{6\cdot(-2/3)}}{5^{3\cdot(-2/3)}}$ Power of a Quotient Rule

$= \dfrac{3^{-4}}{5^{-2}} = \dfrac{5^2}{3^4} = \dfrac{25}{81}$

LEARNING TIP

You can change negative exponents to positive exponents at the beginning or at the end of the problem. You will develop a preference as you gain experience.

The following is an alternative method.

$$\left(\frac{3^6}{5^3}\right)^{-2/3} = \left(\frac{5^3}{3^6}\right)^{2/3} \qquad \left(\frac{a}{b}\right)^{-n} = \left(\frac{b}{a}\right)^n$$

$$= \frac{5^{3\cdot(2/3)}}{3^{6\cdot(2/3)}} \qquad \text{Power of a Quotient Rule}$$

$$= \frac{5^2}{3^4} = \frac{25}{81}$$

(d) $16^{1/3} \cdot 16^{5/12} = 16^{(1/3)+(5/12)}$ Product Rule for Exponents

$$= 16^{(4/12)+(5/12)}$$

$$= 16^{9/12} = 16^{3/4} = \left(\sqrt[4]{16}\right)^3 = 2^3 = 8$$

Recall that $\sqrt[n]{b^n} = |b|$ if n is even and $\sqrt[n]{b^n} = b$ if n is odd. But $\sqrt[n]{b^n} = (b^n)^{1/n}$. Thus we can state the following generalization.

Simplifying Expressions of the Form $(b^n)^{1/n}$

For an even positive integer n, $(b^n)^{1/n} = |b|$.

For an odd integer n, $n > 1$, $(b^n)^{1/n} = b$.

EXAMPLE 2

Simplifying Expressions of the Form $(b^n)^{1/n}$

(a) $[(-3)^2]^{1/2} = |-3| = 3$ Because n is even, absolute value symbols are needed.

(b) $[(-3)^5]^{1/5} = -3$ Because n is odd, absolute value symbols are not used.

(c) $(y^3)^{1/3} = y$ Because n is odd, absolute value symbols are not used.

(d) $(z^8)^{1/8} = |z|$ Because n is even, absolute value symbols are needed.

Unless otherwise indicated, the variables in the remainder of this section will represent positive numbers. This will allow us to perform operations with exponential and radical expressions without regard to absolute value in the results.

EXAMPLE 3

Simplifying Exponential Expressions

Use the properties of exponents to simplify each of the following. Write all results with positive exponents.

(a) $\dfrac{x^{1/3}}{x^{4/3}}$ (b) $y^{1/3} \cdot y^{1/6}$ (c) $(a^3b^{-1})^{-1/2}$

(d) $\left(\dfrac{x^{1/2}}{y^3}\right)^{-1/3}$ (e) $a^{1/3}(a^{5/3} - a^{-2/3})$

Solution

(a) $\dfrac{x^{1/3}}{x^{4/3}} = x^{1/3-4/3}$ Quotient Rule for Exponents

$$= x^{-3/3} = x^{-1} = \frac{1}{x}$$

(b) $y^{1/3} \cdot y^{1/6} = y^{1/3+1/6}$ Product Rule for Exponents

$= y^{2/6+1/6} = y^{3/6} = y^{1/2}$

(c) $(a^3b^{-1})^{-1/2} = a^{3\cdot(-1/2)}b^{-1\cdot(-1/2)}$ Power of a Product Rule

$= a^{-3/2}b^{1/2} = \dfrac{b^{1/2}}{a^{3/2}}$

(d) $\left(\dfrac{x^{1/2}}{y^3}\right)^{-1/3} = \left(\dfrac{y^3}{x^{1/2}}\right)^{1/3}$ $\left(\dfrac{a}{b}\right)^{-n} = \left(\dfrac{b}{a}\right)^{n}$

$= \dfrac{y^{3\cdot(1/3)}}{x^{(1/2)\cdot(1/3)}}$ Power of a Quotient Rule

$= \dfrac{y}{x^{1/6}}$

(e) $a^{1/3}(a^{5/3} - a^{-2/3}) = a^{1/3} \cdot a^{5/3} - a^{1/3} \cdot a^{-2/3}$ Distributive Property

$= a^{1/3+5/3} - a^{1/3-2/3}$ Product Rule for Exponents

$= a^{6/3} - a^{-1/3}$

$= a^2 - \dfrac{1}{a^{1/3}}$ Definition of negative exponent

Think About It

Factor $x - 9$ by writing x as $(x^{1/2})^2$. Use rational exponents to factor $x^{2/3} - 4$.

Simplifying Radical Expressions

If the index of a radical and the exponent of the radicand have a common factor, the expression can be rewritten with a smaller index. We call this process **reducing the index.**

As we will see in Example 4, the technique begins with writing the radical in exponential form.

EXAMPLE 4

Reducing the Index of a Radical

Write the given expression with a smaller index.

(a) $\sqrt[12]{9^6}$ (b) $\sqrt[15]{t^{10}}$ (c) $\sqrt[6]{8y^3}$

Solution

(a) $\sqrt[12]{9^6} = 9^{6/12} = 9^{1/2} = \sqrt{9} = 3$

(b) $\sqrt[15]{t^{10}} = t^{10/15} = t^{2/3} = \sqrt[3]{t^2}$

(c) $\sqrt[6]{8y^3} = \sqrt[6]{(2y)^3} = (2y)^{3/6} = (2y)^{1/2} = \sqrt{2y}$

The properties of exponents can also be used to write a term with more than one radical as a term involving only one radical.

EXAMPLE 5

Reducing the Number of Radicals in a Term

Write each expression with only one radical symbol.

(a) $\sqrt[3]{\sqrt{2}}$ (b) $\sqrt[3]{x} \cdot \sqrt[4]{x}$

Solution

(a) $\sqrt[3]{\sqrt{2}} = (2^{1/2})^{1/3} = 2^{1/6} = \sqrt[6]{2}$

(b) $\sqrt[3]{x} \cdot \sqrt[4]{x} = x^{1/3} \cdot x^{1/4} = x^{4/12} \cdot x^{3/12} = x^{7/12} = \sqrt[12]{x^7}$

8.3 QUICK REFERENCE

Simplifying Exponential Expressions

- The properties of exponents that we stated for integer exponents in Sections 1.6 and 1.7 are valid for rational exponents, provided the expressions are defined.
- Expressions of the form $(b^n)^{1/n}$, where n is an integer, $n > 1$, can be simplified according to the following rules:

  If n is even, $(b^n)^{1/n} = |b|$.

  If n is odd, $(b^n)^{1/n} = b$.

Simplifying Radical Expressions

- If the index of a radical and the exponent of the radicand have a common factor, we can **reduce the index.**
  1. Write the radical in exponential form, and reduce the exponent to lowest terms.
  2. Write the exponential expression as a radical.
- The properties of exponents can be used to write a term with more than one radical as a term involving only one radical.

8.3 EXERCISES

Concepts and Skills

1. If a and b are positive real numbers, only one of the following statements is true.

   (i) $(a^2b^2)^{1/2} = ab$ (ii) $(a^2 + b^2)^{1/2} = a + b$

   Identify the statement that is true, and state a rational exponent rule that supports your claim. For the statement that is not true, select values for a and b for which the left and right sides are not equal.

2. Explain why it is easier to simplify $[(x^{1/3})^{1/5}]^{1/2}$ than it is to simplify $x^{1/3} \cdot x^{1/5} \cdot x^{1/2}$.

In Exercises 3–22, use the properties of exponents to evaluate the given expression.

3. $(6^3)^{2/3}$
4. $(3^{-6})^{1/3}$
5. $(2^{2/7})^{14}$
6. $(3^{2/5})^{10}$
7. $12^{1/2} \cdot 3^{1/2}$
8. $2^{2/3} \cdot 4^{2/3}$
9. $\left(\dfrac{7^3}{2^6}\right)^{-2/3}$
10. $\left(\dfrac{5^6}{3^9}\right)^{1/3}$
11. $65^{1/3} \cdot 65^{2/3}$
12. $7^{1/4} \cdot 7^{-1/4}$
13. $49^{7/10} \cdot 49^{-1/5}$
14. $16^{-1} \cdot 16^{1/2}$
15. $8^{2/9} \cdot 8^{1/9}$
16. $16^{5/8} \cdot 16^{-3/8}$
17. $6^{0.6} \cdot 6^{0.4}$
18. $5^{0.22} \cdot 5^{1.78}$
19. $\dfrac{27^{1/3}}{27^{2/3}}$
20. $\dfrac{16^{-3/8}}{16^{-1/8}}$
21. $8^{1/6} \div 8^{1/2}$
22. $25^{1/4} \div 25^{3/4}$

23. Which of the following is true? Explain your choice.

    (i) $x^{2/5} \cdot x^{5/2} = x^1 = x$

    (ii) $x^{2/5} \cdot x^{5/2} = x^{29/10}$

24. Describe the conditions under which each of the following is true.

    (a) $(x^n)^{1/n} = x$ (b) $(x^n)^{1/n} = |x|$

In Exercises 25–40, use the properties of exponents to simplify the given expression. Write the result with only positive exponents.

25. $x^{2/3} \cdot x^{1/3}$

26. $a^{-2/5} \cdot a^{4/5}$

27. $t \cdot t^{5/7}$

28. $b^{-1/3} \cdot b^{1/4} \cdot b^{5/6}$

29. $(a^{-1/2}b)(a^{3/4}b^{1/2})$

30. $(x^{1/3}y^{-1/3})(x^{2/3}y^{-2/3})$

31. $\dfrac{z^{1/2}}{z^{1/3}}$

32. $\dfrac{m^{1/2}}{m^2}$

33. $\dfrac{a^{1/3}b}{b^{-1/3}}$

34. $\dfrac{x^{1/2}y^{-2}}{xy^{1/2}}$

35. $(a^{-2})^{1/2}$

36. $(y^{-3/4})^{-4}$

37. $(y^{2/3})^{3/2}$

38. $(z^{3/4})^{-4/3}$

39. $(a^6b^4)^{3/2}$

40. $\left(\dfrac{a^6}{b^8}\right)^{-1/2}$

In Exercises 41–44, supply the missing number.

41. $(9x^4)^{(\square)} = 3x^2$

42. $x^{1/2} \cdot x^{(\square)} = x^{3/2}$

43. $(x^{2/3})^{(\square)} = x$

44. $\dfrac{8^{2/3}}{8^{(\square)}} = 32$

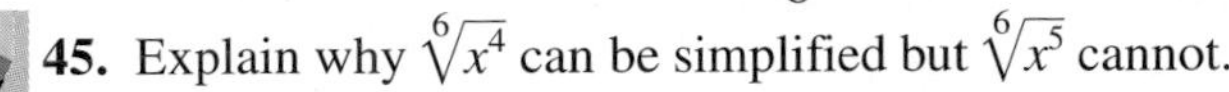
45. Explain why $\sqrt[6]{x^4}$ can be simplified but $\sqrt[6]{x^5}$ cannot.

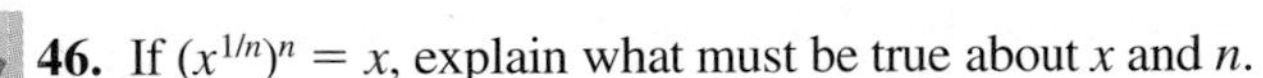
46. If $(x^{1/n})^n = x$, explain what must be true about x and n.

In Exercises 47–54, simplify the expression. All variables represent any real number.

47. $[(-2)^6]^{1/2}$

48. $[(-3)^3]^{1/3}$

49. $(a^{15})^{1/5}$

50. $(x^4)^{1/4}$

51. $(a^4b^2)^{1/2}$

52. $(x^3y^6)^{1/3}$

53. $[(x + y)^2]^{1/2}$

54. $[(2x - 1)^2]^{1/2}$

In Exercises 55–72, use the properties of exponents to simplify the given expression. Write the result with only positive exponents.

55. $(a^{-4}b^{1/3})^{1/2}$

56. $(x^2y^{-1})^{-1/2}$

57. $(a^6b^{-9})^{2/3}$

58. $(25a^4b^6)^{1/2}$

59. $(-3^{2/3}x^{1/2}y^{-1/3})^3$

60. $(x^{3/4}y^{-3/8}z^{-1/4})^8$

61. $\left(\dfrac{a^{4/5}}{b^{8/9}}\right)^{-15/4}$

62. $\left(\dfrac{x^6}{y^{-3/2}}\right)^{-2/3}$

63. $\left(\dfrac{x^{-1/3}}{y^{-1/2}}\right)^{-6}$

64. $\left(\dfrac{a^{-4}}{b^{-2}}\right)^{-5/2}$

65. $\left(\dfrac{16x^{-12}y^8}{z^4}\right)^{1/4}$

66. $\left(\dfrac{27x^{-12}y^9}{z^{-6}}\right)^{1/3}$

67. $(a^{-2/3}b^3)^6(a^{5/7}b^{-2})^7$

68. $(-27x^9)^{1/3}(2x^{-1/2})^2$

69. $\dfrac{(a^{5/6}b^{-2/5}c^{7/60})^{30}}{a^{21}b^{-5}}$

70. $\dfrac{(x^{3/4}y^{-5/6}z^{-5/24})^{12}}{x^7z^{1/2}}$

71. $\dfrac{(9x^4y^{-2})^{-1/2}}{(x^6y^{-3})^{1/3}}$

72. $\left(\dfrac{x^{-2}y^{2/3}}{z^{-1}}\right)\left(\dfrac{8x^3y}{z^6}\right)^{1/3}$

In Exercises 73–80, find the product.

73. $a^{3/4}(a^{5/4} + a^{-3/4})$

74. $y^{-2/3}(y^{5/3} + y^{2/3})$

75. $x^{-1/4}(x^{9/4} - x^{2/3})$

76. $b^{7/5}(b^{-2/5} - b^{4/15})$

77. $(x^{1/2} + y^{1/2})(x^{1/2} - y^{1/2})$

78. $(3^{1/2} + 2^{1/2})(3^{1/2} - 2^{1/2})$

79. $(x^{1/2} - 3^{1/2})^2$

80. $(x^{1/2} + y^{1/2})^2$

In Exercises 81–88, reduce the index.

81. $\sqrt[6]{x^3}$

82. $\sqrt[4]{x^2}$

83. $\sqrt[9]{8x^3}$

84. $\sqrt[8]{81x^4}$

85. $\sqrt[10]{x^4y^6}$

86. $\sqrt[8]{x^2y^4}$

87. $\left(\sqrt[4]{a}\right)^2$

88. $\left(\sqrt[6]{a}\right)^2$

In Exercises 89–96, write the given expression with just one radical symbol.

89. $\sqrt[3]{x^2}\,\sqrt[4]{x}$

90. $\sqrt[3]{x}\,\sqrt[5]{x}$

91. $\dfrac{\sqrt[4]{x}}{\sqrt[3]{x^2}}$

92. $\dfrac{x}{\sqrt[3]{x}}$

93. $\sqrt[4]{\sqrt[3]{7}}$

94. $\sqrt[4]{\sqrt[5]{x^3}}$

95. $\sqrt{2}\,\sqrt[3]{5}$

96. $\sqrt[3]{7}\,\sqrt{3}$

Real-Life Applications

97. A criminal justice expert developed a model to predict the number of crimes in a city for a given year. According to the model, for every 1000 people in a city, $16p^{1/4}$ people would be crime victims, where p is the population of the city (in thousands).

(a) Suppose that a city's population is 90,000. According to the model, for each 1000 people, how many would be crime victims?

(b) Write a simplified expression for the total number of crimes in a given year in a city with population of p thousand.

(c) How many total crimes does the model predict for a city with a population of 1 million?

98. Outdoor Enthusiast is a large mail-order company that sells clothes and equipment for outdoor activities. The company predicts that following the issue of a new catalog, the total number of daily orders (in thousands) will be $3w^{-1/3}$, where w is the num-

ber of weeks since the catalog was issued, and that the total revenue (in thousands of dollars) will be $1200w^{-4/5}$.

(a) Write a simplified expression for the average dollar amount of an order.

(b) For each order, about 1% of the price of the order covers the cost of shipping the items. What is the average shipping cost per order 4 weeks after a catalog is issued?

Group Project: National Park Service

The following table shows the number (in millions) of visitors to all National Park Service sites during the period 1990–1994. Also shown in the table are the expenditures (in millions of dollars) for the National Park Service during the same period. (Source: U.S. National Park Service.)

	1990	1991	1992	1993	1994
Visitors	258.7	267.8	274.7	273.1	277.6
Expenditures	986.1	1104.4	1268.7	1429.4	1404.0

With t representing the number of years since 1985, the function $V(t) = 195t^{1/6}$ models the number of visitors, and the function $E(t) = 340t^{2/3}$ models the expenditures.

99. Write the function $C(t) = \dfrac{E(t)}{V(t)}$, and simplify the expression to the exponential form at^n, where a is rounded to the nearest tenth and n is a fraction.

100. What is your interpretation of $C(t)$?

101. Produce a first-quadrant graph of $C(t)$. Using your interpretation in Exercise 100, what can you conclude?

102. Suppose that a senator sees the data in Exercise 101 and concludes that appropriations for the National Park Service should be cut. How would you support or oppose this decision?

Challenge

Sometimes expressions with rational exponents have a common factor that can be factored out of the expression. Remember to look for the smallest exponent.

Example: $5x^{-2/3} - 8x^{1/3} = x^{-2/3}(5 - 8x)$

In Exercises 103–106, use this technique to factor the expression.

103. $3x^{-1/2} + 5x^{3/2}$

104. $a^{-1/2} + a^{-5/2}$

105. $5a^{1/2} - 10a^{-1/2}$

106. $6x^{1/3} - 3x^{4/3}$

In Exercises 107–110, note that the given expression is in the form $ax^2 + bx + c$, and use the trial-and-check method to factor the expression.

107. $t^{2/5} + 4t^{1/5} - 5$

108. $3x^{4/3} + 2x^{2/3} - 5$

109. $2x^{1/3} - 7x^{1/6} - 15$

110. $6y^{2/3} - 5y^{1/3} + 1$

In Exercises 111 and 112, write the given expression with just one radical symbol.

111. $\sqrt{\sqrt{\sqrt{x}}}$

112. $\sqrt[4]{\sqrt[3]{\sqrt{x}}}$

8.4 THE PRODUCT RULE FOR RADICALS

Multiplying Radicals • Simplifying with the Product Rule

Multiplying Radicals

The easiest operation to perform with radicals is multiplication.

EXPLORING THE CONCEPT

Products of Radicals

We begin our discussion of products of radicals with a product involving two familiar square roots.

$$\sqrt{4} \cdot \sqrt{9} = 2 \cdot 3 = 6$$

Because $\sqrt{36}$ is also 6, we conclude that $\sqrt{4} \cdot \sqrt{9} = \sqrt{36}$. This result suggests that the product of two square roots is the square root of the product of the radicands. In Fig. 8.7 we see further evidence to support this conjecture, and we see that the proposed rule can be extended to cube roots.

Figure 8.7

```
√6√2
          3.464101615
√12
          3.464101615
³√15³√3
          3.556893304
³√45
          3.556893304
```

(i) $\sqrt{6}\sqrt{2} = \sqrt{12}$

(ii) $\sqrt[3]{15}\sqrt[3]{3} = \sqrt[3]{45}$

Finally, we can use the Product to a Power Rule for Exponents to justify the proposed rule.

$$\sqrt[n]{a} \cdot \sqrt[n]{b} = a^{1/n} \cdot b^{1/n} = (ab)^{1/n} = \sqrt[n]{ab}$$

Of course, the usual restrictions apply when the index n is even.

We summarize our findings with the following rule.

The Product Rule for Radicals

For all real numbers a and b for which the operations are defined,

$$\sqrt[n]{a} \cdot \sqrt[n]{b} = \sqrt[n]{ab}$$

Note: The Product Rule for Radicals applies only if the indices are the same.

So that all radicals are defined and results can be expressed without absolute value, we will assume that all variables in this section represent positive numbers.

EXAMPLE 1

Using the Product Rule for Radicals

Use the Product Rule for Radicals to multiply the radical expressions.

(a) $\sqrt{2a} \cdot \sqrt{7b}$ (b) $\sqrt[4]{\frac{1}{x}} \cdot \sqrt[4]{\frac{2}{y}}$ (c) $\sqrt[3]{4y} \cdot \sqrt[3]{3y}$ (d) $\sqrt{2} \cdot \sqrt[3]{3}$

Solution

(a) $\sqrt{2a} \cdot \sqrt{7b} = \sqrt{2a \cdot 7b} = \sqrt{14ab}$

(b) $\sqrt[4]{\frac{1}{x}} \cdot \sqrt[4]{\frac{2}{y}} = \sqrt[4]{\frac{1}{x} \cdot \frac{2}{y}} = \sqrt[4]{\frac{2}{xy}}$

(c) $\sqrt[3]{4y} \cdot \sqrt[3]{3y} = \sqrt[3]{12y^2}$

(d) The Product Rule for Radicals does not apply to $\sqrt{2} \cdot \sqrt[3]{3}$ because the indices are not the same.

Think About It

For what values of n are the expressions in each part equal?
(i) $\sqrt{n}\sqrt{n}, n$ (ii) $\sqrt{-n}\sqrt{n}, -n$
(iii) $\sqrt{-n}\sqrt{-n}, n$

Simplifying with the Product Rule

In Section 8.1 we found that we could determine the indicated root of any radical that can be written in the form $\sqrt[n]{b^n}$.

$$\sqrt{36x^8} = \sqrt{(6x^4)^2} = 6x^4$$
$$\sqrt[3]{8y^9} = \sqrt[3]{(2y^3)^3} = 2y^3$$

The radicand $36x^8$ is a *perfect square* because each factor can be written with an exponent of 2: $36x^8 = 6^2 \cdot (x^4)^2$. Similarly, $8y^9$ is a *perfect cube* because each factor can be written with an exponent of 3: $8y^9 = 2^3 \cdot (y^3)^3$.

LEARNING TIP

Observe that a^m is a perfect nth power if m is divisible by n.

In general, an expression is a **perfect *n*th power** if each of its factors can be written with an exponent of n.

Another method for determining the indicated roots of $\sqrt{36x^8}$ and $\sqrt[3]{8y^9}$ is to use the Product Rule for Radicals.

$$\sqrt{36x^8} = \sqrt{36} \cdot \sqrt{x^8} = 6x^4 \qquad \sqrt{ab} = \sqrt{a}\sqrt{b}$$
$$\sqrt[3]{8y^9} = \sqrt[3]{8} \cdot \sqrt[3]{y^9} = 2y^3 \qquad \sqrt[3]{ab} = \sqrt[3]{a}\sqrt[3]{b}$$

Because the radicand in each case is a perfect nth power, either of the preceding methods can be used to determine the indicated root or write the expression without a radical symbol. This process is an instance of *simplifying* a radical.

We say that a radical is *simplified* if it satisfies three conditions. In this section we consider the first condition.

Simplifying a Radical: Condition 1

The radicand of a simplified nth root radical must not contain a perfect nth power factor.

If the radicand is not a perfect nth power, we may still be able to use the Product Rule for Radicals to satisfy condition 1. If the radicand is an integer, write it as a product of the largest perfect nth power factor of the integer and another factor.

EXAMPLE 2

Using the Product Rule to Simplify a Radical

Simplify each radical. Use a calculator to verify that your result has the same value as the original radical.

(a) $\sqrt{75} = \sqrt{25 \cdot 3}$ — The largest perfect square factor of 75 is 25.

$= \sqrt{25} \cdot \sqrt{3}$ — Product Rule for Radicals

$= 5\sqrt{3}$

(b) $\sqrt{32} = \sqrt{16 \cdot 2}$ — The largest perfect square factor of 32 is 16.

$= \sqrt{16} \cdot \sqrt{2}$ — Product Rule for Radicals

$= 4\sqrt{2}$

(c) $\sqrt[3]{24} = \sqrt[3]{8 \cdot 3}$ — The largest perfect cube factor of 24 is 8.

$= \sqrt[3]{8} \cdot \sqrt[3]{3}$ — Product Rule for Radicals

$= 2\sqrt[3]{3}$

(d) $\sqrt[4]{48} = \sqrt[4]{16 \cdot 3}$ — The largest perfect fourth power factor of 48 is 16.

$= \sqrt[4]{16} \cdot \sqrt[4]{3}$ — Product Rule for Radicals

$= 2\sqrt[4]{3}$

(e) $\sqrt[3]{-54} = \sqrt[3]{-27 \cdot 2}$ — We choose the negative perfect cube factor -27.

$= \sqrt[3]{-27} \cdot \sqrt[3]{2}$ — Product Rule for Radicals

$= -3\sqrt[3]{2}$

A similar procedure is used when the radicand contains a variable raised to a power.

EXAMPLE 3

Using the Product Rule to Simplify a Radical

(a) $\sqrt{y^7} = \sqrt{y^6 \cdot y}$ — The largest perfect square factor of y^7 is y^6.

$= \sqrt{y^6} \cdot \sqrt{y}$ — Product Rule for Radicals

$= y^3\sqrt{y}$

(b) $\sqrt[3]{a^4b^8} = \sqrt[3]{a^3 \cdot a \cdot b^6 \cdot b^2}$ — The largest perfect cube factors are a^3 and b^6.

$= \sqrt[3]{a^3b^6} \cdot \sqrt[3]{ab^2}$ — Product Rule for Radicals

$= ab^2\sqrt[3]{ab^2}$

(c) $\sqrt{x^2 + 12x + 36} = \sqrt{(x + 6)^2}$ — Factor the radicand.

$= x + 6$

To save a step in the simplifying process, we can just write the given radical as a product of two radicals with one of the radicands containing all the perfect *n*th power factors.

EXAMPLE 4

Using the Product Rule to Simplify a Radical

(a) $\sqrt{49a^7} = \sqrt{49a^6} \cdot \sqrt{a} = 7a^3\sqrt{a}$

(b) $\sqrt{300r^{12}s^9} = \sqrt{100r^{12}s^8} \cdot \sqrt{3s} = 10r^6s^4\sqrt{3s}$

(c) $\sqrt[3]{24x^9y^{11}} = \sqrt[3]{8x^9y^9} \cdot \sqrt[3]{3y^2} = 2x^3y^3\sqrt[3]{3y^2}$

(d) $\sqrt[5]{64p^{12}q^{23}} = \sqrt[5]{32p^{10}q^{20}} \cdot \sqrt[5]{2p^2q^3} = 2p^2q^4\sqrt[5]{2p^2q^3}$

(e) $\sqrt{2x^2 + 12x + 18} = \sqrt{2(x^2 + 6x + 9)}$ — Factor out the GCF 2.

$= \sqrt{2(x + 3)^2}$ — Factor the trinomial.

$= \sqrt{(x + 3)^2} \cdot \sqrt{2}$ Product Rule for Radicals

$= (x + 3)\sqrt{2}$ Note that parentheses are needed.

Sometimes we need to use the Product Rule for Radicals twice in the same problem. First, we use it to multiply the radicals. Then we use it again to simplify the result.

Note: Although it is often possible to do some simplifying before multiplying, it is generally less confusing to find the product first and then simplify.

EXAMPLE 5

Multiplying and Simplifying Radicals

(a) $\sqrt{12a^4b^4} \cdot \sqrt{3a^2b^6} = \sqrt{36a^6b^{10}}$ Product Rule for Radicals

$= 6a^3b^5$ The radicand is a perfect square.

(b) $\sqrt{6x^3y^7} \cdot \sqrt{8x^4y^3} = \sqrt{48x^7y^{10}}$ Product Rule for Radicals

$= \sqrt{16x^6y^{10}} \cdot \sqrt{3x}$ Product Rule for Radicals

$= 4x^3y^5\sqrt{3x}$

8.4 QUICK REFERENCE

Multiplying Radicals

- The Product Rule for Radicals

  For all real numbers a and b for which the operations are defined, $\sqrt[n]{a} \cdot \sqrt[n]{b} = \sqrt[n]{ab}$.

Simplifying with the Product Rule

- An expression is a **perfect *n*th power** if each of its factors can be written with an exponent of n.
- Simplifying a Radical: Condition 1

  The radicand of a simplified nth root radical must not contain a perfect nth power factor.
- To use the Product Rule for Radicals to simplify a radical, write the radical as a product of two radicals with one of the radicands containing all the perfect nth power factors.

8.4 EXERCISES

Concepts and Skills

1. Explain why the following is an incorrect application of the Product Rule for Radicals.

   $\sqrt{-3}\sqrt{-12} = \sqrt{36} = 6$

2. Explain whether the Product Rule for Radicals applies to the problem $\sqrt[3]{2}\,\sqrt{2}$.

In Exercises 3–10, use the Product Rule for Radicals to multiply the radicals.

3. $\sqrt{2}\sqrt{3}$

4. $\sqrt{7}\sqrt{x}$

5. $\sqrt{3x}\sqrt{5y}$

6. $\sqrt{2b}\sqrt{ac}$

7. $(-4\sqrt{2})(3\sqrt{7})$

8. $(2\sqrt[5]{6})(\sqrt[5]{3})$

9. $\sqrt[3]{x-2}\sqrt[3]{x+3}$

10. $\sqrt[4]{x}\sqrt[4]{x+2}$

11. What is the first step in simplifying the radical $\sqrt{x^2+10x+25}$? Why is this step necessary? Can this same step be used to simplify the radical $\sqrt{x^2+10x+21}$? Explain.

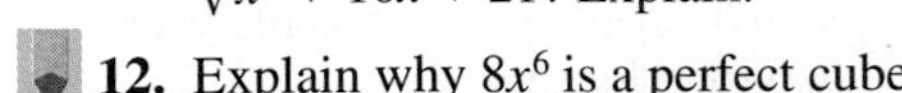

12. Explain why $8x^6$ is a perfect cube.

In Exercises 13–26, determine the indicated root.

13. $\sqrt{25y^2}$

14. $\sqrt{16x^6y^{10}}$

15. $\sqrt{49w^8t^4}$

16. $\sqrt{36x^8y^{12}}$

17. $\sqrt[3]{8t^6}$

18. $\sqrt[4]{16x^8y^{16}}$

19. $\sqrt[4]{81x^{12}y^{20}}$

20. $\sqrt[3]{27x^{27}y^{12}}$

21. $\sqrt{x^2(y-3)^4}$

22. $\sqrt{y^4(2x+7)^2}$

23. $\sqrt{x^2+12x+36}$

24. $\sqrt{y^2+6y+9}$

25. $\sqrt{4y^2+4y+1}$

26. $\sqrt{9t^2+12t+4}$

27. Explain why $\sqrt{32}$ can be simplified but $\sqrt{30}$ cannot.

28. To simplify $\sqrt{16x^{16}}$, we treat the coefficient and the exponent differently even though they are the same number. Explain what to do with each.

In Exercises 29–40, simplify the radical. Use your calculator to verify that the result has the same value as the original radical.

29. $\sqrt{12}$

30. $\sqrt{20}$

31. $\sqrt{50}$

32. $\sqrt{18}$

33. $\sqrt[3]{54}$

34. $\sqrt[3]{192}$

35. $\sqrt[3]{-48}$

36. $\sqrt[3]{-54}$

37. $\sqrt[4]{32}$

38. $\sqrt[4]{162}$

39. $\sqrt[5]{256}$

40. $\sqrt[5]{128}$

In Exercises 41–46, simplify.

41. $\sqrt{a^5b^3}$

42. $\sqrt{x^9y^4z^7}$

43. $\sqrt[3]{a^9b^8}$

44. $\sqrt[3]{x^{17}y^{15}}$

45. $\sqrt[4]{x^{16}y^9}$

46. $\sqrt[4]{a^5b^7}$

In Exercises 47–60, simplify.

47. $\sqrt{60t}$

48. $\sqrt{20x^{10}}$

49. $\sqrt{25x^{25}}$

50. $\sqrt{18a^9}$

51. $\sqrt{500a^7b^{14}}$

52. $\sqrt{100a^3b}$

53. $-\sqrt{49x^6y^7}$

54. $-\sqrt{8r^5t^9}$

55. $\sqrt{28w^6t^9}$

56. $\sqrt{75x^5y^7z^9}$

57. $\sqrt[3]{81x^9y^{11}}$

58. $\sqrt[3]{54x^8}$

59. $\sqrt[5]{-32x^7y^9}$

60. $\sqrt[4]{32x^6y^8}$

61. It would be incorrect to simplify $\sqrt{4x^2+16}$ by taking the square root of each term. What step must first be taken before the radical can be simplified?

62. Use the following methods to simplify $\sqrt{8x}\sqrt{2x^3}$.

(a) Simplify each radical, then multiply.

(b) Multiply, then simplify the result.

Which method seems easier?

In Exercises 63–68, simplify.

63. $\sqrt{12x+8y}$

64. $\sqrt{ab^2+b^3}$

65. $\sqrt{4x^2+16}$

66. $\sqrt{x^4+x^2}$

67. $\sqrt{9x^7y^6+9x^6y^7}$

68. $\sqrt{50a^3+75a^4b}$

In Exercises 69–74, simplify. Assume all variables represent any real number.

69. $\sqrt{9y^2+9y^4}$

70. $\sqrt{t^6+t^4}$

71. $\sqrt{r^4t^2+r^6t^4}$

72. $\sqrt{a^4b^6+a^2b^8}$

73. $\sqrt{y^2-10y+25}$

74. $\sqrt{2t^2-28t+98}$

In Exercises 75–90, multiply and simplify.

75. $(-2\sqrt{5})(2\sqrt{5})$

76. $(4\sqrt{6x})(-2\sqrt{2x})$

77. $\sqrt[3]{3}\sqrt[3]{54}$

78. $\sqrt[3]{2}\sqrt[3]{20}$

79. $\sqrt{7x}\sqrt{14x}$

80. $\sqrt{3y^2}\sqrt{15y}$

81. $\sqrt{3x^3}\sqrt{2x^6}$

82. $(\sqrt{3x^2y})^2$

83. $(2\sqrt{5})^2$

84. $(a\sqrt{3})^2$

85. $\sqrt{6xy^3}\sqrt{12x^3y^2}$

86. $\sqrt{3c^{11}d^{13}}\sqrt{15c^{12}d^{14}}$

87. $\sqrt[3]{25x^2y^2}\sqrt[3]{5x^3y^4}$

88. $(\sqrt[3]{9x^2y^4})^2$

89. $\sqrt[4]{54x^3y}\sqrt[4]{3x^2y^4}$

90. $\sqrt[4]{27x^2y^5}\sqrt[4]{3x^3y^3}$

In Exercises 91–96, supply the missing numbers to make the statement true.

91. $\sqrt{\square\, a^{\square} b^{\square}} = 4a^2b\sqrt{b}$

92. $\square\ \sqrt{4x^{\square}y^{\square}} = -6x^3y^2\sqrt{xy}$

93. $\sqrt{3x^{\square}y^3}\ \sqrt{\square x^2y^{\square}} = 3x^2y^3$

94. $\sqrt{6a^{\square}b^3}\ \sqrt{\square ab^{\square}} = 3a^2b^4\sqrt{2a}$

95. $\sqrt[3]{\square x^{\square}y^{\square}} = 2xy\sqrt[3]{2x}$

96. $\sqrt[3]{\square x^{\square}y^{\square}} = 5x^2y\sqrt[3]{2y^2}$

Modeling with Real Data

97. As shown in the accompanying table, the number of households with computers increased during the period 1990–1995. (Source: Energy Information Administration.)

Year	Percentage of Homes with Computers
1990	16%
1993	23%
1994	36%
1995	47%

The following are functions of the number of years since 1970.

Percentage of households with computers:

$$C(x) = 0.00002x^4\sqrt{x}$$

Number of households (in millions):

$$H(x) = 3.7\sqrt{11x + 410}$$

(a) Write a function that models the actual number of households with computers.

(b) How many households are projected to have computers in the year 2001?

98. In the United States, the number of deaths per 100,000 population due to motor vehicle accidents declined from 23.4 in 1980 to 16.5 in 1994. (Source: National Safety Council.) The following are model functions, where x is the number of years since 1970.

U.S. population (in millions):

$$P(x) = \sqrt{1150x + 41{,}000}$$

Motor vehicle deaths (per 100,000 population):

$$D(x) = \frac{78\sqrt{x}}{x}$$

(a) Write a function N that models the actual number of deaths due to motor vehicle accidents.

(b) Produce the graph of N, and describe the trend in the actual number of deaths.

Group Project: Union Workers

The accompanying bar graph shows that the percentage of unionized workers has declined in recent years. (Source: Bureau of Labor Statistics.)

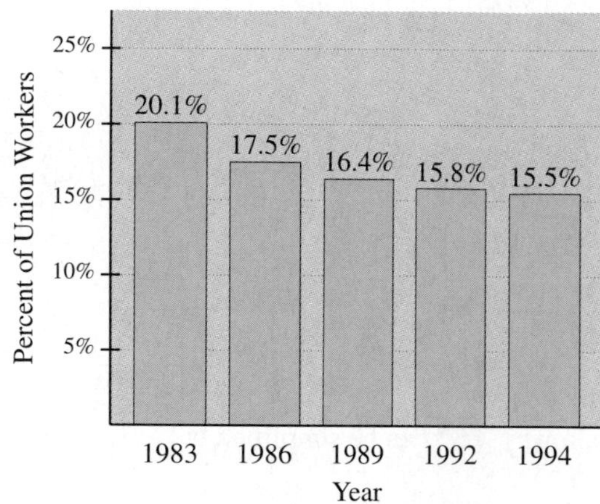

The data in the figure can be modeled by the function $M(t) = 20.13t^{-0.10}$, where M is the percentage of the work force belonging to unions and t is the number of years since 1982.

99. Use the data for 1983, 1989, and 1994 and your calculator's quadratic regression capability to obtain a function of the form $Q(t) = at^2 + bt + c$ that models the data.

100. Use your calculator to produce a scatter plot of the data in the bar graph, the graph of $M(t)$, and the graph of $Q(t)$ on the same screen with a maximum x-value of 15. How well do the two functions model the data?

101. Now extend the maximum x-value to 40 and reproduce the scatter plot and the two graphs. What long-range trends are indicated by the two models? Which model do you think is more realistic?

102. Suppose that the federal government passed legislation banning companies from permanently replacing striking employees. How might this action invalidate $M(t)$ as a model for future years?

Challenge

103. To simplify $\sqrt[6]{z^{10}}$, we can proceed as follows:

$$\sqrt[6]{z^{10}} = z^{10/6} = z^{5/3} = \sqrt[3]{z^5}$$

Is the radical now simplified? Why?

In Exercises 104–108, simplify.

104. $\sqrt[15]{x^{20}}$

105. $\sqrt[8]{16x^{12}y^{20}}$

106. $\sqrt[12]{27x^6y^9}$

107. $\sqrt[21]{64x^{18}y^{15}}$

108. $\sqrt[6]{8x^{15}y^9}$

109. Suppose you want to write a computer program that will factor any given positive integer n. One approach is to successively divide n by 2, then 3, then 4, and so on, until a number is found that divides n. Explain why it would not be necessary to continue testing numbers beyond $\sqrt{n}$.

8.5 THE QUOTIENT RULE FOR RADICALS

Dividing Radicals • Simplifying with the Quotient Rule • Rationalizing Monomial Denominators

Dividing Radicals

The following equalities show how a quotient of two square roots can be written as a single square root.

$$\frac{\sqrt{100}}{\sqrt{25}} = \frac{10}{5} = 2$$

$$\sqrt{\frac{100}{25}} = \sqrt{4} = 2$$

The fact that $\frac{\sqrt{100}}{\sqrt{25}} = \sqrt{\frac{100}{25}}$ suggests that the quotient of two square roots is the square root of the quotient of the radicands. We can use the Quotient to a Power Rule for Exponents to generalize this rule to quotients of nth root radicals.

$$\frac{\sqrt[n]{a}}{\sqrt[n]{b}} = \frac{a^{1/n}}{b^{1/n}} = \left(\frac{a}{b}\right)^{1/n} = \sqrt[n]{\frac{a}{b}}$$

The Quotient Rule for Radicals

For all real numbers a and b for which the operations are defined, $\frac{\sqrt[n]{a}}{\sqrt[n]{b}} = \sqrt[n]{\frac{a}{b}}$.

Note: The Quotient Rule for Radicals applies only if the indices are the same.

In this section we continue to assume that all variables represent positive numbers.

EXAMPLE 1

Using the Quotient Rule for Radicals

(a) $\dfrac{\sqrt[5]{64}}{\sqrt[5]{2}} = \sqrt[5]{\dfrac{64}{2}} = \sqrt[5]{32} = 2$

(b) $\dfrac{\sqrt{x^5}}{\sqrt{x^3}} = \sqrt{\dfrac{x^5}{x^3}} = \sqrt{x^2} = x$

(c) $\dfrac{\sqrt{6x^3}}{\sqrt{2x^2}} = \sqrt{\dfrac{6x^3}{2x^2}} = \sqrt{3x}$

Think About It

The Quotient Rule for Radicals does not apply to $\dfrac{\sqrt{8}}{\sqrt[3]{2}}$. Use another method to write this expression as a single radical.

After we divide radicals, the Product Rule for Radicals may be needed to simplify the result.

EXAMPLE 2

Using the Quotient and Product Rules for Radicals

(a) $\dfrac{\sqrt{a^{11}b^8c^5}}{\sqrt{a^2bc}} = \sqrt{\dfrac{a^{11}b^8c^5}{a^2bc}}$ Quotient Rule for Radicals

$= \sqrt{a^9b^7c^4}$ Divide the numerator by the denominator.

$= \sqrt{a^8b^6c^4}\sqrt{ab}$ Product Rule for Radicals

$= a^4b^3c^2\sqrt{ab}$

(b) $\dfrac{\sqrt[3]{48x^9y^{12}}}{\sqrt[3]{3xy^2}} = \sqrt[3]{\dfrac{48x^9y^{12}}{3xy^2}}$ Quotient Rule for Radicals

$= \sqrt[3]{16x^8y^{10}}$ Divide the numerator by the denominator.

$= \sqrt[3]{8x^6y^9}\sqrt[3]{2x^2y}$ Product Rule for Radicals

$= 2x^2y^3\sqrt[3]{2x^2y}$

After applying the Quotient Rule for Radicals, we may find that the numerator is not divisible by the denominator. However, we may be able to simplify the radicand.

EXAMPLE 3

Using the Quotient Rule for Radicals

(a) $\dfrac{\sqrt{45xy}}{\sqrt{x^3}} = \sqrt{\dfrac{45xy}{x^3}}$ Quotient Rule for Radicals

$= \sqrt{\dfrac{45y}{x^2}}$ Simplify the radicand.

$= \dfrac{\sqrt{9}\sqrt{5y}}{\sqrt{x^2}}$ Product Rule for Radicals

$= \dfrac{3\sqrt{5y}}{x}$

(b) $\dfrac{\sqrt[3]{72x^2y^7}}{\sqrt[3]{x^{11}}} = \sqrt[3]{\dfrac{72x^2y^7}{x^{11}}}$ Quotient Rule for Radicals

$= \sqrt[3]{\dfrac{72y^7}{x^9}}$ Simplify the radicand.

$= \dfrac{\sqrt[3]{8y^6} \cdot \sqrt[3]{9y}}{\sqrt[3]{x^9}}$ Product Rule for Radicals

$= \dfrac{2y^2\sqrt[3]{9y}}{x^3}$

Simplifying with the Quotient Rule

The following is the second condition for a simplified radical.

Simplifying a Radical: Condition 2

The radicand of a simplified radical must not contain a fraction.

The easiest fractional radicand to simplify is one in which the denominator is a perfect power. Then the Quotient Rule for Radicals can be applied directly to simplify the radical.

EXAMPLE 4

Simplifying a Radical with a Fractional Radicand

(a) $\sqrt{\dfrac{7}{9}} = \dfrac{\sqrt{7}}{\sqrt{9}}$ $\sqrt[n]{\dfrac{a}{b}} = \dfrac{\sqrt[n]{a}}{\sqrt[n]{b}}$

$= \dfrac{\sqrt{7}}{3}$ Simplify the denominator.

(b) $\sqrt[3]{\dfrac{9z}{125}} = \dfrac{\sqrt[3]{9z}}{\sqrt[3]{125}}$ $\sqrt[n]{\dfrac{a}{b}} = \dfrac{\sqrt[n]{a}}{\sqrt[n]{b}}$

$= \dfrac{\sqrt[3]{9z}}{5}$ Simplify the denominator.

(c) $\sqrt[4]{\dfrac{a^5}{16a^8}} = \dfrac{\sqrt[4]{a^5}}{\sqrt[4]{16x^8}}$ $\sqrt[n]{\dfrac{a}{b}} = \dfrac{\sqrt[n]{a}}{\sqrt[n]{b}}$

$= \dfrac{\sqrt[4]{a^4} \cdot \sqrt[4]{a}}{2x^2}$ Product Rule for Radicals

$= \dfrac{a\sqrt[4]{a}}{2x^2}$ $\sqrt[4]{a^4} = a,\ a > 0$

If the radicand is a fraction, it may be possible to perform a simple division to eliminate the fraction. If so, then we can use previously discussed procedures to complete the simplification.

EXAMPLE 5

Simplifying a Radical with a Fractional Radicand

$$\sqrt{\frac{45x^9y}{x^3}} = \sqrt{45x^6y}$$ Perform the indicated division in the radicand.

$$= \sqrt{9x^6}\sqrt{5y}$$ Product Rule for Radicals

$$= 3x^3\sqrt{5y}$$

Rationalizing Monomial Denominators

The following is the third condition for a simplified radical.

Simplifying a Radical: Condition 3

A simplified radical must not contain a radical in the denominator.

The expression $\frac{3}{\sqrt{5}}$ is not considered simplified because the radical $\sqrt{5}$ appears in the denominator. If we multiply the numerator and the denominator by $\sqrt{5}$, the resulting denominator will be a rational number, and the expression will be simplified.

$$\frac{3}{\sqrt{5}} = \frac{3 \cdot \sqrt{5}}{\sqrt{5} \cdot \sqrt{5}} = \frac{3\sqrt{5}}{\sqrt{25}} = \frac{3\sqrt{5}}{5}$$

This process is called **rationalizing the denominator.**

Note: We can save a step by recalling the fact that $\sqrt{n}\sqrt{n} = n$ for nonnegative numbers n. Also, remember that the $\sqrt{5}$ in the numerator and the 5 in the denominator are different numbers, and they cannot be divided out.

EXAMPLE 6

Rationalizing a Monomial Denominator

Rationalize the denominator.

(a) $\frac{\sqrt{7}}{\sqrt{3}}$ (b) $\frac{4\sqrt{3}}{\sqrt{2}}$ (c) $\frac{12}{\sqrt{27}}$

LEARNING TIP

When you simplify radicals, often the best approach is to multiply or divide first, then simplify with the Product or Quotient Rule, and finally rationalize the denominator.

Solution

(a) Multiply the numerator and denominator by $\sqrt{3}$.

$$\frac{\sqrt{7}}{\sqrt{3}} = \frac{\sqrt{7} \cdot \sqrt{3}}{\sqrt{3} \cdot \sqrt{3}} = \frac{\sqrt{21}}{3}$$

(b) Multiply the numerator and denominator by $\sqrt{2}$.

$$\frac{4\sqrt{3}}{\sqrt{2}} = \frac{4\sqrt{3} \cdot \sqrt{2}}{\sqrt{2} \cdot \sqrt{2}} = \frac{4\sqrt{6}}{2} = 2\sqrt{6}$$

(c) We can simplify the expression before rationalizing the denominator.

$$\frac{12}{\sqrt{27}} = \frac{12}{\sqrt{9}\sqrt{3}} \quad \text{Product Rule for Radicals}$$

$$= \frac{12}{3\sqrt{3}}$$

$$= \frac{4}{\sqrt{3}} \quad \text{Divide out the common factor 3.}$$

$$= \frac{4 \cdot \sqrt{3}}{\sqrt{3} \cdot \sqrt{3}} \quad \text{Multiply the numerator and denominator by } \sqrt{3}.$$

$$= \frac{4\sqrt{3}}{3}$$

In Example 7, instead of simplifying radicals first, we multiply the numerator and the denominator by a radical selected so that the resulting radicand in the denominator is a perfect square.

EXAMPLE 7

Rationalizing a Monomial Denominator

(a) $\sqrt{\frac{45}{8}} = \frac{\sqrt{45} \cdot \sqrt{2}}{\sqrt{8} \cdot \sqrt{2}} = \frac{\sqrt{90}}{\sqrt{16}} = \frac{\sqrt{9}\sqrt{10}}{4} = \frac{3\sqrt{10}}{4}$

(b) $\frac{3\sqrt{2a}}{\sqrt{x^3}} = \frac{3\sqrt{2a} \cdot \sqrt{x}}{\sqrt{x^3} \cdot \sqrt{x}} = \frac{3\sqrt{2ax}}{\sqrt{x^4}} = \frac{3\sqrt{2ax}}{x^2}$

(c) $\frac{x}{\sqrt{x^2+4}} = \frac{x \cdot \sqrt{x^2+4}}{\sqrt{x^2+4} \cdot \sqrt{x^2+4}} = \frac{x\sqrt{x^2+4}}{x^2+4}$

If the index of the radical in a denominator is 2, we multiply by a number that makes the result a perfect square. If the index is 3, we multiply by a number that makes the result a perfect cube. A similar strategy is used for higher indices.

EXAMPLE 8

Rationalizing Denominators with Higher Indices

Rationalize the denominator.

(a) $\frac{10}{\sqrt[3]{4}}$ (b) $\frac{\sqrt[3]{2}}{\sqrt[3]{r^4}}$ (c) $\frac{-2}{\sqrt[4]{y^7}}$

Solution

(a) Multiplying by $\sqrt[3]{4}$ would result in $\sqrt[3]{16}$ in the denominator, but 16 is not a perfect cube. Instead, we multiply by $\sqrt[3]{2}$ to obtain the perfect cube radicand 8.

$$\frac{10}{\sqrt[3]{4}} = \frac{10 \cdot \sqrt[3]{2}}{\sqrt[3]{4} \cdot \sqrt[3]{2}} = \frac{10\sqrt[3]{2}}{\sqrt[3]{8}} = \frac{10\sqrt[3]{2}}{2} = 5\sqrt[3]{2}$$

(b) Multiply by $\sqrt[3]{r^2}$ to obtain a perfect cube in the denominator.

$$\frac{\sqrt[3]{2}}{\sqrt[3]{r^4}} = \frac{\sqrt[3]{2}\cdot\sqrt[3]{r^2}}{\sqrt[3]{r^4}\cdot\sqrt[3]{r^2}} = \frac{\sqrt[3]{2r^2}}{\sqrt[3]{r^6}} = \frac{\sqrt[3]{2r^2}}{r^2}$$

(c) Multiply by $\sqrt[4]{y}$ to obtain a perfect fourth power in the denominator.

$$\frac{-2}{\sqrt[4]{y^7}} = \frac{-2\cdot\sqrt[4]{y}}{\sqrt[4]{y^7}\cdot\sqrt[4]{y}} = \frac{-2\sqrt[4]{y}}{\sqrt[4]{y^8}} = \frac{-2\sqrt[4]{y}}{y^2}$$

8.5 QUICK REFERENCE

Dividing Radicals

- The Quotient Rule for Radicals
  For all real numbers a and b for which the operations are defined, $\frac{\sqrt[n]{a}}{\sqrt[n]{b}} = \sqrt[n]{\frac{a}{b}}$.

Simplifying with the Quotient Rule

- Simplifying a Radical: Condition 2
  The radicand of a simplified radical must not contain a fraction.
- The Quotient Rule for Radicals can be used to satisfy condition 2 if the denominator is a perfect nth power (possibly after simplifying the fraction) or if the numerator is divisible by the denominator.

Rationalizing Monomial Denominators

- Simplifying a Radical: Condition 3
  A simplified radical must not contain a radical in the denominator.
- We satisfy condition 3 by **rationalizing the denominator.** The method is to multiply the numerator and the denominator by a number or expression selected so that the resulting radicand in the denominator is a perfect nth power.

8.5 EXERCISES

Concepts and Skills

1. Explain whether the Quotient Rule for Radicals applies to the expression $\frac{\sqrt[3]{16}}{\sqrt{2}}$.

2. To simplify $\frac{\sqrt{15}}{\sqrt{9}}$, why is it unnecessary to apply the Quotient Rule for Radicals?

In Exercises 3–12, divide and simplify, if possible.

3. $\frac{\sqrt{30}}{\sqrt{5}}$
4. $\frac{\sqrt{42t}}{\sqrt{7}}$
5. $\frac{\sqrt{48}}{\sqrt{3}}$
6. $\frac{\sqrt[3]{54}}{\sqrt[3]{2}}$
7. $\frac{\sqrt{x^7}}{\sqrt{x}}$
8. $\frac{\sqrt{45a^{11}b^3}}{\sqrt{5ab}}$
9. $\frac{\sqrt{49a^7b^4}}{\sqrt{4a}}$
10. $\frac{\sqrt{81a^5b^7}}{\sqrt{16a^3b^5}}$
11. $\frac{\sqrt[3]{x^{11}}}{\sqrt[3]{x^8}}$
12. $\frac{\sqrt[4]{x^{15}}}{\sqrt[4]{x^{11}}}$

In Exercises 13–20, divide and simplify.

13. $\dfrac{\sqrt{x^5y^{12}}}{\sqrt{xy^3}}$

14. $\dfrac{\sqrt{60a^{10}b^9}}{\sqrt{3a^3b^9}}$

15. $\dfrac{\sqrt{45w^9t^8}}{\sqrt{w^2t^3}}$

16. $\dfrac{\sqrt{40a^{14}b^7}}{\sqrt{5a^9b^4}}$

17. $\dfrac{\sqrt{18x^9d^{13}}}{\sqrt{2x^4d^{10}}}$

18. $\dfrac{\sqrt{144r^{11}s^{12}}}{\sqrt{2r^4}}$

19. $\dfrac{\sqrt[3]{54x^{13}}}{\sqrt[3]{x^2}}$

20. $\dfrac{\sqrt[4]{32a^7b^{15}}}{\sqrt[4]{2a^2b^5}}$

In Exercises 21–30, simplify.

21. $\dfrac{\sqrt{10}}{\sqrt{360}}$

22. $\dfrac{\sqrt[4]{3}}{\sqrt[4]{48}}$

23. $\dfrac{\sqrt{t^3}}{\sqrt{t^9}}$

24. $\dfrac{\sqrt{12a^4b^7}}{\sqrt{3a^{10}b^5}}$

25. $\dfrac{\sqrt{3a^3}}{\sqrt{75a^5}}$

26. $\dfrac{\sqrt{18t^7}}{\sqrt{50t^{13}}}$

27. $\dfrac{\sqrt{16xy}}{\sqrt{9x^3y^5}}$

28. $\dfrac{\sqrt{ab^3}}{\sqrt{a^7b^5}}$

29. $\dfrac{\sqrt[4]{x^6}}{\sqrt[4]{16x^2}}$

30. $\dfrac{\sqrt[3]{24x^5}}{\sqrt[3]{3x^{20}}}$

31. If P and Q are monomials, describe two circumstances in which simplifying $\sqrt{\dfrac{P}{Q}}$ would be possible without rationalizing the denominator.

32. We can satisfy condition 2 by writing $\sqrt{\dfrac{5x^4}{x}}$ as $\sqrt{5x^3}$. Explain why this result is still not simplified.

In Exercises 33–40, simplify.

33. $\sqrt{\dfrac{5}{49}}$

34. $\sqrt{\dfrac{17x}{100}}$

35. $\sqrt[3]{\dfrac{16x^2}{2x^{14}}}$

36. $\sqrt[4]{\dfrac{y^{11}}{81y^3}}$

37. $\sqrt{\dfrac{28}{9}}$

38. $\sqrt{\dfrac{125a}{144}}$

39. $\sqrt[3]{\dfrac{-16a}{1000}}$

40. $\sqrt[4]{\dfrac{243w}{16}}$

In Exercises 41–54, simplify.

41. $\sqrt{\dfrac{60x^{11}y^3}{3y}}$

42. $\sqrt{\dfrac{24x^{13}y^5}{2x^3}}$

43. $\sqrt{\dfrac{75x^7y^{13}}{x^3y^5}}$

44. $\sqrt{\dfrac{150x^{15}y^7}{2y^3}}$

45. $\sqrt{\dfrac{245x^7y^9z^{12}}{5x^3y^2z^5}}$

46. $\sqrt{\dfrac{54r^7s^5w^9}{3rw}}$

47. $\sqrt[3]{\dfrac{72a^9b^{10}}{3a^2b}}$

48. $\sqrt[3]{\dfrac{108a^5b^7c^9}{2abc}}$

49. $\sqrt{\dfrac{x^3y^5}{x^5y}}$

50. $\sqrt{\dfrac{ab^5}{a^5b^4}}$

51. $\sqrt{\dfrac{2a^{12}b^3}{18a^9b^9}}$

52. $\sqrt{\dfrac{8b^5c^4}{2b^7c^3}}$

53. $\sqrt[4]{\dfrac{x^7}{16x^2}}$

54. $\sqrt[3]{\dfrac{40x^{10}y^5}{5xy^{11}}}$

55. To simplify $\dfrac{5}{2\sqrt{3}}$, should you multiply the numerator and denominator by $2\sqrt{3}$ or by just $\sqrt{3}$? Will you obtain the correct result either way? If so, which method do you prefer?

56. To simplify $\dfrac{1}{\sqrt[n]{x^m}}$, we must multiply the numerator and denominator by a radical of the form $\sqrt[n]{x^k}$. For the method to succeed, what must be the relationship among m, k, and n?

In Exercises 57–66, rationalize the denominator. Use your calculator to verify that the result and the original expression have the same value.

57. $\dfrac{5}{\sqrt{5}}$

58. $\dfrac{6}{\sqrt{3}}$

59. $\dfrac{-\sqrt{5}}{\sqrt{3}}$

60. $\dfrac{1}{3\sqrt{2}}$

61. $\dfrac{15}{\sqrt{10}}$

62. $\dfrac{\sqrt{6}}{-3\sqrt{12}}$

63. $\dfrac{7}{\sqrt[3]{3}}$

64. $\dfrac{9}{\sqrt[3]{25}}$

65. $\dfrac{9}{\sqrt[4]{8}}$

66. $\dfrac{2}{\sqrt[5]{8}}$

In Exercises 67–82, simplify.

67. $\dfrac{6}{\sqrt{3x}}$

68. $\dfrac{3}{5\sqrt{t}}$

69. $\dfrac{2}{\sqrt{x+2}}$

70. $\dfrac{4}{\sqrt{4+y^2}}$

71. $\sqrt{\dfrac{a^3}{8}}$

72. $\sqrt{\dfrac{x^4}{20}}$

73. $\sqrt{\dfrac{5}{y^5}}$

74. $\dfrac{x^3}{\sqrt{x^3}}$

75. $\dfrac{\sqrt{5x^2}}{\sqrt{12x^7}}$

76. $\dfrac{\sqrt{a^4}}{\sqrt{18a^5}}$

77. $\dfrac{18x^2y^3}{\sqrt{3x^2y^5}}$

78. $\dfrac{\sqrt{8a^6b^4}}{\sqrt{32a^4b^5}}$

79. $\dfrac{9x}{\sqrt{12x^5}}$

80. $\dfrac{\sqrt{8xy^4}}{\sqrt{6x^2}}$

81. $\sqrt{\dfrac{x^5y^2}{20y^3}}$

82. $\sqrt{\dfrac{4xy}{16x^3y^2}}$

In Exercises 83–88, supply the missing number to make the statement true.

83. $\dfrac{\sqrt{3}}{\sqrt{5}} = \dfrac{\sqrt{15}}{\square}$

84. $\sqrt{\dfrac{3}{8}} = \dfrac{\sqrt{6}}{\square}$

85. $\sqrt{\dfrac{2}{7}} = \dfrac{\square}{7}$

86. $\dfrac{\sqrt{63}}{\sqrt{5}} = \dfrac{\square}{5}$

87. $\dfrac{3}{\square} = \sqrt{3}$

88. $\dfrac{5}{\square} = \dfrac{\sqrt{5x}}{x}$

In Exercises 89–96, simplify.

89. $\dfrac{2b}{\sqrt[3]{b^2}}$

90. $\dfrac{7}{\sqrt[4]{27}}$

91. $\sqrt[3]{\dfrac{5}{4}}$

92. $\dfrac{5}{\sqrt[4]{125}}$

93. $\dfrac{x}{\sqrt[4]{xy^3}}$

94. $\dfrac{5x}{\sqrt[5]{x^2}}$

95. $\sqrt[4]{\dfrac{4x^5}{64xy^3}}$

96. $\sqrt[5]{\dfrac{64a^4b^2}{a^6b^6}}$

In Exercises 97–102, determine the value of a.

97. $4\sqrt{x} = \sqrt{ax}$

98. $10\sqrt{x^2y} = \sqrt{ax^2y}$

99. $\dfrac{6}{\sqrt{x}} = \sqrt{\dfrac{a}{x}}$

100. $\dfrac{2}{\sqrt{y}} = \sqrt{\dfrac{a}{y}}$

101. $\sqrt{x^3} = \sqrt[a]{x^6}$

102. $\sqrt[3]{x^5} = \sqrt[a]{x^{10}}$

Real-Life Applications

103. A used car dealer planned an extensive advertising campaign for a Labor Day weekend sale. The dealer estimated that d dollars (in thousands) of advertising would result in $30\sqrt{d^3}$ sales (number of sales) with a total revenue of $90{,}000d^2$ (in dollars).

(a) Write a simplified expression for the estimated average price of each car sold.

(b) If the dealer spends \$4000 on advertising, what is the average price of a car?

104. A naturalist estimates that the number of birds that can be attracted to a preserve is $120\sqrt[5]{m^2}$, where m is the number of months after the preserve is established. The number of pounds of bird seed that must be provided per month per bird to supplement natural sources is predicted to be $\dfrac{1}{3\sqrt[5]{m}}$, where m is the number of months after the establishment of the preserve.

(a) What is the estimated number of birds attracted to the preserve after 6 months?

(b) Write a simplified expression for the number of pounds per month of bird seed required after m months.

(c) Does the seed requirement per bird increase or decrease as the preserve becomes more established? Does the total seed requirement increase or decrease?

Group Project: Tax Payments and Refunds

In 1996 a survey showed that 59% of all taxpayers expected a tax refund, whereas 41% expected to pay more taxes than were withheld. However, the percentages varied according to age groups. (Source: Bruskin/Goldring Research Survey for Turbo Tax.)

The function $R(x) = \dfrac{734}{x^{0.7}}$ can be used to model the percentage $R(x)$ of taxpayers expecting a refund, and the function $P(x) = \dfrac{0.2x(\sqrt{x}+12)}{\sqrt{x}}$ can be used to model the percentage $P(x)$ of taxpayers expecting to pay more taxes. In both models, x represents the age of the taxpayer.

105. Produce the graph of $R(x)$, and describe the trend in refund expectations as taxpayers grow older.

106. Trace the graph of $P(x)$, where $18 \le x \le 70$, to estimate the age at which 20% of the taxpayers expect to pay more taxes.

107. Using 0 to 100 for both axes, produce the graphs of both model functions. Estimate the age at which the percentage of people expecting refunds equals the percentage of people expecting to pay more taxes. Based on this estimate, what percentage of taxpayers at that age expect neither a refund nor a payment?

108. Suppose that instead of withholding taxes throughout the year, a new system were instituted whereby a flat tax on all earned income became due at the end of the year. What is the likelihood that the federal government's revenue projections would be met?

Challenge

109. Use your calculator to evaluate each of the following expressions.

$$\sqrt{2} \qquad \sqrt{2+\sqrt{2}} \qquad \sqrt{2+\sqrt{2+\sqrt{2}}}$$

$$\sqrt{2+\sqrt{2+\sqrt{2+\sqrt{2}}}}$$

$$\sqrt{2+\sqrt{2+\sqrt{2+\sqrt{2+\sqrt{2}}}}}$$

What number do the values of these expressions seem to be approaching? Do you think that this pattern holds true for any positive integer n? If so, repeat the experiment for $n = 3$.

In Exercises 110 and 111, rationalize the denominator.

110. $\dfrac{1}{\sqrt{\sqrt{x}}}$ **111.** $\dfrac{1}{\sqrt{\sqrt[3]{x}}}$

112. Consider three nonzero numbers x, y, and z. How many of these numbers must be positive for the following to be defined in the real number system?

(a) $\sqrt{x}\sqrt{y}\sqrt{z}$ (b) $\sqrt{xyz}$

113. Use your calculator to show that $\dfrac{7}{\sqrt{7}} = \sqrt{7}$. From this, you may conclude that dividing a number by its square root always results in the square root of the number. Use rational exponents to prove that your conclusion is true for any positive number.

8.6 OPERATIONS WITH RADICALS

Addition and Subtraction • Multiplication • Rationalizing Binomial Denominators

Addition and Subtraction

Radicals with the same radicand and the same index are **like radicals**. In a manner similar to adding like terms, we use the Distributive Property to add or subtract like radicals.

$$3x + 5x = (3+5)x = 8x$$

$$3\sqrt{7} + 5\sqrt{7} = (3+5)\sqrt{7} = 8\sqrt{7}$$

Sometimes one or both of two unlike radicals can be simplified so that they become like radicals and can then be combined.

LEARNING TIP

The familiar steps of removing parentheses and combining like terms apply to simplifying radical expressions. There is nothing new about the techniques used in this section.

EXAMPLE 1

Adding and Subtracting Radicals

(a) $\sqrt{3} + 5\sqrt{3} = 1\sqrt{3} + 5\sqrt{3} = (1 + 5)\sqrt{3} = 6\sqrt{3}$ Distributive Property

(b) $\sqrt{45} - \sqrt{80} = \sqrt{9}\sqrt{5} - \sqrt{16}\sqrt{5}$ Product Rule for Radicals

$= 3\sqrt{5} - 4\sqrt{5}$

$= (3 - 4)\sqrt{5}$ Distributive Property

$= -\sqrt{5}$

(c) $3\sqrt{75} - 2\sqrt{48} = 3\sqrt{25}\sqrt{3} - 2\sqrt{16}\sqrt{3}$ Product Rule for Radicals

$= 3 \cdot 5\sqrt{3} - 2 \cdot 4\sqrt{3}$

$= 15\sqrt{3} - 8\sqrt{3}$

$= 7\sqrt{3}$ Combine the coefficients and retain the radical.

(d) $\sqrt[3]{32} + \sqrt[3]{4} = \sqrt[3]{8}\sqrt[3]{4} + \sqrt[3]{4}$ Product Rule for Radicals

$= 2\sqrt[3]{4} + 1\sqrt[3]{4}$

$= 3\sqrt[3]{4}$ Combine the coefficients and retain the radical.

Note: Radicals can be combined only if the radicands and indices are the same. Thus the terms cannot be combined in either $4\sqrt{2} + 3\sqrt[3]{2}$ or $\sqrt{3} + \sqrt{2}$.

So that all radicals are defined and results can be expressed without regard to absolute value, we will assume that all variables in this section represent positive numbers.

EXAMPLE 2

Adding and Subtracting Radical Expressions

(a) $\sqrt{48x} - \sqrt{27x} = \sqrt{16} \cdot \sqrt{3x} - \sqrt{9} \cdot \sqrt{3x}$ Product Rule for Radicals

$= 4\sqrt{3x} - 3\sqrt{3x}$

$= \sqrt{3x}$ Combine the terms.

(b) $4\sqrt{a^3} + 2a\sqrt{9a} = 4\sqrt{a^2} \cdot \sqrt{a} + 2a\sqrt{9} \cdot \sqrt{a}$ Product Rule for Radicals

$= 4a\sqrt{a} + 2a \cdot 3\sqrt{a}$

$= 4a\sqrt{a} + 6a\sqrt{a}$

$= 10a\sqrt{a}$ Combine the terms.

(c) $3\sqrt[3]{16t^7} + 4t\sqrt[3]{2t^4} = 3\sqrt[3]{8t^6} \cdot \sqrt[3]{2t} + 4t\sqrt[3]{t^3} \cdot \sqrt[3]{2t}$ Product Rule for Radicals

$= 3 \cdot 2t^2\sqrt[3]{2t} + 4t \cdot t\sqrt[3]{2t}$

$= 6t^2\sqrt[3]{2t} + 4t^2\sqrt[3]{2t}$

$= 10t^2\sqrt[3]{2t}$ Combine the terms.

Note: To combine terms, the radical factors *and the variable factors* must be identical. [See parts (b) and (c) of Example 2.] For $3\sqrt{4t} + 5t\sqrt{4t}$, the Distributive Property can be used to factor the expression as $(3 + 5t)\sqrt{4t}$, but the terms cannot be combined.

The following example illustrates the method for combining radicals whose radicands are fractions or rational expressions.

EXAMPLE 3

Adding and Subtracting Radical Expressions

(a) $\sqrt{\frac{8}{9}} - \sqrt{\frac{18}{49}} = \frac{\sqrt{8}}{\sqrt{9}} - \frac{\sqrt{18}}{\sqrt{49}}$ Quotient Rule for Radicals

$= \frac{\sqrt{4}\cdot\sqrt{2}}{3} - \frac{\sqrt{9}\cdot\sqrt{2}}{7}$ Product Rule for Radicals

$= \frac{2\sqrt{2}}{3} - \frac{3\sqrt{2}}{7}$

$= \frac{14\sqrt{2}}{21} - \frac{9\sqrt{2}}{21}$ Rewrite the fractions with 21 as the LCD.

$= \frac{5\sqrt{2}}{21}$ Subtract the numerators.

(b) $\sqrt{\frac{81}{a^6}} + \sqrt{\frac{49}{a^{10}}} = \frac{9}{a^3} + \frac{7}{a^5}$ Both radicands are perfect squares.

$= \frac{9a^2}{a^5} + \frac{7}{a^5}$ Rewrite the fractions with a^5 as the LCD.

$= \frac{9a^2 + 7}{a^5}$ Add the numerators.

Multiplication

We can use the Product Rule for Radicals along with the Distributive Property to multiply radical expressions that have more than one term.

EXAMPLE 4

Multiplying Radical Expressions

(a) $3\sqrt{5}(2\sqrt{5} - \sqrt{2}) = (3\sqrt{5})(2\sqrt{5}) - (3\sqrt{5})(\sqrt{2})$ Distributive Property

$= 6 \cdot 5 - 3\sqrt{10}$ Product Rule for Radicals

$= 30 - 3\sqrt{10}$

(b) $\sqrt{6}(\sqrt{8} - \sqrt{2}) = \sqrt{6}\sqrt{8} - \sqrt{6}\sqrt{2}$ Distributive Property

$= \sqrt{48} - \sqrt{12}$ Product Rule for Radicals

$= \sqrt{16}\sqrt{3} - \sqrt{4}\sqrt{3}$ Product Rule for Radicals

$= 4\sqrt{3} - 2\sqrt{3}$ Combine the terms.

$= 2\sqrt{3}$

(c) $(\sqrt{2} - 3)(\sqrt{3} + \sqrt{2}) = \sqrt{2}\sqrt{3} + \sqrt{2}\sqrt{2} - 3\sqrt{3} - 3\sqrt{2}$ FOIL

$= \sqrt{6} + 2 - 3\sqrt{3} - 3\sqrt{2}$ Product Rule for Radicals

None of the four terms are like terms, so the result cannot be simplified further.

(d) $\left(\sqrt{3}+\sqrt{6}\right)^2 = \left(\sqrt{3}\right)^2 + 2\cdot\sqrt{3}\sqrt{6} + \left(\sqrt{6}\right)^2$ $(A+B)^2 = A^2 + 2AB + B^2$

$= 3 + 2\sqrt{18} + 6$ Product Rule for Radicals

$= 9 + 2\sqrt{18}$

$= 9 + 2\sqrt{9}\sqrt{2}$ Product Rule for Radicals

$= 9 + 2\cdot 3\sqrt{2}$

$= 9 + 6\sqrt{2}$

Rationalizing Binomial Denominators

Recall that the product of a sum of two terms and the difference of the same two terms is the difference of their squares.

$$(A+B)(A-B) = A^2 - B^2$$

The expressions $A + B$ and $A - B$ are called **conjugates** of each other.

EXPLORING THE CONCEPT

Square Root Conjugates

Consider the following products of square root conjugates.

$$\left(4-\sqrt{5}\right)\left(4+\sqrt{5}\right) = (4)^2 - \left(\sqrt{5}\right)^2 = 16 - 5 = 11$$
$$\left(x+3\sqrt{2}\right)\left(x-3\sqrt{2}\right) = (x)^2 - \left(3\sqrt{2}\right)^2 = x^2 - 18$$
$$\left(\sqrt{x}-2\sqrt{y}\right)\left(\sqrt{x}+2\sqrt{y}\right) = \left(\sqrt{x}\right)^2 - \left(2\sqrt{y}\right)^2 = x - 4y$$

In each case, multiplying square root conjugates results in a product that has no square root.

Note: The preceding conclusion holds only for square root conjugates. For example, in the following product of cube root conjugates, the radical is not eliminated.

$$\left(\sqrt[3]{5}-1\right)\left(\sqrt[3]{5}+1\right) = \sqrt[3]{25} - 1$$

If the denominator of a fraction is a binomial with at least one square root term, sometimes we can rationalize the denominator by multiplying the numerator and the denominator by the conjugate of the denominator.

EXAMPLE 5

Rationalizing a Binomial Denominator

(a) $\dfrac{10}{\sqrt{6}-2} = \dfrac{10\cdot\left(\sqrt{6}+2\right)}{\left(\sqrt{6}-2\right)\cdot\left(\sqrt{6}+2\right)}$ Multiply the numerator and denominator by the conjugate of $\sqrt{6}-2$.

$= \dfrac{10\left(\sqrt{6}+2\right)}{\left(\sqrt{6}\right)^2 - 2^2}$ $(A+B)(A-B) = A^2 - B^2$

$= \dfrac{10\left(\sqrt{6}+2\right)}{6-4}$

$$= \frac{5 \cdot 2 \cdot (\sqrt{6} + 2)}{2}$$ Factor the numerator.

$$= 5(\sqrt{6} + 2)$$ Divide out the common factor 2.

(b) $$\frac{\sqrt{3} - \sqrt{2}}{\sqrt{5} + \sqrt{3}} = \frac{(\sqrt{3} - \sqrt{2})(\sqrt{5} - \sqrt{3})}{(\sqrt{5} + \sqrt{3})(\sqrt{5} - \sqrt{3})}$$ Multiply the numerator and denominator by the conjugate of $\sqrt{5} + \sqrt{3}$.

$$= \frac{\sqrt{15} - \sqrt{9} - \sqrt{10} + \sqrt{6}}{(\sqrt{5})^2 - (\sqrt{3})^2}$$ FOIL

$$= \frac{\sqrt{15} - 3 - \sqrt{10} + \sqrt{6}}{5 - 3}$$

$$= \frac{\sqrt{15} - 3 - \sqrt{10} + \sqrt{6}}{2}$$

(c) $$\frac{\sqrt{x}}{\sqrt{x} + \sqrt{2y}} = \frac{\sqrt{x}(\sqrt{x} - \sqrt{2y})}{(\sqrt{x} + \sqrt{2y})(\sqrt{x} - \sqrt{2y})}$$ Multiply the numerator and denominator by the conjugate of $\sqrt{x} + \sqrt{2y}$.

$$= \frac{\sqrt{x^2} - \sqrt{2xy}}{(\sqrt{x})^2 - (\sqrt{2y})^2}$$ $(A + B)(A - B) = A^2 - B^2$

$$= \frac{x - \sqrt{2xy}}{x - 2y}$$ We cannot simplify further because the numerator and denominator cannot be factored.

Think About It

Rationalize the denominator of $\frac{1}{\sqrt{\sqrt{x} + 3}}$.

8.6 QUICK REFERENCE

Addition and Subtraction

- **Like radicals** have the same radicand and the same index. Only like radicals can be combined by addition or subtraction.
- Simplify all radicals before attempting to add or subtract them.
- Use the Distributive Property to add or subtract like radicals.

Multiplication

- To multiply radical expressions that have more than one term, use the Distributive Property, the FOIL method, and other familiar multiplication techniques that we used with polynomials.
- The special product patterns for polynomials also apply to multiplication of radicals.

  $(A + B)^2 = A^2 + 2AB + B^2$

  $(A - B)^2 = A^2 - 2AB + B^2$

  $(A + B)(A - B) = A^2 - B^2$

Rationalizing Binomial Denominators

- Expressions of the form $A + B$ and $A - B$ are said to be **conjugates** of one another.
- The product of square root conjugates does not contain a square root.
- If the denominator of a fraction is a binomial with at least one square root term, sometimes we can rationalize the denominator by multiplying the numerator and the denominator by the conjugate of the denominator.

8.6 EXERCISES

Concepts and Skills

1. Explain how the Distributive Property is used to add or subtract radical expressions. Demonstrate with $3\sqrt{5} + 4\sqrt{5}$.

2. Consider the following two uses of the Distributive Property.

$$3\sqrt{5} + 7\sqrt{5} = (3 + 7)\sqrt{5} = 10\sqrt{5}$$

$$3\sqrt{5} + x\sqrt{5} = (3 + x)\sqrt{5}$$

Explain why parentheses are needed in the second result, but not in the first.

In Exercises 3–10, perform the indicated operation.

3. $5\sqrt{13} + \sqrt{13}$

4. $\sqrt{7} - 2\sqrt{7}$

5. $6\sqrt{x} + 10\sqrt{x}$

6. $5\sqrt{ax} - 6\sqrt{ax}$

7. $9\sqrt[3]{y} - 12\sqrt[3]{y}$

8. $3\sqrt[3]{x^2} + 7\sqrt[3]{x^2}$

9. $\sqrt{2x} + 4\sqrt{3} - 5\sqrt{2x} - 3\sqrt{3}$

10. $ab\sqrt{a} + 2\sqrt{b} - 2ab\sqrt{a} + \sqrt{b}$

11. Explain why the terms of $\sqrt{18} + \sqrt{27}$ cannot be combined with the expression in its present form. Use the Distributive Property in your explanation. What step should you take before deciding whether the two terms can be combined?

12. Explain how the Commutative and Associative Properties of Multiplication are used to determine the product $(3\sqrt{2})(5\sqrt{7})$.

In Exercises 13–22, perform the indicated operation. Use your calculator to verify the result.

13. $\sqrt{27} - \sqrt{12}$

14. $\sqrt{45} + \sqrt{80}$

15. $\sqrt[3]{250} + \sqrt[3]{54}$

16. $\sqrt[3]{24} - \sqrt[3]{192}$

17. $\sqrt[4]{162} - \sqrt[4]{32}$

18. $\sqrt[4]{48} + \sqrt[4]{243}$

19. $5\sqrt{32} + 7\sqrt{72}$

20. $10\sqrt{20} - 3\sqrt{45}$

21. $4\sqrt{12} - 2\sqrt{27} + 5\sqrt{8}$

22. $3\sqrt{50} - 2\sqrt{20} + 6\sqrt{18}$

In Exercises 23–36, perform the indicated operation.

23. $\sqrt{72x} - \sqrt{50x}$

24. $\sqrt{80c} + \sqrt{20c}$

25. $x\sqrt{48} + 3\sqrt{27x^2}$

26. $3\sqrt{48x^2y} - 2x\sqrt{75y}$

27. $4a\sqrt{20a^2b} + a^2\sqrt{45b}$

28. $5\sqrt{a^5} + 7a\sqrt{4a^3}$

29. $x\sqrt{2x} - 3\sqrt{8x^2} + \sqrt{50x^3}$

30. $\sqrt{9x^4} + x\sqrt{12x} - \sqrt{27x^3}$

31. $\sqrt{9xy^3} + 4\sqrt{x^3y} - 5y\sqrt{4xy}$

32. $3\sqrt{x^2} + \sqrt{x^9} - x^3\sqrt{x^3}$

33. $2b\sqrt[3]{54b^5} - 3\sqrt[3]{16b^8}$

34. $\sqrt[4]{32x^3} + \sqrt[4]{2x^3}$

35. $\sqrt[3]{-54x^3} - \sqrt[3]{-16x^3}$

36. $x\sqrt[4]{xy^4} - y\sqrt[4]{x^5}$

In Exercises 37–44, multiply and simplify the result.

37. $4\sqrt{3}(2\sqrt{3} - \sqrt{7})$

38. $2\sqrt{2}(3\sqrt{8} + \sqrt{5})$

39. $5\sqrt{7}(4\sqrt{3} + 7)$

40. $6\sqrt{11}(5\sqrt{5} - 8)$

41. $\sqrt{5}(\sqrt{35} - \sqrt{15})$

42. $\sqrt{7x}(\sqrt{28} + \sqrt{63x})$

43. $\sqrt{3}(\sqrt{y} - \sqrt{3y})$

44. $\sqrt{5}(\sqrt{10y} - 2y)$

In Exercises 45–48, supply the missing number to make each statement true.

45. ▒ $(3 - \sqrt{2}) = 3\sqrt{7} - \sqrt{14}$

46. ▒ $(2 - \sqrt{3y}) = 2y\sqrt{3} - 4\sqrt{y}$

47. $4($ ▒ $) = 4x + 8\sqrt{3}$

48. $\sqrt{3}($ ▒ $) = 3\sqrt{5} - 5\sqrt{3}$

49. According to the Product Rule for Radicals, the following is true: $\sqrt{x + 2}\sqrt{x - 2} = \sqrt{x^2 - 4}$. Produce the graphs of

$$y_1 = \sqrt{x + 2}\sqrt{x - 2} \quad \text{and} \quad y_2 = \sqrt{x^2 - 4}$$

Explain why they are different.

50. Which of the following is not a rational number? Explain your choice.

(i) The square of $2\sqrt{5}$.

(ii) The square of $2 + \sqrt{5}$.

(iii) The product of $2 + \sqrt{5}$ and its conjugate.

In Exercises 51–60, multiply and simplify the result.

51. $(\sqrt{3} - 2)(\sqrt{2} + \sqrt{3})$

52. $(3\sqrt{5} - 2\sqrt{3})(\sqrt{3} - 5)$

53. $(\sqrt{7x} - 3)(\sqrt{3} + \sqrt{7x})$

54. $(\sqrt{3t} - 2)(\sqrt{3} + 4t)$

55. $(2\sqrt{10} + \sqrt{5})(2\sqrt{2} - 5)$

56. $(\sqrt{x} - 5)(2\sqrt{x} + 3)$

57. $(\sqrt{2} - \sqrt{6})^2$ **58.** $(\sqrt{3} + \sqrt{15})^2$

59. $(\sqrt{5} + \sqrt{10})^2$ **60.** $(\sqrt{10} - \sqrt{2})^2$

In Exercises 61–66, determine the product of the expression and its conjugate.

61. $2\sqrt{2} - 3$ **62.** $3\sqrt{2} + 2$

63. $3\sqrt{2} + 4\sqrt{6}$ **64.** $4\sqrt{3} - 3\sqrt{5}$

65. $10 - 5t\sqrt{2}$ **66.** $7w + 3\sqrt{3}$

In Exercises 67–70, determine the product.

67. $\dfrac{\sqrt{3}+2}{\sqrt{3}-2} \cdot \dfrac{\sqrt{3}+2}{\sqrt{3}+2}$

68. $\dfrac{\sqrt{5}-\sqrt{3}}{\sqrt{3}+\sqrt{2}} \cdot \dfrac{\sqrt{3}-\sqrt{2}}{\sqrt{3}-\sqrt{2}}$

69. $\dfrac{\sqrt{7}}{\sqrt{3}+\sqrt{7}} \cdot \dfrac{\sqrt{3}-\sqrt{7}}{\sqrt{3}-\sqrt{7}}$

70. $\dfrac{5}{\sqrt{3}-1} \cdot \dfrac{\sqrt{3}+1}{\sqrt{3}+1}$

In Exercises 71–74, show that the given number and its conjugate are both solutions of the equation.

71. $x^2 - 4x + 1 = 0;\quad 2 + \sqrt{3}$

72. $x^2 - 2x - 4 = 0;\quad 1 - \sqrt{5}$

73. $4x^2 - 12x + 7 = 0;\quad \dfrac{3 - \sqrt{2}}{2}$

74. $9x^2 - 12x - 41 = 0;\quad \dfrac{2 + 3\sqrt{5}}{3}$

In Exercises 75–86, rationalize the denominator.

75. $\dfrac{3}{3 - \sqrt{6}}$ **76.** $\dfrac{12}{\sqrt{5} - \sqrt{2}}$

77. $\dfrac{\sqrt{5} - \sqrt{3}}{\sqrt{5} + \sqrt{3}}$ **78.** $\dfrac{\sqrt{5} - \sqrt{2}}{\sqrt{7} + \sqrt{3}}$

79. $\dfrac{\sqrt{y} + 5}{\sqrt{y} - 4}$ **80.** $\dfrac{5z}{\sqrt{z} + z}$

81. $\dfrac{\sqrt{x}}{\sqrt{x} + 2}$ **82.** $\dfrac{\sqrt{3}}{\sqrt{3} - \sqrt{x}}$

83. $\dfrac{1}{\sqrt{x+1} - 2}$ **84.** $\dfrac{2}{3 + \sqrt{x+2}}$

85. $\dfrac{\sqrt{x} - \sqrt{y}}{\sqrt{x} + \sqrt{y}}$ **86.** $\dfrac{\sqrt{xy}}{\sqrt{x} + \sqrt{y}}$

87. Explain why $\dfrac{1}{\sqrt{5}} + \dfrac{\sqrt{2}}{\sqrt{10}}$ cannot be performed with the fractions in their given form. Must the denominators of the two fractions be rationalized before the fractions can be added? Why?

88. The following shows the start of two different approaches to performing the given addition.

$$\frac{1}{\sqrt{2}} + \frac{1}{\sqrt{3}} = \frac{1}{\sqrt{2}} \cdot \frac{\sqrt{2}}{\sqrt{2}} + \frac{1}{\sqrt{3}} \cdot \frac{\sqrt{3}}{\sqrt{3}} = \cdots$$

$$\frac{1}{\sqrt{2}} + \frac{1}{\sqrt{3}} = \frac{1}{\sqrt{2}} \cdot \frac{\sqrt{3}}{\sqrt{3}} + \frac{1}{\sqrt{3}} \cdot \frac{\sqrt{2}}{\sqrt{2}} = \cdots$$

Complete the two problems and verify that the results are the same. Then comment on the approach that you believe to be more efficient.

In Exercises 89–98, simplify.

89. $\dfrac{\sqrt{x} - 3}{\sqrt{x}}$ **90.** $\dfrac{5 + \sqrt{2t}}{2\sqrt{t}}$

91. $\dfrac{1}{\sqrt{5}} + \sqrt{45}$ **92.** $\dfrac{\sqrt{2}}{\sqrt{3}} + \dfrac{1}{\sqrt{6}}$

93. $1 - \dfrac{3}{\sqrt{2}}$ **94.** $\dfrac{1}{\sqrt{2}} - 2\sqrt{8}$

95. $2\sqrt{x+1} - \dfrac{2x}{\sqrt{x+1}}$

96. $\dfrac{x}{\sqrt{x+3}} - \sqrt{x+3}$

97. $\dfrac{2}{\sqrt{3}+2} - \dfrac{1}{\sqrt{3}-2}$

98. $\dfrac{\sqrt{x}}{\sqrt{x}+5}+\dfrac{2\sqrt{x}}{\sqrt{x}-5}$

A rationalized *numerator* is not one of the conditions for a simplified radical. However, in more advanced studies, there are occasions when we must rationalize the numerator. The technique is the same as that for rationalizing denominators.

In Exercises 99–104, rationalize the *numerator*.

99. $\dfrac{\sqrt{5}+\sqrt{7}}{4}$

100. $\dfrac{\sqrt{11}-\sqrt{3}}{8}$

101. $\dfrac{\sqrt{x}+3}{\sqrt{x}-3}$

102. $\dfrac{5-\sqrt{a}}{5+\sqrt{a}}$

103. $\dfrac{\sqrt{x+2}-\sqrt{x}}{2}$

104. $\dfrac{\sqrt{x+h}-\sqrt{x}}{h}$

105. Determine two consecutive integers whose square roots are also consecutive integers.

106. Consider two positive consecutive integers. Show that the reciprocal of the sum of their square roots is equal to the difference of their square roots. Also, explain how you know that both expressions are positive.

Geometric Models

107. A 50-foot pole is supported by two guy wires, one attached to the top of the pole and the other attached below the top of the pole. The two guy wires, one 10 feet shorter than the other, are attached at the same point on the ground. (See figure.)

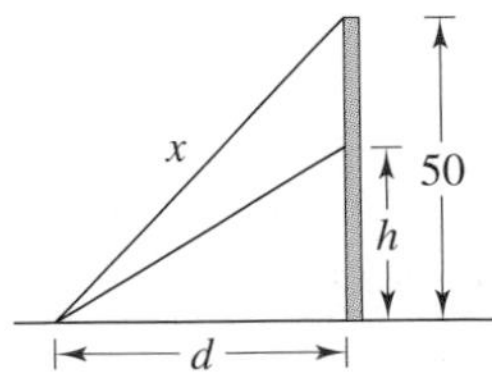

(a) Use the Pythagorean Theorem to write an expression in x for the distance d from the base of the pole to the point at which the wires are attached to the ground.

(b) Write an expression in x for the height h at which the shorter wire is attached to the pole.

108. A rectangular piece of paper is folded so that the top edge is along the side edge. (See figure.) Show that the length d of the diagonal is a multiple of $\sqrt{2}$ and that its value depends only on the width of the paper, not the length.

Figure for 108

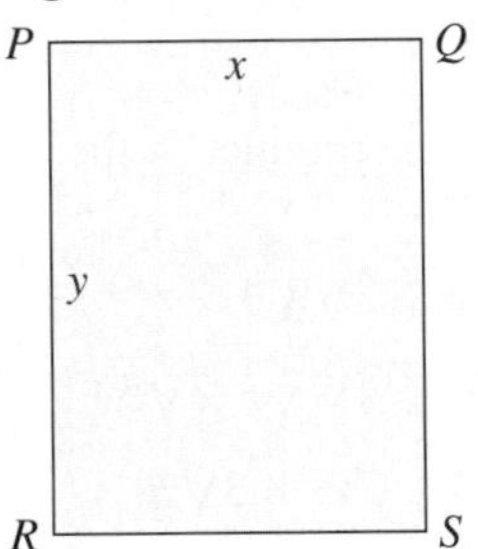

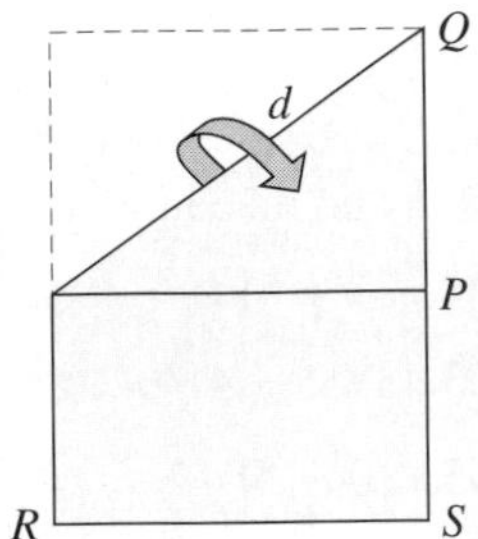

Real-Life Applications

109. Using T to represent the water temperature (in °F), a commercial fisherman determines that for each 100 pounds of fish in the nets, $\sqrt{T}$ pounds are not marketable fish and are returned to the water.

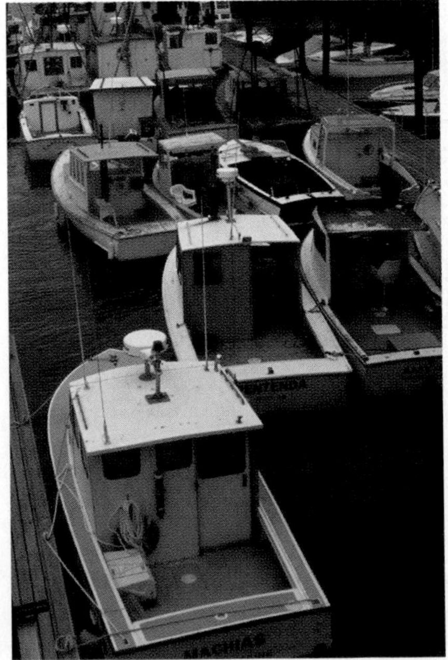

(a) Suppose that the water temperature is 60°F. What percentage of the catch is marketable?

(b) Write a simplified expression for the ratio of the number of pounds of fish that are not marketable to the number of pounds that are marketable.

110. A computer firm sells antivirus protection software. The probability that the consumer is not fully protected against a virus is $\dfrac{0.01\sqrt{D}}{0.01\sqrt{D}+1}$, where D is the number of days after the software product is developed.

(a) Write a simplified expression for the probability expression.

(b) What is the probability that a person is not protected after 30 days?

Group Project: Employee Drug Testing

As more companies began testing potential employees for drugs, the percentage of candidates testing positive declined. The accompanying bar graph shows the trend for the period 1989–1994. (Source: American Management Association.)

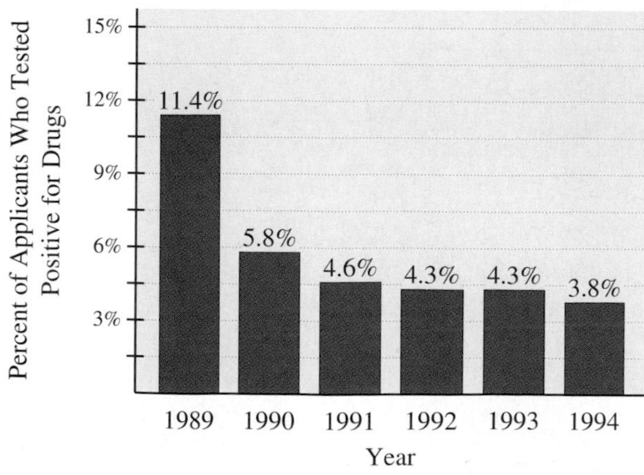

The percentage $P(t)$ of potential new hires testing positive for drugs can be modeled with the function $P(t) = \dfrac{72 - 28\sqrt[3]{t}}{\sqrt[3]{t^2}}$, where t represents the number of years since 1985.

111. Write $P(t)$ with a rationalized denominator. Is the resulting expression now a rational function?

112. Produce and trace the graph of $P(t)$ to estimate the projected year in which no new hires would test positive for drugs.

113. Suppose that companies had stopped drug testing after 1994 and that the graph for the next 6 years was a mirror image of the graph of $P(t)$. What kind of expression do you think might model the data for the 12-year period?

114. Some argue that drug usage off the job is a personal matter and that drug testing of this kind is a violation of privacy rights. What is your position on this issue?

Challenge

In Exercises 115 and 116, evaluate the expression.

115. $\sqrt[3]{3\sqrt[3]{3\sqrt[3]{3}}}$ **116.** $\sqrt{3\sqrt{3\sqrt{3}}}$

In Exercises 117–120, perform the indicated operation.

117. $\sqrt[4]{x^2} + \sqrt{x}$ **118.** $\sqrt{12} + \sqrt[4]{9}$

119. $\sqrt[6]{8} - 2\sqrt[4]{4}$ **120.** $\sqrt[6]{x^2} - \sqrt[9]{x^3}$

121. Rationalize the denominator of

$$\frac{\sqrt{2}-\sqrt{3}}{\sqrt{2}+\sqrt{3}+\sqrt{5}}$$

122. Write the special factoring pattern for the difference of two cubes. Then explain how this pattern can be used to rationalize the denominator of

$$\frac{1}{\sqrt[3]{x}-\sqrt[3]{y}}$$

In Exercises 123 and 124, rationalize the denominator.

123. $\dfrac{3}{\sqrt[3]{x}-\sqrt[3]{2}}$ **124.** $\dfrac{5}{\sqrt[3]{x}-\sqrt[3]{3}}$

In Exercises 125 and 126, simplify the complex fraction and express the result with a rationalized denominator.

125. $\dfrac{\dfrac{1}{\sqrt{3}}+\dfrac{\sqrt{3}}{\sqrt{x}}}{\dfrac{\sqrt{x}}{\sqrt{3}}+\dfrac{3}{\sqrt{3x}}}$ **126.** $\dfrac{\dfrac{3}{\sqrt{y}}+\dfrac{3}{\sqrt{x}}}{\dfrac{1}{\sqrt{y}}+\dfrac{1}{\sqrt{x}}}$

8.7 EQUATIONS WITH RADICALS AND EXPONENTS

Domain of a Radical Function • Radical Equations • Equations with Exponential Expressions

Domain of a Radical Function

A **radical function** is a function defined by a radical expression with a variable in the radicand. The domain of a radical function is the set of values of the variable for which the expression is defined.

EXAMPLE 1

Determining the Domain of a Radical Function

Use a graph to estimate the domain of each function. Then determine the domain algebraically.

(a) $y = \sqrt{1 - 2x}$ (b) $y = \sqrt[3]{x + 3}$ (c) $y = \sqrt[4]{x^2 + 1}$

Solution

(a) In Fig. 8.8 the graph of $y = \sqrt{1 - 2x}$ appears to have an x-intercept $\left(\frac{1}{2}, 0\right)$, and the rest of the graph is to the left of that point. We estimate the domain to be $\{x \mid x \le \frac{1}{2}\}$.

Figure 8.8

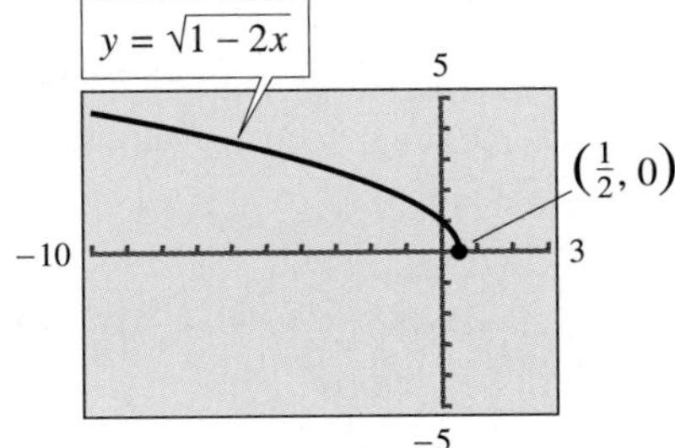

The expression $\sqrt{1 - 2x}$ is defined for all numbers for which $1 - 2x \ge 0$. To determine the domain, we solve the following inequality.

$$
\begin{aligned}
1 - 2x &\ge 0 \\
-2x &\ge -1 \\
x &\le \frac{1}{2}
\end{aligned}
$$

The domain of the function is $\{x \mid x \le \frac{1}{2}\}$.

(b) In Fig. 8.9 the graph of $y = \sqrt[3]{x + 3}$ appears to extend forever to the left and right. This suggests that the domain is **R.**

Because the index is odd, $\sqrt[3]{x + 3}$ is defined for all real numbers. Therefore, the domain of the function is **R.**

Figure 8.9

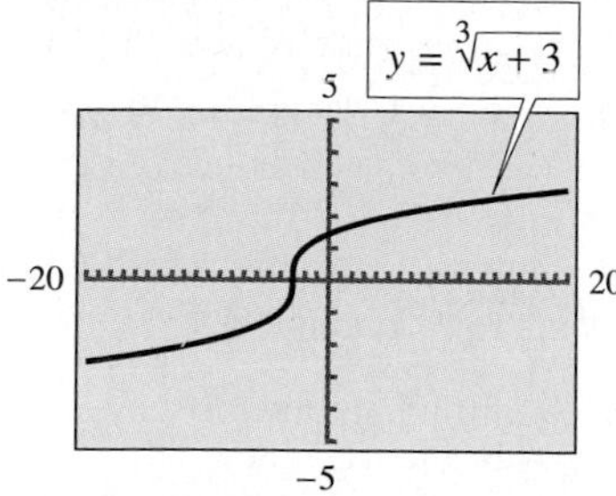

Figure 8.10

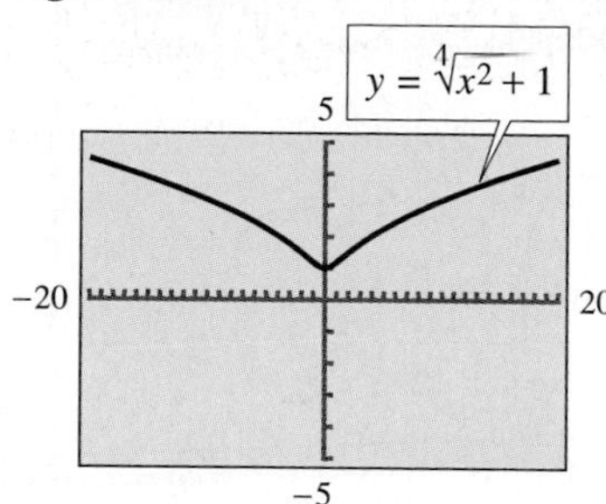

(c) In Fig. 8.10 the graph of $y = \sqrt[4]{x^2 + 1}$ apparently extends to the left and right forever. Thus the domain appears to be **R.**

Although the index is even, $x^2 + 1$ will be positive for any replacement for x. Therefore, the domain of the function is **R.**

Radical Equations

An equation that contains a radical with a variable in the radicand is called a **radical equation.** The following are radical equations that we will solve in subsequent examples.

$$\sqrt{1 - 2x} = 3 \qquad \sqrt{6x + 13} = 2x + 1 \qquad \sqrt[3]{x + 1} - 3 = 1$$

The following property can be used to solve a radical equation that contains a square root radical.

Squaring Property of Equality

For real numbers a and b, if $a = b$, then $a^2 = b^2$.

The Squaring Property of Equality can be used to eliminate a square root radical in a radical equation. For example, if $\sqrt{x} = 4$, then squaring both sides produces $x = 16$.

In general, if both sides of an equation are squared, all solutions of the original equation will be solutions of the new equation. However, the new equation may have solutions that are not solutions of the original equation. For instance, the equation $x = 5$ has just one solution, 5. Squaring both sides produces $x^2 = 25$, which has two solutions, 5 and -5. In this case, -5 is an **extraneous solution.**

LEARNING TIP

A negative number may be a solution to a radical equation. However, we must reject any value that makes the radicand negative.

Note: Because extraneous solutions can occur when we solve a radical equation, we must check all solutions in the original equation.

As with any equation, we can graph the left and right sides of a radical equation to estimate the solution. In the next two examples we use both graphing and algebraic methods.

EXAMPLE 2

Solving an Equation with One Radical

Solve the equation $\sqrt{1 - 2x} = 3$.

Solution

Figure 8.11 shows the graphs of the left and right sides of the equation. The solution appears to be about -4.

Figure 8.11

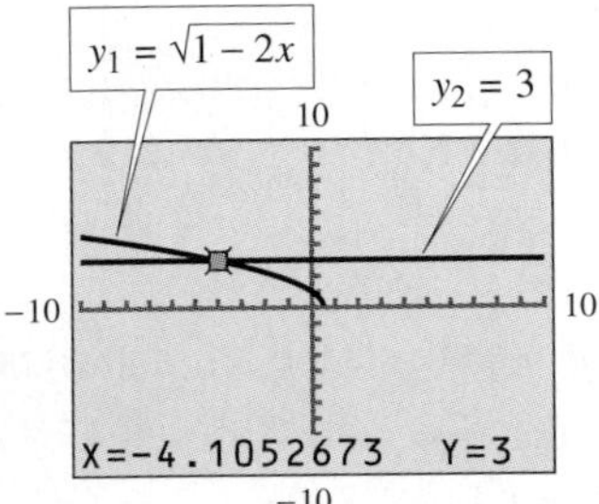

To solve the equation algebraically, we use the Squaring Property of Equality.

$$\sqrt{1 - 2x} = 3$$
$$\left(\sqrt{1 - 2x}\right)^2 = 3^2 \quad \text{Square both sides of the equation.}$$
$$1 - 2x = 9 \quad \text{The radical is eliminated.}$$
$$-2x = 8$$
$$x = -4$$

The solution confirms our graphing estimate. We can check the solution by substitution. When $x = -4$, $\sqrt{1 - 2x} = \sqrt{1 - 2(-4)} = \sqrt{9} = 3$.

EXAMPLE 3

Solving an Equation with One Radical

Solve the equation $\sqrt{6x + 13} = 2x + 1$.

Solution

Figure 8.12 shows the graphs of the left and right sides of the equation. The solution appears to be about 2.

Figure 8.12

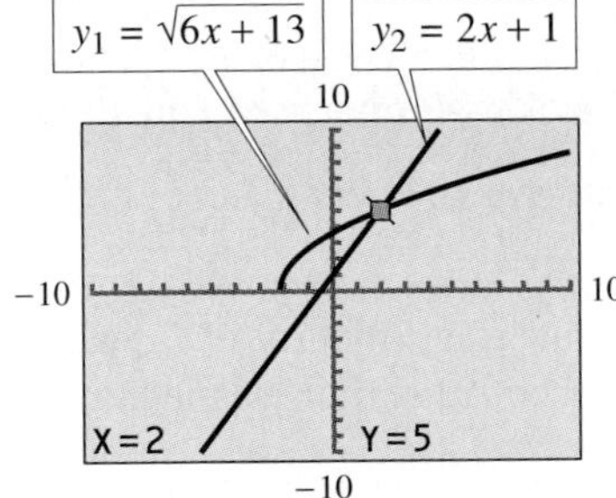

Now we solve the equation algebraically.

$$\sqrt{6x + 13} = 2x + 1$$
$$\left(\sqrt{6x + 13}\right)^2 = (2x + 1)^2 \quad \text{Square both sides.}$$
$$6x + 13 = 4x^2 + 4x + 1 \quad \text{The radical is eliminated.}$$
$$0 = 4x^2 - 2x - 12 \quad \text{Write the resulting quadratic equation in standard form and factor.}$$
$$0 = 2(2x^2 - x - 6)$$
$$0 = 2(2x + 3)(x - 2)$$

Figure 8.13

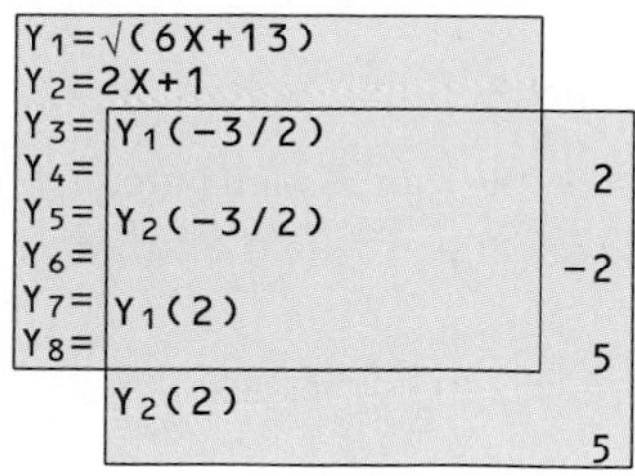

$$2x + 3 = 0 \quad \text{or} \quad x - 2 = 0 \qquad \text{Zero Factor Property}$$
$$2x = -3 \quad \text{or} \quad x = 2 \qquad \text{Solve each case for } x.$$
$$x = -\frac{3}{2} \quad \text{or} \quad x = 2$$

From the graph we expected only *one* solution. Solving algebraically produced *two* values of x.

This time we use a calculator to check the solutions. In Fig. 8.13 we see that $y_1 \neq y_2$ when $x = -\frac{3}{2}$, which means that $-\frac{3}{2}$ is an extraneous solution. When $x = 2$, $y_1 = y_2$. Thus 2 is the only solution of the equation.

Example 3 shows that a graph can be very helpful in detecting extraneous solutions. Although we will not always show a graph in the examples that follow, combining graphing and algebraic methods continues to be an excellent approach.

The Squaring Property of Equality can be extended to any power. If $a = b$, then $a^n = b^n$. In words, both sides of an equation can be raised to the nth power. Remember, however, that doing this may produce extraneous solutions.

As illustrated in Example 4, an initial step in solving radical equations is to isolate the radical before raising both sides of the equation to a power.

EXAMPLE 4

Solving an Equation with a Third Root Radical

Solve the equation $\sqrt[3]{x + 1} - 3 = 1$.

Solution

$$\sqrt[3]{x + 1} - 3 = 1$$
$$\sqrt[3]{x + 1} = 4 \qquad \text{Add 3 to both sides to isolate the radical.}$$
$$\left(\sqrt[3]{x + 1}\right)^3 = 4^3 \qquad \text{Raise both sides to the third power.}$$
$$x + 1 = 64 \qquad \text{The radical is eliminated.}$$
$$x = 63$$

You can use your calculator to check that 63 is the solution.

To solve equations with two radicals, it may be necessary to square both sides more than once.

EXAMPLE 5

Solving an Equation with Two Radicals

Solve the equation $\sqrt{2a} = 2 + \sqrt{a - 2}$.

Solution

We need to isolate one of the radicals. It is usually better to isolate the one with the greater number of terms in the radicand.

$$\sqrt{2a} = 2 + \sqrt{a-2}$$

$$\sqrt{2a} - 2 = \sqrt{a-2}$$ Isolate the radical on the right.

$$\left(\sqrt{2a} - 2\right)^2 = \left(\sqrt{a-2}\right)^2$$ Square both sides.

$$2a - 4\sqrt{2a} + 4 = a - 2$$ The radical on the right is eliminated.

$$-4\sqrt{2a} = -a - 6$$ Subtract $2a$ and 4 to isolate the radical term.

$$\left(-4\sqrt{2a}\right)^2 = (-a-6)^2$$ Square both sides.

$$16(2a) = a^2 + 12a + 36$$ The radical is eliminated.

$$32a = a^2 + 12a + 36$$

$$0 = a^2 - 20a + 36$$ Write the quadratic equation in standard form.

$$0 = (a-18)(a-2)$$ Factor.

$$a - 18 = 0 \quad \text{or} \quad a - 2 = 0$$ Zero Factor Property

$$a = 18 \quad \text{or} \quad a = 2$$

Check that 18 and 2 are both solutions.

EXAMPLE 6

Solving an Equation with Two Radicals

For the equation $1 = \sqrt{t-3} - \sqrt{2t-4}$, estimate the solution graphically. Then solve algebraically.

Solution

Figure 8.14

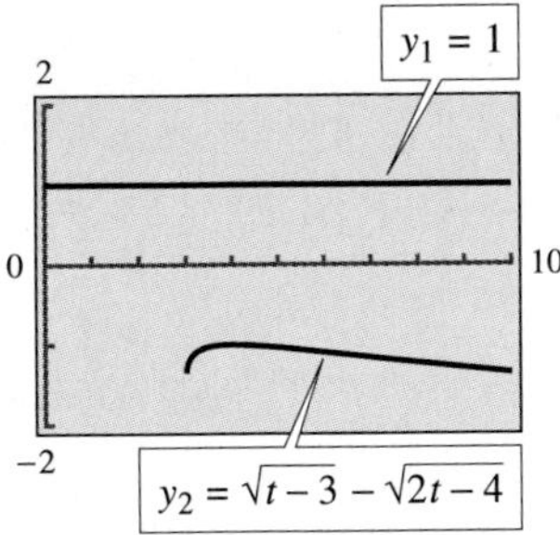

Figure 8.14 shows the graph of each side of the equation.

The graph of the left side is a horizontal line *above* the x-axis. The graph of the right side appears to lie entirely *below* the x-axis. From the graph we conclude that there is no solution.

$$1 = \sqrt{t-3} - \sqrt{2t-4}$$

$$1 + \sqrt{2t-4} = \sqrt{t-3}$$ Isolate one radical.

$$\left(1 + \sqrt{2t-4}\right)^2 = \left(\sqrt{t-3}\right)^2$$ Square both sides.

$$1 + 2\sqrt{2t-4} + 2t - 4 = t - 3$$

$$2\sqrt{2t-4} + 2t - 3 = t - 3$$

$$2\sqrt{2t-4} = -t$$ Isolate the radical term.

$$\left(2\sqrt{2t-4}\right)^2 = (-t)^2$$ Square both sides.

$$4(2t-4) = t^2$$

$$8t - 16 = t^2$$

$$0 = t^2 - 8t + 16$$ Quadratic equation in standard form

$$0 = (t-4)^2$$ Factor.

$$0 = t - 4$$ Zero Factor Property

$$4 = t$$

When $t = 4$, the left side of the equation has a value of 1, but the right side has a value of -1. Thus 4 is an extraneous solution. This equation has no solution.

Equations with Exponential Expressions

We now expand our equation-solving routines to include methods for solving equations that contain exponential expressions.

There are two properties that we use to solve equations in the form $A^n = B$. The first is the Odd Root Property.

The Odd Root Property

Suppose that n is an odd integer, $n > 1$, and c is a real number. If $x^n = c$, then $x = \sqrt[n]{c}$.

EXAMPLE 7

Solving Equations with the Odd Root Property

Solve each equation.

(a) $y^5 = -32$ (b) $(x - 3)^7 = 1$

Solution

(a) $y^5 = -32$

$y = \sqrt[5]{-32}$ Odd Root Property

$y = -2$

(b) $(x - 3)^7 = 1$

$x - 3 = \sqrt[7]{1}$ Odd Root Property

$x - 3 = 1$

$x = 4$

The other property that we use for solving equations that contain exponential expressions is the Even Root Property.

The Even Root Property

Suppose that n is a positive even integer and c is a real number.

1. If $c > 0$ and $x^n = c$, then $x = \sqrt[n]{c}$ or $x = -\sqrt[n]{c}$.
2. If $c = 0$ and $x^n = c$, then $x = 0$.
3. If $c < 0$, then $x^n = c$ has no real number solution.

Note: The symbol $\pm$ indicates two values, one positive and the other negative. Thus part 1 of the Even Root Property can be stated in a more compact form: If $c > 0$ and $x^n = c$, then $x = \pm\sqrt[n]{c}$.

EXAMPLE 8

Solving Equations with the Even Root Property

Solve each equation.

(a) $x^4 = 81$ (b) $(x + 5)^6 = 0$

(c) $(x - 1)^2 - 4 = -1$ (d) $2(t + 1)^2 + 3 = -3$

Solution

(a) $x^4 = 81$

$x = \pm\sqrt[4]{81}$ Even Root Property with $c > 0$

$x = \pm 3$ Both 3 and -3 are solutions.

(b) $(x + 5)^6 = 0$

$x + 5 = 0$ Even Root Property with $c = 0$

$x = -5$

(c) $(x - 1)^2 - 4 = -1$ Isolate the variable term.

$(x - 1)^2 = 3$ Add 4 to both sides.

$x - 1 = \pm\sqrt{3}$ Even Root Property with $c > 0$

$x = 1 \pm \sqrt{3}$ Add 1 to both sides.

$x \approx 2.73$ or $x \approx -0.73$

(d) $2(t + 1)^2 + 3 = -3$

$2(t + 1)^2 = -6$ Subtract 3 from both sides.

$(t + 1)^2 = -3$ Divide both sides by 2.

Because $c < 0$, the Even Root Property states that the equation has no real number solution.

To solve equations of the form $A^{m/n} = B$ for A, raise both sides to the nth power to obtain an integer exponent on A.

EXAMPLE 9

Solving an Equation with a Rational Exponent

Solve the equation $(x - 2)^{2/3} = 4$.

Solution

The graph in Fig. 8.15 suggests that we can anticipate two solutions.

Figure 8.15

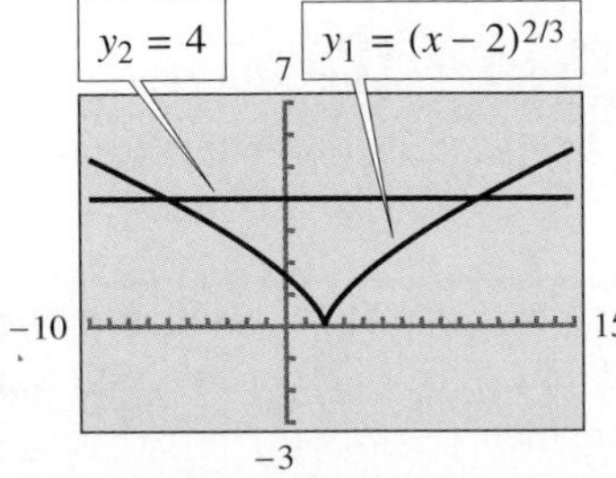

Think About It

In Example 9, why not just raise both sides to the $\frac{3}{2}$ power and apply the properties of exponents?

$$(x - 2)^{2/3} = 4$$

$$[(x - 2)^{2/3}]^3 = 4^3 \qquad \text{Raise both sides to the third power.}$$

$$(x - 2)^2 = 64 \qquad \text{Now the exponent is an integer.}$$

$$x - 2 = \pm 8 \qquad \text{Even Root Property}$$

$$x = 2 + 8 = 10 \quad \text{or} \quad x = 2 - 8 = -6$$

Use a calculator to check the solutions.

8.7 QUICK REFERENCE

Domain of a Radical Function

- A **radical function** is a function defined by a radical expression with a variable in the radicand. The domain of a radical function is the set of values of the variable for which the expression is defined.
- The domain of an odd root radical function is the set of all values of the variable for which the radicand is defined. The domain of an even root radical function is the set of values of the variable for which the radicand is nonnegative.

Radical Equations

- Squaring Property of Equality

  For real numbers a and b, if $a = b$, then $a^2 = b^2$.
- The Squaring Property of Equality can be extended to any power. If $a = b$, then $a^n = b^n$.
- All solutions of a radical equation must be checked because raising both sides of the equation to a power may produce extraneous solutions.

Equations with Exponential Expressions

- The Odd Root Property

  Suppose that n is an odd integer, $n > 1$, and c is a real number. If $x^n = c$, then $x = \sqrt[n]{c}$.
- The Even Root Property

  Suppose that n is an even positive integer and c is a real number.

  1. If $c > 0$ and $x^n = c$, then $x = \pm\sqrt[n]{c}$.
  2. If $c = 0$ and $x^n = c$, then $x = 0$.
  3. If $c < 0$, then $x^n = c$ has no real number solution.
- To solve equations of the form $A^{m/n} = B$ for A, raise both sides to the nth power to obtain an integer exponent on A. Then apply the Odd or Even Root Property.

8.7 EXERCISES

Concepts and Skills

1. Consider the radical function $y = \sqrt{R}$, where R represents the radicand. Explain how to determine the domain of the function.

2. Explain how you can tell that $\sqrt{x - 3} = \sqrt{2 - x}$ has no solution without solving it. (Hint: Consider the domains of the radical expressions.)

In Exercises 3–14, determine the domain of the given radical expression.

3. $\sqrt{x-2}$ **4.** $\sqrt{x+7}$

5. $\sqrt{2x-3}$ **6.** $\sqrt{2x}$

7. $\sqrt{2-x}$ **8.** $\sqrt{1-2x}$

9. $\sqrt{1+3x^2}$ **10.** $\sqrt{x^2+5}$

11. $\sqrt[3]{3-2x}$ **12.** $\sqrt[5]{5x+7}$

13. $\sqrt[4]{x+3}$ **14.** $\sqrt[6]{5-x}$

15. If we square both sides of the radical equation $\sqrt{x} = -5$, we obtain $x = 25$. Are the two equations equivalent? Why? Explain why it is essential to check the values obtained when solving a radical equation.

16. Suppose that you are solving the radical equation $\sqrt{x} + 1 = 5$ and you begin by squaring both sides. What is the resulting equation? Have you met your goal of eliminating the radical from the equation? If not, what should your first step have been?

In Exercises 17–42, solve the equation.

17. $\sqrt{5x+1} = 4$ **18.** $\sqrt{2x+1} = \frac{1}{2}$

19. $\sqrt[3]{3x-1} = 2$ **20.** $\sqrt[3]{4x+1} + 3 = 0$

21. $\sqrt{1-5x} + 6 = 3$ **22.** $\sqrt{x^2} = 3$

23. $\sqrt[4]{x+1} = 2$

24. $\sqrt[4]{2x+1} = \sqrt[4]{x+6}$

25. $\sqrt{x^2-3x} = 2$

26. $\sqrt{3x^2+7x-5} = 1$

27. $\sqrt{x}\sqrt{x-5} = 6$ **28.** $x = 4 + \sqrt{32-2x}$

29. $x\sqrt{2} = \sqrt{5x-2}$ **30.** $x\sqrt{3} = \sqrt{9x+30}$

31. $\sqrt{4x+13} = 2x-1$ **32.** $\sqrt{2x-3} = 5-2x$

33. $\sqrt[3]{x^2+x+2} = 2$ **34.** $\sqrt[3]{x^2+6x+9} = 1$

35. $\sqrt[4]{x+1} + \sqrt[4]{2x-3} = 0$ **36.** $\sqrt[5]{x^2-7x+9} = -1$

37. $\sqrt{x(x-2)} - x = -10$ **38.** $\sqrt{x}\sqrt{x-3} = x - 12$

39. $x - 9 = \sqrt{x^2-x-4}$ **40.** $x = \sqrt{45-11x} + 5$

41. $\sqrt{\frac{t}{3}} = \sqrt{\frac{t+4}{2+t}}$ **42.** $\sqrt{t+\frac{1}{t}} = \frac{\sqrt{17}}{2}$

43. In which of the following equations can we eliminate the radical by squaring both sides once?

(i) $\sqrt{2x} = x$ (ii) $\sqrt{2x} + 1 = x$

(iii) $\sqrt{2x} = \sqrt{x+1}$

Describe the resulting equations in (i) and (ii).

44. For which one of the following equations will it be necessary to square both sides twice in order to solve it?

(i) $\sqrt{x+1} = \sqrt{x} + 1$ (ii) $x + 1 = \sqrt{x+1}$

Explain how the other equation can be solved by squaring both sides just once.

In Exercises 45–60, solve the radical equation.

45. $\sqrt{x-3} + 1 = \sqrt{x}$

46. $\sqrt{x+5} = 1 + \sqrt{x}$

47. $\sqrt{4x-3} - \sqrt{3x-5} = 1$

48. $\sqrt{5x+1} - \sqrt{3x-5} = 2$

49. $\sqrt{5x-9} + 3 = \sqrt{x}$

50. $\sqrt{x-8} + \sqrt{x} = 2$

51. $\sqrt{7x+4} - \sqrt{x+1} = 3$

52. $\sqrt{7-2x} - 1 = \sqrt{2}\sqrt{x-1}$

53. $\sqrt{x+3} = 1 + \sqrt{x-2}$

54. $\sqrt{x^2+3x-1} - 1 = \sqrt{x^2+x-2}$

55. $\sqrt{5x-6} = 1 + \sqrt{3x-5}$

56. $\sqrt{2x-1} - 1 = \sqrt{x-1}$

57. $\sqrt{4x+1} - \sqrt{3x-2} = 1$

58. $\sqrt{2x+3} - \sqrt{x-2} = 2$

59. $\sqrt{7x-3} - \sqrt{2x+1} = 2$

60. $\sqrt{5x} - \sqrt{x+4} = 2$

61. Suppose $x^{a/b} = k$. What is the first step in solving the equation for x? What is the purpose of this step?

62. Without solving, explain why the following equation has no solution: $\sqrt{x+1} + 1 = 0$.

In Exercises 63–74, solve the given equation.

63. $x^2 = 6$ **64.** $\frac{1}{3}x^2 = 15$

65. $t^2 + 9 = 0$ **66.** $3x^2 - 27 = 0$

67. $(2x - 5)^2 = 16$

68. $25(x + 2)^2 = 16$

69. $(3x - 1)^2 = 3$

70. $(x + 2)^2 = 18$

71. $(x - 5)^5 = 0$

72. $(x - 3)^3 = -125$

73. $(2x - 3)^5 = -32$

74. $(2x + 3)^4 = -2$

In Exercises 75–82, solve the equations involving rational exponents.

75. $x^{2/3} = 9$

76. $y^{3/4} = -8$

77. $x^{-3/4} = 27$

78. $y^{-2/3} = 25$

79. $(x - 3)^{-2/3} = 4$

80. $(x + 5)^{-3/5} = 8$

81. $(3x - 2)^{3/4} = 8$

82. $(2x - 3)^{2/3} = 16$

83. The square root of the product of a positive number and its square is twice the number. What is the number?

84. The fourth root of the cube root of the square root of a number is 2. What is the number?

In Exercises 85–88, determine the value of x.

85.

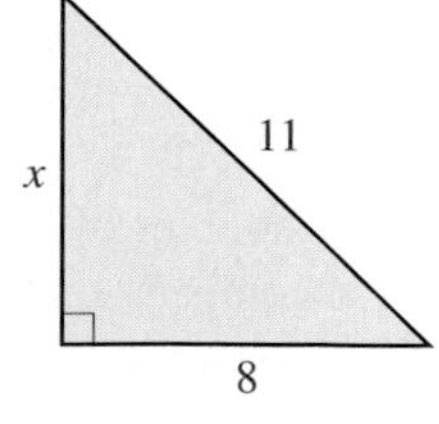

86.

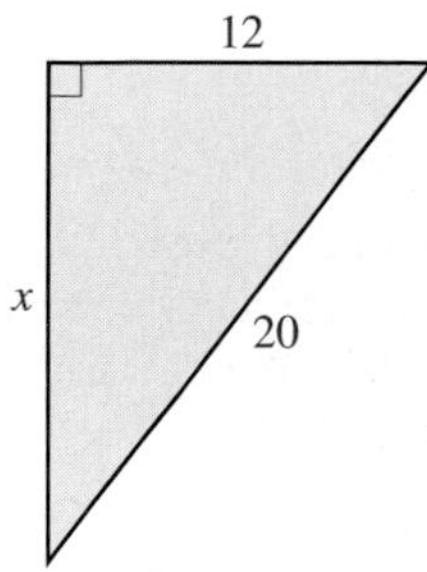

87.

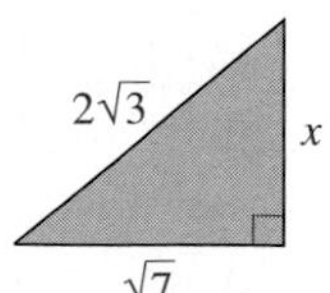

88.

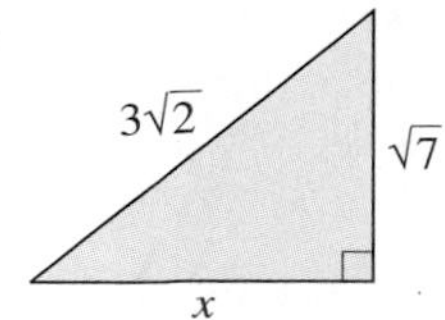

Geometric Models

89. A 26-foot ladder rests against a wall with the bottom of the ladder 7 feet from the base of the wall. How far up the wall does the top of the ladder rest? (See figure.)

Figure for 89

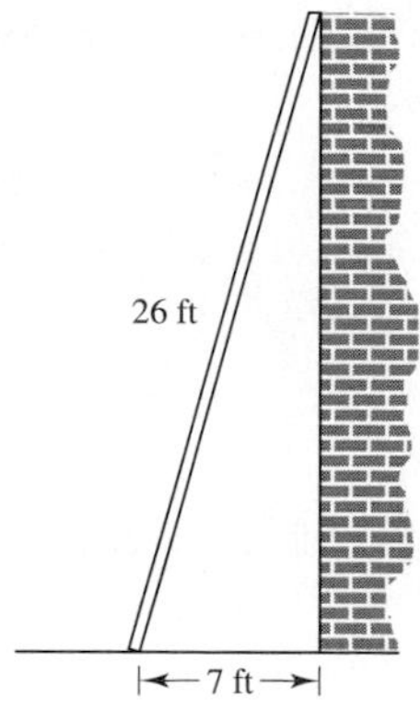

90. A guy wire 44 feet long reaches from the top of a telephone pole to a point on the ground 14 feet from the base of the pole. How tall is the telephone pole? (See figure.)

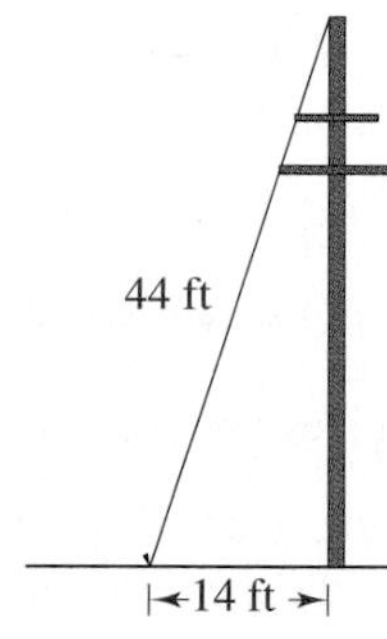

91. A billboard that is 30 feet high is supported by a brace that is anchored to the ground at a point that is 10 feet from the base of the sign. How long is the brace?

92. A 32-foot ramp connects a platform with a sidewalk that is at ground level. If the platform is 6 feet above ground level, what is the distance from the base of the platform to the sidewalk?

93. A surveyor needs to know the distance BC across the pond in the figure on the following page. She places a stake at point C so that the line segments $\overline{AC}$ and $\overline{BC}$ are at right angles. The trees are 150 feet apart, and the distance from the stake to the tree at point A is 125 feet. What is the distance from the stake to the tree at point B?

Figure for 93

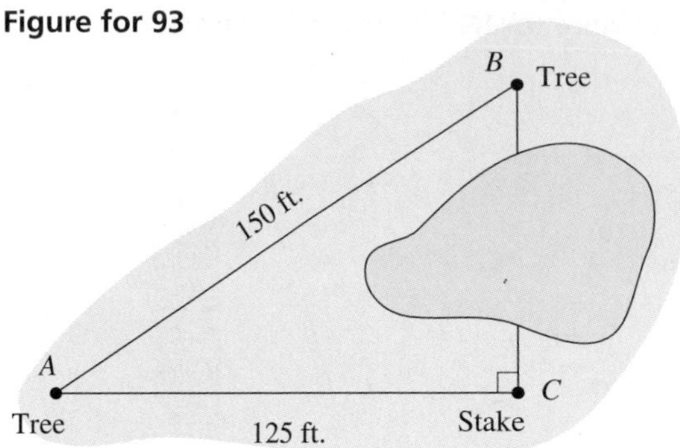

94. A car traveled south from Peoria at 46 miles per hour (mph). A second car left the same point at the same time and traveled west at 54 mph. How far apart were the cars after 1.5 hours?

Real-Life Applications

95. A circular rug was advertised as having an area of 95 square feet. Will the rug fit in a square room 11 feet on each side? Show how you know.

96. Clocks with pendulums can be adjusted to run faster or slower by changing the position of the weight at the end of the pendulum. The time t (in seconds) that it takes a pendulum to swing through one cycle is given by the formula $t = 2\pi\sqrt{\frac{L}{980}}$, where L is the length (in centimeters) of the pendulum. To the nearest tenth of a centimeter, what is the length of a pendulum that takes 1 second to swing through one cycle?

97. The price of plastic rain ponchos at an amusement park depends on demand d. As demand d (in hundreds) increases, the price P also increases and is given by the function $P(d) = 3d - 3\sqrt{d + 2}$. For what demand is the price $12?

98. A weight-loss clinic estimates that the number of people (in hundreds) who have completed its program successfully is $\sqrt{m}$ women and $\sqrt{m - 7}$ men, where m is the number of months since the clinic opened. After how many months have 700 people completed the program?

99. A dolphin trainer at Marine World estimated that the number of commands C that a dolphin learns to understand is a function of the number w of weeks of instruction: $C(w) = w - \sqrt{w}$. After how many weeks will the dolphin understand two commands?

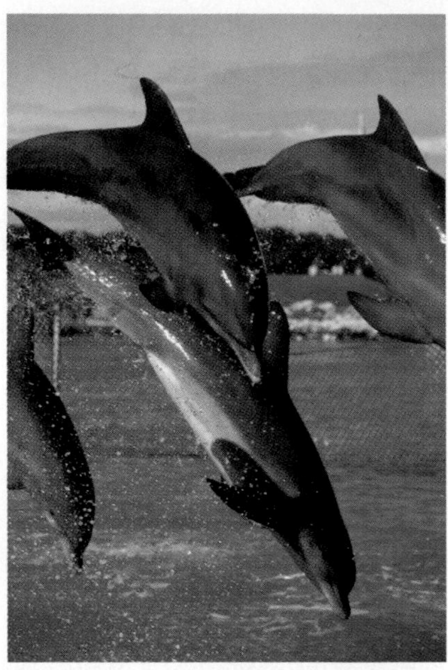

100. The wildlife population of a certain area began to decrease as a result of development of the land. The population is modeled by the expression $270\sqrt{2y + 1} - 270\sqrt{3y - 5}$, where y is the number of years after development began. After how many years will the population reach 0?

Group Project: College Financial Aid

Loans have grown faster than grants as sources of financial aid for a college education. The accompanying table compares the data (in billions of dollars) over a 20-year period. (Source: The College Board.)

	1974–1975	1994–1995
Grants	19.225	20.484
Loans	5.146	25.683

The following functions model the amount of grants and loans (in billions of dollars), where x is the number of years since 1970.

Grants: $G(x) = \sqrt{2.5x + 357}$

Loans: $L(x) = 7.43\sqrt{x} - 11.5$

101. Write an equation to determine the year in which the amount from loans was the same as the amount from grants.

102. Use a graph to estimate the solution of the equation in Exercise 101.

103. Write an equation to determine the year in which the amount from loans is projected to be twice the amount from grants.

104. Use a graph to estimate the solution of the equation in Exercise 103.

Challenge

In Exercises 105 and 106, use a graph to estimate the domain of the given function.

105. $y = \sqrt{x^2 + x - 12}$ **106.** $y = \sqrt{-x^2 - 6x - 5}$

In Exercises 107–110, solve the equation.

107. $\sqrt[6]{3 + t - t^2} = \sqrt[3]{2t + 3}$

108. $\sqrt{\sqrt{t^2 + 3t + 12} + 3} = \sqrt{t + 5}$

109. $\sqrt{x + 2} - \sqrt{x - 3} = \sqrt{4x - 1}$

110. $\sqrt[4]{x + 3} = \sqrt{x + 1}$

111. If we square both sides of each of the following equations, the solution set of each appears to be the set of all real numbers. However, closer examination shows that the solution sets are all different. Solve the equations and compare solution sets.

(a) $\sqrt{x^2} = x$ (b) $\sqrt{x^2} = -x$ (c) $\sqrt{x^4} = x^2$

8.8 COMPLEX NUMBERS

Imaginary Numbers • Complex Number System • Addition and Subtraction • Multiplication and Division

Imaginary Numbers

The equation $x^2 = -1$ has no solution in the set of real numbers. To be able to solve such equations, we begin expanding the number system by defining an **imaginary number** i.

Definition of the Imaginary Number *i*

The symbol i represents an **imaginary number** with the properties $i = \sqrt{-1}$ and $i^2 = -1$.

Using this new number i, we can define the square root of any real number.

Definition of $\sqrt{-n}$

For any positive real number n, $\sqrt{-n} = i\sqrt{n}$.

Note: The word *imaginary* is used to distinguish these new numbers from real numbers. It should not be inferred from their name that imaginary numbers do not really exist. Like all numbers in mathematics, imaginary numbers exist because we *define* them to exist.

EXAMPLE 1

Using the Definition of $\sqrt{-n}$

Write each square root in terms of i.

(a) $\sqrt{-49} = i\sqrt{49} = 7i$

(b) $\sqrt{-7} = i\sqrt{7}$

(c) $\sqrt{-\frac{4}{9}} = i\sqrt{\frac{4}{9}} = \frac{2}{3}i$

Note: In part (b) of Example 1, to avoid confusing $\sqrt{7}i$ with $\sqrt{7i}$, we usually write $i\sqrt{7}$.

Imaginary numbers are not real numbers, and some properties of real numbers do not apply to imaginary numbers. For example, we cannot apply the Product Rule for Radicals to $\sqrt{-12}\sqrt{-3}$ to obtain $\sqrt{36}$. However, some properties, such as the Distributive Property, do apply.

As we perform operations with imaginary numbers, we must begin with the definition of $\sqrt{-n}$.

EXAMPLE 2

Products of Imaginary Numbers

Determine the products.

(a) $\sqrt{-12}\sqrt{-3}$ (b) $\sqrt{-2}(\sqrt{-8} - \sqrt{-6})$

Solution

(a)
$$\begin{aligned} \sqrt{-12}\sqrt{-3} &= i\sqrt{12} \cdot i\sqrt{3} && \text{Definition of } \sqrt{-n} \\ &= i^2 \cdot \sqrt{36} && \text{Product Rule for Radicals} \\ &= -1 \cdot 6 && i^2 = -1 \\ &= -6 \end{aligned}$$

(b)
$$\begin{aligned} \sqrt{-2}(\sqrt{-8} - \sqrt{-6}) &= i\sqrt{2}(i\sqrt{8} - i\sqrt{6}) && \text{Definition of } \sqrt{-n} \\ &= i^2\sqrt{16} - i^2\sqrt{12} && \text{Distributive Property} \\ &= -1 \cdot 4 - (-1)\sqrt{4}\sqrt{3} && \text{Simplify the radicals.} \\ &= -4 + 2\sqrt{3} \end{aligned}$$

We have defined $i = \sqrt{-1}$ and $i^2 = -1$. To facilitate computations, we need to investigate higher powers of i.

EXPLORING THE CONCEPT

Powers of i

Consider the following successive powers of i.

$$\begin{aligned}
i &= & i\\
i^2 &= & -1\\
i^3 &= i^2 \cdot i = -1 \cdot i = & -i\\
i^4 &= (i^2)^2 = (-1)^2 = & 1\\
i^5 &= (i^2)^2 \cdot i = (-1)^2 \cdot i = 1 \cdot i = & i\\
i^6 &= (i^2)^3 = (-1)^3 = & -1\\
i^7 &= (i^2)^3 \cdot i = (-1)^3 \cdot i = -1 \cdot i = & -i\\
i^8 &= (i^2)^4 = (-1)^4 = & 1
\end{aligned}$$

By continuing these computations, we can see that all powers of i cycle through the numbers i, -1, $-i$, and 1. Furthermore by writing powers of i as $(i^2)^n$ or $(i^2)^n \cdot i$, the computations reduce to raising -1 to an integer power.

EXAMPLE 3

Evaluating Powers of *i*

Evaluate.

(a) i^{11} (b) i^{201} (c) i^{48}

Solution

(a) $i^{11} = (i^2)^5 \cdot i = (-1)^5 \cdot i = -1 \cdot i = -i$

(b) $i^{201} = (i^2)^{100} \cdot i = (-1)^{100} \cdot i = 1 \cdot i = i$

(c) $i^{48} = (i^2)^{24} = (-1)^{24} = 1$

Complex Number System

Think About It

For what values of n is i^n a real number? an imaginary number? a complex number?

We define an expanded number system, called the **complex number system,** to be numbers that can be written as the sum of a real number and a multiple of i.

The word *complex* means consisting of more than one part. In the definition of a complex number, we refer to numbers consisting of a *real part* and an *imaginary part.*

Definition of a Complex Number

If a and b are real numbers, then $a + bi$ is the standard form of a **complex number.** If $b \neq 0$, then $a + bi$ is an **imaginary number.** If $b = 0$, then $a + bi$ is simply the real number a. For $a + bi$, a is called the **real part** and b is called the **imaginary part.**

Figure 8.16 shows the structure of the complex number system.

Figure 8.16

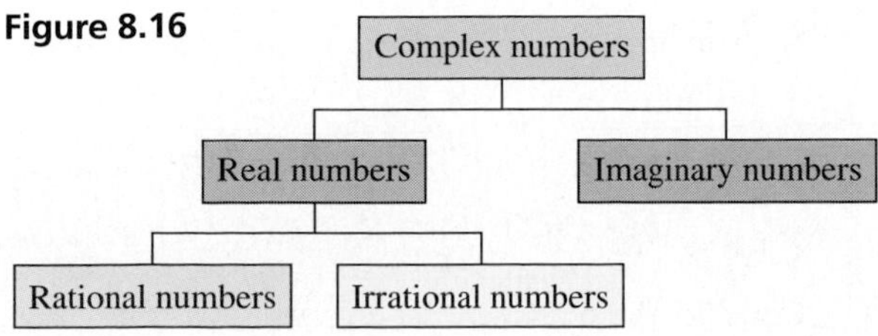

Just as real numbers are either rational or irrational, but not both, complex numbers are either real or imaginary, but not both.

The real number 5 is a complex number because it can be written $5 + 0i$. Similarly, the imaginary number $3i$ is a complex number because it can be written $0 + 3i$.

Note: Although numbers such as $3i$ are called **pure imaginary numbers,** they are still complex numbers.

Figure 8.17

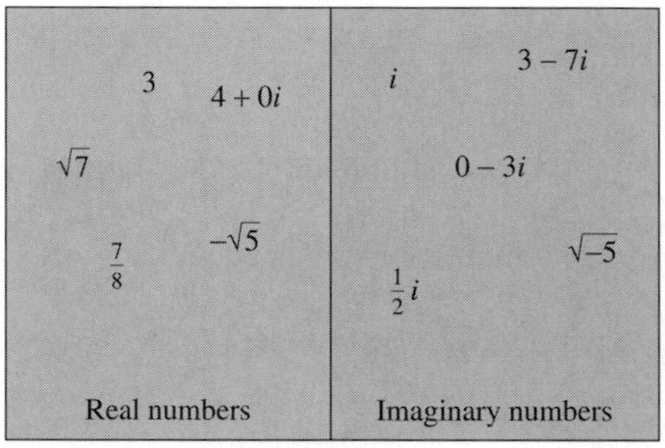

Real numbers	Imaginary numbers
3, $4 + 0i$, $\sqrt{7}$, $-\sqrt{5}$, $\frac{7}{8}$	i, $3 - 7i$, $0 - 3i$, $\sqrt{-5}$, $\frac{1}{2}i$

Complex Numbers

In Fig. 8.17 we see that 3, $\sqrt{7}$, $4 + 0i$, $-\sqrt{5}$, and $\frac{7}{8}$ are real numbers. The numbers i, $0 - 3i$, $3 - 7i$, $\sqrt{-5}$, and $\frac{1}{2}i$ are imaginary numbers. Every number in the figure is a complex number.

Two complex numbers are defined to be equal if their real parts are equal and their imaginary parts are equal.

Definition of Equality of Complex Numbers

$a + bi = c + di$ if and only if $a = c$ and $b = d$.

LEARNING TIP

Although i is not a variable, treating it as a variable in operations is often helpful.

In the complex number system, addition and multiplication are commutative and associative. The Distributive Property also holds for complex numbers.

When we add, subtract, and multiply, it is helpful to treat complex numbers as though they were binomials.

Addition and Subtraction

To add (or subtract) complex numbers, we add (or subtract) the real and imaginary parts.

Definitions of Addition and Subtraction

The sum and difference of two complex numbers $a + bi$ and $c + di$ are defined as follows:

$$(a + bi) + (c + di) = (a + c) + (b + d)i$$

$$(a + bi) - (c + di) = (a - c) + (b - d)i$$

EXAMPLE 4

Adding and Subtracting Complex Numbers

(a) $(3 - 2i) + (-4 + 7i) = (3 - 4) + (-2 + 7)i = -1 + 5i$

(b) $8 - (5 - 2i) = 8 - 5 + 2i = 3 + 2i$

(c) $(1 + i) - (1 - i) = 1 + i - 1 + i = 2i$

Multiplication and Division

To find the product of two complex numbers, use the Distributive Property or the FOIL method just as you did for binomials. Remember that $i \cdot i = i^2 = -1$.

Definition of Multiplication

The product of two complex numbers $a + bi$ and $c + di$ is defined as follows:

$$(a + bi)(c + di) = (ac - bd) + (bc + ad)i$$

EXAMPLE 5

Multiplying Complex Numbers

(a) $5i(4 - 3i) = 20i - 15i^2$ Distributive Property

$= 20i - 15(-1)$

$= 15 + 20i$

(b) $(3 + 2i)(-1 + 4i) = -3 + 12i - 2i + 8i^2$ FOIL

$= -3 + 12i - 2i + 8(-1)$

$= -3 + 12i - 2i - 8$

$= -11 + 10i$

(c) $(2 + i)^2 = 2^2 + 4i + i^2$ Square of a binomial

$= 4 + 4i - 1$

$= 3 + 4i$

(d) $(5 + 4i)(5 - 4i) = 25 - 20i + 20i - 16i^2$ FOIL

$= 25 - 16(-1)$

$= 41$

In part (d) of Example 5 we call the pair of complex numbers $5 + 4i$ and $5 - 4i$ **complex conjugates.** Note that the product of these complex conjugates is the real number 41.

In general, the product of complex conjugates is the sum of the squares of the real parts and the imaginary parts, and the result is always a real number.

LEARNING TIP

Be careful not to confuse $(a + b)(a - b) = a^2 - b^2$ with $(a + bi)(a - bi) = a^2 + b^2$.

Product of Complex Conjugates

For any two complex conjugates $a + bi$ and $a - bi$,

$$(a + bi)(a - bi) = a^2 + b^2$$

To divide a complex number $a + bi$ by a nonzero real number c, we divide the real part and the imaginary part by c.

$$\frac{a + bi}{c} = \frac{a}{c} + \frac{b}{c}i$$

To divide a complex number by an imaginary number, we simply multiply the numerator and the denominator by the conjugate of the denominator.

EXAMPLE 6

Dividing Complex Numbers

(a) $\dfrac{4 + 6i}{2} = \dfrac{4}{2} + \dfrac{6}{2}i = 2 + 3i$

(b) $\dfrac{2 - i}{1 + 2i} = \dfrac{(2 - i)(1 - 2i)}{(1 + 2i)(1 - 2i)}$ Multiply the numerator and the denominator by the conjugate of the denominator.

$= \dfrac{2 - 4i - i - 2}{1 + 4}$ $(a + bi)(a - bi) = a^2 + b^2$

$= -\dfrac{5i}{5}$

$= -i$

(c) $\dfrac{5 + 2i}{2i} = \dfrac{(5 + 2i)(-2i)}{(2i)(-2i)}$ The conjugate of the denominator, $0 + 2i$, is $0 - 2i$, or $-2i$.

$= \dfrac{-10i + 4}{4}$ $(-2i)(2i) = -4i^2 = -4(-1) = 4$

$= 1 - \dfrac{5}{2}i$

8.8 QUICK REFERENCE

Imaginary Numbers

- The symbol i represents an **imaginary number** with the properties $i = \sqrt{-1}$ and $i^2 = -1$.
- For any positive real number n, $\sqrt{-n} = i\sqrt{n}$.
- To perform operations with radicals having negative radicands, use the definition of i to rewrite the radicals in the form of imaginary numbers.
- Powers of i can be evaluated by writing the expression in the form $(i^2)^n$ or $(i^2)^n \cdot i$, where n is an integer.

Complex Number System

- If a and b are real numbers, then $a + bi$ is the standard form of a **complex number.** If $b \neq 0$, then $a + bi$ is an **imaginary number.** If $b = 0$, then $a + bi$ is the real number a.
- For $a + bi$, a is called the **real part** and b is called the **imaginary part.**
- Imaginary numbers and real numbers are also complex numbers.
- The complex numbers $a + bi$ and $c + di$ are defined to be equal if and only if $a = c$ and $b = d$.

Addition and Subtraction

- The sum of two complex numbers is defined as
$$(a + bi) + (c + di) = (a + c) + (b + d)i$$
- The difference of two complex numbers is defined as
$$(a + bi) - (c + di) = (a - c) + (b - d)i$$

Multiplication and Division

- The product of two complex numbers can be found by treating the numbers as binomials and using the FOIL method. The definition is
$$(a + bi)(c + di) = (ac - bd) + (bc + ad)i$$
- The complex numbers $a + bi$ and $a - bi$ are called **complex conjugates.**
- In general, $(a + bi)(a - bi) = a^2 + b^2$.
- If c is a nonzero real number, then $\frac{a + bi}{c} = \frac{a}{c} + \frac{b}{c}i$.
- To find the quotient of $a + bi$ and $c + di$, multiply the numerator and denominator by the conjugate of the divisor.

8.8 EXERCISES

Concepts and Skills

1. In order to solve $x^2 = -1$, it is necessary to define a new type of number. What is the number, and what two properties does it possess according to its definition?

2. To perform $\sqrt{-3}\sqrt{-5}$, we begin by writing $\sqrt{-3}\sqrt{-5} = i\sqrt{3} \cdot i\sqrt{5} = (i \cdot i) \cdot (\sqrt{3} \cdot \sqrt{5})$. What two properties are used and for what purpose?

In Exercises 3–10, write the radical expression in the complex number form $a + bi$.

3. $\sqrt{-9}$
4. $\sqrt{-5}$
5. $\sqrt{-18}$
6. $\sqrt{-\frac{16}{25}}$
7. $\sqrt{25} + \sqrt{-36}$
8. $\sqrt{80} + \sqrt{-27}$
9. $\frac{-4 - \sqrt{-12}}{2}$
10. $\frac{-3 + \sqrt{-27}}{-3}$

11. Describe a method for simplifying i^{34}.

12. Explain how to add two complex numbers.

In Exercises 13–16, determine a and b.

13. $(a + 1) + 2bi = 3 - 2i$
14. $(2a + 1) + 3bi = 7 + 6i$
15. $2a + bi = \sqrt{-25} - 2$
16. $4a + 2bi = 3\sqrt{-4}$

17. Which of the following statements is true?
    (a) The set of real numbers is a subset of the set of imaginary numbers.
    (b) If a and b are nonzero real numbers, then $a + bi$ is an imaginary number.
    (c) The number 0 is both a real number and an imaginary number.

18. Classify the following as real numbers or imaginary numbers.
    (a) $\sqrt{7}$
    (b) $i\sqrt{7}$
    (c) $\sqrt{-7}$
    (d) $\sqrt{7} + \sqrt{-7}$
    (e) $-\sqrt{7} + \sqrt{7}$
    (f) $\sqrt{-7} - \sqrt{-7}$

In Exercises 19–30, perform the indicated operation.

19. $(5 + 6i) + (-4 - 5i)$
20. $(-8 + 4i) - (9 - 4i)$
21. $(7 + 3i) - (-8 - 7i)$
22. $(-5 + 2i) + (5 - 3i)$
23. $7i - 2(-3 - 7i)$
24. $8 - 3(5 + 2i)$
25. $(7 - 2i) - (3 - 4i) + 2(5 - i)$
26. $(3 - 4i) + 3(2 - 6i) - (3 + 2i)$

27. $\sqrt{-36} + \sqrt{-49}$

28. $\sqrt{-32} - \sqrt{-72}$

29. $3 - i - \sqrt{4} - \sqrt{-9}$

30. $\left(4 + \sqrt{-50}\right) - 4\left(3 - \sqrt{-72}\right)$

In Exercises 31–34, determine a and b.

31. $(a + 3i) + (2 + bi) = i$

32. $(a - 2i) - (7 + bi) = -4 - 8i$

33. $\left(5 - \sqrt{-4}\right) - (a + bi) = 6$

34. $(a + bi) + \left(\sqrt{4} - \sqrt{-16}\right) = -1 - 6i$

In Exercises 35–46, determine the product.

35. $(2i)(5i)$

36. $(-6i)(2i)$

37. $7i(3 - 4i)$

38. $-2i(-3 - 5i)$

39. $(2 - 3i)(3 - 4i)$

40. $(1 + i)(-3 + 5i)$

41. $(3 + i)^2$

42. $(2 - i)(-4 + 3i)$

43. $\sqrt{-8}\sqrt{-2}$

44. $\sqrt{-10}\sqrt{2}$

45. $\sqrt{-3}\left(\sqrt{-6} + \sqrt{-27}\right)$

46. $\left(\sqrt{-25} - \sqrt{4}\right)\left(\sqrt{-4} + \sqrt{25}\right)$

In Exercises 47–54, evaluate the given power of i.

47. i^{15}

48. i^{45}

49. i^{52}

50. i^{22}

51. i^{30}

52. i^{39}

53. i^{13}

54. i^{24}

55. Suppose that a and b are nonzero real numbers. What is the difference in the results of the following operations?

(a) $(a + b)(a - b)$ (b) $(a + bi)(a - bi)$

56. To find the quotient $\dfrac{a + bi}{c + di}$, where a, b, c, and d are nonzero real numbers, what step must we take first?

In Exercises 57–62, write the complex conjugate of the given complex number. Then find the product of the complex number and its conjugate.

57. $7i$

58. $-4i$

59. $2 + 3i$

60. $3 - 4i$

61. $\sqrt{5} - 4i$

62. $7 - i\sqrt{2}$

In Exercises 63–74, determine the quotient.

63. $\dfrac{-3}{2i}$

64. $\dfrac{4}{5i}$

65. $\dfrac{9 + 6i}{3i}$

66. $\dfrac{-4 + 10i}{-2i}$

67. $\dfrac{3 - i}{1 + 2i}$

68. $\dfrac{4 + 5i}{3 - 2i}$

69. $\dfrac{3 - 4i}{-7 - 5i}$

70. $\dfrac{5 - 2i}{3 + 5i}$

71. $\dfrac{3}{2 + \sqrt{-3}}$

72. $\dfrac{-4}{1 - \sqrt{-2}}$

73. $\dfrac{\sqrt{-5} - \sqrt{3}}{\sqrt{-3} - \sqrt{9}}$

74. $\dfrac{\sqrt{6} + \sqrt{-3}}{3 - \sqrt{-2}}$

In Exercises 75–92, perform the indicated operations for each expression.

75. $(3 - 4i) + (5 + i)$

76. $(-2 + 3i) - (2 - 7i)$

77. $(4i)(-7i)$

78. $(-3i)(-i)$

79. $\dfrac{1 + 2i}{3i}$

80. $\dfrac{i - 4}{-2i}$

81. $3(2 + i) + 3i(1 - i)$

82. $5 - (4 - 5i)$

83. $(-6 + 5i)^2$

84. $(-2 - i)(3 + 2i)$

85. $\dfrac{3 + 4i}{3 - 4i}$

86. $\dfrac{2i}{5 - i}$

87. $7\sqrt{-25} - 4\sqrt{-49}$

88. $5\sqrt{-28} + 2\sqrt{-63}$

89. $\left(1 + \sqrt{-3}\right)\left(-1 - \sqrt{-12}\right)$

90. $\sqrt{-2}\left(\sqrt{18} - \sqrt{-8}\right)$

91. $\dfrac{\sqrt{-75}}{\sqrt{-3}}$

92. $\dfrac{\sqrt{6}\sqrt{-10}}{\sqrt{-2}}$

In Exercises 93–96, verify that the given number and its conjugate are both solutions of the given equation.

93. $x^2 - 4x + 13 = 0;\quad 2 + 3i$

94. $x^2 + 9 = 0;\quad -3i$

95. $x^2 - 4x + 5 = 0;\quad 2 - i$

96. $4x^2 + 3 = 0;\quad \dfrac{\sqrt{3}}{2}i$

The rule for the product of complex conjugates states that $(a + bi)(a - bi) = a^2 + b^2$. We can use this rule to factor expressions such as $x^2 + 4$ *over the complex numbers.*

Example: $x^2 + 4 = (x + 2i)(x - 2i)$

In Exercises 97–100, use the rule stated on page 544 to factor the given expression over the complex numbers.

97. $x^2 + 9$

98. $x^2 + 25$

99. $4x^2 + 1$

100. $49x^2 + 16$

Challenge

101. For a nonzero complex number $a + bi$ and a positive integer n, $(a + bi)^{-n}$ is defined as

$$(a + bi)^{-n} = \frac{1}{(a + bi)^n}$$

Use this definition to evaluate the following.

(a) i^{-2}

(b) i^{-3}

(c) $(2 + 3i)^{-1}$

(d) $(1 - 2i)^{-2}$

102. Verify that $\frac{\sqrt{2}}{2}(1 + i)$ is a square root of i. What is your conjecture about another square root of i? Show that your conjecture is correct.

103. In the complex number system, a nonzero number has two square roots. For example, the square roots of 4 are 2 and -2; the square roots of -4 are $2i$ and $-2i$.

One cube root of -8 is -2. What is your conjecture about the total number of cube roots of a number in the complex number system? To support your conjecture, verify the following.

(a) $\left(1 + i\sqrt{3}\right)^3 = -8$

(b) $\left(1 - i\sqrt{3}\right)^3 = -8$

In Exercises 104 and 105, perform the indicated operations and express the result in the form $a + bi$.

104. $\dfrac{2i + 1}{i} - \dfrac{1 - 3i}{4 - i}$

105. $\dfrac{2 + 3i}{2 - 3i} + \dfrac{i}{1 - i}$

8. CHAPTER PROJECT

Visitors to National Parks and Monuments

James Bryce, a British ambassador to the United States, observed that the national parks are "the best idea America ever had." Theodore Roosevelt, who was instrumental in preserving the natural wonders in the United States, once said, "Leave it as it is. You cannot improve on it. The ages have been at work on it, and only man can mar it." Roosevelt's philosophy led to the establishment of the National Park Service in 1916 to preserve our natural and historic treasures. Today there are 357 sites and 80.7 million acres under the administration of the National Park Service.

The accompanying table shows the number of visitors (in millions) to national parks and to national monuments during the period 1990–1994.

	Visitors (millions)	
Year	**National Parks**	**National Monuments**
1990	57.7	23.9
1991	57.4	25.8
1992	58.7	26.6
1993	59.8	26.5
1994	60.1	26.5

The following functions model the number (in millions) of visitors to national parks and monuments as functions of the number of years since 1983.

National parks: $P(x) = 46.5x^{1/10} + 1$

National monuments: $M(x) = 16.1x^{1/5}$

1. Overcrowding in national parks has become such a serious problem that the number of visitors is sometimes restricted. If the National Park Service were to determine that 62 million is the maximum number of visitors that the national parks can absorb, in what year would that maximum be reached?
2. Our national monuments occupy 4.8 million acres. Write an expression in x that models the number of visitors per acre. Use this expression to determine the percentage increase in visitors per acre from 1990 to 1994.
3. Write a simplified expression (ratio) that compares the number of visitors to national parks with the number of visitors to national monuments. What is this ratio projected to be in the year 2000?
4. Wrangell–St. Elias National Park and Preserve in Alaska is the park with the largest land area, representing 13.36% of all the land administered by the National Park Service. Determine the number of acres in this park.
5. The smallest unit in the National Park Service system is Thaddeus Kosciuszko National Memorial in Pennsylvania. This unit occupies just 0.02 acre. If the land for this memorial were a square, what would be the length (in feet) of each side of the square? (1 acre = 43,560 square feet.)
6. State parks in the United States consist of 11,831,000 acres and had 724.8 million visitors in 1995. After determining the total number N of acres of state parks in your state, calculate the ratio of your state's population to N for each of the last four census dates.
   (a) What do these data suggest regarding the need for more state park land in your state?
   (b) If the population of your state is increasing, how does that fact work against the establishment of new state park areas?

8. CHAPTER REVIEW EXERCISES

Section 8.1

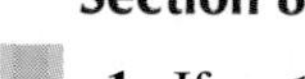

1. If $x < 0$, then $\sqrt{x^2} = |x|$ and $\sqrt[3]{x^3} = x$. Explain why absolute value symbols are used in one result but not the other.

2. Because $3^2 = 9$ and $(-3)^2 = 9$, we say that 9 has two square roots, 3 and -3. Why, then, do we write $\sqrt{9} = 3$?

In Exercises 3–6, evaluate, if possible.

3. $-\sqrt{64}$

4. $-\sqrt[3]{-64}$

5. $-\sqrt[3]{64}$

6. $\sqrt[4]{-64}$

In Exercises 7–10, simplify the given expression.

7. $\sqrt[4]{x^8}$

8. $\sqrt{(3x)^4}$

9. $\sqrt{(x+4)^2}$

10. $7\sqrt[3]{x^9}$

11. Translate the given phrase into a radical expression.

(a) The square root of 10 minus a.

(b) The square root of the quantity 10 minus a.

12. The side of a cube is given by $s = \sqrt[3]{V}$, where V is the volume of the cube. What is the side of a cube whose volume is 4.096 cubic inches?

Section 8.2

13. Explain why $(-4)^{1/2}$ is not a real number.

14. When converting $b^{m/n}$ into a radical, what are the interpretations of m and n?

In Exercises 15–19, evaluate the given expression, if possible.

15. $16^{3/2}$

16. $\left(-\frac{8}{27}\right)^{1/3}$

17. $(-27)^{-2/3}$

18. $-9^{-1/2}$

19. $(-9)^{-1/2}$

20. Write $4x^{6/7}$ and $(4x)^{6/7}$ as radical expressions.

21. Show how to evaluate $25^{1.5}$ without a calculator.

22. The length d of a diagonal of a rectangle is given by $d = (L^2 + W^2)^{1/2}$, where L and W are the length and width of the rectangle, respectively. To the nearest hundredth of a foot, how long is the diagonal of a rectangle whose width is 7 feet and whose length is 5 yards?

Section 8.3

23. Explain why the result of simplifying $(x^4)^{1/4}$ depends on whether x is nonnegative or is any real number.

24. Which of the following is a correct way to begin simplifying the given expression? Explain.

(i) $\left(\frac{4}{9}\right)^{-1/2} = \left(\frac{9}{4}\right)^{1/2} = \cdots$

(ii) $\left(\frac{4}{9}\right)^{-1/2} = \frac{4^{-1/2}}{9^{-1/2}} = \cdots$

25. Use the properties of exponents to evaluate each expression.

(a) $18^{1/2} \cdot 2^{1/2}$

(b) $8^{1/6} \cdot 8^{1/2}$

In Exercises 26–31, simplify the expression and write the result with positive exponents.

26. $\frac{f^{5/7}}{f^{3/7}}$

27. $(b^{-4/3}c^2)^9$

28. $(a^{-2})^{-1/4}$

29. $\left(\frac{27x^6z^{-9}}{y^{-3}}\right)^{1/3}$

30. $(2x^{-3/2})^{-1}$

31. $\frac{(3x^{-1})^{-2}}{x^{1/2}}$

32. Determine the values of a and b that make the statement true.

$(x^a y^{12})^{1/4} = x^2 y^b$

33. Simplify $x^{2/3}(x^{-5/3} - x^{7/3})$, $x \neq 0$. Write the result with positive exponents.

34. (a) Write $\sqrt[3]{\sqrt[5]{x}}$ with just one radical symbol.

(b) Write $\sqrt[10]{x^8y^6}$ with a smaller index.

Section 8.4

35. Explain why the Product Rule for Radicals cannot be used to multiply $\sqrt[3]{5}\sqrt{5}$. Show a method that can be used to perform the operation.

36. Explain why $\sqrt[3]{24}$ can be simplified but $\sqrt[3]{25}$ cannot.

In Exercises 37–44, simplify, if possible.

37. $(3\sqrt{5x})^2$

38. $\sqrt[3]{27a^6b^9}$

39. $\sqrt{6}\sqrt{7x}$

40. $\sqrt{a}\sqrt{a+1}$

41. $\sqrt{x^5}$

42. $\sqrt{45}$

43. $\sqrt{c^{17}d^{20}}$

44. $\sqrt{x^2+9}$

45. Assuming that x represents any real number, show that $\sqrt{x^2 - 6x + 9}$ represents the distance between x and 3 on a number line.

46. If the length and width of a rectangle are $2\sqrt{x}$ and $5\sqrt{x+3}$, find a simplified expression for the area.

Section 8.5

47. State the three conditions under which a radical is simplified.

48. Explain why we cannot rationalize the denominator of $\frac{1}{\sqrt[3]{x}}$ by multiplying the numerator and denominator by $\sqrt[3]{x}$.

In Exercises 49–52, simplify.

49. $\frac{\sqrt[3]{x^5}}{\sqrt[3]{x^2}}$

50. $\frac{\sqrt{9x^3y^4}}{\sqrt{4xy^2}}$

51. $\sqrt{\frac{5a^7}{9b^4}}$

52. $\sqrt{\frac{3x^{15}}{192x^5}}$

In Exercises 53–57, rationalize the denominator.

53. $\frac{6}{\sqrt{6}}$

54. $\sqrt{\frac{13}{3}}$

55. $\frac{3x}{\sqrt{x^5}}$

56. $\frac{1}{\sqrt[4]{8}}$

57. $\frac{\sqrt{18xy^5}}{\sqrt{2x^2y^3}}$

58. Show that $\frac{\sqrt[3]{x^2}}{\sqrt{x}}$ and $\frac{\sqrt{x}}{\sqrt[3]{x}}$ both have the same simplified form.

59. Use the Quotient Rule for Radicals to show that the square root of the reciprocal of a positive number is the reciprocal of the square root of the number.

60. One of the following expressions *cannot* be simplified with just one operation. Identify this expression and then simplify the other two expressions.

(a) $\frac{\sqrt{15}}{\sqrt{3}}$ (b) $\frac{\sqrt{16}}{\sqrt{3}}$ (c) $\frac{\sqrt{15}}{\sqrt{9}}$

Section 8.6

61. Explain why the terms in the given expression cannot be combined.

(a) $3\sqrt{x} + 3\sqrt{y}$ (b) $2\sqrt{x} + 2\sqrt[3]{x}$

62. Explain why we cannot rationalize the denominator of $\frac{1}{3 + \sqrt{2}}$ by multiplying the numerator and denominator by $\sqrt{2}$. Describe the correct method.

In Exercises 63–65, perform the indicated operations.

63. $3\sqrt{45x} + 2\sqrt{20x}$

64. $\sqrt[3]{54} - \sqrt[3]{16}$

65. $(\sqrt{3} + \sqrt{5})^2$

66. Find the product of $\sqrt{x} + \sqrt{2}$ and its conjugate.

67. Supply the missing number to make the statement true.

$$\blacksquare(\sqrt{3} + \sqrt{5x}) = \sqrt{6} + \sqrt{10x}$$

68. Determine which of the following is a rational number by evaluating each expression.

(a) $(3\sqrt{5})^2$ (b) $(3 + \sqrt{5})(3 - \sqrt{5})$

(c) $(3 + \sqrt{5})^2$

In Exercises 69 and 70, rationalize the denominator.

69. $\frac{5}{5 - \sqrt{3}}$ 70. $\frac{\sqrt{x} + 7}{\sqrt{x} + 3}$

In Exercises 71 and 72, simplify.

71. $\frac{1}{\sqrt{3}} + \frac{5\sqrt{3}}{3}$ 72. $\frac{1}{\sqrt{5} + a} - \frac{1}{\sqrt{5} - a}$

73. The lengths of the three sides of a triangle are $\sqrt{20}$, $\sqrt{125}$, and $\sqrt{180}$. Express the perimeter of the triangle as a single simplified radical.

74. The length of a rectangle is one more than the square root of a number x. If the area of the rectangle is $x + \sqrt{x}$, find a simplified expression for the width of the rectangle.

Section 8.7

75. What is an extraneous solution?

76. Describe the first step in solving an equation of the form $A^{m/n} = B$.

In Exercises 77–80, solve.

77. $(2x - 3)^3 = 27$

78. $\sqrt{6x + 1} = 2x - 3$

79. $\sqrt{5x - 4} - \sqrt{x + 3} = -1$

80. $(x + 2)^{2/3} = 9$

81. An old totem pole is supported by a 20-foot wire extending from the top of the pole to a point 13 feet from the base of the pole. How tall is the totem pole?

82. The square root of the sum of a number and 4 is 2 more than the square root of the difference of the number and 4. What is the number?

83. One number is 5 more than another number. If the product of their square roots is 6, what are the numbers?

84. A Texas ranch is in the shape of a square 60 miles on each side. If the ranch house is at the exact center of the square, how far is it from the house to any corner of the property?

Section 8.8

85. In the real number system, $\sqrt{-4}\sqrt{-9}$ cannot be performed. Why? What is the result in the complex number system?

86. What is the difference, if any, between the results of $(a + b)(a - b)$ and $(a + bi)(a - bi)$?

In Exercises 87–94, perform the indicated operations for each expression.

87. $(3 - 5i) - (5 - 7i)$ 88. $\sqrt{-9} + (4 - 5i)$

89. $(6 - 7i)(2 + 3i)$ 90. $(5 + i)^2$

91. $\dfrac{8-4i}{2}$

92. $(4+\sqrt{-7})(4-\sqrt{-7})$

93. i^{57}

94. $\dfrac{3+4i}{1-2i}$

95. If $i(a+bi)=6$, determine a and b.

96. Which of the following statements is false?

(a) The number $a+bi$ is an imaginary number if b is not 0.

(b) The number $a+bi$ is a real number if b is 0.

(c) The number $a+bi$ is a complex number if b is 0.

(d) The number $a+bi$ is a real number, an imaginary number, and a complex number if neither a nor b is 0.

LOOKING AHEAD

The following exercises review concepts and skills that you will need in Chapter 9.

1. Use a graph to estimate the x- and y-intercepts of the graph of the function $f(x)=6-x-x^2$.

In Exercises 2–5, solve the given equation by factoring.

2. $2x^2-3x-5=0$
3. $9x^2+4=12x$
4. $5x-15x^2=0$
5. $x^3+3x^2-4x-12=0$
6. Use the Even Root Property to solve $(2x-1)^2=16$.

In Exercises 7 and 8, evaluate $\dfrac{-b+\sqrt{b^2-4ac}}{2a}$ for the given values of a, b, and c.

7. $a=2, b=-7, c=3$
8. $a=4, b=-9, c=-3$

In Exercises 9 and 10, solve the given equation.

9. $\dfrac{12}{t^2-9}=\dfrac{2}{t-3}-1$
10. $x=2+\sqrt{2x-1}$
11. Explain how to graph the inequality $y\le x-6$.
12. Working together, an office worker and his assistant can complete the job of copying department personnel manuals in $2\frac{2}{5}$ hours. Working alone, the office worker could complete the job 2 hours sooner than the assistant. If the assistant began working alone at 8:00 A.M. and did not take a break, at what time would he complete the job?

8. CHAPTER TEST

1. Determine the domain of $f(x)=\sqrt{3-x}$.
2. Simplify $(-8x^{-6}y^{12})^{1/3}$.

3. Explain the difference between $-25^{1/2}$ and $(-25)^{1/2}$.
4. Supply the missing number to make the statement true.

$$\sqrt{\frac{\square}{49}}=\frac{2\sqrt{5}}{7}$$

5. Write $\sqrt[3]{\sqrt[4]{x^7}}$ as a single radical.
6. Write $\sqrt[3]{x}\cdot\sqrt{x}$ as a single radical expression.
7. Add $\dfrac{\sqrt{12}}{4}+\dfrac{\sqrt{27}}{6}$.
8. Multiply $(3\sqrt{5}-4)(2\sqrt{5}+1)$.

9. For the expression $\dfrac{\sqrt{b}}{\sqrt{a}}$, describe two conditions under which the denominator can be rationalized without multiplying the numerator and denominator by $\sqrt{a}$.
10. Rationalize the denominator of $\dfrac{3\sqrt{2}-1}{\sqrt{2}+3}$.
11. For the triangle in the accompanying figure, show that the length of the hypotenuse is $x+2$.

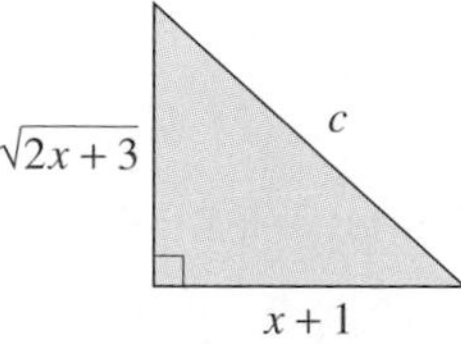

12. Simplify $\sqrt{98x^{11}y^{16}}$.
13. Evaluate $\left(\dfrac{16}{25}\right)^{-3/2}$.
14. Solve $(3t-1)^{2/3}-4=0$.

15. Explain why it is essential to check the numbers obtained when solving a radical equation.

16. Solve $3\sqrt{x} + 4 = 13$.

17. An **isosceles triangle** is a triangle with two sides of the same length. If the lengths of these two sides are $\sqrt{1 - 3x}$ and $\sqrt{3x} + 1$, how long are the two sides?

18. Subtract $(6 - 4i) - (7 - 10i)$.

19. What is the square of the complex number $3 + 2i$?

20. Divide $\frac{1 - 2i}{3 - i}$.

21. In what form must the numbers $\sqrt{-7}$ and $\sqrt{-3}$ be written before they can be multiplied?

22. Which *two* of the following statements are true?

    (i) Every imaginary number is a complex number, but not every complex number is an imaginary number.

    (ii) The product of complex conjugates is an imaginary number.

    (iii) The radicals $\sqrt{4x^2}$, $\sqrt{x^2 + 4}$, and $\sqrt{x^2 + 4x + 4}$ can all be simplified.

    (iv) The expressions $3 + \sqrt{5}$ and $3 - \sqrt{5}$ are irrational numbers, but their product is a rational number.

23. The volume V of a cube is given by $V = s^3$, where s is the length of every side of the cube. Use a rational exponent to write a formula for s. Then use your calculator to determine the side of a cube whose volume is 200 cubic inches. (Round to two decimal places.)

24. Simplify $\sqrt{3} + \sqrt{-3} + \sqrt{-3}(\sqrt{3} + \sqrt{-3})$.

25. In each part, complete the sentence by stating the conditions under which the statement is true.

    (a) If $A^n = B$ and n is an even positive integer, then the equation has no real number solution if ______.

    (b) For a positive integer n, $\sqrt[n]{a^n} = a$ if ______ or if ______.

Chapter 9

Quadratic Equations

Medical advances and attention to good health practices brought about a dramatic decrease in deaths due to heart disease from 1985 to 1990. However, the accompanying bar graph indicates that heart disease–related deaths have been on the rise in the 1990s.

By using a **quadratic function** to model the data in the bar graph, we can graphically estimate the years in which the number of deaths due to heart disease was above or below a specified level. And we can determine these years exactly by solving **quadratic inequalities.** (For more on this topic, see Exercises 95–98 at the end of Section 9.6, and see the Chapter Project.)

In Chapter 9 we add the Quadratic Formula and other methods to our techniques for solving quadratic equations, equations that are quadratic in form, and applications. We also learn how to solve quadratic inequalities and inequalities with rational expressions. The topic of quadratic functions and properties of their graphs concludes the chapter.

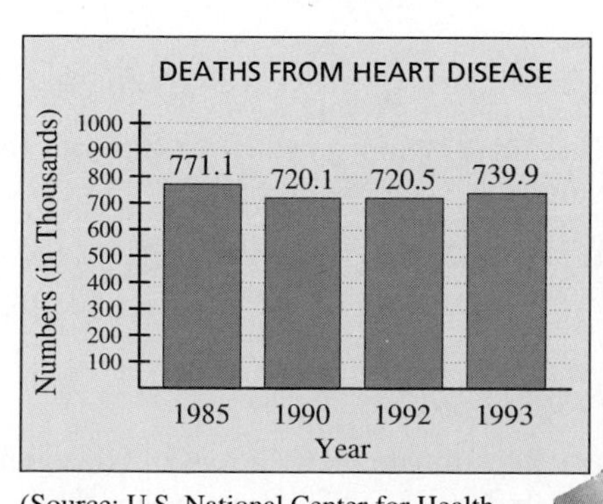

(Source: U.S. National Center for Health Statistics.)

9.1 SPECIAL METHODS

Real Number Solutions • Complex Number Solutions

Real Number Solutions

In Section 6.6 we had our first experience with a class of equations known as **quadratic equations.** The following is a restatement of the definition of a quadratic equation.

Definition of a Quadratic Equation

A **quadratic equation** is an equation that can be written in the *standard form* $ax^2 + bx + c = 0$, where a, b, and c are real numbers and $a \neq 0$.

Associated with a quadratic equation is a **quadratic function.**

Definition of a Quadratic Function

A **quadratic function** is a function of the form

$$f(x) = ax^2 + bx + c$$

where a, b, and c are real numbers and $a \neq 0$.

Note: Observe the difference between a quadratic *equation*, which can be written in the form $ax^2 + bx + c = 0$, and a quadratic *function*, which is written in the form $f(x) = ax^2 + bx + c$. The quadratic equation is the special case $f(x) = 0$.

The following example provides a review of three previously discussed methods for solving a quadratic equation.

EXAMPLE 1

Quadratic Equations: Graphing and Algebraic Methods

Use the graphing method to estimate the solution(s) of the given equation. Then solve the equation algebraically.

(a) $x^2 - 2x - 8 = 0$ (b) $9x^2 - 12x + 4 = 0$

(c) $5 - x^2 = 0$ (d) $x^2 + 4 = 0$

Figure 9.1

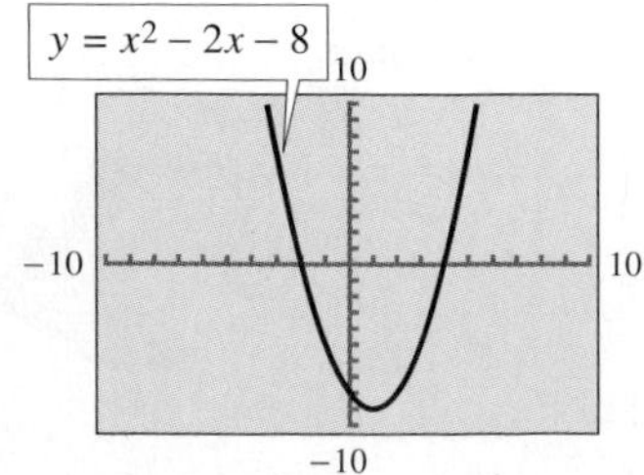

Solution

(a) To estimate the solutions of the equation, produce the graph of $y = x^2 - 2x - 8$ and trace to the points where the y-coordinate is 0, that is, to the x-intercepts. From the graph in Fig. 9.1, we see that the equation has two solutions, approximately -2 and 4.

To solve algebraically, we use the factoring method.

$$x^2 - 2x - 8 = 0$$
$$(x - 4)(x + 2) = 0 \quad \text{Factor the trinomial.}$$
$$x - 4 = 0 \quad \text{or} \quad x + 2 = 0 \quad \text{Zero Factor Property}$$
$$x = 4 \quad \text{or} \quad x = -2 \quad \text{Solve each case for } x.$$

Figure 9.2

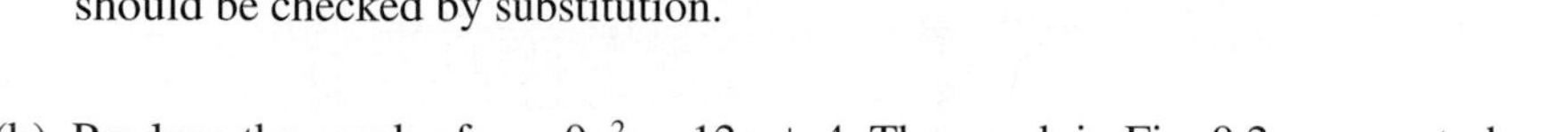

This confirms the solutions suggested by the graphing method. The solutions should be checked by substitution.

(b) Produce the graph of $y = 9x^2 - 12x + 4$. The graph in Fig. 9.2 appears to have only one x-intercept, which suggests that the equation has only one solution. In this case, the approximate solution, 0.66, is called a **double root.**

We solve algebraically by factoring.

$$9x^2 - 12x + 4 = 0$$
$$(3x - 2)(3x - 2) = 0 \quad \text{Factor the trinomial.}$$
$$3x - 2 = 0 \quad \text{Zero Factor Property}$$
$$x = \frac{2}{3}$$

Because $\frac{2}{3} = 0.\overline{6}$, the exact solution and the estimated solution are very close. To verify, check the *exact* solution.

(c) The graph in Fig. 9.3 indicates that the solutions are approximately 2.2 and -2.2.

Figure 9.3

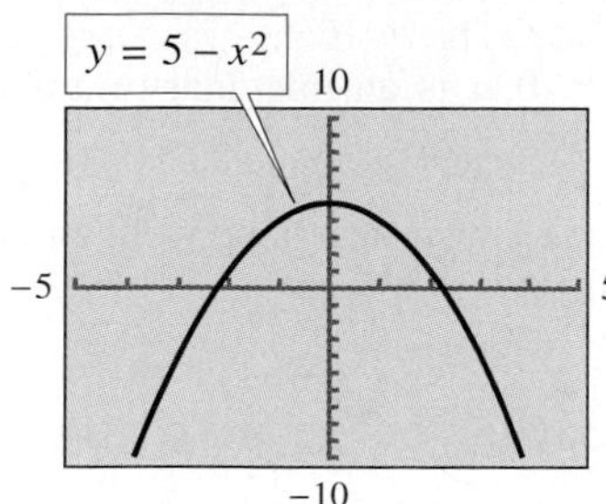

To solve algebraically, we use the Even Root Property:

If $x^2 = c$ and $c > 0$, then $x = \sqrt{c}$ or $x = -\sqrt{c}$.

$$5 - x^2 = 0$$
$$x^2 = 5 \quad \text{Write in the form } x^2 = c.$$
$$x = \sqrt{5} \approx 2.24 \quad \text{or} \quad x = -\sqrt{5} \approx -2.24 \quad \text{Even Root Property}$$

These solutions are consistent with those indicated in Fig. 9.3.

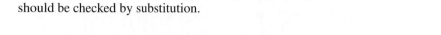

(d) Because the graph of $y = x^2 + 4$ has no x-intercepts (see Fig. 9.4), the equation has no real number solutions.

Figure 9.4

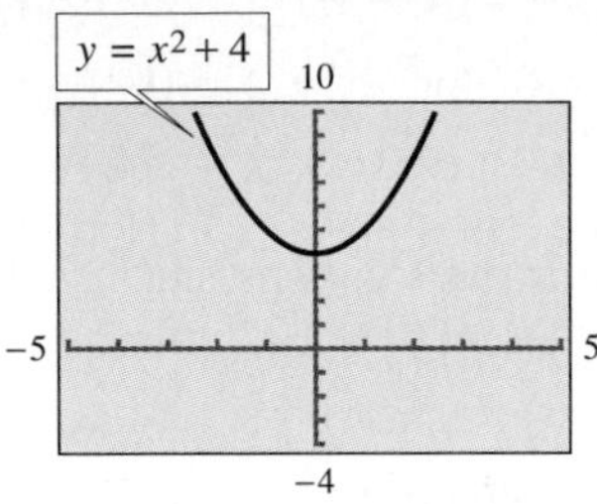

The Even Root Property states that an equation of the form $x^2 = c$ has no real number solution if $c < 0$. Thus $x^2 = -4$ has no real number solution.

Think About It

Experiment with the graph of $y = x^2 + k$, where k is an integer of your choice. For what values of k does $x^2 + k = 0$ have two real number solutions? a double root? no real number solution?

Because a graph provides a picture of all the real number solutions of $y = ax^2 + bx + c$, we can use it to anticipate the number of solutions a quadratic equation will have, and we can obtain estimates of the solutions. For complete accuracy, we use an algebraic method for solving the equation. As always, the best approach is a combination of the two methods.

Complex Number Solutions

In Example 1 both the graphing method and the Even Root Property showed that the equation $x^2 = -4$ has no real number solution. An extension of the Even Root Property for equations of the form $x^2 = c$ enables us to determine complex number solutions of quadratic equations.

The Square Root Property

For any real number c, if $x^2 = c$, then $x = \sqrt{c}$ or $x = -\sqrt{c}$.

Note that for $c < 0$, x is an imaginary number. Thus the equation $x^2 = -4$ has two solutions: $\pm\sqrt{-4} = \pm 2i$.

Our next example illustrates how we can use the Square Root Property to solve quadratic equations.

EXAMPLE 2

LEARNING TIP

The standard form $ax^2 + bx + c = 0$ is required for the factoring method, but the form $A^2 = c$, where A is an algebraic expression, is needed to apply the Square Root Property.

Using the Square Root Property to Solve Quadratic Equations

Use the Square Root Property to solve the given equation.

(a) $(2x + 1)^2 = 5$ (b) $y^2 + 9 = 0$ (c) $(t - 3)^2 = -5$ (d) $7 + 25(2x + 3)^2 = 0$

Solution

(a) $(2x + 1)^2 = 5$

$2x + 1 = \pm\sqrt{5}$ Square Root Property with $c > 0$

$2x = -1 \pm \sqrt{5}$ Solve for x.

$$x = \frac{-1 \pm \sqrt{5}}{2}$$

Recall that the symbol $\pm$ indicates two solutions—one evaluated with the plus symbol, the other evaluated with the minus symbol. When using your calculator to evaluate the two solutions, be sure to enclose the numerator in parentheses. Rounded to two decimal places, the solutions are 0.62 and -1.62.

(b) $y^2 + 9 = 0$

$y^2 = -9$ — Write in the form $A^2 = c$.

$y = \pm\sqrt{-9} = \pm i\sqrt{9} = \pm 3i$ — Square Root Property with $c < 0$

(c) $(t - 3)^2 = -5$

$t - 3 = \pm\sqrt{-5}$ — Square Root Property with $c < 0$

$t - 3 = \pm i\sqrt{5}$

$t = 3 \pm i\sqrt{5} \approx 3 \pm 2.24i$ — Add 3 to both sides to solve for t.

(d) $7 + 25(2x + 3)^2 = 0$ — Isolate the squared quantity.

$25(2x + 3)^2 = -7$

$(2x + 3)^2 = -\frac{7}{25}$

$2x + 3 = \pm\sqrt{-\frac{7}{25}}$ — Square Root Property with $c < 0$

$2x + 3 = \pm\frac{i\sqrt{7}}{5}$ — Subtract 3 and divide by 2 to solve for x.

$2x = -3 \pm \frac{i\sqrt{7}}{5}$

$x = -\frac{3}{2} \pm \frac{\sqrt{7}}{10}i \approx -1.5 \pm 0.26i$

9.1 QUICK REFERENCE

Real Number Solutions

- A **quadratic equation** is an equation that can be written in the *standard form* $ax^2 + bx + c = 0$, where a, b, and c are real numbers and $a \neq 0$.
- Three previously discussed methods for solving a quadratic equation are as follows:
  1. Graphing method
     (a) The graphing method can be used to learn how many solutions a quadratic equation has and to estimate what the solutions are.
     (b) To use the graphing method, produce the graph of the associated function $f(x) = ax^2 + bx + c$. The x-intercepts of the graph of the quadratic function represent the solutions of the quadratic equation.
  2. Factoring method
     (a) Write the quadratic equation in standard form and factor the polynomial.
     (b) Apply the Zero Factor Property and set each factor equal to zero.
     (c) Solve the resulting equations.

3. Even Root Property
If $x^2 = c$ and $c > 0$, then $x = \sqrt{c}$ or $x = -\sqrt{c}$.

Complex Number Solutions

- Square Root Property
For any real number c, if $x^2 = c$, then $x = \sqrt{c}$ or $x = -\sqrt{c}$.
- In the Square Root Property, if $c < 0$, then x is an imaginary number.
- The Square Root Property enables us to determine complex number solutions of a quadratic equation.

9.1 EXERCISES

Concepts and Skills

1. When using the graphing method to estimate the solutions of a quadratic equation in standard form, what feature of the graph is relevant? Why?

2. Describe how to use the graphing method to solve the quadratic equation $x^2 - 2x = 15$ without writing the equation in standard form.

In Exercises 3–14, use the graphing method to estimate the solution(s) of the given equation.

3. $2x^2 + 5x - 3 = 0$
4. $3x^2 + 10x = 8$
5. $3 - 8x = 3x^2$
6. $6 + x - x^2 = 0$
7. $x^2 - 6x + 9 = 0$
8. $x^2 + 8x + 16 = 0$
9. $10x = 25 + x^2$
10. $14x = 49 + x^2$
11. $x^2 + 3x + 5 = 0$
12. $x^2 - 4x + 6 = 0$
13. $10x - 27 - x^2 = 0$
14. $8x - 18 - x^2 = 0$

In Exercises 15–24, use the factoring method to solve.

15. $x^2 + 7x + 6 = 0$
16. $4x^2 + 3 = 8x$
17. $x^2 = x$
18. $25 = 10x - x^2$
19. $\frac{3x}{4} = \frac{1}{2} + \frac{2}{x}$
20. $x = \frac{28}{x+3}$
21. $x(x - 3) = x - 3$
22. $(3x - 1)^2 - 25x^2 = 0$
23. $8x^2 + x - 1 = 3x^2 - 2x + 1$
24. $2x(x + 3) = -5(3 + x)$

25. The equations $6 - x^2 = 0$ and $6 + x^2 = 0$ are very similar, but their solution sets are quite different. Use the Even Root Property to explain why one equation has two solutions and the other equation has no real number solution.

26. Suppose that $x^2 = c$, where $c < 0$. Describe the difference between the Even Root Property and the Square Root Property as each property applies to the solution(s) of the equation.

In Exercises 27–34, solve by using the Square Root Property.

27. $(5x - 4)^2 = 9$
28. $(3x - 2)^2 = 15$
29. $16(x + 5)^2 - 5 = 0$
30. $9(x + 3)^2 - 4 = 0$
31. $\left(x - \frac{3}{2}\right)^2 = \frac{9}{4}$
32. $(4 - x)^2 = \frac{36}{25}$
33. $x(x - 6) = 3(3 - 2x)$
34. $x(x + 12) = 12(3 + x)$

In Exercises 35–52, determine the complex number solutions of the given equation.

35. $t^2 + 16 = 0$
36. $25 + t^2 = 0$
37. $24 + y^2 = 0$
38. $y^2 + 18 = 0$
39. $(x - 3)^2 = -4$
40. $(x + 4)^2 = -9$
41. $(t - 1)^2 + 169 = 144$
42. $(2t + 1)^2 + 25 = 16$
43. $(2x + 5)^2 + 6 = 0$
44. $(3x + 7)^2 + 5 = 0$
45. $8(x + 1)^2 + 21 = 3$
46. $25(x + 2)^2 + 11 = 0$
47. $10 + \frac{4}{y^2} = 1$
48. $40 + \frac{25}{y^2} = 4$
49. $4 = \frac{9}{(1 - x)^2}$
50. $\frac{16}{(2 - x)^2} = 49$

51. $x(x + 4) = -4(1 - x)$

52. $x(8 - x) = 8(2 + x)$

53. The accompanying figure shows the graph of $y = x^2 + 3x + 4$. According to the graph, how many real number solutions are there for the quadratic equation $x^2 + 3x + 4 = 0$? Why? Does this mean that the equation has no solution? Explain.

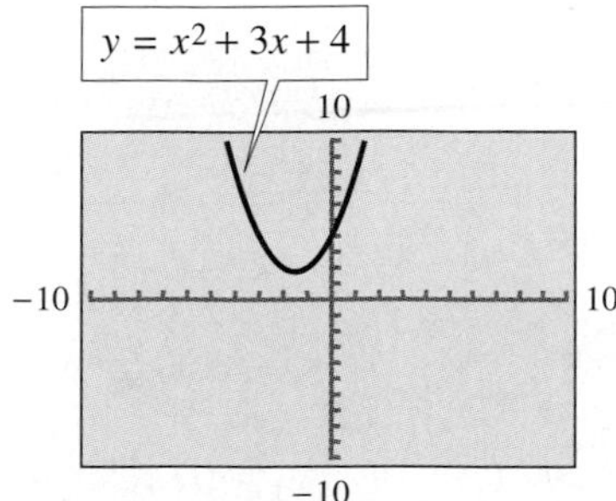

54. When solving a quadratic equation, how can we recognize when to use the factoring method and when to use the Square Root Property?

In Exercises 55–76, solve.

55. $3x = x^2$

56. $12x^2 = 21 + 4x$

57. $t^2 + 45 = 0$

58. $x^2 + 20 = 0$

59. $36x^2 + 12x + 1 = 0$

60. $9 - 6t + t^2 = 0$

61. $y^2 = -1.44$

62. $x^2 = 1.21$

63. $x(x - 5) = 5(2 - x)$

64. $x^2 + 6x = 6(3 + x)$

65. $x - 2 = \dfrac{35}{x}$

66. $\dfrac{7}{x} + 4 = \dfrac{15}{x^2}$

67. $(4x + 5)^2 + 49 = 0$

68. $(3x - 4)^2 + 36 = 0$

69. $\left(\dfrac{3x + 2}{4}\right)^2 = 12$

70. $\left(\dfrac{x - 1}{2}\right)^2 = 6$

71. $t(t + 3) = 3(t - 2)$

72. $5(x^2 + 15) = 2x^2$

73. $x(x - 4) = x - 4$

74. $6x^2 + 9x + 3 = 1 - 4x^2$

75. $(4x - 1)(x - 1) = 7x - 8$

76. $(4x + 1)(5x + 1) = x(-5x - 1)$

Given two numbers, we can write a quadratic equation for which the numbers are solutions. For example, if 2 and −5 are solutions, then

$$x = 2 \quad \text{or} \quad x = -5$$
$$x - 2 = 0 \quad \text{or} \quad x + 5 = 0$$
$$(x - 2)(x + 5) = 0$$
$$x^2 + 3x - 10 = 0$$

Note that the equation is not unique. Multiplying both sides by any nonzero number produces an equivalent equation.

In Exercises 77–84, write a quadratic equation with integer coefficients that has the given pair of numbers as solutions.

77. 3, 5

78. −1, −2

79. 3, 3

80. 0, 0

81. $3, \dfrac{4}{3}$

82. $\dfrac{2}{3}, \dfrac{3}{4}$

83. $-4i, 4i$

84. $i\sqrt{3}, -i\sqrt{3}$

85. The product of two consecutive negative integers is 72. What are the two integers?

86. A book lies open so that the product of the facing page numbers is 1806. To what possible pages is the book open?

Geometric Models

87. Suppose that a rectangle is described as having an area of 24 square units and dimensions that are consecutive odd integers. Try to determine the dimensions and interpret the result.

88. Two less than the area of a square is numerically equal to 5 less than the length of a side. Use the graphing method to estimate the dimensions of the square.

89. The length of a rectangle is 1 inch less than twice the width. If the area is 45 square inches, what are the dimensions of the rectangle?

90. The height of a triangle is one-third the length of the base. If the area of the triangle is 24 square feet, what is the height?

Real-Life Applications

91. A physical education major is doing his student teaching at an elementary school. One day he lines up his fifth-grade students and asks them to count off by even numbers beginning with 2. Standing side by side, two students have numbers whose product is 360. What possible numbers could they have?

92. As a project for an elementary science class, a teacher launches a model rocket, and the students use a stop watch to time the flight. The teacher knows that the height h after t seconds is given by $h = 96t - 16t^2$. What flight time does the teacher anticipate that the students will record?

93. A sentry walks back and forth from his guard house to the end of a building. His distance d from the guard house at any given time t (in minutes) is given by $d = 18t^2 - 90t$. How long does the sentry take to complete one round trip?

94. Two trucks left a truck stop together. One traveled north and the other traveled east. When the trucks were 50 miles apart, the eastbound truck had traveled 10 miles farther than the northbound truck. How far had each truck traveled?

Modeling with Real Data

95. The accompanying line graph shows the number of cellular phone customers (in thousands) served by Southwestern Bell for selected years in the period 1984–1990. (Source: Southwestern Bell Corporation.)

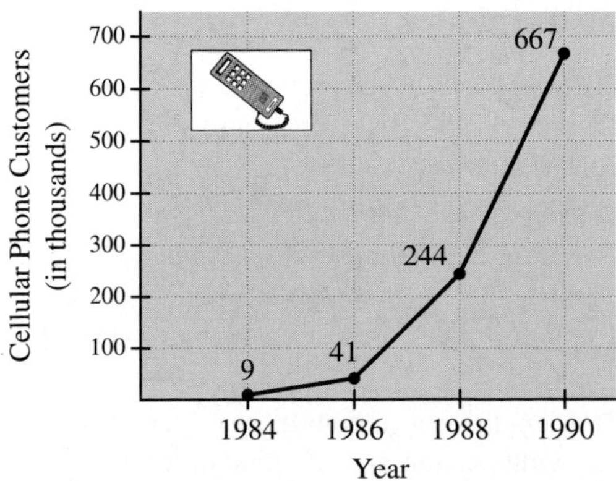

The function $C(t) = 24.4t^2 - 233.3t + 554$ models the number C of customers (in thousands), where t is the number of years since 1980.

(a) Write an equation that you can use to estimate the year in which the company served 1 million customers.

(b) Use the graphing method to estimate the solution of your equation. Round to the nearest integer.

96. The accompanying table shows that U.S. citizenship applications have risen steadily in the 1990s. (Source: Immigration and Naturalization Service.)

Year	Approximate Number of Applications
1990	234,000
1991	210,000
1992	300,000
1993	450,000
1994	500,000
1995	1,000,000

The function $f(x) = 46x^2 - 88x + 251$ models the number of applications f (in thousands) for each year x after 1990.

(a) Write an equation to determine the year in which the number of applications will be three times the number in 1995.

(b) Use the graphing method to estimate the solution of your equation. Round to the nearest integer.

Challenge

In Exercises 97–100, solve for x. Assume that all variables (other than x) represent positive numbers.

97. $(x + 2)^2 = -y$

98. $(x + a)^2 = 9$

99. $(x + c)^2 = -16$

100. $(ax + 1)^2 + 25 = 9$

101. Show that $x - 1 + i$ and $x - 1 - i$ are factors of $x^2 - 2x + 2$. Then use the Zero Factor Property to solve $x^2 - 2x + 2 = 0$.

102. Show that $x - 2 - 3i$ and $x - 2 + 3i$ are factors of $x^2 - 4x + 13$. Then use the Zero Factor Property to solve $x^2 - 4x + 13 = 0$.

In Exercises 103 and 104, use the Square Root Property to solve.

103. $(2x + 5)^2 = (x - 1)^2$

104. $(3 - x)^2 = (1 - 3x)^2$

9.2 COMPLETING THE SQUARE

Perfect Square Trinomial • Solving Quadratic Equations • Complex Number Solutions

Perfect Square Trinomial

In the preceding section we used the Square Root Property to solve equations such as $(x - 3)^2 = 5$ and $(x + 2)^2 = -4$. Using a method called **completing the square,** we can write any quadratic equation in the form $(x \pm h)^2 = k$. The method is based on the special products for the square of a binomial.

$$(A \pm B)^2 = A^2 \pm 2AB + B^2$$

EXPLORING THE CONCEPT

Completing a Perfect Square Trinomial

Consider the expression $x^2 + 8x + c$. To determine the value of c for which this expression is a perfect square trinomial, we compare the expression to the special product form:

$$x^2 + \quad 8x + c$$
$$x^2 + 2Bx + B^2$$

For the coefficients of x to be equal, $2B = 8$, and so $B = 4$. Thus $B^2 = c = 16$. The perfect square trinomial is $x^2 + 8x + 16$. In general, we can determine the constant term by squaring half the coefficient of x.

Completing a Perfect Square Trinomial

If the first part of a perfect square trinomial is $x^2 + bx$, then the entire perfect square trinomial is $x^2 + bx + \left(\frac{b}{2}\right)^2$.

In words, the constant term is the square of half the coefficient of x.

EXAMPLE 1

Completing a Perfect Square Trinomial

Write a perfect square trinomial whose first two terms are given.

(a) $x^2 + 12x + \square$ (b) $x^2 - 5x + \square$

Solution

(a) Half the coefficient of x is 6 and $6^2 = 36$. Thus the constant term is 36, and the trinomial is $x^2 + 12x + 36$.

(b) Half the coefficient of x is $-\frac{5}{2}$ and $\left(-\frac{5}{2}\right)^2 = \frac{25}{4}$. The constant term is $\frac{25}{4}$, and the trinomial is $x^2 - 5x + \frac{25}{4}$.

Think About It

To complete the square for $x^2 + ix$, would the constant term be a real number or an imaginary number?

Solving Quadratic Equations

By completing the square, we can write any quadratic equation in the form $(x \pm h)^2 = k$. Then we can use the Square Root Property to solve the equation.

Solving a Quadratic Equation by Completing the Square

1. If necessary, divide both sides by the leading coefficient to obtain a coefficient of 1 for the quadratic term.
2. Write the equation in the form $x^2 + bx = k$.
3. Add $\left(\frac{b}{2}\right)^2$ to both sides to complete the square.
4. Factor the resulting perfect square trinomial.
5. Solve with the Square Root Property.

EXAMPLE 2

LEARNING TIP

Keep in mind that we complete the square in order to take advantage of the Square Root Property. However, if the expression can be factored, the factoring method is usually easier.

Solving Quadratic Equations by Completing the Square

(a) $x^2 - 10x - 1 = 0$

$x^2 - 10x = 1$ — Add 1 to both sides.

$x^2 - 10x + 25 = 1 + 25$ — Add $\left(\frac{-10}{2}\right)^2 = (-5)^2 = 25$ to both sides to complete the square.

$(x - 5)^2 = 26$ — Factor the perfect square trinomial.

$x - 5 = \pm\sqrt{26}$ — Square Root Property

$x = 5 \pm \sqrt{26}$ — Add 5 to both sides to solve for x.

$x \approx 10.10$ or $x \approx -0.10$

(b) $x(2x + 3) = 6$

$2x^2 + 3x = 6$ — Remove parentheses.

$x^2 + \frac{3}{2}x = 3$ — Divide both sides by 2.

$x^2 + \frac{3}{2}x + \frac{9}{16} = \frac{3}{1} + \frac{9}{16}$ — Add $\left(\frac{1}{2} \cdot \frac{3}{2}\right)^2 = \left(\frac{3}{4}\right)^2 = \frac{9}{16}$ to both sides to complete the square.

$\left(x + \frac{3}{4}\right)^2 = \frac{57}{16}$ — Add the fractions and factor.

$x + \frac{3}{4} = \pm\frac{\sqrt{57}}{4}$ — Square Root Property

$x = -\frac{3}{4} \pm \frac{\sqrt{57}}{4}$ — Subtract $\frac{3}{4}$ from both sides to solve for x.

$x \approx 1.14$ or $x \approx -2.64$

(c) $x^2 + 6x + 8 = 0$

$x^2 + 6x = -8$ — Subtract 8 from both sides.

$x^2 + 6x + 9 = -8 + 9$ — Add $\left(\frac{6}{2}\right)^2 = 3^2 = 9$ to both sides to complete the square.

$(x + 3)^2 = 1$ Factor the perfect square trinomial.

$x + 3 = \pm\sqrt{1}$ Square Root Property

$x = -3 \pm 1$ Subtract 3 from both sides to solve for x.

$x = -2$ or $x = -4$

Note: The equation in part (c) of Example 2 could have been solved by factoring. We used the method of completing the square to illustrate that the method is applicable to any quadratic equation.

Complex Number Solutions

Although imaginary number solutions of a quadratic equation cannot be found with the graphing method, they can be determined with the method of completing the square.

EXAMPLE 3

Quadratic Equations with Complex Number Solutions

Determine the solutions of each equation.

(a) $x^2 + 4x + 5 = 0$ (b) $3x^2 - 2x + 1 = 0$

Solution

Figure 9.5

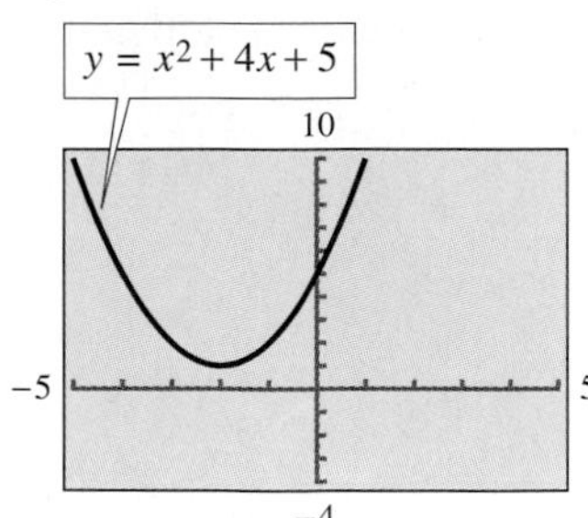

(a) Figure 9.5 shows the graph of $f(x) = x^2 + 4x + 5$. Because the graph has no x-intercepts, we know that the equation has no real number solutions.

Now we solve algebraically by completing the square.

$x^2 + 4x + 5 = 0$

$x^2 + 4x = -5$ Subtract 5 from both sides.

$x^2 + 4x + 4 = -5 + 4$ Add $\left(\frac{1}{2} \cdot 4\right)^2 = 2^2 = 4$ to both sides.

$x^2 + 4x + 4 = -1$

$(x + 2)^2 = -1$ Factor and write in exponential form.

$x + 2 = \pm\sqrt{-1}$ Square Root Property

$x + 2 = \pm i$

$x = -2 \pm i$ Subtract 2 from both sides to solve for x.

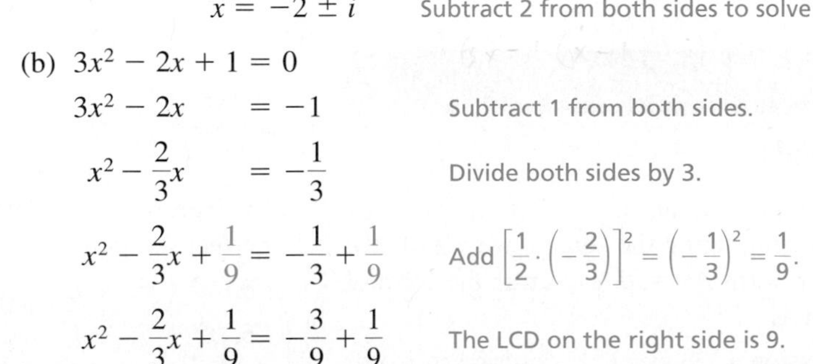

(b) $3x^2 - 2x + 1 = 0$

$3x^2 - 2x = -1$ Subtract 1 from both sides.

$x^2 - \frac{2}{3}x = -\frac{1}{3}$ Divide both sides by 3.

$x^2 - \frac{2}{3}x + \frac{1}{9} = -\frac{1}{3} + \frac{1}{9}$ Add $\left[\frac{1}{2} \cdot \left(-\frac{2}{3}\right)\right]^2 = \left(-\frac{1}{3}\right)^2 = \frac{1}{9}$.

$x^2 - \frac{2}{3}x + \frac{1}{9} = -\frac{3}{9} + \frac{1}{9}$ The LCD on the right side is 9.

$$\left(x - \frac{1}{3}\right)^2 = -\frac{2}{9}$$ Add the fractions and factor.

$$x - \frac{1}{3} = \pm\sqrt{-\frac{2}{9}}$$ Square Root Property

$$x - \frac{1}{3} = \pm\frac{i\sqrt{2}}{3}$$

$$x = \frac{1}{3} \pm \frac{\sqrt{2}}{3}i \approx 0.33 \pm 0.47i$$

9.2 QUICK REFERENCE

Perfect Square Trinomial

- If the first part of a quadratic trinomial is $x^2 + bx$, then the entire perfect square trinomial is $x^2 + bx + \left(\frac{b}{2}\right)^2$.
- In words, to determine the constant needed to form a perfect square trinomial (with 1 as the leading coefficient), take half the coefficient of x and square the result. This process is called **completing the square.**

Solving Quadratic Equations

- To solve a quadratic equation $ax^2 + bx + c = 0$ by completing the square, follow these steps:
  1. If necessary, divide both sides by the leading coefficient to obtain a coefficient of 1 for the quadratic term.
  2. Write the equation in the form $x^2 + bx = k$.
  3. Add $\left(\frac{b}{2}\right)^2$ to both sides to complete the square.
  4. Factor the resulting perfect square trinomial.
  5. Solve with the Square Root Property.

Complex Number Solutions

- The method of completing the square could result in an equation of the form $(x \pm h)^2 = k$, where $k < 0$. In this case, the solutions will be imaginary numbers.

9.2 EXERCISES

Concepts and Skills

1. To solve a quadratic equation $ax^2 + bx + c = 0$ by the method of completing the square, we rewrite the equation in the form $A^2 = B$. Describe the form of the expression A.

2. For the expression $x^2 + bx + k$, describe in words how to determine k in order to complete the square.

In Exercises 3–8, determine k so that the trinomial is a perfect square.

3. $x^2 + 6x + k$

4. $x^2 - 8x + k$

5. $x^2 + 7x + k$

6. $x^2 - 9x + k$

7. $x^2 + \frac{4}{3}x + k$

8. $x^2 - \frac{5}{2}x + k$

9. Two students are solving $x^2 - 8x = -12$. One student rewrites the equation in standard form and solves with the factoring method. The other student adds 16 to both sides to complete the square and then solves with the Square Root Property. Which method is correct? Why?

10. Two students are solving $x^2 + x = -1$. One student rewrites the equation in standard form and tries to solve with the factoring method. The other student tries to solve with the method of completing the square. Which student will be successful? Why?

In Exercises 11–18, solve by factoring and then solve again by completing the square.

11. $x^2 + 4x + 3 = 0$

12. $x^2 + 6x + 8 = 0$

13. $t^2 + 2t = 15$ **14.** $t^2 = 6t + 55$

15. $x^2 - x - 12 = 0$ **16.** $x^2 + 5x - 6 = 0$

17. $6 = x^2 + x$ **18.** $x^2 = 5x + 14$

In Exercises 19–24, solve by completing the square.

19. $x^2 - 6x + 4 = 0$ **20.** $x^2 + 2x = 5$

21. $x^2 + 2x = 24$ **22.** $x^2 - 4x = 5$

23. $x^2 = 4x + 4$ **24.** $x^2 = 4x + 8$

25. If you want to solve the equation $3x^2 - 5x = 7$ with the method of completing the square, what is the first step you must take?

26. Although we regard the factoring method and the method of completing the square as two different methods for solving a quadratic equation, the method of completing the square does include factoring. At what step of the procedure do we need to factor?

In Exercises 27–30, solve by factoring and then solve again by completing the square.

27. $3y^2 + 4y + 1 = 0$

28. $5y^2 + 6y = 8$

29. $15 = 2x^2 + 7x$

30. $6x^2 + 3 = 11x$

In Exercises 31–38, solve by completing the square.

31. $4x^2 = 8x + 21$ **32.** $4x^2 = 27 + 12x$

33. $4x^2 + 25 = 20x$ **34.** $16x^2 + 9 = 24x$

35. $x(2x - 3) = 1$ **36.** $4x^2 = 3(x + 1)$

37. $2x^2 + 7x - 1 = 0$ **38.** $4x^2 + 5x - 3 = 0$

39. Determine k so that $kx^2 + 6x + 1$ is a perfect square.

40. Determine k so that $4x^2 + kx + 25$ is a perfect square.

In Exercises 41–48, determine the imaginary number solutions by completing the square.

41. $x^2 - 4x + 8 = 0$ **42.** $x^2 + 6x + 10 = 0$

43. $x^2 + 6 = 4x$ **44.** $x^2 + 7 = 2x$

45. $x^2 + 3x + 5 = 0$ **46.** $x^2 + 5x + 8 = 0$

47. $2x^2 + x + 3 = 0$ **48.** $3x^2 + 2x + 7 = 0$

49. Show that the solutions of $x^2 + bx = k$ are imaginary numbers if $k < -\left(\frac{b}{2}\right)^2$.

50. For what value of k does $x^2 + bx = k$ have only one solution?

In Exercises 51–74, solve the equation.

51. $x^2 + 16 = 10x$ **52.** $2x^2 + x - 6 = 0$

53. $36x^2 - 49 = 0$ **54.** $\frac{1}{7}y^2 = 49$

55. $x(x - 3) = 10$ **56.** $b(b + 5) = 36$

57. $4(x + 3)^2 - 7 = 0$ **58.** $(3x - 2)^2 = 16$

59. $x(x + 1) + 9 = 0$

60. $2a(a + 2) = 3$

61. $(x - 3)(x - 1) = 5x - 6$

62. $(2x - 1)(8x - 1) = -1 - 2x$

63. $x^2 + \frac{3}{2}x = \frac{3}{4}$ **64.** $x^2 + \frac{4}{3}x = \frac{5}{9}$

65. $x^2 = x - 9$ **66.** $x^2 = x + 4$

67. $4x^2 + 7x = 3$ **68.** $3x^2 - x + 8 = 0$

69. $3x^2 + x = \frac{2}{3}$

70. $0.5x^2 - 2x = 1.75$

71. $(x - 2)^2 = -3(x - 2)$

72. $(x + 1)^2 = 3(x + 2)$

73. $4x^2 + 9 = 2x$ **74.** $2x^2 + 9 = 5x$

In Exercises 75–80, one solution is given. Determine a, b, or c and then determine another solution.

75. $ax^2 + x - 5 = 0;\ \ 1$

76. $ax^2 + 9x - 5 = 0;\ \ \frac{1}{3}$

77. $6x^2 + bx - 3 = 0;\ \ \frac{1}{3}$

78. $5x^2 + bx - 3 = 0;\ \ 1$

79. $x^2 + 2x + c = 0$; 6

80. $2x^2 + 7x + c = 0$; $-\frac{1}{2}$

81. The sum of the reciprocal of an integer and the square of the reciprocal is $\frac{6}{25}$. What is the integer?

82. The product of the reciprocal of a number and 20 is 8 less than the number. What two numbers satisfy this condition?

83. The square of a positive number is 3 more than 5 times the number. What is the number? (Round to two decimal places.)

84. Seven times a negative number is 1 less than the square of the number. What is the number? (Round to two decimal places.)

Real-Life Applications

85. A construction crew built a platform in the shape of a right triangle for a political campaign rally. If the lengths of the legs of the triangle differ by 8 feet and the length of the longest side of the platform is 50 feet, how long is the shortest side of the platform? (Round to the nearest foot.)

86. Taxes on commercial property have increased by \$1.00 per frontage foot along Hayes Street. A store is on a lot in the shape of a right triangle with one leg along Hayes Street. The frontage along Minnow Road is 100 feet. (See figure.)

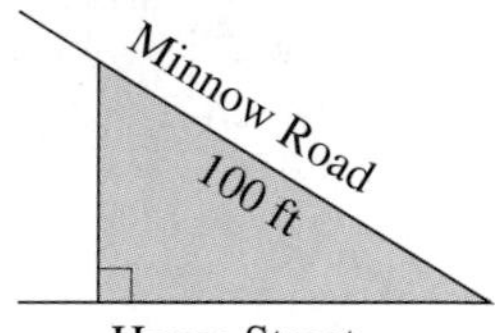

If the Hayes Street frontage is 30 feet more than twice the length of the other leg of the lot, what is the amount of the tax increase? (Round to the nearest dollar.)

87. A woman uses a rototiller to till her garden. Her neighbor has a larger tiller and could do the job in 30 minutes less time. How long does the woman take to till the garden if she and her neighbor, working together, can complete the job in 1 hour? (Round to the nearest minute.)

88. A bricklayer can complete a job in 2 hours less time than his apprentice takes. Working together, the two of them can complete the job in 5 hours. How long would the apprentice take working alone? (Round to the nearest hour.)

Group Project: Nuclear Testing

The decline in nuclear testing began with the Nuclear Non-Proliferation Treaty, which went into effect in 1970. The accompanying table shows the number of nuclear tests in the decades following the 1940s. (Source: *Nuclear Notebook*.)

Years	Number of Nuclear Tests
1945–1949	7
1950–1959	292
1960–1969	692
1970–1979	535
1980–1989	452
1990–1995	41

The function $T(x) = -96x^2 + 1248x - 3552$ models the number T of nuclear tests in decade x, where $x = 5$ for the 1950s, $x = 6$ for the 1960s, and so on.

89. Use the method of completing the square to solve the equation $T(x) = 0$. Round results to the nearest integer.

90. Interpret your solutions in Exercise 89.

91. Write the equation $T(x) = 0$ in standard form with the GCF divided out. Explain why the factoring method would not have been convenient for solving the equation.

92. Not all countries are signatories of the treaty, and some are intent on developing nuclear capabilities. Under what conditions might the model function T cease to be valid in the future?

Challenge

In Exercises 93 and 94, use the method of completing the square to solve for x.

93. $x^2 - 2cx + 4c^2 = 0$

94. $ax^2 + 2x + a = 0, a \neq 0$

In Exercises 95 and 96, solve by completing the square.

95. $x^2 + 2\sqrt{5}x - 4 = 0$

96. $x^2 - 6\sqrt{3}x + 23 = 0$

In Exercises 97 and 98, solve for x. Determine the values of k for which (a) the solutions will be real numbers and (b) the solutions will be imaginary numbers.

97. $x^2 + kx + 1 = 0$

98. $x^2 + 2x + k = 0$

In Exercises 99 and 100, use the method of completing the square to solve the equation.

99. $x^2 + 4ix - 8 = 0$

100. $x^2 - 8ix - 25 = 0$

101. The method of completing the square can be illustrated geometrically. Consider the square (with a missing corner) in the accompanying figure.

(a) Find the total area of the figure by adding the areas of the pieces. Simplify the result.

Figure for 101

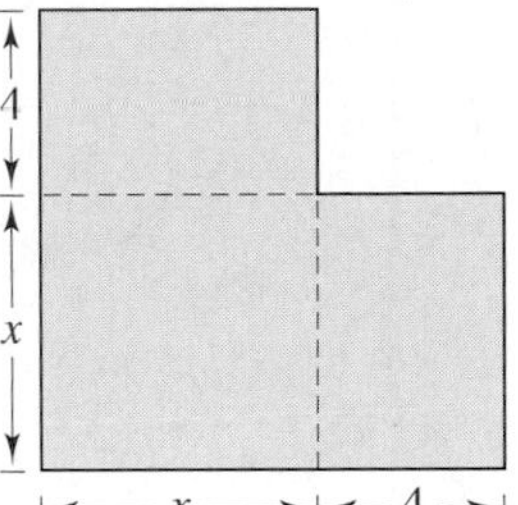

(b) What area needs to be added to "complete the square"? That is, what is the area of the missing corner in the figure?

(c) Find the area of the complete square by adding the areas found in parts (a) and (b).

(d) Use the fact that the lengths of the sides of the complete square are $x + 4$ to calculate the area of the complete square. Do not simplify.

(e) Compare the results in parts (c) and (d).

9.3 THE QUADRATIC FORMULA

Derivation and Use • Complex Number Solutions • The Discriminant

Derivation and Use

Each time we solve a quadratic equation with the method of completing the square, we follow the same steps. By performing the steps just once for the general equation $ax^2 + bx + c = 0$, we can derive a formula for finding the solutions of any quadratic equation.

EXPLORING THE CONCEPT

The Quadratic Formula

We begin with the quadratic equation written in standard form, and we complete the square.

$ax^2 + bx + c = 0$ Standard form of a quadratic equation

$ax^2 + bx = -c$ Subtract c from both sides.

$x^2 + \frac{b}{a}x = -\frac{c}{a}$ Leading coefficient must be 1. Divide both sides by a.

$x^2 + \frac{b}{a}x + \frac{b^2}{4a^2} = -\frac{c}{a} + \frac{b^2}{4a^2}$ To complete the square, add $\left(\frac{1}{2} \cdot \frac{b}{a}\right)^2 = \frac{b^2}{4a^2}$ to both sides.

$$x^2 + \frac{b}{a}x + \frac{b^2}{4a^2} = \frac{-4ac}{4a^2} + \frac{b^2}{4a^2}$$ The LCD on the right side is $4a^2$.

$$x^2 + \frac{b}{a}x + \frac{b^2}{4a^2} = \frac{b^2 - 4ac}{4a^2}$$ Combine the fractions on the right side.

$$\left(x + \frac{b}{2a}\right)^2 = \frac{b^2 - 4ac}{4a^2}$$ Factor and write in exponential form.

$$x + \frac{b}{2a} = \frac{\pm\sqrt{b^2 - 4ac}}{2a}$$ Square Root Property

$$x = \frac{-b}{2a} \pm \frac{\sqrt{b^2 - 4ac}}{2a}$$ Subtract $\frac{b}{2a}$ to solve for x.

$$x = \frac{-b \pm \sqrt{b^2 - 4ac}}{2a}$$ Combine the fractions.

This formula is called the **Quadratic Formula.**

Think About It

In the Quadratic Formula, either b or c can be 0, but a cannot be 0. Why would this problem never arise?

The Quadratic Formula

For real numbers a, b, and c, with $a \neq 0$, the solutions of the quadratic equation $ax^2 + bx + c = 0$ are given by the **Quadratic Formula.**

$$x = \frac{-b \pm \sqrt{b^2 - 4ac}}{2a}$$

Note: A quadratic equation must be written in standard form before the Quadratic Formula can be used to solve it.

EXAMPLE 1

Using the Quadratic Formula: Two Rational Number Solutions

Use the Quadratic Formula to solve $3x^2 - 13x = 30$.

Solution

$$3x^2 - 13x = 30$$

$$3x^2 - 13x - 30 = 0$$ Write the equation in standard form.

Substitute $a = 3$, $b = -13$, and $c = -30$ into the Quadratic Formula.

$$x = \frac{-b \pm \sqrt{b^2 - 4ac}}{2a}$$

$$x = \frac{-(-13) \pm \sqrt{(-13)^2 - 4(3)(-30)}}{2(3)}$$ Note that the opposite of b is 13.

$$x = \frac{13 \pm \sqrt{169 + 360}}{6}$$ Simplify the radicand first.

$$x = \frac{13 \pm \sqrt{529}}{6}$$ Note that 529 is a perfect square.

LEARNING TIP

Writing the Quadratic Formula each time you solve a problem will help you to learn the formula. List the numbers a, b, and c, and substitute these values into the formula before you attempt to perform operations.

$$x = \frac{13 \pm 23}{6}$$ Use the symbol + for one solution and the symbol − for the other solution.

$$x = 6 \quad \text{or} \quad x = -\frac{5}{3}$$

The equation in Example 1 has two real number solutions, and the equation could have been solved by factoring. Note that after substituting into the Quadratic Formula and simplifying, the radicand is a perfect square.

EXAMPLE 2

Using the Quadratic Formula: Double Root

Use the Quadratic Formula to solve $16t^2 + 9 = 24t$.

Solution

$$16t^2 + 9 = 24t$$

$$16t^2 - 24t + 9 = 0$$ Write the equation in standard form.

Substitute $a = 16$, $b = -24$, and $c = 9$ into the Quadratic Formula.

$$t = \frac{24 \pm \sqrt{(-24)^2 - 4(16)(9)}}{2(16)}$$

$$t = \frac{24 \pm \sqrt{576 - 576}}{32}$$ Simplify the radicand first.

$$t = \frac{24 \pm 0}{32} = \frac{24}{32} = \frac{3}{4}$$ Note that the radicand is 0.

The equation in Example 2 has only one distinct solution. The equation could have been solved by factoring, which would have revealed $\frac{3}{4}$ to be a double root. Because the radicand was 0, the ± symbol instructed us to add 0 for one root and subtract 0 for the other root. Of course, the result is the same in both cases.

EXAMPLE 3

Using the Quadratic Formula: Two Irrational Number Solutions

Use the Quadratic Formula to solve $3x^2 - 10 = 0$.

Solution

$$3x^2 - 10 = 0$$ Note that $b = 0$.

$$x = \frac{0 \pm \sqrt{0^2 - 4(3)(-10)}}{2(3)}$$ Substitute $a = 3$, $b = 0$, and $c = -10$ into the Quadratic Formula.

$$x = \frac{0 \pm \sqrt{0 + 120}}{6}$$

$$x = \pm\frac{\sqrt{120}}{6} = \pm\frac{\sqrt{30}}{3}$$ Note that 120 is not a perfect square.

$$x \approx 1.83 \quad \text{or} \quad x \approx -1.83$$

The equation in Example 3 has two irrational number solutions and could have been solved with the Square Root Property. The fact that the radicand is not a perfect square indicates that the factoring method could not have been used.

EXAMPLE 4

Using the Quadratic Formula: Two Irrational Number Solutions

Use the Quadratic Formula to solve $4x = x^2 + 1$.

Solution

$$4x = x^2 + 1$$

$$-x^2 + 4x - 1 = 0$$ Write the equation in standard form.

$$x = \frac{-4 \pm \sqrt{4^2 - 4(-1)(-1)}}{2(-1)}$$ Substitute $a = -1$, $b = 4$, and $c = -1$ into the Quadratic Formula.

$$x = \frac{-4 \pm \sqrt{16 - 4}}{-2}$$

$$x = \frac{-4 \pm \sqrt{12}}{-2} = 2 \pm \sqrt{3}$$

$$x \approx 0.27 \quad \text{or} \quad x \approx 3.73$$

The equation in Example 4 has two irrational number solutions. The factoring method could not have been used. The equation could have been solved by completing the square, but the Quadratic Formula was derived from that method.

Complex Number Solutions

In each of the preceding examples, after we substituted a, b, and c into the Quadratic Formula, the radicand was never negative. Therefore, the solutions were always real numbers.

For other equations, it is possible for the radicand in the Quadratic Formula to be negative. In such cases the solutions are imaginary numbers.

The Quadratic Formula can be used to determine all complex number solutions of any quadratic equation.

EXAMPLE 5

Using the Quadratic Formula: Complex Number Solutions

Use the Quadratic Formula to solve $3x^2 + 2x + 1 = 0$.

Solution

$$3x^2 + 2x + 1 = 0$$

$$x = \frac{-2 \pm \sqrt{2^2 - 4(3)(1)}}{2(3)}$$ Substitute $a = 3$, $b = 2$, and $c = 1$ into the Quadratic Formula.

$$x = \frac{-2 \pm \sqrt{4 - 12}}{6}$$

$$x = \frac{-2 \pm \sqrt{-8}}{6}$$ The radicand is negative.

$$x = \frac{-2 \pm i\sqrt{8}}{6}$$

$$x = \frac{-2}{6} \pm \frac{\sqrt{8}}{6}i = -\frac{1}{3} \pm \frac{\sqrt{2}}{3}i \approx -0.33 \pm 0.47i$$

In Example 5 the radicand is negative, and the solutions are complex conjugates.

The Discriminant

In the Quadratic Formula, the radicand is $b^2 - 4ac$. This expression is called the **discriminant.** The discriminant provides information about the number and type of solutions of a quadratic equation with rational coefficients.

Information Obtained from the Discriminant $b^2 - 4ac$

If a, b, and c are rational numbers, the value of the discriminant determines the number and type of solutions of a quadratic equation as follows:

1. If the discriminant is positive and
   (a) a perfect square, there are two rational solutions.
   (b) not a perfect square, there are two irrational solutions.
2. If the discriminant is zero, there is one rational solution (double root).
3. If the discriminant is negative, the solutions are complex conjugates.

EXAMPLE 6

Using the Discriminant to Characterize Solutions of Quadratic Equations

Use the discriminant to determine the number and type of the solutions of each equation.

(a) $3x^2 + 1 = 2x$ (b) $x(5x + 1) = 12$

(c) $2 + 3x + x^2 = 0$ (d) $25x^2 - 20x + 4 = 0$

Solution

(a) $3x^2 + 1 = 2x$

$3x^2 - 2x + 1 = 0$ Write the equation in standard form.

$b^2 - 4ac = (-2)^2 - 4(3)(1) = 4 - 12 = -8$

Because the discriminant is negative, the solutions of the equation are complex conjugates.

(b) $x(5x + 1) = 12$

$5x^2 + x - 12 = 0$ Simplify and write the equation in standard form.

$b^2 - 4ac = 1^2 - 4(5)(-12) = 1 + 240 = 241$

The discriminant is positive, but it is not a perfect square. Thus the equation has two irrational solutions.

(c) $2 + 3x + x^2 = 0$

$x^2 + 3x + 2 = 0$

$b^2 - 4ac = 3^2 - 4(1)(2) = 9 - 8 = 1$

Because the discriminant is positive and a perfect square, the equation has two rational solutions.

(d) $25x^2 - 20x + 4 = 0$

$b^2 - 4ac = (-20)^2 - 4(25)(4) = 400 - 400 = 0$

The discriminant is zero. Thus the equation has one rational solution (double root).

We now have five methods available for solving a quadratic equation: the graphing method, the factoring method, the Square Root Property, the method of completing the square, and the Quadratic Formula.

Graphing is an excellent method for estimating solutions and for determining the number of solutions. When exact solutions are required, an algebraic method is necessary, but producing the graph is still a useful visual aid.

When solving a quadratic equation algebraically, choose the method that is easiest for that particular equation. Calculating the discriminant can assist in selecting a solution method. If the discriminant is a perfect square, the equation can be solved by factoring. Otherwise, our other techniques must be used. The following are some other criteria.

If	**Use**
$b = 0$	Square Root Property
$c = 0$	Factoring method
$b, c \neq 0$	Factoring method if it is easy; otherwise use the Quadratic Formula

The Quadratic Formula can be used to solve any quadratic equation, but other methods are usually more efficient when they can be used. Having a calculator to perform the arithmetic, however, makes the Quadratic Formula an appealing choice.

9.3 QUICK REFERENCE

Derivation and Use

- The **Quadratic Formula** is derived by applying the method of completing the square to the general quadratic equation $ax^2 + bx + c = 0$.
- The solutions of the quadratic equation $ax^2 + bx + c = 0$ are given by the Quadratic Formula.

$$x = \frac{-b \pm \sqrt{b^2 - 4ac}}{2a}$$

- To use the Quadratic Formula, write the quadratic equation in standard form and substitute the values of a, b, and c into the formula.

Complex Number Solutions

- When the Quadratic Formula is used to solve a quadratic equation, the radicand may be negative. In such cases, the solutions are imaginary numbers.

The Discriminant

- In the Quadratic Formula, the expression $b^2 - 4ac$ is called the **discriminant.**
- If a, b, and c are rational numbers, then the value of the discriminant determines the number and type of solutions as follows:
  1. If the discriminant is positive and
     (a) a perfect square, there are two rational solutions.
     (b) not a perfect square, there are two irrational solutions.
  2. If the discriminant is zero, there is one rational solution (double root).
  3. If the discriminant is negative, the solutions are complex conjugates.

9.3 EXERCISES

Concepts and Skills

1. To use the Quadratic Formula to solve a quadratic equation, we must know the numbers a, b, and c. What must be true about the equation before these numbers can be determined?

2. What is your opinion about the best method for solving the equation $x^2 - 2x - 1088 = 0$? Can the factoring method be used? If so, what are the factors?

In Exercises 3–6, write the equation in standard form and then state the values of a, b, and c.

3. $x(3x - 2) = 7x + 6$ **4.** $(x - 2)^2 = -4x$

5. $(x + 1)^2 = 3x - 1$ **6.** $1 = 4x(2x - 5)$

In Exercises 7–14, solve with the factoring method and then solve again with the Quadratic Formula.

7. $x^2 - x - 2 = 0$ **8.** $x^2 - 3x + 2 = 0$

9. $x^2 + 16 = 8x$ **10.** $x^2 + 1 = 2x$

11. $2x^2 + 7x = 4$ **12.** $2x^2 = 7x - 3$

13. $x^2 = 5x$ **14.** $4x^2 - 3x = 3x$

In Exercises 15–22, use the Quadratic Formula to solve.

15. $4x^2 = 3 + 9x$ **16.** $4x^2 + 7x = 3$

17. $x^2 - 2x + 9 = 0$ **18.** $x^2 + 4x + 10 = 0$

19. $x(x + 2) = 5$ **20.** $x^2 = 2(2 - x)$

21. $x + 2x^2 + 4 = 0$ **22.** $9 + 3x^2 + 5x = 0$

23. One way to use your calculator with the Quadratic Formula is to store the values of a, b, and c, and then enter the expressions

$$\frac{-b + \sqrt{b^2 - 4ac}}{2a} \quad \text{and} \quad \frac{-b - \sqrt{b^2 - 4ac}}{2a}$$

Write the sequence of key strokes for entering the expressions correctly. What expressions will the calculator evaluate if you forget to enclose the numerators in parentheses?

24. When you have a number of quadratic equations to solve with the Quadratic Formula, a better method than that described in Exercise 23 is to enter the two expressions as Y_1 and Y_2 on the function screen. Then, after storing the values of a, b, and c, simply evaluate Y_1 and Y_2. Use this method for the following equations and interpret the results.
   (a) $2x^2 + 5x - 1 = 0$
   (b) $9x^2 - 6x + 1 = 0$
   (c) $4x^2 + x + 3 = 0$

In Exercises 25–50, solve by any method.

25. $6x - 9x^2 = 0$

26. $5 - 13x - 6x^2 = 0$

27. $9 - 49x^2 = 0$ **28.** $3 - 4x^2 = 0$

29. $49 = 14x - x^2$

30. $9x^2 - 6x = -1$

31. $(5x - 4)^2 = 10$

32. $(4x + 5)^2 = 49$

33. $0.25x^2 - 0.5x - 2 = 0$

34. $0.5x^2 + x = 0.875$

35. $x(x + 4) = 5$

36. $3 = x(x - 6)$

37. $5x^2 + 2x + 7 = 0$

38. $41 + 8x - 4x^2 = 0$

39. $x^2 - 3x + 8 = 0$

40. $x^2 - 5x - 5 = 0$

41. $6x - x^2 - 4 = 0$

42. $x^2 = 4 - 2x$

43. $x(5 + x) + 9 = 0$

44. $5 = 2x(1 - 3x)$

45. $\frac{1}{2}x^2 - \frac{2}{3}x + \frac{4}{3} = 0$

46. $\frac{1}{6}x^2 + \frac{3}{2} = \frac{1}{2}x$

47. $x^2 = \frac{5x}{2} + \frac{1}{2}$

48. $\frac{2}{3}x^2 + 2x = \frac{5}{9}$

49. $(x - 3)(2x + 1) = x(x - 4)$

50. $(2x + 1)^2 = 2(3x + 1)$

In Exercises 51–58, use the discriminant to determine the number and the nature of the solutions of the given equation.

51. $4x^2 + 12x = 7$

52. $x^2 + 12 = 8x$

53. $2x^2 + 5x + 7 = 0$

54. $4x^2 + x + 9 = 0$

55. $36x^2 + 1 = 12x$

56. $4x(x + 1) + 1 = 0$

57. $3x^2 + 4x - 8 = 0$

58. $1 + 6x - 9x^2 = 0$

In Exercises 59–62, determine the values of k so that the equation has two unequal real roots.

59. $x^2 + 8x + k = 0$

60. $3x^2 + 2x + k = 0$

61. $kx^2 - 4x + 1 = 0$

62. $kx^2 - 3x + 2 = 0$

In Exercises 63–66, determine the values of k so that the equation has two imaginary solutions.

63. $x^2 + 6x + k = 0$

64. $5x^2 + 4x + k = 0$

65. $kx^2 - 3x + 1 = 0$

66. $kx^2 - 5x + 4 = 0$

In Exercises 67–70, determine the values of k so that the equation has a double root.

67. $x^2 + kx + 9 = 0$

68. $x^2 + kx + 3 = 0$

69. $x^2 + kx + 2k = 0$

70. $kx^2 + (2k + 4)x + 9 = 0$

71. (a) Explain why $x^2 + ax - 5 = 0$ always has real solutions, regardless of the value of the real number a.

(b) For any positive real number k, does the equation $x^2 + ax - k = 0$ always have real solutions, regardless of the value of the real number a? Explain.

72. Explain how to use the discriminant to determine whether the factoring method can be used to solve the given equation.

(a) $2x^2 + 5x = 7$ (b) $2x^2 + 3x = 8$

In Exercises 73–76, use the Quadratic Formula to factor the polynomial.

73. $x^2 + 4x - 480$

74. $x^2 + 32x + 240$

75. $x^2 + 7x - 450$

76. $x^2 + 4x - 672$

77. Use the Quadratic Formula to show that the sum of the roots of $ax^2 + bx + c = 0$ is $-b/a$. Test the conclusion by solving the equation $2x^2 - 7x - 15 = 0$ and showing that the sum of the two roots is $-b/a$.

78. Use the conclusion in Exercise 77 to explain why 5 and -4 cannot possibly be solutions of the equation $2x^2 - 3x - 20 = 0$ even though the numbers are factors of -20.

79. Use the Quadratic Formula to show that the product of the roots of $ax^2 + bx + c = 0$ is c/a.

80. Use the conclusion in Exercise 79 to explain why 5 and -4 cannot possibly be solutions of the equation $2x^2 - 3x - 20 = 0$ even though the numbers are factors of -20.

In Exercises 81–84, determine the sum and the product of the roots of the given equation without solving the equation.

81. $3x^2 - 2x - 1 = 0$

82. $2x^2 - 2x + 1 = 0$

83. $x^2 + x + 1 = 0$

84. $x^2 - 4x - 1 = 0$

85. Show that $x^2 + kx + (k - 1) = 0$ has two real number solutions for any value of k except 2.

86. Show that $k^2x^2 - 4kx + 4 = 0$ has a double root for any nonzero value of k.

Real-Life Applications

87. A softball field should be laid out so that the distance between bases is 60 feet. Suppose the infield of a certain softball field is perfectly square and the distance between first and third base is 84.9 feet. To the nearest foot, is this a regulation softball field?

88. An auto dealer wants to string a wire from the top of a pole to the ground and hang colorful pennants from it. The distance from the base of the pole to the point where the wire is attached to the ground is 3 times the height of the pole. If the wire is 63 feet long, how tall is the pole?

89. A firm that manufactures ready-to-assemble computer desks can sell 200 desks per day at a wholesale price of \$90. However, if the company can cut costs and reduce the price, sales will increase by 10 desks per day for each \$2 reduction in price. For what prices will the total revenue be \$20,000 per day?

90. A white water rafting outfitter provides equipment and pickup service for trips down the Ocoee River. When the price is \$40 per person, the outfitter can anticipate 140 customers per day on the weekend. For each \$5 increase in price, 12 customers will go to a competitor. For what price will the total revenue be \$4200 per weekend day?

Group Project: Post-GED Education

The general equivalency diploma (GED) is given to high school dropouts who pass the GED test. The accompanying bar graph shows a rise in the percentage of GED recipients who planned additional schooling. (Source: American Council on Education.)

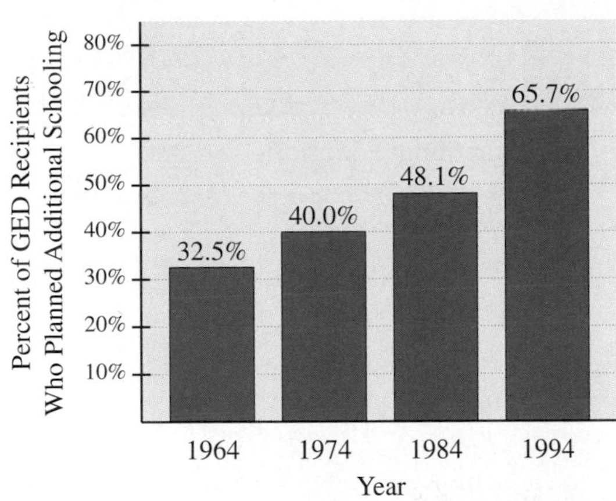

The function $p(x) = 0.22(0.1x^2 + x + 145)$ models the data, where $p(x)$ is the percentage who planned more schooling and x is the number of years since 1960.

91. Use the Quadratic Formula to determine the first-quadrant point of the graph of the function where the y-coordinate is 55.

92. Interpret each of the coordinates of the point in Exercise 91.

93. Determine and explain the significance of the sign of the discriminant in $p(x) = 0$.

94. Many colleges have programs to help otherwise underprepared applicants to prepare for college study. Should state-supported schools offer such programs to anyone who wishes to attend college, or should there be minimum standards?

Challenge

In Exercises 95 and 96, solve for x.

95. $x^2 + ax - 2a = 0$ **96.** $x^2 - yx - 2 = 0$

In Exercises 97 and 98, solve for y.

97. $x = y^2 + 2y - 8$ **98.** $x = 2y - 5 - y^2$

In Exercises 99 and 100, write a quadratic equation with the given solutions.

99. $-5 \pm 2\sqrt{5}$ **100.** $-2 \pm 3i$

In Exercises 101 and 102, use the Quadratic Formula to solve each equation.

101. $x^2 + \sqrt{3}x - 1 = 0$ **102.** $\sqrt{2}x^2 + \sqrt{5}x - \sqrt{8} = 0$

It can be shown that the Quadratic Formula holds for quadratic equations with imaginary coefficients. In Exercises 103 and 104, solve the equations and express the solutions in the complex number form $a + bi$.

103. $x^2 - 2ix - 3 = 0$ **104.** $ix^2 + 3x - 2i = 0$

9.4 EQUATIONS IN QUADRATIC FORM

Fractional Equations • Radical Equations • Quadratic Forms • Complex Number Solutions

Fractional Equations

Clearing fractions in an equation with rational expressions sometimes leads to a quadratic equation.

EXAMPLE 1

Solving Equations with Rational Expressions

Solve the equation $\frac{2}{t} - \frac{t}{t-2} = 5$.

Solution

Figure 9.6

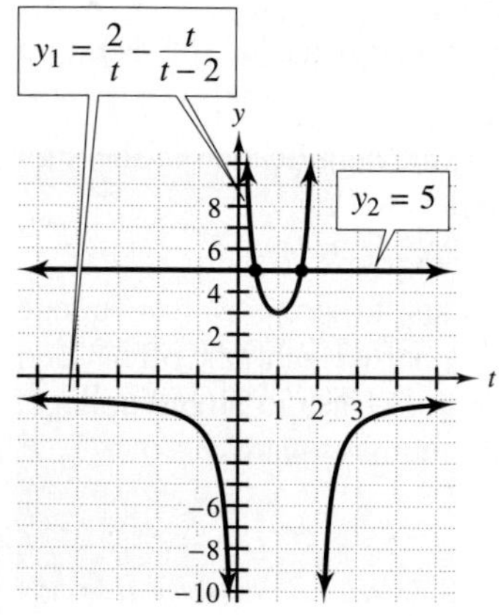

Figure 9.6 shows the graphs of $y_1 = \frac{2}{t} - \frac{t}{t-2}$ and $y_2 = 5$.

The two points of intersection indicate that the equation has two solutions.

$$\frac{2}{t} - \frac{t}{t-2} = 5$$ Note that 0 and 2 are restricted values.

$$\frac{t(t-2)}{1} \cdot \frac{2}{t} - \frac{t(t-2)}{1} \cdot \frac{t}{t-2} = 5t(t-2)$$ Multiply by the LCD $t(t-2)$.

$$2(t-2) - t^2 = 5t^2 - 10t$$

$$2t - 4 - t^2 = 5t^2 - 10t$$ Simplify.

$$-6t^2 + 12t - 4 = 0$$ Write the equation in standard form.

Now use the Quadratic Formula to solve for t.

$$t = \frac{-12 \pm \sqrt{12^2 - 4(-6)(-4)}}{2(-6)}$$

$$t = \frac{-12 \pm \sqrt{144 - 96}}{-12}$$

$$t = \frac{-12 \pm \sqrt{48}}{-12}$$

$t \approx 1.58$ or $t \approx 0.42$ Neither solution is restricted.

Note that these solutions are consistent with those suggested by the graph in Fig. 9.6.

Radical Equations

To solve equations involving radicals, we raise both sides of the equation to a power equal to the index of the radical. The result may be a quadratic equation that we can solve with our present methods.

EXAMPLE 2

Solving Equations with Radical Expressions

Solve the equation $\sqrt{x+1} = x - 2$.

Solution

Figure 9.7

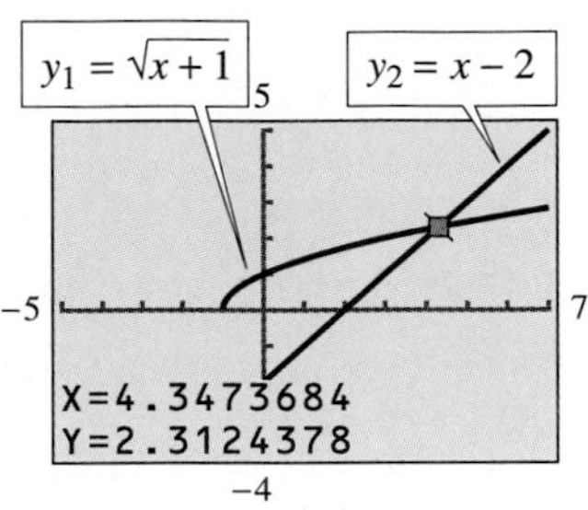

Figure 9.7 shows the graphs of $y_1 = \sqrt{x+1}$ and $y_2 = x - 2$.

The one point of intersection represents the one real number solution of the equation. The solution is approximately 4.3.

$$\sqrt{x+1} = x - 2$$

$$\left(\sqrt{x+1}\right)^2 = (x-2)^2 \qquad \text{Square both sides of the equation.}$$

$$x + 1 = x^2 - 4x + 4 \qquad \text{Square the binomial.}$$

$$0 = x^2 - 5x + 3 \qquad \text{Write the equation in standard form.}$$

Use the Quadratic Formula to solve for x.

$$x = \frac{5 \pm \sqrt{(-5)^2 - 4(1)(3)}}{2(1)} = \frac{5 \pm \sqrt{13}}{2}$$

$$x \approx 4.30 \quad \text{or} \quad x \approx 0.70$$

The solution 4.30 is consistent with the solution that we estimated from the graph. However, when we check the value 0.70, we find that it is an extraneous solution.

Quadratic Forms

An equation is in quadratic form if it can be written in the form $au^2 + bu + c = 0$, where u is an algebraic expression. To solve such equations, we make a substitution to write the equation in terms of u and then apply our methods for solving quadratic equations. The method is illustrated in Example 3.

EXAMPLE 3

LEARNING TIP

To identify an appropriate substitution, look for a variable expression and its square, and let u represent the expression.

Solving Equations That Are in Quadratic Form

Solve each equation.

(a) $x^4 - 6x^2 + 8 = 0$

(b) $16(z-1)^2 - 8(z-1) - 2 = 0$

(c) $28x^{-2} + x^{-1} - 2 = 0$

Solution

Figure 9.8

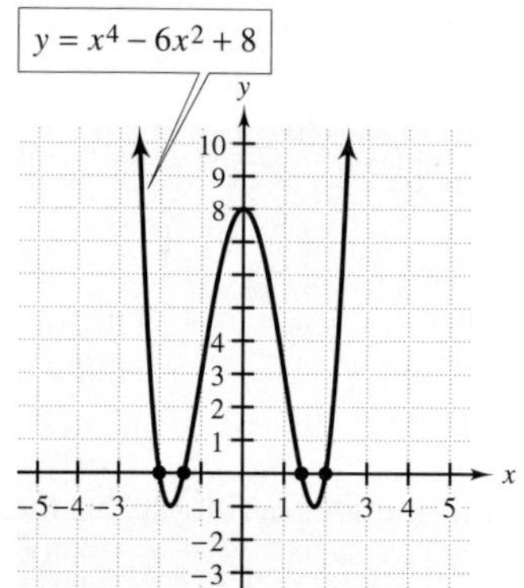

(a) In Figure 9.8 the graph of $y = x^4 - 6x^2 + 8$ suggests two positive solutions and two negative solutions.

The first term x^4 is the square of the variable factor x^2 in the second term.

$$x^4 - 6x^2 + 8 = 0$$
$$(x^2)^2 - 6x^2 + 8 = 0$$
$$u^2 - 6u + 8 = 0 \qquad \text{Let } u = x^2. \text{ Note that } u^2 = x^4.$$
$$(u - 2)(u - 4) = 0 \qquad \text{Factor.}$$
$$u - 2 = 0 \quad \text{or} \quad u - 4 = 0 \qquad \text{Zero Factor Property}$$
$$u = 2 \quad \text{or} \quad u = 4 \qquad \text{Solve each case for } u.$$

Now replace u with x^2.

$$x^2 = 2 \quad \text{or} \quad x^2 = 4$$
$$x = \pm\sqrt{2} \approx \pm 1.41 \quad \text{or} \quad x = \pm 2 \qquad \text{Square Root Property}$$

As indicated by the graph, the equation has two positive solutions and two negative solutions.

Figure 9.9

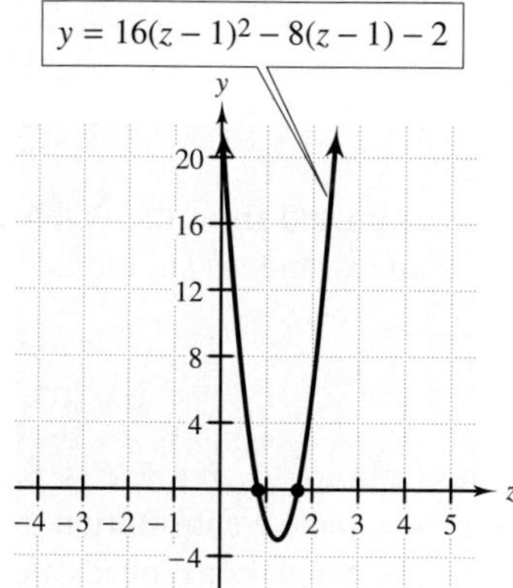

(b) Figure 9.9 shows the graph of $y = 16(z - 1)^2 - 8(z - 1) - 2$. From the graph, we anticipate two positive solutions between 0 and 2.

$$16(z - 1)^2 - 8(z - 1) - 2 = 0$$
$$16u^2 - 8u - 2 = 0 \qquad \text{Let } u = z - 1. \text{ Then } u^2 = (z - 1)^2.$$

Use the Quadratic Formula to solve for u.

$$u = \frac{8 \pm \sqrt{(-8)^2 - 4(16)(-2)}}{2(16)}$$
$$u = \frac{8 \pm \sqrt{64 + 128}}{32}$$
$$u = \frac{8 \pm \sqrt{192}}{32}$$
$$u \approx 0.68 \quad \text{or} \quad u \approx -0.18$$

To solve for z, replace u with the expression it represents.

$$z - 1 \approx 0.68 \quad \text{or} \quad z - 1 \approx -0.18$$
$$z \approx 1.68 \quad \text{or} \quad z \approx 0.82$$

(c)

$$28x^{-2} + x^{-1} - 2 = 0$$
$$28(x^{-1})^2 + x^{-1} - 2 = 0$$
$$28u^2 + u - 2 = 0 \qquad \text{Let } u = x^{-1}. \text{ Then } u^2 = x^{-2}.$$
$$(4u - 1)(7u + 2) = 0 \qquad \text{Factor.}$$
$$4u - 1 = 0 \quad \text{or} \quad 7u + 2 = 0 \qquad \text{Zero Factor Property}$$
$$4u = 1 \quad \text{or} \quad 7u = -2 \qquad \text{Solve each case for } u.$$
$$u = \frac{1}{4} \quad \text{or} \quad u = -\frac{2}{7}$$

To solve for x, replace u with the expression it represents.

$$x^{-1} = \frac{1}{4} \quad \text{or} \quad x^{-1} = -\frac{2}{7} \qquad \text{Solve for } x.$$

$$x = 4 \quad \text{or} \quad x = -\frac{7}{2}$$

An equation involving rational exponents may be quadratic in form if one of the exponents is twice another exponent. Recognizing such patterns is essential to detecting quadratic forms.

EXAMPLE 4

Solving Equations with Rational Exponents

Solve the equation $x^{2/3} - 2x^{1/3} - 24 = 0$.

Solution

$$x^{2/3} - 2x^{1/3} - 24 = 0$$

$$u^2 - 2u - 24 = 0 \qquad \text{Let } u = x^{1/3}. \text{ Then } u^2 = (x^{1/3})^2 = x^{2/3}.$$

$$(u - 6)(u + 4) = 0 \qquad \text{Factor.}$$

$$u - 6 = 0 \quad \text{or} \quad u + 4 = 0 \qquad \text{Zero Factor Property}$$

$$u = 6 \quad \text{or} \quad u = -4 \qquad \text{Solve each case for } u.$$

Replace u with $x^{1/3}$.

$$x^{1/3} = 6 \quad \text{or} \quad x^{1/3} = -4$$

$$(x^{1/3})^3 = 6^3 \quad \text{or} \quad (x^{1/3})^3 = (-4)^3 \qquad \text{Raise both sides to the third power.}$$

$$x = 216 \quad \text{or} \quad x = -64$$

Think About It

Suppose that to solve the equation $(x + 1) + 5\sqrt{x + 1} + 6 = 0$, you let $u = \sqrt{x + 1}$ and write $u^2 + 5u + 6 = 0$. How can you tell immediately that the original equation has no solution?

Complex Number Solutions

Whenever quadratic equations emerge in the process of solving an equation, the solutions may be imaginary numbers.

EXAMPLE 5

Equations with Complex Number Solutions

Solve the equation $x^4 + x^2 = 12$.

Solution

$$x^4 + x^2 = 12$$

$$x^4 + x^2 - 12 = 0$$

Figure 9.10 shows the graph of $y = x^4 + x^2 - 12$. From the graph we anticipate two real number solutions. An algebraic method of solving is needed to determine any imaginary number solutions.

Figure 9.10

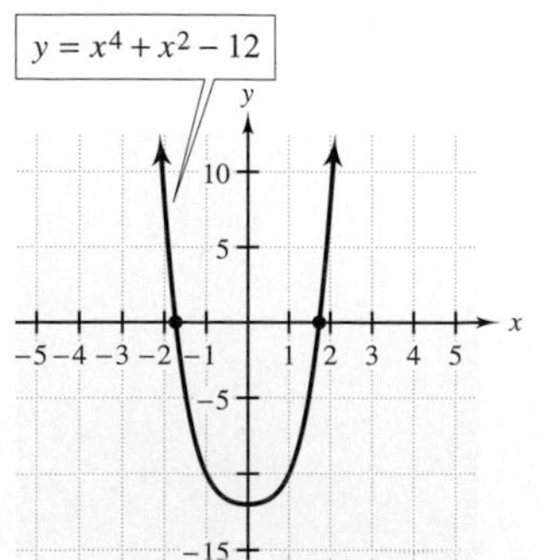

$$x^4 + x^2 = 12$$

$$u^2 + u = 12 \quad \text{Let } u = x^2. \text{ Then } u^2 = x^4.$$

$$u^2 + u - 12 = 0 \quad \text{Write the equation in standard form.}$$

$$(u + 4)(u - 3) = 0 \quad \text{Factor.}$$

$$u + 4 = 0 \quad \text{or} \quad u - 3 = 0 \quad \text{Zero Factor Property}$$

$$u = -4 \quad \text{or} \quad u = 3 \quad \text{Solve each case for } u.$$

The solutions for u are real numbers. To solve for x, replace u with the expression it represents.

$$x^2 = -4 \quad \text{or} \quad x^2 = 3$$

$$x = \pm 2i \quad \text{or} \quad x = \pm\sqrt{3} \approx \pm 1.73$$

Two of the solutions for x are real (irrational) numbers (as the graph suggested), and two solutions are complex conjugates.

In this section we have seen that the ability to solve quadratic equations benefits us well beyond quadratic equations themselves. Solving a quadratic equation may be one part of solving a higher-degree equation; an equation involving rational expressions, radicals, or rational exponents; or any other equation that is quadratic in form.

9.4 QUICK REFERENCE

Fractional Equations

- To solve a fractional equation, clear the fractions by multiplying both sides of the equation by the LCD. This step may lead to a quadratic equation.
- Make sure that any solutions you obtain are not restricted values.

Radical Equations

- To solve a radical equation, raise both sides of the equation to a power equal to the index of the radical. The result may be a quadratic equation.
- Make sure that any solutions you obtain are not extraneous.

Quadratic Forms

- An equation is in quadratic form if it can be written in the form $au^2 + bu + c = 0$, where u is an algebraic expression.
- To solve such equations, we make a substitution to write the equation in terms of u and then apply our methods for solving quadratic equations.

Complex Number Solutions

- Whenever quadratic equations emerge in the process of solving an equation, the solutions may be imaginary numbers.

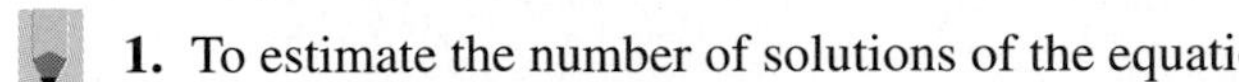

9.4 EXERCISES

Concepts and Skills

1. To estimate the number of solutions of the equation

$$x + \frac{1}{x-1} - \frac{5-4x}{x-1} = 0$$

we can produce the graph of

$$f(x) = x + \frac{1}{x-1} - \frac{5-4x}{x-1}$$

and observe that there is one x-intercept. (See figure.)

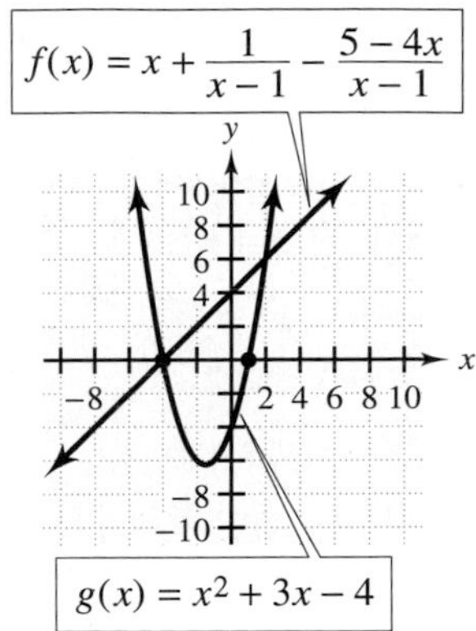

Now clear the fractions and show that the result is the quadratic equation $x^2 + 3x - 4 = 0$. From the figure we can see that the graph of $g(x) = x^2 + 3x - 4$ has two intercepts. Explain why the graphs indicate only one solution for the original equation but two solutions for the resulting quadratic equation.

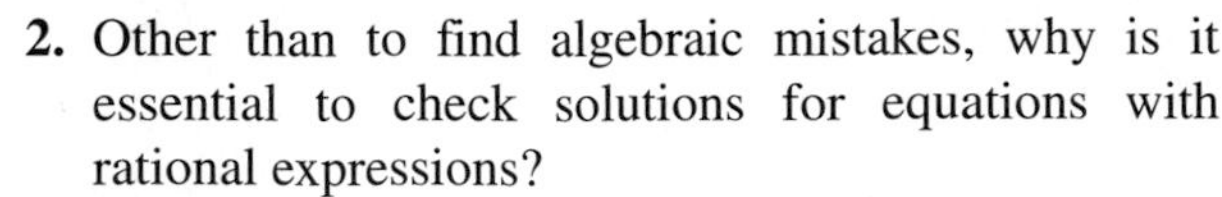

2. Other than to find algebraic mistakes, why is it essential to check solutions for equations with rational expressions?

In Exercises 3–14, solve the given equation.

3. $\frac{3}{x} = 2 + \frac{4}{x^2}$

4. $x + \frac{5}{x} = 3$

5. $\frac{t-3}{t} = \frac{2}{t-3}$

6. $\frac{y}{y+1} = \frac{y+1}{3}$

7. $x + 2 = \frac{6}{1-x}$

8. $x + 1 = \frac{2}{2x+3}$

9. $\frac{x}{x+3} + \frac{3}{x} = 4$

10. $\frac{2}{x-2} + \frac{4}{x+3} = 1$

11. $\frac{x}{x-2} - \frac{5}{x+2} = \frac{12}{x^2-4}$

12. $\frac{x}{x-5} - \frac{16}{x^2-25} = \frac{1}{x+5}$

13. $\frac{x+3}{x-2} - \frac{5}{2x^2-3x-2} = \frac{2}{2x+1}$

14. $\frac{x}{x-3} - \frac{1}{2x-1} = \frac{5x}{2x^2-7x+3}$

15. To estimate the number of solutions of the equation $\sqrt{x^2-5} + 2\sqrt{x} = 0$, we can produce the graph of $f(x) = \sqrt{x^2-5} + 2\sqrt{x}$ and observe that there are no x-intercepts. (See figure.)

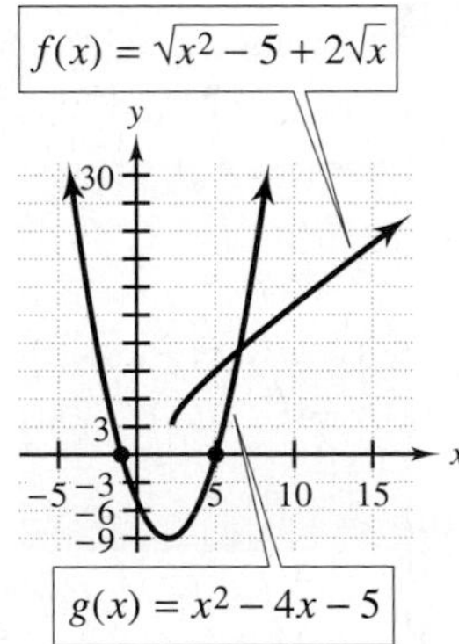

Now isolate the radicals, square both sides, and show that the result is $x^2 - 4x - 5 = 0$. From the figure we can see that the graph of the function $g(x) = x^2 - 4x - 5$ has two x-intercepts. Explain why the graphs indicate that the original equation has no solution but that the resulting quadratic equation has two solutions.

16. Other than to find algebraic mistakes, why is it essential to check your solutions for radical equations?

In Exercises 17–26, solve the radical equation.

17. $5 = 3\sqrt{x} + \frac{2}{\sqrt{x}}$

18. $\sqrt{x+1} - \frac{2}{\sqrt{x+1}} = 1$

19. $2y + 1 = \sqrt{11y - 1}$

20. $2 + x = \sqrt{2x + 3}$

21. $x - 3 = \sqrt{x - 1}$

22. $2x + 5 = \sqrt{4x + 14}$

23. $\sqrt{3x - 2} = \sqrt{x + 2} - 1$

24. $\sqrt{3x + 4} = \sqrt{x - 2} - 3$

25. $\sqrt{x}\sqrt{x + 3} = 1$

26. $\sqrt{2x}\sqrt{x - 3} = 6$

27. Explain the role of substitution in solving equations that are quadratic in form. After you have solved the resulting equation, how do you determine the solution(s) of the original equation?

28. Suppose that an equation is quadratic in form. The two variable terms are $x^{1/6}$ and $x^{1/3}$ and you want to use a substitution to solve the equation. What is an appropriate substitution? Why?

In Exercises 29–48, solve by using an appropriate substitution to rewrite the given equation in quadratic form.

29. $b^4 - 7b^2 + 12 = 0$

30. $x^4 - 14x^2 + 45 = 0$

31. $y^4 + 2y^2 = 63$

32. $x^4 + 17x^2 + 16 = 0$

33. $x - 6x^{1/2} + 8 = 0$

34. $c + 21 = 10c^{1/2}$

35. $y^{-2} - 2 = y^{-1}$

36. $1 - 6x^{-1} + 9x^{-2} = 0$

37. $y^{2/3} + 12 = 7y^{1/3}$

38. $3x^{1/10} + x^{1/5} = 4$

39. $(x^2 - 5x)^2 - 8(x^2 - 5x) = 84$

40. $(x^2 - x)^2 - 26(x^2 - x) + 120 = 0$

41. $(x + 1)^2 + 2(x + 1) = 2$

42. $(y - 2)^2 - 2(y - 2) = 4$

43. $(x^2 - 3)^2 - 5(x^2 - 3) - 6 = 0$

44. $(x^2 + 1)^2 - 2(x^2 + 1) - 15 = 0$

45. $3(\sqrt{x} - 2)^2 - 7(\sqrt{x} - 2) + 2 = 0$

46. $(\sqrt{x} - 8)^2 + 9(\sqrt{x} - 8) + 20 = 0$

47. $x - \sqrt{x} = 6$

48. $x + 3\sqrt{x} - 10 = 0$

In Exercises 49–78, determine all complex number solutions.

49. $x^4 - 7x^2 + 12 = 0$

50. $x^4 - 10x^2 + 16 = 0$

51. $\dfrac{x}{x + 2} + \dfrac{4}{x} = 5$

52. $\dfrac{4}{x} + \dfrac{5x}{x + 3} = 2$

53. $2x + 5 = 2\sqrt{x + 3}$

54. $2x = 1 + \sqrt{x + 7}$

55. $a^4 + a^2 = 20$

56. $x^4 + 5x^2 + 6 = 0$

57. $x - 13x^{1/2} + 42 = 0$

58. $b = 4b^{1/2} + 21$

59. $\dfrac{x + 3}{x - 1} = \dfrac{5}{x}$

60. $x + 2 = \dfrac{6}{1 - x}$

61. $\sqrt{3x + 12} - \sqrt{x + 8} = 2$

62. $\sqrt{3x - 2} = \sqrt{x + 3} + 1$

63. $a^{2/3} = 2a^{1/3} + 15$

64. $2x^{1/6} + x^{1/3} = 8$

65. $\sqrt{3x} = \sqrt{2 - 7x}$

66. $2x = \sqrt{3}\sqrt{2x + 1}$

67. $\dfrac{x + 1}{3 - x} + \dfrac{12}{x^2 - 3x} + \dfrac{5}{x} = 0$

68. $1 = \dfrac{3}{x + 4} - \dfrac{5x}{x^2 + 3x - 4}$

69. $(3 - y)^2 - 3(3 - y) = 6$

70. $(x + 1)^2 + (x + 1) - 3 = 0$

71. $\sqrt{2x - 3} = 4 - \sqrt{x + 2}$

72. $\sqrt{3x - 5} = 2 - \sqrt{x + 1}$

73. $(3 - x)^{-2} - 3(3 - x)^{-1} = 4$

74. $4(x + 1)^{-1} + (x + 1)^{-2} - 21 = 0$

75. $\dfrac{x}{4(x - 4)} = \dfrac{1}{x + 2} + \dfrac{6}{x^2 - 2x - 8}$

76. $\dfrac{x + 4}{2x + 1} + \dfrac{1}{x + 1} = \dfrac{-1}{2x^2 + 3x + 1}$

77. $(x^2 - x)^2 + 3(x^2 - x) + 2 = 0$

78. $(2x^2 - 7x)^2 - 2(2x^2 - 7x) - 8 = 0$

Real-Life Applications

79. A small manufacturer makes motors for yard tractors. The cost to produce x hundred motors is $8000 + 8000x^2 - 80x^4$, and the total revenue from the sale of the motors is $1680x^2 - 4960$. How many motors must the company make and sell to break even?

80. Mail Express provides a mailing service for bulk-mail advertising for small businesses. The company estimates that for x hundred brochures, preparation and mailing cost C (in dollars) will be modeled by $C = 20 + 20\sqrt{2x + 11}$, and the revenue R (in dollars) will be modeled by $R = 20\sqrt{5x + 1}$. For what size order will the company break even?

81. As part of a summer program, a college obtains a low bid of \$195,000 for a group of students to travel and study in South America. After initial registration, spaces for 40 students remain. If the remaining spaces are filled, the per-person cost will be reduced by \$100. How many students can the program accommodate, and what is the cost per person if all spaces are filled?

82. A business traveler left home at 6:00 A.M. and drove 50 miles to the airport for a flight to Pittsburgh. After the first 20 miles, traffic became congested and reduced the motorist's average speed by 24 miles per hour. If the traveler arrived at the airport at 7:10 A.M., how many minutes of the trip were spent in heavy traffic?

Modeling with Real Data

83. A study shows that the percentage of hotels with no vacancy depends on the day of the week. (Source: Horwath U.S. Hotel Industry Study.) The function $v(x) = -0.48x^2 + 3.4x + 10.3$ models the no-vacancy rate for day x, where Monday is $x = 1$, Tuesday is $x = 2$, and so on.

(a) Write and solve an equation to determine the day(s) of the week for which the no-vacancy rate is about 15.3%.

(b) The following function provides a no-vacancy rate model for days 8–14.

$$v_1(x) = -0.48(x - 7)^2 + 3.4(x - 7) + 10.3$$

Write an equation to determine the day(s) of the second week for which the no-vacancy rate is about 15.3%.

(c) Use a substitution to write the equation in part (b) in the standard form of a quadratic equation. Solve the equation and compare your result with the result in part (a).

84. A survey found that having a neat house is not a major concern for the majority of the 30–49 age group. The accompanying table shows the percentages of "yes" responses to the question, "Is a neat house very important?" (Source: Soap and Detergent Association.)

Age Group	Percent
18–29	57%
30–49	48%
50 and over	62%

The following function models the percentages of people who think that a neat house is very important, where x is the number of years over (or under) age 40.

$$h(x) = 0.037(x + 40)^2 - 3.01(x + 40) + 109.14$$

(a) Write an equation to determine the age at which 50% of the respondents answered "Yes."

(b) Use a substitution to write the equation in part (a) in the standard form of a quadratic equation. Then solve the equation and interpret the result.

Group Project: School Lunch Program

The cost of the federally funded National School Lunch Program has risen steadily since its inception. In 1985, the cost was about \$2.6 billion. More recent costs are shown in the accompanying table. (Source: Food and Nutrition Service, USDA.)

Year	Cost (in millions)
1990	3214
1993	4018
1994	4283

With t representing the number of years since 1990, the function $C(t) = 15.5(t + 5)^2 + 52.7(t + 5) + 2576$ models the cost C (in millions of dollars) of the school lunch program.

85. If the model function remains valid, what would be the national per capita cost for the school lunch program in the year 2002? (Assume a U.S. population of 270 million.)

86. Note that the expression for $C(t)$ is quadratic in form. Using a suitable substitution to solve the equation, determine the value of t for which $C(t) = 10{,}000$. Interpret the result.

87. The form of the model function makes it easy to evaluate $C(-5)$ without a calculator. What previously given information does the result represent?

88. In 1994 there were approximately 41 million elementary and secondary students, of whom nearly 26 million participated in the school lunch program. Do you think that 63% of all students require federal lunch subsidies? Why or why not?

Challenge

In Exercises 89 and 90, rewrite the equation so that it is quadratic in form. Then make an appropriate substitution and solve.

89. $t = \sqrt{t+3}$

(Hint: Consider adding 3 to both sides.)

90. $x^2 + 6x - \sqrt{x^2 + 6x - 2} = 22$

In Exercises 91–96, determine all complex number solutions.

91. $x^6 - 7x^3 = 8$

92. $x^6 + 26x^3 = 27$

93. $x^2 + 2x = |2x + 1|$

94. $|x - 3| = x^2 - 3x$

95. $y^{-2} + 1 = y^{-1}$

96. $2\sqrt{x} - \dfrac{1}{\sqrt{x}} = 2$

9.5 APPLICATIONS

Geometric Figures • Ratio and Proportion • Distance, Rate, Time • Work Rate • Investment

This section includes a sampling of applications that can be modeled with a variety of equations. As we have seen, even if the initial equation used to describe conditions is not quadratic, the solving process may lead to a quadratic equation.

Geometric Figures

Problems involving the area of a rectangle often can be modeled with a quadratic equation.

EXAMPLE 1

Figure 9.11

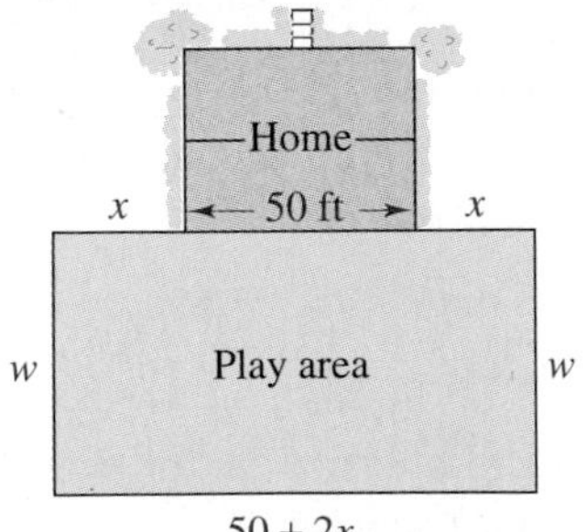

The Area of a Rectangular Play Area

A day-care operator plans to use 170 feet of fencing to enclose a rectangular play area of 2800 square feet. She plans to construct the play area so that it is centered against the back of her home, which is 50 feet in length. Determine the dimensions necessary to enclose the required area.

Solution

Let x = the distance from the corner of the house to the corner of the fence. (See Fig. 9.11.) Then the length of the rectangle is $50 + 2x$.

Let w = the width of the play area. Note that no fencing is needed along the back of the home.

$$x + w + (50 + 2x) + w + x = 170 \qquad \text{Total fencing used is 170 feet.}$$
$$2w + 50 + 4x = 170 \qquad \text{Combine like terms.}$$
$$2w = 120 - 4x \qquad \text{Solve for } w.$$
$$w = 60 - 2x$$

$$L \cdot W = A \qquad \text{Formula for the area of a rectangle}$$
$$(50 + 2x)(60 - 2x) = 2800 \qquad \text{Required area is 2800 square feet.}$$
$$3000 + 20x - 4x^2 = 2800 \qquad \text{FOIL}$$

$$-4x^2 + 20x + 200 = 0 \quad \text{Write the equation in standard form.}$$
$$x^2 - 5x - 50 = 0 \quad \text{Divide both sides by } -4.$$
$$(x - 10)(x + 5) = 0 \quad \text{Factor.}$$
$$x - 10 = 0 \text{ or } x + 5 = 0 \quad \text{Zero Factor Property}$$
$$x = 10 \text{ or } x = -5 \quad \text{Solve each case for } x.$$

Because the distance is not negative, the only applicable solution is 10. The fence extends 10 feet on each side of the house, which means that the length of the play area is 70 feet. The width is $60 - 2(10) = 40$ feet.

Think About It

It can be shown that $1 + 2 + 3 + \cdots + n = \frac{n(n+1)}{2}$. If the sum of n numbers is 946, what is the last number in the sum?

Ratio and Proportion

According to early Greek mathematicians, the most perfect rectangles were those whose width W and length L satisfied the *Golden Ratio.*

$$\frac{W}{L} = \frac{L}{W + L}$$

We can use this formula in application problems in which a rectangle is to have dimensions that satisfy the Golden Ratio.

EXAMPLE 2

Rectangles That Satisfy the Golden Ratio

To create a miniature reproduction of the courtyard of an ancient Greek temple, an architect designed the courtyard so that it satisfied the Golden Ratio. If the perimeter of the miniature courtyard was 20 inches, what were the dimensions of the rectangle?

Solution

Let L = the length of the rectangle, W = the width of the rectangle, and P = the perimeter, which is given as 20 inches.

$$2L + 2W = P \quad \text{Formula for the perimeter of a rectangle}$$
$$2L + 2W = 20 \quad \text{The required perimeter is 20 inches.}$$
$$L + W = 10 \quad \text{Divide both sides by 2.}$$
$$W = 10 - L \quad \text{Solve for } W.$$

The Golden Ratio formula is $\frac{W}{L} = \frac{L}{W + L}$.

$$\frac{10 - L}{L} = \frac{L}{10} \quad \text{Replace } W \text{ with } 10 - L.$$
$$\frac{10L}{1} \cdot \frac{10 - L}{L} = \frac{10L}{1} \cdot \frac{L}{10} \quad \text{Multiply by the LCD } 10L \text{ to clear the fractions.}$$
$$100 - 10L = L^2 \quad \text{Simplify.}$$
$$L^2 + 10L - 100 = 0 \quad \text{Write the equation in standard form.}$$

We use the Quadratic Formula to solve for L.

$$L = \frac{-10 \pm \sqrt{10^2 - 4(1)(-100)}}{2(1)}$$

$$L = \frac{-10 \pm \sqrt{100 + 400}}{2}$$

$$L = \frac{-10 \pm \sqrt{500}}{2}$$

$$L \approx 6.18 \quad \text{or} \quad L \approx -16.18$$

Because length is positive, the negative solution is disqualified. The length is approximately 6.18 inches, and the width is approximately $10 - 6.18 = 3.82$ inches.

Distance, Rate, Time

Recall that distance D, rate R, and time T are related by the formula $D = RT$.

EXAMPLE 3

A Boating and Walking Trip

From a dock, the distance directly across a lake to a marina is 3 miles. An angler travels by boat at 5 mph from the dock to a pier on the opposite shore. Then he walks at 4 mph from the pier to a fishing spot that is 6 miles down the shoreline from the marina. (See Fig. 9.12.) If he made the trip in 1.5 hours, how far did he walk?

Figure 9.12

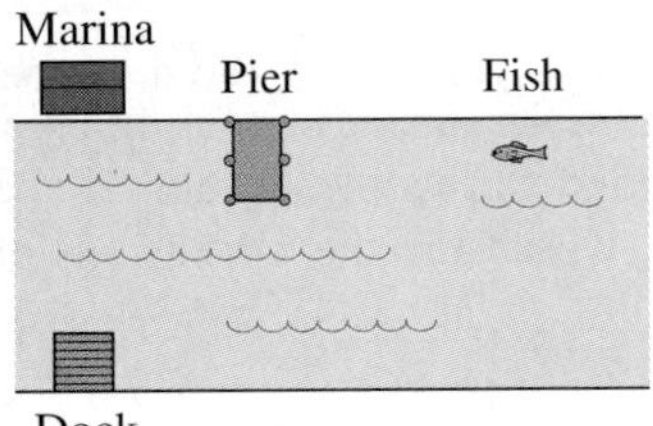

Solution

In Fig. 9.13 we identify locations with letters:

- D: Dock
- M: Marina
- P: Pier
- F: Fishing spot

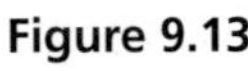

Figure 9.13

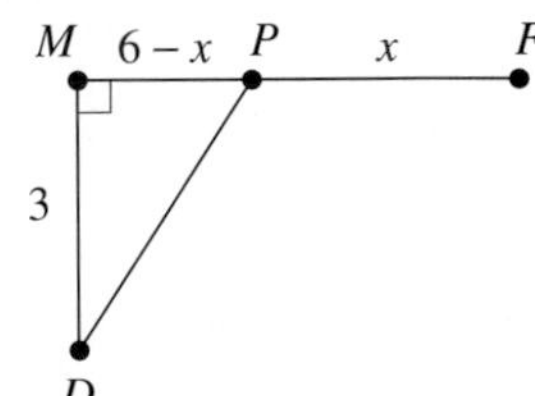

Let $x =$ the distance the angler walked from P to F. The distance from the marina M to the fishing spot F is 6 miles. Therefore, the distance from the marina M to the pier P is $6 - x$.

We use the Pythagorean Theorem on the right triangle DMP to determine the distance DP traveled by boat.

$$(DP)^2 = (DM)^2 + (MP)^2$$

$$(DP)^2 = 3^2 + (6 - x)^2$$

$$DP = \sqrt{9 + (6 - x)^2}$$

	Distance (miles)	Rate (mph)	Time (hours)
Boat	$\sqrt{9 + (6 - x)^2}$	5	$\frac{\sqrt{9 + (6 - x)^2}}{5}$
Walking	x	4	$\frac{x}{4}$

$$\frac{\sqrt{9+(6-x)^2}}{5}+\frac{x}{4}=1.5$$ The total travel time was 1.5 hours.

$$\frac{20}{1}\cdot\frac{\sqrt{9+(6-x)^2}}{5}+\frac{20}{1}\cdot\frac{x}{4}=20\cdot 1.5$$ Clear fractions by multiplying by the LCD 20.

$$4\sqrt{9+(6-x)^2}+5x=30$$ Simplify.

$$4\sqrt{9+(6-x)^2}=30-5x$$ Isolate the radical.

$$\left[4\sqrt{9+(6-x)^2}\right]^2=(30-5x)^2$$ Square both sides.

$$16[9+(6-x)^2]=(30-5x)^2$$ $(ab)^2=a^2b^2$

$$16(9+36-12x+x^2)=(30-5x)^2$$ Square the binomial $6-x$.

$$16(45-12x+x^2)=900-300x+25x^2$$ Square $30-5x$.

$$720-192x+16x^2=900-300x+25x^2$$ Distributive Property

$$9x^2-108x+180=0$$ Write the equation in standard form.

$$x^2-12x+20=0$$ Divide both sides by 9.

$$(x-2)(x-10)=0$$ Factor.

$$x-2=0 \quad\text{or}\quad x-10=0$$ Zero Factor Property

$$x=2 \quad\text{or}\quad x=10$$ Solve each case for x.

By substitution, we find that 10 is an extraneous solution. For $x = 2$, the distance $6 - x$ would be 4, which is an acceptable value. Therefore, the angler walked 2 miles.

Work Rate

If T is the time needed to complete a task, then $1/T$ is the work rate. If two or more persons, groups, or machines work at the task, then the combined work rate is the sum of the individual work rates.

EXAMPLE 4

Combined Work Rates

Working together, two details of Air Force reservists can load a C-130 aircraft with relief supplies for hurricane victims in 6 hours. Working alone, the first detail can load the plane in 2 hours less time than the second detail can when working alone. How long would each detail, working alone, take to load the plane?

Solution

Let t = number of hours for the second detail, working alone, to load the plane. Then $t - 2$ = number of hours for the first detail, working alone, to load the plane.

	Time (hours)	Work Rate (loads/hour)
First detail	$t - 2$	$\frac{1}{t-2}$
Second detail	t	$\frac{1}{t}$
Together	6	$\frac{1}{6}$

$$\frac{1}{t-2} + \frac{1}{t} = \frac{1}{6}$$ Sum of the work rates equals the combined work rate.

$$\frac{6t(t-2)}{1} \cdot \frac{1}{t-2} + \frac{6t(t-2)}{1} \cdot \frac{1}{t} = \frac{6t(t-2)}{1} \cdot \frac{1}{6}$$ Multiply by the LCD $6t(t-2)$.

$$6t + 6(t-2) = t(t-2)$$ Simplify.

$$6t + 6t - 12 = t^2 - 2t$$ Distributive Property

$$t^2 - 14t + 12 = 0$$ Write the equation in standard form.

To solve for t, use the Quadratic Formula.

$$t = \frac{14 \pm \sqrt{(-14)^2 - 4(1)(12)}}{2(1)}$$

$$t = \frac{14 \pm \sqrt{196 - 48}}{2}$$

$$t = \frac{14 \pm \sqrt{148}}{2}$$

$$t \approx 13.08 \quad \text{or} \quad t \approx 0.92$$

Because the solution 0.92 makes the first detail's time negative, the only applicable solution is 13.08. The second detail would take approximately 13.08 hours to load the plane, and the first detail would take approximately 11.08 hours.

Investment

In previous investment problems, we have used this relationship:

Number of shares · price per share = total investment

LEARNING TIP

Consider writing a "word equation" to relate the quantities involved in a problem. For example, (number of items) · (cost per item) = total cost.

If the focus of an investment application is on price per share, then we can rewrite the relationship this way.

$$\text{Price per share} = \frac{\text{total investment}}{\text{number of shares}}$$

EXAMPLE 5

An Investment Application Problem

Six years ago a woman invested \$42,000 in stock. As a result of stock splits and dividend reinvestment, she now has 1100 more shares of the stock than she originally purchased, and the market value is \$73,600. However, the price per share is \$3 per share less than the price per share of the original purchase. How many shares did she purchase originally and how many does she own now?

Solution

Let $x =$ number of shares initially purchased and $x + 1100 =$ number of shares currently owned. The price per share is the market value divided by the number of shares.

$$\frac{42{,}000}{x} = \text{price per share of initial purchase}$$

$$\frac{73{,}600}{x + 1100} = \text{current price per share}$$

The difference between the initial price per share and the current price per share is \$3.

$$\frac{42{,}000}{x} - \frac{73{,}600}{x + 1100} = 3$$

Clear fractions by multiplying by the LCD $x(x + 1100)$.

$$\frac{x(x + 1100)}{1} \cdot \frac{42{,}000}{x} - \frac{x(x + 1100)}{1} \cdot \frac{73{,}600}{x + 1100} = 3x(x + 1100)$$

$$42{,}000(x + 1100) - 73{,}600x = 3x(x + 1100)$$

$$42{,}000x + 46{,}200{,}000 - 73{,}600x = 3x^2 + 3300x$$

$$46{,}200{,}000 - 31{,}600x = 3x^2 + 3300x$$

$$3x^2 + 34{,}900x - 46{,}200{,}000 = 0$$

To solve for x, use the Quadratic Formula.

$$x = \frac{-34{,}900 \pm \sqrt{34{,}900^2 - 4(3)(-46{,}200{,}000)}}{2(3)}$$

$$x = \frac{-34{,}900 \pm \sqrt{1{,}772{,}410{,}000}}{6}$$

$$x = 1200 \quad \text{or} \quad x = -12{,}833.33$$

Because the number of shares must be positive, the only applicable solution is 1200. Thus the woman bought 1200 shares initially, and she now owns 2300 shares.

9.5 EXERCISES

Real-Life Applications

Geometric Figures

1. The dimensions of a rectangular front yard are 60 feet by 80 feet. If you begin mowing around the outside edge, how wide a border around the lawn must you mow to complete three-fourths of the job?
2. The dimensions of a rectangular hay field are 60 yards by 100 yards. A farmer is going to cut the hay around the outside edge. How wide a border around the field must the farmer cut to complete two-thirds of the job?
3. A triangular-shaped fountain area occupies one corner of a rectangular park. A low wall is to be constructed along the longest side, which is 160 feet long. A higher wall is to be constructed along the two perpendicular sides, whose lengths differ by 30 feet. What are the lengths of the two higher walls?
4. When a painter leans a 24-foot ladder against a house, the distance from the base of the ladder to the house is 8 feet less than the distance from the top of the ladder to the ground. How far up the house is the top of the ladder?

Ratio and Proportion

5. A woodworker is asked to build a rectangular shadow box with a perimeter of 56 inches. To make the shape aesthetically appealing, the rectangle should satisfy the Golden Ratio. What are the appropriate dimensions of the shadow box?
6. The design editor for a publisher created a front cover for a book. The cover satisfied the Golden Ratio and had a perimeter of 31 inches. What were the dimensions of the cover?
7. In the accompanying figure, triangles ABC and DEF are similar triangles. Determine the lengths AC and DE.

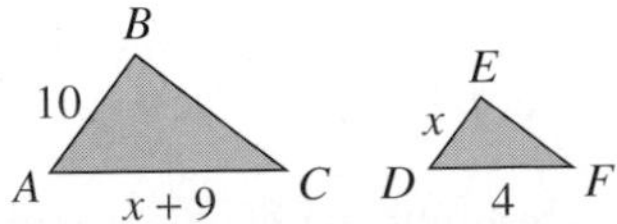

8. In the accompanying figure, triangles ABC and DEF are similar triangles. Determine the lengths AB and DF.

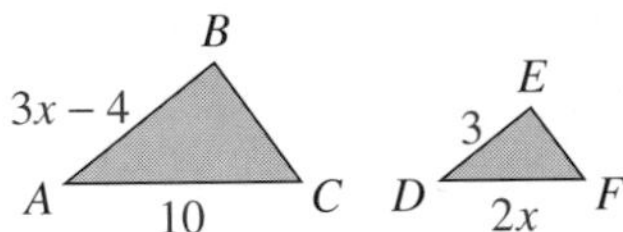

Distance, Rate, Time

9. A boater drove a motorboat upstream a distance of 46 miles and then immediately drove back to the starting point. The total time of the trip was 3.5 hours. If the rate of the current was 4 mph, how fast would the boat have traveled in still water?
10. Two campers left camp at 8:00 A.M. to paddle their canoe upstream 10 miles to a fishing cove. They fished for 3.75 hours and then paddled back to the campsite and arrived at 6:00 P.M. If the rate of the current was 3 mph, how fast can the campers paddle in still water?
11. A family drove 45 miles and then hiked 1 mile to their campsite. The trip took 1 hour and 15 minutes. If they drove 55 mph faster than they hiked, how fast did they drive?
12. Two track team members ran a 200-meter relay in 29 seconds. (Each person ran 100 meters.) One sprinter ran 0.55 meters per second faster than the other. What was the speed of each runner?

Work Rate

13. Working together, two roommates can paint their apartment in 6 hours. Working alone, one person can complete the job in 2 hours less time than the other person working alone. How long would each person take, working alone?
14. Two landscape workers can complete a certain landscaping job in a total of 20 hours. Working alone, one can do this same job in 2 hours less time than the other. How long would the faster worker take to do the job if the slower worker were unable to work?

15. A backyard swimming pool has two fill pipes and a drain pipe. The pool fills in 9 hours with the first fill pipe. Draining the pool with the drain pipe takes 1 hour longer than filling the pool with the second fill pipe. If all three pipes are open at once, the pool can be filled in 8 hours. How many hours would draining the pool take with the two fill pipes closed?

16. Both the inlet and outlet pipes to a cider barrel were mistakenly left open. Nevertheless, the barrel filled in 12 hours. With only the outlet pipe open, a full barrel of cider can be emptied in 1 hour more than the time to fill an empty barrel with only the inlet pipe open. How long would draining a full barrel of cider take with the outlet pipe open and the inlet pipe closed?

Investment

17. Ten years ago a man had less than 400 shares of stock worth \$16,095. Today he has 405 more shares of stock than he had originally, and the investment is worth \$34,875. The price per share today is \$1.50 more than it was 10 years ago. How many shares of stock did the man have originally?

18. A woman invested \$13,500 to buy more than 200 shares of a certain stock. After 5 years she had 22.23 more shares of stock, and her investment was worth \$16,189.13. If each share increased in value by \$4, how many shares did the woman purchase initially?

19. A community service organization sells citrus fruit each year to raise money for the local library. Last year the total sales were \$8568. This year the organization sold 102 more boxes of fruit than last year, and the total sales were \$11,594. This year's selling price was up \$2 from last year's selling price. If fewer than 500 boxes were sold last year, how many boxes of fruit were sold last year and this year, and what was the price per box this year?

20. The community theater group performed a popular musical. The tickets for the Friday night performance were \$2 more than the Thursday night tickets. The total receipts on Thursday night were \$967.50, and the total receipts on Friday night were \$2034.00. On Friday 65 more people attended the play than had attended on Thursday night. How many people attended on Friday night?

Miscellaneous

21. The owner of an apartment complex collected \$24,000 per month in rent when all the apartments were occupied. Then the owner raised the rent \$80 per month, and ten apartments were vacated. However, the owner still collects \$24,000 a month in rent. How many apartments are in the complex?

22. An Olympics vendor purchased a box of souvenir keychains for \$70. When all but 12 were sold at a profit of 75¢ per keychain, he had recovered his original investment. How many keychains were in the box, and what was the selling price?

23. Find two consecutive odd integers such that the difference of their cubes is 866.

24. Find two consecutive integers such that the difference of their cubes is 169.

25. A plot of land next to a river is to be enclosed with 300 feet of fencing to form a rectangular area of 11,500 square feet. The river will serve as the fourth side. An 8-foot-wide gate is to be installed in the fence that is parallel to the river. (See figure.) How far should the fence extend from the river if the minimum distance is 70 feet?

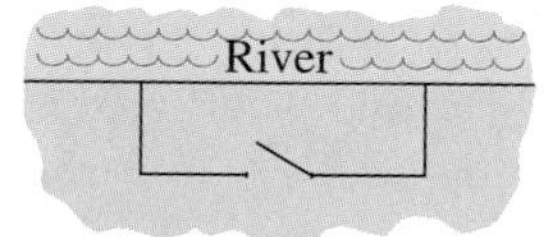

26. A woman bought a 21-square-yard rectangular piece of carpeting for a room that is 12 feet by 18 feet. She chose the dimensions of the carpet so that a uniformly wide strip of uncovered floor would be around the outside of the carpet. How wide will this strip be?

27. If a rectangular sign is 5 feet wide, how long must it be in order to satisfy the Golden Ratio?

28. An architect's plans include a rectangular iron gate with a length that is 3 feet more than the width and that satisfies the Golden Ratio. What are the dimensions of the gate?

29. A polygon of n sides has $\frac{1}{2}n(n - 3)$ diagonals. If a polygon has 27 diagonals, how many sides does it have?

30. The height h of an object projected upward from an initial height h_0 with an initial velocity v_0 is given by $h = -16t^2 + v_0t + h_0$.

The floor of the Golden Gate Bridge is 220 feet above the water. If a rock is tossed upward from the floor of the bridge with an initial velocity of 30 feet per second, how many seconds later will the rock hit the water? (Round your answer to the nearest tenth.)

31. In geometry the altitude h to the hypotenuse of a right triangle is related to the two segments a and b shown in the figure by the equation $\frac{a}{h} = \frac{h}{b}$.

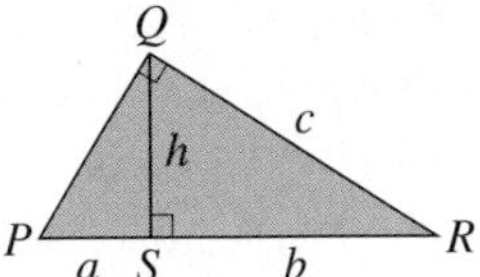

Suppose that a home gardener uses the design shown in the figure to build a cold frame. The front segment b is 2.5 feet longer than the back segment a, and the height of the frame is 3 feet. How long is the front panel c?

32. In the figure, $\overline{AB}$ is a 4-inch mirror with two vertical shields represented by $\overline{AQ}$ and $\overline{BR}$. Light ray L_1 strikes the mirror at point P and is reflected as light ray L_2. By a law of optics, $\angle 1$ (the *angle of incidence*) has the same measure as $\angle 2$ (the *angle of reflection*). Given the relative dimensions in the figure, how tall is the shield $\overline{AQ}$?

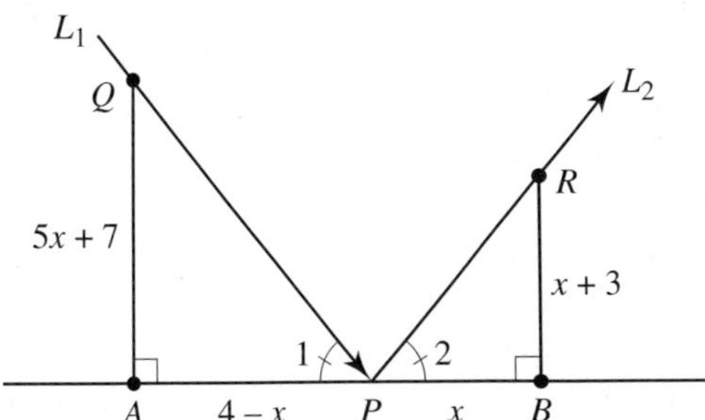

33. A farmer has two irrigation sprinklers that rotate about a fixed point to water circular areas. The circumference of the circle watered by the larger system is 4π yards more than the circumference of the circle covered by the smaller system. The total area irrigated by both systems is 164π square yards. What is the radius of the circular area irrigated by the smaller system?

34. An automobile center has a large metal drum to store used motor oil. The height of the drum is 4 feet, and the total surface area is 24π square feet. What is the diameter of the drum? (Hint: The total surface area S of a right circular cylinder is $S = 2\pi rh + 2\pi r^2$, where r is the radius and h is the height.)

35. The area of a triangular mainsail is 440 square feet. If the height of the sail is 10 feet less than the base, what is the height?

36. A car and a truck left Luckenback, Texas, at noon. The car traveled due north and the truck traveled due east. In 2 hours time, they were 155 miles apart, and the car had traveled 30 miles more than the truck. How far was each vehicle from Luckenback at 2:00 P.M.?

37. A shop for custom-made furniture has an order for a rectangular table that is 2 feet longer than it is wide. The customer wants the table to be designed so that an average adult seated at the table can reach the center from any position at the table. The designer knows that the average person can reach 3 feet. What is the width of the table?

38. To construct a paved walk of uniform width around a rectangular swimming pool, a contractor uses 5.5 cubic yards of concrete. The pool is 30 feet long and 20 feet wide. If the walk is to be 3 inches thick, how wide can it be?

39. When 1 is added to a number, the result is 2 more than the square root of the result. What is the number?

40. The square root of 7 more than a number is one less than twice the number. What is the number?

41. A woman swims from a starting point S to a marker M that is 0.15 mile from the shore. From the marker the woman swims at a rate of 0.5 mph back to the shore at point B, which is a bicycle pickup point. From there she rides the bike at 12 mph to the finish line F, which is 4.2 miles from the starting point. If the total time from the marker to the finish line is 50 minutes, how far did the woman ride the bicycle?

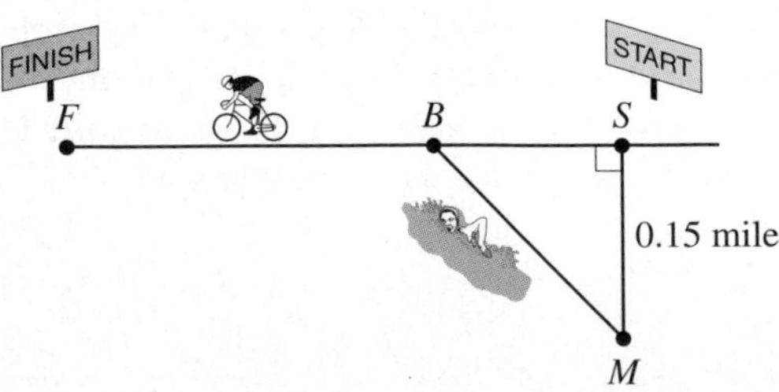

42. A tour bus traveled the first 200 miles of a trip at an average speed that was 6 mph faster than the average speed for the last 180 miles. If the entire trip took 6 hours and 40 minutes, what was the average speed for the first portion of the trip?

43. Working alone, one person can prepare a quarterly bank report in 4 hours less time than a less experienced person. If both begin working at 9:00 A.M. and stop at 5:00 P.M., but the less experienced worker takes a 2-hour lunch break while the experienced worker skips lunch, the job is 65% complete at the end of the day. If both begin working at 9:00 A.M. the next day, at what time will they complete the report?

44. Two air-conditioning units can cool a building to 72°F in 3 hours. Working alone, the older unit takes 5 hours longer than the newer unit to cool the building. How many hours were required to cool the building before the new unit was installed?

45. As a result of a basketball injury, a senior forward played in three fewer games than a junior forward. The senior, who scored a total of 174 points, averaged 5.1 points per game more than the junior, who scored a total of 141 points. How many games did the senior play?

46. At 7:00 A.M. a trucker left Little Rock and traveled north. At 8:00 A.M. a second trucker left the same point and drove east, going 12 mph faster than the first. At noon the two trucks were 310 miles apart. To the nearest tenth, what was the average speed of the second truck?

Group Project: Credit Card Payoffs

A majority of Americans have credit cards, but the percentage of card holders who pay the entire balance due each month varies with age. The accompanying table shows representative data for selected ages of card holders. (Source: Federal Reserve System.)

Age	Percent Having Credit Cards	Percent Who Always Pay the Balance Due
30	57.6	39.6
40	64.2	45.5
50	72.5	52.1
60	68.0	64.8
70	65.0	74.9
80	52.1	78.9

47. The function $N(x) = -0.025x^2 + 2.71x - 1.76$ can be used to model the percentage $N(x)$ of card holders at age x. Determine and interpret the x-intercepts of the graph of $N(x)$.

48. As is true with most models, the validity of $N(x)$ is restricted to the given x-values. To illustrate, use $N(x)$ to determine the percentage of 10-year-olds who have credit cards.

49. Based on the data in the table, would

$$y = ax + b \quad \text{or} \quad y = ax^2 + bx + c$$

be the better model for the percentage of card holders who always pay the balance due? Use the data for ages 30 and 70 to write a model function for these data. How would the graph of this function compare with the graph of a function that models the percentage of card holders who rarely pay the balance due?

50. The table indicates that the percentage of card holders who pay off their balances increases with age. Why do you suppose this is so?

Challenge

51. A campus co-op bought a shipment of calculators for \$10,000. The co-op first sold calculators to all students at a price required to cover the total cost. The remaining 40 calculators were then sold to nonstudents at the same price that the students paid, and the profit was distributed to the students. If all the calculators had been sold to students, the price would have been \$12.50 less than what they paid. What share of the profit did each student receive?

52. A fragile item is packed in a carton in the shape of a cube. This carton is then placed inside a larger carton, which is also in the shape of a cube and which has sides 2 feet longer than the sides of the smaller carton. The larger carton is then filled with 40 cubic feet of packing material around the smaller carton. How long is a side of the smaller carton?

53. A spring is attached at point A [see Fig. (a) on the following page] and is connected to the rim of a wheel at point B so that the spring is tangent to the wheel. In this position the spring is 26 inches long. The wheel is turned clockwise so that the spring stretches, and the end of the spring is now at point D. [See Fig. (b) on the following page.] If the distance between points C and D is 8 inches, how far did the spring stretch from its original position?

(Hint: A theorem from geometry relates AB, AC, and AD as $AB/AD = AC/AB$.)

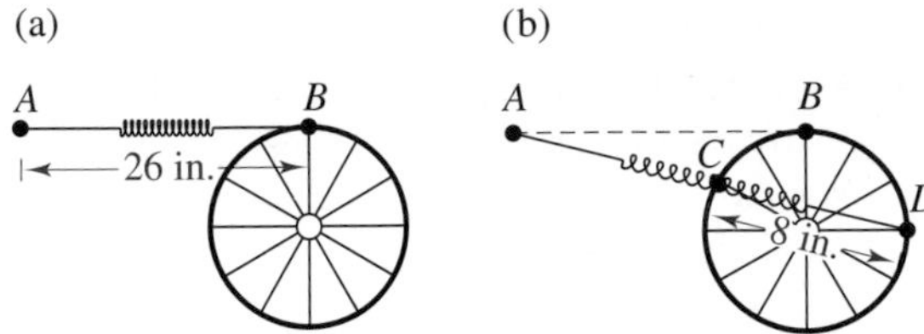

54. A ceremonial cannon, located on the bank of a river, is about to be fired. (See point C in the accompanying figure.) Observer 1 is directly across the river at point A, and Observer 2 is farther down the river at point B. Observer 1 is 2000 feet farther from Observer 2 than he is from the cannon. When the cannon is fired, Observer 2 hears the sound 3 seconds after she sees the flash. How wide is the river? (Assume that the speed of sound is 1100 feet per second.)

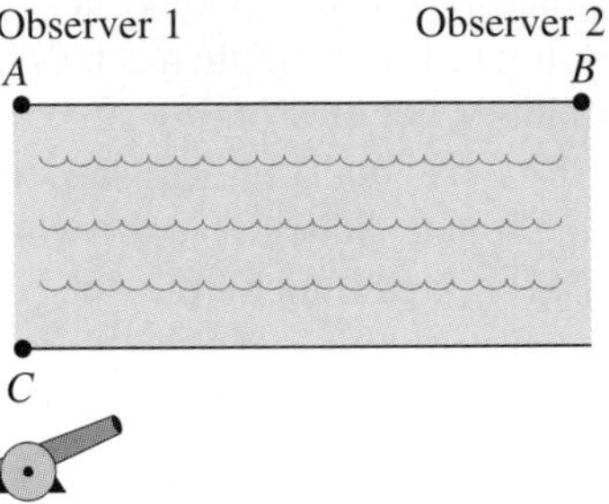

9.6 QUADRATIC INEQUALITIES

Graphing and Algebraic Methods • Special Cases • Real-Life Applications

Graphing and Algebraic Methods

The standard form of a quadratic equation is $ax^2 + bx + c = 0$. When the equality symbol is replaced by an inequality symbol, we obtain a **quadratic inequality.**

Definition of a Quadratic Inequality

A **quadratic inequality** is an inequality that can be written in the standard form $ax^2 + bx + c > 0$, where a, b, and c are real numbers and $a \neq 0$. The symbols $<$, $\leq$, and $\geq$ also can be used.

The **solution set** of a quadratic inequality is the set of all replacements for the variable that make the inequality true.

The graphing method is a visual way of estimating solution sets.

EXPLORING THE CONCEPT

The Graphing Method

Consider the quadratic inequality $x^2 + 2x - 8 < 0$. Figure 9.14 shows the graph of $y = x^2 + 2x - 8$.

Figure 9.14

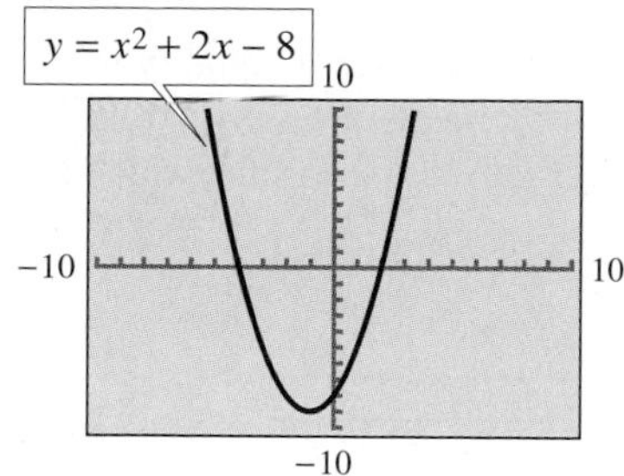

Observe that the x-intercepts appear to be $(-4, 0)$ and $(2, 0)$. These points can be verified algebraically.

$$x^2 + 2x - 8 = 0$$
$$(x + 4)(x - 2) = 0$$
$$x + 4 = 0 \quad \text{or} \quad x - 2 = 0$$
$$x = -4 \quad \text{or} \quad x = 2$$

The solutions of the inequality are the values of x for which $y < 0$, and so the solutions are represented by points of the graph that are below the x-axis.

From Fig. 9.14 we see that $y < 0$ in the interval between the two x-intercepts, that is, for all values of x such that $-4 < x < 2$. Thus the solution set of the inequality is $(-4, 2)$. (See Fig. 9.15.)

Figure 9.15

For the inequality $x^2 + 2x - 8 > 0$, we look for points of the graph that are above the x-axis. Referring again to Fig. 9.14, we see that these points are to the left of $(-4, 0)$ and to the right of $(2, 0)$. Thus the solution set is $(-\infty, -4) \cup (2, \infty)$.

For inequalities involving $\leq$ or $\geq$, the x-intercepts also represent solutions. For example, the solution set of $x^2 + 2x - 8 \leq 0$ is $[-4, 2]$.

The ideal method for solving a quadratic inequality is to determine the x-intercepts algebraically and then to use a graph to assist you in writing the solution set.

LEARNING TIP

The graph of $y = ax^2 + bx + c$ provides valuable assistance in visualizing where the graph is above or below the x-axis. However, algebraic methods are needed to determine the x-intercepts.

Solving a Quadratic Inequality with Graphing and Algebraic Methods

1. Write the inequality in standard form.
2. Produce the graph of $y = ax^2 + bx + c$.
3. Determine the x-intercepts algebraically. For inequalities $y \leq 0$ or $y \geq 0$, these points represent solutions of the inequality.
4. For inequalities $y < 0$ or $y \leq 0$, determine the points where the graph is below the x-axis. For inequalities $y > 0$ or $y \geq 0$, determine the points where the graph is above the x-axis. The x-coordinates of all such points are solutions.

An alternative to the graphing method is the test-point method.

EXPLORING THE CONCEPT

The Test-Point Method

Consider again the quadratic inequality $x^2 + 2x - 8 < 0$. We previously determined that the x-intercepts are $(-4, 0)$ and $(2, 0)$. We can use these intercepts to divide the number line into intervals, which are labeled A, B, and C in Fig. 9.16.

Figure 9.16

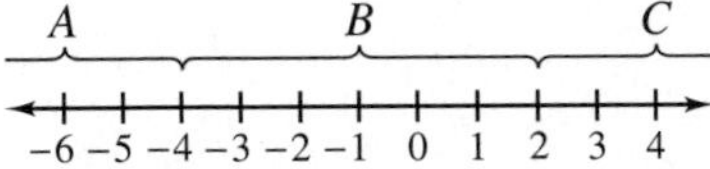

In each interval we select a test value for x (shown in blue in Figure 9.17). If the inequality is true for a certain test value, then the inequality is true for all values of x in that interval.

Figure 9.17

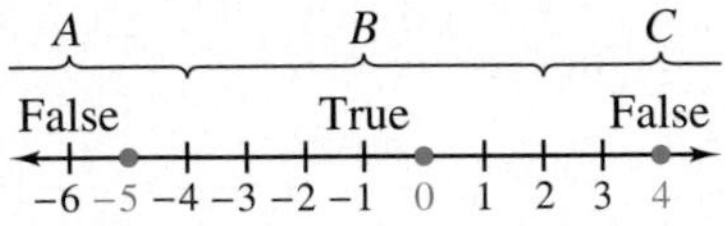

A B C

False True False

−6 −5 −4 −3 −2 −1 0 1 2 3 4

From Fig. 9.17 we see that the inequality is true in interval B. Thus the solution set is $(-4, 2)$.

Note: As with the graphing method, the test-point method only gives us intervals of solutions. Be sure to check the inequality symbol to determine whether the endpoints of those intervals are also solutions.

In Example 1 we illustrate both the graphing method and the test-point method.

EXAMPLE 1

Solving a Quadratic Inequality

Solve the quadratic inequality $3x - x^2 \leq -10$.

Solution

$-x^2 + 3x + 10 \leq 0$ Write the inequality in standard form.

Figure 9.18 shows the graph of $y = -x^2 + 3x + 10$. The x-intercepts appear to be $(-2, 0)$ and $(5, 0)$.

Figure 9.18

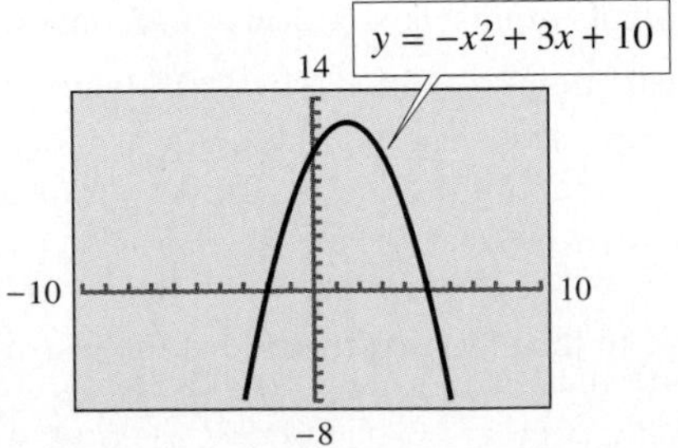

We can determine the x-intercepts algebraically by solving the associated quadratic equation.

$$\begin{aligned} -x^2 + 3x + 10 &= 0 \\ -1(x^2 - 3x - 10) &= 0 && \text{Factor out } -1. \\ -1(x - 5)(x + 2) &= 0 && \text{Factor the trinomial.} \\ x - 5 = 0 \quad \text{or} \quad x + 2 &= 0 && \text{Zero Factor Property} \\ x = 5 \quad \text{or} \quad x &= -2 && \text{Solve each case for } x. \end{aligned}$$

Because the solutions of the inequality are all x-values for which $y \leq 0$, we note the two intervals where the graph is on or *below* the x-axis. One interval is to the *left* of $(-2, 0)$, and the other interval is to the *right* of $(5, 0)$. Note that the x-intercepts also represent solutions.

Figure 9.19

−2 5

The solution set is the union of the two intervals of x-values: $(-\infty, -2] \cup [5, \infty)$. (See Fig. 9.19.)

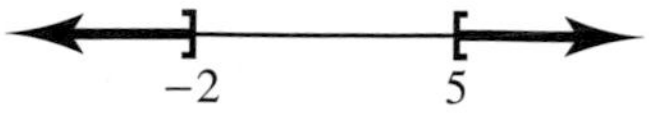

To use the test-point method, we plot the x-intercepts and thereby divide the number line into three intervals.

Think About It

The graphs of $y_1 = x^2 + 2x - 15$ and $y_2 = -x^2 - 2x + 15$ have the same x-intercepts, $(0, -5)$ and $(0, 3)$. Use the expressions for y_1 and y_2 to write inequalities that have the same solution set, $[-5, 3]$.

Figure 9.20

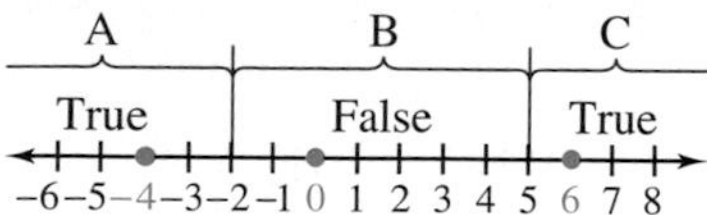

Figure 9.20 shows the results of testing a value (shown in blue) in each interval, and we can see that the solutions are in intervals A and C, including the endpoints of those intervals. Thus the solution set is $(-\infty, -2] \cup [5, \infty)$.

To solve a quadratic inequality, we use the factoring method, the Square Root Property, or the Quadratic Formula to determine the x-intercepts.

EXAMPLE 2

Solving Quadratic Inequalities

Solve the following inequalities.

(a) $1 > x(4 + x)$ (b) $12x < 4x^2 + 7$

Solution

(a)

$$\begin{aligned} 1 &> x(4 + x) \\ 1 &> 4x + x^2 && \text{Distributive Property} \\ -x^2 - 4x + 1 &> 0 && \text{Write the inequality in standard form.} \end{aligned}$$

Figure 9.21

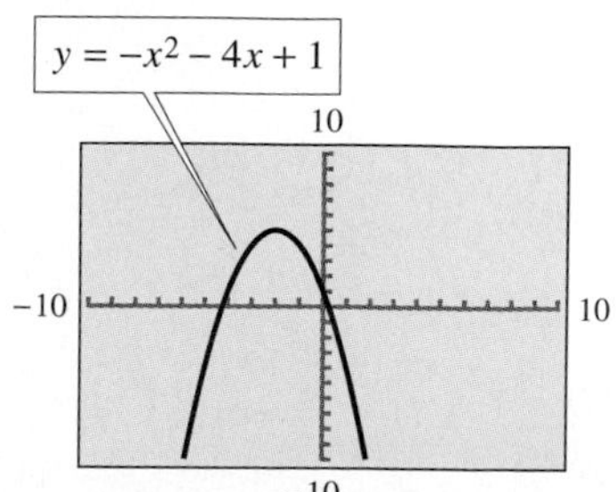

Figure 9.21 shows the graph of $y = -x^2 - 4x + 1$. The estimated x-intercepts are $(-4, 0)$ and $(0, 0)$.

To determine the actual x-intercepts of the graph, we solve the associated quadratic equation $-x^2 - 4x + 1 = 0$ with the Quadratic Formula.

$$x = \frac{4 \pm \sqrt{(-4)^2 - 4(-1)(1)}}{2(-1)}$$

$$x = \frac{4 \pm \sqrt{16 + 4}}{-2}$$

$$x = \frac{4 \pm \sqrt{20}}{-2} = -2 \pm \sqrt{5}$$

$$x \approx -4.24 \quad \text{or} \quad x \approx 0.24$$

Figure 9.22

We observe that the graph of the function is above the x-axis ($y > 0$) for x-values in the interval *between* the two x-intercepts. Therefore, the (approximate) solution set of the inequality is $(-4.24, 0.24)$. (See Fig. 9.22.)

(b) $$12x < 4x^2 + 7$$

$$-4x^2 + 12x - 7 < 0$$ Write the inequality in standard form.

Figure 9.23

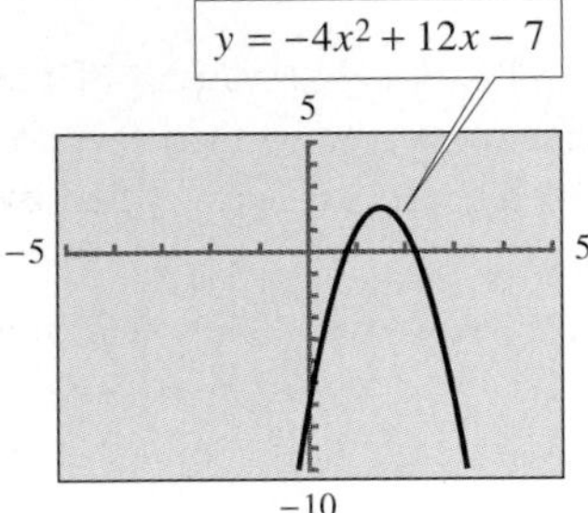

Figure 9.23 shows the graph of the function $y = -4x^2 + 12x - 7$. From the graph, we estimate the x-intercepts: $(1, 0)$ and $(2, 0)$.

To determine the x-intercepts algebraically, we solve the associated quadratic equation $-4x^2 + 12x - 7 = 0$ with the Quadratic Formula.

$$x = \frac{-12 \pm \sqrt{12^2 - 4(-4)(-7)}}{2(-4)}$$

$$x = \frac{-12 \pm \sqrt{144 - 112}}{-8}$$

$$x = \frac{-12 \pm \sqrt{32}}{-8} = \frac{3 \pm \sqrt{2}}{2}$$

$$x \approx 0.79 \quad \text{or} \quad x \approx 2.21$$

Figure 9.24

0.79 2.21

For $y < 0$, we note that the graph is *below* the x-axis to the *left* of $(0.79, 0)$ and to the *right* of $(2.21, 0)$. Therefore, an approximation of the solution set of the inequality is $(-\infty, 0.79) \cup (2.21, \infty)$. (See Fig. 9.24.)

Special Cases

LEARNING TIP

Rather than memorize special cases, use the graph to visualize the situation.

In each of the preceding examples, the solution set was an interval of x-values or the union of two intervals of x-values. The next example illustrates some special cases of quadratic inequalities that lead to different solution sets.

EXAMPLE 3

Special Cases of Quadratic Inequalities

Solve the inequalities.

(a) $x^2 + 3 \geq 0$ (b) $x^2 + 3 < 0$

(c) $x^2 + 6x + 9 \leq 0$ (d) $x^2 + 6x + 9 > 0$

Solution

(a) Figure 9.25 shows the graph of $y = x^2 + 3$. Because the entire graph lies *above* the x-axis, we can see that $x^2 + 3 \geq 0$ for all values of x. The solution set of the inequality is **R**.

(b) For $x^2 + 3 < 0$, Fig. 9.25 shows that there is no part of the graph *below* the x-axis. Therefore, the solution set is Ø.

Figure 9.25

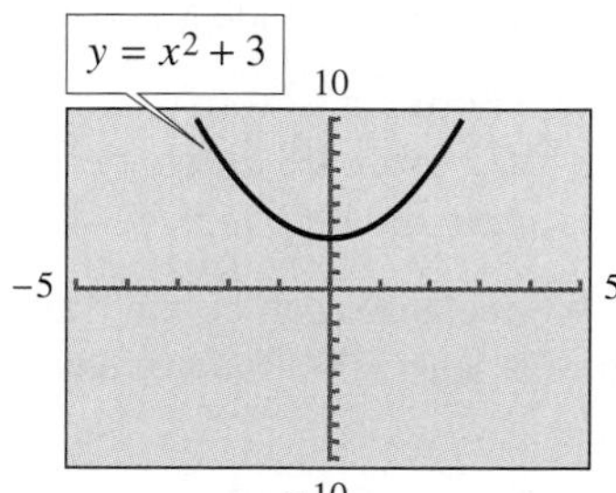

Figure 9.26

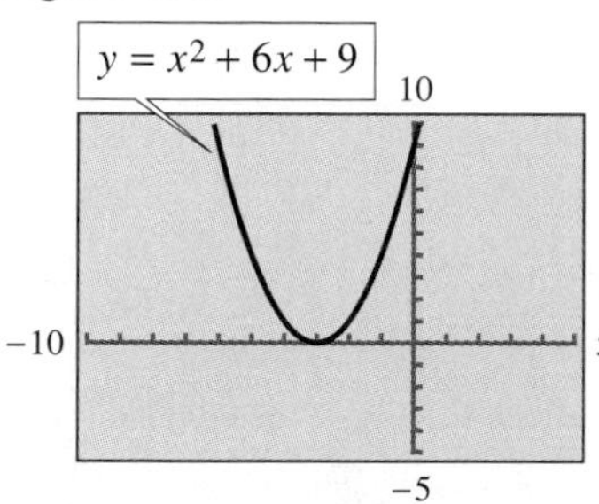

(c) Figure 9.26 shows the graph of $y = x^2 + 6x + 9$. We see that no part of the graph is *below* the x-axis. However, there is one point, the x-intercept, for which $x^2 + 6x + 9 = 0$. This point is $(-3, 0)$, which easily can be verified by substitution. The only solution of $x^2 + 6x + 9 \leq 0$ is -3.

(d) The graph in Fig. 9.26 lies entirely *above* the x-axis except for one point, the x-intercept. Therefore, every point of the graph except $(-3, 0)$ represents a solution of the inequality $x^2 + 6x + 9 > 0$. In interval notation, the solution set is $(-\infty, -3) \cup (-3, \infty)$.

Note: When we solve a quadratic inequality, we may find that the graph has no x-intercept. However, as part (a) of Example 3 shows, this does not imply that the inequality has no solution.

Real-Life Applications

Applications involving quadratic inequalities are characterized by such words and phrases as *less than* and *at least.*

EXAMPLE 4

Minimum Area of a Rectangle

According to a local election code, candidates must be provided with at least 8000 square feet of space outside a polling place for campaigning activities. If the poll manager has 400 feet of rope with which to enclose a rectangular area, what are the possible lengths of the space?

Solution

Let $L =$ the length of the rectangle and $W =$ the width of the rectangle.

$2L + 2W = 400$ — The perimeter of the rectangle is 400 feet.

$2W = 400 - 2L$ — Solve for W.

$W = 200 - L$

Figure 9.27

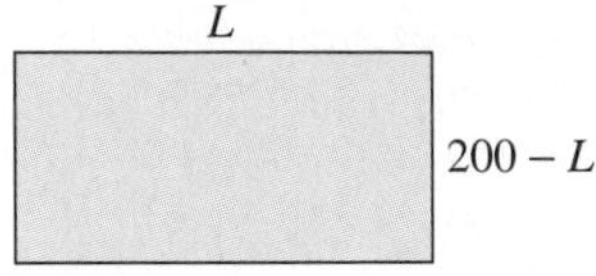

Figure 9.27 shows the rectangle with the dimensions labeled.

The area of the rectangle is given by the formula $A = LW$.

$$A = LW = L(200 - L) = -L^2 + 200L$$

$$-L^2 + 200L \geq 8000 \quad \text{The area must be at least 8000 square feet.}$$

$$-L^2 + 200L - 8000 \geq 0 \quad \text{Write the inequality in standard form.}$$

Figure 9.28

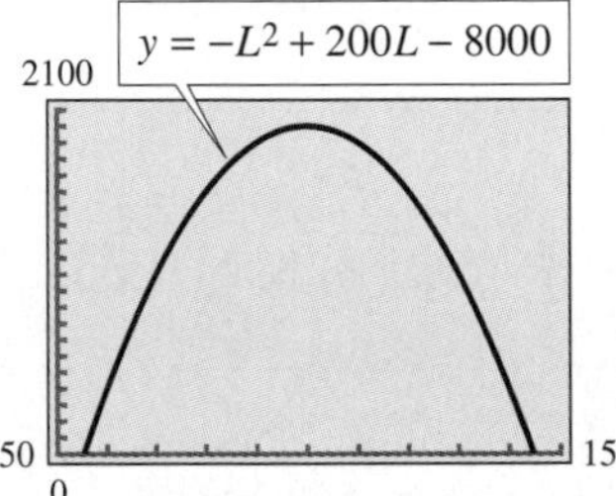

Figure 9.28 shows the graph of $y = -L^2 + 200L - 8000$. From the graph, we see that y is *at least* 0 for that portion of the graph between (and including) the x-intercepts.

Use the Quadratic Formula to verify that the solutions of $-L^2 + 200L - 8000 = 0$ are 55.28 and 144.72 (rounded to two decimal places). Therefore, the solution set of the inequality is approximately [55.28, 144.72].

The length of the rectangle can be any value between 55.28 feet and 144.72 feet to obtain an area of at least 8000 square feet.

9.6 QUICK REFERENCE

Graphing and Algebraic Methods

- A **quadratic inequality** is an inequality that can be written in the standard from $ax^2 + bx + c > 0$ (or $<, \leq, \geq$), where a, b, and c are real numbers and $a \neq 0$.
- The **solution set** of a quadratic inequality is the set of all replacements for the variable that make the inequality true.
- The following is a combined graphing and algebraic method for solving a quadratic inequality.
  1. Write the inequality in standard form.
  2. Produce the graph of $y = ax^2 + bx + c$.
  3. Determine the x-intercepts algebraically. For inequalities $y \leq 0$ or $y \geq 0$, these points represent solutions of the inequality.
  4. For inequalitites $y < 0$ or $y \leq 0$, determine the points where the graph is below the x-axis. For inequalities $y > 0$ or $y \geq 0$, determine the points where the graph is above the x-axis. The x-coordinates of all such points are solutions.
- To use the test-point method, plot the x-intercepts on a number line to divide the line into intervals. In each interval, test an x-value to determine whether the inequality is true or false. The solution set consists of the union of all intervals for which the inequality is true.

Special Cases

- When we solve a quadratic inequality, the graph of the associated quadratic function may have only one x-intercept $(a, 0)$, or it may have no x-intercept.
- In such special cases, the solution set may be **R**, Ø, a set containing just the one number a, or the set of all real numbers except the number a.

9.6 EXERCISES

Concepts and Skills

1. Suppose that you want to use a graph to assist you in solving a quadratic inequality, and you have already determined the x-intercepts. Describe the next step you will take to write the solution set.

2. To solve a quadratic inequality, we begin by determining the x-intercepts of the graph of the associated function. What algebraic methods can be used? Which of these methods will always work?

In Exercises 3–6, for what values of x is the given expression positive?

3. $(3 - x)(1 + x)$
4. $2x^2 + 5x - 3$
5. $5x^2 - 8x$
6. $6 - x^2 + 3x$

In Exercises 7–10, for what values of x is the given expression negative?

7. $x^2 + 10 - 7x$
8. $12 + 4x - x^2$
9. $3 + x - 2x^2$
10. $x^2 - 3x - 6$

11. Suppose that you want to use a graph to assist you in solving $2x^2 + 7 < -3x$. What first step must you take if you want to produce a single graph rather than the graphs of both sides? What is the function whose graph you will produce?

12. For a quadratic inequality, suppose that the graph of the associated function has two x-intercepts. Describe the possible intervals of x-values that comprise the solution set.

In Exercises 13–18, use a graph to estimate the solution set of the given inequality.

13. $9 - x^2 < 0$
14. $x(x + 3) \geq 0$
15. $5 - 4x^2 \geq 8x$
16. $x^2 > 2(x + 3)$
17. $x(7 - x) \geq 8$
18. $4x - 2x^2 + 3 < 0$

In Exercises 19–26, solve the inequality.

19. $(x - 2)(x + 5) > 8$
20. $5 + 4x - x^2 \geq 0$
21. $x^2 \leq 5x$
22. $4x^2 < 7$
23. $x(12x + 5) \geq 2$
24. $x(4x + 9) > 9$
25. $2x(x - 2) < 7$
26. $x^2 + 2x \geq 7$

In Exercises 27–30, match the inequality in column B with the sentence in column A.

	Column A	*Column B*
27.	A number does not exceed 5.	(a) $x > 5$
28.	A number is at least 5.	(b) $x \leq 5$
29.	A number is more than 5.	(c) $x - 5 < 0$
30.	A number is less than 5.	(d) $5 \leq x$

In Exercises 31–38, translate the given information into an inequality and then solve it.

31. The product of x and $x + 1$ is at least 6.
32. The product of $x + 3$ and $x - 1$ is less than 12.
33. The product of x and $x - 1$ is not more than 6.
34. The sum of $3x^2$ and $5x$ is less than 2.
35. Six less than the product of x and $6x - 5$ is negative.
36. The difference of 6 and $5x^2 + 7x$ is positive.
37. Seven more than the product of $2x$ and $2 - x$ is not positive.
38. The difference of $4x + 7$ and $2x^2$ is not negative.

In Exercises 39–42, for what values of x is the function nonnegative?

39. $g(x) = (4 - x)(2 + x)$
40. $f(x) = x^2 + 3 - 4x$
41. $g(x) = 4x^2 - 7x$
42. $h(x) = 8 - 5x^2$

In Exercises 43–46, for what values of x is the function nonpositive?

43. $f(x) = 16 - x^2$
44. $h(x) = 3(1 - x^2) - 8x$
45. $g(x) = x(x - 8) - 5$
46. $f(x) = 9x^2 + 6x - 8$

47. Suppose that you use a graph to solve a quadratic inequality and you find that the graph has no x-intercepts. List all the possible solution sets of the inequality.

48. Suppose that you use a graph to solve a quadratic inequality and you find that the graph has only one x-intercept, (3, 0). List all the possible solution sets of the inequality.

49. Suppose that $x - 2$ is a factor of the expression $ax^2 + bx + c$. Explain whether 2 can be a solution of the inequality $ax^2 + bx + c > 0$.

50. The solution set of $x^2 - x - 6 < 0$ is the interval $(-2, 3)$. Explain how to determine the solution set of $x^2 - x - 6 \geq 0$.

In Exercises 51–76, solve the inequality.

51. $4x^2 + 23x + 15 \geq 0$

52. $x^2 + 7x + 6 < 0$

53. $5x - x^2 < 0$

54. $25 - x^2 \geq 0$

55. $x^2 + 3x + 6 < 0$

56. $3x - 2x^2 > 7$

57. $4x^2 + 20x + 25 \leq 0$

58. $6x^2 - 41x + 70 \leq 0$

59. $x^2 - 4x \geq 4$

60. $x^2 + 2x + 5 \leq 0$

61. $x^2 + 9 \leq 6x$

62. $x^2 + 16 \leq 8x$

63. $2x^2 < 4x + 5$

64. $2x^2 \geq 5x + 4$

65. $x^2 + 5 \geq 6x$

66. $5x - x^2 < 8$

67. $0.5x - 0.25x^2 \leq 0.9$

68. $0.3x^2 - 0.7x + 0.9 \leq 0$

69. $x(6x + 17) \leq -7$

70. $3x(2 - 3x) \geq 1$

71. $x^2 + \frac{3}{4}x > \frac{5}{2}$

72. $\frac{x^2}{2} + x \geq 3$

73. $3x(x + 2) < 4$

74. $2x(3 - x) < 9$

75. $(2x + 1)(x + 2) \leq (x - 3)(x + 2)$

76. $(2x + 3)(x + 1) > 2(2x + 3)(1 - x)$

In Exercises 77–80, use the discriminant to determine all values of k so that the solution set of the quadratic inequality is **R**.

77. $2x^2 - 3x + k > 0$

78. $kx^2 + 1 - 3x > 0$

79. $9 + kx - x^2 > 0$

80. $7x^2 - kx > 0$

In Exercises 81–84, use the discriminant to determine all values of k so that the real number solution set of the quadratic inequality is Ø.

81. $x^2 + 5x + k < 0$

82. $kx^2 + 2x + 1 < 0$

83. $kx^2 + 8x - 5 > 0$

84. $5x^2 - 2x + k < 0$

In Exercises 85–88, use the discriminant to determine all values of k so that the real number solution set of the quadratic inequality is a set with only one element.

85. $x^2 + 12x + k \leq 0$

86. $x^2 + 3x + k \leq 0$

87. $x^2 + kx + 9 \leq 0$

88. $4x^2 + kx + 1 \leq 0$

89. Suppose that two numbers differ by 5. If the product of the numbers is at most 6, what is the largest possible value of the smaller number?

90. Consider a circle with radius r. For what values of r does the numerical value of the circumference of the circle exceed the numerical value of the area?

Real-Life Applications

91. A builder plans to construct a home on a lot that is 100 feet wide. The convenants for the subdivision include the following provisions.
   (a) The front of a home must be at least 35 feet longer than its depth.
   (b) No home can be less than 15 feet from any property line.
   (c) All homes must have more than 2556 square feet of ground-floor living space.

   Show why the builder cannot construct the home.

92. The Woodstock Dairy Company wants to design a holding tank for chocolate milk. The tank is to be in the shape of a right circular cylinder 9.5 feet high. What should the radius of the tank be so that it holds at least 10,000 gallons? (1 gallon = 231 cubic inches.)

93. The Mathis Package Company wants to design a box that is 11 inches high with a base perimeter of 80 inches. The box's volume needs to be at least 2 cubic feet. What should be its length?

94. An army unit fired a mortar shell over a hill 112 feet high. The height h of the shell was $h = -16t^2 + 128t$, where t was the time of flight. To the nearest tenth of a second, during what time interval was the shell at least 30 feet higher than the hill?

Group Project: Heart Disease

The accompanying figure shows deaths (in thousands) due to heart disease in the United States for selected years in the period 1985–1993. (Source: U.S. National Center for Health Statistics.)

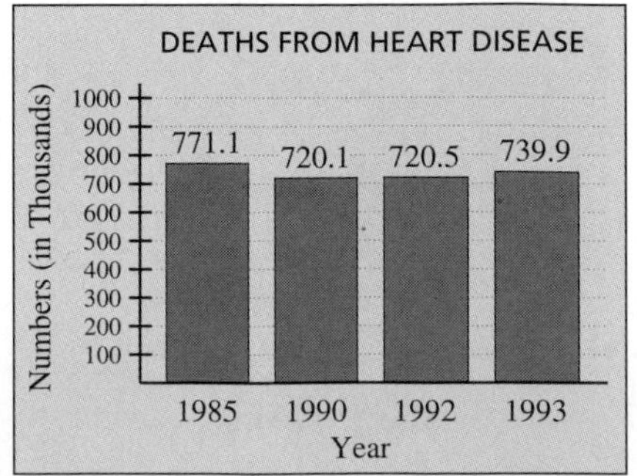

95. Use the quadratic regression capability of your calculator to create a model function $H(x)$ for the data in the bar graph. (Let x represent the number of years since 1985.) Then, using 0 to 20 along the x-axis and 0 to 1000 along the y-axis, produce the graph of your model function.

96. Produce the graph of $y = 750$ in the same graphing window and trace to estimate the x-coordinates of the points of intersection. (Round your estimates to the nearest whole number.)

97. To determine the points of intersection, use the Quadratic Formula to solve the equation $H(x) = 750$. (Round your results to the nearest whole number.) Using these results and the graph, determine the period of years in which, according to the model, the number of deaths due to heart disease was no more than 750,000.

98. In addition to poor health practices, heart disease is often attributed to work-related stress. What are some of the stress factors that have been particularly prevalent in the 1990s?

Challenge

99. Produce the graph of $g(x) = x^2 - 2x - 15$.

(a) By looking at the graph of g, explain how you would estimate the domain of

$$f(x) = \sqrt{x^2 - 2x - 15}$$

(b) By looking at the graph of g, what do you think the graph of $h(x) = |x^2 - 2x - 15|$ would look like? Why?

100. For $ax^2 > 0$ $(a \neq 0)$, use the discriminant to prove that the solution set cannot be **R**.

101. For nonzero real numbers a and b, use the discriminant to prove that the solution set of $ax^2 + bx > 0$ cannot be **R**.

In Exercises 102 and 103, determine all values of k so that the solution set of the quadratic inequality is **R**.

102. $x^2 + kx + 4 > 0$

103. $9x^2 + kx + 1 \geq 0$

In Exercises 104 and 105, determine all values of k so that the real number solution set of the quadratic inequality is $\varnothing$.

104. $3x^2 + kx - 1 > 0$

105. $-2x^2 + kx - 2 > 0$

In Exercises 106 and 107, determine all values of k so that the real number solution set of the quadratic inequality is a set with only one element.

106. $kx^2 + 3x + (k - 1) \geq 0$

107. $2kx^2 + x + (k - 1) \leq 0$

9.7 HIGHER-DEGREE AND RATIONAL INEQUALITIES

Higher-Degree Inequalities • Rational Inequalities • Real-Life Applications

Higher-Degree Inequalities

In general, any polynomial inequality can be solved in a manner similar to that used for solving quadratic inequalities. To solve a polynomial inequality such as $P(x) > 0$, we can produce the graph of $P(x)$, estimate the x-intercepts, if any, and observe the interval(s) of x-values for which the graph lies above the x-axis. Alternatively, we can use the test-point method described in Section 9.6. Similar procedures are used for $\leq$, $<$, and $\geq$ inequalities.

For quadratic inequalities it is always possible to determine the exact x-intercepts by solving the associated equation $P(x) = 0$. For polynomial inequalities of higher degree, this step will not be possible unless the polynomial can be factored.

This is one instance in which the power of the graphing method is evident. In real-data applications, higher-degree polynomials are not often factorable, and graphing emerges as the best method for routine problem solving.

EXAMPLE 1

Solving Higher-Degree Polynomial Inequalities

Solve the following inequalities.

(a) $x^3 + 2x^2 - 15x < 0$

(b) $2x^4 - 5x^3 + x - 1 \leq 0$

Solution

Figure 9.29

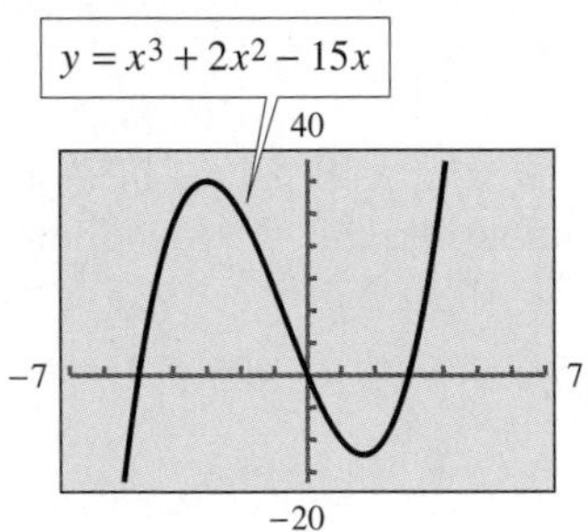

(a) Figure 9.29 shows the graph of the polynomial function $y = x^3 + 2x^2 - 15x$. From the graph, we estimate the x-intercepts to be $(-5, 0)$, $(0, 0)$, and $(3, 0)$.

Because $x^3 + 2x^2 - 15x$ is factorable, we can determine the x-intercepts by solving the associated equation.

$$x^3 + 2x^2 - 15x = 0$$
$$x(x^2 + 2x - 15) = 0$$
$$x(x + 5)(x - 3) = 0$$
$$x = 0 \quad \text{or} \quad x + 5 = 0 \quad \text{or} \quad x - 3 = 0$$
$$x = 0 \quad \text{or} \quad x = -5 \quad \text{or} \quad x = 3$$

Figure 9.30

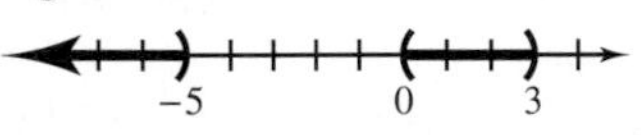

The graph lies *below* the x-axis to the *left* of $(-5, 0)$ and *between* $(0, 0)$ and $(3, 0)$. Therefore, the solution set of the inequality is $(-\infty, -5) \cup (0, 3)$. (See Fig. 9.30.)

To use the test-point method, we plot the x-intercepts, which divide the number line into four intervals. (See Fig. 9.31.)

Think About It

Why is it that the solution set of $x^3(x + 4)^2 \leq 0$ is a single interval, but the solution set of $x^3(x - 4)^2 \leq 0$ is not?

Figure 9.31

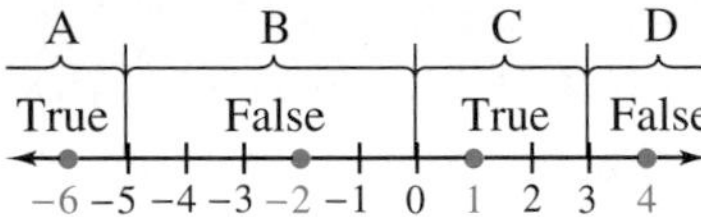

By testing the indicated x-values (shown in blue) in each interval, we find that the solutions of the inequality lie in intervals A and C. Thus the solution set is the union of those intervals: $(-\infty, -5) \cup (0, 3)$.

Figure 9.32

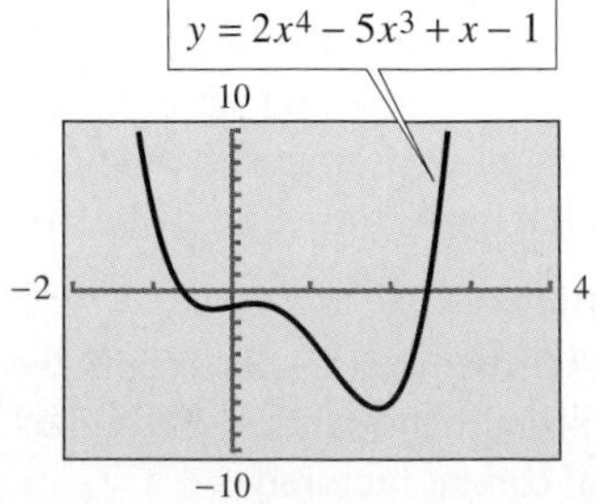

(b) Figure 9.32 shows the graph of the function $y = 2x^4 - 5x^3 + x - 1$. Because $2x^4 - 5x^3 + x - 1$ is not factorable, we will not be able to determine the x-intercepts algebraically. This means we must be more careful in our estimates of the x-intercepts on the graph.

By zooming in on the two x-intercepts and rounding to the nearest hundredth, we obtain the points $(-0.64, 0)$ and $(2.45, 0)$. Note that the x-intercepts represent solutions of the inequality.

Figure 9.33

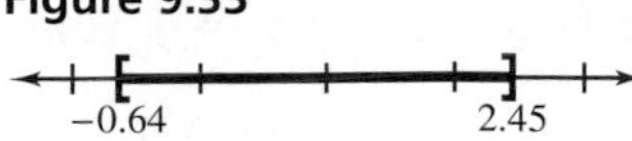

Other solutions are represented by points that are *below* the x-axis, that is, *between* the two x-intercepts. The approximate solution set is $[-0.64, 2.45]$. (See Fig. 9.33.)

Rational Inequalities

We can use similar graphing or test-point procedures to solve inequalities with rational expressions. However, we must remember that any restricted values must be excluded from the solution set.

EXAMPLE 2

Solving Rational Inequalities

Solve the following rational inequalities.

(a) $\dfrac{2}{x-3} \geq 1$ (b) $\dfrac{x}{x-16} < \dfrac{2}{6-x}$

Solution

(a) $\dfrac{2}{x-3} \geq 1$ Note that 3 is a restricted value.

$\dfrac{2}{x-3} - 1 \geq 0$ We write the inequality with 0 on one side.

Figure 9.34 shows the graph of $y = \dfrac{2}{x-3} - 1$. From the graph we estimate the one x-intercept to be $(5, 0)$. We can use substitution to verify this.

Figure 9.34

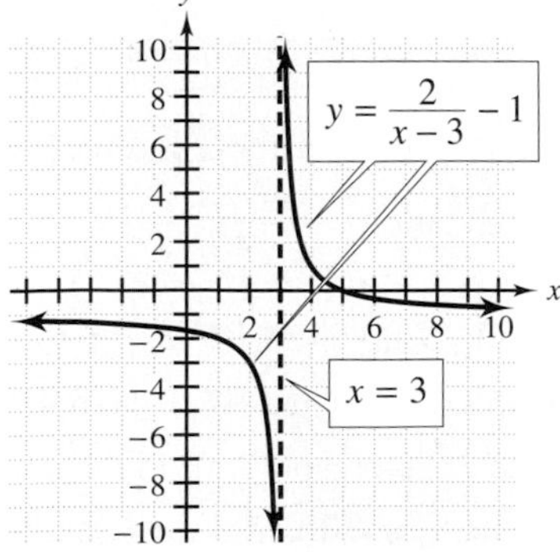

The dashed vertical line in Fig. 9.34 indicates that 3 is a restricted value. The graph lies *above* the x-axis in the interval from the vertical line $x = 3$ to the x-intercept $(5, 0)$. The right endpoint of the interval 5 represents a solution, but the left endpoint 3 does not because it is a restricted value. Therefore, the solution set of the inequality is $(3, 5]$. (See Fig. 9.35.)

Figure 9.35

3 5

LEARNING TIP

Depending on the inequality symbol, x-intercepts may represent solutions. However, restricted values are never solutions.

Figure 9.36

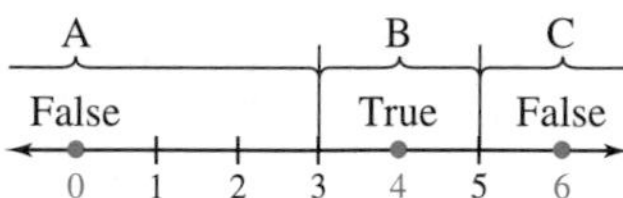

To use the test-point method, we solve the associated equation $\frac{2}{x-3} = 1$, from which we learn that the x-intercept is (5, 0). The x-intercept and the restricted value 3 divide the number line into three intervals, as shown in Fig 9.36.

By testing the indicated x-values (shown in blue) in each interval, we find that the solutions lie in interval B. As before, 5 is a solution, but 3 is not. Thus the solution set is (3, 5].

(b) $\frac{x}{x-16} < \frac{2}{6-x}$ Note that 16 and 6 are restricted values.

$\frac{x}{x-16} - \frac{2}{6-x} < 0$ Write the inequality with 0 on one side.

Figure 9.37 shows the graph of $y = \frac{x}{x-16} - \frac{2}{6-x}$. Note the two dashed vertical lines indicating the restricted values 6 and 16.

Figure 9.37

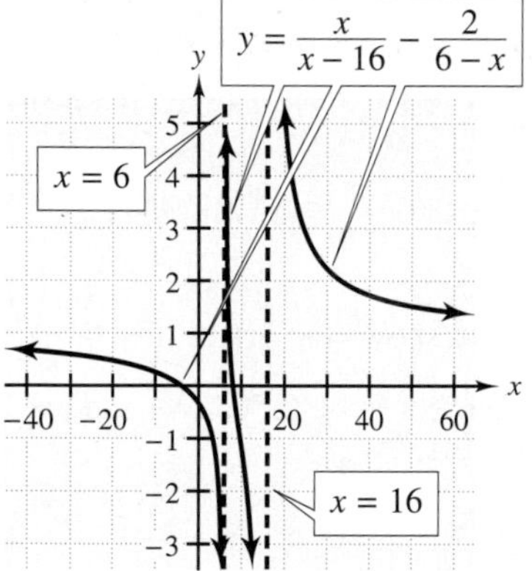

We can determine that the x-intercepts are (−4, 0) and (8, 0) by solving the associated equation $\frac{x}{x-16} = \frac{2}{6-x}$.

All points of the graph lie *below* the x-axis in the intervals

(i) from the x-intercept (−4, 0) to the vertical line at $x = 6$ and

(ii) from the x-intercept (8, 0) to the vertical line at $x = 16$.

Therefore, the solution set of the inequality is $(-4, 6) \cup (8, 16)$. (See Fig. 9.38.)

Figure 9.38

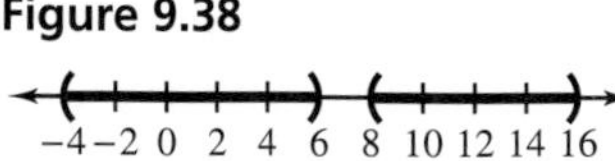

Note: In Example 2 we began each part by writing the inequality with 0 on one side. It is not essential to do this, but instead of producing the graphs of both sides of the inequality, it is often easier to analyze just one graph with respect to the x-axis.

Real-Life Applications

A higher-degree inequality or a rational inequality may be needed to describe the conditions in an application problem.

EXAMPLE 3

Allowable Dimensions of a Box

A wholesale florist ships cut flowers in large, elongated, rectangular cartons. For stability, the carton must be 3 feet longer than the width, and the height must be 1 foot greater than the width. (See Fig. 9.39.) The shipper will not handle boxes of more than 12 cubic feet. Use a graph to estimate the allowable widths of the box.

Figure 9.39

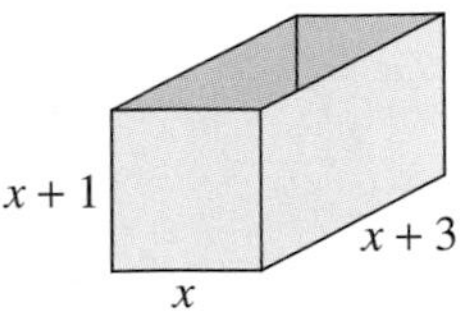

Solution

Let x = width of the box, $x + 3$ = length of the box, and $x + 1$ = height of the box.

The volume V is determined by the formula $V = LWH$.

$$x(x + 3)(x + 1) \leq 12 \quad \text{The volume cannot exceed 12 cubic feet.}$$

$$x(x + 3)(x + 1) - 12 \leq 0 \quad \text{Write the inequality with 0 on one side.}$$

Figure 9.40 shows the graph of the function $y = x(x + 3)(x + 1) - 12$. From the graph we estimate the x-intercept to be (1.25, 0).

Figure 9.40

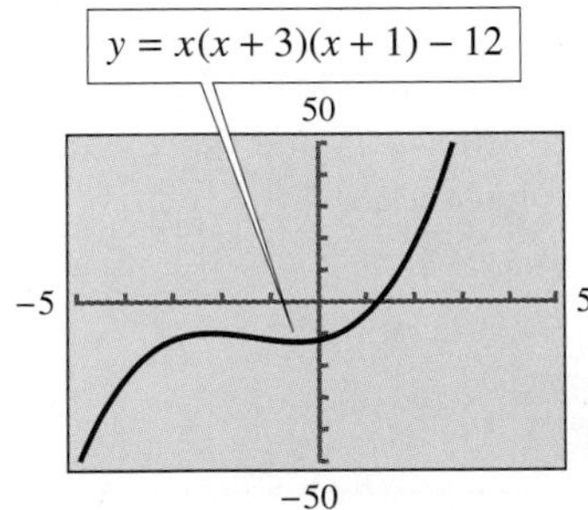

Because the inequality is of the form $y \leq 0$, we look for points of the graph that are *on* or *below* the x-axis. These are the x-intercept and all points to the *left* of the x-intercept. However, because x represents the width of the box, x must be positive. Therefore, the valid interval of x-values is between the y-axis and the x-intercept.

The solution set is (0, 1.25]. The width of the box can be any positive number up to and including 1.25 feet.

EXAMPLE 4

Using a Rational Inequality to Calculate Election Returns

An incumbent county commissioner can count on 1000 votes in the upcoming election. The challenger, who needs more than 50% of the vote to win, can count on 800 votes. Polls show that three-fourths of the remaining uncommitted votes are leaning toward the challenger, and the other one-fourth are leaning toward the incumbent.

For this reason, the challenger hopes that a great many of the currently uncommitted voters show up on election day. If the polls are completely accurate, at least how many of the uncommitted voters must vote to ensure the challenger's election?

Solution

If x = the number of uncommitted voters who do vote, then $800 + \frac{3}{4}x$ = the number of votes the challenger receives.

Because the total number of votes cast is $1800 + x$, the fraction of votes cast for the challenger is given by

$$\frac{\text{Votes for challenger}}{\text{Total votes cast}} = \frac{800 + \frac{3}{4}x}{1800 + x}$$

To win, the challenger's percentage must be more than 50%.

$$\frac{800 + \frac{3}{4}x}{1800 + x} > 0.50$$

$$\frac{800 + \frac{3}{4}x}{1800 + x} - 0.50 > 0$$

Figure 9.41

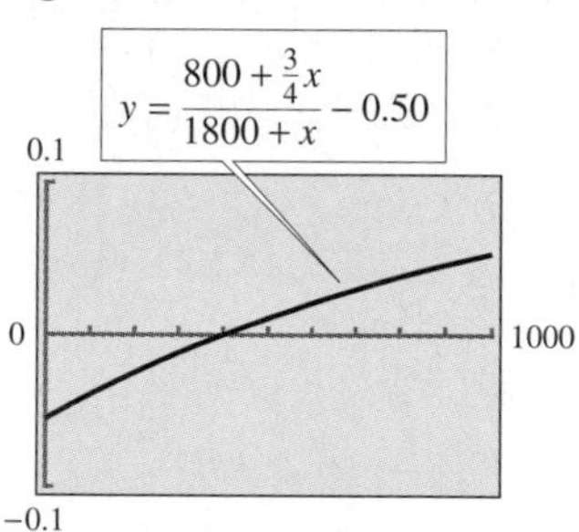

Figure 9.41 shows the graph of $y = \frac{800 + \frac{3}{4}x}{1800 + x} - 0.50$.

We estimate the x-intercept to be (400, 0). This can be easily verified by storing 400 for x and evaluating the function.

All points of the graph lie *above* the x-axis from the x-intercept to the *right*. The solution set is $(400, \infty)$. For the challenger to win, more than 400 of the uncommitted voters must vote.

9.7 QUICK REFERENCE

Higher-Degree Inequalities

- To solve a polynomial inequality $P(x) > 0$, produce the graph of $P(x)$, estimate the x-intercepts, if any, and observe the interval(s) of x-values for which the graph lies above the x-axis. Alternatively, use the test-point method. Similar procedures are used for $\geq$, $<$, and $\leq$ inequalities.
- Unless $P(x)$ is factorable, it will not be possible to solve the equation $P(x) = 0$ in order to determine the x-intercepts algebraically.

Rational Inequalities

- The following is a summary of the graphing and algebraic methods for solving a rational inequality $R(x) > 0$:
  1. Produce the graph of the function $R(x)$, and estimate the x-intercepts, if any. Determine the x-intercepts by solving the associated rational equation $R(x) = 0$, or verify by substitution.
  2. Determine the interval(s) of x-values for which the graph is above the x-axis. Use the interval(s) to write the solution set of the inequality. *Exclude restricted values.*

3. Similar procedures are used for $\geq$, $<$, and $\leq$ inequalities. Except for restricted values, the solution set includes endpoint x-values for $\geq$ and $\leq$ inequalities.

- Alternatively, solve the associated equation to determine the x-intercept(s). Use the x-intercept(s) and any restricted values to divide the number line into intervals. Then use the test-point method.

9.7 EXERCISES

Concepts and Skills

1. Suppose you use a graph to solve the inequality $P(x) < 0$, where $P(x)$ is a polynomial of degree greater than 2. From the graph you can estimate the x-intercepts. Under what conditions would it not be possible to determine the x-intercepts algebraically?

2. Let $P(x) = x^3 - x^2 - 12x + k$, where k is a real number. Use a graph to estimate the solution set of $P(x) > 0$ when $k = 15$. What is the effect on the solution set if k is increased to 25?

In Exercises 3–16, solve the inequality. Determine all interval endpoints algebraically.

3. $(x - 3)(x + 2)(x - 1) > 0$
4. $3x(x - 3)(2x + 9) < 0$
5. $(x + 2)^2(x - 3)(x + 6) \leq 0$
6. $x^2(x + 4)(8 - x) \geq 0$
7. $x^3 - 3x^2 > 0$
8. $2x^7 - x^6 \leq 0$
9. $x^5 + 6x^4 + 5x^3 < 0$
10. $x^6 - 7x^5 + 6x^4 \geq 0$
11. $x^4 - 13x^2 + 36 > 0$
12. $x^4 - 3x^2 - 4 \geq 0$
13. $x^4 - 6x^2 + 5 \leq 0$
14. $x^4 + 2x^2 - 15 < 0$
15. $x^3 + 3x^2 - x - 3 > 0$
16. $9x^3 - 45x^2 - 16x + 80 \leq 0$

In Exercises 17–22, solve the given inequality. Verify all interval endpoints by substitution.

17. $x^3 - 4x^2 + x + 6 > 0$
18. $x^3 + 5x^2 + 2x - 8 < 0$
19. $x^3 + 6x^2 + 3x - 10 \leq 0$
20. $x^3 - 3x^2 - 6x + 8 \geq 0$
21. $x^4 + 6x^3 + 7x^2 - 6x - 8 < 0$
22. $x^4 + 5x^3 + 5x^2 - 5x - 6 \geq 0$

23. Suppose that you are solving the rational inequality $R(x) \geq 0$. Explain how it is possible that every point of the graph of $R(x)$ is above the x-axis between the x-values a and b and yet a and b are not solutions.

24. Suppose that you are solving the rational inequality $R(x) \leq 0$, where m and n are restricted values. You produce the graph of $R(x)$ and find that there is one x-intercept $(a, 0)$ where $m < a < n$. From the following list, identify the two sets that could *not* be the solution set of the inequality and explain why.

(a) (m, n) (b) (a, n) (c) $(m, a]$
(d) $[m, a]$ (e) $\{a\}$ (f) $[a, n)$

In Exercises 25–34, solve the inequality.

25. $\dfrac{-2x}{x + 1} \geq 0$
26. $\dfrac{x - 2}{x + 5} > 0$
27. $\dfrac{x - 6}{x + 1} < 3$
28. $\dfrac{3}{4 - x} \geq 1$
29. $\dfrac{4 - 3x}{4x^2 + 1} > 0$
30. $\dfrac{2x + 5}{(x - 1)^2} \leq 0$
31. $\dfrac{x - 2}{x + 4} > \dfrac{4}{x - 3}$
32. $\dfrac{x + 3}{4 - x} < \dfrac{4}{x + 2}$
33. $\dfrac{x}{x + 3} \geq 2x$
34. $\dfrac{x - 2}{x(3 + x)} \leq 2$

In Exercises 35 and 36, for what values of x is the expression nonnegative?

35. $x(3 - x)(3x + 4)$
36. $\dfrac{x + 1}{(2x - 1)(x - 4)}$

In Exercises 37 and 38, for what values of x is the expression nonpositive?

37. $\dfrac{2x - 5}{x^2 + 9}$
38. $(4 - x)(x + 3)(x - 2)$

In Exercises 39–42, write the inequality and solve.

39. The quotient of $5 - 2x$ and $3x$ is at least $2x - 3$.

40. Three divided by $x - 4$ is not greater than 2.

41. The difference of $\frac{x}{x+2}$ and $\frac{x-1}{x+1}$ is negative.

42. The product of $2x + 1$ and $x - 3$ divided by $x - 5$ is positive.

In Exercises 43 and 44, for what values of x is the function positive?

43. $f(x) = x^2(3 - x)(x + 5)$

44. $g(x) = \frac{x+3}{x(x-1)} - 1$

In Exercises 45 and 46, for what values of x is the function negative?

45. $h(x) = \frac{x^2 - 9x + 20}{x^2 - 9}$

46. $g(x) = (x + 1)(x - 3)^2(x + 5)$

47. Suppose that you are solving the rational inequality $\frac{x-3}{x+1} > 0$. Would it be correct to conclude that $x - 3$ and $x + 1$ must both be positive in order for the fraction to be positive? Why or why not?

48. Suppose you are using a graph to solve a rational inequality. You notice that the displayed graph includes a vertical line at $x = 4$. Is the line part of the graph? Explain the significance of the line.

In Exercises 49–70, solve the inequality.

49. $\frac{1}{x} < \frac{-2}{3}$

50. $\frac{3}{x} \geq \frac{2}{5}$

51. $(x - 5)(3 - x)(2 + x) \geq 0$

52. $5x(2x - 1)(4 - x) \leq 0$

53. $\frac{x+5}{x-2} \leq \frac{x}{x-2} - 3$

54. $\frac{x-6}{x+2} \geq \frac{5}{x+2} - 2$

55. $(2x - 5)(x + 1)^2 > 0$

56. $x^3(x - 1)^2(5 - x) > 0$

57. $\frac{(x+2)(7-x)}{x-5} < 0$

58. $\frac{x+4}{(1-3x)(5-x)} \geq 0$

59. $(x^2 + 1)(2x - 1) \leq 0$

60. $x^4 \leq 1$

61. $\frac{x+2}{x+3} - \frac{x-1}{x-2} \leq 1$

62. $\frac{1-x}{x+3} + \frac{x}{x-1} \geq 2$

63. $x^4 - 5x^2 > 36$

64. $x^3 + 2x^2 < 16x + 32$

65. $\frac{x^2(2x+1)(x-3)}{(x+1)^3(x-1)} \leq 0$

66. $\frac{(3x-2)^2(x+4)(x+1)}{(x-3)(x+2)^3} > 0$

67. $x^4 < 2x^5$

68. $x^5 + 9x^4 + 8x^3 \geq 0$

69. $\frac{8x^2 - 23x - 3}{2x^2 + 1} \geq 1$

70. $\frac{(2x-3)^2}{x^2 + x + 5} \leq 1$

71. For the inequality $\frac{x^2}{(x-2)^2} \geq 0$, would it be correct to state that x^2 and $(x - 2)^2$ are nonnegative for *any* replacement for x; therefore, the solution set is the set of all real numbers? Why or why not?

72. The inequality $x(x - 1)^2 < 0$ states that the product of x and $(x - 1)^2$ is negative. Why does the factor x determine the sign of the product, whereas the factor $(x - 1)^2$ does not?

In Exercises 73–78, solve the given inequality by inspection.

73. $x(x - 5)^2 < 0$

74. $(x - 1)^4(x + 1)^2 > 0$

75. $\frac{x}{(x+2)^4} > 0$

76. $\frac{x^4}{x-1} \leq 0$

77. $\frac{x^2}{x+3} > 0$

78. $\frac{-2x^4}{(x+1)^2} \leq 0$

In Exercises 79–82, solve the inequality.

79. $\frac{2x^2 - 5x - 7}{x^2 - 1} \geq 0$

80. $(x + 1)(2 - 3x)^2 > 0$

81. $x^2(x + 6)(x - 2) \geq 0$

82. $3x^3(x + 2)^4 \geq 0$

83. What are three consecutive odd integers whose product is at least 105?

84. The *successor* of an integer is one more than the integer. For example, the successor of 8 is 9, and the successor of -5 is -4. What is the smallest positive integer for which the quotient of the integer and its successor is at least 2? Explain how a graph can be used to support your conclusion.

Real-Life Applications

85. A candidate for a certain Marine unit must complete a 5-mile course in 93 minutes or less. If the candidate anticipates that the average speed on the second part of the course, a 2-mile obstacle course, will be 1.5 miles per hour less than the speed on the first portion of the course, for what average speed on the first portion of the course will the candidate qualify?

86. The speed of the current in the Hillsborough River is 2 mph. At what speed would a canoeist have to paddle to go upstream 6 miles and back in less than 3 hours?

87. A fish hatchery that produces trout for stocking mountain streams needs a holding tank with a capacity of at least 225 cubic feet of water. The sum of the depth and width must be 10 and the length must be 1 foot less than twice the depth. What are the possible depths of the tank?

88. The publisher of a popular magazine determines that for x thousand copies of the magazine the cost C (in dollars) for publishing is modeled by $C = 600x + 12{,}000$, and the revenue R (in dollars) from sales is modeled by $R = 6x^3 + 120x^2$. For how many copies will the publisher break even or make a profit?

Group Project: Cost of Preparing Tax Returns

In 1996 the average cost to prepare a tax return was \$79.40. However, as the accompanying table shows, the average cost varied with income. (Source: Bruskin/Goldring Research.)

Income Range	Average Cost
Less than \$20,000	\$ 52.40
\$20,000–\$29,999	70.30
\$30,000–\$39,999	60.50
\$40,000–\$49,999	88.30
\$50,000 and over	118.80

The function $c(x) = 0.0025x^3 - 0.22x^2 + 6.82x - 7.2$ models the cost c of preparing a tax return for income x (in thousands of dollars).

89. Write an inequality to determine the income range that would result in an average preparation cost in excess of \$100.

90. Use a graph to estimate the solution of the inequality in Exercise 89.

91. For what incomes is the average cost between \$60 and \$80?

92. Some Libertarians argue that the federal tax code is unconstitutional because no single individual can possibly know all the tax laws. What is your opinion of this? Name two professions that would likely resist simplifying the tax code.

Challenge

93. Produce the graph of $y = (x - 3)(x + 2)(x - 1)$. Explain how you would use the graph to determine the domain of the radical function

$$f(x) = \sqrt{(x - 3)(x + 2)(x - 1)}$$

In Exercises 94–98, solve.

94. $x|x - 4| \geq 0$

95. $|x(x - 4)| \geq 0$

96. $x^2 - 3x \leq |x - 2|$

97. $x^3 - 1 < 0$

98. $\dfrac{x+1}{x-1} \leq \dfrac{x+4}{2x+1} + \dfrac{6x}{2x^2 - x - 1}$

In Exercises 99–102, compare the solution sets of each pair of inequalities.

99. (a) $(2x - 1)^2(x + 4)^4 \geq 0$
(b) $(2x - 1)^2(x + 4)^4 > 0$

100. (a) $(2x - 1)^2(x + 4)^4 \leq 0$
(b) $(2x - 1)^2(x + 4)^4 < 0$

101. (a) $\dfrac{x}{(x-3)^2} < 0$
(b) $\dfrac{x}{(x-3)^2} \leq 0$

102. (a) $\dfrac{x}{(x-3)^2} \geq 0$
(b) $\dfrac{x}{(x-3)^2} > 0$

9.8 QUADRATIC FUNCTIONS

Properties of Graphs • Variations of the Graph of $f(x) = x^2$ • Vertex and Intercepts • Real-Life Applications

Properties of Graphs

Recall that a quadratic function is a function of the form $f(x) = ax^2 + bx + c$, where a, b, and c are real numbers and $a \neq 0$. Throughout this chapter we have been using graphs of quadratic functions to assist us in solving quadratic equations and inequalities. In this section we consider quadratic functions and their graphs in greater detail.

EXPLORING THE CONCEPT

Variations of the Graph of $y = x^2$

Figures 9.42 and 9.43 show the graphs of simple quadratic functions of the form $y = ax^2$.

Figure 9.42

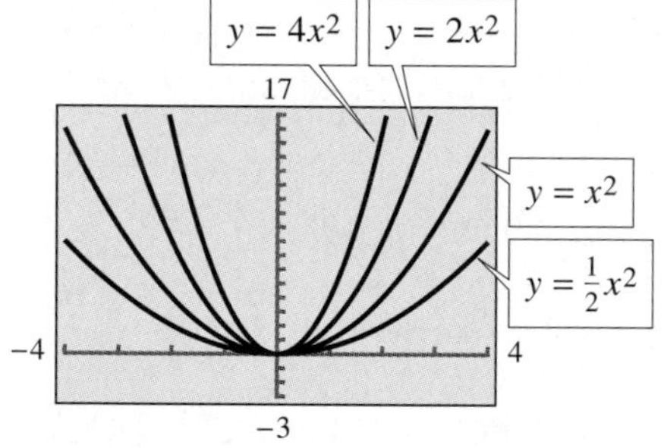

Figure 9.43

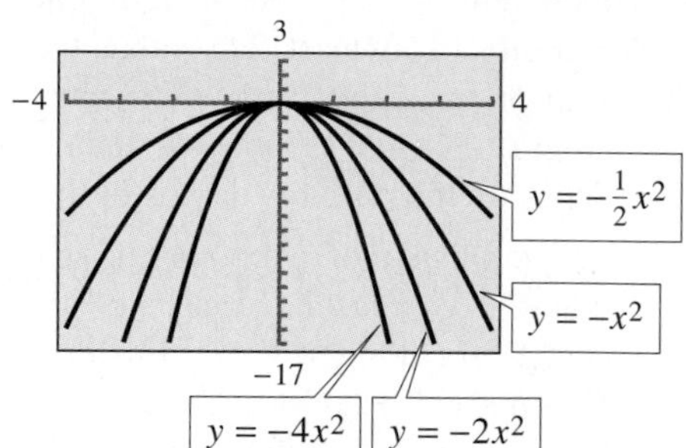

Observe that $a > 0$ for all the graphs in Fig. 9.42, whereas $a < 0$ for all the graphs in Fig. 9.43. We can see that the graph of $y = ax^2$ opens upward when $a > 0$ and downward when $a < 0$.

The figures also show that the graphs become narrower as $|a|$ increases. However, in all cases, the graphs extend without bound to the left and right, and so the domain of the function is **R**. Finally, for $a > 0$, the range is $\{y \mid y \geq 0\}$, and for $a < 0$, the range is $\{y \mid y \leq 0\}$.

The graph of a quadratic function is called a **parabola**. When a parabola opens upward, as in Fig. 9.44, it has a low point corresponding to the minimum value of the function. The parabola in Fig. 9.45 opens downward and has a high point corresponding to the maximum value of the function. The highest or lowest point of a parabola is called the **vertex**.

Figure 9.44

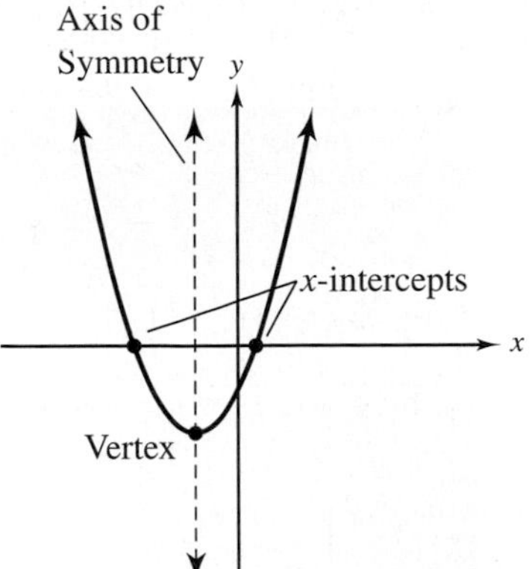

Figure 9.45

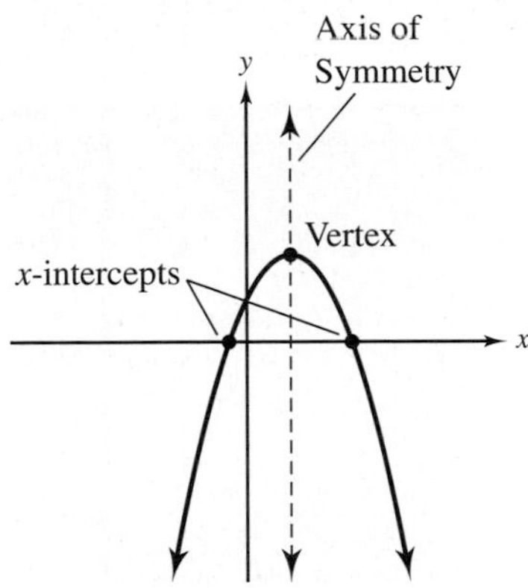

The parabolas in Figs. 9.44 and 9.45 are *symmetric* with respect to a vertical line containing the vertex. This line is called the **axis of symmetry**. If the graph were folded together along the axis of symmetry, the two sides of the parabola would coincide.

Note: Parabolas can be oriented in any direction in the coordinate plane. A parabola with a vertical axis of symmetry is sometimes called a *vertical parabola*.

Variations of the Graph of $f(x) = x^2$

We have seen that the absolute value of a affects the width of a parabola and that the sign of a affects the direction in which it opens. But the vertex for $f(x) = ax^2$ is the origin for all values of a.

Now we consider how the vertex of a parabola can be shifted vertically and horizontally.

EXPLORING THE CONCEPT

Vertical and Horizontal Shifts

Figure 9.46 shows the graph of $y = x^2$ along with the graphs of $f(x) = x^2 + k$ for $k = 5, -4$, and -8. Each graph of $f(x) = x^2 + k$ is the same as the graph of $y = x^2$ but shifted vertically. If $k > 0$, the vertex of the graph is k units above the vertex of $y = x^2$, and if $k < 0$, the vertex of the graph is $|k|$ units below the vertex of $y = x^2$.

Figure 9.46

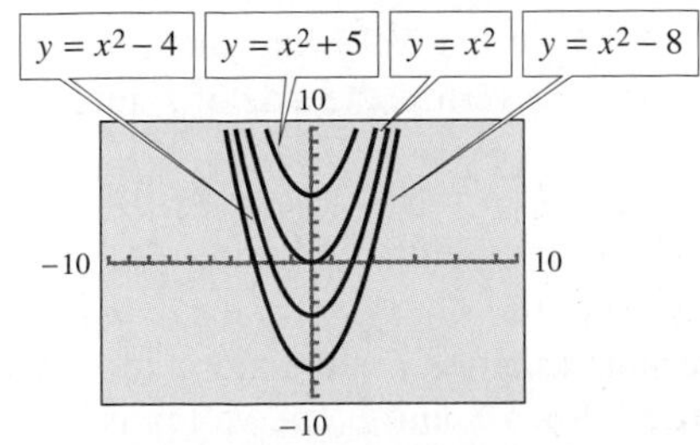

Similarly, Fig. 9.47 shows the graph of $y = x^2$ along with the graphs of $g(x) = (x - h)^2$ for $h = -6$, 2, and 5. Each graph of $g(x) = (x - h)^2$ is the same as the graph of $y = x^2$ but shifted horizontally. If $h > 0$, the vertex of the graph is h units to the right of the vertex of $y = x^2$, and if $h < 0$, the vertex of the graph is $|h|$ units to the left of the vertex of $y = x^2$.

Figure 9.47

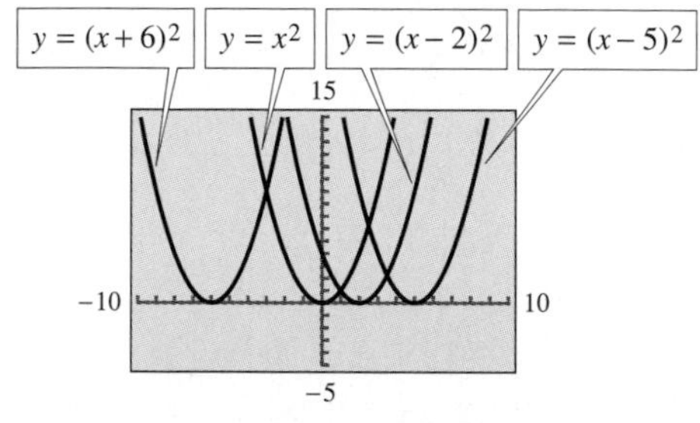

These observations can be generalized as follows:

1. The graph of $f(x) = x^2 + k$ is a vertical shift of the graph of $y = x^2$. If $k > 0$, the graph is shifted upward k units; if $k < 0$, the graph is shifted downward $|k|$ units.
2. The graph of $g(x) = (x - h)^2$ is a horizontal shift of the graph of $y = x^2$. If $h > 0$, the graph is shifted to the right h units; if $h < 0$, the graph is shifted to the left $|h|$ units.

A vertical or horizontal shift of $y = x^2$ does not affect the domain of the function, but it may affect the vertex, the axis of symmetry, and the range.

	Vertex	*Axis of Symmetry*	*Range*
$f(x) = x^2 + k$	$(0, k)$	$x = 0$	$\{y \mid y \geq k\}$
$g(x) = (x - h)^2$	$(h, 0)$	$x = h$	$\{y \mid y \geq 0\}$

Note: Our conclusions about $g(x) = (x - h)^2$ are valid only if the function is written in that form. To analyze the function $g(x) = (x + h)^2$, we first write $g(x) = [x - (-h)]^2$.

In Example 1 we combine the conclusions drawn from our observations.

EXAMPLE 1

Figure 9.48

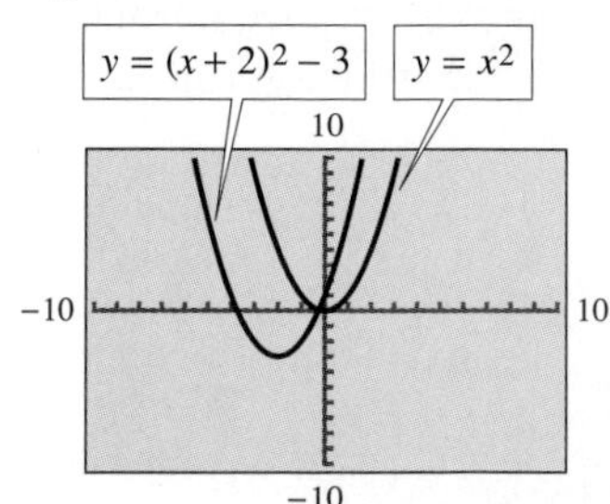

Properties of Graphs of Quadratic Functions

Produce the graph of the following functions. Compare each graph to the graph of $y = x^2$. Describe the vertex, the axis of symmetry, and the range of the function.

(a) $f(x) = (x + 2)^2 - 3$ (b) $g(x) = (x - 4)^2 + 5$

Solution

(a) The graph of function f is the graph of $y = x^2$ with the vertex shifted to the left 2 units and down 3 units. The vertex is $(-2, -3)$, and the axis of symmetry is the line $x = -2$. The range of f is $\{y \mid y \geq -3\}$. (See Fig. 9.48.)

Figure 9.49

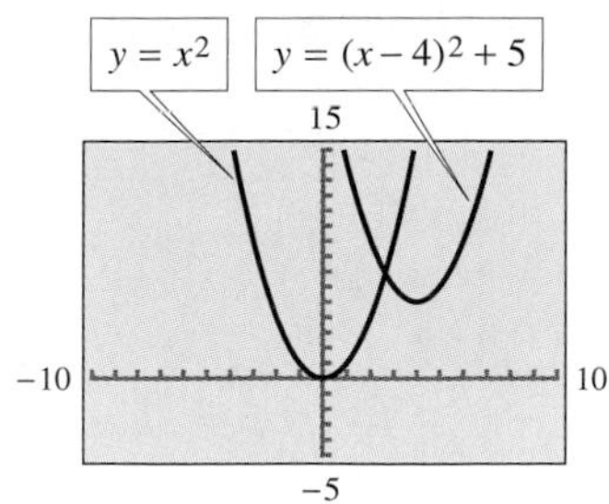

(b) The graph of function g is the graph of $y = x^2$ with the vertex shifted to the right 4 units and up 5 units. The vertex is (4, 5), and the axis of symmetry is the line $x = 4$. The range of g is $\{y \mid y \geq 5\}$. (See Fig. 9.49.)

Example 1 prompts us to generalize as follows.

Properties of the Graph of $f(x) = a(x - h)^2 + k$

The graph of $f(x) = a(x - h)^2 + k$ is the graph of $y = ax^2$ with the vertex shifted to the point (h, k).

Vertex and Intercepts

By producing the graph of a quadratic function, we can estimate the coordinates of the vertex. To determine the exact coordinates, we complete the square to write the function in the form $f(x) = a(x - h)^2 + k$. The process is illustrated in Example 2.

EXAMPLE 2

Determining the Vertex of a Parabola Algebraically

Determine the vertex of the graph of $f(x) = x^2 - 8x + 10$.

Solution

You may want to begin by producing the graph of the function and tracing the graph to estimate the coordinates of the vertex.

By using the method of completing the square, we will write the function in the form $f(x) = (x - h)^2 + k$.

$$f(x) = x^2 - 8x + 10$$

$$f(x) - 10 = x^2 - 8x \qquad \text{Subtract 10 from both sides.}$$

$$f(x) - 10 + 16 = x^2 - 8x + 16 \qquad \text{To complete the square, add } \left(-\frac{8}{2}\right)^2 = 16 \text{ to both sides.}$$

$$f(x) + 6 = (x - 4)^2 \qquad \text{Factor and write in exponential form.}$$

$$f(x) = (x - 4)^2 + (-6) \qquad \text{Add } -6 \text{ to both sides to obtain the function.}$$

In this form we see that $h = 4$ and $k = -6$. Therefore, the vertex is $(4, -6)$.

The procedure illustrated in Example 2 is one that we need to repeat each time we wish to determine the vertex of a parabola. By working through the same routine for the general quadratic function $f(x) = ax^2 + bx + c$, we can derive formulas for determining the coordinates of the vertex $V(h, k)$. (See Exercise 111.)

LEARNING TIP

Observe that the x-coordinate of the vertex is midway between the x-coordinates of the x-intercepts. If you know the x-intercepts, the average of the x-coordinates is the x-coordinate of the vertex.

$$h = -\frac{b}{2a} \quad \text{and} \quad k = \frac{4ac - b^2}{4a}$$

The vertex is $\left(-\frac{b}{2a}, \frac{4ac - b^2}{4a}\right)$.

Note: The coordinates of the vertex can be found by using these two formulas. However, in practice, it is easier to determine the first coordinate by evaluating $-\frac{b}{2a}$. Then the y-coordinate can be determined by evaluating $f\left(-\frac{b}{2a}\right)$.

EXAMPLE 3

Determining the Vertex Algebraically

Determine the vertex of the graph of $f(x) = 2x^2 + 12x + 17$.

Solution

The x-coordinate of the vertex is

$$x = -\frac{b}{2a} = -\frac{12}{2(2)} = -3$$

Because $f(-3) = -1$, the vertex is $(-3, -1)$.

In general, the y-intercept of a graph is a point whose x-coordinate is 0. For a quadratic function $f(x) = ax^2 + bx + c$, $f(0) = c$. Therefore, a vertical parabola always has one y-intercept $(0, c)$.

To determine the x-intercepts, we solve the quadratic equation $ax^2 + bx + c = 0$. We know this equation has zero, one, or two real number solutions. Thus a vertical parabola has zero, one, or two x-intercepts.

The following summarizes the characteristics of the graph of a quadratic function $f(x) = ax^2 + bx + c$.

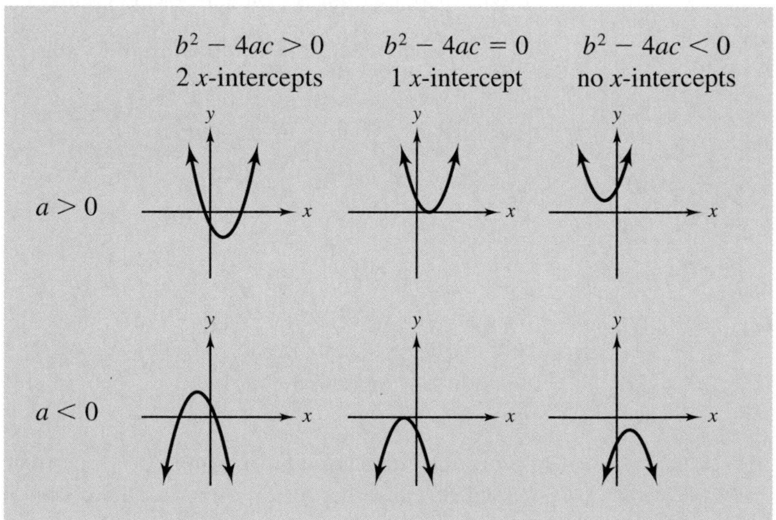

EXAMPLE 4

Determining the Vertices and Intercepts of Parabolas

Produce the graph of each of the following. Determine its vertex and intercepts algebraically.

(a) $h(x) = x^2 - 4x + 1$

(b) $g(x) = 4x^2 - 12x + 9$

(c) $s(x) = -2x^2 + 12x - 19$

Solution

(a) Because $a = 1 > 0$, the graph opens upward. The y-intercept is $(0, 1)$. There appear to be two x-intercepts. (See Fig. 9.50.)

Figure 9.50

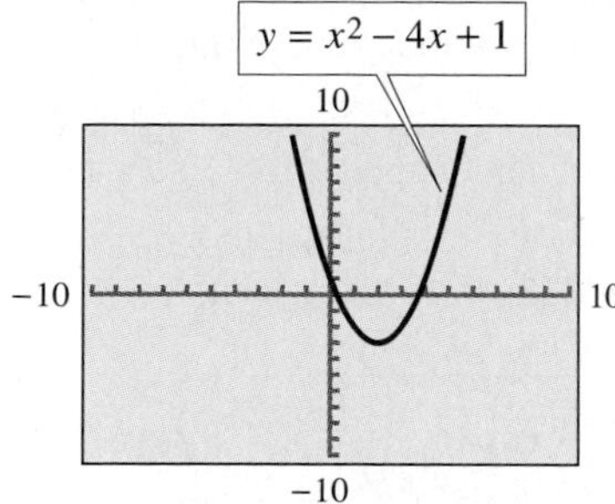

To determine the x-intercepts exactly, use the Quadratic Formula to solve the equation $x^2 - 4x + 1 = 0$.

$$x = \frac{4 \pm \sqrt{(-4)^2 - 4(1)(1)}}{2(1)} = \frac{4 \pm \sqrt{16 - 4}}{2} = \frac{4 \pm \sqrt{12}}{2} = 2 \pm \sqrt{3}$$

$$x \approx 3.73 \quad \text{or} \quad x \approx 0.27$$

The (approximate) x-intercepts are $(3.73, 0)$ and $(0.27, 0)$. Now we determine the coordinates of the vertex.

$$x = -\frac{b}{2a} = -\frac{-4}{2(1)} = 2$$

Because $h(2) = -3$, the vertex is $(2, -3)$.

(b) Because $a = 4 > 0$, the graph opens upward. The y-intercept is $(0, 9)$. There appears to be only one x-intercept. (See Fig. 9.51.)

Figure 9.51

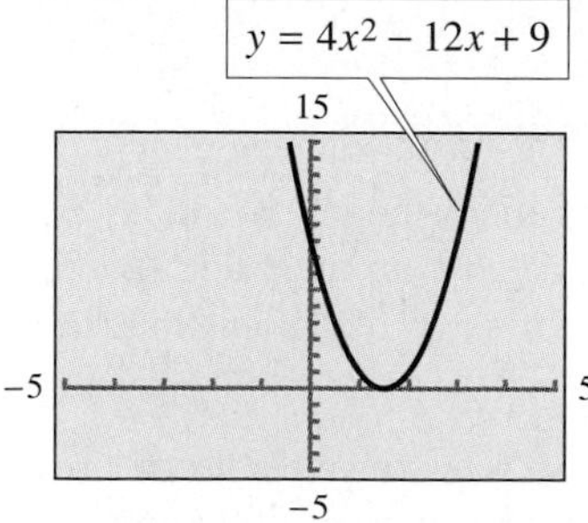

To determine the x-intercept exactly, solve the associated equation.

$$\begin{aligned} 4x^2 - 12x + 9 &= 0 \\ (2x - 3)^2 &= 0 \\ 2x - 3 &= 0 \\ 2x &= 3 \\ x &= \frac{3}{2} \end{aligned}$$

The x-intercept is $\left(\frac{3}{2}, 0\right)$.

Now we determine the coordinates of the vertex.

$$x = -\frac{b}{2a} = -\frac{-12}{2(4)} = \frac{3}{2}$$

Because $g\left(\frac{3}{2}\right) = 0$, the vertex is $\left(\frac{3}{2}, 0\right)$. (Note that the vertex and the x-intercept are the same point.)

(c) Because $a = -2 < 0$, the parabola opens downward. The y-intercept is $(0, -19)$. There appear to be no x-intercepts. (See Fig. 9.52.)

Figure 9.52

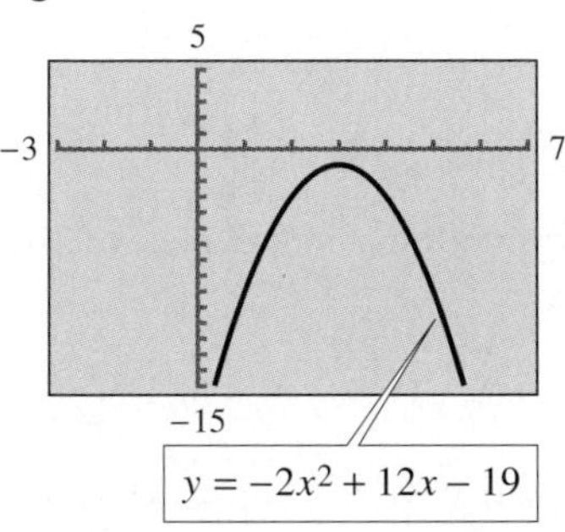

We confirm that there are no x-intercepts by evaluating the discriminant.

$$b^2 - 4ac = 12^2 - 4(-2)(-19) = 144 - 152 = -8$$

Because the discriminant is negative, there are no x-intercepts. Now we determine the coordinates of the vertex.

$$x = -\frac{b}{2a} = -\frac{12}{2(-2)} = 3$$

Because $s(3) = -1$, the vertex is $(3, -1)$.

Real-Life Applications

Think About It

Two points determine a line. What is the minimum number of points needed to determine a parabola? For a particular parabola, if you know that $c = 3$ and that $-\frac{b}{2a} = 6$, can you you write the quadratic function?

Applications in which a quantity is to have an optimum (maximum or minimum) value can often be modeled with quadratic functions. The vertex of the graph is the point of interest because it is at this point that the function has its maximum or minimum value.

In Section 9.5 we saw an example of a day-care operator who wished to enclose a rectangular play yard of a specified area. In Example 5 we see a similar problem, but this time the goal is to *maximize* the area.

EXAMPLE 5

Maximizing a Rectangular Area

A day-care operator wants to use 200 feet of fencing to enclose a rectangular play area for her day-care center. She plans to use an existing fence as one side. The existing fence extends from each back corner of her home to the property line. (See Fig. 9.53.) What dimensions should she use to maximize the rectangular area? What is the maximum area?

Figure 9.53

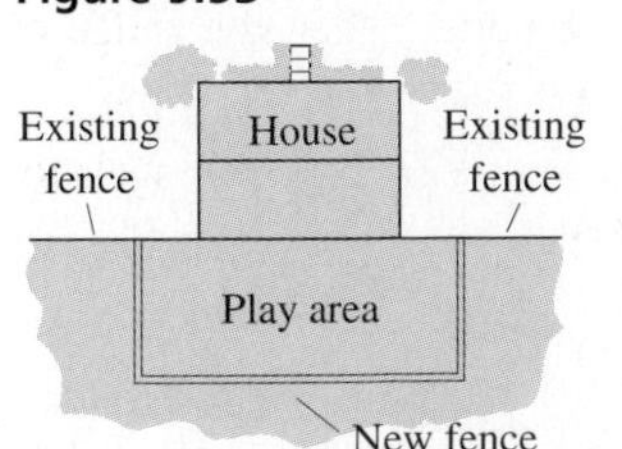

Solution

Figure 9.54

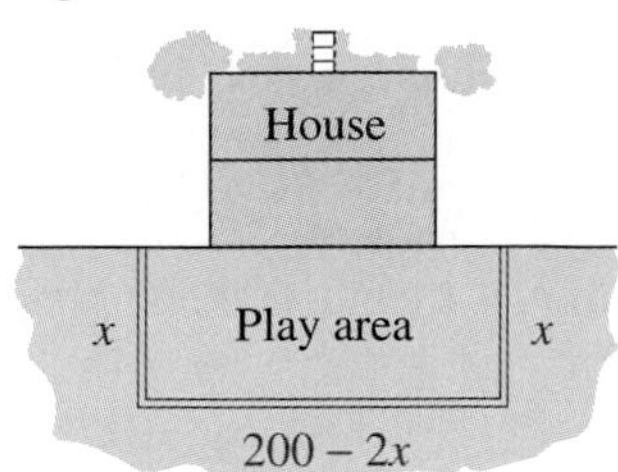

Let $x =$ the width of the rectangle. Because the amount of fencing available is 200 feet, the length of the rectangle is $200 - 2x$. (See Fig. 9.54.) The area is given by the function $A(x) = x(200 - 2x) = -2x^2 + 200x$.

Because $a = -2 < 0$, the parabola opens downward. The maximum value of the function occurs at the vertex, which we estimate as (50, 5000). (See Fig. 9.55.)

Figure 9.55

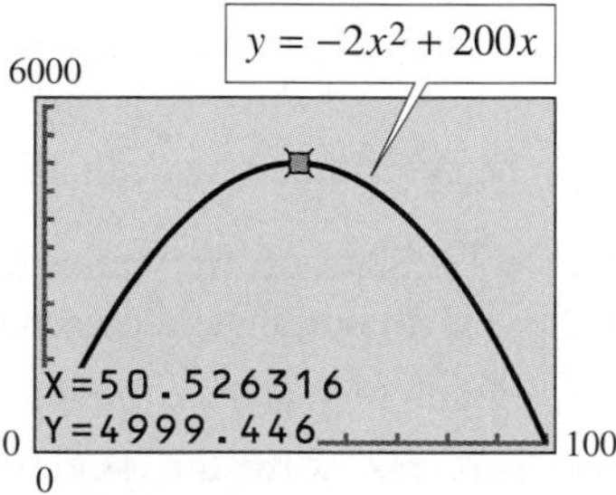

The x-coordinate of the vertex is determined as follows.

$$x = -\frac{b}{2a} = -\frac{200}{2(-2)} = 50$$

Because x represents the width of the play area, the maximum area is achieved when the width is 50 feet. The length is

$$200 - 2x = 200 - 2(50) = 100$$

The area of the yard can be found either by evaluating the function for $x = 50$ or by using the formula $A = LW$.

$$A(50) = 5000$$
$$A = LW = 100 \cdot 50 = 5000$$

The maximum area of 5000 square feet is achieved when the width is 50 feet and the length is 100 feet.

9.8 QUICK REFERENCE

Properties of Graphs

- The graph of a quadratic function $f(x) = ax^2 + bx + c$ is called a **parabola**.
- If $a > 0$, the parabola opens upward; if $a < 0$, the parabola opens downward. As $|a|$ increases, the parabola becomes narrower; as $|a|$ decreases, the parabola becomes wider.
- The lowest point of a parabola (when $a > 0$) or the highest point (when $a < 0$) is called the **vertex**.
- The domain of a quadratic function is **R**. If $V(h, k)$ is the vertex, then the range of the function is $\{y \mid y \geq k\}$ when $a > 0$ or $\{y \mid y \leq k\}$ when $a < 0$.
- The graph of a quadratic function is **symmetric** with respect to a vertical line containing the vertex. This line is called the **axis of symmetry**.

Variations of the Graph of $f(x) = x^2$

- For $f(x) = x^2 + k$, the vertex of the parabola is $V(0, k)$.
- For $f(x) = (x - h)^2$, the vertex of the parabola is $V(h, 0)$.
- The graph of $f(x) = a(x - h)^2 + k$ is the graph of $y = ax^2$ with the vertex shifted to $V(h, k)$.

Vertex and Intercepts

- To determine the vertex of the graph of a quadratic function, we can use the method of completing the square to rewrite the function in the form $f(x) = a(x - h)^2 + k$.
- In general, the x-coordinate of the vertex is $-\dfrac{b}{2a}$.

  The y-coordinate can be determined by evaluating $f\left(-\dfrac{b}{2a}\right)$.
- If $a > 0$, then the y-coordinate of the vertex represents the minimum value of the function; if $a < 0$, then the y-coordinate represents the maximum value of the function.
- The y-intercept of the graph of $f(x) = ax^2 + bx + c$ is $(0, c)$.
- To determine the x-intercepts, solve the quadratic equation $ax^2 + bx + c = 0$. The number of x-intercepts depends on the sign of the discriminant.

9.8 EXERCISES

Concepts and Skills

1. What do we call the graph of a quadratic function $f(x) = ax^2 + bx + c$? Describe how the value of a affects the shape and orientation of the graph.

2. What is the domain of any quadratic function? If you know the vertex $V(h, k)$ of the graph of a quadratic function, can you specify the range of the function? If so, state what the range is. If not, what additional information do you need?

In Exercises 3–6, for $f(x) = ax^2 + bx + c$, the value of a is given along with the vertex V of the graph of the function. What is the range of the function?

3. $a = -2$; $V(1, 3)$

4. $a = 1$; $V(-2, 1)$

5. $a = \dfrac{3}{5}$; $V(0, 2)$

6. $a = -3$; $V(-4, 0)$

7. (a) If the vertex $V(h, k)$ of a vertical parabola is known, explain how to write the equation of the axis of symmetry.

   (b) If the equation of the axis of symmetry of a parabola is $x = a$, what can you conclude about the vertex of the parabola?

8. (a) Describe the graph of $f(x) = x^2 + k$ in comparison with the graph of $y = x^2$.

   (b) Describe the graph of $g(x) = (x - h)^2$ in comparison with the graph of $y = x^2$.

In Exercises 9–12, write the equation of the vertical axis of symmetry for a parabola with vertex V.

9. $V(0, -1)$
10. $V(-1, -5)$
11. $V(3, 0)$
12. $V(0, 4)$

In Exercises 13–18, match the function with its graph.

13. $f(x) = x^2 - 4$
14. $g(x) = x - 4$
15. $h(x) = -2x^2$
16. $f(x) = (x - 3)^2$
17. $g(x) = -\dfrac{1}{2}x - 2$
18. $h(x) = -\dfrac{1}{2}(x + 1)^2 - 2$

(a)

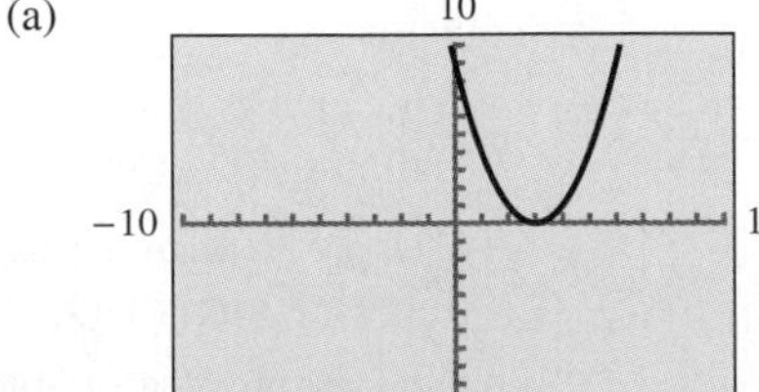

(b)

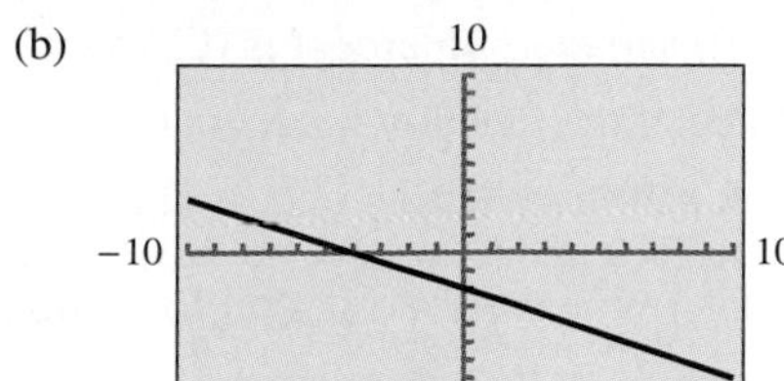

(c)

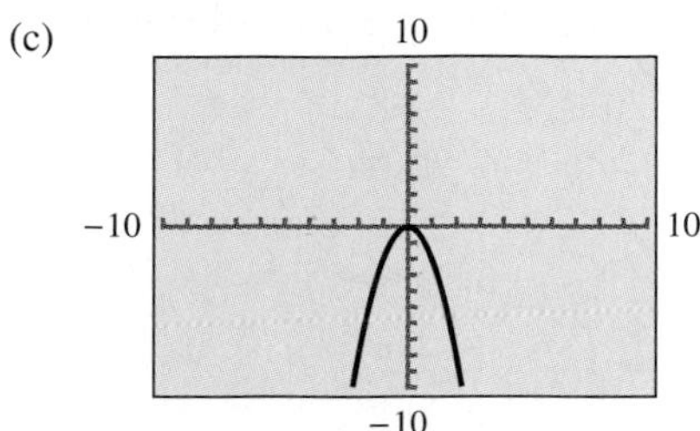

(d)

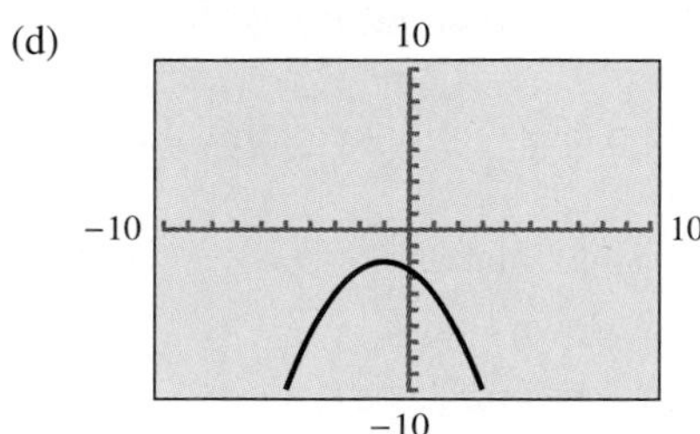

(e)

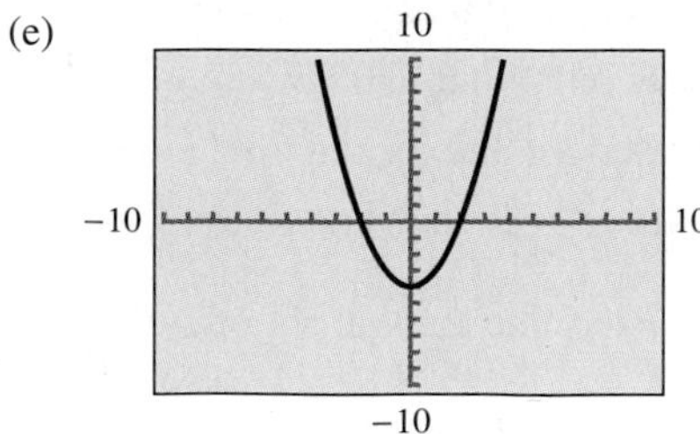

(f)

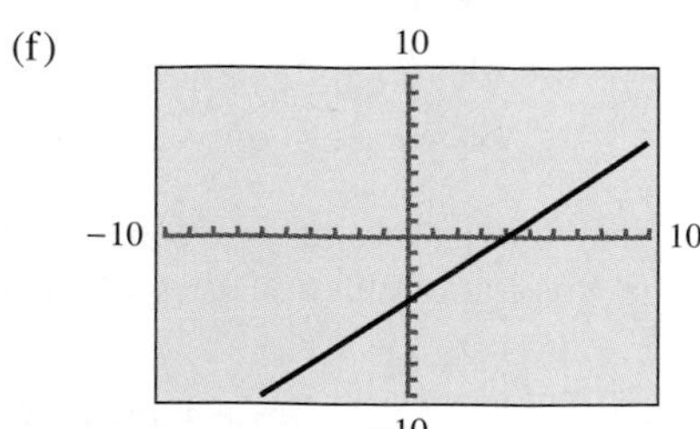

In Exercises 19–24, compare the graphs of the given pair of functions to the graph of $y = x^2$. State whether the graphs are more or less narrow and describe any horizontal or vertical shifts of the vertices.

19. (a) $y = -x^2$ (b) $y = -2x^2$

20. (a) $y = x^2 + 4$ (b) $y = x^2 - 4$

21. (a) $y = (x + 3)^2$ (b) $y = (x - 3)^2$

22. (a) $y = x^2 - 1$ (b) $y = (x - 1)^2$

23. (a) $y = (x - 3)^2 - 2$ (b) $y = (x + 1)^2 + 4$

24. (a) $y = -2(x + 1)^2 - 1$ (b) $y = 3(x - 1)^2 + 2$

25. Given the function $f(x) = ax^2 + bx + c$, we can determine the vertex of the graph by using the method of completing the square. Describe a faster method for determining the x-coordinate of the vertex. How do we then determine the y-coordinate?

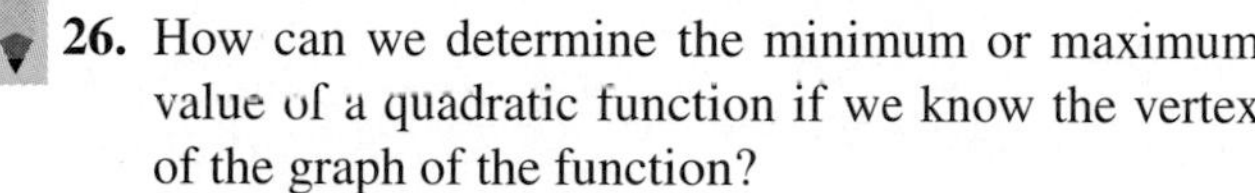

26. How can we determine the minimum or maximum value of a quadratic function if we know the vertex of the graph of the function?

In Exercises 27–32, produce the graph of the given function and estimate the vertex. Then determine the vertex algebraically and state the minimum or maximum value of the function.

27. $f(x) = (x - 3)^2 + 4$

28. $g(x) = (x + 3)^2 - 4$

29. $h(x) = x^2 + 2x - 3$

30. $f(x) = x^2 - 6x + 5$

31. $g(x) = 10 - 6x - 3x^2$

32. $h(x) = -6 + 5x - 3x^2$

In Exercises 33–36, the range of a quadratic function is given along with the equation of the axis of symmetry of the graph of the function. What is the vertex of the graph?

33. $\{y \mid y \geq 3\}$; $x = 0$

34. $\{y \mid y \leq -2\}$; $x = -3$

35. $\{y \mid y \leq -1\}$; $x = 5$

36. $\{y \mid y \geq 0\}$; $x = -2$

In Exercises 37–42, for the given function, determine the vertex of its graph and the equation of the axis of symmetry.

37. $f(x) = x^2 - 6x + 9$

38. $g(x) = x^2 + 4x + 9$

39. $h(x) = 2x^2 + 4x - 7$

40. $g(x) = 3x^2 + 6x + 8$

41. $h(x) = 7 - 4x - x^2$

42. $f(x) = 5x - 2x^2$

43. Use what you know about the domain of a quadratic function to explain why its graph always has a y-intercept. If you know what the function is, what is an easy way to determine the y-intercept?

44. If we know that the solutions of $ax^2 + bx + c = 0$ are m and n, what information do we have about the graph of $f(x) = ax^2 + bx + c$?

In Exercises 45–52, use a graph to estimate the y-intercepts and the x-intercepts, if any. Then verify the intercepts algebraically.

45. $f(x) = x^2 - 3x - 28$

46. $h(x) = 2x^2 - 11x + 12$

47. $g(x) = 4x^2 - 5x + 9$

48. $f(x) = -3 + x - x^2$

49. $f(x) = 5 - 4x - 3x^2$

50. $h(x) = 2x^2 + 5x - 2$

51. $h(x) = 3x^2 + 5x$

52. $g(x) = 0.2x - 0.4x^2$

In Exercises 53–56, $f(x) = ax^2 + bx + c$, where a, b, and c are rational numbers. In each exercise the value of the discriminant is given for the associated equation $f(x) = 0$. How many x-intercepts does the graph of $f(x)$ have? Are the x-coordinates of these points rational or irrational numbers?

53. -4

54. 0

55. 16

56. 20

In Exercises 57–62, $f(x) = ax^2 + bx + c$. In each exercise the value of a is given along with the vertex of the graph of $f(x)$. Determine the number of x-intercepts of the graph.

57. -3, $V(-1, -2)$

58. -1, $V(3, 2)$

59. $\frac{1}{2}$, $V(2, -4)$

60. 2, $V(-2, 1)$

61. 3, $V(2, 0)$

62. -4, $V(0, 1)$

In Exercises 63–68, information is given about the graph of a quadratic function. Determine the number of x-intercepts.

63. The vertex is $(-2, -5)$, and the graph opens upward.

64. The vertex is $(0, 1)$, and the graph opens upward.

65. The vertex is $(-3, 0)$.

66. The vertex is $(1, 2)$, and the y-intercept is $(0, 7)$.

67. The maximum value of the function occurs at $(-4, 3)$.

68. The minimum value of the function occurs at $(3, 0)$.

In Exercises 69–74, determine the following information for each function.

(a) The domain and range of the given function

(b) The x- and y-intercepts of its graph

(c) The vertex and the equation of the axis of symmetry

69. $f(x) = x^2 - 6x + 8$

70. $g(x) = x^2 + 8x + 12$

71. $f(x) = 5 - 4x - x^2$

72. $g(x) = 12 - 4x - x^2$

73. $h(x) = 2x^2 + 4x + 9$

74. $g(x) = 2x - x^2 - 8$

In Exercises 75–82, you are given information about the location of the graph of $f(x) = (x - h)^2 + k$ relative to the location of the graph of $y = x^2$. Determine the values of h and k.

75. The graph of $f(x)$ is up 3 units.

76. The graph of $f(x)$ is down $\frac{1}{2}$ unit.

77. The graph of $f(x)$ is left $\frac{3}{5}$ unit.

78. The graph of $f(x)$ is right 1 unit.

79. The graph of $f(x)$ is up 2 units and left 4 units.

80. The graph of $f(x)$ is left 1 unit and down 4 units.

81. The graph of $f(x)$ is right 3 units and up 2 units.

82. The graph of $f(x)$ is right 2 units and down 3 units.

In Exercises 83–88, determine the values of b and c in the quadratic function $f(x) = x^2 + bx + c$ so that the graph of f has the given x-intercept(s).

83. $(5, 0)$ and $(-3, 0)$

84. $(-2, 0)$ and $(-5, 0)$

85. $(0, 0)$ and $(-2, 0)$

86. $(-4, 0)$ and $(4, 0)$

87. $(4, 0)$ only

88. $(-1, 0)$ only

In Exercises 89–92, use the given information about the graph of the given quadratic function to write the specific function.

89. The graph of $f(x) = x^2 + bx + c$ has a y-intercept $(0, 3)$ and an axis of symmetry whose equation is $x = 2$.

90. The graph of $f(x) = 2x^2 + bx + c$ has the vertex $V(2, -3)$.

91. The graph of $f(x) = ax^2 + x + c$ has an x-intercept $(-1, 0)$, and the first coordinate of the vertex is $-\frac{1}{6}$.

92. The graph of $f(x) = ax^2 - 8x + c$ has the vertex $V(1, -9)$.

93. The sum of two numbers is 30. Determine the numbers such that the product is a maximum.

94. The difference of two numbers is 36. Determine the two numbers such that their product is a minimum.

Real-Life Applications

95. A springboard diver is propelled upward from the board, and after reaching the peak of the dive, she performs a double twist before entering the water. Her height h (in feet) after t seconds is given by $h = 10 + 30t - 16t^2$. After how many seconds will the diver be at the peak of the dive?

96. The Canton Package Co. wants to design a box that is 11 inches high with a perimeter around the base of 80 inches. What should be the length and width of the box to have the maximum volume? What is the maximum volume (in cubic feet)?

97. A convention setup crew had 150 linear feet of dividers to create five adjacent meeting rooms of identical size. For each room, a wall of the building served as one side, and the dividers were used to form the other three sides. What was the largest area that the rooms could have?

98. The manager of a resort condominium complex can rent each of the 80 units for \$140 per night. However, five units will remain vacant for each \$20 per night increase in the rental price. For what price will the total income be the greatest?

99. An engineer wants to design a parabolic arch whose height is 60 feet. At ground level, the span of the arch is 100 feet. If one ground point of the arch is placed at the origin of a coordinate system and the other ground point is placed along the positive x-axis, determine the quadratic function that models the arch.

100. In Exercise 99, suppose the engineer places the vertex of the parabola on the positive y-axis and the ground points on the x-axis. Determine the quadratic function that models the arch and compare it to the function determined in Exercise 99.

Group Project: Mutual Fund Investors

As the accompanying bar graph shows, an increasing number of investors have sought to balance their investments through mutual funds. (Source: Scudder Investor Services, Inc.)

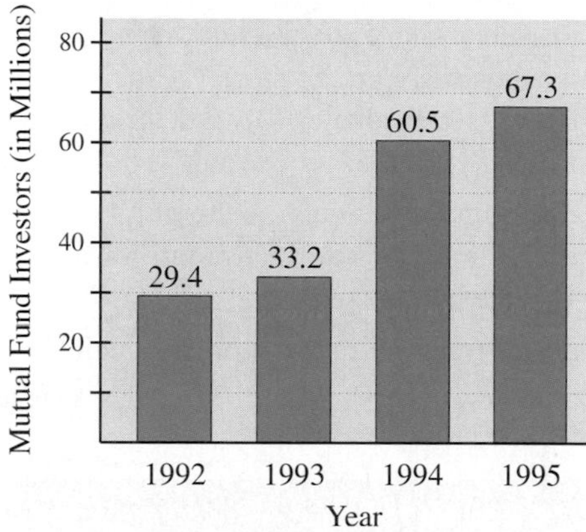

Let x represent the number of years since 1992.

101. Looking at the bar graph, would you expect a parabola that models the data for the years 1992, 1993, and 1995 to open upward or downward? Use the quadratic regression feature of your calculator to write a model quadratic function for those three years. Does the value of a confirm your expectation?

102. Looking at the bar graph, would you expect a parabola that models the data for the years 1992, 1994, and 1995 to open upward or downward? Use the quadratic regression feature of your calculator to write a model quadratic function for those 3 years. Does the value of a confirm your expectation?

103. Determine and interpret the coordinates of the vertices of the two parabolas. Do you think that either function is a reasonable model for future trends in the number of mutual fund investors?

104. Name two advantages and two disadvantages of investing in mutual funds.

Challenge

In Exercises 105–108, the graph of the function $f(x) = ax^2 + bx + c$ is a parabola containing the given points. Determine the vertex of the parabola.

105. $(-1, 0)$, $(1, -6)$, and $(2, -3)$

106. $(1, 6)$, $(-2, -12)$, and $(3, 8)$

107. $(1, 3)$, $(2, 7)$, and $(3, 13)$

108. $(0, 1)$, $\left(-\frac{1}{2}, 0\right)$, and $(1, 9)$

109. A commercial developer finds that the rental income I (in dollars) for a shopping mall is given by $I(x) = (700 + 25x)(30 - x)$, where x is the number of vacant stores.

(a) It seems reasonable to expect a maximum income when there are no store vacancies, that is, when $x = 0$. Evaluate $I(0)$ to determine the income when no stores are vacant.

(b) Now determine the vertex of the graph of the income function. For how many store vacancies is the income maximized?

(c) How many stores are in the shopping mall?

110. For a certain flight, a charter airline charges \$96 per person. Each person is charged an additional \$4 for each unsold seat. If the plane has 60 seats, find the number of unsold seats that produces the maximum revenue.

111. Using the following steps as a guide, derive the formulas for the coordinates of the vertex of a parabola.

(a) Begin with the general quadratic function $f(x) = ax^2 + bx + c$. Subtract c from both sides, and then divide both sides by a.

(b) By completing the square, write the right side as the square of a binomial.

(c) Write the function in the following form.

$$f(x) = a(x - h)^2 + k$$

9. CHAPTER PROJECT Heart Attack Risks

Research indicates that lifestyle has a substantial effect on the risk of a heart attack. Weight, exercise, and blood cholesterol levels are three of the many factors on which medical research focuses in the prevention of heart disease. The *New England Journal of Medicine* reported the following factors and the amount by which heart attack risk is reduced.

Factor	Risk Reduction
Exercising regularly	45%
Maintaining ideal weight	35–55%
Decreasing cholesterol (per 1% decrease in cholesterol)	2–3%

1. One recommendation for exercise is to maintain a target heart rate for 20 minutes each day. The following formulas give the target heart rate T for a person age A. (Source: *The Aerobics Program for Total Well-Being*, by Kenneth H. Cooper.)

   Women: $T = 0.8(220 - A)$

   Men: $T = 0.8(205 - 0.5A)$

   Using the formulas, make a table showing the target heart rates for men and women ages 20 through 70 in increments of 10 years.

2. Use algebraic methods to determine the age at which the target heart rates for men and women are the same.

3. The ideal weight W (in pounds) of a person with a medium frame can be modeled as a function of height h (in inches).

   Women: $W(h) = 2.95h - 57.32$

   Men: $W(h) = 0.077h^2 - 7.61h + 310.11$

(a) Use the appropriate model to estimate the ideal weight for a man whose height is 5 feet 11 inches.

(b) For women, what is the approximate increase in weight for each additional inch in height?

4. Exercise is also a factor in weight control. The number of calories C burned per hour by a 120-pound person walking at rate r in miles per hour is modeled by $C(r) = 25.3r^{1.78}$.

   (a) How many calories are burned per hour by a 120-pound person walking 5 miles per hour? (Round to the nearest whole number.)

   (b) Use a graph to estimate the walking rate required to burn 700 calories per hour. (Round to the nearest tenth.)

5. A person's cholesterol level C increases with age a. Using the following model functions, what can you conclude about men and women age 60?

   Women: $C(a) = 1.17a + 147.15$

   Men: $C(a) = 0.76a + 171.4$

6. In groups, choose and investigate another factor in heart attack risk, such as alcohol consumption, smoking, blood pressure, or use of aspirin. Perhaps a representative of the nursing faculty, college wellness center, or college health education department can provide additional information and discuss efforts toward education to promote a healthy lifestyle on your campus or in your community. Report your findings and support your information with model functions and graphs.

9. CHAPTER REVIEW EXERCISES

Section 9.1

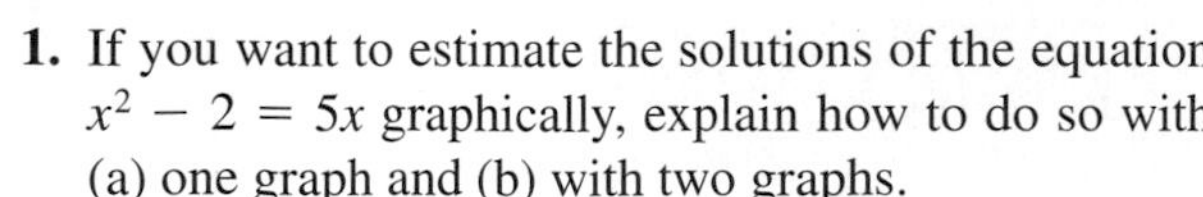

1. If you want to estimate the solutions of the equation $x^2 - 2 = 5x$ graphically, explain how to do so with (a) one graph and (b) with two graphs.

In Exercises 2 and 3, use the graphing method to estimate the solution(s) of the given equation.

2. $a^2 - 7a + 10 = 0$

3. $13 - 3z^2 = 0$

In Exercises 4 and 5, use the Zero Factor Property to solve.

4. $\left(4x - \frac{2}{7}\right)\left(6x - \frac{3}{5}\right) = 0$

5. $(d - 3)(d + 2) = 36$

In Exercises 6 and 7, use the factoring method to solve.

6. $16 + 9c^2 = 24c$

7. $x(x + 2) = 15$

In Exercises 8 and 9, solve by using the Square Root Property.

8. $25 - y^2 = 0$

9. $25(b + 4)^2 - 16 = 0$

In Exercises 10 and 11, determine the complex number solutions.

10. $a^2 + 36 = 0$

11. $(b - 2)^2 + 37 = 0$

12. The sum of the squares of two consecutive odd integers is 130. What are the integers?

Section 9.2

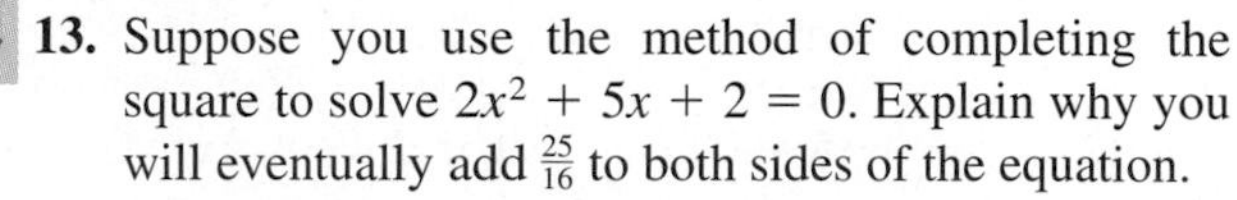

13. Suppose you use the method of completing the square to solve $2x^2 + 5x + 2 = 0$. Explain why you will eventually add $\frac{25}{16}$ to both sides of the equation.

14. Determine the value of k so that $x^2 - 7x + k$ is a perfect square.

In Exercises 15 and 16, solve by completing the square.

15. $x^2 + 10x - 1 = 0$

16. $3y(y - 5) = 12$

In Exercises 17 and 18, determine the imaginary number solutions by completing the square.

17. $x^2 - 6x + 10 = 0$ **18.** $3x^2 - 2x + 4 = 0$

In Exercises 19–23, determine the complex number solutions.

19. $x^2 - 5x + 3 = 0$ **20.** $x^2 + 7 = 3x$

21. $3 - x - 2x^2 = 0$ **22.** $2x^2 + x + 1 = 0$

23. $1 = 2x^2 + 6x$

24. The sum of the reciprocals of two consecutive integers is $\frac{7}{12}$. What are the integers?

Section 9.3

25. Which of the following equations can be solved with the Quadratic Formula?

(i) $3x^2 = 12$ (ii) $(x + 1)(x - 5) = 0$

(iii) $x^2 + x + 1 = 0$ (iv) $(x + 3)^2 = 5$

Identify those equations that can be solved with an alternative method and state the method.

In Exercises 26–29, solve by using the Quadratic Formula.

26. $x(x + 3) = 7$ **27.** $5x^2 - 16 = 0$

28. $3 + \frac{4}{x^2} = \frac{7}{x}$ **29.** $4x^2 = 5x + 3$

In Exercises 30 and 31, determine the complex number solutions by using the Quadratic Formula.

30. $x^2 + 4x + 7 = 0$ **31.** $2x^2 + 9 = x$

In Exercises 32–35, use the discriminant to determine the number and the nature of the solutions of the equation.

32. $8x = x^2 + 16$ **33.** $4x = x^2 - 12$

34. $8x = 2x^2 - 7$ **35.** $6x^2 - 2x + 5 = 0$

36. The sum of the square of a number and 5 times the number is 7. What is the number?

Section 9.4

37. If $x > 0$, which of the following equations are quadratic in form?

(a) $\frac{2}{x^2} + \frac{3}{x} - 5 = 0$ (b) $2x + \sqrt{2x} = 3$

(c) $(x + 1)^2 + 2(x + 1) + 1 = 0$

(d) $x^{1/4} - x^{1/2} + 3 = 0$

For those you selected, state the substitution you would use; rewrite the equation as a quadratic equation.

In Exercises 38–45, determine the real number solutions of the equation.

38. $\frac{3x}{x - 4} + \frac{3}{x + 3} = 4$ **39.** $\frac{x}{x + 1} = 2 + \frac{2}{x - 2}$

40. $\sqrt{3x - 4} = \sqrt{x + 1} - 2$

41. $\sqrt{1 - 2x} = x + 1$

42. $a^4 - 5a^2 + 6 = 0$

43. $x - 3x^{1/2} = 10$

44. $x^{1/6} + x^{1/3} = 2$

45. $(x^2 + x)^2 + 72 = 18(x^2 + x)$

In Exercises 46 and 47, determine all complex number solutions.

46. $t^4 + 8t^2 - 9 = 0$ **47.** $\frac{x}{x - 2} = \frac{2}{x + 5}$

48. Use an appropriate substitution to write the equation $3 + 10x^{-1} + 8x^{-2} = 0$ as a quadratic equation. Then determine the solutions of the original equation.

Section 9.5

49. Two tablecloths are to be cut from a rectangular piece of material. One tablecloth is a circle and the other is a rectangle whose length is 3 feet longer than its width. (See figure.) The total area of the tablecloths is 90 square feet. How many square yards of material are needed?

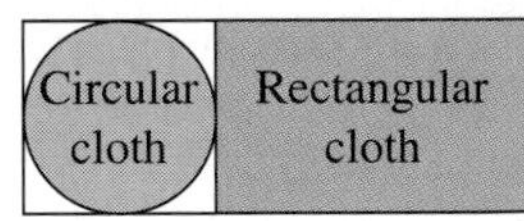

50. A fraternity sold tickets for its annual dinner-dance. Ten couples could not attend the dinner but did attend the dance. The fraternity collected \$1680 for dinner tickets and \$1300 for dance tickets. If the dinner tickets cost \$15 more than the dance tickets, how many couples attended the dance?

51. For Hanukkah an aunt sent \$6 to each child and a grandmother sent \$60 to be divided equally among the children of the family. An uncle sent \$56 to be divided among the children and three other relatives. A grandfather sent each child \$5. If each child received a total of \$30, how many children were in the family?

52. A boat traveled up a river and back, a total distance of 40 miles, in 6 hours. If the rate of the current was 5 mph, what would the rate of the boat have been in still water?

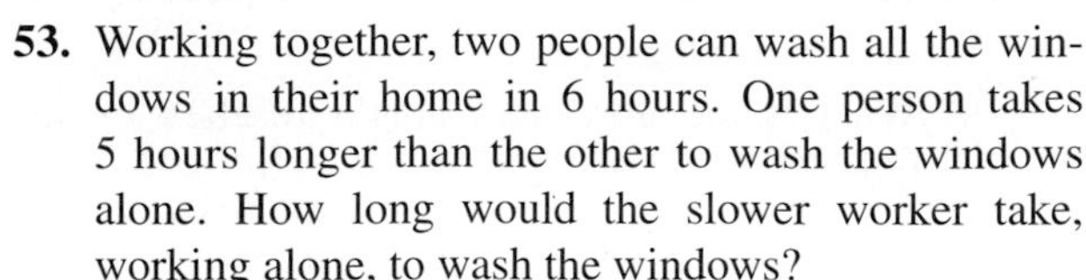

53. Working together, two people can wash all the windows in their home in 6 hours. One person takes 5 hours longer than the other to wash the windows alone. How long would the slower worker take, working alone, to wash the windows?

54. A metal fabricator needs a piece of aluminum in the shape of a right triangle with an area of 150 square inches. If the length of one leg is 1 inch more than twice the length of the other leg, what is the length of the hypotenuse? (Round the result to the nearest tenth.)

Section 9.6

55. The solution set of $x^2 + 5x + 6 < 0$ is $(-3, -2)$. Explain how to determine the solution set of the inequality $x^2 + 5x + 6 \geq 0$.

In Exercises 56–61, solve the inequality.

56. $2x^2 + 6 \geq 7x$ 57. $x(x - 6) + 9 \leq 0$

58. $4x(x + 5) + 9 \leq 0$ 59. $x^2 + 5 \geq 3x$

60. $2x - x^2 + 2 > 0$ 61. $x - x^2 - 1 > 0$

62. A dog owner has 250 feet of fencing to enclose a rectangular run area for his dogs. If he wants the area to be at least 3500 square feet, what can the length of the rectangle be?

In Exercises 63 and 64, determine the values of x for which the function satisfies the given condition.

63. $f(x) = 8 + 3x - 2x^2$; $f(x) > 0$

64. $g(x) = x^2 + 8x + 16$; $g(x) \geq 0$

In Exercises 65 and 66, use the discriminant to determine all values of k so that the quadratic inequality has the given solution set.

65. $3x^2 + 2x + k > 0$; **R**

66. $kx^2 + 3x + 2 < 0$; $\emptyset$

Section 9.7

67. Suppose that you use the accompanying figure to solve the inequality $(x + 5)(x - 3)^2 > 0$.

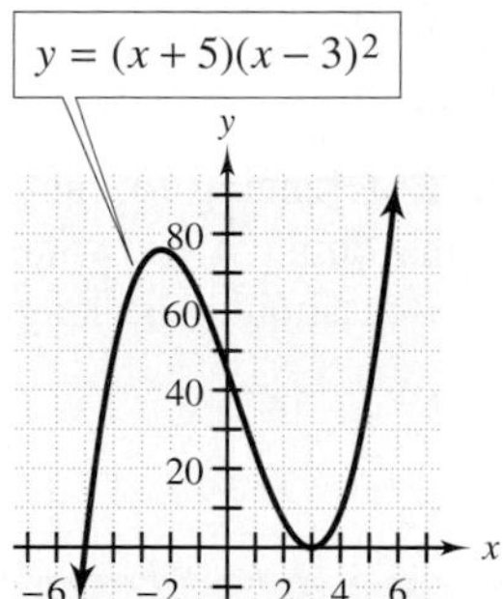

From the following list, identify the correct solution set and explain why it is correct.

(i) $(-5, 3)$ (ii) $(-5, \infty)$

(iii) $(-5, 3) \cup (3, \infty)$ (iv) $(-5, 3]$

In Exercises 68–75, use a graph to estimate the x-intercepts. Then determine the x-intercepts algebraically and write the solution set of the inequality.

68. $(x - 2)(x + 1)(x + 3) > 0$

69. $(x - 3)^2(x + 1) \geq 0$

70. $x^4 - 2x^3 - 3x^2 < 0$

71. $x^3 + x^2 - 4x - 4 \leq 0$

72. $\dfrac{x + 3}{x - 4} < 0$ 73. $\dfrac{3 - 4x}{3x - 4} \geq 1$

74. $\dfrac{(x + 1)(x - 5)}{x + 3} \geq 0$ 75. $\dfrac{x}{x + 2} > 2x$

76. Solve the inequality $5x^3(x + 6)^4 \leq 0$ by inspection.

Section 9.8

77. Explain how you can use a graph to estimate the domain and range of a quadratic function.

78. Explain how you can use the graph of the function $f(x) = ax^2 + bx + c$ to estimate the solutions of the equation $ax^2 + bx + c = 0$.

In Exercises 79 and 80, determine the vertex and the intercepts of the graph of the given function.

79. $f(x) = x^2 - 5x + 9$

80. $g(x) = 3x^2 - 4x - 7$

In Exercises 81–84, determine whether the graph of the given function opens upward or downward. Then find the vertex of the graph and the range of the function.

81. $f(x) = 17 - x^2$

82. $g(x) = 2 - \dfrac{4}{5}x - \dfrac{2}{3}x^2$

83. $h(x) = 2x^2 - 5x$

84. $f(x) = 4x^2 - 6x + 7$

85. A dog owner has 250 feet of fencing to enclose a rectangular run area for his dogs. If he wants the maximum possible area, what should the length of the rectangle be?

86. If an object is thrown vertically upward, its height (in feet) after t seconds is given by the function $h(t) = 64t - 16t^2$.

(a) In how many seconds will the object be at its maximum height?

(b) What is the maximum height reached by the object?

LOOKING AHEAD

The following exercises review concepts and skills that you will need in Chapter 10.

In Exercises 1 and 2, determine the domain of the given function.

1. $f(x) = \dfrac{x+2}{x-3}$ **2.** $g(x) = \sqrt{x+5}$

3. Explain how the Vertical Line Test is used to determine if a graph represents a function.

4. Determine whether the given relation represents a function.

(a) $y = x^2$ (b) $x = |y|$

5. Perform the indicated operations.

(a) Add the polynomials $x + 3$ and $x^2 + 4x - 3$.

(b) Subtract $4 - 6t$ from $3t + 1$.

6. Perform the indicated operations.

(a) $(x + 3)(x - 2)$

(b) $(x^2 - 5x + 6) \div (x - 2)$

In Exercises 7 and 8, evaluate the given function as indicated.

7. $f(x) = \sqrt{5 - x};\ f(-4)$ **8.** $g(x) = \dfrac{x-1}{x+1};\ g(2)$

In Exercises 9 and 10, evaluate $f(x) = 3x - 7$ as indicated.

9. $f(t + 4)$ **10.** $f(t^2)$

11. Use the integer setting to graph the function $g(x) = 0.01x^3 + x$. Then use your graph to estimate the values in parts (a)–(c).

(a) $g(3)$

(b) x, if $g(x) = -10.43$ (c) x, if $g(x) = 8.16$

12. Evaluate the exponential expressions.

(a) 2^{-4} (b) $\left(\dfrac{1}{3}\right)^{-3}$

(c) 15^0 (d) $4^{3/2}$

(e) $125^{-2/3}$

9. CHAPTER TEST

1. Solve $3t^2 + 4t = 1$ by completing the square.

In Questions 2–8, solve the given equation.

2. $x(6x - 5) = 6$ **3.** $(2x - 1)^2 - 5 = 0$

4. $2 = \dfrac{2}{t} + \dfrac{1}{t^2}$

5. $(2y + 1)(y + 3) = 2$

6. $3x^2 - x + 2 = 0$

7. $9x^4 + 7x^2 = 2$

8. $\dfrac{2}{x+7} + \dfrac{16}{x^2 + 6x - 7} = \dfrac{x}{1-x}$

9. The area A of a triangle is 20 square feet. The base b is 3 feet longer than the height h. What are the dimensions of the triangle? $\left(A = \frac{1}{2}bh\right)$

10. At a dairy a milk storage tank has an inlet pipe for receiving milk from the milking station. An outlet pipe takes the milk to the bottling room. The tank fills in 8 hours if both pipes are open. Emptying the tank takes 2 hours longer with the outlet pipe open than filling the tank with the inlet pipe. How long does emptying a full tank take?

11. A gutter is formed from a 100-foot-long sheet of metal that is 12 inches wide. To make the gutter, the sides are folded up the same amount on each side. (See figure.) What should the height of the gutter be for maximum capacity?

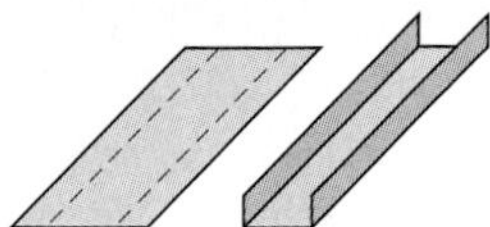

12. Suppose that you use a graph to assist you in solving a quadratic inequality. Explain how the graph can have no x-intercept but the solution set of the inequality is **R**.

In Questions 13–15, solve the given inequality.

13. $6 + x - x^2 \leq 0$

14. $\dfrac{2}{x-3} < \dfrac{1}{x+2}$

15. $t(2t + 1)(t - 5) \geq 0$

16. Suppose $f(x) = x^2 - 6x + 11$. Write the function in the form $f(x) = (x - h)^2 + k$. Then give the coordinates of the vertex of the graph of f and write the equation of the axis of symmetry.

17. The figure shows the graph of $f(x) = ax^2 + bx + c$.

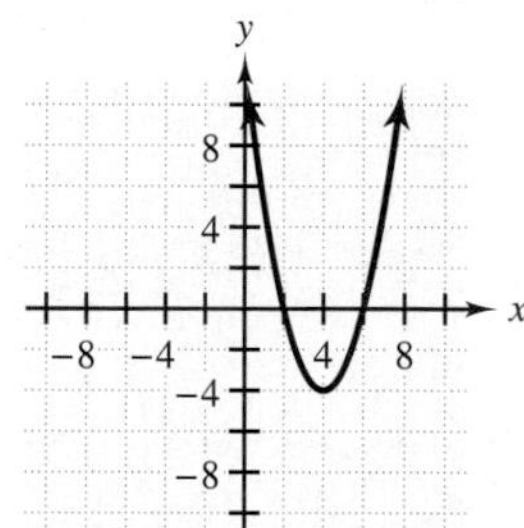

(a) What are the domain and range of the function?

(b) What do you know about the coefficient a?

(c) What are the estimated solutions of the equation $ax^2 + bx + c = 0$?

18. To the nearest hundredth, determine the x-intercepts of the graph of $f(x) = x^2 + 6x - 20$.

Chapter 10

Exponential and Logarithmic Functions

Work at home by way of computers, faxes, E-mail, and other forms of communications technology is becoming a reality for millions of workers.

The accompanying bar graph suggests that the number of *telecommuters* increased linearly from 1989 to 1993, but studies indicate that a better model for the future is an **exponential function.** If the benefits of telecommuting are outweighed by its disadvantages, the growth may be slower and might be modeled with a **logarithmic function.** (For more on this topic, see Exercises 97–100 at the end of Section 10.2, and see the Chapter Project.)

In Chapter 10 we discuss the algebra of functions and their inverses. In particular, we study the properties and graphs of two inverse functions called *exponential* and *logarithmic functions.* Finally, we learn how to solve equations and applications that involve exponential and logarithmic expressions.

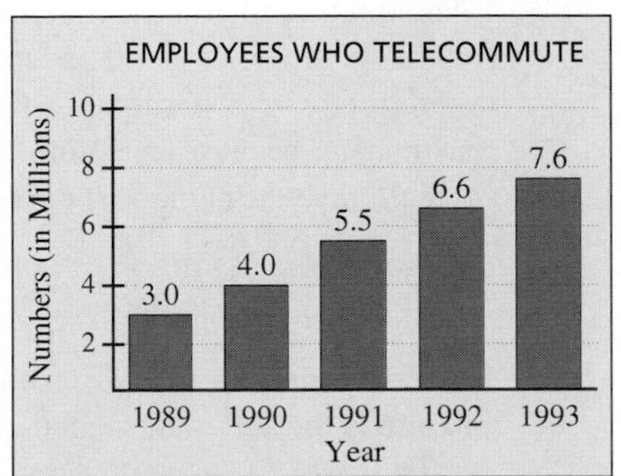

(Source: LINK Resources Corporation.)

10.1 ALGEBRA OF FUNCTIONS

Review of Functions • Basic Operations on Functions • Composition of Functions • Real-Life Applications

Review of Functions

In Chapter 2 we defined a function, introduced function notation, and discussed methods for evaluating functions. We begin with a brief review of these topics.

Definition of a Function

A **function** is a set of ordered pairs in which no two different ordered pairs have the same first coordinate.

We can use the Vertical Line Test to determine whether a graph represents a function. Because no two ordered pairs of a function can have the same first coordinate, any vertical line will intersect the graph of a function at most once. For example, the graph in Fig. 10.1 is the graph of a function; the graph in Fig. 10.2 is not the graph of a function.

Figure 10.1

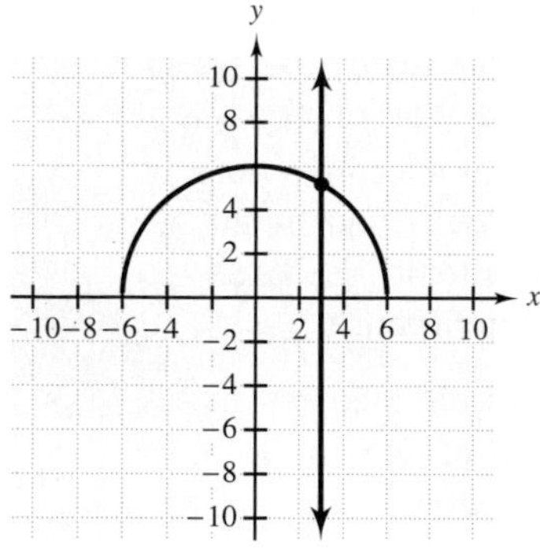

Figure 10.2

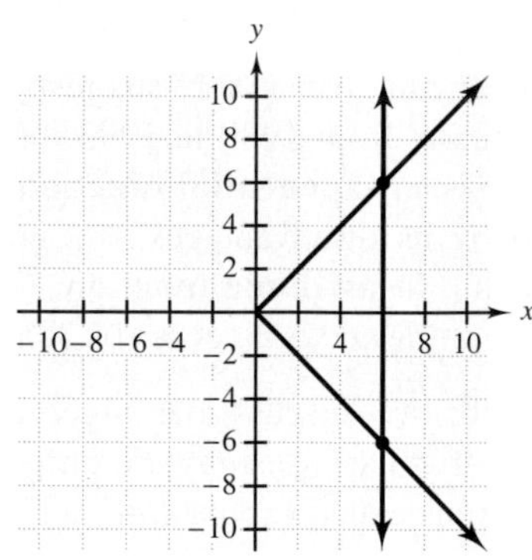

An equation in x and y describes a function if each value of x corresponds to a unique value of y. For example, because every real number has only one absolute value, the equation $y = |x|$ describes a function. However, the equation $x = |y|$ does not describe a function because its solution set includes, for instance, the pairs (4, 4) and (4, −4), which have the same first coordinate.

Although the definition of *function* refers specifically to a set of ordered pairs, we usually call equations such as $y = |x|$ functions because their solution sets are functions.

The **domain** of a function is the set of all first coordinates of the ordered pairs, and the **range** is the set of all second coordinates of the ordered pairs.

Note: Recall that a graph can be used to estimate the domain and range of a function.

Basic Operations on Functions

In previous chapters we have added, subtracted, multiplied, and divided algebraic expressions. Functions can also be combined with these operations.

Definitions of Basic Operations on Functions

If functions f and g have domains with at least one element in common, then the following operations are defined for each x in the domains of both f and g.

$(f + g)(x) = f(x) + g(x)$ Sum

$(f - g)(x) = f(x) - g(x)$ Difference

$(fg)(x) = f(x)g(x)$ Product

$\left(\frac{f}{g}\right)(x) = \frac{f(x)}{g(x)}$, for $g(x) \neq 0$ Quotient

EXAMPLE 1

Performing Basic Operations on Functions

Let $f(x) = x^2 + 2$ and $g(x) = |x - 1|$. Evaluate each of the following.

(a) $(f + g)(1)$

(b) $(f - g)(-1)$

(c) $(fg)(2)$

(d) $\left(\frac{f}{g}\right)(0)$

Solution

(a) Both f and g are defined at 1.

$$\begin{aligned}(f + g)(1) &= f(1) + g(1) && \text{Definition of sum}\\ &= (1^2 + 2) + |1 - 1| && \text{Evaluate each function.}\\ &= 3\end{aligned}$$

(b) Both f and g are defined at -1.

$$\begin{aligned}(f - g)(-1) &= f(-1) - g(-1) && \text{Definition of difference}\\ &= [(-1)^2 + 2] - |-1 - 1| && \text{Evaluate each function.}\\ &= [1 + 2] - |-2|\\ &= 1\end{aligned}$$

(c) Both f and g are defined at 2.

$$\begin{aligned}(fg)(2) &= f(2)g(2) && \text{Definition of product}\\ &= (2^2 + 2)\,|2 - 1| && \text{Evaluate each function.}\\ &= (4 + 2)\,|1|\\ &= 6\end{aligned}$$

(d) Both f and g are defined at 0 and $g(0) \neq 0$.

$$\left(\frac{f}{g}\right)(0) = \frac{f(0)}{g(0)} \quad \text{Definition of quotient}$$

$$= \frac{0^2 + 2}{|0 - 1|} \quad \text{Evaluate each function.}$$

$$= \frac{2}{1}$$

$$= 2$$

EXAMPLE 2

Performing Basic Operations on Functions

Let $f(x) = 2x + 3$ and $g(x) = x - 5$. Determine each of the following.

(a) $(f + g)(x)$ (b) $(f - g)(x)$

(c) $(fg)(x)$ (d) $\left(\frac{f}{g}\right)(x)$

Solution

Note that the domain for both functions f and g is **R**. Therefore, the operations can be performed for all values of x except as indicated in part (d).

(a) $(f + g)(x) = f(x) + g(x) = (2x + 3) + (x - 5) = 3x - 2$

(b) $(f - g)(x) = f(x) - g(x) = (2x + 3) - (x - 5) = x + 8$

(c) $(fg)(x) = f(x)g(x) = (2x + 3)(x - 5) = 2x^2 - 7x - 15$

(d) $\left(\frac{f}{g}\right)(x) = \frac{f(x)}{g(x)} = \frac{2x + 3}{x - 5} \quad (x \neq 5)$

Note that 5 is in the domain of both functions f and g, but because $g(5) = 0$, the domain of f/g does not include 5.

EXAMPLE 3

Determining the Domain of a Combined Function

Let $f(x) = \sqrt{x + 3}$ and $g(x) = \sqrt{2 - x}$. What is the domain of $(f + g)$?

Solution

$$(f + g)(x) = f(x) + g(x) = \sqrt{x + 3} + \sqrt{2 - x}$$

The domain of $(f + g)$ is the intersection of the domains of f and g. From the graphs of f and g in Fig. 10.3, we can estimate that the intersection of the domains of f and g is the interval of x-values between -3 and 2, inclusive.

Figure 10.3 also shows the graph of $f + g$, whose domain appears to be the same interval.

Figure 10.3

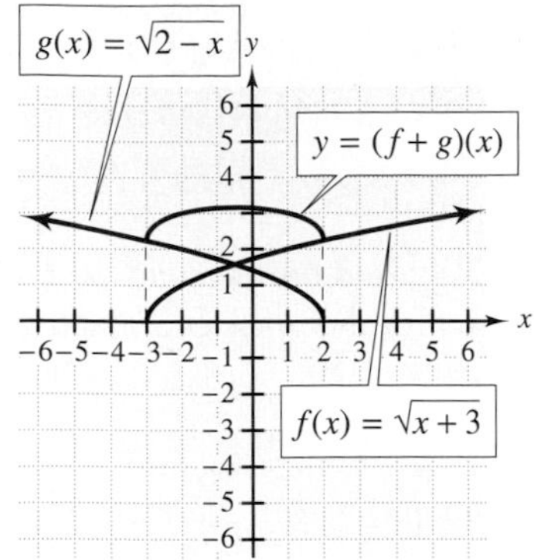

These estimates can be verified algebraically.

Function	*Requirement*	*Domain*
f	$x + 3 \geq 0$	$x \geq -3$
g	$2 - x \geq 0$	$x \leq 2$

Because the intersection of the domains is $-3 \leq x \leq 2$, the domain of $f + g$ is $[-3, 2]$.

EXAMPLE 4

Determining the Domain of a Combined Function

Let $f(x) = x^2 - 16$ and $g(x) = 2x - 8$. What is the domain of f/g?

Solution

Although the domain of both functions f and g is the set of all real numbers, f/g is defined only if $g(x) \neq 0$. Because $g(x) = 0$ if $x = 4$, the domain of f/g is the set of all real numbers *except* 4.

The process of determining f/g will also reveal the domain of the function.

$$\left(\frac{f}{g}\right)(x) = \frac{f(x)}{g(x)}$$

$$= \frac{x^2 - 16}{2x - 8}$$

$$= \frac{(x + 4)(x - 4)}{2(x - 4)} \qquad \text{Factor.}$$

$$= \frac{x + 4}{2} \qquad \text{Divide out the common factor } x - 4.$$

When we divide out the common factor $x - 4$, we are assuming that $x - 4$ is not 0, and so $x \neq 4$. That is why 4 must be excluded from the domain of f/g, even though the simplified expression for f/g appears to have no restricted values.

Composition of Functions

Another way to combine functions is to *compose* the functions.

Definition of Composition

The **composition** of functions f and g is a function $f \circ g$, where $(f \circ g)(x) = f(g(x))$. The domain of $f \circ g$ is the set of all x in the domain of g for which $g(x)$ is in the domain of f.

Note: The notation $f \circ g$, used for the *composition* of f and g, does not mean the same thing as fg, which is used for the *product* of f and g.

EXAMPLE 5

Composition of Two Functions

Let $f(x) = x^2 - 1$ and $g(x) = x + 3$. Determine each of the following.

(a) $(f \circ g)(-2)$ (b) $(g \circ f)(-2)$

(c) $(f \circ g)(x)$ (d) $(g \circ f)(x)$

LEARNING TIP

To evaluate the composite function $(f \circ g)(x)$, write $f(g(x))$, and begin by evaluating $g(x)$. Then substitute the value or expression for $g(x)$ for the variable in the definition of f.

Solution

(a) Because $g(-2) = -2 + 3 = 1$,

$$(f \circ g)(-2) = f(g(-2)) = f(1) = 1^2 - 1 = 0$$

(b) Because $f(-2) = (-2)^2 - 1 = 4 - 1 = 3$,

$$(g \circ f)(-2) = g(f(-2)) = g(3) = 3 + 3 = 6$$

(c) We determine $(f \circ g)(x)$ by substituting $g(x)$ for all x in $f(x)$.

$$\begin{aligned}(f \circ g)(x) &= f(g(x)) \\ &= f(x + 3) \\ &= (x + 3)^2 - 1 \\ &= x^2 + 6x + 9 - 1 \\ &= x^2 + 6x + 8\end{aligned}$$

(d) $(g \circ f)(x) = g(f(x)) = g(x^2 - 1) = (x^2 - 1) + 3 = x^2 + 2$

From Example 5 we observe that the operation of composition is not commutative. In general, $f \circ g \neq g \circ f$.

Sometimes we need to regard a function as the composition of simpler functions. For instance, $F(x) = (x + 1)^2$ consists of two operations (adding 1 and squaring), so F can be written as the composition of $g(x) = x + 1$ (adding 1) and $h(x) = x^2$ (squaring). We can write $F(x) = (h \circ g)(x)$ because

$$(h \circ g)(x) = h(g(x)) = h(x + 1) = (x + 1)^2 = F(x)$$

A given function can be written as the composition of two other functions.

EXAMPLE 6

Writing a Function as a Composite Function

Use a composition of $f(x) = 2x - 1$ and $g(x) = \dfrac{1}{x}$ to write the given function h.

(a) $h(x) = \dfrac{2}{x} - 1$ (b) $h(x) = 4x - 3$

Think About It

Suppose that the rule for function f is take three steps forward and that the rule for function g is turn right 90°. If you follow the rule for $f \circ g$, will you arrive at the same point as if you follow the rule for $g \circ f$?

Solution

(a) $h(x) = \frac{2}{x} - 1 = 2 \cdot \frac{1}{x} - 1 = 2 \cdot g(x) - 1 = f(g(x)) = (f \circ g)(x)$

(b) $h(x) = 4x - 3 = 4x - 2 - 1 = 2(2x - 1) - 1$
$= 2 \cdot f(x) - 1 = f(f(x)) = (f \circ f)(x)$

Real-Life Applications

In certain applications one variable may be a function of a second variable, which, in turn, is a function of a third variable. To describe how the first and third variables are related, a composite function can be used.

EXAMPLE 7

Distance, Time, and Fuel Consumption

The distance d (in miles) a race car has traveled after t minutes is given by $d(t) = 2.5t$. The number f of gallons of fuel remaining after d miles is given by $f(d) = 20 - 0.125d$.

(a) How far has the car traveled after 20 minutes?

(b) How much fuel remains after 20 minutes?

(c) Write a function to describe the amount of fuel remaining after t minutes.

(d) Use the function in part (c) to calculate the amount of fuel remaining after 20 minutes. Compare the result with that of part (b).

Solution

(a) $d(20) = 2.5(20) = 50$ Evaluate $d(t)$ for $t = 20$.

After 20 minutes, the car has traveled 50 miles.

(b) $f(50) = 20 - 0.125(50)$ From part (a), when $t = 20$, $d = 50$.
$= 13.75$ Evaluate $f(d)$ for $d = 50$.

After 20 minutes, 13.75 gallons of fuel remain.

(c) Fuel f is a function of distance d, which is a function of time t. To relate fuel and time directly, we compose functions f and d.

$$(f \circ d)(t) = f(d(t)) = f(2.5t) = 20 - 0.125(2.5t) = 20 - 0.3125t$$

(d) $(f \circ d)(20) = 20 - 0.3125(20) = 20 - 6.25 = 13.75$

As in part (b), 13.75 gallons of fuel remain after 20 minutes.

10.1 QUICK REFERENCE

Review of Functions

- A **function** is a set of ordered pairs in which no two different ordered pairs have the same first coordinate.
- The **domain** of a function is the set of all first coordinates of the ordered pairs; the **range** is the set of all second coordinates.

- A relation is a function if
  1. the graph of the relation passes the Vertical Line Test.
  2. each element of the domain corresponds to a unique element of the range.

Basic Operations on Functions

- If functions f and g have domains with at least one element in common, then the following operations are defined for each x in the domains of both f and g:
  1. $(f + g)(x) = f(x) + g(x)$ — Sum
  2. $(f - g)(x) = f(x) - g(x)$ — Difference
  3. $(fg)(x) = f(x)g(x)$ — Product
  4. $\left(\frac{f}{g}\right)(x) = \frac{f(x)}{g(x)}$, for $g(x) \neq 0$ — Quotient
- The domain of $f + g$, $f - g$, and fg is the intersection of the domains of f and g. The domain of f/g is the intersection of the domains of f and g, excluding any x-values for which $g(x) = 0$.

Composition of Functions

- The **composition** of functions f and g is a function $f \circ g$ where $(f \circ g)(x) = f(g(x))$. The domain of $f \circ g$ is the set of all x in the domain of g for which $g(x)$ is in the domain of f.
- In certain special cases, $(f \circ g)(x) = (g \circ f)(x)$, but in general, $f \circ g$ and $g \circ f$ are not equal.

10.1 EXERCISES

Concepts and Skills

1. Suppose that $f(x) = x - 4$ and $g(x) = 3x^2$. Describe two methods for determining $(f + g)(1)$.

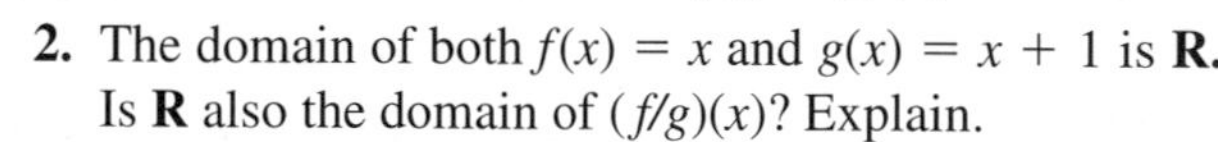

2. The domain of both $f(x) = x$ and $g(x) = x + 1$ is **R.** Is **R** also the domain of $(f/g)(x)$? Explain.

In Exercises 3–10, determine the domain of the function.

3. $g(x) = |2x^2 - x - 1|$

4. $h(x) = 4 - 3|2 - x|$

5. $f(x) = \sqrt{1 - 2x}$

6. $g(x) = \dfrac{x}{\sqrt{x + 5}}$

7. $h(x) = \dfrac{x + 4}{x^3 + 3x^2 - 4x}$

8. $f(x) = \dfrac{x^2 - 9}{x^2 + 25}$

9. $g(x) = \sqrt{x^2 - 16}$

10. $f(x) = \sqrt{x^2 + 2x - 8}$

In Exercises 11–16, for the given functions f and g, determine each of the following.

(a) $(f + g)(-1)$ (b) $(f - g)(1)$

(c) $(fg)(0)$ (d) $\left(\dfrac{f}{g}\right)(0)$

11. $f(x) = 3x$, $g(x) = x - 1$

12. $f(x) = \sqrt{x + 1}$, $g(x) = 3$

13. $f(x) = |2x + 1|$, $g(x) = x^2$

14. $f(x) = \dfrac{1}{x + 2}$, $g(x) = \dfrac{1}{(x + 2)^2}$

15. $f(x) = 2^x$, $g(x) = 2^{-x}$

16. $f(x) = 3^{x-1}$, $g(x) = 3^{1-x}$

In Exercises 17–22, use the given functions f and g to evaluate the combined function in each part.

17. Let $f(x) = -4x$ and $g(x) = 3 - 2x$.

(a) $(f + g)(-3)$ (b) $(f - g)(4)$

(c) $(fg)(0)$ (d) $\left(\dfrac{f}{g}\right)(1)$

18. Let $f(x) = 3x + 1$ and $g(x) = x - 6$.

(a) $(f - g)(-1.2)$ (b) $(f + g)\left(-\dfrac{1}{2}\right)$

(c) $\left(\dfrac{f}{g}\right)(2.5)$ (d) $(fg)\left(\dfrac{2}{3}\right)$

19. Let $f(x) = |x - 3|$ and $g(x) = 2x - 3$.
(a) $(f + g)(2)$ (b) $(f - g)(-2)$
(c) $(fg)(0)$ (d) $\left(\dfrac{f}{g}\right)(-1)$

20. Let $f(x) = \sqrt{5 - x}$ and $g(x) = x - 5$.
(a) $(f + g)(4)$ (b) $(f - g)(-4)$
(c) $(fg)(5)$ (d) $\left(\dfrac{f}{g}\right)(1)$

21. Let $f(x) = x^2 - 2$ and $g(x) = \sqrt[3]{x}$.
(a) $(f + g)(0)$ (b) $(f - g)(1)$
(c) $\left(\dfrac{f}{g}\right)(27)$ (d) $(fg)(8)$

22. Let $f(x) = (x + 1)^{1/2}$ and $g(x) = x^3 + 2x$
(a) $(f + g)(3)$ (b) $(f - g)(-1)$
(c) $(fg)(1)$ (d) $\left(\dfrac{g}{f}\right)(0)$

23. Suppose that $f(x) = x^2 + 3x$, $g(x) = x$, and $h(x) = x + 3$. Show that $(f/g)(x) = x + 3$. Is the domain of f/g the same as the domain of h? Explain.

24. Suppose that $A = \{x \mid x < 4\}$ is the domain of function f and $B = \{x \mid x > -3\}$ is the domain of function g. Use interval notation to express the domain of $f + g$, $f - g$, and fg.

In Exercises 25–30, functions f and g are given. Determine the simplified combined function in each part and state its domain.

25. Let $f(x) = x^2 - 9$ and $g(x) = x + 3$.
(a) $(f - g)(x)$ (b) $(f + g)(x)$
(c) $\left(\dfrac{f}{g}\right)(x)$ (d) $(fg)(x)$

26. Let $f(x) = x + 3$ and $g(x) = x^2 + 4x + 3$.
(a) $(f + g)(x)$ (b) $(f - g)(x)$
(c) $(fg)(x)$ (d) $\left(\dfrac{f}{g}\right)(x)$

27. Let $f(x) = x^2 + 4$ and $g(x) = x^2 - 4$.
(a) $\left(\dfrac{f}{g}\right)(x)$ (b) $\left(\dfrac{g}{f}\right)(x)$
(c) $\left(\dfrac{f}{f}\right)(x)$ (d) $\left(\dfrac{g}{g}\right)(x)$

28. Let $f(x) = x^2 - 5x + 6$ and $g(x) = x - 2$.
(a) $\left(\dfrac{g}{f}\right)(x)$ (b) $\left(\dfrac{f}{g}\right)(x)$ (c) $\left(\dfrac{g}{g}\right)(x)$ (d) $\left(\dfrac{f}{f}\right)(x)$

29. Let $f(x) = \dfrac{1}{x + 2}$ and $g(x) = \dfrac{1}{x - 3}$.
(a) $(f \quad g)(x)$ (b) $(f + g)(x)$
(c) $\left(\dfrac{f}{g}\right)(x)$ (d) $(fg)(x)$

30. Let $f(x) = \dfrac{x + 1}{x + 3}$ and $g(x) = \dfrac{x + 3}{2 - x}$.
(a) $(f + g)(x)$ (b) $(f - g)(x)$
(c) $\left(\dfrac{g}{f}\right)(x)$ (d) $\left(\dfrac{f}{g}\right)(x)$

In Exercises 31–34, let $f(x) = x^2 - 4$ and $g(x) = x - 1$. Determine the given combined function.

31. $(f - g)(2t)$ **32.** $(f + g)(-t)$

33. $\left(\dfrac{f}{g}\right)\left(\dfrac{t}{2}\right)$ **34.** $(fg)(t + 2)$

In Exercises 35–38, let $f(x) = x + 5$ and $g(x) = x^2 - 3$. Determine the given combined function.

35. $(f + g)(3t)$ **36.** $(f - g)(t + 1)$

37. $(fg)(-2t)$ **38.** $\left(\dfrac{f}{g}\right)\left(\dfrac{t}{3}\right)$

39. Explain how you can use the graphs of functions f and g to estimate the domain of $f + g$.

40. Let $f(x) = \sqrt{x + 3}$ and $g(x) = \sqrt{4 - x}$.
(a) Estimate the domains of f and g by producing their graphs.
(b) Use the results of part (a) to estimate the domain of $f + g$.
(c) Explain why the graph of $f + g$ confirms your estimate in part (b).

In Exercises 41–44, functions f and g are given. Determine the domain of the combined function given in each part.

41. Let $f(x) = \sqrt{2x - 3}$ and $g(x) = \sqrt{5 - x}$.
(a) $(f + g)(x)$
(b) $(f - g)(x)$

42. Let $f(x) = \sqrt{2 - x}$ and $g(x) = \sqrt{3 - x}$.
(a) $(f - g)(x)$ (b) $(f + g)(x)$

43. Let $f(x) = \sqrt{8 - x}$ and $g(x) = \sqrt{x - 1}$.
(a) $\left(\dfrac{f}{g}\right)(x)$ (b) $(fg)(x)$

44. Let $f(x) = \sqrt{2 - x}$ and $g(x) = \sqrt{x - 3}$.

(a) $(fg)(x)$ (b) $\left(\frac{g}{f}\right)(x)$

45. For two given functions f and g, determining the value of $(f \circ g)(5)$ requires two calculations. Describe these calculations in the order in which they are performed.

46. Suppose that $f = \{(1, 2), (2, 5), (4, -1)\}$ and that $g = \{(1, 3), (2, 5)\}$. Explain why $(g \circ f)(1)$ is defined but $(f \circ g)(1)$ is not.

In Exercises 47–50, functions f and g are given. Evaluate the composite function given in each part.

47. Let $f(x) = x - 3$ and $g(x) = 2 - x^2$.

(a) $(f \circ g)(1)$ (b) $(f \circ g)(0)$

(c) $(g \circ f)(3)$ (d) $(g \circ f)(-1)$

48. Let $f(x) = 2x + 1$ and $g(x) = x^3$.

(a) $(f \circ g)(-1)$ (b) $(f \circ g)(2)$

(c) $(g \circ f)(0)$ (d) $(g \circ f)(-1)$

49. Let $f(x) = x^2$ and $g(x) = 2x - 3$.

(a) $(g \circ f)(-3)$ (b) $(f \circ g)\left(-\frac{1}{2}\right)$

(c) $(g \circ g)\left(\frac{3}{2}\right)$ (d) $(f \circ f)(-1)$

50. Let $f(x) = x^2 - 2x + 3$ and $g(x) = x + 4$.

(a) $(f \circ g)(-5)$ (b) $(g \circ f)(3)$

(c) $(f \circ f)(0)$ (d) $(g \circ g)(2)$

In Exercises 51 and 52, for the given functions f and g, determine each of the following.

(a) $(f \circ g)(2)$ (b) $(g \circ f)(-1)$ (c) $(f \circ g)(-3)$

51. $f = \{(-2, 3), (2, -4), (1, 5), (-3, -2), (-1, 3)\}$

$g = \{(-2, 5), (2, 1), (-3, 2), (3, -2)\}$

52. $f = \{(2, 5), (3, 6), (4, 7), (-2, 1), (-1, 2)\}$

$g = \{(-5, -1), (-3, -2), (0, 4), (-1, 3), (2, 3)\}$

In Exercises 53–60, for the given functions f and g, determine $(f \circ g)(x)$ and $(g \circ f)(x)$.

53. $f(x) = 3 - x^2, \quad g(x) = 2x + 1$

54. $f(x) = 3x - 2, \quad g(x) = 4x - x^2$

55. $f(x) = x^2 - 3x + 5, \quad g(x) = 3x - 2$

56. $f(x) = 2x - 5, \quad g(x) = x^2 - 4x - 5$

57. $f(x) = x^3, \quad g(x) = \sqrt[3]{x - 2}$

58. $f(x) = \sqrt[3]{2x - 3}, \quad g(x) = -x^3$

59. $f(x) = 5, \quad g(x) = 2x - 3$

60. $f(x) = x^2 - 3, \quad g(x) = 7$

In Exercises 61–68, for the given functions f and g, determine $(f \circ f)(x)$ and $(g \circ g)(x)$.

61. $f(x) = 3x - 4, \quad g(x) = x + 5$

62. $f(x) = x - 2, \quad g(x) = 4 - 3x$

63. $f(x) = |x - 3|, \quad g(x) = x^2 + 3$

64. $f(x) = x^2 - 2, \quad g(x) = |x + 3|$

65. $f(x) = 2x^2 - 3x, \quad g(x) = 2 - x$

66. $f(x) = x - 3, \quad g(x) = 3x^2 - 2x$

67. $f(x) = 3, \quad g(x) = 2$

68. $f(x) = -5, \quad g(x) = -2$

In Exercises 69–72, show that $(f \circ g)(x) = x$ for each given pair of functions.

69. $f(x) = 2x + 3, \quad g(x) = \dfrac{x - 3}{2}$

70. $f(x) = 2x - 5, \quad g(x) = \dfrac{x + 5}{2}$

71. $f(x) = \dfrac{x}{3} + 2, \quad g(x) = 3x - 6$

72. $f(x) = \dfrac{x}{4} - 2, \quad g(x) = 4x + 8$

In Exercises 73–80, let $f(x) = x + 3$, $g(x) = 2x$, and $h(x) = x^2$. Use f, g, and h to write the function as a composition of two functions.

73. $F(x) = 2x + 6$

74. $H(x) = 4x^2$

75. $H(x) = 4x$

76. $G(x) = x + 6$

77. $G(x) = x^2 + 3$

78. $F(x) = (x + 3)^2$

79. $F(x) = 2x^2$

80. $G(x) = 2x + 3$

Real-Life Applications

81. On interstate trips, a driver averages 54 mph. The distance d (in miles) traveled in t hours is given by $d(t) = 54t$. Because the driver averages 25 miles per gallon, the number of gallons g used is given by $g(d) = d/25$. The cost per gallon is \$1.03, so the total fuel cost c is given by $c(g) = 1.03g$.

(a) Write a function describing the number of gallons used in t hours of travel.

(b) Write a function describing the total fuel cost in t hours of travel.

(c) Determine the total fuel cost of a 12-hour trip.

82. An oil tanker runs aground and springs a leak. The oil spreads out in a semicircular pattern from the shoreline. The distance r (in feet) from the tanker to the edge of the oil spill at time t (in minutes) is given by the function $r(t) = 20t$. If the area of the semicircle is given by $A(x) = \frac{1}{2}\pi x^2$, where x is the radius, write a function $A \circ r$ for the area covered by the oil at time t. What is the area of the oil spill after 5 minutes?

83. A party caterer charges \$5.00 per person for 100 people. For a gathering of less than 100 people, the caterer increases the price per person by \$0.05 for each person under 100 people. For example, 90 people is 10 less than 100, so the additional price per person is $10(0.05) = \$0.50$. Then the total cost is $(90)(5.50) = \$495.00$.

(a) Write a function $f(x)$ to represent the number of people who attend the gathering.

(b) Write a function $c(x)$ to represent the per-person cost.

(c) Write a function $T(x) = (fc)(x)$ to represent the total cost of the gathering.

(d) Use the function in part (c) to determine the total cost for 84 people.

84. The distance between the bases of a baseball diamond (a square) is 90 feet. A ball is hit directly down the third base line at a speed of 60 feet per second. If x is the distance from home plate to the ball, then x is a function of time t in seconds. (See figure.) Let $f(x)$ represent the distance from the ball to first base.

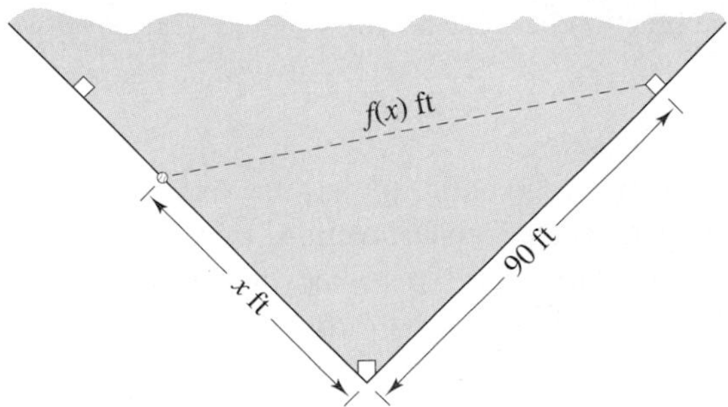

(a) Use the Pythagorean Theorem to write the function f.

(b) Write the function $x(t)$.

(c) Write the function $(f \circ x)(t)$. What does this function represent?

Group Project: ATM Terminals

The accompanying table shows that the number of automatic teller machine (ATM) terminals and the number of transactions at ATMs increased from 1985 to 1994. (Source: *Bank Network News*.)

Year	ATM Terminals (in thousands)	ATM Transactions (in millions)
1985	60.0	3565
1988	72.5	4480
1989	75.6	5116
1990	80.2	5751
1991	83.5	6418
1992	87.3	7206
1993	94.8	7705
1994	109.1	8334

The number of terminals (in thousands) can be modeled as a function of the number of years since 1980.

$$G(t) = 0.13t^3 - 3.4t^2 + 32t - 31.3$$

The number of transactions (in millions) can be modeled as a function of the number of terminals x (in thousands).

$$F(x) = 107x - 2840$$

85. Interpret $(F \circ G)(t)$.

86. Evaluate $G(14)$. Identify and describe the actual entry in the table for which $G(14)$ is an estimate.

87. Evaluate $(F \circ G)(14)$. Identify and describe the actual entry in the table for which $(F \circ G)(14)$ is an estimate.

88. Banks have steadily increased their fees for the use of ATM terminals by people who are not bank customers. What effect might this have on functions F and $F \circ G$?

Challenge

In Exercises 89–93, write each function $f(x)$ as a composition of two simpler functions g and h. State specifically what the functions g and h are.

89. $f(x) = \sqrt{x^2 - 5x + 6}$

90. $f(x) = \dfrac{1}{x^2 + 4x}$

91. $f(x) = (3x + 2)^4$

92. $f(x) = 3x^{1/2} - 5$

93. $f(x) = |2x + 5| + 8$

94. Let $f(x) = 3$ and $g(x) = \dfrac{1}{x - 3}$. Determine $(f \circ g)(x)$ and $(g \circ f)(x)$.

10.2 INVERSE FUNCTIONS

One-to-One Functions • Inverse Functions • Determining the Inverse

One-to-One Functions

We have defined a function as a set of ordered pairs in which no two ordered pairs have the same *first* coordinate. If, in addition, we require that no two ordered pairs have the same *second* coordinate, then we have the special case of a **one-to-one function.**

Think About It

Suppose a function associates each person with that person's mother. Is the function one-to-one?

Definition of a One-to-One Function

A function is a **one-to-one function** if no two different ordered pairs have the same second coordinate.

This means that for any value of y in the range of the function f, there is a *unique* value of x such that $f(x) = y$.

EXAMPLE 1

One-to-One Functions

Determine whether each function is a one-to-one function.

(a) $f = \{(2, 3), (4, 5), (6, 3)\}$

(b) $g = \{(1, 2), (2, 3), (3, 4), (4, 5), (5, 6)\}$

Solution

(a) Function f is not a one-to-one function because the ordered pairs (2, 3) and (6, 3) are different ordered pairs with the same second coordinate.

(b) Function g is a one-to-one function because no two ordered pairs have the same second coordinate.

To determine whether a graph represents a function, we use the Vertical Line Test. To determine whether a graph represents a one-to-one function, we use the Horizontal Line Test. Because no two ordered pairs of a one-to-one function have the same second coordinate, a horizontal line cannot intersect the graph of the function at more than one point.

The Horizontal Line Test for a One-to-One Function

A function is a one-to-one function if and only if no horizontal line intersects the graph of the function at more than one point.

The graphs in Figs. 10.4 and 10.5 both pass the Vertical Line Test, so both represent functions.

Figure 10.4

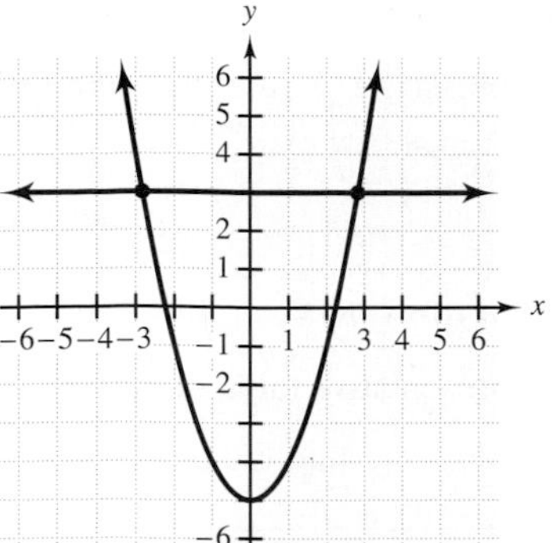

Figure 10.5

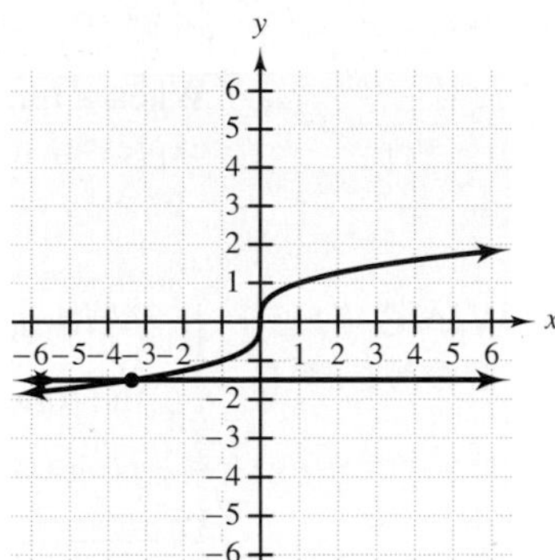

LEARNING TIP

Remember that a one-to-one function must pass both the Vertical Line Test and the Horizontal Line Test.

Using the Horizontal Line Test, we find that the graph in Fig. 10.4 does not represent a one-to-one function, but the graph in Fig. 10.5 does.

Inverse Functions

If the corresponding x- and y-coordinates of a one-to-one function f are interchanged, the resulting set of ordered pairs is also a function. This new function is called the **inverse function** and is written f^{-1} (read "f inverse").

Note: In the notation f^{-1}, the -1 is not an exponent. We read f^{-1} as a single symbol that represents an inverse function.

EXPLORING THE CONCEPT

Inverse Functions

Consider the function

$$f = \{(-3, 5), (-1, -4), (3, -2), (6, 1)\}$$

To write the inverse function, interchange the x- and y-coordinates of each ordered pair.

$$f^{-1} = \{(5, -3), (-4, -1), (-2, 3), (1, 6)\}$$

Note the domains and ranges of f and f^{-1}.

Domain of f: $\{-3, -1, 3, 6\}$ Range of f: $\{-4, -2, 1, 5\}$

Range of f^{-1}: $\{-3, -1, 3, 6\}$ Domain of f^{-1}: $\{-4, -2, 1, 5\}$

The domain of f is the same as the range of f^{-1}, and the range of f is the same as the domain of f^{-1}.

If we sketch the graphs of f and f^{-1} on the same coordinate system (see Fig. 10.6), we observe that the points lie the same distance on either side of the diagonal line whose equation is $y = x$. This line is called the **line of symmetry.**

Figure 10.6

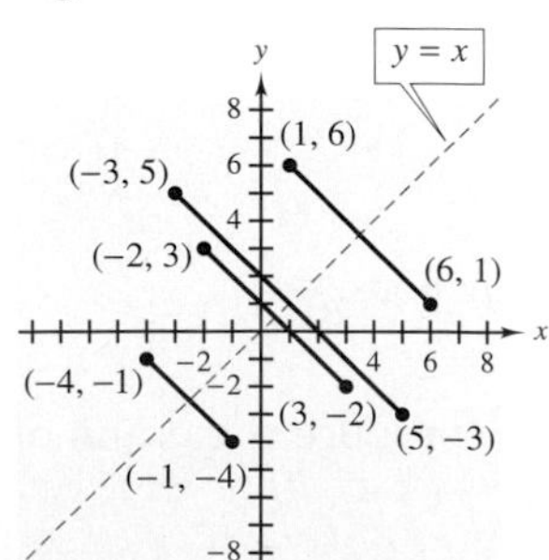

Note: If we fold the graph along the line of symmetry, the points coincide.

When a one-to-one function is described by a set of ordered pairs, forming the inverse function is simply a matter of exchanging coordinates.

When a function f is expressed as a rule, then the inverse function g often also can be expressed as a rule.

EXPLORING THE CONCEPT

Writing an Inverse Function

The following is a representative table of values for $f(x) = x + 12$.

x	-3	-2	-1	0	1	2	3
$f(x)$	9	10	11	12	13	14	15

To write a table of values for the inverse function g, we interchange the x- and y-coordinates in the table for f.

x	9	10	11	12	13	14	15
$g(x)$	-3	-2	-1	0	1	2	3

From the table for g, we see that each y-coordinate is 12 less than the corresponding x-coordinate. Thus the inverse function can be expressed as the rule $g(x) = x - 12$.

Note that function f adds 12 to each value of x, whereas the inverse function g subtracts 12 from each value of x. This suggests that an inverse function f^{-1} reverses the operations indicated in the rule for f. For example, if $f(x) = 3x$, then $f^{-1}(x) = \frac{x}{3}$; if $f(x) = x^3$, then $f^{-1}(x) = \sqrt[3]{x}$.

The definition of inverse function provides a way to verify that our conjecture is correct.

Definition of an Inverse Function

Functions f and g are **inverse functions** if $(f \circ g)(x) = x$ for each x in the domain of g and $(g \circ f)(x) = x$ for each x in the domain of f. We write $g = f^{-1}$ and $f = g^{-1}$.

We can use this definition to verify that $f(x) = x + 12$ and $g(x) = x - 12$ are inverse functions.

$$(f \circ g)(x) = f(g(x)) = f(x - 12) = (x - 12) + 12 = x$$
$$(g \circ f)(x) = g(f(x)) = g(x + 12) = (x + 12) - 12 = x$$

Because $(f \circ g)(x) = x$ and $(g \circ f)(x) = x$, f and g are inverse functions.

Draw Inv

By reversing the coordinates of a function f, a calculator can produce the graph of the inverse function g.

EXAMPLE 2

Verifying Inverse Functions

Verify that the given pair of functions are inverses. Produce the graphs of each pair and observe that the graphs are symmetric with respect to the line $y = x$.

(a) $f(x) = \dfrac{x - 2}{3}$ and $g(x) = 3x + 2$

(b) $f(x) = \sqrt[3]{x + 2}$ and $g(x) = x^3 - 2$

Solution

(a) We apply the definition of inverse functions.

$$(f \circ g)(x) = f(g(x)) = f(3x + 2) = \frac{(3x + 2) - 2}{3} = \frac{3x}{3} = x$$

$$(g \circ f)(x) = g(f(x)) = g\left(\frac{x - 2}{3}\right) = 3 \cdot \frac{x - 2}{3} + 2 = (x - 2) + 2 = x$$

Because $(f \circ g)(x) = x$ and $(g \circ f)(x) = x$, f and g are inverse functions. Figure 10.7 shows their graphs and the line of symmetry.

LEARNING TIP

Rotating a figure or viewing window so that the line $y = x$ is vertical may help you to see the symmetry of the graphs of a function and its inverse.

Figure 10.7

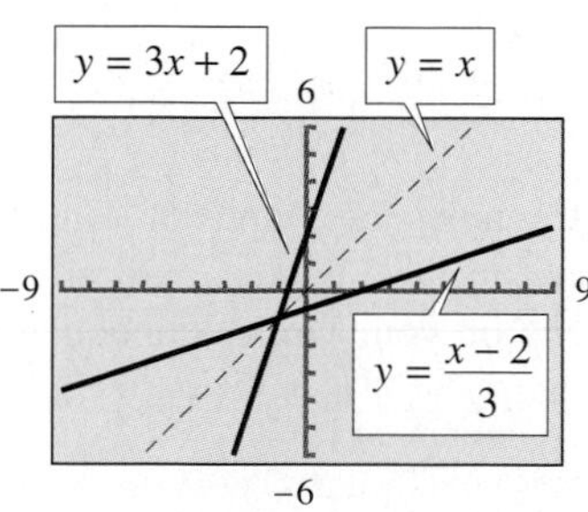

Figure 10.8

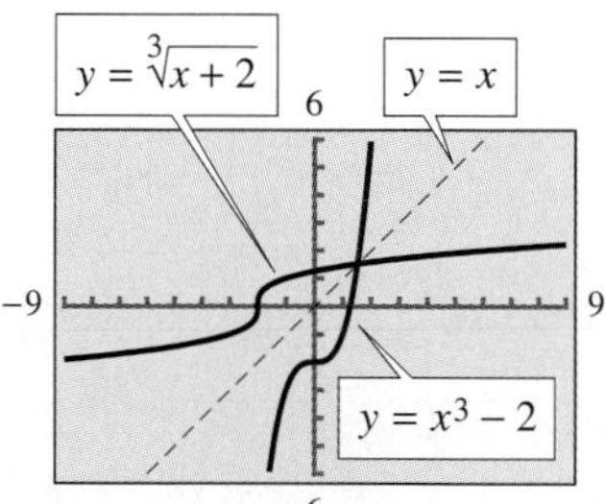

(b) $(f \circ g)(x) = f(x^3 - 2) = \sqrt[3]{(x^3 - 2) + 2} = \sqrt[3]{x^3} = x$

$(g \circ f)(x) = g\left(\sqrt[3]{x + 2}\right) = \left(\sqrt[3]{x + 2}\right)^3 - 2 = (x + 2) - 2 = x$

Because $(f \circ g)(x) = x$ and $(g \circ f)(x) = x$, f and g are inverse functions. Figure 10.8 shows their graphs and the line of symmetry.

In Example 2 the graphs of the inverse functions would coincide if they were folded along the diagonal line shown in the figures. In general, the graphs of f and f^{-1} are symmetric with respect to the line of symmetry $y = x$.

Determining the Inverse

If a one-to-one function is defined by a rule, we can often determine a rule for the inverse function by interchanging the roles of x and y and solving for y.

Determining the Inverse of a Function

To determine the inverse of a function f, perform the following steps:

1. Replace the function notation $f(x)$ with y.
2. Interchange x and y.
3. Solve for y, and replace y with $f^{-1}(x)$.

EXAMPLE 3

Determining an Inverse Function

Determine the inverse of the given function.

(a) $f(x) = 2x - 3$ (b) $g(x) = \sqrt[3]{x + 5}$

Solution

(a)

$$f(x) = 2x - 3$$

$$y = 2x - 3 \quad \text{Replace } f(x) \text{ with } y.$$

$$x = 2y - 3 \quad \text{Interchange } x \text{ and } y.$$

$$x + 3 = 2y \quad \text{Solve for } y.$$

$$\frac{x + 3}{2} = y$$

$$f^{-1}(x) = \frac{x + 3}{2} \quad \text{Replace } y \text{ with } f^{-1}(x).$$

Figure 10.9

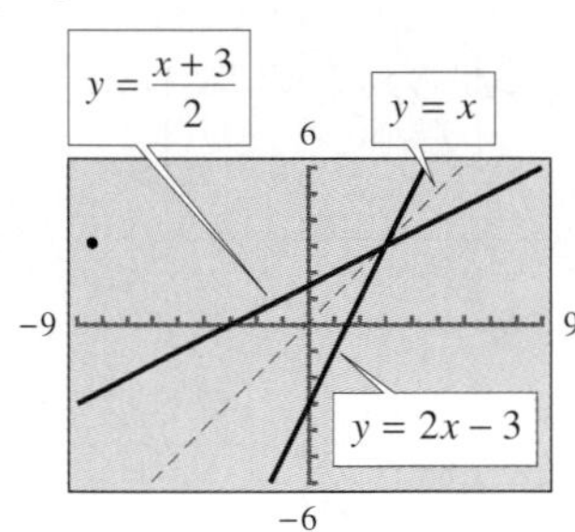

Figure 10.9 shows the graphs of f and f^{-1}. Notice that the graphs are symmetric with respect to the line $y = x$. Producing the graphs of f and f^{-1} gives you a visual check on your work. You can also use the definition of inverse functions to check that $(f \circ f^{-1})(x) = (f^{-1} \circ f)(x) = x$.

(b)

$$g(x) = \sqrt[3]{x + 5}$$

$$y = \sqrt[3]{x + 5} \quad \text{Replace } g(x) \text{ with } y.$$

$$x = \sqrt[3]{y + 5} \quad \text{Interchange } x \text{ and } y.$$

$$x^3 = y + 5 \quad \text{Cube both sides of the equation.}$$

$$x^3 - 5 = y \quad \text{Solve for } y.$$

$$g^{-1}(x) = x^3 - 5 \quad \text{Replace } y \text{ with } g^{-1}(x).$$

10.2 QUICK REFERENCE

One-to-One Functions

- A function is a **one-to-one function** if no two different ordered pairs have the same second coordinate.
- The graph of a one-to-one function passes the Horizontal Line Test: No horizontal line intersects the graph more than once.

Inverse Functions

- If the x- and y-coordinates of a one-to-one function f are interchanged, the resulting function is the **inverse function** f^{-1}.
- The domain of f is the same as the range of f^{-1}, and the range of f is the same as the domain of f^{-1}.
- The graphs of f and f^{-1} are symmetric with respect to the line $y = x$.
- An inverse function f^{-1} reverses the rule of function f.
- Functions f and g are inverse functions if $(f \circ g)(x) = x$ for each x in the domain of g and $(g \circ f)(x) = x$ for each x in the domain of f.
- This definition can be used to determine whether two functions are inverse functions.

Determining the Inverse

- To determine the inverse of a function f, perform the following steps:
  1. Replace the function notation $f(x)$ with y.
  2. Interchange x and y.
  3. Solve for y and replace y with $f^{-1}(x)$.

10.2 EXERCISES

Concepts and Skills

1. Suppose that no two ordered pairs of a set have the same second coordinate. Is the set a one-to-one function? Explain.
2. Describe how the Horizontal Line Test is used to determine whether a function is a one-to-one function.

In Exercises 3–6, determine whether the given function is a one-to-one function.

3. $\{(5, 2), (2, 5), (3, 2), (6, -4)\}$
4. $\{(5, 2), (3, 0), (1, -2), (7, 4)\}$
5. $\{(-3, -1), (-1, 1), (3, 5), (1, 3)\}$
6. $\{(-3, -1), (-1, -3), (3, -2), (2, -1)\}$

In Exercises 7–12, determine whether the given graph represents a one-to-one function.

7.
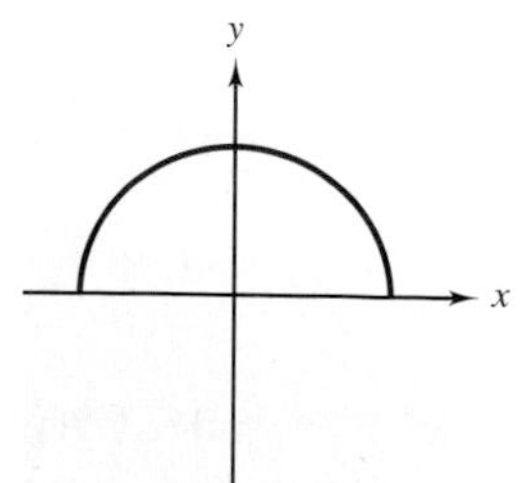

8.
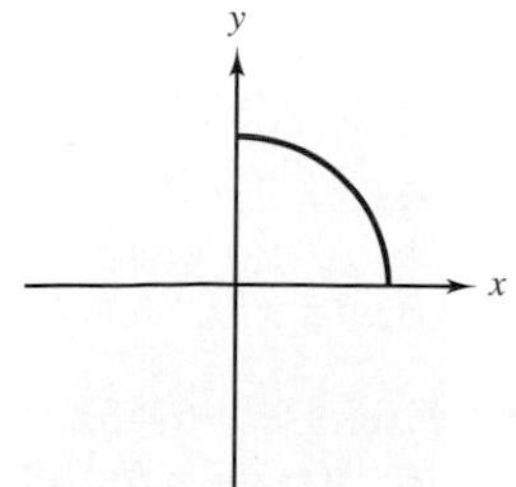

9. 10.
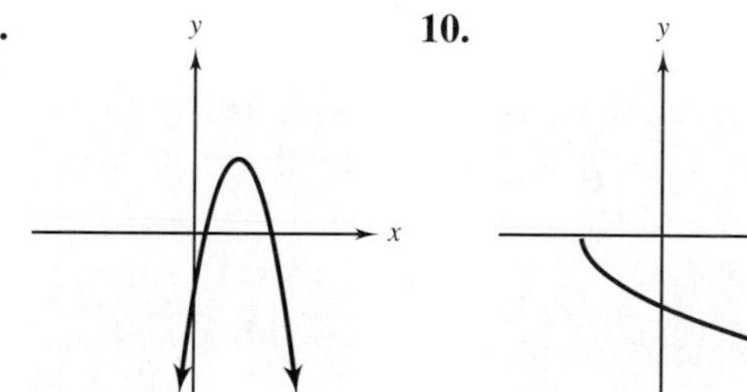

11.
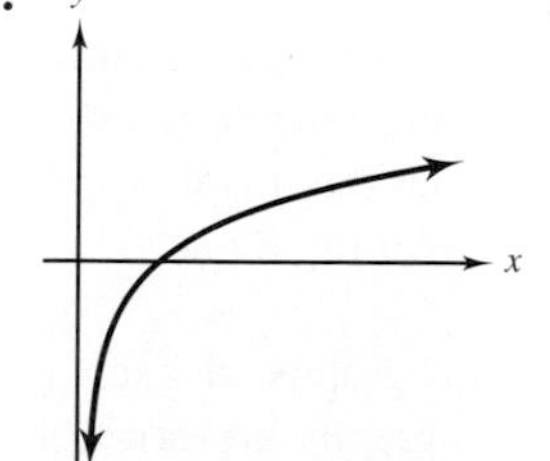

12.
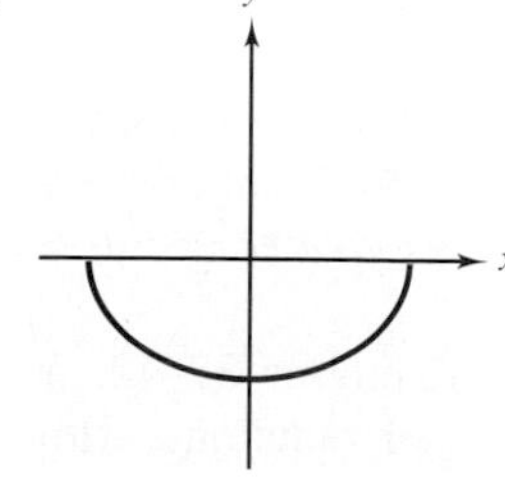

In Exercises 13–18, produce the graph of the given function. From the graph determine whether the function is one-to-one.

13. $f(x) = x^3 - 3x^2 + 6x + 4$
14. $f(x) = x^3 - 2x^2 - 7x + 2$
15. $f(x) = |x - 7| - |x + 5|$

16. $f(x) = |x| + 3$

17. $f(x) = x^3 2^x$

18. $f(x) = x\sqrt{25 - x^2}$

In Exercises 19–22, determine f^{-1}. State the domain and range of f^{-1}.

19. $f = \{(1, 2), (3, 4), (5, 6), (7, -1)\}$

20. $f = \{(-1, 2), (-3, 4), (-5, -7), (5, 8)\}$

21. $f = \{(3, 2), (4, 3), (-5, 1), (-7, -2)\}$

22. $f = \{(5, -6), (7, -8), (-8, 5), (-6, 7)\}$

In Exercises 23–26, f is a one-to-one function. Suppose that $f(3) = 4$, $f(5) = 7$, $f(-2) = -3$, and $f(-4) = -5$. Evaluate f^{-1} as indicated.

23. $f^{-1}(-3)$ **24.** $f^{-1}(-5)$

25. $f^{-1}(4)$ **26.** $f^{-1}(7)$

In Exercises 27–30, use the integer setting to produce the graph of $f(x) = 2^x$. Then trace the graph to evaluate f^{-1} as indicated.

27. $f^{-1}(64)$ **28.** $f^{-1}(8)$

29. $f^{-1}(0.0625)$ **30.** $f^{-1}(0.25)$

In Exercises 31–34, use the integer setting to produce the graph of $f(x) = 2^{-x}$. Then trace the graph to evaluate f^{-1} as indicated.

31. $f^{-1}(32)$ **32.** $f^{-1}(2)$

33. $f^{-1}(0.125)$ **34.** $f^{-1}(0.5)$

35. Suppose that $f(x) = x$. If you press the x^{-1} key on your calculator, will you obtain $f^{-1}(x)$? Explain.

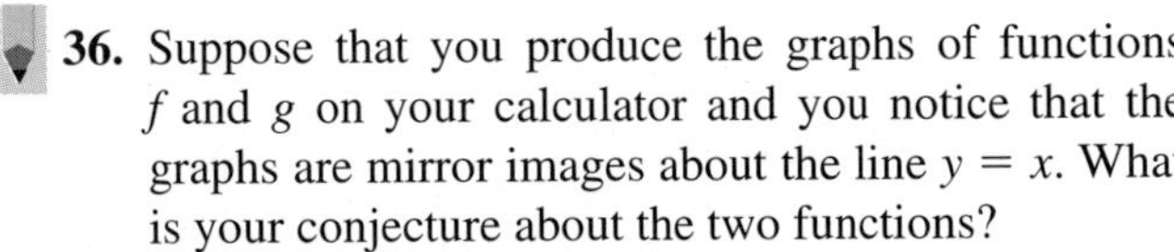

36. Suppose that you produce the graphs of functions f and g on your calculator and you notice that the graphs are mirror images about the line $y = x$. What is your conjecture about the two functions?

In Exercises 37–42, produce the graphs of each given pair of functions. Although the graphs are not conclusive, decide whether it is reasonable to believe that the functions may be inverses.

37. $f(x) = 2x + 1, \quad g(x) = \dfrac{x - 1}{2}$

38. $f(x) = 3x + 1, \quad g(x) = 3x - 2$

39. $f(x) = x^2 + 1, \quad g(x) = 1 - x^2$

40. $f(x) = |x - 1|, \quad g(x) = x + 1$

41. $f(x) = x^3 - 2, \quad g(x) = \sqrt[3]{x + 2}$

42. $f(x) = \sqrt[3]{x} - 1, \quad g(x) = x^3 + 1$

You may be able to use your calculator to produce the graph of the composition of two functions entered as Y_1 and Y_2. Enter $Y_1(Y_2)$ as Y_3 and produce only the graph of Y_3.

In Exercises 43–48, use this method and the integer setting to produce the graph of the composition of f and g. Trace the graph to decide whether f and g are inverse functions.

43. $f(x) = \dfrac{3}{4}x$, $g(x) = \dfrac{4}{3}x$ **44.** $f(x) = 2 - 3x$, $g(x) = \dfrac{1}{3}(2 - x)$

45. $f(x) = x - 1$, $g(x) = \dfrac{1}{x - 1}$ **46.** $f(x) = 2 - x$, $g(x) = x - 2$

47. $f(x) = x^{1/5}$, $g(x) = x^5$ **48.** $f(x) = -3(x - 1)$, $g(x) = 1 - \dfrac{x}{3}$

In Exercises 49–52, use your calculator to produce the graph of the given function. Then sketch the graph of the inverse function.

49. $f(x) = 2^x$ **50.** $f(x) = (0.7)^x$

51. $f(x) = x^3 - 1$ **52.** $f(x) = 0.2x^3 + 1$

53. Explain why a linear function always has an inverse.

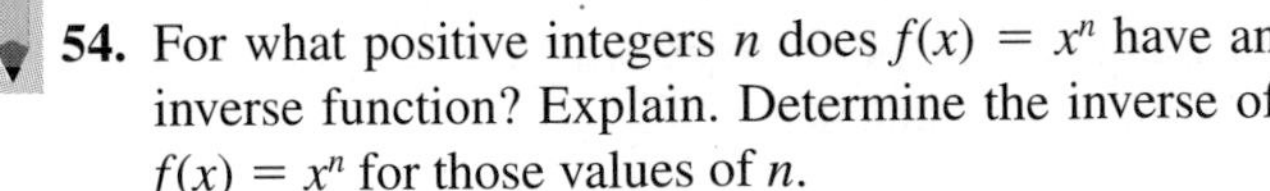

54. For what positive integers n does $f(x) = x^n$ have an inverse function? Explain. Determine the inverse of $f(x) = x^n$ for those values of n.

In Exercises 55–60, determine whether the pairs are inverses by showing that $(f \circ g)(x) = x$ and $(g \circ f)(x) = x$.

55. $f(x) = 2x, \quad g(x) = -2x$

56. $f(x) = x + 2, \quad g(x) = x - 2$

57. $f(x) = \sqrt[3]{x + 1}, \quad g(x) = x^3 - 1$

58. $f(x) = \dfrac{1}{4}x^2, x \geq 0; \quad g(x) = 2\sqrt{x}, x \geq 0$

59. $f(x) = \dfrac{3}{x}, \quad g(x) = \dfrac{3}{x}$

60. $f(x) = x^{1/3}, \quad g(x) = x^{-1/3}$

In Exercises 61–82, determine f^{-1} and verify that $(f \circ f^{-1})(x) = x$ for all x in the domain of f^{-1} and that $(f^{-1} \circ f)(x) = x$ for all x in the domain of f.

61. $f(x) = 3x + 5$

62. $f(x) = 5x - 7$

63. $f(x) = -2x$

64. $f(x) = 2(x - 1)$

65. $f(x) = \frac{x}{3} + 5$

66. $f(x) = \frac{2x - 1}{3}$

67. $f(x) = \frac{1}{x}$

68. $f(x) = \frac{1}{x} - 3$

69. $f(x) = \frac{1}{x + 2}$

70. $f(x) = \frac{3}{1 - x}$

71. $f(x) = \frac{x + 1}{x - 2}$

72. $f(x) = \frac{x + 3}{2x + 1}$

73. $f(x) = x^3 + 1$

74. $f(x) = -\frac{1}{8}x^3$

75. $f(x) = x^{-3}$

76. $f(x) = x^{-5} + 1$

77. $f(x) = \frac{\sqrt{x}}{2}$

78. $f(x) = \sqrt{3x}$

79. $f(x) = \sqrt[3]{2 - x}$

80. $f(x) = \sqrt[5]{3 - x}$

81. $f(x) = \sqrt[5]{x} + 3$

82. $f(x) = 1 - \sqrt[3]{x}$

In Exercises 83–86, a verbal description of function f is given. Write a verbal description of the inverse function f^{-1}.

83. Function f adds 2 to x and divides the result by 3.

84. Function f multiplies x by -2 and then subtracts 1 from the result.

85. Function f cubes 1 more than x.

86. Function f adds 4 to x and then takes half the result.

87. From the following list, identify the pairs of functions that are inverse functions.

(i) $f(x) = \frac{1}{3}$, $g(x) = 3$

(ii) $f(x) = 5$, $g(x) = -5$

(iii) $f(x) = \frac{5}{x}$, $g(x) = \frac{5}{x}$

88. Suppose that the graph of function f lies entirely in the given quadrants. In what quadrants does the graph of function f^{-1} lie?

(a) I and II (b) II and IV

(c) I and III (d) II and III

89. Explain why $y = \sqrt{9 - x^2}$, $x \geq 0$, has an inverse, but $y = \sqrt{9 - x^2}$ does not.

90. Let $f(x) = 2x - 1$. Determine $f^{-1}(x)$ by evaluating f at $f^{-1}(x)$ and then solving for $f^{-1}(x)$.

Real-Life Applications

91. The function $C(t) = \frac{5}{9}(t - 32)$ converts Fahrenheit temperatures to Celsius. Determine the inverse function and describe what it does.

92. Write a function that converts hours to minutes. Find the inverse function, and describe what it does.

93. Suppose that a British pound is worth \$1.50. Write a function that gives the value d in dollars of p British pounds. Then write the inverse of the function, and describe what it does.

94. One meter is 1.094 yards. Write a function that gives the number y of yards in m meters. Then write the inverse of the function, and describe what it does.

Modeling with Real Data

95. The accompanying table shows that the number of cellular phone subscribers has risen along with the number of cell sites. (Source: Cellular Telecommunications Industry Association.)

Year	Cell Sites	Subscribers (in thousands)
1990	5,616	5,283
1991	7,847	7,557
1992	10,307	11,033
1993	12,805	16,009

The number of cell sites C can be written as a function of the number of subscribers s (in thousands): $C(s) = 0.66s + 2588$.

(a) Produce the graph of C to determine whether C is a one-to-one function.

(b) Solve $C(s)$ for s. What does the resulting function represent?

96. Consider the relation R represented by the following mapping diagram.

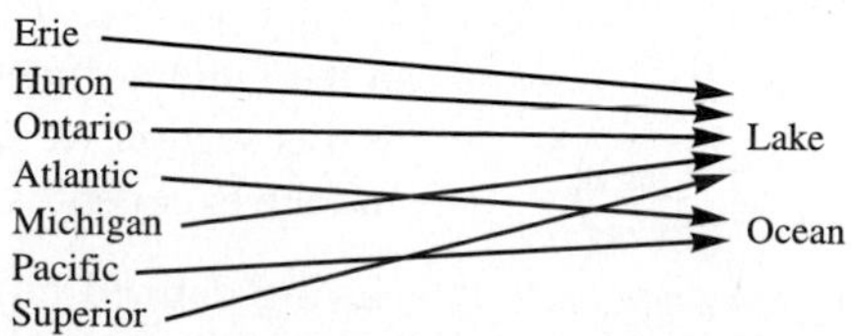

(a) Is R a function? Why or why not?

(b) If R is a function, does the function have an inverse? Why or why not?

Group Project: Telecommuting

A developing trend for American workers is *telecommuting*—using a computer or phone technology to work from home. The accompanying figure shows the number of employees (in millions) involved in telecommuting from 1989 through 1993. (Source: LINK Resources Corporation.)

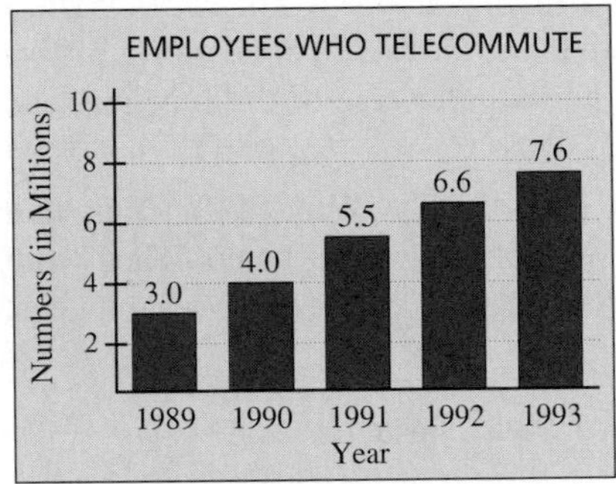

A function that models the data in the figure is

$$N(x) = 1.18x - 1.74$$

where N is the number of employees (in millions) and x is the number of years since 1985.

97. Produce the graph of the model function and the graph of $y = x$ on the same coordinate system. Trace the graph to estimate the point at which the graphs of function N and its inverse would intersect.

98. Determine $N^{-1}(x)$, and produce its graph. Do the graphs of $y_1 = N(x)$ and $y_2 = N^{-1}(x)$ intersect at the point you predicted in Exercise 97?

99. Solve the equation $N(x) = N^{-1}(x)$ algebraically, and compare the solution for x to the first coordinate of the point of intersection in Exercise 98.

100. Working at home sometimes causes social isolation. Think of a compromise solution to this potential problem.

Challenge

In Exercises 101 and 102, produce the graph of the function. Then sketch the graph of the inverse function.

101. $f(x) = x^2 - 4x, \quad x \leq 2$

102. $f(x) = x^2 - 4x + 3, \quad x \leq 2$

In Exercises 103–106, determine f^{-1} and verify that $(f \circ f^{-1})(x) = x$ for all x in the domain of f^{-1} and that $(f^{-1} \circ f)(x) = x$ for all x in the domain of f.

103. $f(x) = x^2 + 1, \quad x \geq 0$

104. $f(x) = |x - 4|, \quad x \leq 4$

105. $f(x) = -|x + 5|, \quad x \leq -5$

106. $f(x) = x^2 - 6x + 5, \quad x \leq 3$

10.3 EXPONENTIAL FUNCTIONS

Definition • Natural Base e • Graphs of Exponential Functions • Exponential Equations • Real-Life Applications

Definition

In the familiar function $g(x) = x^2$, the base is a variable and the exponent is a constant. In this section we consider functions such as $f(x) = 2^x$ in which the base is a constant and the exponent is a variable.

Having defined rational exponents, we know that the expression 2^x is defined for all rational numbers. Although the details are beyond the scope of this book, the expression 2^x also can be defined for any irrational number. Thus the function $f(x) = 2^x$ is defined for all real numbers.

Definition of an Exponential Function

An **exponential function** can be written as $f(x) = b^x$, where $b > 0$, $b \neq 1$, and x is any real number.

The two restrictions on b in the definition are important. First, the definition excludes $b = 1$ because 1^x has a value of 1 for all values of x and the function would simply be the constant function $f(x) = 1$.

The definition also requires b to be positive so that the function can be defined for all real numbers. For example, if $b = -4$ and $x = \frac{1}{2}$, then evaluating the function would result in the expression $(-4)^{1/2}$, which is not a real number.

EXAMPLE 1

Evaluating an Exponential Expression

(a) If $f(x) = 3^x$, then $f(2) = 3^2 = 9$.

(b) If $f(x) = -5^x$, then $f(2) = -5^2 = -1 \cdot 5^2 = -25$.

(c) If $g(x) = 2^x$, then $g(-4) = 2^{-4} = \dfrac{1}{2^4} = \dfrac{1}{16}$.

(d) If $g(x) = \left(\dfrac{1}{4}\right)^x$, then $g(-2) = \left(\dfrac{1}{4}\right)^{-2} = 4^2 = 16$.

(e) If $h(x) = 9^x$, then $h\left(\dfrac{1}{2}\right) = 9^{1/2} = \sqrt{9} = 3$.

Note: In part (b) of Example 1, $f(x) = -5^x$ is an exponential function because the base is 5, which is positive. However, $(-5)^x$ cannot be used to define an exponential function because the base is negative.

Natural Base *e*

An irrational number approximately equal to 2.718281828 occurs frequently in applications in science, engineering, business, and other areas. Because it occurs so frequently, the letter e is used to represent it, just as the Greek letter π is used to represent the irrational number approximately equal to 3.14159.

The number e is often the base in an exponential function.

$$f(x) = e^x$$

This function is called the **natural exponential function,** but more often it is called simply *the* exponential function.

Base e

Most calculators have a special key for the base e.

EXAMPLE 2

Evaluating e^x

Let $f(x) = e^x$. Use your calculator to evaluate the function at the indicated values. Round results to three decimal places.

(a) $f(1)$ (b) $f(-1)$ (c) $f(2)$

Solution

Figure 10.10

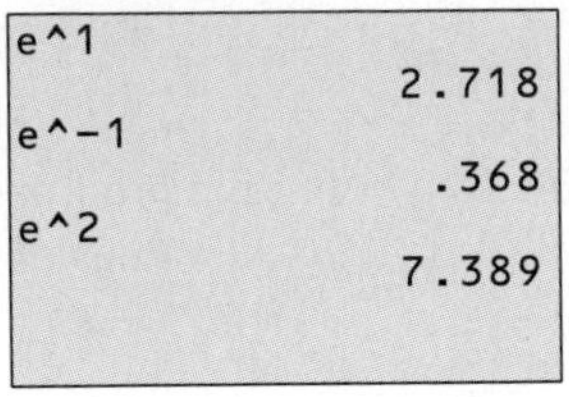

```
e^1
                 2.718
e^-1
                  .368
e^2
                 7.389
```

Graphs of Exponential Functions

We can learn more about the behavior of exponential functions by observing their graphs.

EXPLORING THE CONCEPT

Graphs of Exponential Functions

Figures 10.11 and 10.12 show the graphs of functions of the form $y = b^x$. In Fig. 10.11, $b > 1$, whereas in Fig. 10.12, $0 < b < 1$.

Figure 10.11

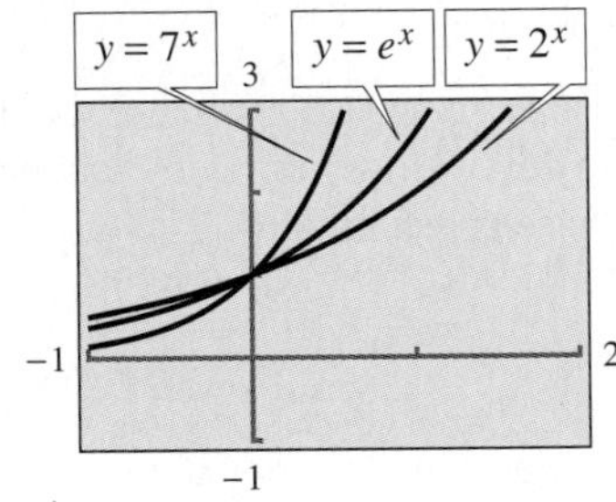

Figure 10.12

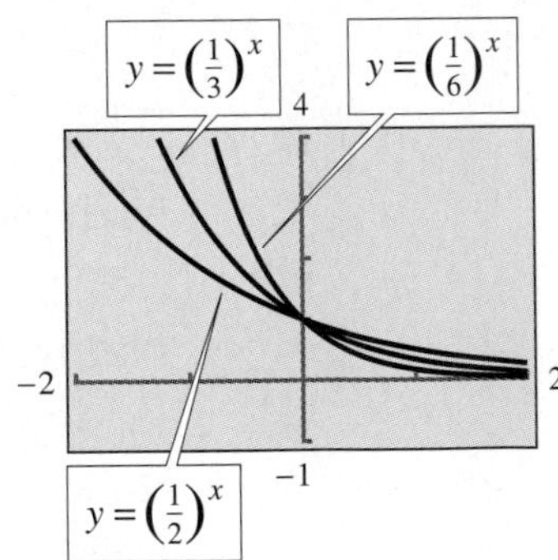

In both figures the domain of each function is **R.** Because $b > 0$, b^x cannot be 0, and so the graphs approach but do not reach the x-axis. Thus the range of each function is $\{y \mid y > 0\}$.

The graphs in the two figures have other characteristics in common. For example, because $b^0 = 1$, the y-intercept of each graph is (0, 1). Moreover, because all the graphs pass the Horizontal Line Test, all the functions are one-to-one functions.

The graphs differ in one important way. In Fig. 10.11 the graphs rise from left to right, and they become steeper as b increases. In Fig. 10.12 the graphs fall from left to right, and they become steeper as b decreases.

Here is a summary of our findings about exponential functions.

Properties of Exponential Functions and Their Graphs

Let $f(x) = b^x$, $b > 0$ and $b \neq 1$.

1. The y-intercept is (0, 1); there is no x-intercept.
2. The function f is a one-to-one function.
3. The domain is the set of all real numbers; the range is $\{y \mid y > 0\}$.
4. The graph approaches but does not reach the x-axis.
5. For $b > 1$, the graph rises from left to right; for $0 < b < 1$, the graph falls from left to right.

Think About It

Describe the graph of the exponential function $f(x) = -b^x$. What are the domain, the range, and the intercepts?

Exponential Equations

Equations that involve a variable in the exponent are called **exponential equations.** The following are some examples of exponential equations.

$$2^x = 8 \qquad 2^{3x} = 16 \qquad 27^x = 81 \qquad \left(\frac{1}{3}\right)^x = 9$$

As we have done before, we can estimate the solutions of such equations by graphing.

EXAMPLE 3

Solving an Exponential Equation by Graphing

Use a graph to estimate the solution of $e^x = 2$.

Solution

Produce the graphs of $y_1 = e^x$ and $y_2 = 2$. (See Fig. 10.13.) After one or two zoom-and-trace cycles, we estimate the solution to be $x \approx 0.68$.

Figure 10.13

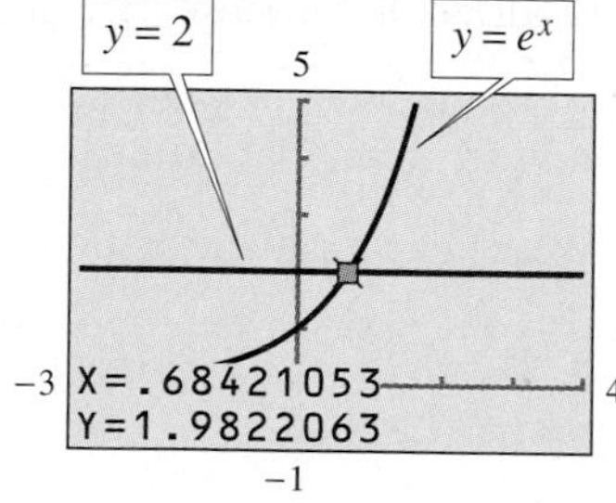

We can verify by evaluating y_1 for $x = 0.68$ or by calculating $e^{0.68}$ directly. In either case, the result is close to 2, and the solution is approximately 0.68.

To solve exponential equations algebraically requires new techniques.

We observed that an exponential function is a one-to-one function. Therefore, if $b^m = b^n$, then m must equal n.

Property of Equivalent Exponents

For $b > 0$ and $b \neq 1$, if $b^m = b^n$, then $m = n$.

We can use this property to solve an exponential equation. If we can write each side of an exponential equation as an exponential function with the same base, then we can equate the exponents.

EXAMPLE 4

Equating Exponents to Solve an Exponential Equation

Use a graph to estimate the solution of the equation $4^x = 8$. Then solve the equation algebraically.

Solution

Figure 10.14

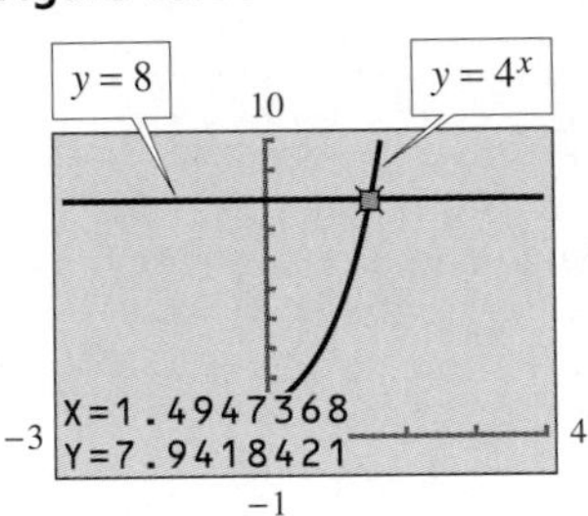

Figure 10.14 shows the graphs of $y_1 = 4^x$ and $y_2 = 8$. We estimate the solution to be about 1.5.

Now we solve algebraically by equating exponents.

$$4^x = 8$$
$$(2^2)^x = 2^3 \quad \text{Write each side with a base of 2.}$$
$$2^{2x} = 2^3 \quad \text{Power to a Power Rule}$$
$$2x = 3 \quad \text{Equate the exponents.}$$
$$x = \frac{3}{2}$$

You should verify the solution by substitution.

As we will see in the next section, it is always possible to write exponential expressions with the same base. However, the method of equating exponents is recommended only for those special cases in which changing the base is easy to do.

EXAMPLE 5

LEARNING TIP

If you are unsuccessful in writing both sides of an exponential equation with the same base, your only recourse so far is to use the graphing method to estimate the solution.

Solving Exponential Equations

Solve each exponential equation.

(a) $4^{2x+1} = 32$ (b) $\left(\frac{2}{3}\right)^{1-x} = \left(\frac{9}{4}\right)^{2x}$ (c) $e^{x^2-1} = 1$

Solution

(a)
$$4^{2x+1} = 32$$
$$(2^2)^{2x+1} = 2^5 \quad \text{Write both sides with base 2.}$$
$$2^{4x+2} = 2^5 \quad \text{Power to a Power Rule}$$
$$4x + 2 = 5 \quad \text{Equate exponents.}$$
$$4x = 3 \quad \text{Solve for } x.$$
$$x = \frac{3}{4}$$

(b) $\left(\frac{2}{3}\right)^{1-x} = \left(\frac{9}{4}\right)^{2x}$

$\left(\frac{2}{3}\right)^{1-x} = \left[\left(\frac{3}{2}\right)^{2}\right]^{2x}$ Write both sides with base $\frac{2}{3}$.

$\left(\frac{2}{3}\right)^{1-x} = \left[\left(\frac{2}{3}\right)^{-2}\right]^{2x}$ $\left(\frac{a}{b}\right)^{-n} = \left(\frac{b}{a}\right)^{n}$

$\left(\frac{2}{3}\right)^{1-x} = \left(\frac{2}{3}\right)^{-4x}$ Power to a Power Rule

$1 - x = -4x$ Equate exponents.

$1 = -3x$ Solve for x.

$x = -\frac{1}{3}$

(c) $e^{x^2-1} = 1$

$e^{x^2-1} = e^0$ Write both sides with base e.

$x^2 - 1 = 0$ Equate exponents.

$(x + 1)(x - 1) = 0$ Factor.

$x + 1 = 0$ or $x - 1 = 0$ Zero Factor Property

$x = -1$ or $x = 1$

Note that we would not be able to use this procedure to solve the equation $e^x = 2$ because we presently have no way of expressing 2 as a power of e. The techniques for solving this type of equation will be introduced later.

Real-Life Applications

Exponential equations play a major role in describing the conditions of certain application problems. Sometimes the exponential equation is a formula that has been derived to relate the variables of the problem.

For example, the formula to compute the value A of an investment of P dollars at an annual interest rate r compounded n times per year for t years is

$$A = P\left(1 + \frac{r}{n}\right)^{nt}$$

EXAMPLE 6

Compound Interest

If an investment earns 8% interest, compounded monthly, in how many years will the investment double?

Solution

If P is the original investment, then it must double to $2P$.

$A = 2P$ $n = 12$ $r = 0.08$

$A = P\left(1 + \frac{r}{n}\right)^{nt}$ Compound interest formula

$$2P = P\left(1 + \frac{0.08}{12}\right)^{12t} \quad \text{Substitute for } A, n, \text{ and } r.$$

$$2 = \left(1 + \frac{0.08}{12}\right)^{12t} \quad \text{Divide both sides by } P.$$

$$2 \approx (1.0067)^{12t}$$

Figure 10.15

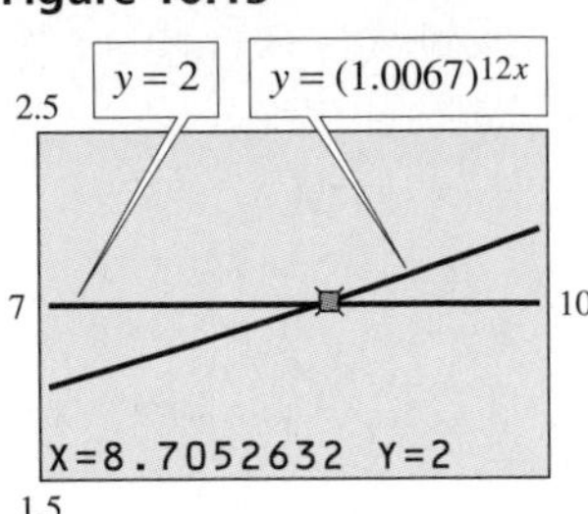

Because we cannot conveniently express both sides with the same base, we will use a graph (with x instead of t) to approximate the solution.

Figure 10.15 shows the graphs of $y_1 = 2$ and $y_2 = (1.0067)^{12x}$. The approximate solution for x is 8.7. It will take about 8.7 years for the investment to double. Note that the amount of the investment has no bearing on the result.

10.3 QUICK REFERENCE

Definition

- An **exponential function** can be written $f(x) = b^x$, where $b > 0$, $b \neq 1$, and x is any real number.

Natural Base *e*

- The irrational number 2.71828 . . . is represented by the letter e.
- The function $f(x) = e^x$ is called the **natural exponential function.**

Graphs of Exponential Functions

- The following are properties of the graph of an exponential function $f(x) = b^x$, $b > 0$ and $b \neq 1$:
  1. The y-intercept is (0, 1); there is no x-intercept.
  2. The function f is a one-to-one function.
  3. The domain is the set of all real numbers; the range is $\{y \mid y > 0\}$.
  4. The graph approaches but does not reach the x-axis.
  5. For $b > 1$, the graph rises from left to right; for $0 < b < 1$, the graph falls from left to right.

Exponential Equations

- An equation in which the variable appears in the exponent is an **exponential equation.**
- Solutions of exponential equations can be estimated by graphing the two sides of the equation and determining the point of intersection.
- Because an exponential function is a one-to-one function, if $b^m = b^n$, then $m = n$. This rule can be used to solve an exponential equation as follows:
  1. Write each side of the equation as a single exponential expression with the same base.
  2. Use the rules of exponents to simplify both sides.
  3. Equate the exponents.
  4. Solve the resulting equation.

10.3 EXERCISES

Concepts and Skills

1. Explain the conditions under which the graph of $f(x) = b^x$
   (a) rises from left to right.
   (b) falls from left to right.

2. If $f(x) = (1/b)^x$ with $b > 1$, as b increases, what is the effect on the graph of f?

In Exercises 3–10, let $f(x) = 2^{3x-1}$ and $g(x) = \left(\frac{1}{3}\right)^{2x}$. Evaluate each of the following.

3. $f(2)$
4. $g(-1)$
5. $g(1)$
6. $f(-1)$
7. $g(0)$
8. $f(0)$
9. $f\left(\frac{2}{3}\right)$
10. $g\left(-\frac{3}{2}\right)$

In Exercises 11–18, let $g(x) = e^{-2x}$ and $h(x) = \left(\frac{1}{5}\right)^x$. Determine each of the following values to the nearest hundredth.

11. $g(2)$
12. $h(-2)$
13. $g(-3)$
14. $h(0)$
15. $h\left(-\frac{3}{2}\right)$
16. $h\left(\frac{1}{2}\right)$
17. $g(-2)$
18. $g(1)$

In Exercises 19–26, evaluate the given function as indicated. Round results to the nearest hundredth.

19. $f(x) = 4e^{3x}$, $f(2)$
20. $g(x) = 100e^{-2x}$, $g(3)$
21. $g(x) = \left(\sqrt{3}\right)^{2x}$, $g(5)$
22. $f(x) = \left(\sqrt{5}\right)^{-3x}$, $f(-3)$
23. $f(x) = (0.23)^{3x-1}$, $f(-2)$
24. $g(x) = (1.45)^{1-2x}$, $g(3)$
25. $g(x) = -3e^{2x} + 4$, $g(-2)$
26. $f(x) = 5 + 3e^{-x+1}$, $f(5)$

27. From the following list, identify the functions that have the same graph and explain why.
   (i) $f(x) = 3^x$ (ii) $g(x) = 3^{-x}$
   (iii) $h(x) = \left(\frac{1}{3}\right)^x$ (iv) $s(x) = \frac{1}{3^x}$

28. If $f(x) = b^x$ with $b > 1$, as b increases, what is the effect on the graph of f?

In Exercises 29–34, use a graph to estimate the range and domain of the given function.

29. $f(x) = -e^x - 2$
30. $g(x) = 2^{x-2} + 4$
31. $f(x) = 2 - 3^{-x}$
32. $h(x) = e^{x^2}$
33. $h(x) = e^{-|x|}$
34. $g(x) = e^x + e^{-x}$

35. Use your calculator to produce the graphs of the following three functions.
   $f(x) = 3^x$ $g(x) = 3^x + 2$ $h(x) = 3^x - 4$
   In general, compare the graphs of b^x and $b^x + c$ when $c > 0$ and when $c < 0$.

36. Use your calculator to produce the graphs of the following three functions.
   $f(x) = 3^x$ $g(x) = 3^{x+2}$ $h(x) = 3^{x-4}$
   In general, compare the graphs of b^x and b^{x+k} when $k > 0$ and when $k < 0$.

In Exercises 37 and 38, compare the graphs of the given pair of functions to the graph of $h(x) = 2^x$.

37. (a) $f(x) = 2^x + 1$ (b) $g(x) = 2^x - 3$
38. (a) $f(x) = 2^{x+3}$ (b) $g(x) = 2^{x-4}$

In Exercises 39 and 40, compare the graphs of the given pair of functions to the graph of $h(x) = \left(\frac{1}{3}\right)^x$.

39. (a) $f(x) = \left(\frac{1}{3}\right)^x + 3$ (b) $g(x) = \left(\frac{1}{3}\right)^{x+3}$
40. (a) $f(x) = \left(\frac{1}{3}\right)^{-x}$ (b) $g(x) = 3^{-x}$

The following figure shows the graph of a function $f(x) = b^x$. In Exercises 41 and 42, use this graph to assist you in matching the given functions to graph (A), (B), (C), or (D).

41. (a) $g(x) = b^x + 3$ (b) $h(x) = 4 - b^x$

42. (a) $g(x) = b^{-x}$ (b) $h(x) = b^{-x} - 5$

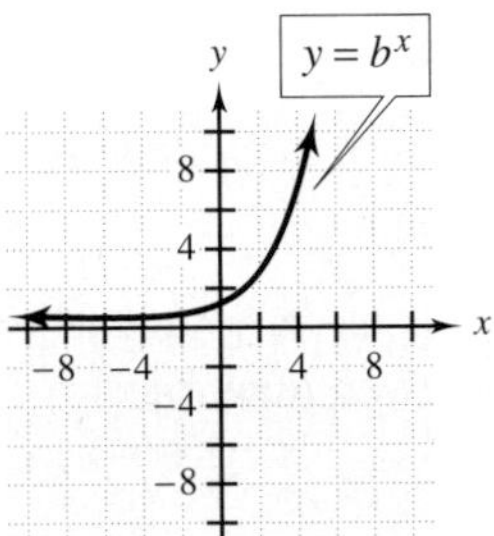

(A)

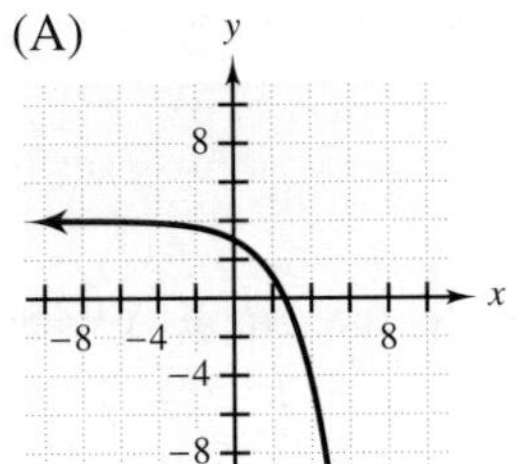

(B)

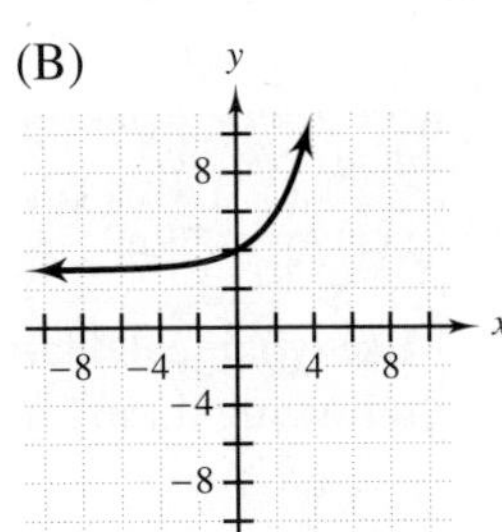

(C)

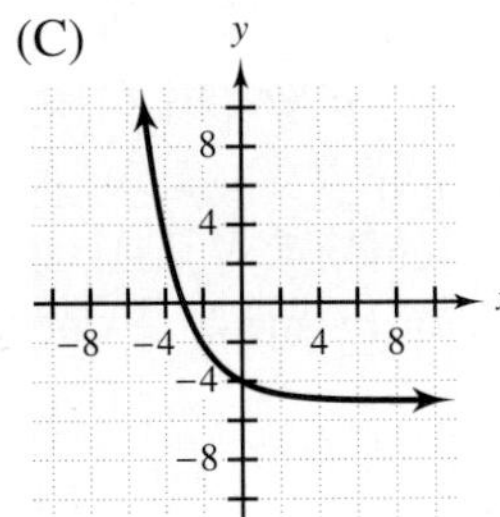

(D)

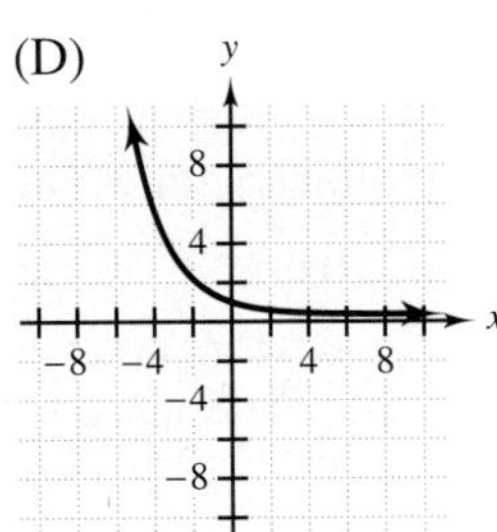

In Exercises 43–48, let $f(x) = e^x$ and $h(x) = (2.3)^x$. Use a graph to estimate the value of x (to the nearest hundredth) for which the function has the given value.

43. $h(x) = 0.76$ **44.** $f(x) = 0.59$

45. $h(x) = 7.51$ **46.** $f(x) = 4.85$

47. $f(x) = -3.45$ **48.** $h(x) = 0$

In Exercises 49–56, use a graph to estimate the solution of the given exponential equation.

49. $e^x = 4$ **50.** $2^{-x} = 7$

51. $(0.27)^x = 6$ **52.** $(1.56)^x = 8$

53. $e^x = -2x + 1$ **54.** $x + 4^x = -3$

55. $e^x = x$ **56.** $-4^x = 2 - x$

57. Why can we solve $4^x = 8$ without using a calculator but cannot use the same method to solve $4^x = 12$?

58. Explain why the graphs of $f(x) = 2^{2x}$ and $g(x) = 4^x$ are the same.

In Exercises 59–66, solve the exponential equation.

59. $5^{-x} = 125$ **60.** $2^{x-1} = 16$

61. $\left(\frac{2}{5}\right)^x = \frac{25}{4}$ **62.** $\left(\frac{3}{2}\right)^x = \frac{8}{27}$

63. $e^{3x+1} = e^{x-2}$ **64.** $e^{x+1} = e^{2x}$

65. $\frac{1}{9^x} = 27$ **66.** $25^x = \frac{1}{5}$

In Exercises 67–90, solve the exponential equation.

67. $49^x = 343$ **68.** $9^{-2x} = 27$

69. $16 = 8^{x-2}$ **70.** $e^{x+4} = \frac{1}{e^{3x}}$

71. $3 \cdot 8^x = 12$ **72.** $-3^x = -9$

73. $9 \cdot 3^x = \frac{1}{3}$ **74.** $4^x \cdot 16^{2-3x} = 8$

75. $2^{-x-3} = -4$ **76.** $-4^{2x} = 2$

77. $\frac{5^{x+1}}{5^{1-x}} = \frac{1}{25^x}$ **78.** $\frac{3^{x+1}}{27^x} = \frac{1}{9^x}$

79. $e^x + 4 = 3$ **80.** $7 - e^{3-2x} = 6$

81. $\left(\frac{1}{3}\right)^x 27^{x+1} = 9^{3-2x}$

82. $\left(\frac{1}{3}\right)^{x^2} = 27^{(2/3)x-1}$

83. $32^{1-x} = 8^{x^2-x}$ **84.** $125^{2-3x} = 25^{3x-5}$

85. $(0.25)^{2x+1} = 8^{2-x}$ **86.** $(0.2)^{x-3} = 25^{1-x}$

87. $e^{x^2-1} = 7^0$

88. $(e^{x+2})^x = (e^{x+2})^2$

89. $\left(\sqrt{3}\right)^{2-2x} = 9^{x^2}$

90. $\left(\sqrt{2}\right)^{4-2x} = \left(\sqrt[3]{2}\right)^{6x}$

Real-Life Applications

91. A couple plans to invest $1000 each year to save for a vacation home. They find two banks that pay the same interest rate of 6%. However, one bank pays only simple interest, whereas the other compounds the interest daily (365 days). What would be the difference in the amount of interest they would earn at the two banks after the first year?

92. Two sisters each received a gift of \$2000 from their aunt with the stipulation that they invest the money for at least 1 year. The older sister invested the money in an account that paid 7% compounded quarterly, and the younger sister invested her money in another bank that paid 6% compounded daily (365 days). What is the difference in the amount of interest that the two sisters earned?

93. When a company downsized 2 years ago, an employee received a lump-sum severance pay that he invested in a growth fund that yielded an annual rate of 9% in dividends, compounded quarterly. The account is now worth \$5974.16. How much severance pay did the person receive?

94. At the beginning of the year an employee invested in the company savings plan as well as in the company retirement plan. The savings plan paid 11% compounded daily (365 days), and the retirement fund paid 7% compounded quarterly. At the end of 1 year, the savings plan had a balance of \$3348.79, and the retirement account balance was \$3322.76. What was the total original investment in both plans?

In Exercises 95 and 96, use the graphing method to estimate the solution.

95. One alumnus of a college set up two small scholarship funds for the students at the college. One, designated for nursing students, was \$15,000 and was invested at 10% compounded daily (365 days). The other, designated for math or science majors, was \$20,000 and was invested at 6% compounded quarterly. However, the alumnus stipulated that no scholarships could be awarded until the two funds had the same amount of money. How many years after the initial gift can the college award a scholarship?

96. When a child was born, a grandmother invested \$3000 in a fund guaranteed to pay 7% compounded quarterly. The purpose was to be able to contribute at least \$10,000 to the child's college education. Approximately how many years will it take to achieve this goal?

Modeling with Real Data

97. In 1995, the average annual per-capita expenditure for books was \$84.20. The amount was expected to increase by 6% annually for the next 5 years. (Source: Standard & Poor's.)

(a) Complete the accompanying table by writing exponential expressions for the expenditures in each year through 2000.

Year	n	Expenditure
1995	0	84.20
1996	1	$84.20(1.06)^1$
1997	2	$84.20(1.06)^2$
1998	3	
1999	4	
2000	5	

(b) Write an exponential function that models the expenditure E in year n, where n is the number of years since 1995.

(c) What expenditure does the model project for 2000? for 2003?

(d) Use a graph to estimate the year in which the average annual expenditure is projected to reach \$150.

98. The number of households using personal computers to conduct their banking is projected to increase exponentially. (Source: *Home Banking Report*.)

Year	Households
1995	754,000
1996	1,884,000
1997	3,297,000
1998	5,769,000
1999	8,653,000
2000	12,980,000

The exponential function $B(x) = 0.95e^{0.55x}$ models the number of households B (in millions) subscribing to on-line banking, where x is the number of years since 1995.

(a) According to the model, how many households are projected to use the on-line banking service in 2002?

(b) Use the graph of B to estimate the year in which 25 million households are expected to have on-line banking services.

Group Project: Cost of Medicare

Not until the mid-1990s was any serious effort made to curtail the cost of health care in America. The accompanying figure shows the rapid increase in federal spending for Medicare during the period 1980–1993. (Source: U.S. Health Care Financing Administration.)

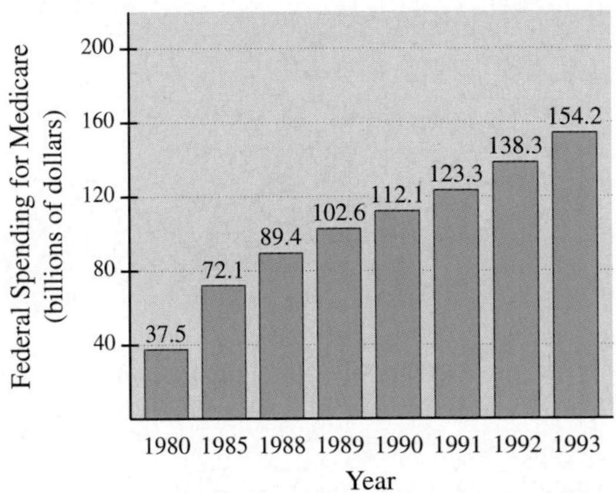

The data in the bar graph can be modeled by $M(t) = 39.126(1.11)^t$, where M is the federal spending (in billions) for Medicare and t is the number of years since 1980.

99. Use a suitable window setting to produce the graph of the function M. How does the appearance of the graph support the claim that Medicare costs were rising sharply during the period 1980–1993?

100. Assuming that the model function remains valid, trace the graph to estimate the cost of Medicare in the year 2000.

101. Assuming a budget of $1 trillion, by what year would the cost of Medicare represent half the federal budget?

102. It has been estimated that 5% of Medicare costs are fraudulent. If so, by how much were American taxpayers cheated in 1993? Compare this figure with the approximately $14.5 billion of federal spending for higher education that year.

Challenge

In Exercises 103–105, use a graph to estimate the solution(s) of the given equation.

103. $6x \cdot 3^x = x + 1$

104. $\dfrac{e^x - e^{-x}}{2} = 2.3$

105. $5xe^x = -1$

In Exercises 106 and 107, use a graph to estimate the solution set of the given inequality.

106. $2^x < 3^x$

107. $e^x < e^{-x}$

In Exercises 108–111, use a graph of the given function to estimate any local minimum or maximum.

108. $y = xe^x$

109. $y = x^2e^x$

110. $y = 7x^2(0.5)^x$

111. $y = e^{|x|} + 3$

10.4 LOGARITHMIC FUNCTIONS

Definition • Evaluation of Logarithms • Logarithmic Equations • Graphs of Logarithmic Functions

Definition

In the preceding section we learned that the exponential function $f(x) = b^x$ is a one-to-one function. This means that an exponential function has an inverse.

Because the graph of a function and its inverse are symmetric with respect to the line $y = x$, we can sketch the graph of the inverse of a function by drawing the mirror image of the function on the other side of the line of symmetry.

EXPLORING THE CONCEPT

Inverses of Exponential Functions

Consider the exponential functions $f(x) = 2^x$ and $g(x) = \left(\frac{1}{2}\right)^x$. Figure 10.16 shows the graphs of f and f^{-1}, and Fig. 10.17 shows the graphs of g and g^{-1}.

Figure 10.16

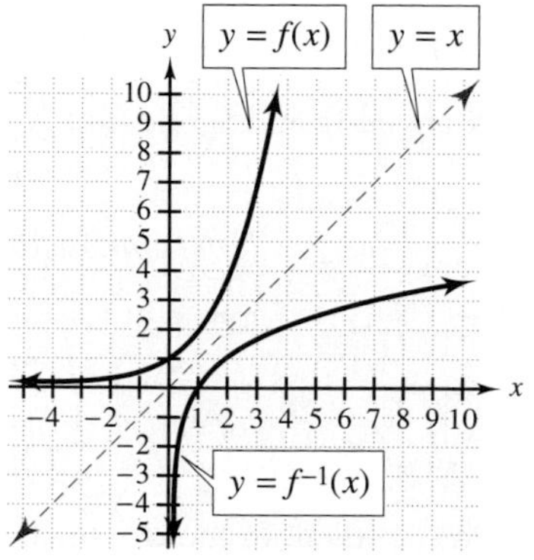

Figure 10.17

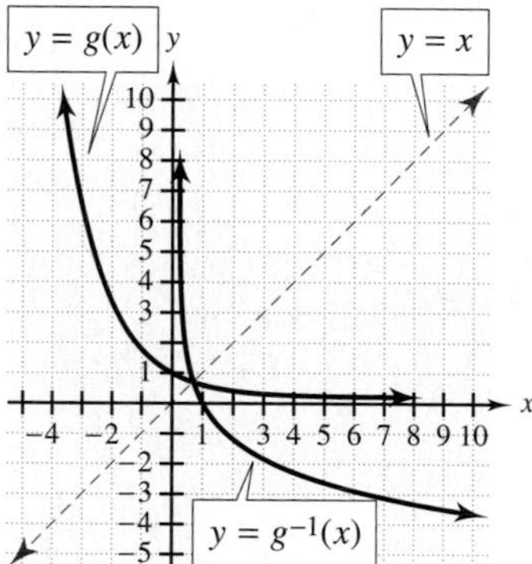

The graphs of f^{-1} and g^{-1} share some common characteristics. For example, the x-intercept in both cases is (1, 0). The graphs approach but never reach the y-axis, so there is no y-intercept. The graphs of f^{-1} and g^{-1} lie entirely to the right of the y-axis, so the domain is $\{x \mid x > 0\}$. The range in both cases is **R.**

These observations are true for the inverse of any exponential function $y = b^x$. Each exponential function has an inverse. We can sketch the graph of the inverse, and we can discover properties of the inverse. However, unlike the functions that we studied in Section 10.2, we cannot interchange the variables x and y and solve for y to obtain an algebraic description of the function. We simply call the inverse function the **logarithmic function with base *b*.**

Definition of Logarithmic Function

For $x > 0$, $b > 0$, and $b \neq 1$, the **logarithm of *x* with base *b*,** represented by $\log_b x$, is defined by the following:

$\log_b x = y$ if and only if $x = b^y$

The function $g(x) = \log_b x$ is called the **logarithmic function with base *b*** and is the inverse of the exponential function $f(x) = b^x$.

The notation $\log_b x$ is often read more simply as the "log base b of x." The number x is called the **argument of the logarithm.**

Natural and Common Logarithms

The logarithmic function with base e is called the **natural logarithmic function** and is written $y = \ln x$ rather than $y = \log_e x$.

The logarithmic function with base 10 is called the **common logarithmic function** and is written $y = \log x$ rather than $y = \log_{10} x$.

Evaluation of Logarithms

Expressions can be changed from exponential to logarithmic form or from logarithmic to exponential form.

Exponential form *Logarithmic form*

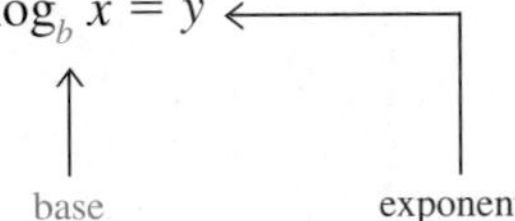

Observe that if $y = \log_b x$, then, in the exponential form, $x = b^y = b^{\log_b x}$. Thus $\log_b x$ is an exponent.

EXAMPLE 1

Exponential Form to Logarithmic Form

Write in logarithmic form.

(a) $3^4 = 81$ (b) $5^{-2} = 0.04$ (c) $10^3 = 1000$

Solution

	Exponential form	*Logarithmic form*	
(a)	$3^4 = 81$	$\log_3 81 = 4$	
(b)	$5^{-2} = 0.04$	$\log_5 0.04 = -2$	
(c)	$10^3 = 1000$	$\log 1000 = 3$	Recall that log 1000 means $\log_{10} 1000$.

EXAMPLE 2

Logarithmic Form to Exponential Form

Write in exponential form.

(a) $\log_4 2 = \frac{1}{2}$ (b) $\ln e = 1$ (c) $\log_5 \frac{1}{5} = -1$

Solution

	Logarithmic form	*Exponential form*	
(a)	$\log_4 2 = \frac{1}{2}$	$4^{1/2} = 2$	
(b)	$\ln e = 1$	$e^1 = e$	Recall that ln e means $\log_e e$.
(c)	$\log_5 \frac{1}{5} = -1$	$5^{-1} = \frac{1}{5}$	

One method for evaluating a logarithmic expression begins with writing the expression in exponential form.

EXAMPLE 3

Evaluating Logarithmic Expressions

Evaluate the logarithms.

(a) $\log_6 6$ (b) $\log_{1/4} 2$ (c) $\log_5 125$

Solution

(a) Let $\log_6 6 = y$.

$6^y = 6$ Write in exponential form.

$6^y = 6^1$

$y = 1$ Equate exponents.

Therefore, $\log_6 6 = 1$.

(b) Let $\log_{1/4} 2 = y$.

$\left(\frac{1}{4}\right)^y = 2$ Write in exponential form.

$\left(\frac{1}{2^2}\right)^y = 2$ Write both sides with the same base.

$(2^{-2})^y = 2$ Definition of negative exponent

$2^{-2y} = 2^1$ Power to a Power Rule

$-2y = 1$ Equate exponents.

$y = -\frac{1}{2}$

Therefore, $\log_{1/4} 2 = -\frac{1}{2}$.

(c) Let $\log_5 125 = y$.

$5^y = 125$ Write in exponential form.

$5^y = 5^3$ Write both sides with the same base.

$y = 3$ Equate exponents.

Therefore, $\log_5 125 = 3$.

Log, Ln

Most calculators have keys for both common logarithms and natural logarithms.

EXAMPLE 4

Evaluating Logarithms with a Calculator

Use a calculator to evaluate the following logarithms. Round results to two decimal places.

(a) $\ln 8$ (b) $\log 20$

Solution

Figure 10.18

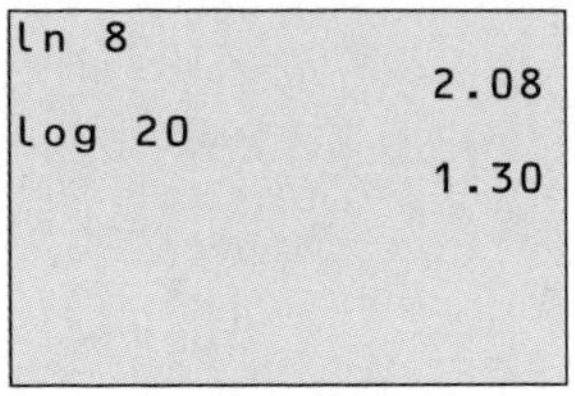

```
ln 8
                2.08
log 20
                1.30
```

Example 4 shows how easy it is to use a calculator to evaluate common logarithms and natural logarithms. However, most calculators do not have keys for evaluating logarithms with bases other than 10 or e.

It can be shown that a logarithm with any base b can be evaluated in terms of logarithms with any other base a.

LEARNING TIP

Noting that x is above b in the expression $\log_b x$ might help you to remember that x is in the numerator and b is in the denominator in the Change of Base Formula.

Change of Base Formula

For $x > 0$ and for any positive bases a and b, where $a \neq 1$ and $b \neq 1$,

$$\log_b x = \frac{\log_a x}{\log_a b}$$

One use of this formula is for evaluating a logarithm with any base in terms of natural or common logarithms.

EXAMPLE 5

Evaluating Logarithms with Bases Other Than 10 or e

(a) $\log_3 5 = \dfrac{\log 5}{\log 3} = 1.46$ Use $\log_b x = \dfrac{\log_a x}{\log_a b}$ with $a = 10$.

(b) $\log_{2.3} 4.72 = \dfrac{\ln 4.72}{\ln 2.3} = 1.86$ Use $a = e$.

Logarithmic Equations

We can often solve equations involving logarithms by converting logarithmic forms to exponential forms.

EXAMPLE 6

Solving Logarithmic Equations

Solve each equation for x.

(a) $\log_x 27 = 3$ (b) $\log_{2/3} x = 2$ (c) $\log x = -1$ (d) $\ln x = 0$

Solution

(a) $\log_x 27 = 3$

$x^3 = 27$ Write in exponential form.

$x = 3$ Odd Root Property

Think About It

Show how to evaluate the following expressions without a calculator.

(a) $(\log_2 3)(\log_3 4)(\log_4 5)(\log_5 6)(\log_6 7)(\log_7 8)$

(b) $(\log_2 4)(\log_4 6)(\log_6 8)$

(b) $\log_{2/3} x = 2$

$\left(\frac{2}{3}\right)^2 = x$ Write in exponential form.

$x = \frac{4}{9}$

(c) $\log x = -1$

$10^{-1} = x$ Common logarithm base is 10.

$x = \frac{1}{10}$

(d) $\ln x = 0$

$e^0 = x$ Natural logarithm base is e.

$x = 1$

Graphs of Logarithmic Functions

We can learn more about the behavior of logarithmic functions by examining their graphs. To use a calculator to produce the graphs of logarithmic functions other than common or natural logarithms, we use the Change of Base Formula.

EXAMPLE 7

Graphs of Logarithmic Functions

Produce the graphs of the following logarithmic functions.

(a) $y = \log_3 x$ (b) $y = \log_{1/3} x$

Solution

Use the Change of Base Formula for each function.

(a) $y = \log_3 x = \dfrac{\log x}{\log 3}$ (b) $y = \log_{1/3} x = \dfrac{\log x}{\log \frac{1}{3}}$

Figure 10.19

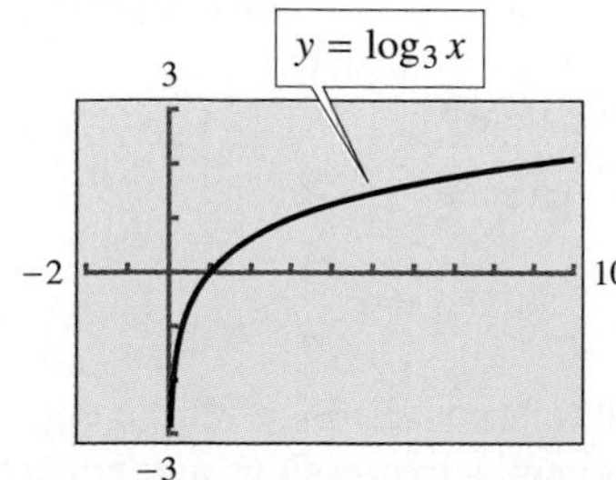

Figure 10.20

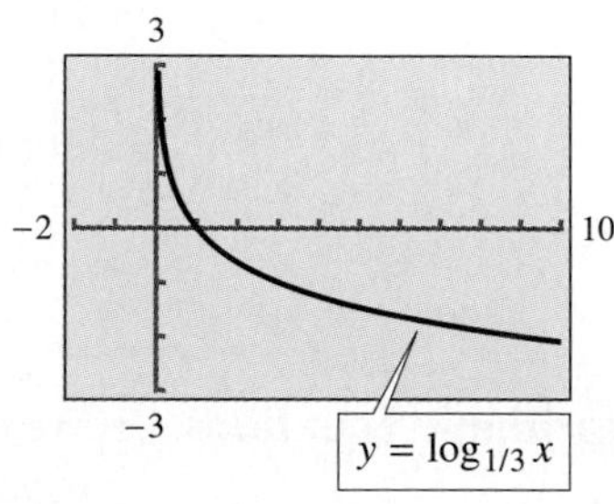

Observe that in part (a) of Example 7, $b > 1$, and the graph rises from left to right. In part (b) of Example 7, $0 < b < 1$, and the graph falls from left to right.

In Example 7 the choice of common logarithms to change the base was arbitrary. For $y = \log_3 x$, we could have used $y = \log_3 x = \dfrac{\ln x}{\ln 3}$, and the graph would have been the same.

Here is a summary of properties of $f(x) = \log_b x$.

Properties of $f(x) = \log_b x$ and Its Graph

1. The domain of f is $\{x \mid x > 0\}$.
2. The range of f is the set of all real numbers.
3. The x-intercept is (1, 0).
4. The graph approaches but does not reach the y-axis; there is no y-intercept.
5. For $b > 1$, the graph rises from left to right.
6. For $0 < b < 1$, the graph falls from left to right.

10.4 QUICK REFERENCE

Definition

- For $x > 0$, $b > 0$, and $b \neq 1$, the **logarithm of x with base b,** represented by $\log_b x$, is defined by the following:

  $$\log_b x = y \quad \text{if and only if} \quad x = b^y$$

  The function $f(x) = \log_b x$ is called the **logarithmic function with base b** and is the inverse of the exponential function $f(x) = b^x$.
- The logarithmic function with base e is called the **natural logarithmic function** and is written $y = \ln x$.
- The logarithmic function with base 10 is called the **common logarithmic function** and is written $y = \log x$.

Evaluation of Logarithms

- Exponential expressions can be changed to logarithmic form, and logarithmic expressions can be changed to exponential form:

  $$x = b^y \longleftrightarrow \log_b x = y$$
- Logarithmic expressions can sometimes be evaluated without a calculator if they are first converted to exponential forms.
- In order to use a calculator to evaluate a logarithm with a base other than 10 or e, the Change of Base Formula is needed:

  For $x > 0$, and for any positive bases a and b, where $a \neq 1$ and $b \neq 1$,

  $$\log_b x = \frac{\log_a x}{\log_a b}$$

Logarithmic Equations

- We can solve some (but not all) logarithmic equations by converting the logarithmic expressions to exponential expressions and equating exponents.

Graphs of Logarithmic Functions

- The following are properties of $f(x) = \log_b x$ and its graph:
  1. The domain of f is $\{x \mid x > 0\}$.
  2. The range of f is the set of all real numbers.
  3. The x-intercept is (1, 0).
  4. The graph approaches but does not reach the y-axis; there is no y-intercept.

5. For $b > 1$, the graph rises from left to right.
6. For $0 < b < 1$, the graph falls from left to right.

- To use a calculator to produce graphs of logarithmic functions with bases other than 10 or e, use the Change of Base Formula.

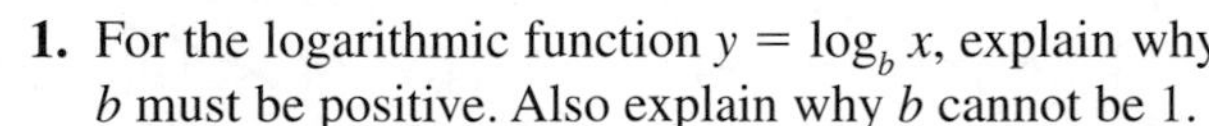

10.4 EXERCISES

Concepts and Skills

1. For the logarithmic function $y = \log_b x$, explain why b must be positive. Also explain why b cannot be 1.

2. Explain why $\log 100$ can be evaluated without a calculator but $\ln 100$ cannot.

In Exercises 3–10, write the given logarithmic equation in exponential form.

3. $\log_7 49 = 2$ **4.** $\log_4 8 = \frac{3}{2}$

5. $\log 0.001 = -3$ **6.** $\log_{1/2} 8 = -3$

7. $\ln \sqrt[3]{e} = \frac{1}{3}$ **8.** $\log_8 \frac{1}{4} = -\frac{2}{3}$

9. $\log_m n = 2$ **10.** $\ln 3 = t$

In Exercises 11–18, write the given exponential equation in logarithmic form.

11. $4^3 = 64$ **12.** $27^{2/3} = 9$

13. $10^{-5} = 0.00001$ **14.** $8^{-2/3} = \frac{1}{4}$

15. $\left(\frac{1}{2}\right)^{-3} = 8$ **16.** $5^{-3} = \frac{1}{125}$

17. $P = e^{rt}$ **18.** $2^x = y$

In Exercises 19–26, for each function, determine f^{-1}.

19. $f(x) = e^x$

20. $f(x) = 3^x$

21. $f(x) = 4^{-x}$

22. $f(x) = \left(\frac{3}{4}\right)^x$

23. $f(x) = \log x$ **24.** $f(x) = \log_5 x$

25. $f(x) = \log_{1.5} x$ **26.** $f(x) = \ln x$

In Exercises 27–50, without using a calculator, evaluate the given logarithms, if possible. Assume that all variables represent positive numbers.

27. $\log_4 16$ **28.** $\log_3 81$

29. $\ln 1$ **30.** $\log_{11} 11$

31. $\log_9 243$ **32.** $\log_9 \frac{1}{27}$

33. $\log 1000$ **34.** $\ln e^5$

35. $\ln(-3)$ **36.** $\log_4(-1)$

37. $\log_{1/9} 3$ **38.** $\log_{2/3}\left(\frac{9}{4}\right)$

39. $\log_2 8^{1/3}$ **40.** $\log_5 25^{-1/2}$

41. $\log_3 \sqrt{3^5}$ **42.** $\log \sqrt[3]{10^2}$

43. $\log 0$ **44.** $\log_7(-7)$

45. $\log_{2/3}\left(\frac{3}{2}\right)$ **46.** $\log_{3/4}\left(\frac{4}{3}\right)$

47. $\log_b b^3$ **48.** $\log_b b$

49. $\log_a \sqrt{a}$ **50.** $\log_b 1$

In Exercises 51–60, evaluate the given logarithm. Round the results to two decimal places.

51. $\ln 4$ **52.** $\log 40$

53. $\log_4 7$ **54.** $\log_2 5$

55. $\ln\left(\frac{\sqrt{3}}{2}\right)$ **56.** $\frac{\ln \sqrt{5}}{\ln 3}$

57. $\log\left(7 - \sqrt{2}\right)$ **58.** $\log_7(\log 2)$

59. $\log_5 7^3$

60. $\log \sqrt{5}$

61. What is the first step in solving a logarithmic equation $c = \log_b x$?

62. If $\log_b(x + 2) = c$, explain how you know that the solution of the equation is greater than -2.

In Exercises 63–82, solve the given equation.

63. $\log_3 x = 5$ **64.** $\log_3 x = -2$

65. $\ln x = 1$ **66.** $\ln t = 0$

67. $\log_8 t = -\frac{2}{3}$ **68.** $\log_{27} x = \frac{4}{3}$

69. $\log_x 81 = 2$

70. $\log_x e^{17} = 17$

71. $\log_x 25 = \frac{2}{3}$

72. $\log_x 0.001 = -\frac{3}{2}$

73. $\log_x 1 = 0$

74. $\log_x x = 1$

75. $\log_2 \sqrt{x} = \frac{1}{2}$

76. $\log_2 \sqrt[3]{x} = \frac{2}{3}$

77. $\log_6 6^x = 3$

78. $\log_3 9^x = 2$

79. $\log_4 (x + 1) = 2$

80. $\log_5 (2x - 1) = 1$

81. $\log_5 x^2 = 4$

82. $\log_3 (x^2 - 1) = 1$

83. Under what conditions will the graph of $\log_b x$

(a) rise from left to right?

(b) fall from left to right?

84. If $f(x) = 7^x$ and $g(x) = \log_7 x$, what is $(f \circ g)(x)$? Explain how you know.

In Exercises 85–88, use the graph of the given function to estimate the domain of the function.

85. $\ln (5 - 2x)$

86. $\log x^2$

87. $\ln |x - 3|$

88. $\ln (x^2 + 4)$

In Exercises 89–92, let $f(x) = \ln x$ and $g(x) = \log x$. Use a graph to determine the value of x corresponding to the given functional value. Round results to the nearest hundredth.

89. $g(x) = 1.751$

90. $g(x) = -1.622$

91. $f(x) = 3.807$

92. $f(x) = -1.328$

In Exercises 93–96, compare the graphs of each pair of functions to the graph of $f(x) = \ln x$.

93. (a) $g(x) = 3 + \ln x$ (b) $h(x) = \ln (3 + x)$

94. (a) $g(x) = \ln (x - 5)$ (b) $h(x) = -5 + \ln x$

95. (a) $g(x) = \ln (-x)$ (b) $h(x) = -\ln x$

96. (a) $g(x) = \log_2 x$ (b) $h(x) = \log_3 x$

The following figure shows the graph of a function $f(x) = \log_b x$. In Exercises 97 and 98, use this graph to assist you in matching the given functions to graph (A), (B), (C), or (D).

97. (a) $g(x) = \log_b (x + 3)$ (b) $h(x) = 3 + \log_b x$

98. (a) $h(x) = 1 - \log_b x$ (b) $g(x) = \log_b (x - 5)$

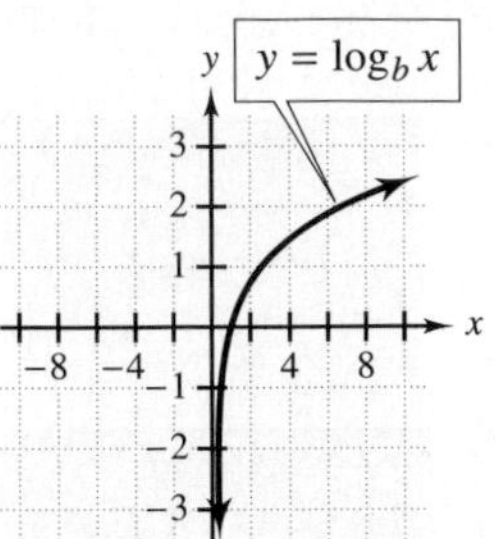

(A)

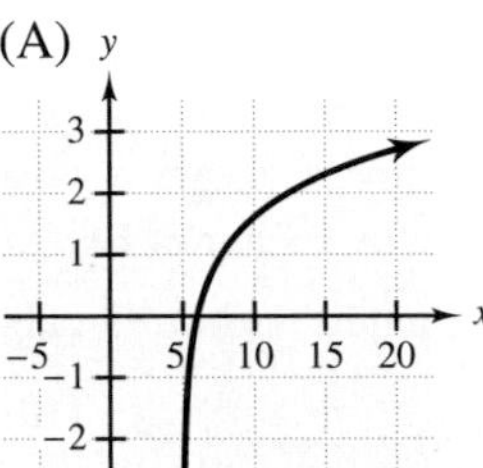

(B)

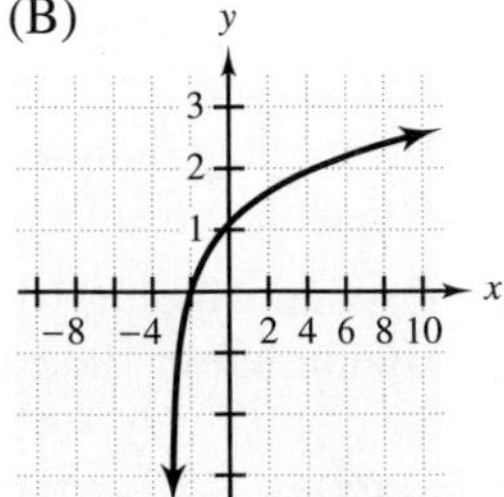

(C)

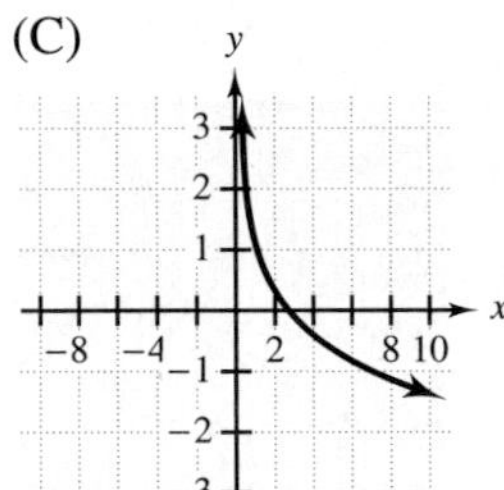

(D)

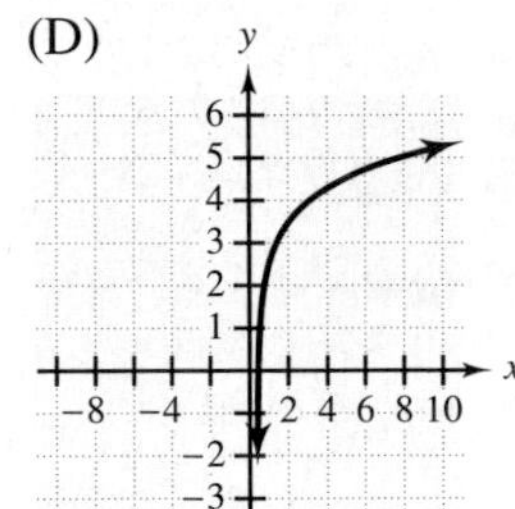

Real-Life Applications

99. An amplifier produces a power gain G (in decibels) given by $G = 10 \log (P_o/P_i)$, where P_o is the power output and P_i is the power input.

(a) To the nearest hundredth, what is the power gain when $P_o = 16$ and $P_i = 0.004$?

(b) What is the power output for an amplifier that produces a 50-decibel gain when the power input is 0.006?

100. The number of cell divisions d that occur in a given time is given by $d = \log_2 (n/n_0)$, where n_0 is the initial cell count and n is the final cell count.

(a) Approximately how many cell divisions occurred if the initial cell count was 50 and the final cell count was 100,000?

(b) Determine the final cell count if an initial cell count of 100 cells undergoes 12 cell divisions.

101. The acidity of a substance is measured by the pH value. To determine the acidity of a substance, chemists measure the concentration $[H^+]$ of hydrogen ions in the substance and then calculate the pH with the formula $pH = -\log [H^+]$.

(a) The hydrogen ion concentration of lemon juice is about 0.005. What is the pH value?

(b) The water in a pond is best for fish if the pH is 7. Determine the hydrogen ion concentration.

102. The air pressure P is related to the altitude y (in kilometers) by the formula $\ln (P/P_0) = -0.116y$, where P_0 is the air pressure at sea level.

(a) To the nearest tenth, at what altitude is the air pressure 75% of the air pressure at sea level?

(b) The air pressure at sea level is 760 millimeters. To the nearest whole number, what is the air pressure at an altitude of 5 kilometers?

Modeling with Real Data

103. The accompanying table shows the average number of home runs per game for the National and American Leagues in the period 1992–1996. (Source: Elias Sports Bureau.)

	Home Runs	
Year	**National**	**American**
1992	1.30	1.57
1993	1.72	1.83
1994	1.91	2.23
1995	1.90	2.14
1996	2.11	2.45

The following natural logarithmic functions model the number of home runs, where x is the number of years since 1990.

National: $N(x) = 0.89 + 0.68 \ln x$

American: $A(x) = 1.04 + 0.77 \ln x$

(a) According to the models, how many home runs per game will each league average in 2001?

(b) Produce the graphs of the two functions in the same coordinate system. What are two conclusions that can be drawn from the graphs?

104. Recycling takes 95% less energy than making new metal from ore. The accompanying table shows the percentage of aluminum cans that were recycled for selected years in the period 1974–1994. (Source: The Aluminum Association.)

Year	Percent
1974	17%
1982	56%
1990	64%
1994	65%

The percentage R of recycled cans x years after 1970 is given by $R(x) = -18.9 + 63.55 \log x$.

(a) According to the model, what percentage of cans is projected to be recycled in 1999?

(b) In what year does the model predict 75% of all aluminum cans will be recycled?

Group Project: NFL Salaries

Each spring the National Football League drafts college players. As the bar graph on the following page shows, the median salaries of draftees who make the roster depend on the round in which the players were drafted. (Source: *USA Today*.)

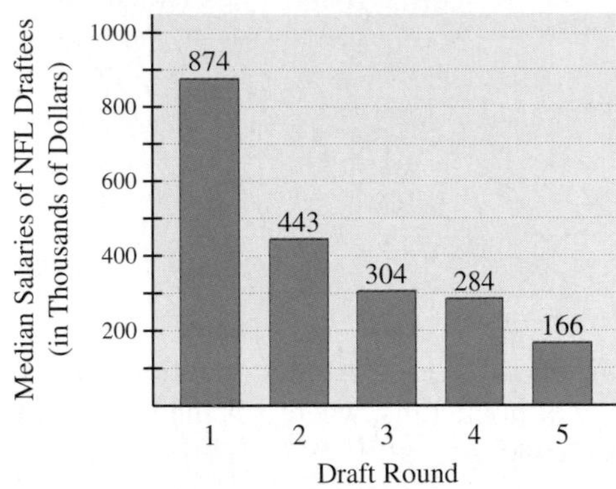

If r represents the round number, the median salary $S(r)$ (in thousands) can be modeled by either of the following functions:

$$S_1(r) = 818 - 422 \ln r$$

$$S_2(r) = 1095e^{-0.377r}$$

105. Using a suitable window, produce both graphs in the same coordinate system. Which function underestimates the actual median salary for round 1?

106. What property of the graph of S_1 makes this model function clearly unrealistic for round 7 and beyond? Why?

107. Suppose that the median salary for every round of the draft increased by \$50,000. Which of the two constants in the model function S_1 would be affected? How?

108. An increasing number of athletes leave college before completing their degree to enter the NFL draft. What factors would you weigh in making such a decision?

Challenge

109. Suppose $a > b$. If you multiply both sides of the inequality by $\ln \frac{1}{2}$, what is the result? Why?

110. Show that $e^{x-\ln x} = x^{-1}e^{x}$.

In Exercises 111 and 112, evaluate the given expression.

111. $e^{\ln 5 - \ln 2}$

112. $\ln e^{\ln e}$

In Exercises 113 and 114, for what values of x is the function positive?

113. $\ln |x|$

114. $\ln (x^2 - 16)$

In Exercises 115 and 116, solve the given inequalities.

115. $\log_3 x < \log_4 x$

116. $\ln x > \log x$

10.5 PROPERTIES OF LOGARITHMS

Basic Properties • Product, Quotient, and Power Rules for Logarithms • Combining the Properties

Basic Properties

Because logarithms are defined in terms of exponents, the properties of logarithms are closely related to the properties of exponents.

The first group of properties follows directly from the definition of logarithm.

Basic Properties of Logarithms

1. $\log_b 1 = 0$
2. $\log_b b = 1$
3. $\log_b b^x = x$
4. $b^{\log_b x} = x$

Think About It

To solve exponential equations, we wrote both sides with the same base and equated the exponents. For the equation $3^x = 6$, use property 4 to write 6 with a base of 3. Then solve the equation.

For each property an equivalent exponential form can be written. In the following table, the logarithmic properties in the left column are justified by the corresponding exponential forms in the center column.

Logarithmic form	*Exponential form*	*Example*
$\log_b 1 = 0$	$b^0 = 1$	$\log_7 1 = 0$
$\log_b b = 1$	$b^1 = b$	$\log_{3/4} \frac{3}{4} = 1$
$\log_b b^x = x$	$b^x = b^x$	$\log_8 8^5 = 5$
$\log_b x = \log_b x$	$b^{\log_b x} = x$	$5^{\log_5 3} = 3$

Product, Quotient, and Power Rules for Logarithms

Some logarithmic expressions can be evaluated in two different ways.

EXPLORING THE CONCEPT

Rules for Logarithms

Figures 10.21, 10.22, and 10.23 show the results of evaluating some logarithmic expressions.

Figure 10.21

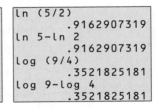

```
ln (5*2)
         2.302585093
ln 5+ln 2
         2.302585093
log (7*3)
         1.322219295
log 7+log 3
         1.322219295
```

Figure 10.22

```
ln (5/2)
         .9162907319
ln 5-ln 2
         .9162907319
log (9/4)
         .3521825181
log 9-log 4
         .3521825181
```

Figure 10.23

```
ln 2^3
         2.079441542
3ln 2
         2.079441542
log 5^4
         2.795880017
4log 5
         2.795880017
```

We summarize these results in the following table.

Figure	*Results*	*General rule*
10.21	$\ln(5 \cdot 2) = \ln 5 + \ln 2$ $\log(7 \cdot 3) = \log 7 + \log 3$	The logarithm of a product of two numbers is equal to the sum of the logarithms of the numbers.
10.22	$\ln \frac{5}{2} = \ln 5 - \ln 2$ $\log \frac{9}{4} = \log 9 - \log 4$	The logarithm of a quotient of two numbers is equal to the difference of the logarithms of the two numbers.
10.23	$\ln 2^3 = 3 \cdot \ln 2$ $\log 5^4 = 4 \cdot \log 5$	The logarithm of a number raised to a power is equal to the power times the logarithm of the number.

The generalizations that are informally given in the table can be stated formally as rules of logarithms.

LEARNING TIP

To help remember the properties of logarithms, compare them with the corresponding properties of exponents. For example:

Product *Sum*
$b^m b^n = b^{m+n}$
$\log(MN) = \log M + \log N$

Product, Quotient, and Power Rules for Logarithms

For $b > 0$, $b \neq 1$, positive real numbers M and N, and any real number c, the following are properties of logarithms:

1. $\log_b MN = \log_b M + \log_b N$ Product Rule for Logarithms
2. $\log_b \dfrac{M}{N} = \log_b M - \log_b N$ Quotient Rule for Logarithms
3. $\log_b M^c = c(\log_b M)$ Power Rule for Logarithms

The following demonstrates why the Product Rule for Logarithms is true.

Let $\log_b M = r$ and $\log_b N = s$. Then $b^r = M$ and $b^s = N$. Therefore, the product $MN = b^r b^s = b^{r+s}$. But $b^{r+s} = MN$ implies that

$$\log_b MN = r + s = \log_b M + \log_b N$$

Justifications for the Quotient and Power Rules for Logarithms are similar to that used for the Product Rule for Logarithms. The proofs are left as exercises.

Note: There is no property for the logarithm of a *sum* of two numbers or a *difference* of two numbers. In particular, $\log_b (x + y)$ is not equal to $\log_b x + \log_b y$, and $\log_b (x - y)$ is not equal to $\log_b x - \log_b y$.

In the remainder of this section we assume that all logarithmic expressions are defined.

EXAMPLE 1

Rewriting Logarithms Whose Arguments are Products, Quotients, or Exponential Expressions

(a) $\log_4 5(t + 2) = \log_4 5 + \log_4 (t + 2)$ Product Rule for Logarithms

(b) $\ln e(e - 1) = \ln e + \ln (e - 1)$ Product Rule for Logarithms

$= 1 + \ln (e - 1)$ $\ln e = 1$

(c) $\ln \dfrac{1}{y} = \ln 1 - \ln y$ Quotient Rule for Logarithms

$= 0 - \ln y$ $\ln 1 = 0$

$= -\ln y$

(d) $\log_6 \dfrac{y + 4}{y} = \log_6 (y + 4) - \log_6 y$ Quotient Rule for Logarithms

(e) $\log z^4 = 4 \log z$ Power Rule for Logarithms

(f) $\ln \dfrac{1}{x^3} = \ln x^{-3} = -3 \ln x$ Power Rule for Logarithms

EXAMPLE 2

Writing a Logarithmic Expression as a Single Logarithm

(a) $\ln 3 + \ln x = \ln (3x)$ Product Rule for Logarithms

(b) $\log (x + 1) + \log (x - 1) = \log [(x + 1)(x - 1)]$

$= \log (x^2 - 1)$

(c) $\ln (2x) - \ln 3 = \ln \dfrac{2x}{3}$ Quotient Rule for Logarithms

(d) $\log(x^2 + 4) - \log(x + 2) = \log \dfrac{x^2 + 4}{x + 2}$

(e) $\dfrac{1}{2} \ln t = \ln t^{1/2} = \ln \sqrt{t}$ Power Rule for Logarithms

(f) $-2 \log_7 t = \log_7 t^{-2} = \log_7 \dfrac{1}{t^2}$

Combining the Properties

When we need to expand a single logarithm into two or more logarithms or to write a logarithmic expression with a single logarithm, we will sometimes need to use more than one of the rules for logarithms.

EXAMPLE 3

Expanding a Logarithmic Expression

(a) $\log_5 3(x + 4)^2 = \log_5 3 + \log_5 (x + 4)^2$ Product Rule for Logarithms

$= \log_5 3 + 2 \log_5 (x + 4)$ Power Rule for Logarithms

(b) $\ln \left(\dfrac{x + 2}{x + 1}\right)^2 = 2 \ln \dfrac{x + 2}{x + 1}$ Power Rule for Logarithms

$= 2[\ln(x + 2) - \ln(x + 1)]$ Quotient Rule for Logarithms

$= 2 \ln(x + 2) - 2 \ln(x + 1)$ Distributive Property

(c) $\log \sqrt{\dfrac{x^3}{x + 2}} = \log \left(\dfrac{x^3}{x + 2}\right)^{1/2}$ Definition of rational exponent

$= \dfrac{1}{2} \log \dfrac{x^3}{x + 2}$ Power Rule for Logarithms

$= \dfrac{1}{2} [\log x^3 - \log(x + 2)]$ Quotient Rule for Logarithms

$= \dfrac{1}{2} [3 \log x - \log(x + 2)]$ Power Rule for Logarithms

$= \dfrac{3}{2} \log x - \dfrac{1}{2} \log(x + 2)$ Distributive Property

EXAMPLE 4

Combining Logarithms

(a) $2 \ln t + 3 \ln(t + 2) = \ln t^2 + \ln(t + 2)^3$ Power Rule for Logarithms

$= \ln t^2(t + 2)^3$ Product Rule for Logarithms

(b) $4[\log_6 (y + 1) - \log_6 y] = 4 \log_6 \dfrac{y + 1}{y}$ Quotient Rule for Logarithms

$= \log_6 \left(\dfrac{y + 1}{y}\right)^4$ Power Rule for Logarithms

(c) $2 \log(2x + 1) + \log x - 4 \log(x + 3)$

$= \log(2x + 1)^2 + \log x - \log(x + 3)^4$ Power Rule for Logarithms

$= \log x(2x + 1)^2 - \log(x + 3)^4$ Product Rule for Logarithms

$= \log \dfrac{x(2x + 1)^2}{(x + 3)^4}$ Quotient Rule for Logarithms

10.5 QUICK REFERENCE

Basic Properties

- Certain properties of logarithms follow directly from the definition of logarithm and can be used to evaluate a logarithmic expression.

1. $\log_b 1 = 0$
2. $\log_b b = 1$
3. $\log_b b^x = x$
4. $b^{\log_b x} = x$

Product, Quotient, and Power Rules for Logarithms

- For $b > 0$, $b \neq 1$, positive real numbers M and N, and any real number c, the following are properties of logarithms:

1. $\log_b MN = \log_b M + \log_b N$ Product Rule for Logarithms
2. $\log_b \frac{M}{N} = \log_b M - \log_b N$ Quotient Rule for Logarithms
3. $\log_b M^c = c(\log_b M)$ Power Rule for Logarithms

Combining the Properties

- Logarithmic expressions may involve combinations of products, quotients, and powers. The Product, Quotient, and Power Rules for Logarithms can be used to expand a single logarithm into two or more logarithmic expressions.
- These same rules can be used to combine two or more logarithms into a single logarithmic expression.

10.5 EXERCISES

Concepts and Skills

1. Treating a logarithm as an exponent, we can evaluate $\log_4 16$ by asking, "What exponent on 4 results in 16?" Because the answer is 2, $\log_4 16 = 2$. What question would you ask in order to evaluate $\log_9 3$? What is the answer?

2. What rule of logarithms are we using if we write $\log x^2 = \log (x \cdot x) = \log x + \log x = 2 \log x$? What rule of logarithms would be more efficient in obtaining the same result?

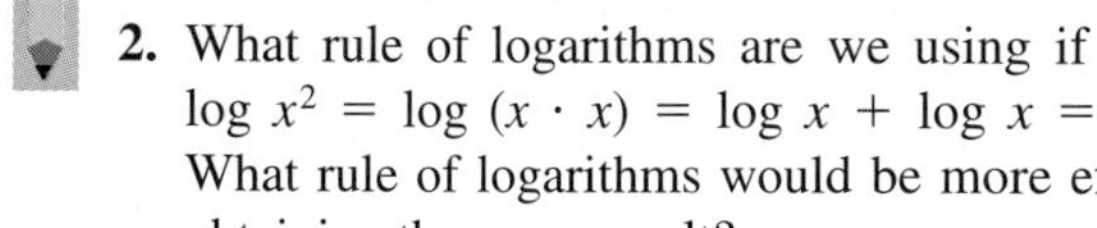

In Exercises 3–18, evaluate the given expression.

3. $\log_5 5^7$ **4.** $\ln e^2$

5. $\log_{1.2} 1$ **6.** $\log_7 7$

7. $7^{\log_7 2}$ **8.** $3.5^{\log_{3.5} 10}$

9. $\log_9 \sqrt{3}$ **10.** $\log_4 \sqrt{8}$

11. $9^{-\log_3 2}$ **12.** $125^{-\log_5 4}$

13. $e^{2 \ln 3}$ **14.** $10^{3 \log 2}$

15. $\log_3 (\log_4 4)$ **16.** $\ln (\log 10)$

17. $\log_3 (\log_2 8)$ **18.** $\log_2 (\log_2 16)$

In Exercises 19–22, write the given logarithm as a sum or difference of logarithms.

19. $\log_7 3x$ **20.** $\log_8 5mn$

21. $\log_7 \frac{5}{y}$ **22.** $\log_6 \frac{y}{t}$

In Exercises 23–26, write the given logarithm with no exponents or radicals in the argument.

23. $\log x^2$ **24.** $\log_3 7^{-4}$

25. $\log_9 \sqrt{n}$ **26.** $\ln \sqrt[3]{x}$

In Exercises 27–34, use the rules of logarithms to expand the logarithm in column A. Then, letting $\log_3 2 = A$ and $\log_3 5 = B$, match the result with one of the forms in column B.

Column A	*Column B*
27. $\log_3 \left(\frac{5}{6}\right)^3$	(a) $A - \frac{1}{2}B + 1$
28. $\log_3 \frac{45}{8}$	(b) $-\frac{2}{3}A + \frac{1}{3}B - \frac{2}{3}$
29. $\log_3 \frac{6\sqrt{5}}{5}$	(c) $-3A + 3B - 3$

Column A	*Column B*
30. $\log_3 \dfrac{36\sqrt{2}}{125}$	(d) $-\dfrac{3}{4}A - \dfrac{3}{4}$
31. $\log_3 \dfrac{5\sqrt{10}}{9}$	(e) $\dfrac{1}{5}A - \dfrac{8}{5}B - 1$
32. $\log_3 \dfrac{\sqrt[3]{30}}{6}$	(f) $-3A + B + 2$
33. $\log_3 \dfrac{\sqrt[4]{96}}{12}$	(g) $\dfrac{1}{2}A + \dfrac{3}{2}B - 2$
34. $\log_3 \dfrac{\sqrt[5]{50}}{75}$	(h) $\dfrac{5}{2}A - 3B + 2$

35. To evaluate $\log (AB)^n$, should you first apply the Product Rule for Logarithms or the Power Rule for Logarithms? Why?

36. From the following list, identify two logarithms that cannot be expanded and explain why.

(i) $\log (AB)$ (ii) $\log (A - B)$

(iii) $\log (A + B)$ (iv) $\log \dfrac{A}{B}$

In Exercises 37–58, if possible, use the properties of logarithms to expand the given expression.

37. $\log_9 x^2y^{-3}$ **38.** $\log_3 27(x + 5)^3$

39. $\log \dfrac{(x + 1)^2}{x + 2}$ **40.** $\log \dfrac{x^2y^3}{z^4}$

41. $\ln \left(\dfrac{2x + 5}{x + 7}\right)^5$ **42.** $\log_3 \dfrac{(x^2z)^3}{(y^2z)^5}$

43. $\log \dfrac{\sqrt{10}}{x^2}$ **44.** $\ln \sqrt[4]{\dfrac{x^3y}{z^2}}$

45. $\log_5 \sqrt[3]{xy^2}$ **46.** $\log_3 81x^2\sqrt{y}$

47. $\log_2 \dfrac{\sqrt{x + 1}}{32}$ **48.** $\log_8 \sqrt{x(x + 2)}$

49. $\log_2 \dfrac{x^2\sqrt{y}}{\sqrt[3]{x}}$ **50.** $\ln \dfrac{(x + 2)\sqrt{x}}{x^2}$

51. $\log (4x^2 + 4x + 1)$ **52.** $\log (x^2 + 3x + 2)$

53. $\ln 1000e^{0.08t}$ **54.** $\ln (e^2 - e)$

55. $\log_4 (x + 4)$ **56.** $\ln (e^2 + 1)$

57. $\ln Pe^{rt}$ **58.** $\ln P\left(1 + \dfrac{r}{t}\right)^{nt}$

59. Explain how to combine $2 \log 3 + \log 3$ into a single logarithm without the Product Rule for Logarithms. Can you combine $2 \log x + \log y$ into a single logarithm without the Product Rule for Logarithms if $x \neq y$? Explain.

60. Which rule of logarithms would you use to show that $-\log x = \log (1/x)$? Would you use the same rule to show that $\log (1/x) = -\log x$? Write the steps to show that both equations are true.

In Exercises 61–64, write the given logarithmic expression as a single logarithm.

61. $\log_4 3 + \log_4 5$ **62.** $\ln x^2 + \ln x^{-1}$

63. $\log_3 10 - \log_3 5$ **64.** $\log_7 u^5 - \log_7 u^2$

In Exercises 65–68, write the given logarithmic expression as a single logarithm with a coefficient of 1.

65. $3 \log_2 y$ **66.** $-\log_5 x$

67. $\dfrac{1}{2} \log x$ **68.** $\dfrac{2}{3} \ln t$

In Exercises 69–72, write the given expression as a single logarithm and evaluate the result. Use a calculator to verify that the single logarithm has the same value as the original expression.

69. $\log_4 60 - \log_4 15$

70. $\log_3 4 + \log_3 6 - \log_3 8$

71. $\log_5 50 + 2 \log_5 10 - \log_5 40$

72. $\dfrac{1}{3} \log_6 27 - \log_6 8 - \log_6 81$

In Exercises 73–88, use the rules of logarithms to combine the given expression into a single logarithmic expression with a coefficient of 1.

73. $\log_8 3 + \log_8 x - \log_8 y$

74. $\ln x - \ln 3 - \ln 2$

75. $2 \log_7 x - \log_7 y$

76. $\log_8 5 + 2 \log_8 t$

77. $\dfrac{1}{2} \log x - \dfrac{2}{3} \log y$

78. $\dfrac{3}{2} \log_6 r + \dfrac{1}{2} \log_6 s$

79. $3 \log y - 2 \log x - 4 \log z$

80. $\log y + 2 \log (x + 2) - 5 \log z$

81. $2 \log (x + 3) - 3 \log (x + 1)$

82. $\dfrac{1}{2} \log (t + 1) - 3 \log t$

83. $\frac{1}{2}[\log x - \log(x+1)]$

84. $\frac{1}{3}[2\log(x+2) - \log(x-5)]$

85. $5 + 2\log_3 x$

86. $1 + 4\log_2 \sqrt{x}$

87. $rt + \ln P_0$

88. $0.06t + \ln 1000$

In Exercises 89–94, determine whether the equation is true.

89. $\frac{\log_8 3}{\log_8 5} = \log_8 \frac{3}{5}$ **90.** $\frac{\log_3 25}{\log_3 5} = 2$

91. $\frac{\log_3 81}{\log_3 9} = \log_3 9$ **92.** $\frac{\log_2 12}{\log_2 4} = \log_2 3$

93. $(\log 3)(\log 2) = \log 6$

94. $-2 = (\log_7 9)(\log_{1/3} 7)$

In Exercises 95–100, determine whether the given equation is always true.

95. $\log_2 x^x = x \log_2 x$ **96.** $\log 10^{x+1} = x + 1$

97. $\log_5 x^2y = 2\log_5 x + \log_5 y$

98. $\ln \frac{e}{x} = 1 - \ln x$

99. $\log_3 (x + 10) = \log_3 x + \log_3 10$

100. $\log x^2 = (\log x)^2$

Real-Life Applications

101. A consultant measured the effect of travel fatigue on productivity of business travelers. The productivity was modeled by $P(t) = \frac{10}{\log 5 + \log(t+2)}$, where t is the number of hours that the person spent in transit and waiting for transit (such as sitting in airports).

(a) By what percentage did the productivity decline with 4 hours of travel time?

(b) Write the function with a single logarithm in the denominator.

102. Suppose that the sales (in thousands of units) of a certain new brand of cereal is given by the function $S(t) = 100\log_2(t+2)$, where t is the number of months since the brand was introduced. How many units of the product were sold 6 months after its introduction? Show that an equivalent formula can be written with natural logarithms:

$$S(t) = \frac{\ln(t+2)^{100}}{\ln 2}$$

Group Project: College Enrollment

The enrollment in public and private colleges and universities has increased steadily since 1970. The accompanying table shows total and projected enrollment figures (in thousands) for selected years in the period 1970–2004. (Source: U.S. National Center for Educational Statistics.)

Year	Enrollment (in thousands)
1970	8,581
1980	12,097
1985	12,247
1990	13,820
1992	14,191
1993	14,600
1994	14,700
1995	14,946
2000	15,462
2003	15,802
2004	15,892

The enrollment $N(t)$ (in thousands) can be modeled by $N(t) = 3969 + 3336 \ln t$, where t is the number of years since 1970.

103. Using 0 to 35 for x and 0 to 18,000 for y, produce a scatter plot of the data points and the graph of function N in the same coordinate system. What do you observe when you trace to the point representing the data for 1970? Why is this so?

104. Use the Power Rule for Logarithms to rewrite $N(t)$. Even though the expressions are equivalent, why do you think that it is impossible to produce the graph of this revised function on some calculators?

105. Using the data for 1980 and 2004, write a linear function to model enrollment for the period 1980–2004. Add the graph of this function to the graphing screen. By extending the viewing window for both axes, compare the two models with respect to projections beyond 2004.

106. If admissions standards were significantly raised or lowered, what effect do you think there would be on the graph of $N(t)$?

Challenge

107. Let $\log_b M = r$ and $\log_b N = s$. Using steps similar to those used to justify the Product Rule for Logarithms, show that the Quotient Rule for Logarithms is true.

108. Let $\log_b M = r$. Show that the Power Rule for Logarithms is true.

109. Show that $2 + \log_3 x = \log_3 9x$.

110. Show that

$$\log_2 x + \log_3 x = \frac{(\log 2 + \log 3) \log x}{(\log 2)(\log 3)}$$

111. Show that $\log_2 (xy) = \dfrac{\ln x + \ln y}{\ln 2}$.

112. Suppose $a^2 + b^2 = 7ab$. Show that the following is true.

$$\log (a + b) - \log 3 = \frac{1}{2}(\log a + \log b)$$

10.6 EXPONENTIAL AND LOGARITHMIC EQUATIONS

Definitions and Rules • Exponential Equations • Logarithmic Equations • Real-Life Applications

Definitions and Rules

An equation in which a variable appears in the argument of a logarithm is called a **logarithmic equation.** An equation in which a variable appears as an exponent is called an **exponential equation.**

Because both exponential and logarithmic functions are one-to-one functions, the following properties can be used to solve exponential and logarithmic equations.

Properties of Exponential and Logarithmic Equations

Assuming that all expressions are defined, the following properties hold for exponential and logarithmic equations:

1. If $x = y$, then $b^x = b^y$. — Equating exponential expressions
2. If $b^x = b^y$, then $x = y$. — Equating exponents
3. If $x = y$, then $\log_b x = \log_b y$. — Equating logarithmic expressions
4. If $\log_b x = \log_b y$, then $x = y$. — Equating arguments of logarithms

Exponential Equations

Recall that when an exponential equation can be written with both sides having the same base, we can equate the exponents and solve the resulting equation. However, if both sides of an equation cannot be written with the same base, other algebraic techniques are needed to solve the equation.

EXPLORING THE CONCEPT

Solving Exponential Equations

Consider the equation $3^{2x} = 5$. We can use a graph to estimate that the solution is 0.7. (See Fig. 10.24 on the following page.)

Figure 10.24

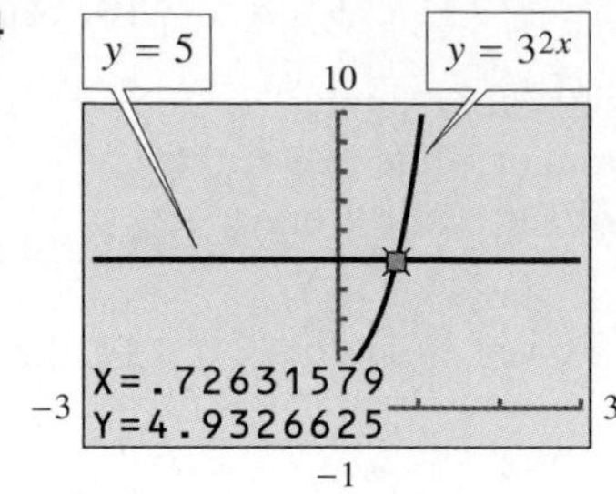

To solve the equation algebraically, we choose a logarithm with a convenient base, usually ln or log, and equate the logarithms of both sides.

$$\begin{aligned} 3^{2x} &= 5 && \\ \ln 3^{2x} &= \ln 5 && \text{Equate the logarithms of both sides.} \\ 2x \cdot \ln 3 &= \ln 5 && \text{Power Rule for Logarithms} \\ (2 \ln 3)\, x &= \ln 5 && \\ x &= \frac{\ln 5}{2 \ln 3} && \text{Divide both sides by 2 ln 3.} \\ x &\approx 0.73 && \end{aligned}$$

Although the solutions of the equations in Example 1 can be estimated by graphing, we focus on the algebraic method of equating logarithms.

EXAMPLE 1

Solving Exponential Equations

(a)

$$\begin{aligned} 7^{x+1} &= 14 && \\ \log 7^{x+1} &= \log 14 && \text{Equate the logarithms of both sides.} \\ (x + 1) \log 7 &= \log 14 && \text{Power Rule for Logarithms} \\ x \log 7 + \log 7 &= \log 14 && \text{Distributive Property} \\ x \log 7 &= \log 14 - \log 7 && \text{Isolate the variable term.} \\ x &= \frac{\log 14 - \log 7}{\log 7} && \\ x &\approx 0.36 && \end{aligned}$$

(b) When an equation involves the number e, we often choose to use natural logarithms to solve.

$$\begin{aligned} e^x &= 2 && \\ \ln e^x &= \ln 2 && \text{Equate the logarithms of both sides.} \\ x \ln e &= \ln 2 && \text{Power Rule for Logarithms} \\ x \cdot 1 &= \ln 2 && \ln e = 1 \\ x &= \ln 2 && \\ x &\approx 0.69 && \end{aligned}$$

(c)

$$8 - 3^{2x-1} = 1 \quad \text{Isolate the exponential term.}$$
$$-3^{2x-1} = -7 \quad \text{Subtract 8 from both sides.}$$
$$3^{2x-1} = 7 \quad \text{Multiply both sides by } -1.$$
$$\log 3^{2x-1} = \log 7 \quad \text{Equate the logarithms of both sides.}$$
$$(2x - 1)\log 3 = \log 7 \quad \text{Power Rule for Logarithms}$$
$$2x - 1 = \frac{\log 7}{\log 3} \quad \text{Divide both sides by log 3.}$$
$$2x = \frac{\log 7}{\log 3} + 1 \quad \text{Solve for } x.$$
$$x = \frac{1}{2}\left(\frac{\log 7}{\log 3} + 1\right)$$
$$x \approx 1.39$$

Logarithmic Equations

We can also use graphs to estimate the solutions of logarithmic equations.

EXPLORING THE CONCEPT

Solving Logarithmic Equations

Consider the equation $\ln(x - 3) + \ln(x + 1) = \ln 5$.

Figure 10.25 shows the graphs of $y_1 = \ln(x - 3) + \ln(x + 1)$ and $y_2 = \ln 5$.

Figure 10.25

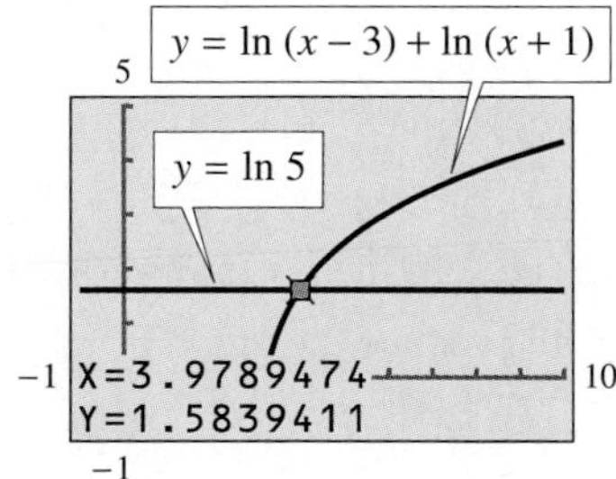

The solution is approximately 4. (In fact, it is easy to verify by substitution that the solution is *exactly* 4.)

To solve the equation algebraically, we write the left side as a single logarithm and then equate the arguments of the logarithms.

$$\ln(x - 3) + \ln(x + 1) = \ln 5$$
$$\ln(x - 3)(x + 1) = \ln 5 \quad \text{Product Rule for Logarithms}$$
$$(x - 3)(x + 1) = 5 \quad \text{Equate the arguments of the logarithms.}$$
$$x^2 - 2x - 3 = 5 \quad \text{FOIL}$$
$$x^2 - 2x - 8 = 0 \quad \text{Write the quadratic equation in standard form.}$$
$$(x - 4)(x + 2) = 0 \quad \text{Factor.}$$
$$x - 4 = 0 \quad \text{or} \quad x + 2 = 0 \quad \text{Zero Factor Property}$$
$$x = 4 \quad \text{or} \quad x = -2$$

LEARNING TIP

For a logarithmic equation, use the properties of logarithms to write each side of the equation as a single term.

The graph in Fig. 10.25 suggests that the equation has only one solution, but the algebraic approach produces two solutions.

Replacing x with -2 gives $\ln(-5) + \ln(-1) = \ln 5$. But logarithmic functions are defined only for positive numbers. Therefore, -2 is an extraneous solution. The only solution of the equation is 4.

In Example 2 we use both graphing and algebraic methods.

EXAMPLE 2

Solving Logarithmic Equations

Use a graph to estimate the solution(s) of the given equation. Then solve algebraically.

(a) $\ln(1 - 2x) = \ln 5$

(b) $\log_3 x + \log_3 (x - 2) = 1$

(c) $\ln(4x - 5) - \ln(x - 2) = \ln(2x + 1)$

Solution

Figure 10.26

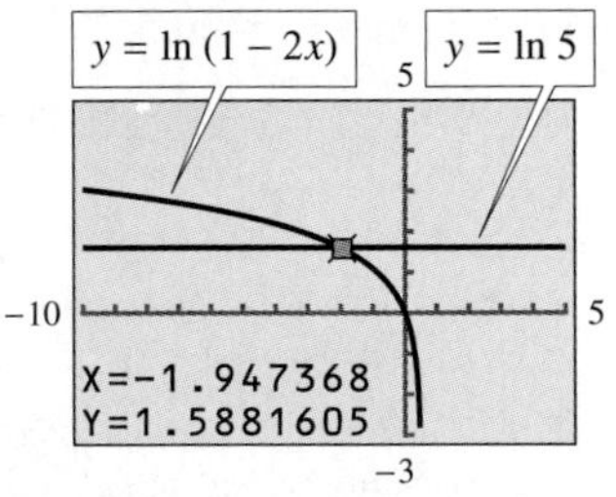

(a) Figure 10.26 shows the graphs of $y_1 = \ln(1 - 2x)$ and $y_2 = \ln 5$. We estimate the one solution to be -1.95.

$$\ln(1 - 2x) = \ln 5$$
$$1 - 2x = 5 \qquad \text{Equate arguments.}$$
$$-2x = 4$$
$$x = -2$$

The solution is easily verified by substitution.

(b) To produce the graph of $y_1 = \log_3 x + \log_3 (x - 2)$, we use the Change of Base Formula.

$$\log_3 x + \log_3 (x - 2) = \frac{\log x}{\log 3} + \frac{\log (x - 2)}{\log 3}$$

Figure 10.27

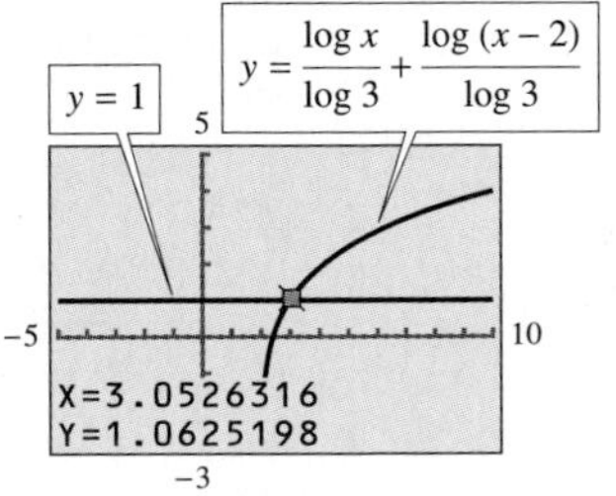

From Fig. 10.27 we estimate the solution to be approximately 3.

$$\log_3 x + \log_3 (x - 2) = 1$$
$$\log_3 x(x - 2) = 1 \qquad \text{Product Rule for Logarithms}$$
$$x(x - 2) = 3^1 \qquad \text{Exponential form}$$
$$x^2 - 2x = 3 \qquad \text{Distributive Property}$$
$$x^2 - 2x - 3 = 0 \qquad \text{Quadratic Equation in standard form}$$
$$(x - 3)(x + 1) = 0 \qquad \text{Factor.}$$
$$x - 3 = 0 \quad \text{or} \quad x + 1 = 0 \qquad \text{Zero Factor Property}$$
$$x = 3 \quad \text{or} \quad x = -1$$

The value -1 is not in the domain of $\log_3 (x - 2)$, so it is an extraneous solution. The only solution is $x = 3$.

Figure 10.28

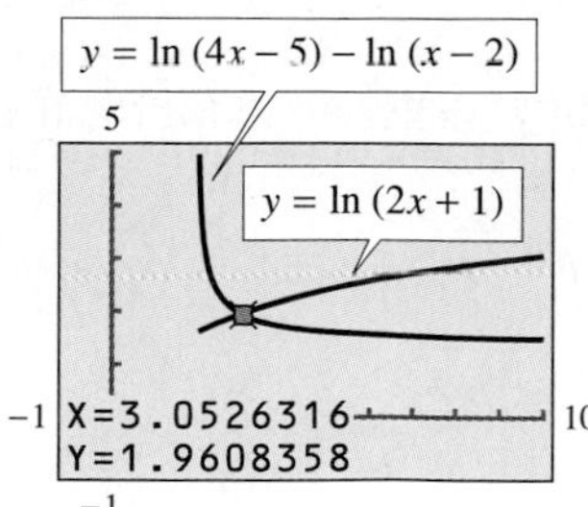

(c) We produce the graphs of $y_1 = \ln(4x - 5) - \ln(x - 2)$ and $y_2 = \ln(2x + 1)$. (See Fig. 10.28.) The estimated solution is 3.

$$\ln(4x - 5) - \ln(x - 2) = \ln(2x + 1)$$

$$\ln\frac{4x - 5}{x - 2} = \ln(2x + 1) \qquad \text{Quotient Rule for Logarithms}$$

$$\frac{4x - 5}{x - 2} = 2x + 1 \qquad \text{Equate arguments.}$$

$$4x - 5 = (2x + 1)(x - 2) \qquad \text{Clear the fraction.}$$

$$4x - 5 = 2x^2 - 3x - 2 \qquad \text{FOIL}$$

$$0 = 2x^2 - 7x + 3 \qquad \text{Standard form}$$

$$0 = (2x - 1)(x - 3) \qquad \text{Factor.}$$

$$2x - 1 = 0 \quad \text{or} \quad x - 3 = 0 \qquad \text{Zero Factor Property}$$

$$2x = 1 \quad \text{or} \quad x = 3$$

$$x = \frac{1}{2} \quad \text{or} \quad x = 3$$

When x is $\frac{1}{2}$, the left side of the equation is not defined. Therefore, $\frac{1}{2}$ is an extraneous solution, and 3 is the only solution.

Think About It

Show that $\log 10^{\ln e} = \ln e^{\log 10}$.

Real-Life Applications

When interest is *compounded continuously*, the amount A on deposit after t years is given by $A = Pe^{rt}$, where r is the interest rate and P is the initial investment.

EXAMPLE 3

Continuously Compounded Interest

At age 24, a person has dreams of becoming a millionaire. If this person's current net worth is \$150,000, what growth rate is required to reach the \$1,000,000 goal by age 50?

Solution

$A = 1{,}000{,}000$ Desired net worth at age 50

$P = 150{,}000$ Current net worth

$t = 26$ Number of years until the age is 50

$$A = Pe^{rt} \qquad \text{Model for continuous compounding}$$

$$1{,}000{,}000 = 150{,}000e^{26r} \qquad \text{Substitute } A, P, \text{ and } t.$$

$$6.667 \approx e^{26r} \qquad \text{Divide both sides by 150,000.}$$

$$\ln 6.667 \approx 26r \qquad \text{Convert to logarithmic form.}$$

$$r \approx \frac{\ln 6.667}{26}$$

$$r \approx 0.073$$

For the person to be a millionaire by age 50, the current assets must be invested at an interest rate of 7.3%, compounded continuously.

If a quantity increases (growth) or decreases (decay) over a period of time, the formula $P = P_0e^{kt}$ can sometimes be used to describe the amount P of the quantity after a certain time t. The initial amount of the quantity (at time $t = 0$) is P_0. The constant k is called the *growth* or *decay constant.* A positive value of k indicates growth and a negative value of k indicates decay.

Growth or decay that can be approximated by the model $P = P_0e^{kt}$ is called *exponential growth* or *decay.*

EXAMPLE 4

Exponential Decay of Toxic Waste

Containers of toxic waste totaling 500 pounds are buried in a landfill. After 30 years, 400 pounds of toxic waste still remain. If the decay is exponential, how much will remain after 100 years?

Solution

$P_0 = 500$ Initial amount of toxic waste

$P = 500e^{kt}$ Exponential decay model with $P_0 = 500$

$400 = 500e^{30k}$ When $t = 30$, $P = 400$.

$0.8 = e^{30k}$ Divide both sides by 500.

$\ln 0.8 = 30k$ Convert to logarithmic form.

$$k = \frac{\ln 0.8}{30}$$

$k \approx -0.0074$ Decay constant is negative.

Now we use the formula again, this time with the known decay constant.

$P \approx 500e^{-0.0074t}$

$P \approx 500e^{-0.0074(100)}$ Time t is 100 years.

$P \approx 500e^{-0.74}$

$P \approx 238.56$

After 100 years, approximately 238.56 pounds of toxic waste will remain in the landfill.

10.6 QUICK REFERENCE

Definitions and Rules

- An equation in which a variable appears in the argument of a logarithm is a **logarithmic equation.** An equation in which a variable appears as an exponent is an **exponential equation.**
- Assuming that all expressions are defined, the following properties hold for exponential and logarithmic equations:

  1. If $x = y$, then $b^x = b^y$.
  2. If $b^x = b^y$, then $x = y$.
  3. If $x = y$, then $\log_b x = \log_b y$.
  4. If $\log_b x = \log_b y$, then $x = y$.

Exponential Equations

- To solve an exponential equation that can be written with both sides having the same base, equate the exponents and solve the resulting equation.
- To solve an exponential equation for which the bases are different, take the logarithm (with a convenient base) of both sides, simplify, and solve the resulting equation.

Logarithmic Equations

- To solve a logarithmic equation algebraically, follow these steps:
  1. Write one side of the equation as a single logarithm and the other as a number or as a single logarithm. The logarithms on each side must have the same base.
  2. If one side of the equation is a number, convert the equation to an exponential equation and solve.
  3. If both sides involve a logarithmic expression with the same base, equate the arguments and solve the resulting equation.
  4. Check all solutions in the original equation. Using the properties of logarithms can introduce extraneous solutions.

10.6 EXERCISES

Concepts and Skills

1. To solve the equation $b^x = 3$, under what conditions is it not necessary to write the equation in logarithmic form?

2. To solve the equation $e^x = 5$, why is it not possible to use the method of equating exponents?

In Exercises 3–6, solve the given equation.

3. $3^{5x} = \frac{1}{27}$ **4.** $5^{2x} = \frac{1}{25}$

5. $7^{1-x} = 49$ **6.** $\left(\frac{1}{9}\right)^{x-3} = 27$

In Exercises 7–10, solve for x. Express the result to the nearest hundredth.

7. $3^x = 5$ **8.** $e^{3x} = 5$

9. $3^{x-3} = 2^{x+2}$ **10.** $4^{2x-1} = 10^{1-x}$

In Exercises 11–28, solve the exponential equation. Express the result to the nearest hundredth.

11. $e^{20k} = 0.6$ **12.** $e^{0.25t} = 10$

13. $4e^{2x-1} = 5$ **14.** $\frac{2}{3}e^{-x/2} = 10$

15. $\left(\frac{5}{3}\right)^{-x} = 20$ **16.** $5 \cdot 4^{1-3x} = 12$

17. $3 + e^x = 7$ **18.** $2e^x - 5 = 19$

19. $5^x + 7 = 2$ **20.** $15 - 2^x = 12$

21. $600(1 + 0.02)^{4t} = 1000$

22. $750\left(1 + \frac{0.06}{12}\right)^{12t} = 1200$

23. $5^{-x^2} = 0.2$ **24.** $3^{x^2} = 5$

25. $5^{|x+1|} = 2$ **26.** $3^{|x|} = 0.5$

27. $|2 - 3^x| = 1$ **28.** $|3e^x + 1| = 2$

29. To solve the equation $\log_5 x = \log 8$, can you use the method of equating the arguments of the logarithms? Explain.

30. Solving the equation $\log x + \log (x + 3) = \log 18$ leads to two values for x: 3 and -6. Assuming that all steps of the solving process are correct, what is the solution of the equation? Why?

In Exercises 31–44, solve for x.

31. $\log_5 x = \log_5 (2x - 3)$

32. $\log (3 - x) = \log (2x)$

33. $\log_5 (x - 1) = 2$

34. $7 + \log_2 (x + 1) = 10$

35. $\log_3 \sqrt[3]{3x - 5} = 1$

36. $1 + \log_2 \sqrt{x} = 5$

37. $\frac{1}{3}\log_8(x - 7) = \log_8 3$

38. $\frac{1}{2}\ln(3x - 1) = \ln 5$

39. $\log_2(x^2 - 7x) = 3$

40. $\log_6 x(x + 1) = 1$

41. $\log_4(5 - 3x)^3 = 6$

42. $\log_2(x^2 + 4x)^2 = 10$

43. $\frac{1}{2}\log_9(6 - x) = \log_9 x$

44. $\frac{1}{2}\log_2(x - 1) = \log_2(x - 3)$

In Exercises 45–62, solve the given equation.

45. $2 - \log_9 x^2 = 1$

46. $1 + \log_4(x + 1)^2 = 4$

47. $\log_4 6x - \log_4(x - 5) = 2$

48. $\log(3t - 1) = 1 + \log(5t - 2)$

49. $\log x + \log(7 - x) = 1$

50. $2\log_3 x = \log_3(5x - 4)$

51. $1 - \log(x - 5) = \log\left(\frac{x}{5}\right)$

52. $\log t = 2 - \log(t + 21)$

53. $\log_5(2x - 3) - \log_5 x = 1$

54. $\log_2(4x + 5) - \log_2(x - 7) = 2$

55. $\log_5 x(x + 2) = 0$

56. $\log_5 x(x + 3) = 1$

57. $\log x + \log(4 + x) = \log 5$

58. $\ln(2x + 1) = \ln 14 - \ln(x + 2)$

59. $\log(x + 6) - 2\log x = \log 12$

60. $\ln 3 = \ln(6 - 7x) - 2\ln x$

61. $|2 - \log_2(x - 1)| = 3$

62. $\log_6|2x + 1| = 1$

63. The following is one of the basic exponent properties.

For $b > 0$ and $b \neq 1$, if $b^x = b^y$, then $x = y$.

Using an example, explain why $b \neq 1$ is a necessary condition.

64. Use what you know about the graphs of the functions $f(x) = e^x$ and $g(x) = \ln x$ to explain why the equation $e^x = \ln x$ has no solution.

In Exercises 65–72, each equation is quadratic in form. Use the substitution methods discussed in Section 9.4 to solve the equation.

65. $(\log x)^2 + \log x^3 - 4 = 0$

66. $(\ln x)^2 - \ln x - 6 = 0$

67. $\log_4 x - 3\sqrt{\log_4 x} + 2 = 0$

68. $\log_3 x - 4\sqrt{\log_3 x} + 3 = 0$

69. $e^{2x} + 6 = 5e^x$

70. $e^{2x} - 4e^x + 3 = 0$

71. $5^{2x} = 5^x + 42$

72. $7^{2x} + 7^x - 2 = 0$

In Exercises 73–80, solve for the indicated variable.

73. $R = \log I$ for I

74. $P = \log\left(\frac{10P_I}{P_0}\right)$ for P_0

75. $L = 10\log\left(\frac{I}{I_0}\right)$ for I

76. $\ln M = \ln Q - \ln(1 - Q)$ for Q

77. $P = P_0e^{rt}$ for r

78. $A = A_0e^{-kt}$ for k

79. $2 = \left(1 + \frac{r}{n}\right)^{nt}$ for t

80. $A = P(1 + r)^t$ for t

Real-Life Applications

Compound Interest

If P dollars is invested at an interest rate r for a period of t years, and if the interest is compounded continuously, then the value A of the investment is given by $A = Pe^{rt}$.

81. A certain bank advertises that customers can double their investments in 14 years because interest is compounded continuously. What is the annual interest rate offered by the bank?

82. At the time of their first child's birth, a couple invested \$5000 at 5.5% compounded continuously. Their plan was to give the money to the child when the value of the investment reached \$20,000. At what age will the child receive the money?

Exponential Growth

During certain periods of time, populations of people or other life forms may grow exponentially. Therefore, models describing such growth may be exponential equations.

83. A planning study for a small country predicted that the country's population P (in millions) could be modeled as a function of the number t of years since 1990: $P(t) = 14e^{0.03t}$.

(a) What was the population in 1990?

(b) What will the population be in 2010?

(c) What will the population be in 2020?

84. In an industrial city the number of toxic particles in the atmosphere is found to be increasing over time according to the formula $P(t) = 400 \cdot 2^{0.4t}$, where P is measured in standard units. Letting $t = 0$ for the year 1988, estimate the number of toxic particles in the year 2000.

Exponential Decay

Certain radioactive substances decay according to the formula $P(t) = P_0e^{rt}$, where P_0 is the original amount of the substance, $P(t)$ is the amount of substance remaining after time t, and r is the rate of decay.

The *half-life* of a substance is the time t_h it takes for half the original amount of the substance to remain: $P(t_h) = \frac{1}{2}P_0$.

85. The half-life of carbon-14 is 5730 years. In how many years will only 30% of an original amount of carbon-14 remain?

86. What is the half-life of a substance when only 30% of the substance remains after 5 days?

Depreciation

Accountants refer to the decrease in the value of an item over time as *depreciation*. Depreciation is sometimes modeled with an exponential equation.

87. A bulldozer depreciates so that the value after t years is given by the function $V(t) = V_0e^{-0.19t}$, where V_0 is the initial cost. After 7 years the bulldozer is worth \$22,480.57. What was the original cost?

88. A front-end loader depreciates so that the value after t years is given by the function $V(t) = V_0e^{-0.27t}$, where V_0 is the initial cost. After 6 years the front-end loader is worth \$9894.94. What was the original cost?

Law of Cooling

The temperature T of an object after an elapsed period of time t is given by the formula $T = T_m + D_0e^{kt}$, where T_m is the constant temperature of the surrounding medium, D_0 is the difference between the initial temperature of the object and the temperature of the surrounding medium, and k is a cooling constant that depends on the composition of the object.

89. The coroner arrived at a murder scene at 8:00 A.M. The body was in a 60°F room and the body's temperature was 88°F. After 45 minutes, the body's temperature was 85°F. If you assume that a normal body temperature is 98.6°F, at what time did the murder occur?

90. A cake is removed from a 350°F oven and placed in a room whose temperature is 60°F. How long will the cake take to cool to 75°F? (If t is in minutes, $k = -0.1$.)

Richter Scale

A Richter scale reading is given by $R = \log I$, where I is the number of times more intense an earthquake is than a very small quake.

91. The 1988 earthquake in Armenia registered 6.9 on the Richter scale. What would be the Richter scale reading for an earthquake that is 10 times as intense?

92. The Richter scale reading for the 1964 Alaskan earthquake was 8.5. For the 1985 Mexico City earthquake, the Richter scale reading was 7.8. How much more intense was the Alaskan earthquake?

Modeling with Real Data

93. The number of wild horses on Cumberland Island has increased dramatically since 1960 to the point where they endanger island beaches by eating sea oats that hold the sand dunes together and keep the beach from eroding. (Source: National Park Service.)

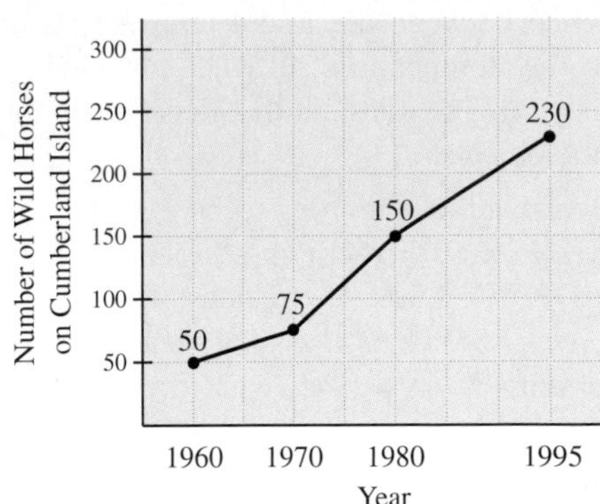

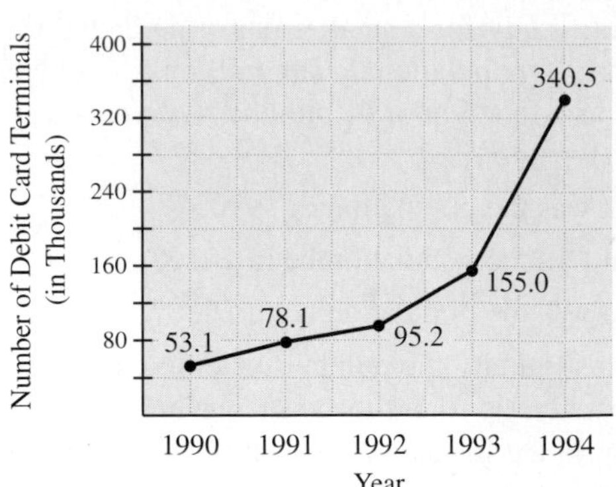

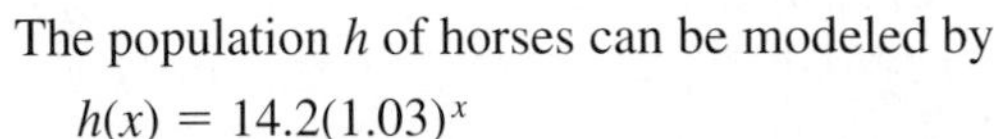

The population h of horses can be modeled by

$$h(x) = 14.2(1.03)^x$$

where x is the number of years since 1900.

(a) Write an equation to estimate the year in which the horse population is projected to reach 300.

(b) Solve the equation in part (a).

94. School districts are spending more each year on technology such as hardware, software, supplies, and training. (Source: Quality Education Data.)

School Year Beginning	Technology Spending (billions of dollars)
1991	2.1
1993	2.8
1995	4.0

The function

$$S(t) = 1.77e^{0.16t}$$

models the spending S (in billions of dollars), where t is the number of years since 1990.

(a) Write an equation to determine the year when spending will double from the 1995–1996 level.

(b) Solve the equation in part (a).

Group Project: Debit Cards

An increasing number of shoppers use bank cards that can be used at the point of sale (POS) to debit their checking accounts. The number of such transactions has increased with the increasing number of terminals. (Source: *Bank Network News*.)

A function that models the number $T(x)$ of terminals is $T(x) = 48.02(1.56)^x$, where x is the number of years since 1990 and $T(x)$ is in thousands.

95. Use a suitable window setting and the graphing method to estimate the projected year when there will be 1.5 million POS terminals.

96. Write and solve an equation to answer the question in Exercise 95.

97. The function $G(x) = 18.91(1.68)^x$ models the number (in thousands) of POS terminals located in grocery stores. Write and solve an equation to determine the year when half of all POS terminals were in grocery stores.

98. If you always pay off your credit card balance, will you save money or lose money by using a debit card? Why?

Challenge

In Exercises 99 and 100, $f(x) = \log x$ and $g(x) = \ln x$. Solve the given equation.

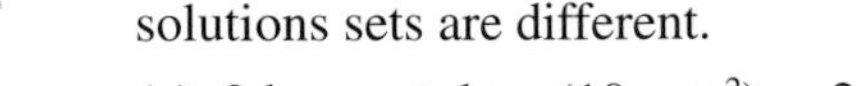

99. $(g \circ f)(x) = 1$ **100.** $(f \circ g)(x) = 1$

101. Solve the following equations and explain why the solutions sets are different.

(a) $2 \log_3 x + \log_3 (10 - x^2) = 2$

(b) $\log_3 x^2 + \log_3 (10 - x^2) = 2$

102. The following is a proof of the Change of Base Formula. Fill in the blanks with a rule or definition that justifies each step.

$x = b^{\log_b x}$ (a) ______

$\log_a x = \log_a b^{\log_b x}$ (b) ______

$\log_a x = (\log_b x)(\log_a b)$ (c) ______

$\dfrac{\log_a x}{\log_a b} = \log_b x$ (d) ______

103. Consider the following equation.

$$\ln(3 - 4x) - \ln(x - 2) = \ln(2x + 3)$$

Produce the graph of the left side of the equation. What is your conclusion about the solution set of the equation? Support your conclusion by determining the domains of the logarithmic expressions.

In Exercises 104 and 105, use the graphing method to estimate the solution(s) of the given equations.

104. $e^x - x = -\ln x$

105. $2^x = x^2$

10. CHAPTER PROJECT Telecommuting

The increasingly widespread use of personal computers, fax machines, E-mail, and other communications technology has made telecommuting a viable work option for many companies and their employees.

Studies show greater employee productivity and job satisfaction. Employees are relieved of the time, cost, and stress of commuting, and they enjoy a greater work-home quality of life. Societal benefits include less traffic congestion, cleaner air, and better land use.

The state of Oregon has been particularly aggressive in promoting telecommuting. Financial grants, tax incentives, formal research, and educational programs for employers, unions, and employees are just a few of Oregon's attempts to solve emerging and related problems of population growth, land use, and infrastructure.

The following information will be useful in answering Questions 1–3 in this project. (Source: Oregon Department of Energy.)

(a) A research report showed that the average total miles driven by a worker on a nontelecommuting day was 43.2 miles, whereas the average total miles driven on a telecommuting day was 8.0 miles.

(b) The number of miles driven in Oregon doubled between 1970 and 1990.

(c) Oregon's work force consists of approximately 1.3 million people.

1. About 46% of Oregon's work force hold positions that are amenable to telecommuting. If all these employees telecommuted on a single day, what would be the difference in the total number of miles driven in Oregon?
2. Between 1970 and 1990, the average auto fuel efficiency for new cars increased from 14 miles per gallon to 28 miles per gallon. What gain was made in reducing the total gasoline use in that period?
3. A study showed that 86% of Oregon's work force drove to work. Of these, 949,000 drove alone. Approximately what percent of Oregon's work force carpooled?
4. A 1996 national study predicts that the number of employees who will telecommute at least 1 day per week will increase by 15% each year for the next 7 years. What kind of function—linear, exponential, or logarithmic—would be best for modeling the projected number of telecommuters in this period? Why?
5. The number of telecommuting residents of San Diego was estimated at 60,000 in 1996. Assuming that the information in Question 4 holds for this city, write a function that would model the number of telecommuters in years after 1996. Then use your model to estimate the number of telecommuters in 2001.

6. Investigate the efforts being made in your city or state to promote telecommuting. Present your findings with graphs and tables, and discuss the advantages and problems associated with this emerging work style.

10. CHAPTER REVIEW EXERCISES

Section 10.1

In Exercises 1–8, let $f(x) = x^2 - 3x$, $g(x) = 3x + 4$, and $h(x) = \sqrt{x + 1}$. Determine each of the following.

1. $(f + g)(x)$

2. $(f - g)(2.4)$

3. $\left(\frac{f}{g}\right)(3.7)$

4. $(fg)(x)$

5. $(f \circ g)(x)$

6. $(g \circ h)(x)$

7. $(f \circ h)(3.56)$

8. $(h \circ f)(4.23)$

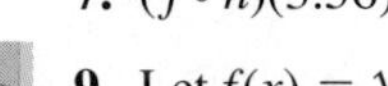

9. Let $f(x) = \sqrt{x - 1}$ and $g(x) = x + 1$. Since $g(-1)$ is defined, can we conclude that -1 is in the domain of $f \circ g$? Explain.

In Exercises 10 and 11, for the given functions f and g, determine $(f \circ g)(x)$ and $(g \circ f)(x)$.

10. $f(x) = 8, \quad g(x) = 2x^2 - 3x + 4$

11. $f(x) = \dfrac{1}{x + 2}, \quad g(x) = |2x + 5|$

12. For a 50-mile ride, a cyclist's time t (in minutes) increases as the temperature x increases. Suppose the following function describes the relationship between t and x, where $70 < x < 90$.

$$t(x) = 2x + 160$$

If the cyclist's rate is $r = 50/t$, write the function $(r \circ t)(x)$. What is your interpretation of this function?

Section 10.2

13. Explain the difference between the Vertical Line Test and the Horizontal Line Test.

In Exercises 14 and 15, produce the graph of the function. From the graph determine whether the function is one-to-one.

14. $f(x) = x^3 - 2x^2 + 3x - 4$

15. $f(x) = x^3 + x^2 - 4x - 3$

16. Determine whether the function is a one-to-one function.

(a) $f = \{(3, 6), (-1, 5), (5, 6), (1, 4)\}$

(b) $g = \{(7, 2), (-3, 6), (-2, -1), (0, 4)\}$

17. For each function in Exercise 16 that is one-to one, determine the inverse function. State the domain and range of the inverse function.

In Exercises 18 and 19, determine whether the pairs of functions are inverses by comparing $f \circ g$ and $g \circ f$.

18. $f(x) = 1 - 2x, \quad g(x) = 2x + 1$

19. $f(x) = \dfrac{1}{x} + 1, \quad g(x) = \dfrac{1}{x - 1}$

In Exercises 20 and 21, determine f^{-1}. Then verify that $(f \circ f^{-1})(x) = x$ for all x in the domain of f^{-1}.

20. $f(x) = 5x + 8$

21. $f(x) = x^3 - 5$

22. What is the equation of the line of symmetry for the graphs of inverse functions? Describe the graphs of f and f^{-1} if the equation $f(x) = f^{-1}(x)$ has no solution.

Section 10.3

23. For the exponential function $f(x) = b^x$, what restrictions, if any, are placed on b and x?

24. For the graph of the exponential function $f(x) = b^x$, state each of the following.

(a) x-intercept

(b) y-intercept

(c) domain

(d) range

25. Evaluate the given function as indicated.

(a) $f(x) = -8e^{-2x}, \quad f(1)$

(b) $g(x) = \left(\dfrac{9}{16}\right)^{3x}, \quad g\left(\dfrac{1}{2}\right)$

26. Use a graph to estimate the domain and range of $f(x) = 5 - 2^{-x}$.

In Exercises 27 and 28, use a graph to estimate the solution(s) of the given equation.

27. $\pi^x = 5$

28. $(0.57)^x = 6$

In Exercises 29–32, solve the given exponential equation algebraically. Verify the solution with your calculator.

29. $4^x = 2048$ **30.** $16^{2-3x} = 8^{5-6x}$

31. $\left(\frac{1}{27}\right)^{3x} = 9^{4x-5}$ **32.** $(\sqrt{5})^{6-2x} = 25^{x^2}$

33. Suppose that you want to model the growth behavior of a certain quantity over time t, and you decide to use the function $f(t) = b^t$. Would the base b be greater than 1 or would it be between 0 and 1? Why?

34. Five years ago, some money was invested in an account at 7.5% interest compounded monthly. Today the account is worth $7629.80. How much was originally invested?

Section 10.4

In Exercises 35 and 36, write the given exponential equation in logarithmic form.

35. $8^{2/3} = 4$ **36.** $49^{-1/2} = \frac{1}{7}$

In Exercises 37 and 38, write the given logarithmic equation in exponential form.

37. $\log_{25} \frac{1}{5} = -\frac{1}{2}$ **38.** $\log_{1/2} 16 = -4$

39. For the definition of the logarithmic function, explain why $b = 1$ is excluded.

40. What are the bases of the following logarithms?

(a) $\log_3 x$ (b) $\ln x$ (c) $\log x$

In Exercises 41–44, evaluate the logarithms without a calculator.

41. $\ln e^{3.23}$ **42.** $\log_{27} \frac{1}{9}$

43. $\log_{1/4} 32$ **44.** $\log 100^{2.5}$

In Exercises 45 and 46, use your calculator to solve for x. Round the results to two decimal places.

45. $\ln x = 5.2$ **46.** $\log x = 5.2$

In Exercises 47–50, solve for x.

47. $\log_{343} x = \frac{2}{3}$ **48.** $\log_x 676 = 2$

49. $\log_4 32 = x$ **50.** $\log_x 128 = \frac{7}{3}$

51. For the graph of the function $f(x) = \log_b x$, state each of the following.

(a) x-intercept (b) y-intercept

(c) domain (d) range

52. Suppose $g(t) = \log_b t$, where $0 < b < 1$. In the 20th century, and for appropriate values of time t, would g be more likely to model the change in the cost of living or the incidence of polio? Why?

Section 10.5

In Exercises 53–56, use the basic properties of logarithms to simplify the given expression.

53. $\log_{21} 21^{41}$ **54.** $21^{\log_{21} 43}$

55. $\log_{21} 21$ **56.** $\log_b 1$

57. Describe how you can use the fact that $\log 7 \approx 0.845$ to estimate $\log 49$ without a calculator.

58. Let $\log_2 3 = C$ and $\log_2 5 = D$. Write

$$\log_2 \frac{5\sqrt[3]{60}}{24}$$

as an expression involving C and D.

In Exercises 59 and 60, use the properties of logarithms to expand the given expression. Write your answer in terms of logarithms of x, y, and z. Assume that x, y, and z represent positive numbers.

59. $\log \frac{\sqrt[4]{x^2y^3}}{x^3z^4}$ **60.** $\ln \sqrt[6]{\frac{x^5y^4}{z^3}}$

In Exercises 61 and 62, use the properties of logarithms to combine the given expression into a single logarithmic expression. Assume that x, y, and z represent positive numbers.

61. $\frac{1}{4}[3 \log (x + 4) - \log y - \log 2z]$

62. $3 \log (x + 2) - 5 \log (x + 1)$

63. Write $\log_2 24 + 2 \log_2 5 - \log_2 75$ as a single logarithm. Then evaluate it.

64. Write $\log_3 (x + 5)$ in terms of natural logarithms.

Section 10.6

65. Explain why $3^x = \frac{1}{9}$ can be solved by equating exponents, but $3^x = 4$ cannot. For the latter equation, what method should you use to solve it?

66. If you wished to solve $\log_2 x = 4$ by equating arguments of logarithms, you could replace 4 with $\log_2 16$. What is an easier method for solving the equation?

In Exercises 67–72, solve for x. Round your solutions to two decimal places.

67. $3^x = 7$

68. $e^{-x} = 0.78$

69. $\log_4 (x + 3) + \log_4 (x - 3) = 2$

70. $2^{x-1} = 3^{x+2}$

71. $\ln (x - 4) - \ln (3x - 10) = \ln \frac{1}{x}$

72. $\log_2 (x + 1) + \log_2 (3x - 5) = \log_2 (5x - 3) + 2$

73. If $a = 6.9e^{-2.3b}$, find b when $a = 0.45$.

74. Suppose $7000 is invested in a savings fund in which interest is compounded continuously at the rate of 12% per year. How long will it take for the money to double in value?

75. Suppose the number N of bacteria present in a certain culture after t minutes is $N = ke^{0.04t}$, where k is a constant. If 5500 bacteria are present after 11 minutes, how many bacteria were present initially?

76. An amount of $800 is deposited in a savings account and earns interest for 8 years at 5.25% compounded monthly. If there were no withdrawals or additional deposits, how much is in the account at the end of the 8 years?

LOOKING AHEAD

The following exercises review concepts and skills that you will need in Chapter 11.

1. Write the quadratic function $y = x^2 + 6x + 8$ in the form $y = (x - h)^2 + k$.

2. Determine the vertex and the equation of the axis of symmetry for the graph of the quadratic function in Exercise 1.

3. Determine the intercepts and vertex of the graph of the quadratic function $f(x) = 21 - 4x - x^2$.

4. Determine the domain and range of the function $g(x) = 15 - 2x - x^2$.

5. Suppose that the graph of $f(x) = (x - h)^2 + k$ is the graph of $y = x^2$ shifted left 2 units and down 3 units. What are the values of h and k?

6. Suppose that the graph of $y = x^2 + bx + c$ has x-intercepts $(-1, 0)$ and $(5, 0)$. What are the values of b and c?

7. Calculate the distance between the points $(3, -1)$ and $(-5, 14)$.

8. What is the radius of a circle?

9. Remove the parentheses and simplify the expression $(y - 4)^2 + (x + 3)^2$.

10. Use the graphing method to estimate the solution of the following system of equations.

$$2x + 5y + 1 = 0$$
$$3y = x - 5$$

11. Use the substitution method to solve the following system of equations.

$$2x = 3y + 3$$
$$-2x + 3y = 15$$

12. Compare the graphs of the inequalities $y \leq x + 6$ and $y < x + 6$.

10. CHAPTER TEST

For Questions 1–4, let $f(x) = x^2 - 4$, $g(x) = x + 2$, and $h(x) = \sqrt{x}$.

1. Find $(f + g)(x)$.

2. Find $\left(\frac{f}{g}\right)(x)$.

3. Find $(f \circ g)(x)$.

4. Find $(h \circ g)(x)$.

5. Verify that $f(x) = \frac{x - 5}{3}$ and $g(x) = 3x + 5$ are inverse functions.

6. For the function $f(x) = \sqrt[3]{x + 2}$, determine f^{-1}.

7. The graph of a function is given in the figure. Sketch the graph of the inverse function.

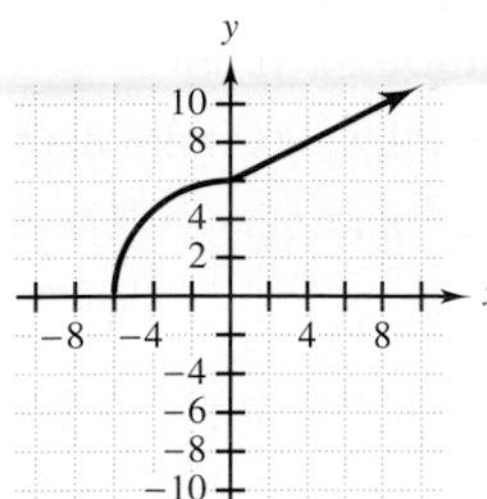

For Questions 8 and 9, evaluate the expression.

8. $\log_{25} 5$

9. $\log_2 4^{3/4}$

For Questions 10–13, solve the equation.

10. $5^{2-3x} = 25$

11. $\left(\frac{1}{16}\right)^{x-1} = 8$

12. $8^x = 3^{x+1}$

13. $e^{2x} = 10$

14. Explain how to evaluate $\log_5 13$ with your calculator.

15. Write the expression $\log\left(\frac{A^2}{\sqrt{B}}\right)$, where A and B represent positive numbers, in terms of $\log A$ and $\log B$.

For Questions 16 and 17, write the given expression as a single logarithmic expression.

16. $\log(x^2 - x - 2) - \log(x - 2)$

17. $2\log_6 x + \log_6(x - 1) - 1$

For Questions 18–21, solve the given equation.

18. $\log(2x + 1) = \log 5$

19. $\log x + \log(x + 3) = 1$

20. $\ln(2t + 1) - \ln t = \ln 3$

21. $2 + \ln x = 0$

22. At the birth of a child, \$5000 is invested in a college fund. What rate of interest, compounded continuously, is required for the investment to be worth \$23,000 eighteen years later?

23. The demand (in thousands of cases) for a certain brand of soap t weeks after a major advertising campaign is given by $D(t) = 2 + 5(0.89)^t$.

(a) What was the demand when the ad campaign began?

(b) Using a graph of $y = D(t)$, state your opinion about the effectiveness of the ad campaign.

(c) After approximately how many weeks had the demand for this brand of soap fallen to half the initial demand?

24. The Richter scale rating of an earthquake is given by the formula $R = \log I$, where I is the number of times more intense the earthquake is than a very small quake. The San Francisco earthquake that occurred during the World Series in 1989 registered 7.1 on the Richter scale. How much more intense was the 1906 San Francisco earthquake, which registered 8.3 on the Richter scale?

25. The temperature T of an object after an elapsed period of time t is given by $T = T_m + D_0e^{kt}$, where T_m is the constant temperature of the surrounding medium, D_0 is the difference between the initial temperature of the object and the temperature of the surrounding medium, and k is a cooling constant.

A can of soda had been left in a hot car where the soda reached a temperature of 90°F. If the can was placed in a cooler of ice at 32°F, how long did it take to chill the beverage to 40°F? (With time measured in hours, the cooling constant is $k = -1.32$.)

8–10 CUMULATIVE TEST

1. Simplify. Assume that y represents any real number.

(a) $\sqrt{y^8}$ (b) $\sqrt{y^{10}}$

2. Determine the real number value, if any.

(a) $(-27)^{2/3}$ (b) $(-27)^{-2/3}$

(c) $(-16)^{3/4}$ (d) $-16^{-3/4}$

3. Simplify and write the result with positive exponents. Assume that all variables represent positive numbers.

(a) $\frac{a^{-2}b^{1/3}}{a^{1/2}b}$ (b) $(x^{-1}y^2)^{-1/2}$ (c) $\left(\frac{x^{-1}}{y^2}\right)^{-3}$

4. Simplify.

(a) $\sqrt[3]{\frac{x^6}{8y^3}}$ (b) $\sqrt[5]{3x}\sqrt[5]{2x^2y}$ (c) $\frac{\sqrt{x^3y}}{\sqrt{xy^3}}$

5. Simplify.

(a) $\sqrt{20x^9}$ (b) $\sqrt[14]{y^{21}}$

6. Perform the indicated operations.

(a) $\sqrt{12x^3} - x\sqrt{3x}$ (b) $(\sqrt{a} + \sqrt{2})^2$

7. What is the product of $3 - \sqrt{5}$ and its conjugate?

8. Rationalize the denominator.

(a) $\dfrac{5}{\sqrt[3]{2}}$ (b) $\dfrac{\sqrt{x}}{\sqrt{2x} + \sqrt{y}}$

9. Solve.

(a) $\sqrt{2x + 1} = x\sqrt{3}$ (b) $(2x - 3)^{2/3} = 9$

10. Simplify.

(a) i^{19} (b) $(3 - i)^2$ (c) $\dfrac{i}{2 + i}$

11. In addition to the method of completing the square, name three methods for solving a quadratic equation. Which of these three methods works for all quadratic equations?

12. Solve $(3x - 1)(2x + 3) = x(x - 2)$. Round your solution(s) to the nearest hundredth.

13. Six less than a number is 7 times the reciprocal of the number. What is the number?

14. If $ax^2 + bx + c = 0$ has two imaginary solutions, what is the relationship between b^2 and $4ac$? Why?

15. Solve $a + a^{1/2} = 12$.

16. A person jogged 2 miles and then walked back to the starting point. He jogged 3 mph faster than he walked. If the total time was 39 minutes, how fast did he jog?

17. Suppose $p(x) = x^2 + 2x - 24$. Solve each of the following.

(a) $p(x) > 0$

(b) $p(x) < -25$

(c) $p(x) > -26$

18. Solve.

(a) $x^3 - 3x^2 + x - 1 \leq 0$ (b) $\dfrac{x}{x - 3} > 0$

19. If $f(x) = ax^2 - 6x + c$, tell what you know about a or c for each of the following descriptions of the graph.

(a) The y-intercept is 3.

(b) The parabola opens downward.

(c) The x-coordinate of the vertex is -1.

20. Suppose that the daily profit P from producing x items is given by $P(x) = 8x - x^2$.

(a) How many items should be produced each day to maximize the profit?

(b) What will the maximum daily profit be?

21. For $f(x) = x^2$ and $g(x) = 2x$, evaluate each of the following.

(a) $(f + g)(0)$

(b) $(f - g)(2)$

(c) $(fg)(-2)$

22. For $f(x) = |x - 1|$ and $g(x) = 2x + 1$, evaluate each of the following, if possible.

(a) $(f \circ g)\left(-\dfrac{1}{2}\right)$

(b) $\left(\dfrac{f}{g}\right)\left(-\dfrac{1}{2}\right)$

23. If $f(x) = \dfrac{x}{x - 3}$, determine f^{-1}. Then show that $(f \circ f^{-1})(x) = x$ for all x in the domain of f^{-1} and that $(f^{-1} \circ f)(x) = x$ for all x in the domain of f.

24. What properties do the graphs of $f(x) = 5^x$ and $g(x) = \left(\frac{1}{5}\right)^x$ have in common? Describe how the graphs are different.

25. Solve each of the following equations.

(a) $\log_2 32 = x$

(b) $\log_3 (x - 1) = 2$

(c) $\log_x 7 = 2$

26. (a) Expand $\ln \dfrac{x^2 y}{\sqrt{z}}$.

(b) Write $3 \log (x - 1) - \log (x + 1)$ as a single logarithm.

27. Solve each of the following equations. Round your results to the nearest hundredth.

(a) $\log_3 x(x - 6) = 3$

(b) $3^{x+1} = 5$

(c) $4e^{2x} = 1$

28. For continuously compounded interest, the value A of an investment of P dollars after t years is given by $A = Pe^{rt}$, where r is the interest rate. Approximately what interest rate is needed for a \$1200 investment to be worth \$2187 after 10 years?

Chapter 11

Conic Sections

The National Endowment for the Arts (NEA) is a federally funded agency that provides grants and support for the arts in the United States. The accompanying bar graph shows the trends in appropriations for the NEA in the 1990s.

The data in the graph can be modeled by a function whose graph is a **parabola,** which is a curve that is called a **conic section.** The highest level of funding is represented by the **vertex** of the parabola. The model function and its graph could be used, for example, to project trends in public funding for the arts. (For more on this topic, see Exercises 73–76 at the end of Section 11.1, and see the Chapter Project.)

Most of this chapter is devoted to a discussion of the four conic sections: the parabola, the circle, the ellipse, and the hyperbola. We also learn how to solve systems of nonlinear equations and inequalities.

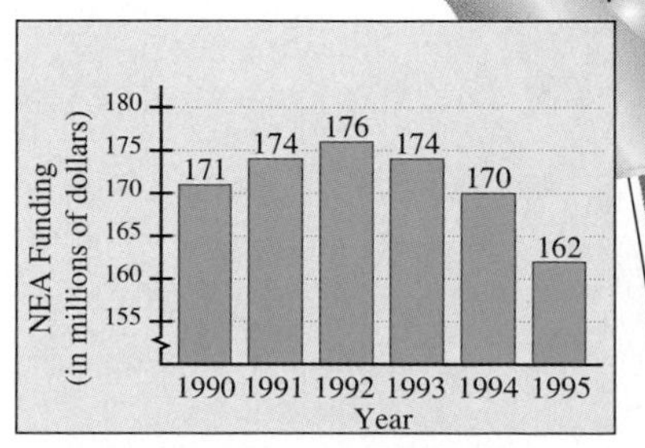

(Source: National Endowment for the Arts.)

11.1 THE PARABOLA

Definition of a Parabola • Vertex and Intercepts • Writing the Equation of a Parabola

Definition of a Parabola

In this chapter we study **conic sections,** which are the graphs of second-degree equations in two variables.

Conic sections are the curves obtained by intersecting a plane with an **infinite right circular cone.** As shown in Fig. 11.1, altering the angle of the plane produces four types of curves: parabola, circle, ellipse, and hyperbola.

Figure 11.1

The geometric definitions of conic sections are stated in terms of distances. We begin our discussion with parabolas.

Figure 11.2

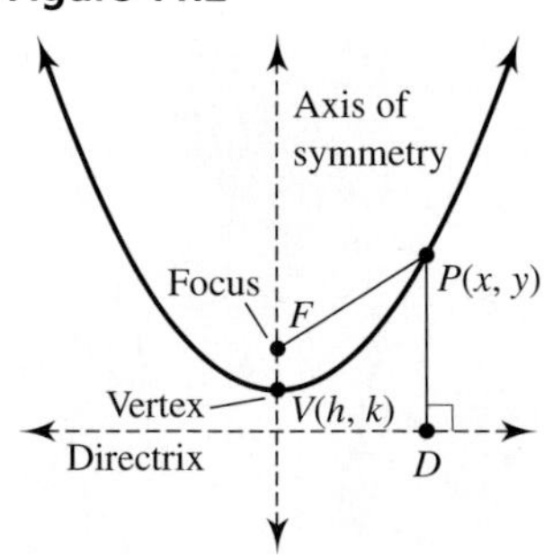

> **Definition of a Parabola**
>
> A **parabola** is the set of all points that are equidistant from a fixed point called the **focus** and a fixed line called the **directrix.**

Figure 11.2 shows a *vertical parabola* whose directrix is horizontal and whose axis of symmetry is vertical.

The axis of symmetry contains the vertex and the focus, and it is perpendicular to the directrix. Note that the vertex is the midpoint of the line segment that joins the focus and the directrix.

In Fig. 11.2 the definition of parabola requires that $PF = PD$. This geometric requirement leads to the equation $y = a(x - h)^2 + k$, where $V(h, k)$ is the vertex of the parabola. Note that this equation for a vertical parabola is the same as that discussed in Chapter 9.

EXAMPLE 1

Graphing a Vertical Parabola

Sketch the graph of $y = \frac{1}{2}(x - 6)^2 + 3$.

Solution

The graph of $y = \frac{1}{2}(x - 6)^2 + 3$ is the graph of $y = \frac{1}{2}x^2$ with the vertex shifted right 6 units and up 3 units. The vertex is (6, 3), and the axis of symmetry is $x = 6$. (See Fig. 11.3.)

Figure 11.3

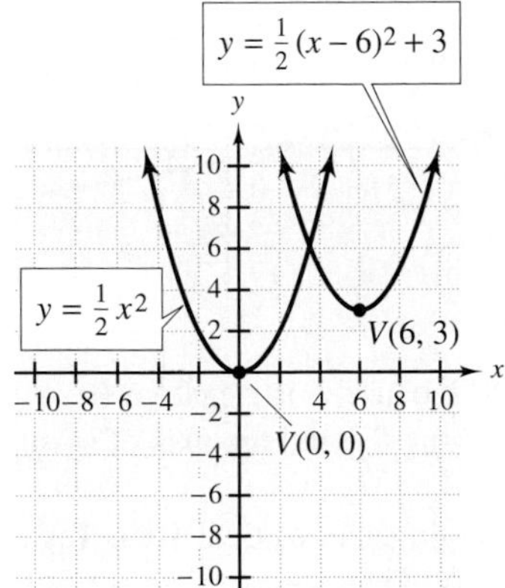

LEARNING TIP

Associating h with x and k with y will help you to remember the different equation forms for vertical and horizontal parabolas. In both cases, however, the vertex is $V(h, k)$.

Parabolas also can open to the right or left. The graph of the second-degree equation $x = a(y - k)^2 + h$ is a parabola with a vertex at $V(h, k)$ and a horizontal axis of symmetry. We refer to such parabolas as *horizontal parabolas*.

For a vertical parabola $y = a(x - h)^2 + k$, the parabola opens upward if a is positive and downward if a is negative. Similarly, for a horizontal parabola $x = a(y - k)^2 + h$, the parabola opens to the right if a is positive and to the left if a is negative. (See Fig. 11.4.)

Figure 11.4

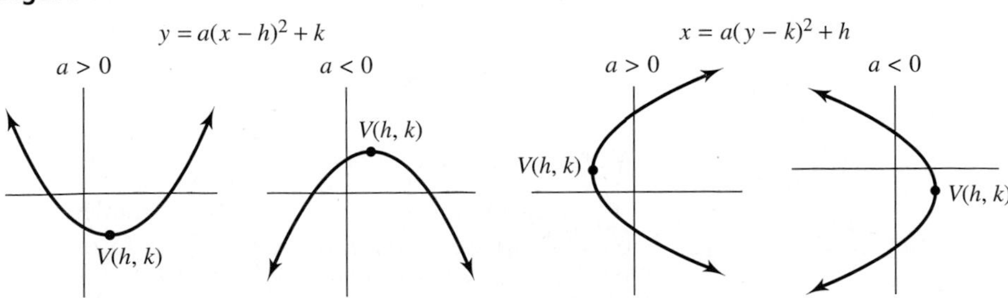

EXAMPLE 2

Graphs of Horizontal Parabolas

Sketch the graph of each of the following equations.

(a) $x = (y - 2)^2 - 5$ (b) $x = -(y + 1)^2 + 4$

Solution

(a) Because a is positive ($a = 1$), the parabola opens to the right. The vertex is $(-5, 2)$, and the axis of symmetry is $y = 2$. (See Fig. 11.5.)

Figure 11.5

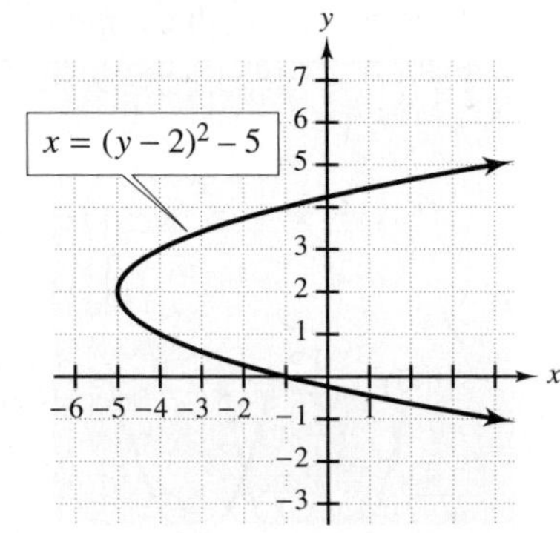

Figure 11.6

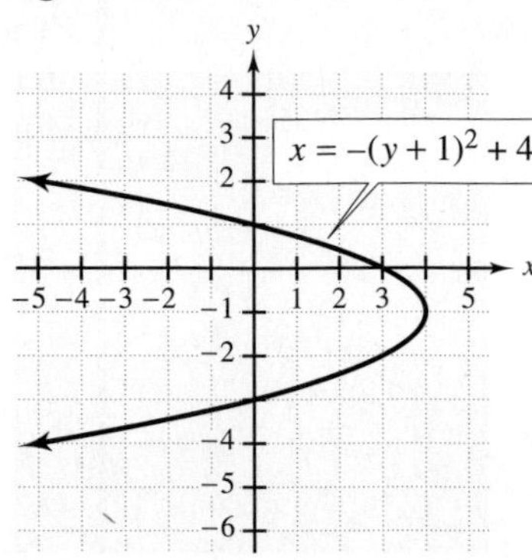

(b) Because a is negative ($a = -1$), the parabola opens to the left. The vertex is $(4, -1)$, and the axis of symmetry is $y = -1$. (See Fig. 11.6.)

By applying the Vertical Line Test, we see that horizontal parabolas do not represent functions. Thus we cannot enter a single function in a calculator to produce the graph of such equations. However, we can produce the graph by entering two related functions. For instance, to graph $x = y^2$, we solve for y to obtain the two functions $y = \sqrt{x}$ and $y = -\sqrt{x}$. The two graphs together make up the graph of $x = y^2$. (See Fig. 11.7.)

Figure 11.7

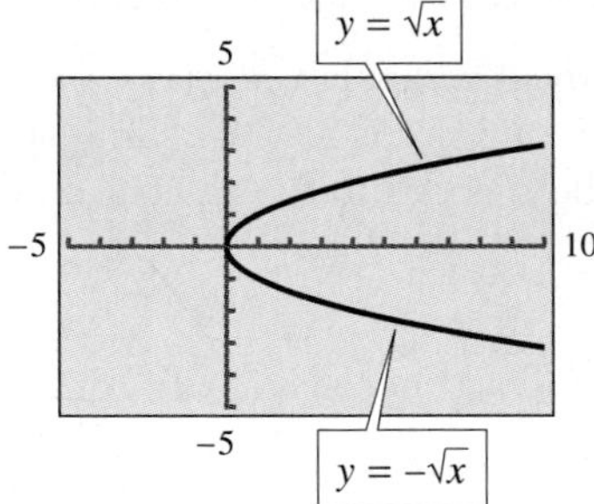

Think About It

Solve each equation in Example 2 for y, and produce the graphs of the related functions. Compare your results with Figs. 11.5 and 11.6.

Vertex and Intercepts

In Chapter 9 we used the method of completing the square to find the vertex of a parabola whose equation was given in the form $y = ax^2 + bx + c$. Then, by generalizing the method, we obtained a formula for the x-coordinate of the vertex: $x = -\dfrac{b}{2a}$.

We obtain similar results for a horizontal parabola. The equation can be written in the form $x = ay^2 + by + c$. We can complete the square to write the equation in the form $x = a(y - k)^2 + h$ to determine the vertex. The method of completing the square also can be used to derive a formula for the y-coordinate of the vertex: $y = -\dfrac{b}{2a}$.

EXAMPLE 3

Determining the Vertex of a Parabola

Dctermine the vertex of the graph of each equation.

(a) $y = 3x^2 + 12x - 3$ (b) $x = y^2 - 6y + 5$

Solution

(a) The x-coordinate of the vertex is $x = -\dfrac{b}{2a} = -\dfrac{12}{2(3)} = -2.$

$$y = 3(-2)^2 + 12(-2) - 3 = -15$$ Replace x with -2 to obtain the y-coordinate.

The vertex is $(-2, -15)$.

(b) The y-coordinate of the vertex is $y = -\dfrac{b}{2a} = -\dfrac{-6}{2(1)} = 3.$

$$x = 3^2 - 6(3) + 5 = -4$$ Replace y with -3 to obtain the x-coordinate.

The vertex is $(-4, 3)$.

A vertical parabola $y = ax^2 + bx + c$ has one y-intercept $(0, c)$. The graph has zero, one, or two x-intercepts that we find by solving the quadratic equation $ax^2 + bx + c = 0$. A horizontal parabola $x = ay^2 + by + c$ has one x-intercept $(c, 0)$ and zero, one, or two y-intercepts that can be found by solving the quadratic equation $ay^2 + by + c = 0$.

EXAMPLE 4

Analyzing a Horizontal Parabola

Determine the vertex and intercepts of each parabola.

(a) $x = 2y^2 + 8y - 7$ (b) $x = -y^2 + 3y - 10$

Solution

(a) Because a is positive $(a = 2)$, the parabola opens to the right. Because $c = -7$, the x-intercept is $(-7, 0)$. To determine the y-intercepts, we use the Quadratic Formula to solve the equation $2y^2 + 8y - 7 = 0$. Verify that the solutions are $y \approx 0.74$ and $y \approx -4.74$. The (approximate) y-intercepts are $(0, -4.74)$ and $(0, 0.74)$.

Now determine the coordinates of the vertex.

$$y = -\frac{b}{2a} = -\frac{8}{2(2)} = -2$$ Determine the y-coordinate of the vertex.

$$x = 2(-2)^2 + 8(-2) - 7 = -15$$ Replace y with -2.

The vertex is $(-15, -2)$.

(b) Because a is negative $(a = -1)$, the parabola opens to the left. Because $c = -10$, the x-intercept is $(-10, 0)$. Use the Quadratic Formula to verify that the equation $-y^2 + 3y - 10 = 0$ has no solution. Thus the parabola has no y-intercepts.

Now determine the coordinates of the vertex.

$$y = -\frac{b}{2a} = -\frac{3}{2(-1)} = \frac{3}{2}$$ Determine the y-coordinate of the vertex.

$$x = -\left(\frac{3}{2}\right)^2 + 3\left(\frac{3}{2}\right) - 10 = -\frac{31}{4}$$ Replace y with $\frac{3}{2}$.

The vertex is $\left(-\frac{31}{4}, \frac{3}{2}\right)$.

Writing the Equation of a Parabola

Given sufficient information about a parabola, we can write its equation.

EXAMPLE 5

Writing an Equation of a Vertical Parabola

Determine the equation of the vertical parabola whose x-intercepts are (2, 0) and (−3, 0) and whose y-intercept is (0, 1).

Solution

As we have noted before, if $(x_1, 0)$ is an x-intercept of the graph of a quadratic function, then $(x - x_1)$ is a factor of the quadratic expression. Because (2, 0) and (−3, 0) are x-intercepts, the equation can be written as follows.

$$y = a(x - 2)(x + 3)$$
$$y = a(x^2 + x - 6)$$

Because $y = 1$ when $x = 0$, $1 = a(-6)$, and so $a = -\frac{1}{6}$.

The complete equation is $y = -\frac{1}{6}(x^2 + x - 6)$. The parabola is shown in Fig. 11.8.

Figure 11.8

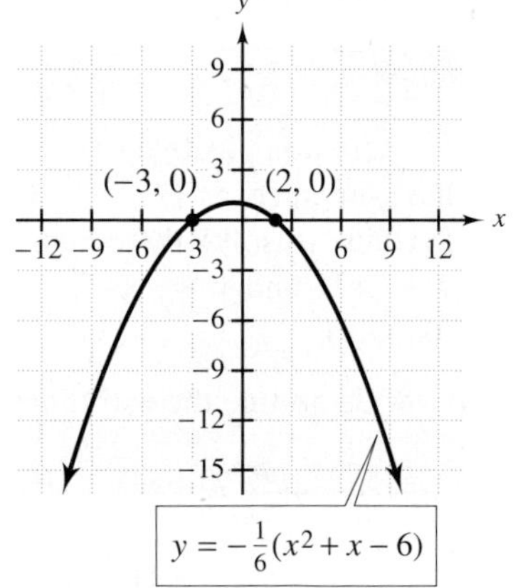

EXAMPLE 6

Writing an Equation of a Horizontal Parabola

Write an equation of a horizontal parabola whose vertex is (−1, 4) and whose x-intercept is (−3, 0).

Solution

Because the vertex is known, we can use the form $x = a(y - k)^2 + h$. The vertex is $(-1, 4)$, so the equation is

$$x = a(y - 4)^2 - 1$$

This equation is satisfied by the coordinates of the x-intercept.

$$-3 = a(0 - 4)^2 - 1 \quad \text{Replace } x \text{ with } -3 \text{ and } y \text{ with } 0.$$

$$-2 = 16a \quad \text{Simplify.}$$

$$a = -\frac{1}{8} \quad \text{Solve for } a.$$

The equation of the parabola is $x = -\frac{1}{8}(y - 4)^2 - 1$. Figure 11.9 shows the parabola.

Figure 11.9

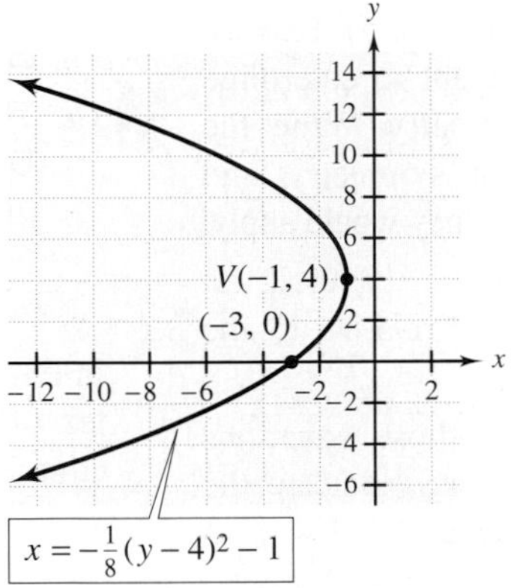

11.1 QUICK REFERENCE

Definition of a Parabola

- **Conic sections** are the curves obtained by intersecting a plane with an **infinite right circular cone.**
- The conic sections are the parabola, the circle, the ellipse, and the hyperbola.
- A **parabola** is the set of all points that are equidistant from a fixed point called the **focus** and a fixed line called the **directrix.**
- For a vertical parabola $y = a(x - h)^2 + k$, the parabola opens upward if a is positive and downward if a is negative. Similarly, for a horizontal parabola $x = a(y - k)^2 + h$, the parabola opens to the right if a is positive and to the left if a is negative.
- A vertical parabola represents a function, but a horizontal parabola does not.

Vertex and Intercepts

- The equation of a vertical parabola can be written in the form $y = ax^2 + bx + c$. The x-coordinate of the vertex is $-\dfrac{b}{2a}$.
- The equation of a horizontal parabola can be written in the form $x = ay^2 + by + c$. The y-coordinate of the vertex is $-\dfrac{b}{2a}$.

- A vertical parabola has one y-intercept, $(0, c)$ and zero, one, or two x-intercepts.
- A horizontal parabola has one x-intercept, $(c, 0)$ and zero, one, or two y-intercepts.

Writing the Equation of a Parabola

- The equation forms $y = a(x - h)^2 + k$ and $y = ax^2 + bx + c$ are convenient models for writing the equation of a vertical parabola.
- For horizontal parabolas, the equation forms $x = a(y - k)^2 + h$ and $x = ay^2 + by + c$ can be used to write the equation.

11.1 EXERCISES

Concepts and Skills

1. Suppose the vertex of a parabola is $V(3, 5)$. Explain why this information is not sufficient to determine the equation of the axis of symmetry. Write the *possible* equations of the axis of symmetry, and describe the conditions under which they would apply.
2. Use the geometric definition of a parabola to describe the location of the vertex relative to the focus and directrix.
3. From the following list, identify those equations that represent functions. For all parts, indicate the direction in which the parabola opens.

   (i) $y = -3x^2$ (ii) $x = 3y^2$

   (iii) $x = 5 - y^2$ (iv) $y = x^2 + 1$
4. In each part, information is given about a certain parabola. Indicate whether the parabola is the graph of a function.

   (a) The parabola has two x-intercepts.

   (b) The parabola has a horizontal axis of symmetry.

   (c) The vertex is $V(7, 3)$, and the focus is $F(-1, 3)$.

   (d) The equation of the axis of symmetry is $x = -3$.

In Exercises 5–8, determine the orientation, vertex, and axis of symmetry. Produce the graph on your calculator to determine whether the displayed parabola is consistent with your answers.

5. $y + 16 = (x + 2)^2$
6. $y = 5 - x^2$
7. $y = -(x - 3)^2$
8. $y - 3 = 2(x - 5)^2$

In Exercises 9–14, determine the orientation, vertex, and axis of symmetry.

9. $x = 6 - y^2$
10. $x + 1 = (y - 3)^2$
11. $x = (y + 4)^2 - 16$
12. $x + (y - 1)^2 = 7$
13. $(y - 8)^2 = x - 2$
14. $x = -(y + 1)^2$

15. Consider the equation $x = ay^2 + by + c$. What information does the number c provide about the graph of the equation?
16. The graph of $x = ay^2 + by + c$ is a parabola. Describe two ways to determine the y-coordinate of the vertex.

In Exercises 17–32, determine the intercepts and the vertex of the graph of the given equation. Then write the equation of the axis of symmetry and sketch the graph.

17. $x + 9 = y^2$
18. $y + 7 = x^2$
19. $y + 16 = (x + 1)^2$
20. $x + 25 = (y + 2)^2$
21. $(x - 5)^2 = y + 1$
22. $y^2 + 24 = 10y + x$
23. $y + 4x = 12 - x^2$
24. $x + 16 = 8y - y^2$
25. $6y - x = y^2$
26. $x^2 - y = 4x + 12$
27. $y = 10x - 25 - x^2$
28. $y + x^2 = 6x$
29. $x + 16 = (2y + 1)^2$
30. $y + 5x + 2x^2 = 12$
31. $y^2 + x + 3y = 4$
32. $x + 3y^2 = 6 - 7y$

In Exercises 33–36, determine the domain and range of the given relation.

33. $x + 3y = y^2 + 2$
34. $x^2 + x = y + 12$
35. $3 - 2x = y + x^2$
36. $x + y^2 - y = 6$

37. Describe the steps you must take to produce the graph of $x = 4y^2$ on your calculator.
38. If a parabola's equation is $y = ax^2 + bx + c$ and the x-intercepts are $(x_1, 0)$ and $(x_2, 0)$, explain how to determine the equation of the axis of symmetry.

In Exercises 39–44, information about a parabola is given. Determine the number of x- and y-intercepts.

39. The vertex is $V(3, -5)$, $a > 0$, and the equation of the axis of symmetry is $x = 3$.

40. The axis of symmetry is horizontal, the vertex is $V(-3, 0)$, and $a < 0$.

41. Two points of the graph are $A(-1, 2)$ and $B(-1, 12)$, and the vertex is $V(-4, 7)$.

42. The vertex is in Quadrant III, $a < 0$, and the axis of symmetry is vertical.

43. The parabola contains $P(0, 3)$, $a > 0$, and the y-axis is the axis of symmetry.

44. The equation of the axis of symmetry is $y = 3$, $a > 0$, and the vertex is $V(-1, 3)$.

In Exercises 45–52, determine the orientation of the parabola.

45. The parabola contains the point (2, 6) and has an x-intercept (5, 0). The range is $\{y \mid y \geq 0\}$.

46. The intercepts are $(-3, 0)$, $(-6, 0)$, and $(0, -5)$.

47. The intercepts are $(-8, 0)$ and (0, 6), and the domain is $\{x \mid x \leq 0\}$.

48. The axis is horizontal, the vertex is $(2, -3)$, and the parabola contains $(1, -1)$.

49. The intercepts are $(-4, 0)$ and $(0, -8)$, and the range is $\{y \mid y \leq 0\}$.

50. The axis is vertical, the vertex is $(-4, -2)$, and the parabola contains $(2, -5)$.

51. The parabola contains the point (6, 3), the y-intercept is (0, 0), and the domain is $\{x \mid x \geq 0\}$.

52. The intercepts are (0, 5), (0, 1), and $(-5, 0)$.

In Exercises 53–56, determine the equation of the axis of symmetry of the parabola with the given intercepts.

53. (0, 5), (6, 0), and $(0, -3)$

54. $(-5, 0)$, (0, 7), and (3, 0)

55. $(-1, 0)$, (0, 7), and (8, 0)

56. (0, 2), $(0, -3)$, and $(-4, 0)$

In Exercises 57–60, determine the equation of the parabola with the given intercepts.

57. (4, 0), $(-2, 0)$, and (0, 5)

58. $(-1, 0)$, (2, 0), and $(0, -3)$

59. $(0, -1)$, (0, 3), and (4, 0)

60. $(0, -2)$, $(0, -4)$, and (8, 0)

In Exercises 61–68, determine the equation of the parabola that satisfies the given conditions.

61. The parabola contains the point $(-2, 8)$ and has x-intercepts (3, 0) and $(-5, 0)$.

62. The intercepts are (2, 0) and (0, 3), and the range is $\{y \mid y \geq 0\}$.

63. The parabola contains the point $(3, -2)$ and has y-intercepts (0, 4) and $(0, -3)$.

64. The intercepts are (5, 0) and (0, 3), and the domain is $\{x \mid x \geq 0\}$.

65. The axis is horizontal, the vertex is (2, 5), and the parabola contains (6, 1).

66. The parabola contains the point $(-1, 2)$ and has a y-intercept $(0, -1)$. The domain is $\{x \mid x \leq 0\}$.

67. The axis is vertical, the vertex is $(2, -3)$, and the parabola contains $(1, -1)$.

68. The parabola contains the point $(-1, -3)$ and has an x-intercept $(-3, 0)$. The range of the parabola is $\{y \mid y \leq 0\}$.

Real-Life Applications

69. A *learning curve* is a graph that represents the proficiency or the amount of knowledge a person acquires over time. If y is a person's proficiency rating and t represents time, what is your opinion of $t = \frac{1}{9}y^2$ as a possible learning-curve model for estimating the time it takes to achieve a given proficiency rating? What restriction would you need to place on y in order for the model to be acceptable?

70. Suppose that an engineer has designed a radar antenna with the model $y = 0.05x^2$. For a revision of the design, she decides to have the antenna open to the right with the vertex on the x-axis and three units to the right of the origin. If the antenna has the same shape as it had originally, what equation should the engineer use?

71. The coordinate-system diagram for a parabolic arch of a bridge contains the point (3, 6) and shows the base of the arch at the points $(-3, 0)$ and (5, 0).

(a) Write an equation to model the arch.

(b) What is the height of the arch?

72. The cross-sectional design of a parabolic mirror for a certain telescope is drawn on a coordinate system. As shown in the figure on page 700, the ends of the curve are the points A and B of the y-axis. The parabola contains the point $P(-1, 0)$, and its vertex is $V\left(-\frac{16}{7}, -3\right)$.

(a) Write an equation to model the cross section of the mirror.

(b) What is the distance AB across the top of the mirror?

Figure for 72

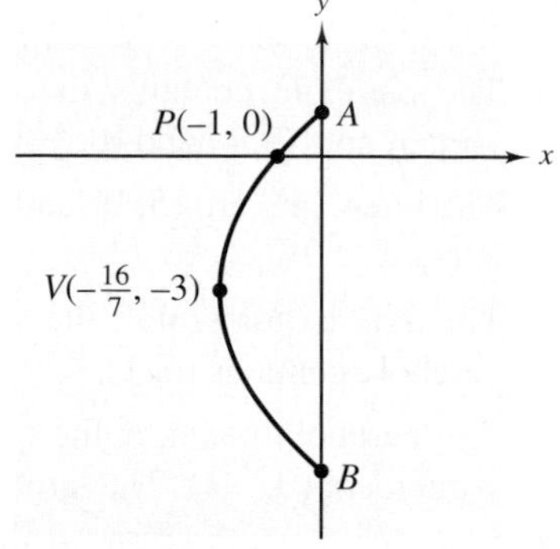

Group Project: Funding for the Arts

The National Endowment for the Arts (NEA) is one of many federal agencies that has had sharp reductions in funding as a result of increased pressures to balance the federal budget. The accompanying bar graph shows the trends in the 1990s.

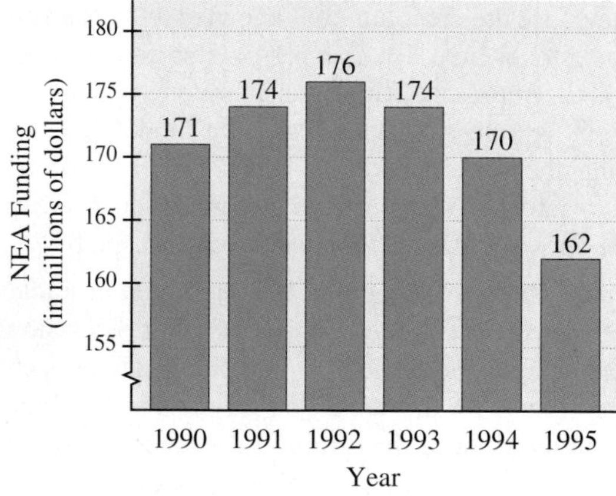

The NEA's annual appropriation A (in millions of dollars) can be modeled by the quadratic function $A(x) = -1.41x^2 + 5.37x + 170.68$, where x is the number of years since 1990.

73. Rounding to integer coordinates, estimate the y-intercept of the graph of A. What does this point represent? How accurately does it model the actual data?

74. Rounding to integer coordinates, estimate the vertex of the graph of A. What is the significance of this point?

75. For the graph of A, estimate the x-intercept that has a positive x-coordinate. What is the significance of this point?

76. If Congress succeeds in creating a balanced budget by the year 2003, do you think that the result in Exercise 75 could become a reality? Do you think that it should?

Challenge

77. Explain why the points $A(2, -1)$, $B(-1, 1)$, and $C(-4, 3)$ could not be points of a parabola.

78. Consider any parabola whose axis of symmetry is either horizontal or vertical. Compare the equation of the axis of symmetry with the equation of the directrix.

79. An object thrown directly upward reaches a height h (in feet) after t seconds according to the function $h(t) = 32t - 16t^2$.

(a) What is the maximum height of the object?

(b) Solve the equation $h = 32t - 16t^2$ for t.

(c) Explain how the equation obtained in part (b) can be used to confirm your answer in part (a).

80. Suppose that the vertex of a parabola with a horizontal axis of symmetry is $V(3, 4)$. If the equation of the directrix is $x = 1$, what are the coordinates of the focus?

81. The accompanying figure shows a parabola whose vertex is the origin. The focus is d units above the vertex, and the directrix is a horizontal line d units below the vertex. The point $P(x, y)$ is any point of the parabola. Point Q is a point of the directrix and is directly below point P.

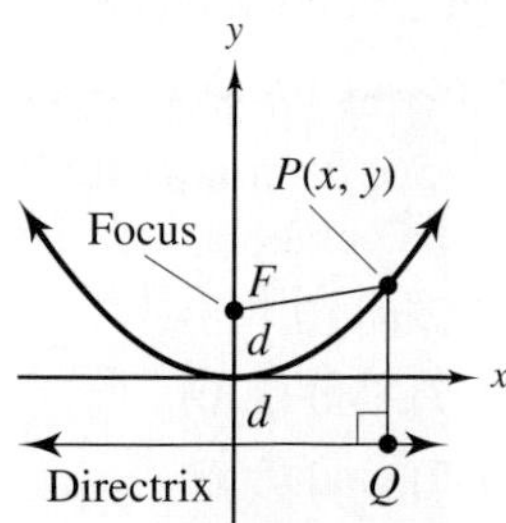

(a) What are the coordinates of the focus?

(b) What is the equation of the directrix?

(c) What are the coordinates of point Q?

(d) What is the distance PF?

(e) What is the distance PQ?

Using the fact that, by the geometric definition of a parabola, $PF = PQ$, and using parts (d) and (e), show that the equation of the parabola is $4dy = x^2$.

11.2 THE CIRCLE

Center at the Origin • Center Not at the Origin • Writing an Equation of a Circle

Center at the Origin

Probably the most familiar of the conic sections is a circle.

Definition of a Circle

A **circle** is the set of all points in a plane that are at a fixed distance (called the **radius**) from a fixed point (called the **center**).

The simplest location for the center is the origin.

EXPLORING THE CONCEPT

A Circle with Center (0, 0)

To write the equation of such a circle, let (x, y) represent any point of the circle (see Fig. 11.10), and apply the Distance Formula to write an expression for r.

Figure 11.10

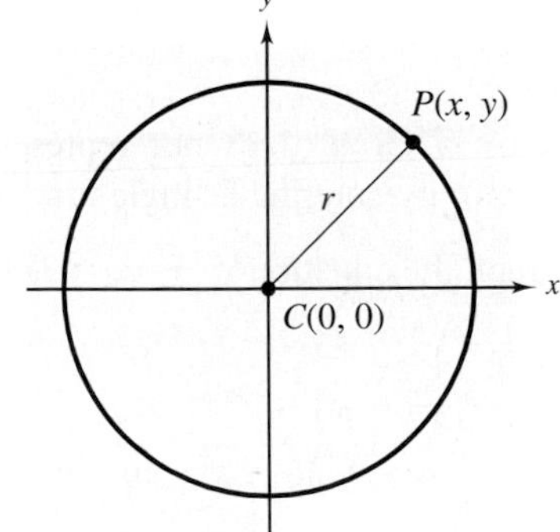

$$\sqrt{(x_2 - x_1)^2 + (y_2 - y_1)^2} = d \quad \text{Distance Formula}$$

$$\sqrt{(x - 0)^2 + (y - 0)^2} = r \quad \text{Distance from } P(x, y) \text{ to } C(0, 0) \text{ is } r.$$

$$(x - 0)^2 + (y - 0)^2 = r^2 \quad \text{Square both sides to eliminate the radical.}$$

$$x^2 + y^2 = r^2 \quad \text{Simplify.}$$

Circle with Center at the Origin

The graph of the equation $x^2 + y^2 = r^2$, $r > 0$, is a circle with center $C(0, 0)$ and radius r.

EXAMPLE 1

Sketching a Circle

Sketch the graph of the equation $x^2 + y^2 = 36$.

Solution

$$x^2 + y^2 = 36$$
$$x^2 + y^2 = 6^2$$

The equation describes a circle with a center at (0, 0). Since $r^2 = 36$, $r = 6$ and the radius of the circle is 6. (See Fig. 11.11.)

Figure 11.11

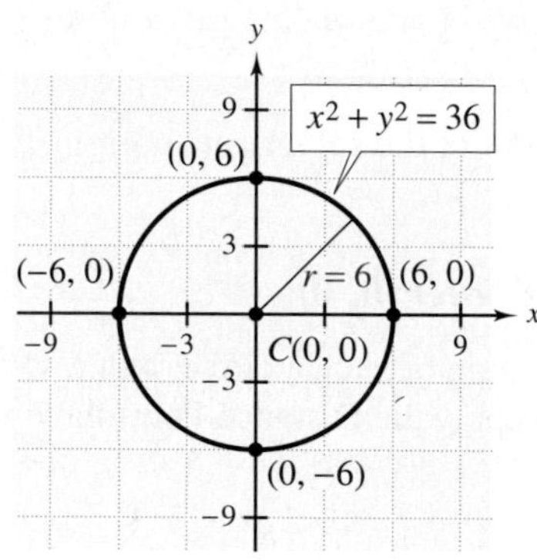

Think About It

The set of points whose coordinates satisfy $x^2 + y^2 = r^2$, where $r = 0$, is called a *degenerate circle*. Describe this degenerate circle.

Because a circle does not represent a function, we cannot produce a circle on a calculator by entering a single function.

To graph the equation $x^2 + y^2 = 36$, we first solve for y.

$$x^2 + y^2 = 36$$
$$y^2 = 36 - x^2$$
$$y = \sqrt{36 - x^2} \quad \text{or} \quad y = -\sqrt{36 - x^2} \qquad \text{Square Root Property}$$

The graph of $y = \sqrt{36 - x^2}$ is a semicircle *above* the x-axis, and the graph of $y = -\sqrt{36 - x^2}$ is a semicircle *below* the x-axis. To display the full circle, we produce the graphs of both functions on the same coordinate axes.

Center Not at the Origin

The equation of a circle must be modified if the center is not the origin.

EXPLORING THE CONCEPT

A Circle with Center (h, k)

To write the equation of a circle with center at $C(h, k)$ and radius r, let $P(x, y)$ represent a point of the circle, and again, apply the Distance Formula. (See Fig. 11.12.)

Figure 11.12

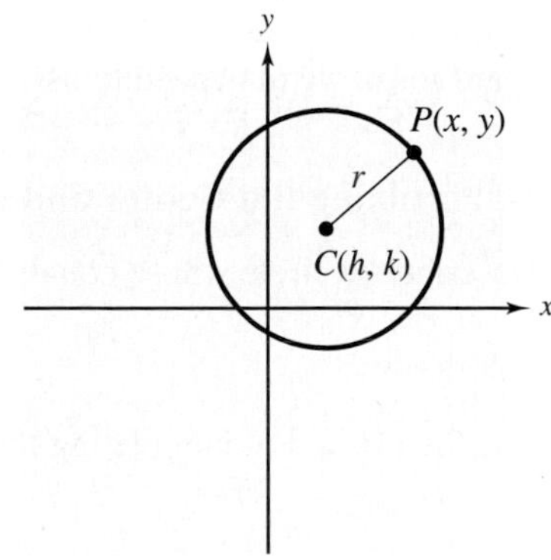

$\sqrt{(x_2 - x_1)^2 + (y_2 - y_1)^2} = d$	Distance Formula
$\sqrt{(x - h)^2 + (y - k)^2} = r$	Distance from $P(x, y)$ to $C(h, k)$ is r.
$(x - h)^2 + (y - k)^2 = r^2$	Square both sides to eliminate the radical.

Equation of a Circle with Center $C(h, k)$

The **standard form** of the equation of a circle with center $C(h, k)$ and radius r is $(x - h)^2 + (y - k)^2 = r^2$.

EXAMPLE 2

Graphing a Circle

Sketch the graph of $(x + 2)^2 + (y - 1)^2 = 10$.

Solution

$$(x + 2)^2 + (y - 1)^2 = 10$$

$$[x - (-2)]^2 + (y - 1)^2 = \left(\sqrt{10}\right)^2 \qquad \text{Standard form.}$$

The graph is a circle whose center is $C(-2, 1)$ and whose radius is $\sqrt{10} \approx 3.16$. (See Fig. 11.13.)

LEARNING TIP

A common error is to interpret $x^2 + y^2 = 10$ as a circle whose radius is 10. Be aware that the constant is the value of r^2, not r.

Figure 11.13

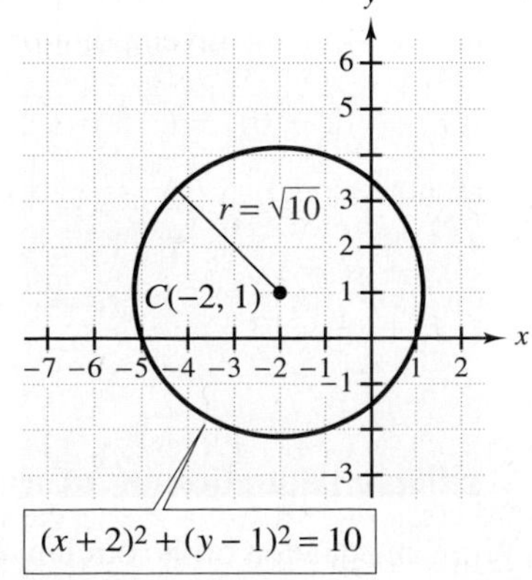

Although equations of circles can be written in many forms, when the equation is written in the standard form, the center and radius are easily identified. To obtain the standard form, we may need to use the method of completing the square.

EXAMPLE 3

Determining the Center and Radius of a Circle

Describe the circle whose equation is $x^2 + y^2 + 6x - 5y = -1$.

Solution

Use the method of completing the square to write the equation in the standard form $(x - h)^2 + (y - k)^2 = r^2$.

$$x^2 + y^2 + 6x - 5y = -1$$

$$(x^2 + 6x + \quad) + (y^2 - 5y + \quad) = -1 \qquad \text{Complete the square for each group.}$$

$$(x^2 + 6x + 9) + \left(y^2 - 5y + \frac{25}{4}\right) = -1 + 9 + \frac{25}{4} \qquad \text{Add 9 and } \tfrac{25}{4} \text{ to both sides of the equation.}$$

$$(x + 3)^2 + \left(y - \frac{5}{2}\right)^2 = \frac{57}{4} \qquad \text{Factor.}$$

$$[x - (-3)]^2 + \left(y - \frac{5}{2}\right)^2 = \left(\frac{\sqrt{57}}{2}\right)^2$$

The center is $(-3, 2.5)$, and the radius is $\frac{\sqrt{57}}{2} \approx 3.77$. (See Fig. 11.14.)

Figure 11.14

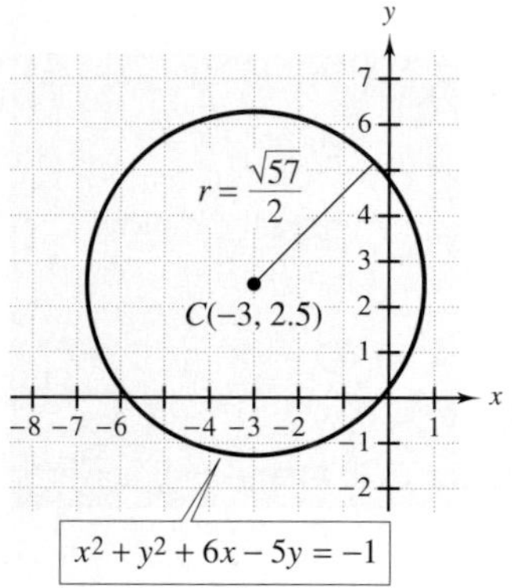

Writing an Equation of a Circle

If we know (or can determine) the center and radius of a circle, then we can write its equation.

EXAMPLE 4

Writing an Equation of a Circle

Write an equation of the circle whose center is the origin and that contains the point $(2, -3)$.

Solution

Figure 11.15 shows the circle. The radius of the circle is the distance from the center $(0, 0)$ to the given point on the circle $(2, -3)$.

$$r = \sqrt{(2 - 0)^2 + (-3 - 0)^2} = \sqrt{4 + 9} = \sqrt{13}$$

Using the model $x^2 + y^2 = r^2$, we write the equation.

$$x^2 + y^2 = \left(\sqrt{13}\right)^2$$

$$x^2 + y^2 = 13$$

Figure 11.15

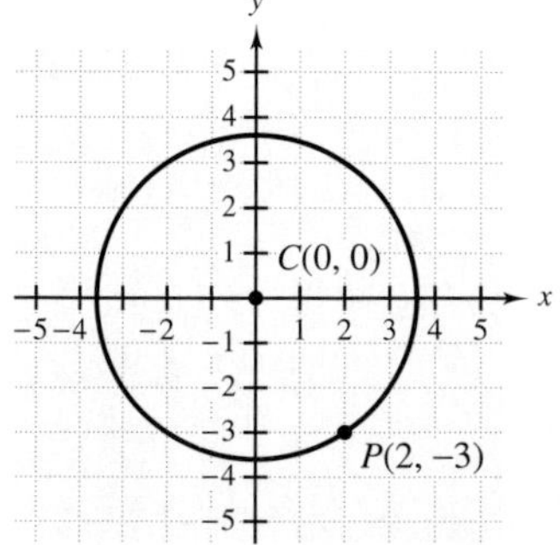

EXAMPLE 5

Writing an Equation of a Circle

Write an equation of the circle with center $(-3, 2)$ and radius 5.

Solution

Figure 11.16 shows the circle.

Figure 11.16

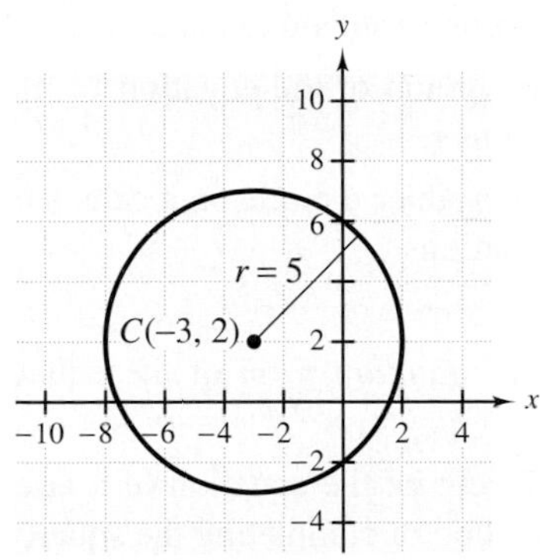

Because the center is not the origin, we use the model $(x - h)^2 + (y - k)^2 = r^2$.

$$[x - (-3)]^2 + (y - 2)^2 = 5^2 \quad h = -3, k = 2, r = 5$$

$$(x + 3)^2 + (y - 2)^2 = 25$$

EXAMPLE 6

Writing an Equation of a Circle

Write an equation of the circle with center $(3, -1)$ and containing the point $(7, 1)$.

Solution

Figure 11.17 shows the circle.

Figure 11.17

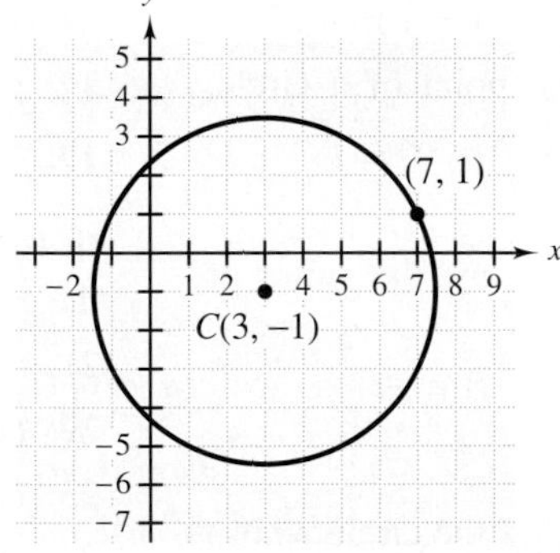

The radius is the distance from $(3, -1)$ to $(7, 1)$.

$$r = \sqrt{(7 - 3)^2 + [1 - (-1)]^2} = \sqrt{4^2 + 2^2} = \sqrt{16 + 4} = \sqrt{20}$$

We use the model $(x - h)^2 + (y - k)^2 = r^2$.

$$(x - 3)^2 + [y - (-1)]^2 = \left(\sqrt{20}\right)^2 \quad h = 3, k = -1, r = \sqrt{20}$$

$$(x - 3)^2 + (y + 1)^2 = 20$$

11.2 QUICK REFERENCE

Center at the Origin

- A **circle** is the set of all points in a plane that are at a fixed distance (called the **radius**) from a fixed point (called the **center**).
- The graph of the equation $x^2 + y^2 = r^2$, $r > 0$, is a circle with center $C(0, 0)$ and radius r.
- To produce a circle on a calculator, solve its equation for y and graph the two related functions.

Center Not at the Origin

- The *standard form* of the equation of a circle with center $C(h, k)$ and radius r is $(x - h)^2 + (y - k)^2 = r^2$.
- To rewrite the equation of a circle in standard form, it may be necessary to use the method of completing the square.

Writing an Equation of a Circle

- If we know (or can determine) the center and radius of a circle, then we can write an equation of the circle.
- Determining the radius of a circle may require the Distance Formula.

11.2 EXERCISES

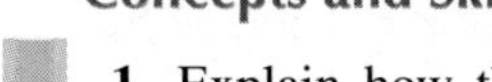

Concepts and Skills

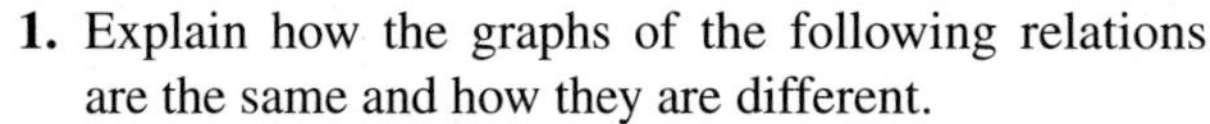

1. Explain how the graphs of the following relations are the same and how they are different.

$$x^2 + y^2 = 25 \qquad (x + 2)^2 + (y - 3)^2 = 25$$

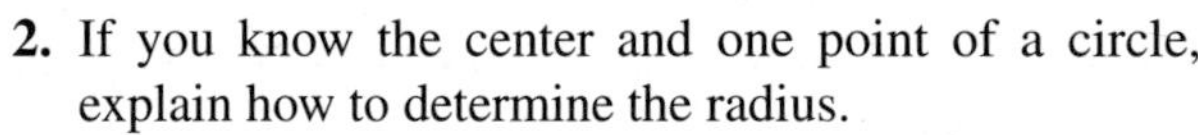

2. If you know the center and one point of a circle, explain how to determine the radius.

In Exercises 3–6, determine the center and the radius of the graph of the given relation.

3. $x^2 + y^2 = 16$
4. $x^2 + y^2 = 1$
5. $9x^2 + 9y^2 = 1$
6. $4x^2 + 4y^2 = 36$

In Exercises 7–10, write an equation of the circle whose center is the origin and whose radius is given. Write the equation without fractions.

7. 4
8. $\frac{4}{3}$
9. $\sqrt{15}$
10. $3\sqrt{2}$

In Exercises 11–16, sketch the graph of the given relation.

11. $x^2 + y^2 = 25$
12. $x^2 + y^2 = 49$
13. $16x^2 + 16y^2 = 25$
14. $2x^2 + 2y^2 = 8$
15. $x^2 + y^2 = 6$
16. $\frac{x^2}{2} + \frac{y^2}{2} = 9$

In Exercises 17–20, write an equation of the circle whose center is (0, 0) and that contains the given point.

17. $(5, 0)$
18. $(-3, -2)$
19. $(6, -1)$
20. $(0, -4)$

In Exercises 21–26, determine the center and radius of the graph of the given relation.

21. $x^2 + (y + 2)^2 = 3$
22. $(x - 7)^2 + (y + 8)^2 - 9 = 0$
23. $(x + 3)^2 + (y - 2)^2 - 25 = 0$
24. $(x - 7)^2 + (y + 4)^2 + 7 = 11$
25. $4(x - 2)^2 + 4(y + 3)^2 = 11$
26. $4(y - 5)^2 + 4(x + 2)^2 = 25$

In Exercises 27–32, write an equation of a circle with the given center and radius.

27. $(4, 2)$, $r = 3$
28. $(-3, 7)$, $r = 5$
29. $(0, 4)$, $r = 2$
30. $(-6, 0)$, $r = 10$
31. $(5, 6)$, $r = \sqrt{11}$
32. $(2, -7)$, $r = \sqrt{2}$

In Exercises 33–36, write an equation of a circle whose center is C and that contains the point P.

33. $C(2, -4), P(5, -4)$

34. $C(4, -3), P(4, 2)$

35. $C(5, 7), P(4, 2)$

36. $C(3, -6), P(-1, -2)$

In Exercises 37–46, sketch the graph of the given equation.

37. $(x - 2)^2 + y^2 = 36$

38. $(x + 2)^2 + (y - 3)^2 = 49$

39. $(y - 4)^2 + (x + 3)^2 = 64$

40. $y^2 + (x - 5)^2 = 25$

41. $(x + 4)^2 + (y + 3)^2 - 25 = 75$

42. $(x - 3)^2 + (y - 5)^2 - 6 = 30$

43. $(y + 5)^2 + (x + 3)^2 + 5 = 30$

44. $(y + 3)^2 + (x + 4)^2 + 10 = 2$

45. $4x^2 + 4(y + 2)^2 = 64$

46. $9(x + 5)^2 + 9y^2 = 25$

In Exercises 47–56, determine the center and the radius of the circle.

47. $x^2 + y^2 + 8y + 3 = 0$

48. $x^2 + y^2 + 4x - 1 = 0$

49. $x^2 + y^2 - 6x + 4y - 12 = 0$

50. $x^2 + y^2 - 2x - 6y + 1 = 0$

51. $x^2 + y^2 + 8x - 4y - 16 = 0$

52. $x^2 + y^2 + 6y - 10x - 15 = 0$

53. $x^2 + y^2 - 4y + 2x + 2 = 0$

54. $x^2 + y^2 + 8x + 2y + 10 = 0$

55. $x^2 + y^2 - 7x + 5y + \frac{5}{2} = 0$

56. $x^2 + y^2 + 3x - 5y - \frac{1}{2} = 0$

57. A line that is tangent to a circle is a line that intersects the circle in only one point. What is the relationship between a tangent line and a radius that is drawn to the point of tangency?

58. If you know the center and diameter of a circle, explain how to write the equation of the circle.

In Exercises 59–64, write an equation of a circle with the given conditions.

59. The center is $(-2, 5)$, and the circle is tangent to the x-axis.

60. The center is $(3, -6)$, and the circle is tangent to the line $x + 4 = 0$.

61. The center is a point of the line $x = 2$, and the circle contains the points $(2, 5)$ and $(2, -3)$.

62. The center is on the x-axis, and the x-intercepts are $(1, 0)$ and $(-5, 0)$.

63. The endpoints of the diameter are $(7, -10)$ and $(-3, 2)$.

64. The circle contains $(3, 2)$ and $(7, 10)$, and these are the endpoints of a line segment that passes through the center of the circle.

In Exercises 65–68, determine the domain and range of the given relation.

65. $x^2 + y^2 = 100$

66. $x^2 + y^2 = 32$

67. $(x - 5)^2 + (y + 2)^2 = 16$

68. $(x + 1)^2 + y^2 = 81$

In Exercises 69–72, the graph of the given relation is a half-circle. Determine whether the graph is the upper, lower, right, or left half of the circle. When possible, use your calculator to produce the graph. Otherwise, draw a sketch of the graph.

69. $y = \sqrt{9 - x^2}$

70. $x = -\sqrt{25 - y^2}$

71. $y = -\sqrt{9 - (x + 3)^2}$

72. $x = -2 + \sqrt{49 - (y - 4)^2}$

In Exercises 73–78, write the name of the graph of the equation.

73. $y^2 = 2 + 2x - x^2$

74. $x^2 = 2 - 4y - y^2$

75. $2x + 3y = 15$

76. $y = 4x$

77. $y^2 = x - 6y - 9$

78. $x^2 + 5x - y = 2$

79. If the phrase *in a plane* were not included in the definition of a circle, the definition would describe a different geometric figure. What do you think it would be and why?

80. If the radius of a circle is positive, explain why the center is not a point of the graph.

Real-Life Applications

81. **Concentric circles** are circles with the same center but with different radii. Suppose that a food court is in the shape of a circle whose equation is $x^2 + y^2 = 9$. Surrounding the food court is a circular walkway whose outer edge has the equation $x^2 + y^2 = 25$. If the variables represent yards, what is the area of the walkway?

82. A draftsman has drawn the cross section of a circular rod with the center at the origin. If the equation of the circle is $x^2 + y^2 = 0.6$, what is the circumference of the rod to the nearest hundredth?

83. A mechanical engineer's design of a clutch system shows two circular disks whose centers are 12 units apart. The disks are in the same plane and their radii are r_1 and r_2, where $r_1 < 12$ and $r_2 < 12$. What are the conditions on r_1 and r_2 if the disks

(a) do not touch?

(b) touch at exactly one point?

(c) overlap?

84. On a fleet deployment map, a carrier is located at the origin of a coordinate system whose positive y-axis represents north. Six destroyers are each positioned 10 miles from the carrier in the northwest and southwest quadrants. Write an equation to model the possible coordinates of any one of the destroyers.,

Challenge

In Exercises 85 and 86, write an equation of the line that is tangent to the given circle at point P.

85. $x^2 + y^2 = 10$; $P(-1, 3)$

86. $(x + 5)^2 + (y + 2)^2 = 41$; $P(-10, -6)$

In Exercises 87 and 88, the equation of a circle is given. Write an equation that describes the indicated half-circle.

87. $x^2 + y^2 = 8$; left half

88. $(x - 2)^2 + (y + 4)^2 = 12$; top half

89. A **chord** of a circle is a line segment whose endpoints are points of the circle. The word *radius* is often used to mean the line segment whose endpoints are the center and one point of the circle.

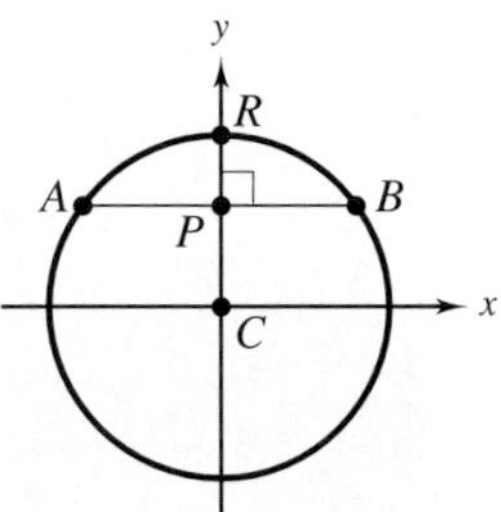

In the accompanying figure, $\overline{AB}$ is a chord, $\overline{CR}$ is the radius that is perpendicular to $\overline{AB}$, and P is their point of intersection.

(a) What name do we give to the longest chord of *any* circle?

(b) Suppose the equation of the circle in the figure is $x^2 + y^2 = 100$. If chord $\overline{AB}$ is parallel to the x-axis and $AB = 6$, what is the distance from point P to point C?

11.3 THE ELLIPSE

Center at the Origin • Center Not at the Origin • Writing an Equation of an Ellipse

Center at the Origin

Another type of conic section is an **ellipse**, which is described by the following geometric definition.

Definition of an Ellipse

An **ellipse** is the set of all points in a plane such that the sum of the distances from each point to two fixed points (called **foci**) is constant.

Figure 11.18 shows an ellipse with two foci $F_1(-c, 0)$ and $F_2(c, 0)$. Figure 11.19 shows an ellipse with two foci $F_1(0, -c)$ and $F_2(0, c)$. Point $P(x, y)$ is any point of the ellipse. By definition, $PF_1 + PF_2 = k$, where k is a constant.

Figure 11.18

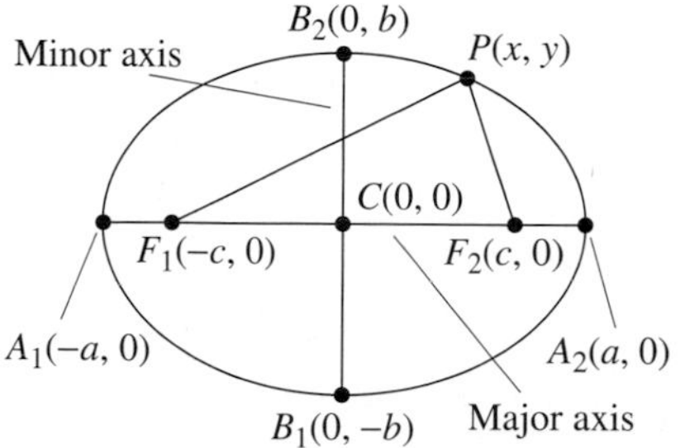

Figure 11.19

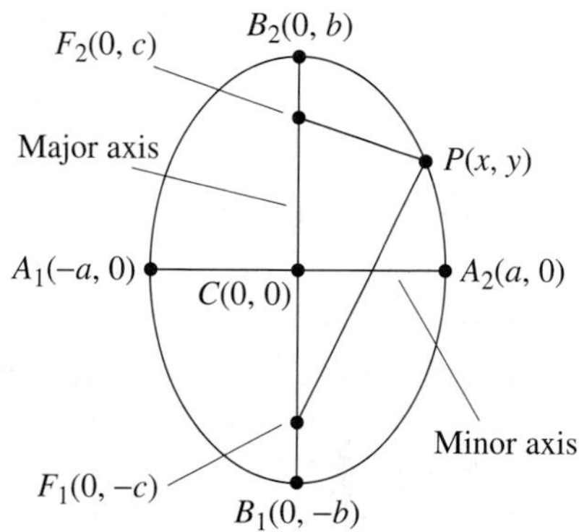

Think About It

Using only the definition of an ellipse, explain how you know that the following is true.

$A_1F_1 + A_1F_2 = B_1F_1 + B_1F_2$

The line containing the foci intersects the ellipse at points called the **vertices** of the ellipse. The vertices are $A_1(-a, 0)$ and $A_2(a, 0)$ in Fig. 11.18 and $B_1(0, -b)$ and $B_2(0, b)$ in Fig. 11.19. The line segment connecting the vertices is the **major axis**, and its midpoint is the **center** $C(0, 0)$ of the ellipse.

A line perpendicular to the major axis at the center of the ellipse intersects the ellipse at points sometimes called the **co-vertices**. The co-vertices are $B_1(0, -b)$ and $B_2(0, b)$ in Fig. 11.18 and $A_1(-a, 0)$ and $A_2(a, 0)$ in Fig. 11.19. The line segment connecting the co-vertices is the **minor axis**.

We can use the definition of ellipse to derive an equation of an ellipse. (See Exercise 67.)

Ellipse with Center at the Origin

The graph of the equation

$$\frac{x^2}{a^2} + \frac{y^2}{b^2} = 1, \qquad a > 0 \text{ and } b > 0$$

is an ellipse whose center is the origin.

For an ellipse whose center is the origin, $(a, 0)$ and $(-a, 0)$ are the x-intercepts, and $(0, b)$ and $(0, -b)$ are the y-intercepts. These are the key points to plot when we need to sketch the graph of the ellipse.

EXAMPLE 1

Sketching an Ellipse Whose Center Is the Origin

Sketch the ellipse whose equation is $\dfrac{x^2}{9} + \dfrac{y^2}{16} = 1$.

Solution

The equation $\frac{x^2}{9} + \frac{y^2}{16} = 1$ describes an ellipse whose center is the origin.

Because $a^2 = 9$, $a = 3$ and the x-intercepts are $(3, 0)$ and $(-3, 0)$. Because $b^2 = 16$, $b = 4$ and the y-intercepts are $(0, 4)$ and $(0, -4)$. Figure 11.20 shows how these points are used to sketch the ellipse.

Figure 11.20

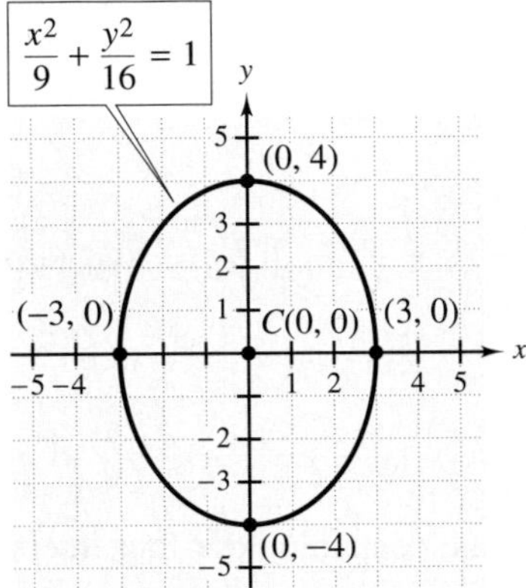

An ellipse does not represent a function because it does not pass the Vertical Line Test. Therefore, to produce an ellipse on a calculator, we must solve the equation for y and then graph the two related functions. You should verify that solving for y in the equation in Example 1 gives the following related functions.

$$y = \frac{\sqrt{144 - 16x^2}}{3} \quad \text{and} \quad y = -\frac{\sqrt{144 - 16x^2}}{3}$$

Center Not at the Origin

As is true for circles, an ellipse can be shifted so that its center is $C(h, k)$. (See Fig. 11.21.)

Figure 11.21

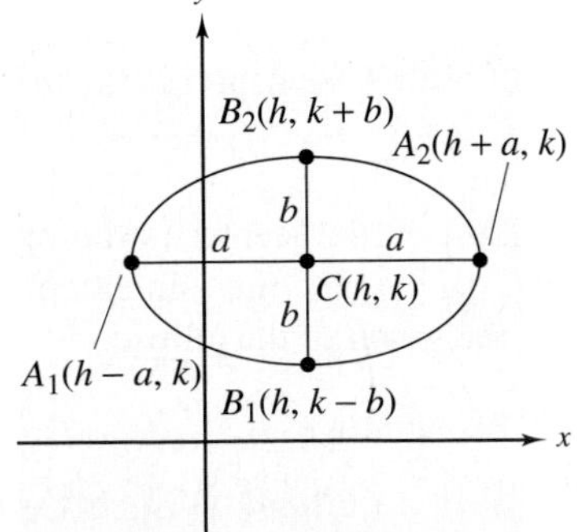

Notice that the numbers a and b indicate the distances from the center of the ellipse to the endpoints of the axes.

Equation of an Ellipse with Center $C(h, k)$

The *standard form* of the equation of an ellipse with center $C(h, k)$ is

$$\frac{(x-h)^2}{a^2} + \frac{(y-k)^2}{b^2} = 1, \qquad a > 0 \text{ and } b > 0$$

EXAMPLE 2

Sketching an Ellipse with Center $C(h, k)$

Sketch the graph of each equation.

(a) $\dfrac{(x-4)^2}{9} + \dfrac{(y-2)^2}{36} = 1$ (b) $4(x+3)^2 + 25(y-1)^2 = 100$

Solution

(a) $\dfrac{(x-4)^2}{9} + \dfrac{(y-2)^2}{36} = 1$

Because $h = 4$ and $k = 2$, the center of the ellipse is $(4, 2)$. The values of a $(a = 3)$ and b $(b = 6)$ give the distances from the center to the endpoints of the axes. As in Fig. 11.22, we use these values to locate the endpoints of the axes and then sketch the graph.

Figure 11.22

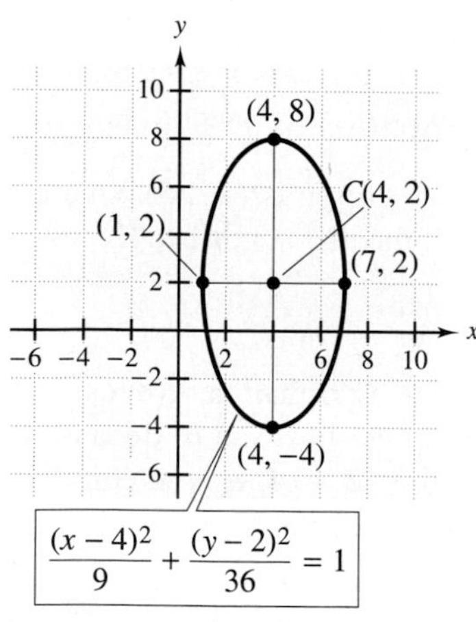

Figure 11.23

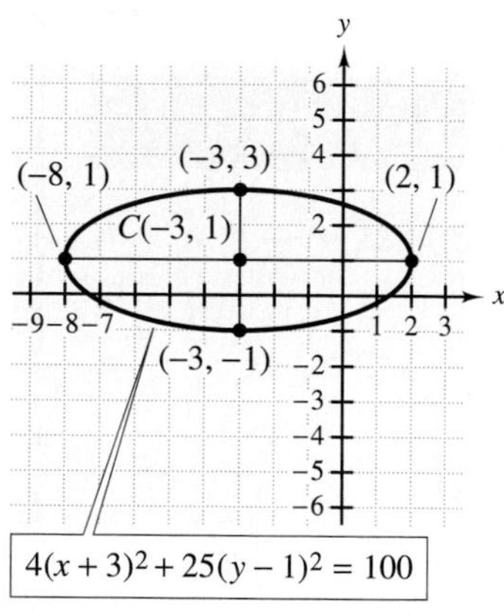

(b) $4(x+3)^2 + 25(y-1)^2 = 100$ Divide both sides of the equation by 100.

$\dfrac{(x+3)^2}{25} + \dfrac{(y-1)^2}{4} = 1$ Standard form

The center is $(-3, 1)$. Use the values of a $(a = 5)$ and b $(b = 2)$ to locate the endpoints of the axes. Figure 11.23 shows the graph.

LEARNING TIP

Whether you are sketching an ellipse or writing its equation, you will need to know the numbers a, b, h, and k. Determining these numbers is usually your first priority.

Writing an Equation of an Ellipse

Given sufficient information about an ellipse, we can write an equation of the ellipse.

EXAMPLE 3

Writing an Equation of an Ellipse with Center $C(0, 0)$

Write an equation of the ellipse whose center is the origin, whose x-intercepts are $(-2, 0)$ and $(2, 0)$, and whose y-intercepts are $(0, -5)$ and $(0, 5)$.

Solution

By plotting the intercepts, we obtain the ellipse shown in Fig. 11.24.

Figure 11.24

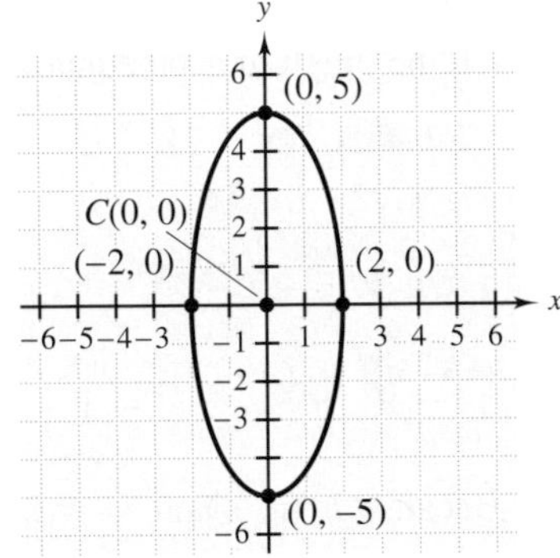

$$\frac{x^2}{a^2} + \frac{y^2}{b^2} = 1 \quad \text{Model for equation of an ellipse with center } C(0, 0).$$

$$\frac{x^2}{4} + \frac{y^2}{25} = 1 \quad a = 2 \text{ and } b = 5$$

EXAMPLE 4

Writing an Equation of an Ellipse with Center $C(h, k)$

The center of an ellipse is $C(2, 5)$. One endpoint of the minor axis is $P(6, 5)$, and one endpoint of the major axis is $Q(2, -2)$. Write an equation of the ellipse.

Figure 11.25

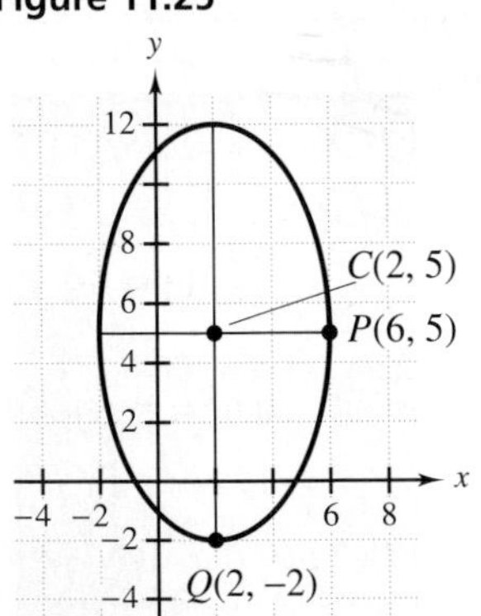

Solution

Using the given information, we can sketch the ellipse (see Fig. 11.25) and determine the values of a and b. Because the horizontal distance from the center C to point P is $6 - 2 = 4$, a is 4. Similarly, the vertical distance from C to Q is $5 - (-2) = 7$, so $b = 7$.

$$\frac{(x-h)^2}{a^2} + \frac{(y-k)^2}{b^2} = 1 \quad \text{Model for equation of an ellipse with center } C(h, k).$$

$$\frac{(x-2)^2}{16} + \frac{(y-5)^2}{49} = 1 \quad \text{Center } C(2, 5),\ a = 4,\ b = 7$$

11.3 QUICK REFERENCE

Center at the Origin

- An **ellipse** is the set of all points in a plane such that the sum of the distances from each point to two fixed points (called **foci**) is constant.

- The line containing the foci intersects the ellipse at points called the **vertices**. The line segment connecting the vertices is the **major axis**, and its midpoint is the **center** of the ellipse.
- The line perpendicular to the major axis at the center of an ellipse intersects the ellipse at points called **co-vertices**. The line segment connecting the co-vertices is the **minor axis**.
- The equation $\dfrac{x^2}{a^2} + \dfrac{y^2}{b^2} = 1$, $a > 0$ and $b > 0$, describes an ellipse whose center is the origin.

Center Not at the Origin

- The *standard form* of the equation of an ellipse with center $C(h, k)$ is $\dfrac{(x-h)^2}{a^2} + \dfrac{(y-k)^2}{b^2} = 1$, $a > 0$ and $b > 0$.

Writing an Equation of an Ellipse

- To write an equation of an ellipse, we must know the center and at least one endpoint of each axis.

11.3 EXERCISES

Concepts and Skills

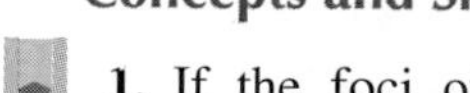

1. If the foci of an ellipse were moved closer and closer together until they became the same point, how would you describe the resulting figure?
2. The vertices of an ellipse are the endpoints of the major axis. Write a similar sentence describing the co-vertices of an ellipse.

In Exercises 3–8, determine the intercepts of the graph of the given relation.

3. $\dfrac{x^2}{25} + \dfrac{y^2}{16} = 1$
4. $\dfrac{x^2}{36} + \dfrac{y^2}{25} = 1$
5. $\dfrac{x^2}{9} + y^2 = 1$
6. $4x^2 + 9y^2 = 36$
7. $9x^2 + 4y^2 = 1$
8. $5x^2 + 8y^2 = 40$

In Exercises 9–14, sketch the graph of the given relation.

9. $\dfrac{x^2}{81} + \dfrac{y^2}{100} = 1$
10. $x^2 + \dfrac{y^2}{4} = 4$
11. $25x^2 + 4y^2 = 100$
12. $4x^2 + y^2 = 4$
13. $\dfrac{x^2}{25} + \dfrac{4y^2}{9} = 1$
14. $\dfrac{x^2}{4} + \dfrac{y^2}{25} = \dfrac{1}{100}$

In Exercises 15–20, write an equation of an ellipse whose center is the origin and that satisfies the given conditions.

15. The intercepts are $(\pm 3, 0)$ and $(0, \pm 5)$.
16. The intercepts are $(\pm 7, 0)$ and $(0, \pm 6)$.
17. Two intercepts are $(9, 0)$ and $(0, 4)$.
18. Two intercepts are $(-6, 0)$ and $(0, -8)$.
19. The length of the horizontal axis is 8, and the length of the vertical axis is 12.
20. The length of the vertical axis is 2, and the length of the horizontal axis is 6.

In Exercises 21–26, determine the center and the endpoints of the major and minor axes for the graph of the given relation.

21. $\dfrac{(x-3)^2}{16} + \dfrac{(y-4)^2}{25} = 1$
22. $\dfrac{(x+5)^2}{49} + \dfrac{(y+2)^2}{9} = 1$

23. $\frac{(x-5)^2}{4} + (y+1)^2 = 4$

24. $(x+2)^2 + 9y^2 = 9$

25. $25(x-1)^2 + 16(y+3)^2 = 400$

26. $9(x+3)^2 + 5(y-2)^2 = 45$

In Exercises 27–32, sketch the graph of the given relation.

27. $\frac{(x-2)^2}{36} + \frac{(y+3)^2}{25} = 1$

28. $(x+5)^2 + \frac{y^2}{4} = 1$

29. $\frac{9(x-1)^2}{16} + \frac{(y+4)^2}{25} = 1$

30. $25(x-1)^2 + 36(y-2)^2 = 1$

31. $\frac{(x-1)^2}{9} + \frac{(y+1)^2}{16} = \frac{1}{144}$

32. $5(x-1)^2 + 3(y-4)^2 = 15$

In Exercises 33–40, write an equation of the ellipse that satisfies the given conditions.

33. The center is (4, 5), an endpoint of one axis is (4, 9), and an endpoint of the other axis is (7, 5).

34. The center is (−1, 2), an endpoint of one axis is (−7, 2), and an endpoint of the other axis is (−1, −2).

35. The center is (−2, −3), a vertex is (6, −3), and a co-vertex is (−2, 0).

36. The center is (5, −6), a vertex is (5, −13), and a co-vertex is (0, −6).

37. The endpoints of the axes are (3, 9), (3, −1), (1, 4), and (5, 4).

38. The endpoints of the axes are (−7, 5), (5, 5), (−1, 9), and (−1, 1).

39. The vertices are (2, −2) and (−8, −2); the co-vertices are (−3, 0) and (−3, −4).

40. The vertices are (1, 3) and (1, −11); the co-vertices are (5, −4) and (−3, −4).

41. If the graph of a relation is an ellipse whose center is (0, 0), describe the domain and range of the relation in terms of the endpoints of the axes.

42. For the relation $\frac{x^2}{a^2} + \frac{y^2}{b^2} = 1$, under what conditions is the graph of the relation a circle? an ellipse?

In Exercises 43–48, determine the domain and range of the relation.

43. $9x^2 + 4y^2 = 36$

44. $4x^2 + 25y^2 = 100$

45. $4x^2 + 9y^2 = 1$

46. $16(x+5)^2 + 25(y-2)^2 = 400$

47. $36(x+3)^2 + 4(y+5)^2 = 144$

48. $36(x+2)^2 + 25(y+1)^2 = 1$

In Exercises 49–56, identify the graph.

49. $y^2 - x = 2y$

50. $y - x^2 = 2$

51. $y + x = 9$

52. $2x + 3y = 12$

53. $3y^2 + 3x^2 = 27$

54. $x^2 - 4x + y^2 = 0$

55. $4x^2 + 9y^2 = 36$

56. $x^2 + 4y^2 = 9$

57. Push two pins or thumbtacks into a soft board. Attach a string to the pins so that the string is somewhat loose. Pull the string taut with a pencil and, keeping the string tight, move the pencil wherever the string will allow it to go. (See figure.)

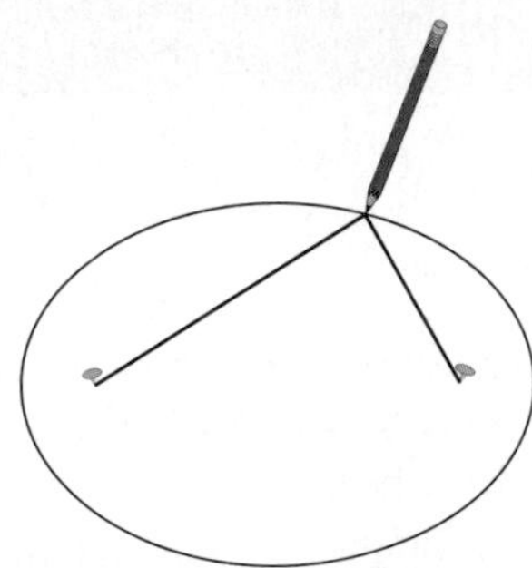

(a) Explain why the resulting curve is an ellipse.

(b) How does the shape of the ellipse change when the tacks are moved farther apart?

(c) How does the shape of the ellipse change when the tacks are moved closer together?

(d) What is the resulting curve when the two ends of the string are attached to just one tack?

58. For an ellipse whose center is the origin, the equation is $\frac{x^2}{a^2} + \frac{y^2}{b^2} = 1$.

How do the relative sizes of a and b influence the shape of the ellipse? To explore this question, sketch the following ellipses. What conclusions can you draw from this experiment?

(a) $\frac{x^2}{100} + \frac{y^2}{49} = 1, \quad a > b$

(b) $\dfrac{x^2}{49} + \dfrac{y^2}{100} = 1, \quad a < b$

(c) $\dfrac{x^2}{49} + \dfrac{y^2}{49} = 1, \quad a = b$

Real-Life Applications

59. A civil engineer had sketched a *semielliptical* arch (the upper half of an ellipse) designed to support a bridge over a small river. The equation of the full ellipse was $\dfrac{x^2}{625} + \dfrac{y^2}{324} = 1$, where x and y are in feet. The accompanying figure shows the semiellipse for which $y \geq 0$. Later, the engineer learned that the height of the arch needed to be increased by 2 feet. What equation should be used for the full ellipse? What is the equation for the semiellipse?

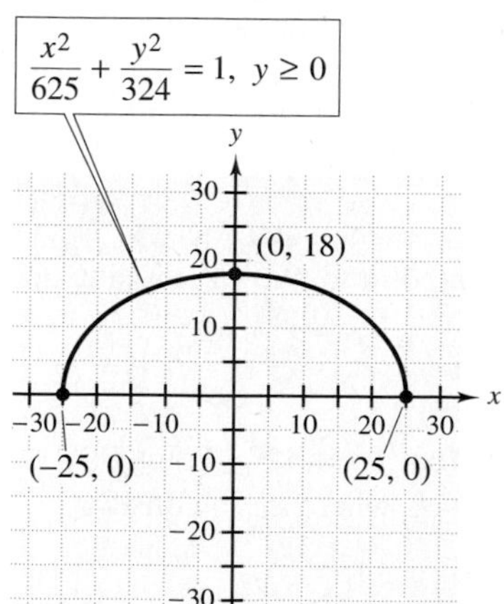

60. The earth travels in an elliptical orbit about the sun with the sun located at a focus $F(c, 0)$ of the ellipse. (See figure.) If the equation of the ellipse is $\dfrac{x^2}{a^2} + \dfrac{y^2}{b^2} = 1$, write expressions in terms of a and c that represent the greatest and least distances from the earth to the sun.

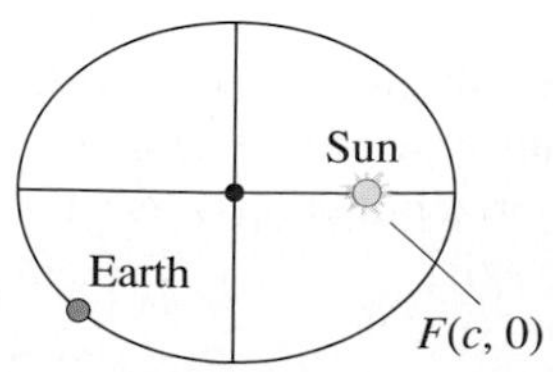

Challenge

In Exercises 61 and 62, use the method of completing the square to write the equation in standard form. Then, sketch the graph.

61. $4x^2 + 8x + y^2 - 2y = 11$

62. $x^2 - 6x + 4y^2 + 16y = 0$

In Exercises 63 and 64, write the equation for the specified part of the ellipse.

63. $16x^2 + y^2 = 9$; top half

64. $4x^2 + 16y^2 = 16$; left half

65. Consider the ellipse $9x^2 + 16y^2 = 144$. If P_1 and P_2 are points of the ellipse and the x-coordinate of both points is 2, what is the distance between the two points? Round to the nearest hundredth.

66. Consider the ellipse $5x^2 + 9y^2 = 45$ and a circle whose center is the origin. If both the ellipse and the circle have the same y-intercepts, what is the distance between their positive x-intercepts? Round your answer to the nearest hundredth.

67. The accompanying figure shows an ellipse whose center is the origin and with foci $F_1(-c, 0)$ and $F_2(c, 0)$. One x-intercept is $A(a, 0)$ and one y-intercept is $B(0, b)$. Point $P(x, y)$ is any point of the ellipse. Let d represent the sum of the distances from any point of the ellipse to F_1 and F_2.

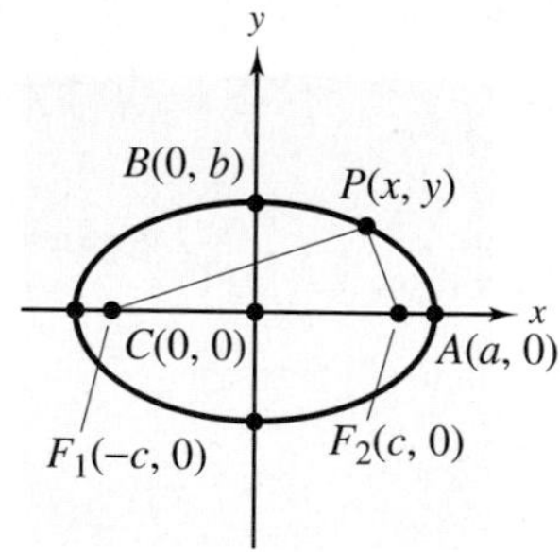

(a) Point B is a point of the ellipse. Use the Distance Formula to show that

$$d = BF_1 + BF_2 = 2\sqrt{b^2 + c^2}$$

(b) Point A is also a point of the ellipse. Show that $d = AF_1 + AF_2 = 2a$.

(c) From parts (a) and (b), $2\sqrt{b^2 + c^2} = 2a$. From this, show that $a^2 - c^2 = b^2$.

(d) Point P is also a point of the ellipse. Use the Distance Formula to write an equation reflecting the fact that $PF_1 + PF_2 = d$, but use $2a$ instead of d.

(e) Eliminate the radicals from the equation written in part (d) by squaring both sides. It will be necessary to do this twice.

(f) Simplify the resulting equation, and show that $(a^2 - c^2)x^2 + a^2y^2 = a^2(a^2 - c^2)$.

(g) Substitute b^2 for $a^2 - c^2$ [see part (c)], and divide both sides of the equation by a^2b^2. The result is the standard form for an ellipse whose center is the origin.

$$\frac{x^2}{a^2} + \frac{y^2}{b^2} = 1$$

11.4 THE HYPERBOLA

Center at the Origin • Center Not at the Origin • Nonstandard Hyperbolas

Center at the Origin

The last conic section we will consider is the **hyperbola.**

Definition of a Hyperbola

A **hyperbola** is the set of all points in a plane such that the absolute value of the difference of the distances from each point to two fixed points (called **foci**) is a constant.

Figure 11.26 shows a hyperbola with two foci F_1 and F_2. Point $P(x, y)$ is any point of the hyperbola. By definition, $|PF_1 - PF_2| = k$, where k is a constant.

Figure 11.26

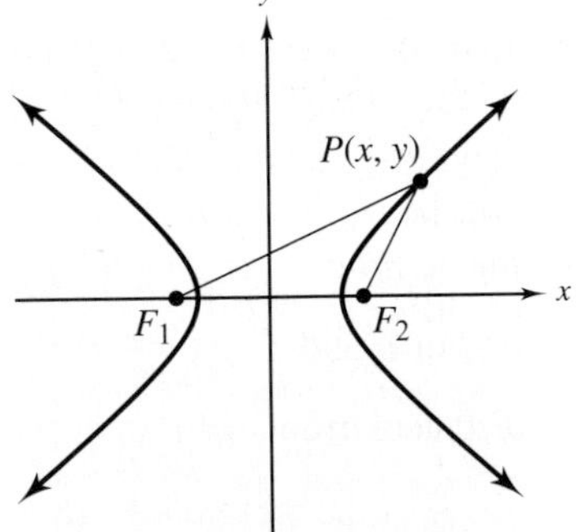

The midpoint of the line segment connecting the foci is the **center** of the hyperbola. The line containing the foci (called the **transverse axis**) can be horizontal or vertical. Figure 11.27 shows a hyperbola with a horizontal transverse axis, and Fig. 11.28 shows a hyperbola with a vertical transverse axis.

Figure 11.27

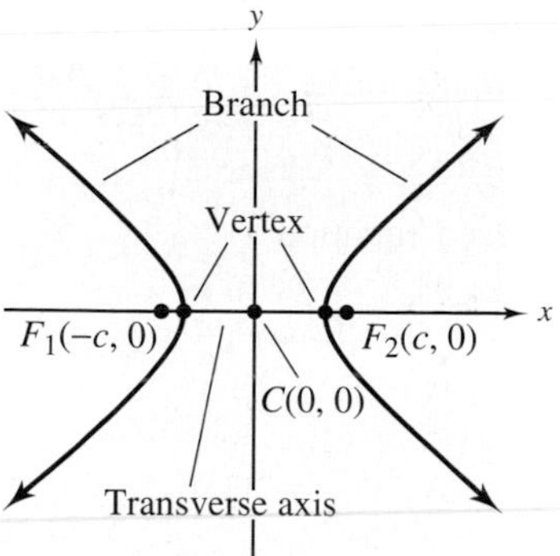

Figure 11.28

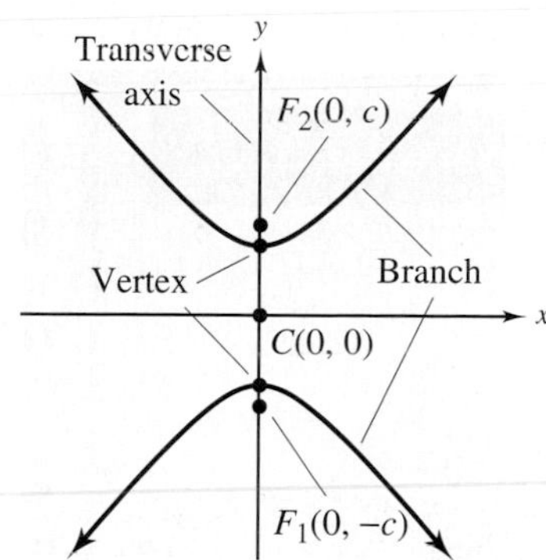

Using the definition of a hyperbola, we can derive an equation for a hyperbola whose center is the origin.

Equation of a Hyperbola Whose Center is the Origin

The standard form of the equation of a hyperbola with center $C(0, 0)$ is

$$\frac{x^2}{a^2} - \frac{y^2}{b^2} = 1 \qquad \text{Horizontal transverse axis}$$

and

$$\frac{y^2}{b^2} - \frac{x^2}{a^2} = 1 \qquad \text{Vertical transverse axis}$$

where $a > 0$ and $b > 0$.

The two separate parts of the graph are called **branches.** Each branch intersects the transverse axis at a point called the **vertex.** As the two branches are extended, they become closer to (but never touch) two lines called **asymptotes.** The asymptotes are the extended diagonals of the **central rectangle.** Observe that the midpoints of the sides of the central rectangle are $(\pm a, 0)$ and $(0, \pm b)$. (See Figs. 11.29 and 11.30.)

Think About It

Consider a hyperbola whose center is the origin and whose transverse axis is horizontal. What are the slopes of the asymptotes?

Figure 11.29

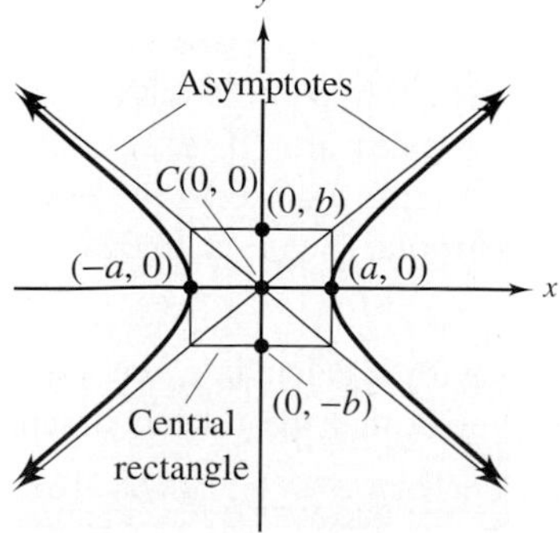

Figure 11.30

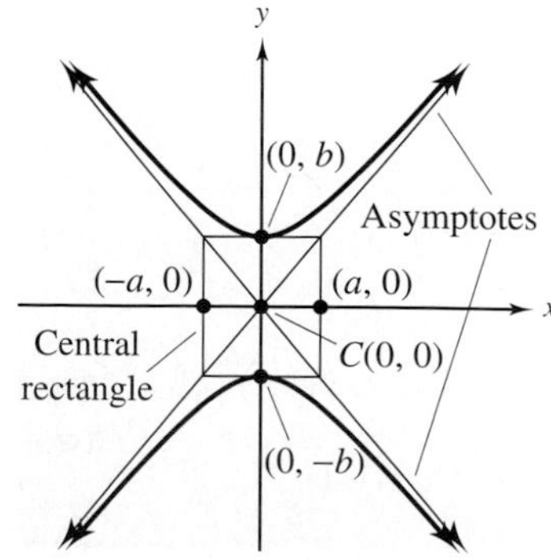

Note: The two asymptotes and the central rectangle are *not* part of the hyperbola. They are simply aids in sketching the graph and in studying its behavior.

EXAMPLE 1

Sketching the Graph of a Hyperbola

Sketch the graph of each equation.

(a) $\frac{x^2}{9} - \frac{y^2}{25} = 1$ (b) $4y^2 - x^2 = 16$

Solution

(a) $\frac{x^2}{9} - \frac{y^2}{25} = 1$

The graph is a horizontal hyperbola with its center at the origin. Because $a = 3$ and $b = 5$, the vertices are $(\pm 3, 0)$, and we draw the central rectangle through the points $(\pm 3, 0)$ and $(0, \pm 5)$. Next, draw the asymptotes through the rectangle, and then sketch the hyperbola. (See Fig. 11.31.)

Figure 11.31

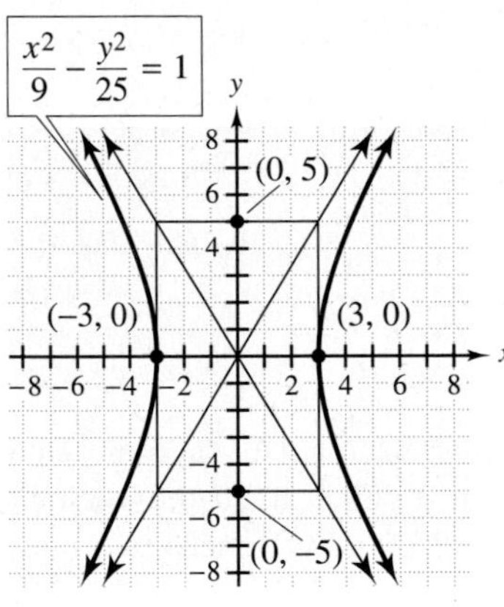

Figure 11.32

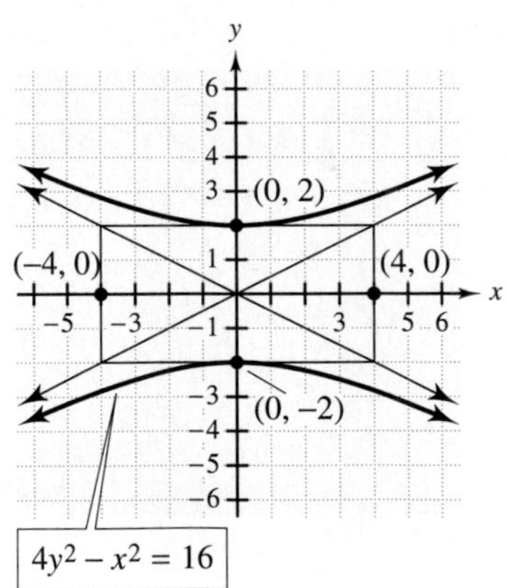

(b) $4y^2 - x^2 = 16$ Divide both sides of the equation by 16.

$\frac{y^2}{4} - \frac{x^2}{16} = 1$ Standard form

The graph is a vertical hyperbola with its center at the origin. Because $a = 4$ and $b = 2$, the vertices are $(0, \pm 2)$, and the central rectangle contains the points $(0, \pm 2)$ and $(\pm 4, 0)$. Sketch the asymptotes and use them as guides for sketching the hyperbola. (See Fig. 11.32.)

To display a hyperbola with a calculator, we solve its equation for y and graph the two related functions. Verify that the related functions for the equation in part (b) of Example 1 are $y = \frac{\sqrt{x^2 + 16}}{2}$ and $y = -\frac{\sqrt{x^2 + 16}}{2}$.

Center Not at the Origin

As with circles and ellipses, the center of a hyperbola can be any point $C(h, k)$, as in Fig. 11.33.

LEARNING TIP

The numbers a, b, h, and k used to draw the central rectangle are the same as those used to sketch an ellipse. Visualize the ellipse inside the central rectangle and the hyperbola outside it.

Figure 11.33

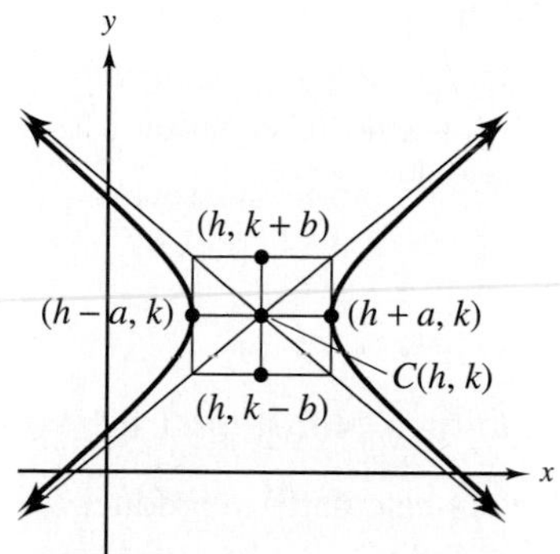

Equation of a Hyperbola with Center $C(h, k)$

The *standard forms* of the equation of a hyperbola with center $C(h, k)$ are

$$\frac{(x-h)^2}{a^2} - \frac{(y-k)^2}{b^2} = 1 \quad \text{Horizontal transverse axis}$$

and

$$\frac{(y-k)^2}{b^2} - \frac{(x-h)^2}{a^2} = 1 \quad \text{Vertical transverse axis}$$

where $a > 0$ and $b > 0$.

EXAMPLE 2

Sketching a Hyperbola with Center $C(h, k)$

Sketch the graph of $\dfrac{(x-3)^2}{4} - \dfrac{(y+1)^2}{9} = 1$.

Solution

$$\frac{(x-3)^2}{2^2} - \frac{[y-(-1)]^2}{3^2} = 1 \quad \text{Standard form}$$

From the equation, the center of the hyperbola is $(3, -1)$, $a = 2$ and $b = 3$. Because $a = 2$, the vertices are two units to the left and right of the center: $(1, -1)$ and $(5, -1)$. Because $b = 3$, the midpoints of the other two sides of the rectangle are three units above and below the center: $(3, -4)$ and $(3, 2)$. Construct the central rectangle, plot the vertices, and draw the asymptotes. Then sketch the hyperbola. (See Fig. 11.34.)

Figure 11.34

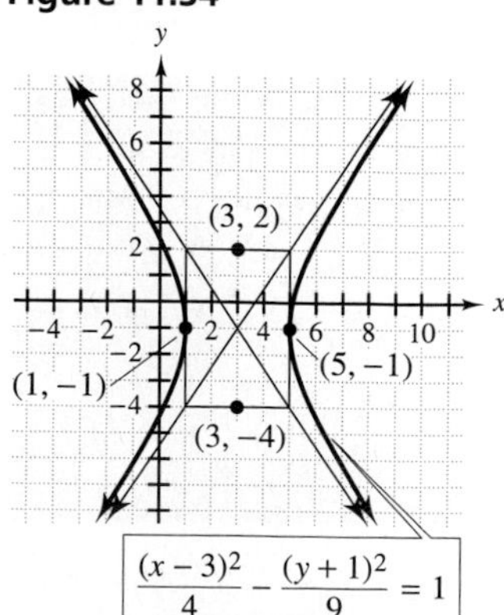

Nonstandard Hyperbolas

A **nonstandard hyperbola** is a hyperbola whose equation can be written $xy = c$, where c is a nonzero constant. Two examples of such equations are $xy = 4$ and $x = -\frac{1}{y}$.

Because this kind of equation is easy to solve for y, the graph is easy to produce on a calculator.

EXAMPLE 3

Graphing a Nonstandard Hyperbola

Use your calculator to produce the graph of $xy = 4$.

Solution

$xy = 4$ Solve for y.

$y = \frac{4}{x}$ Enter the function in your calculator.

Figure 11.35 shows the graph of the hyperbola. Note that unlike other hyperbolas we have studied, the asymptotes of the graph of a nonstandard hyperbola are the x- and y-axes.

Figure 11.35

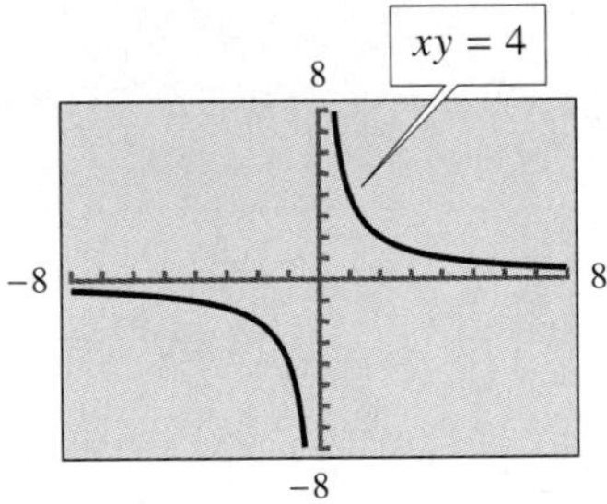

11.4 QUICK REFERENCE

Center at the Origin

- A **hyperbola** is the set of all points in a plane such that the absolute value of the difference of the distances from each point to two fixed points (called **foci**) is a constant.
- The midpoint of the line segment connecting the foci is the **center** of the hyperbola. The line containing the foci (called the **transverse axis**) can be horizontal or vertical.
- The *standard form* of the equation of a hyperbola with center $C(0, 0)$ is

  $$\frac{x^2}{a^2} - \frac{y^2}{b^2} = 1 \quad \text{Horizontal transverse axis}$$

  and

  $$\frac{y^2}{b^2} - \frac{x^2}{a^2} = 1 \quad \text{Vertical transverse axis}$$

  where $a > 0$ and $b > 0$.
- The two separate parts of the graph are called **branches.** Each branch intersects the transverse axis at a point called the **vertex.**
- The **asymptotes** are the extended diagonals of the **central rectangle.** The midpoints of the sides of the central rectangle are $(\pm a, 0)$ and $(0, \pm b)$.

Center Not at the Origin

- The *standard forms* of the equation of a hyperbola with center $C(h, k)$ are

  $$\frac{(x-h)^2}{a^2} - \frac{(y-k)^2}{b^2} = 1 \quad \text{Horizontal transverse axis}$$

  and

  $$\frac{(y-k)^2}{b^2} - \frac{(x-h)^2}{a^2} = 1 \quad \text{Vertical transverse axis}$$

  where $a > 0$ and $b > 0$.

Nonstandard Hyperbolas

- A **nonstandard hyperbola** is a hyperbola whose equation can be written $xy = c$, where c is a nonzero constant.
- The asymptotes of the graph of a nonstandard hyperbola are the x- and y-axes.

11.4 EXERCISES

Concepts and Skills

1. Explain how to construct the central rectangle for a hyperbola whose center is the origin and whose transverse axis is horizontal.

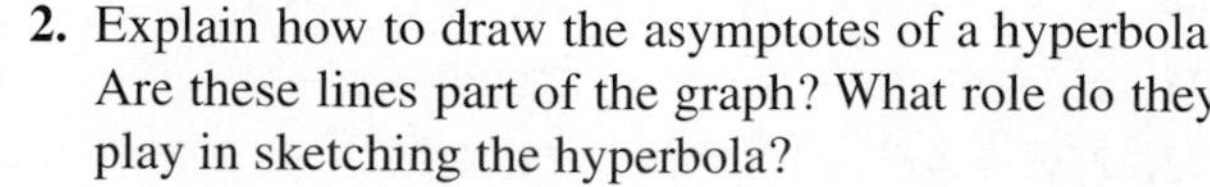

2. Explain how to draw the asymptotes of a hyperbola. Are these lines part of the graph? What role do they play in sketching the hyperbola?

In Exercises 3–6, sketch the graph of the given relation.

3. $\frac{x^2}{25} - \frac{y^2}{16} = 1$

4. $x^2 - \frac{y^2}{4} = 1$

5. $4x^2 - 9y^2 = 36$

6. $x^2 - 9y^2 = 9$

In Exercises 7–10, sketch the graph of the given relation.

7. $\frac{y^2}{25} - \frac{x^2}{36} = 1$

8. $\frac{y^2}{25} - \frac{x^2}{25} = 1$

9. $5y^2 - 5x^2 = 45$ **10.** $25y^2 - x^2 = 25$

11. Explain how to determine the orientation of a hyperbola from its equation.

12. Consider the equation $xy = 1$, where $x > 0$ and $y > 0$. Describe how the y-values are affected by increases or decreases in the x-values. How does the graph of the equation support your description?

In Exercises 13–20, determine the orientation of the hyperbola by indicating whether the transverse axis is horizontal or vertical.

13. $\frac{9x^2}{4} - 4y^2 = 1$ **14.** $x^2 - \frac{y^2}{36} = 1$

15. $\frac{4y^2}{9} - \frac{9x^2}{4} = 1$ **16.** $y^2 - x^2 = 1$

17. $9y^2 - x^2 = 9$ **18.** $x^2 = y^2 - 25$

19. $-4x^2 + y^2 = -36$ **20.** $9x^2 - 5y^2 - 45 = 0$

In Exercises 21–32, sketch the graph.

21. $\frac{x^2}{6} - y^2 = 1$ **22.** $9x^2 - 4y^2 = 4$

23. $y^2 - \frac{x^2}{25} = 1$ **24.** $y^2 = x^2 + 121$

25. $4x^2 - y^2 = 4$ **26.** $y^2 - 4x^2 = 16$

27. $\frac{y^2}{8} - x^2 = 1$ **28.** $2y^2 - 25x^2 = 50$

29. $x^2 = 12 + 3y^2$ **30.** $\frac{x^2}{9} - \frac{5y^2}{9} = 1$

31. $3y^2 - x^2 = 16$ **32.** $x^2 - 4y^2 = 20$

In Exercises 33–36, sketch the graph of the given relation.

33. $\frac{(x-3)^2}{16} - \frac{(y-4)^2}{25} = 1$

34. $\frac{(x+5)^2}{49} - \frac{(y+2)^2}{9} = 1$

35. $25(x-1)^2 - 16(y+3)^2 = 400$

36. $16(x+5)^2 - 25(y-2)^2 = 100$

In Exercises 37–40, sketch the graph of the given relation.

37. $\frac{(y+5)^2}{9} - \frac{(x-6)^2}{4} = 1$

38. $\frac{(y-1)^2}{100} - \frac{(x-3)^2}{64} = 1$

39. $4(y-4)^2 - 9(x+2)^2 = 36$

40. $16(y+3)^2 - 25(x-5)^2 = 4$

In Exercises 41–48, determine the orientation of the hyperbola by indicating whether the transverse axis is horizontal or vertical.

41. $\frac{(y-4)^2}{9} - \frac{x^2}{4} = 1$

42. $\frac{(y+2)^2}{36} - \frac{(x-2)^2}{36} = 1$

43. $\frac{x^2}{36} - \frac{(y+2)^2}{16} = 1$

44. $\frac{(x+1)^2}{16} - \frac{(y-2)^2}{16} = 1$

45. $25(x+2)^2 - 9y^2 = 225$

46. $4(x-2)^2 - 36(y-3)^2 = 9$

47. $25x^2 - 25(y+1)^2 + 36 = 0$

48. $3(x+1)^2 = (y+4)^2 - 12$

In Exercises 49–56, sketch the graph.

49. $\frac{(x+4)^2}{169} - \frac{(y-5)^2}{144} = 1$

50. $\frac{(x+4)^2}{9} - \frac{(y-5)^2}{16} = 1$

51. $4(x+2)^2 - 9y^2 - 36 = 0$

52. $16x^2 - 9(y+2)^2 = 144$

53. $(y-1)^2 - 36(x-4)^2 = 36$

54. $9(y-2)^2 - 4x^2 = 1$

55. $3(y+1)^2 = (x-3)^2 + 9$

56. $36(y+6)^2 - 16(x+3)^2 = 25$

In Exercises 57–64, graph the nonstandard hyperbola.

57. $y = \frac{3}{x}$ **58.** $y = -\frac{1}{x}$

59. $xy = 7$ **60.** $xy = -3$

61. $2xy = 5$ **62.** $3xy = -4$

63. $x = -\frac{8}{y}$ **64.** $x = \frac{2}{y}$

In Exercises 65–74, identify the graph.

65. $y + 4x - 3 = 0$ **66.** $3x = 5y + 30$

67. $x^2 + 4y^2 - 1 = 0$

68. $2(x+3)^2 = 8 - y^2$

69. $x^2 + y^2 + 4y = 0$ **70.** $y^2 = 5 - x^2$

71. $x^2 = y - 6x + 3$ **72.** $y^2 + x - 4 = 0$

73. $25y^2 = 4x^2 + 100$ **74.** $16x^2 - y^2 = 25$

75. The models

$$\frac{x^2}{a^2} + \frac{y^2}{b^2} = 1 \quad \text{and} \quad \frac{x^2}{a^2} - \frac{y^2}{b^2} = 1$$

are quite similar. Explain the difference in their graphs.

76. Suppose the equations of two hyperbolas are

$$\frac{x^2}{36} - \frac{y^2}{9} = 1 \quad \text{and} \quad \frac{y^2}{9} - \frac{x^2}{36} = 1.$$

Think about the central rectangles and asymptotes of the hyperbolas. Without sketching the graphs, explain how you can tell that the hyperbolas do not intersect.

Real-Life Applications

77. To design a "sweep" for the lawn area in a park, a landscape architect uses the right branch of a hyperbola whose center is the origin of a rectangular coordinate system. The central rectangle of the hyperbola is 30 feet wide and 20 feet high. The architect knows that the focus $F(c, 0)$ of this branch of the hyperbola is located so that $c = \sqrt{a^2 + b^2}$. If he wants to place a historical marker at the focus of the hyperbola, how far (to the nearest foot) will it be from the origin?

78. An engineer uses the following functions to design two possible arches for a bridge.

$$y = \frac{2}{3}\sqrt{9 - x^2} \quad \text{and} \quad y = -\frac{2}{3}\sqrt{9 + x^2} + 4$$

All values for x and y are measured in tens of feet. Use your calculator to produce the graphs of the functions. What conic sections do these graphs represent? Compare the heights of the two arches. One arch is wider at the ground level (x-axis) than the other. Calculate the difference in their widths.

Challenge

79. For each of the following relations, predict the quadrants in which the branches of the graph are located. Verify your predictions by producing the graphs on a calculator.

(a) $xy = 1$ (b) $x^2y = 1$

(c) $xy^2 = 1$ (d) $x^2y^2 = 1$

80. Explain your reasoning in sketching the graph of $xy = 0$. Describe the graph.

In Exercises 81 and 82, use the method of completing the square to sketch the graph of the given relation.

81. $x^2 - y^2 + 4x + 3 = 0$

82. $y^2 - x^2 + 2y + 4x - 7 = 0$

11.5 SYSTEMS OF NONLINEAR EQUATIONS

The Substitution Method • The Addition Method • Real-Life Applications

The Substitution Method

In Chapter 5 we discussed methods for solving systems of linear equations. In this section we examine systems of *nonlinear* equations.

Definition of System of Nonlinear Equations

A **system of nonlinear equations** is a system of equations in which at least one equation is nonlinear.

We will focus on systems with two variables in which one of the equations is the equation of a conic section. A **solution** of such a system is an ordered pair of numbers that satisfies all equations simultaneously.

Algebraic methods for solving nonlinear systems are similar to the methods for solving linear systems. In the examples we illustrate the graphing method, the substitution method, and the addition method.

EXAMPLE 1

Solving a System of Nonlinear Equations with the Substitution Method

Use the substitution method to solve the following system of equations.

$$x^2 + y^2 = 25$$
$$4x - 3y = 0$$

Solution

By examining the graphs, we see that there are two points of intersection. (See Figure 11.36.) The estimated solutions are $(-3, -4)$ and $(3, 4)$.

Figure 11.36

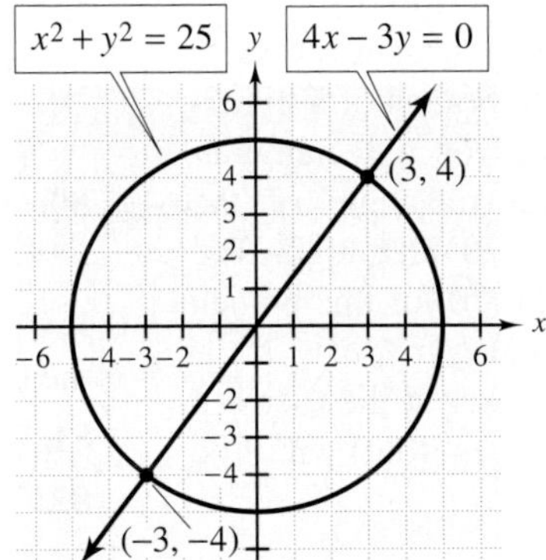

Think About It

Solve the equation $x^2 + y^2 = 25$, and use your calculator's default setting to graph the related functions. Does the graph look like a circle? If not, what can you do?

To solve the system algebraically, we solve the linear equation for y and substitute the resulting expression for y in the first equation.

$$x^2 + y^2 = 25$$
$$x^2 + \left(\frac{4}{3}x\right)^2 = 25 \qquad \text{Substitute } \tfrac{4}{3}x \text{ for } y.$$
$$x^2 + \frac{16}{9}x^2 = 25 \qquad \text{Simplify.}$$
$$9x^2 + 16x^2 = 225 \qquad \text{Clear the fraction.}$$
$$25x^2 = 225$$
$$x^2 = 9$$
$$x = \pm 3 \qquad \text{Square Root Property}$$

Substituting 3 and -3 for x in the second equation, we obtain corresponding y-values of 4 and -4, respectively. The two solutions of the system of equations are $(3, 4)$ and $(-3, -4)$.

Note: When writing solutions, be sure to match the x- and y-values that correspond to each other.

EXAMPLE 2

Solving a System of Nonlinear Equations with the Substitution Method

Solve the following system of equations.

$$\frac{x^2}{4} + \frac{y^2}{36} = 1$$
$$x = y^2 - 2$$

Solution

The graph of the first equation is an ellipse whose center is the origin, and the graph of the second equation is a horizontal parabola. (See Fig. 11.37.)

The graph suggests that the system has three solutions.

Figure 11.37

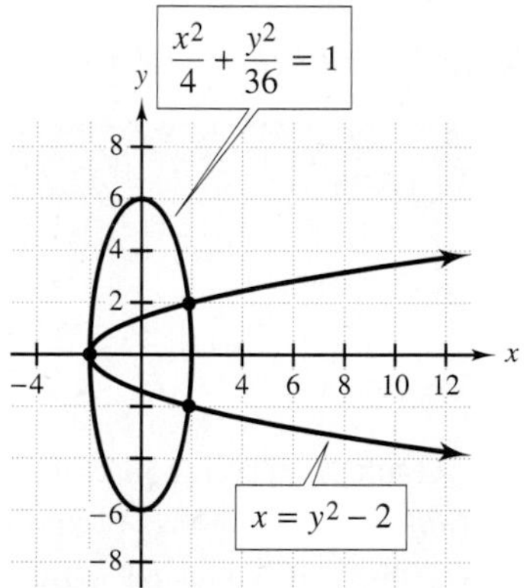

LEARNING TIP

The possible number of solutions of a nonlinear system can vary greatly. Even a rough sketch of the graphs will help you at least to anticipate the number of solutions that you should obtain.

The substitution method works well because the second equation is already solved for x. Replace x in the first equation with $y^2 - 2$.

$$\frac{x^2}{4} + \frac{y^2}{36} = 1$$

$$\frac{(y^2-2)^2}{4} + \frac{y^2}{36} = 1 \qquad \text{Replace } x \text{ with } y^2 - 2.$$

$$9(y^2-2)^2 + y^2 = 36 \qquad \text{Clear the fractions.}$$

$$9(y^4 - 4y^2 + 4) + y^2 = 36 \qquad \text{Square the binomial.}$$

$$9y^4 - 36y^2 + 36 + y^2 = 36 \qquad \text{Distributive Property}$$

$$9y^4 - 35y^2 = 0 \qquad \text{Simplify.}$$

$$y^2(9y^2 - 35) = 0 \qquad \text{Factor.}$$

$$y^2 = 0 \quad \text{or} \quad 9y^2 - 35 = 0 \qquad \text{Zero Factor Property}$$

$$y^2 = 0 \quad \text{or} \quad 9y^2 = 35$$

$$y^2 = 0 \quad \text{or} \quad y^2 = \frac{35}{9}$$

$$y = 0 \quad \text{or} \quad y = \pm\frac{\sqrt{35}}{3} \approx \pm 1.97 \qquad \text{Square Root Property}$$

To determine x, substitute these y-values into the second equation.

For $y = 0$:

$$x = y^2 - 2$$
$$x = 0^2 - 2 = -2$$

For $y \approx \pm 1.97$:

$$x = y^2 - 2$$
$$x \approx (\pm 1.97)^2 - 2 \approx 1.88$$

One exact solution is $(-2, 0)$. Two approximate solutions are $(1.88, -1.97)$ and $(1.88, 1.97)$.

The Addition Method

Sometimes systems cannot be solved quickly with substitution. Systems in which both equations appear in a similar form can sometimes be solved with the addition method.

EXAMPLE 3

Solving a System of Nonlinear Equations with the Addition Method

Solve the following system of equations.

$$\frac{x^2}{4} - \frac{y^2}{25} = 1$$
$$\frac{y^2}{9} - \frac{x^2}{16} = 1$$

Solution

The graph of each equation is a hyperbola whose center is the origin. The transverse axis of the first hyperbola is horizontal; the transverse axis of the second hyperbola is vertical. (See Fig. 11.38.)

Figure 11.38

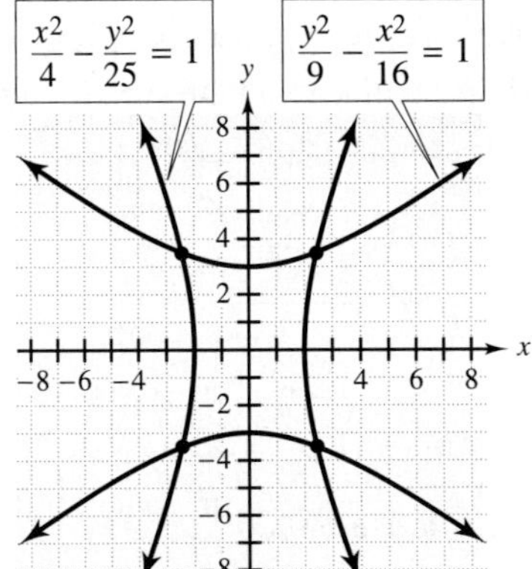

From the graph it appears that the system has four solutions.

To clear the fractions, multiply the first equation by 100 and the second equation by 144.

$$25x^2 - 4y^2 = 100$$
$$-9x^2 + 16y^2 = 144$$

To use the addition method for eliminating y, multiply the first equation by 4, and then add the equations.

$$\begin{aligned} 100x^2 - 16y^2 &= 400 \\ -9x^2 + 16y^2 &= 144 \\ \hline 91x^2 \qquad\quad &= 544 \\ x^2 &= \frac{544}{91} \\ x &= \pm\sqrt{\frac{544}{91}} \approx \pm 2.44 \end{aligned}$$

Square Root Property

Substitute these values for x in either equation and solve for y. Verify that $y \approx \pm 3.49$. The four approximate solutions are $(2.44, 3.49)$, $(2.44, -3.49)$, $(-2.44, 3.49)$, and $(-2.44, -3.49)$.

EXAMPLE 4

Solving a System of Nonlinear Equations with the Addition Method

Solve the following system of equations.

$$\begin{aligned} x^2 + y^2 &= 81 \\ 9x^2 + 4y^2 &= 36 \end{aligned}$$

Solution

The graph of the first equation is a circle, and the graph of the second equation is an ellipse. Each is centered at the origin. (See Fig. 11.39.) The graph indicates that the circle and ellipse have no intersection, and so we conclude that the system has no real number solution.

Figure 11.39

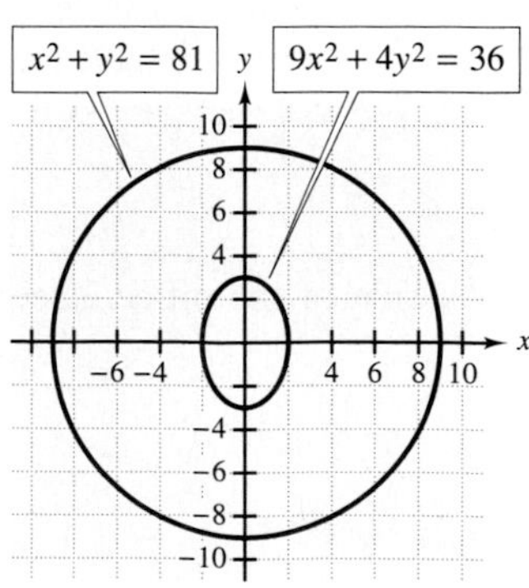

We can verify our conclusion algebraically. To solve with the addition method, we multiply the first equation by -4 and add the equations.

$$\begin{aligned} -4x^2 - 4y^2 &= -324 \\ 9x^2 + 4y^2 &= \quad 36 \\ \hline 5x^2 \qquad\quad &= -288 \\ x^2 &= -\frac{288}{5} \end{aligned}$$

Because this equation has no real number solution, the system of equations has no real number solution.

Although the system of equations in Example 4 has no real number solutions, the system does have complex number solutions. We can continue the solving process to determine the complex number solutions for x.

$$x^2 = -\frac{288}{5}$$

$$x = \pm i\sqrt{\frac{288}{5}} \approx \pm 7.59i$$

Now we substitute $-\frac{288}{5}$ for x^2 in the equation $x^2 + y^2 = 81$.

$$-\frac{288}{5} + y^2 = 81$$

$$y^2 = \frac{405}{5} + \frac{288}{5} = \frac{693}{5}$$

$$y = \pm\sqrt{\frac{693}{5}} \approx \pm 11.77$$

Thus the complex number solutions of the system are as follows:

$(7.59i, 11.77)$, $(7.59i, -11.77)$, $(-7.59i, 11.77)$, $(-7.59i, -11.77)$

Real-Life Applications

Systems of nonlinear equations may be used to model applications that involve two or more variables. In order to solve such application problems, sufficient information must be known to write two or more different equations in those variables.

EXAMPLE 5

Determining the Optimal Area of a Floor Plan

A small store is designed in the shape of a square. The store is to have an adjoining square office. The total area of the store and office is to be 1744 square feet. The perimeter of the combined rooms, not counting the interior wall, is to be 184 feet. What should be the dimensions of the store and office so that the store will have the largest possible area?

Solution

Figure 11.40

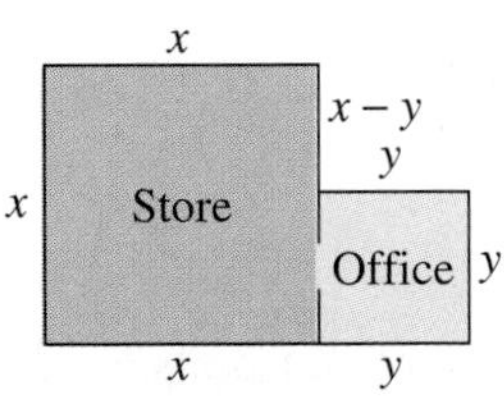

Let x = length of a side of the store and y = length of a side of the office. (See Fig. 11.40.)

The following system of equations models the information.

$x^2 + y^2 = 1744$ — The combined area must be 1744 square feet.

$2x + y = 92$ — The total distance around the outside walls must be 184 feet. Simplify $x + x + x + y + y + y + (x - y) = 184$.

Use the substitution method to solve the system.

$x^2 + (92 - 2x)^2 = 1744$ — Solve the second equation for y to obtain $y = 92 - 2x$, and substitute into the first equation.

$5x^2 - 368x + 6720 = 0$ — Simplify and write in standard form.

Use the Quadratic Formula to verify that the solutions for x are 33.6 and 40.

Although either value for x is permissible, we use $x = 40$ to make the store area as large as possible. The store dimensions are 40 feet on each side, and the office dimensions are $y = 92 - 2(40)$, or 12 feet on each side.

11.5 QUICK REFERENCE

The Substitution Method

- A **system of nonlinear equations** is a system of equations in which at least one equation is nonlinear.
- A **solution** of such a system is an ordered pair of numbers that satisfies all equations simultaneously.
- The graphing method suggests the number of real number solutions of a system and gives an estimate of the solutions.
- The substitution method for solving a nonlinear system of equations works best when one of the equations can be conveniently solved for one of the variables.

The Addition Method

- When the substitution method is not convenient, consider using the addition method to eliminate one of the variables.
- The addition method is particularly appropriate when no equation of a system contains a linear term.

11.5 EXERCISES

Concepts and Skills

1. Given the equation $ax^2 + by^2 = c^2$, what must be true about a and b in order for the graph of the equation to be
   (a) a circle?
   (b) an ellipse?
   (c) a hyperbola?

2. Describe the difference between the graphs of $y = ax + b$ and $y = ax^2 + b$. What information can we obtain from the value of a in each case?

In Exercises 3–16, identify the graph that is described by the given equation.

3. $9(x + 5)^2 + (y - 6)^2 = 9$
4. $2x^2 + 4x = 12y - 3y^2 + 17$
5. $x - 9y^2 = 9$
6. $y^2 + x - 2y + 3 = 0$
7. $9y^2 - 25x^2 = 225$
8. $4(y - 2)^2 - 9(x + 3)^2 = 144$
9. $3x + 25 - 5y = 0$
10. $\frac{x}{4} + \frac{y}{9} = 1$
11. $8(x - 6)^2 + 8(y + 3)^2 = 72$
12. $x^2 + y^2 - 4x + 6y - 12 = 0$
13. $x^2 - 4y^2 = 10$
14. $2x^2 + 4x = 3y^2 - 12y + 16$
15. $4(y + 4) - 25(x + 1)^2 = 100$
16. $x^2 + y - x + 2 = 0$

In Exercises 17–26, sketch an example of a graph of two conic sections with the indicated number of points of intersection.

17. two parabolas, 4 points
18. circle, ellipse, 4 points
19. ellipse, parabola, 3 points
20. ellipse, hyperbola, 0 points
21. two ellipses, 2 points
22. parabola, hyperbola, 3 points
23. circle, line, 0 points
24. parabola, line, 1 point
25. two circles, 2 points
26. ellipse, hyperbola, 4 points

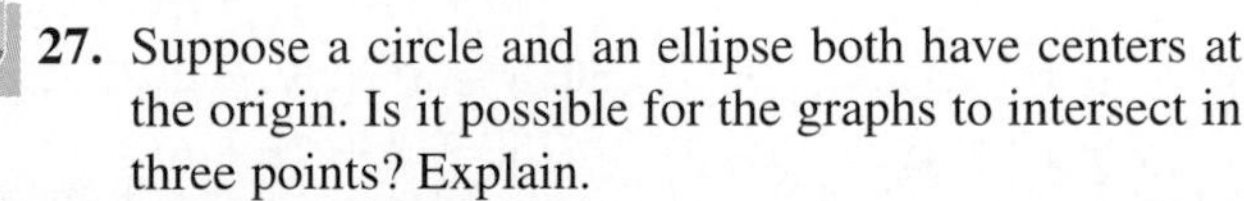

27. Suppose a circle and an ellipse both have centers at the origin. Is it possible for the graphs to intersect in three points? Explain.
28. What are two benefits from graphing a system of nonlinear equations?

In Exercises 29–36, use a graph to estimate the real number solution(s) of the given system of equations.

29. $x^2 + y^2 = 36$
$y - x = 6$

30. $x^2 + y^2 = 45$
$y = 2x$

31. $x^2 + y^2 = 16$
$y + x = 6$

32. $y = x^2 - 3x - 4$
$y + 2x = 2$

33. $xy = 6$
$2y = x + 4$

34. $\dfrac{x^2}{4} + \dfrac{y^2}{64} = 1$
$x^2 + y^2 = 36$

35. $4x^2 + 16y^2 = 64$
$9x^2 + 16y^2 = 1$

36. $9x^2 - 4y^2 = 36$
$4x^2 + 9y^2 = 36$

37. Under what conditions is the substitution method a convenient method for solving a system of nonlinear equations?

38. Suppose that you use the addition method to solve a system of nonlinear equations. After eliminating one variable, you find that the resulting equation has no real number solution. Does this necessarily imply that the system has no solution? Why?

In Exercises 39–46, use the substitution method to solve the given system of equations. Round decimal solutions to the nearest hundredth.

39. $xy + 6 = 0$
$3y + x + 3 = 0$

40. $xy = 8$
$y - x = 2$

41. $y + x^2 + 4x + 3 = 0$
$y = 3x + 7$

42. $x + y^2 = 6y - 5$
$2y - x = 2$

43. $(x - 2)^2 + y^2 = 4$
$y + x = 4$

44. $x^2 - 2y^2 = -9$
$x^2 + y^2 = 18$

45. $y^2 = 25 - x^2$
$3x - 4y = 25$

46. $2y = x^2 + 6$
$x^2 + y^2 = 9$

In Exercises 47–62, solve the given system of equations. Round any decimal solutions to the nearest hundredth.

47. $x^2 + y^2 - 6 = 0$
$y = 2$

48. $4x^2 - y^2 = 20$
$x = 3$

49. $y = x^2 - 4$
$y = x^2 + 3x$

50. $y = x^2 + 3x - 2$
$y = x^2 + 7x + 6$

51. $x^2 + y^2 = 9$
$4x^2 - 9y^2 = 36$

52. $(x - 1)^2 + y^2 = 26$
$x^2 + y^2 = 25$

53. $y^2 + 4x^2 = 4$
$y = 2x^2 - 2$

54. $4xy = 5$
$y = 5x$

55. $x^2 - y^2 = 7$
$y = x^2 - 9$

56. $x^2 + y^2 = 36$
$y = 6 - x^2$

57. $y + 2x = 3$
$y = x^2 - 3x - 4$

58. $xy = 6$
$y = x^2 + x - 6$

59. $xy = 3$
$y = -3x$

60. $y = 2x^2 + 3$
$y = x^2 - 5$

61. $4y^2 - x^2 = 4$
$y + x^2 = 2$

62. $x^2 + y^2 = 49$
$36y^2 - 25x^2 = 900$

63. The sum of two numbers is 4, and the product is -21. Find the sum of the squares of the numbers.

64. A positive number exceeds a negative number by 3, and it exceeds the square of the negative number by 1. What are the two numbers?

65. A 56-yard wire is cut into two pieces. Each piece is bent into the shape of a square. The sum of the areas of the squares is 106 square yards. How long is each wire piece?

66. A farmer has 1000 yards of fencing to make a rectangular pasture for horses. The area of the pasture is 52,500 square yards. What is the length of the pasture?

Real-Life Applications

67. A motorist traveled from Boise to Idaho Falls, a distance of 260 miles. The motorist's average speed for that trip was reduced by 13 mph on the return trip, and so the time to make the return trip increased by 1 hour. What was the motorist's average speed and travel time on the initial trip?

68. A student borrowed money from a college student loan program and paid \$44.80 in simple interest for 1 year. In the following year, the interest rate rose by 2%. The student borrowed \$112 less that year, but she paid the same amount of interest. What was the original amount of the loan and the original rate of interest?

69. A manufacturer of deck chairs models revenue with $y = 26x - 0.2x^2$ and cost with $y = 4x + 174$, where x is number of chairs manufactured and y represents dollars. For what number of chairs does the manufacturer break even, and what is the revenue at the break-even point? (Round your answers to the nearest whole number.)

70. A surveyor has plats of two 30-acre parcels of land. One parcel is a triangle, and the other is a parallelogram. (See figure.) If one square inch on the plats is equivalent to one acre, determine the dimensions b and h on the plats.

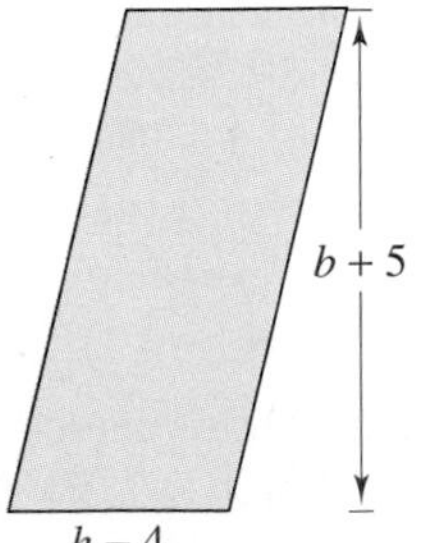

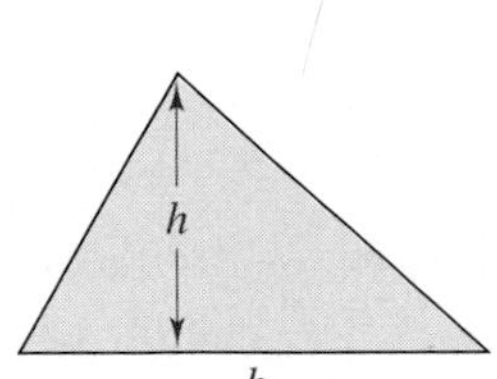

Group Project: Dressing for Work

The accompanying table shows the time spent by white-collar and professional workers to dress for work. (Source: NPD Group.)

Time (in minutes)	Men	Women
Less than 5	21%	6%
5–9	25%	18%
10–19	37%	40%
20 or more	18%	35%

The percentages P of men and women who spend t minutes dressing for work can be modeled by the following system of nonlinear equations.

Men: $P = -0.13t^2 + 3.73t + 9.05$

Women: $P = -0.14t^2 + 5.38t - 10.24$

71. Use the graphing method to estimate the solution of the system of equations.
72. Solve the system algebraically.
73. Interpret the solution.
74. Use the graph to estimate the vertex of each parabola. What does the vertex represent?

Challenge

75. The only point of intersection of a line and the parabola $y = x^2$ is $P(-2, 4)$. Determine the equation of the line.
76. A line whose equation is $2x + y = c$ intersects the circle $x^2 + y^2 = 16$ at just one point. Determine the value of c.
77. Use graphs to verify that neither of the following two systems of equations has a real number solution.

    (i) $x^2 + y^2 = 4$, $x^2 + y^2 = 1$ (ii) $x^2 + y^2 = 4$, $x = 3$

    Now use algebraic methods to show that one system also has no complex number solutions but that the other system has two complex number solutions.
78. A person uses a total of 38.85 feet of low picket fencing to edge a circular pond and a small, square patio next to it. The total area of the pond and the patio is 53.27 square feet. What is the radius of the pond? What is the area of the patio?

11.6 SYSTEMS OF INEQUALITIES

Nonlinear Inequalities • Systems of Inequalities • Real-Life Applications

Nonlinear Inequalities

In this section we will consider *nonlinear* inequalities in two variables. The methods for graphing their solution sets are similar to those used for linear inequalities.

LEARNING TIP

There is nothing in this section that is conceptually or procedurally different from systems of linear inequalities. We still graph the solid or dashed boundary lines and use test points to determine the region(s) to shade.

Procedure for Graphing a Nonlinear Inequality

1. Graph the corresponding equation.
2. Use a solid line for $\leq$ or $\geq$ and a broken line for $<$ or $>$.
3. Test points to determine the region or regions that contain the solutions.
4. Shade the region or regions that contain the solutions.

EXAMPLE 1

Graphing the Solution Set of a Nonlinear Inequality

Sketch the graph of $x^2 + y^2 < 25$.

Solution

The graph of $x^2 + y^2 = 25$ is a circle whose radius is 5 and whose center is the origin. We test the point $A(0, 0)$ in the interior of the circle and $B(7, 3)$ in the exterior of the circle.

$x^2 + y^2 < 25$		$x^2 + y^2 < 25$	
$0^2 + 0^2 < 25$	Test (0, 0).	$7^2 + 3^2 < 25$	Test (7, 3).
$0 < 25$	True	$58 < 25$	False

The graph of the solution set consists of all points inside the circle. (See Fig. 11.41.)

Figure 11.41

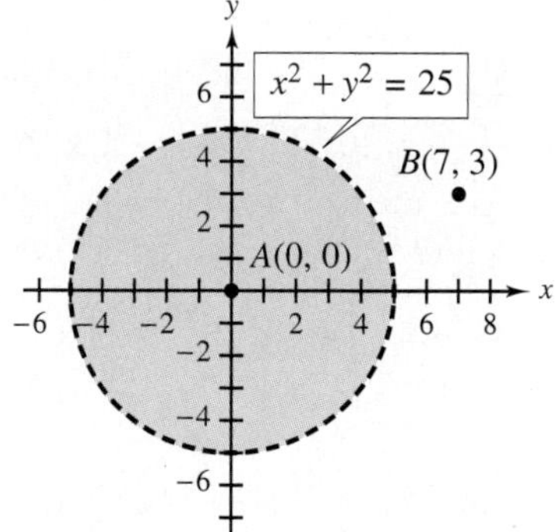

EXAMPLE 2

Graphing the Solution Set of a Nonlinear Inequality

Graph the inequality $4y^2 - 9x^2 \geq 36$.

Solution

The graph of $4y^2 - 9x^2 = 36$ is given in Fig. 11.42. Also shown are points in the three regions that are created by the hyperbola. We will test these points to determine whether they represent solutions of the inequality.

Figure 11.42

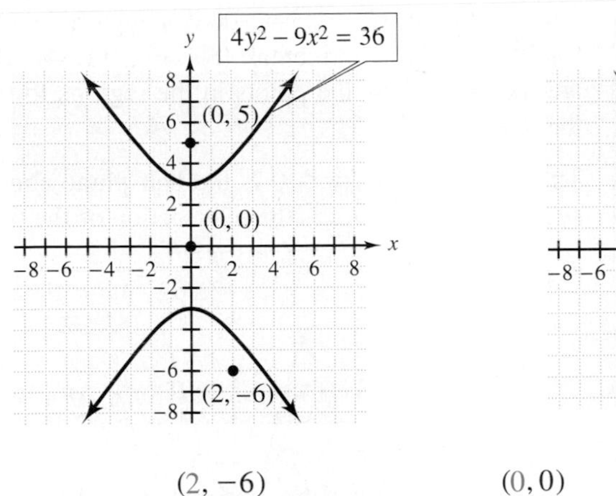

Figure 11.43

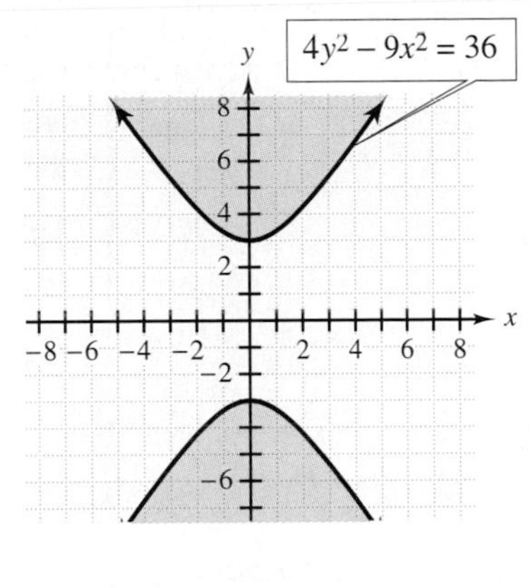

(2, −6)	(0, 0)	(0, 5)
$4y^2 - 9x^2 \geq 36$	$4y^2 - 9x^2 \geq 36$	$4y^2 - 9x^2 \geq 36$
$4(-6)^2 - 9(2)^2 \geq 36$	$4(0)^2 - 9(0)^2 \geq 36$	$4(5)^2 - 9(0)^2 \geq 36$
$108 \geq 36$	$0 \geq 36$	$100 \geq 36$
True	False	True

The test points indicate that the solutions are above and below the branches of the hyperbola. The points of the hyperbola also represent solutions. (See Fig. 11.43.)

Systems of Inequalities

When a problem involving inequalities requires that two or more conditions must be satisfied, a system of inequalities is needed.

If at least one inequality of a system is nonlinear, then we call the system a **system of nonlinear inequalities.** The **solution set** of such a system is the set of all ordered pairs of numbers that satisfy each inequality.

Methods for graphing solution sets of systems of nonlinear inequalities are similar to those used for systems of linear inequalities.

To Graph a System of Nonlinear Inequalities

1. Graph each inequality.
2. Determine the intersection of the graphs of the individual inequalities.

EXAMPLE 3

Graphing a System of Nonlinear Inequalities

Graph the following system of inequalities.

$$25x^2 - 4y^2 \geq 100$$
$$x - 2y \leq 0$$

Solution

The graph of $25x^2 - 4y^2 = 100$ is a hyperbola whose center is the origin and whose transverse axis is horizontal. (See Fig. 11.44.) The solutions of $25x^2 - 4y^2 \geq 100$ are represented by the points in the regions to the left and right of each branch of the hyperbola.

The graph of $x - 2y \leq 0$ is the half-plane above the line. The solution set of the system of inequalities is the intersection of the two shaded regions.

Figure 11.44

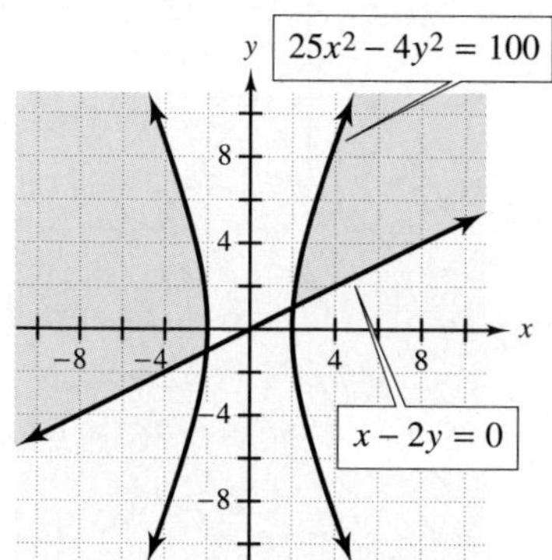

Think About It

Consider the following system of inequalities.

$(x - h)^2 + (y - k)^2 \leq 4$
$kx - hy \leq 0$

What is the area of the graph of the solution set?

Real-Life Applications

An application problem may require two or more inequalities to describe the conditions of the problem.

EXAMPLE 4

Possible Dimensions That Meet Fabrication Specifications

A rectangular piece of stainless steel is to be fabricated according to the following specifications.

(i) The area of the piece cannot exceed 50 square feet.
(ii) The length must be no more than 4 feet longer than the width.
(iii) The width can be no more than 7 feet.

Write a system of inequalities that describes the specifications. Then use a graph to represent the solution set of the system, that is, all possible dimensions that meet the specifications.

Solution

Let $x =$ the width of the rectangle and $y =$ the length of the rectangle. We translate the given specifications into a system of inequalities.

$xy \leq 50$	The area of the piece cannot exceed 50 square feet.
$y \leq x + 4$	Length is not more than 4 feet longer than the width.
$x \leq 7$	The width is no more than 7 feet.
$x \geq 0$	The width is nonnegative.
$y \geq 0$	The length is nonnegative.

The last two inequalities imply that we limit our region of interest to the first quadrant.

Figure 11.45

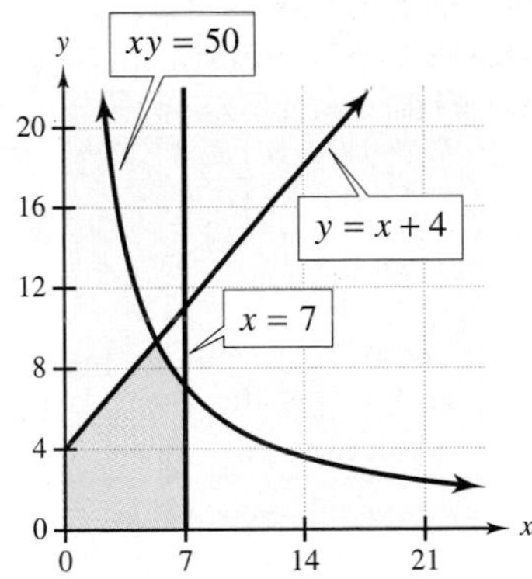

The solution set shown in Fig. 11.45 represents all the possible dimensions of the sheet of steel. For example, the point (4, 6) is in the shaded region. Therefore, a rectangular piece 4 feet wide and 6 feet long satisfies every condition.

11.6 QUICK REFERENCE

Nonlinear Inequalities

- To graph a nonlinear inequality, follow these steps:
  1. Graph the corresponding equation.
  2. Use a solid line for $\leq$ or $\geq$ and a broken line for $<$ or $>$.
  3. Test points to determine the region or regions that represent the solutions.
  4. Shade the region or regions that represent the solutions.

Systems of Inequalities

- Two or more inequalities, at least one of which is nonlinear, form a **system of nonlinear inequalities.** A **solution** of such a system is a pair of numbers that satisfies each inequality.
- To graph the solution set of a system of inequalities, follow these steps:
  1. Graph each inequality.
  2. Determine the intersection of the graphs of the individual inequalities.

11.6 EXERCISES

Concepts and Skills

1. Generally describe the graph of the solution set of the inequality

$$\frac{x^2}{a^2} + \frac{y^2}{b^2} < 1, \qquad a \neq b$$

2. Describe the difference between the graphs of the inequalities $y > x^2$ and $y \geq x^2$.

In Exercises 3–24, graph the solution set of the inequality.

3. $x^2 + 9y^2 > 36$

4. $9x^2 \geq 4 - y^2$

5. $x^2 < 49 - y^2$

6. $4x^2 + 4y^2 \geq 25$

7. $4x^2 - y^2 > 16$

8. $\frac{x^2}{25} - \frac{y^2}{9} \leq 1$

9. $y + 4x > x^2 + 3$

10. $x + 8 \leq y^2 + 2y$

11. $x^2 + y^2 \leq 36$

12. $3x^2 + 3y^2 < 27$

13. $x + y^2 \geq 4y$

14. $y < 3 - x^2$

15. $16y^2 > 64 - 4x^2$

16. $12x^2 + 6y^2 > 48$

17. $y^2 - x^2 > 1$

18. $9y^2 \leq 36 + 4x^2$

19. $y \geq 6 + 5x - x^2$

20. $x - y^2 > 0$

21. $(x + 4)^2 + y^2 \leq 16$

22. $(x - 3)^2 > 1 - (y + 1)^2$

23. $xy > 3$

24. $y \leq -\dfrac{2}{x}$

25. Without graphing, explain how you can tell that the following system of inequalities has no solution.

$x^2 + y^2 \leq 9$
$y \geq 4$

26. Consider the following system of inequalities.

$y \leq x^2 + 1$
$x < 0$
$y > 0$

Without graphing, explain how you can tell that the solution set (if it is not empty) is represented by points in Quadrant II.

In Exercises 27–38, graph the system of inequalities.

27. $x^2 + y^2 \geq 25$
$3y + 4x + 25 \geq 0$

28. $16y^2 \leq 144 - 9x^2$
$3y \leq x + 4$

29. $4y^2 - 9x^2 \geq 36$
$y \leq 3x - 1$

30. $x > y^2 - 2y - 8$
$y < 2x - 5$

31. $x^2 + y^2 \geq 1$
$x^2 + y^2 \leq 25$

32. $x^2 + y^2 > 10$
$x^2 + 9y^2 < 4$

33. $2x^2 + y^2 \geq 8$
$y^2 - x^2 \leq 1$

34. $y \geq 2x^2 - 11$
$x^2 + y^2 < 64$

35. $y \leq x^2 + x - 1$
$y \geq x^2 + 4x + 3$

36. $x^2 + y^2 \geq 9$
$x \leq 20 + y - y^2$

37. $y^2 \geq x^2 + 16$
$\dfrac{y^2}{25} + \dfrac{x^2}{4} \leq 1$

38. $xy < -5$
$4y^2 - x^2 < 16$

In Exercises 39–44, graph the system of inequalities.

39. $x^2 + y^2 \geq 9$
$x \geq 0$
$y \geq 0$

40. $\dfrac{x^2}{5} + \dfrac{y^2}{25} \geq 1$
$x \leq 0$
$y \geq 0$

41. $x + y^2 < 3y - 4$
$x > 0$
$y \leq 0$

42. $\dfrac{x^2}{9} - y^2 > 1$
$x \geq 0$
$y \leq 0$

43. $x^2 + y^2 < 49$
$y + x > 0$
$y - x > 0$

44. $9x^2 - y^2 \leq 36$
$y < 3x$
$y \geq 0$

In Exercises 45–48, write an inequality or a system of inequalities to describe the given set of points.

45.

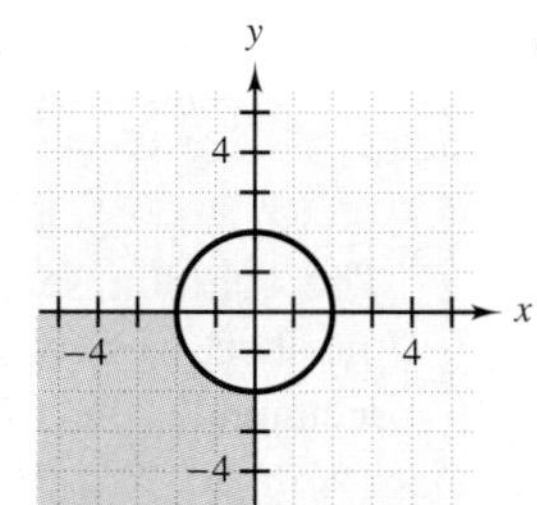

46.

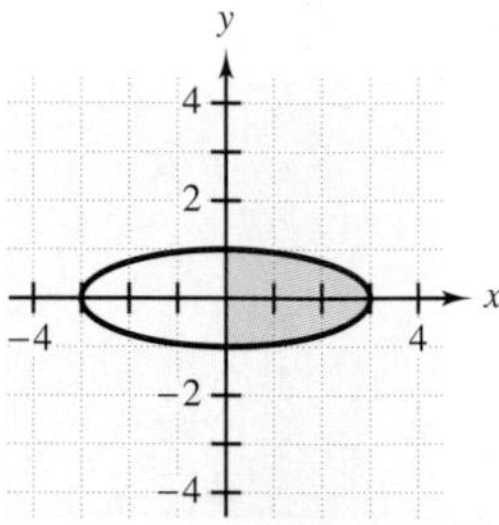

47. The points are in the interior of a parabola whose vertex is $V(0, 4)$. One of the two x-intercepts is $(16, 0)$.

48. The points are between the branches of a hyperbola with a vertical transverse axis. The midpoints of the central rectangle are $(0, \pm 5)$ and $(\pm 2, 0)$.

Real-Life Applications

In Exercises 49 and 50, write a system of inequalities that describes the given conditions. Then use a graph to represent the full set of possible solutions.

49. A drafting specialist is designing a square warehouse adjoined by four square offices. (See figure.) The total area of the warehouse and the offices cannot exceed 10,900 square feet. The area of each office must be at least 225 square feet.

50. A cabinetmaker is making tops for two square tables. The length of the larger walnut top is y and the length of the smaller oak top is x. The difference of the areas of the tops must exceed 10, and the sum of the areas must exceed 25.

51. An architect designed an art museum and performing arts center as two adjacent buildings bound by the curves $\frac{y^2}{4} - \frac{x^2}{9} \geq 1$, $\frac{y^2}{25} - \frac{x^2}{36} \leq 1$, $x \leq 8$, and $x \geq -8$. Sketch the plan for the two buildings.

52. The floor plan for an amphitheater is bounded by the curves $\frac{x^2}{16} - \frac{y^2}{49} \leq 1$, $x \leq 100 - y^2$, and $x \geq 0$. Sketch the floor plan.

Challenge

In Exercises 53 and 54, graph the solution set of the given system of inequalities.

53. $x^2 + y^2 \leq 25$
$|x| \geq 3$

54. $\frac{y^2}{4} - \frac{x^2}{4} \leq 1$
$y \geq |x|$

In Exercises 55 and 56, write a system of inequalities whose solution set is represented by the given graph.

55.

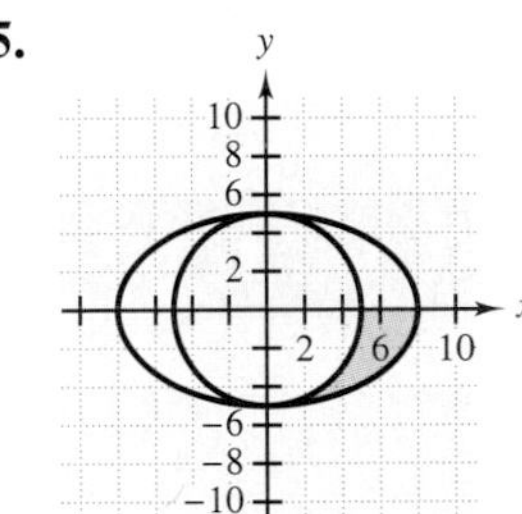

56.

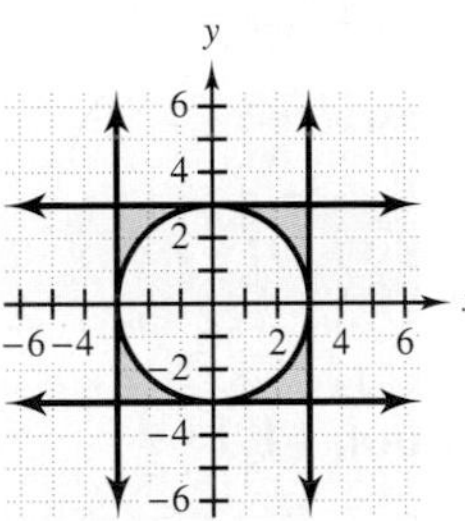

11. CHAPTER PROJECT Funding for the Arts

In the Group Project for Section 11.1 we saw that the annual appropriation A (in millions of dollars) for the National Endowment for the Arts (NEA) can be modeled by $A(x) = -1.41x^2 + 5.37x + 170.68$, where x is the number of years since 1990.

A companion federal agency, the National Endowment for the Humanities (NEH), has received appropriations H (in millions of dollars) that can be modeled by $H_1(x) = 5.9x + 143.4$, where x is the number of years since 1990.

1. For function A, what does the coefficient of x^2 indicate, and how can that information be interpreted with regard to NEA funding?
2. For function H_1, what does the coefficient of x indicate, and how can that information be interpreted with regard to NEH funding?
3. Using a suitable window, display the graphs of A and H_1 in the same coordinate system. Interpret the point of intersection of the graphs in Quadrant I.
4. Suppose that Congress had decided to model NEH funding with a parabola whose vertex was the point in Exercise 3 and whose y-intercept was the same as that for function H_1. Write the new model function in the form $H_2(x) = a(x - h)^2 + k$.
5. Add the graph of H_2 to the graphs of A and H_1. Verify that the graph of H_2 contains the point of intersection in Exercise 3 and the y-intercept of H_1. If Congress were to allow funding for the NEA and the NEH to expire completely, according to the model functions A and H_2, which of the two agencies would close its doors first? Which points of the graphs support your conclusion?
6. Opponents of the NEA argue that a federal subsidy of the arts is nothing more than a welfare program for artists. However, a 1994 economic impact study estimates that the nonprofit arts industry generates:

   1.3 million jobs

   $36.8 billion in expenditures

   $790 million in local tax revenues

$1.2 billion in state tax revenues

$3.4 billion in federal tax revenues

(*Source*: National Assembly of Local Arts Agencies)

Would you say that the federal government is subsidizing the arts, or are the revenues from the arts subsidizing other federal programs?

11. CHAPTER REVIEW EXERCISES

Section 11.1

In Exercises 1–6, determine the intercepts, the vertex, and the equation of the axis of symmetry. Sketch the graph.

1. $y = x^2 - 7x$ **2.** $x = y^2 - 7y$

3. $x = 6 + y - 2y^2$ **4.** $y = 6 + x - 2x^2$

5. $x = 2y^2 + 6y + 7$ **6.** $y = 6x - 7 - 2x^2$

In Exercises 7–10, determine the equation of the parabola that satisfies the given conditions.

7. The intercepts are $(0, -10)$, $(2, 0)$, and $(5, 0)$.

8. The intercepts are $(10, 0)$, $(0, 2)$, and $(0, 5)$.

9. The range is $\{y \mid y \geq 0\}$, the x-intercept is $(4, 0)$, and the parabola contains the point $(5, 4)$.

10. The domain is $\{x \mid x \leq 0\}$, the y-intercept is $(0, 5)$, and the parabola contains the point $(-2, 4)$.

11. If a parabola has intercepts $(0, 5)$, $(5, 0)$, and $(0, -11)$, determine the equation of the axis of symmetry.

In Exercises 12 and 13, determine the domain and range of the relation.

12. $y^2 + 5y = 6 + x$

13. $2x^2 + 7x = 15 + y$

Section 11.2

In Exercises 14–17, determine the center and radius of the circle.

14. $x^2 + (y - 2)^2 = 81$

15. $x^2 + y^2 - 8x + 10y + 5 = 0$

16. $x^2 + 4x + y^2 + 2y + 5 = 0$

17. $x^2 + y^2 + 2x - 6y + 14 = 0$

In Exercises 18–20, write an equation of a circle with the given conditions.

18. Center is $(-3, 5)$; radius is 5.

19. Center is $(3, -6)$; radius is $\sqrt{6}$.

20. Center is the origin; circle contains $(-3, 5)$.

Section 11.3

In Exercises 21 and 22, determine the intercepts of the ellipse.

21. $\dfrac{x^2}{49} + \dfrac{y^2}{64} = 1$ **22.** $25x^2 + 36y^2 = 900$

In Exercises 23 and 24, determine the endpoints of the axes of the ellipse and sketch the graph.

23. $\dfrac{(x + 4)^2}{81} + \dfrac{(y - 3)^2}{64} = 1$

24. $25(x - 2)^2 + 9(y + 5)^2 = 225$

In Exercises 25 and 26, write an equation of an ellipse with the following conditions.

25. Center is the origin, and intercepts are $(\pm 5, 0)$ and $(0, \pm 8)$.

26. Center is $(-4, -3)$, a vertex is $(1, -3)$, and a co-vertex is $(-4, 1)$.

Section 11.4

In Exercises 27 and 28, determine the intercepts and sketch the central rectangle, the asymptotes, and the hyperbola.

27. $\dfrac{x^2}{36} - \dfrac{y^2}{49} = 1$ **28.** $9y^2 - 25x^2 = 225$

In Exercises 29 and 30, determine the center and sketch the central rectangle, the asymptotes, and the hyperbola.

29. $\dfrac{(y + 6)^2}{49} - \dfrac{(x - 7)^2}{36} = 1$

30. $36(x - 5)^2 - 9(y - 4)^2 = 324$

Section 11.5

In Exercises 31–34, identify the conic section.

31. $x^2 - 4y^2 = 64$

32. $16x^2 + 4y^2 = 64$

33. $4x^2 + 4y^2 = 64$

34. $4x + 4y^2 = 64$

In Exercises 35–37, solve the system of equations.

35. $x^2 - y^2 = 5$
$y = x^2 - 7$

36. $y^2 + 5x^2 = 8$
$x^2 + y^2 = 11$

37. $y = x^2 - 5x - 6$
$y = x - 2$

Section 11.6

In Exercises 38 and 39, graph the solution set for the inequality.

38. $9x^2 - y^2 \geq 36$

39. $x > y^2 - 5y - 6$

In Exercises 40 and 41, graph the given system of inequalities.

40. $y^2 + 5x^2 \geq 8$
$x^2 + y^2 \leq 11$

41. $x \geq y^2 - y - 12$
$y + 2x < 1$

LOOKING AHEAD

The following exercises review concepts and skills that you will need in Chapter 12.

In Exercises 1 and 2, evaluate the expression for the indicated values of the variable.

1. n^{-2}; 3, 5

2. $\dfrac{1 + (-1)^n}{n}$; 4, 7

In Exercises 3 and 4, use the formula $b = a + (n - 1)d$ to determine the indicated value.

3. b if $a = -2$, $n = 7$, and $d = -2$

4. d if $a = 12$, $b = -13$, and $n = 6$

In Exercises 5–7, evaluate each formula as indicated.

5. $b = ar^{n-1}$ for $a = 4$, $r = -3$, and $n = 5$

6. $S = \dfrac{n(a + b)}{2}$ for $n = 50$, $a = 7$, and $b = 35$

7. $S = \dfrac{a(1 - r^n)}{1 - r}$ for $a = 5$, $r = \dfrac{1}{2}$, and $n = 3$

8. Solve the equation $r^5 = -243$.

9. Solve the following system of equations.

$13 = a + 2d$
$33 = a + 7d$

In Exercises 10–12, multiply and simplify.

10. $(c + 4)^2$

11. $(x - 2)^3$

12. $(a + b)^4$

11. CHAPTER TEST

In Questions 1–4, the graph of each equation is a conic section. Identify the conic section.

1. $x - y^2 - 3y + 2 = 0$

2. $x^2 - 3x + y^2 = 0$

3. $9x^2 - 25y^2 - 225 = 0$

4. $9x^2 + 16y^2 - 144 = 0$

In Questions 5 and 6, determine the vertex and axis of symmetry of the parabola.

5. $x = (y + 2)^2 + 1$

6. $y = -x^2 - 3x + 2$

In Questions 7 and 8, determine the intercepts of the parabola and sketch the graph.

7. $x = 8 - 2y^2$

8. $y = 3x^2 - 9x$

In Questions 9 and 10, determine the equation of the parabola that satisfies the given conditions.

9. vertex $(2, -1)$, x-intercept $(-1, 0)$, horizontal axis of symmetry

10. vertex $(-3, 2)$, contains point $(-2, 5)$, no x-intercept

In Questions 11 and 12, graph the circle.

11. $9x^2 + 9y^2 = 49$

12. $(x + 2)^2 + (y - 6)^2 = 25$

In Questions 13 and 14, write the equation of the circle that satisfies the given information.

13. center $(0, 0)$, contains the point $(-1, 3)$

14. center $(1, -3)$, radius 4

In Questions 15 and 16, graph the ellipse.

15. $x^2 + 9y^2 = 9$

16. $\dfrac{(x-5)^2}{49} + \dfrac{(y+3)^2}{25} = 1$

17. Write an equation of the ellipse whose center is (0, 0) and whose intercepts are (−4, 0) and (0, 2).

In Questions 18 and 19, graph the hyperbola.

18. $x^2 - 25y^2 = 25$

19. $\dfrac{y^2}{9} - \dfrac{x^2}{4} = 1$

In Questions 20 and 21, solve the system of equations.

20. $x^2 + y^2 = 9$
$y = x^2 - 3$

21. $2x^2 + y^2 = 16$
$x^2 - y^2 = -4$

22. Graph the inequality $4x^2 - 9y^2 \geq 36$.

23. Graph the solution set of the following system of inequalities.

$$y \leq 4 - x^2$$
$$4x^2 + 9y^2 \geq 36$$

Write a system of equations describing the following information. Then solve the system and answer the question.

24. A used-car lot is in the shape of a right triangle. If 300 feet of fencing is required to enclose the lot and the length of the hypotenuse is 130 feet, what is the area of the lot?

Chapter 12

Sequences and Series

A **sequence** is a list of numbers (called **terms**), and a **series** is the sum of the terms of a sequence. Some sequences, such as an **arithmetic sequence** and a **geometric sequence,** follow a pattern. A well-known sequence is the *Fibonacci sequence* 1, 1, 2, 3, 5, 8, 13, . . . , in which the first two terms are 1 and each succeeding term is the sum of the two previous terms. Fibonnaci numbers often occur in nature. For example, pine cones, daisies, and ears of corn exhibit growth patterns that can be modeled by the Fibonacci sequence. In this chapter, we consider the reproductive patterns of rabbits and bees as other examples of the Fibonacci sequence. (For more on this topic, see Exercises 83–86 at the end of Section 12.1, and see the Chapter Project.)

In our final chapter we study the general topics of sequences and series, and then we focus on two particular types: arithmetic and geometric sequences and series. We conclude the chapter with a discussion of expanding binomials with the Binomial Theorem.

12.1 SEQUENCES

Terms of a Sequence • Factorials • Increasing, Decreasing, and Alternating Sequences • The General Term of a Sequence

Terms of a Sequence

In mathematics, a **sequence** is a list of numbers called **terms.** A sequence is arranged in a particular order with a first term, a second term, and so on. For instance, the daily high temperature in Cleveland for a 4-day period is an example of a finite sequence with four terms: 12, 8, 10, 9. The positive odd integers constitute an infinite sequence: 1, 3, 5, 7, ...

More precisely, a sequence is defined as a special type of function.

Infinite and Finite Sequences

An **infinite sequence** is a function whose domain is the set of natural numbers. A **finite sequence** is a function whose domain is the first n natural numbers.

Whether a sequence is infinite or finite is usually clear from the context. Thus we usually refer to either type as simply a *sequence*.

EXPLORING THE CONCEPT

The Terms of a Sequence

For the sequence 4, 8, 12, 16, . . . , the elements of the domain and range are as follows:

Domain:	1	2	3	4	5	. . .	n	. . .
	↓	↓	↓	↓	↓		↓	
Range:	4	8	12	16	20	. . .	$4n$	. . .

LEARNING TIP

A *general term* is simply a formula for calculating any term of a sequence.

Because a sequence is a function, it can be described with function notation. For example, the preceding sequence can be defined by $a(n) = 4n$. We call $4n$ the **general term** or ***n*th term.** Thus the first three terms of the sequence $a(n) = 4n$ are as follows:

$$a(1) = 4(1) = 4 \qquad a(2) = 4(2) = 8 \qquad a(3) = 4(3) = 12$$

Conventionally, the terms of a sequence are identified with subscripts rather than function notation:

$$a_1 = 4, \quad a_2 = 8, \quad \ldots, \quad a_{10} = 40, \quad \ldots, \quad a_n = 4n$$

For a finite sequence, such as 6, 7, 8, 9, we simply restrict n: $b_n = n + 5$, where $1 \leq n \leq 4$.

EXAMPLE 1

Writing the Terms of a Finite Sequence

Write all the terms of the sequence $a_n = \frac{n}{n+1}, 1 \leq n \leq 6$.

Solution

The sequence has six terms.

$$a_1 = \frac{1}{1+1} = \frac{1}{2} \qquad a_2 = \frac{2}{2+1} = \frac{2}{3} \qquad a_3 = \frac{3}{3+1} = \frac{3}{4}$$

$$a_4 = \frac{4}{4+1} = \frac{4}{5} \qquad a_5 = \frac{5}{5+1} = \frac{5}{6} \qquad a_6 = \frac{6}{6+1} = \frac{6}{7}$$

The sequence is $\frac{1}{2}, \frac{2}{3}, \frac{3}{4}, \frac{4}{5}, \frac{5}{6}, \frac{6}{7}$.

EXAMPLE 2

Writing the Terms of an Infinite Sequence

(a) Write the first four terms of the sequence $a_n = 2n - 7$.
(b) Write the fifth term of $b_n = (-2)^n$.

Solution

(a) $a_1 = 2(1) - 7 = -5 \qquad a_2 = 2(2) - 7 = -3$ Replace n with 1, 2, 3, and 4.

$a_3 = 2(3) - 7 = -1 \qquad a_4 = 2(4) - 7 = 1$

The first four terms are $-5, -3, -1$, and 1.

(b) $b_5 = (-2)^5 = -32$ For the fifth term, $n = 5$.

Factorials

Many important sequences involve **factorials.**

Definition of Factorial

For a positive integer n, the number $\boldsymbol{n!}$ (which is read $\boldsymbol{n}$ **factorial**) is the product of the integers 1 through n: $n! = n(n-1)(n-2) \cdot \cdot \cdot 2 \cdot 1$. The factorial 0! is defined to be 1.

 Factorial

We can use a calculator to evaluate factorials.

Here are some examples of factorials.

$$2! = 2 \cdot 1 = 2$$

$$3! = 3 \cdot 2 \cdot 1 = 6$$

$$4! = 4 \cdot 3 \cdot 2 \cdot 1 = 24$$

$$7! = 5040$$

$$10! = 3{,}628{,}800$$

Figure 12.1

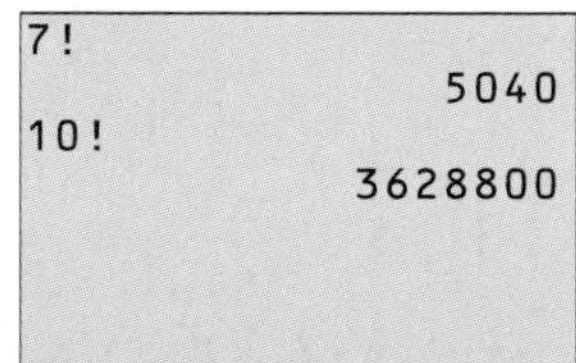

EXAMPLE 3

Writing the Terms of a Sequence Involving Factorials

Write the first four terms of each sequence.

(a) $a_n = (n - 1)!$ (b) $b_n = \dfrac{(n + 1)!}{n!}$

Solution

(a) $a_n = (n - 1)!$

$a_1 = (1 - 1)! = 0! = 1$ $a_2 = (2 - 1)! = 1! = 1$

$a_3 = (3 - 1)! = 2! = 2$ $a_4 = (4 - 1)! = 3! = 6$

The first four terms of the sequence are 1, 1, 2, and 6.

(b) $b_n = \dfrac{(n + 1)!}{n!}$

$b_1 = \dfrac{2!}{1!} = \dfrac{2 \cdot 1}{1} = 2$ $b_2 = \dfrac{3!}{2!} = \dfrac{3 \cdot 2 \cdot 1}{2 \cdot 1} = 3$

$b_3 = \dfrac{4!}{3!} = \dfrac{4 \cdot 3 \cdot 2 \cdot 1}{3 \cdot 2 \cdot 1} = 4$ $b_4 = \dfrac{5!}{4!} = \dfrac{5 \cdot 4 \cdot 3 \cdot 2 \cdot 1}{4 \cdot 3 \cdot 2 \cdot 1} = 5$

The first four terms of the sequence are 2, 3, 4, and 5.

Increasing, Decreasing, and Alternating Sequences

The terms of a sequence may increase, decrease, or alternate in sign.

Definitions of Increasing, Decreasing, and Alternating Sequences

A sequence is an **increasing sequence** if $a_n \le a_{n+1}$ for each n. A sequence is a **decreasing sequence** if $a_n \ge a_{n+1}$ for each n. A sequence whose terms alternate in sign is an **alternating sequence.**

In words, a sequence is increasing if every term is greater than or equal to the preceding term. Similarly, a sequence is decreasing if every term is less than or equal to the preceding term.

Note: If for each n, $a_n < a_{n+1}$, we say that the sequence is *strictly increasing*. A sequence is *strictly decreasing* if $a_n > a_{n+1}$ for each n.

EXAMPLE 4

Identifying Increasing, Decreasing, and Alternating Sequences

(a) 3, 8, 13, 18, . . .	An increasing sequence
(b) 1, 2, −3, 4, 5, −6, 7, 8, . . .	The sequence is neither increasing, decreasing, nor alternating.
(c) 1, −3, 9, −27, . . .	An alternating sequence
(d) 4, 2, 0, −2, −4, −6, . . .	A decreasing sequence

The General Term of a Sequence

To determine an expression for the general term of a sequence, we must recognize the pattern of the terms of the sequence and express that pattern algebraically.

EXAMPLE 5

Writing the General Term of a Sequence

Write an expression for the general term of each sequence.

(a) 4, 5, 6, 7, 8, . . .

(b) $\frac{1}{3}, \frac{1}{9}, \frac{1}{27}, \frac{1}{81}, \frac{1}{243}, \ldots$

(c) 1, 0, −1, −2, −3, . . .

Solution

(a) $a_1 = 4 = 1 + 3 \qquad a_2 = 5 = 2 + 3 \qquad a_3 = 6 = 3 + 3$

The pattern suggests that the general term may be $a_n = n + 3$.

(b) $b_1 = \frac{1}{3} = \frac{1}{3^1} \qquad b_2 = \frac{1}{9} = \frac{1}{3^2} \qquad b_3 = \frac{1}{27} = \frac{1}{3^3}$

The pattern suggests that the general term may be $b_n = \frac{1}{3^n}$.

(c) $c_1 = 1 = 2 - 1 \qquad c_2 = 0 = 2 - 2 \qquad c_3 = -1 = 2 - 3$

The pattern suggests that the general term may be $c_n = 2 - n$.

Think About It

In words, what is the meaning of $a_n = a_{n-1} + a_{n-2}$?

12.1 QUICK REFERENCE

Terms of a Sequence

- A **sequence** is a list of numbers called **terms.**
- An **infinite sequence** is a function whose domain is the set of natural numbers.

A **finite sequence** is a function whose domain is the first n natural numbers.

- The **general** or ***n*th term** is a formula for determining any term of a sequence.

Factorials

- For a positive integer n, the number $n!$ (which is read ***n* factorial**) is the product of the integers 1 through n. The factorial 0! is defined to be 1.

Increasing, Decreasing, and Alternating Sequences

- The following are special sequences:
  1. **Increasing sequence:** $a_n \le a_{n+1}$ for each n
  2. **Decreasing sequence:** $a_n \ge a_{n+1}$ for each n
  3. **Alternating sequence:** terms alternate in sign

The General Term of a Sequence

- To write an expression for the general term of a sequence, determine the pattern of the terms, and express that pattern as a formula.

12.1 EXERCISES

Concepts and Skills

1. What names are given to the following two sequences? Describe the difference between these sequences.

(a) 0, 1, 2, 3, . . .

(b) 0, 1, 2, 3

2. Explain how to determine the domain of a finite sequence of n terms.

In Exercises 3–8, predict the next two terms.

3. 5, 8, 11, 14, . . .

4. 5, 3, 1, −1, −3, . . .

5. 5, 7, 6, 8, 7, 9, 8, 10, . . .

6. 2, −1, 3, −2, 4, −3, 5, −4, . . .

7. 3, 9, 27, 81, . . .

8. 3, 4, 7, 11, 18, . . .

In Exercises 9–18, write the first four terms of the sequence.

9. $a_n = 3^n$ **10.** $c_n = 1 - 2n$

11. $b_n = n^2$ **12.** $k_n = n^n$

13. $c_n = 1 + (-1)^n$ **14.** $a_n = (-1)^{n-1}n$

15. $a_n = (n + 2)!$

16. $b_n = (n - 1)(n - 2)(n - 3)$

17. $b_n = \dfrac{(-1)^n}{n}$ **18.** $k_n = \dfrac{n^2}{n!}$

19. Suppose the general term of a sequence is represented by the expression a_n. What is the meaning of the letter n?

20. What do we call the sequence $a_n = (-1)^n$? Why?

In Exercises 21–28, write all the terms in the sequence.

21. $a_n = \dfrac{n}{2 + n}, \quad 1 \le n \le 5$

22. $c_n = n^{-2}, \quad 1 \le n \le 5$

23. $a_n = \sqrt{n - 1}, \quad 1 \le n \le 7$

24. $a_n = (-1)^{2n}, \quad 1 \le n \le 7$

25. $k_n = \dfrac{3^n}{n^3}, \quad 1 \le n \le 4$

26. $c_n = n^2 + (n - 1)^n, \quad 1 \le n \le 4$

27. $k_n = (-1)^n n^2, \quad 1 \le n \le 6$

28. $b_n = \dfrac{(n + 1)!}{n!}, \quad 1 \le n \le 4$

In Exercises 29–36, for the sequence whose nth term is given, determine the indicated term.

29. $a_n = 5 - 3n; \quad a_{25}$

30. $b_n = 2(n + 1)^2 + 3n; \quad b_{14}$

31. $k_n = \dfrac{1 + (-1)^n}{n}; \quad k_5$ **32.** $c_n = \dfrac{n^2 - 4}{n + 3}; \quad c_7$

33. $c_n = \frac{\sqrt{n}}{n}$; c_{100}

34. $k_n = 8\left(\frac{1}{2}\right)^{n-1}$; k_5

35. $a_n = \frac{(n+5)!}{n!4!}$; a_3

36. $b_n = \frac{2^n}{1+3^n}$; b_5

In Exercises 37–44, identify the sequence as increasing, decreasing, alternating, or none of these.

37. $-7, -3, 1, 5, 9, 13, \ldots$

38. $81, 27, 9, 3, 1, \frac{1}{3}, \frac{1}{9}, \ldots$

39. $81, -27, 9, -3, 1, -\frac{1}{3}, \frac{1}{9}, \ldots$

40. $7, 9, 6, 8, 5, 7, \ldots$

41. $2, 1, \frac{1}{2}, \frac{1}{4}, \frac{1}{8}, \ldots$

42. $1, 1, 2, 6, 24, 120, \ldots$

43. $1, 2, -3, 4, 5, -6, \ldots$

44. $1, -1, 2, -6, 24, -120, \ldots$

In Exercises 45–52, identify the sequence as increasing, decreasing, alternating, or none of these.

45. $k_n = 3n + 2$

46. $b_n = 1 - 5n$

47. $c_n = (-2)^{n-1}$

48. $a_n = (-1)^n n^2$

49. $k_n = \frac{12}{n}$

50. $b_n = n^2$

51. $k_n = 2 + (-1)^n$

52. $c_n = 1 - (-2)^n$

In Exercises 53–64, write an expression for the general term of the sequence.

53. $5, 6, 7, 8, 9, \ldots$

54. $9, 8, 7, 6, 5, \ldots$

55. $\frac{1}{2}, \frac{2}{3}, \frac{3}{4}, \frac{4}{5}, \ldots$

56. $\frac{\ln 2}{2}, \frac{\ln 3}{3}, \frac{\ln 4}{4}, \ldots$

57. $5, 7, 9, 11, 13, \ldots$

58. $15, 12, 9, 6, 3, \ldots$

59. $\frac{1}{2 \cdot 1}, \frac{2}{3 \cdot 2 \cdot 1}, \frac{3}{4 \cdot 3 \cdot 2 \cdot 1}, \ldots$

60. $e, \frac{e^2}{2}, \frac{e^3}{3 \cdot 2 \cdot 1}, \frac{e^4}{4 \cdot 3 \cdot 2 \cdot 1}, \ldots$

61. $1, -2, 4, -8, 16, \ldots$

62. $2, -2, 2, -2, 2, \ldots$

63. $-1, 4, -9, 16, \ldots$

64. $12, -4, \frac{4}{3}, -\frac{4}{9}, \ldots$

If we describe a sequence by giving the first term and stating the relationship between each term and its successor, the sequence is said to be defined **recursively.**

Example: Let $a_1 = 2$ and $a_n = a_{n-1} + 5$. In words, each term of the sequence (except the first) is 5 more than the previous term. The first five terms of the sequence are 2, 7, 12, 17, and 22.

In Exercises 65–72, for the sequence that has been defined recursively, write the first four terms.

65. $a_1 = 5, \quad a_n = a_{n-1} + 3$

66. $a_1 = 3, \quad a_n = a_{n-1} - 2$

67. $a_1 = 5, \quad a_n = 2a_{n-1}$

68. $c_1 = 8, \quad c_n = -\frac{1}{2}c_{n-1}$

69. $b_1 = 0, \quad b_n = 2b_{n-1} + 3$

70. $b_1 = 7, \quad b_n = (-1)^{n-1}b_{n-1}$

71. $a_1 = 1, \quad a_2 = 2, \quad a_n = a_{n-1}a_{n-2}$

72. $b_1 = 2, \quad b_2 = 4, \quad b_n = \frac{b_{n-1} + b_{n-2}}{2}$

73. (a) Show that the following are true.

$6! = 6 \cdot 5!$

$10! = 10 \cdot 9!$

(b) Use this information as a guide for writing an alternative expression for $n!$.

(c) What is $123!/122!$?

74. Simplify the given expressions.

(a) $\frac{(n+1)!}{n!}$ (b) $\frac{(n-1)!}{(n+1)!}$

75. Is the sequence 3, 3, 3, 3, . . . an increasing sequence, a decreasing sequence, or neither? Why?

76. Explain why an alternating sequence cannot be increasing or decreasing.

77. An expression for the general term of the sequence of positive odd integers is $a_n = 2n - 1$, where $n \geq 1$. What is the 53rd odd integer in the sequence?

78. Write an expression for the general term of the sequence of positive multiples of 7. Then find the term number of the term 581.

Real-Life Applications

79. A tourist accidentally dropped a camera from the top of the Hancock Center. The camera drops a total of $16t^2$ feet after t seconds (ignoring air resistance). Calculate the distance that the camera drops during each second, where $1 \leq t \leq 4$, and write the distances as a sequence. Then write an expression for the general term of the sequence and use it to determine the distance that the camera falls during the seventh second.

80. A school system estimates that its textbooks depreciate in value by 25% each year. For a book whose original value is \$40, write a sequence that lists the value of the book at the end of each of the first 4 years of use. Then write an expression for the general term of the sequence and use it to determine the number of years after which the value of the book will be less than \$5.

81. A person borrowed \$2000 and agreed to pay it back in monthly payments of \$200 plus interest of 1% of the unpaid balance. Write a sequence for the first five payments and write a simplified expression for the general term of the sequence. After 7 months how much of the loan remains to be paid?

82. According to an economist, 60% of all money that is spent in a certain community is respent in that community within a week. Suppose that a special event results in \$150,000 of tourist income for the community. Write the general term of a sequence that describes the amount of the tourist income that is respent at the end of week n.

Group Project: Birth Patterns of Rabbits

Suppose that a pair of rabbits will produce a new pair of rabbits in their second month and will continue to produce a new pair each month thereafter. Assuming that all the rabbits will do the same and that none of the rabbits die, we can determine the number of rabbits at the end of a given number of months.

In the accompanying table, N means newborn and R means ready to reproduce.

Number of Months	Status of Rabbit Pairs	Number of Pairs
Start	N	1
1	R	1
2	RN	2
3	RRN	3
4	$RRRNN$	5
5	$RRRRRNNN$	8
6		

83. Each R-pair in the fifth month produces an N-pair in the sixth month, and each N-pair in the fifth month becomes an R-pair in the sixth month. Complete the entries for the sixth month.

84. Extend the table to 11 months, and complete the entries for the added months.

85. The **Fibonacci sequence** is 1, 1, 2, 3, 5, 8, ... What evidence do you see in the table that the birth patterns of rabbits can be modeled by the Fibonacci sequence?

86. In the Fibonacci sequence, $a_1 = 1$ and $a_2 = 1$. Write a formula for a_n, where $n \geq 3$.

Challenge

87. Consider the following recursively defined sequence: $b_1 = 1, b_n = nb_{n-1}$. Generate terms of the sequence until you can detect its special pattern. Use factorial notation to write an alternative expression for the general term of the sequence.

88. In the following sequence, the first two terms are given, and the general term is defined recursively as the difference in the two previous terms.

$$a_1 = 5, \quad a_2 = 2, \quad a_n = a_{n-2} - a_{n-1}$$

Show that the sequence alternates after three terms.

89. Suppose $\log 10! = a$. Without using a calculator, show how to evaluate $\log 9!$ in terms of a.

90. Let $a_1 = 1$ and $a_2 = 1$. Write the recursive formula for the term a_n in the Fibonacci sequence.

91. Use the recursive formula found in Exercise 90 to determine the first ten terms of the Fibonacci sequence.

12.2 SERIES

Definition • Summation Notation • Writing a Series

Definition

A **series** is the sum of the terms of a sequence. Like a sequence, a series can be finite or infinite.

Definitions of Finite and Infinite Series

A **finite series** is the sum of the terms of a finite sequence. An **infinite series** is the sum of the terms of an infinite sequence.

The following are examples of sequences and their related series.

Sequence	*Series*	
2, 5, 8, 11	$2 + 5 + 8 + 11$	Finite
$1, \frac{1}{2}, \frac{1}{4}, \frac{1}{8}, \frac{1}{16}, \ldots$	$1 + \frac{1}{2} + \frac{1}{4} + \frac{1}{8} + \frac{1}{16} + \cdots$	Infinite

In this section we restrict our attention to finite series.

Summation Notation

To write the sum of a sequence, we use a notation called **summation notation.** The notation uses the Greek letter Σ (**sigma**) to represent the word *sum.*

The sum of the terms of the sequence

$$a_n = 3n - 1, \quad 1 \leq n \leq 4$$

is written with summation notation as

$$\sum_{i=1}^{4} (3i - 1)$$

This notation is read, "the sum of $(3i - 1)$ from $i = 1$ to 4."

The letter i is the **index** of the summation. The lower number 1 is the **lower limit,** and the higher number 4 is the **upper limit.** Letters other than i can also be used.

LEARNING TIP

The symbol $\sum_{i=1}^{4} (3i - 1)$ looks complicated, but it only means that we are to add the terms obtained by replacing i with 1, 2, 3, and 4.

To **expand** the series, we replace i with 1, 2, 3, and 4 to obtain the following:

$$\begin{aligned}\sum_{i=1}^{4} (3i - 1) &= (3 \cdot 1 - 1) + (3 \cdot 2 - 1) + (3 \cdot 3 - 1) + (3 \cdot 4 - 1) \\ &= 2 + 5 + 8 + 11 \\ &= 26\end{aligned}$$

EXAMPLE 1

Expanding a Series

(a) $\sum_{i=1}^{7} (2i)^i = (2 \cdot 1)^1 + (2 \cdot 2)^2 + (2 \cdot 3)^3 + \cdots + (2 \cdot 7)^7$
$= 2^1 + 4^2 + 6^3 + 8^4 + 10^5 + 12^6 + 14^7$

(b) $\sum_{i=1}^{5} \ln(i+2) = \ln 3 + \ln 4 + \ln 5 + \ln 6 + \ln 7$

To **evaluate** a series, we simply perform the indicated addition.

EXAMPLE 2

Evaluating a Series

(a) $\sum_{j=0}^{5} (2j) = 0 + 2 + 4 + 6 + 8 + 10 = 30$

(b) $\sum_{i=2}^{5} (-1)^i(i^2 - 2) = 2 - 7 + 14 - 23 = -14$

Think About It
If k is a constant, explain why $\sum_{i=1}^{n} k = nk$.

 Summation

By entering the general term and limits of the summation, you can use a calculator to evaluate a series. Figure 12.2 shows the results for the series in Example 2.

Figure 12.2

```
sum(seq(2J,J,0,5,
1))
                30
sum(seq((-1)^I(I²
-2),I,2,5,1))
               -14
```

Writing a Series

To write a series with summation notation, we determine the general term and the domain of the sequence.

EXAMPLE 3

Writing a Series in Summation Notation

Write each series in summation notation.

(a) $\frac{1}{4} + \frac{1}{5} + \frac{1}{6} + \cdots + \frac{1}{10}$

(b) $e + e^2 + e^3 + e^4 + e^5$

Solution

(a) $\sum_{n=1}^{7} \frac{1}{n+3}$ The general term is $a_n = \frac{1}{n+3}$, $1 \le n \le 7$.

(b) $\sum_{n=1}^{5} e^n$ The general term is $a_n = e^n$, $1 \le n \le 5$.

The same series can be represented in different ways. For example, the series $2 + 5 + 8 + 11$ can be described by both the following summation expressions.

$$\sum_{i=1}^{4} (3i - 1) = 2 + 5 + 8 + 11 = 26$$

$$\sum_{j=0}^{3} (3j + 2) = 2 + 5 + 8 + 11 = 26$$

When we change the lower limit of a sum, we must also change the expression for the general term. The variable in the expression must be *increased* by the same amount that the lower limit is *decreased*; the variable must be *decreased* by the same amount that the lower limit is *increased*. The following illustrates this process.

For the series $\sum_{i=1}^{4} (3i - 1)$, the index i has values from 1 to 4. To rewrite the series with index j having values from 0 to 3, note that the values of the index are decreased by 1 so that $j = i - 1$ or $i = j + 1$. To write the general term, replace i with $j + 1$.

$$3i - 1 = 3(j + 1) - 1 = 3j + 2$$

The series can be written $\sum_{j=0}^{3} (3j + 2)$.

EXAMPLE 4

Changing the Index of Summation

Rewrite the series $\sum_{i=0}^{4} \frac{i}{i + 1}$ with index j and lower limit 1.

Solution

Note that $i = j - 1$. Write the general term.

$$\frac{i}{i+1} = \frac{j-1}{(j-1)+1} = \frac{j-1}{j} \qquad \text{Replace } i \text{ with } j - 1.$$

Rewrite the series as $\sum_{j=1}^{5} \frac{j-1}{j}$.

12.2 QUICK REFERENCE

Definition

- A **series** is the sum of the terms of a sequence.
- A **finite series** is the sum of the terms of a finite sequence; an **infinite series** is the sum of the terms of an infinite sequence.

Summation Notation

- We use the Greek letter Σ (**sigma**) to represent a series.
- In the **summation notation** $\sum_{i=m}^{n} a_i$, the letter i is the **index** of the summation, and the numbers m and n are the **lower** and **upper limits.**
- To **expand** a series $\sum_{i=m}^{n} a_i$, we replace i with integers from m to n. To **evaluate** a series, we perform the indicated addition.

Writing a Series
- To write a series with summation notation, determine the general term and the domain of the sequence.
- The indices of a series can be changed to write the series in an alternative way. The variable in the general expression must be increased (decreased) by the same amount that the lower index is decreased (increased).

12.2 EXERCISES

Concepts and Skills

1. What is the difference between a *sequence* and a *series*?

2. What is the difference between *expanding* a series and *evaluating* a series?

In Exercises 3–8, expand the series.

3. $\sum_{k=0}^{4} (-2)^k$

4. $\sum_{i=3}^{7} \frac{1}{i+2}$

5. $\sum_{k=3}^{9} k(k-1)$

6. $\sum_{i=1}^{5} (-1)^{i+1} i^2$

7. $\sum_{k=1}^{5} \frac{k^2}{2k-1}$

8. $\sum_{k=0}^{5} \frac{(-1)^k}{k!}$

In Exercises 9–24, evaluate the series.

9. $\sum_{k=1}^{6} 5$

10. $\sum_{k=3}^{7} -3k^0$

11. $\sum_{j=1}^{6} j^2$

12. $\sum_{j=1}^{6} 2j$

13. $\sum_{k=3}^{5} \frac{1}{k}$

14. $\sum_{k=1}^{3} k^{-1}$

15. $\sum_{i=1}^{5} (2i-3)$

16. $\sum_{i=3}^{7} (3i+4)$

17. $\sum_{k=1}^{4} (k-1)(k-2)(k-3)$

18. $\sum_{k=1}^{5} (k-2)(k+1)$

19. $\sum_{k=0}^{3} 2\left(\frac{1}{3}\right)^k$

20. $\sum_{j=1}^{4} \frac{1}{3j}$

21. $\sum_{i=1}^{8} (-1)^{i+1}(i+1)$

22. $\sum_{k=0}^{5} \frac{(-1)^k}{k+1}$

23. $\sum_{j=3}^{6} (j^2+2j)$

24. $\sum_{j=1}^{4} \frac{j+1}{j}$

25. To write a series in summation notation, what two facts must be known about the related sequence?

26. For $\sum_{i=1}^{n} x^{2i} = x^2 + x^4 + \cdots + x^{18}$, explain how to determine the upper limit n.

In Exercises 27–36, write the series in summation notation.

27. $5 + 6 + 7 + \cdots + 15$

28. $3 + 8 + 13 + 18 + 23$

29. $1 + \sqrt{2} + \sqrt{3} + 2 + \sqrt{5} + \cdots + 3$

30. $1 + e + e^2 + e^3$

31. $11 + 13 + 15 + 17 + \cdots + 27$

32. $\frac{1}{2} + \frac{2}{5} + \frac{3}{8} + \frac{4}{11} + \frac{5}{14}$

33. $-1 + \frac{1}{2} - \frac{1}{3} + \frac{1}{4} - \cdots - \frac{1}{9}$

34. $\frac{3}{1+\pi} + \frac{4}{2+\pi} + \frac{5}{3+\pi} + \cdots + \frac{10}{8+\pi}$

35. $\frac{1}{3} - \frac{1}{2} + \frac{3}{5} - \frac{2}{3} + \frac{5}{7} - \frac{3}{4}$

36. $24 - 26 + 28 - 30 + 32 - 34$

37. Suppose you wish to change the lower limit of the series $\sum_{i=1}^{10} i$ to 3. State what the revised general term must be, and explain how you determined it.

38. Give an example in which $\sum_{i=1}^{3} a_i = \sum_{i=1}^{3} b_i$, but $a_i \neq b_i$.

In Exercises 39–44, rewrite the given series by changing the lower or upper limit as indicated.

39. $\sum_{k=4}^{7} 3k$; lower limit 1

40. $\sum_{i=0}^{6} \frac{i-3}{2}$; upper limit 8

41. $\sum_{i=2}^{7} (2i + 3)$; lower limit 4

42. $\sum_{j=4}^{7} \frac{1}{2j+1}$; lower limit 1

43. $\sum_{k=1}^{5} (-2)^{-k}$; upper limit 12

44. $\sum_{k=1}^{6} (-1)^k(k-1)^{k+1}$; upper limit 10

45. Consider the series $\sum_{j=2}^{6} \frac{j^2+j}{j+1}$.

(a) Evaluate the series.

(b) Simplify the general expression, and then evaluate the series again.

(c) What do you observe about the results in parts (a) and (b)?

46. Show that $\sum_{i=1}^{6} \ln i = \ln 6!$.

Real-Life Applications

47. Each year n after a textile mill closed, the population of a community changed by $10n - 130$. Write and evaluate a series to model the total change in population for the first 5 years after the mill closed.

48. An orange grower has a potential harvest of 5000 bushels of oranges. On consecutive weeks, a crew picks $\frac{2}{5}$ of the remaining oranges. Write and evaluate a series to determine the total harvest at the end of 3 weeks.

49. A certain beetle destroys $\frac{1}{10}$ of the remaining pines in an area each year. If an area initially has 800 pines, write and evaluate a series to model the total number of pines destroyed after 3 years.

50. An exterminator claims that for the first week after treating a house, each day 60% of the remaining insects will be killed. If the house has 240 insects, write and evaluate a series to determine the total number of insects killed after 3 days.

Modeling with Real Data

51. The table shows the number of computers sold in the United States in the period 1993–1996. (Source: Dataquest, Inc.)

Year	Units sold (in millions)
1993	5.8
1994	6.7
1995	8.2
1996	9.5

(a) Write the information as a sequence C_n with n representing the number of years since 1990.

(b) Evaluate the series $\sum_{n=3}^{6} C_n$, and interpret the result.

52. Refer to the data in Exercise 51.

(a) Write the information as a sequence C_j with j representing the number of years since 1993.

(b) Evaluate the series $\sum_{j=0}^{3} C_j$, and compare the result with part (b) of Exercise 51.

53. The table shows the number of airline passengers (in millions) during the period 1992–1994. (Source: Air Transportation Association of America.)

Year	Passengers enplaned (in millions)
1992	475
1993	488
1994	528

(a) Write the information as a sequence P_n with n representing the number of years since 1990.

(b) Evaluate the series $\sum_{n=2}^{4} P_n$, and interpret the result.

54. Refer to the data in Exercise 53.

(a) Write the information as a sequence P_i with i representing the number of years since 1992.

(b) Evaluate the series $\sum_{i=0}^{2} P_i$, and compare the result with part (b) of Exercise 53.

Group Project: Marriages and Divorces

The table shows the number (in millions) of marriages and divorces in the United States during the period 1990–1993. (Source: Vital Statistics of the United States.)

Year	Marriages (in millions)	Divorces (in millions)
1990	2.45	1.18
1991	2.37	1.18
1992	2.36	1.22
1993	2.33	1.19

55. Let n represent the number of years since 1990. Write the information as sequences m_n (marriages) and d_n (divorces).

56. What do the series $\sum_{n=0}^{3} m_n$ and $\sum_{n=0}^{3} d_n$ represent? Evaluate each series.

57. What does the sequence $C_n = m_n - d_n$ represent? Is the sequence increasing or decreasing?

58. What factors contribute to the trend indicated by the sequence in Exercise 57?

Challenge

In Exercises 59 and 60, without simplifying the expression for the general term, evaluate the given series.

59. $\sum_{k=1}^{10} [(k-1)^2 - (k+1)^2]$

60. $\sum_{k=1}^{15} \left[\frac{1}{k+1} - \frac{1}{k}\right]$

61. If c is a constant, show that $\sum_{k=1}^{n} ca_k = c\sum_{k=1}^{n} a_k$.

62. Show that $\sum_{k=1}^{n} (a_k + b_k) = \sum_{k=1}^{n} a_k + \sum_{k=1}^{n} b_k$.

In Exercises 63 and 64, expand both sides of the given equation to determine if the equation is true or false.

63. $\sum_{k=1}^{2} a_k^2 = \left(\sum_{k=1}^{2} a_k\right)^2$

64. $\left(\sum_{k=1}^{2} a_k\right)^2 = 2a_1a_2 + \sum_{k=1}^{2} a_k^2$

12.3 ARITHMETIC SEQUENCES AND SERIES

Arithmetic Sequences • Arithmetic Series • Real-Life Applications

Arithmetic Sequences

Suppose a fence post is placed at the corner of a field. Then additional posts are placed 20 feet apart in a row along the edge of the field. The location of the posts can be described by the sequence 0, 20, 40, 60, 80, ... Because the difference between any two consecutive terms is always 20, the sequence is an example of an **arithmetic sequence.**

Definition of an Arithmetic Sequence

An **arithmetic sequence** is a sequence in which the difference between any term and the preceding term is a constant called the **common difference.**

LEARNING TIP

The symbol a_n refers to any term of a sequence, and the symbol a_{n-1} refers to the preceding term.

For any arithmetic sequence, the common difference d is the difference between any term and the preceding term: $a_n - a_{n-1} = d$.

Sequence	*Common difference*	
$-1, 1, 3, 5, \ldots$	2	$1-(-1)=2, 3-1=2, 5-3=2, \ldots$
$7, 3, -1, -5, \ldots$	-4	$3-7=-4, -1-3=-4, -5-(-1)=-4, \ldots$

If we know the first term and the common difference of an arithmetic sequence, generating the terms is easy.

EXAMPLE 1

Writing the Terms of an Arithmetic Sequence

Write the first five terms of an arithmetic sequence whose first term is 2 and whose common difference is -3.

Solution

Each term is found by adding -3 to the preceding term. The first five terms of the sequence are $2, -1, -4, -7, -10$.

In general, the terms of an arithmetic sequence with common difference d may be written as follows:

$$a_1, \quad a_2 = a_1 + d, \quad a_3 = a_1 + 2d, \quad a_4 = a_1 + 3d, \quad \ldots$$

Note that the coefficient of d is always 1 less than the term number.

The General Term of an Arithmetic Sequence

The general term of an arithmetic sequence with first term a_1 and common difference d is $a_n = a_1 + (n - 1)d$.

EXAMPLE 2

Writing the General Term of an Arithmetic Sequence

Write the general term for the sequence $3, \frac{9}{2}, 6, \frac{15}{2}, \ldots$.

Solution

$$d = \frac{9}{2} - 3 = \frac{3}{2}$$ To find the common difference, subtract any two adjacent terms.

Because we know that $a_1 = 3$ and $d = \frac{3}{2}$, we can write the general term.

$$a_n = a_1 + (n - 1)d$$

$$a_n = 3 + (n - 1)\left(\frac{3}{2}\right) = 3 + \frac{3}{2}n - \frac{3}{2} = \frac{3}{2}n + \frac{3}{2}$$

EXAMPLE 3

Writing the General Term of an Arithmetic Sequence

If two terms of an arithmetic sequence are $a_1 = -1$ and $a_5 = 7$, write the general term.

Solution

We can use a_1 and a_5 to determine d.

$$a_n = a_1 + (n-1)d$$
$$7 = -1 + (5-1)d \qquad a_1 = -1,\ n = 5, \text{ and } a_5 = 7$$
$$7 = -1 + 4d$$
$$8 = 4d$$
$$d = 2$$

Think About It

If 6 fenceposts are used to string 30 feet of fencing, is the distance (common difference) between the posts 30/6 = 5 feet? Why?

Now we know a_1 and d, so we can write the general term.

$$a_n = -1 + (n-1)(2) = -1 + 2n - 2 = 2n - 3$$

EXAMPLE 4

Finding a Specific Term of an Arithmetic Sequence

Find a_5 of an arithmetic sequence with $a_3 = -5$ and $a_{10} = -19$.

Solution

We use the given information in the formula $a_n = a_1 + (n-1)d$.

$$n = 3,\ a_3 = -5 \qquad\qquad n = 10,\ a_{10} = -19$$
$$-5 = a_1 + (3-1)d \qquad\qquad -19 = a_1 + (10-1)d$$
$$-5 = a_1 + 2d \qquad\qquad -19 = a_1 + 9d$$

These two equations form a system of equations. We multiply both sides of $a_1 + 2d = -5$ by -1. Then we add the result to $a_1 + 9d = -19$.

$$\begin{aligned} -a_1 - 2d &= 5 \\ a_1 + 9d &= -19 \\ \hline 7d &= -14 \\ d &= -2 \end{aligned}$$

To determine a_1, substitute $d = -2$ into either equation.

$$a_1 + 9(-2) = -19$$
$$a_1 = -1$$

Knowing a_1 and d, we can write the general term.

$$a_n = -1 + (n-1)(-2)$$
$$a_n = 1 - 2n$$

Therefore, $a_5 = 1 - 2(5) = -9$.

EXAMPLE 5

Finding the Number of Terms in an Arithmetic Sequence

Determine the number of terms in the finite sequence

$$6, 2, -2, -6, \ldots, -70$$

Solution

The common difference is -4 and $a_1 = 6$. Let n represent the number of terms.

$$a_n = a_1 + (n - 1)d$$
$$-70 = 6 + (n - 1)(-4)$$
$$-70 = 6 - 4n + 4$$
$$-70 = 10 - 4n$$
$$-80 = -4n$$
$$n = 20$$

There are 20 terms in the sequence.

Arithmetic Series

Associated with an arithmetic sequence is an **arithmetic series.**

Definition of an Arithmetic Series

The sum of the terms of an arithmetic sequence is an **arithmetic series.** The sum of the first n terms of a sequence is called the ***n*th partial sum.**

EXPLORING THE CONCEPT

The nth Partial Sum of an Arithmetic Sequence

Let S_n represent the sum of the first n terms of an arithmetic sequence. We can write the sum in both increasing and decreasing order and add the terms.

$$S_n = a_1 + (a_1 + d) + (a_1 + 2d) + \cdots + [a_1 + (n - 1)d]$$
$$S_n = a_n + (a_n - d) + (a_n - 2d) + \cdots + [a_n - (n - 1)d]$$

$$2S_n = (a_1 + a_n) + (a_1 + a_n) + (a_1 + a_n) + \cdots + (a_1 + a_n)$$
$$2S_n = n(a_1 + a_n) \quad \text{The sum contains } n \text{ terms } (a_1 + a_n).$$
$$S_n = \frac{n(a_1 + a_n)}{2} \quad \text{Divide both sides by 2.}$$

The result is a formula for the nth partial sum of an arithmetic sequence.

The *n*th Partial Sum of an Arithmetic Sequence

For an arithmetic sequence whose first term is a_1 and whose nth term is a_n, the nth partial sum S_n is given by

$$S_n = \frac{n(a_1 + a_n)}{2}$$

In words, the nth partial sum can be found by averaging the first and last terms of the sequence and multiplying the result by n.

EXAMPLE 6

Finding the Sum of an Arithmetic Sequence

Determine the sum $3 + 6 + 9 + 12 + \cdots + 90$.

Solution

Since $a_1 = 3$ and $d = 3$, the number of terms can be determined as follows:

$$a_n = a_1 + (n - 1)d$$
$$90 = 3 + (n - 1)(3)$$
$$90 = 3 + 3n - 3$$
$$90 = 3n$$
$$30 = n$$

The sum can now be determined as follows:

$$S_n = \frac{n(a_1 + a_n)}{2}$$
$$S_{30} = \frac{30(3 + 90)}{2} = 1395$$

EXAMPLE 7

Finding the Sum of an Arithmetic Sequence

Determine the sum of the first 100 terms of an arithmetic sequence with $a_1 = 3$ and $d = -2$.

Solution

First, we determine the 100th term.

$$a_n = a_1 + (n - 1)d$$
$$a_{100} = 3 + (100 - 1)(-2) \qquad a_1 = 3,\ d = -2,\ n = 100.$$
$$a_{100} = -195$$

Now we can determine the sum of the first 100 terms.

$$S_n = \frac{n(a_1 + a_n)}{2} \qquad n = 100,\ a_1 = 3,\ a_{100} = -195$$
$$S_{100} = \frac{100\,[3 + (-195)]}{2} = -9600$$

EXAMPLE 8

Evaluating an Arithmetic Series

Evaluate the series $\sum_{i=1}^{50} (2 - 3i)$.

Solution

We are finding the sum of 50 terms ($n = 50$) with $a_1 = 2 - 3(1) = -1$ and $a_{50} = 2 - 3(50) = -148$.

$$S_n = \frac{n(a_1 + a_n)}{2} = \frac{50(-1 - 148)}{2} = -3725$$

Real-Life Applications

Arithmetic sequences and series may be models for the conditions of certain application problems.

EXAMPLE 9

Well-Drilling Costs as an Arithmetic Sequence

Welltech drills oil wells. The company charges \$0.75 for the first foot and for each additional foot, 6 cents more than the charge for the previous foot.

(a) What is the charge for drilling the last foot of a well that is 325 feet deep?

(b) What is the total cost for drilling the well?

Solution

(a) The charge for each foot is described by an arithmetic sequence whose first term $a_1 = 0.75$ and whose common difference $d = 0.06$.

$$a_1 = 0.75, \quad a_2 = 0.81, \quad a_3 = 0.87, \quad \ldots$$

The general term is given by

$$a_n = a_1 + (n - 1)d$$
$$a_n = 0.75 + (n - 1)(0.06) = 0.69 + 0.06n$$

Because the well is 325 feet deep, the charge for the last foot is determined by $a_{325} = 0.69 + 0.06(325) = 20.19$. The last foot costs \$20.19.

(b) The total cost is given by

$$S_n = \frac{n(a_1 + a_n)}{2}$$
$$S_n = \frac{325(0.75 + 20.19)}{2} = 3402.75$$

The total cost is \$3402.75.

12.3 QUICK REFERENCE

Arithmetic Sequences

- An **arithmetic sequence** is a sequence in which the difference between any term and the preceding term is a constant called the **common difference.**
- The **general term** of an arithmetic sequence with first term a_1 and common difference d is $a_n = a_1 + (n - 1)d$.

Arithmetic Series

- The sum of the terms of an arithmetic sequence is an **arithmetic series.** The sum of the first n terms is called the ***n*th partial sum** of the sequence.
- The nth partial sum of an arithmetic sequence is given by $S_n = \dfrac{n(a_1 + a_n)}{2}$.

12.3 EXERCISES

Concepts and Skills

1. Describe the property that distinguishes arithmetic sequences from other kinds of sequences.
2. In the following list, only one entry contains sufficient information to generate the terms of an arithmetic sequence. Explain why the information in the other two entries is not sufficient.

   (i) $a_1 = 3, n = 20$ (ii) $a_4 = 12, d = 5$

   (iii) $d = -2, n = 6$

In Exercises 3–8, determine whether the given sequence is an arithmetic sequence.

3. $1, 3, 5, 7, 9, \ldots$
4. $1, 4, 9, 16, 25, \ldots$
5. $-8, -8, -8, -8, \ldots$
6. $4\pi, 7\pi, 10\pi, 13\pi, \ldots$
7. $\frac{1}{2}, \frac{5}{4}, \frac{25}{8}, \frac{125}{16}, \frac{625}{32}, \ldots$
8. $3, 3.3, 3.33, 3.333, \ldots$

In Exercises 9–14, determine whether the given sequence is an arithmetic sequence.

9. $k_n = \frac{1}{n}$
10. $a_n = \frac{n}{3} - 2$
11. $k_n = \frac{n+3}{4}$
12. $a_n = \frac{n-1}{n}$
13. $b_n = (-1)^{n+1}(n-2)$
14. $b_n = 1.2n$

In Exercises 15–22, use the given information to write the first five terms of the arithmetic sequence.

15. $a_1 = 2, \quad d = 3$
16. $a_1 = -5, \quad d = 2$
17. $b_1 = \frac{2}{3}, \quad d = \frac{1}{6}$
18. $c_1 = 1.2, \quad d = 0.3$
19. $a_n = 5n + 8$
20. $a_n = \frac{3-n}{2}$
21. $c_2 = 7, \quad d = 3$
22. $k_3 = 4, \quad d = \frac{1}{2}$

In Exercises 23–30, write the general term for the arithmetic sequence.

23. $a_1 = 3, \quad d = 5$
24. $c_1 = 25, \quad d = -6$
25. $a_1 = -2, \quad d = -2$
26. $b_1 = \frac{1}{4}, \quad d = \frac{1}{2}$
27. $k_5 = 20, \quad d = -2$
28. $a_3 = -4, \quad d = 5$
29. $a_4 = -7, \quad d = 3$
30. $k_3 = 2, \quad d = -0.4$

In Exercises 31–36, write the general term for the arithmetic sequence.

31. $2, 4, 6, 8, 10, \ldots$
32. $9, 6, 3, 0, -3, \ldots$
33. $5, 2, -1, -4, -7, \ldots$
34. $-10, -3, 4, 11, \ldots$
35. $\frac{1}{3}, 0, -\frac{1}{3}, -\frac{2}{3}, -1, \ldots$
36. $\frac{2}{3}, 2, \frac{10}{3}, \frac{14}{3}, 6, \ldots$

In Exercises 37–44, two terms of an arithmetic sequence are given. Write the general term.

37. $c_1 = 7, \quad c_4 = 19$
38. $a_1 = 5, \quad a_5 = -7$
39. $a_1 = 12, \quad a_6 = -13$
40. $b_1 = 8, \quad b_7 = 20$
41. $a_3 = 13, \quad a_8 = 33$
42. $b_4 = 16, \quad b_{10} = 34$
43. $b_7 = 3, \quad b_{15} = 4$
44. $c_5 = -3, \quad c_8 = -5$

In Exercises 45–52, for the given arithmetic sequence, find the indicated term.

45. $a_1 = 5, \quad d = 4; \quad a_{12}$
46. $b_1 = -2, \quad d = -3; \quad b_{14}$
47. $3, \frac{7}{2}, 4, \frac{9}{2}, 5, \ldots; \quad c_{10}$
48. $-\frac{3}{2}, -3, -\frac{9}{2}, -6, -\frac{15}{2}, \ldots; \quad a_8$
49. $c_7 = -7, \quad d = 3; \quad c_{13}$
50. $b_3 = 17, \quad d = -4; \quad b_{10}$
51. $c_3 = 14, \quad d = -2; \quad c_8$
52. $a_7 = -3, \quad d = -5; \quad a_9$

In Exercises 53–56, for the given arithmetic sequence, find the indicated term.

53. $c_8 = 11, \quad c_4 = 23; \quad c_1$
54. $c_{11} = 9, \quad c_5 = 39; \quad c_1$
55. $a_5 = 19, \quad a_9 = 7; \quad a_{15}$
56. $a_6 = 17, \quad a_{10} = 29; \quad a_{21}$

In Exercises 57–62, determine the number of terms in the given finite arithmetic sequence.

57. $1, 3, 5, 7, 9, \ldots, 211$ **58.** $-5, -2, 1, \ldots, 73$

59. $171, 167, 163, \ldots, 19$ **60.** $1, 5, 9, \ldots, 93$

61. $7, 3, -1, \ldots, -61$

62. $-5.1, -4.9, -4.7, \ldots, 1.9$

63. In words, describe a method for finding the nth partial sum of a finite arithmetic sequence.

64. Explain why it is not possible to determine S_{15} for the sequence $1, 3, 5, 7, \ldots, 27$.

In Exercises 65–70, for the given sequence, find the sum of the given number of terms.

65. $-6, -6, -6, -6, \ldots; \quad n = 46$

66. $8, 10, 12, 14, 16, \ldots; \quad n = 45$

67. $9, 11, 13, 15, 17, \ldots; \quad n = 30$

68. $1, 5, 9, 13, 17, \ldots; \quad n = 25$

69. $5, 2, -1, -4, -7, \ldots; \quad n = 20$

70. $50, 45, 40, 35, 30, \ldots; \quad n = 50$

In Exercises 71–74, evaluate the given series.

71. $51 + 52 + 53 + \cdots + 100$

72. $-23 + (-17) + (-11) + \cdots + 79$

73. $79 + 72 + 65 + \cdots + (-33)$

74. $6 + 2 - 2 - 6 - 10 - \cdots - 66$

In Exercises 75–78, determine whether the given series is an arithmetic series.

75. $\sum_{k=1}^{18} 3(k - 4)$ **76.** $\sum_{k=1}^{15} (k + 1)^2$

77. $\sum_{k=0}^{9} (-1)^k k$ **78.** $\sum_{k=1}^{14} \frac{3 - k}{2}$

In Exercises 79–90, find the sum of the arithmetic series.

79. $\sum_{i=1}^{20} (2i - 3)$ **80.** $\sum_{k=1}^{16} (3 - 5k)$

81. $\sum_{i=1}^{15} (3i + 4)$ **82.** $\sum_{j=1}^{24} (5 - j)$

83. $\sum_{j=1}^{15} \left(\frac{1}{2}j + 7\right)$ **84.** $\sum_{k=1}^{15} \frac{5}{2}k$

85. $\sum_{k=1}^{14} \frac{2}{3}(k - 1)$ **86.** $\sum_{k=1}^{18} \frac{3 - 2k}{6}$

87. $\sum_{k=10}^{20} 2k$

88. $\sum_{k=4}^{12} \frac{k + 3}{2}$

89. $\sum_{k=15}^{25} (k + 3)$

90. $\sum_{k=12}^{30} 3(k - 2)$

91. Explain why an arithmetic sequence is increasing if $d > 0$ and decreasing if $d < 0$.

92. For a certain arithmetic sequence, $a_1 = 1$ and $S_{20} = -30$. What can you conclude about d? Why?

93. Find the sum of the first 50 odd natural numbers.

94. Find the sum of the first 50 even natural numbers.

95. In the arithmetic sequence $-4, -1, 2, 5, 8, \ldots,$ how many terms are needed for the sequence to have a sum of 255?

96. In the arithmetic sequence $25, 21, 17, 13, 9, \ldots,$ how many terms are needed for the sequence to have a sum of -260?

97. A student received money for college for 10 months. She received \$225 the first month, \$210 the second month, \$195 the third month, and so on. What was the total amount she received?

98. The salary schedule for a bricklayer's apprentice is \$500 for the first month and a monthly increase of \$35 for the rest of the year. What is the salary of the apprentice in the 12th month? What is his total salary for the year?

99. A neat pile of bricks has 45 bricks in the first row, 43 in the second row, 41 in the third row, and so on. There is one brick on the top row. How many bricks are in the 13th row? How many bricks are in the pile?

100. A ladder has 11 rungs that uniformly increase in length from 20 inches at the top to 26 inches at the base. Determine the lengths of the other 9 rungs.

Real-Life Applications

101. The contract for renovating an office building required completion of the work by the end of April and imposed a penalty for noncompletion of \$300 for the first day, \$450 for the second day, \$600 for the third day, and so on. Suppose that the contractor did not complete the work until May 10. What was the penalty for that day, and what was the total penalty?

102. A neighborhood association held a Labor Day weekend yard sale from Thursday through Monday and attracted 88 customers the first day, 76 the second day, 64 the third day, and so on. Each customer spent an average of \$7.40. How much money did the sale generate?

103. A harvesting crew gathers an average of 17 bushels of tomatoes per person each day. On Monday, a crew of 14 reported for work. However each following day, 2 more people were absent than the day before.

(a) How many bushels of tomatoes were harvested on Friday?

(b) How many bushels were harvested in 5 days?

104. In a rapidly growing county, 410 new students entered the school system on the first day of school. Then each day during the first 2 weeks of the school year, 28 fewer students enter the school system than entered the previous school day.

(a) How many students entered the system on the seventh school day?

(b) How many new students entered during the first 10 school days?

Modeling with Real Data

105. The table shows the number (in millions) of bank credit cards in circulation during the period 1992–1996. (Source: RAM Research Group.)

Year	Number (in millions)
1992	240
1993	270
1994	290
1995	320
1996	360

The data can be modeled by $C_n = 29n + 180$, where n is the number of years since 1990.

(a) What is the common difference for the arithmetic sequence C_n?

(b) Evaluate C_0. What does C_0 represent?

106. Refer to the data in Exercise 105.

(a) Use the data to determine the average number of credit cards in circulation per year from 1992 to 1996.

(b) Evaluate the arithmetic series $\sum_{n=2}^{6} (29n + 180)$. Use the result to find the average number of credit cards in circulation per year.

(c) Compare the results in parts (a) and (b).

107. The table shows parcel post rates for local packages in 1997. (Source: U.S. Postal Service.)

Maximum Weight (in pounds)	Cost
2	\$2.56
3	\$2.63
4	\$2.71
5	\$2.77
6	\$2.84

(a) Compute the average change in rates for the data given in the table.

(b) Use the average change in rates as the common difference to write the general term C_w of an arithmetic sequence to model the data. Let w represent the weight of the package.

108. Refer to the model in Exercise 107.

(a) Suppose that a mailing service ships 5 packages weighing 2, 3, 4, 5, and 6 pounds, respectively. Write a series for the total shipping cost.

(b) Evaluate the series in part (a).

Challenge

109. For a given arithmetic sequence, $a_1 = \frac{1}{25}$ and $a_2 = \frac{1}{15}$. What is a_5?

110. For a given arithmetic sequence, $a_{12} = 84.26$ and $S_{12} = 550.08$. What is the first term of the sequence?

111. Show whether 3000 is a term in the following arithmetic sequence.

5, 8, 11, 14, . . .

If not, what term of the sequence is closest to 3000?

112. Eight days after a cyclist began a trip along the Geauga Trail at a constant rate of 15 miles per day, a second cyclist began a trip along the same route. The second cyclist covered 14 miles the first day and increased the mileage by 2 miles each day thereafter. How many days does it take for the second cyclist to overtake the first cyclist?

12.4 GEOMETRIC SEQUENCES AND SERIES

Geometric Sequences • Geometric Series • Real-Life Applications

Geometric Sequences

Consider the terms of the sequence 2, 4, 8, 16, 32, ... Clearly, each term is two times the previous term. In this case we say that the ratio of any term to the preceding term is 2.

Definition of a Geometric Sequence

A **geometric sequence** is a sequence in which the ratio of consecutive terms is a constant called the **common ratio.**

Symbolically, there is a constant ratio r such that $\dfrac{a_n}{a_{n-1}} = r$, or $a_n = ra_{n-1}$ for each term number n.

EXAMPLE 1

LEARNING TIP

Try to develop a sense of the effect of r when $a_1 > 0$. For example, if $r > 1$, then the terms ofthe sequence become greaterand greater. If $0 < r < 1$, the terms become smaller and smaller. If $r < 0$, the signs of the terms alternate.

Finding the Common Ratio

Sequence	*Ratio*	
(a) $2, 4, 8, 16, 32, \ldots$	2	$\frac{4}{2} = 2, \frac{8}{4} = 2, \frac{16}{8} = 2, \frac{32}{16} = 2, \ldots$
(b) $4, -2, 1, -\frac{1}{2}, \frac{1}{4}, \ldots$	$-\frac{1}{2}$	$\frac{-2}{4} = -\frac{1}{2}, \frac{1}{-2} = -\frac{1}{2}, \frac{-1/2}{1} = -\frac{1}{2}, \frac{1/4}{-1/2} = -\frac{1}{2}, \ldots$

EXAMPLE 2

Writing the Terms of a Geometric Sequence

Write the first four terms of the geometric sequence whose first term is 12 and whose common ratio is $\frac{1}{3}$.

Solution

Because $r = \frac{1}{3}$, $a_n = \frac{1}{3}a_{n-1}$.

$$a_1 = 12 \qquad a_2 = \frac{1}{3}(12) = 4 \qquad a_3 = \frac{1}{3}(4) = \frac{4}{3} \qquad a_4 = \frac{1}{3}\left(\frac{4}{3}\right) = \frac{4}{9}$$

The terms of a geometric sequence can be written as follows:

$$a_1, \quad a_2 = a_1r, \quad a_3 = a_1r^2, \quad a_4 = a_1r^3, \quad \ldots$$

General Term of a Geometric Sequence

The **general term** of a geometric sequence with first term a_1 and common ratio r is $a_n = a_1r^{n-1}$.

EXAMPLE 3

Finding the General Term of a Geometric Sequence

Write the general term for the geometric sequence whose first term is 3 and whose common ratio is -2. Then determine the 10th term.

Solution

$$a_n = a_1 r^{n-1} = 3(-2)^{n-1} \qquad a_1 = 3 \text{ and } r = -2$$

For the 10th term, $a_{10} = 3(-2)^9 = -1536$

EXAMPLE 4

Finding the General Term of a Geometric Sequence

Write the general term for the geometric sequence

$$6, 4, \frac{8}{3}, \frac{16}{9}, \ldots$$

Solution

The common ratio is $r = \dfrac{a_2}{a_1} = \dfrac{4}{6} = \dfrac{2}{3}$.

$$a_n = a_1 r^{n-1} = 6\left(\frac{2}{3}\right)^{n-1} \qquad a_1 = 6 \text{ and } r = \frac{2}{3}$$

Think About It

Describe a geometric sequence in which $-1 < r < 0$.

EXAMPLE 5

Finding the General Term of a Geometric Sequence

Write the general term for a geometric sequence whose third term is 18 and whose eighth term is -4374.

Solution

$$a_3 = a_1 r^2 = 18 \qquad a_8 = a_1 r^7 = -4374$$

$$\frac{a_8}{a_3} = \frac{a_1 r^7}{a_1 r^2} = \frac{-4374}{18} \qquad \text{Divide } a_8 \text{ by } a_3 \text{ and simplify.}$$

$$r^5 = -243 \qquad \text{Divide out } a_1 \text{ and subtract exponents.}$$

$$r = -3 \qquad \text{Odd Root Property}$$

Now use $r = -3$ and $a_3 = 18$ to solve for a_1.

$$a_3 = a_1 r^2 = a_1(-3)^2 = 18$$

$$9a_1 = 18$$

$$a_1 = 2$$

The general term is $a_n = 2(-3)^{n-1}$.

Geometric Series

A **geometric series** is the sum of the terms of a geometric sequence. The sum of the first n terms of a geometric sequence is called the ***n*th partial sum.**

EXPLORING THE CONCEPT

The nth Partial Sum of a Geometric Sequence

To obtain a formula for the nth partial sum of a geometric sequence, let S_n represent the nth partial sum.

$$S_n = a_1 + ra_1 + r^2a_1 + r^3a_1 + \cdots + r^{n-1}a_1$$

If we multiply S_n by $-r$, we obtain the following:

$$-rS_n = -ra_1 - r^2a_1 - r^3a_1 - \cdots - r^{n-1}a_1 - r^na_1$$

Now we add the two sums.

$$\begin{aligned} S_n - rS_n &= a_1 + ra_1 + r^2a_1 + r^3a_1 + \cdots + r^{n-1}a_1 \\ & - ra_1 - r^2a_1 - r^3a_1 - \cdots - r^{n-1}a_1 - r^na_1 \\ &= a_1 - r^na_1 \\ (1-r)S_n &= a_1(1-r^n) \qquad \text{Factor both sides.} \\ S_n &= \frac{a_1(1-r^n)}{1-r} \qquad \text{Divide both sides by } 1-r,\ r \neq 1. \end{aligned}$$

The *n*th Partial Sum of a Geometric Sequence

The sum of the first n terms of a geometric sequence whose first term is a_1 and whose common ratio is r is given by

$$S_n = \frac{a_1(1-r^n)}{1-r}, \qquad r \neq 1$$

Note: In the formula, r cannot equal 1. If $r = 1$, then

$$S_n = \underbrace{a_1 + a_1 + a_1 + \cdots + a_1}_{n \text{ terms}} = na_1$$

EXAMPLE 6

Finding the *n*th Partial Sum of a Geometric Sequence

Determine the sum of the first ten terms of the sequence

$2, -4, 8, -16, \ldots$

Solution

Because $\frac{a_2}{a_1} = \frac{-4}{2} = -2$, $r = -2$.

We use $a_1 = 2$, $r = -2$, and $n = 10$ to determine the sum.

$$S_n = \frac{a_1(1-r^n)}{1-r}$$

$$S_{10} = \frac{2[1-(-2)^{10}]}{1-(-2)} = \frac{2(1-1024)}{3} = -682$$

EXAMPLE 7

Finding the *n*th Partial Sum of a Geometric Sequence

Evaluate the series $\sum_{i=0}^{6} 4(3)^i$.

Solution

$$\sum_{i=0}^{6} 4(3)^i = 4 + 4(3) + 4(3)^2 + \cdots + 4(3)^6$$

Using $a_1 = 4$, $r = 3$, and $n = 7$, the sum is as follows:

$$S_n = \frac{a_1(1 - r^n)}{1 - r}$$

$$S_7 = \frac{4(1 - 3^7)}{1 - 3} = 4372$$

In certain cases the nth partial sums of a geometric sequence become closer and closer to a limiting value as n increases.

EXPLORING THE CONCEPT

The Sum of an Infinite Geometric Sequence

For the infinite geometric sequence $1, \frac{1}{2}, \frac{1}{4}, \frac{1}{8}, \frac{1}{16}, \ldots, a_1 = 1$ and the common ratio r is 0.5. As n becomes larger and larger, the term r^n approaches 0. (See Exercise 59.) Thus r^n "vanishes" in the formula $S_n = \frac{a_1(1 - r^n)}{1 - r}$, and so the partial sums approach $\frac{a_1}{1 - r} = \frac{1}{1 - 0.5} = 2$ as n becomes larger. (See Exercise 60.) For this reason, we *define* the sum of this infinite series to be 2.

Note that these results are obtained only if $|r| < 1$. If $r = 1$, then $r^n = 1$ no matter how large n becomes. If $r > 1$, then as n becomes larger, the term r^n also becomes larger.

An infinite series $a_1 + a_2 + a_3 + \cdots$ can be written $\sum_{i=1}^{\infty} a_i$. If this sum exists, then we represent it with the symbol S.

Definition of the Sum of an Infinite Geometric Series

The sum of an infinite geometric series is

$$S = \frac{a_1}{1 - r}, \qquad |r| < 1$$

For $|r| \geq 1$, the sum does not exist.

EXAMPLE 8

Finding the Sum of an Infinite Series

Determine the sum of $4 - 2 + 1 - \frac{1}{2} + \frac{1}{4} - \cdots$.

Solution

We see that $a_1 = 4$ and $r = -\frac{1}{2}$. Because $\left|-\frac{1}{2}\right| < 1$, it is possible to determine the sum.

$$S = \frac{a_1}{1 - r} = \frac{4}{1 - \left(-\frac{1}{2}\right)} = \frac{4}{\frac{3}{2}} = \frac{8}{3}$$

EXAMPLE 9

Evaluating the Sum of an Infinite Series

Evaluate $\sum_{i=1}^{\infty} \left(\frac{2}{3}\right)^i$.

Solution

The series is $\frac{2}{3} + \frac{4}{9} + \frac{8}{27} + \cdots$ with $a_1 = \frac{2}{3}$ and $r = \frac{2}{3}$.

$$S = \frac{a_1}{1 - r} = \frac{\frac{2}{3}}{1 - \frac{2}{3}} = \frac{\frac{2}{3}}{\frac{1}{3}} = 2$$

Real-Life Applications

Geometric sequences and series are sometimes needed to describe the conditions of application problems.

EXAMPLE 10

A Sequence of Withdrawals from a College Fund

At the beginning of her first semester in college, a student began withdrawing money from her college fund. She continued to make semiannual withdrawals for the 4 years she was in college. However, each withdrawal was 25% larger than the preceding withdrawal. At the end of 4 years, she had withdrawn a total of \$10,913. How much were her first and last withdrawals?

Solution

Let a_1 = the amount of the first withdrawal. Because the student made eight withdrawals, the amounts are represented by the geometric sequence

$$a_1, \quad a_1(1.25), \quad a_1(1.25)^2, \quad \ldots, \quad a_1(1.25)^7$$

The total amount withdrawn is given by the series $\sum_{i=1}^{8} a_1(1.25)^{i-1}$. The sum of the withdrawals is the eighth partial sum.

$$S_n = \frac{a_1(1 - r^n)}{1 - r}$$

$$S_8 = 10{,}913$$

$$\frac{a_1[1 - (1.25)^8]}{1 - 1.25} = 10{,}913$$

$$\frac{a_1[1 - (1.25)^8]}{-0.25} = 10{,}913$$

$$a_1[1 - (1.25)^8] = 10{,}913(-0.25)$$

$$a_1 = \frac{10{,}913(-0.25)}{1 - (1.25)^8}$$

$$a_1 = 550$$

The first withdrawal was \$550. The last withdrawal was $550(1.25)^7$, or \$2622.60.

EXAMPLE 11

Total Distance Traveled by a Pendulum

Air resistance reduces the length of the path of a swing of a pendulum to 95% of the length of the preceding swing. If the first swing is 10 feet, what is the total distance the pendulum swings before it comes to rest?

Solution

The distance of each swing of the pendulum is described by a geometric sequence.

$$a_1 = 10, \quad a_2 = 10(0.95), \quad a_3 = 10(0.95)^2, \quad \ldots$$

The general term is $a_n = 10(0.95)^{n-1}$.

The total distance is given by the series $\sum_{i=1}^{\infty} 10(0.95)^{i-1}$.

Using $a_1 = 10$ and $r = 0.95$, the total distance is found as follows:

$$S = \frac{a_1}{1 - r} = \frac{10}{1 - 0.95} = 200$$

The pendulum travels a total distance of 200 feet.

12.4 QUICK REFERENCE

Geometric Sequences

- A **geometric sequence** is a sequence in which the ratio of consecutive terms is a constant called the **common ratio.**
- The **general term** of a geometric sequence with first term a_1 and common ratio r is $a_n = a_1 r^{n-1}$.

Geometric Series

- A **geometric series** is the sum of the terms of a geometric sequence.
- The sum of the first n terms of a geometric sequence is called the ***n*th partial sum.** For a sequence whose first term is a_1 and whose common ratio is r, $r \neq 1$, the formula for the nth partial sum is

$$S_n = \frac{a_1(1 - r^n)}{1 - r}$$

If $r = 1$, $S_n = na_1$.

- The sum of an infinite geometric series is defined only if $|r| < 1$. The formula for the sum is $S = \frac{a_1}{1 - r}$.

12.4 EXERCISES

Concepts and Skills

1. Describe the property that distinguishes a geometric sequence from other kinds of sequences.
2. What information must be known (or can be determined) in order to generate the terms of a geometric sequence?

In Exercises 3–6, determine whether the given sequence is a geometric sequence.

3. $2, 1, \frac{1}{2}, \frac{1}{4}, \frac{1}{8}, \ldots$
4. $1, \frac{1}{2!}, \frac{1}{3!}, \frac{1}{4!}, \ldots$
5. $1, 3, 5, 7, 9, \ldots$
6. $6, -2, \frac{2}{3}, -\frac{2}{9}, \ldots$

In Exercises 7–14, write the first five terms of the geometric sequence.

7. $a_1 = -3, \quad r = 2$
8. $a_1 = 5, \quad r = 0.3$
9. $k_1 = 64, \quad r = -\frac{1}{2}$
10. $a_1 = 6, \quad r = \sqrt{3}$
11. $a_n = \frac{1}{3^{n-4}}$
12. $a_n = (-3)^{n+1}$
13. $b_n = -4\left(\frac{2}{3}\right)^{n-1}$
14. $c_n = 10(2)^{n-3}$

In Exercises 15–22, write the general term for the geometric sequence.

15. $a_1 = -27, \quad r = \frac{1}{3}$
16. $a_1 = 2, \quad r = 3$
17. $c_1 = \frac{1}{8}, \quad r = -2$
18. $b_1 = 2, \quad r = \sqrt{5}$
19. $a_3 = 0.0008, \quad r = 0.2$
20. $b_4 = 64, \quad r = -2$
21. $k_4 = -1, \quad r = -\frac{1}{2}$
22. $a_4 = \frac{81}{256}, \quad r = -\frac{3}{4}$

In Exercises 23–26, write the general term for the geometric sequence.

23. $5, 10, 20, 40, 80, \ldots$
24. $24, 18, \frac{27}{2}, \frac{81}{8}, \frac{243}{32}, \ldots$
25. $-\frac{1}{9}, \frac{1}{3}, -1, 3, \ldots$
26. $e^5, e^4, e^3, e^2, e, \ldots$

In Exercises 27–32, write the general term of the geometric sequence.

27. $a_1 = -1, \quad a_4 = -8$
28. $a_1 = 6, \quad a_4 = -\frac{16}{9}$
29. $a_1 = \frac{5}{8}, \quad a_4 = \frac{1}{25}$
30. $a_1 = \frac{\sqrt{3}}{2}, \quad a_5 = 2\sqrt{3}$
31. $a_3 = 18, \quad a_6 = -486$
32. $a_4 = 64, \quad a_6 = 1024$

33. Describe the terms of a geometric sequence whose common ratio is 1.
34. What condition must be satisfied in order for an infinite geometric series to have a sum?

In Exercises 35–44, for each geometric sequence, determine the indicated term.

35. $a_1 = 0.3, \quad r = 0.1; \quad a_5$
36. $a_1 = \frac{4}{9}, \quad r = \frac{3}{2}; \quad a_8$
37. $32, 16, 8, 4, \ldots; \quad a_{10}$
38. $16, -8, 4, -2, \ldots; \quad a_{12}$
39. $a_1 = 2, \quad a_5 = 162; \quad a_8$
40. $a_1 = 16, \quad a_6 = -\frac{1}{2}; \quad a_{12}$
41. $a_3 = 1166.40, \quad r = 1.08; \quad a_6$
42. $a_4 = \frac{81}{16}, \quad r = \frac{3}{2}; \quad a_2$
43. $a_3 = 2, \quad a_7 = 32; \quad a_{12}$
44. $a_5 = 18, \quad a_7 = 54; \quad a_{10}$

In Exercises 45–48, find the nth partial sum of the geometric sequence for the given value of n.

45. $2, 4, 8, 16, 32, \ldots; \quad n = 12$

46. $1, -3, 9, -27, 81, \ldots; \quad n = 11$

47. $3, -2, \frac{4}{3}, -\frac{8}{9}, \ldots; \quad n = 8$

48. $1, \frac{2}{3}, \frac{4}{9}, \frac{8}{27}, \ldots; \quad n = 10$

In Exercises 49–52, determine the sum.

49. $2 - 4 + 8 - 16 + 32 - \cdots - 4096$

50. $-2 + 6 - 18 + 54 - \cdots + 4374$

51. $4 + 2 + 1 + \cdots + \frac{1}{64}$

52. $81 + 27 + 9 + 3 + \cdots + \frac{1}{81}$

In Exercises 53–58, evaluate the geometric series.

53. $\sum_{i=1}^{12} 5(2^{i-1})$

54. $\sum_{i=1}^{8} 3^i$

55. $\sum_{i=1}^{8} (-2)5^i$

56. $\sum_{i=0}^{10} \left(\frac{3}{4}\right)^i$

57. $\sum_{i=1}^{12} 9\left(\frac{2}{3}\right)^{i-1}$

58. $\sum_{k=1}^{10} 4\left(-\frac{3}{2}\right)^{k-1}$

59. Consider the sequence $1, \frac{1}{2}, \frac{1}{4}, \frac{1}{8}, \frac{1}{16}, \ldots$. Use a calculator to compute r^n for $1 \le n \le 8$. Based on your calculations, what value does the term r^n approach as n becomes larger and larger?

60. Consider the sequence $1, \frac{1}{2}, \frac{1}{4}, \frac{1}{8}, \frac{1}{16}, \ldots$. Use a calculator to compute S_n for $1 \le n \le 8$. Based on your calculations, what value does S_n approach as n becomes larger and larger?

In Exercises 61–68, evaluate the sum.

61. $\sum_{i=1}^{\infty} \left(\frac{1}{2}\right)^i$

62. $\sum_{i=1}^{\infty} \frac{3}{5}\left(\frac{5}{9}\right)^i$

63. $\sum_{k=0}^{\infty} 4^{1-k}$

64. $\sum_{k=1}^{\infty} 10^{3-k}$

65. $\sum_{k=1}^{\infty} \frac{4}{2^{k-1}}$

66. $\sum_{k=0}^{\infty} \frac{3}{(-5)^{k+1}}$

67. $\sum_{k=1}^{\infty} -6\left(-\frac{2}{3}\right)^{k-1}$

68. $\sum_{k=0}^{\infty} (-0.3)^{k+1}$

In Exercises 69–74, evaluate the sum.

69. $4 - 2 + 1 - \frac{1}{2} + \cdots$

70. $-81 + 27 - 9 + 3 - \cdots$

71. $3 + \frac{3}{4} + \frac{3}{16} + \frac{3}{64} + \cdots$

72. $15 + 5 + \frac{5}{3} + \frac{5}{9} + \cdots$

73. $1 + 3^{-2} + 3^{-4} + 3^{-6} + \cdots$

74. $8 + 4\sqrt{2} + 4 + 2\sqrt{2} + \cdots$

75. Explain how to determine whether a sequence is arithmetic or geometric.

76. Write a sequence that is both arithmetic and geometric. Explain your reasoning.

In Exercises 77–82, state whether the given series is geometric, arithmetic, or neither.

77. $\sum_{i=1}^{15} \left(-\frac{2}{3}\right)^i$

78. $\sum_{i=1}^{40} -\frac{3^{2i}}{4^i}$

79. $\sum_{k=1}^{12} \frac{4}{5}k$

80. $\sum_{k=0}^{20} \frac{2k+3}{5}$

81. $\sum_{k=1}^{5} \frac{k}{k+2}$

82. $\sum_{j=3}^{7} (2j)^2$

A repeating decimal can be written as a geometric series.

Example: $0.\overline{3} = 0.3 + 0.03 + 0.003 + \cdots$
Here, $a_1 = 0.3$ and $r = 0.1$. Because $|r| < 1$,

$$S = \frac{a_1}{1-r} = \frac{0.3}{1-0.1} = \frac{0.3}{0.9} = \frac{1}{3}$$

Therefore, $0.\overline{3} = \frac{1}{3}$.

In Exercises 83–88, use the sum of an infinite geometric series to write the repeating decimal as a fraction.

83. $0.\overline{4}$ **84.** $0.\overline{7}$ **85.** $0.\overline{19}$

86. $0.\overline{35}$ **87.** $0.\overline{123}$ **88.** $0.\overline{251}$

Real-Life Applications

89. The temperature in a newly filled hot tub is 65°F. If the temperature rises 10% each hour, how warm is the water after 4 hours?

90. Suppose you start a savings program by saving one penny on January 1. The next day you will save two pennies. Each day for the remainder of the month you will double the number of pennies you saved on the day before. How many pennies will you need to save on January 31 in order to complete the program?

91. A major manufacturer has a monthly payroll of \$250,000. If 75% of the money is respent in the community, what is the total economic impact of the monthly payroll?

92. For an address to the Chamber of Commerce, a college president released a report on the economic impact of student employees. If the payroll for 1 week was \$12,000 and 80% was respent in the community, what was the total economic impact?

Group Project: Annual Earnings for Graduates

In 1996, the average annual earnings for a high school graduate was \$18,737, and the average annual earnings for a college graduate (bachelor's degree) was \$32,629. (Source: Department of Commerce.)

93. Suppose a person receives a 5% raise each year. Write the general terms for sequences h_n and c_n that model the yearly earnings of high school and college graduates, respectively.

94. What are the projected annual earnings for high school and college graduates after 10 years of employment? after 20 years?

95. What are the projected total earnings for high school and college graduates after 10 years of employment?

96. Determine the projected difference in total earnings after 20 years of employment for a college graduate and a high school graduate.

Challenge

97. Show that $0.\overline{9} = 1$.

98. Determine whether the given sequence is arithmetic, geometric, or neither.
    (a) ln 3, ln 6, ln 9, ln 12, . . .
    (b) ln 3, ln 9, ln 27, ln 81, . . .
    (c) ln 3, ln 9, ln 81, ln 6561, . . .

99. For the sequence $-10, 20, -40, 80, \ldots,$
    (a) is 20,480 a term of the sequence?
    (b) is 2560 a term of the sequence?

100. Determine three numbers that meet all the following conditions:
    (i) Their sum is 9.
    (ii) The numbers form an arithmetic sequence.
    (iii) The squares of the numbers form a geometric sequence.

12.5 THE BINOMIAL THEOREM

Pascal's Triangle • The Binomial Theorem • Specified Terms

Pascal's Triangle

By observing the patterns that result from raising a binomial to a power (sometimes called *expanding* the binomial), we can determine a general method for expanding $(a + b)^n$ for any nonnegative integer n.

EXPLORING THE CONCEPT

Pascal's Triangle

Consider the following patterns:

$$(a + b)^0 = 1$$
$$(a + b)^1 = a + b$$
$$(a + b)^2 = a^2 + 2ab + b^2$$
$$(a + b)^3 = a^3 + 3a^2b + 3ab^2 + b^3$$
$$(a + b)^4 = a^4 + 4a^3b + 6a^2b^2 + 4ab^3 + b^4$$

Observe that the exponents on a begin with n and are decreasing, the exponents on b are increasing to n, and the sum of the exponents in each term is n. Furthermore, the coefficients follow a definite pattern.

1	$(a + b)^0 = 1$
1 1	$(a + b)^1 = a + b$
1 2 1	$(a + b)^2 = a^2 + 2ab + b^2$
1 3 3 1	$(a + b)^3 = a^3 + 3a^2b + 3ab^2 + b^3$
1 4 6 4 1	$(a + b)^4 = a^4 + 4a^3b + 6a^2b^2 + 4ab^3 + b^4$
1 5 10 10 5 1	Coefficients of $(a + b)^5$

Note that each entry in the triangle is the sum of the two numbers above it. By continuing this pattern, we can write additional lines that contain the coefficients of expansions of $(a + b)^n$ for higher values of n.

Think About It

Add the entries in each row of Pascal's triangle. Compare your results to the value of 2^n, where n is a whole number.

The triangular array of numbers is called **Pascal's triangle.** We can refer to the appropriate line in Pascal's triangle to determine the coefficients for the expansion of $(a + b)^n$ for any positive integer n.

Suppose that we wish to expand $(a + b)^5$. As we observed, the exponents on a begin with 5 and decrease, and the exponents on b increase to 5.

$$a^5 \qquad a^4b \qquad a^3b^2 \qquad a^2b^3 \qquad ab^4 \qquad b^5$$

To determine the coefficients, we can refer to the appropriate line in Pascal's triangle.

$$(a + b)^5 = 1a^5 + 5a^4b + 10a^3b^2 + 10a^2b^3 + 5ab^4 + 1b^5$$

EXAMPLE 1

Using Pascal's Triangle to Expand a Binomial

Expand $(x - 2)^4$.

Solution

From Pascal's triangle, we know the coefficients of $(a + b)^4$ are

$$1 \qquad 4 \qquad 6 \qquad 4 \qquad 1$$

and so $(a + b)^4 = a^4 + 4a^3b + 6a^2b^2 + 4ab^3 + b^4$.

Now replace a with x and b with -2.

$$\begin{aligned}(a + b)^4 &= a^4 + 4a^3b + 6a^2b^2 + 4ab^3 + b^4 \\ (x - 2)^4 &= x^4 + 4x^3(-2) + 6x^2(-2)^2 + 4x(-2)^3 + (-2)^4 \\ &= x^4 + 4x^3(-2) + 6x^2(4) + 4x(-8) + 16 \\ &= x^4 - 8x^3 + 24x^2 - 32x + 16\end{aligned}$$

In Example 1 the binomial $x - 2$ is a difference. Therefore, the terms in the expansion of $(x - 2)^4$ alternate in sign. This is true for all expansions of the form $(a - b)^n$.

The Binomial Theorem

When we expand $(a + b)^n$, the simple exponent patterns on a and b make it easy to determine the variable factors of each term.

$$a^n \qquad a^{n-1}b \qquad a^{n-2}b^2 \qquad \cdots \qquad ab^{n-1} \qquad b^n$$

The coefficients (called **binomial coefficients**) of the terms can be found with Pascal's triangle, but this is not a practical method when n is large. The Binomial Theorem offers an alternative method for determining the coefficients.

LEARNING TIP

Memorizing the notation for the Binomial Theorem is unrealistic. Instead, be aware of how the exponents behave and how the coefficients can be determined.

The Binomial Theorem

For positive integer n,

$$(a + b)^n = {}_nC_0a^n + {}_nC_1a^{n-1}b + {}_nC_2a^{n-2}b^2 + \cdots + {}_nC_{n-1}ab^{n-1} + {}_nC_nb^n$$

$$= \sum_{r=0}^{n} {}_nC_ra^{n-r}b^r$$

where ${}_nC_r = \dfrac{n!}{r!\,(n-r)!}$.

The proofs that ${}_nC_n = 1$ and ${}_nC_0 = 0$ are easy and are left as exercises.

Coefficient

As Example 2 shows, we can use either of two methods for evaluating ${}_nC_r$ with a calculator.

EXAMPLE 2

Evaluating a Binomial Coefficient

Evaluate ${}_{10}C_4$ in each of the following ways.

(a) Without a calculator

(b) With factorials on a calculator

(c) With ${}_nC_r$ on a calculator

Solution

(a) $${}_{10}C_4 = \frac{10!}{4!(10-4)!} = \frac{10 \cdot 9 \cdot 8 \cdot 7 \cdot 6 \cdot 5 \cdot 4 \cdot 3 \cdot 2 \cdot 1}{4 \cdot 3 \cdot 2 \cdot 1 \cdot 6 \cdot 5 \cdot 4 \cdot 3 \cdot 2 \cdot 1}$$

$$= \frac{10 \cdot 9 \cdot 8 \cdot 7}{4 \cdot 3 \cdot 2 \cdot 1} = 210$$

(b) **Figure 12.3**

```
10!/(4!6!)
                210
```

(c) **Figure 12.4**

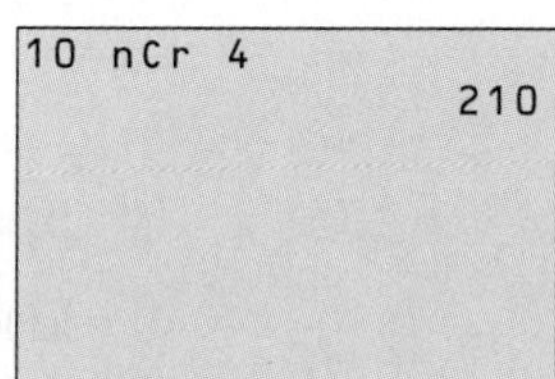

Now that we know how to evaluate ${}_nC_r$, we can use the Binomial Theorem to expand any binomial.

EXAMPLE 3

Using the Binomial Theorem to Expand a Binomial

Expand $(x + 3)^6$.

Solution

First, we determine the coefficients.

$${}_6C_0 = 1 \qquad {}_6C_1 = 6 \qquad {}_6C_2 = 15 \qquad {}_6C_3 = 20$$
$${}_6C_4 = 15 \qquad {}_6C_5 = 6 \qquad {}_6C_6 = 1$$

Now we know that the expansion has the following form.

$$(a + b)^6 = a^6 + 6a^5b + 15a^4b^2 + 20a^3b^3 + 15a^2b^4 + 6ab^5 + b^6$$

Replace a with x and b with 3.

$$\begin{aligned}(x + 3)^6 &= x^6 + 6x^5(3) + 15x^4(3)^2 + 20x^3(3)^3 + 15x^2(3)^4 + 6x(3)^5 + (3)^6 \\ &= x^6 + 18x^5 + 135x^4 + 540x^3 + 1215x^2 + 1458x + 729\end{aligned}$$

EXAMPLE 4

Using the Binomial Theorem to Expand a Binomial

Expand $(2y - 3)^5$.

Solution

First, we determine the coefficients.

$${}_5C_0 = 1 \qquad {}_5C_1 = 5 \qquad {}_5C_2 = 10$$
$${}_5C_3 = 10 \qquad {}_5C_4 = 5 \qquad {}_5C_5 = 1$$

The expansion has the following form.

$$(a + b)^5 = a^5 + 5a^4b + 10a^3b^2 + 10a^2b^3 + 5ab^4 + b^5$$

Replace a with $2y$ and b with -3.

$$\begin{aligned}[2y + (-3)]^5 &= (2y)^5 + 5(2y)^4(-3) + 10(2y)^3(-3)^2 \\ &\quad + 10(2y)^2(-3)^3 + 5(2y)(-3)^4 + (-3)^5 \\ &= 32y^5 - 240y^4 + 720y^3 - 1080y^2 + 810y - 243\end{aligned}$$

Note that the signs alternate.

Specified Terms

The coefficient of the first term of a binomial expansion is ${}_nC_0$, for the second term, ${}_nC_1$, and so on. In general, for the kth term, the coefficient is ${}_nC_{k-1}$.

The kth Term of a Binomial Expansion

When $(a + b)^n$ is expanded, the kth term, where $k \le n + 1$, is given by

$$ {}_nC_{k-1}a^{n-(k-1)}b^{k-1} $$

EXAMPLE 5

Finding a Specified Term of a Binomial Expansion

Write the eighth term of the expansion of $(x - 2)^{10}$.

Solution

We use the formula for the kth term of a binomial expansion. Substitute $n = 10$, $k = 8$, $a = x$, and $b = -2$.

$$ \begin{aligned} {}_nC_{k-1}a^{n-(k-1)}b^{k-1} &= {}_{10}C_7x^3(-2)^7 \\ &= 120x^3(-128) \\ &= -15{,}360x^3 \end{aligned} $$

This chapter contains formulas for performing a variety of tasks. Some of the formulas are quite detailed and, because they are not used frequently, difficult to remember for very long.

It is not as important to memorize formulas as it is to know that the formulas exist and to be able to retrieve and use them when they are needed.

12.5 QUICK REFERENCE

Pascal's Triangle

- **Pascal's triangle** is a triangular arrangement of numbers with each entry being the sum of the two numbers above it.
- To expand $(a + b)^n$, we use the following properties:
  1. The terms all contain the factors a and b. The exponents on a decrease from n to 0, while the exponents on b increase from 0 to n.

     $$ a^n \qquad a^{n-1}b \qquad a^{n-2}b^2 \qquad \cdots \qquad ab^{n-1} \qquad b^n $$

  2. The coefficients of the terms can be found in the appropriate line of Pascal's triangle.
  3. For $(a + b)^n$, the signs are all positive; for $(a - b)^n$, the signs alternate.

The Binomial Theorem

- The Binomial Theorem summarizes the known information about the expansion of the special product $(a + b)^n$. For a positive integer n, $(a + b)^n$ can be found by

$$ {}_nC_0a^n + {}_nC_1a^{n-1}b + {}_nC_2a^{n-2}b^2 + \cdots + {}_nC_{n-1}ab^{n-1} + {}_nC_nb^n $$

$$ = \sum_{r=0}^{n} {}_nC_ra^{n-r}b^r $$

- The coefficients of a binomial expansion can be found by

$$_nC_r = \frac{n!}{r!(n-r)!}$$

for integers r, $0 \le r \le n$.

Specified Terms

- When $(a + b)^n$ is expanded, the kth term, where $k \le n + 1$, is given by

$$_nC_{k-1}a^{n-(k-1)}b^{k-1}$$

12.5 EXERCISES

Concepts and Skills

1. Explain how each entry in Pascal's triangle is determined.
2. Suppose you were given the first 15 lines of Pascal's triangle. What is the greatest value of n for which you could use the triangle to expand $(a + b)^n$? Explain your answer.

In Exercises 3–6, use Pascal's triangle to expand the binomial.

3. $(2x - 3y)^3$
4. $(2x - 3y)^4$
5. $(x + 2y)^4$
6. $(x + 2y)^5$
7. For the symbol $_nC_r$, what are the meanings of n and r?
8. To which line in Pascal's triangle would the values of $_6C_r$, $0 \le r \le 6$, correspond? How do you know?

In Exercises 9–18, use a calculator to evaluate the binomial coefficient.

9. $_8C_4$
10. $_{10}C_7$
11. $_7C_5$
12. $_9C_4$
13. $_{15}C_0$
14. $_{17}C_{17}$
15. $_{10}C_1$
16. $_{12}C_{11}$
17. $_{20}C_{18}$
18. $_{16}C_2$

In Exercises 19–28, use the Binomial Theorem to expand the binomial.

19. $(x + 2)^5$
20. $(x - 3)^6$
21. $(x - 4)^5$
22. $(t + 5)^4$
23. $(3x - 1)^5$
24. $(2x + y)^4$
25. $(1 - 2a^2)^5$
26. $(t^2 + 2)^6$
27. $\left(x + \frac{1}{2}\right)^6$
28. $(2x^2 - y)^6$

In Exercises 29–36, write the first four terms in the expansion of the given expression.

29. $(x^3 - y^2)^{12}$
30. $(r^2 + 2t)^9$
31. $(x^2 - 2y)^{10}$
32. $(4r + t^3)^6$
33. $(3 - 2b)^{12}$
34. $(3x - y)^8$
35. $\left(\frac{s}{5} - \frac{t}{2}\right)^5$
36. $\left(\frac{1}{x} - 2\right)^8$

In Exercises 37–40, write the first two terms in the expansion of the given expression.

37. $(a + b)^{200}$
38. $(x - y)^{300}$
39. $(x + 2)^{100}$
40. $(x - 3)^{50}$
41. Suppose that you want to find the sixth term in the expansion of $(a + b)^9$. Explain why the coefficient of the term is $_9C_5$ rather than $_9C_6$.
42. For each term in the expansion of $(c + d)^n$, explain why the sum of the exponents on c and d is always n.

In Exercises 43–50, for the expansion of the given expression, find the indicated term.

43. $(2x - 1)^9$; last term
44. $(2s + 3t^2)^5$; last term
45. $(x^2 + y^3)^6$; 4th term
46. $(t^3 + s)^7$; 6th term
47. $(6 - x)^9$; 2nd term
48. $(x - 2)^{11}$; 7th term
49. $(x + 3y)^8$; 3rd term
50. $(a^2 - 2b^3)^9$; 3rd term

In Exercises 51–58, the expansion of the given expression will include a term whose variable part is given. Write the coefficient of the term.

51. $(x - 2)^{11}$; x^7
52. $(5 + y)^8$; y^2

53. $(2x + 5y)^9$; x^6y^3

54. $(3x - 4y)^5$; x^4y

55. $(x^3 + 2)^{11}$; x^{12}

56. $(3 + y^2)^6$; y^6

57. $(2a^2 - b)^8$; a^8b^4

58. $(a^3 + 3b^2)^{10}$; $a^{21}b^6$

59. Show that ${}_nC_n = 1$ for a positive integer n.

60. Show that ${}_nC_0 = 1$ for a positive integer n.

61. Use a calculator to write the 10th row of Pascal's triangle.

62. Use a calculator to write the 11th row of Pascal's triangle.

63. Before automatic pinsetters were invented, owners of bowling alleys hired young men to set bowling pins by hand. The figure shows the patterns in which the pins might be set.

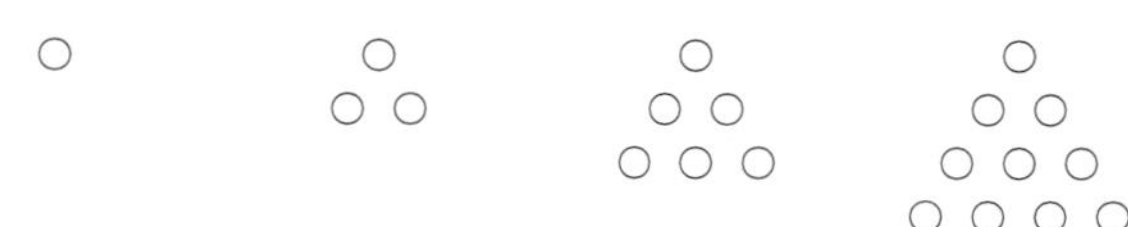

(a) Draw what the pattern would look like if there were one more row of bowling pins.

(b) Write a sequence indicating the number of pins in each of the five patterns.

(c) Predict how many pins there would be if the pattern contained six rows.

(d) The numbers in this sequence are called *triangular numbers*. Write the first seven rows of Pascal's triangle and see if you can locate the sequence of triangular numbers along any row, column, or diagonal.

64. The accompanying figure shows the first six rows of Pascal's triangle. Add the numbers along the indicated diagonal lines and write the results as a sequence. What have you discovered?

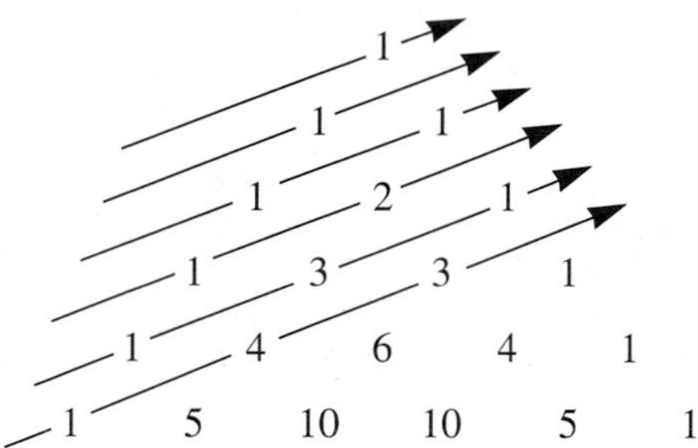

Challenge

In Exercises 65 and 66, use the Binomial Theorem to raise the given complex number to the indicated power.

65. $(1 + i)^6$

66. $(2 - i)^4$

67. Show that the expansion of $(x - 3)^6$ is

$$\sum_{k=0}^{6} \frac{6!}{k!(6-k)!} x^{6-k}(-3)^k$$

68. Draw a vertical line through the center of Pascal's triangle in the figure for Exercise 64. Note that the entries on the left and right of the line are symmetric. Here is the last row in the figure.

1	5	10	10	5	1
${}_5C_0$	${}_5C_1$	${}_5C_2$	${}_5C_3$	${}_5C_4$	${}_5C_5$
${}_5C_{5-0}$	${}_5C_{5-1}$	${}_5C_{5-2}$	${}_5C_{5-3}$	${}_5C_{5-4}$	${}_5C_{5-5}$

In general, show that the binomial coefficients are symmetric by showing that ${}_nC_j = {}_nC_{n-j}$.

69. Use your calculator to compute 11^n for $n = 0, 1, 2, 3$, and 4. What do you observe about the digits in your results? Can your observations be generalized for all n?

12. CHAPTER PROJECT Generations of Bees

A male honey bee has no "father" because it comes from an unfertilized egg. A female honey bee comes from a fertilized egg, so it has a "mother" and a "father."

The accompanying figure shows a *tree diagram* of six generations for a male honey bee and its ancestors.

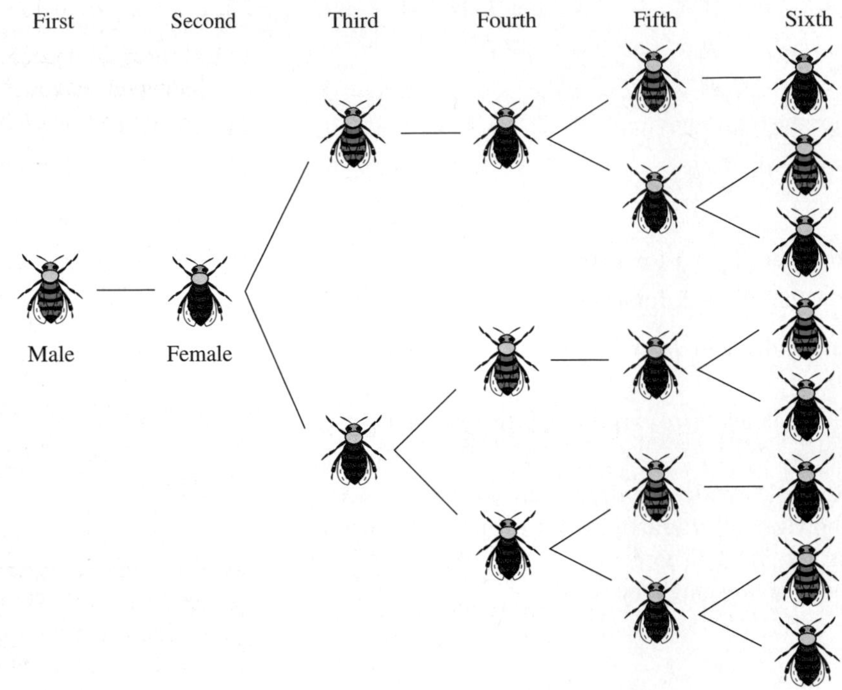

Observe that the male bee (first generation) has only a "mother" (second generation), which in turn has both a "father" and a "mother" (third generation).

1. Extend the tree diagram to a seventh generation. Then, using M (male) and F (female), list the entries in the seventh column.
2. Write a sequence in which each term a_n is the number of bees in generation n, where $1 \leq n \leq 7$.
3. The Fibonacci sequence is 1, 1, 2, 3, 5, 8, ... Compare your sequence in Exercise 2 with the Fibonacci sequence. What can you conclude?
4. In what generation would a single male bee have 144 ancestors?
5. Suppose that the diagram in the figure had begun with a female bee in the first generation. Then $a_1 = 1$ and $a_2 = 2$. Write a formula for a_n, where $n \geq 3$. Is this sequence the same as the Fibonacci sequence?
6. Research and report on other natural phenomena that exhibit Fibonacci number patterns.

12. CHAPTER REVIEW EXERCISES

Section 12.1

1. Find the next four terms in the following sequence.

5, 1, 4, 0, 3, −1, 2, . . .

In Exercises 2 and 3, write the first four terms and the eighth term of the sequence whose nth term is given.

2. $a_n = (-1)^n 2^{n-2}$

3. $b_n = 17 - 4n$

4. Write all the terms in the following sequence.

$$a_n = (n - 1)! - 7, \qquad 1 \le n \le 5$$

In Exercises 5 and 6, for the sequence whose nth term is given, find the indicated term.

5. $b_n = \dfrac{n^2 - 3n}{n + 2}$; b_{10}

6. $c_n = (-1)^n[(n + 2)^2 - 2n]$; c_9

In Exercises 7 and 8, write the general term for each sequence.

7. 2, 5, 8, 11, 14, . . .

8. 2, 4, 8, 16, 32, . . .

9. Write the first five terms of the sequence whose recursive definition is as follows.

$$a_1 = 4; \qquad a_{n+1} = 2a_n - 3$$

Section 12.2

In Exercises 10 and 11, expand the series but do not evaluate it.

10. $\sum_{i=1}^{5} (-1)^i(i^2 - 3i)$ **11.** $\sum_{i=1}^{6} \ln(i + 1)$

In Exercises 12 and 13, evaluate the series.

12. $\sum_{i=1}^{4} \frac{1}{2}i$ **13.** $\sum_{k=0}^{5} (-1)^k(k + 3)$

In Exercises 14 and 15, write the series expression in summation notation.

14. $e^2 + e^4 + e^6 + e^8 + e^{10}$

15. $\frac{1}{4} + \frac{3}{5} + \frac{5}{6} + 1 + \frac{9}{8} + \frac{11}{9}$

In Exercises 16 and 17, change the index as indicated.

16. $\sum_{i=3}^{8} (i + 2)$; lower limit 1

17. $\sum_{i=0}^{6} \frac{1}{2i - 1}$; upper limit 8

Section 12.3

In Exercises 18 and 19, write the first five terms of the arithmetic sequence with the given first term a_1 and the common difference d.

18. $a_1 = -7$, $d = 3$ **19.** $a_1 = 7$, $d = -3$

In Exercises 20–23, write the general term for the arithmetic sequence.

20. $a_1 = 9$, $d = 3$ **21.** 4, 7, 10, 13, 16, . . .

22. $a_1 = 5$, $a_7 = 17$ **23.** $a_5 = 2$, $a_9 = -14$

In Exercises 24 and 25, determine the indicated term for the arithmetic sequence.

24. 27, 23, 19, . . . ; a_{12}

25. $a_4 = 13$, $a_9 = 33$; a_{15}

In Exercises 26 and 27, determine the number of terms in the finite sequence.

26. $-\frac{13}{3}, -\frac{10}{3}, -\frac{7}{3}, \ldots, \frac{50}{3}$

27. 79, 73, 67, . . . , −59

In Exercises 28 and 29, determine the sum.

28. 4 + 9 + 14 + 19 + · · · + 124

29. $\sum_{i=1}^{25} (4i - 7)$

30. On March 1, a person agreed to work 2 hours for a total of $8.54. He agrees to work 2 hours each day during the month of March with the understanding that he will receive an increase of 62 cents each day. How much will he be paid for March 31?

31. The rows of seats in an outdoor amphitheater are curved with 52 seats in the front row. Each row thereafter contains 4 more seats than the row directly in front. What is the total number of seats in the first 30 rows?

Section 12.4

In Exercises 32 and 33, write the first five terms of the geometric sequence with the given first term a_1 and the common ratio r.

32. $a_1 = -\frac{1}{32}$, $r = -2$ **33.** $a_1 = 243$, $r = \frac{1}{3}$

In Exercises 34–37, write the general term for the geometric sequence.

34. $a_1 = 10$, $r = 0.2$

35. $e^4, e^7, e^{10}, e^{13}, \ldots$

36. $a_1 = -2$, $a_6 = 64$

37. $a_4 = 1$, $a_7 = 27$

In Exercises 38 and 39, determine the indicated term for the geometric sequence.

38. $5, 10\sqrt{5}, 100, \ldots$; a_7

39. 1024, 512, 256, . . . ; a_{14}

40. Find the sum of the first 12 terms of the following geometric sequence.

$3, -9, 27, \ldots$

41. Evaluate the following geometric series.

$$\sum_{i=0}^{10} (-2)^i$$

42. Evaluate the following infinite geometric series.

$$\sum_{i=1}^{\infty} \left(-\frac{1}{2}\right)^i$$

43. Use the sum of an infinite geometric series to change 0.222 . . . to a fraction.

44. In January 1995, a person made \$1000. In each succeeding month, her earnings increased by 3% of her previous month's earnings. How much did she earn in 1995?

Section 12.5

In Exercises 45 and 46, use a calculator to evaluate the binomial coefficient.

45. ${}_{13}C_7$ **46.** ${}_{11}C_6$

In Exercises 47 and 48, use the Binomial Theorem to expand the binomial.

47. $(x - 3)^6$ **48.** $(2x - 3y)^4$

In Exercises 49 and 50, find the indicated term in the expansion.

49. $(x + y)^{13}$; 7th term

50. $(2x + 3y)^6$; 4th term

12. CHAPTER TEST

1. Write the first four terms of the sequence whose general term is $a_n = \dfrac{n-1}{n!}$.

2. Write all the terms of the following sequence.

$$b_n = \frac{n-2}{n^2}, \qquad 1 \le n \le 5$$

3. Determine b_{12}, where $b_n = n(n - 2)$.

4. Determine the first term of a geometric sequence whose fourth term $c_4 = \frac{27}{5}$ and whose common ratio $r = \frac{3}{5}$.

5. Determine the first four terms of the geometric sequence whose first term $b_1 = \frac{7}{12}$ and whose common ratio $r = \frac{3}{2}$.

6. Write the next three terms of the following sequence.

$4, 7, 6, 9, 8, 11, 10, \ldots$

In Questions 7–12, write an expression for the general term of the given sequence.

7. $\dfrac{1}{2^2}, \dfrac{1}{3^2}, \dfrac{1}{4^2}, \dfrac{1}{5^2}, \ldots$ **8.** $1, \dfrac{1}{3}, -\dfrac{1}{3}, -1, -\dfrac{5}{3}, \ldots$

9. $-\dfrac{1}{2}, \dfrac{1}{4}, -\dfrac{1}{8}, \dfrac{1}{16}, \ldots$

10. The arithmetic sequence with $a_1 = 3$ and $a_9 = 7$

11. The arithmetic sequence with $a_1 = \frac{1}{2}$ and whose seventh partial sum is $S_7 = \frac{49}{2}$

12. The geometric sequence with $a_3 = 12$ and $a_6 = 96$

13. Determine the number of terms in the following arithmetic sequence.

$2, -3, -8, -13, \ldots, -58$

14. Write the following series with summation notation.

$$\frac{1}{2} + \frac{2}{3} + \frac{3}{4} + \frac{4}{5} + \frac{5}{6}$$

15. Expand and evaluate the following series.

$$\sum_{n=0}^{5} (n-2)^2$$

16. Evaluate the following arithmetic series.

$$\sum_{n=0}^{20} (2n-3)$$

17. Evaluate the following geometric series.

$$\sum_{n=0}^{20} 4\left(\frac{1}{3}\right)^{n-2}$$

18. Evaluate the following infinite geometric series.

$$3 - 1 + \frac{1}{3} - \frac{1}{9} + \cdots$$

In Questions 19 and 20, use the Binomial Theorem to expand the given binomial.

19. $(x + 3)^5$

20. $(2x - 3y)^4$

21. Write the fourth term of the expansion of $(2x - 1)^{10}$.

22. Prior to the opening of a new industrial complex, the population of Gober was 1000. Projections indicated that the population would increase by 25% the first year after the complex opened. After the first year, the population would continue to increase by 10% each year. What is the projected population after 5 years?

23. Determine the sum of the odd integers between 200 and 300.

24. At the end of the first year, the trade-in value of a certain car was \$2500 less than the original cost. Thereafter, the trade-in value decreased by \$1200 per year. If the original cost was \$20,000, write an expression for the general term of the sequence of trade-in values for the first 5 years. What was the trade-in value after 4 years?

25. A genealogist listed all her ancestors for 12 generations. Assuming there were no intermarriages or second marriages, how many people are in the list?

11–12 CUMULATIVE TEST

1. Write $x = y^2 + 6y - 1$ in the form $x = a(y - k)^2 + h$. Then determine the intercepts and the vertex of the graph.

2. Write an equation of a circle whose center is $C(-3, 5)$ and that contains the point $P(1, -2)$.

3. Determine the center and the endpoints of the major and minor axes of the ellipse whose equation is $(x - 3)^2 + 4y^2 = 16$.

4. Sketch the graph of

$$16(x - 4)^2 - 25(y + 2)^2 = 100$$

5. A wire is bent into the shape of a right triangle whose hypotenuse is 26 inches and whose area is 120 square inches. How long are the legs of the triangle?

6. Graph the following system of inequalities.

$$\frac{x^2}{16} + \frac{y^2}{25} \leq 1$$

$$x - y \leq -4$$

$$x \leq 0$$

$$y \geq 0$$

7. For each part, state whether the given sequence is increasing, decreasing, or alternating. Then give the fifth term of the sequence.

(a) $a_n = 1 - 3n$

(b) $b_n = (-2)^{n+1}$

(c) $c_n = -\dfrac{15}{n}$

8. Expand the series $\displaystyle\sum_{i=0}^{3} \frac{x^i}{i + 1}$.

9. How many terms are in the following sequence?

$$-1, 2, 5, \ldots, 59$$

10. A company is organized so that exactly 6 people report to each manager at each managerial level. The president is at the top of the organization chart, and there are 1296 employees at the lowest level. How many people work at the company?

11. Evaluate ${}_{15}C_{11}$.

12. Use the Binomial Theorem to expand $(2x - 3)^5$.

Answers to Exercises

CHAPTER 1

Section 1.1 *(page 8)*

1. The decimal name of a rational number is a terminating or repeating decimal.

3. Terminating **5.** Repeating **7.** False **9.** True

11. True **13.** False **15.** $\{0, 1, 2, 3, 4, 5, 6\}$

17. $\varnothing$ **19.** Negative integers

21. Numbers that can be written as p/q such that $q = 1$

23. False **25.** True **27.** False

29. The number 1.75 is a terminating decimal, and the digit 5 is followed by zeros. The number $1.\overline{75}$ is a repeating decimal, and the block of digits 75 repeats without end.

31. Rational **33.** Irrational **35.** Rational **37.** Irrational

39. **(a)** $\sqrt{16}$ **(b)** $0, \sqrt{16}$ **(c)** $-4, 0, -17, \sqrt{16}$
(d) $-4, 0, \frac{3}{5}, 0.25, -17, 0.\overline{63}, \sqrt{16}$ **(e)** $\sqrt{7}, \pi$

41. 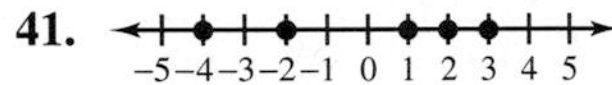

43. $-\frac{4}{3}$ −1.2 $\frac{1}{2}$ 0.75
−3 −2 −1 0 1 2

45. $-\pi$ $-\sqrt{4}$ $-\frac{1}{2}$ $\sqrt{4}$ 4.3
−5 −4 −3 −2 −1 0 1 2 3 4 5
$-2\frac{1}{3}$

47. $<$ **49.** $<$ **51.** $>$ **53.** $>$ **55.** $>$ **57.** $10 > x$

59. $-14.2 < -14$ **61.** $3 < y$ **63.** $-2 \le x \le 3$

65. $a < 0$ 6 **67.** $-2 < x < 2$

69. $(-\infty, -4)$ $x < -4$
−6 −5 −4 −3 −2 −1 0

71. $[-5, \infty)$ $x \ge -5$
−7 −6 −5 −4 −3 −2 −1 0

73. $(-3, -1)$ $-3 < x < -1$
−5 −4 −3 −2 −1 0 1 2

75. $[2, 5]$ $2 \le x \le 5$
0 1 2 3 4 5 6 7

77. $(-3, 7]$ $-3 < x \le 7$
−3 −1 1 3 5 7

79. $\{x \mid -3 \le x < 7\}$ **81.** $\{x \mid x \le 0\}$ **83.** 6 **85.** 0

87. The symbols $|7|$ and $|-7|$ represent the distances from 0 to 7 and from 0 to -7, respectively. Because these distances are the same, $|7| = |-7|$.

89. $-5, 5$ **91.** $-6, 2$ **93.** 6 **95.** -7 **97.** 0 **99.** 0

101. 1 **103.** \$450 **105.** \$1.90 **107.** 3.15 gallons

109. 19.2 ounces **111.** $>$ **113.** $>$ **115.** $\frac{27}{10}$ **117.** $\frac{1}{3}$

Section 1.2 *(page 16)*

1. Every pair of negative factors results in a positive product. If the number of negative factors is even, they can all be paired, and the complete product is positive. If the number of negative factors is odd, all but one of the factors can be paired, and the remaining negative factor makes the complete product negative.

3. -8 **5.** 8 **7.** 6 **9.** -11 **11.** $-\frac{1}{15}$ **13.** -6

15. -11.4 **17.** 7 **19.** 6 **21.** $\frac{11}{4}$ **23.** 60 **25.** 30

27. $\frac{1}{4}$ **29.** -4 **31.** Undefined **33.** $-\frac{3}{32}$ **35.** 16

37. -27 **39.** -49 **41.** $-\frac{27}{64}$

43. **(a)** If $b < 0$, then b^n is positive if n is an even integer.
(b) If $b < 0$, then b^n is negative if n is an odd integer.

45. 2 **47.** -6 **49.** Not a real number

51. $\frac{2}{3}$ **53.** 10 **55.** 13 **57.** 7 **59.** 70 **61.** 54

63. -36 **65.** 1 **67.** 14 **69.** -14 **71.** -36

73. -7 **75.** $-\frac{1}{2}$ **77.** 4 **79.** \$8.41 **81.** 14,775 feet

83. \$51.95 **85.** 13

87. **(a)** $+8.98$ **(b)** 4.29 **(c)** 19.61 **(d)** 2.59 **(e)** -5.23

89. **(a)** Houston and Memphis **(b)** 20,440

91. 41% **93.** The total number of ATM cards is not given.

95. If $a > 1$, then $a^2 - a$ is positive. However, if $a = 1$, then $a^2 - a = 0$, and if $0 < a < 1$, then $a^2 - a$ is negative.

97. If $a > 1$, then $a - \sqrt{a}$ is positive. However, if $a = 1$, then $a - \sqrt{a} = 0$, and if $0 < a < 1$, then $a - \sqrt{a}$ is negative.

99. $1 - 2 + 3 - (4 + 5 - 6) + 7 - 8 + 9 = 7$

101. **(a)** No value of x **(b)** $x = 0$ **(c)** $x \ne 0$

Section 1.3 *(page 25)*

1. When $-9 - 4x + x^2$ is written as $-9 + (-4x) + x^2$, the Commutative Property of Addition can be applied to write the expression as $x^2 + (-4x) + (-9) = x^2 - 4x - 9$.

3. Associative Property of Addition

5. Additive Identity Property

7. Distributive Property **9.** Distributive Property

11. Multiplicative Identity Property

13. Multiplication Property of -1

15. Distributive Property

17. (i) **19.** (b) **21.** (c) **23.** (f) **25.** $x + (4 + 3)$

27. $(6 \cdot 2)x$ **29.** $x + 6$ **31.** $1x$ **33.** $3(4z + 5)$

35. Suppose that 0 had a reciprocal r. Then, according to the Property of Multiplicative Inverses, $0 \cdot r = 1$. But the Multiplication Property of 0 states that $0 \cdot r = 0$. Therefore, 0 does not have a reciprocal.

37. $\frac{1}{5}$; Property of Multiplicative Inverses

39. 1; Multiplicative Identity Property

41. -1; Distributive Property

43. **(a)** Commutative Property of Addition
(b) Associative Property of Addition
(c) Property of Additive Inverses
(d) Additive Identity Property

45. **(a)** Definition of Subtraction
(b) Associative Property of Addition
(c) Property of Additive Inverses
(d) Additive Identity Property

47. $x + (3 + 2) = x + 5$ **49.** $2z + (7 - 2) = 2z + 5$

51. $(-5 \cdot 3)x = -15x$ **53.** $\left[-\frac{5}{6}\left(-\frac{9}{10}\right)\right]x = \frac{3}{4}x$

55. $20 + 15b$ **57.** $-3x - 12$ **59.** $-15 + 6x$

61. $xy + xz$ **63.** $5(x + y)$ **65.** $3(x + 4)$ **67.** $5(y + 1)$

69. $3(2 - x)$ **71.** $-x + 4$ **73.** $-2x - 5y + 3$

75. $12x$ **77.** $-12x^2$ **79.** $x + 8$ **81.** 3

83. $\left(\frac{3}{7} + \frac{4}{7}\right) + \left(\frac{2}{9} + \frac{5}{9} + \frac{2}{9}\right) = 2$

85. $(2 \cdot 5)[-87(-1)] = 870$

87. $7(100 + 8) = 700 + 56 = 756$

89. $15(100 - 2) = 1500 - 30 = 1470$

91. If n is the number, $9n = (10 - 1)n = 10n - n$.

93. $-\frac{5}{7}, \frac{7}{5}$ **95.** $-\frac{2}{3}x, \frac{3}{2x}$ **97.** $-x - 3, \frac{1}{x + 3}$

99. $3x - 4, \frac{1}{4 - 3x}$

101. Property of Additive Inverses

103. Commutative Property of Addition

105. **(a)** 10:47 A.M. **(b)** 11 hours per week

107. **(a)** 14% **(b)** \$350

109. Sometimes: True if $a = b$; otherwise, false.

111. Sometimes: True if a and b have like signs; otherwise, false.

113. False **115.** $\frac{a + b}{3} = \frac{1}{3}(a + b) = \frac{1}{3}a + \frac{1}{3}b = \frac{a}{3} + \frac{b}{3}$

Section 1.4 *(page 33)*

1. Both types of expressions contain numbers, grouping symbols, and operations symbols. However, algebraic expressions contain variables, whereas numerical expressions do not.

3. -11 **5.** -3 **7.** 2 **9.** 16 **11.** 14 **13.** -6

15. -18 **17.** 6

19. **(a)** $-3(4 - 5)$ **(b)** $(-3 \cdot 4) - 5$ **(c)** $-(3 \cdot 4 - 5)$

21. 24 **23.** 25 **25.** 17 **27.** 0.22

29. If $x = 2$, the denominator is 0, and so the expression is undefined.

31. 10 **33.** -3 **35.** 3 **37.** -1 **39.** 5 **41.** 10

43. Undefined **45.** 0 **47.** 2 **49.** $-\frac{2}{3}$ **51.** 2 **53.** $-\frac{2}{5}$

55. -9.41 **57.** -13.20 **59.** -246.84

61. The result is $\sqrt{-1}$, which is not a real number.

63. 4 **65.** $\frac{9}{5}$ **67.** 14 or -4

69. In both cases we replace the variable with a given value and perform the indicated operations. For an expression, the result is the value of the expression. For an equation, the given number is a solution only if the expressions on the right and left sides of the equation have the same value.

71. Yes **73.** No **75.** Yes **77.** No **79.** No **81.** Yes

83. Yes **85.** Yes **87.** 84.9 feet

89. No. The ladder, building, and ground form a triangle whose sides are 6, 22, and 24. Because $6^2 + 22^2 \neq 24^2$, the wall is not perfectly vertical.

91. 10°C **93.** \$1300

95. **(a)**

Year	x	A	y	$A - y$
1980	0	27	22.4	+4.6
1984	4	36	33.8	−2.8
1988	8	50	55.2	−5.2
1992	12	72	71.6	+0.4
1996	16	91	88.0	+3.0

(b) The $A - y$ column shows the difference between the actual data and the data estimated from the model expression. The model expression is most accurate for 1992 and the least accurate for 1988.

(c) The fact that the average is 0 tells us nothing about the accuracy of the model. Very large discrepancies could have an average of 0.

97.

t	E_1	E_2
0	3.3	3.3
5	3.7	3.7
7	3.9	3.9
10	4.3	4.1

Because the values of the expressions are approximately equal to the data, the models are reasonably accurate.

99. For values of t between 0 and 10, the term $0.003t^2$ is small.

101. **(a)** Positive **(b)** Either **(c)** Positive **(d)** Negative **(e)** Negative **(f)** Negative **(g)** Positive **(h)** Negative

103. $3^3 + 3$

105. Suppose that n is the line number in the table. Then, for $a^2 + b^2 = c^2$, $a = 2n + 1$, $b = n(a + 1)$, and $c = b + 1$. The next two lines are $9^2 + 40^2 = 41^2$ and $11^2 + 60^2 = 61^2$.

Section 1.5 *(page 41)*

1. Suppose that A is an algebraic expression with no explicitly written constant term. Because $A = A + 0$, the constant term of the expression is 0.

3. Terms: $2x, -y, 5$ Coefficients: $2, -1, 5$

5. Terms: $3a^3, -4b^2, 5c, -6d$ Coefficients: $3, -4, 5, -6$

7. Terms: $7y^2, -2(3x - 4), \frac{2}{3}$ Coefficients: $7, -2, \frac{2}{3}$

9. Terms: $\frac{x+4}{5}, -5x, 2y^3$ Coefficients: $\frac{1}{5}, -5, 2$

11. The expression $2x + 3$ contains one plus sign that separates the terms $2x$ and 3. The plus sign in the expression $2(x + 3)$ is inside parentheses and is not considered when identifying terms. Thus this expression has only one term.

13. $-x + 3$ **15.** $3x - 3y - 5$ **17.** $6x - 7y$

19. 4 **21.** $-x + y$ **23.** $17x^2 + 3x$ **25.** $10ac^2 - 13ac$

27. $2.75x^3 - 4.40x^2$

29. We can write $-(2x - 5)$ as $-1(2x - 5)$, and we can write $3 - (4a + 1)$ as $3 + (-1)(4a + 1)$. In both cases, distributing -1 changes the signs of the terms inside the grouping symbols.

31. $-a + b$ **33.** $10x - 15$ **35.** $-2a + 3b - 7d$

37. $3x - 6y - z + 2$ **39.** (d) **41.** (f) **43.** (a)

45. $-x - 1$ **47.** $x - 12$ **49.** $-7a - 2b$ **51.** -8

53. $-3x + 17$ **55.** $x + 2$ **57.** $3x + 5$ **59.** $8t + 13$

61. $2x - 4$ **63.** $c - 125$ **65.** $-14x - 37$ **67.** 4

69. $x^2y^3 + 3x^3y^2$ **71.** 0 **73.** $\frac{2n}{12} = \frac{n}{6}$

75. $3n + n = 4n$ **77.** (b) **79.** (b) **81.** $x - 5 = 3$

83. $8x = 11$ **85.** $2x = x - 3$ **87.** $x - 0.4x = 50$

89. $x - 2y = -5$ **91.** $3L$ **93.** $25q + 10d$

95. $2.5x + 15$ **97.** $3n + 3$ **99.** $0.08d$

101. $\sqrt{a^2 + (a - 2)^2}$ **103.** $1.05s$

105. The phrase is ambiguous because we can read it as "(the product of 3 and a number) increased by 7" or as "the product of 3 and (a number increased by 7)." The corresponding translations are $3x + 7$ and $3(x + 7)$.

107. **(a)**

Age	10	12	14	16
Amount spent per trip	18.50	23.58	28.66	33.74

(b) When $A = 13$, the value of the model expression is 26.12, which exceeds the average of \$25. This suggests that spending exceeds the average for the 13–17 age group.

(c)

Age	10	12	14	16
Amount spent per month	222.00	282.96	343.92	404.88

109. -127°F **111.** -17°F **113.** (a) **115.** (a) **117.** (b)

119. $2x^2$

Section 1.6 *(page 49)*

1. For $2x^0$, the base is x. We evaluate x^0 and then multiply by 2. For $(2x)^0$, the base is $2x$. We multiply 2 and x and then raise the result to the 0 power.

3. 1 **5.** -1 **7.** 3 **9.** 3 **11.** $-\frac{1}{8}$ **13.** 15 **15.** 4

17. $\frac{29}{18}$ **19.** $\frac{1}{8}$ **21.** -7 **23.** 4

25. The base for the exponent -5 is x, not $3x$. Therefore, $3x^{-5} = 3 \cdot \frac{1}{x^5} = \frac{3}{x^5}$.

27. $\frac{1}{x^7}$ **29.** $\frac{4}{x^4}$ **31.** $\frac{x^3}{32}$ **33.** $-\frac{1}{x^3y^4}$ **35.** $\frac{1}{y^2}$ **37.** $-\frac{k^6}{5}$

39. $\frac{1}{(2+y)^3}$ **41.** $-\frac{3b}{2a}$ **43.** 2^{11} **45.** $(6 + t)^9$ **47.** 1

49. $\frac{3}{x^3}$ **51.** $\frac{14y}{x^7}$

53. The first approach uses the definition of a negative exponent, whereas the second approach uses the Quotient Rule for Exponents.

55. 2^3 **57.** x^{11} **59.** $\frac{1}{y^5}$ **61.** $(x + 1)^2$ **63.** $-\frac{x}{2}$

65. $-2t^{12}$ **67.** x^{-8} **69.** x^4

71. $-1 - 4^0, -|8|^0, 0, (-10)^0, 4^0 + x^0$

73. $(4 - 3)^{-9}$ **75.** $(2^5 + 1)^0$ **77.** $(8 \cdot 5)^2$ **79.** $(5 \div 3)^{-1}$

81. $\frac{3}{2z^3}$ **83.** $\frac{d^{10}}{c^6}$ **85.** $\frac{x}{y^2}$ **87.** $\frac{y^2}{3}$ **89.** $\frac{2x^2}{y^4}$ **91.** $\frac{y^{10}}{x^5}$

93. $\frac{y^5}{3x^5}$ **95.** $40t$ **97.** 85

99. (a) 58 billion (b) $0.00125t^2$

101. (a) $\frac{1553}{a}$ (b) Answers will vary.

103. Answers will vary. **105.** $m = 2, n = -6$

107. Answers will vary.

Section 1.7 *(page 56)*

1. For $x^3 \cdot x^3$, we add exponents to obtain x^6 (Product Rule for Exponents); for $x^3 + x^3$, we combine the coefficients, but do nothing with the exponents, to obtain $2x^3$ (Distributive Property); for $(x^3)^3$, we multiply exponents to obtain x^9 (Power to a Power Rule).

3. x^{12} **5.** $\frac{a^{20}}{16}$ **7.** $(x - 3)^{12}$ **9.** $a^{30}b^{54}$ **11.** $\frac{a^{24}}{b^{30}}$

13. $\frac{16y^8}{x^6}$ **15.** $x^4(x + 2)^{12}$ **17.** $\frac{49x^6}{25y^{10}}$ **19.** $-\frac{k^5}{32}$

21. $\frac{x^{12}}{16y^{10}}$ **23.** $\frac{9x^6y^4}{25}$

25. Select values for c and d and substitute. For $c = 1$ and $d = 2$, $(1 + 2)^2 \neq 1^2 + 2^2$.

27. $-\frac{x^2}{5y}$ **29.** $-\frac{x^2}{2y^7}$ **31.** $24x^{14}$ **33.** $\frac{1}{x^8}$ **35.** $48x^{17}$

37. y^8 **39.** $\frac{2}{5x^4}$ **41.** $\frac{b^2}{a^2}$ **43.** $\frac{1}{25}$ **45.** $\frac{3}{x^3}$ **47.** $2(xy)^{-1}$

49. $(x^{-1} + y)^{-1}$

51. The number n is such that $1 \le n \le 10$, and p is an integer.

53. $1.25 \cdot 10^5$ **55.** $-5.376 \cdot 10^8$ **57.** $2.5 \cdot 10^{-2}$

59. $6.45 \cdot 10^{-6}$ **61.** 1,340,000 **63.** -0.0000004214

65. $-572{,}000{,}000$ **67.** 0.00089 **69.** $4.3 \cdot 10^{52}$

71. $1.2 \cdot 10^{-11}$ **73.** 6000 **75.** 200 **77.** 5

79. $3.65 \cdot 10^{11}$ **81.** $1.06 \cdot 10^{15}$

83. 0.000000000053 meter **85.** 5,000,000

87. \$2,280,000 or $\$2.28 \cdot 10^6$ **89.** $7.2 \cdot 10^6$ deaths

91. $1.1 \cdot 10^8$ person-hours **93.** $8.0 \cdot 10^{-4}$ pound

95. $\$1.43 \cdot 10^{10}$, $\$3.32 \cdot 10^9$ **97.** \$17,500,000,000

99. Answers will vary. **101.** $n = -1, m = 5$

103. $n = -6$

Chapter Review Exercises *(page 60)*

1. The decimal names of the rational numbers are terminating or repeating. The decimal names of the irrational numbers are nonterminating and nonrepeating.

3. True **5.** False

7. $0 \le x < 3$

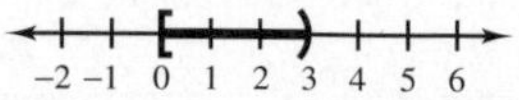

9. -3 **11.** $\frac{11}{8}$ **13.** -120 **15.** -27 **17.** 11

19. \$866.28

21. Associative Property of Addition

23. Property of Additive Inverses

25. Distributive Property **27.** $x^2 + 8x - 7$

29. 1 **31.** 24 **33.** -1 **35.** 25 **37.** 8

39. They are all grouping symbols.

41. There are two terms. The coefficients are 3 and -5.

43. $3x^2 + 2x - 4y$ **45.** $6x + 11$ **47.** $xy^2z + xyz^2 + x^2yz$

49. $1.2w$

51. $b \neq 0$ **53.** $\frac{2y^4}{x^4}$ **55.** 1 **57.** xyz

59. The base does not change sign when applying the definition of a negative exponent.

61. To apply the Product Rule, write $(x^5)^2 = x^5 \cdot x^5$ and add the exponents. To apply the Power to a Power Rule, multiply exponents.

63. $\frac{a^8}{b^{10}}$ **65.** $-\frac{9}{2x^7}$ **67.** $2.9 \cdot 10^{-5}$ **69.** $1.41 \cdot 10^{87}$

Chapter Test *(page 62)*

1. [*1.1*] (a) $-2, 0, 6, \sqrt{9}$
(b) $\pi, \sqrt{7}$
(c) $-3.7, -2, -0.\overline{14}, 0, \frac{2}{3}, 6, \sqrt{9}$

2. [*1.1*] $\{-2, -1, 0, 1, 2, 3, 4, 5\}$

3. [*1.1*] $-2 < x \le 3$

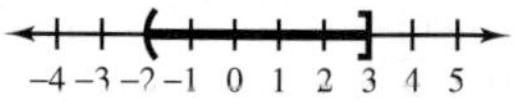

4. [*1.2*] (a) -4 (b) $-\frac{1}{15}$ (c) $\frac{5}{37}$ (d) -7 (e) -1
(f) 4 (g) $-\frac{5}{4}$ (h) -9.2 (i) 0 (j) 8

5. [*1.3*] (a) $3x - 12$ (b) $-x + 4$ (c) $3 + (t - 3r)$

6. [*1.4*] $\frac{[2(-9 + 4)]^2}{-17 - 3} = -5$ **7.** [*1.4*] $-3.49°$ **8.** [*1.4*] -2

9. [*1.4*] Terms: $5x^2, -4x, 2$; coefficients: $5, -4, 2$

10. [*1.4*] For -2^2, the base is 2: $-2^2 = -4$
For $(-2)^2$, the base is -2: $(-2)^2 = 4$

11. [*1.6, 1.7*] (a) x^4y^6 (b) t^2 (c) $\frac{4b^8}{a^{10}}$ (d) $\frac{1}{a^4b^6}$ (e) c^2d^6

12. [*1.6*] (a) $-\frac{1}{32}$ (b) 3 (c) $-\frac{1}{8}$

13. [*1.7*] (a) 27,000,000 (b) 0.0000456

14. [*1.7*] $5.7 \cdot 10^{-7}$ **15.** [*1.4*] (a) -4 (b) -3 (c) 20.216

16. [*1.5*] (a) $3x + 5y$ (b) $4s - t$ (c) $-4x + 2$

17. [*1.4*] Yes **18.** [*1.4*] $\frac{1}{2}L - 3$

19. [1.4] $n + 3n + 2(n - 1)$ **20.** [1.7] 3.7×10^9 miles

CHAPTER 2

Section 2.1 *(page 73)*

1. Starting at the origin, move 2 units to the left along the x-axis and then 4 units vertically upward. The destination is the point $A(-2, 4)$

3.

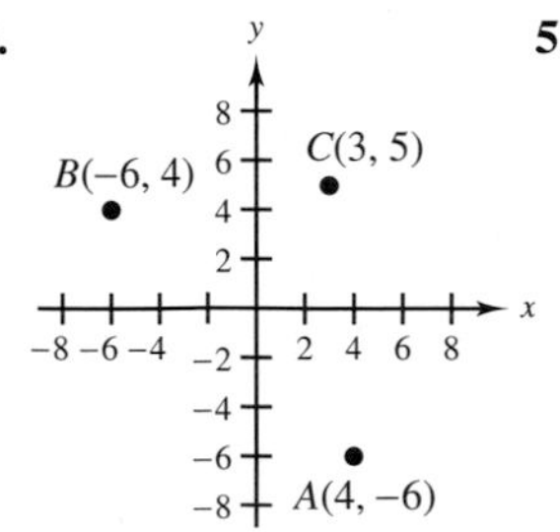

5.

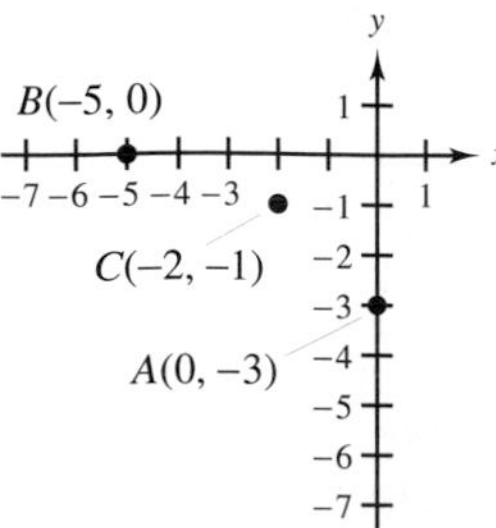

7. (8, 3)

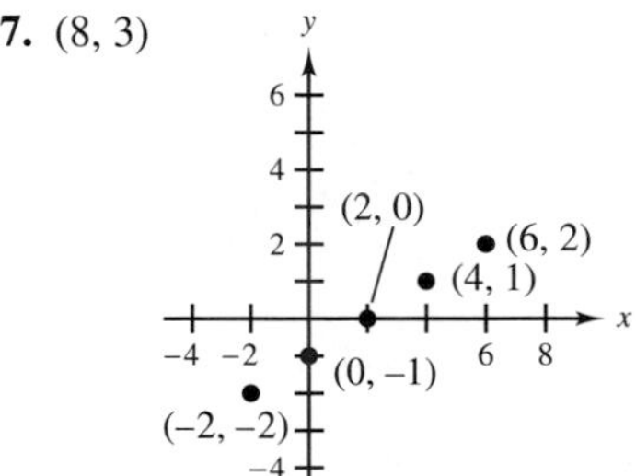

9. (2, 5)

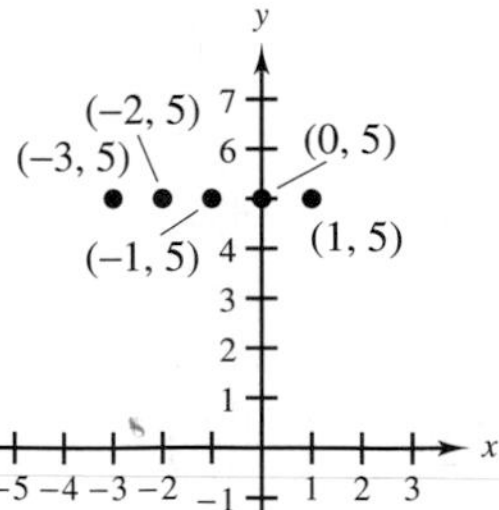

11. (6, 3)

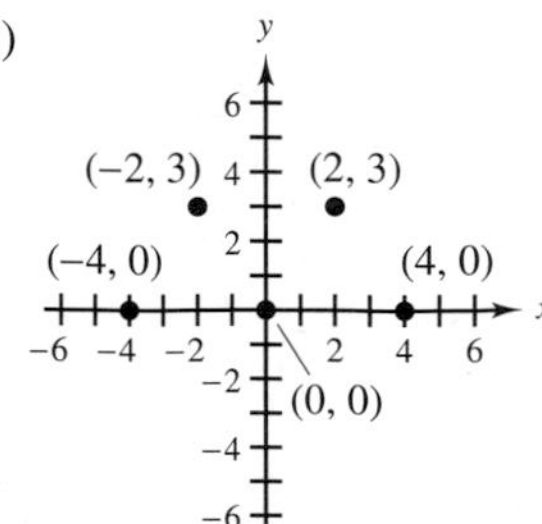

13. III **15.** I **17.** IV **19.** (−6, 4) **21.** (0, 0)

23. (5, 0)

25. Points whose x-coordinates are 0 belong to the y-axis.

27.

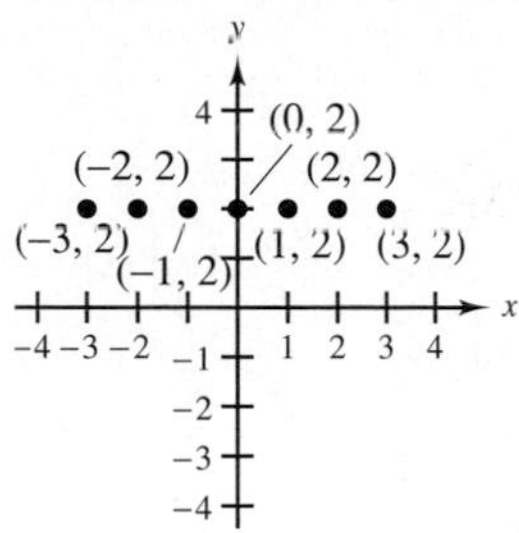

29.

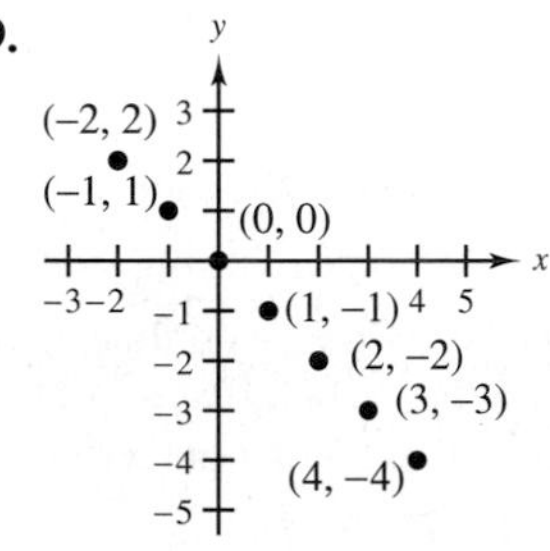

31.

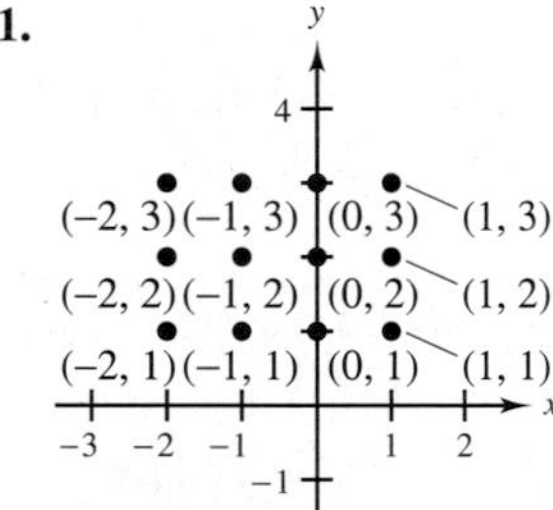

33. I, IV, or x-axis **35.** III, IV, or y-axis **37.** x-axis

39. II, III, or x-axis **41.** II, IV

43. **(a)** A point of the y-axis below the origin

(b) A point of the x-axis to the right of the origin

(c) A point of the y-axis above the origin

45. **(a)** (2, −1), (3, −7), (5, −4)

(b) (−2, 1), (−3, 7), (−5, 4)

47. **(a)** (1, 5), (5, 1), (1, 11), (11, 1)

(b) (−1, −5), (−5, −1), (−1, −11), (−11, −1)

49.

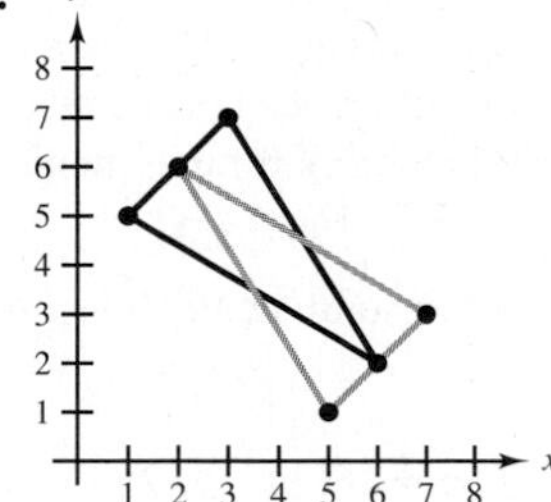

51.

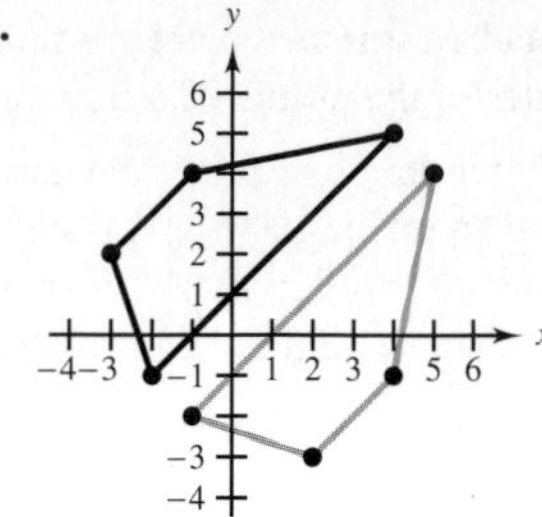

53. (0, 0), (5, −3), (−2, −6)

55. (0, 6), (6, 3), (5, −3), (−2, 0)

57. $A(0, 0)$, (1, 3), (4, 3), (5, 0)

59. $A(0, 0)$, (9, 2), (9, −5)

61. Find the square of the difference of the x-coordinates and the square of the difference of the y-coordinates. Add these quantities and take the square root of the result.

63. 17 **65.** 16 **67.** 10 **69.** 5 **71.** 10.20

73. Collinear

75. Noncollinear **77.** (4, −2) **79.** (1, 5) **81.** (0, 1.75)

83. **(a)** 40 **(b)** 25.30 **85.** **(a)** 25 **(b)** 20.65

91. Yes **93.** No

95. **(a)** 28 **(b)** 26.25 **97.** No **99.** 21.6 inches

101. **(a)** Employed full time with one child

(b) Employed full time with four or more children

103. **(a)**

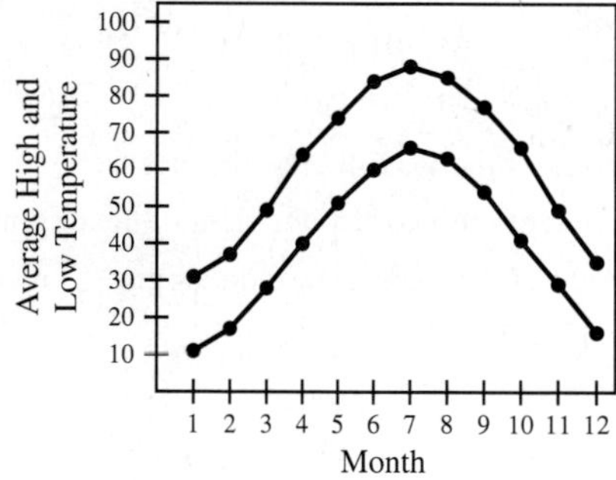

(b) Choose the two points representing the average high and the average low for any month. If the y-coordinates of those points are y_H and y_L, then the distance $y_H - y_L$ is the difference between the average high and low temperatures for that month.

105. (1960, 55), (1970, 63), (1980, 75), (1990, 85)

107. The estimated population for the South in 1975 was 70 million.

109.

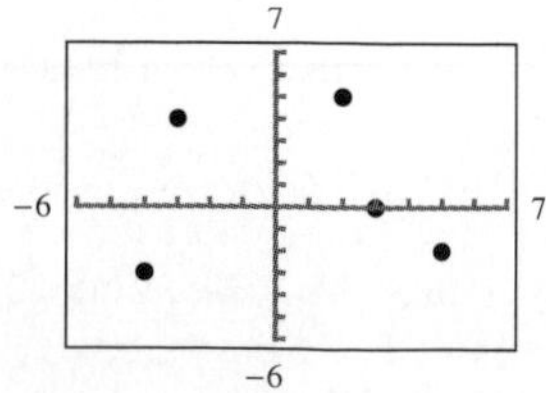

111.

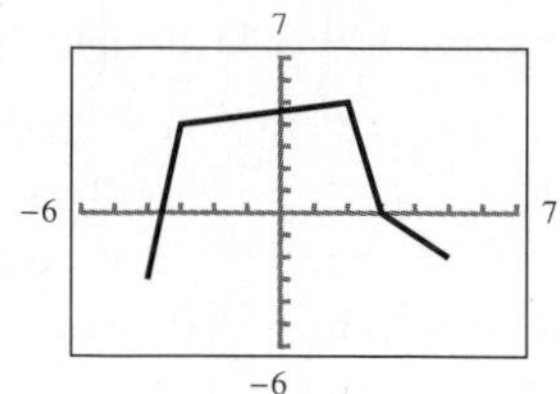

Section 2.2 *(page 82)*

1. On the home screen, store 2 in X and retrieve the value of Y_1.

3. −4, −13, 5, −11 **5.** 3, 5, −22, 6 **7.** 1, 2, 0, $\sqrt{5}$

9. $-4, \frac{1}{5}, -\frac{6}{5}, 0$ **11.** 2, 6.5, 8.75 **13.** 21, 9, 9

15. 4, 5, 13 **17.** 0, undefined, 0.375

19. The second coordinate is 7 because the value of the expression $2x + 1$ is 7 when $x = 3$.

21.

x	−2	1	−4	0
$3x$	−6	3	−12	0
$(x, 3x)$	(−2, −6)	(1, 3)	(−4, −12)	(0, 0)

23.

x	0	−2	2	−4
$2x^2$	0	8	8	32
$(x, 2x^2)$	(0, 0)	(−2, 8)	(2, 8)	(−4, 32)

25. (ii) **27.** $\frac{5}{3}, 7, -\frac{17}{3}, -3$ **29.** −7, 1, 1, −7

31. $-\frac{1}{5}, \frac{1}{3}, \frac{1}{4}, -\frac{1}{2}$

33. 11, 3, 5, 11 **35.** 10, −12, −7, −2

37. 1, 4, −3, −5 **39.** 0, −8, 27, none

41. −6, 0; −3; none; −8, 2

43. −8, 7; none; −1, 0; −0.5

45. You would trace to the y-axis because every point of the y-axis has an x-coordinate of 0.

47. **(a)** 12 **(b)** 4 **49.** **(a)** 9 **(b)** −3 or 3

51. **(a)** 2 **(b)** −2 **53.** **(a)** −16 **(b)** −2 or 2

55. 8 **57.** −5, 0 **59.** −12, 0 **61.** 1 time **63.** 3 times

65. (i), (ii) **67.** (i), (iii) **69.** 2 **71.** 0.6 **73.** $B(15, 17)$

75. $B(18, -6)$

77. **(a)** $1{,}000{,}000 + 40{,}000x$, where x is the number of additional workers **(b)** 4

79. **(a)** $12 + \frac{4x}{5000}$ **(b)** 28 calls **(c)** \$60,000

81. 1985 and 1995

83. Favorable: 1996; lack credibility: 1994

85. The sales of real trees are rising, whereas the sales of artificial trees are declining.

87. 1997 **89.** −2, 6; $0 \le x \le 4$; none

91. −4; none; 4 **93.** 46, 286

Section 2.3 *(page 91)*

1. A relation is a set of ordered pairs. A function is a relation in which each first coordinate is paired with exactly one second coordinate. Every function is a relation, but not every relation is a function.

3. $\{(-3, 9), (-2, 4), (-1, 1), (0, 0), (1, 1), (2, 4), (3, 9)\}$
Domain: $\{-3, -2, -1, 0, 1, 2, 3\}$
Range: $\{0, 1, 4, 9\}$

5. $\{(1, 4), (2, 8), (3, 12), (4, 16), (5, 20)\}$
Domain: $\{1, 2, 3, 4, 5\}$
Range: $\{4, 8, 12, 16, 20\}$

7. $\{(10, 70), (20, 140), (30, 210), (40, 280)\}$
Domain: $\{10, 20, 30, 40\}$
Range: $\{70, 140, 210, 280\}$

9. **11.**

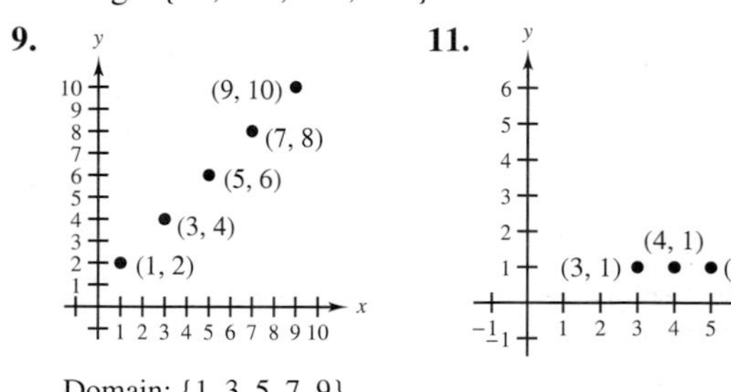

9. Domain: $\{1, 3, 5, 7, 9\}$
Range: $\{2, 4, 6, 8, 10\}$
Function

11. Domain: $\{3, 4, 5\}$
Range: $\{1\}$
Function

13. **15.**

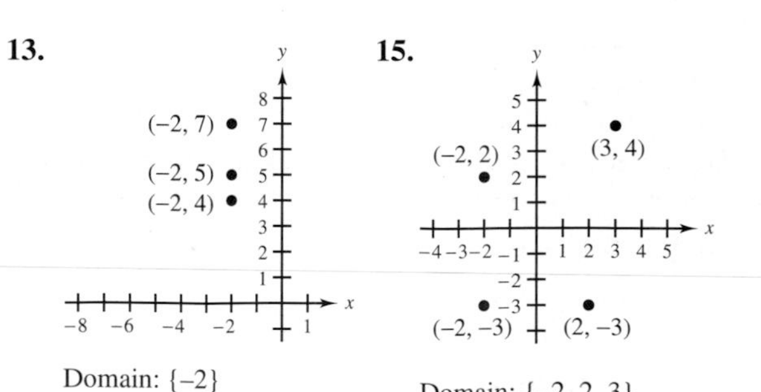

13. Domain: $\{-2\}$
Range: $\{4, 5, 7\}$
Not a function

15. Domain: $\{-2, 2, 3\}$
Range: $\{-3, 2, 4\}$
Not a function

17.

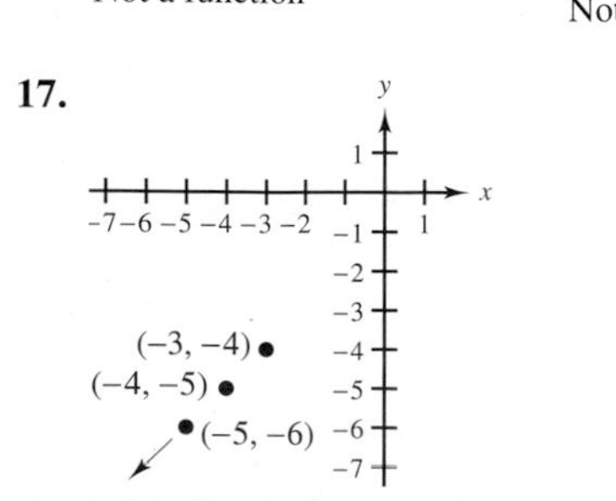

Domain: $\{\ldots, -5, -4, -3\}$
Range: $\{\ldots, -6, -5, -4\}$
Function

19. The domain of a relation can be estimated from a graph by determining the x-coordinates of the points of the graph.

21. For a graph to represent a function, no vertical line can intersect the graph in more than one point. If a vertical line intersects a graph in more than one point, there are two ordered pairs with the same first coordinate. This violates the definition of a function.

23. Domain: $\{3.00, 4.50, 6.00\}$
Range: $\{1.2, 2.7, 3.8, 4.1, 5.3\}$
Not a function

24. Domain: {Washington, Dallas, San Francisco}
Range: $\{1992, 1993, 1994, 1995, 1996\}$
Not a function

25. Domain: {Maine, Michigan, Montana, Texas, Oregon}
Range: {Augusta, Lansing, Helena, Austin, Salem}
Function

27. Domain: $[-7, 1]$
Range: $[-6, 2]$
Not a function

29. Domain: $\{-5, -4, -3, -2, -1, 0, 1, 2, 3\}$
Range: $\{3\}$
Function

31. Domain: $[-1, \infty)$
Range: $\mathbf{R}$
Not a function

33. Domain: $\mathbf{R}$
Range: $(-\infty, -3] \cup [3, \infty)$
Not a function

35. Domain: $\mathbf{R}$
Range: $\mathbf{R}$
Function

37. Domain: $\mathbf{R}$
Range: $[-7.5, \infty)$
Function

39. Yes **41.** No **43.** Yes **45.** No **47.** Yes

49. No **51.** No **53.** Yes **55.** Yes

57. Domain: $\mathbf{R}$
Range: $\mathbf{R}$

59. Domain: $\mathbf{R}$
Range: $[-4, \infty)$

61. Domain: $\mathbf{R}$
Range: $(-\infty, 4]$

63. Domain: $[-9, \infty)$
Range: $[-5, \infty)$

65. Domain: $\mathbf{R}$
Range: $[0, \infty)$

67. Domain: $\mathbf{R}$
Range: $[0, \infty)$

69. **(a)** $\{(1, 190), (2, 330), (3, 470), (4, 610), (5, 750), (6, 890), (7, 1030)\}$
(b) Domain: $\{1, 2, 3, 4, 5, 6, 7\}$
Range: $\{190, 330, 470, 610, 750, 890, 1030\}$
(c) Yes

71. **(a)** $I = x(12 - x)$ **(b)** \$36 **(c)** 6 **(d)** $\{x \mid 0 \le x \le 12\}$
(e) $\{y \mid 0 \le y \le 36\}$ **(f)** 4 or 8

73. **(a)** The percentage of movie-goers decreases with age.
(b) 34 **(c)** Answers will vary.

75. **(a)** 1992 **(b)** 1985

77. $A =$ {(New Orleans, 256), (Tampa, 246), (Jackson, 231), (Mobile, 217), (Casper, 217)} Set A is a function.

79. No **81.** No **83.** Yes **85.** No

87. Domain: **R**, $x \neq -3$
Range: **R**, $y \neq -6$

89. Domain: **R**
Range: (0, 1]

Section 2.4 *(page 99)*

1. To evaluate $f(3)$, replace x in the expression $2x + 1$ with 3 and perform the indicated operations. $f(3) = 2(3) + 1 = 7$.

3. (a) $-2, -2$ **(b)** $2t, 2t$ **(c)** $a + 1, a + 1$

5. $4, -14$ **7.** $-10, -4$ **9.** $-8, 2$ **11.** $-1, \frac{3}{2}$

13. (a) -3 **(b)** -9 **15. (a)** 4 **(b)** 3

17. One point of the graph is $(4, -7)$. The x-coordinate of each point of the graph is the value of the variable, and the y-coordinate is the value of the function.

19. (a) -13 **(b)** -23 **(c)** -11 **(d)** -5

21. (a) 2 **(b)** 2 **(c)** 18 **(d)** 0

23. (a) 1 **(b)** 4 **(c)** 7 **(d)** 1

25. Enter the function on the Y screen as Y_1. Then go to the home screen and store the value for which the function is to be evaluated. Retrieving Y_1 displays the value of the function for the stored value of the variable.

27. (a) -32.9 **(b)** -2.75 **(c)** 2 **(d)** 397

29. (a) -4.44 **(b)** -559.04 **(c)** -4165.54 **(d)** 4.02

31. (a) $h(5) = -2, k(5) = -\frac{3}{4}$

(b) The parentheses in function k require the subtraction to be performed before the division.

33. -2 **35.** 5 **37.** -59 **39.** 26

41. (a) $G(1) = G(-4) = G(2.3) = -3$

(b) The results in part (a) suggest that $G(x) = -3$ for all x.

(c) $G(x) = -3$

43. (a) -1 **(b)** 2X − 3(X − √(X − 1))

45. (a) 3 − 5 * ABS (X − 3) **(b)** -27

47. (a) 8.15 **(b)** 2.59 **49. (a)** 9.49 **(b)** 3.77

51. (a) 11,441 **(b)** 7736 **53. (a)** 0 **(b)** 1.87

55. (a) $f(7) = -49, g(7) = 49, h(7) = 49$

(b) $f(-7) = -49, g(-7) = 49, h(-7) = 49$

(c) f: B, A; g: B; h: A, B

(d) Functions g and h have the same graph because $x^2 = (-x)^2$.

57. (a) $B(0) = C(0) = 5$ **(b)** $B(3) = \sqrt{34}, C(3) = 8$

(c) $B(-6) = \sqrt{61}, C(-6) = -1$

(d) The two functions do not have the same graph because $B(x) \neq C(x)$ for all x.

59. (a) Yes **(b)** No **61. (a)** $10a - 7$ **(b)** $5a + 3$

63. (a) $t^2 - 5t + 6$ **(b)** $4t^2 - 10t + 6$

65. (a) $\sqrt{a^2 + 5}$ **(b)** $\sqrt{4a^2 + 5}$

67. (a) $|-s^3 - 2|$ **(b)** $|27s^3 - 2|$

69. (a) 1 **(b)** $2h + 1$ **(c)** $2h$ **(d)** 2

71. (a) -25 **(b)** $4h - 25$ **(c)** $4h$ **(d)** 4

73. 55 paces per minute

75. (a) 5 days, 75,000 **(b)** 10 days

77.

Year	1992	1995	1998	2001
x	2	5	8	11
Semiprivate	7157	8142	9127	10,112
Private	5204	4759	4314	3870

79. 0, 2, 3, 5

81. For 1988, $t = -2$. From the model, helmet sales in 1988 would be a negative number. The model appears not to be valid prior to 1990.

83. (a) 3, 30, 300, 3000, 30,000

(b) As x becomes smaller, $\frac{3}{x}$ becomes larger.

(c) The result is an error message because division by 0 is undefined.

Section 2.5 *(page 109)*

1. (a) Trace to the point whose x-coordinate is 0 and read the corresponding y-coordinate.

(b) Evaluate the function for $x = 0$.

3. If the graph of a relation has two y-intercepts, the relation contains two ordered pairs with the same first coordinate. The graph would not pass the Vertical Line Test. Thus the relation is not a function.

5. (a) 10 **(b)** $(4, 0), (0, -4)$

7. (a) 9 **(b)** $(24, 0), (0, -8)$

9. (a) -3.45 **(b)** $(0, -3), (-3, 0), (1, 0)$

11. (a) 1.45 **(b)** $(0, 8), (-4, 0), (2, 0)$

13. (a) -15 **(b)** 10 **(c)** 50

15. (a) 27 **(b)** 5 or -3 **(c)** None

17. (a) 2 **(b)** 18 **(c)** 0

19. (a) 11 **(b)** 14 or -14 **(c)** None

21. None; 0 **23.** 9; none **25.** None; 6 **26.** 6; none

27. (a) $(0.82, -1.09)$ **(b)** $(-0.82, 1.09)$

29. (a) $(2.63, -9.71)$ **(b)** $(-0.63, 7.71)$

31. Local maximum: $(-1.43, 16.90)$
Local minimum: $(2.10, -5.05)$

33. Because $f(x) = -(2 + |x|)$ is always negative, the graph of f has no x-intercepts.

35.

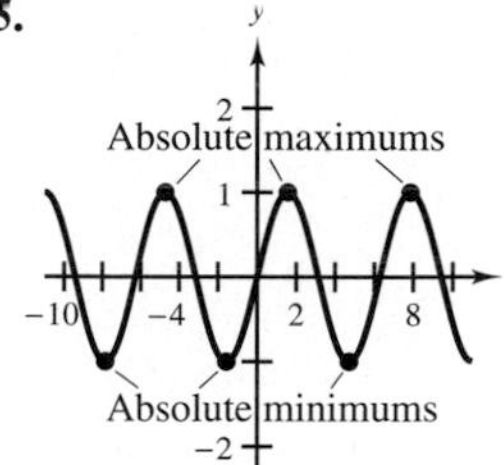

37.

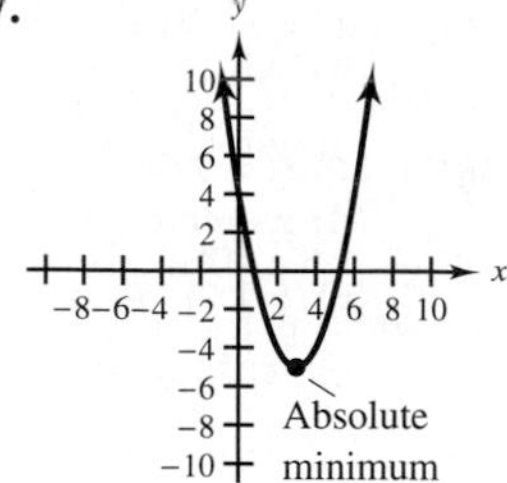

39.

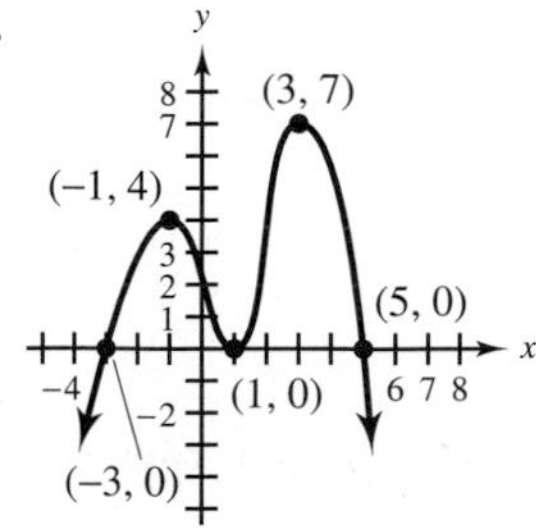

41. **(a)** 4 x-intercepts and 1 y-intercept

(b) 2 local minimums and 1 local maximum

(c) Domain: **R**; range: $[-4.84, \infty)$

(d) $(1, 0)$

43. If $x = -5$, then $g(-5) = \sqrt{-7}$, which is not a real number.

45. **(a)** No x-intercepts; y-intercept is $(0, 5)$

(b) Absolute minimum: 3; no absolute maximum

(c) Domain: **R**; range: $[3, \infty)$

47. **(a)** x-intercept is $(1, 0)$; no y-intercept

(b) Absolute minimum: 0; no absolute maximum

(c) Domain: $[1, \infty)$; range: $[0, \infty)$

49. **(a)** f **(b)** g **(c)** f and g

51. **(a)** f **(b)** g **(c)** f and g

53.

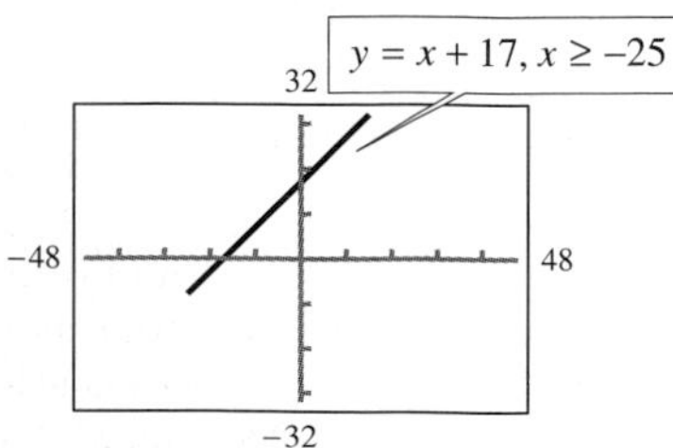

55.

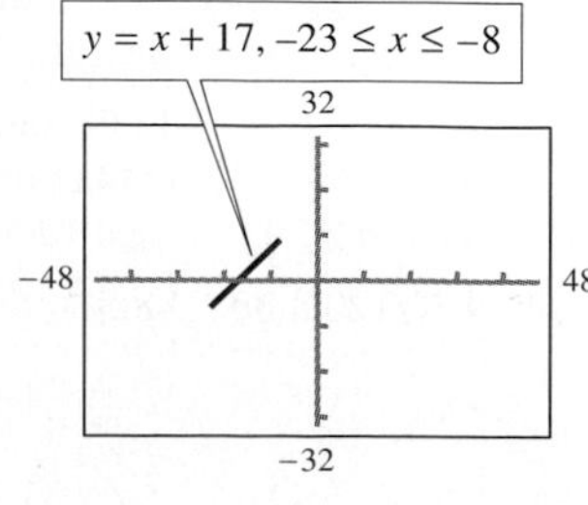

57.

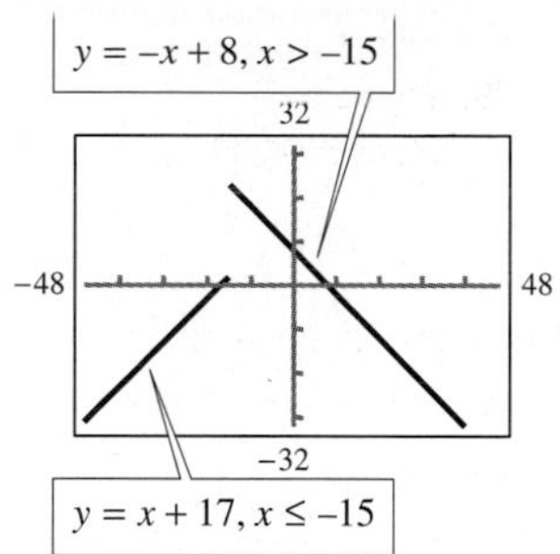

59. **(a)** The pair (2, 8) belongs to the area graph. When the width is 2 feet, the area is 8 square feet.

(b) The pair (2, 12) belongs to the perimeter graph. When the width is 2 feet, the perimeter is 12 feet.

(c) (3, 18)

(d) The board should be 3 feet wide and 6 feet long.

(e) The area would be 18 square feet, and the perimeter would be 18 feet.

61. 1.25 seconds and 5 seconds

63. **(a)** 1991 **(b)** \$27.9 billion

65. Opal: 4; Erin: 1; Roxanne: 3; Camille: 5; David: 1; Humberto: 2; Carla: 4

67. 1960; 800 **69.** 2294 **71.** **(a)** $x \le 3$ **(b)** $x \ge 3$

73. **(a)** $-3 \le x \le 3$ **(b)** $x \le -3$ or $x \ge 3$

75. **(a)** $x = 0$ **(b)** $x \ge 0$

77. **(a)**

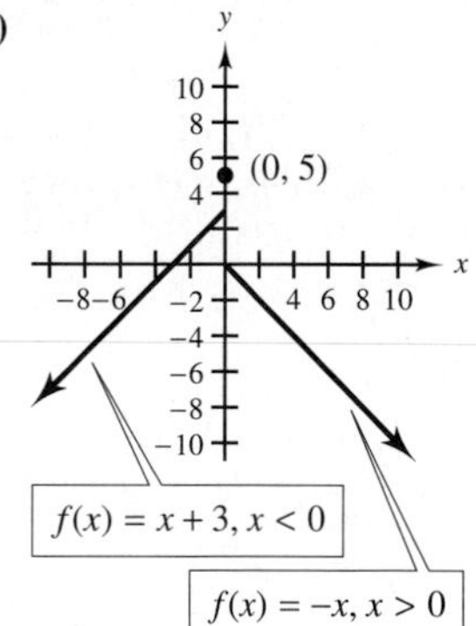

(b) 5

(c) The range is the set containing 5 and all real numbers less than 3.

Chapter Review Exercises *(page 114)*

1. **(a)** $(-4, 3)$ **(b)** $(5, 0)$ **(c)** $(-2, -3)$ **(d)** $(1, 1)$ **(e)** $(3, -4)$ **(f)** $(0, -2)$

3. IV **5.** y-axis

7.

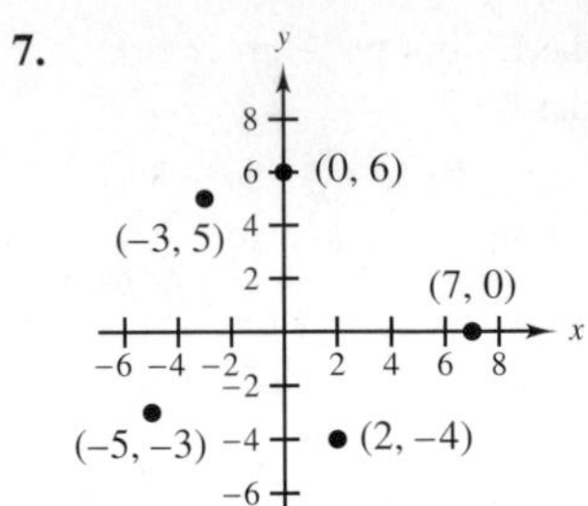

9. (a) 9.49 (b) $(-6.5, -0.5)$

11. 36 13. $-32, 0, 3, -21$ 15. -4 17. -6.5

19. The second coordinate of each point of the graph is the value of the expression when the value of the variable is the first coordinate. Because the value of the expression is 0, the desired point has the form $(x, 0)$, which is a point of the x-axis.

21. (i), (ii), (iii)

23. Domain: $\{-4, 0, 3, 5\}$
Range: $\{-6, 1, 5, 7\}$
Function

25. Yes 27. No

29. Domain: $(-\infty, 9]$ 31. (ii) 33. 8 35. $\sqrt{21}$ 37. $\frac{3}{76}$

39. -4 41. (a) 3 (b) $-\frac{15}{2}$

43. (√(X^2−3)−4)÷(ABS(X^−1−2))

45. (a) Domain: $[-2, \infty)$
Range: $(-\infty, 20]$
(b) 14 (c) 6

47. (a) $(-3, 0)$; $(0, 3)$ (b) 5 (c) 4 49. x-intercepts

Chapter Test *(page 116)*

1. [*2.1*] $A(-3, 6)$, $B(0, 3)$, $C(3, 0)$, $D(6, -3)$
2. [*2.1*] (a) I, IV, or x-axis (b) II
3. [*2.1*] (a) (2.5, 1) (b) $\sqrt{34}$
4. [*2.1*] (a) About 76 miles; no
   (b) The locations are not points of a line.
5. [*2.2*] The second coordinate is the value of the expression when the variable is 0.
6. [*2.2*] 7, 0, 3.75, 4.375
7. [*2.2*] (a) 2 (b) -8 or 8 8. [*2.2*] 2 or -5
9. [*2.4*] $\frac{1}{4}$ 10. [*2.4*] 3 11. [*2.4*] $3t$
12. [*2.3*] (a) Domain: $[-4, 4]$; range: $[-5, 5]$
   (b) Domain: $\{1, 2, 3, 5\}$; range: $\{-1, 3, 7\}$
13. [*2.3*] (a) Yes. The graph passes the Vertical Line Test.
   (b) No. The graph does not pass the Vertical Line Test.
14. [*2.5*] (a) $[-30.25, \infty)$ (b) $(-1, 0)$, $(10, 0)$
15. [*2.5*] (a) -5 (b) $(-\infty, 4]$
16. [*2.5*] (a) $(-4, 6)$, $(4, 2)$ (b) $(0, 2)$
17. [*2.5*] (a) Local maximum
   (b) Absolute maximum
   (c) y-intercept
   (d) x-intercept
   (e) Local minimum
   (f) Absolute maximum
18. [*2.5*] (a) The absolute maximum is 800. A maximum daily profit of \$800 is earned when the price of admission is \$5.
   (b) 2.55 or 7.45; Ticket prices at which the profit is \$500.

CHAPTER 3

Section 3.1 *(page 124)*

1. An equation is a statement that two algebraic expressions have the same value.

3. Yes 5. No 7. Yes 9. No 11. Yes 13. Yes

15. No 17. Yes 19. No

21. The equation is true for some values of x ($x \geq 0$), but it is false for other values of x ($x < 0$).

33. 5 35. -5 37. $-3, 0, 2$ 39. -11

41. The graphs of the left and right sides of an inconsistent linear equation are parallel lines. The solution set is Ø.

43. 3 45. 3 47. 2 49. 2 51. **R** 53. -2 55. 0

57. Ø 59. -18 61. **R** 63. 6 65. 6

67. (a) The number a is the solution of the equation.
   (b) The number b is the value of both the left and right sides when the variable is replaced with a.

69. 11; (11, 25) 71. -5; $(-5, 3)$ 73. 15; $(15, -3.75)$

75. -12; $(-12, -7)$

77. (a) $20 + 2h$, $10 + 2.25h$ (b) $h = 40$
   (c) For 40 on-line hours, the two plans cost the same.
   (d) For both plans, the cost is \$100 for 40 on-line hours.

79. (a) 4.8
   (b) The first coordinate represents the number of years since 1990 when the number of new applications and the number of renewals were approximately equal. The year is about 1995. The second coordinate represents the equal number of new applications and renewals. The number is approximately 2.58 million.
   (c) The line graph for renewals eventually will fall below the x-axis, which would mean a negative number of renewals. Thus the model is unreliable beyond 1996.

81. $1.16a + 9.27 = 40$; $a \approx 26$ 83. 120 85. -8.5

87. $-6, 10$ 89. Ø

91. The equation $0 \cdot x + B = C$ is not a linear equation in one variable because it cannot be written as $Ax + B = 0$, where $A \neq 0$. The equation is an identity if $B = C$.

Section 3.2 *(page 134)*

1. Equivalent equations are equations that have exactly the same solution sets.

3. For $x + 2 = 3$, 2 is added to x, so we add -2 to both sides to isolate the variable. For $-2x = 6$, x is multiplied by -2, so we divide both sides by -2 to isolate the variable.

5. 3 **7.** -12 **9.** -6 **11.** **R** **13.** -8 **15.** 17 **17.** $\emptyset$

19. 6

21. The Multiplication Property of Equations permits us to multiply both *sides* by 15: $15(\frac{1}{3} - 2x) = 15(\frac{2}{5})$. By the Distributive Property, we obtain $15 \cdot \frac{1}{3} - 15 \cdot 2x = 15 \cdot \frac{2}{5}$.

23. $-\frac{1}{24}$ **25.** 1 **27.** $-\frac{10}{3}$ **29.** $\frac{40}{3}$ **31.** 5 **33.** 4

35. -100 **37.** Yes **39.** Yes **41.** No **43.** 6; conditional

45. **R**; identity **47.** $\emptyset$; inconsistent **49.** 15; conditional

51. -15 **53.** 8 **55.** 3 **57.** 1 **59.** 3 **61.** 5 **63.** 16

65. -22 **67.** $\frac{5}{2}$ **69.** -5 **71.** **R** **73.** $\frac{1}{3}$ **75.** $\emptyset$ **77.** $\frac{21}{2}$

79. $\frac{23}{5}$ **81.** 3 **83.** 5

85. **(a)** $30 + 8(s - 1)$ **(b)** $8s + 22 = 62$; $s = 5$

87. **(a)** 1989 **(b)** 2015

89. **(a)** $0.28x - 1.07 = 20$

(b) $x \approx 75$; the tax rate will reach 20% by the year 2010.

91. The points represent the years 1979 ($x = 0$) and 1995 ($x = 16$) and the number of production-line jobs in each of those years. The information from the model agrees with the data.

93. The coordinates (12, 18.85) indicate that in 1991 there were 18.75 million production-line jobs.

95. $\frac{b-c}{a}$, $a \neq 0$ **97.** $\frac{d-b}{a-c}$, $a \neq c$ **99.** $-\frac{26}{77}$ **101.** No

Section 3.3 *(page 141)*

1. To solve a formula for a given variable, treat all other variables as constants and use normal equation-solving procedures to isolate the given variable.

3. $m = \dfrac{F}{a}$ **5.** $r = \dfrac{I}{Pt}$ **7.** $t = \dfrac{s}{v}$ **9.** $b = \dfrac{2A}{h}$

11. $a = \dfrac{2A - bh}{h}$ **13.** $\pi = \dfrac{A}{r^2}$ **15.** $m = \dfrac{E}{c^2}$

17. $v = at + w$ **19.** $r = \dfrac{A - P}{Pt}$ **21.** $R = \dfrac{PV}{nT}$

23. $R_1 = 2R_T - R_2$ **25.** $y = -\dfrac{ax + c}{b}$ **27.** $v = \dfrac{2s - gt^2}{2t}$

29. $n = \dfrac{A - a + d}{d}$ **31.** $a = \dfrac{2S - n(n-1)d}{2n}$

33. To isolate y, we can divide both sides by 3, or we can divide each term of both sides by 3.

35. $y = -\frac{2}{5}x + 2$ **37.** $y = x + 5$ **39.** $y = -\frac{2}{3}x + 3$

41. $y = \frac{3}{4}x - \frac{11}{4}$ **43.** $y = -\frac{1}{2}x$ **45.** $y = \frac{4}{3}x + 40$

47. $y = \frac{1}{2}x + 7$ **49.** $y = \frac{6}{5}x + 12$ **51.** $y = \frac{3}{4}x + \frac{17}{4}$

53. **(a)** 144.71 square inches

(b) $h = \dfrac{A - 2\pi r^2}{2\pi r}$ **(c)** 22 centimeters

55. All except (*v*) **57.** 672.27 feet **59.** 67 feet

61. The depth H of the water in the second tank is calculated by $H = \frac{V}{LW} = \frac{15}{6(1.5)} = \frac{15}{9} \approx 1.7$ feet. Thus the water in the second tank will be deeper.

63. 555 feet **65.** $S = 2WH + 2LH + 2WL$

67. **(a)** 160 **(b)** 130 **(c)** 190 **(d)** 180

69. $D = (1 - r)R$ **(a)** \$102.17 **(b)** 24%

(c) \$84.20 **(d)** \$220.20 **(e)** 44%

71. **(a)** 127 miles **(b)** 159 miles **(c)** 239 miles

73. 20.56 meters

75. The model appears to be a good approximation of the data.

77. The average of the differences is -0.83. Although this difference is small, the model is not a good approximation of the data because the magnitude of the differences is very large.

79. $x = \dfrac{ac + bc}{b - a}$ **81.** $r = \dfrac{S - a}{S}$ **83.** $V_2 = \dfrac{P_1V_1T_2}{P_2T_1}$

85. $L = \dfrac{A - 2WH}{2W + 2H}$

Section 3.4 *(page 151)*

1. The other number is $T - n$. The sum of the numbers is $n + (T - n) = T$.

3. -3 **5.** 13 **7.** 36

9. Both consecutive even integers and consecutive odd integers are 2 units apart on the number line. Whether the integers are even or odd depends on whether x represents an even integer or an odd integer. The representations are the same in both cases.

11. 47, 48 **13.** 13 **15.** 4 feet, 8 feet, 33 feet

17. 30 yards, 25 yards, 50 yards **19.** 40°, 60°, 80°

21. Yes; 30°, 60°, 90° **23.** 20 feet, 12 feet

25. 13 yards, 17 yards **27.** 21 **29.** 19

31. 80° **33.** No, the book is 14 inches high.

35. 9 inches **37.** No, the window is 45 inches high.

39. Yes, the angles are 26°, 50°, and 104°.

41. 56.4° **43.** 12° **45.** Answers will vary.

47. Because the sum of even integers is always even, the sum cannot be 19.

49. 160 miles **51.** \$81.36 **53.** 30 feet

55. Base: 9 inches; sides: 6 inches

57. 640 **59.** 26 **61.** \$400, \$800 **63.** 4 **65.** 2 points

67. The sum of an odd integer and an even integer cannot be an even integer.

69. 49 **71.** 8 meters, 10 meters, 22 meters

73. 12 **75.** 58 **77.** \$40,000

79. 38 children, 152 men, 76 women

81. She cannot know for sure. The solution of $x + (x + 2) + (x + 4) = 597$ is 197. Therefore, the highest score was $x + 4 = 201$, but it was not necessarily the husband's score.

83. **(a)** $37{,}700 + 6000y$

(b) $37{,}700 + 6000y = 61{,}700$; $y = 4$ (1999)

85. $-3x + 127 = 50$; $x \approx 25$ (1995)

87. 3 metric tons per year

Section 3.5 *(page 164)*

1. In mathematics, *of* means *times*: 0.15(30).

3. \$205.33 **5.** \$350, lost money **7.** \$2000 **9.** 20%

11. \$25 **13.** \$299

15. The costs in (a), (c), and (d) are variable costs; the costs in (b) and (e) are fixed costs.

17. 7.6 hours **19.** 200

21. The value is the principal P plus 7% of the principal: (iii).

23. \$571.43 **25.** \$1100 **27.** \$20,000

29. \$2500 at 9%, \$4000 at 6%

31. \$4000 **33.** 4 mph **35.** bus: 56 mph; car: 62 mph

37. 2:24 P.M., 168 miles **39.** 10 pounds

41. 21 pounds **43.** 3 pounds

45. 50 gallons of 3%, 100 gallons of 4.5%

47. 2 ounces of each **49.** 64 ounces **51.** \$1475

53. English: \$33; biology: \$37; math: \$49 **55.** 4 pounds

57. 20° **59.** 1.54 cups of 12%, 0.46 cup of 25%

61. 12:00 noon **63.** \$82,000

65. Rabbit: 5.84 seconds; cat: 6.82 seconds; elephant: 8.18 seconds; coyote: 4.76 seconds; tortoise: 20.06 minutes; snail: 1.89 hours

66. \$560

67. Traffic: 61.3 million; civil: 11.5 million; criminal: 10.0 million; domestic: 3.5 million; juvenile: 1.2 million

69. About a 1% decrease

71. You should not agree to the plan. If x = wholesale cost, then $1.25x$ = retail value. The buyer's offer is $0.75(1.25x) = 0.9375x$. Because this is less than the amount you paid for the inventory, you would lose money.

Section 3.6 *(page 177)*

1. The interval $[a, b]$ contains a, but the interval $(a, b]$ does not.

3. $(-\infty, 2)$, $\{x \mid x < 2\}$

5. $[3, \infty)$, $\{x \mid x \geq 3\}$

7. $(-3, 2]$, $\{x \mid -3 < x \leq 2\}$

9. In both cases we graph the left and right sides. For an equation, the solution is the x-coordinate of the point of intersection. For an inequality, the solution set is the set of x-values for which the graph of one side is above (or below) the graph of the other side.

11. **(a)** $[-14, \infty)$ **(b)** $(-\infty, -14)$

13. $(-\infty, 8)$ **15.** $[-5, \infty)$ **17.** **R** **19.** $(-\infty, 4]$ **21.** Ø

23. **(a)** The graphs are parallel (or coincide for $<$ or $>$).

(b) The graphs are parallel (or coincide for $\leq$ or $\geq$).

25. $<$ **27.** $>$ **29.** $>$ **31.** $<$ **33.** $c < 0$

35. Any real number **37.** $c > 0$

39. The methods are the same, except that we must reverse the inequality symbol if we multiply or divide both sides of an inequality by a negative number.

41. $(-\infty, 6)$ **43.** $[5, \infty)$ **45.** **R** **47.** $[3, \infty)$ **49.** $(-\infty, 8)$

51. $\left(-\infty, \frac{2}{5}\right]$ **53.** Ø **55.** $[0, \infty)$ **57.** Ø **59.** $\left(-\infty, \frac{5}{2}\right)$

61. $(-\infty, -8]$ **63.** $\left(-\infty, \frac{35}{4}\right)$ **65.** $\left(-\frac{19}{2}, \infty\right)$ **67.** $\left(-\infty, \frac{5}{29}\right]$

69. $\left(-\infty, -\frac{5}{11}\right)$ **71.** $[0.23, \infty)$ **73.** $(-\infty, 0.27]$

75. The sum of the lengths of any two sides must be greater than the length of the third side: $a + b > c$, $a + c > b$, $b + c > a$.

77. 4 feet **79.** Third place; you need at least 4 points.

81. 790 miles **83.** \$6000

85. **(a)** $\frac{105}{2}t - \frac{79}{3} \geq 870$

(b) The solutions are $t \geq 17.07$, which means that there would be an average of at least two nonincumbent candidates for each seat starting in about 1997.

87. **(a)** $0.408x + 7.348 > 0.549x + 7.734$

(b) The solutions are $x < -2.74$, which indicates that the National League averaged more runs prior to about 1987, but it will not do so after 1987.

89. $\left(-\infty, \dfrac{b-a}{5}\right]$ **91.** $\left[\dfrac{8}{a^2}, \infty\right)$

93. The first number is at most -2.

Section 3.7 *(page 186)*

1. The solution set is the intersection of the solution sets of the individual inequalities.

3. Yes, no **5.** Yes, yes **7.** Yes, yes **9.** Yes, no

11. A conjunction is a compound inequality in which the connective is *and*. A disjunction is a compound inequality in which the connective is *or*.

13. 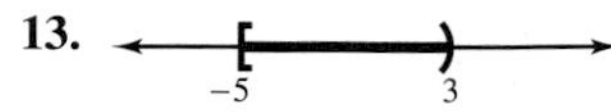

15.

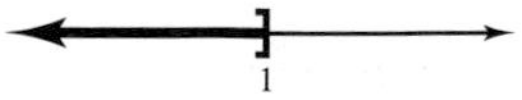

17. Ø

19.

21.

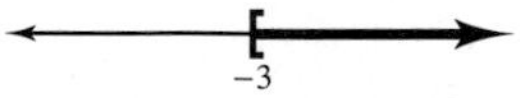

23.

25. **(a)** The graph of y_2 intersects or is above the graph of y_1 at and to the right of the point (−4, 8). The graph of y_2 intersects or is below the graph of y_3 at and to the left of the point (8, 20). Points that satisfy both conditions have x-coordinates in the interval [−4, 8].
    (b) The graph of y_2 intersects or is below the graph of y_1 at and to the left of the point (−4, 8). The graph of y_2 intersects or is above the graph of y_3 at and to the right of the point (8, 20). Points that satisfy at least one condition have x-coordinates in the interval $(-\infty, 4] \cup [8, \infty)$.

27. $\left[-\frac{10}{3}, \frac{10}{3}\right]$ **29.** (−5, 7)

31. **(a)** The only solution is the number c: $\{c\}$.
    (b) No number can be less than c and also greater than c: Ø.
    (c) Every real number is either at most c or at least c: **R**.
    (d) The only number that does not satisfy at least one of the conditions is c: $\{x \mid x \neq c\}$.

33. $-7 < x < 4$ **35.** $-6 \le t \le -\frac{2}{3}$ **37.** $-1.5 \le x \le 4$

39. $-4.5 < x \le -1$ **41.** $-1 \le x < 4$

43. $-4 < x \le 5.75$ **45.** $-4 < x \le 8$

47. The solution set of a disjunction is the union of intervals, whereas the solution set of a conjunction is the intersection of intervals. Thus, for example, the union of $(-\infty, 1)$ and $[1, \infty)$ is **R**, but the intersection of those intervals is Ø.

49. [−5, 7] **51.** [4, 14) **53.** (−11, 6]

55. $(-\infty, -4] \cup [3, \infty)$ **57.** $(-\infty, -7) \cup (-1, \infty)$

59. $(-\infty, -1.25) \cup [1, \infty)$ **61.** (−9, −6]

63. $(-\infty, -1) \cup (5, \infty)$ **65.** $(-\infty, -3] \cup (4, \infty)$ **67.** Ø

69. $[4, \infty)$ **71.** [0, 2] **73.** $(-\infty, 0) \cup (6, \infty)$ **75.** (−5, 1)

77. Ø **79.** $[3, \infty)$ **81.** $(2, \infty)$ **83.** 4 **85.** $(-2, \infty)$

87. Ø **89.** 10 **91.** **R**

93. Because $3 < x > 8$ is incorrect notation, it has no valid interpretation.

95. 45, 40

97. Between 2.5 hours and 3.75 hours, inclusive

99. Between 29 and 35, inclusive

101. Between 3 and 6, inclusive

103. Between 80 and 100, inclusive

105. **(a)** $4486 + 0.093m$
    (b) $5000 \le 4486 + 0.093m \le 7000$
    (c) $5527 \le m \le 27{,}032$
    If the company car is driven between 5527 and 27,032 miles, the cost will be within the budget.

107. **(a)** $0.17t + 1.35 = 0.21t + 1.20$; $t = 3.75$ (about 1994)
    (b) $0.17t + 1.35 < 0.21t + 1.20$; $t > 3.75$ (after about 1994)

109. $0.93x + 5.55 \le 16$

111. $0.93x + 5.55 \le 16$ and $0.40x + 0.72 \ge 4$; $8.2 \le x \le 11.2$ (between about 1998 and 2001)

113. $k \ge 3$

115. **(a)** $y_1 = 0.5x - 15$, $y_2 = x - 5$, $y_3 = -x + 21$
    (b) The points of intersection are the points for which $y_1 = y_2$ and $y_2 = y_3$.
    (c) The solution set is represented by that portion of the graph of y_2 that lies between the graphs of y_1 and y_3. The solution set is (−20, 13).

117. [−7, 3] **119.** **R**

Section 3.8 *(page 197)*

1. The number represented by x is 5 units from 2.

3. −12, 12 **5.** 5 **7.** −25, 5 **9.** −2.5, 7.5

11. 0.4 **13.** −14, 6

15. **(a)** $(-\infty, -14) \cup (8, \infty)$ **(b)** (−14, 8) **(c)** $\{-14, 8\}$

17. (−12, 12) **19.** (−3, 15) **21.** **R**

23. The inequality $|x| \ge 5$ is equivalent to the disjunction $x \le -5$ or $x \ge 5$ and has the solution set $(-\infty, -5] \cup [5, \infty)$. The inequality $|x| \le 5$ is equivalent to the conjunction $-5 \le x \le 5$ and has the solution set [−5, 5].

25. (−4, 5) **27.** −7 **29.** [−3, 11] **31.** Ø

33. $(-\infty, -3] \cup \left[\frac{5}{3}, \infty\right)$ **35.** $(-\infty, 5) \cup (5, \infty)$

37. **R** **38.** **R** **39.** $\left(-\infty, -\frac{11}{4}\right) \cup \left(\frac{5}{4}, \infty\right)$

41. $|x - 5| < 7$ **43.** $|x + 7| \ge 10$ **45.** $|x + 1| < 3$

47. $|x| \ge 2$ **49.** −2.5, 5.5 **51.** −11, 9 **53.** Ø

55. $-1, 11$ **57.** $\emptyset$ **59.** $\frac{4}{3}$ **61.** $-3, 2$ **63.** $-\frac{7}{3}, \frac{17}{3}$

65. 10 **67.** $\frac{5}{3}, 5$ **69.** **R**

71. Divide both sides by -3 to obtain $|x - 2| \le -4$.

73. $(-\infty, -5) \cup (2, \infty)$ **75.** $\left(\frac{1}{3}, \frac{7}{3}\right)$ **77.** **R**

79. $\left(-\infty, -\frac{1}{3}\right] \cup \left[\frac{11}{3}, \infty\right)$ **81.** $\frac{5}{3}$ **83.** $(-\infty, 1) \cup \left(\frac{7}{3}, \infty\right)$

85. $\emptyset$ **87.** $(-1, 2)$ **89.** $\left(-\infty, -\frac{5}{2}\right) \cup \left(\frac{3}{2}, \infty\right)$

91. $\left(-\infty, -\frac{3}{2}\right]$ **93.** $(-\infty, -2)$ **95.** $(-\infty, -1)$

97. $[5, 15]$

99. By taking the absolute value, we guarantee that the distance is not negative.

101. $|L - 12| \le 0.24$; $-0.24 \le L - 12 \le 0.24$

103. $|d - 3| \le \frac{1}{16}$; $-\frac{1}{16} \le d - 3 \le \frac{1}{16}$

105. $|50 - s| \le 6$; between 44 and 56, inclusive

107. $|98.6 - t| \le 0.5$; between 98.1°F and 99.1°F, inclusive

109. $|37.7t - 1804| \ge 400$

111. $|37.7t - 1804| < 326.8$; $39.2 < t < 56.5$; medical costs for those between the ages of 39 and 56 years differ from the average by less than 20%.

113. **(a)** By the definition of absolute value, $|x - 2| = x - 2$ only if $x - 2 \ge 0$ or $x \ge 2$.

(b) By the definition of absolute value, $|x - 2| = -(x - 2)$ only if $x - 2 \le 0$ or $x \le 2$.

(c) The absolute values of a number and its opposite are equal. Therefore, the equation is true for all real numbers.

115. 0 **117.** -2.5 **119.** $x \le 0$

Chapter Review Exercises *(page 201)*

1. No

5. To estimate the solution, produce the graph of each side of the equation and trace to estimate the point of intersection. The x-coordinate of that point corresponds to the solution of the equation. To check the solution, store the value of the estimated solution in X and evaluate each expression.

7. 6 **9.** All real numbers **11.** -30 **13.** 55

15. 0 **17.** -7 **19.** 1.4; conditional **21.** **R**; identity

23. -5 **25.** $H = \dfrac{V}{LW}$ **27.** $d = \dfrac{6xy - acy}{cx}$

29. $I = \dfrac{E}{R}$ **31.** $y = \dfrac{3}{4}x - \dfrac{25}{4}$ **33.** $r = \dfrac{A - P}{Pt}$; 6.5%

35. 2 feet **37.** 32 feet, 33 feet, 34 feet **39.** 225

41. \$155 **43.** 9 hours **45.** 70 mph, 64 mph

47. 15°, 75° **49.** $[2, \infty)$ **51.** $(-\infty, 4)$

53. 

55. $(4, \infty)$ **57.** **R** **59.** $(-\infty, 6)$ **61.** $[3, \infty)$

63. $(-2.5, -0.25]$ **65.** $[11.5, \infty)$ **67.** $(-7, 4)$

69. $(-5, 2]$ **71.** 60 meters **73.** 0 to 600 miles **75.** **R**

77. $[-6, 2]$ **79.** **R** **81.** $(-\infty, -2] \cup [7, \infty)$

83. $(-\infty, -2) \cup (3, \infty)$ **85.** $(-\infty, -1)$

87. Between \$1875 and \$2500, inclusive

89. $-8, 8$ **91.** $-4.5, 7.5$ **93.** $1.5, 2.5$ **95.** $4, \frac{8}{3}$

97. $\left(-\frac{19}{3}, 3\right)$ **99.** $(-\infty, 2] \cup [8, \infty)$

101. $(-\infty, -7] \cup [7, \infty)$ **103.** $-5 \le x - 3 \le 5$

105. $(-\infty, 2)$ **107.** $(-3.5, \infty)$ **109.** 6 or 12

111. Answers will vary.

Chapter Test *(page 205)*

1. [*3.2*] **(a)** Addition Property of Equations

(b) Property of Additive Inverses

(c) Multiplication Property of Equations

2. [*3.1*] 3; conditional **3.** [*3.1*] **R**; identity

4. [*3.1*] $\emptyset$; inconsistent **5.** [*3.3*] $t = \dfrac{A - P}{Pr}$

6. [*3.3*] $y = -\frac{2}{3}x + 4$ **7.** [*3.5*] \$2300

8. [*3.4*] 62.5 meters by 37.5 meters **9.** [*3.5*] 1.2 gallons

10. [*3.5*] 42 mph, 54 mph **11.** [*3.6*] $[-5, 2)$

12. [*3.6*] $(4, \infty)$ **13.** [*3.6*] $(-\infty, -3)$

14. [*3.6*] $(-\infty, -7.5]$ **15.** [*3.6*] $(-\infty, 1.5)$

16. [*3.7*] $[-2, 2]$ **17.** [*3.7*] $(-2, 3)$ **18.** [*3.7*] $[-3, \infty)$

19. [*3.7*] **R** **20.** [*3.8*] $-4, 0$ **21.** [*3.8*] $(-\infty, -1) \cup (3, \infty)$

22. [*3.8*] $[-2, 3]$ **23.** [*3.8*] $\emptyset$

24. [*3.7*] Between 12 minutes and 42 minutes, inclusive

25. [*3.6*] At most 1.35 inches

Cumulative Test: Chapters 1–3 *(page 205)*

1. [*1.1*] **(a)** Rational numbers

(b) Whole numbers

(c) Irrational numbers

(d) Real numbers

2. [*1.2*] **(a)** $\frac{7}{16}$ **(b)** -3 **(c)** 20 **(d)** -12

3. [*1.3*] **(a)** Distributive Property

(b) Associative Property of Multiplication

(c) Property of Additive Inverses

(d) Distributive Property

4. [*1.4*] **(a)** 0 **(b)** 1 **5.** [*1.5*] **(a)** $-9 - x$ **(b)** 0

6. [*1.6*] **(a)** -7 **(b)** $\dfrac{y^4}{x^5}$ **7.** [*1.7*] **(a)** x^{12} **(b)** $\dfrac{x^2y^6}{4}$

8. [2.1] **(a)** 17.20 **(b)** (1, −2)

9. [2.2] For the graph to represent a function, no vertical line can be drawn that intersects the graph at more than one point.

10. [2.2] 2, −18 **11.** [2.2] −1.4

12. [2.4] **(a)** −19 **(b)** −5.39

13. [2.5] **(a)** (0, 0), (2.2, 0), (−2.2, 0)
(b) Local maximum (−1.3, 4.3); local minimum (1.3, −4.3)

14. [3.1] The solution is 4. When x is replaced with 4, the value of each side of the equation is 17.

15. [3.2] **(a)** $\frac{6}{5}$ **(b)** $\frac{49}{26}$ **16.** [3.3] 72

17. [3.4] 22°, 78°, 80° **18.** [3.5] 1:45 P.M.

19. [3.6] **(a)** $\left(-\frac{2}{3}, \infty\right)$ **(b)** $(-\infty, -1]$

20. [3.7] **(a)** $(-\infty, -1) \cup [3, \infty)$ **(b)** (3, 6)

21. [3.8] **(a)** (−5, 11) **(b)** 3, −5
(c) $(-\infty, -5] \cup [-1, \infty)$

22. [3.6] 15 inches

CHAPTER 4

Section 4.1 *(page 215)*

1. The equation $x = -5$ can be written as $1x + 0y = -5$, which is the standard form of a linear equation in two variables.

3. Yes **5.** No **7.** No **9.** Yes **11.** No **13.** No

15. 5 **17.** 5 **19.** −0.5 **21.** $a = 11, b = -3, c = 6$

23. $a = 3, b = 3, c = 3$ **25.** $a = -5, b = -1, c = 16$

27. For $|x| = 2$, there are two solutions, −2 and 2. For $|x| < 2$, the solutions are all numbers in the open interval from −2 to 2. For $y = x + 4$, one solution is the ordered pair (−2, 2).

29.

x	10	−6	7
y	−6	10	−3

31.

x	0	3	−5
$f(x)$	1	10	−14
(x, y)	(0, 1)	(3, 10)	(−5, −14)

33. **(a)** $2x - 3y - 6 = 0$ **(b)** $c = -1.5, d = -\frac{2}{3}$

35.

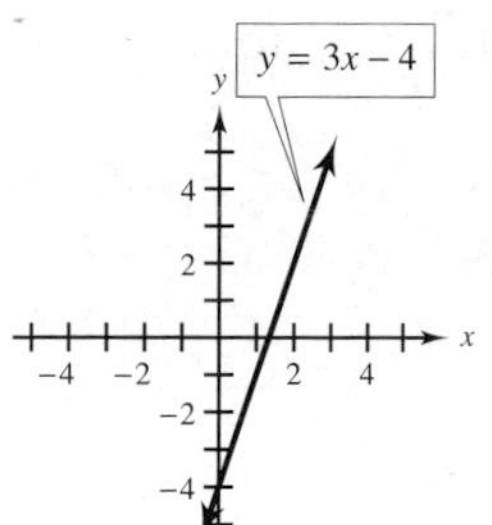

37.

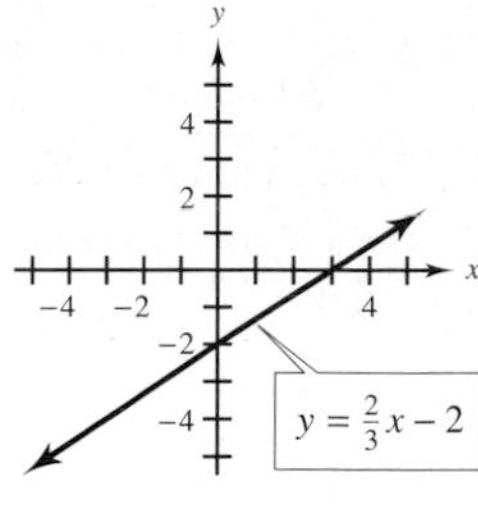

39.

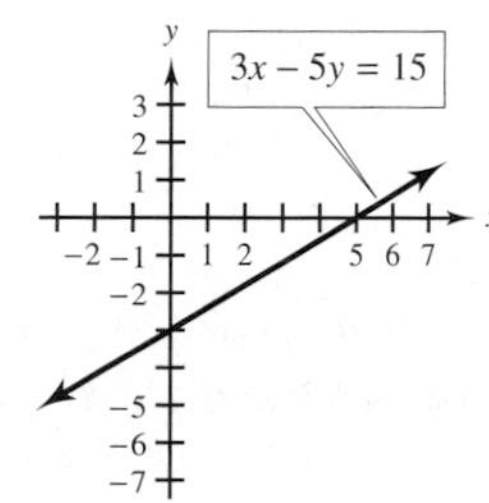

41.

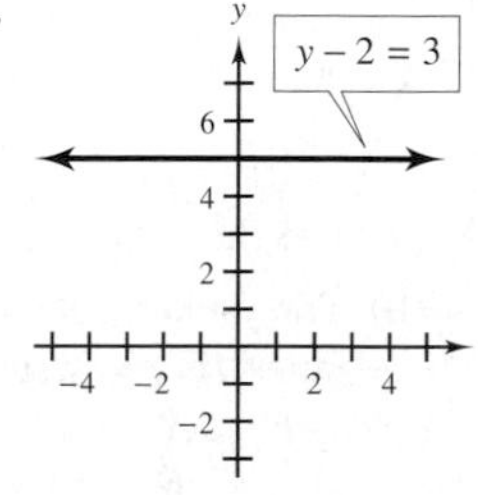

43.

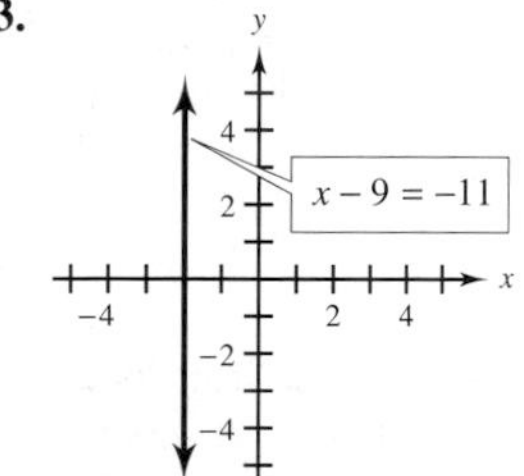

45.

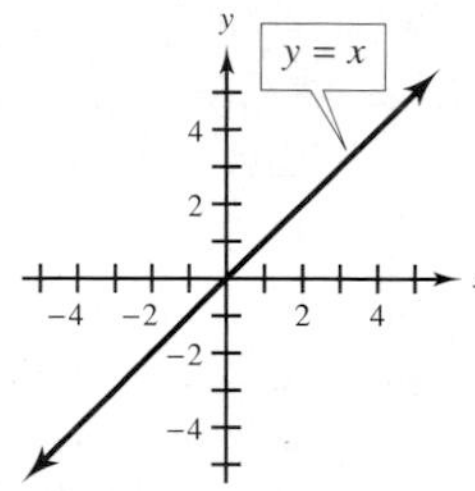

47.

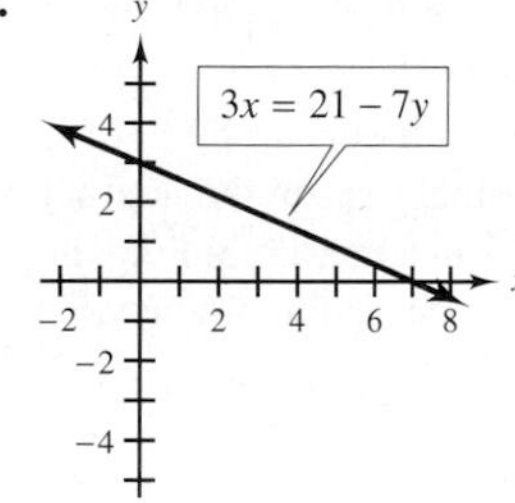

49. (0, 12), (4, 0) **51.** (0, −10), (30, 0)

53. (0, −9), (12, 0)

55. $5x + 3y - 9 = 0$ and $6x + 2y - 6 = 0$

57. $y = -\frac{2}{5}x + \frac{14}{5}$; $\left(0, \frac{14}{5}\right)$ **59.** $y = -4x - 10$; (0, −10)

61. (0, 0) **63.** (0, −3), (4, 0) **65.** (0, 4), (3, 0)

67. (0, 15)

69. If the x- and y-intercepts are the same, the line contains the origin. Therefore, $b = 0$.

71. (a) $y = x + 4$

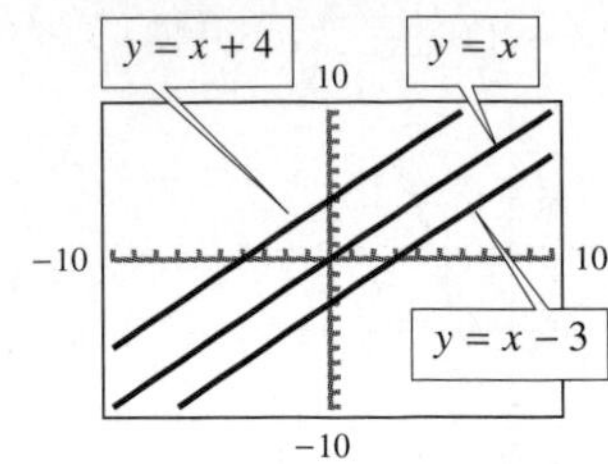

(b) $y = x - 3$

73. $a = -2, c = 8$ **75.** $a = 5, b = -3$

77. Vertical **79.** Neither **81.** Horizontal

83. $x = -y$ **85.** $y = \frac{1}{2}x$ **87.** $x + y = 5$

89. (a) $y = 32 - x$

(b) The y-intercept (0, 32) represents the start of the test when all questions remain to be answered. The x-intercept (32, 0) represents the end of the test when all questions have been answered.

91. \$7.00, \$7.15, \$7.30, \$7.45, \$7.60, \$7.75

93. (a) The y-intercept (0, 676) indicates that 676 million books were sold in 1990.

(b) 2002

95. (a) 0.625 loaf

(b) The x-intercept is (1910, 0), which indicates that no bread was consumed in 1910. The model function is not reliable outside the years 1988–1992.

97. The x-intercept, which is approximately (1988, 0), indicates that in 1988 no E85 was used.

99. The linear model indicates that the consumption of E85 is increasing. However, large increases in the number of cars that use E85 would result in a more rapid rise in consumption, and the linear model might not be valid.

101. The x-intercept is $(g, 0)$, and the y-intercept is $(0, h)$.

103.

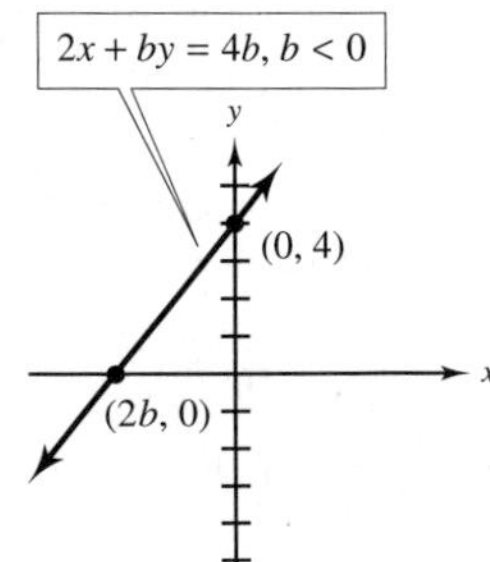

105.

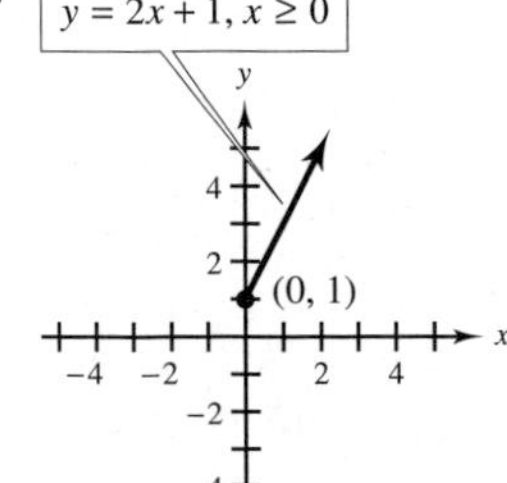

107.

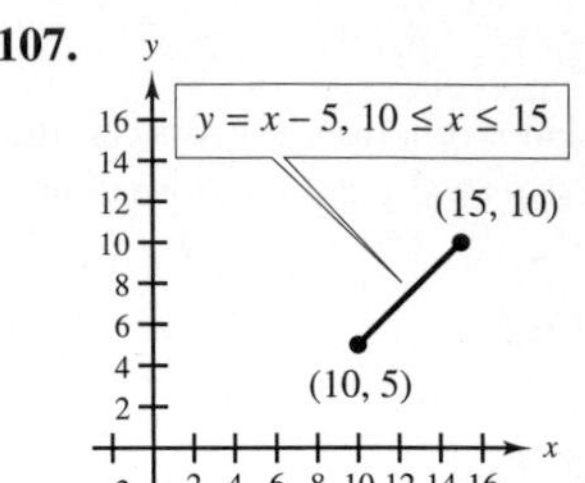

Section 4.2 *(page 227)*

1. The geometric interpretation is the rise divided by the run.

3. 2 **5.** −1 **7.** 0 **9.** Undefined **11.** 5 **13.** $\frac{7}{2}$

15. Undefined **17.** −2.825 **19.** $-\frac{4}{5}$ **21.** 0 **23.** $\frac{5}{2}$

25. Procedures (i) and (iii) are correct. Procedure (ii) refers to a rise and a run that are both negative. Thus the slope is +3.

27. $y = 3$ **29.** $x = 2$ **31.** $y = 4$

33. Method (ii) is the correct use of the slope formula. In method (i), the coordinates are not subtracted in the same order in the numerator and the denominator.

35. $-\frac{3}{5}$ **37.** 0 **39.** $\frac{3}{4}$ **41.** $y = -2x + 17; m = -2$

43. $y = 2x; m = 2$ **45.** $y = 15; m = 0$

47. $y = -\frac{1}{3}x + 2; m = -\frac{1}{3}$ **49.** Undefined

51. $y = \frac{5}{3}x - \frac{7}{3}; m = \frac{5}{3}$ **53.** $y = \frac{9}{8}x - 18; m = \frac{9}{8}$

57. (i), (iii) **59.** 0

61.

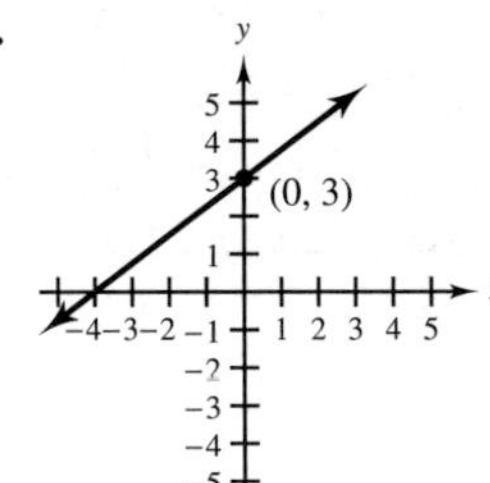

63.

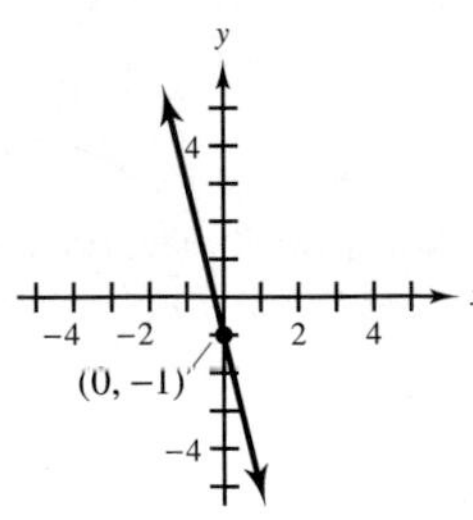

65.

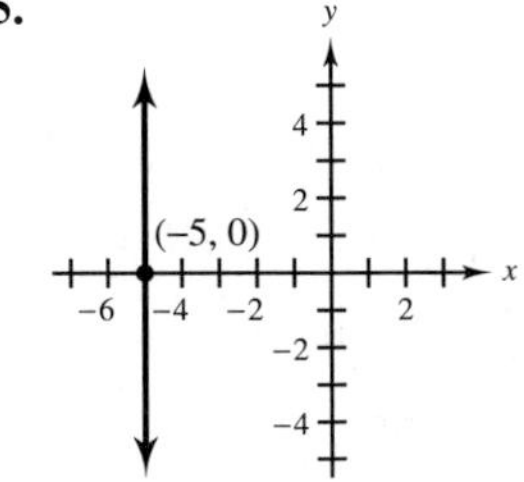

67.

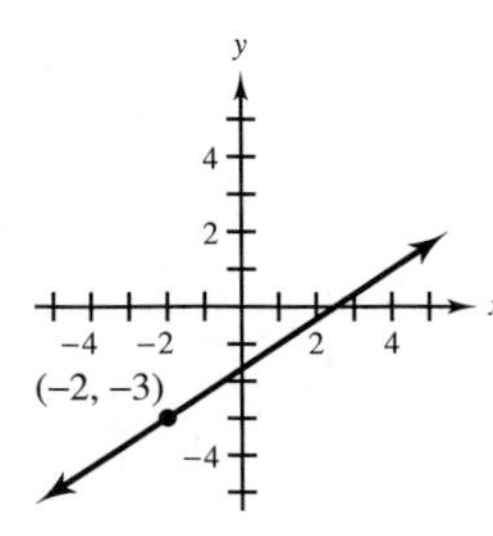

69.

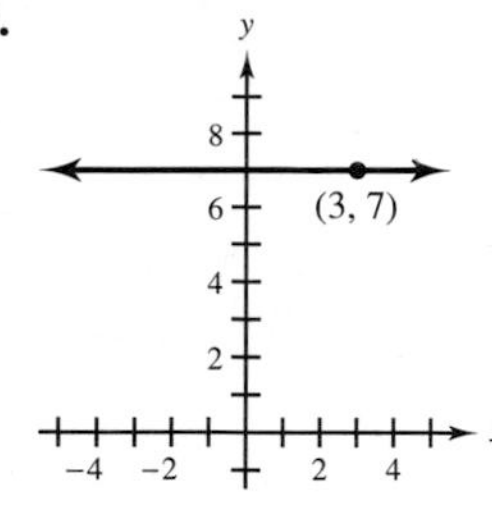

71.

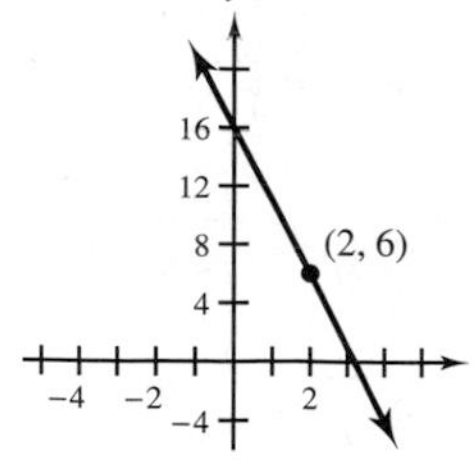

73.

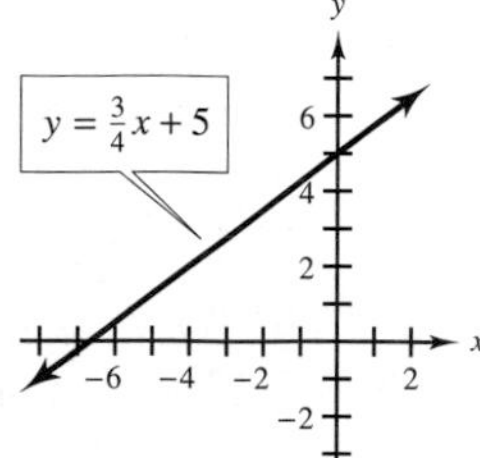

75.

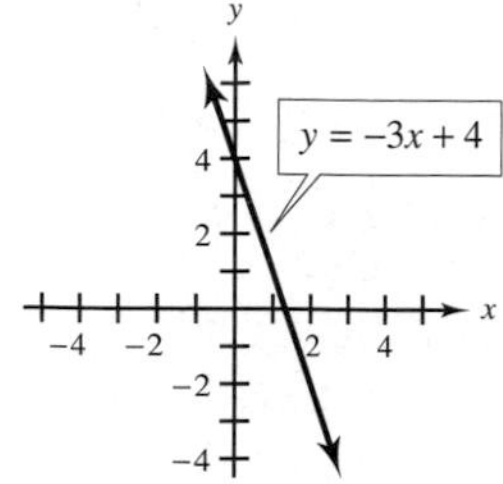

77.

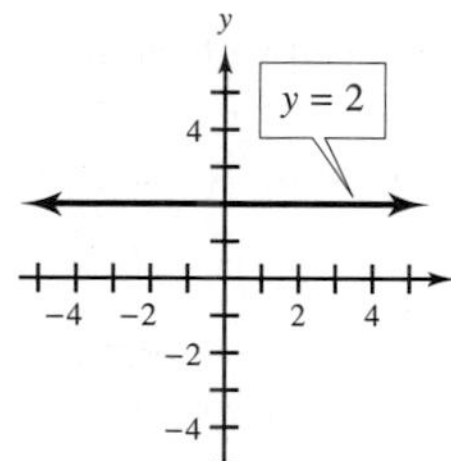

79.

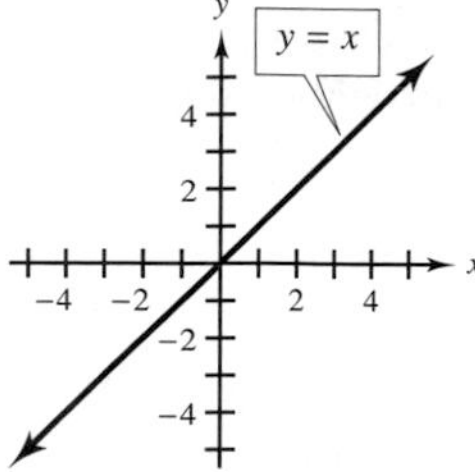

81.

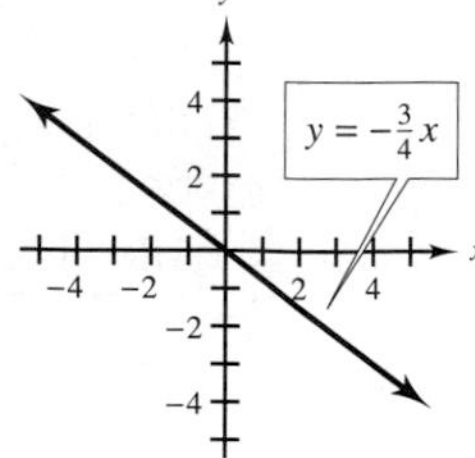

83.

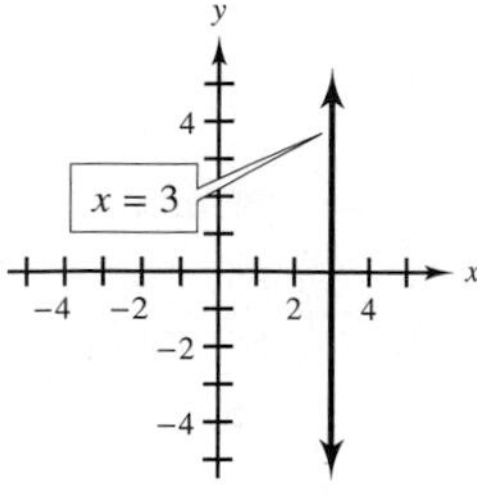

85. The graph retains the same slope, but it is shifted upward.

87. $y = -\frac{2}{5}x + 2$

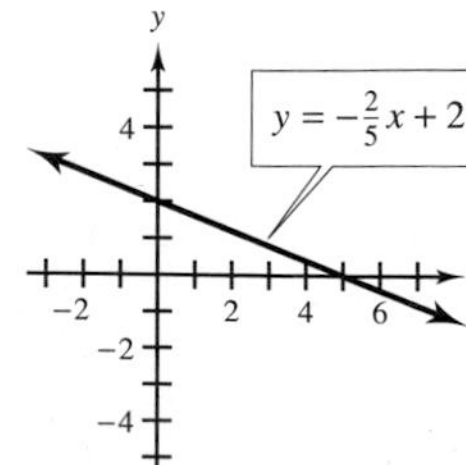

89. $x = 7$

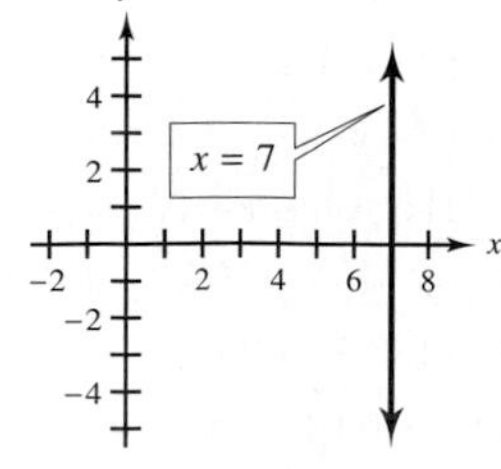

91. $y = 2x - 8$

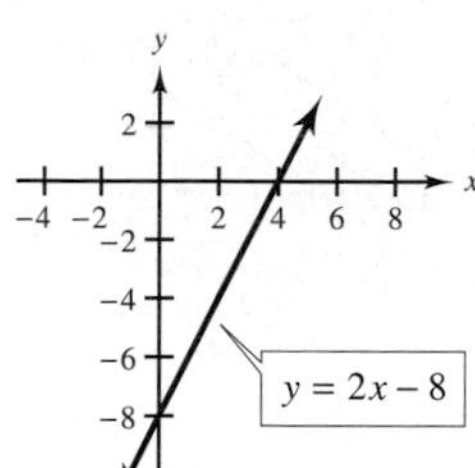

93. $y = 7$

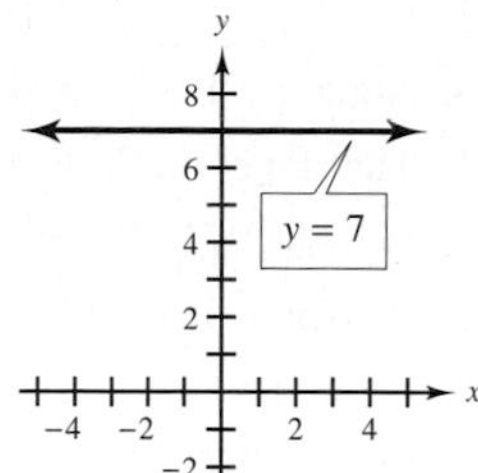

95. $y = 2x$

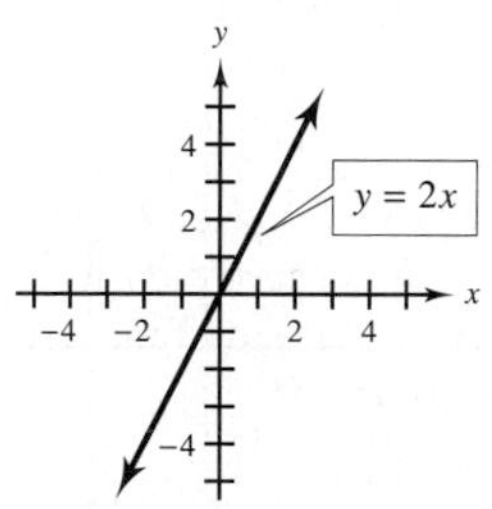

97. **(a)** 5 feet 1 inch **(b)** 2
(c) The slope of the graph is the same as the average annual growth rate.
(d) The age is 10 and the height is 4 feet 5 inches.

99. **(a)** -75 **(b)** 1200 **(c)** Rise: -525; run: 7
(d) 2006

101. **(a)** 5.6 **(b)** $5.\overline{3}$

103. **(a)** Semiprivate: $m = 328.4$
Private: $m = -148.2$
(b) The positive slope indicates an increase in the number of semiprivate golf courses, whereas the negative slope indicates a decrease in the number of private courses.

105. Because the coefficient of x is positive, the graph of the model function rises from left to right, which is consistent with the trend shown in the bar graph.

107. It would be negative. The percentage of men decreases with increasing percentages of women.

109. The settings for X must be made small or the settings for Y must be made large. The ratio of Y to X must be large.

111. For example, Xmin = -5, Xmax = 5, Ymin = -100, Ymax = 500

113. For example, Xmin = -1000, Xmax = 3000, Ymin = -10, Ymax = 50

Section 4.3 *(page 238)*

1. Perpendicular lines are lines that intersect to form a right angle.

3. Neither **5.** Parallel **7.** Perpendicular

9. **(a)** -3 **(b)** $\frac{1}{3}$ **11.** **(a)** $-\frac{3}{4}$ **(b)** $\frac{4}{3}$

13. **(a)** 0 **(b)** Undefined **15.** Parallel **16.** Parallel

17. Neither **19.** Perpendicular

21. The slope of a vertical line is undefined.

23. Parallel **25.** Perpendicular **27.** Neither

29. Perpendicular **31.** Parallel **33.** Neither

35. Undefined **37.** 0

39. **(a)** 2 **(b)** $-\frac{1}{2}$ **41.** **(a)** $-\frac{4}{3}$ **(b)** $\frac{3}{4}$

43. **(a)** 0 **(b)** Undefined

45. $-\frac{2}{3}$ **47.** Undefined

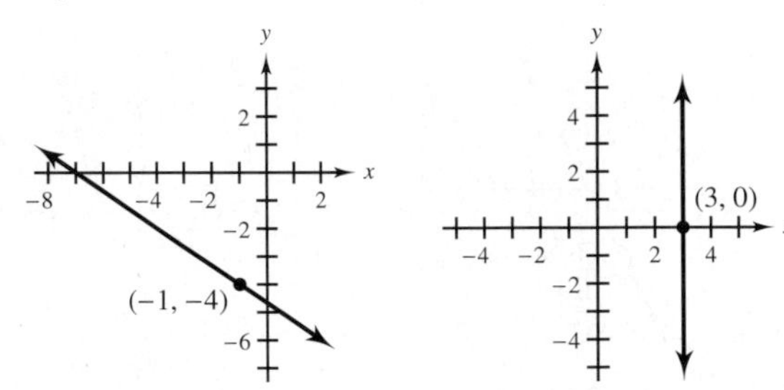

49. **(a)** 10.5 **(b)** $-\frac{14}{3}$ **53.** Collinear

59. The slope of a line represents the rate at which y changes with respect to a change in x.

61. $\frac{1}{2}$ **63.** -4 **65.** $\frac{3}{2}$ **67.** $-\frac{7}{3}$ **69.** y increased by 6

71. x decreased by 4 **73.** y decreased by 10

75. x increased by 4

77. There exists a constant k such that $A = kB$.

79. Yes **81.** Yes **83.** No

85. Increases, decreases

87. **(a)** $c = 0.1m + 29$ **(b)** 0.1

89. 1.25 cents **91.** 2820 pounds **93.** 247.5 square feet

95. $156.25

97. **(a)** The cost C is $C(d) = 233d$, where d is the number of days of vacation travel.

(b) The function is in the form $y = kx$. **(c)** 233

99. **(a)** The slope is 1.3, which is the average rate of change of the pole speeds.

(b) The average rate of change of pole speeds from 1980 to 1995 is 1.2, which is slightly lower than the rate of change over the entire period.

(c) The predicted speed is 204 mph, which seems possible.

101.

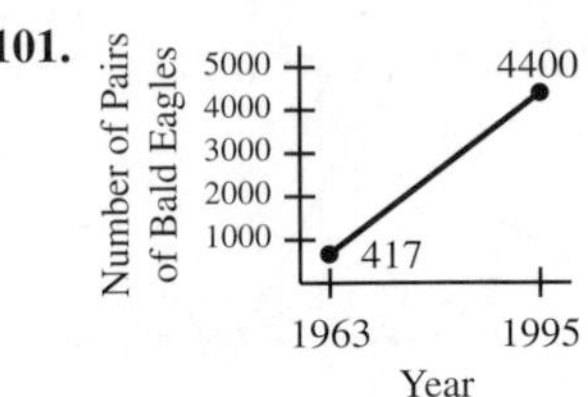

103. For just the two given data points, the rate of change is about 124 pairs of eagles per year. This projects to 5646 pairs of eagles in the year 2005.

105. 6 **107.** 1 **109.** 0

Section 4.4 *(page 250)*

1. We must know the slope and the y-intercept.

3. $y = -4x - 3$ **5.** $y = \frac{1}{2}x + 3$ **7.** $y = -5x + 1$

9. $y = -\frac{3}{4}x - 5$ **11.** $y = 7x - \frac{2}{3}$ **13.** $y = \frac{3}{4}x + 2$

15. $x = 4$ **17.** $y = -\frac{5}{2}x - 3$

19. A horizontal line has a slope of 0, but a vertical line has no defined slope.

21. $y = 2x + 1$ **23.** $y = \frac{2}{5}x - 8$ **25.** $y = -\frac{2}{3}x - \frac{19}{3}$

27. $y = -5$ **29.** $x = -3$ **31.** $x - y = 3$

33. $5x - 3y = 10$ **35.** $2x + 8y = -7$

37. Use the slope formula to determine the slope.

39. $y = -2x + 10$ **41.** $y = -2x + 3$ **43.** $y = x + 1$

45. $y = -\frac{2}{3}x - \frac{5}{3}$ **47.** $x = -4$ **49.** $y = \frac{1}{2}x - \frac{5}{2}$

51. $y = 5$

53. **(a)** A line that is perpendicular to a vertical line is a horizontal line. The equation is in the form $y = b$.

(b) A line that is parallel to the y-axis is a vertical line. The equation is in the form $x = c$.

55. $y = -4$ **57.** $y = 5$ **59.** $y = -\frac{2}{3}x - \frac{13}{3}$ **61.** $x = -2$

63. $y = 3x + 2$ **65.** $x = -2$ **67.** $y = 4x + 2$

69. $x = -2$ **71.** $L_1\colon y = -\frac{4}{3}x + \frac{11}{3}$ $L_2\colon y = \frac{3}{4}x - \frac{1}{2}$

73. $y = -x - 7$ **75.** **(a)** $y = -x + 3$ **(b)** $y = -\frac{3}{4}x - 3$

77. 6 **79.** 4 **81.** $-\frac{1}{6}$ **83.** 1 **85.** $y = 0.5x - 7.5$

87. $y = -1.25x - 428.25$

89. $\overline{PQ}\colon y = \frac{1}{3}x + 2$; $\overline{PR}\colon y = \frac{1}{3}x + 2$

The points are collinear because the equations are the same.

91. **(a)** $v = -200t + 1500$ **(b)** $700

93. $153.3 million; $6.8 million per year

95. **(a)** 0.01 **(b)** $y = 0.01x + 1.17$

(c) The value of b would be 1.19.

97. **(a)** 9.6%

(b) Let x represent the number of years since 1990.

(c) $y = 1.2x$

99. $C = 2.51x + 41.21$ **101.** 7 **105.** (3, 0), (0, 2)

107. $\left(\frac{5}{2}, 0\right), \left(0, -\frac{4}{3}\right)$ **109.** $bx - ay = 0$

Section 4.5 *(page 263)*

1. The solution set of $x + 2y \leq 6$ contains ordered pairs that satisfy $x + 2y = 6$, but the solution set of $x + 2y < 6$ does not.

3. No, yes, no **5.** Yes, yes, no **7.** No, yes, no

9. In the first case, we are assuming that $x + 3 > 7$ is an inequality in one variable. In the second case, we are assuming that $x + 3 > 7$ is an inequality in two variables.

11.

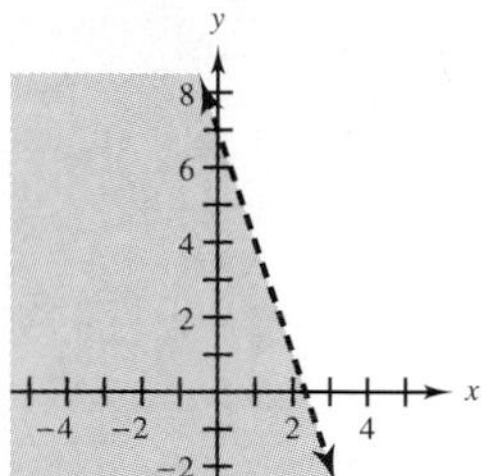

13.

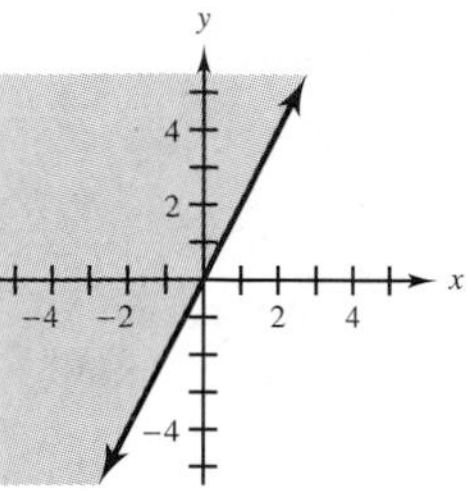

15.

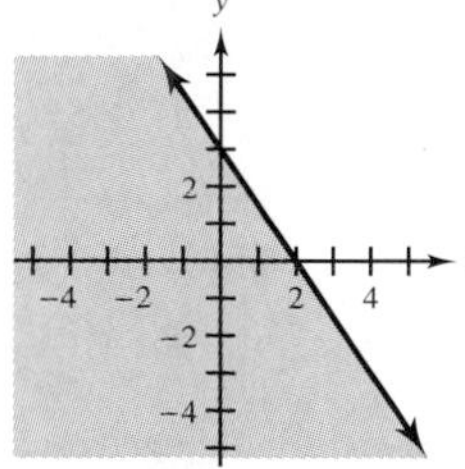

17.

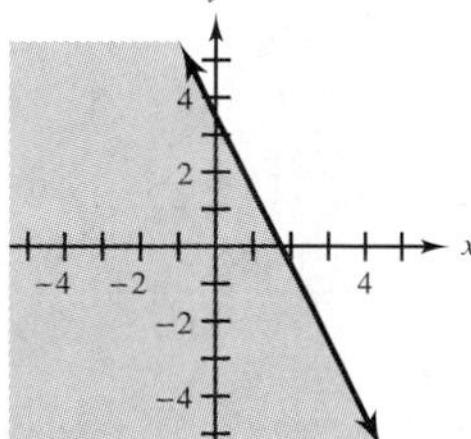

19.

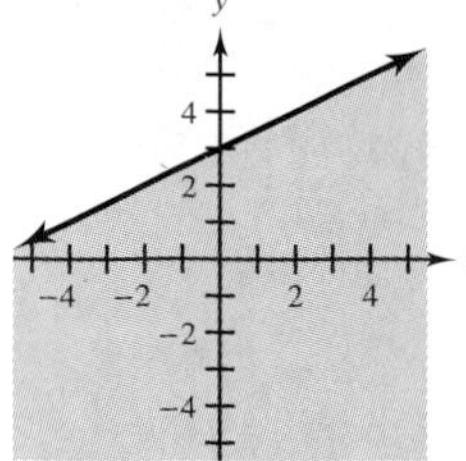

21.

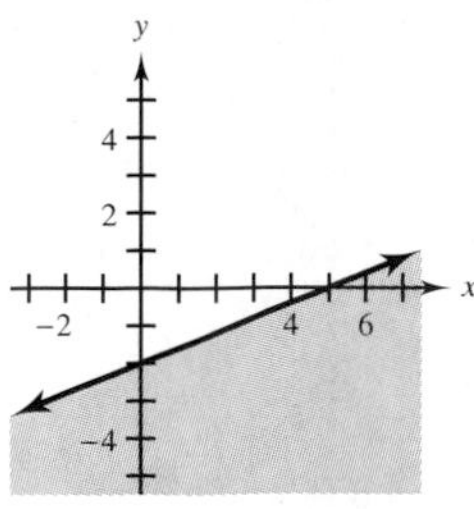

23.

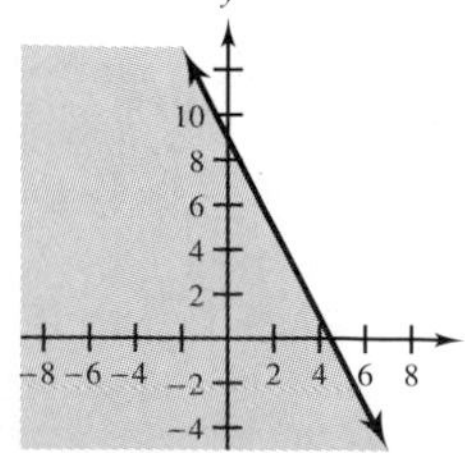

25.

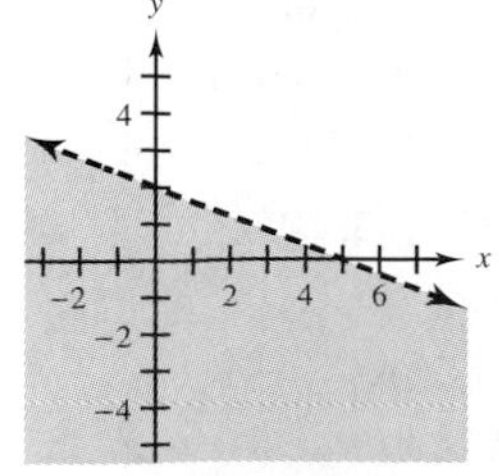

27.

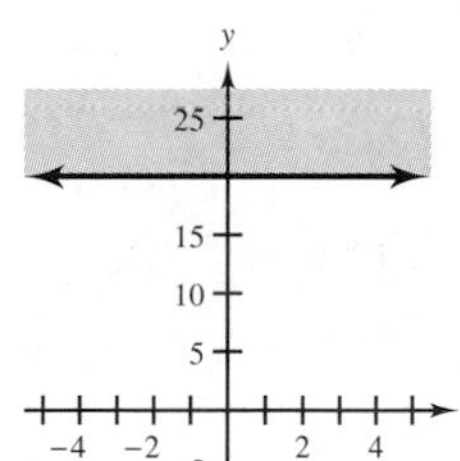

29.

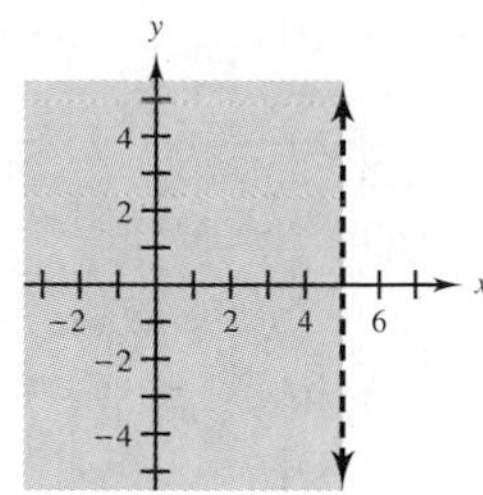

31. $y \leq 3$ **33.** $y < 0.5x + 1$ **35.** $x \leq 2$

37. $x \geq y + 1$ **39.** $y \geq 3x - 4$

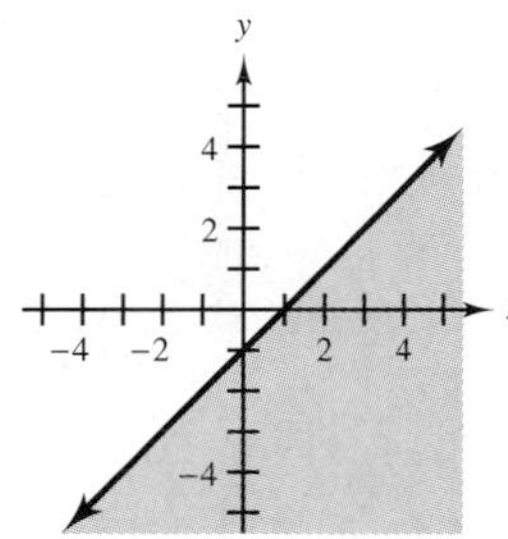

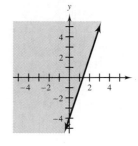

41. $x + 2y > 10$

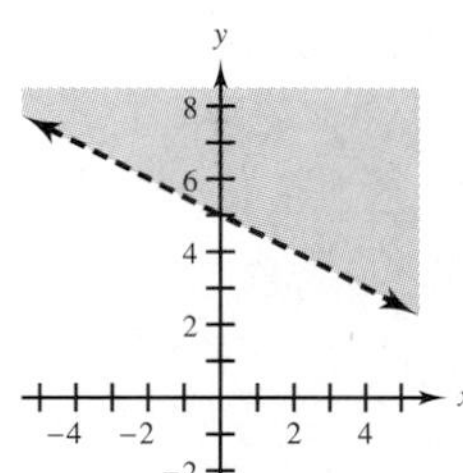

43.

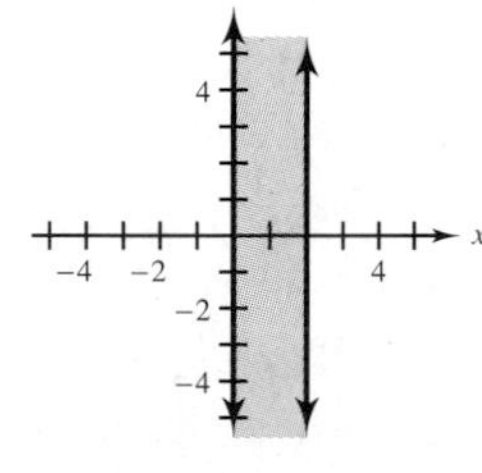

45.

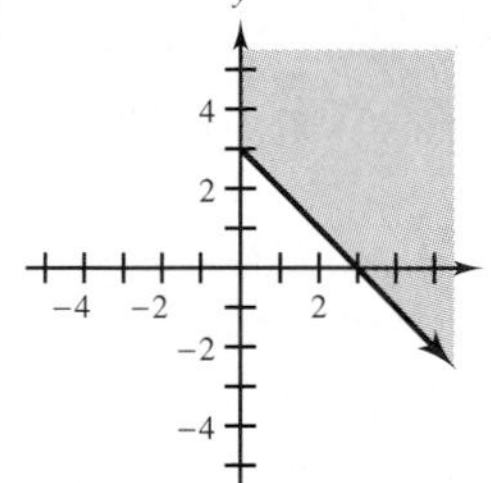

47.

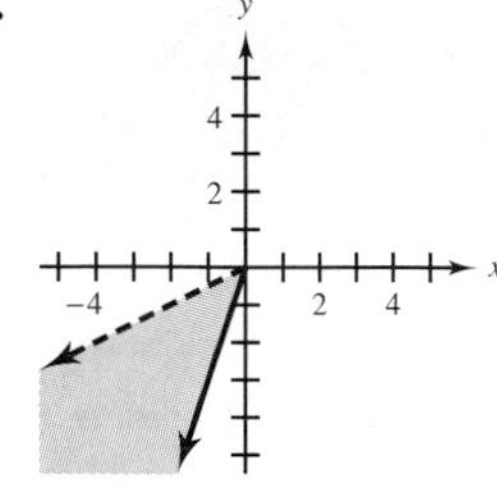

49.

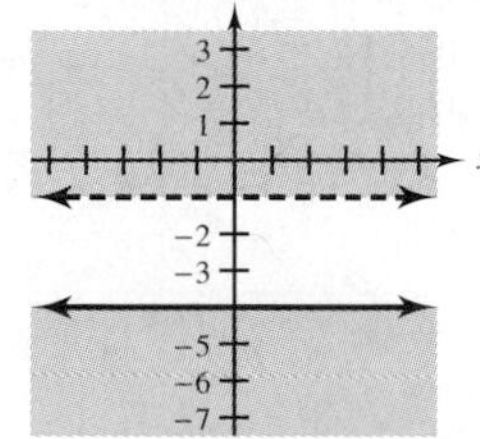

51.

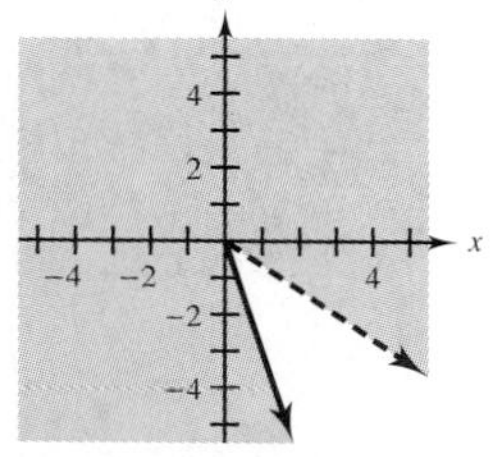

53.

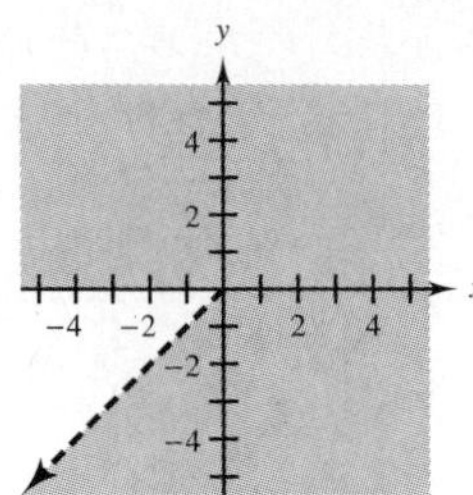

55.

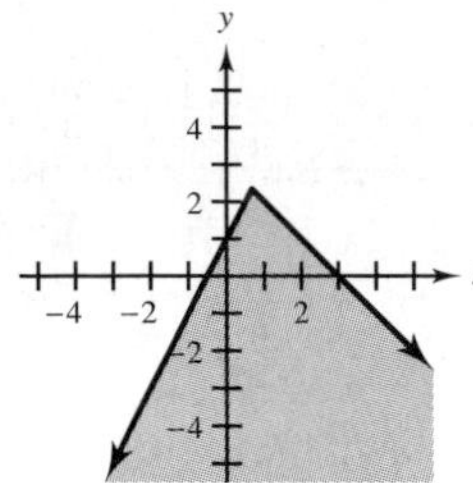

57.

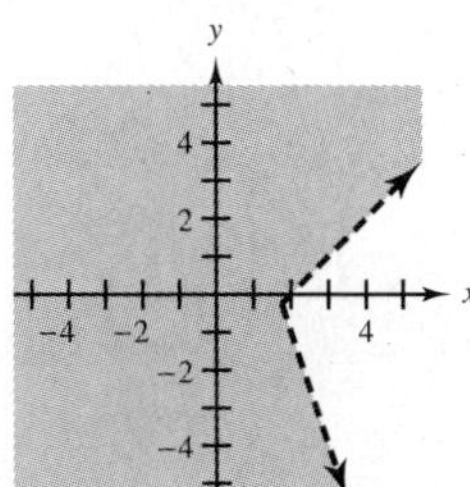

59.

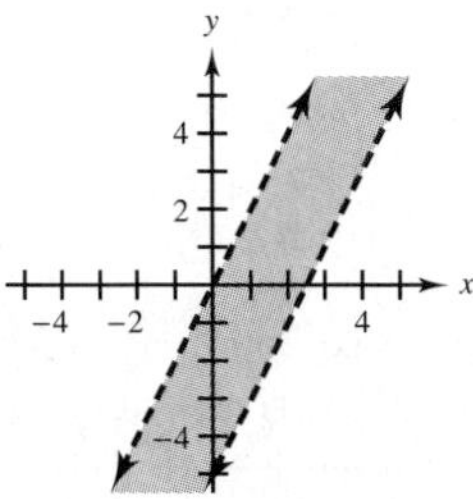

61.

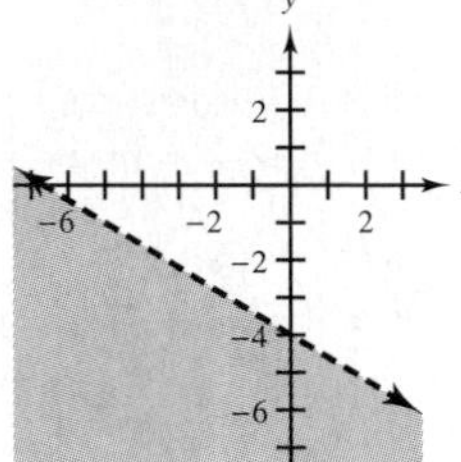

63.

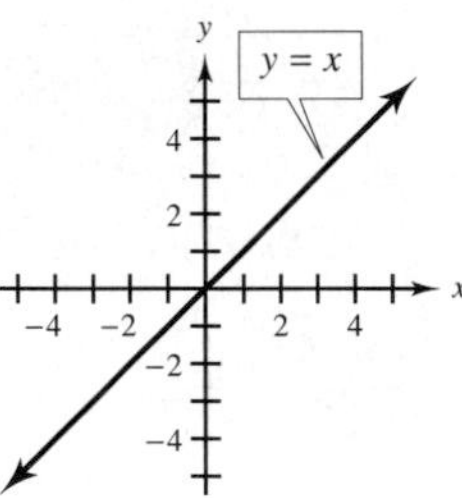

65.

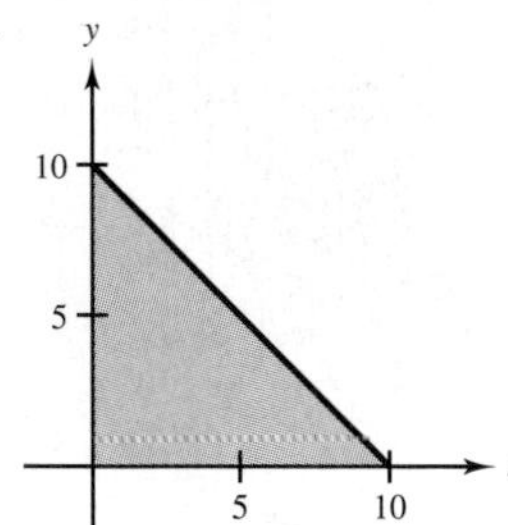

67. $y < 2$ and $y \geq -3$ **69.** $x \leq -2$ or $x > 1$

71. $y \geq x + 3$ or $y \leq x - 2$

73.

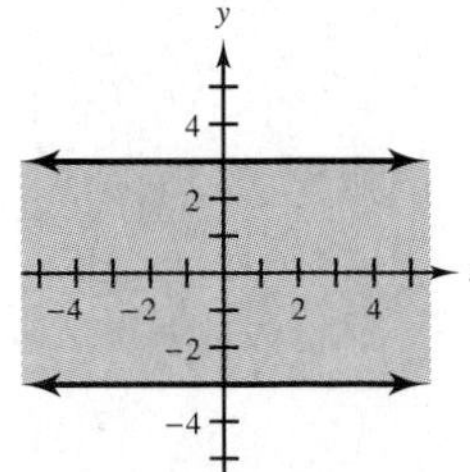

75.

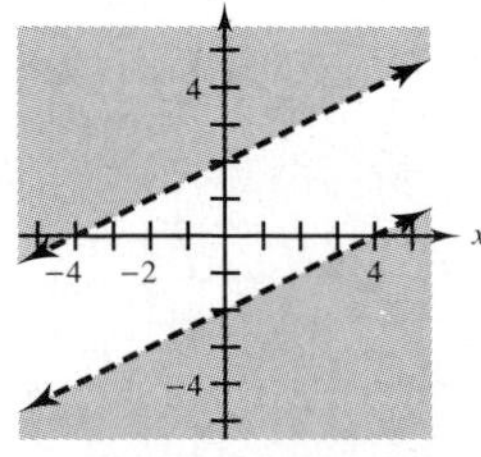

77.

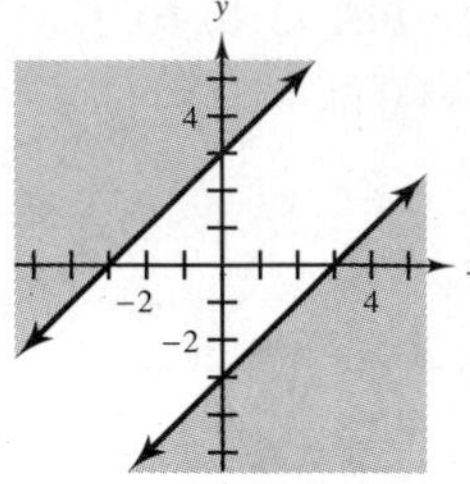

79. $\emptyset$

81.

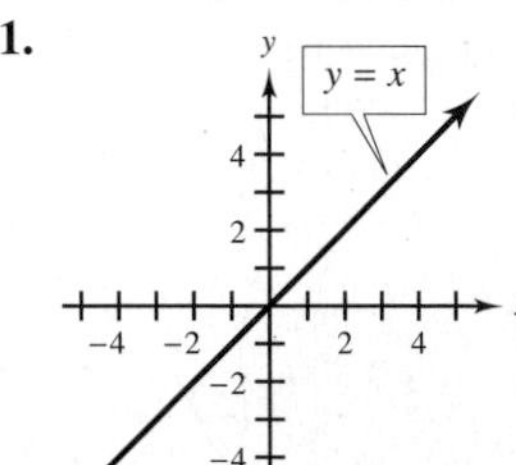

83.

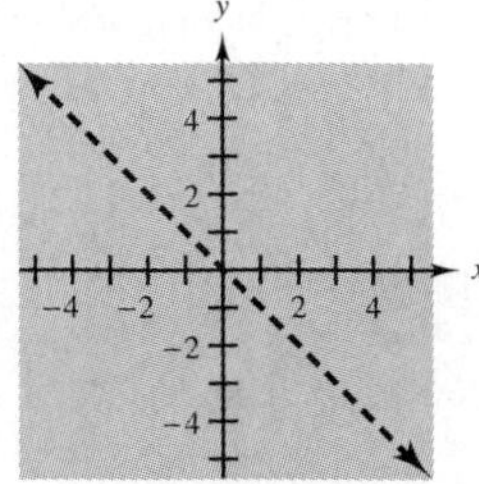

85. x = number of 4×4 pallets; y = number of 3×5 pallets
$16x + 15y \leq 6000$, $x \geq 0$, $y \geq 0$

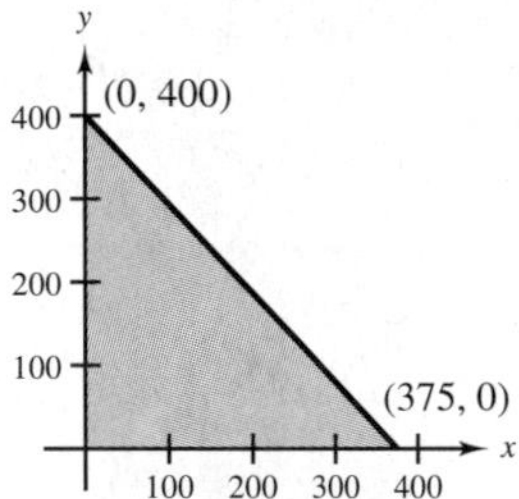

87. x = score of one team, y = score of the other team
$|x - y| \leq 5$, $x \geq 0$, $y \geq 0$

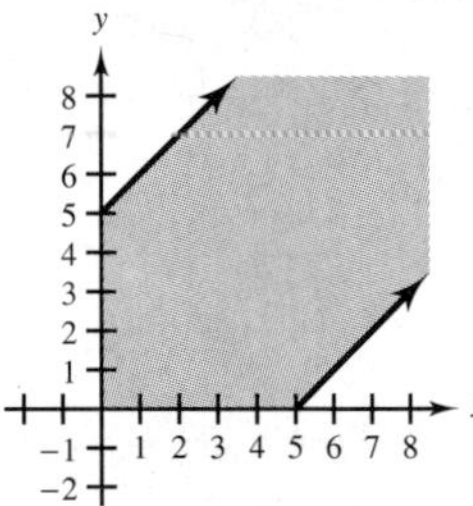

89. c = number of cats; d = number of dogs
$20 \leq c + d \leq 100$, $c \geq 0$, $d \geq 0$

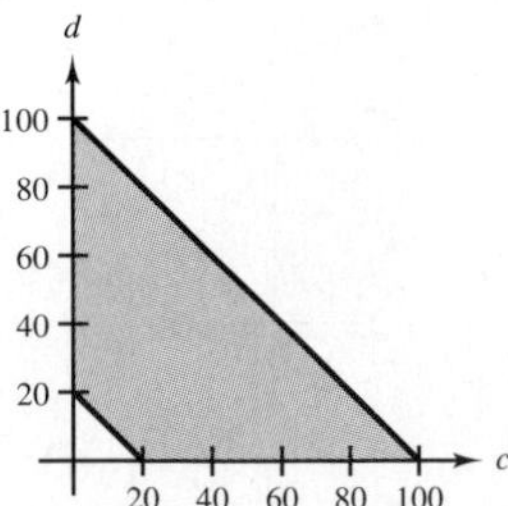

91. **(a)** $R + D \le 101.84$

(b) The coordinates represent the number of votes for the two candidates.

93. **(a)** $y \le 5850 + 0.28(x - 39{,}000)$ **(b)** Below

95. 3 **97.** Below

99.

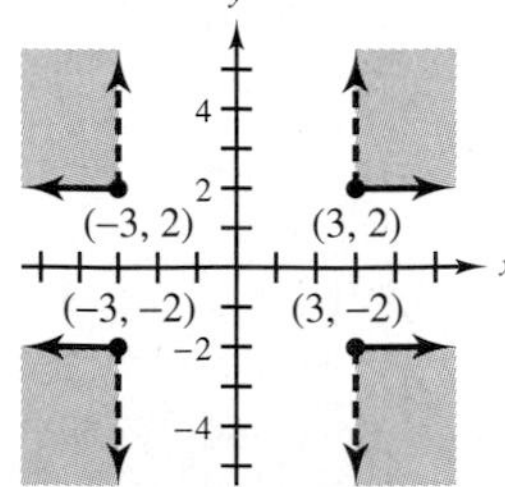

101. Entire plane

103.

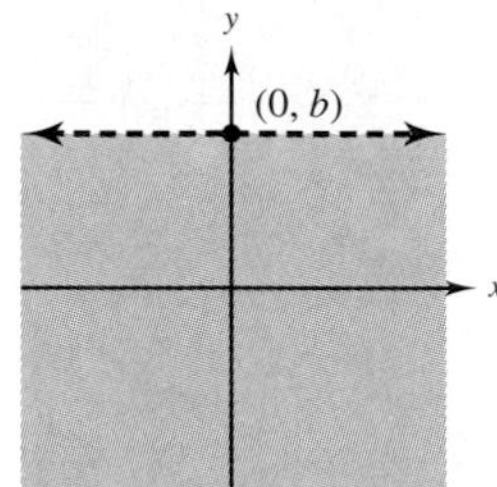

105.

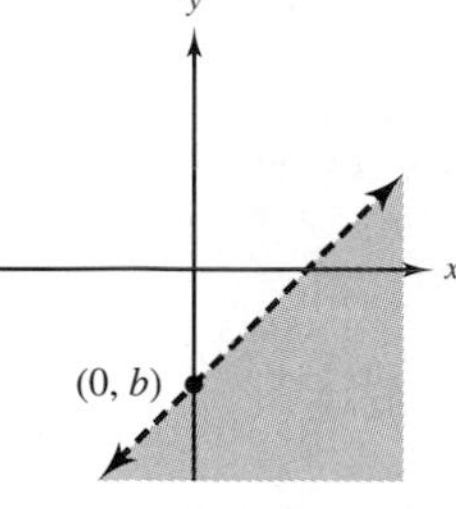

Chapter Review Exercises *(page 267)*

1. No **3.** Yes **7.** $a = 9, b = -7, c = 12$

9.

x	3	7	−7
y	9	15	−6

11. $(0, -7), (3, 0)$ **13.** $(0, 20)$; no x-intercept

15. $y = \frac{1}{2}x - \frac{5}{2}; \left(0, -\frac{5}{2}\right)$ **17.** $y = 2x - 8; (0, -8)$

19.

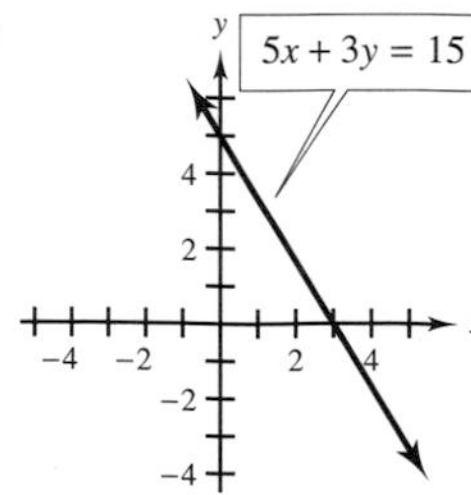

21.

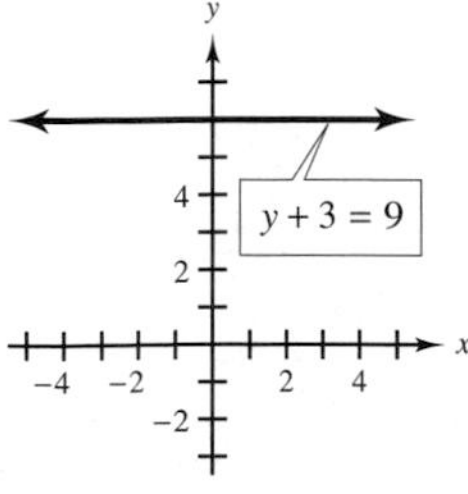

23. (ii) **25.** $\frac{5}{7}$ **27.** $-\frac{10}{9}$ **29.** $y = \frac{3}{4}x - \frac{15}{4}; \frac{3}{4}$

31.

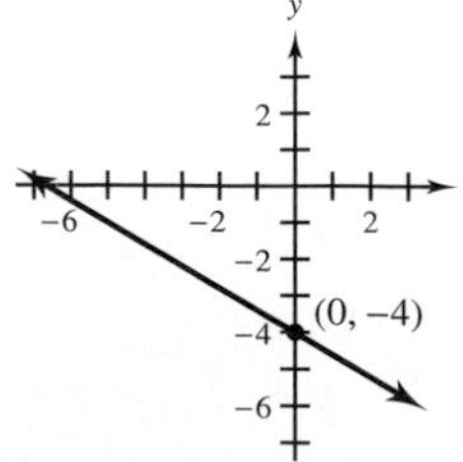

33.

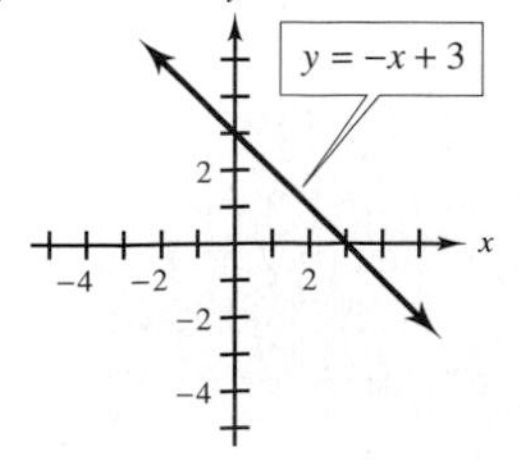

35.

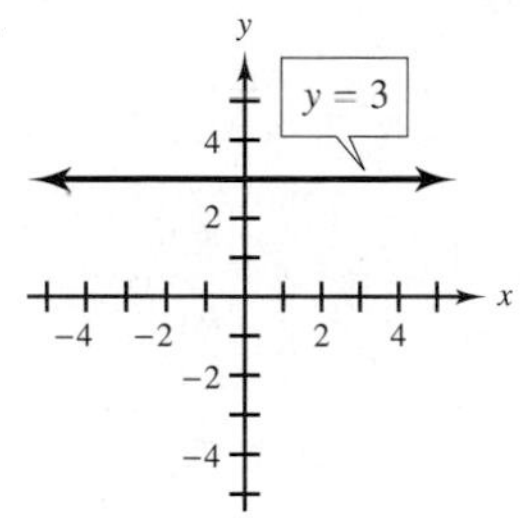

37. Parallel **39.** Perpendicular **41.** Perpendicular

43. Yes **45.** −3 **47.** y is decreased by 15

49. \$1.25 **51.** No **53.** Yes **55.** 112,500

57. $y = -3x - 1$ **59.** $y = -x - 2$ **61.** $y = \frac{5}{4}x - \frac{9}{4}$

63. $y = 6$ **65.** $y = -\frac{2}{3}x + \frac{17}{3}$ **67.** $y = -0.4x + 2$

69. 1999 **71.** Yes, no, yes **73.** No, yes, no

75.

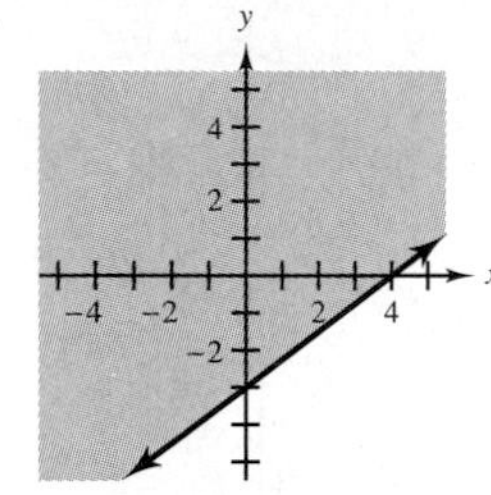

77.

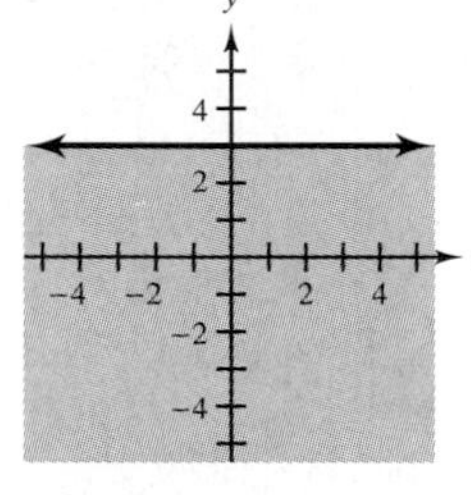

79.

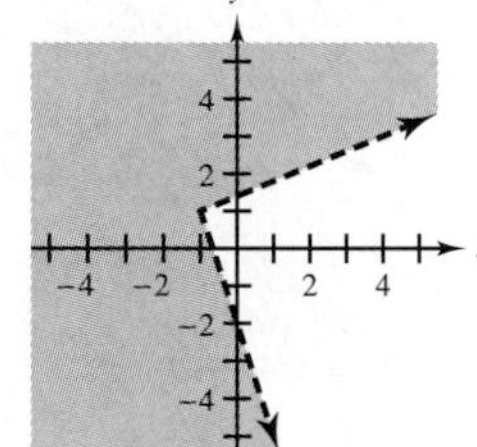

81.

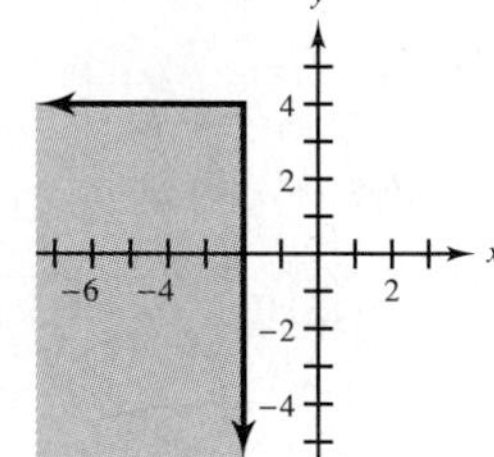

83.

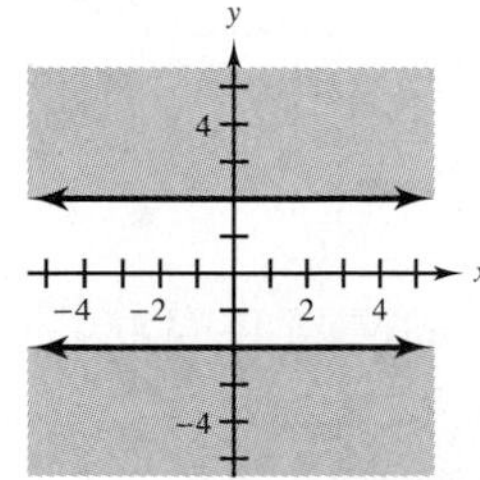

85.

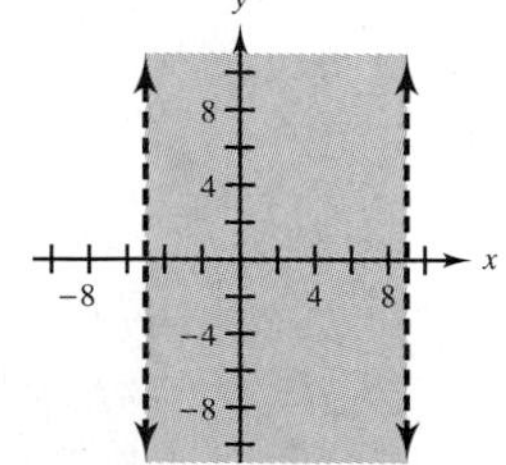

87. x = number of grandstand tickets sold
y = number of bleacher tickets sold
$6x + 4y < 11{,}400, x \ge 0, y \ge 0$

89. x = number of girls, y = number of boys
$x + y \le 400$ and $x \ge 100, y \ge 0$

Chapter Test *(page 270)*

1. [*4.1*] **(a)** $(-1, 0)$ **(b)** $(0, 2)$ **(c)** $(-2, -2)$ **(d)** $(0.5, 3)$

2. [*4.2*] $(0, 0)$; $m = 1$

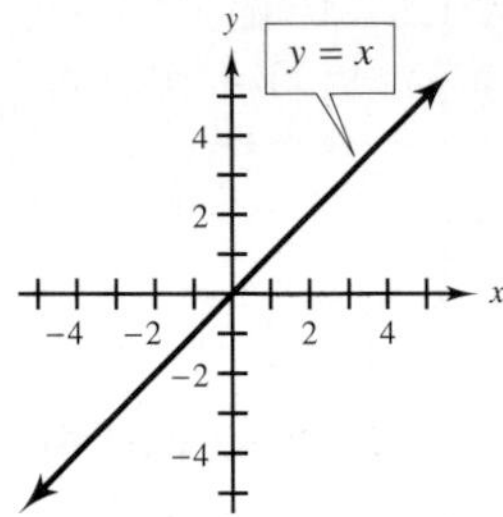

3. [*4.2*] $(0, -2)$, $(-2, 0)$; $m = -1$

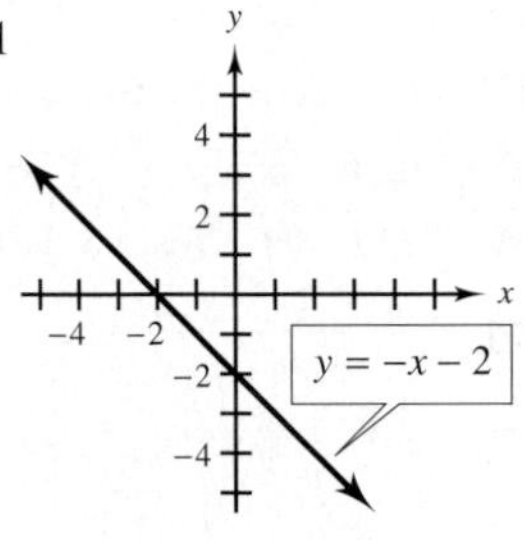

4. [*4.2*] $(0, -3)$, $(4.5, 0)$; $m = \frac{2}{3}$

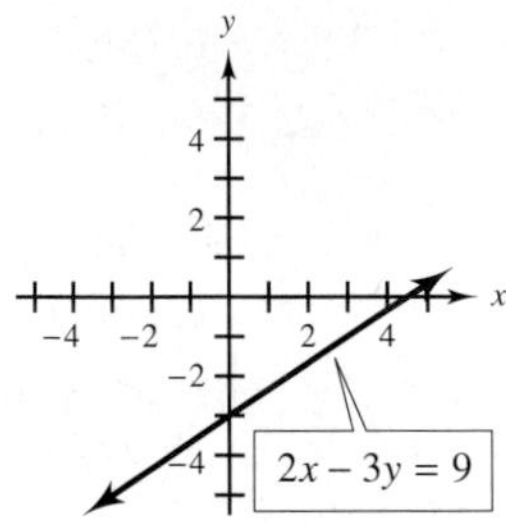

5. [*4.2*] $(5, 0)$; slope undefined

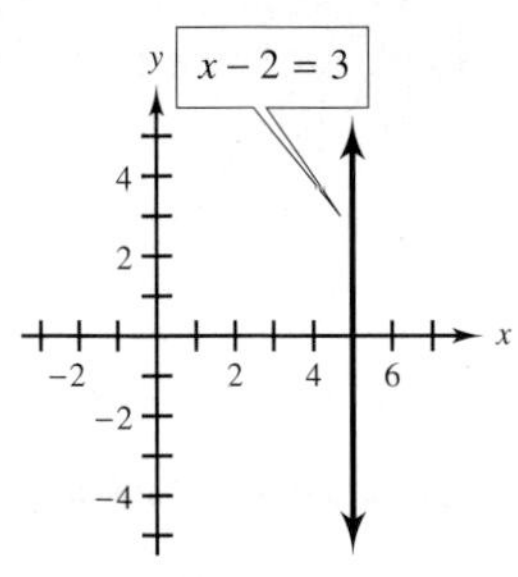

6. [*4.2*] $m < 0, b < 0$ **7.** [*4.4*] $y = \frac{2}{3}x - \frac{13}{3}$

8. [*4.4*] $y = 2x + 5$ **9.** [*4.4*] $x = 2$ **10.** [*4.4*] $y = 3$

11. [*4.4*] $y = 2x + 2$ **12.** [*4.4*] $y = 0.75x + 3$

13. [*4.5*]

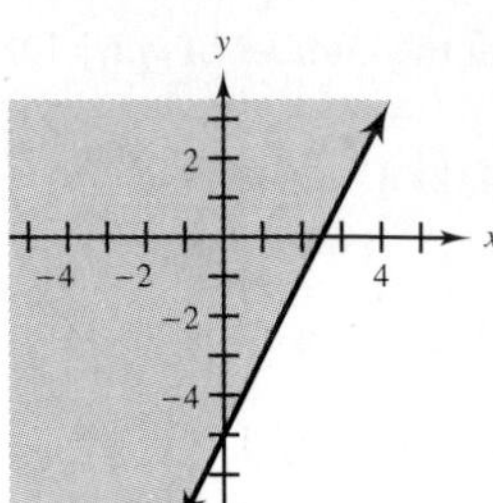

14. [*4.5*]

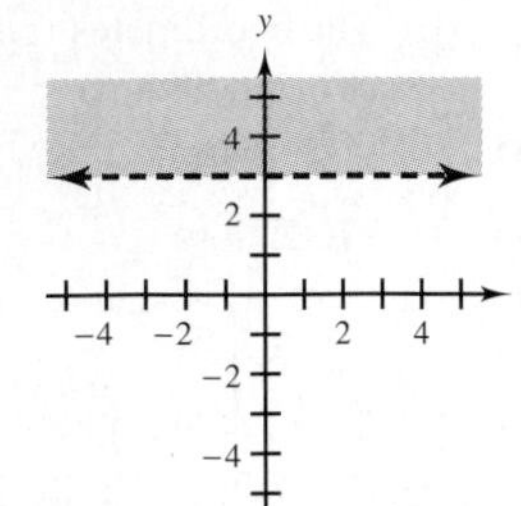

15. [*4.5*]

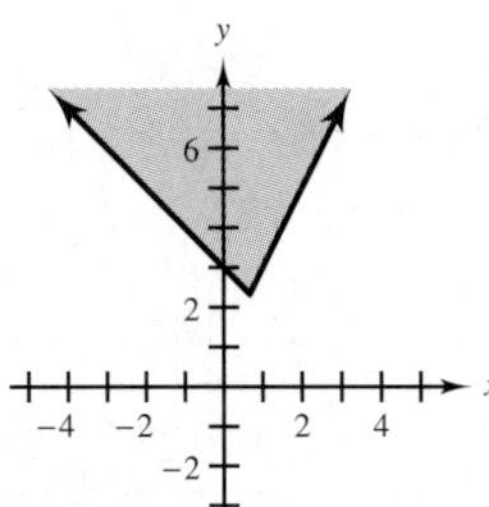

16. [*4.5*]

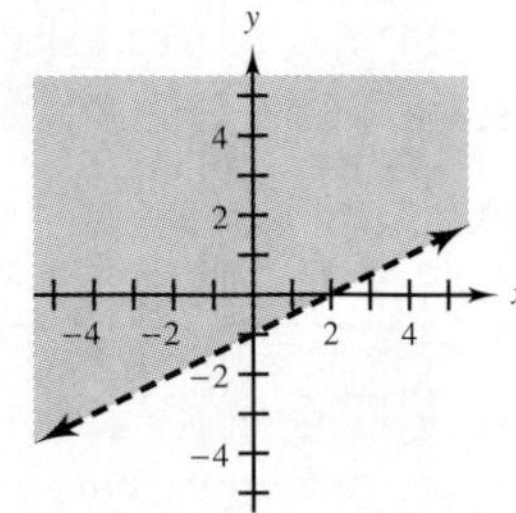

17. [*4.5*]

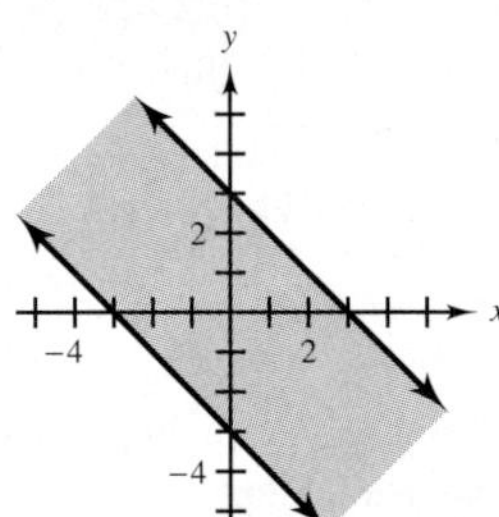

18. [*4.5*]

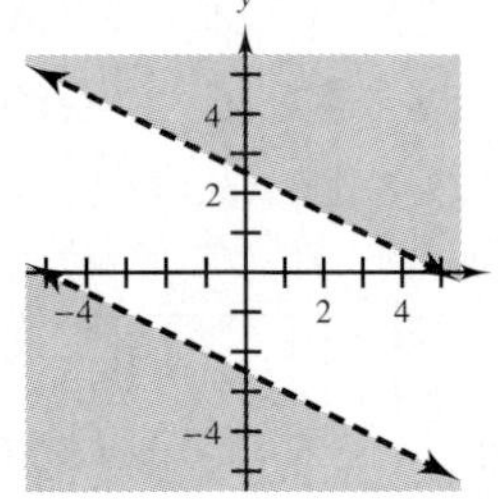

19. [*4.3*] Answers will vary.

20. [*4.2*]

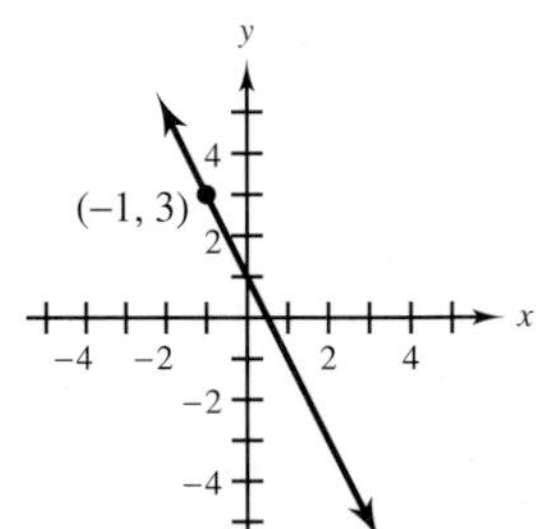

21. [*4.3*] Perpendicular **22.** [*4.3*] Neither

23. [*4.3*] $v = 32t$; 64 feet per second

24. [*4.3*] $-\$0.45$

25. [*4.3*] t = number of hours since 6:00 A.M.
$T = 2.5t + 88$
At 3:00 P.M., $t = 9$ and $T = 110.5°F$.

CHAPTER 5

Section 5.1 *(page 281)*

1. Substitute the numbers into each equation. If both equations are satisfied, the ordered pair is a solution.

3. No, yes, no **5.** No, yes, yes **7.** $a = 1, b = 2$

9. $a = 4, c = 7$

11. Two lines cannot intersect at exactly two points.

13. None **15.** 1 **17.** Infinitely many

19. (a) (i) The graphs do not coincide.
(ii) The graphs coincide.
(b) Yes, the lines could be parallel.

21. (0, 6) **23.** (0, 0) **25.** (3, 4) **27.** No solution

29. (5, −2) **31.** (11, 17)

33. Infinitely many solutions **35.** (27, −18) **37.** (−3, 4)

39. (−4, 7)

41. (a) Yes, either equation can be solved for either variable.
(b) Solving the second equation for y is easier.

43. (−22, −27) **45.** Dependent **47.** Inconsistent

49. Dependent **51.** (2, −3) **53.** Inconsistent **55.** (0, 5)

57. (80, 80) **59.** −2 **61.** 5 **63.** −2 **65.** −4 **67.** 10, 6

69. 20, 12 **71.** 67°, 23° **73.** 112°, 68°

75. 27 feet, 10 feet **77.** 10 inches, 30 inches

79. 3500 **81.** 6

83. (a) $y = -1.16x + 90.73$ **(b)** Approximately (35.1, 50)
(c) At about age 35, 50% pay off their credit card bills each month, and 50% do not.

85. $y_1 = 0.32x + 10.9$ and $y_2 = 0.68x + 1.7$

87. The solution (25.56, 19.08) indicates that in 2015 all discarded paper will be recycled.

89. (−2, 5), (2, 5)

91. (2, −10), (8, −4), (−5, 4), (1, 10); parallelogram

93. (−3, 2)

95. No; there are either infinitely many solutions or no solution.

Section 5.2 *(page 290)*

1. (a) Multiply the second equation by −3, and add the equations.
(b) Multiply the first equation by 4, and add the equations.

3. (6, 4) **5.** (−10, 7) **7.** (−2, 5) **9.** $\left(2, \frac{2}{3}\right)$

11. No, the result means that there are infinitely many solutions.

13. (10, −7) **15.** (5, −3) **17.** $\left(\frac{1}{2}, \frac{2}{3}\right)$ **19.** (5, 4))

21. The lines intersect at (3, 2). Many other pairs of lines also intersect at this point.

23. (1, 1) **25.** (0, 2) **27.** (−2, 1) **29.** (0, 1)

31. Both variables are eliminated, and the resulting equation is true.

33. (−3, −4) **35.** No solution

37. Infinitely many solutions

39. No solution **41.** (4, −2)

43. Infinitely many solutions **45.** $\left(\frac{41}{7}, \frac{23}{7}\right)$ **47.** No solution

49. Infinitely many solutions **51.** (3, −3) **53.** (1, 2)

55. Infinitely many solutions **57.** $a = 3, b = 2$

59. $a = 1, b = 3$ **61.** −4 **63.** −6 **65.** $c = 4$

67. $a = 3, b = -3$ **69.** $\left(\frac{b+c}{2}, \frac{b-c}{2}\right)$

71. $\left(\frac{c+3}{a+1}, \frac{3-ac}{a+1}\right), a \neq -1$ **73.** $\left(\frac{1}{a}, 1\right), a \neq 0$

75. $(1, 0), a \neq -b$

77. Hamburger: $1.90; milk shake: $1.20

79. 27 **81.** 43 nickels, 21 dimes **83.** 6 **85.** 15%

87. (a) France: $y = 500$, China: $y = 450$, United Kingdom: $y = 200$. The lines are all horizontal and distinct, and thus parallel. Therefore, there is no point of intersection.
(b) Approximately (8.3, 6394)
(c) The United States and Russia are projected to have an equal number of warheads, 6394, in 1998.

89. (7.9, 44.2))

91. There are other forms of communication such as telephone, voice mail, and personal conversations.

93. (2, 4) **95.** (−5, 2.5)

97. The system has no solution because $\frac{1}{x} = 0$ has no solution.

99. $b \neq -4$ **101.** $c = 5$ **103.** $a \neq 1$ **105.** $c \neq 10$

Section 5.3 *(page 302)*

1. No, not all coefficients of a linear equation can be zero.

3. Yes, no **5.** No, yes

7. $a = 3, b = 0, c = -2$ **9.** $b = 3, c = 0, d = 1$

11. The graph of a linear equation in two variables is a line in a plane. The graph of a linear equation in three variables is a plane in three-dimensional space.

13. (a) At least two planes are parallel or a plane is parallel to the line of intersection of the other two planes.
(b) The three planes coincide or intersect in a common line.

15. (−1, 3, −2) **17.** (7, 5, 4) **19.** (3, −2, 1)

21. (4, −3, −1) **23.** (1, 3, −2) **25.** $\left(3, \frac{1}{2}, -\frac{1}{3}\right)$

27. $\left(1, -\frac{1}{3}, \frac{1}{2}\right)$ **29.** $\left(\frac{1}{2}, \frac{1}{3}, -\frac{2}{3}\right)$

31. **(a)** After eliminating one variable, the resulting system of two equations in two variables is inconsistent.

(b) After eliminating one variable, the two equations in two variables in the resulting system are dependent.

33. Ø **34.** Ø

35. The set of all solutions to $x + 2y - z = 3$

37. (3, −2, 1) **39.** (1, 2, 3) **41.** (5, −4, −1)

43. (5, −4, −3) **45.** (−7, −4, 3) **47.** Ø

49. Set of all solutions to $4x - 3y + 6z = 12$

51. (−3, −6, 4) **53.** (−3, 2, 5) **55.** (−1, −3, 2)

57. 23°, 64°, 93° **59.** 27 yards, 52 yards, 21 yards

61. 27 nickels, 56 dimes, 72 quarters

63. **(a)** $R + D = 0.08(1200)$
$R + D + I = 1200$
$R = D + 4$

(b) Republican: 50, Democrat: 46, Independent: 1104

65. **(a)** $b + w + p = 100$
$b = 2w$
$w = 3p$

(b) Brass: 60; woodwind: 30; percussion: 10

67. (30, 11.6), (70, 4.8), (94, 8.7)

69. $p(t) = 0.005t^2 - 0.690t + 27.610$

71. (1, 2, −1, 3) **73.** (1, −1, 0.5)

75. $\left(\frac{a-b+c}{2}, a - \frac{a-b+c}{2}, c - \frac{a-b+c}{2}\right)$

Section 5.4 *(page 312)*

1. A 2 × 4 matrix is an array of numbers with two rows and four columns.

3. 2 × 4 **5.** 2 × 2 **7.** 3 × 4

9. **(a)** A coefficient matrix is a matrix whose elements are the coefficients of the variables of the equations in a system of equations.

(b) A constant matrix is a matrix whose elements are the constants of the equations in a system of equations.

(c) An augmented matrix is a matrix whose elements are the coefficients and the constants of the equations in a system of equations.

11. $\left[\begin{array}{cc|c} 3 & 2 & 6 \\ 1 & -4 & 9 \end{array}\right]$ **13.** $\left[\begin{array}{ccc|c} 1 & 2 & -3 & 5 \\ 1 & 0 & 2 & 15 \\ 0 & 2 & -1 & 6 \end{array}\right]$

15. $2x + y = 1$
$3x - 2y = 12$

17. $x + y - z = 2$
$2x - 3y + z = 5$
$3x + 2y - 4z = 3$

19. The goal is to have 1's along the main diagonal of the matrix and 0's elsewhere.

21. **(a)** $R_1 \div 3$ **(b)** $-2R_1 + R_2 \rightarrow R_2$ **(c)** $R_2 \div (-7)$ **(d)** $-2R_2 + R_1 \rightarrow R_1$ **(e)** (2, 1)

23. **(a)** $-2R_1 + R_2 \rightarrow R_2$ **(b)** $-1R_1 + R_3 \rightarrow R_3$ **(c)** $R_2 \div (-3)$ **(d)** $-1R_3$ **(e)** $R_3 + R_1 \rightarrow R_1$ **(f)** $R_3 + R_2 \rightarrow R_2$ **(g)** $-1R_2 + R_1 \rightarrow R_1$ **(h)** (1, 2, −1)

25. **(a)** 2 **(b)** 2 **(c)** −56 **(d)** 2 **(e)** 1 **(f)** −3 **(g)** 1 **(h)** (−3, 8)

27. **(a)** 3 **(b)** −3 **(c)** 3 **(d)** −2 **(e)** 1 **(f)** −2 **(g)** −6 **(h)** −2 **(i)** 1 **(j)** 0 **(k)** 1 **(l)** 0 **(m)** 0 **(n)** 0 **(o)** (1, 0, −2)

29. The equations must be written in standard form.

31. (3, −4) **33.** $\left(\frac{1}{2}, -\frac{1}{4}\right)$ **35.** Inconsistent **37.** Dependent

39. (4, 2, −3) **41.** (1, −3, 2)

43. $\left[\begin{array}{cc|c} 2 & -5 & -5 \\ 5 & 2 & 2 \end{array}\right]$ **45.** $\left[\begin{array}{ccc|c} 3 & -4 & 1 & 2 \\ 1 & 1 & 3 & 1 \\ 2 & 3 & -1 & 0 \end{array}\right]$

47. (−3, 1) **49.** (3, −2, −2) **51.** (0, −1, 2)

53. (1, −2, −1)

55. **(a)** $\begin{bmatrix} 428 & 453 \\ 408 & 430 \\ 483 & 544 \end{bmatrix}$

The average number of minutes in a woman's work day in developing countries

(b) $\begin{bmatrix} 428 & 408 & 483 \\ 453 & 430 & 544 \end{bmatrix}$; column 1

57. **(a)** $z = -3, x = 4, y = -5$ **(b)** Yes

59. (−2, −3, 1, 2) **61.** (1, −1, −1, 2)

63. If $a \neq \frac{1}{3}$, then the solution is unique; if $a = \frac{1}{3}$, then there is no solution.

Section 5.5 *(page 322)*

1. A matrix is an array of numbers. A determinant is a number associated with a matrix.

3. 7 **5.** 0 **7.** 0 **9.** 1 **11.** −1 **13.** 0 **15.** −51

17. −8 **19.** 0 **21.** 16 **23.** 0 **25.** $x^2 - 9$ **27.** $x - 5y$

29. 7 **31.** 1 **33.** 4

35. Division by zero is not defined.

37. No **39.** No **41.** Yes **43.** (2, 5) **45.** (−1, 2)

47. (5, 3) **49.** (3, 2) **51.** (5, 2) **53.** (3, 0.5)

55. Cramer's Rule does not apply.

57. $\left(-\frac{1}{4}, \frac{2}{3}\right)$ **59.** (1, 3, 2) **61.** (−2, −3, 4)

63. Cramer's Rule does not apply. **65.** (2, −1, −2)

67. (3, −1, 0) **69.** $\left(-1, 7, \frac{1}{3}\right)$ **73.** −9 **75.** 16

77. −2.5 **79.** −4

81. Grocery store, 14 years; pharmacy, 10 years

83. $1600 in checking, $800 in savings

85. 27 nickels, 9 dimes, 5 quarters **87.** $(2\sqrt{3}, -\sqrt{2})$

89. $\begin{vmatrix} 2 & y \\ 3 & x \end{vmatrix}$ **91.** $\begin{vmatrix} a & y \\ -b & x \end{vmatrix}$ **93.** Answers will vary.

Section 5.6 *(page 332)*

1. 12 pounds **3.** 5 pounds

5. 42 cakes, 48 pies, 50 dozen cookies

7. 237 patrons, 219 others **9.** 14,261

11. 1450 orchestra, 1550 loge, 2500 balcony

13. $3490 at 8.5%; $6510 at 6.9%

15. $3800 at 6%; $3200 at 7%

17. Stock fund: $5400; bond fund: $4100; mutual fund: $7000

19. The number of ounces of acid is found by multiplying the total volume of solution (20 ounces) by the concentration (0.10).

21. 34%: 50 liters; 55%: 20 liters **23.** 200 gallons

25. 10%: 6 liters; 20%: 4 liters; 50%: 8 liters

27. Because the boat is pushed by the current, the effective rate is the sum of the rates of the boat and the current.

29. Canoe: 5 mph; current: 2 mph

31. Airplane: 450 mph; wind: 30 mph

33. Canoe: 4 mph; current: 2 mph

35. $y = -2x + 4$ **37.** $y = 2x^2 - 3x - 5$

39. $y = -4x^2 + x + 7$ **41.** Father: $34; son $17

43. 39 five-dollar bills, 31 ten-dollar bills

45. Heavy-duty: 12; homeowner: 19

47. Gloss: 32 gallons; flat: 43 gallons

49. Orchids: 37; carnations: 58 **51.** 140

53. 8-mile: 22; 6-mile: 10; 5.5-mile: 8 **55.** A: 5; B: 8; C: 4

57. First set: 15 volumes; second set: 14 volumes; third set: 6 volumes

59. Hamburger: $1.95; shake: $1.50; fries: $0.95

61. **(a)** $y = 0.11x + 4.9$ **(b)** $y = 0.07x + 5.9$ **(c)** $(24, 7.5)$

63. The solution is approximately $(-8.71, 4.25)$, which represents 1965, when the costs of attending both public and private colleges were 4.25% of the median family income.

65.
$$\begin{aligned} 4a + 2b + c &= 473 \\ 16a + 4b + c &= 808 \\ 36a + 6b + c &= 1328 \end{aligned}$$

67. The models are the same. **69.** 3.94 ounces

71. Supervisor: 5 mph; inspector: 7 mph

Section 5.7 *(page 343)*

1. The graph of each linear inequality is a half-plane. The graph of the solution set of the system is the intersection of the half-planes.

3.

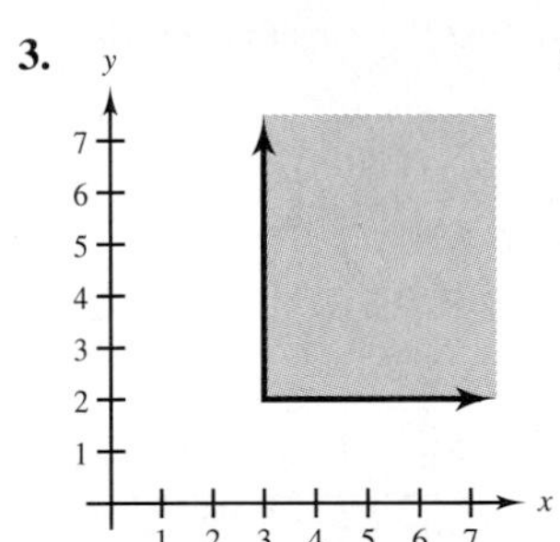

5.

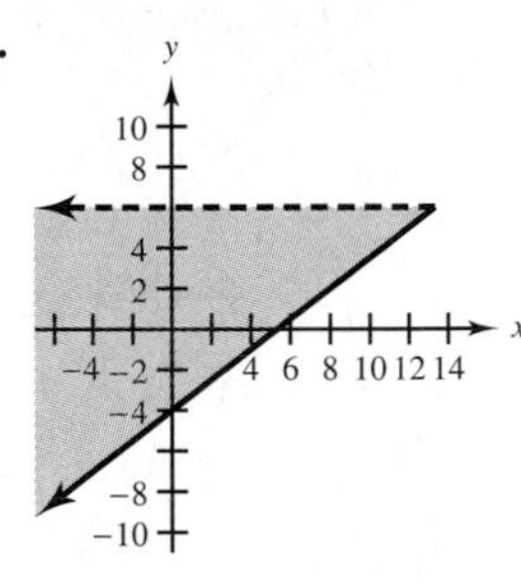

7.

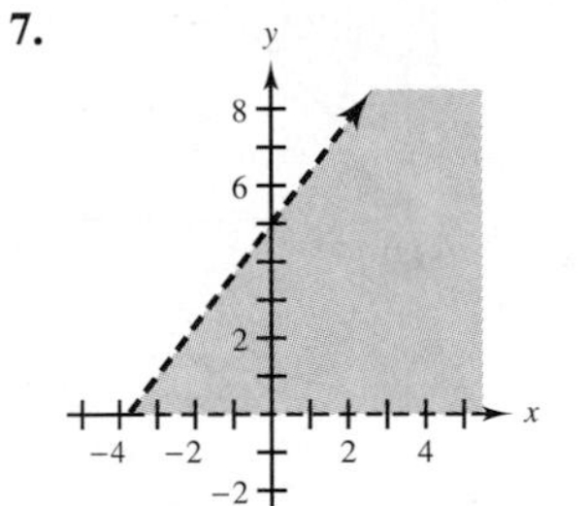

9.

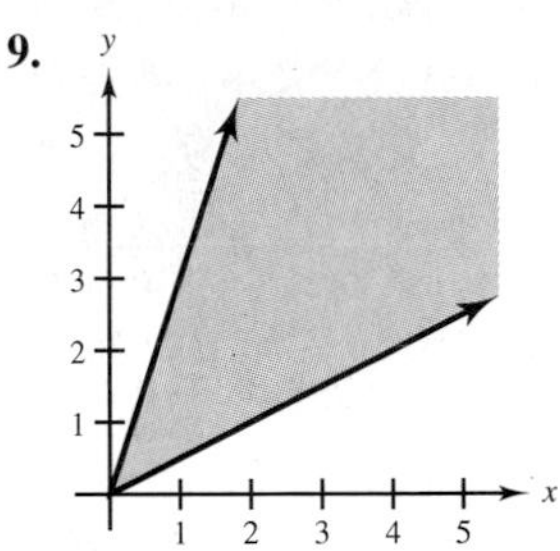

11.

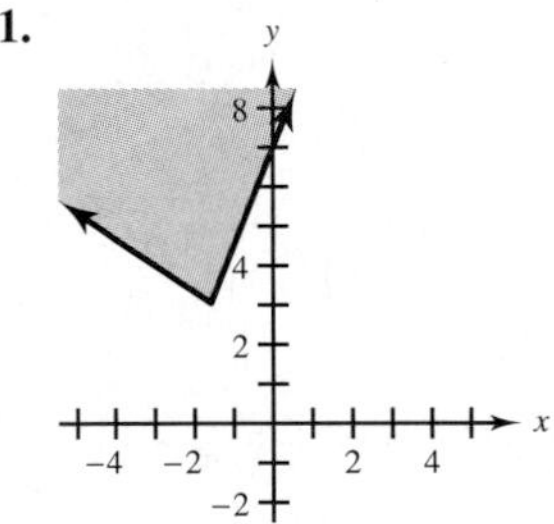

13. $\varnothing$ **15.** $y \le 4, x \le 5, x \ge 0, y \ge 0$

17. $y \le -8x + 8, y > -\frac{1}{3}x - 2$

19. The graph is a square, centered at the origin, and its interior. The sides are parallel to the axes, and the length of each side is 10.

21. **23.**

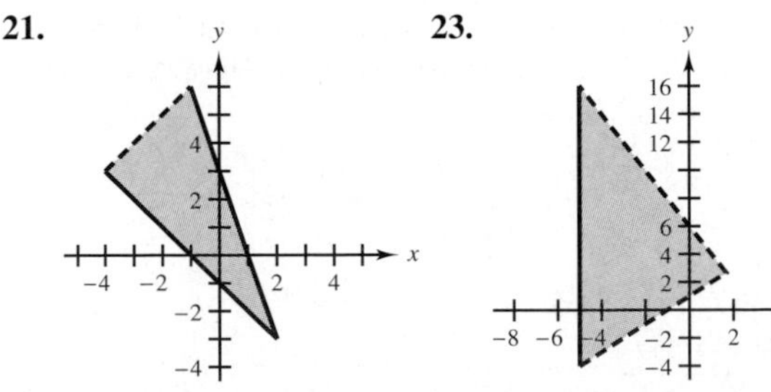

25.

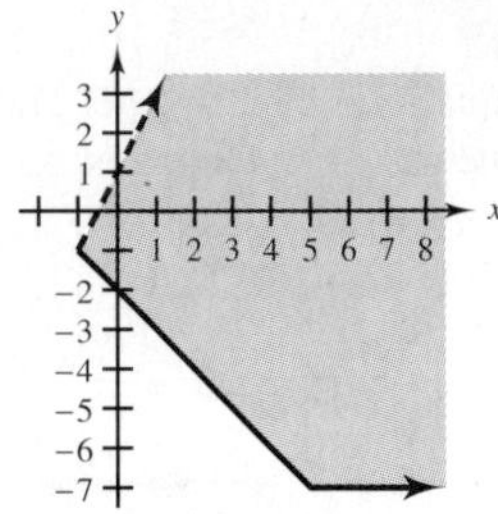

27. The number of people cannot be negative: $x \geq 0$ and $y \geq 0$.

29.

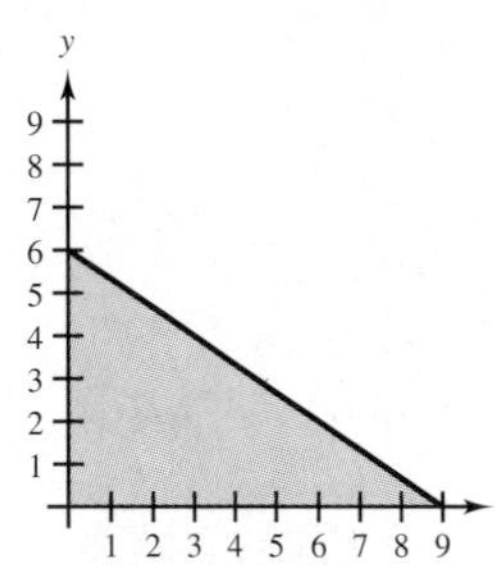

31.

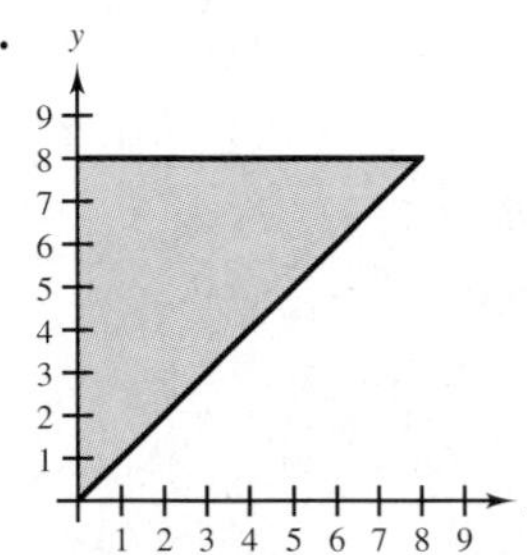

33.

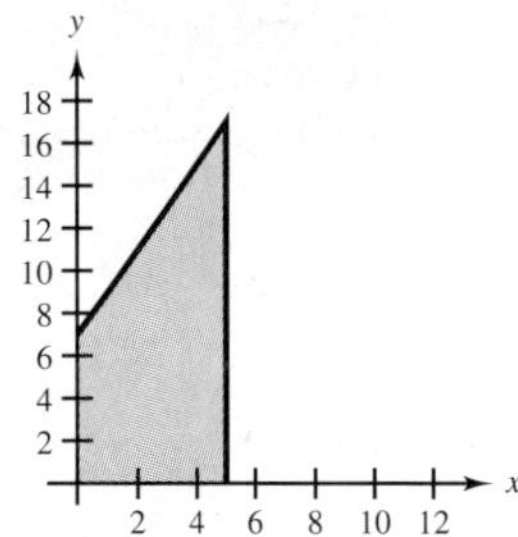

35.

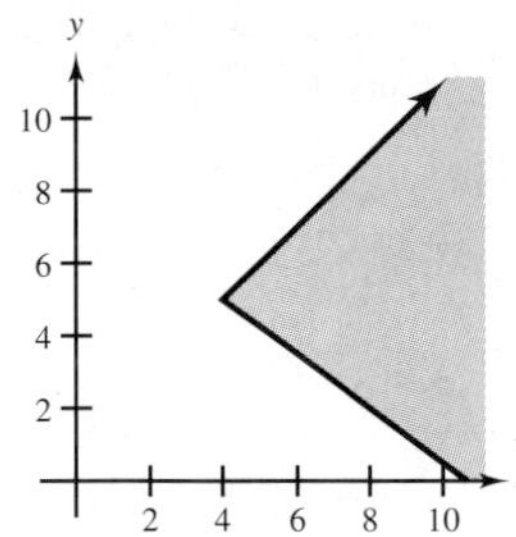

37.

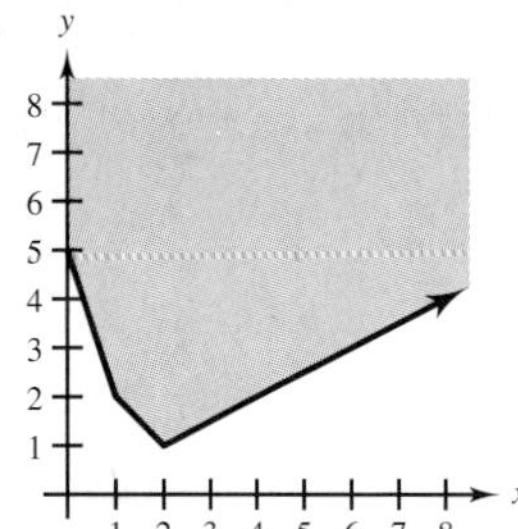

39. $\varnothing$

41. x = number of millions of dollars in short-term notes
y = number of millions of dollars in mortgages
$x + y \leq 10, y \leq 7, x \geq 0, y \geq 0$
\$7 million, \$4 million

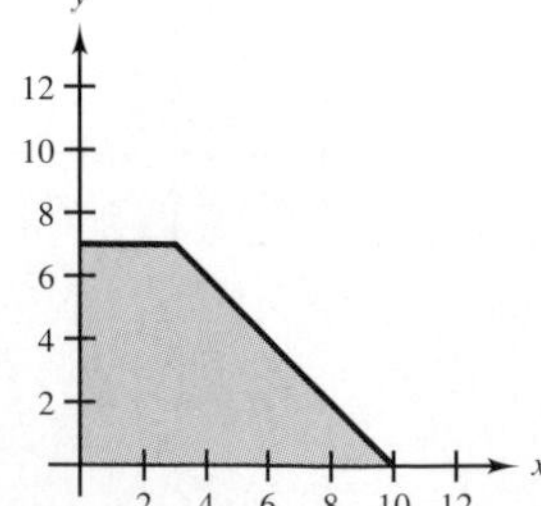

43. x = number of tuna fruit salads,
y = number of chicken Caesar salads
$x \leq \frac{2}{3}y, x + y \geq 50, x \geq 0, y \geq 0$
From 0 to 40 tuna salads can be prepared.

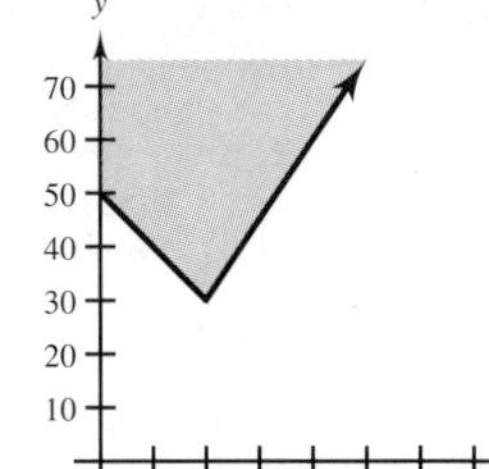

45. **(a)** x = number of barns, y = number of tool buildings
$15x + 5y \leq 90, 7x + 6y \leq 84, x \geq 0, y \geq 0$

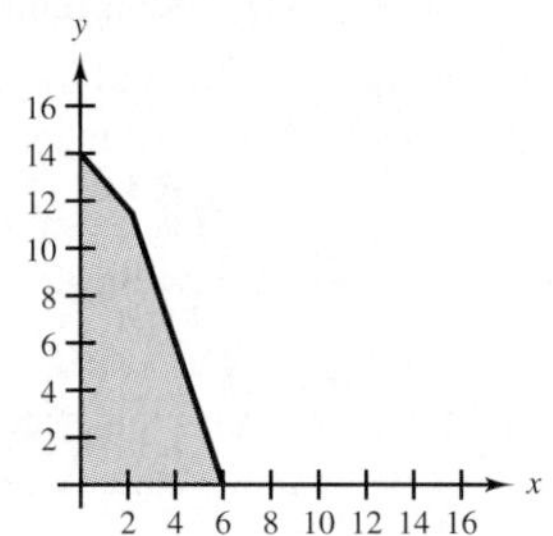

(b) Yes **(c)** No

47. Short-term notes: \$3 million; mortgages: \$7 million

Chapter Review Exercises *(page 346)*

1. The point represents a solution of both equations.

3. Yes **5.** (3, 1) **7.** Infinitely many solutions

9. 1 **11.** 1 **13.** (0.6, 1.6)

15. Infinitely many solutions **17.** (0.75, −0.25)

19. Return to the original system and use the addition method again, but this time to eliminate x.

21. (2, −4) **23.** Infinitely many solutions **25.** (8, −6)

27. No solution **29.** −12 **31.** −3

33. The solution set is represented by a plane drawn in a three-dimensional coordinate space.

35. No **37.** (−2, 3, 1) **39.** (2, −3, −5) **41.** No solution

43. (−8, 1, 3) **45.** (1, 2, −5)

47. A coefficient matrix consists of the coefficients of the variables of a system of equations. An augmented matrix is the combination of a coefficient matrix and a constant matrix.

49. $\left[\begin{array}{cc|c} 4 & -3 & 12 \\ 1 & -2 & -3 \end{array}\right]$ **51.** $\begin{aligned} x + 2y - 3z &= 4 \\ 3x + y - 2z &= 5 \\ x + y &= 0 \end{aligned}$

53. (2, −1) **55.** (−22, −27) **57.** (0, 0, 1) **59.** (−10, 9)

61. (−1, 4, 0)

63. Because the value of the determinant is 0, the expression is undefined. Thus the system does not have a unique solution.

65. 10 **67.** −6 **69.** (−1, 1) **71.** 33 fives, 27 twenties

73. Sledge: 15; claw: 27 **75.** 435

77.

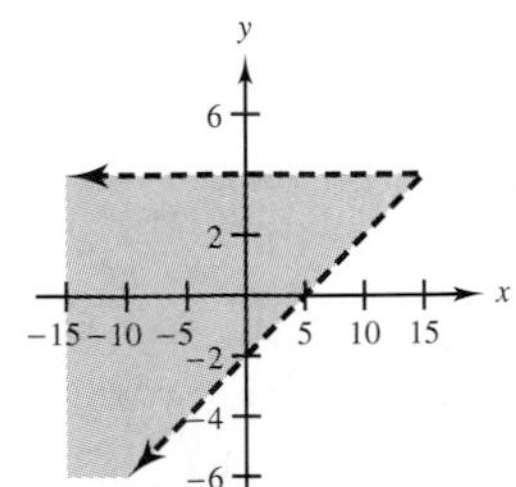

79.

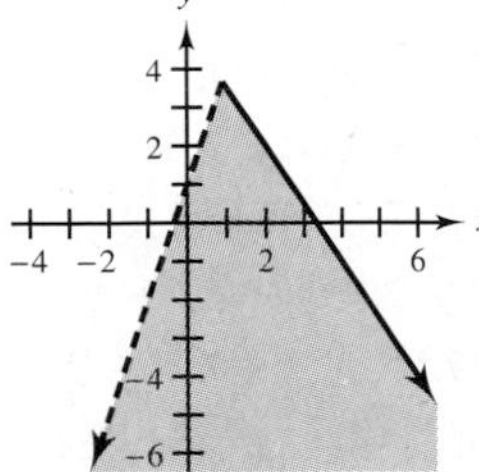

81.

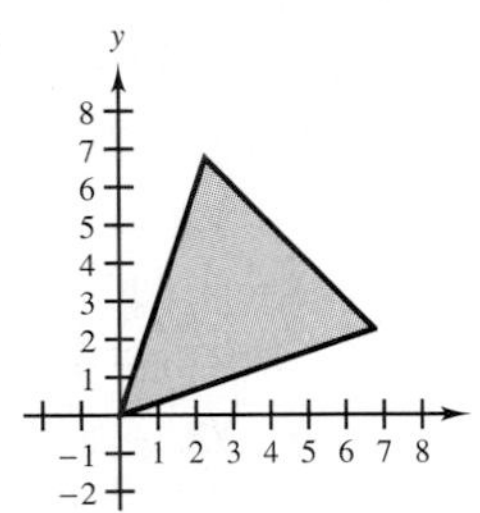

83.

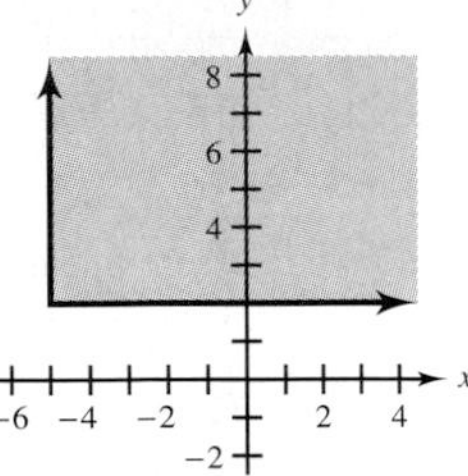

85.

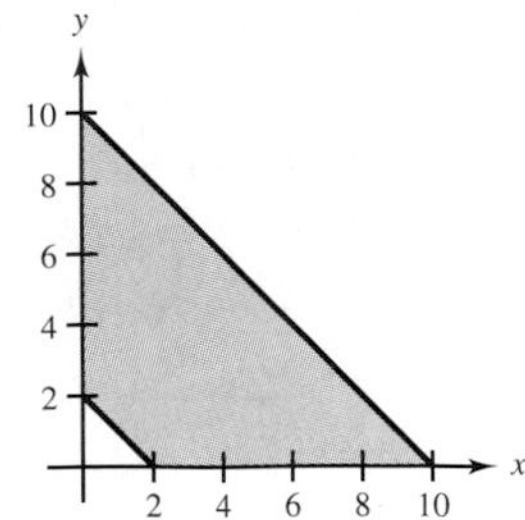

87. x = number of bags of mortar, y = number of bags of lime
$x + y \le 20$, $80x + 40y \le 1000$, $x \ge 0$, $y \ge 0$

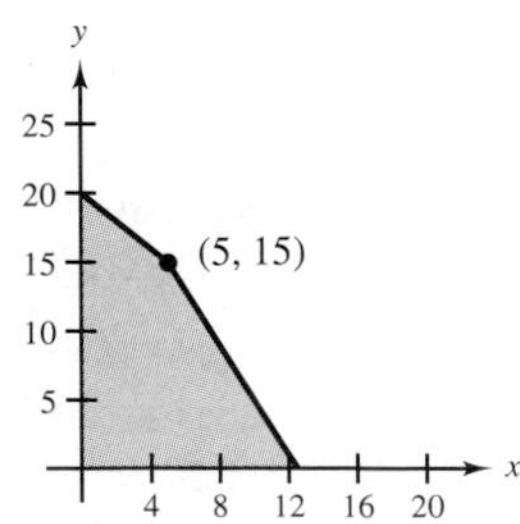

Chapter Test *(page 350)*

1. [*5.1*] (2, −1) **2.** [*5.1*] (−12, 11)

3. [*5.1*] (5, −3) **4.** [*5.2*] Infinitely many solutions

5. [*5.2*] (10, −2) **6.** [*5.3*] (2, −1, 1)

7. [*5.4*] (4, −2) **8.** [*5.4*] (1, 3, −2)

9. [*5.4*] (−3, 0.5, 1) **10.** [*5.1*] Dependent

11. [*5.1*] Inconsistent **12.** [*5.1*] Unique solution

13. [*5.1*] **(a)** The graphs coincide.
(b) The graphs intersect at one point.
(c) The graphs are parallel.

14. [*5.5*] 1 **15.** [*5.5*] (−5, 1)

16. [*5.7*]

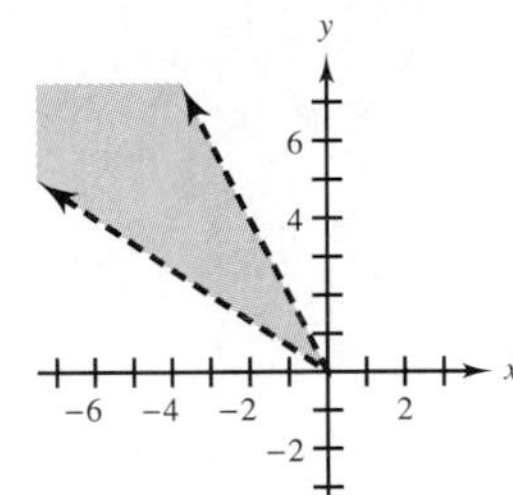

17. [*5.7*]

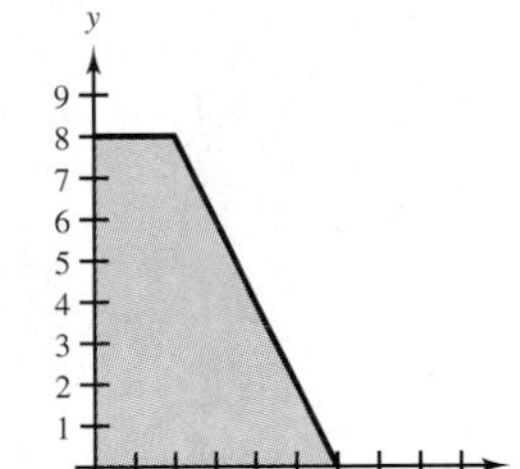

18. [*5.6*] Wind speed, 25 mph; airspeed of plane, 175 mph

19. [*5.6*] $31\frac{2}{3}°$, $58\frac{1}{3}°$

20. [*5.6*] 40 liters of the 15% solution, 60 liters of the 40% solution

21. [*5.6*] Shrimp, 45.5 pounds; crab, 43 pounds; lobster, 48 pounds

22. [*5.7*] p = number of pines, d = number of dogwoods
$p + d \le 20$, $p \ge 2d$, $p \ge 0$, $d \ge 0$

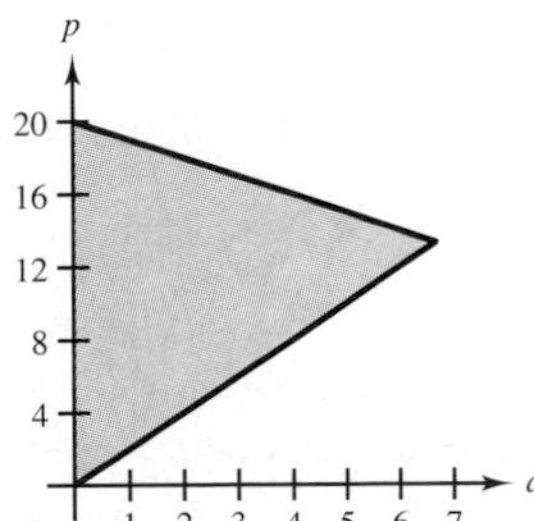

Cumulative Test: Chapters 4–5 *(page 351)*

1. [*4.1*] $a = \frac{1}{3}$, $b = 2$

2. [*4.1*] (0, −10) (−2, 0)

3. [*4.2*] **(a)** Negative **(b)** Undefined
(c) Zero **(d)** Positive

4. [*4.2*] $y = \frac{1}{2}x + 2$

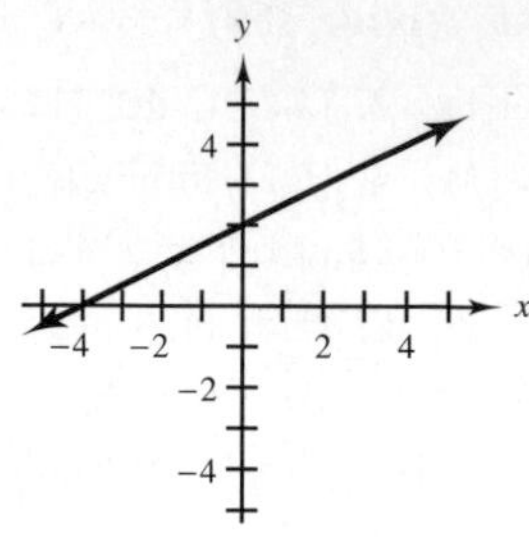

5. [*4.3*] The rate of change of y with respect to x is the slope of the line: $\frac{2}{3}$.

6. [*4.3*] $\frac{7}{5}$ **7.** [*4.4*] $x - y = 3$

8. [*4.4*] $y = -\frac{2}{3}x - \frac{10}{3}$

9. [*4.5*] **(a)** The boundary line is the dashed line $y = x - 7$.

(b) Either test a point or write the inequality as $y > x - 7$ and shade above the line.

10. [*4.5*]

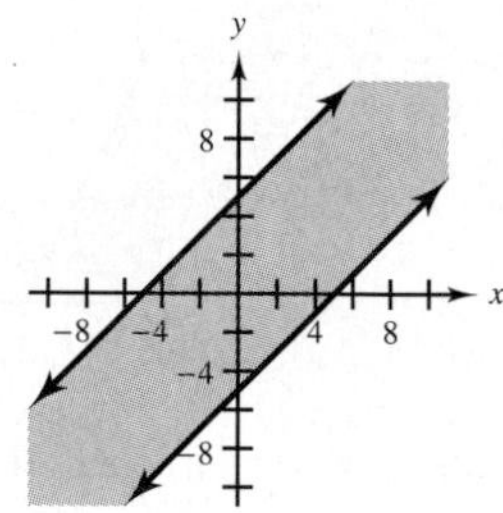

11. [*5.1*] **(a)** Empty set **(b)** One point

(c) All points on the line

12. [*5.1*] $(-2, 6)$

13. [*5.2*] **(a)** Dependent **(b)** $(-1, 5)$ **(c)** Inconsistent

14. [*5.2*] 200 mathematics books, 220 English books

15. [*5.3*] $(-1, 0, 4)$ **16.** [*5.3*] 40°, 70°, 140°

17. [*5.4*] $(3, -2, 1)$ **18.** [*5.5*] -3

19. [*5.5*] The determinant of the coefficient matrix is zero.

20. [*5.6*] 4 mph **21.** [*5.6*] 150

22. [*5.7*]

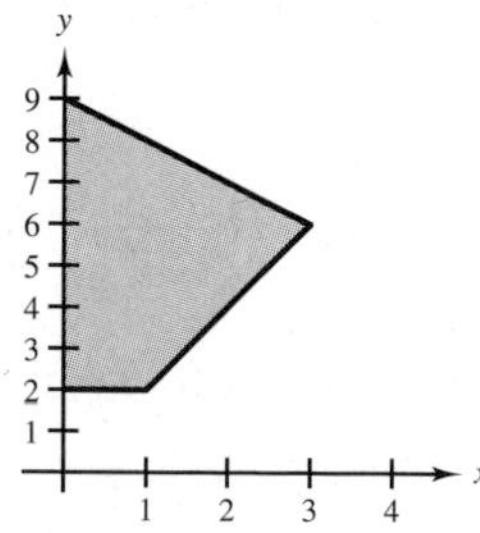

23. [*5.7*] $2L + 2W \leq 40$

$W \leq \frac{1}{2}L$

$L \geq 0$

$W \geq 0$

CHAPTER 6

Section 6.1 *(page 359)*

1. (a) The variable is in the denominator.

(b) The exponent is negative.

3. Yes **5.** Yes **7.** 5, 0 **9.** $-4, 3$ **11.** 1, 2

13. The degree of a term is the value of the exponent on the variable (or the sum of the exponents if there is more than one variable). The degree of a polynomial is the degree of the term with the highest degree.

15. 1 **17.** 5 **19.** 5, trinomial **21.** 3, neither

23. 5, binomial **25.** 2, $-3y^2 + 6y + 6$, $6 + 6y - 3y^2$

27. 9, $4x^9 - 7x^8 - 5x^6 + x^3$, $x^3 - 5x^6 - 7x^8 + 4x^9$

29. 12, $n^{12} + 4n^8 - 2n^6 - 3n^4$, $-3n^4 - 2n^6 + 4n^8 + n^{12}$

31. -4 **33.** 15 **35.** -24 **37.** $4a$ **39.** $3xy^2$

41. $-7x + 6$ **43.** $4x^2 - x + 14$ **45.** $-2t^2 - 3t + 1$

47. $n^2 - m^2 + 7mn$ **49.** $-2xy + 5y^2$ **51.** $-2a^2 + 5a^2b$

53. $-2x^3 - 8x$ **55.** $-4x^2 + 5x - 8$ **57.** $x^2 - 12$

59. $2x^2 + 2x + 2$

61. (ii) The minuend is $3x + 4$, and the subtrahend is $x - 2$.

63. $3x^2 - 2x + 7$ **65.** $-9x^2 - 3x - 9$ **67.** $-x + 7$

69. $x - 3$ **71.** $x^2 - 5x + 2$ **73.** $-x - 1$ **75.** -2

77. $-3y + 3$ **79.** $-2x^2 - 6xy + 2y^2 + 4xy^2$

81. 5, 0 **83.** $-1, -11$ **85.** -130.05, 68.33

87. $x^2 + 4x - 2$ **89.** $3x^2 - 11x - 16$

91. (a) 1600 **(b)** 1000 **(c)** 232 **(d)** 0 **(e)** 9

(f) $40 \leq t \leq 100$

93. (a) \$2012 **(b)** \$9710 **(c)** $T(x) = x^2 + 17x + 22$

(d) \$11,722 **(e)** Answers will vary. **(f)** \$20,800

(g) $P(x) = x^2 - 9x - 22$ **(h)** \$9078

(i) Answers will vary.

95. (a) Lowest: 1983; highest: 1997

(b) A downward trend in construction is predicted after 2000.

97. $a = -1, b = 1, c = 4$ **99.** $a = 0, b = 6, c = -9$

101. When we simplify the left side, we obtain $(a + b)x^2 - 2 = 4x^2 - 2$, which means that $a + b = 4$. There are infinitely many values of a and b such that $a + b = 4$.

103. 2, 3, 5

Section 6.2 *(page 368)*

1. Statement (ii) is false. For example, $(x + 2)(x - 2) = x^2 - 4$, which is a binomial.

3. $6x^3 - 9x^2$ **5.** $-8x^5 - 20x^3$ **7.** $-x^3y^2 + x^2y^3 + 2xy^2$

9. $x^2 - 2x - 24$

11. $-y^2 - 2y + 15$ **13.** $9x^2 + 29x + 6$

15. $x^2 - 13xy + 36y^2$ **17.** $12x^2 - 28xy + 15y^2$

19. $x^3 - 2x^2 - 3x + 10$ **21.** $6x^3 - 7x^2 - 5x + 3$

23. $x^3 + y^3$ **25.** $4x^4 - 4x^3 + 7x^2 - 11x + 4$

27. $x^4 - 2x^2 + 1$ **29.** $x^3 - 8$

31. $2x^4 - 8x^3 + 19x^2 - 22x + 15$

33. Yes, but recognizing the special pattern saves the step of determining the inner and outer products.

35. $x^2 + 6x + 9$ **37.** $49x^2 - 42x + 9$

39. $4x^2 + 20xy + 25y^2$ **41.** $x^2 - 16$ **43.** $4x^2 - 9$

45. $81n^2 - 16m^2$

47. The Associative Property of Multiplication guarantees that either grouping is correct. Method (i) is easier because it can be done in one step.

49. $a^6 - b^6$ **51.** $16 + 4p^2q - 4pq^2 - p^3q^3$

53. $x^4y^4 + 4x^2y^2 + 4$ **55.** $12x^5 - 26x^4 - 10x^3$

57. $-19x - 18$ **59.** $2b^3 - 5b^2 - 13b + 30$

61. $16x^4 - 1$ **63.** $-20x - 50$ **65.** $-x^2 + x - 18$

67. $12x$ **69.** $32x^5 - 48x^4 + 18x^3$ **71.** $5y^2 + 20y - 29$

73. $a = 4, b = -2$ **75.** $a = 3, b = 2$

77. $2rt - r + 5t - 5$ **79.** $300w - 2w^2$

81. The changes from 1985 to 1987 and from 1987 to 1989 are large compared with the changes from 1989 to 1991 and from 1991 to 1993. Because the rate of change is not approximately constant, a linear model does not appear to be appropriate.

83. $E(x) = -0.33x^3 - 4.54x^2 + 178.4x + 1531.46$ represents the total expenditure for textbooks. The graph indicates that future total expenditures will decline.

85. $x^{2n} + 6x^n + 8$ **87.** $x^{2n} - 16$

91. Write the expression as $[(x + 3y) - 4][(x + 3y) + 4]$ and use the special product form $(A - B)(A + B)$. The result is $x^2 + 6xy + 9y^2 - 16$.

93. $$[(x^2 + 5x) + 3][(x^2 + 5x) - 3] = (x^2 + 5x)^2 - 9$$
$$= x^4 + 10x^3 + 25x^2 - 9$$

95. $x^2 - 6x + 9 - 25y^2$

Section 6.3 *(page 376)*

1. Choose the smallest exponent for each variable.

3. $3(x + 4)$ **5.** $5(x + 2y - 6)$ **7.** $4x(3x - 1)$

9. Prime **11.** $x^2(x - 1)$ **13.** $3x(x^2 - 3x + 4)$

15. $xy^2(x^4y^2 - x^3y + 1)$ **17.** $(a + by)(3x - 2y)$

19. $(2x + 3)(4x^2 - 7y^2)$

21. All are correct because the product in each case is $4x - 12$.

23. $4(x - 3), -4(3 - x)$

25. $y(6 + 3x - x^2), -y(x^2 - 3x - 6)$

27. $(y - 3)(y - 2)$ **29.** $(3 - z)(5 + z)$

31. $(x - 4)(2x + 5)$

33. The answer is obtained from multiplying the factors. $(2x - 5)(x + 3) = 2x^2 + 6x - 5x - 15$.

35. $(b + y)(a + x)$ **37.** $(d - 4)(c + 3)$ **39.** Prime

41. $(x^2 + y)(a + 3)$ **43.** $(x^2 + 3)(x + 2)$

45. $3(y^2 + 4)(y + 3)$ **47.** $(a^2 + 1)(b + y)$

49. $(x^2 + 3)(x - 5)$ **51.** $(3x^2 + 4)(2x - 3)$

53. $3(2x^2 - 5)(3x + 1)$ **55.** $x(3y^2 + 7)(y - 2)$ **57.** $4y - 5$

59. $3x^2 + 2x + 6$ **61.** $\frac{1}{3}(x + 15)$ **63.** $\frac{1}{12}(2x^2 + 3x + 36)$

65. $d^{15}(d^5 - 1)$ **67.** $11x^{48}(8 + x^3)$

69. $6ab^2(4a^4b - b^2 + 2a^2)$

71. $4a^2b^2(x - 4)^2(b + 3ax - 12a)$

73. $A = P(1 + r)$ **75.** $A = 2\pi r(h + r)$

77. Use the Distributive Property to multiply $x^{-6}(3x^2 + 2)$.

79. $2x - 5$ **81.** $2x^{-4}(x^7 - 5)$ **83.** $5x^{-4}y^{-2}(y^9 - 3x^7)$

85. $(x + 3)(a^2 + b - 5)$

Section 6.4 *(page 380)*

1. A difference of two squares consists of two perfect square terms separated by a minus sign.

3. $(x + 3)(x - 3)$ **5.** $(1 + 2x)(1 - 2x)$

7. $(4x + 5y)(4x - 5y)$ **9.** $\left(\frac{1}{2}x + 1\right)\left(\frac{1}{2}x - 1\right)$

11. $(a^3 + b^8)(a^3 - b^8)$ **13.** $(a + b + 2)(a + b - 2)$

15. $(16 + x - y)(16 - x + y)$

17. A perfect square trinomial is a trinomial with two terms that are perfect squares and a third term that is twice the product of the square roots of the other two terms.

19. $(a - 2)^2$ **21.** $(2x + 3)^2$ **23.** $(x - 7y)^2$

25. $\left(x + \frac{1}{2}\right)^2$ **27.** $(x^2 - 5)^2$ **29.** $(x + 1)(x^2 - x + 1)$

31. $(3y - 2)(9y^2 + 6y + 4)$ **33.** $(5 - 2x)(25 + 10x + 4x^2)$

35. $(3x + y)(9x^2 - 3xy + y^2)$

37. $(x^2 + 1)(x^4 - x^2 + 1)$ **39.** $(x + 1)^2$

41. $(0.5x + 0.7)(0.5x - 0.7)$ **43.** $(x - 1)(x^2 + x + 1)$

45. $(c - 10)^2$ **47.** $2(x + 3)(x^2 - 3x + 9)$

49. Prime **51.** $(6 - 7y)^2$ **53.** $4(x + 4)(x - 4)$

55. $(5x - 4y)^2$ **57.** $(x^2 + 4)(x + 2)(x - 2)$

59. $(2 + 5x^2)(4 - 10x^2 + 25x^4)$

61. $(y^4 - 3)^2$ **63.** $(6x - y)(36x^2 + 6xy + y^2)$

65. $(x - 2)^2(x^2 + 2x + 4)^2$ **67.** $(x^6 + 2y^{10})(x^6 - 2y^{10})$

69. $(x^3 - 2)^2$ **71.** $2(7 + y^3)(7 - y^3)$

73. $x^2(5xy - 1)(25x^2y^2 + 5xy + 1)$ **75.** $x^3(7x^2 - 5)^2$

77. Prime **79.** $(4 + 3xy^3)(16 - 12xy^3 + 9x^2y^6)$
81. $(x + 2)^2(x - 2)^2$ **83.** $(4x^2 + 5y)(4x^2 - 5y)$
85. $(8x^2 + 5y)^2$ **87.** $(4a^2 + 9d^2)(2a + 3d)(2a - 3d)$
89. $2x(x + 2)(x^2 - 2x + 4)$
91. $(5x^2 - 3y^4)(25x^4 + 15x^2y^4 + 9y^8)$
93. $4(a + 2b)(x + y)(x - y)$ **95.** $(a + b)(x + y)(x - y)$
97. $4(x - 2)^2$ **99.** $(x + 2)^2(x - 2)^2$
101. Because $x^6 - y^6 = (x^3)^2 - (y^3)^2$, the expression is a difference of two squares. Because $x^6 - y^6 = (x^2)^3 - (y^2)^3$, the expression is a difference of two cubes.
103. $-6, 6$ **105.** 1 **107.** 16
111. $\frac{1}{2}mv_1^2 - \frac{1}{2}mv_2^2, \frac{1}{2}m(v_1 + v_2)(v_1 - v_2)$
115. $C(x) = -1020(4x + 87)(x + 12)(x - 12)$
117. $(x + 2)(x - 2)(x^2 - 2x + 4)(x^2 + 2x + 4)$
119. $(a + b)(a - b)(a^2 - ab + b^2)(a^2 + ab + b^2)$
121. $(2x + 1)(2x - 1)(x - 2)(x^2 + 2x + 4)$
123. $(x + 1)(x - 1)(y - z)(y^2 + yz + z^2)$

Section 6.5 *(page 390)*

1. The Commutative Property of Multiplication allows us to write the factors in either order.
3. $(x + 1)(x + 2)$ **5.** $(x - 3)(x - 6)$ **7.** $(x + 2)(x + 7)$
9. Prime **11.** $(x - 8y)(x + 2y)$ **13.** $(2x + 1)(x + 2)$
15. $(2 + x)(6 - x)$ **17.** $(3x + 1)(x - 3)$
19. $(x - 5)(4x + 3)$ **21.** Prime **23.** $(c + 2)(c - 1)$
25. $(6x - 5)(x + 3)$ **27.** $(x + 8)(x - 3)$
29. $(4 + x)(6 - x)$ **31.** $(b - 5)(b + 3)$
33. $(3x + 2)(x + 1)$ **35.** $(3x - 5)(2x - 3)$
37. $(x + 5)^2$ **39.** $(2 - 5x)(1 + 3x)$ **41.** $(9 - 10x)(2 + x)$
43. $(a - 6)(a - 4)$ **45.** $(3x + 2)(3x - 5)$
47. $(x - 7)(x - 9)$ **49.** $9(x - 2)(x + 1)$
51. $(2x + 3)(6x - 1)$ **53.** $(4x - 7y)(2x + 3y)$
55. $2(x + 3)(x - 2)$ **57.** $4(x + 3)^2$
59. There are fewer combinations to try because 17 and 23 have fewer factors than 24 and -18.
61. $8xy(x - 5y)(x + 2y)$ **63.** $-3x(2x - 5)(2x + 3)$
65. $2y(2x - 3)(7x + 3)$ **67.** $(ab - 5)(ab + 2)$
69. $xy(x + 6)^2$ **71.** $(xy - 8)(xy + 6)$ **73.** $2(3x - y)^2$
75. $(6m - n)(m + n)$ **77.** $2xy^2(4x - y)(2x - y)$
79. $x(x - 1)(x - 3)(x + 14)$
81. $(x + 5)(x - 5)(x + 3)(x - 3)(x + 1)(x - 1)$
83. $(xy + 1)(2xy - 3)(xy + 3)$
85. No, the factor $x^2 - 4$ can be factored as $(x + 2)(x - 2)$.
87. $(x^2 + 30)(x^2 - 2)$ **89.** $(x^3 + 5)(x^3 + 12)$
91. $(x + 3)(x - 3)(x + 2)(x - 2)$
93. $(5x^4 - 7)(x^4 + 2)$ **95.** $(2x + 1)(2x - 1)(x + 3)(x - 3)$
97. $(3x + 2)(3x - 2)(x + 1)(x - 1)$
99. $(x + 3)(x + 6)$ **101.** $2(x - 3)(2x - 7)$
103. $(2x^2 + 4x - 3)(x^2 + 2x + 2)$ **105.** $(x + 8)(x - 3)$
107. $f(t) = 52(2.5t + 60)$
109. The coefficient of t indicates that food costs were increasing at an annual rate of \$130 during this period. In the factored form, the coefficient of t indicates that food costs were increasing at a weekly rate of \$2.50.
111. $-7, 7, -11, 11$ **113.** $-8, 8, -16, 16$
115. 2 **117.** $(x^n + 8)(x^n + 3)$
119. $(y^{2n} - 11)(y^n + 3)(y^n - 3)$ **121.** $(5x^n - 3)(x^n + 1)$
123. $x^{2n}(3x^n + 2)(x^n + 1)$

Section 6.5 Supplementary *(page 392)*

1. $(5x + 3)(x - 2)$ **3.** $(5 + 9y)(5 - 9y)$
5. $(x^2 - 3)(2x + 1)$ **7.** $(x - 2)(3x + 5y)$
9. $(5x + 1)(25x^2 - 5x + 1)$ **11.** $(x + 2)(x + 4)$
13. Prime **15.** $(2x + 5)^2$ **17.** $(a - b)(x - 3)$
19. $25(y - 1)(y^2 + y + 1)$
21. $(y^2 + 4x^2)(y + 2x)(y - 2x)$
23. $a^2b^4(b^3 + 3ab - 1)$ **25.** $(3x + 1)(7x - 2)$
27. $(4 - a)^2$ **29.** $2(5 + 3x^2)(5 - 3x^2)$
31. $(y + 5)(2 - 3x)$ **33.** $4x^2(3x - 2)(2x + 3)$
35. $(c^{10} + d^4)(c^5 + d^2)(c^5 - d^2)$ **37.** Prime
39. $(a + 2)(a - 2)(x - y)(x^2 + xy + y^2)$
41. $a^5b^3(b - a^2b^5 + a)$ **43.** $(m - 8)(m - 2)$
45. $4(x^2 + 4y^2)(x + 2y)(x - 2y)$
47. $(a + b - 2)(a^2 + 2ab + b^2 + 2a + 2b + 4)$
51. $2a^4(3a - 5)(5a - 9)$
53. $(c - 1)(c^2 + c + 1)(c + 2)(c - 2)$
55. $(x^2 + 11y^2)(x + 3y)(x - 3y)$

Section 6.6 *(page 398)*

1. Produce the graph and estimate the x-intercepts.
3. **(a)** $-2, 6$ **(b)** No solution **(c)** 2 **(d)** $-4, 8$
5. $-3, 4$ **7.** 0, 6 **9.** 5 **11.** $-5, -1, 3$
13. Assuming that the equation has only one variable, the largest exponent on the variable is 1 for a first-degree equation and 2 for a second-degree equation.
15. $-7, 0$ **17.** $-3, 3$ **19.** 2, 3 **21.** $-1, 2.5$ **23.** 0, 5
25. $-3, 7$ **27.** $-6, 1$ **29.** 5, 7 **31.** 1.5, 4 **33.** $-2, \frac{4}{9}$
35. $-6, 5$ **37.** 2, 30 **39.** $-4, 2$ **41.** 2.5, 2 **43.** $-3, 4$

45. $-4, 1.5$ **47.** $0, 1.25$ **49.** $-6, 3, 5$ 2.5,

51. $-3, 0, 3$ **53.** $-1, 0, \frac{2}{3}$ **55.** $-1, 1, 5$ **57.** $-2, 1.5, 2$

59. $-3, -2, 2, 3$ **61.** $2, \frac{5}{2}, \frac{5}{3}$ **63.** $-3, 2, 3$

65. $b = 5; -0.5$ **67.** $c = 6; -3.5$ **69.** $x^2 + x - 12 = 0$

71. $x^2 - 8x + 16 = 0$ **73.** $2x^2 - x = 0$ **75.** $-c, 2c$

77. $-\frac{3y}{2}, \frac{y}{2}$ **79.** $-3a, a$ **81.** $-0.65, 4.65, 1, 3$

83. $-0.5, 2$ **85.** $-0.5, 0, 0.83, 1.33$ **87.** $-6, -3, -2, 1$

89. $-9, -7, -3, -1$ **91.** $-1, 1, 2, 4$

93. -9 or 8 **95.** 3, 4, 5 or $-5, -4, -3$

97. 0, 2 or 3 **99.** -2 and 5

101. 22 meters, 27 meters

103. 8 inches, 15 inches, 17 inches

105. 2 feet **107.** 15 inches, 20 inches

109. **(a)** $C(x) = -0.05(x^2 - 88x + 640)$
(b) $C(x) = -0.05(x - 80)(x - 8)$ **(c)** 8, 80
(d) No person of age 8 or age 80 becomes upset about not having coffee at a regular time.

111. The value of c for which $p(c) = 0$ is -1. The quadratic factor cannot be factored. The graph has no x-intercepts.

113. $x = 1$, y is any number **115.** $-2, 1$

Section 6.7 *(page 408)*

1. Dividend: $x + 7$; divisor: x; quotient: 1; remainder: 7

3. $2x^2 + 4x$ **5.** $2x^3 + 3x + \frac{1}{2}$ **7.** $8a^3 + 6a^2 - 3a + 4$

9. $5x^3 - 3x^2 + \dfrac{3}{2} - \dfrac{2}{x}$ **11.** $5c^3d^5 + 2c^2d^3 - \dfrac{3d}{c}$

13. $9x^2 + 7x - 6$ **15.** 8

17. We stop dividing when the remainder is 0 or when the degree of the remainder is less than the degree of the divisor.

19. $Q(x) = 2, R(x) = 7$ **21.** $Q(x) = x + 2, R(x) = -1$

23. $Q(x) = 3x + 3, R(x) = 5$

25. $Q(a) = a^2 - a + 2, R(a) = -2$

27. $Q(x) = 3x^2 + 4x + 1, R(x) = 0$

29. $Q(x) = 1, R(x) = 4$ **31.** $Q(x) = 1, R(x) = x - 4$

33. $Q(a) = a^2 + 2a, R(a) = a - 3$

35. Q: $x + 6y$; R: $15y^2$ **37.** Q: $4x^2 - 4xy + y^2$; R: $-3y^3$

39. Write the divisor as $x - (-3)$. Then the divider is -3.

41. $Q(x) = 3x + 2, R(x) = 0$ **43.** $Q(x) = 6x + 4, R(x) = 3$

45. $Q(x) = 2x^3 + x^2 + 3x + 3, R(x) = 10$

47. $Q(x) = x^3 - 2x^2 + 5x - 10, R(x) = 6$

49. $Q(x) = x^2 + 4x + 5, R(x) = 4$

51. No, the divisor cannot be second degree. It must be in the form $x - c$.

53. $a = 1, b = -2, c = 0, d = 1, e = -2$

55. Dividend: $x^3 + 3x^2 - 2x - 4$; divisor: $x + 1$

57. $x + 4 + \frac{15}{x - 6}$ **59.** $3x + 14 + \frac{44}{x - 3}$ **61.** $x + 5$

63. $2a^2 + 3a + 2 + \frac{7}{3a - 2}$ **65.** $2x^2 + x - 5 + \frac{12}{x + 2}$

67. $x^4 - 2x^3 + 2x^2 - 4x + 9 - \frac{13}{x + 2}$

69. $(x - 1)(x + 4)(x - 3)$ **71.** $(x + 3)(x - 4)(x - 2)$

73. $(2x + 1)(x - 1)^2$ **75.** $2x - 13 + \frac{21}{x + 2}$

77. $9x + 15 + \frac{16}{x - 1}$ **79.** $2x^2 + 5x - 1$

81. $x + 2$ **83.** $Q(x) = x^4 + 3x^3 + 6x^2 + x + 4, R(x) = 0$

85. $Q(x) = x + 3, R(x) = 11x - 3$

87. $9x^5 + 12x^4 + 3x^3 - 12x^2$ **89.** $x + 8$

91. **(a)** $x^2 - 8x - 1$ **(b)** 2 feet

93. -8 **95.** 4 **97.** $4x^{2n} - 16$

99. $Q(x) = x^{3n} - x^n + 5, R(x) = -14$

101. 2

Section 6.8 *(page 415)*

1. According to the Remainder Theorem, the remainder is 5.

3. -2 **5.** -15 **7.** 9 **9.** -331 **11.** -1.125

13. We say that c is a root or solution of the equation $P(x) = 0$ because the equation is true when x is replaced with c.

27. $(x^2 - 3x + 4)(x + 3)$ **29.** $(2x^2 + 3x + 6)(x + 2)$

31. $(2x^2 - 4x - 6)\left(x - \frac{1}{2}\right)$ **33.** $(x^3 - x + 2)(x - 4)$

35. Divide the polynomial by $x - 1$, and factor the resulting second-degree polynomial.

37. $(x + 3)^2(x + 2)$ **39.** $(x + 5)(x - 4)(x - 1)$

41. $(x - 1)(2x - 1)(x + 2)$ **43.** $(x - 3)(x + 2)(4x - 3)$

45. $(x + 2)(4x - 3)(2x - 1)$ **47.** $(3x - 2)(x - 3)(x + 1)$

49. $(x + 1)(x - 2)(x - 3)$ **51.** $(x - 3)(x^2 - x + 2)$

53. $(x + 1)(x - 2)(2x - 3)$ **55.** $(x + 1)(3x - 2)(2x - 3)$

57. $(x + 3)(x - 1)(3x + 2)$

59. $(x + 2)(x + 1)(x - 3)(x + 3)$

61. $(x + 2)(x - 1)(x - 2)(x^2 + 1)$

63. Three factors are $(x - a)$, $(x - b)$, and $(x - c)$. This does not imply that $P(x)$ is of degree 3 because one or more of the factors might be repeated.

65. $x^2 - x - 6$ **67.** $x^3 + x^2 - 12x$ **69.** $x - 3$

71. A solution is c because $x - c$ is a factor of $P(x)$ if and only if $P(c) = 0$.

73. $-3, -2, 3$ **75.** $-3, 1, 3$

77. No, $c^4 + c^2 + 1$ is never 0 for any real number c.

79. $F(x) = -0.0075(x^2 - 50x - 2400)$
$F(x) = -0.0075(x - 80)(x + 30)$

81. Because x represents age, only $x = 80$ is meaningful. However, this solution indicates that no people of age 80 fished in 1996, which is not true. Because $F(0) = 18$, the model indicates that 18% of people of age 0 fished. The model is not valid for age 0. We should probably judge the model to be reasonably valid only for the age span given in the table: $7 \le x \le 65$.

83. -1 **85.** 1 or 6 **87.** -2 or 2

89. Positive odd integers n

Chapter Review Exercises *(page 418)*

1. **(a)** Binomial **(b)** 2 **(c)** $-2x^2 + 3x$ **(d)** -2

3. **(a)** Trinomial **(b)** 5 **(c)** $9x^5 + x^2 - 3$ **(d)** 1

5. 6 **7.** 2 **9.** -55 **11.** $2x^2 - 8x + 13$

13. $3x^2 - 2x + 7$ **15.** $-8x^8$ **17.** $10a^2b - 2a^2b^3 + 6a^2b^2$

19. $-6x^2 + 7x + 24$ **21.** $-x^2 - 12x - 19$

23. $8x^2 - 22x + 15$

25. A difference of two squares consists of two perfect square terms separated by a minus sign.

27. $9x^2 - 4$ **29.** $9y^2 - 12y + 4$ **31.** $x^2 - x - 6$

33. $x^3 - x$ **35.** Choose the common factor $-x^2y^2$.

37. $x^{25}(1 + x^{10})$ **39.** $x^3y^3(x^3 - x^4y - y^2)$

41. $(x^2 - 2)(x + 3)$ **43.** $7x^3(2x^4 - 5x^2 + 6)$

45. Answers will vary.

47. The constant term is not a perfect square.

49. $(7x + 4y)(7x - 4y)$ **51.** $(4 + a^2)(2 + a)(2 - a)$

53. $(3b^2 + 2a)(9b^4 - 6ab^2 + 4a^2)$ **55.** $(c + 8)^2$

57. Answers will vary. **59.** $(x - 6)(x + 1)$

61. $2(x + 7)(x - 1)$

63. Because the numbers 6 and 24 have several factors, they lead to a large number of possible factorizations to try.

65. $(3x + 5)(x + 3)$ **67.** $(6x - 5)(2x + 3)$

69. $(x + 4)(x - 2)$ **71.** $(5x - 4)(25x^2 + 20x + 16)$

73. No, but you can factor out the 2 to obtain $[2(x + 2y)]^2 = 4(x + 2y)^2$.

75. $(3x - 5)(x + 3)$ **77.** $(x^3 + 3y^2)(x^6 - 3x^3y^2 + 9y^4)$

79. $x^3y^3(y - x)$ **81.** $x(x + 3)(6x + 7)$

83. In $a + b = 0$, we know that a and b are opposites, but we do not know what the numbers are. In $ab = 0$, either a or b must be 0.

85. 0, 8 **87.** $-7, 6$ **89.** $-6, 4$ **91.** $-3, -2, 2$

93. $-3, -2, -\frac{1}{3}, \frac{2}{3}$ **95.** -9 or 7

97. Multiplying the quotient times the divisor and adding the remainder should result in the dividend.

99. $3x^3 - 4x$ **101.** $Q(x) = 3x^2 - 4x - 5, R(x) = -5$

103. Q: $x^2 + 3y^2$; R: $8y^3$

105. $P(x) = 2x^3 + x^2 + 4$

107. $Q(x) = 2x^2 - x - 1, R(x) = 4$

109. $Q(x) = x^4 + 2x^3 + 4x^2 + 8x + 16, R(x) = 0$

111. The value of $P(2)$ is the remainder when $P(x)$ is divided by $x - 2$. Therefore, $P(2) = 3$.

113. 3

115. The graph of P has at least three x-intercepts: $(-4, 0)$, $(2, 0)$, and $(5, 0)$. If $P(x)$ is divided by $x - c$ and the remainder is 0, then $P(c) = 0$ and $(c, 0)$ is an x-intercept.

117. $(x + 1)(x - 3)(x - 2)$

Chapter Test *(page 421)*

1. [*6.1*] -16

2. [*6.1*] **(a)** $(p + q)(x) = 3x^4 + x^3 - 5x^2 - 2x - 5$
$(p + q)(1) = -8$
$p(1) + q(1) = -7 + (-1) = -8$
(b) $(p - q)(x) = 3x^4 - 9x^3 - 5x^2 + 14x - 9$
$(p - q)(-1) = -16$
$p(-1) - q(-1) = -11 - 5 = -16$

3. [*6.2*] $-2x^5 + 10x^4 - 6x^3$ **4.** [*6.2*] $6t^2 + 11st + 4s^2$

5. [*6.2*] $x^3 - 11x + 6$ **6.** [*6.2*] $x^4 + 6x^2y + 9y^2$

7. [*6.2*] $25x^2 - 49z^2$ **8.** [*6.2*] $-6a^5b^6$

9. [*6.2*] $P(x) = x^5 + x^3 - x^2 - 1$; no portion of the graph is in Quadrant II.

10. [*6.7*] $P(x) = 2x^3 + 5x^2 - 1$
$2x^3 + 5x^2 - 1 = (2x^3 + 5x^2 - 1)(x + 3) - 10$

11. [*6.7*] $Q(x) = 3x^2 - 8x + 21$
$R(x) = -49$

12. [*6.7*] $Q(x) = -5x^5y^2 + 7x^4y^6$
$R(x) = 0$

13. [*6.7*] $Q(x) = x + 3$
$R(x) = -2x + 4$

14. [*6.7*] $Q(x) = x^4 - 2x^3 + 4x^2 - 8x + 16$
$R(x) = -33$

15. [*6.7*] The number k is the remainder after dividing $P(x) = ax^3 + bx^2 + cx + d$ by $x + 2$.

17. [*6.3*] $6r^3s^4(3s - 2r^2)$ **18.** [*6.3*] $(y - 1)(x + 3)$

19. [*6.4*] $3(3 + 2x)(3 - 2x)$ **20.** [*6.4*] $(2y - 5x)^2$

21. [*6.4*] $(1 - 2x)(1 + 2x + 4x^2)$

22. [*6.5*] $(3x + 4)(2x + 1)$

23. [*6.8*] $(x + 3)(x - 2)(x - 4)$

24. [*6.8*] $(x - 2)(x - 7)(x + 5)$ **25.** [*6.6*] $0, \frac{4}{3}$

26. [*6.6*] $-3, 0.4$ **27.** [*6.6*] $-2, 0.5, 2$

28. [*6.6*] $-1, 0.5, 1, 2.5$ **30.** [*6.6*] 7 feet, 10 feet

31. [*6.6*] 100 feet, 120 feet

CHAPTER 7

Section 7.1 *(page 430)*

1. Any number that makes the denominator zero is excluded from the domain.

3. $x \neq 5$ **5.** 0.5, 0, undefined **7.** -0.875, 0, undefined

9. $\{x \mid x \neq 4\}$ **11.** $\{x \mid x \neq -2\}$ **13.** $\{x \mid x \neq -3, 3\}$

15. $\{x \mid x \neq -4, 2\}$ **17.** **R** **19.** + **21.** -25

23. The expression in (i) has a common factor $(x + 3)$ that can be divided out. In (ii), there is no common factor in the numerator and the denominator.

25. $\frac{2x-1}{3}$ **27.** 1 **29.** $x - 9$ **31.** Cannot be simplified

33. $\frac{1}{1+5x}$ **35.** $\frac{x+4}{x+3}$ **37.** $\frac{a(a+1)}{6}$ **39.** $\frac{1}{2x+1}$

41. $\frac{x-5}{x-8}$ **43.** 1 **45.** $\frac{a}{b+3}$ **47.** $\frac{1}{x-y}$

49. $\frac{c-2d}{2c+d}$ **51.** -2 **53.** $3 - x$

55. **(a)** Neither **(b)** 1 **(c)** -1

57. **(a)** -1 **(b)** 1 **(c)** Neither

59. -1 **61.** $-\frac{1}{3}$ **63.** $-(a + 7)$ **65.** $\frac{-1}{5w+z}$

67. $-\frac{x+4}{x+3}$ **69.** $x - 3$ **71.** $\frac{x^2-2x+4}{x-2}$

73. $\frac{3x(3x-2)}{2(2x+3)}$ **75.** $12x^2y$ **77.** 25 **79.** $(x + 5)^2$

81. $5(x - 3)$ **83.** $-3(x + 4)$ **85.** $9(x - 5)$ **87.** $-12t$

89. $-1(1 + x)(1 - x)$

91. **(a)** $2 million

(b) No, the function is not defined for $x = 100$.

(c) 90%, $10 million

93. $-0.293x + 28.873$

95. $N(x) = (x - 28)(-0.293x + 28.873)$; $x = 28, 98.5$. The result indicates that people age 28 and age 98.5 have a net worth of zero.

97. $-\frac{1}{x^3}$ **99.** x^5

101. For even n, $\{x \mid x \neq -1, 1\}$; for odd n, $\{x \mid x \neq 1\}$

105. $\frac{(x+1)^2-(x-2)^2}{x(x+4)-(x+1)^2} = 3$

Section 7.2 *(page 437)*

1. An easier method is to factor first and then divide out common factors.

3. $\frac{1}{5a}$ **5.** $\frac{a-4}{3}$ **7.** 1 **9.** $\frac{c-6}{20c}$ **11.** $\frac{1}{3}$ **13.** $\frac{y+1}{y+8}$

15. $3 - x$ **17.** $2x^2 - x - 1$

19. The opposite of $x + 3$ is $-(x + 3) = -x - 3$.

The reciprocal of $x + 3$ is $\frac{1}{x+3}$.

21. $\frac{x^3}{6y^2}$ **23.** $\frac{6}{x}$ **25.** $\frac{-x(x-2)}{x+2}$ **27.** $\frac{x-1}{8x}$ **29.** $\frac{x-6}{x-2}$

31. $\frac{1}{2x(3x-2)}$ **33.** $6x^2 - 11x + 4$ **35.** $x^3 - 8$

37. $\frac{-20x^3}{3y}$ **39.** $\frac{2(x+3)}{x-4}$ **41.** $\frac{x+2}{3x(x-1)}$ **43.** $\frac{1}{2}$

45. $\frac{a+9}{a-9}$ **47.** $\frac{x+1}{x+4}$ **49.** $\frac{x}{(2x-3)(2x-5)}$

51. 1 **53.** $x + 5$ **55.** $-\frac{9}{10}$ **57.** 4 **59.** $3x^2(x - 1)$

61. $\frac{x-3}{x(x+1)}$

63. **(a)** $\frac{B(x)+G(x)}{2} = \frac{5}{4}x + \frac{11}{6}$

(b) The graph for the combined data lies between the graphs of B and G.

65. $A(t) = \left(\frac{26t+20}{t+1.5} \cdot 10^9\right) \div \left(\frac{32t+52}{t+2.6} \cdot 10^6\right)$

67. $A(4) - A(0) = 826.67 - 666.67 = 160$. After peaking in 1995, the cost per person is projected to decrease slowly thereafter.

69. $\frac{x-2}{x+3}$ **71.** $\frac{2x-3}{x+1}$

73. As x approaches 0, the y-values become larger. By selecting x-values closer and closer to 0, we can make the y-value as large as we want.

Section 7.3 *(page 446)*

1. Add the numerators and retain the common denominator.

3. $\frac{8}{a}$ **5.** $\frac{-3x-9}{x}$ **7.** 6 **9.** -6 **11.** $\frac{3}{x-4}$ **13.** $x + 7$

15. $\frac{1}{x+2}$ **17.** $\frac{9-2x}{2x+3}$ **19.** $\frac{1}{x-2}$ **21.** $x + 6$ **23.** $3t - 3$

25. $2x^2 - x$

27. Multiply the numerator and the denominator of the second fraction by -1.

29. 5; $a + 5$ **31.** $8x$; $11x + 9$ **33.** $\frac{9x+3}{7x-8}$ **35.** $x + 4$

37. $\frac{a^2+b^2}{a-b}$ **39.** $30x^3y^2$ **41.** $6x(x - 4)$ **43.** $-30t(t + 5)$

45. $(3x - 2)(x + 1)(4x + 1)$ **47.** $3(x + 1)(x + 2)(x - 2)$

49. Multiply the numerator and the denominator by x.

51. $\frac{5b + 7a^2}{a^3b^2}$ **53.** $\frac{x^2 - 3}{x}$ **55.** $\frac{11x - 12}{x(x - 4)}$

57. $\frac{12}{(x - 5)(x + 7)}$ **59.** $\frac{t^2 + 2t + 3}{t + 1}$ **61.** $\frac{2t + 5}{18(t - 5)}$

63. $\frac{2t - 1}{t(t + 1)^2}$ **65.** $\frac{13t^2 + 2t - 2}{3t^2(2t - 1)(t + 1)}$ **67.** $\frac{1}{2x + 1}$

69. $\frac{x + 5}{4(x - 9)}$ **71.** $\frac{8x - 49}{(x + 7)(x + 8)(x - 8)}$

73. $\frac{x - 12}{x - 6}$ **75.** $\frac{x - 6}{(x + 3)(x - 3)}$ **77.** $\frac{x + 9}{(x + 3)(x - 1)}$

79. $(x + y)^{-1} = \frac{1}{x + y}$ and $x^{-1} + y^{-1} = \frac{1}{x} + \frac{1}{y} = \frac{y + x}{xy}$.
The expressions are not equivalent.

81. $\frac{y - x}{xy}$ **83.** $\frac{x^2 + x + 9}{(x + 3)(x - 2)}$

85. (a) The graph suggests that increasing the hours of training will reduce errors, but errors can never be reduced to zero.

(b) $E_1(x) = \frac{5}{x(x + 1)}$; the graph suggests that the incentive plan will reduce errors even further.

87. $V_1(t) = \frac{200(10t + 77)}{t + 25}$

89. $\frac{200(65t^2 + 1692t + 3578)}{(t + 25)(t + 14)}$. The expression represents the total annual fixed and variable cost (in dollars) for 10,000 miles of driving.

91. $\frac{-1}{x(x + h)}$ **93.** $\frac{1 - x^2}{1 + x^2}$

Section 7.4 *(page 453)*

1. The expressions do not mean the same thing. The first expression means $\frac{2}{3}$ divided by 4, and its value is $\frac{1}{6}$. The second expression means 2 divided by $\frac{3}{4}$, and its value is $\frac{8}{3}$.

3. $\frac{x^2}{3}$ **5.** $\frac{1}{3x}$ **7.** $\frac{x - 5}{2}$ **9.** $\frac{x - 9}{48x}$ **11.** $\frac{x - 4}{2x}$

13. $\frac{x - 3}{7x(x - 2)}$ **15.** $-\frac{x + 2}{x - 3}$

17. Add or subtract the fractions in the numerator and the denominator. Then multiply the numerator by the reciprocal of the denominator.

19. $\frac{4(t + 6)}{3(t + 2)}$ **21.** $\frac{1}{y}$ **23.** $\frac{2x - 1}{x}$ **25.** -1 **27.** $\frac{1}{x}$

29. xy **31.** $\frac{-x - 15}{2x^2 - 17}$ **33.** $-\frac{2x^2 + 3x - 2}{2x^2 - 5x - 5}$ **35.** $\frac{1}{2}$

37. $\frac{x - 2}{2x - 1}$ **39.** $\frac{x + 2y}{x - 2y}$ **41.** $x + 2$

43. Exclude values for which Q, R, or S is 0.

45. $\frac{x + y}{x - 3y}$ **47.** $\frac{-1}{x(x + h)}$ **49.** $a - b$ **51.** $\frac{x - 1}{(x - 2)(x + 1)}$

53. $\frac{a + b}{a^3b^3}$

55. Multiplying the given expression by the LCD would change the value of the expression.

57. $\frac{1}{a + b}$ **59.** $\frac{x^3 + x}{x^4 + 1}$ **61.** $\frac{x - 1}{3x - 4}$ **63.** $\frac{x^3 + y^4}{x^2y^3(x - y)}$

65. $\frac{x + 1}{x^2 + x + 5}$ **67. (a)** $\frac{\frac{x}{x + 1}}{\frac{x + 1}{x + 2}}$ **(b)** $\frac{x^2 + 2x}{x^2 + 2x + 1}$

69. (a) $\frac{x - m}{x}, \frac{y - n}{y}$

(b) $\frac{(x - m) + (y - n)}{x + y}$. This is not the sum of the fractions in part (a).

(c) $\frac{\frac{x - m}{x}}{\frac{y - n}{y}} = \frac{y(x - m)}{x(y - n)}$. This is a comparison of the ratio of accepted males who enrolled with the ratio of accepted females who enrolled.

71. $\frac{1}{(x + 3)(x^2 + 1)}$

73. $p(t) \cdot \frac{r(t)}{100} = \frac{-0.0026t^3 + 0.1755t^2 - 0.312t}{0.7t + 10}$

75. $\frac{13(2049t - 4890)}{20(2.8t^3 - 34t^2 + 189t - 5)}$

77. Attendance increased by about 26%, whereas salaries increased by about 5%.

79. $\frac{-2}{(2x + 2h - 1)(2x - 1)}$ **81.** $\frac{-3}{(x + h - 2)(x - 2)}$

83. $\frac{14x^2 - 9x - 3}{7x - 2}$ **85.** $\frac{8}{5}$

Section 7.5 *(page 463)*

1. We call -3 and 0 restricted values because each one makes a denominator 0.

3. $-3, 0, 4$ **5.** $-3, 0, 3$

7. No, because 2 is a restricted value.

9. $-3, 1$ **11.** $-3, -1$ **13.** $-3, -2, 2$ **15.** 20

17. 4 **19.** 7 **21.** $-\frac{3}{2}$ **23.** $\frac{2}{3}$ **25.** 8 **27.** $-1.5, 5$

29. $-4, 5$ **31.** 3 **33.** $-4, 1$

35. In (i), 9 is an extraneous solution. In (ii), clearing fractions leads to $-4 = 4$, which is false, and the equation has no solution.

37. 4 **39.** -5 **41.** No solution **43.** $\frac{9}{8}, -\frac{1}{2}$ **45.** $-6, \frac{1}{2}$

47. $-8, \frac{1}{2}$ **49.** -2 **51.** $-\frac{121}{32}$ **53.** $-\frac{10}{7}$ **55.** 2

57. No solution **59.** No solution

61. To solve the equation, clear the fractions by multiplying both sides by the LCD. To perform the addition, rewrite each expression with the LCD as the denominator.

63. $\dfrac{x^2 - 5x - 9}{(x - 8)(x - 3)}$ **65.** -1 **67.** $\dfrac{y}{8}$ **69.** $\dfrac{-2t^2 - 7t + 3}{t + 3}$

71. 0, 6 **73.** $\dfrac{4}{x + 2}$ **75.** $R = \dfrac{E}{I}$ **77.** $T_1 = \dfrac{P_1V_1T_2}{P_2V_2}$

79. $r = \dfrac{I}{Pt}$ **81.** $m_2 = \dfrac{F\mu r^2}{m_1}$ **83.** $a = \dfrac{S(1 - r)}{1 - r^n}$

85. $a = \dfrac{bx}{b - y}$ **87.** $y = -2x - 2$ **89.** $y = -\dfrac{4}{5}(x + 4)$

91. The greatest frequency of DUI arrests per 100,000 drivers occurs among younger drivers.

93. The highest frequency of DUI arrests per 100,000 drivers is 356, which occurs at age 25.6.

95. 7 **97.** $-3, -2, 3$ **99.** $y = m(x - x_1) + y_1$

101. $-\frac{5}{4}, 0$

Section 7.6 *(page 471)*

1. In (i) we are asked to find the sum of two fractions. Write the fractions with the LCD and add. In (ii) we are asked to solve an equation. Clear the fractions by multiplying both sides by the LCD.

3. 4 inches **5.** 540 **7.** 9 **9.** 1200 **11.** 15

13. 6 **15.** 7.2 hours **17.** $19.50 **19.** 15 minutes

21. 6 hours **23.** Plane: 150 mph; train: 60 mph

25. 12 mph

27. Choices (c) and (e) illustrate inverse relations.

29. 60 feet **31.** 24 minutes **33.** 30° **35.** 6 seconds

37. 200 mph **39.** 5.5 hours **41.** 28 days

43. (a) $\dfrac{100(0.050x - 0.06)}{0.095x + 0.80}$ **(b)** $\dfrac{100(0.006x^2 + 0.2)}{-0.020x^2 + 0.43x + 12.7}$

45. 47

47. Small: 8 pounds; medium: 12 pounds; large: 5 pounds

49. 4

Chapter Review Exercises *(page 477)*

1. The numbers are 5 and -3, which are called restricted values.

3. $\{x \mid x \neq 2.5\}$

5. In the first expression, 3 is a factor in both the numerator and the denominator. There is no common factor in the numerator and denominator of the second expression.

7. $\dfrac{x + 4}{2x - 3}$ **9.** $\dfrac{b + 4}{b + 2}$ **11.** $5(x - 4)$ **13.** $\dfrac{8}{x^2}$

15. $\dfrac{x + 1}{x - 2}$

17. (a) $\dfrac{1}{xy}$ **(b)** $\dfrac{1}{x} \cdot \dfrac{1}{y}$

(c) The expressions are equivalent because $\dfrac{1}{x} \cdot \dfrac{1}{y} = \dfrac{1}{xy}$.

19. $\dfrac{6}{x + 2}$ **21.** $\dfrac{x - 1}{x - 4}$ **23.** $\dfrac{2}{x - 2}$ **25.** $9(x + 7)(x - 7)$

27. If the denominators were not the same, the terms would not have a common factor, and the Distributive Property would not apply.

29. $\dfrac{-4x + 5}{5x - 4}$ **31.** $\dfrac{10x - 37}{(x + 4)(x - 4)(x - 3)}$ **33.** $\dfrac{5x + 4}{x(x + 2)}$

35. $\dfrac{6x + 27}{(x + 8)(x - 6)(x + 5)}$

37. Multiply the numerator by the reciprocal of the denominator, or multiply the numerator and the denominator by the LCD of all the fractions.

39. $\dfrac{a - 3}{12}$ **41.** Answers will vary. **43.** $\dfrac{x^2 + 16}{x}$

45. An apparent solution does not check if it makes a denominator zero. Such a solution is called an extraneous solution.

47. $\dfrac{1}{2}, -\dfrac{2}{5}$ **49.** No solution **51.** 2 **53.** No solution

55. 1 solution **57.** 24 hours **59.** 15 hours **61.** 54 mph

Chapter Test *(page 479)*

1. [*7.1*] When $x = 3$, the denominator is zero.

2. [*7.1*] $\{x \mid x \neq 0, -3\}$ **3.** [*7.1*] $\dfrac{2(x + 2)}{x(x - 2)}$ **4.** [*7.1*] $\dfrac{x + 5}{x - 9}$

5. [*7.1*] $-\dfrac{x + 2y}{x + y}$ **6.** [*7.2*] $\dfrac{t + 2}{t + 3}$ **7.** [*7.2*] $\dfrac{1}{2(y - 5)}$

8. [*7.3*] $3x(x - 2)$

9. [*7.3*] The expression can be written $\dfrac{r}{m + n} - \dfrac{r - 1}{m + n}$. Because the denominators are the same, the subtraction is easy to perform. The result is $\dfrac{1}{m + n}$.

10. [*7.3*] 1 **11.** [*7.3*] $\dfrac{3 - 5t}{t^2}$

12. [*7.3*] $\dfrac{2x^2 - 16x - 25}{(2x - 1)(x + 5)(4x + 3)}$

13. [*7.4*] All are complex fractions. The last two are equivalent.

14. [7.4] $\frac{6}{x-3y}$ **15.** [7.4] $\frac{x-3}{x+2}$ **16.** [7.4] $\frac{b^2}{b^2-a^2}$

17. [7.5] 12 **18.** [7.5] $-2, 5$ **19.** [7.5] No solution

20. [7.5] -3.2

21. [7.5] In (a) the apparent solution is 1, but it is an extraneous solution. In (b) the equation leads to $-2 = 3$, so the equation has no solution.

22. [7.5] $B = \frac{A}{2A-1}$ **23.** [7.6] 40% **24.** [7.6] 2.4 hours

25. [7.6] 5 mph

Cumulative Test: Chapters 6–7 *(page 480)*

1. [6.1] $x^3 + x^2 + x - 1$ **2.** [6.2] $-6x + 18$

3. [6.3] **(a)** $x^3y^3(x - y)$
(b) $(x + 3y)(a - 7b)$

4. [6.4] **(a)** $(2x + 5)(2x - 3)$
(b) $2(x - 3)^2$
(c) $(y + 2)(y^2 - 2y + 4)$

5. [6.5] **(a)** $(z + 9)(z - 7)$
(b) $(a - 9b)(a + 5b)$
(c) $3(4 - x)(1 + x)$

6. [6.6] After factoring the left side, set each factor equal to 0. Then solve each case for x.

7. [6.6] **(a)** 0, 7 **(b)** $-1, 1, -5, 5$

8. [6.6] 8 or -3

9. [6.7] $Q(c) = 2c^3 + c^2 + 3c - 2$, $R(c) = -2$

10. [6.7] $Q(x) = x^4 + 3x^3 + 6x^2 + x + 4$, $R(x) = -1$

11. [6.8] Divide $P(x)$ by $x - c$. If the remainder is 0, then $x - c$ is a factor of $P(x)$.

12. [6.8] -3 **13.** [7.1] $\frac{x+3}{4x+3}$ **14.** [7.2] $\frac{a+4}{a-4}$

15. [7.3] $\frac{y+12}{(y+10)(y+4)}$ **16.** [7.4] $\frac{x}{3}$ **17.** [7.5] -4

18. [7.6] $3\frac{3}{7}$ hours **19.** [7.6] 55 mph

CHAPTER 8

Section 8.1 *(page 488)*

1. The number 9 has two square roots, -3 and 3, because squaring either results in 9. The expression $\sqrt{9}$ means the principal or positive square root of 9, which is 3.

3. 6, -6 **5.** Not a real number **7.** 3 **9.** 5, -5

11. -3 **13.** Not a real number **15.** 4 **17.** -4 **19.** $\frac{2}{3}$

21. Not a real number **23.** -4 **25.** 20 **27.** 4 **29.** 2

31. 7 **33.** 2.69 **35.** -0.92

37. The expression x^3 could be positive or negative depending on the value of x. The expression x^2 is nonnegative for all x.

39. 4 **41.** 7 **43.** 10 **45.** 2 **47.** 21 **49.** $2\sqrt{x}$

51. $\sqrt[5]{a} - 9$ **53.** $\sqrt{c+b}$ **55.** x^5 **57.** $4y^3$ **59.** x^7

61. $-y^5$ **63.** x^7 **65.** $-x^2$ **67.** $(3x)^3$ **69.** $(x + 3)^5$

71. **(a)** The graph of y_1 is the same as the graph of $|x|$, which is not the same as the graph of y_2.
(b) The graphs are the same.

73. $|x^3|$ **75.** $-x^4$ **77.** x^3 **79.** $|a|$ **81.** $|x + 2|$

83. $-2|x - 1|$ **85.** $1 - x$ **87.** $>$ **89.** $>$ **91.** 4 **93.** 8

95. -3

97. **(a)** \$103 million **(b)** 2002

99. Cheyenne: $-10.1°$; Boston: $-3.6°$; Seattle: 21.8°; Norfolk: 24.1°

101. 37.5° **103.** -1 **105.** 2 **107.** 2

109. **(a)** If we did not exclude $n = 1$, then $\sqrt[1]{b} = a$ would imply that $a^1 = b$, which would mean that $\sqrt[1]{b} = b$. Thus nothing is gained by including $n = 1$ in the definition.
(b) For $b = 1$, $\sqrt[0]{b} = a$ implies that $a^0 = b$ or $a^0 = 1$. But this statement is true for all $a \neq 0$. Thus $\sqrt[0]{b}$ would not have a unique value if $b = 1$.
For $b \neq 1$, $\sqrt[0]{b} = a$ implies that $a^0 = b$, which implies that $a^0 \neq 1$. Because this statement is false, $\sqrt[0]{b}$ has no value.

Section 8.2 *(page 494)*

1. Choice (iii) is false because $16^{1/2}$ is the principal square root 4.

3. 7 **5.** -4 **7.** Not a real number **9.** $\frac{2}{3}$ **11.** -3

13. -2 **15.** Not a real number **17.** -5 **19.** 1.97

21. Not a real number **23.** -1.41

25. The numerator is the exponent, and the denominator is the index.

27. $3\sqrt[3]{x^2}$ **29.** $\sqrt[3]{(3x)^2}$ **31.** $\sqrt[3]{(2x+3)^2}$ **33.** $3\sqrt{x} + \sqrt{2y}$

35. $y^{2/3}$ **37.** $(3x^3)^{1/4}$ **39.** $(x^2 + 4)^{1/2}$ **41.** $t^{1/2} - 5^{1/2}$

43. $(2x - 1)^{2/3}$ **45.** -2.46 **47.** -3.01 **49.** 29.93

51. Use the $\sqrt[x]{\ }$ key with $x = 5$ or evaluate $470^{1/5}$.

53. $\frac{1}{3}$ **55.** 3 **57.** 9 **59.** $\frac{1}{9}$ **61.** 8

63. Not a real number **65.** $\frac{1}{4}$ **67.** $\frac{9}{4}$ **69.** $-\frac{1}{16}$

71. Not a real number **73.** $-\frac{1}{32}$ **75.** 11 **77.** $\frac{7}{2}$ **79.** 8

81. 2 **83.** $\frac{1}{3}$ **85.** 6.35 **87.** 0.72 **89.** 0.53 **91.** 16

93. -8 **95.** $\frac{2}{3}$

97. **(a)** The graphs intersect at (0, 0) and (1, 1). If $x = 0$, then $x^{1/n} = 0^{1/n} = 0$ for all $n > 1$. If $x = 1$, then $x^{1/n} = 1^{1/n} = 1$ for all $n > 1$.
(b) $0 < x < 1$ **(c)** $x > 1$

99. **(a)** 2 hours **(b)** 4.2 additional hours
(c) 0; yes. Even if a student did not study for the final, it is unlikely that the exam grade would be 0. The model indicates that a perfect score can be earned with about 10 hours of study. This seems realistic.

101. **(a)** 81% **(b)** 37 **103.** $\frac{1}{2}$ **105.** $\frac{1}{2}$

107. The function is not defined if x is an even number.

Section 8.3 *(page 500)*

1. Choice (i) is true according to the Power of a Product Rule and the Power to a Power Rule for Exponents. Choice (ii) is untrue for $a = 3$ and $b = 4$, for example.

3. 36 **5.** 16 **7.** 6 **9.** $\frac{16}{49}$ **11.** 65 **13.** 7 **15.** 2

17. 6 **19.** $\frac{1}{3}$ **21.** $\frac{1}{2}$

23. Choice (i) is false because the exponents were multiplied when they should have been added. Choice (ii) is true because the Product Rule for Exponents was applied properly.

25. x **27.** $t^{12/7}$ **29.** $a^{1/4}b^{3/2}$ **31.** $z^{1/6}$ **33.** $a^{1/3}b^{4/3}$

35. $\dfrac{1}{a}$ **37.** y **39.** a^9b^6 **41.** $\dfrac{1}{2}$ **43.** $\dfrac{3}{2}$

45. In the first expression, the index and the exponent have a common factor. Thus, when the expression is written with a rational exponent, the exponent can be reduced. None of this is true for the second expression.

47. 8 **49.** a^3 **51.** $a^2|b|$ **53.** $|x + y|$ **55.** $\dfrac{b^{1/6}}{a^2}$

57. $\dfrac{a^4}{b^6}$ **59.** $\dfrac{-9x^{3/2}}{y}$ **61.** $\dfrac{b^{10/3}}{a^3}$ **63.** $\dfrac{x^2}{y^3}$ **65.** $\dfrac{2y^2}{x^3z}$

67. ab^4 **69.** $\dfrac{a^4c^{7/2}}{b^7}$ **71.** $\dfrac{y^2}{3x^4}$ **73.** $a^2 + 1$ **75.** $x^2 - x^{5/12}$

77. $x - y$ **79.** $x + 3 - 2(3x)^{1/2}$ **81.** $\sqrt{x}$ **83.** $\sqrt[3]{2x}$

85. $\sqrt[5]{x^2y^3}$ **87.** $\sqrt{a}$ **89.** $\sqrt[12]{x^{11}}$ **91.** $\dfrac{1}{\sqrt[12]{x^5}}$ **93.** $\sqrt[12]{7}$

95. $\sqrt[6]{200}$

97. **(a)** 49 **(b)** $16p^{5/4}$ **(c)** 89,975

99. $C(t) = 1.7t^{1/2}$

101. The trend in expenditures per visitor shows slight increases.

103. $x^{-1/2}(3 + 5x^2)$ **105.** $5a^{-1/2}(a - 2)$

107. $(t^{1/5} + 5)(t^{1/5} - 1)$ **109.** $(2x^{1/6} + 3)(x^{1/6} - 5)$

111. $\sqrt[8]{x}$

Section 8.4 *(page 506)*

1. The Product Rule for Radicals applies to products of real numbers. The factors $\sqrt{-3}$ and $\sqrt{-12}$ are not real numbers.

3. $\sqrt{6}$ **5.** $\sqrt{15xy}$ **7.** $-12\sqrt{14}$ **9.** $\sqrt[3]{x^2 + x - 6}$

11. The first step is to factor the trinomial: $\sqrt{(x + 5)^2} = x + 5$. This step is necessary because there is no rule for simplifying a radical whose radicand is a sum. Although $x^2 + 10x + 21$ can be factored, the result is not a perfect square. Therefore, the radical cannot be simplified.

13. $5y$ **15.** $7w^4t^2$ **17.** $2t^2$ **19.** $3x^3y^5$ **21.** $x(y - 3)^2$

23. $x + 6$ **25.** $2y + 1$

27. The radicand 32 has a perfect square factor, 16. The radicand 30 does not have a perfect square factor.

29. $2\sqrt{3}$ **31.** $5\sqrt{2}$ **33.** $3\sqrt[3]{2}$ **35.** $-2\sqrt[3]{6}$ **37.** $2\sqrt[4]{2}$

39. $2\sqrt[5]{8}$ **41.** $a^2b\sqrt{ab}$ **43.** $a^3b^2\sqrt[3]{b^2}$ **45.** $x^4y^2\sqrt[4]{y}$

47. $2\sqrt{15t}$ **49.** $5x^{12}\sqrt{x}$ **51.** $10a^3b^7\sqrt{5a}$ **53.** $-7x^3y^3\sqrt{y}$

55. $2w^3t^4\sqrt{7t}$ **57.** $3x^3y^3\sqrt[3]{3y^2}$ **59.** $-2xy\sqrt[5]{x^2y^4}$

61. Factor out the common factor 4 from the radicand. Then apply the Product Rule for Radicals.

63. $2\sqrt{3x + 2y}$ **65.** $2\sqrt{x^2 + 4}$ **67.** $3x^3y^3\sqrt{x + y}$

69. $3|y|\sqrt{1 + y^2}$ **71.** $r^2|t|\sqrt{1 + r^2t^2}$ **73.** $|y - 5|$ **75.** -20

77. $3\sqrt[3]{6}$ **79.** $7x\sqrt{2}$ **81.** $x^4\sqrt{6x}$ **83.** 20 **85.** $6x^2y^2\sqrt{2y}$

87. $5xy^2\sqrt[3]{x^2}$ **89.** $3xy\sqrt[4]{2xy}$ **91.** $\sqrt{16a^4b^3}$

93. $\sqrt{3x^2y^3}\sqrt{3x^2y^3}$ **95.** $\sqrt[3]{16x^4y^3}$

97. **(a)** $N(x) = 0.74x^4\sqrt{11x^2 + 410x}$ **(b)** 104,274,787

99. $Q(t) = 0.04t^2 - 0.93t + 20.99$

101. Function M projects a continued slight decline in union membership, while function Q projects a sharp increase. In the long run, function M is probably the more realistic model.

103. The radical is not simplified because the radicand has a perfect cube factor.

105. $xy^2\sqrt{2xy}$ **107.** $\sqrt[7]{4x^6y^5}$

109. If n were divisible by a number larger than $\sqrt{n}$, then the quotient would be a number less than $\sqrt{n}$, and that number would have been tested already. Thus only numbers that do not exceed $\sqrt{n}$ need to be tested.

Section 8.5 *(page 514)*

1. The rule does not apply because the indices are different.

3. $\sqrt{6}$ **5.** 4 **7.** x^3 **9.** $\dfrac{7a^3b^2}{2}$ **11.** x **13.** $x^2y^4\sqrt{y}$

15. $3w^3t^2\sqrt{5wt}$ **17.** $3x^2d\sqrt{xd}$ **19.** $3x^3\sqrt[3]{2x^2}$ **21.** $\dfrac{1}{6}$

23. $\frac{1}{t^3}$ **25.** $\frac{1}{5a}$ **27.** $\frac{4}{3xy^2}$ **29.** $\frac{x}{2}$

31. The radical could be simplified if Q is a perfect square or if Q divides P.

33. $\frac{\sqrt{5}}{7}$ **35.** $\frac{2}{x^4}$ **37.** $\frac{2\sqrt{7}}{3}$ **39.** $-\frac{\sqrt[3]{2a}}{5}$ **41.** $2x^5y\sqrt{5x}$

43. $5x^2y^4\sqrt{3}$ **45.** $7x^2y^3z^3\sqrt{yz}$ **47.** $2a^2b^3\sqrt[3]{3a}$ **49.** $\frac{y^2}{x}$

51. $\frac{a\sqrt{a}}{3b^3}$ **53.** $\frac{x\sqrt[4]{x}}{2}$

55. Either method is correct, but multiplying by $\sqrt{3}$ requires fewer steps.

57. $\sqrt{5}$ **59.** $-\frac{\sqrt{15}}{3}$ **61.** $\frac{3\sqrt{10}}{2}$ **63.** $\frac{7\sqrt[3]{9}}{3}$ **65.** $\frac{9\sqrt[4]{2}}{2}$

67. $\frac{2\sqrt{3x}}{x}$ **69.** $\frac{2\sqrt{x+2}}{x+2}$ **71.** $\frac{a\sqrt{2a}}{4}$ **73.** $\frac{\sqrt{5y}}{y^3}$

75. $\frac{\sqrt{15x}}{6x^3}$ **77.** $6x\sqrt{3y}$ **79.** $\frac{3\sqrt{3x}}{2x^2}$ **81.** $\frac{x^2\sqrt{5xy}}{10y}$

83. 5 **85.** $\sqrt{14}$ **87.** $\sqrt{3}$ **89.** $2\sqrt[3]{b}$ **91.** $\frac{\sqrt[3]{10}}{2}$

93. $\frac{\sqrt[4]{x^3y}}{y}$ **95.** $\frac{x\sqrt[4]{y}}{2y}$ **97.** 16 **99.** 36 **101.** 4

103. (a) $3000\sqrt{d}$ (b) \$6000

105. The graph suggests that the percentage of taxpayers who expect a refund decreases with the age of the taxpayer.

107. 75; about 28% .

109. 1.414, 1.848, 1.962, 1.990, 1.998. The values appear to be approaching 2. The pattern does not hold for $n = 3$.

111. $\frac{\sqrt[6]{x^5}}{x}$

113. Answers will vary.

Section 8.6 *(page 522)*

1. The Distributive Property is used to factor out the common radical factor: $3\sqrt{5} + 4\sqrt{5} = (3 + 4)\sqrt{5} = 7\sqrt{5}$.

3. $6\sqrt{13}$ **5.** $16\sqrt{x}$ **7.** $-3\sqrt[3]{y}$ **9.** $-4\sqrt{2x} + \sqrt{3}$

11. The terms must have a common factor for the Distributive Property to apply. The radicals should be simplified before we can decide whether the terms can be combined.

13. $\sqrt{3}$ **15.** $8\sqrt[3]{2}$ **17.** $\sqrt[4]{2}$ **19.** $62\sqrt{2}$

21. $2\sqrt{3} + 10\sqrt{2}$ **23.** $\sqrt{2x}$ **25.** $13x\sqrt{3}$ **27.** $11a^2\sqrt{5b}$

29. $6x\sqrt{2x} - 6x\sqrt{2}$ **31.** $4x\sqrt{xy} - 7y\sqrt{xy}$ **33.** 0

35. $-x\sqrt[3]{2}$ **37.** $24 - 4\sqrt{21}$ **39.** $20\sqrt{21} + 35\sqrt{7}$

41. $5\sqrt{7} - 5\sqrt{3}$ **43.** $\sqrt{3y} - 3\sqrt{y}$ **45.** $\sqrt{7}$ **47.** $x + 2\sqrt{3}$

49. The domain of y_1 is $[2, \infty)$, whereas the domain of y_2 is $(-\infty, -2] \cup [2, \infty)$.

51. $\sqrt{6} - 2\sqrt{2} + 3 - 2\sqrt{3}$

53. $\sqrt{21x} + 7x - 3\sqrt{3} - 3\sqrt{7x}$ **55.** $3\sqrt{5} - 8\sqrt{10}$

57. $8 - 4\sqrt{3}$ **59.** $15 + 10\sqrt{2}$ **61.** -1 **63.** -78

65. $100 - 50t^2$ **67.** $-7 - 4\sqrt{3}$

69. $\frac{7 - \sqrt{21}}{4}$ **75.** $3 + \sqrt{6}$ **77.** $4 - \sqrt{15}$

79. $\frac{y + 9\sqrt{y} + 20}{y - 16}$ **81.** $\frac{x - 2\sqrt{x}}{x - 4}$ **83.** $\frac{\sqrt{x+1} + 2}{x - 3}$

85. $\frac{x - 2\sqrt{xy} + y}{x - y}$

87. The operation cannot be performed because the denominators are not the same. It is not necessary to rationalize the denominators in order to add. It is only necessary that the fractions have the same denominator.

89. $\frac{x - 3\sqrt{x}}{x}$ **91.** $\frac{16\sqrt{5}}{5}$ **93.** $\frac{2 - 3\sqrt{2}}{2}$ **95.** $\frac{2\sqrt{x} + 1}{x + 1}$

97. $6 - \sqrt{3}$ **99.** $\frac{-1}{2(\sqrt{5} - \sqrt{7})}$ **101.** $\frac{x - 9}{x - 6\sqrt{x} + 9}$

103. $\frac{1}{\sqrt{x+2} + \sqrt{x}}$ **105.** 0 and 1

107. (a) $d = \sqrt{x^2 - 2500}$ (b) $h = \sqrt{2600 - 20x}$

109. (a) 92% (b) $\frac{T + 100\sqrt{T}}{10{,}000 - T}$

111. $P(t) = \frac{72\sqrt[3]{t} - 28\sqrt[3]{t^2}}{t}$. This is not a rational function because the numerator is not a polynomial.

113. A quadratic expression **115.** 1.63 **117.** $2\sqrt{x}$

119. $-\sqrt{2}$ **121.** $\frac{-\sqrt{6} - 2\sqrt{15} + 3\sqrt{10}}{12}$

123. $\frac{3(\sqrt[3]{x^2} + \sqrt[3]{2x} + \sqrt[3]{4})}{x - 2}$ **125.** $\frac{\sqrt{x} + 3}{x + 3}$

Section 8.7 *(page 533)*

1. To determine the domain of the function, solve the inequality $R \geq 0$.

3. $\{x \mid x \geq 2\}$ **5.** $\{x \mid x \geq 1.5\}$ **7.** $\{x \mid x \leq 2\}$ **9.** **R**

11. **R** **13.** $\{x \mid x \geq -3\}$

15. The equations are not equivalent because the first has no solution, whereas the second has the solution 25. The methods used to solve a radical equation may introduce an extraneous solution.

17. 3 **19.** 3 **21.** No solution **23.** 15 **25.** 4, -1 **27.** 9

29. $\frac{1}{2}$, 2 **31.** 3 **33.** -3, 2 **35.** No solution

37. No solution **39.** No solution **41.** 4

43. The radical in each equation can be eliminated by squaring once. The first two resulting equations are second degree.

45. 4 **47.** 3, 7 **49.** No solution **51.** 3 **53.** 6

55. 2, 3 **57.** 2, 6 **59.** 4

61. Raise both sides to the bth power to obtain an integer exponent on x.

63. $\pm\sqrt{6}$ **65.** No solution **67.** 4.5, 0.5 **69.** $\frac{1 \pm \sqrt{3}}{3}$

71. 5 **73.** $\frac{1}{2}$ **75.** ± 27 **77.** $\frac{1}{81}$ **79.** $\frac{25}{8}, \frac{23}{8}$ **81.** 6

83. 4 **85.** $\sqrt{57} \approx 7.55$ **87.** $\sqrt{5} \approx 2.24$

89. 25.04 feet **91.** 31.62 feet **93.** 82.92 feet

95. Yes, $r = 5.499$ feet **97.** 700 ponchos **99.** 4 weeks

101. $7.43\sqrt{x} - 11.5 = \sqrt{2.5x + 357}$

103. $7.43\sqrt{x} - 11.5 = 2\sqrt{2.5x + 357}$

105. $\{x \mid x \le 4 \text{ or } x \ge 3\}$ **107.** $-\frac{6}{5}, -1$ **109.** No solution

111. **(a)** $\{x \mid x \ge 0\}$ **(b)** $\{x \mid x \le 0\}$ **(c)** **R**

Section 8.8 *(page 543)*

1. The number is i, and the properties are $i^2 = -1$ and $i = \sqrt{-1}$.

3. $3i$ **5.** $3i\sqrt{2}$ **7.** $5 + 6i$ **9.** $-2 - i\sqrt{3}$

11. Write i^{34} as a power of i^2: $i^{34} = (i^2)^{17} = (-1)^{17} = -1$.

13. $a = 2, b = -1$ **15.** $a = -1, b = 5$ **17.** (b)

19. $1 + i$ **21.** $15 + 10i$ **23.** $6 + 21i$ **25.** 14

27. $13i$ **29.** $1 - 4i$ **31.** $a = -2, b = -2$

33. $a = -1, b = -2$ **35.** -10 **37.** $28 + 21i$

39. $-6 - 17i$ **41.** $8 + 6i$ **43.** -4 **45.** $-9 - 3\sqrt{2}$

47. $-i$ **49.** 1 **51.** -1 **53.** i

55. In part (a) the result is $a^2 - b^2$, but in part (b) the result is $a^2 + b^2$.

57. $-7i$, 49 **59.** $2 - 3i$, 13 **61.** $\sqrt{5} + 4i$, 21

63. $\frac{3}{2}i$ **65.** $2 - 3i$ **67.** $\frac{1 - 7i}{5}$ **69.** $\frac{-1 + 43i}{74}$

71. $\frac{6 - 3i\sqrt{3}}{7}$ **73.** $\frac{\sqrt{15} + 3\sqrt{3} + 3i - 3i\sqrt{5}}{12}$

75. $8 - 3i$ **77.** 28 **79.** $\frac{2 - i}{3}$ **81.** $9 + 6i$ **83.** $11 - 60i$

85. $\frac{-7 + 24i}{25}$ **87.** $7i$ **89.** $5 - 3i\sqrt{3}$ **91.** 5

97. $(x + 3i)(x - 3i)$ **99.** $(2x + i)(2x - i)$

101. **(a)** -1 **(b)** i **(c)** $\frac{2 - 3i}{13}$ **(d)** $\frac{-3 + 4i}{25}$ **105.** $-\frac{23}{26} + \frac{37}{26}i$

Chapter Review Exercises *(page 546)*

1. The number x^2 has two square roots, x and $-x$, but $\sqrt{x^2}$ is the principal (nonnegative) square root. Thus we use absolute value symbols to ensure that the result is nonnegative. The number x^3 has only one cube root, and so absolute value symbols are not used.

3. -8 **5.** -4 **7.** x^2 **9.** $x + 4$

11. **(a)** $\sqrt{10} - a$ **(b)** $\sqrt{10 - a}$

13. The expression $(-4)^{1/2}$ means $\sqrt{-4}$, but there is no real number whose square is -4.

15. 64 **17.** $\frac{1}{9}$ **19.** Not a real number

21. $25^{1.5} = 25^{3/2} = \left(\sqrt{25}\right)^3 = 5^3 = 125$

23. When n is even, the initial result is always $|x|$. However, if x is nonnegative, then $|x| = x$.

25. **(a)** 6 **(b)** 4 **27.** $\frac{c^{18}}{b^{12}}$ **29.** $\frac{3x^2y}{z^3}$ **31.** $\frac{x^{3/2}}{9}$

33. $\frac{1}{x} - x^3$

35. The Product Rule cannot be used because the indices are different. We can use rational exponents to write

$$\sqrt[3]{5} \cdot \sqrt{5} = 5^{1/3} \cdot 5^{1/2} = 5^{2/6} \cdot 5^{3/6} = 5^{5/6} = \sqrt[6]{5^5}$$

37. $45x$ **39.** $\sqrt{42x}$ **41.** $x^2\sqrt{x}$ **43.** $c^8d^{10}\sqrt{c}$

45. Answers will vary.

47. The radicand must contain no perfect nth power factors. The radicand must not contain a fraction. There must be no radical in the denominator of a fraction.

49. x **51.** $\frac{a^3\sqrt{5a}}{3b^2}$ **53.** $\sqrt{6}$ **55.** $\frac{3\sqrt{x}}{x^2}$ **57.** $\frac{3y\sqrt{x}}{x}$

59. Answers will vary.

61. **(a)** The indices are the same, but the radicands are different.

(b) The radicands are the same, but the indices are different.

63. $13\sqrt{5x}$ **65.** $8 + 2\sqrt{15}$ **67.** $\sqrt{2}$

69. $\frac{5\left(5 + \sqrt{3}\right)}{22}$ **71.** $2\sqrt{3}$ **73.** $13\sqrt{5}$

75. When an algebraic method is used correctly to solve an equation, an apparent solution may not satisfy the equation. Such a number is an extraneous solution.

77. 3 **79.** 1 **81.** 15.20 feet **83.** 4 and 9

85. The square root of a negative number is not a real number.

$$\sqrt{-4}\sqrt{-9} = (2i)(3i) = 6i^2 = -6$$

87. $-2 + 2i$ **89.** $33 + 4i$ **91.** $4 - 2i$ **93.** i
95. $a = 0, b = -6$

Chapter Test (*page 549*)

1. [*8.1*] $\{x \mid x \le 3\}$ **2.** [*8.2*] $\frac{-2y^4}{x^2}$

3. [*8.2*] $-25^{1/2} = -\sqrt{25} = -5$. $(-25)^{1/2} = \sqrt{-25}$, which is not a real number.

4. [*8.5*] 20 **5.** [*8.3*] $\sqrt[12]{x^7}$ **6.** [*8.3*] $\sqrt[6]{x^5}$

7. [*8.6*] $\sqrt{3}$ **8.** [*8.6*] $26 - 5\sqrt{5}$

9. [*8.5*] The conditions are (i) b is divisible by a, or (ii) a is a perfect square.

10. [*8.6*] $\frac{10\sqrt{2} - 9}{7}$ **11.** [*8.4*] Answers will vary.

12. [*8.4*] $7x^5y^8\sqrt{2x}$ **13.** [*8.3*] $\frac{125}{64}$ **14.** [*8.7*] $-\frac{7}{3}, 3$

15. [*8.7*] Extraneous solutions can occur.

16. [*8.7*] 9 **17.** [*8.7*] 1 **18.** [*8.8*] $-1 + 6i$

19. [*8.8*] $5 + 12i$ **20.** [*8.8*] $\frac{1-i}{2}$ **21.** [*8.8*] $i\sqrt{7}, i\sqrt{3}$

22. [*8.8*] (i) and (iv) **23.** [*8.2*] $s = V^{1/3}$, 5.85 inches

24. [*8.8*] $\sqrt{3} - 3 + i\sqrt{3} + 3i$

25. [*8.7*] **(a)** $B < 0$ [8.1] **(b)** $a > 0$ or n is odd.

CHAPTER 9

Section 9.1 (*page 556*)

1. The x-intercepts correspond to the solutions of the equation.

3. $\frac{1}{2}, -3$ **5.** $\frac{1}{3}, -3$ **7.** 3 **9.** 5

11. No real number solution **13.** No real number solution

15. $-6, -1$ **17.** 0, 1 **19.** $-\frac{4}{3}, 2$ **21.** 1, 3 **23.** $-1, \frac{2}{5}$

25. If $A^2 = c$ and $c > 0$, then $A = \pm\sqrt{c}$. Thus the equation $x^2 = 6$ has two solutions, $\pm\sqrt{6}$. If $A^2 = c$ and $c < 0$, then there are no real number solutions. Thus $x^2 = -6$ has no real number solution.

27. 0.2, 1.4 **29.** $-5 \pm \frac{\sqrt{5}}{4} \approx -4.44, -5.56$

31. 0, 3 **33.** $-3, 3$ **35.** $\pm 4i$ **37.** $\pm 2i\sqrt{6} \approx \pm 4.90i$

39. $3 \pm 2i$ **41.** $1 \pm 5i$ **43.** $\frac{-5 \pm i\sqrt{6}}{2} \approx -2.5 \pm 1.22i$

45. $-1 \pm \frac{3}{2}i$ **47.** $\pm\frac{2}{3}i$ **49.** $-\frac{1}{2}, \frac{5}{2}$ **51.** $\pm 2i$

53. Because there are no x-intercepts, the equation has no real number solution. There are two imaginary number solutions.

55. 0, 3 **57.** $\pm 3i\sqrt{5} \approx \pm 6.71i$ **59.** $-\frac{1}{6}$ **61.** $\pm 1.2i$

63. $\pm\sqrt{10} \approx \pm 3.16$ **65.** $-5, 7$ **67.** $-\frac{5}{4} \pm \frac{7}{4}i$

69. $\frac{-2 \pm 8\sqrt{3}}{3} \approx 3.95, -5.29$ **71.** $\pm i\sqrt{6} \approx \pm 2.45i$

73. 1, 4 **75.** $\frac{3}{2}$ **77.** $x^2 - 8x + 15 = 0$

79. $x^2 - 6x + 9 = 0$ **81.** $3x^2 - 13x + 12 = 0$

83. $x^2 + 16 = 0$ **85.** $-8, -9$

87. The resulting integers are even, not odd.

89. 5 inches, 9 inches **91.** 18, 20 **93.** 5 minutes

95. (a) $24.4t^2 - 233.3t + 554 = 1000$ **(b)** $t \approx 11$ (1991)

97. $-2 \pm i\sqrt{y}$ **99.** $-c \pm 4i$ **101.** $1 \pm i$ **103.** $-6, -\frac{4}{3}$

Section 9.2 (*page 562*)

1. The expression A is a first-degree binomial.

3. 9 **5.** $\frac{49}{4}$ **7.** $\frac{4}{9}$

9. Both methods are correct. The factoring method can be used because $x^2 - 8x + 12$ is factorable. The method of completing the square can be used for any quadratic equation.

11. $-3, -1$ **13.** $-5, 3$ **15.** $-3, 4$ **17.** $-3, 2$

19. $3 \pm \sqrt{5} \approx 0.76, 5.24$ **21.** $-6, 4$

23. $2 \pm 2\sqrt{2} \approx -0.83, 4.83$

25. Divide both sides by 3 to make the leading coefficient 1.

27. $-1, -\frac{1}{3}$ **29.** $-5, \frac{3}{2}$ **31.** $-\frac{3}{2}, \frac{7}{2}$ **33.** $\frac{5}{2}$

35. $\frac{3 \pm \sqrt{17}}{4} \approx -0.28$ or 1.78

37. $\frac{-7 \pm \sqrt{57}}{4} \approx -3.64$ or 0.14

39. 9 **41.** $2 \pm 2i$ **43.** $2 \pm i\sqrt{2} \approx 2 \pm 1.41i$

45. $\frac{-3 \pm i\sqrt{11}}{2} \approx -1.5 \pm 1.66i$

47. $\frac{-1 \pm i\sqrt{23}}{4} \approx -0.25 \pm 1.20i$

49. Answers will vary. **51.** 2, 8 **53.** $\pm\frac{7}{6}$ **55.** $-2, 5$

57. $-3 \pm \frac{\sqrt{7}}{2} \approx -4.32, -1.68$

59. $\frac{-1 \pm i\sqrt{35}}{2} \approx -0.5 \pm 2.96i$

61. $\frac{9 \pm 3\sqrt{5}}{2} \approx 1.15, 7.85$ **63.** $\frac{-3 \pm \sqrt{21}}{4} \approx -1.90, 0.40$

65. $\frac{1 \pm i\sqrt{35}}{2} \approx 0.5 \pm 2.96i$

67. $\frac{-7 \pm \sqrt{97}}{8} \approx -2.11$ or 0.36 **69.** $-\frac{2}{3}, \frac{1}{3}$ **71.** $-1, 2$

73. $\frac{1 \pm i\sqrt{35}}{4} \approx 0.25 \pm 1.48i$

75. $a = 4; -\frac{5}{4}$ **77.** $b = 7; -\frac{3}{2}$ **79.** $c = -48; -8$

81. 5 **83.** 5.54 **85.** 31 feet **87.** 2 hours and 17 minutes

89. 4, 9 **91.** $x^2 - 13x + 37 = 0$. The polynomial is prime.

93. $c \pm ci\sqrt{3}$ **95.** $-\sqrt{5} \pm 3 \approx -5.24, 0.76$

97. **(a)** $k \geq 2$ or $k \leq -2$ **(b)** $-2 < k < 2$ **99.** $-2i \pm 2$

101. **(a)** $x^2 + 8x$ **(b)** 16 **(c)** $x^2 + 8x + 16$ **(d)** $(x + 4)^2$
(e) $(x + 4)^2 = x^2 + 8x + 16$

Section 9.3 *(page 571)*

1. The equation must be in standard form.

3. 3, −9, −6 **5.** 1, −1, 2 **7.** −1, 2 **9.** 4

11. $-4, \frac{1}{2}$ **13.** 0, 5 **15.** $\dfrac{9 \pm \sqrt{129}}{8} \approx -0.29, 2.54$

17. $1 \pm 2i\sqrt{2} \approx 1 \pm 2.83i$

19. $-1 \pm \sqrt{6} \approx -3.45, 1.45$

21. $\dfrac{-1 \pm i\sqrt{31}}{4} \approx -0.25 \pm 1.39i$

23. (−B+√(B^2−4AC))/(2A); (−B−√(B^2−4AC))/(2A)
If parentheses in the numerators are omitted, only the radical term will be divided by 2A.

25. $0, \frac{2}{3}$ **27.** $\pm\frac{3}{7}$ **29.** 7 **31.** $\dfrac{4 \pm \sqrt{10}}{5} \approx 0.17$ or 1.43

33. −2, 4 **35.** −5, 1 **37.** $\dfrac{-1 \pm i\sqrt{34}}{5} \approx -0.2 \pm 1.17i$

39. $\dfrac{3 \pm i\sqrt{23}}{2} \approx 1.5 \pm 2.40i$ **41.** $3 \pm \sqrt{5} \approx 0.76$ or 5.24

43. $\dfrac{-5 \pm i\sqrt{11}}{2} \approx -2.5 \pm 1.66i$

45. $\dfrac{2 \pm 2i\sqrt{5}}{3} \approx 0.67 \pm 1.49i$

47. $\dfrac{5 \pm \sqrt{33}}{4} \approx -0.19$ or 2.69

49. $\dfrac{1 \pm \sqrt{13}}{2} \approx -1.30$ or 2.30 **51.** 2 rational

53. 2 imaginary **55.** 1 rational (double root)

57. 2 irrational **59.** $k < 16$ **61.** $k < 4$ **63.** $k > 9$

65. $k > \frac{9}{4}$ **67.** ±6 **69.** 0, 8

71. **(a)** The discriminant is $a^2 + 20$, which is positive for all values of a. Therefore, the solutions are always real numbers.
(b) The discriminant is $a^2 + 4k$. Because k is positive, the discriminant is always positive, and the solutions are always real numbers.

73. $(x - 20)(x + 24)$ **75.** $(x - 18)(x + 25)$

77. Answers will vary. **79.** Answers will vary.

81. $\frac{2}{3}, -\frac{1}{3}$ **83.** −1, 1 **87.** Yes **89.** \$50 or \$80

91. Approximately (28, 55)

93. The sign is negative, which indicates that the graph of y has no x-intercept. The model indicates that there is no year in which no GED recipient planned additional schooling.

95. $x = \dfrac{-a \pm \sqrt{a^2 + 8a}}{2}$ **97.** $-1 \pm \sqrt{9 + x}$

99. $x^2 + 10x + 5 = 0$

101. $\dfrac{-\sqrt{3} \pm \sqrt{7}}{2} \approx 0.46$ or −2.19 **103.** $i \pm \sqrt{2} \approx i \pm 1.41$

Section 9.4 *(page 579)*

1. Multiplying both sides of the original equation by $x - 1$ introduces an extraneous solution.

3. $\dfrac{3 \pm i\sqrt{23}}{4} \approx 0.75 \pm 1.20i$ **5.** $4 \pm \sqrt{7} \approx 6.65, 1.35$

7. $\dfrac{-1 \pm i\sqrt{15}}{2} \approx -0.5 \pm 1.94i$

9. $\dfrac{-3 \pm \sqrt{21}}{2} \approx -3.79, 0.79$

11. $\dfrac{3 \pm \sqrt{17}}{2} \approx -0.56, 3.56$ **13.** −2

15. Squaring both sides of the original equation introduces extraneous solutions.

17. $1, \frac{4}{9}$ **19.** $\dfrac{7 \pm \sqrt{17}}{8} \approx 1.39$ or 0.36 **21.** 5

23. $\dfrac{6 - \sqrt{19}}{2} \approx 0.82$ **25.** $\dfrac{-3 + \sqrt{13}}{2} \approx 0.30$

27. By making an appropriate substitution, we write the given equation in the form $au^2 + bu + c = 0$, where u represents an expression. After solving this equation for u, we replace u with the expression that it represents and solve for the original variable.

29. $\pm 2, \pm\sqrt{3} \approx \pm 1.73$ **31.** $\pm 3i, \pm\sqrt{7} \approx \pm 2.65$

33. 4, 16 **35.** $-1, \frac{1}{2}$ **37.** 64, 27 **39.** 7, −2, 3, 2

41. $-2 \pm \sqrt{3} \approx -0.27, -3.73$ **43.** $\pm 3, \pm\sqrt{2} \approx \pm 1.41$

45. $16, \frac{49}{9}$ **47.** 9 **49.** $\pm 2, \pm\sqrt{3} \approx \pm 1.73$

51. $\dfrac{-3 \pm \sqrt{41}}{4} \approx 0.85, -2.35$

53. $\dfrac{-4 + \sqrt{3}}{2} \approx -1.13$ **55.** $\pm 2, \pm i\sqrt{5} \approx \pm 2.24i$

57. 49, 36 **59.** $1 \pm 2i$ **61.** 8 **63.** −27, 125

65. $\dfrac{-7 + \sqrt{73}}{6} \approx 0.26$ **67.** 1 **69.** $\dfrac{3 \pm \sqrt{33}}{2} \approx -1.37, 4.37$

71. $53 - 8\sqrt{39} \approx 3.04$ **73.** $\frac{11}{4}, 4$ **75.** No solution

77. $\dfrac{1 \pm i\sqrt{7}}{2} \approx 0.5 \pm 1.32i, \dfrac{1 \pm i\sqrt{3}}{2} \approx 0.5 \pm 0.87i$

79. 900 motors **81.** 300 students, $650 per person

83. **(a)** $-0.48x^2 + 3.4x + 10.3 = 15.3$; $x \approx 2.1$ (Tuesday) or $x = 5$ (Friday)

(b) $-0.48(x - 7)^2 + 3.4(x - 7) + 10.3 = 15.3$

(c) Letting $u = x - 7$, $-0.48u^2 + 3.4u + 10.3 = 15.3$; $u \approx 21$ or $u = 5$; $x \approx 9.1$ (Tuesday) or $x = 12$ (Friday)

85. About $29.45 per person

87. $C(-5) = 2576$, which is the cost (in millions of dollars) of the school lunch program in 1985.

89. $\dfrac{1 + \sqrt{13}}{2} \approx 2.30$

91. $2, -1, -1 \pm i\sqrt{3} \approx -1 \pm 1.73i, \dfrac{1 \pm i\sqrt{3}}{2} \approx 0.5 \pm 0.87i$

93. $1, -2 - \sqrt{3} \approx -3.73$ **95.** $\dfrac{1 \pm i\sqrt{3}}{2} \approx 0.5 \pm 0.87i$

Section 9.5 *(page 588)*

1. 16.97 feet **3.** 97.14 feet, 127.14 feet

5. 17.3 inches, 10.7 inches **7.** $AC = 12.26$, $DE = 3.26$

9. 26.88 mph **11.** 57.16 mph

13. 11.08 hours, 13.08 hours **15.** 9 hours **17.** 370 shares

19. Last year: 459 boxes at $18.67; this year: 561 boxes at $20.67

21. 60 **23.** 11 and 13 or −13 and −11

25. 90.38 feet **27.** 8 feet **29.** 9 **31.** 5.41 feet

33. 8 yards **35.** 25.08 feet **37.** 3.12 feet **39.** 3

41. 4 miles **43.** About 12:49 **45.** 12

47. The x-intercepts are about (1, 0) and (108, 0). The interpretation is that no one aged 1 or 108 has a credit card.

49. Because the percentage of card holders who always pay the balance due increases with age, the linear model $y = ax + b$ is the better model. A model is $y = 0.88x + 13.13$. Because the slope is positive, the graph rises from left to right. A function that models the percentage of card holders who do not pay their balances would have a graph that falls from left to right.

51. $15.63 **53.** 4.31 inches

Section 9.6 *(page 599)*

1. For $\le$ or $\ge$, the x-intercepts represent solutions. Then look for the x-interval(s) in which the graph is above the x-axis (for $>$ or $\ge$) or below the x-axis (for $<$ or $\le$).

3. $(-1, 3)$ **5.** $(-\infty, 0) \cup (\frac{8}{5}, \infty)$ **7.** $(2, 5)$

9. $(-\infty, -1) \cup (\frac{3}{2}, \infty)$

11. Write the inequality in standard form. The function is $f(x) = 2x^2 + 3x + 7$.

13. $(-\infty, -3) \cup (3, \infty)$ **15.** $[-2.5, 0.5]$

17. $[1.44, 5.56]$ **19.** $(-\infty, -6) \cup (3, \infty)$ **21.** $[0, 5]$

23. $\left(-\infty, -\frac{2}{3}\right] \cup \left[\frac{1}{4}, \infty\right)$ **25.** $(-1.12, 3.12)$

27. (b) **29.** (a) **31.** $(-\infty, -3] \cup [2, \infty)$ **33.** $[-2, 3]$

35. $\left(-\frac{2}{3}, \frac{3}{2}\right)$ **37.** $(-\infty, -1.12] \cup [3.12, \infty)$

39. $[-2, 4]$ **41.** $\left(-\infty, 0\right] \cup \left[\frac{7}{4}, \infty\right)$ **43.** $(-\infty, -4] \cup [4, \infty)$

45. $[-0.58, 8.58]$ **47.** **R** or Ø

49. Because $x - 2$ is a factor of $ax^2 + bx + c$, (2, 0) is an x-intercept of the graph of $f(x) = ax^2 + bx + c$. But the solutions of $ax^2 + bx + c > 0$ are represented by points of the graph that are above the x-axis. Thus 2 is not a solution.

51. $(-\infty, -5) \cup \left[-\frac{3}{4}, \infty\right)$ **53.** $(-\infty, 0) \cup (5, \infty)$ **55.** Ø

57. $\{-2.5\}$ **59.** $(-\infty, -0.83] \cup [4.83, \infty)$ **61.** $\{3\}$

63. $(-0.87, 2.87)$ **65.** $(-\infty, 1] \cup [5, \infty]$ **67.** **R**

69. $\left[-\frac{7}{3}, -\frac{1}{2}\right]$ **71.** $(-\infty, -2) \cup (\frac{5}{4}, \infty)$

73. $(-2.53, 0.53)$ **75.** $[-4, -2]$ **77.** $k > \frac{9}{8}$

79. Not possible **81.** $k \ge \frac{25}{4}$ **83.** $k \le -3.2$ **85.** 36

87. ±6 **89.** 1

91. If x and $x + 35$ represent the width and length of the home, then $x(x + 35) > 2556$, from which $x > 35$. Thus the length of the home is more than 70 feet, which violates provision (b).

93. Between 10.74 inches and 29.26 inches

95. With x representing the number of years since 1985, $H(x) = 2.14x^2 - 21.49x + 771.38$.

97. 1, 9; 1986–1994

99. **(a)** The domain of f is the set of all x for which the radicand is nonnegative. Therefore, we estimate the intervals for which the graph of g is on or above the x-axis: $(-\infty, -3] \cup [5, \infty)$.

(b) Because $h(x)$ is nonnegative for all x, the portion of the graph of g that is below the x-axis is inverted to the other side of the x-axis.

103. $-6 \le k \le 6$ **105.** $-4 \le k \le 4$ **107.** About 1.11

Section 9.7 *(page 607)*

1. If the polynomial cannot be factored, we cannot determine the x-intercepts with algebraic techniques presented to this point.

3. $(-2, 1) \cup (3, \infty)$ **5.** $[-6, 3]$ **7.** $(3, \infty)$

9. $(-\infty, -5) \cup (-1, 0)$ **11.** $(-\infty, -3) \cup (-2, 2) \cup (3, \infty)$

13. $[-2.24, -1] \cup [1, 2.24]$ **15.** $(-3, -1) \cup (1, \infty)$

17. $(-1, 2) \cup (3, \infty)$ **19.** $(-\infty, -5] \cup [-2, 1]$

21. $(-4, -2) \cup (-1, 1)$

23. If a and b are restricted values, they are not included in the solution set.

25. $(-1, 0]$ **27.** $\left(-\infty, -\frac{9}{2}\right) \cup (-1, \infty)$ **29.** $\left(-\infty, \frac{4}{3}\right)$

31. $(-\infty, -4) \cup (-1, 3) \cup (10, \infty)$

33. $(-\infty, -3) \cup \left[-\frac{5}{2}, 0\right]$

35. $\left(-\infty, -\frac{4}{3}\right] \cup [0, 3]$ **37.** $\left(-\infty, \frac{5}{2}\right]$

39. $\left(-\infty, -\frac{1}{2}\right] \cup \left(0, \frac{5}{3}\right]$ **41.** $(-2, -1)$

43. $(-5, 0) \cup (0, 3)$ **45.** $(-3, 3) \cup (4, 5)$

47. No, both expressions could be negative.

49. $\left(-\frac{3}{2}, 0\right)$ **51.** $(-\infty, -2] \cup [3, 5]$ **53.** $\left[\frac{1}{3}, 2\right)$

55. $\left(\frac{5}{2}, \infty\right)$ **57.** $(-2, 5) \cup (7, \infty)$ **59.** $\left(-\infty, \frac{1}{2}\right]$

61. $(-\infty, -4.19] \cup (-3, 1.19] \cup (2, \infty)$

63. $(-\infty, -3) \cup (3, \infty)$ **65.** $\left(-1, -\frac{1}{2}\right] \cup (1, 3]$ **67.** $\left(\frac{1}{2}, \infty\right)$

69. $(-\infty, -0.17] \cup [4, \infty)$

71. Both are nonnegative for all x. However, the quotient is not defined for $x = 2$. The solution set is **R**, $x \neq 2$.

73. $(-\infty, 0)$ **75.** $(0, \infty)$ **77.** $(-3, 0) \cup (0, \infty)$

79. $(-\infty, -1) \cup (-1, 1) \cup \left[\frac{7}{2}, \infty\right)$

81. $(-\infty, -6] \cup \{0\} \cup [2, \infty)$

83. Any three consecutive odd integers where the smallest integer is at least 3

85. At least 4 mph **87.** Between 5 and 8.23 feet

89. $0.0025x^3 - 0.22x^2 + 6.82x - 7.2 > 100$

91. About \$19,000–\$44,000

93. Because the radicand must be nonnegative, the domain of f is represented by portions of the graph that are on or above the x-axis.

95. **R** **97.** $(-\infty, 1)$ **99.** **(a)** **R** **(b)** **R**, $x \neq -4, 0.5$

101. **(a)** $(-\infty, 0)$ **(b)** $(-\infty, 0]$

Section 9.8 *(page 618)*

1. The graph is a parabola. For positive a, the graph opens upward, and for negative a, the graph opens downward. The graph becomes narrower as $|a|$ increases.

3. $\{y \mid y \leq 3\}$ **5.** $\{y \mid y \geq 2\}$

7. **(a)** Because the axis of symmetry contains the vertex, the equation is $x = h$.

(b) The first coordinate of the vertex is a.

9. $x = 0$ **11.** $x = 3$ **13.** (e) **15.** (c) **17.** (b)

19. **(a)** Reflected in the x-axis

(b) Reflected in the x-axis and narrower

21. **(a)** Shifted left 3 units

(b) Shifted right 3 units

23. **(a)** Shifted right 3 units and down 2 units

(b) Shifted left 1 unit and up 4 units

25. The x-coordinate is $-\frac{b}{2a}$. Determine the y-coordinate by evaluating the function at $-\frac{b}{2a}$.

27. $(3, 4)$; minimum: 4 **29.** $(-1, -4)$; minimum: -4

31. $(-1, 13)$; maximum: 13 **33.** $(0, 3)$ **35.** $(5, -1)$

37. $(3, 0)$, $x = 3$ **39.** $(-1, -9)$, $x = -1$

41. $(-2, 11)$, $x = -2$

43. Because the domain is **R**, the function is defined for $x = 0$. The y-intercept is $(0, c)$, where c is the constant term of the function.

45. $(0, -28)$, $(7, 0)$, $(-4, 0)$ **47.** $(0, 9)$, no x-intercept

49. $(0, 5)$, $(-2.12, 0)$, $(0.79, 0)$ **51.** $(0, 0)$, $\left(-\frac{5}{3}, 0\right)$

53. None **55.** Two, rational **57.** None **59.** Two

61. One **63.** Two **65.** One **67.** Two

69. **(a)** Domain: **R**; range: $\{y \mid y \geq -1\}$

(b) $(0, 8)$, $(4, 0)$, $(2, 0)$

(c) $(3, -1)$, $x = 3$

71. **(a)** Domain: **R**; range: $\{y \mid y \leq 9\}$

(b) $(0, 5)$, $(-5, 0)$, $(1, 0)$

(c) $(-2, 9)$, $x = -2$

73. **(a)** Domain: **R**; range: $\{y \mid y \geq 7\}$

(b) $(0, 9)$

(c) $(-1, 7)$, $x = -1$

75. $h = 0, k = 3$ **77.** $h = -\frac{3}{5}, k = 0$ **79.** $h = -4, k = 2$

81. $h = 3, k = 2$ **83.** $b = -2, c = -15$

85. $b = 2, c = 0$ **87.** $b = -8, c = 16$

89. $f(x) = x^2 - 4x + 3$ **91.** $f(x) = 3x^2 + x - 2$

93. 15, 15 **95.** $\frac{15}{16}$ second **97.** 187.5 square feet

99. $y = -0.024x^2 + 2.4x$

101. Upward: $f(x) = 4.42x^2 - 0.62x + 29.4$

103. For the graph of f, the vertex is about $(0.1, 29.4)$. In 1992, the number of investors was 29.4 million, the lowest number in the period. For the graph of g, the vertex is about $(3.7, 68.5)$. In 1996, the number of investors was 68.5 million, the highest number in the period. Neither model is likely to be valid beyond the given period.

105. $\left(\frac{3}{4}, -\frac{49}{8}\right)$ **107.** $\left(-\frac{1}{2}, \frac{3}{4}\right)$

109. **(a)** \$21,000

(b) $V(1, 21025)$; the income is maximized if one store is vacant. **(c)** 30

111. $f(x) = ax^2 + bx + c$

$$f(x) - c = ax^2 + bx$$

$$\frac{f(x)}{a} - \frac{c}{a} = x^2 + \frac{b}{a}x$$

$$\frac{f(x)}{a} - \frac{c}{a} + \frac{b^2}{4a^2} = x^2 + \frac{b}{a}x + \frac{b^2}{4a^2}$$

$$\frac{f(x)}{a} - \frac{4ac - b^2}{4a^2} = \left(x + \frac{b}{2a}\right)^2$$

$$\frac{f(x)}{a} = \left(x + \frac{b}{2a}\right)^2 + \frac{4ac - b^2}{4a^2}$$

$$f(x) = \left(x + \frac{b}{2a}\right)^2 + \frac{4ac - b^2}{4a}$$

With the function written in this form, we see that

$h = -\dfrac{b}{2a}$ and $k = \dfrac{4ac - b^2}{4a}$.

Chapter Review Exercises *(page 623)*

1. (a) For one graph, write the equation in standard form, graph, and estimate the x-coordinates of the x-intercepts.

(b) For two graphs, graph the left and right sides, and estimate the x-coordinates of the points of intersection.

3. $\pm\dfrac{\sqrt{39}}{3} \approx \pm 2.08$ **5.** $-6, 7$ **7.** $-5, 3$ **9.** $-\frac{24}{5}, -\frac{16}{5}$

11. $2 \pm i\sqrt{37} \approx 2 \pm 6.08i$

13. After dividing both sides by 2, we obtain an x-coefficient of $\frac{5}{2}$. To complete the square, add the square of half the x-coefficient to both sides. This number is $\frac{25}{16}$.

15. $-5 \pm \sqrt{26} \approx 0.10$ or -10.10 **17.** $3 \pm i$

19. $\dfrac{5 \pm \sqrt{13}}{2} \approx 4.30$ or 0.70 **21.** $-\frac{3}{2}, 1$

23. $\dfrac{-3 \pm \sqrt{11}}{2} \approx 0.16$ or -3.16

25. All can be solved with the Quadratic Formula or by completing the square. Parts (i) and (iv) can be solved with the Square Root Property. Parts (i) and (ii) can be solved by factoring.

27. $\pm\dfrac{4\sqrt{5}}{5} \approx \pm 1.79$ **29.** $\dfrac{5 \pm \sqrt{73}}{8} \approx -0.44$ or 1.69

31. $\dfrac{1 \pm i\sqrt{71}}{4} \approx 0.25 \pm 2.11i$ **33.** 2, rational

35. 2, imaginary

37. All are quadratic in form.

(a) $u = \frac{1}{x}$; $2u^2 + 3u - 5 = 0$

(b) $u = \sqrt{2x}$; $u^2 + u - 3 = 0$

(c) $u = x + 1$; $u^2 + 2u + 1 = 0$

(d) $u = x^{1/4}$; $u - u^2 + 3 = 0$

39. $-1 \pm \sqrt{3} \approx 0.73, -2.73$ **41.** 0 **43.** 25

45. $-4, -3, 2, 3$

47. $\dfrac{-3 \pm i\sqrt{7}}{2} \approx -1.5 \pm 1.32i$ **49.** 10.95 square yards

51. 5 **53.** 15 hours

55. The graph of $f(x) = x^2 + 5x + 6$ is below the x-axis in the x-interval $(-3, -2)$. Therefore, the graph of f is on or above the x-axis for $x \le -3$ and for $x \ge -2$. Thus the solution set is $(-\infty, -3] \cup [-2, \infty)$.

57. $\{3\}$ **59. R** **61.** Ø **63.** $(-1.39, 2.89)$

65. $k > \frac{1}{3}$

67. The solutions are represented by those points that are above the x-axis. Therefore, the solutions are all numbers greater than -5 except 3. Thus (iii) is correct.

69. $[-1, \infty)$ **71.** $(-\infty, -2] \cup [-1, 2]$

73. $\left[1, \frac{4}{3}\right)$ **75.** $(-\infty, -2) \cup \left(-\frac{3}{2}, 0\right)$

77. The graph shows that the domain is **R**. The range can be estimated by first locating the vertex (h, k). If the parabola opens upward, the range is $\{y \mid y \ge k\}$; if the parabola opens downward, the range is $\{y \mid y \le k\}$.

79. $V\left(\frac{5}{2}, \frac{11}{4}\right)$; $(0, 9)$ **81.** Downward; $V(0, 17)$; $\{y \mid y \le 17\}$

83. Upward; $V\left(\frac{5}{4}, -\frac{25}{8}\right)$; $\left\{y \mid y \ge -\frac{25}{8}\right\}$

85. 62.5 feet

Chapter Test *(page 626)*

1. [*9.2*] $\dfrac{-2 \pm \sqrt{7}}{3} \approx -1.55$ or 0.22

2. [*9.1*] $-\frac{2}{3}$ or $\frac{3}{2}$

3. [*9.1*] $\dfrac{1 \pm \sqrt{5}}{2} \approx -0.62$ or 1.62

4. [*9.3*] $\dfrac{1 \pm \sqrt{3}}{2} \approx -0.37$ or 1.37

5. [*9.3*] $\dfrac{-7 \pm \sqrt{41}}{4} \approx -0.15$ or -3.35

6. [*9.3*] $\dfrac{1 \pm i\sqrt{23}}{6} \approx 0.17 \pm 0.80i$

7. [*9.3*] $\pm i, \pm\dfrac{\sqrt{2}}{3} \approx \pm 0.47$

8. [*9.1*] -2 **9.** [*9.5*] $h = 5$ feet, $b = 8$ feet

10. [*9.5*] 5.12 hours **11.** [*9.5*] 3 inches

12. [*9.6*] If the graph has no x-intercept, then it lies entirely above or entirely below the x-axis. If, for example, the graph is entirely above the x-axis and the inequality symbol is $>$ or $\geq$, then the solution set is **R**.

13. [*9.6*] $(-\infty, -2] \cup [3, \infty)$

14. [*9.7*] $(-\infty, -7) \cup (-2, 3)$

15. [*9.7*] $\left[-\frac{1}{2}, 0\right] \cup [5, \infty)$

16. [*9.8*] $f(x) = (x - 3)^2 + 2$; $V(3, 2)$; $x = 3$

17. [*9.8*] **(a)** Domain: **R**; range: $\{y \mid y \geq -4\}$

(b) $a > 0$ **(c)** 2, 6

18. [*9.8*] $(2.39, 0)$, $(-8.39, 0)$

CHAPTER 10

Section 10.1 *(page 636)*

1. Evaluate f and g at 1 and add the results, or algebraically add f and g and then evaluate the resulting function at 1.

3. **R** **5.** $(-\infty, 0.5]$ **7.** **R**, $x \neq -4, 0, 1$

9. $(-\infty, -4] \cup [4, \infty)$

11. **(a)** -5 **(b)** 3 **(c)** 0 **(d)** 0

13. **(a)** 2 **(b)** 2 **(c)** 0 **(d)** Undefined

15. **(a)** 2.5 **(b)** 1.5 **(c)** 1 **(d)** 1

17. **(a)** 21 **(b)** -11 **(c)** 0 **(d)** -4

19. **(a)** 2 **(b)** 12 **(c)** -9 **(d)** -0.8

21. **(a)** -2 **(b)** -2 **(c)** $\frac{727}{3}$ **(d)** 124

23. No; 0 is in the domain of h, but 0 is not in the domain of f/g because $g(0) = 0$.

25. **(a)** $x^2 - x - 12$; **R** **(b)** $x^2 + x - 6$; **R**

(c) $x - 3$; **R**, $x \neq -3$ **(d)** $x^3 + 3x^2 - 9x - 27$; **R**

27. **(a)** $\frac{x^2 + 4}{x^2 - 4}$; **R**, $x \neq -2, 2$ **(b)** $\frac{x^2 - 4}{x^2 + 4}$; **R**

(c) 1; **R** **(d)** 1; **R**, $x \neq -2, 2$

29. **(a)** $\frac{-5}{(x + 2)(x - 3)}$; **R**, $x \neq -2, 3$

(b) $\frac{2x - 1}{(x + 2)(x - 3)}$; **R**, $x \neq -2, 3$

(c) $\frac{x - 3}{x + 2}$; **R**, $x \neq -2, 3$

(d) $\frac{1}{(x + 2)(x - 3)}$; **R**, $x \neq -2, 3$

31. $4t^2 - 2t - 3$ **33.** $\frac{t^2 - 16}{2t - 4}$ **35.** $9t^2 + 3t + 2$

37. $-8t^3 + 20t^2 + 6t - 15$

39. Produce the graphs of f and g and note the intersection of the x-values for which f and g are defined.

41. **(a)** $[1.5, 5]$ **(b)** $[1.5, 5]$ **43.** **(a)** $(1, 8]$ **(b)** $[1, 8]$

45. First evaluate $g(5)$, and then evaluate $f(g(5))$.

47. **(a)** -2 **(b)** -1 **(c)** 2 **(d)** -14

49. **(a)** 15 **(b)** 16 **(c)** -3 **(d)** 1

51. **(a)** 5 **(b)** -2 **(c)** -4

53. $-4x^2 - 4x + 2$; $-2x^2 + 7$

55. $9x^2 - 21x + 15$; $3x^2 - 9x + 13$ **57.** $x - 2$; $\sqrt[3]{x^3 - 2}$

59. 5; 7 **61.** $9x - 16$; $x + 10$

63. $\left||x - 3| - 3\right|$; $x^4 + 6x^2 + 12$

65. $8x^4 - 24x^3 + 12x^2 + 9x$; x **67.** 3; 2 **73.** $(g \circ f)(x)$

75. $(g \circ g)(x)$ **77.** $(f \circ h)(x)$ **79.** $(g \circ h)(x)$

81. **(a)** $g(d(t)) = \frac{54t}{25}$ **(b)** $c(g(d(t))) = 1.03\left(\frac{54t}{25}\right)$ **(c)** \$26.70

83. **(a)** $f(x) = x$ **(b)** $c(x) = 10 - 0.05x$

(c) $T(x) = 10x - 0.05x^2$ **(d)** \$487.20

85. The composition of F and G is the number of transactions as a function of the number of years since 1980.

87. $(F \circ G)(14) = 8611.14$

The actual entry is 8334, which is the number (in millions) of ATM transactions in 1994.

89. $f = g \circ h$, where $g(x) = \sqrt{x}$ and $h(x) = x^2 - 5x + 6$

91. $f = g \circ h$, where $g(x) = x^4$ and $h(x) = 3x + 2$

93. $f = g \circ h$, where $g(x) = x + 8$ and $h(x) = |2x + 5|$

Section 10.2 *(page 645)*

1. The set is not necessarily a one-to-one function. If the set has any ordered pairs with the same first coordinate, then the set is not even a function.

3. No **5.** Yes **7.** No **9.** No **11.** Yes **13.** Yes

15. No **17.** No

19. $f^{-1} = \{(2, 1), (4, 3), (6, 5), (-1, 7)\}$
Domain: $\{2, 4, 6, -1\}$; range: $\{1, 3, 5, 7\}$

21. $f^{-1} = \{(2, 3), (3, 4), (1, -5), (-2, -7)\}$
Domain: $\{2, 3, 1, -2\}$; range: $\{3, 4, -5, -7\}$

23. -2 **25.** 3 **27.** 6 **29.** -4 **31.** -5 **33.** 3

35. No; the x^{-1} key returns the reciprocal of x, not the inverse function. In this case, $f(x) = f^{-1}(x) = x$.

37. Yes **39.** No **41.** Yes **43.** Yes **45.** No **47.** Yes

49.

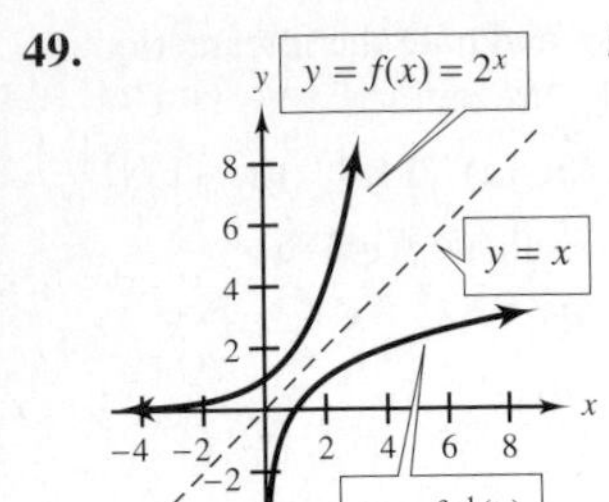

51.

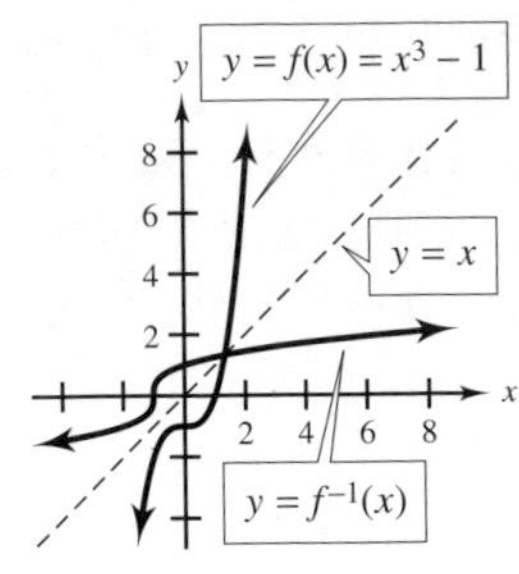

53. The graph of a linear function is a nonvertical line that passes the Horizontal Line Test. Therefore, the function has an inverse.

55. No **57.** Yes **59.** Yes **61.** $\dfrac{x-5}{3}$ **63.** $-0.5x$

65. $3(x-5)$ **67.** $\dfrac{1}{x}$ **69.** $\dfrac{1}{x} - 2$ **71.** $\dfrac{2x+1}{x-1}$

73. $\sqrt[3]{x-1}$ **75.** $x^{-1/3}$ **77.** $4x^2, x \geq 0$ **79.** $2 - x^3$

81. $(x-3)^5$ **83.** Multiply x by 3 and then subtract 2.

85. Take the cube root of x and then subtract 1. **87.** (iii)

89. The first is a one-to-one function, but the second is not.

91. $\dfrac{9}{5}t + 32$; converts Celsius temperatures to Fahrenheit

93. $d = 1.50p$; the inverse function is $p = \dfrac{d}{1.50}$, which gives the value p in pounds of d dollars.

95. **(a)** C is a one-to-one function.

(b) $s = \dfrac{C - 2588}{0.66}$. This inverse function represents the number of subscribers (in thousands) as a function of the number of cell sites.

97. Approximately (9.6, 9.6)

99. $x = \dfrac{29}{3} = 9.\overline{6}$

101.

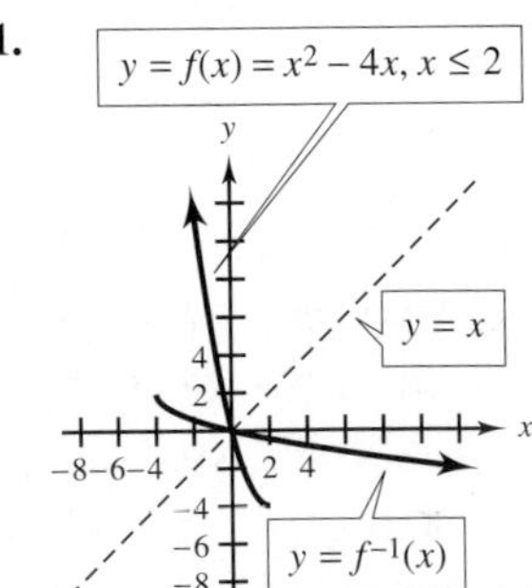

103. $\sqrt{x-1}, x \geq 1$ **105.** $x - 5, x \leq 0$

Section 10.3 *(page 655)*

1. **(a)** The graph rises from left to right if $b > 1$.

(b) The graph falls from left to right if $0 < b < 1$.

3. 32 **5.** $\frac{1}{9}$ **7.** 1 **9.** 2 **11.** 0.02 **13.** 403.43

15. 11.18 **17.** 54.60 **19.** 1613.72 **21.** 243

23. 29,370.08 **25.** 3.95

27. Because $3^{-x} = \left(\dfrac{1}{3}\right)^x = \dfrac{1}{3^x}$, functions (ii), (iii), and (iv) all have the same graph.

29. Domain: **R**; range: $\{y \mid y < -2\}$

31. Domain: **R**; range: $\{y \mid y < 2\}$

33. Domain: **R**; range: $\{y \mid 0 < y \leq 1\}$

35. For $c > 0$, the graph is shifted up c units, and for $c < 0$, the graph is shifted down $|c|$ units.

37. **(a)** Up 1 unit **(b)** Down 3 units

39. **(a)** Up 3 units **(b)** Left 3 units

41. **(a)** B **(b)** A **43.** -0.33 **45.** 2.42 **47.** None

49. 1.39 **51.** -1.37 **53.** 0 **55.** No solution

57. Both sides of the equation $4^x = 8$ can be written with the same base, and then the exponents can be equated. The two sides of $4x = 12$ cannot be written with the same integer base.

59. -3 **61.** -2 **63.** -1.5 **65.** -1.5 **67.** 1.5 **69.** $\frac{10}{3}$

71. $\frac{2}{3}$ **73.** -3 **75.** No solution **77.** 0 **79.** No solution

81. 0.5 **83.** $1, -\frac{5}{3}$ **85.** -8 **87.** ± 1 **89.** $-1, 0.5$

91. \$1.83 **93.** \$5000 **95.** 7.12 years

97. **(a)**

Year	n	Expenditure
1998	3	$84.20(1.06)^3$
1999	4	$84.20(1.06)^4$
2000	5	$84.20(1.06)^5$

(b) $E(n) = 84.20(1.06)^n$ **(c)** \$112.68, \$134.20

(d) About 2005

99. The graph rises rapidly from left to right.

101. 2004 **103.** $0.16, -2.19$ **105.** $-0.26, -2.54$

107. $(-\infty, 0)$ **109.** Minimum: (0, 0); maximum: $(-2, 0.54)$

111. Minimum: (0, 4)

Section 10.4 *(page 665)*

1. The logarithmic function is the inverse of the exponential function, which is not defined for $b \leq 0$ or for $b = 1$.

3. $7^2 = 49$ **5.** $10^{-3} = 0.001$ **7.** $e^{1/3} = \sqrt[3]{e}$ **9.** $m^2 = n$

11. $\log_4 64 = 3$ **13.** $\log 0.00001 = -5$

15. $\log_{1/2} 8 = -3$ **17.** $\ln P = rt$ **19.** $\ln x$ **21.** $\log_{1/4} x$

23. 10^x **25.** 1.5^x **27.** 2 **29.** 0 **31.** 2.5 **33.** 3

35. Not defined **37.** −0.5 **39.** 1 **41.** 2.5

43. Not defined **45.** −1 **47.** 3 **49.** 0.5 **51.** 1.39

53. 1.40 **55.** −0.14 **57.** 0.75 **59.** 3.63

61. Write the equation in exponential form.

63. 243 **65.** e **67.** 0.25 **69.** 9 **71.** 125

73. $x > 0, x \neq 1$ **75.** 2 **77.** 3 **79.** 15 **81.** ± 25

83. **(a)** The graph rises from left to right if $b > 1$.
(b) The graph falls from left to right if $0 < b < 1$.

85. $(-\infty, 2.5)$ **87.** $\mathbf{R}, x \neq 3$ **89.** 56.36 **91.** 45.02

93. **(a)** Up 3 units **(b)** Left 3 units

95. **(a)** Reflected in the y-axis **(b)** Reflected in the x-axis

97. **(a)** B **(b)** D **99.** **(a)** 36.02 decibels **(b)** 600

101. **(a)** 2.3 **(b)** 10^{-7}

103. **(a)** National: 2.52; American: 2.89
(b) The American League average will always be higher than the National League average, and the difference in the averages will increase.

105. Both functions

107. Increasing 818 to 868 would shift the graph upward to reflect the $50,000 increase.

109. Because $\ln \frac{1}{2} < 0$, reverse the inequality symbol to obtain $a(\ln \frac{1}{2}) < b(\ln \frac{1}{2})$.

111. 2.5 **113.** $(-\infty, -1) \cup (1, \infty)$ **115.** $(0, 1)$

Section 10.5 *(page 672)*

1. What exponent on 9 results in 3? The answer is $\frac{1}{2}$.

3. 7 **5.** 0 **7.** 2 **9.** 0.25 **11.** 0.25 **13.** 9 **15.** 0

17. 1 **19.** $\log_7 3 + \log_7 x$ **21.** $\log_7 5 - \log_7 y$

23. $2 \log x$ **25.** $\frac{1}{2} \log_9 n$ **27.** (c) **29.** (a) **31.** (g)

33. (d)

35. Begin with the Power Rule for Logarithms to obtain $n \log (AB)$. In the Order of Operations, exponents have priority over products.

37. $2 \log_9 x - 3 \log_9 y$ **39.** $2 \log (x + 1) - \log (x + 2)$

41. $5 \ln (2x + 5) - 5 \ln (x + 7)$

43. $\frac{1}{2} - 2 \log x$ **45.** $\frac{1}{3} \log_5 x + \frac{2}{3} \log_5 y$

47. $-5 + \frac{1}{2} \log_2 (x + 1)$ **49.** $\frac{5}{3} \log_2 x + \frac{1}{2} \log_2 y$

51. $2 \log (2x + 1)$ **53.** $0.08t + 3 \ln 10$

55. Cannot be expanded **57.** $rt + \ln P$

59. Use the Distributive Property to obtain $(2 + 1) \log 3 = 3 \log 3$. For $2 \log x + \log y, x \neq y$, the Distributive Property does not apply.

61. $\log_4 15$ **63.** $\log_3 2$ **65.** $\log_2 y^3$ **67.** $\log \sqrt{x}$

69. 1 **71.** 3 **73.** $\log_8 \dfrac{3x}{y}$ **75.** $\log_7 \dfrac{x^2}{y}$ **77.** $\log \dfrac{\sqrt{x}}{\sqrt[3]{y^2}}$

79. $\log \dfrac{y^3}{x^2 z^4}$ **81.** $\log \dfrac{(x + 3)^2}{(x + 1)^3}$ **83.** $\log \sqrt{\dfrac{x}{x + 1}}$

85. $\log_3 243x^2$ **87.** $\ln P_0 e^{rt}$ **89.** False **91.** True

93. False **95.** True **97.** True **99.** False

101. **(a)** 32.3% **(b)** $P(t) = \dfrac{10}{\log (5t + 10)}$

103. The graph of N does not contain a point for 1970 because ln 0 is not defined.

105. $L(t) = 158t + 10{,}515$. Beyond 2004, function L projects higher enrollments than does function N.

Section 10.6 *(page 681)*

1. If b can be written as a power of 3, then we can equate exponents instead of writing the equation in logarithmic form.

3. −0.6 **5.** −1 **7.** 1.46 **9.** 11.55 **11.** −0.03

13. 0.61 **15.** −5.86 **17.** 1.39 **19.** No solution

21. 6.45 **23.** ±1 **25.** −0.57, −1.43 **27.** 0, 1

29. No; the base must be the same to equate arguments.

31. 3 **33.** 26 **35.** $\frac{32}{3}$ **37.** 34 **39.** 8, −1 **41.** $-\frac{11}{3}$

43. 2 **45.** ±3 **47.** 8 **49.** 2, 5 **51.** 10

53. No solution **55.** −2.41, 0.41 **57.** 1 **59.** 0.75

61. 33, 1.5

63. If b were permitted to be 1, then $1^3 = 1^8$, but $3 \neq 8$.

65. $10, 10^{-4}$ **67.** 4, 256 **69.** ln 3, ln 2 **71.** $\dfrac{\ln 7}{\ln 5}$

73. $I = 10^R$ **75.** $I = I_0 \sqrt[10]{10^L}$ **77.** $r = \dfrac{1}{t} \ln \left(\dfrac{P}{P_0}\right)$

79. $t = \dfrac{\ln 2}{n \ln \left(1 + \dfrac{r}{n}\right)}$ **81.** 4.95%

83. **(a)** 14 million **(b)** 25.51 million **(c)** 34.43 million

85. 9953 years **87.** $85,000 **89.** 5:53 A.M. **91.** 7.9

93. **(a)** $14.2(1.03)^x = 300$ **(b)** $x \approx 103$ (about 2003)

95. About 1998

97. $18.91(1.68)^x = 0.5[48.02(1.56)^x]$; $x \approx 3.22$ (about 1993)

99. 10^e **100.** e^{10}

101. **(a)** 1, 3 **(b)** ±1, ±3
In (a), negative numbers are not in the domain of $\log_3 x$, but in (b), negative numbers are in the domain of both logarithmic functions.

103. The graph of the left side cannot be produced, which implies that the equation has no solution. The domain of $\ln(3 - 4x)$ is $(-\infty, 0.75)$, and the domain of $\ln(x - 2)$ is $(2, \infty)$. The intersection of these two domains is $\emptyset$.

105. $-0.77, 2, 4$

Chapter Review Exercises *(page 686)*

1. $x^2 + 4$ **3.** 0.17 **5.** $9x^2 + 15x + 4$ **7.** -1.85

9. No; $g(-1) = 0$, but $f(0)$ is not defined.

11. $\dfrac{1}{|2x + 5| + 2}, \left|\dfrac{2}{x + 2} + 5\right|$

13. Use the Vertical Line Test to determine if a relation is a function. Use the Horizontal Line Test to determine if a function is one-to-one.

15. Not one-to-one

17. $g^{-1} = \{(2, 7), (6, -3), (-1, -2), (4, 0)\}$
Domain: $\{2, 6, -1, 4\}$; range: $\{7, -3, -2, 0\}$

19. Inverses **21.** $\sqrt[3]{x + 5}$

23. The exponent x can be any real number, but $b > 0$ and $b \neq 1$.

25. **(a)** $-\dfrac{8}{e^2}$ **(b)** $\frac{27}{64}$ **27.** 1.41 **29.** 5.5 **31.** $\frac{10}{17}$

33. For growth, the function should be increasing: $b > 1$.

35. $\log_8 4 = \frac{2}{3}$ **37.** $25^{-1/2} = \frac{1}{5}$

39. The function $y = 1^x$ is not one-to-one. Thus it does not have an inverse.

41. 3.23 **43.** -2.5 **45.** 181.27 **47.** 49 **49.** 2.5

51. **(a)** $(1, 0)$ **(b)** None **(c)** $(0, \infty)$ **(d)** **R**

53. 41 **55.** 1

57. Because $\log 49 = 2 \log 7$, we estimate $\log 49$ as $2(0.845) = 1.69$.

59. $\frac{3}{4}\log y - \frac{5}{2}\log x - 4\log z$ **61.** $\log\sqrt[4]{\dfrac{(x + 4)^3}{2yz}}$

63. 3

65. In the first equation, both sides can be written as powers of 3. In the second equation, 4 cannot easily be written as a power of 3. The equation can be solved with logarithms.

67. 1.77 **69.** 5 **71.** 5 **73.** 1.19 **75.** 3542

Chapter Test *(page 688)*

1. *[10.1]* $x^2 + x - 2$ **2.** *[10.1]* $x - 2, x \neq -2$

3. *[10.1]* $x^2 + 4x$ **4.** *[10.1]* $\sqrt{x + 2}$

5. *[10.2]* $(f \circ g)(x) = (g \circ f)(x) = x$ **6.** *[10.2]* $x^3 - 2$

7. *[10.2]*

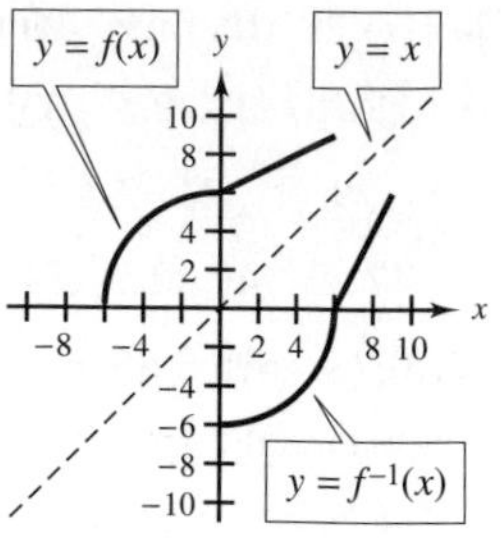

y = f(x)
y = x
y = f^{-1}(x)

8. *[10.4]* 0.5 **9.** *[10.4]* 1.5 **10.** *[10.3]* 0

11. *[10.3]* 0.25 **12.** *[10.6]* 1.12 **13.** *[10.6]* 1.15

14. *[10.4]* Use the Change of Base Formula to obtain
$$\log_5 13 = \frac{\ln 13}{\ln 5}.$$

15. *[10.5]* $2\log A - \frac{1}{2}\log B$ **16.** *[10.5]* $\log(x + 1)$

17. *[10.5]* $\log_6\left[\dfrac{x^2(x - 1)}{6}\right]$ **18.** *[10.6]* 2 **19.** *[10.6]* 2

20. *[10.6]* 1 **21.** *[10.6]* $e^{-2} \approx 0.14$ **22.** *[10.6]* 8.48%

23. *[10.5]* **(a)** 7000
(b) As time t increases, demand D decreases. The ad campaign is ineffective.
(c) 10.3 weeks

24. *[10.6]* 15.85 times as intense **25.** *[10.6]* 1.5 hours

Cumulative Test: Chapters 8–10 *(page 689)*

1. *[8.1]* **(a)** y^4 **(b)** $|y^5|$

2. *[8.2]* **(a)** 9 **(b)** $\frac{1}{9}$ **(c)** Not a real number **(d)** $-\frac{1}{8}$

3. *[8.3]* **(a)** $\dfrac{1}{a^{5/2}b^{2/3}}$ **(b)** $\dfrac{x^{1/2}}{y}$ **(c)** x^3y^6

4. *[8.4]* **(a)** $\dfrac{x^2}{2y}$ **(b)** $\sqrt[5]{6x^3y}$ **(c)** $\dfrac{x}{y}$

5. *[8.5]* **(a)** $2x^4\sqrt{5x}$ **(b)** $y\sqrt{y}$

6. *[8.6]* **(a)** $x\sqrt{3x}$ **(b)** $a + 2\sqrt{2a} + 2$

7. *[8.6]* 4 **8.** *[8.7]* **(a)** $\dfrac{5\sqrt[3]{4}}{2}$ **(b)** $\dfrac{x\sqrt{2} - \sqrt{xy}}{2x - y}$

9. *[8.8]* **(a)** 1 **(b)** $-12, 15$

10. *[8.9]* **(a)** $-i$ **(b)** $8 - 6i$ **(c)** $\frac{1 + 2i}{5}$

11. *[9.3]* Graphing can be used to estimate solutions. Algebraic methods include factoring, using the Square Root Property, and using the Quadratic Formula. The Quadratic Formula always works.

12. *[9.3]* $\dfrac{-9 \pm \sqrt{141}}{10} \approx -2.09, 0.29$ **13.** *[9.5]* $-1, 7$

14. [*9.3*] If the equation has complex number solutions, then the discriminant is negative. Thus $b^2 - 4ac < 0$, and so $b^2 < 4ac$.

15. [*9.4*] 9 **16.** [*9.5*] 8 mph

17. [*9.6*] **(a)** $(-\infty, -6) \cup (4, \infty)$ **(b)** $\varnothing$ **(c)** $\mathbf{R}$

18. [*9.7*] **(a)** $(-\infty, 2.77]$ **(b)** $(-\infty, 0) \cup (3, \infty)$

19. [*9.8*] **(a)** $c = 3$ **(b)** $a < 0$ **(c)** $a = -3$

20. [*9.8*] **(a)** 4 **(b)** 16

21. [*10.1*] **(a)** 0 **(b)** 0 **(c)** -16

22. [*10.1*] **(a)** 1 **(b)** Not defined

23. [*10.2*] $f^{-1}(x) = \dfrac{3x}{x - 1}$

24. [*10.3*] Both have a y-intercept of $(0, 1)$ and no x-intercept. The graph of f rises and the graph of g falls from left to right.

25. [*10.4*] **(a)** 5 **(b)** 10 **(c)** $\sqrt{7}$

26. [*10.5*] **(a)** $2 \ln x + \ln y - \frac{1}{2} \ln z$ **(b)** $\log \dfrac{(x - 1)^3}{x + 1}$

27. [*10.6*] **(a)** 9, -3 **(b)** 0.46 **(c)** -0.69

28. [*10.6*] 6%

CHAPTER 11

Section 11.1 *(page 698)*

1. The orientation of the parabola is not known. For a vertical parabola, the equation of the axis of symmetry would be $x = 3$. For a horizontal parabola, the equation would be $y = 5$.

3. **(a)** Function; down **(b)** Not a function; right **(c)** Not a function; left **(d)** Function; up

5. Vertical; $V(-2, -16)$; $x = -2$

7. Vertical; $V(3, 0)$; $x = 3$ **9.** Horizontal; $V(6, 0)$; $y = 0$

11. Horizontal; $V(-16, -4)$; $y = -4$

13. Horizontal; $V(2, 8)$; $y = 8$

15. The x-intercept of the graph is $(c, 0)$.

17. Axis: $y = 0$ **19.** Axis: $x = -1$

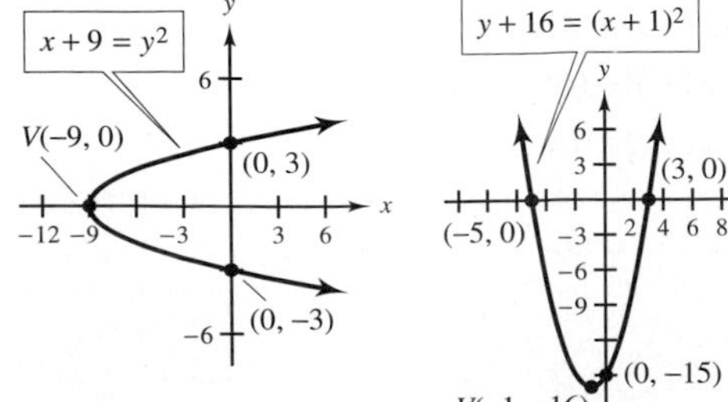

21. Axis: $x = 5$

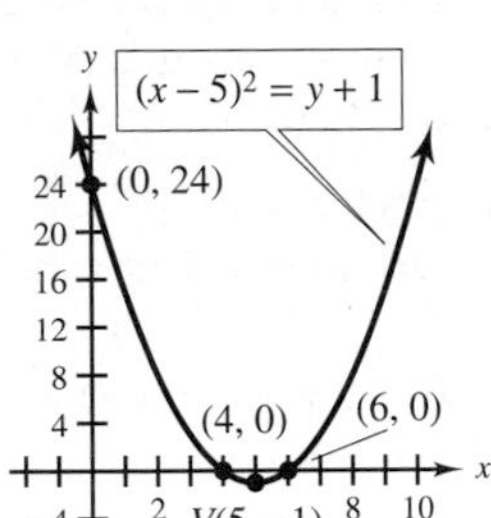

23. Axis: $x = -2$

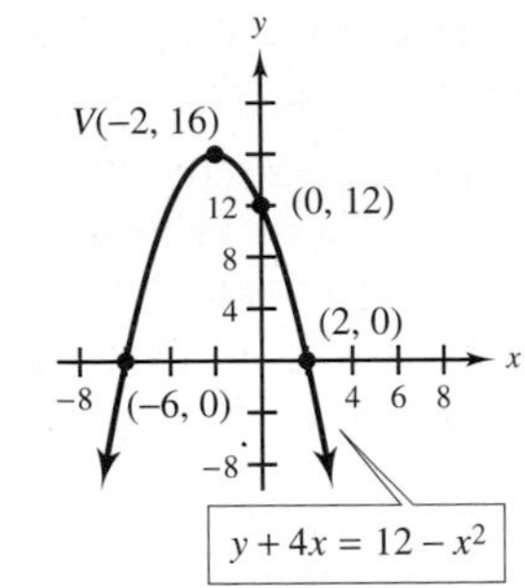

25. Axis: $y = 3$

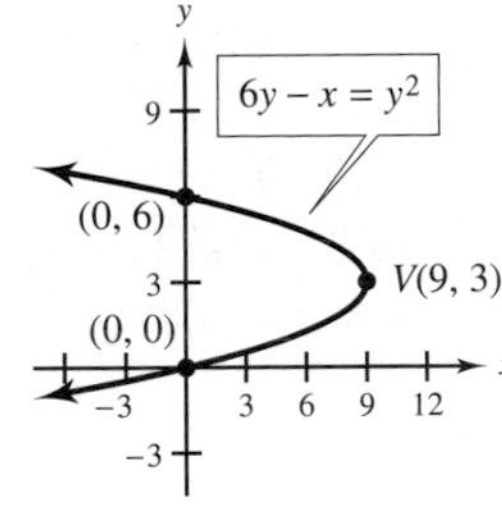

27. Axis: $x = 5$

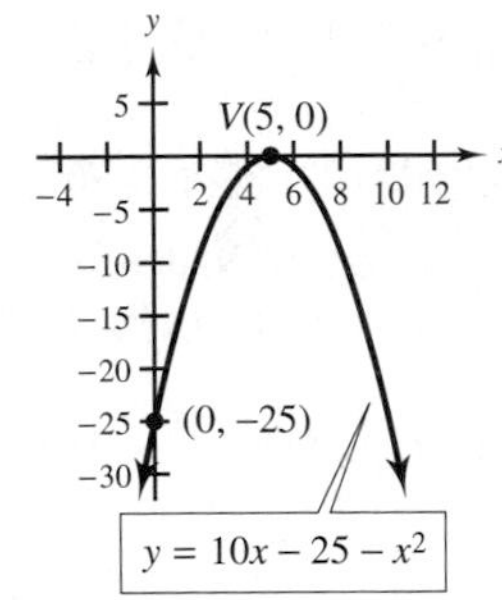

29. Axis: $y = -0.5$

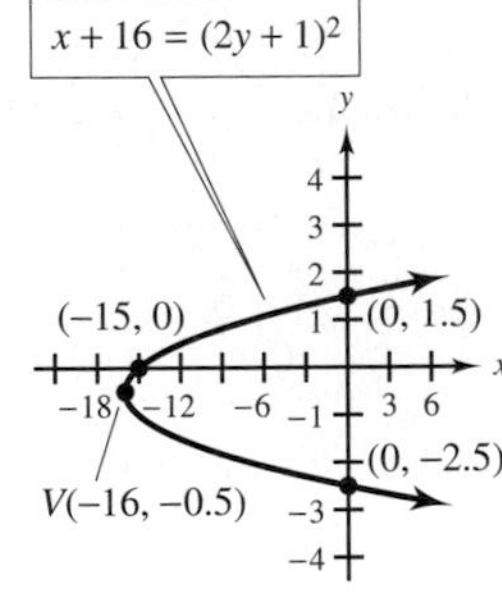

31. Axis: $y = -1.5$

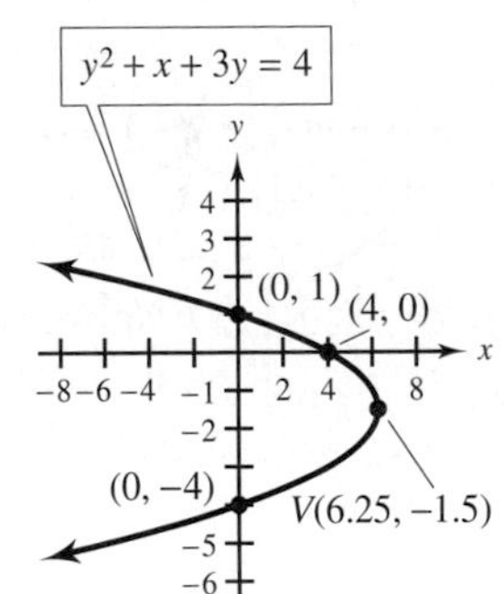

33. Domain: $\left[-\frac{1}{4}, \infty\right)$; range: $\mathbf{R}$

35. Domain: $\mathbf{R}$; range: $(-\infty, 4]$

37. Solve the equation for y, enter the two resulting functions, and graph both of them.

39. x-intercepts: 2; y-intercepts: 1

41. x-intercepts: 1; y-intercepts: 2

43. x-intercepts: none; y-intercepts: 1

45. Opens up **47.** Opens left **49.** Opens down

51. Opens right **53.** $y = 1$ **55.** $x = \frac{7}{2}$

57. $y = -\frac{5}{8}(x^2 - 2x - 8)$ **59.** $x = -\frac{4}{3}(y^2 - 2y - 3)$

61. $y = -\frac{8}{15}(x^2 + 2x - 15)$ **63.** $x = -\frac{1}{2}(y^2 - y - 12)$

65. $x = \frac{1}{4}(y - 5)^2 + 2$ **67.** $y = 2(x - 2)^2 - 3$

69. The graph models a rapid initial learning rate, but with a decreasing rate over time. The y-values must be nonnegative.

71. **(a)** $y = -0.5x^2 + x + 7.5$ **(b)** 8

73. The y-intercept (0, 171) represents the NEA's appropriation for 1990. The model value and the actual data are the same.

75. The x-intercept (13, 0) represents the year 2003 when the model projects no funding for the NEA.

77. Because $m_{\overline{AB}} = m_{\overline{BC}}$, the three points are points of a line.

79. **(a)** 16 feet **(b)** $t = 1 \pm \frac{1}{4}\sqrt{16 - h}$

(c) Time t is defined only if $16 - h \geq 0$, that is, if $h \leq 16$.

81. **(a)** $(0, d)$ **(b)** $y = -d$ **(c)** $(x, -d)$

(d) $\sqrt{x^2 + (y - d)^2}$ **(e)** $y + d$

Section 11.2 *(page 706)*

1. The graphs of both equations are circles with radius 5. The center of the first circle is (0, 0); the center of the second circle is $(-2, 3)$.

3. $C(0, 0)$; $r = 4$ **5.** $C(0, 0)$; $r = \frac{1}{3}$ **7.** $x^2 + y^2 = 16$

9. $x^2 + y^2 = 15$

11.

(−5, 0) (0, 5) C(0, 0) (5, 0) (0, −5)

$x^2 + y^2 = 25$

13.

(−1.25, 0) (0, 1.25) (1.25, 0) C(0, 0) (0, −1.25)

$16x^2 + 16y^2 = 25$

15.

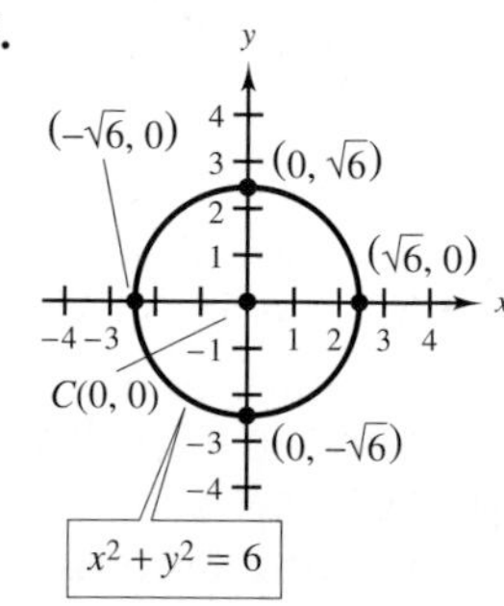

17. $x^2 + y^2 = 25$ **19.** $x^2 + y^2 = 37$

21. $C(0, -2)$; $r = \sqrt{3} \approx 1.73$ **23.** $C(-3, 2)$; $r = 5$

25. $C(2, -3)$; $r = \dfrac{\sqrt{11}}{2} \approx 1.66$ **27.** $(x - 4)^2 + (y - 2)^2 = 9$

29. $x^2 + (y - 4)^2 = 4$ **31.** $(x - 5)^2 + (y - 6)^2 = 11$

33. $(x - 2)^2 + (y + 4)^2 = 9$ **35.** $(x - 5)^2 + (y - 7)^2 = 26$

37.

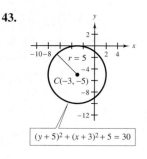

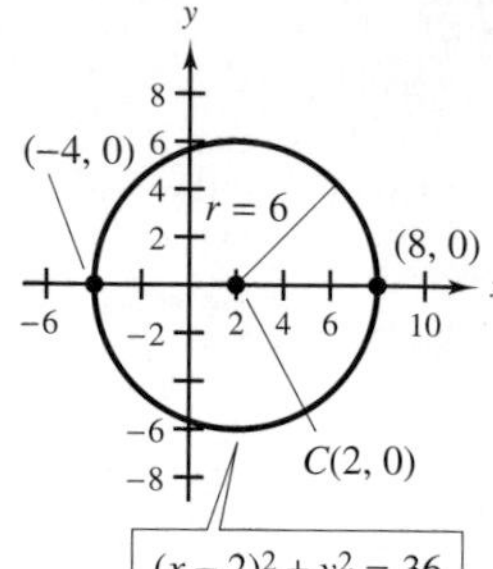

39.

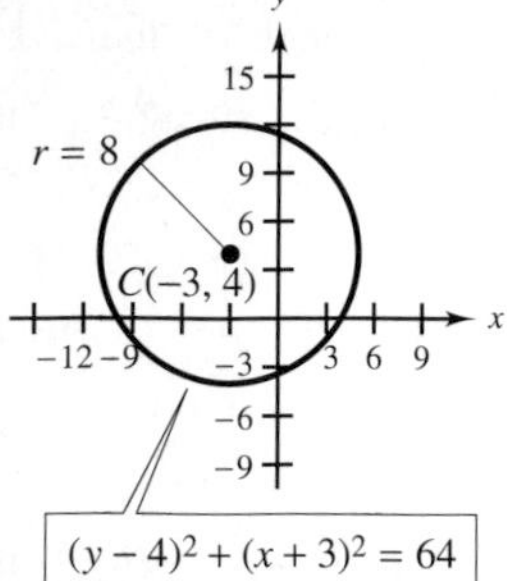

41.

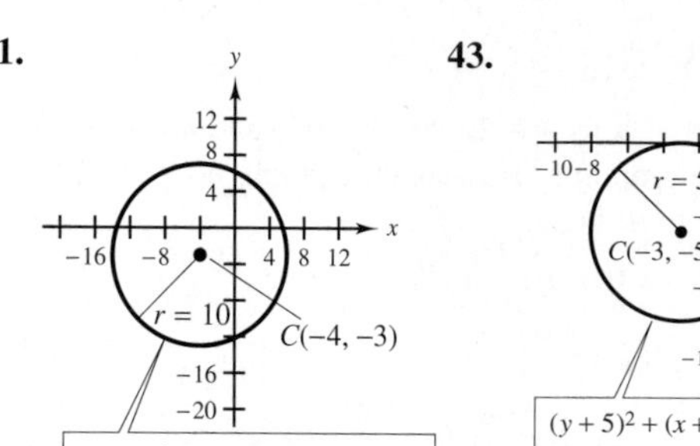

43.

r = 5 C(−3, −5)

$(y + 5)^2 + (x + 3)^2 + 5 = 30$

45.

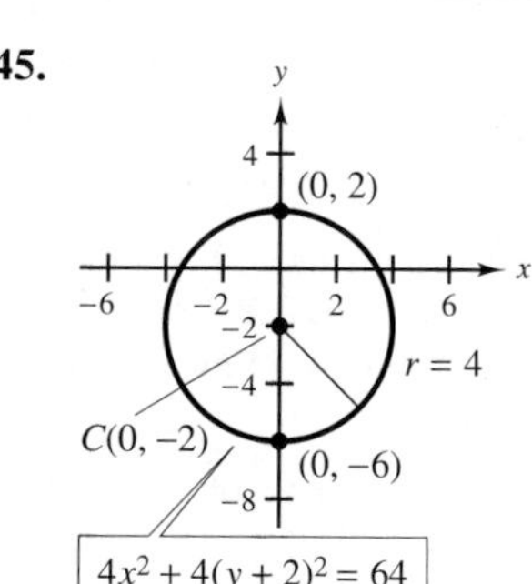

47. $C(0, -4)$; $r = \sqrt{13} \approx 3.61$

49. $C(3, -2)$; $r = 5$ **51.** $C(-4, 2)$; $r = 6$

53. $C(-1, 2)$; $r = \sqrt{3} \approx 1.73$ **55.** $C(3.5, -2.5)$; $r = 4$

57. A tangent line to a circle is perpendicular to the radius that is drawn to the point of tangency.

59. $(x + 2)^2 + (y - 5)^2 = 25$ **61.** $(x - 2)^2 + (y - 1)^2 = 16$

63. $(x - 2)^2 + (y + 4)^2 = 61$

65. Domain: $[-10, 10]$; range: $[-10, 10]$

67. Domain: $[1, 9]$; range: $[-6, 2]$

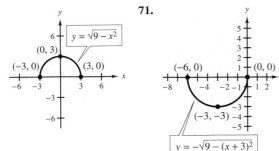

69. **71.**

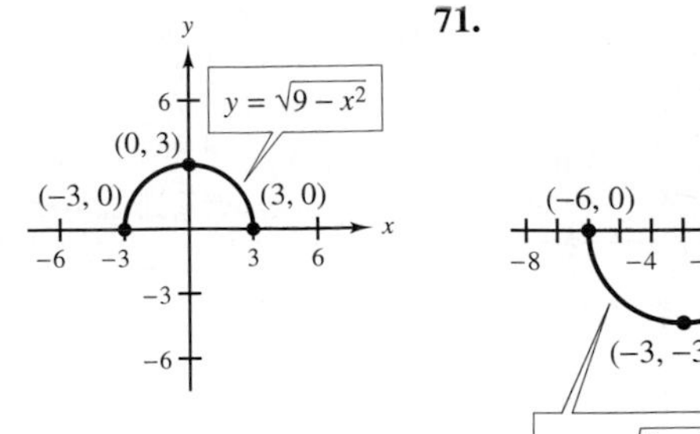

73. Circle **75.** Line **77.** Parabola **79.** A sphere

81. 16π square yards

83. **(a)** $r_1 + r_2 < 12$ **(b)** $r_1 + r_2 = 12$ **(c)** $r_1 + r_2 > 12$

85. $x - 3y = -10$ **87.** $x = -\sqrt{8 - y^2}$

89. **(a)** Diameter **(b)** $\sqrt{91} \approx 9.54$

Section 11.3 *(page 713)*

1. The resulting figure would be a circle.

3. $(\pm 5, 0), (0, \pm 4)$ **5.** $(\pm 3, 0), (0, \pm 1)$

7. $\left(\pm\frac{1}{3}, 0\right), \left(0, \pm\frac{1}{2}\right)$

9.

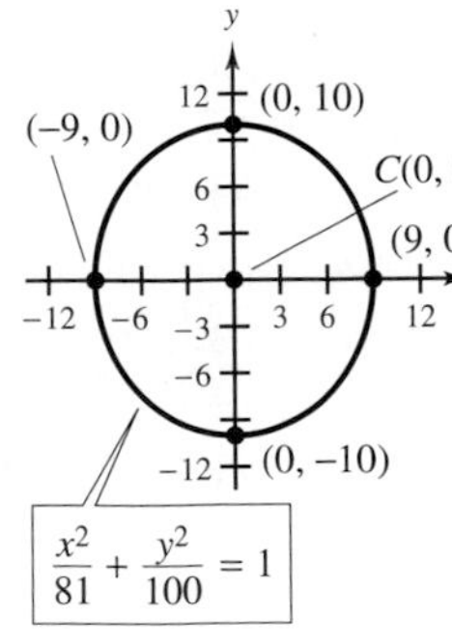

11.

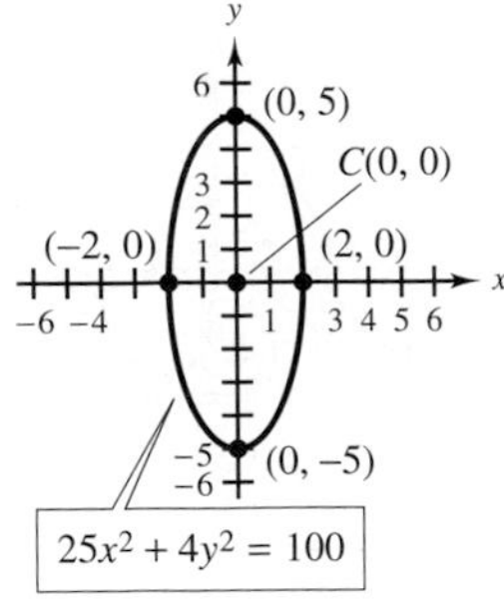

13.

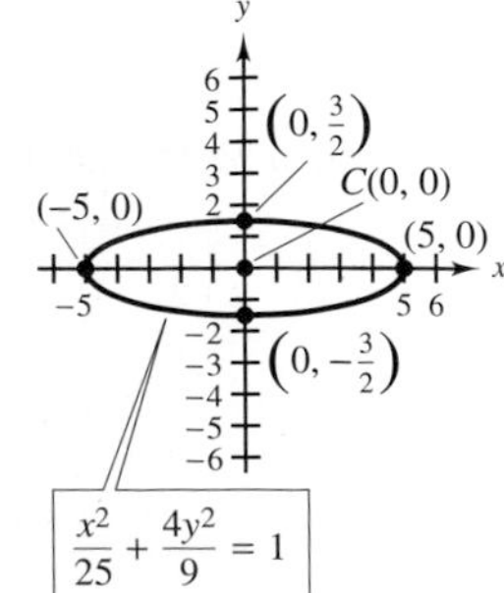

15. $\frac{x^2}{9} + \frac{y^2}{25} = 1$ **17.** $\frac{x^2}{81} + \frac{y^2}{16} = 1$ **19.** $\frac{x^2}{16} + \frac{y^2}{36} = 1$

21. $C(3, 4)$; $(-1, 4), (7, 4), (3, 9), (3, -1)$

23. $C(5, -1)$; $(9, -1), (1, -1), (5, 1), (5, -3)$

25. $C(1, -3)$; $(5, -3), (-3, -3), (1, 2), (1, -8)$

27. **29.**

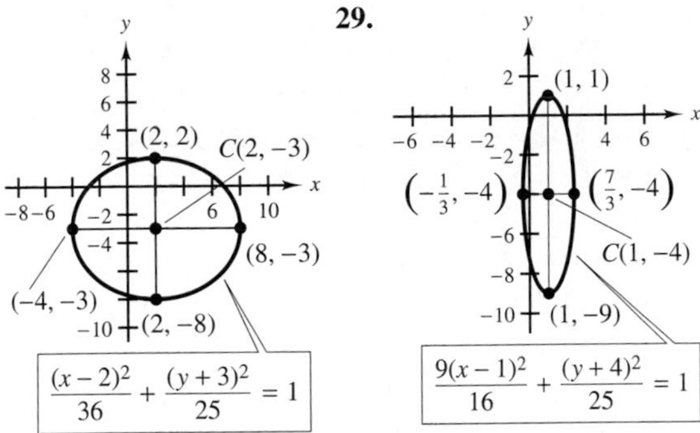

31.

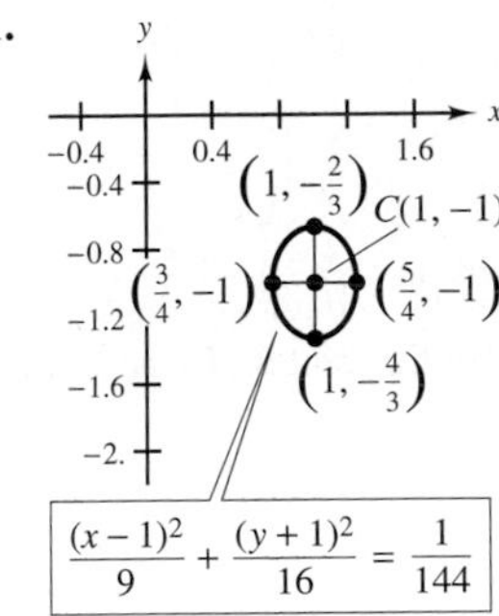

33. $\frac{(x-4)^2}{9} + \frac{(y-5)^2}{16} = 1$ **35.** $\frac{(x+2)^2}{64} + \frac{(y+3)^2}{9} = 1$

37. $\frac{(x-3)^2}{4} + \frac{(y-4)^2}{25} = 1$ **39.** $\frac{(x+3)^2}{25} + \frac{(y+2)^2}{4} = 1$

41. If the endpoints of the horizontal axis are $(\pm a, 0)$ and the endpoints of the vertical axis are $(0, \pm b)$, then the domain of the relation is $[-a, a]$ and the range is $[-b, b]$.

43. Domain: $[-2, 2]$; range: $[-3, 3]$

45. Domain: $\left[-\frac{1}{2}, \frac{1}{2}\right]$; range: $\left[-\frac{1}{3}, \frac{1}{3}\right]$

47. Domain: $[-5, -1]$; range: $[-11, 1]$

49. Parabola **51.** Line **53.** Circle **55.** Ellipse

57. **(a)** The curve is an ellipse because the sum of the distances from the pencil to the two tacks (foci) is constant.

(b) If the tacks are moved farther apart horizontally, the ellipse becomes flatter and longer.

(c) As the tacks are moved closer together, the ellipse becomes closer to a circle.

(d) The resulting curve is a circle with the one tack as the center.

59. Full ellipse: $\frac{x^2}{625} + \frac{y^2}{400} = 1$; semiellipse: $y = 0.8\sqrt{625 - x^2}$

61. $\dfrac{(x+1)^2}{4} + \dfrac{(y-1)^2}{16} = 1$

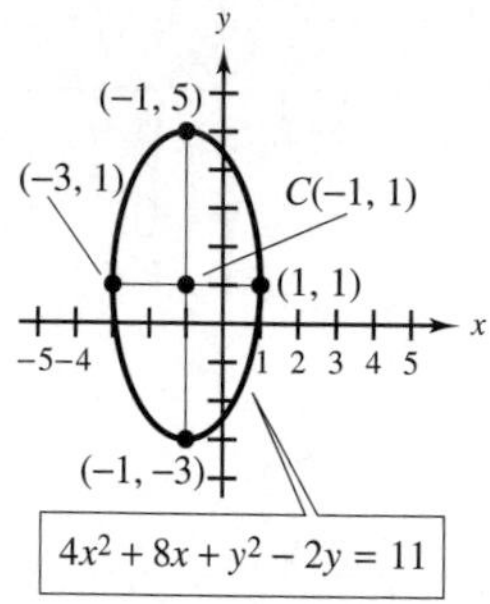

63. $y = \sqrt{9 - 16x^2}$ **65.** 5.20

Section 11.4 *(page721)*

1. The equation of the hyperbola is $\dfrac{x^2}{a^2} - \dfrac{y^2}{b^2} = 1$. The midpoints of the sides of the central rectangle are $(-a, 0)$, $(a, 0)$, $(0, -b)$, and $(0, b)$.

3.

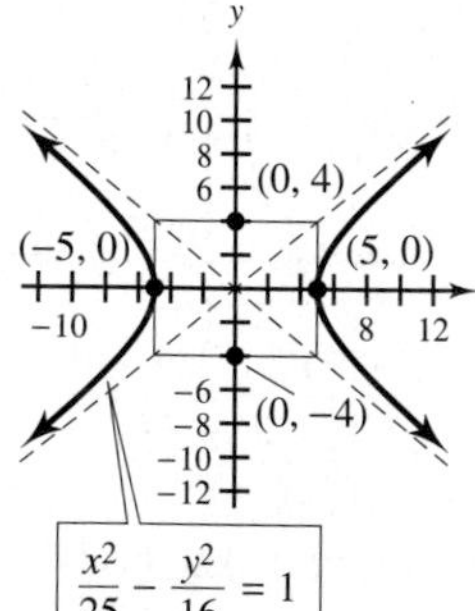

5.

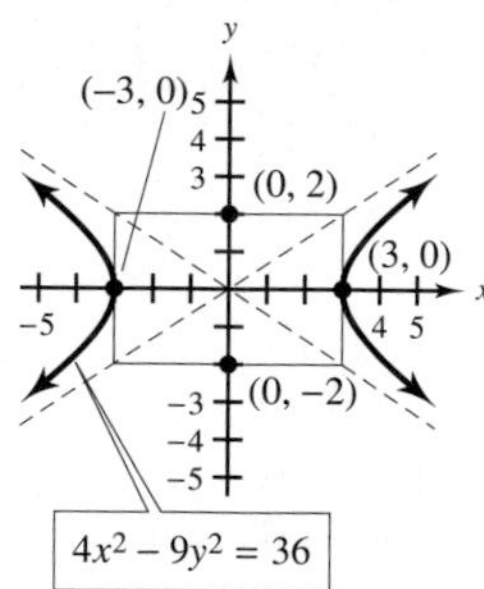

7.

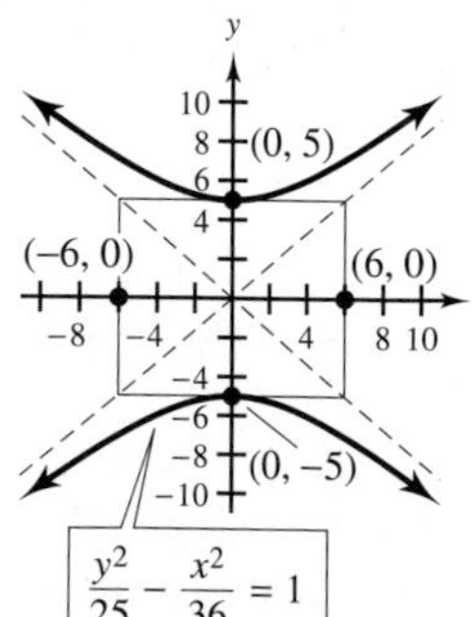

9.

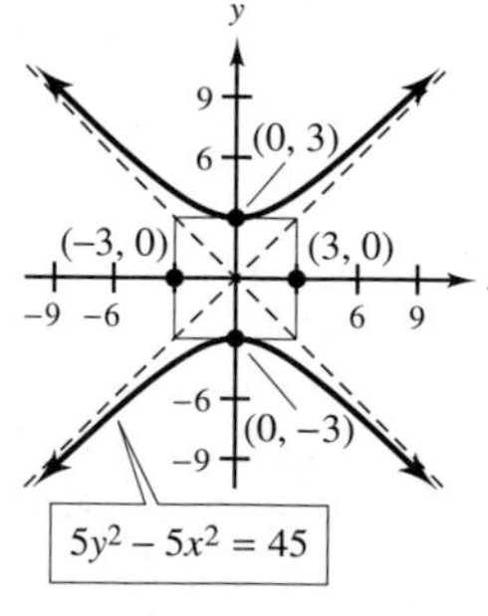

11. If the minus sign precedes the y^2 term of the equation, the transverse axis is horizontal. If the minus sign precedes the x^2 term of the equation, the transverse axis is vertical.

13. Horizontal **15.** Vertical **17.** Vertical **19.** Horizontal

21.

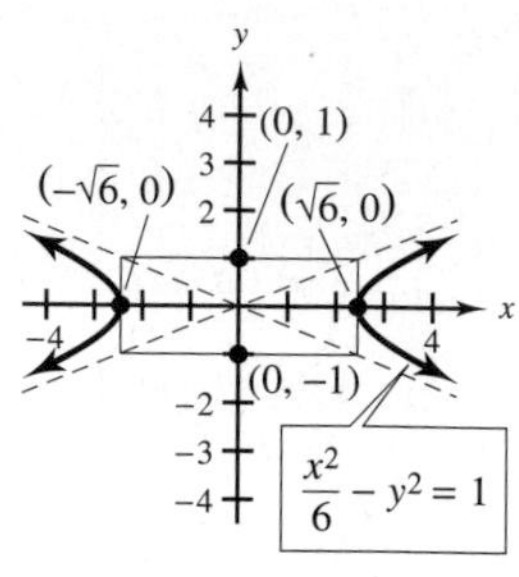

23.

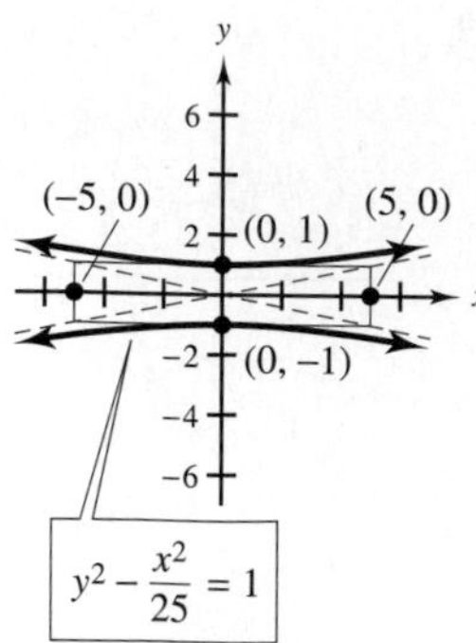

25.

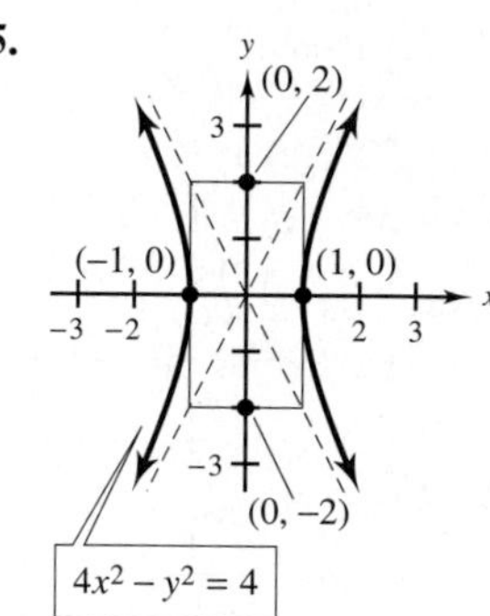

27.

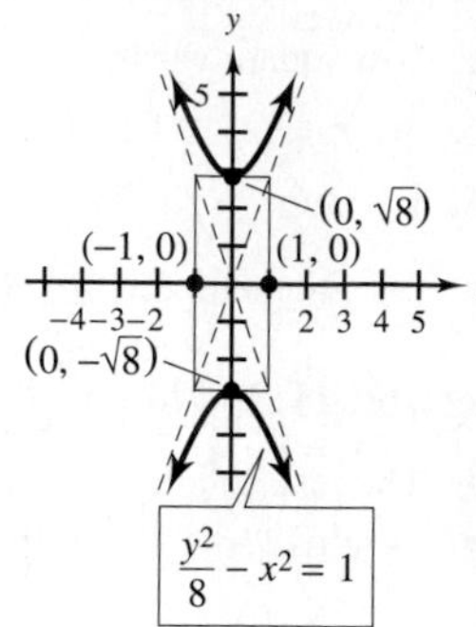

29.

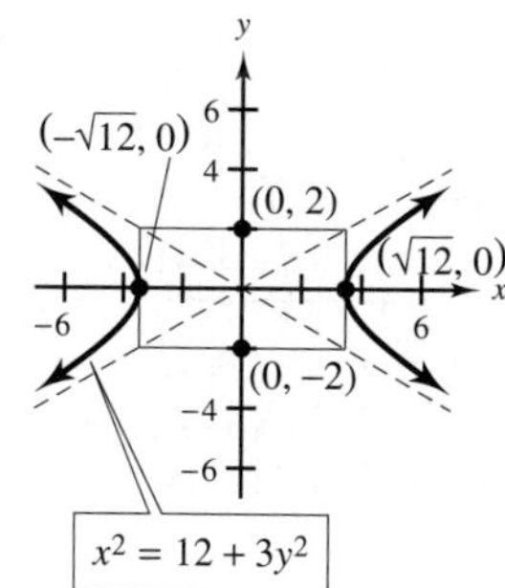

31.

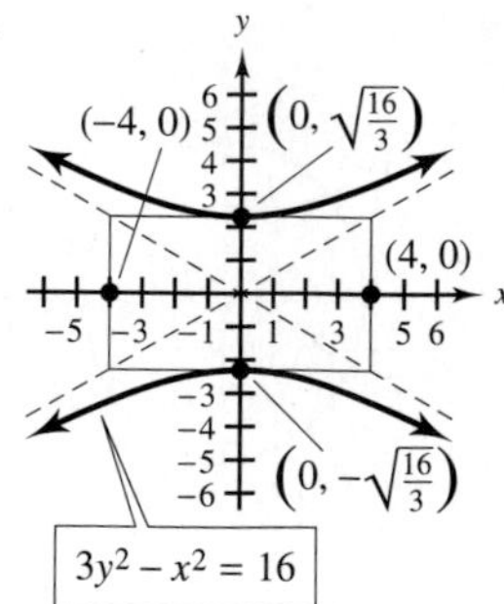

33.

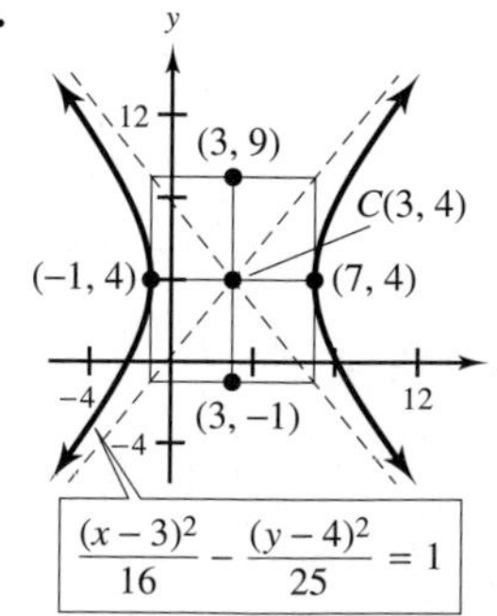

35.

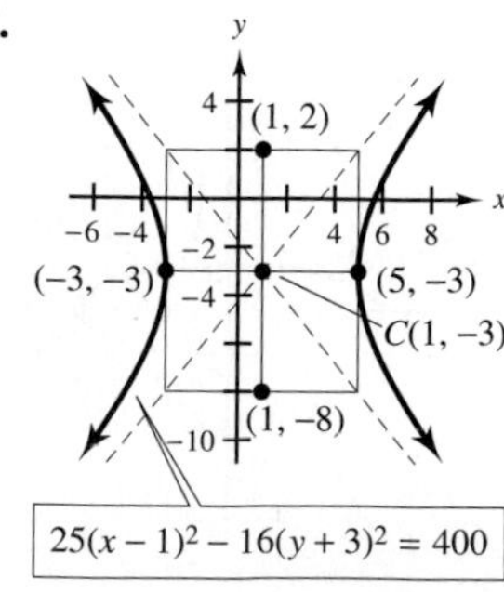

37.

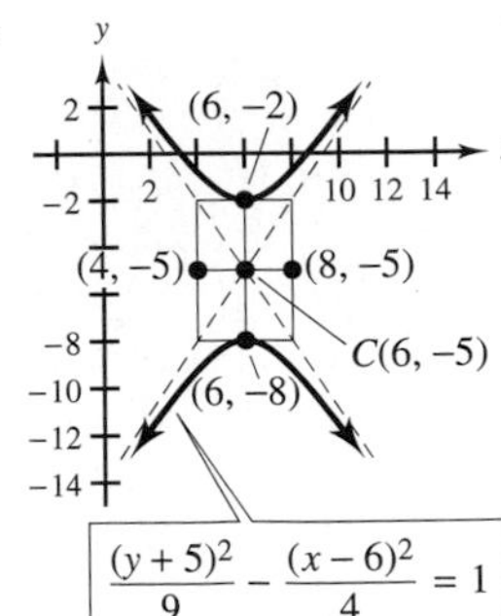

39.

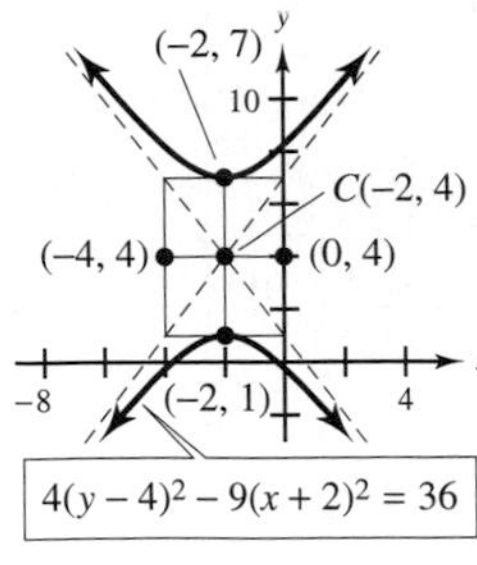

41. Vertical **43.** Horizontal **45.** Horizontal **47.** Vertical

49.

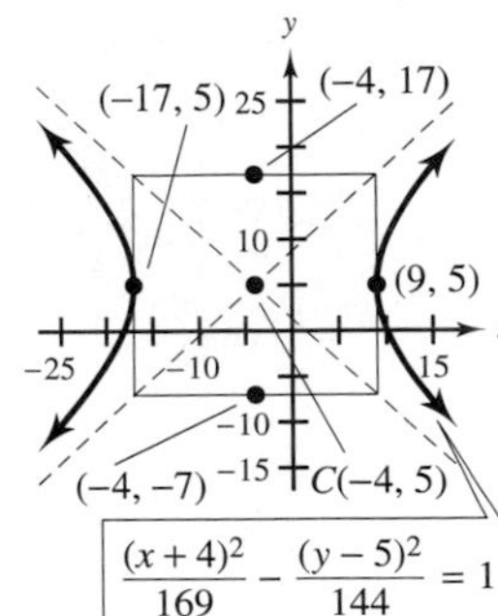

51.

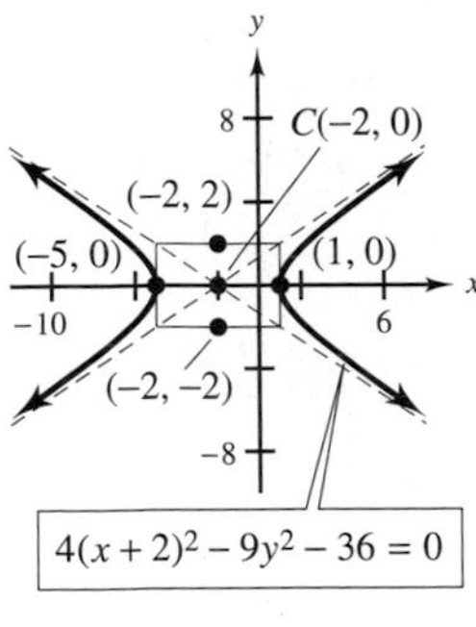

53.

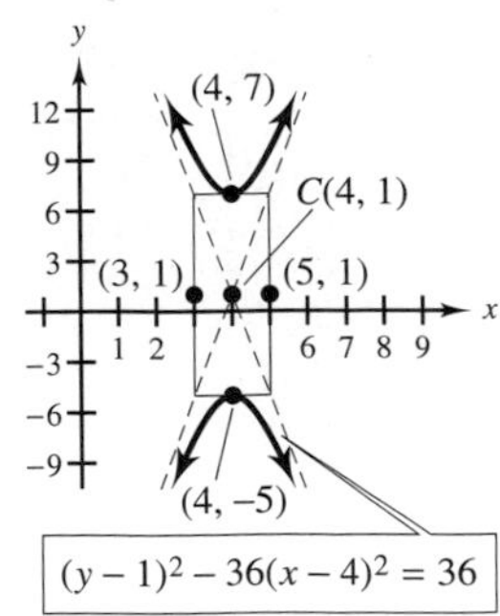

55.

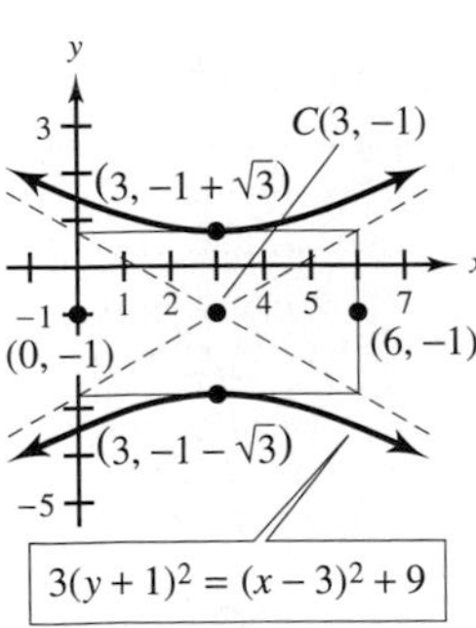

57.

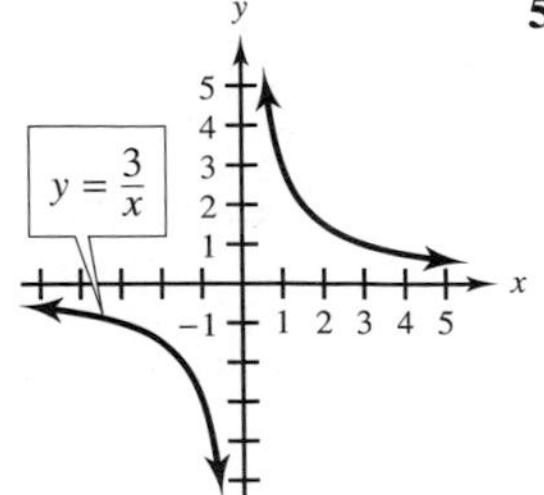

59.

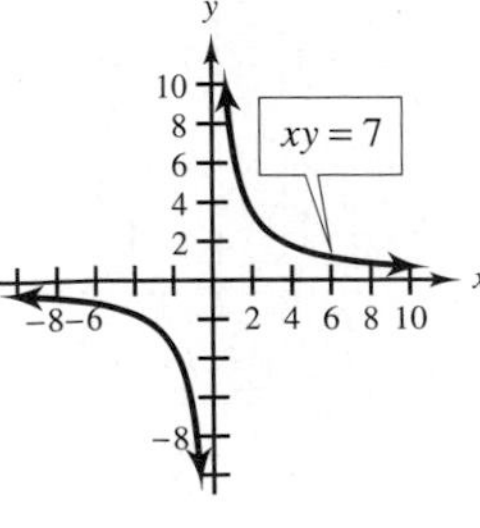

61. **63.**

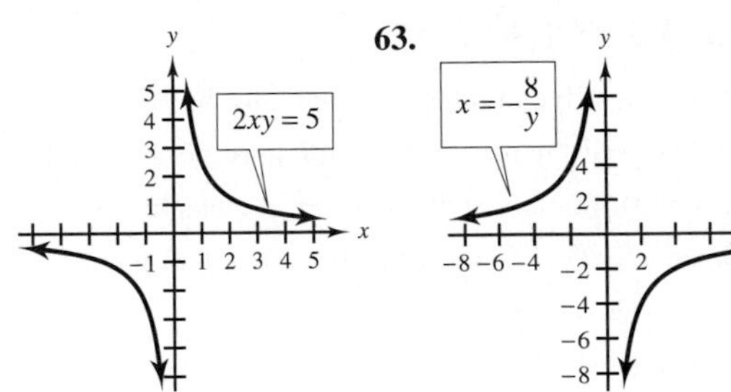

65. Line **67.** Ellipse **69.** Circle **71.** Parabola

73. Hyperbola

75. The equation $\frac{x^2}{a^2} + \frac{y^2}{b^2} = 1$ is the model for an ellipse whose center is the origin. The equation $\frac{x^2}{a^2} - \frac{y^2}{b^2} = 1$ is the model for a hyperbola whose center is the origin.

77. 18 feet

79. **(a)** **(b)**

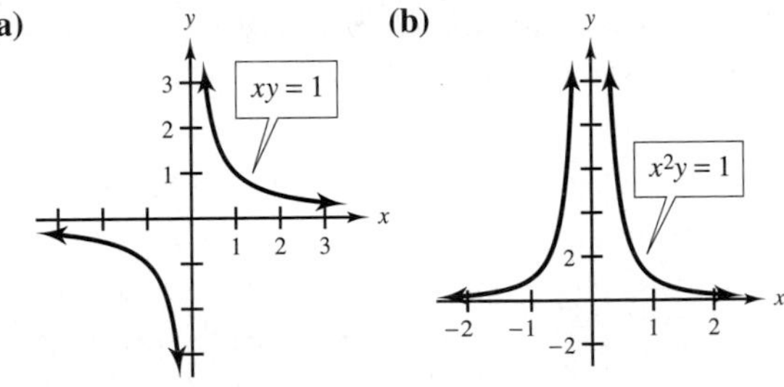

(c)

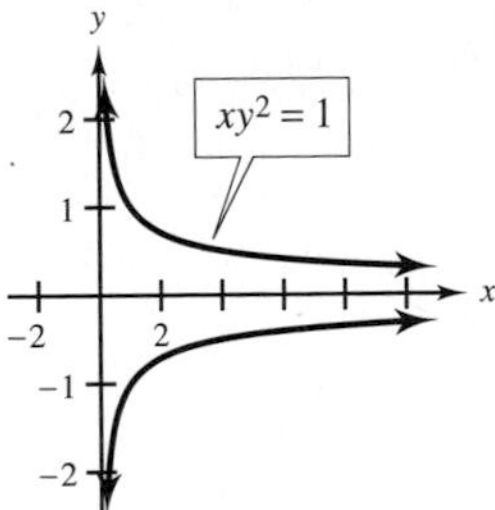

(d)

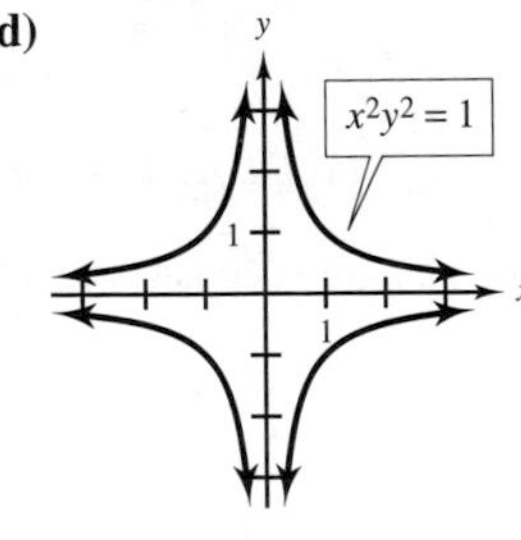

81.

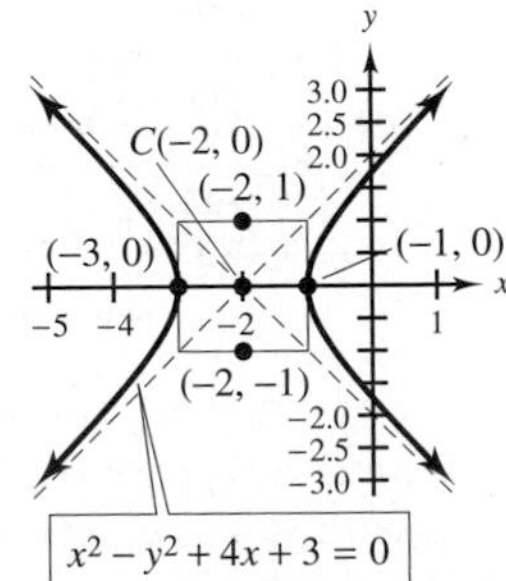

Section 11.5 *(page 729)*

1. (a) The positive numbers a and b must be equal.
   (b) The positive numbers a and b must not be equal.
   (c) The numbers a and b must have opposite signs.

3. Ellipse **5.** Parabola **7.** Hyperbola **9.** Line

11. Circle **13.** Hyperbola **15.** Parabola

27. The circle and ellipse can intersect in 0, 2, or 4 points. In order to intersect in 3 points, the centers of the two figures must be different.

29. $(0, 6), (-6, 0)$ **31.** No real number solutions

33. $(-6, -1), (2, 3)$ **35.** No real number solutions

37. The substitution method is a convenient method for solving a system of equations if one of the equations can be easily solved for one of the variables.

39. $(3, -2), (-6, 1)$ **41.** $(-5, -8), (-2, 1)$

43. $(4, 0), (2, 2)$ **45.** $(3, -4)$ **47.** $(1.41, 2), (-1.41, 2)$

49. $(-1.33, -2.22)$ **51.** $(3, 0), (-3, 0)$

53. $(1, 0), (-1, 0), (0, -2)$

55. $(3.32, 2), (-3.32, 2), (2.83, -1), (-2.83, -1)$

57. $(3.19, -3.39), (-2.19, 7.39)$ **59.** $(i, -3i), (-i, 3i)$

61. $(0.95, 1.11), (-0.95, 1.11), (1.83, -1.36), (-1.83, -1.36)$

63. 58 **65.** 20 yards and 36 yards **67.** 65 mph, 4 hours

69. 9 chairs, \$208; 101 chairs, \$580 **71.** $(12.66, 35.4)$

73. According to the model, the same percentage (35.4%) of men and women spend 12.66 minutes dressing for work.

75. $y = -4x - 4$ **77.** Answers will vary.

Section 11.6 *(page 735)*

1. The graph consists of all points in the interior of the ellipse whose equation is $\frac{x^2}{a^2} + \frac{y^2}{b^2} = 1$.

3.

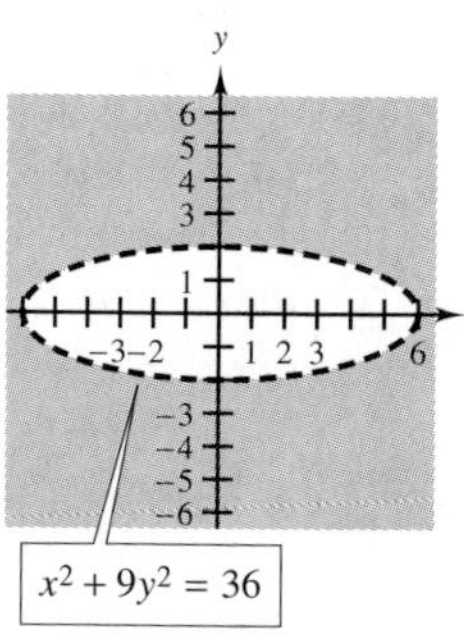

5.

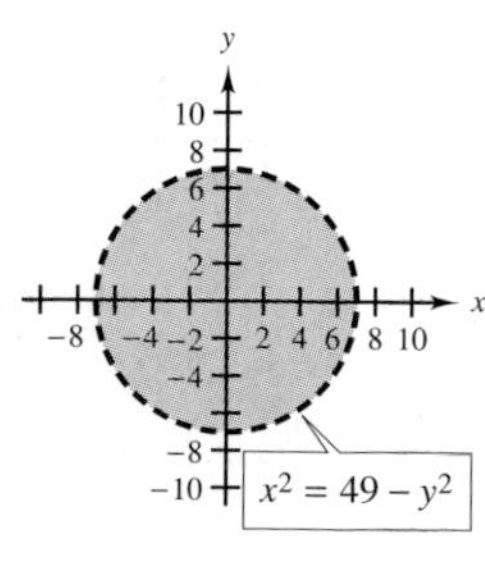

7.

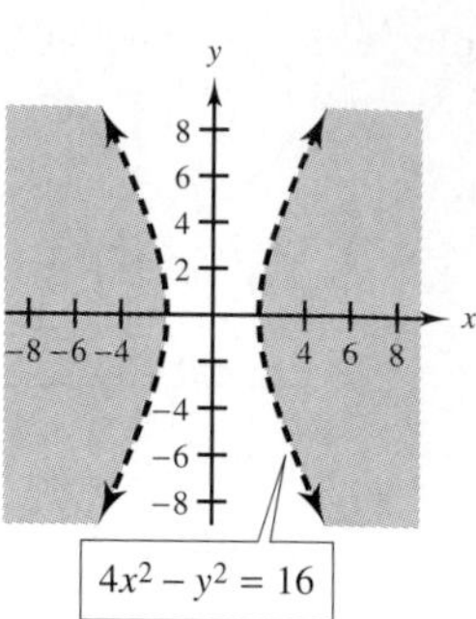

9.

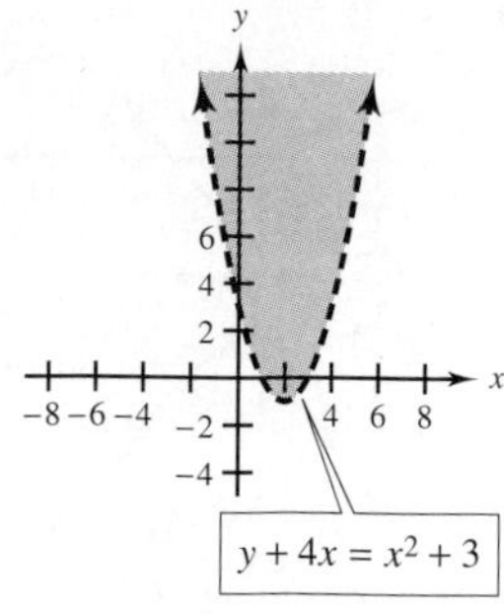

11.

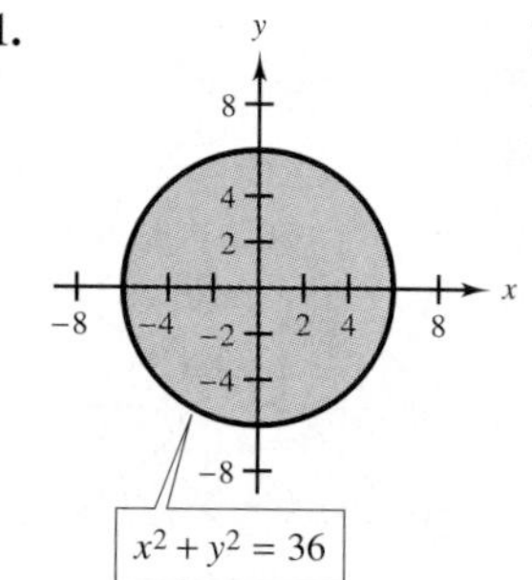

13.

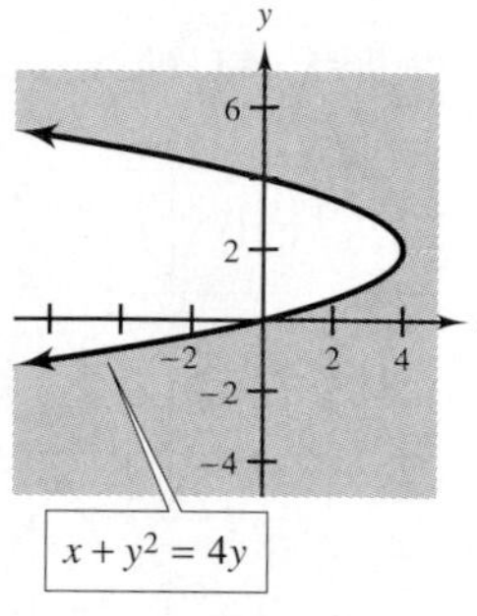

15.

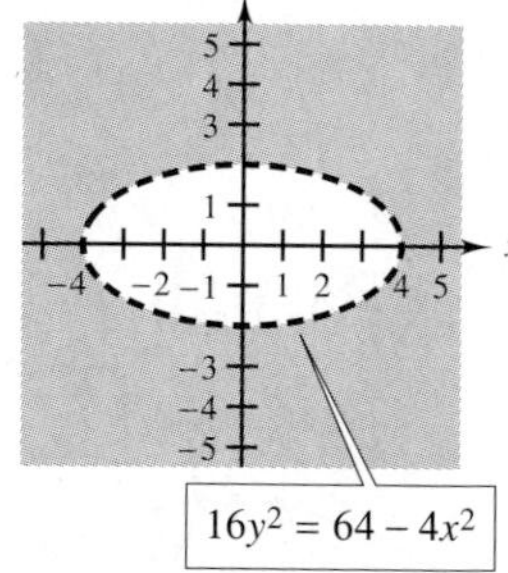

17.

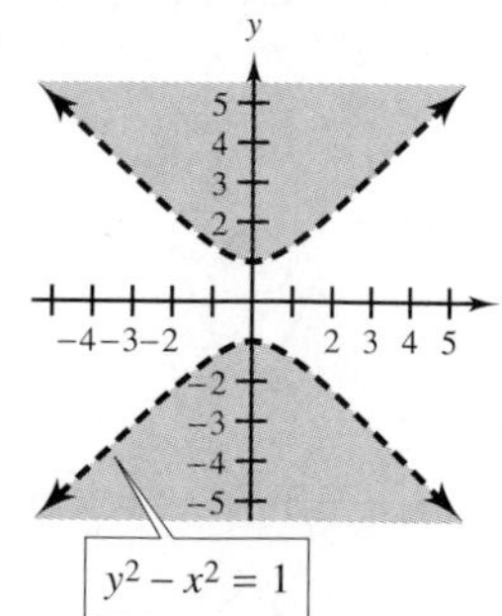

19.

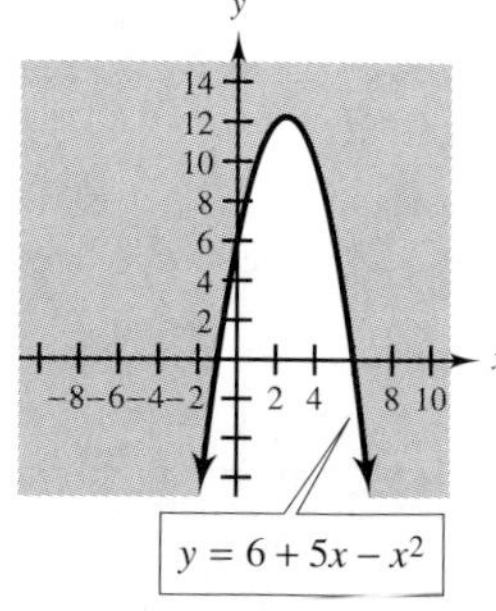

21.

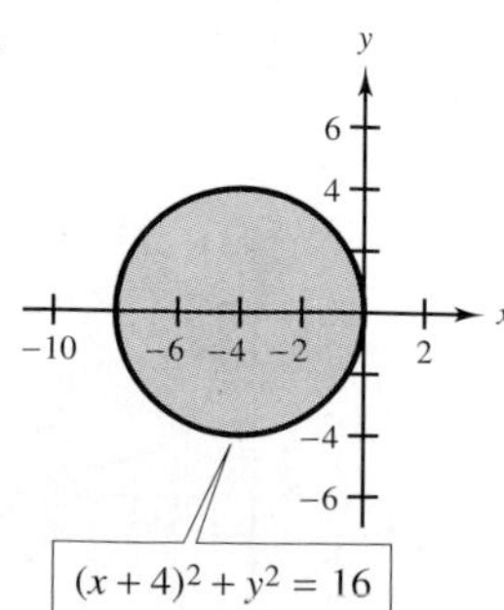

23.

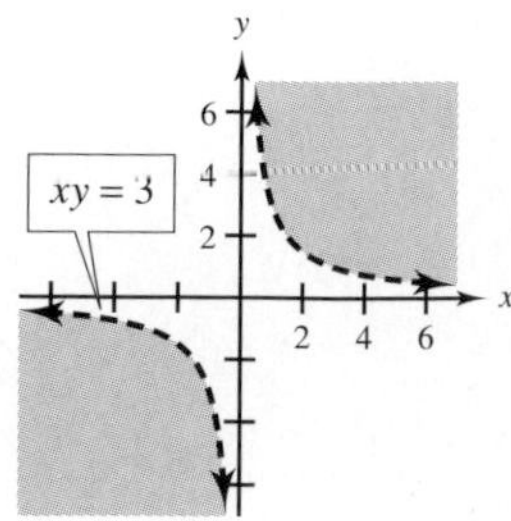

25. The interior of a circle whose radius is 3 does not intersect with the half-plane above the line $y = 4$.

27.

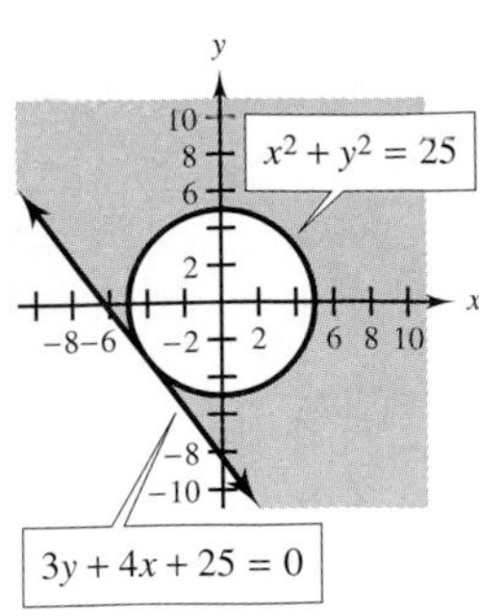

29.

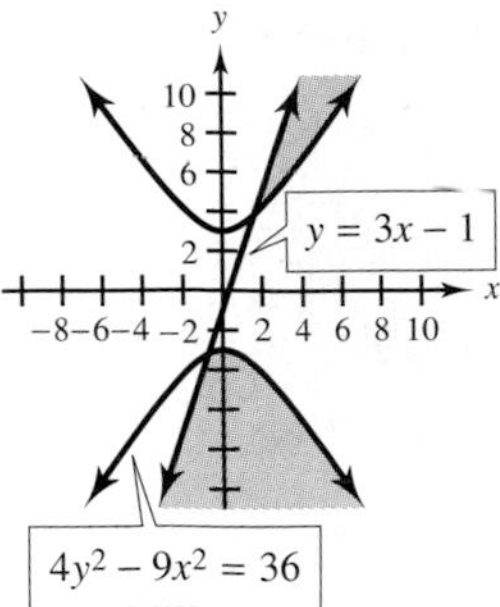

31.

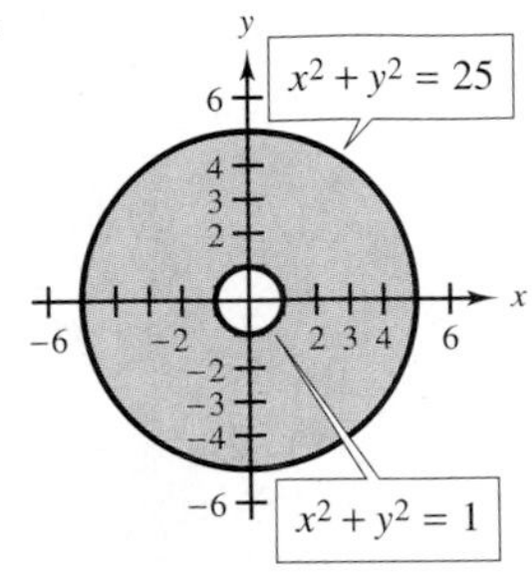

33.

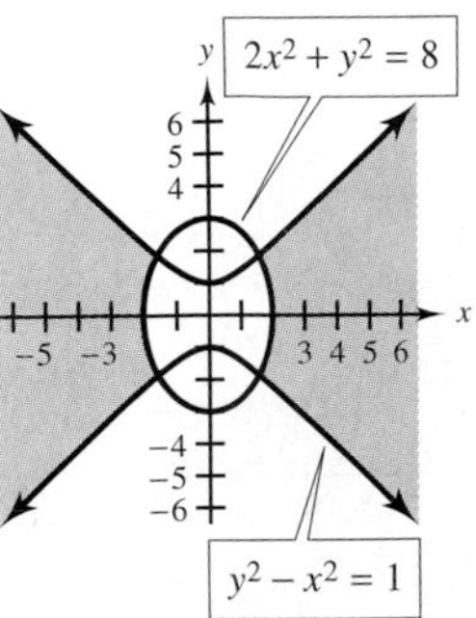

35.

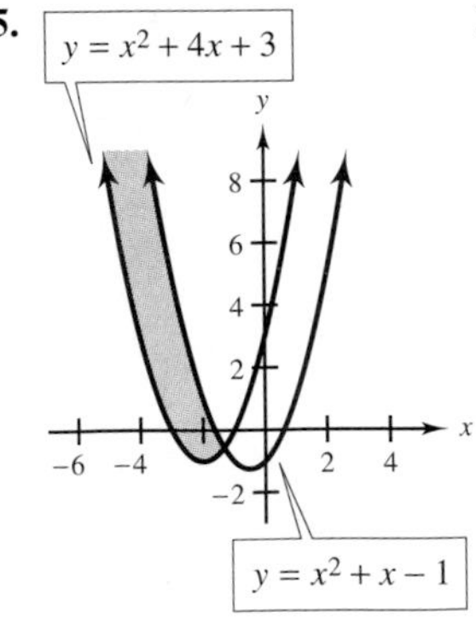

37.

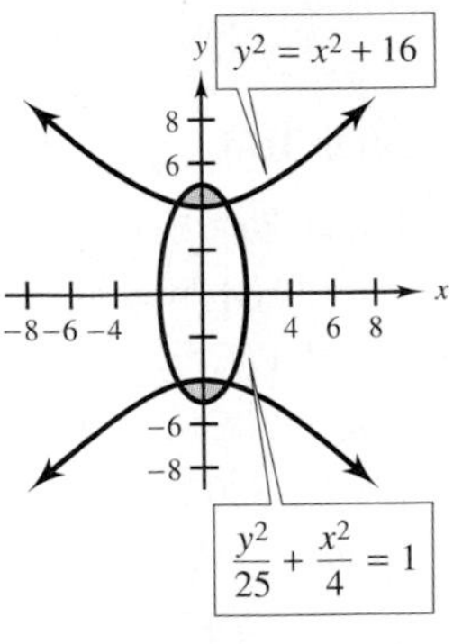

39.

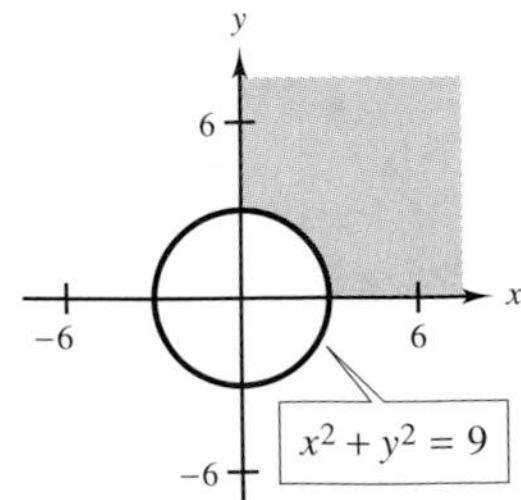

41. No solution

43.

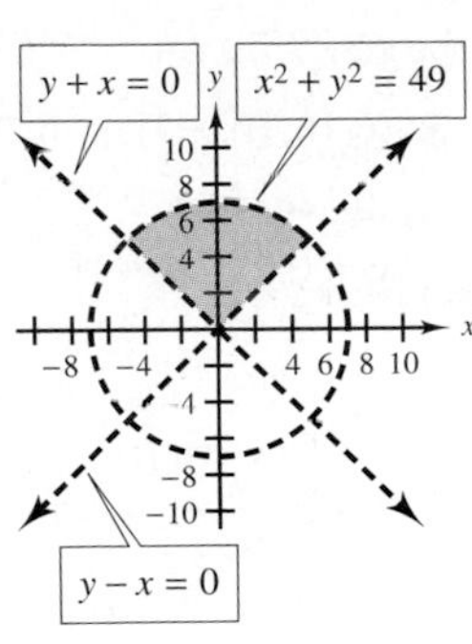

45. $x^2 + y^2 \geq 4$, $x \leq 0$, $y \leq 0$ **47.** $y < 4 - \frac{1}{64}x^2$

49. If x is the length of a side of the warehouse and y is the length of a side of an office, $x^2 + 4y^2 \leq 10{,}900$ and $y^2 \geq 225$ and $x \geq 0$ and $y \geq 0$.

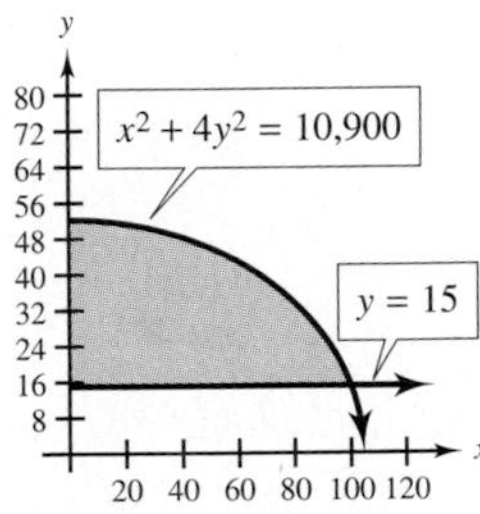

51.

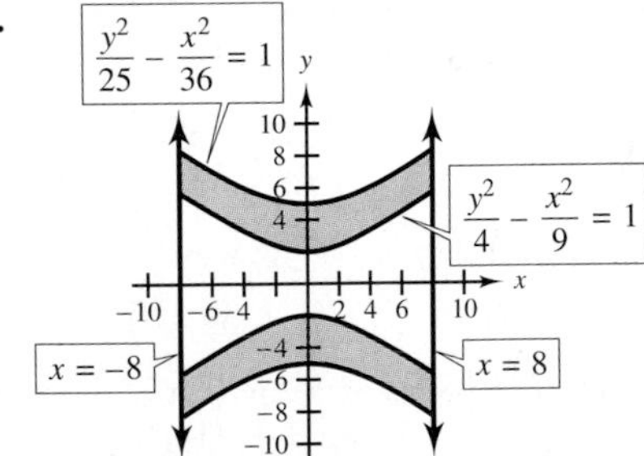

53.

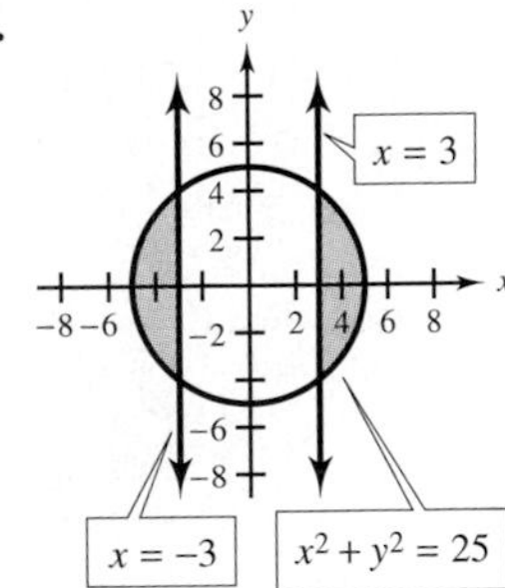

55. $x^2 + y^2 \geq 25$ and $25x^2 + 64y^2 \leq 1600$ and $x \geq 0$ and $y \leq 0$

Chapter Review Exercises *(page 738)*

1. (0, 0), (7, 0)
(3.5, −12.25)
$x = 3.5$

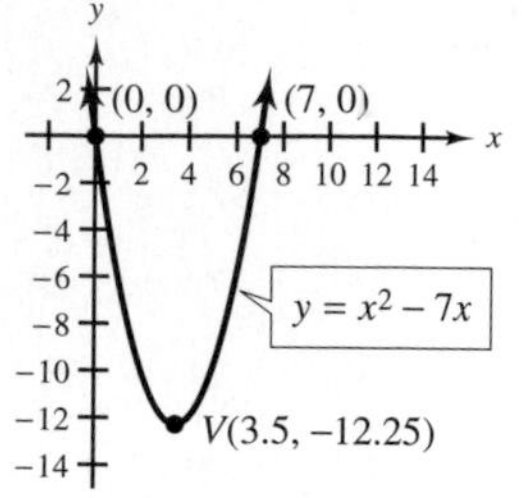

3. (6, 0), (0, −1.5), (0, 2)
(6.125, 0.25)
$y = 0.25$

V(6.125, 0.25)
(0, 2)
(0, −1.5)
$x = 6 + y - 2y^2$

5. (7, 0)
(2.5, −1.5)
$y = -1.5$

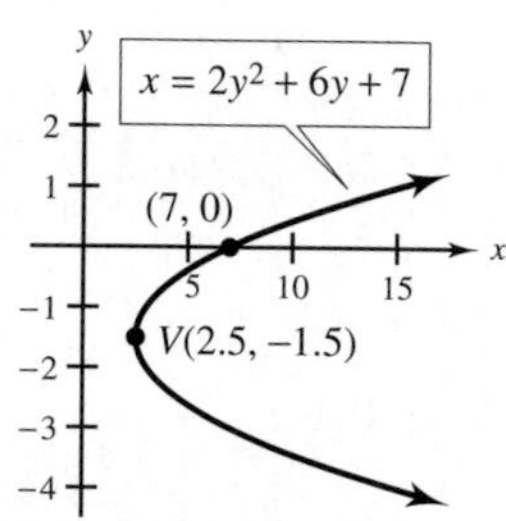

7. $y = -x^2 + 7x - 10$ **9.** $y = 4x^2 - 32x + 64$
11. $y = -3$ **13.** Domain: **R**; range: $\{y \mid y \geq -21.125\}$
15. $C(4, -5)$; $r = 6$ **17.** Ø **19.** $(x - 3)^2 + (y + 6)^2 = 6$
21. (0, ±8), (±7, 0)
23. (− , 3), (5, 3), (−4, −5), (−4, 11)

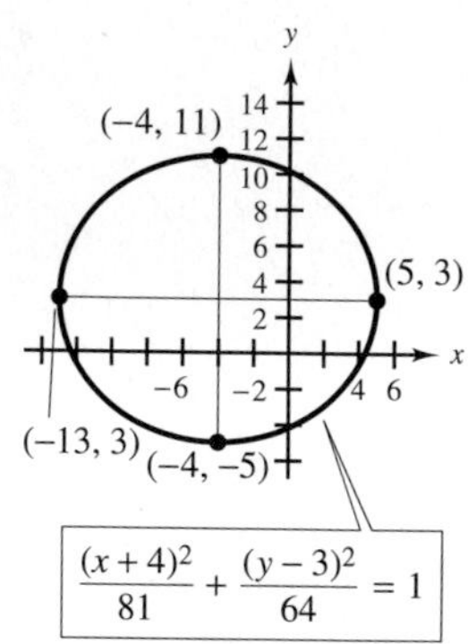

25. $\dfrac{x^2}{25} + \dfrac{y^2}{64} = 1$

27. (±6, 0)

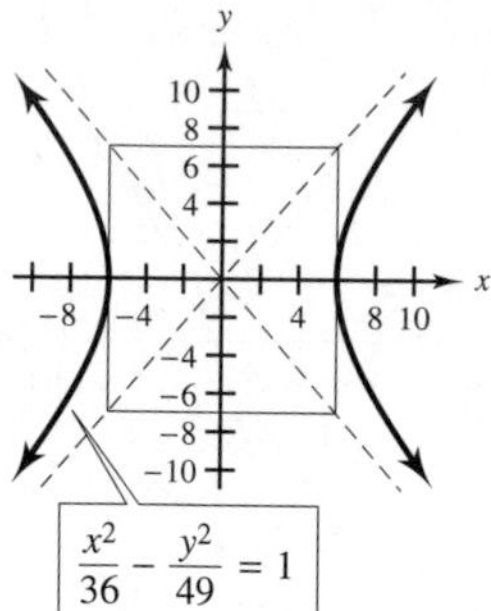

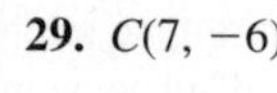
29. $C(7, -6)$

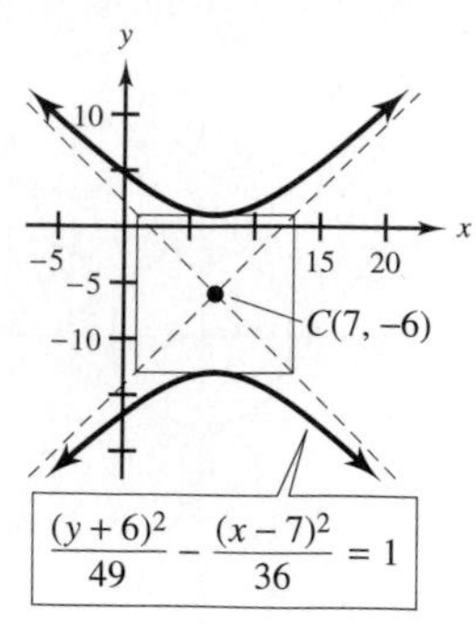

31. Hyperbola **33.** Circle **35.** (±2.45, −1), (±3, 2)
37. (6.61, 4.61), (−0.61, −2.61)
39.

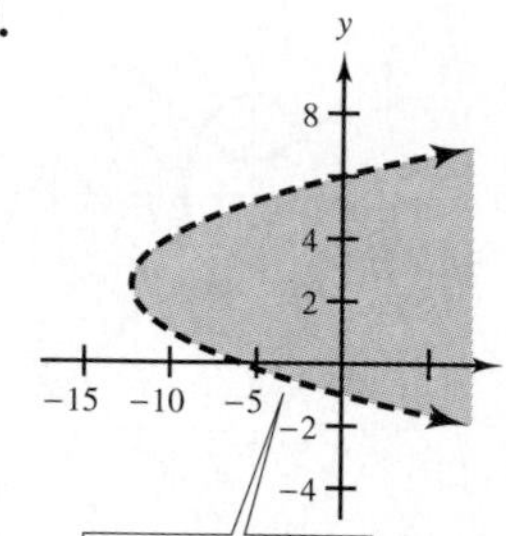

41.

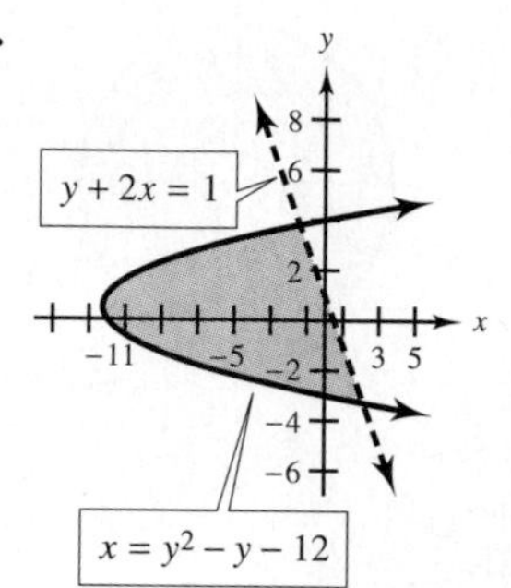

Chapter Test *(page 739)*

1. [*11.1*] Parabola **2.** [*11.2*] Circle
3. [*11.4*] Hyperbola **4.** [*11.3*] Ellipse
5. [*11.1*] $V(1, -2)$, $y = -2$
6. [*11.1*] $V(-1.5, 4.25)$, $x = -1.5$
7. [*11.1*] (8, 0), (0, ±2)

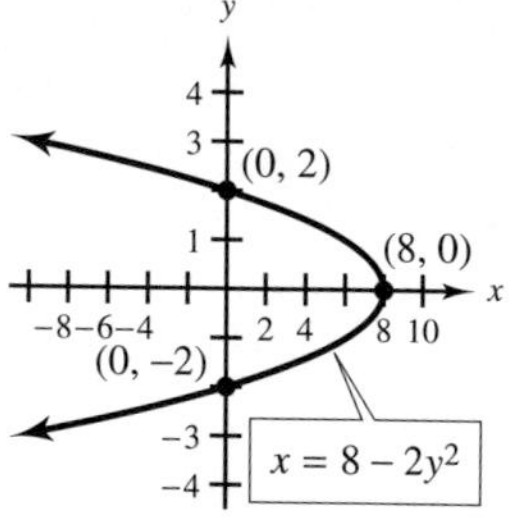

8. [*11.1*] (0, 0), (3, 0)

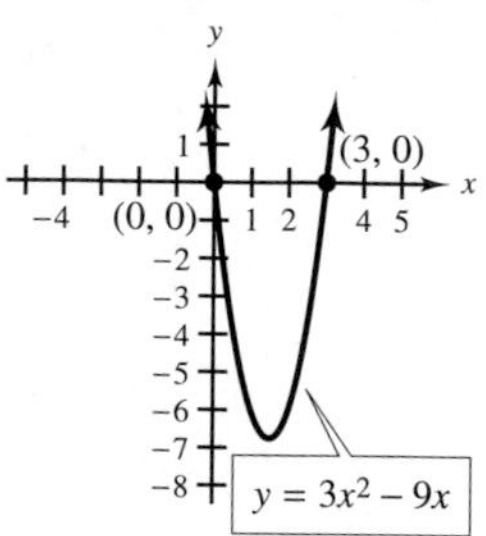

9. [*11.1*] $x = -3(y + 1)^2 + 2$
10. [*11.1*] $y = 3(x + 3)^2 + 2$

11. [*11.2*]

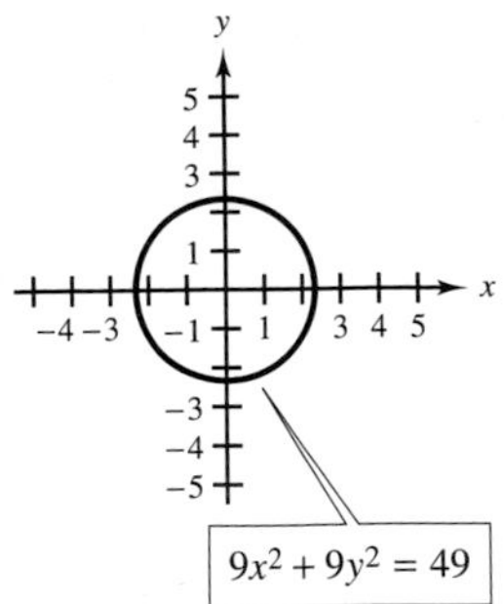

12. [*11.2*]

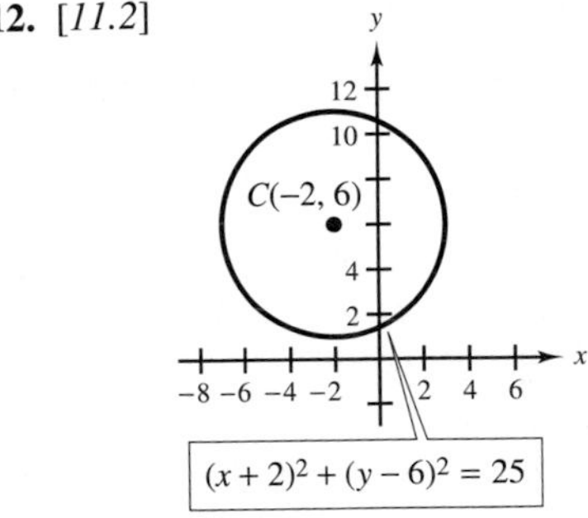

13. [*11.2*] $x^2 + y^2 = 10$

14. [*11.2*] $(x - 1)^2 + (y + 3)^2 = 16$

15. [*11.3*]

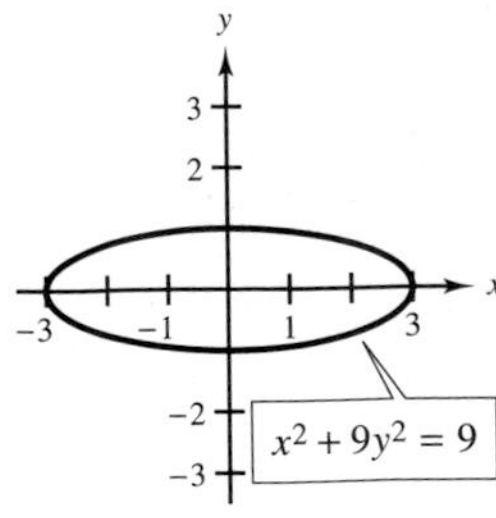

16. [*11.3*]

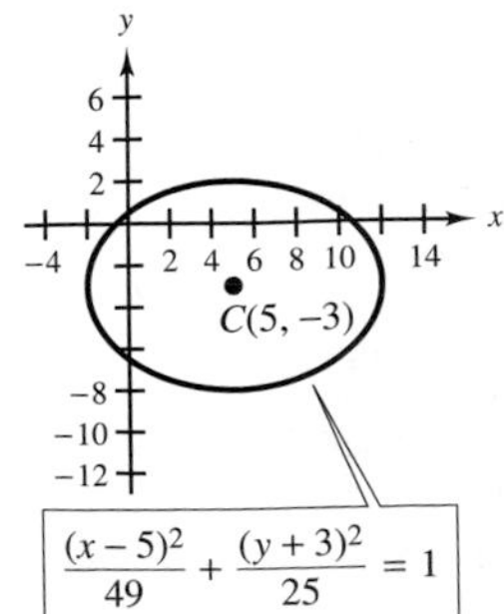

17. [*11.3*] $\frac{x^2}{16} + \frac{y^2}{4} = 1$

18. [*11.4*]

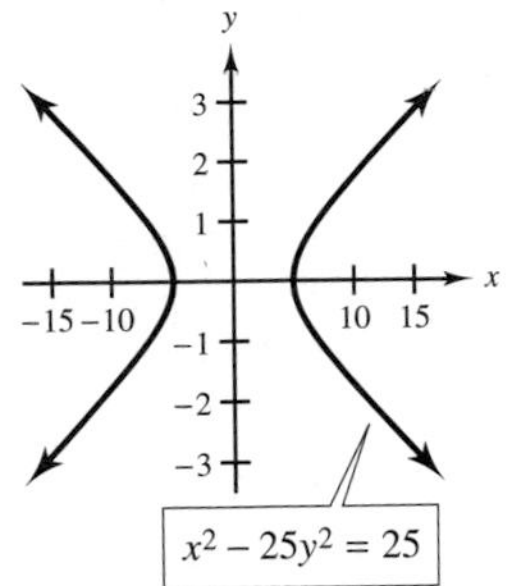

19. [*11.4*]

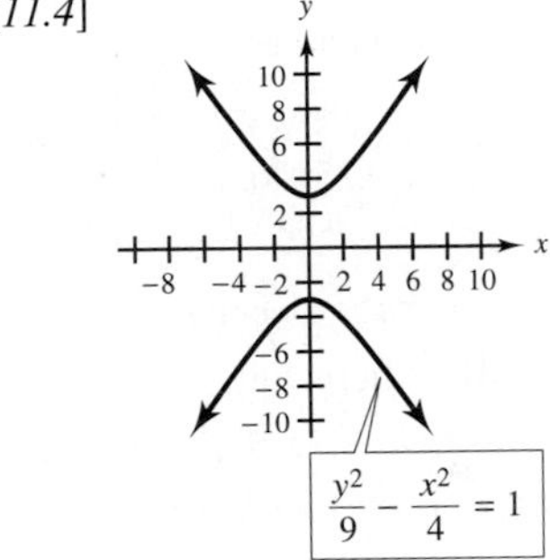

20. [*11.5*] $(0, -3)$, $(\pm 2.24, 2)$

21. [*11.5*] $(\pm 2, 2.83)$, $(\pm 2, -2.83)$

22. [*11.6*]

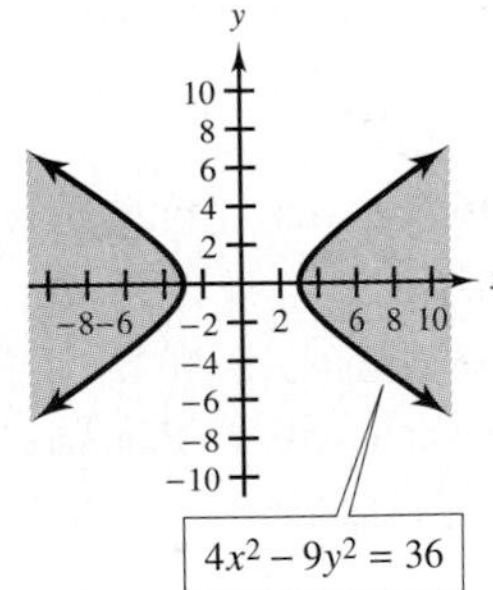

23. [*11.6*]

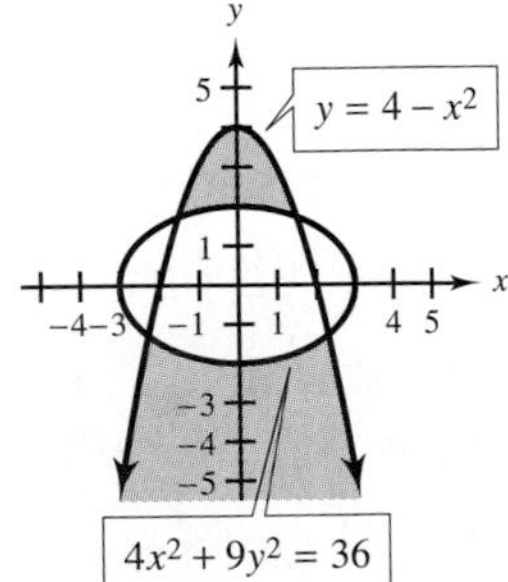

24. [*11.5*] 3000 square feet

CHAPTER 12

Section 12.1 *(page 746)*

1. (a) Infinite sequence

(b) Finite sequence

An infinite sequence has infinitely many terms; a finite sequence has a finite number of terms.

3. 17, 20 **5.** 9, 11 **7.** 243, 729 **9.** 3, 9, 27, 81

11. 1, 4, 9, 16 **13.** 0, 2, 0, 2 **15.** 6, 24, 120, 720

17. $-1, \frac{1}{2}, -\frac{1}{3}, \frac{1}{4}$

19. The letter n represents the term number.

21. $\frac{1}{3}, \frac{1}{2}, \frac{3}{5}, \frac{2}{3}, \frac{5}{7}$ **23.** $0, 1, \sqrt{2}, \sqrt{3}, 2, \sqrt{5}, \sqrt{6}$

25. $3, \frac{9}{8}, 1, \frac{81}{64}$ **27.** $-1, 4, -9, 16, -25, 36$

29. -70 **31.** 0 **33.** $\frac{1}{10} = 0.1$ **35.** 280

37. Increasing **39.** Alternating **41.** Decreasing

43. None **45.** Increasing **47.** Alternating

49. Decreasing **51.** None **53.** $n + 4$ **55.** $\frac{n}{n+1}$

57. $2n + 3$ **59.** $\frac{n}{(n+1)!}$ **61.** $(-2)^{n-1}$ **63.** $(-1)^n n^2$

65. 5, 8, 11, 14 **67.** 5, 10, 20, 40 **69.** 0, 3, 9, 21

71. 1, 2, 2, 4 **73. (b)** $n! = n(n-1)!$ **(c)** 123

75. By definition, the sequence is both increasing and decreasing because $a_n \geq a_{n-1}$ and $a_n \leq a_{n-1}$ for each term.

77. 105

79. 16, 48, 80, 112; $a_n = 16(2n - 1)$; $a_7 = 208$ (feet)

81. \$220, \$218, \$216, \$214, \$212; $a_n = 222 - 2n$; \$600

83. *RRRRRRRRNNNNN*; 13

85. The entries in the third column form a Fibonacci sequence.

87. The sequence is 1, 2, 6, 24, 120, ... An alternative expression for the general term is $n!$.

89. $\log 9! = a - 1$ **91.** 1, 1, 2, 3, 5, 8, 13, 21, 34, 55

Section 12.2 *(page 752)*

1. A *sequence* is a list of terms. A *series* is the sum of the terms of a sequence.

3. $1 - 2 + 4 - 8 + 16$

5. $6 + 12 + 20 + 30 + 42 + 56 + 72$

7. $1 + \frac{4}{3} + \frac{9}{5} + \frac{16}{7} + \frac{25}{9}$ **9.** 30 **11.** 91 **13.** $\frac{47}{60}$ **15.** 15

17. 6 **19.** $\frac{80}{27}$ **21.** -4 **23.** 122

25. To write a series in summation notation, the general term and the domain of the sequence must be known.

27. $\sum_{n=1}^{11}(n + 4)$ **29.** $\sum_{n=1}^{9}\sqrt{n}$ **31.** $\sum_{n=1}^{9}(2n + 9)$

33. $\sum_{n=1}^{9}\frac{(-1)^n}{n}$ **35.** $\sum_{n=1}^{6}(-1)^{n-1}\frac{n}{n+2}$

37. Because the lower limit is being increased from 1 to 3, the index of the general term must be decreased by 2. The revised general term is $i - 2$.

39. $\sum_{k=1}^{4}(3k + 9)$ **41.** $\sum_{i=4}^{9}(2i - 1)$ **43.** $\sum_{k=8}^{12}(-2)^{7-k}$

45. (a) 20 **(b)** $\sum_{j=2}^{6} j = 20$

(c) The results in parts (a) and (b) are the same.

47. $\sum_{n=1}^{5}(10n - 130) = -500$ **49.** $\sum_{n=1}^{3}\frac{1}{10}\left(\frac{9}{10}\right)^{n-1}(800) = 216.8$

51. (a) $C_3 = 5.8, C_4 = 6.7, C_5 = 8.2, C_6 = 9.5$

(b) 30.2 (million); the total number of computers sold from 1993 to 1996

53. (a) $P_2 = 475, P_3 = 488, P_4 = 528$

(b) 1491 million passengers from 1992 to 1994

55. $m_0 = 2.45, m_1 = 2.37, m_2 = 2.36, m_3 = 2.33$
$d_0 = 1.18, d_1 = 1.18, d_2 = 1.22, d_3 = 1.19$

57. The amount by which the number of marriages exceeds the number of divorces; decreasing.

59. -220 **63.** False

Section 12.3 *(page 760)*

1. Every term of an arithmetic sequence differs from the preceding term by the same amount.

3. Yes **5.** Yes **7.** No **9.** No **11.** Yes **13.** No

15. 2, 5, 8, 11, 14 **17.** $\frac{2}{3}, \frac{5}{6}, 1, \frac{7}{6}, \frac{4}{3}$ **19.** 13, 18, 23, 28, 33

21. 4, 7, 10, 13, 16 **23.** $5n - 2$ **25.** $-2n$ **27.** $30 - 2n$

29. $3n - 19$ **31.** $2n$ **33.** $8 - 3n$ **35.** $\frac{2-n}{3}$

37. $4n + 3$ **39.** $17 - 5n$ **41.** $4n + 1$ **43.** $\frac{n+17}{8}$

45. 49 **47.** $\frac{15}{2}$ **49.** 11 **51.** 4 **53.** 32 **55.** -11

57. 106 **59.** 39 **61.** 18

63. Find the average of the first and last terms, and multiply the result by the number of terms.

65. -276 **67.** 1140 **69.** -470 **71.** 3775

73. 391 **75.** Yes **77.** No **79.** 360 **81.** 420

83. 165 **85.** $\frac{182}{3}$ **87.** 330 **89.** 253

91. If $d > 0$, then each term is greater than the preceding term, and the sequence is increasing. If $d < 0$, then each term is less than the preceding term, and the sequence is decreasing.

93. 2500 **95.** 15 **97.** \$1575 **99.** 21; 529

101. \$1650, \$9750

103. (a) 102 bushels **(b)** 850 bushels

105. (a) 29

(b) 180; the number (in millions) of credit cards in 1990

107. **(a)** \$0.07 **(b)** $C_w = 2.42 + 0.07w$

109. $\frac{11}{75}$ **111.** No; 2999

Section 12.4 *(page 769)*

1. The ratio of any term to its predecessor is always the same.

3. Yes **5.** No **7.** $-3, -6, -12, -24, -48$

9. $64, -32, 16, -8, 4$ **11.** $27, 9, 3, 1, \frac{1}{3}$

13. $-4, -\frac{8}{3}, -\frac{16}{9}, -\frac{32}{27}, -\frac{64}{81}$ **15.** $a_n = -27\left(\frac{1}{3}\right)^{n-1}$

17. $c_n = \frac{1}{8}(-2)^{n-1}$ **19.** $a_n = 0.02(0.2)^{n-1}$

21. $k_n = 8\left(-\frac{1}{2}\right)^{n-1}$ **23.** $a_n = 5(2)^{n-1}$

25. $a_n = -\frac{1}{9}(-3)^{n-1}$ **27.** $a_n = -1(2)^{n-1}$

29. $a_n = \frac{5}{8}\left(\frac{2}{5}\right)^{n-1}$ **31.** $a_n = 2(-3)^{n-1}$

33. All the terms are the same as the first term.

35. $a_5 = 0.00003$ **37.** $a_{10} = \frac{1}{16} = 0.0625$

39. $a_8 = \pm 4374$ **41.** $a_6 \approx 1469.33$ **43.** $a_{12} = \pm 1024$

45. $S_{12} = 8190$ **47.** $S_8 = \frac{1261}{729} \approx 1.73$ **49.** $S_{12} = -2730$

51. $S_9 = \frac{511}{64} \approx 7.98$ **53.** $S_{12} = 20{,}475$

55. $S_8 = -976{,}560$ **57.** $S_{12} \approx 26.79$ **59.** 0 **61.** $S = 1$

63. $S = \frac{16}{3} \approx 5.33$ **65.** $S = 8$ **67.** $S = -\frac{18}{5} = -3.6$

69. $S = \frac{8}{3} \approx 2.67$ **71.** $S = 4$ **73.** $S = \frac{9}{8} = 1.125$

75. If the difference between any term and its predecessor is always the same, the sequence is arithmetic. If the ratio of any term to its predecessor is always the same, the sequence is arithmetic.

77. Geometric **79.** Arithmetic **81.** Neither **83.** $\frac{4}{9}$

85. $\frac{19}{99}$ **87.** $\frac{41}{333}$ **89.** $a_4 \approx 86.5$°F. **91.** \$1 million

93. $h_n = 18{,}737(1.05)^{n-1}$; $c_n = 32{,}629(1.05)^{n-1}$

95. \$235,671.97, \$410,404.06 **97.** Answers will vary.

99. **(a)** Yes; 20,480 is the 12th term. **(b)** No.

Section 12.5 *(page 776)*

1. Each entry is the sum of the two entries that are diagonally above it.

3. $8x^3 - 36x^2y + 54xy^2 - 27y^3$

5. $x^4 + 8x^3y + 24x^2y^2 + 32xy^3 + 16y^4$

7. For the binomial coefficient ${}_nC_r$, n is the power in $(a + b)^n$. The number r is an integer between 0 and n, and $r + 1$ is the term number in the expansion.

9. 70 **11.** 21 **13.** 1 **15.** 10 **17.** 190

19. $x^5 + 10x^4 + 40x^3 + 80x^2 + 80x + 32$

21. $x^5 - 20x^4 + 160x^3 - 640x^2 + 1280x - 1024$

23. $243x^5 - 405x^4 + 270x^3 - 90x^2 + 15x - 1$

25. $1 - 10a^2 + 40a^4 - 80a^6 + 80a^8 - 32a^{10}$

27. $x^6 + 3x^5 + \frac{15}{4}x^4 + \frac{5}{2}x^3 + \frac{15}{16}x^2 + \frac{3}{16}x + \frac{1}{64}$

29. $x^{36} - 12x^{33}y^2 + 66x^{30}y^4 - 220x^{27}y^6 + \cdots$

31. $x^{20} - 20x^{18}y + 180x^{16}y^2 - 960x^{14}y^3 + \cdots$

33. $531{,}441 - 4{,}251{,}528b + 15{,}588{,}936b^2 - 34{,}642{,}080b^3 + \cdots$

35. $\frac{1}{3125}s^5 - \frac{1}{250}s^4t + \frac{1}{50}s^3t^2 - \frac{1}{20}s^2t^3 + \cdots$

37. $a^{200} + 200a^{199}b + \cdots$ **39.** $x^{100} + 200x^{99} + \cdots$

41. Because $0 \le r \le n$, the value of r is always 1 less than the term number.

43. -1 **45.** $20x^6y^9$ **47.** $-15{,}116{,}544x$ **49.** $252x^6y^2$

51. 5280 **53.** 672,000 **55.** 42,240 **57.** 1120

61. 1 9 36 84 126 126 84 36 9 1

63. **(a)** The next row would contain 5 pins.

(b) 1, 3, 6, 10, 15 **(c)** 21

(d) The sequence in part (b) appears in the third diagonal.

65. $-8i$ i

69. $11^0 = 1$; $11^1 = 11$; $11^2 = 121$; $11^3 = 1331$; $11^4 = 14{,}641$. The digits are the entries in the first five lines of Pascal's triangle. This pattern does not continue beyond 11^4.

Chapter Review Exercises *(page 778)*

1. $-2, 1, -3, 0$ **3.** $13, 9, 5, 1$ $b_8 = -15$

5. $b_{10} = \frac{35}{6}$ **7.** $a_n = 3n - 1$ **9.** 4, 5, 7, 11, 19

11. $\ln 2 + \ln 3 + \ln 4 + \ln 5 + \ln 6 + \ln 7$ **13.** -3

15. $\displaystyle\sum_{n=1}^{6} \frac{2n-1}{n+3}$ **17.** $\displaystyle\sum_{i=2}^{8} \frac{1}{2i-5}$

19. $7, 4, 1, -2, -5$ **21.** $a_n = 3n + 1$ **23.** $a_n = 22 - 4n$

25. 57 **27.** 24 **29.** 1125 **31.** 3300 **33.** 243, 81, 27, 9, 3

35. $a_n = e^{3n+1}$ **37.** $a_n = 3^{n-4}$ **39.** 0.125 **41.** 683

43. $\frac{2}{9}$ **45.** 1716

47. $x^6 - 18x^5 + 135x^4 - 540x^3 + 1215x^2 - 1458x + 729$

49. $1716x^7y^6$

Chapter Test *(page 780)*

1. [*12.1*] $0, \frac{1}{2}, \frac{1}{3}, \frac{1}{8}$ **2.** [*12.1*] $-1, 0, \frac{1}{9}, \frac{1}{8}, \frac{3}{25}$

3. [*12.1*] 120 **4.** [*12.4*] 25

5. [*12.4*] $\frac{7}{12}, \frac{7}{8}, \frac{21}{16}, \frac{63}{32}$ **7.** $a_n = 3n - 1$ **6.** [*12.1*] 13, 12, 15

7. [*12.1*] $a_n = \dfrac{1}{(n+1)^2}$ **8.** [*12.3*] $a_n = \dfrac{5 - 2n}{3}$

9. [*12.4*] $a_n = \left(-\dfrac{1}{2}\right)^n$ **10.** [*12.3*] $a_n = \dfrac{n+5}{2}$

11. [*12.3*] $a_n = n - \frac{1}{2}$ **12.** [*12.4*] $a_n = 3(2)^{n-1}$

13. [*12.3*] 13 **14.** [*12.2*] $\displaystyle\sum_{n=1}^{5} \frac{n}{n+1}$ **15.** [*12.2*] 19

16. [*12.3*] 357 **17.** [*12.4*] 54 **18.** [*12.4*] 2.25

19. [*12.5*] $x^5 + 15x^4 + 90x^3 + 270x^2 + 405x + 243$
20. [*12.5*] $16x^4 - 96x^3y + 216x^2y^2 - 216xy^3 + 81y^4$
21. [*12.5*] $-15360x^7$ **22.** [*12.4*] 1830 **23.** [*12.3*] 12,500
24. [*12.3*] $a_n = 18{,}700 - 1200n$; \$13,900 **25.** [*12.4*] 8190

Cumulative Test: Chapters 11–12 *(page 781)*

1. [*11.1*] $(-1, 0), (0, 0.16), (0, -6.16); V(-10, -3)$
2. [*11.2*] $(x + 3)^2 + (y - 5)^2 = 65$
3. [*11.3*] $C(3, 0); (7, 0), (-1, 0), (3, 2), (3, -2)$
4. [*11.4*]

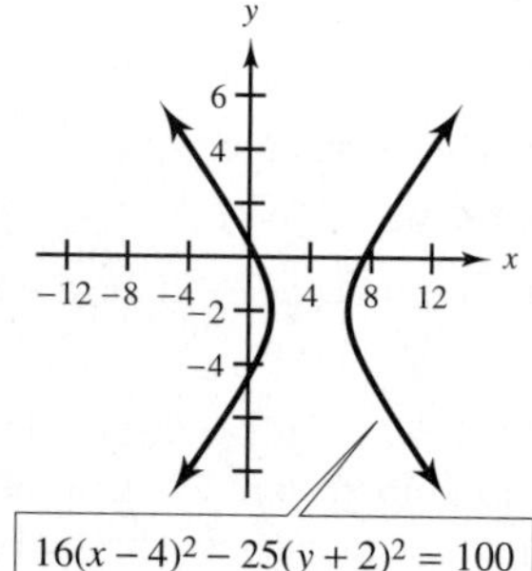

5. [*11.5*] 24 inches, 10 inches
6. [*11.6*]

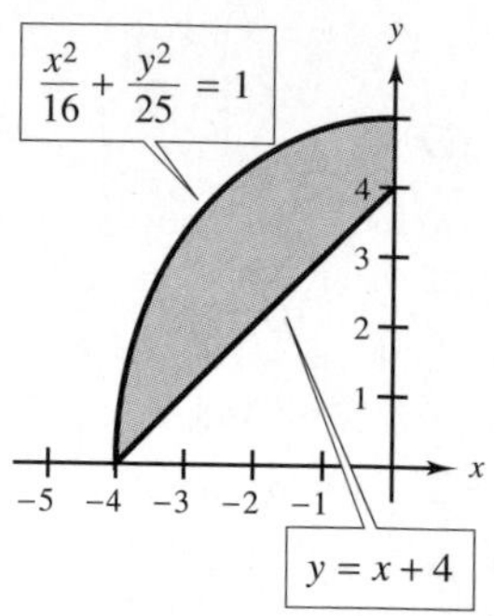

7. [*12.1*] **(a)** Decreasing; -14 **(b)** Alternating; 64
(c) Increasing; -3

8. [*12.2*] $1 + \frac{x}{2} + \frac{x^2}{3} + \frac{x^3}{4}$

9. [*12.3*] 21 **10.** [*12.4*] 1555 **11.** [*12.5*] 1365
12. [*12.5*] $32x^5 - 240x^4 + 720x^3 - 1080x^2 + 810x - 243$

Indexes

Index of Real-Life Applications

Agriculture and Land Use

Business

Economics

Engineering and Technology

Geometry

Home and Community

Life Sciences

Miscellaneous

Physical Sciences

Social Sciences

Sports and Recreation

Index of Real-Data Applications

Business and Finance

Education

Family and Community

Government and Military

Life Sciences and Health

Physical Sciences

Social Sciences

Sports and Recreation

Technology

Transportation and Energy

Index

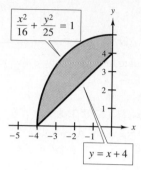

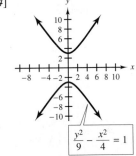

Government and Military

Life Sciences and Health

Physical Sciences

Social Sciences

Sports and Recreation

Technology

Transportation and Energy

Index

Selected Keys from Typical Graphing Calculators

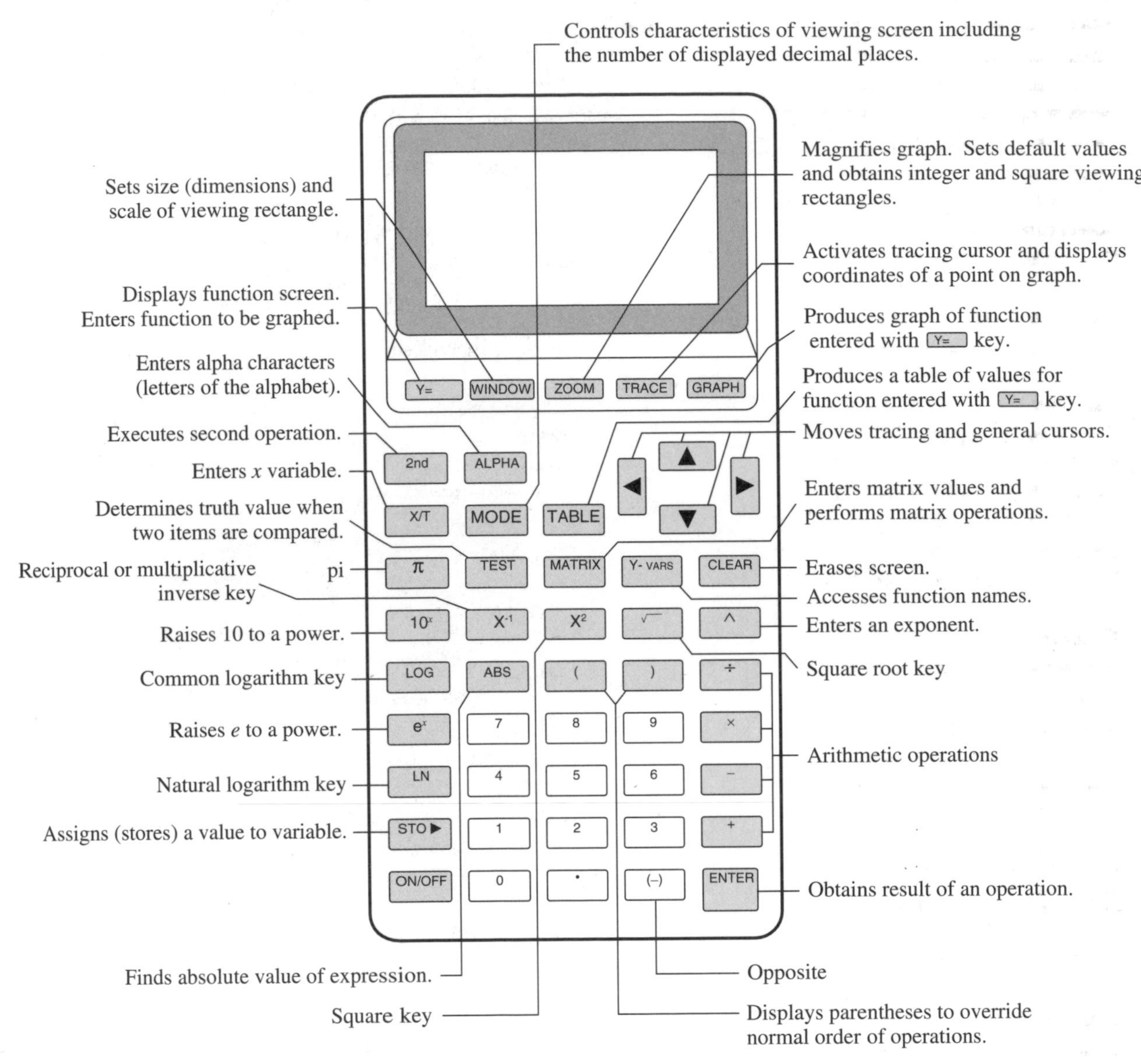

Controls characteristics of viewing screen including the number of displayed decimal places.
Sets size (dimensions) and scale of viewing rectangle.
Magnifies graph. Sets default values and obtains integer and square viewing rectangles.
Activates tracing cursor and displays coordinates of a point on graph.
Displays function screen. Enters function to be graphed.
Produces graph of function entered with Y= key.
Enters alpha characters (letters of the alphabet).
Produces a table of values for function entered with Y= key.
Y=
WINDOW
ZOOM
TRACE
GRAPH
Executes second operation.
Moves tracing and general cursors.
2nd
ALPHA
Enters x variable.
X/T
MODE
TABLE
Enters matrix values and performs matrix operations.
Determines truth value when two items are compared.
TEST
MATRIX
Y- VARS
CLEAR
Erases screen.
pi
π
Reciprocal or multiplicative inverse key
Accesses function names.
Raises 10 to a power.
10ˣ
X⁻¹
X²
√
^
Enters an exponent.
Common logarithm key
LOG
ABS
(
)
÷
Square root key
Raises e to a power.
eˣ
7
8
9
×
Arithmetic operations
Natural logarithm key
LN
4
5
6
−
Assigns (stores) a value to variable.
STO▶
1
2
3
+
ON/OFF
0
.
(−)
ENTER
Obtains result of an operation.
Finds absolute value of expression.
Opposite
Square key
Displays parentheses to override normal order of operations.